Information Science and Statistics

Information Science and Statistics

Akaike and Kitagawa: The Practice of Time Series Analysis.
Bishop: Pattern Recognition and Machine Learning.
Cowell, Dawid, Lauritzen, and Spiegelhalter: Probabilistic Networks and Expert Systems.
Doucet, de Freitas, and Gordon: Sequential Monte Carlo Methods in Practice.
Fine: Feedforward Neural Network Methodology.
Hawkins and Olwell: Cumulative Sum Charts and Charting for Quality Improvement.
Jensen: Bayesian Networks and Decision Graphs.
Marchette: Computer Intrusion Detection and Network Monitoring: A Statistical Viewpoint.
Rubinstein and Kroese: The Cross-Entropy Method: A Unified Approach to Combinatorial Optimization, Monte Carlo Simulation, and Machine Learning.
Studený: Probabilistic Conditional Independence Structures.
Vapnik: The Nature of Statistical Learning Theory, Second Edition.
Wallace: Statistical and Inductive Inference by Minimum Massage Length.

Christopher M. Bishop

Pattern Recognition and Machine Learning

Springer

Christopher M. Bishop F.R.Eng.
Assistant Director
Microsoft Research Ltd
Cambridge CB3 0FB, U.K.
cmbishop@microsoft.com
http://research.microsoft.com/~cmbishop

Library of Congress Control Number: 2006922522

ISBN 978-0387-31073-2

Printed on acid-free paper.

9 8 (corrected at 8th printing 2009)

springer.com

This book is dedicated to my family:

Jenna, Mark, and Hugh

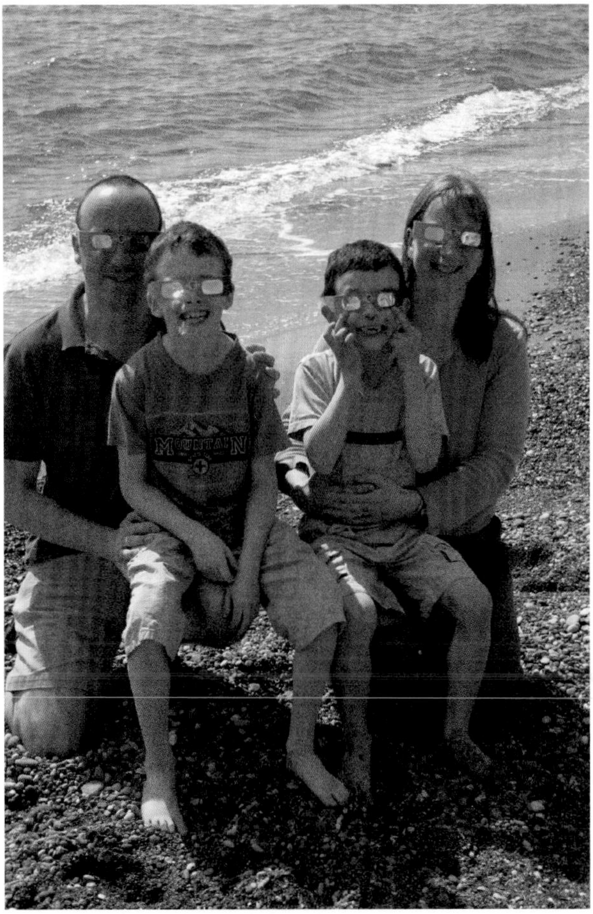

Total eclipse of the sun, Antalya,Turkey, 29 March 2006.

Preface

Pattern recognition has its origins in engineering, whereas machine learning grew out of computer science. However, these activities can be viewed as two facets of the same field, and together they have undergone substantial development over the past ten years. In particular, Bayesian methods have grown from a specialist niche to become mainstream, while graphical models have emerged as a general framework for describing and applying probabilistic models. Also, the practical applicability of Bayesian methods has been greatly enhanced through the development of a range of approximate inference algorithms such as variational Bayes and expectation propagation. Similarly, new models based on kernels have had significant impact on both algorithms and applications.

This new textbook reflects these recent developments while providing a comprehensive introduction to the fields of pattern recognition and machine learning. It is aimed at advanced undergraduates or first year PhD students, as well as researchers and practitioners, and assumes no previous knowledge of pattern recognition or machine learning concepts. Knowledge of multivariate calculus and basic linear algebra is required, and some familiarity with probabilities would be helpful though not essential as the book includes a self-contained introduction to basic probability theory.

Because this book has broad scope, it is impossible to provide a complete list of references, and in particular no attempt has been made to provide accurate historical attribution of ideas. Instead, the aim has been to give references that offer greater detail than is possible here and that hopefully provide entry points into what, in some cases, is a very extensive literature. For this reason, the references are often to more recent textbooks and review articles rather than to original sources.

The book is supported by a great deal of additional material, including lecture slides as well as the complete set of figures used in the book, and the reader is encouraged to visit the book web site for the latest information:

http://research.microsoft.com/∼cmbishop/PRML

Exercises

The exercises that appear at the end of every chapter form an important component of the book. Each exercise has been carefully chosen to reinforce concepts explained in the text or to develop and generalize them in significant ways, and each is graded according to difficulty ranging from $(\star)$, which denotes a simple exercise taking a few minutes to complete, through to $(\star\star\star)$, which denotes a significantly more complex exercise.

It has been difficult to know to what extent worked solutions should be made widely available. Those engaged in self study will find worked solutions very beneficial, whereas many course tutors request that solutions be available only via the publisher so that the exercises may be used in class. In order to try to meet these conflicting requirements, those exercises that help amplify key points in the text, or that fill in important details, have solutions that are available as a PDF file from the book web site. Such exercises are denoted by **WWW**. Solutions for the remaining exercises are available to course tutors by contacting the publisher (contact details are given on the book web site). Readers are strongly encouraged to work through the exercises unaided, and to turn to the solutions only as required.

Although this book focuses on concepts and principles, in a taught course the students should ideally have the opportunity to experiment with some of the key algorithms using appropriate data sets. Matlab software implementing many of the algorithms discussed in this book, together with example data sets, will be available through the book web site, along with a companion tutorial (Bishop and Nabney, 2008) describing practical algorithms for solving the optimization problems which arise in machine learning.

Acknowledgements

First of all I would like to express my sincere thanks to Markus Svensén who has provided immense help with preparation of figures and with the typesetting of the book in LaTeX. His assistance has been invaluable.

I am very grateful to Microsoft Research for providing a highly stimulating research environment and for giving me the freedom to write this book (the views and opinions expressed in this book, however, are my own and are therefore not necessarily the same as those of Microsoft or its affiliates).

Springer has provided excellent support throughout the final stages of preparation of this book, and I would like to thank my commissioning editor John Kimmel for his support and professionalism, as well as Joseph Piliero for his help in designing the cover and the text format and MaryAnn Brickner for her numerous contributions during the production phase. The inspiration for the cover design came from a discussion with Antonio Criminisi.

I also wish to thank Oxford University Press for permission to reproduce excerpts from an earlier textbook, *Neural Networks for Pattern Recognition* (Bishop, 1995a). The images of the Mark 1 perceptron and of Frank Rosenblatt are reproduced with the permission of Arvin Calspan Advanced Technology Center. I would also like to thank Asela Gunawardana for plotting the spectrogram in Figure 13.1, and Bernhard Schölkopf for permission to use his kernel PCA code to plot Figure 12.17.

Many people have helped by proofreading draft material and providing comments and suggestions, including Shivani Agarwal, Cédric Archambeau, Arik Azran, Andrew Blake, Hakan Cevikalp, Michael Fourman, Brendan Frey, Zoubin Ghahramani, Thore Graepel, Katherine Heller, Ralf Herbrich, Geoffrey Hinton, Adam Johansen, Matthew Johnson, Michael Jordan, Eva Kalyvianaki, Anitha Kannan, Julia Lasserre, David Liu, Tom Minka, Ian Nabney, Tonatiuh Pena, Yuan Qi, Sam Roweis, Balaji Sanjiya, Toby Sharp, Ana Costa e Silva, David Spiegelhalter, Jay Stokes, Tara Symeonides, Martin Szummer, Marshall Tappen, Ilkay Ulusoy, Chris Williams, John Winn, and Andrew Zisserman.

Finally, I would like to thank my wife Jenna who has been hugely supportive throughout the several years it has taken to write this book.

Chris Bishop
Cambridge
February 2006

Mathematical notation

I have tried to keep the mathematical content of the book to the minimum necessary to achieve a proper understanding of the field. However, this minimum level is nonzero, and it should be emphasized that a good grasp of calculus, linear algebra, and probability theory is essential for a clear understanding of modern pattern recognition and machine learning techniques. Nevertheless, the emphasis in this book is on conveying the underlying concepts rather than on mathematical rigour.

I have tried to use a consistent notation throughout the book, although at times this means departing from some of the conventions used in the corresponding research literature. Vectors are denoted by lower case bold Roman letters such as $\mathbf{x}$, and all vectors are assumed to be column vectors. A superscript T denotes the transpose of a matrix or vector, so that $\mathbf{x}^{\mathrm{T}}$ will be a row vector. Uppercase bold Roman letters, such as $\mathbf{M}$, denote matrices. The notation $(w_1, \ldots, w_M)$ denotes a row vector with M elements, while the corresponding column vector is written as $\mathbf{w} = (w_1, \ldots, w_M)^{\mathrm{T}}$.

The notation $[a, b]$ is used to denote the *closed* interval from a to b, that is the interval including the values a and b themselves, while (a, b) denotes the corresponding *open* interval, that is the interval excluding a and b. Similarly, $[a, b)$ denotes an interval that includes a but excludes b. For the most part, however, there will be little need to dwell on such refinements as whether the end points of an interval are included or not.

The $M \times M$ identity matrix (also known as the unit matrix) is denoted $\mathbf{I}_M$, which will be abbreviated to $\mathbf{I}$ where there is no ambiguity about its dimensionality. It has elements I_{ij} that equal 1 if $i = j$ and 0 if $i \neq j$.

A functional is denoted $f[y]$ where $y(x)$ is some function. The concept of a functional is discussed in Appendix D.

The notation $g(x) = O(f(x))$ denotes that $|f(x)/g(x)|$ is bounded as $x \to \infty$. For instance if $g(x) = 3x^2 + 2$, then $g(x) = O(x^2)$.

The expectation of a function $f(x, y)$ with respect to a random variable x is denoted by $\mathbb{E}_x[f(x, y)]$. In situations where there is no ambiguity as to which variable is being averaged over, this will be simplified by omitting the suffix, for instance

$\mathbb{E}[x]$. If the distribution of x is conditioned on another variable z, then the corresponding conditional expectation will be written $\mathbb{E}_x[f(x)|z]$. Similarly, the variance is denoted $\mathrm{var}[f(x)]$, and for vector variables the covariance is written $\mathrm{cov}[\mathbf{x}, \mathbf{y}]$. We shall also use $\mathrm{cov}[\mathbf{x}]$ as a shorthand notation for $\mathrm{cov}[\mathbf{x}, \mathbf{x}]$. The concepts of expectations and covariances are introduced in Section 1.2.2.

If we have N values $\mathbf{x}_1, \ldots, \mathbf{x}_N$ of a D-dimensional vector $\mathbf{x} = (x_1, \ldots, x_D)^{\mathrm{T}}$, we can combine the observations into a data matrix $\mathbf{X}$ in which the nth row of $\mathbf{X}$ corresponds to the row vector $\mathbf{x}_n^{\mathrm{T}}$. Thus the n, i element of $\mathbf{X}$ corresponds to the ith element of the nth observation $\mathbf{x}_n$. For the case of one-dimensional variables we shall denote such a matrix by x, which is a column vector whose n^{th} element is x_n. Note that x (which has dimensionality N) uses a different typeface to distinguish it from $\mathbf{x}$ (which has dimensionality D).

Contents

1 Introduction

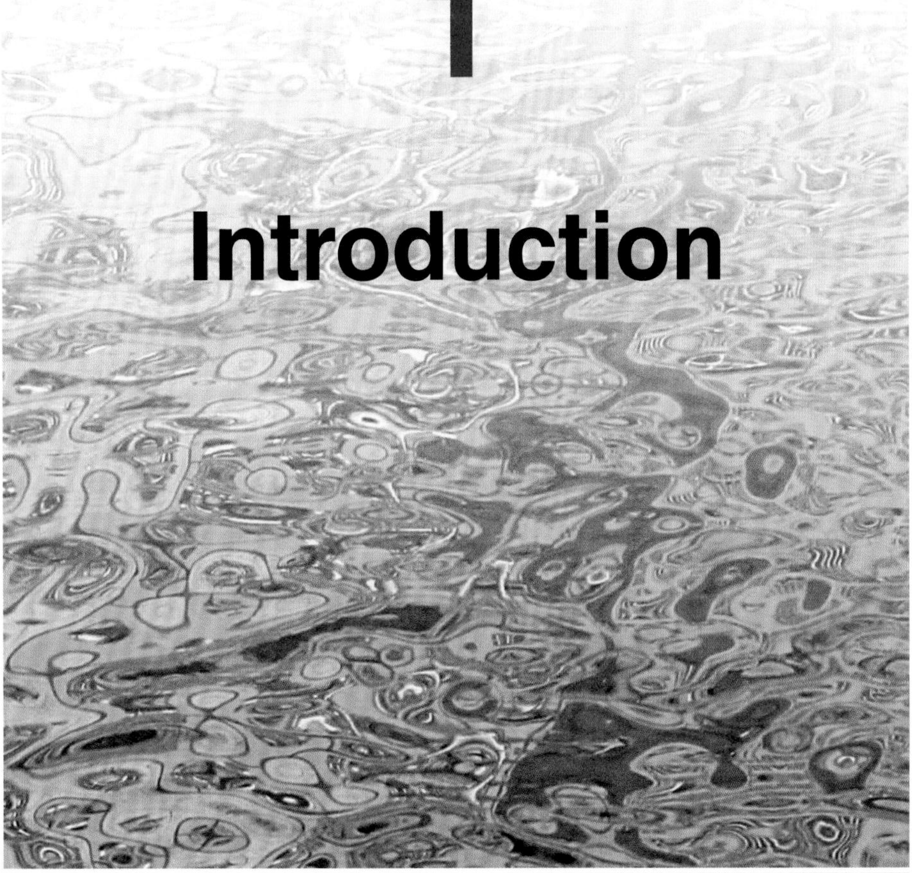

The problem of searching for patterns in data is a fundamental one and has a long and successful history. For instance, the extensive astronomical observations of Tycho Brahe in the 16th century allowed Johannes Kepler to discover the empirical laws of planetary motion, which in turn provided a springboard for the development of classical mechanics. Similarly, the discovery of regularities in atomic spectra played a key role in the development and verification of quantum physics in the early twentieth century. The field of pattern recognition is concerned with the automatic discovery of regularities in data through the use of computer algorithms and with the use of these regularities to take actions such as classifying the data into different categories.

Consider the example of recognizing handwritten digits, illustrated in Figure 1.1. Each digit corresponds to a 28×28 pixel image and so can be represented by a vector $\mathbf{x}$ comprising 784 real numbers. The goal is to build a machine that will take such a vector $\mathbf{x}$ as input and that will produce the identity of the digit $0, \ldots, 9$ as the output. This is a nontrivial problem due to the wide variability of handwriting. It could be

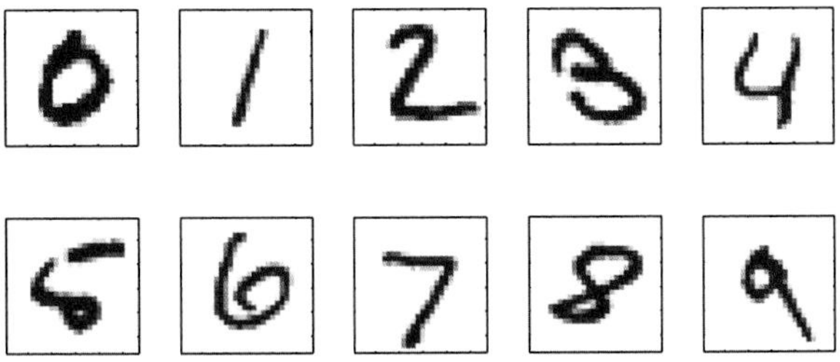

Figure 1.1 Examples of hand-written digits taken from US zip codes.

tackled using handcrafted rules or heuristics for distinguishing the digits based on the shapes of the strokes, but in practice such an approach leads to a proliferation of rules and of exceptions to the rules and so on, and invariably gives poor results.

Far better results can be obtained by adopting a machine learning approach in which a large set of N digits $\{\mathbf{x}_1, \ldots, \mathbf{x}_N\}$ called a *training set* is used to tune the parameters of an adaptive model. The categories of the digits in the training set are known in advance, typically by inspecting them individually and hand-labelling them. We can express the category of a digit using *target vector* $\mathbf{t}$, which represents the identity of the corresponding digit. Suitable techniques for representing categories in terms of vectors will be discussed later. Note that there is one such target vector $\mathbf{t}$ for each digit image $\mathbf{x}$.

The result of running the machine learning algorithm can be expressed as a function $\mathbf{y}(\mathbf{x})$ which takes a new digit image $\mathbf{x}$ as input and that generates an output vector $\mathbf{y}$, encoded in the same way as the target vectors. The precise form of the function $\mathbf{y}(\mathbf{x})$ is determined during the *training* phase, also known as the *learning* phase, on the basis of the training data. Once the model is trained it can then determine the identity of new digit images, which are said to comprise a *test set*. The ability to categorize correctly new examples that differ from those used for training is known as *generalization*. In practical applications, the variability of the input vectors will be such that the training data can comprise only a tiny fraction of all possible input vectors, and so generalization is a central goal in pattern recognition.

For most practical applications, the original input variables are typically *pre-processed* to transform them into some new space of variables where, it is hoped, the pattern recognition problem will be easier to solve. For instance, in the digit recognition problem, the images of the digits are typically translated and scaled so that each digit is contained within a box of a fixed size. This greatly reduces the variability within each digit class, because the location and scale of all the digits are now the same, which makes it much easier for a subsequent pattern recognition algorithm to distinguish between the different classes. This pre-processing stage is sometimes also called *feature extraction*. Note that new test data must be pre-processed using the same steps as the training data.

Pre-processing might also be performed in order to speed up computation. For example, if the goal is real-time face detection in a high-resolution video stream, the computer must handle huge numbers of pixels per second, and presenting these directly to a complex pattern recognition algorithm may be computationally infeasible. Instead, the aim is to find useful features that are fast to compute, and yet that also preserve useful discriminatory information enabling faces to be distinguished from non-faces. These features are then used as the inputs to the pattern recognition algorithm. For instance, the average value of the image intensity over a rectangular subregion can be evaluated extremely efficiently (Viola and Jones, 2004), and a set of such features can prove very effective in fast face detection. Because the number of such features is smaller than the number of pixels, this kind of pre-processing represents a form of dimensionality reduction. Care must be taken during pre-processing because often information is discarded, and if this information is important to the solution of the problem then the overall accuracy of the system can suffer.

Applications in which the training data comprises examples of the input vectors along with their corresponding target vectors are known as *supervised learning* problems. Cases such as the digit recognition example, in which the aim is to assign each input vector to one of a finite number of discrete categories, are called *classification* problems. If the desired output consists of one or more continuous variables, then the task is called *regression*. An example of a regression problem would be the prediction of the yield in a chemical manufacturing process in which the inputs consist of the concentrations of reactants, the temperature, and the pressure.

In other pattern recognition problems, the training data consists of a set of input vectors $\mathbf{x}$ without any corresponding target values. The goal in such *unsupervised learning* problems may be to discover groups of similar examples within the data, where it is called *clustering*, or to determine the distribution of data within the input space, known as *density estimation*, or to project the data from a high-dimensional space down to two or three dimensions for the purpose of *visualization*.

Finally, the technique of *reinforcement learning* (Sutton and Barto, 1998) is concerned with the problem of finding suitable actions to take in a given situation in order to maximize a reward. Here the learning algorithm is not given examples of optimal outputs, in contrast to supervised learning, but must instead discover them by a process of trial and error. Typically there is a sequence of states and actions in which the learning algorithm is interacting with its environment. In many cases, the current action not only affects the immediate reward but also has an impact on the reward at all subsequent time steps. For example, by using appropriate reinforcement learning techniques a neural network can learn to play the game of backgammon to a high standard (Tesauro, 1994). Here the network must learn to take a board position as input, along with the result of a dice throw, and produce a strong move as the output. This is done by having the network play against a copy of itself for perhaps a million games. A major challenge is that a game of backgammon can involve dozens of moves, and yet it is only at the end of the game that the reward, in the form of victory, is achieved. The reward must then be attributed appropriately to all of the moves that led to it, even though some moves will have been good ones and others less so. This is an example of a *credit assignment* problem. A general feature of reinforcement learning is the trade-off between *exploration*, in which the system tries out new kinds of actions to see how effective they are, and *exploitation*, in which the system makes use of actions that are known to yield a high reward. Too strong a focus on either exploration or exploitation will yield poor results. Reinforcement learning continues to be an active area of machine learning research. However, a

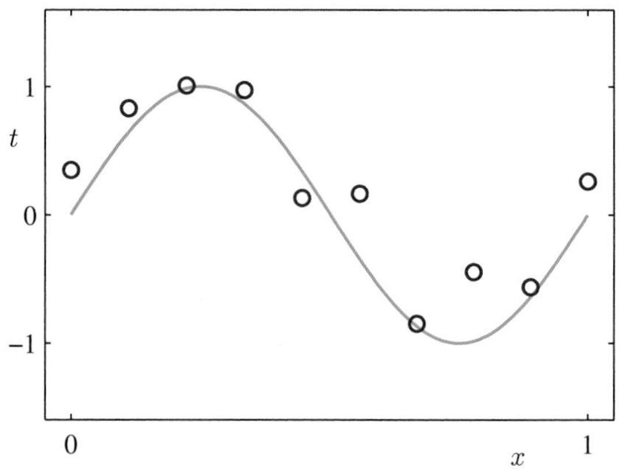

Figure 1.2 Plot of a training data set of $N = 10$ points, shown as blue circles, each comprising an observation of the input variable x along with the corresponding target variable t. The green curve shows the function $\sin(2\pi x)$ used to generate the data. Our goal is to predict the value of t for some new value of x, without knowledge of the green curve.

detailed treatment lies beyond the scope of this book.

Although each of these tasks needs its own tools and techniques, many of the key ideas that underpin them are common to all such problems. One of the main goals of this chapter is to introduce, in a relatively informal way, several of the most important of these concepts and to illustrate them using simple examples. Later in the book we shall see these same ideas re-emerge in the context of more sophisticated models that are applicable to real-world pattern recognition applications. This chapter also provides a self-contained introduction to three important tools that will be used throughout the book, namely probability theory, decision theory, and information theory. Although these might sound like daunting topics, they are in fact straightforward, and a clear understanding of them is essential if machine learning techniques are to be used to best effect in practical applications.

1.1. Example: Polynomial Curve Fitting

We begin by introducing a simple regression problem, which we shall use as a running example throughout this chapter to motivate a number of key concepts. Suppose we observe a real-valued input variable x and we wish to use this observation to predict the value of a real-valued target variable t. For the present purposes, it is instructive to consider an artificial example using synthetically generated data because we then know the precise process that generated the data for comparison against any learned model. The data for this example is generated from the function $\sin(2\pi x)$ with random noise included in the target values, as described in detail in Appendix A.

Now suppose that we are given a training set comprising N observations of x, written $\mathbf{x} \equiv (x_1, \ldots, x_N)^{\mathrm{T}}$, together with corresponding observations of the values of t, denoted $\mathbf{t} \equiv (t_1, \ldots, t_N)^{\mathrm{T}}$. Figure 1.2 shows a plot of a training set comprising $N = 10$ data points. The input data set $\mathbf{x}$ in Figure 1.2 was generated by choosing values of x_n, for $n = 1, \ldots, N$, spaced uniformly in range $[0, 1]$, and the target data set $\mathbf{t}$ was obtained by first computing the corresponding values of the function

$\sin(2\pi x)$ and then adding a small level of random noise having a Gaussian distribution (the Gaussian distribution is discussed in Section 1.2.4) to each such point in order to obtain the corresponding value t_n. By generating data in this way, we are capturing a property of many real data sets, namely that they possess an underlying regularity, which we wish to learn, but that individual observations are corrupted by random noise. This noise might arise from intrinsically stochastic (i.e. random) processes such as radioactive decay but more typically is due to there being sources of variability that are themselves unobserved.

Our goal is to exploit this training set in order to make predictions of the value $\widehat{t}$ of the target variable for some new value $\widehat{x}$ of the input variable. As we shall see later, this involves implicitly trying to discover the underlying function $\sin(2\pi x)$. This is intrinsically a difficult problem as we have to generalize from a finite data set. Furthermore the observed data are corrupted with noise, and so for a given $\widehat{x}$ there is uncertainty as to the appropriate value for $\widehat{t}$. Probability theory, discussed in Section 1.2, provides a framework for expressing such uncertainty in a precise and quantitative manner, and decision theory, discussed in Section 1.5, allows us to exploit this probabilistic representation in order to make predictions that are optimal according to appropriate criteria.

For the moment, however, we shall proceed rather informally and consider a simple approach based on curve fitting. In particular, we shall fit the data using a polynomial function of the form

$$y(x, \mathbf{w}) = w_0 + w_1 x + w_2 x^2 + \ldots + w_M x^M = \sum_{j=0}^{M} w_j x^j \tag{1.1}$$

where M is the *order* of the polynomial, and x^j denotes x raised to the power of j. The polynomial coefficients $w_0, \ldots, w_M$ are collectively denoted by the vector $\mathbf{w}$. Note that, although the polynomial function $y(x, \mathbf{w})$ is a nonlinear function of x, it is a linear function of the coefficients $\mathbf{w}$. Functions, such as the polynomial, which are linear in the unknown parameters have important properties and are called *linear models* and will be discussed extensively in Chapters 3 and 4.

The values of the coefficients will be determined by fitting the polynomial to the training data. This can be done by minimizing an *error function* that measures the misfit between the function $y(x, \mathbf{w})$, for any given value of $\mathbf{w}$, and the training set data points. One simple choice of error function, which is widely used, is given by the sum of the squares of the errors between the predictions $y(x_n, \mathbf{w})$ for each data point x_n and the corresponding target values t_n, so that we minimize

$$E(\mathbf{w}) = \frac{1}{2} \sum_{n=1}^{N} \{y(x_n, \mathbf{w}) - t_n\}^2 \tag{1.2}$$

where the factor of $1/2$ is included for later convenience. We shall discuss the motivation for this choice of error function later in this chapter. For the moment we simply note that it is a nonnegative quantity that would be zero if, and only if, the

Figure 1.3 The error function (1.2) corresponds to (one half of) the sum of the squares of the displacements (shown by the vertical green bars) of each data point from the function $y(x, \mathbf{w})$.

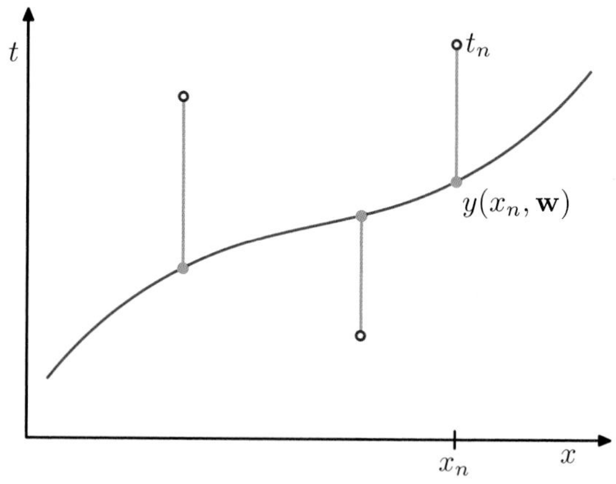

function $y(x, \mathbf{w})$ were to pass exactly through each training data point. The geometrical interpretation of the sum-of-squares error function is illustrated in Figure 1.3.

We can solve the curve fitting problem by choosing the value of $\mathbf{w}$ for which $E(\mathbf{w})$ is as small as possible. Because the error function is a quadratic function of the coefficients $\mathbf{w}$, its derivatives with respect to the coefficients will be linear in the elements of $\mathbf{w}$, and so the minimization of the error function has a unique solution, denoted by $\mathbf{w}^\star$, which can be found in closed form. The resulting polynomial is given by the function $y(x, \mathbf{w}^\star)$.

Exercise 1.1

There remains the problem of choosing the order M of the polynomial, and as we shall see this will turn out to be an example of an important concept called *model comparison* or *model selection*. In Figure 1.4, we show four examples of the results of fitting polynomials having orders $M = 0, 1, 3$, and 9 to the data set shown in Figure 1.2.

We notice that the constant ($M = 0$) and first order ($M = 1$) polynomials give rather poor fits to the data and consequently rather poor representations of the function $\sin(2\pi x)$. The third order ($M = 3$) polynomial seems to give the best fit to the function $\sin(2\pi x)$ of the examples shown in Figure 1.4. When we go to a much higher order polynomial ($M = 9$), we obtain an excellent fit to the training data. In fact, the polynomial passes exactly through each data point and $E(\mathbf{w}^\star) = 0$. However, the fitted curve oscillates wildly and gives a very poor representation of the function $\sin(2\pi x)$. This latter behaviour is known as *over-fitting*.

As we have noted earlier, the goal is to achieve good generalization by making accurate predictions for new data. We can obtain some quantitative insight into the dependence of the generalization performance on M by considering a separate test set comprising 100 data points generated using exactly the same procedure used to generate the training set points but with new choices for the random noise values included in the target values. For each choice of M, we can then evaluate the residual value of $E(\mathbf{w}^\star)$ given by (1.2) for the training data, and we can also evaluate $E(\mathbf{w}^\star)$ for the test data set. It is sometimes more convenient to use the root-mean-square

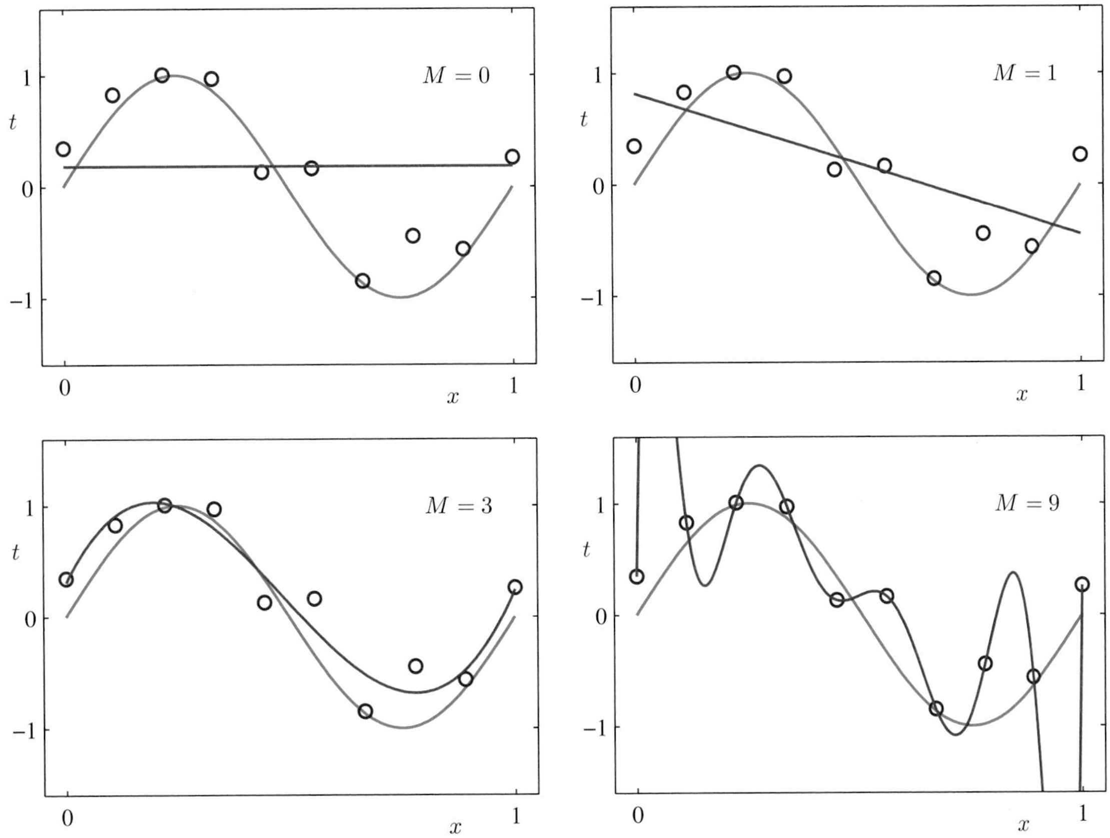

Figure 1.4 Plots of polynomials having various orders M, shown as red curves, fitted to the data set shown in Figure 1.2.

(RMS) error defined by

$$E_{\mathrm{RMS}} = \sqrt{2E(\mathbf{w}^\star)/N} \tag{1.3}$$

in which the division by N allows us to compare different sizes of data sets on an equal footing, and the square root ensures that E_{RMS} is measured on the same scale (and in the same units) as the target variable t. Graphs of the training and test set RMS errors are shown, for various values of M, in Figure 1.5. The test set error is a measure of how well we are doing in predicting the values of t for new data observations of x. We note from Figure 1.5 that small values of M give relatively large values of the test set error, and this can be attributed to the fact that the corresponding polynomials are rather inflexible and are incapable of capturing the oscillations in the function $\sin(2\pi x)$. Values of M in the range $3 \leqslant M \leqslant 8$ give small values for the test set error, and these also give reasonable representations of the generating function $\sin(2\pi x)$, as can be seen, for the case of $M = 3$, from Figure 1.4.

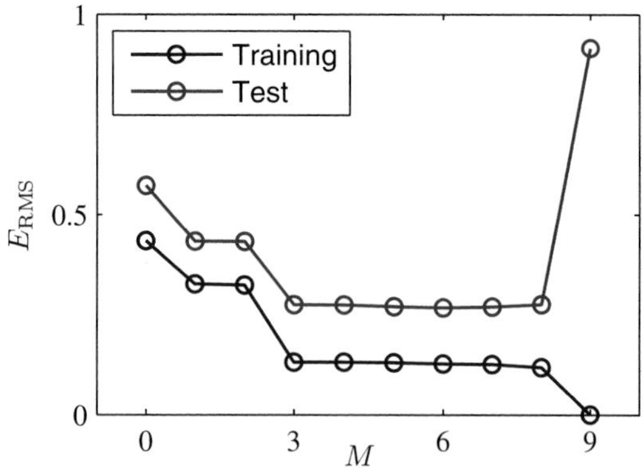

Figure 1.5 Graphs of the root-mean-square error, defined by (1.3), evaluated on the training set and on an independent test set for various values of M.

For $M = 9$, the training set error goes to zero, as we might expect because this polynomial contains 10 degrees of freedom corresponding to the 10 coefficients $w_0, \ldots, w_9$, and so can be tuned exactly to the 10 data points in the training set. However, the test set error has become very large and, as we saw in Figure 1.4, the corresponding function $y(x, \mathbf{w}^\star)$ exhibits wild oscillations.

This may seem paradoxical because a polynomial of given order contains all lower order polynomials as special cases. The $M = 9$ polynomial is therefore capable of generating results at least as good as the $M = 3$ polynomial. Furthermore, we might suppose that the best predictor of new data would be the function $\sin(2\pi x)$ from which the data was generated (and we shall see later that this is indeed the case). We know that a power series expansion of the function $\sin(2\pi x)$ contains terms of all orders, so we might expect that results should improve monotonically as we increase M.

We can gain some insight into the problem by examining the values of the coefficients $\mathbf{w}^\star$ obtained from polynomials of various order, as shown in Table 1.1. We see that, as M increases, the magnitude of the coefficients typically gets larger. In particular for the $M = 9$ polynomial, the coefficients have become finely tuned to the data by developing large positive and negative values so that the correspond-

Table 1.1 Table of the coefficients $\mathbf{w}^\star$ for polynomials of various order. Observe how the typical magnitude of the coefficients increases dramatically as the order of the polynomial increases.

	$M = 0$	$M = 1$	$M = 3$	$M = 9$
$w_0^\star$	0.19	0.82	0.31	0.35
$w_1^\star$		-1.27	7.99	232.37
$w_2^\star$			-25.43	-5321.83
$w_3^\star$			17.37	48568.31
$w_4^\star$				-231639.30
$w_5^\star$				640042.26
$w_6^\star$				-1061800.52
$w_7^\star$				1042400.18
$w_8^\star$				-557682.99
$w_9^\star$				125201.43

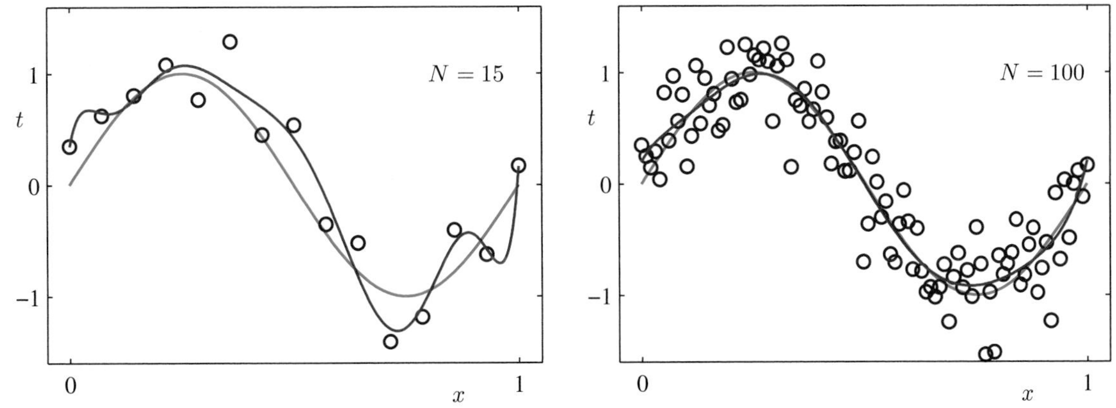

Figure 1.6 Plots of the solutions obtained by minimizing the sum-of-squares error function using the $M = 9$ polynomial for $N = 15$ data points (left plot) and $N = 100$ data points (right plot). We see that increasing the size of the data set reduces the over-fitting problem.

ing polynomial function matches each of the data points exactly, but between data points (particularly near the ends of the range) the function exhibits the large oscillations observed in Figure 1.4. Intuitively, what is happening is that the more flexible polynomials with larger values of M are becoming increasingly tuned to the random noise on the target values.

It is also interesting to examine the behaviour of a given model as the size of the data set is varied, as shown in Figure 1.6. We see that, for a given model complexity, the over-fitting problem become less severe as the size of the data set increases. Another way to say this is that the larger the data set, the more complex (in other words more flexible) the model that we can afford to fit to the data. One rough heuristic that is sometimes advocated is that the number of data points should be no less than some multiple (say 5 or 10) of the number of adaptive parameters in the model. However, as we shall see in Chapter 3, the number of parameters is not necessarily the most appropriate measure of model complexity.

Also, there is something rather unsatisfying about having to limit the number of parameters in a model according to the size of the available training set. It would seem more reasonable to choose the complexity of the model according to the complexity of the problem being solved. We shall see that the least squares approach to finding the model parameters represents a specific case of *maximum likelihood* (discussed in Section 1.2.5), and that the over-fitting problem can be understood as a general property of maximum likelihood. By adopting a *Bayesian* approach, the over-fitting problem can be avoided. We shall see that there is no difficulty from a Bayesian perspective in employing models for which the number of parameters greatly exceeds the number of data points. Indeed, in a Bayesian model the *effective* number of parameters adapts automatically to the size of the data set.

Section 3.4

For the moment, however, it is instructive to continue with the current approach and to consider how in practice we can apply it to data sets of limited size where we

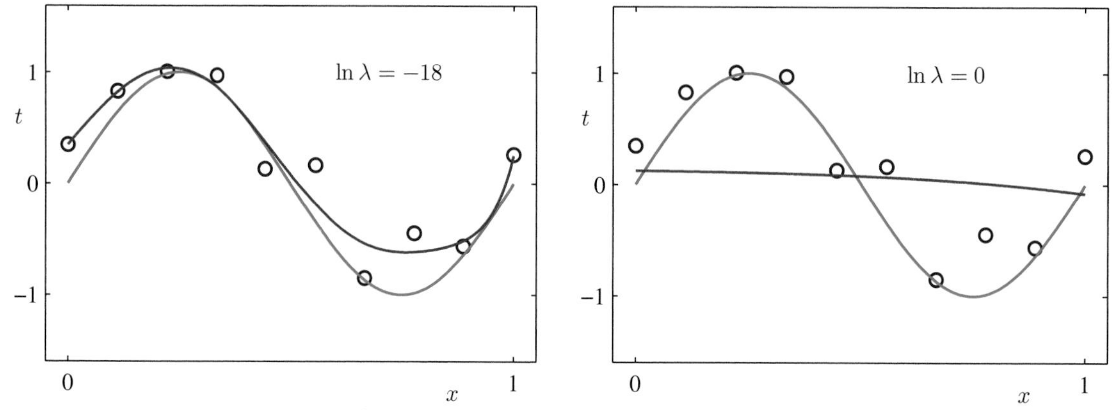

Figure 1.7 Plots of $M = 9$ polynomials fitted to the data set shown in Figure 1.2 using the regularized error function (1.4) for two values of the regularization parameter λ corresponding to $\ln \lambda = -18$ and $\ln \lambda = 0$. The case of no regularizer, i.e., $\lambda = 0$, corresponding to $\ln \lambda = -\infty$, is shown at the bottom right of Figure 1.4.

may wish to use relatively complex and flexible models. One technique that is often used to control the over-fitting phenomenon in such cases is that of *regularization*, which involves adding a penalty term to the error function (1.2) in order to discourage the coefficients from reaching large values. The simplest such penalty term takes the form of a sum of squares of all of the coefficients, leading to a modified error function of the form

$$\widetilde{E}(\mathbf{w}) = \frac{1}{2}\sum_{n=1}^{N}\{y(x_n, \mathbf{w}) - t_n\}^2 + \frac{\lambda}{2}\|\mathbf{w}\|^2 \tag{1.4}$$

where $\|\mathbf{w}\|^2 \equiv \mathbf{w}^{\mathrm{T}}\mathbf{w} = w_0^2 + w_1^2 + \ldots + w_M^2$, and the coefficient λ governs the relative importance of the regularization term compared with the sum-of-squares error term. Note that often the coefficient w_0 is omitted from the regularizer because its inclusion causes the results to depend on the choice of origin for the target variable (Hastie *et al.*, 2001), or it may be included but with its own regularization coefficient (we shall discuss this topic in more detail in Section 5.5.1). Again, the error function in (1.4) can be minimized exactly in closed form. Techniques such as this are known in the statistics literature as *shrinkage* methods because they reduce the value of the coefficients. The particular case of a quadratic regularizer is called *ridge regression* (Hoerl and Kennard, 1970). In the context of neural networks, this approach is known as *weight decay*.

Exercise 1.2

Figure 1.7 shows the results of fitting the polynomial of order $M = 9$ to the same data set as before but now using the regularized error function given by (1.4). We see that, for a value of $\ln \lambda = -18$, the over-fitting has been suppressed and we now obtain a much closer representation of the underlying function $\sin(2\pi x)$. If, however, we use too large a value for λ then we again obtain a poor fit, as shown in Figure 1.7 for $\ln \lambda = 0$. The corresponding coefficients from the fitted polynomials are given in Table 1.2, showing that regularization has the desired effect of reducing

Table 1.2 Table of the coefficients $\mathbf{w}^\star$ for $M = 9$ polynomials with various values for the regularization parameter λ. Note that $\ln \lambda = -\infty$ corresponds to a model with no regularization, i.e., to the graph at the bottom right in Figure 1.4. We see that, as the value of λ increases, the typical magnitude of the coefficients gets smaller.

	$\ln \lambda = -\infty$	$\ln \lambda = -18$	$\ln \lambda = 0$
$w_0^\star$	0.35	0.35	0.13
$w_1^\star$	232.37	4.74	-0.05
$w_2^\star$	-5321.83	-0.77	-0.06
$w_3^\star$	48568.31	-31.97	-0.05
$w_4^\star$	-231639.30	-3.89	-0.03
$w_5^\star$	640042.26	55.28	-0.02
$w_6^\star$	-1061800.52	41.32	-0.01
$w_7^\star$	1042400.18	-45.95	-0.00
$w_8^\star$	-557682.99	-91.53	0.00
$w_9^\star$	125201.43	72.68	0.01

the magnitude of the coefficients.

The impact of the regularization term on the generalization error can be seen by plotting the value of the RMS error (1.3) for both training and test sets against $\ln \lambda$, as shown in Figure 1.8. We see that in effect λ now controls the effective complexity of the model and hence determines the degree of over-fitting.

The issue of model complexity is an important one and will be discussed at length in Section 1.3. Here we simply note that, if we were trying to solve a practical application using this approach of minimizing an error function, we would have to find a way to determine a suitable value for the model complexity. The results above suggest a simple way of achieving this, namely by taking the available data and partitioning it into a training set, used to determine the coefficients $\mathbf{w}$, and a separate *validation* set, also called a *hold-out* set, used to optimize the model complexity (either M or λ). In many cases, however, this will prove to be too wasteful of valuable training data, and we have to seek more sophisticated approaches.

Section 1.3

So far our discussion of polynomial curve fitting has appealed largely to intuition. We now seek a more principled approach to solving problems in pattern recognition by turning to a discussion of probability theory. As well as providing the foundation for nearly all of the subsequent developments in this book, it will also

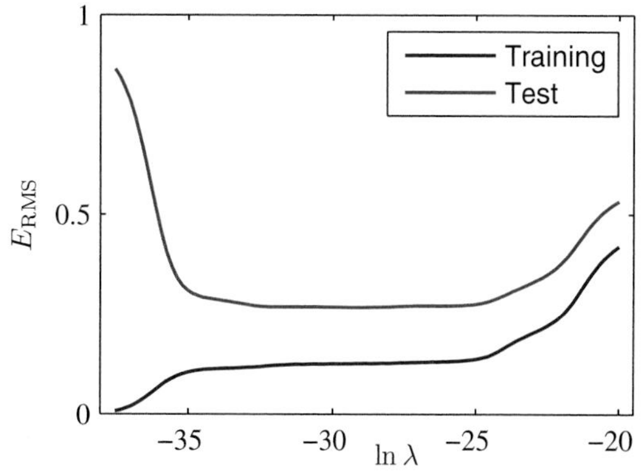

Figure 1.8 Graph of the root-mean-square error (1.3) versus $\ln \lambda$ for the $M = 9$ polynomial.

give us some important insights into the concepts we have introduced in the context of polynomial curve fitting and will allow us to extend these to more complex situations.

1.2. Probability Theory

A key concept in the field of pattern recognition is that of uncertainty. It arises both through noise on measurements, as well as through the finite size of data sets. Probability theory provides a consistent framework for the quantification and manipulation of uncertainty and forms one of the central foundations for pattern recognition. When combined with decision theory, discussed in Section 1.5, it allows us to make optimal predictions given all the information available to us, even though that information may be incomplete or ambiguous.

We will introduce the basic concepts of probability theory by considering a simple example. Imagine we have two boxes, one red and one blue, and in the red box we have 2 apples and 6 oranges, and in the blue box we have 3 apples and 1 orange. This is illustrated in Figure 1.9. Now suppose we randomly pick one of the boxes and from that box we randomly select an item of fruit, and having observed which sort of fruit it is we replace it in the box from which it came. We could imagine repeating this process many times. Let us suppose that in so doing we pick the red box 40% of the time and we pick the blue box 60% of the time, and that when we remove an item of fruit from a box we are equally likely to select any of the pieces of fruit in the box.

In this example, the identity of the box that will be chosen is a random variable, which we shall denote by B. This random variable can take one of two possible values, namely r (corresponding to the red box) or b (corresponding to the blue box). Similarly, the identity of the fruit is also a random variable and will be denoted by F. It can take either of the values a (for apple) or o (for orange).

To begin with, we shall define the probability of an event to be the fraction of times that event occurs out of the total number of trials, in the limit that the total number of trials goes to infinity. Thus the probability of selecting the red box is $4/10$

Figure 1.9 We use a simple example of two coloured boxes each containing fruit (apples shown in green and oranges shown in orange) to introduce the basic ideas of probability.

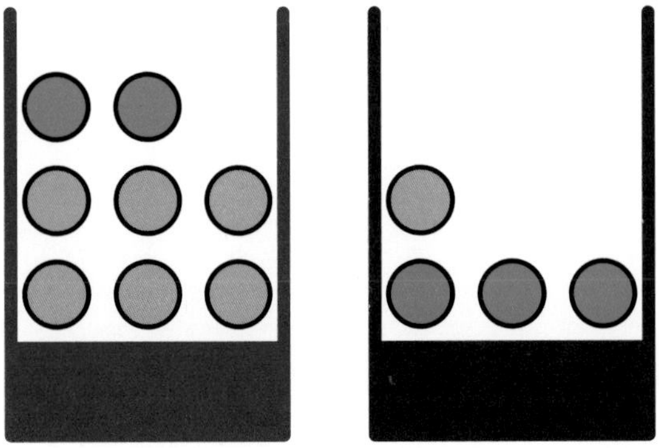

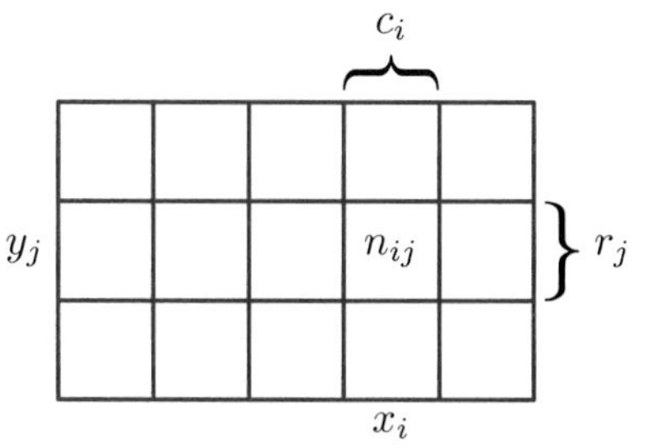

Figure 1.10 We can derive the sum and product rules of probability by considering two random variables, X, which takes the values $\{x_i\}$ where $i = 1, \ldots, M$, and Y, which takes the values $\{y_j\}$ where $j = 1, \ldots, L$. In this illustration we have $M = 5$ and $L = 3$. If we consider a total number N of instances of these variables, then we denote the number of instances where $X = x_i$ and $Y = y_j$ by n_{ij}, which is the number of points in the corresponding cell of the array. The number of points in column i, corresponding to $X = x_i$, is denoted by c_i, and the number of points in row j, corresponding to $Y = y_j$, is denoted by r_j.

and the probability of selecting the blue box is $6/10$. We write these probabilities as $p(B = r) = 4/10$ and $p(B = b) = 6/10$. Note that, by definition, probabilities must lie in the interval $[0, 1]$. Also, if the events are mutually exclusive and if they include all possible outcomes (for instance, in this example the box must be either red or blue), then we see that the probabilities for those events must sum to one.

We can now ask questions such as: "what is the overall probability that the selection procedure will pick an apple?", or "given that we have chosen an orange, what is the probability that the box we chose was the blue one?". We can answer questions such as these, and indeed much more complex questions associated with problems in pattern recognition, once we have equipped ourselves with the two elementary rules of probability, known as the *sum rule* and the *product rule*. Having obtained these rules, we shall then return to our boxes of fruit example.

In order to derive the rules of probability, consider the slightly more general example shown in Figure 1.10 involving two random variables X and Y (which could for instance be the Box and Fruit variables considered above). We shall suppose that X can take any of the values x_i where $i = 1, \ldots, M$, and Y can take the values y_j where $j = 1, \ldots, L$. Consider a total of N trials in which we sample both of the variables X and Y, and let the number of such trials in which $X = x_i$ and $Y = y_j$ be n_{ij}. Also, let the number of trials in which X takes the value x_i (irrespective of the value that Y takes) be denoted by c_i, and similarly let the number of trials in which Y takes the value y_j be denoted by r_j.

The probability that X will take the value x_i and Y will take the value y_j is written $p(X = x_i, Y = y_j)$ and is called the *joint* probability of $X = x_i$ and $Y = y_j$. It is given by the number of points falling in the cell i,j as a fraction of the total number of points, and hence

$$p(X = x_i, Y = y_j) = \frac{n_{ij}}{N}. \tag{1.5}$$

Here we are implicitly considering the limit $N \to \infty$. Similarly, the probability that X takes the value x_i irrespective of the value of Y is written as $p(X = x_i)$ and is given by the fraction of the total number of points that fall in column i, so that

$$p(X = x_i) = \frac{c_i}{N}. \tag{1.6}$$

Because the number of instances in column i in Figure 1.10 is just the sum of the number of instances in each cell of that column, we have $c_i = \sum_j n_{ij}$ and therefore, from (1.5) and (1.6), we have

$$p(X = x_i) = \sum_{j=1}^{L} p(X = x_i, Y = y_j) \tag{1.7}$$

which is the *sum rule* of probability. Note that $p(X = x_i)$ is sometimes called the *marginal* probability, because it is obtained by marginalizing, or summing out, the other variables (in this case Y).

If we consider only those instances for which $X = x_i$, then the fraction of such instances for which $Y = y_j$ is written $p(Y = y_j|X = x_i)$ and is called the *conditional* probability of $Y = y_j$ given $X = x_i$. It is obtained by finding the fraction of those points in column i that fall in cell i,j and hence is given by

$$p(Y = y_j|X = x_i) = \frac{n_{ij}}{c_i}. \tag{1.8}$$

From (1.5), (1.6), and (1.8), we can then derive the following relationship

$$\begin{aligned} p(X = x_i, Y = y_j) &= \frac{n_{ij}}{N} = \frac{n_{ij}}{c_i} \cdot \frac{c_i}{N} \\ &= p(Y = y_j|X = x_i)p(X = x_i) \end{aligned} \tag{1.9}$$

which is the *product rule* of probability.

So far we have been quite careful to make a distinction between a random variable, such as the box B in the fruit example, and the values that the random variable can take, for example r if the box were the red one. Thus the probability that B takes the value r is denoted $p(B = r)$. Although this helps to avoid ambiguity, it leads to a rather cumbersome notation, and in many cases there will be no need for such pedantry. Instead, we may simply write $p(B)$ to denote a distribution over the random variable B, or $p(r)$ to denote the distribution evaluated for the particular value r, provided that the interpretation is clear from the context.

With this more compact notation, we can write the two fundamental rules of probability theory in the following form.

The Rules of Probability

sum rule $$p(X) = \sum_{Y} p(X, Y) \tag{1.10}$$

product rule $$p(X, Y) = p(Y|X)p(X). \tag{1.11}$$

Here $p(X, Y)$ is a joint probability and is verbalized as "the probability of X *and* Y". Similarly, the quantity $p(Y|X)$ is a conditional probability and is verbalized as "the probability of Y *given* X", whereas the quantity $p(X)$ is a marginal probability and is simply "the probability of X". These two simple rules form the basis for all of the probabilistic machinery that we use throughout this book.

From the product rule, together with the symmetry property $p(X, Y) = p(Y, X)$, we immediately obtain the following relationship between conditional probabilities

$$p(Y|X) = \frac{p(X|Y)p(Y)}{p(X)} \tag{1.12}$$

which is called *Bayes' theorem* and which plays a central role in pattern recognition and machine learning. Using the sum rule, the denominator in Bayes' theorem can be expressed in terms of the quantities appearing in the numerator

$$p(X) = \sum_Y p(X|Y)p(Y). \tag{1.13}$$

We can view the denominator in Bayes' theorem as being the normalization constant required to ensure that the sum of the conditional probability on the left-hand side of (1.12) over all values of Y equals one.

In Figure 1.11, we show a simple example involving a joint distribution over two variables to illustrate the concept of marginal and conditional distributions. Here a finite sample of $N = 60$ data points has been drawn from the joint distribution and is shown in the top left. In the top right is a histogram of the fractions of data points having each of the two values of Y. From the definition of probability, these fractions would equal the corresponding probabilities $p(Y)$ in the limit $N \to \infty$. We can view the histogram as a simple way to model a probability distribution given only a finite number of points drawn from that distribution. Modelling distributions from data lies at the heart of statistical pattern recognition and will be explored in great detail in this book. The remaining two plots in Figure 1.11 show the corresponding histogram estimates of $p(X)$ and $p(X|Y = 1)$.

Let us now return to our example involving boxes of fruit. For the moment, we shall once again be explicit about distinguishing between the random variables and their instantiations. We have seen that the probabilities of selecting either the red or the blue boxes are given by

$$p(B = r) = 4/10 \tag{1.14}$$
$$p(B = b) = 6/10 \tag{1.15}$$

respectively. Note that these satisfy $p(B = r) + p(B = b) = 1$.

Now suppose that we pick a box at random, and it turns out to be the blue box. Then the probability of selecting an apple is just the fraction of apples in the blue box which is $3/4$, and so $p(F = a|B = b) = 3/4$. In fact, we can write out all four conditional probabilities for the type of fruit, given the selected box

$$p(F = a|B = r) = 1/4 \tag{1.16}$$
$$p(F = o|B = r) = 3/4 \tag{1.17}$$
$$p(F = a|B = b) = 3/4 \tag{1.18}$$
$$p(F = o|B = b) = 1/4. \tag{1.19}$$

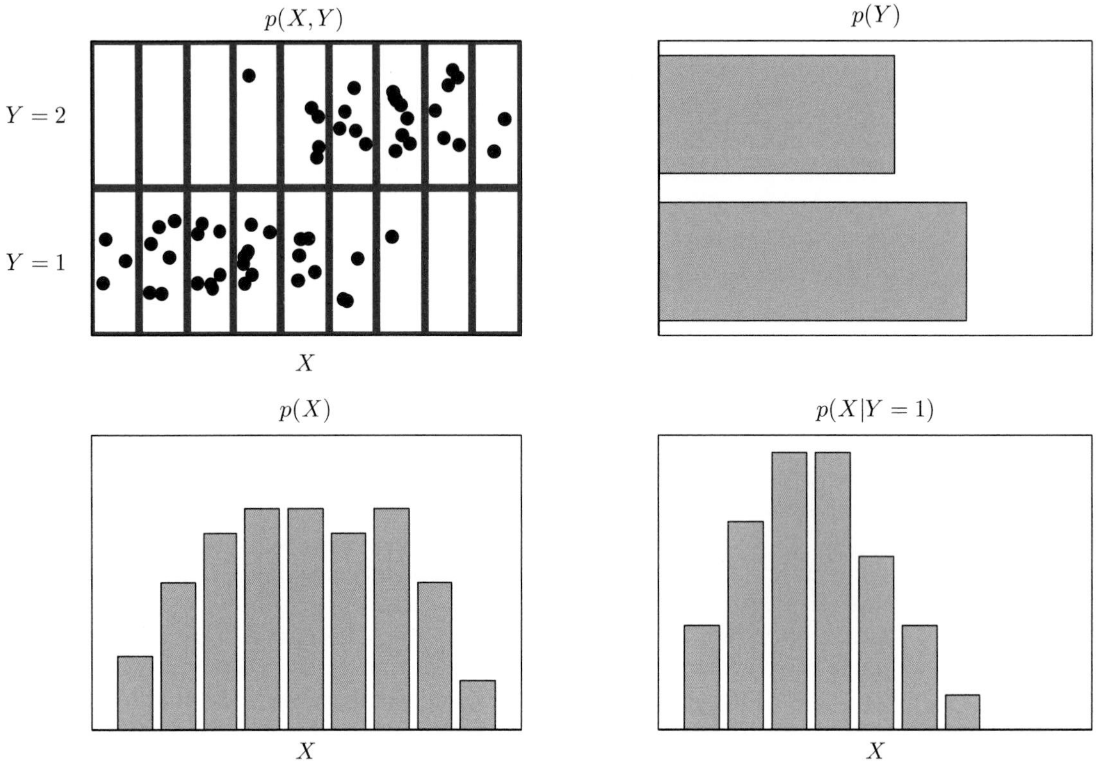

Figure 1.11 An illustration of a distribution over two variables, X, which takes 9 possible values, and Y, which takes two possible values. The top left figure shows a sample of 60 points drawn from a joint probability distribution over these variables. The remaining figures show histogram estimates of the marginal distributions $p(X)$ and $p(Y)$, as well as the conditional distribution $p(X|Y = 1)$ corresponding to the bottom row in the top left figure.

Again, note that these probabilities are normalized so that

$$p(F = a|B = r) + p(F = o|B = r) = 1 \tag{1.20}$$

and similarly

$$p(F = a|B = b) + p(F = o|B = b) = 1. \tag{1.21}$$

We can now use the sum and product rules of probability to evaluate the overall probability of choosing an apple

$$\begin{aligned} p(F = a) &= p(F = a|B = r)p(B = r) + p(F = a|B = b)p(B = b) \\ &= \frac{1}{4} \times \frac{4}{10} + \frac{3}{4} \times \frac{6}{10} = \frac{11}{20} \end{aligned} \tag{1.22}$$

from which it follows, using the sum rule, that $p(F = o) = 1 - 11/20 = 9/20$.

Suppose instead we are told that a piece of fruit has been selected and it is an orange, and we would like to know which box it came from. This requires that we evaluate the probability distribution over boxes conditioned on the identity of the fruit, whereas the probabilities in (1.16)–(1.19) give the probability distribution over the fruit conditioned on the identity of the box. We can solve the problem of reversing the conditional probability by using Bayes' theorem to give

$$p(B = r|F = o) = \frac{p(F = o|B = r)p(B = r)}{p(F = o)} = \frac{3}{4} \times \frac{4}{10} \times \frac{20}{9} = \frac{2}{3}. \tag{1.23}$$

From the sum rule, it then follows that $p(B = b|F = o) = 1 - 2/3 = 1/3$.

We can provide an important interpretation of Bayes' theorem as follows. If we had been asked which box had been chosen before being told the identity of the selected item of fruit, then the most complete information we have available is provided by the probability $p(B)$. We call this the *prior probability* because it is the probability available *before* we observe the identity of the fruit. Once we are told that the fruit is an orange, we can then use Bayes' theorem to compute the probability $p(B|F)$, which we shall call the *posterior probability* because it is the probability obtained *after* we have observed F. Note that in this example, the prior probability of selecting the red box was $4/10$, so that we were more likely to select the blue box than the red one. However, once we have observed that the piece of selected fruit is an orange, we find that the posterior probability of the red box is now $2/3$, so that it is now more likely that the box we selected was in fact the red one. This result accords with our intuition, as the proportion of oranges is much higher in the red box than it is in the blue box, and so the observation that the fruit was an orange provides significant evidence favouring the red box. In fact, the evidence is sufficiently strong that it outweighs the prior and makes it more likely that the red box was chosen rather than the blue one.

Finally, we note that if the joint distribution of two variables factorizes into the product of the marginals, so that $p(X, Y) = p(X)p(Y)$, then X and Y are said to be *independent*. From the product rule, we see that $p(Y|X) = p(Y)$, and so the conditional distribution of Y given X is indeed independent of the value of X. For instance, in our boxes of fruit example, if each box contained the same fraction of apples and oranges, then $p(F|B) = P(F)$, so that the probability of selecting, say, an apple is independent of which box is chosen.

1.2.1 Probability densities

As well as considering probabilities defined over discrete sets of events, we also wish to consider probabilities with respect to continuous variables. We shall limit ourselves to a relatively informal discussion. If the probability of a real-valued variable x falling in the interval $(x, x + \delta x)$ is given by $p(x)\delta x$ for $\delta x \to 0$, then $p(x)$ is called the *probability density* over x. This is illustrated in Figure 1.12. The probability that x will lie in an interval (a, b) is then given by

$$p(x \in (a, b)) = \int_a^b p(x)\,\mathrm{d}x. \tag{1.24}$$

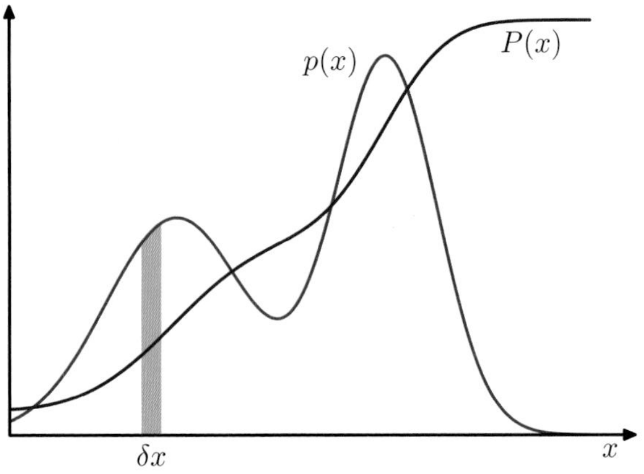

Figure 1.12 The concept of probability for discrete variables can be extended to that of a probability density $p(x)$ over a continuous variable x and is such that the probability of x lying in the interval $(x, x + \delta x)$ is given by $p(x)\delta x$ for $\delta x \rightarrow 0$. The probability density can be expressed as the derivative of a cumulative distribution function $P(x)$.

Because probabilities are nonnegative, and because the value of x must lie somewhere on the real axis, the probability density $p(x)$ must satisfy the two conditions

$$p(x) \geqslant 0 \tag{1.25}$$

$$\int_{-\infty}^{\infty} p(x)\,\mathrm{d}x = 1. \tag{1.26}$$

Under a nonlinear change of variable, a probability density transforms differently from a simple function, due to the Jacobian factor. For instance, if we consider a change of variables $x = g(y)$, then a function $f(x)$ becomes $\widetilde{f}(y) = f(g(y))$. Now consider a probability density $p_x(x)$ that corresponds to a density $p_y(y)$ with respect to the new variable y, where the suffixes denote the fact that $p_x(x)$ and $p_y(y)$ are different densities. Observations falling in the range $(x, x + \delta x)$ will, for small values of δx, be transformed into the range $(y, y + \delta y)$ where $p_x(x)\delta x \simeq p_y(y)\delta y$, and hence

$$\begin{aligned} p_y(y) &= p_x(x)\left|\frac{\mathrm{d}x}{\mathrm{d}y}\right| \\ &= p_x(g(y))\,|g'(y)|\,. \end{aligned} \tag{1.27}$$

One consequence of this property is that the concept of the maximum of a probability density is dependent on the choice of variable.

Exercise 1.4

The probability that x lies in the interval $(-\infty, z)$ is given by the *cumulative distribution function* defined by

$$P(z) = \int_{-\infty}^{z} p(x)\,\mathrm{d}x \tag{1.28}$$

which satisfies $P'(x) = p(x)$, as shown in Figure 1.12.

If we have several continuous variables $x_1, \ldots, x_D$, denoted collectively by the vector $\mathbf{x}$, then we can define a joint probability density $p(\mathbf{x}) = p(x_1, \ldots, x_D)$ such that the probability of $\mathbf{x}$ falling in an infinitesimal volume $\delta\mathbf{x}$ containing the point $\mathbf{x}$ is given by $p(\mathbf{x})\delta\mathbf{x}$. This multivariate probability density must satisfy

$$p(\mathbf{x}) \geqslant 0 \tag{1.29}$$

$$\int p(\mathbf{x})\,\mathrm{d}\mathbf{x} = 1 \tag{1.30}$$

in which the integral is taken over the whole of $\mathbf{x}$ space. We can also consider joint probability distributions over a combination of discrete and continuous variables.

Note that if x is a discrete variable, then $p(x)$ is sometimes called a *probability mass function* because it can be regarded as a set of 'probability masses' concentrated at the allowed values of x.

The sum and product rules of probability, as well as Bayes' theorem, apply equally to the case of probability densities, or to combinations of discrete and continuous variables. For instance, if x and y are two real variables, then the sum and product rules take the form

$$p(x) = \int p(x,y)\,\mathrm{d}y \tag{1.31}$$

$$p(x,y) = p(y|x)p(x). \tag{1.32}$$

A formal justification of the sum and product rules for continuous variables (Feller, 1966) requires a branch of mathematics called *measure theory* and lies outside the scope of this book. Its validity can be seen informally, however, by dividing each real variable into intervals of width Δ and considering the discrete probability distribution over these intervals. Taking the limit $\Delta \to 0$ then turns sums into integrals and gives the desired result.

1.2.2 Expectations and covariances

One of the most important operations involving probabilities is that of finding weighted averages of functions. The average value of some function $f(x)$ under a probability distribution $p(x)$ is called the *expectation* of $f(x)$ and will be denoted by $\mathbb{E}[f]$. For a discrete distribution, it is given by

$$\mathbb{E}[f] = \sum_x p(x)f(x) \tag{1.33}$$

so that the average is weighted by the relative probabilities of the different values of x. In the case of continuous variables, expectations are expressed in terms of an integration with respect to the corresponding probability density

$$\mathbb{E}[f] = \int p(x)f(x)\,\mathrm{d}x. \tag{1.34}$$

In either case, if we are given a finite number N of points drawn from the probability distribution or probability density, then the expectation can be approximated as a finite sum over these points

$$\mathbb{E}[f] \simeq \frac{1}{N} \sum_{n=1}^{N} f(x_n). \tag{1.35}$$

We shall make extensive use of this result when we discuss sampling methods in Chapter 11. The approximation in (1.35) becomes exact in the limit $N \to \infty$.

Sometimes we will be considering expectations of functions of several variables, in which case we can use a subscript to indicate which variable is being averaged over, so that for instance

$$\mathbb{E}_x[f(x, y)] \tag{1.36}$$

denotes the average of the function $f(x, y)$ with respect to the distribution of x. Note that $\mathbb{E}_x[f(x, y)]$ will be a function of y.

We can also consider a *conditional expectation* with respect to a conditional distribution, so that

$$\mathbb{E}_x[f|y] = \sum_x p(x|y) f(x) \tag{1.37}$$

with an analogous definition for continuous variables.

The *variance* of $f(x)$ is defined by

$$\text{var}[f] = \mathbb{E}\left[(f(x) - \mathbb{E}[f(x)])^2\right] \tag{1.38}$$

and provides a measure of how much variability there is in $f(x)$ around its mean value $\mathbb{E}[f(x)]$. Expanding out the square, we see that the variance can also be written in terms of the expectations of $f(x)$ and $f(x)^2$

Exercise 1.5

$$\text{var}[f] = \mathbb{E}[f(x)^2] - \mathbb{E}[f(x)]^2. \tag{1.39}$$

In particular, we can consider the variance of the variable x itself, which is given by

$$\text{var}[x] = \mathbb{E}[x^2] - \mathbb{E}[x]^2. \tag{1.40}$$

For two random variables x and y, the *covariance* is defined by

$$\begin{aligned} \text{cov}[x, y] &= \mathbb{E}_{x,y}\left[\{x - \mathbb{E}[x]\}\{y - \mathbb{E}[y]\}\right] \\ &= \mathbb{E}_{x,y}[xy] - \mathbb{E}[x]\mathbb{E}[y] \end{aligned} \tag{1.41}$$

which expresses the extent to which x and y vary together. If x and y are independent, then their covariance vanishes.

Exercise 1.6

In the case of two vectors of random variables $\mathbf{x}$ and $\mathbf{y}$, the covariance is a matrix

$$\begin{aligned} \text{cov}[\mathbf{x}, \mathbf{y}] &= \mathbb{E}_{\mathbf{x},\mathbf{y}}\left[\{\mathbf{x} - \mathbb{E}[\mathbf{x}]\}\{\mathbf{y}^{\mathrm{T}} - \mathbb{E}[\mathbf{y}^{\mathrm{T}}]\}\right] \\ &= \mathbb{E}_{\mathbf{x},\mathbf{y}}[\mathbf{x}\mathbf{y}^{\mathrm{T}}] - \mathbb{E}[\mathbf{x}]\mathbb{E}[\mathbf{y}^{\mathrm{T}}]. \end{aligned} \tag{1.42}$$

If we consider the covariance of the components of a vector $\mathbf{x}$ with each other, then we use a slightly simpler notation $\text{cov}[\mathbf{x}] \equiv \text{cov}[\mathbf{x}, \mathbf{x}]$.

1.2.3 Bayesian probabilities

So far in this chapter, we have viewed probabilities in terms of the frequencies of random, repeatable events. We shall refer to this as the *classical* or *frequentist* interpretation of probability. Now we turn to the more general *Bayesian* view, in which probabilities provide a quantification of uncertainty.

Consider an uncertain event, for example whether the moon was once in its own orbit around the sun, or whether the Arctic ice cap will have disappeared by the end of the century. These are not events that can be repeated numerous times in order to define a notion of probability as we did earlier in the context of boxes of fruit. Nevertheless, we will generally have some idea, for example, of how quickly we think the polar ice is melting. If we now obtain fresh evidence, for instance from a new Earth observation satellite gathering novel forms of diagnostic information, we may revise our opinion on the rate of ice loss. Our assessment of such matters will affect the actions we take, for instance the extent to which we endeavour to reduce the emission of greenhouse gasses. In such circumstances, we would like to be able to quantify our expression of uncertainty and make precise revisions of uncertainty in the light of new evidence, as well as subsequently to be able to take optimal actions or decisions as a consequence. This can all be achieved through the elegant, and very general, Bayesian interpretation of probability.

The use of probability to represent uncertainty, however, is not an ad-hoc choice, but is inevitable if we are to respect common sense while making rational coherent inferences. For instance, Cox (1946) showed that if numerical values are used to represent degrees of belief, then a simple set of axioms encoding common sense properties of such beliefs leads uniquely to a set of rules for manipulating degrees of belief that are equivalent to the sum and product rules of probability. This provided the first rigorous proof that probability theory could be regarded as an extension of Boolean logic to situations involving uncertainty (Jaynes, 2003). Numerous other authors have proposed different sets of properties or axioms that such measures of uncertainty should satisfy (Ramsey, 1931; Good, 1950; Savage, 1961; deFinetti, 1970; Lindley, 1982). In each case, the resulting numerical quantities behave precisely according to the rules of probability. It is therefore natural to refer to these quantities as (Bayesian) probabilities.

In the field of pattern recognition, too, it is helpful to have a more general no-

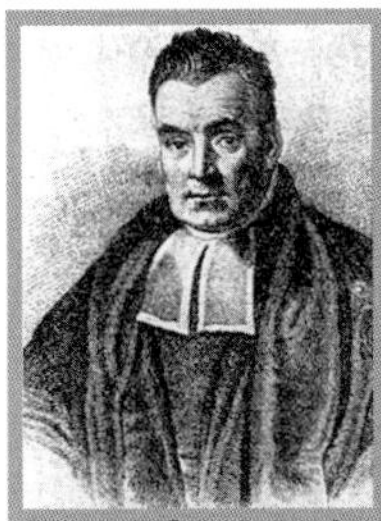

Thomas Bayes
1701–1761

Thomas Bayes was born in Tunbridge Wells and was a clergyman as well as an amateur scientist and a mathematician. He studied logic and theology at Edinburgh University and was elected Fellow of the Royal Society in 1742. During the 18th century, issues regarding probability arose in connection with gambling and with the new concept of insurance. One particularly important problem concerned so-called inverse probability. A solution was proposed by Thomas Bayes in his paper 'Essay towards solving a problem in the doctrine of chances', which was published in 1764, some three years after his death, in the *Philosophical Transactions of the Royal Society*. In fact, Bayes only formulated his theory for the case of a uniform prior, and it was Pierre-Simon Laplace who independently rediscovered the theory in general form and who demonstrated its broad applicability.

tion of probability. Consider the example of polynomial curve fitting discussed in Section 1.1. It seems reasonable to apply the frequentist notion of probability to the random values of the observed variables t_n. However, we would like to address and quantify the uncertainty that surrounds the appropriate choice for the model parameters $\mathbf{w}$. We shall see that, from a Bayesian perspective, we can use the machinery of probability theory to describe the uncertainty in model parameters such as $\mathbf{w}$, or indeed in the choice of model itself.

Bayes' theorem now acquires a new significance. Recall that in the boxes of fruit example, the observation of the identity of the fruit provided relevant information that altered the probability that the chosen box was the red one. In that example, Bayes' theorem was used to convert a prior probability into a posterior probability by incorporating the evidence provided by the observed data. As we shall see in detail later, we can adopt a similar approach when making inferences about quantities such as the parameters $\mathbf{w}$ in the polynomial curve fitting example. We capture our assumptions about $\mathbf{w}$, before observing the data, in the form of a prior probability distribution $p(\mathbf{w})$. The effect of the observed data $\mathcal{D} = \{t_1, \ldots, t_N\}$ is expressed through the conditional probability $p(\mathcal{D}|\mathbf{w})$, and we shall see later, in Section 1.2.5, how this can be represented explicitly. Bayes' theorem, which takes the form

$$p(\mathbf{w}|\mathcal{D}) = \frac{p(\mathcal{D}|\mathbf{w})p(\mathbf{w})}{p(\mathcal{D})} \tag{1.43}$$

then allows us to evaluate the uncertainty in $\mathbf{w}$ *after* we have observed $\mathcal{D}$ in the form of the posterior probability $p(\mathbf{w}|\mathcal{D})$.

The quantity $p(\mathcal{D}|\mathbf{w})$ on the right-hand side of Bayes' theorem is evaluated for the observed data set $\mathcal{D}$ and can be viewed as a function of the parameter vector $\mathbf{w}$, in which case it is called the *likelihood function*. It expresses how probable the observed data set is for different settings of the parameter vector $\mathbf{w}$. Note that the likelihood is not a probability distribution over $\mathbf{w}$, and its integral with respect to $\mathbf{w}$ does not (necessarily) equal one.

Given this definition of likelihood, we can state Bayes' theorem in words

$$\text{posterior} \propto \text{likelihood} \times \text{prior} \tag{1.44}$$

where all of these quantities are viewed as functions of $\mathbf{w}$. The denominator in (1.43) is the normalization constant, which ensures that the posterior distribution on the left-hand side is a valid probability density and integrates to one. Indeed, integrating both sides of (1.43) with respect to $\mathbf{w}$, we can express the denominator in Bayes' theorem in terms of the prior distribution and the likelihood function

$$p(\mathcal{D}) = \int p(\mathcal{D}|\mathbf{w})p(\mathbf{w})\,\mathrm{d}\mathbf{w}. \tag{1.45}$$

In both the Bayesian and frequentist paradigms, the likelihood function $p(\mathcal{D}|\mathbf{w})$ plays a central role. However, the manner in which it is used is fundamentally different in the two approaches. In a frequentist setting, $\mathbf{w}$ is considered to be a fixed parameter, whose value is determined by some form of 'estimator', and error bars on this estimate are obtained by considering the distribution of possible data sets $\mathcal{D}$. By contrast, from the Bayesian viewpoint there is only a single data set $\mathcal{D}$ (namely the one that is actually observed), and the uncertainty in the parameters is expressed through a probability distribution over $\mathbf{w}$.

A widely used frequentist estimator is *maximum likelihood*, in which $\mathbf{w}$ is set to the value that maximizes the likelihood function $p(\mathcal{D}|\mathbf{w})$. This corresponds to choosing the value of $\mathbf{w}$ for which the probability of the observed data set is maximized. In the machine learning literature, the negative log of the likelihood function is called an *error function*. Because the negative logarithm is a monotonically decreasing function, maximizing the likelihood is equivalent to minimizing the error.

One approach to determining frequentist error bars is the *bootstrap* (Efron, 1979; Hastie *et al.*, 2001), in which multiple data sets are created as follows. Suppose our original data set consists of N data points $\mathbf{X} = \{\mathbf{x}_1, \ldots, \mathbf{x}_N\}$. We can create a new data set $\mathbf{X}_{\mathrm{B}}$ by drawing N points at random from $\mathbf{X}$, with replacement, so that some points in $\mathbf{X}$ may be replicated in $\mathbf{X}_{\mathrm{B}}$, whereas other points in $\mathbf{X}$ may be absent from $\mathbf{X}_{\mathrm{B}}$. This process can be repeated L times to generate L data sets each of size N and each obtained by sampling from the original data set $\mathbf{X}$. The statistical accuracy of parameter estimates can then be evaluated by looking at the variability of predictions between the different bootstrap data sets.

One advantage of the Bayesian viewpoint is that the inclusion of prior knowledge arises naturally. Suppose, for instance, that a fair-looking coin is tossed three times and lands heads each time. A classical maximum likelihood estimate of the probability of landing heads would give 1, implying that all future tosses will land heads! By contrast, a Bayesian approach with any reasonable prior will lead to a much less extreme conclusion.

Section 2.1

There has been much controversy and debate associated with the relative merits of the frequentist and Bayesian paradigms, which have not been helped by the fact that there is no unique frequentist, or even Bayesian, viewpoint. For instance, one common criticism of the Bayesian approach is that the prior distribution is often selected on the basis of mathematical convenience rather than as a reflection of any prior beliefs. Even the subjective nature of the conclusions through their dependence on the choice of prior is seen by some as a source of difficulty. Reducing the dependence on the prior is one motivation for so-called *noninformative* priors. However, these lead to difficulties when comparing different models, and indeed Bayesian methods based on poor choices of prior can give poor results with high confidence. Frequentist evaluation methods offer some protection from such problems, and techniques such as cross-validation remain useful in areas such as model comparison.

Section 2.4.3

Section 1.3

This book places a strong emphasis on the Bayesian viewpoint, reflecting the huge growth in the practical importance of Bayesian methods in the past few years, while also discussing useful frequentist concepts as required.

Although the Bayesian framework has its origins in the 18th century, the practical application of Bayesian methods was for a long time severely limited by the difficulties in carrying through the full Bayesian procedure, particularly the need to marginalize (sum or integrate) over the whole of parameter space, which, as we shall see, is required in order to make predictions or to compare different models. The development of sampling methods, such as Markov chain Monte Carlo (discussed in Chapter 11) along with dramatic improvements in the speed and memory capacity of computers, opened the door to the practical use of Bayesian techniques in an impressive range of problem domains. Monte Carlo methods are very flexible and can be applied to a wide range of models. However, they are computationally intensive and have mainly been used for small-scale problems.

More recently, highly efficient deterministic approximation schemes such as variational Bayes and expectation propagation (discussed in Chapter 10) have been developed. These offer a complementary alternative to sampling methods and have allowed Bayesian techniques to be used in large-scale applications (Blei *et al.*, 2003).

1.2.4 The Gaussian distribution

We shall devote the whole of Chapter 2 to a study of various probability distributions and their key properties. It is convenient, however, to introduce here one of the most important probability distributions for continuous variables, called the *normal* or *Gaussian* distribution. We shall make extensive use of this distribution in the remainder of this chapter and indeed throughout much of the book.

For the case of a single real-valued variable x, the Gaussian distribution is defined by

$$\mathcal{N}\left(x|\mu,\sigma^2\right) = \frac{1}{(2\pi\sigma^2)^{1/2}} \exp\left\{-\frac{1}{2\sigma^2}(x-\mu)^2\right\} \tag{1.46}$$

which is governed by two parameters: μ, called the *mean*, and σ^2, called the *variance*. The square root of the variance, given by σ, is called the *standard deviation*, and the reciprocal of the variance, written as $\beta = 1/\sigma^2$, is called the *precision*. We shall see the motivation for these terms shortly. Figure 1.13 shows a plot of the Gaussian distribution.

From the form of (1.46) we see that the Gaussian distribution satisfies

$$\mathcal{N}(x|\mu,\sigma^2) > 0. \tag{1.47}$$

Exercise 1.7

Also it is straightforward to show that the Gaussian is normalized, so that

Pierre-Simon Laplace
1749–1827

It is said that Laplace was seriously lacking in modesty and at one point declared himself to be the best mathematician in France at the time, a claim that was arguably true. As well as being prolific in mathematics, he also made numerous contributions to astronomy, including the nebular hypothesis by which the earth is thought to have formed from the condensation and cooling of a large rotating disk of gas and dust. In 1812 he published the first edition of *Théorie Analytique des Probabilités*, in which Laplace states that "probability theory is nothing but common sense reduced to calculation". This work included a discussion of the inverse probability calculation (later termed Bayes' theorem by Poincaré), which he used to solve problems in life expectancy, jurisprudence, planetary masses, triangulation, and error estimation.

Figure 1.13 Plot of the univariate Gaussian showing the mean μ and the standard deviation σ.

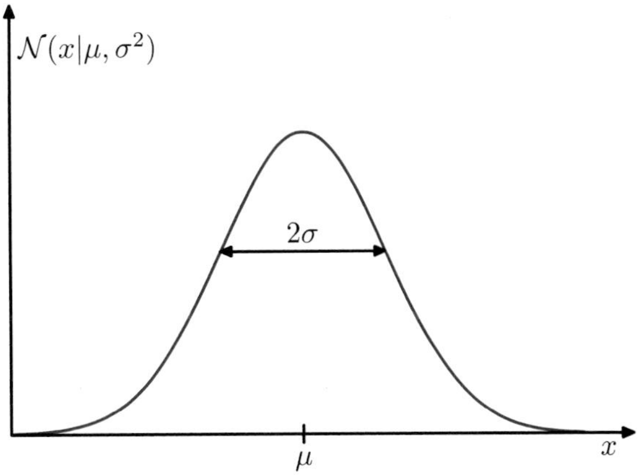

$$\int_{-\infty}^{\infty} \mathcal{N}\left(x|\mu,\sigma^2\right) \mathrm{d}x = 1. \tag{1.48}$$

Thus (1.46) satisfies the two requirements for a valid probability density.

Exercise 1.8

We can readily find expectations of functions of x under the Gaussian distribution. In particular, the average value of x is given by

$$\mathbb{E}[x] = \int_{-\infty}^{\infty} \mathcal{N}\left(x|\mu,\sigma^2\right) x \,\mathrm{d}x = \mu. \tag{1.49}$$

Because the parameter μ represents the average value of x under the distribution, it is referred to as the mean. Similarly, for the second order moment

$$\mathbb{E}[x^2] = \int_{-\infty}^{\infty} \mathcal{N}\left(x|\mu,\sigma^2\right) x^2 \,\mathrm{d}x = \mu^2 + \sigma^2. \tag{1.50}$$

From (1.49) and (1.50), it follows that the variance of x is given by

$$\mathrm{var}[x] = \mathbb{E}[x^2] - \mathbb{E}[x]^2 = \sigma^2 \tag{1.51}$$

Exercise 1.9

and hence σ^2 is referred to as the variance parameter. The maximum of a distribution is known as its mode. For a Gaussian, the mode coincides with the mean.

We are also interested in the Gaussian distribution defined over a D-dimensional vector $\mathbf{x}$ of continuous variables, which is given by

$$\mathcal{N}(\mathbf{x}|\boldsymbol{\mu},\boldsymbol{\Sigma}) = \frac{1}{(2\pi)^{D/2}} \frac{1}{|\boldsymbol{\Sigma}|^{1/2}} \exp\left\{-\frac{1}{2}(\mathbf{x}-\boldsymbol{\mu})^{\mathrm{T}}\boldsymbol{\Sigma}^{-1}(\mathbf{x}-\boldsymbol{\mu})\right\} \tag{1.52}$$

where the D-dimensional vector $\boldsymbol{\mu}$ is called the mean, the $D \times D$ matrix $\boldsymbol{\Sigma}$ is called the covariance, and $|\boldsymbol{\Sigma}|$ denotes the determinant of $\boldsymbol{\Sigma}$. We shall make use of the multivariate Gaussian distribution briefly in this chapter, although its properties will be studied in detail in Section 2.3.

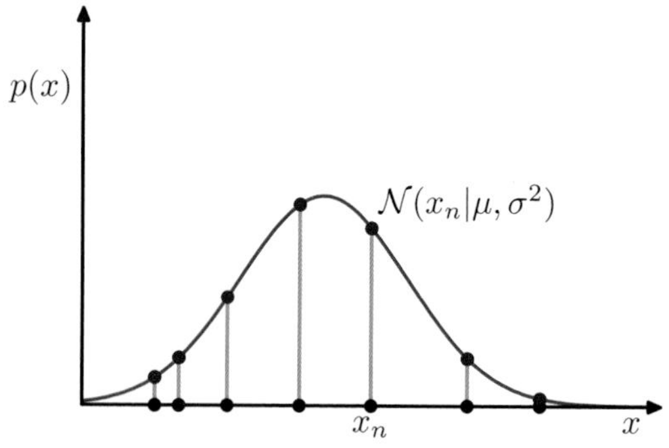

Figure 1.14 Illustration of the likelihood function for a Gaussian distribution, shown by the red curve. Here the black points denote a data set of values $\{x_n\}$, and the likelihood function given by (1.53) corresponds to the product of the blue values. Maximizing the likelihood involves adjusting the mean and variance of the Gaussian so as to maximize this product.

Now suppose that we have a data set of observations $\mathbf{x} = (x_1, \ldots, x_N)^{\mathrm{T}}$, representing N observations of the scalar variable x. Note that we are using the typeface $\mathbf{x}$ to distinguish this from a single observation of the vector-valued variable $(x_1, \ldots, x_D)^{\mathrm{T}}$, which we denote by $\mathbf{x}$. We shall suppose that the observations are drawn independently from a Gaussian distribution whose mean μ and variance σ^2 are unknown, and we would like to determine these parameters from the data set. Data points that are drawn independently from the same distribution are said to be *independent and identically distributed*, which is often abbreviated to i.i.d. We have seen that the joint probability of two independent events is given by the product of the marginal probabilities for each event separately. Because our data set $\mathbf{x}$ is i.i.d., we can therefore write the probability of the data set, given μ and σ^2, in the form

$$p(\mathbf{x}|\mu, \sigma^2) = \prod_{n=1}^{N} \mathcal{N}\left(x_n|\mu, \sigma^2\right). \tag{1.53}$$

When viewed as a function of μ and σ^2, this is the likelihood function for the Gaussian and is interpreted diagrammatically in Figure 1.14.

One common criterion for determining the parameters in a probability distribution using an observed data set is to find the parameter values that maximize the likelihood function. This might seem like a strange criterion because, from our foregoing discussion of probability theory, it would seem more natural to maximize the probability of the parameters given the data, not the probability of the data given the parameters. In fact, these two criteria are related, as we shall discuss in the context of curve fitting.

Section 1.2.5

For the moment, however, we shall determine values for the unknown parameters μ and σ^2 in the Gaussian by maximizing the likelihood function (1.53). In practice, it is more convenient to maximize the log of the likelihood function. Because the logarithm is a monotonically increasing function of its argument, maximization of the log of a function is equivalent to maximization of the function itself. Taking the log not only simplifies the subsequent mathematical analysis, but it also helps numerically because the product of a large number of small probabilities can easily underflow the numerical precision of the computer, and this is resolved by computing instead the sum of the log probabilities. From (1.46) and (1.53), the log likelihood function can be written in the form

$$\ln p\left(\mathbf{x}|\mu, \sigma^2\right) = -\frac{1}{2\sigma^2}\sum_{n=1}^{N}(x_n - \mu)^2 - \frac{N}{2}\ln\sigma^2 - \frac{N}{2}\ln(2\pi). \tag{1.54}$$

Exercise 1.11

Maximizing (1.54) with respect to μ, we obtain the maximum likelihood solution given by

$$\mu_{\mathrm{ML}} = \frac{1}{N}\sum_{n=1}^{N} x_n \tag{1.55}$$

which is the *sample mean*, i.e., the mean of the observed values $\{x_n\}$. Similarly, maximizing (1.54) with respect to σ^2, we obtain the maximum likelihood solution for the variance in the form

$$\sigma^2_{\mathrm{ML}} = \frac{1}{N}\sum_{n=1}^{N}(x_n - \mu_{\mathrm{ML}})^2 \tag{1.56}$$

which is the *sample variance* measured with respect to the sample mean μ_{ML}. Note that we are performing a joint maximization of (1.54) with respect to μ and σ^2, but in the case of the Gaussian distribution the solution for μ decouples from that for σ^2 so that we can first evaluate (1.55) and then subsequently use this result to evaluate (1.56).

Later in this chapter, and also in subsequent chapters, we shall highlight the significant limitations of the maximum likelihood approach. Here we give an indication of the problem in the context of our solutions for the maximum likelihood parameter settings for the univariate Gaussian distribution. In particular, we shall show that the maximum likelihood approach systematically underestimates the variance of the distribution. This is an example of a phenomenon called *bias* and is related to the problem of over-fitting encountered in the context of polynomial curve fitting. We first note that the maximum likelihood solutions μ_{ML} and σ^2_{ML} are functions of the data set values $x_1, \ldots, x_N$. Consider the expectations of these quantities with respect to the data set values, which themselves come from a Gaussian distribution with parameters μ and σ^2. It is straightforward to show that

Section 1.1

Exercise 1.12

$$\mathbb{E}[\mu_{\mathrm{ML}}] = \mu \tag{1.57}$$

$$\mathbb{E}[\sigma^2_{\mathrm{ML}}] = \left(\frac{N-1}{N}\right)\sigma^2 \tag{1.58}$$

so that on average the maximum likelihood estimate will obtain the correct mean but will underestimate the true variance by a factor $(N-1)/N$. The intuition behind this result is given by Figure 1.15.

From (1.58) it follows that the following estimate for the variance parameter is unbiased

$$\widetilde{\sigma}^2 = \frac{N}{N-1}\sigma^2_{\mathrm{ML}} = \frac{1}{N-1}\sum_{n=1}^{N}(x_n - \mu_{\mathrm{ML}})^2. \tag{1.59}$$

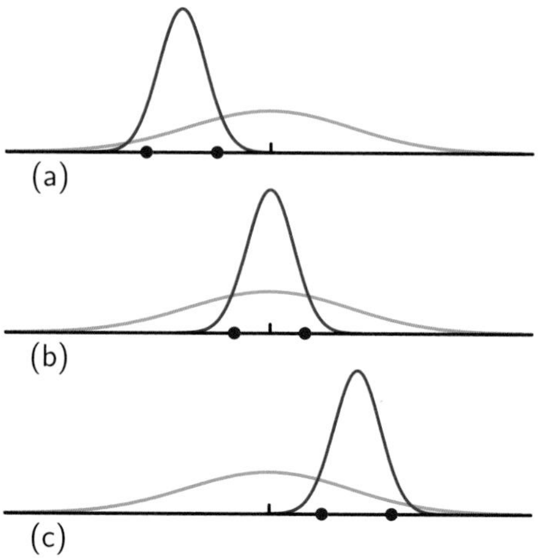

Figure 1.15 Illustration of how bias arises in using maximum likelihood to determine the variance of a Gaussian. The green curve shows the true Gaussian distribution from which data is generated, and the three red curves show the Gaussian distributions obtained by fitting to three data sets, each consisting of two data points shown in blue, using the maximum likelihood results (1.55) and (1.56). Averaged across the three data sets, the mean is correct, but the variance is systematically under-estimated because it is measured relative to the sample mean and not relative to the true mean.

Note that the bias of the maximum likelihood solution becomes less significant as the number N of data points increases, and in the limit $N \to \infty$ the maximum likelihood solution for the variance equals the true variance of the distribution that generated the data. In practice, for anything other than small N, this bias will not prove to be a serious problem. However, throughout this book we shall be interested in more complex models with many parameters, for which the bias problems associated with maximum likelihood will be much more severe. In fact, as we shall see, the issue of bias in maximum likelihood lies at the root of the over-fitting problem that we encountered earlier in the context of polynomial curve fitting.

1.2.5 Curve fitting re-visited

Section 1.1

We have seen how the problem of polynomial curve fitting can be expressed in terms of error minimization. Here we return to the curve fitting example and view it from a probabilistic perspective, thereby gaining some insights into error functions and regularization, as well as taking us towards a full Bayesian treatment.

The goal in the curve fitting problem is to be able to make predictions for the target variable t given some new value of the input variable x on the basis of a set of training data comprising N input values $\mathbf{x} = (x_1, \ldots, x_N)^{\mathrm{T}}$ and their corresponding target values $\mathbf{t} = (t_1, \ldots, t_N)^{\mathrm{T}}$. We can express our uncertainty over the value of the target variable using a probability distribution. For this purpose, we shall assume that, given the value of x, the corresponding value of t has a Gaussian distribution with a mean equal to the value $y(x, \mathbf{w})$ of the polynomial curve given by (1.1). Thus we have

$$p(t|x, \mathbf{w}, \beta) = \mathcal{N}\left(t|y(x, \mathbf{w}), \beta^{-1}\right) \tag{1.60}$$

where, for consistency with the notation in later chapters, we have defined a precision parameter β corresponding to the inverse variance of the distribution. This is illustrated schematically in Figure 1.16.

Figure 1.16 Schematic illustration of a Gaussian conditional distribution for t given x given by (1.60), in which the mean is given by the polynomial function $y(x, \mathbf{w})$, and the precision is given by the parameter β, which is related to the variance by $\beta^{-1} = \sigma^2$.

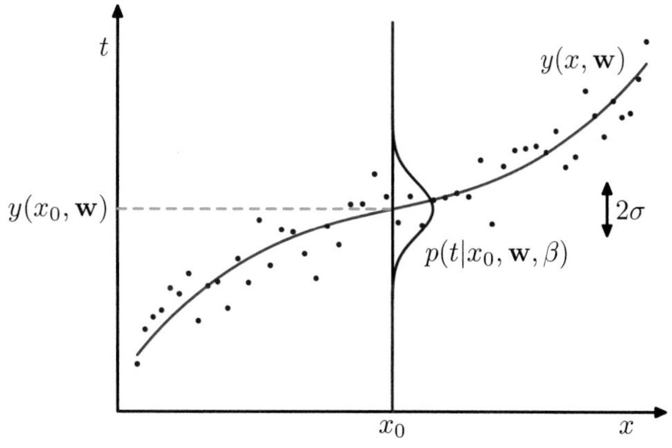

We now use the training data $\{\mathbf{x}, \mathbf{t}\}$ to determine the values of the unknown parameters $\mathbf{w}$ and β by maximum likelihood. If the data are assumed to be drawn independently from the distribution (1.60), then the likelihood function is given by

$$p(\mathbf{t}|\mathbf{x}, \mathbf{w}, \beta) = \prod_{n=1}^{N} \mathcal{N}\left(t_n | y(x_n, \mathbf{w}), \beta^{-1}\right). \tag{1.61}$$

As we did in the case of the simple Gaussian distribution earlier, it is convenient to maximize the logarithm of the likelihood function. Substituting for the form of the Gaussian distribution, given by (1.46), we obtain the log likelihood function in the form

$$\ln p(\mathbf{t}|\mathbf{x}, \mathbf{w}, \beta) = -\frac{\beta}{2} \sum_{n=1}^{N} \{y(x_n, \mathbf{w}) - t_n\}^2 + \frac{N}{2} \ln \beta - \frac{N}{2} \ln(2\pi). \tag{1.62}$$

Consider first the determination of the maximum likelihood solution for the polynomial coefficients, which will be denoted by $\mathbf{w}_{\mathrm{ML}}$. These are determined by maximizing (1.62) with respect to $\mathbf{w}$. For this purpose, we can omit the last two terms on the right-hand side of (1.62) because they do not depend on $\mathbf{w}$. Also, we note that scaling the log likelihood by a positive constant coefficient does not alter the location of the maximum with respect to $\mathbf{w}$, and so we can replace the coefficient $\beta/2$ with $1/2$. Finally, instead of maximizing the log likelihood, we can equivalently minimize the negative log likelihood. We therefore see that maximizing likelihood is equivalent, so far as determining $\mathbf{w}$ is concerned, to minimizing the *sum-of-squares error function* defined by (1.2). Thus the sum-of-squares error function has arisen as a consequence of maximizing likelihood under the assumption of a Gaussian noise distribution.

We can also use maximum likelihood to determine the precision parameter β of the Gaussian conditional distribution. Maximizing (1.62) with respect to β gives

$$\frac{1}{\beta_{\mathrm{ML}}} = \frac{1}{N} \sum_{n=1}^{N} \{y(x_n, \mathbf{w}_{\mathrm{ML}}) - t_n\}^2. \tag{1.63}$$

Again we can first determine the parameter vector $\mathbf{w}_{\rm ML}$ governing the mean and subsequently use this to find the precision $\beta_{\rm ML}$ as was the case for the simple Gaussian distribution.

Section 1.2.4

Having determined the parameters $\mathbf{w}$ and β, we can now make predictions for new values of x. Because we now have a probabilistic model, these are expressed in terms of the *predictive distribution* that gives the probability distribution over t, rather than simply a point estimate, and is obtained by substituting the maximum likelihood parameters into (1.60) to give

$$p(t|x, \mathbf{w}_{\rm ML}, \beta_{\rm ML}) = \mathcal{N}\left(t|y(x, \mathbf{w}_{\rm ML}), \beta_{\rm ML}^{-1}\right). \tag{1.64}$$

Now let us take a step towards a more Bayesian approach and introduce a prior distribution over the polynomial coefficients $\mathbf{w}$. For simplicity, let us consider a Gaussian distribution of the form

$$p(\mathbf{w}|\alpha) = \mathcal{N}(\mathbf{w}|\mathbf{0}, \alpha^{-1}\mathbf{I}) = \left(\frac{\alpha}{2\pi}\right)^{(M+1)/2} \exp\left\{-\frac{\alpha}{2}\mathbf{w}^{\rm T}\mathbf{w}\right\} \tag{1.65}$$

where α is the precision of the distribution, and $M+1$ is the total number of elements in the vector $\mathbf{w}$ for an $M^{\rm th}$ order polynomial. Variables such as α, which control the distribution of model parameters, are called *hyperparameters*. Using Bayes' theorem, the posterior distribution for $\mathbf{w}$ is proportional to the product of the prior distribution and the likelihood function

$$p(\mathbf{w}|\mathsf{x}, \mathsf{t}, \alpha, \beta) \propto p(\mathsf{t}|\mathsf{x}, \mathbf{w}, \beta)p(\mathbf{w}|\alpha). \tag{1.66}$$

We can now determine $\mathbf{w}$ by finding the most probable value of $\mathbf{w}$ given the data, in other words by maximizing the posterior distribution. This technique is called *maximum posterior*, or simply *MAP*. Taking the negative logarithm of (1.66) and combining with (1.62) and (1.65), we find that the maximum of the posterior is given by the minimum of

$$\frac{\beta}{2}\sum_{n=1}^{N}\{y(x_n, \mathbf{w}) - t_n\}^2 + \frac{\alpha}{2}\mathbf{w}^{\rm T}\mathbf{w}. \tag{1.67}$$

Thus we see that maximizing the posterior distribution is equivalent to minimizing the regularized sum-of-squares error function encountered earlier in the form (1.4), with a regularization parameter given by $\lambda = \alpha/\beta$.

1.2.6 Bayesian curve fitting

Although we have included a prior distribution $p(\mathbf{w}|\alpha)$, we are so far still making a point estimate of $\mathbf{w}$ and so this does not yet amount to a Bayesian treatment. In a fully Bayesian approach, we should consistently apply the sum and product rules of probability, which requires, as we shall see shortly, that we integrate over all values of $\mathbf{w}$. Such marginalizations lie at the heart of Bayesian methods for pattern recognition.

In the curve fitting problem, we are given the training data x and t, along with a new test point x, and our goal is to predict the value of t. We therefore wish to evaluate the predictive distribution $p(t|x, \mathsf{x}, \mathsf{t})$. Here we shall assume that the parameters α and β are fixed and known in advance (in later chapters we shall discuss how such parameters can be inferred from data in a Bayesian setting).

A Bayesian treatment simply corresponds to a consistent application of the sum and product rules of probability, which allow the predictive distribution to be written in the form

$$p(t|x, \mathsf{x}, \mathsf{t}) = \int p(t|x, \mathbf{w})p(\mathbf{w}|\mathsf{x}, \mathsf{t})\,\mathrm{d}\mathbf{w}. \tag{1.68}$$

Here $p(t|x, \mathbf{w})$ is given by (1.60), and we have omitted the dependence on α and β to simplify the notation. Here $p(\mathbf{w}|\mathsf{x}, \mathsf{t})$ is the posterior distribution over parameters, and can be found by normalizing the right-hand side of (1.66). We shall see in Section 3.3 that, for problems such as the curve-fitting example, this posterior distribution is a Gaussian and can be evaluated analytically. Similarly, the integration in (1.68) can also be performed analytically with the result that the predictive distribution is given by a Gaussian of the form

$$p(t|x, \mathsf{x}, \mathsf{t}) = \mathcal{N}\left(t|m(x), s^2(x)\right) \tag{1.69}$$

where the mean and variance are given by

$$m(x) = \beta\boldsymbol{\phi}(x)^{\mathrm{T}}\mathbf{S}\sum_{n=1}^{N}\boldsymbol{\phi}(x_n)t_n \tag{1.70}$$

$$s^2(x) = \beta^{-1} + \boldsymbol{\phi}(x)^{\mathrm{T}}\mathbf{S}\boldsymbol{\phi}(x). \tag{1.71}$$

Here the matrix $\mathbf{S}$ is given by

$$\mathbf{S}^{-1} = \alpha\mathbf{I} + \beta\sum_{n=1}^{N}\boldsymbol{\phi}(x_n)\boldsymbol{\phi}(x_n)^{\mathrm{T}} \tag{1.72}$$

where $\mathbf{I}$ is the unit matrix, and we have defined the vector $\boldsymbol{\phi}(x)$ with elements $\phi_i(x) = x^i$ for $i = 0, \ldots, M$.

We see that the variance, as well as the mean, of the predictive distribution in (1.69) is dependent on x. The first term in (1.71) represents the uncertainty in the predicted value of t due to the noise on the target variables and was expressed already in the maximum likelihood predictive distribution (1.64) through β_{ML}^{-1}. However, the second term arises from the uncertainty in the parameters $\mathbf{w}$ and is a consequence of the Bayesian treatment. The predictive distribution for the synthetic sinusoidal regression problem is illustrated in Figure 1.17.

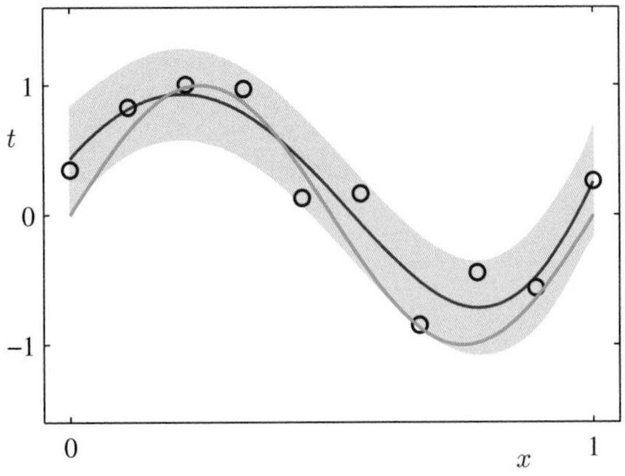

Figure 1.17 The predictive distribution resulting from a Bayesian treatment of polynomial curve fitting using an $M = 9$ polynomial, with the fixed parameters $\alpha = 5 \times 10^{-3}$ and $\beta = 11.1$ (corresponding to the known noise variance), in which the red curve denotes the mean of the predictive distribution and the red region corresponds to ± 1 standard deviation around the mean.

1.3. Model Selection

In our example of polynomial curve fitting using least squares, we saw that there was an optimal order of polynomial that gave the best generalization. The order of the polynomial controls the number of free parameters in the model and thereby governs the model complexity. With regularized least squares, the regularization coefficient λ also controls the effective complexity of the model, whereas for more complex models, such as mixture distributions or neural networks there may be multiple parameters governing complexity. In a practical application, we need to determine the values of such parameters, and the principal objective in doing so is usually to achieve the best predictive performance on new data. Furthermore, as well as finding the appropriate values for complexity parameters within a given model, we may wish to consider a range of different types of model in order to find the best one for our particular application.

We have already seen that, in the maximum likelihood approach, the performance on the training set is not a good indicator of predictive performance on unseen data due to the problem of over-fitting. If data is plentiful, then one approach is simply to use some of the available data to train a range of models, or a given model with a range of values for its complexity parameters, and then to compare them on independent data, sometimes called a *validation set*, and select the one having the best predictive performance. If the model design is iterated many times using a limited size data set, then some over-fitting to the validation data can occur and so it may be necessary to keep aside a third *test set* on which the performance of the selected model is finally evaluated.

In many applications, however, the supply of data for training and testing will be limited, and in order to build good models, we wish to use as much of the available data as possible for training. However, if the validation set is small, it will give a relatively noisy estimate of predictive performance. One solution to this dilemma is to use *cross-validation*, which is illustrated in Figure 1.18. This allows a proportion $(S-1)/S$ of the available data to be used for training while making use of all of the

Figure 1.18 The technique of S-fold cross-validation, illustrated here for the case of $S = 4$, involves taking the available data and partitioning it into S groups (in the simplest case these are of equal size). Then $S - 1$ of the groups are used to train a set of models that are then evaluated on the remaining group. This procedure is then repeated for all S possible choices for the held-out group, indicated here by the red blocks, and the performance scores from the S runs are then averaged.

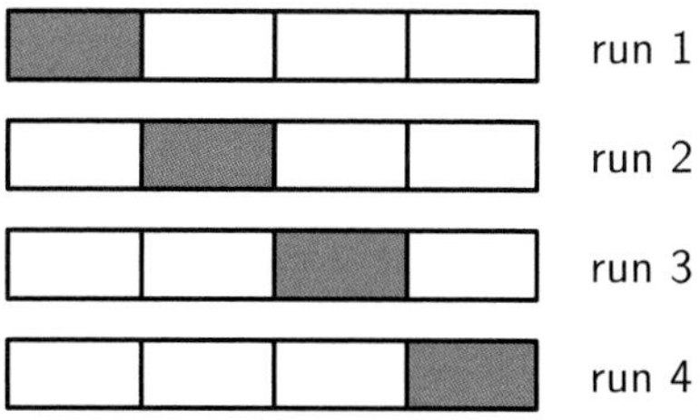

data to assess performance. When data is particularly scarce, it may be appropriate to consider the case $S = N$, where N is the total number of data points, which gives the *leave-one-out* technique.

One major drawback of cross-validation is that the number of training runs that must be performed is increased by a factor of S, and this can prove problematic for models in which the training is itself computationally expensive. A further problem with techniques such as cross-validation that use separate data to assess performance is that we might have multiple complexity parameters for a single model (for instance, there might be several regularization parameters). Exploring combinations of settings for such parameters could, in the worst case, require a number of training runs that is exponential in the number of parameters. Clearly, we need a better approach. Ideally, this should rely only on the training data and should allow multiple hyperparameters and model types to be compared in a single training run. We therefore need to find a measure of performance which depends only on the training data and which does not suffer from bias due to over-fitting.

Historically various 'information criteria' have been proposed that attempt to correct for the bias of maximum likelihood by the addition of a penalty term to compensate for the over-fitting of more complex models. For example, the *Akaike information criterion*, or AIC (Akaike, 1974), chooses the model for which the quantity

$$\ln p(\mathcal{D}|\mathbf{w}_{\mathrm{ML}}) - M \tag{1.73}$$

is largest. Here $p(\mathcal{D}|\mathbf{w}_{\mathrm{ML}})$ is the best-fit log likelihood, and M is the number of adjustable parameters in the model. A variant of this quantity, called the *Bayesian information criterion*, or *BIC*, will be discussed in Section 4.4.1. Such criteria do not take account of the uncertainty in the model parameters, however, and in practice they tend to favour overly simple models. We therefore turn in Section 3.4 to a fully Bayesian approach where we shall see how complexity penalties arise in a natural and principled way.

1.4. The Curse of Dimensionality

In the polynomial curve fitting example we had just one input variable x. For practical applications of pattern recognition, however, we will have to deal with spaces

Figure 1.19 Scatter plot of the oil flow data for input variables x_6 and x_7, in which red denotes the 'homogenous' class, green denotes the 'annular' class, and blue denotes the 'laminar' class. Our goal is to classify the new test point denoted by '$\times$'.

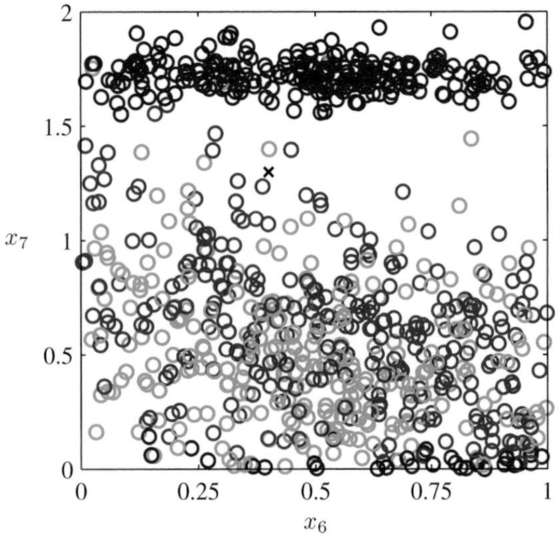

of high dimensionality comprising many input variables. As we now discuss, this poses some serious challenges and is an important factor influencing the design of pattern recognition techniques.

In order to illustrate the problem we consider a synthetically generated data set representing measurements taken from a pipeline containing a mixture of oil, water, and gas (Bishop and James, 1993). These three materials can be present in one of three different geometrical configurations known as 'homogenous', 'annular', and 'laminar', and the fractions of the three materials can also vary. Each data point comprises a 12-dimensional input vector consisting of measurements taken with gamma ray densitometers that measure the attenuation of gamma rays passing along narrow beams through the pipe. This data set is described in detail in Appendix A. Figure 1.19 shows 100 points from this data set on a plot showing two of the measurements x_6 and x_7 (the remaining ten input values are ignored for the purposes of this illustration). Each data point is labelled according to which of the three geometrical classes it belongs to, and our goal is to use this data as a training set in order to be able to classify a new observation (x_6, x_7), such as the one denoted by the cross in Figure 1.19. We observe that the cross is surrounded by numerous red points, and so we might suppose that it belongs to the red class. However, there are also plenty of green points nearby, so we might think that it could instead belong to the green class. It seems unlikely that it belongs to the blue class. The intuition here is that the identity of the cross should be determined more strongly by nearby points from the training set and less strongly by more distant points. In fact, this intuition turns out to be reasonable and will be discussed more fully in later chapters.

How can we turn this intuition into a learning algorithm? One very simple approach would be to divide the input space into regular cells, as indicated in Figure 1.20. When we are given a test point and we wish to predict its class, we first decide which cell it belongs to, and we then find all of the training data points that

Figure 1.20 Illustration of a simple approach to the solution of a classification problem in which the input space is divided into cells and any new test point is assigned to the class that has a majority number of representatives in the same cell as the test point. As we shall see shortly, this simplistic approach has some severe shortcomings.

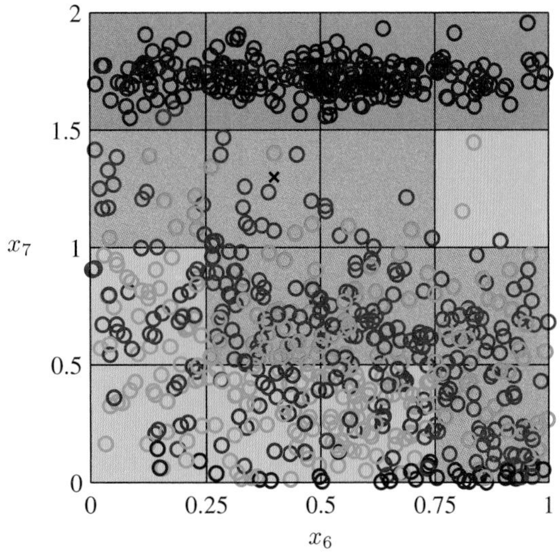

fall in the same cell. The identity of the test point is predicted as being the same as the class having the largest number of training points in the same cell as the test point (with ties being broken at random).

There are numerous problems with this naive approach, but one of the most severe becomes apparent when we consider its extension to problems having larger numbers of input variables, corresponding to input spaces of higher dimensionality. The origin of the problem is illustrated in Figure 1.21, which shows that, if we divide a region of a space into regular cells, then the number of such cells grows exponentially with the dimensionality of the space. The problem with an exponentially large number of cells is that we would need an exponentially large quantity of training data in order to ensure that the cells are not empty. Clearly, we have no hope of applying such a technique in a space of more than a few variables, and so we need to find a more sophisticated approach.

We can gain further insight into the problems of high-dimensional spaces by returning to the example of polynomial curve fitting and considering how we would

Section 1.1

Figure 1.21 Illustration of the curse of dimensionality, showing how the number of regions of a regular grid grows exponentially with the dimensionality D of the space. For clarity, only a subset of the cubical regions are shown for $D = 3$.

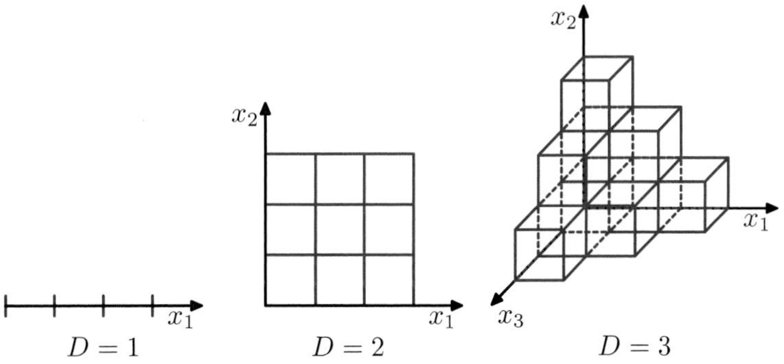

extend this approach to deal with input spaces having several variables. If we have D input variables, then a general polynomial with coefficients up to order 3 would take the form

$$y(\mathbf{x}, \mathbf{w}) = w_0 + \sum_{i=1}^{D} w_i x_i + \sum_{i=1}^{D}\sum_{j=1}^{D} w_{ij} x_i x_j + \sum_{i=1}^{D}\sum_{j=1}^{D}\sum_{k=1}^{D} w_{ijk} x_i x_j x_k. \quad (1.74)$$

As D increases, so the number of independent coefficients (not all of the coefficients are independent due to interchange symmetries amongst the x variables) grows proportionally to D^3. In practice, to capture complex dependencies in the data, we may need to use a higher-order polynomial. For a polynomial of order M, the growth in the number of coefficients is like D^M. Although this is now a power law growth, rather than an exponential growth, it still points to the method becoming rapidly unwieldy and of limited practical utility.

Exercise 1.16

Our geometrical intuitions, formed through a life spent in a space of three dimensions, can fail badly when we consider spaces of higher dimensionality. As a simple example, consider a sphere of radius $r = 1$ in a space of D dimensions, and ask what is the fraction of the volume of the sphere that lies between radius $r = 1-\epsilon$ and $r = 1$. We can evaluate this fraction by noting that the volume of a sphere of radius r in D dimensions must scale as r^D, and so we write

$$V_D(r) = K_D r^D \quad (1.75)$$

Exercise 1.18

where the constant K_D depends only on D. Thus the required fraction is given by

$$\frac{V_D(1) - V_D(1-\epsilon)}{V_D(1)} = 1 - (1-\epsilon)^D \quad (1.76)$$

which is plotted as a function of ϵ for various values of D in Figure 1.22. We see that, for large D, this fraction tends to 1 even for small values of ϵ. Thus, in spaces of high dimensionality, most of the volume of a sphere is concentrated in a thin shell near the surface!

As a further example, of direct relevance to pattern recognition, consider the behaviour of a Gaussian distribution in a high-dimensional space. If we transform from Cartesian to polar coordinates, and then integrate out the directional variables, we obtain an expression for the density $p(r)$ as a function of radius r from the origin. Thus $p(r)\delta r$ is the probability mass inside a thin shell of thickness δr located at radius r. This distribution is plotted, for various values of D, in Figure 1.23, and we see that for large D the probability mass of the Gaussian is concentrated in a thin shell.

Exercise 1.20

The severe difficulty that can arise in spaces of many dimensions is sometimes called the *curse of dimensionality* (Bellman, 1961). In this book, we shall make extensive use of illustrative examples involving input spaces of one or two dimensions, because this makes it particularly easy to illustrate the techniques graphically. The reader should be warned, however, that not all intuitions developed in spaces of low dimensionality will generalize to spaces of many dimensions.

Figure 1.22 Plot of the fraction of the volume of a sphere lying in the range $r = 1-\epsilon$ to $r = 1$ for various values of the dimensionality D.

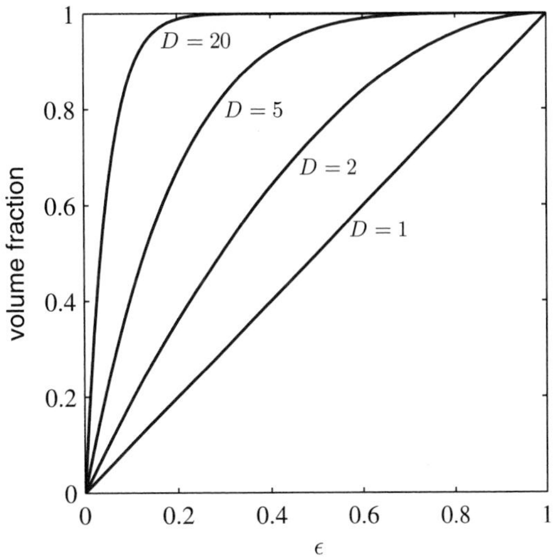

Although the curse of dimensionality certainly raises important issues for pattern recognition applications, it does not prevent us from finding effective techniques applicable to high-dimensional spaces. The reasons for this are twofold. First, real data will often be confined to a region of the space having lower effective dimensionality, and in particular the directions over which important variations in the target variables occur may be so confined. Second, real data will typically exhibit some smoothness properties (at least locally) so that for the most part small changes in the input variables will produce small changes in the target variables, and so we can exploit local interpolation-like techniques to allow us to make predictions of the target variables for new values of the input variables. Successful pattern recognition techniques exploit one or both of these properties. Consider, for example, an application in manufacturing in which images are captured of identical planar objects on a conveyor belt, in which the goal is to determine their orientation. Each image is a point

Figure 1.23 Plot of the probability density with respect to radius r of a Gaussian distribution for various values of the dimensionality D. In a high-dimensional space, most of the probability mass of a Gaussian is located within a thin shell at a specific radius.

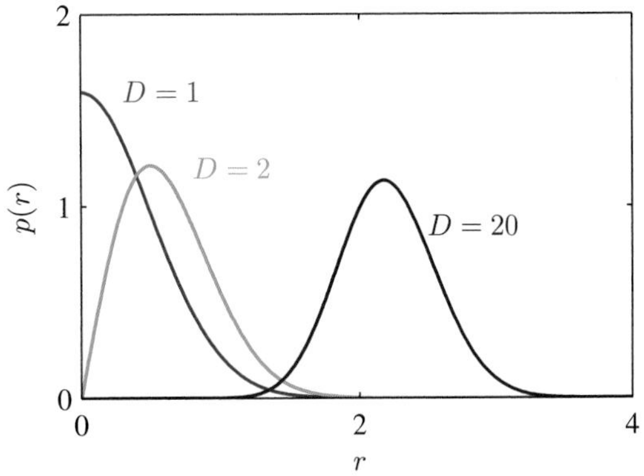

in a high-dimensional space whose dimensionality is determined by the number of pixels. Because the objects can occur at different positions within the image and in different orientations, there are three degrees of freedom of variability between images, and a set of images will live on a three dimensional *manifold* embedded within the high-dimensional space. Due to the complex relationships between the object position or orientation and the pixel intensities, this manifold will be highly nonlinear. If the goal is to learn a model that can take an input image and output the orientation of the object irrespective of its position, then there is only one degree of freedom of variability within the manifold that is significant.

1.5. Decision Theory

We have seen in Section 1.2 how probability theory provides us with a consistent mathematical framework for quantifying and manipulating uncertainty. Here we turn to a discussion of decision theory that, when combined with probability theory, allows us to make optimal decisions in situations involving uncertainty such as those encountered in pattern recognition.

Suppose we have an input vector $\mathbf{x}$ together with a corresponding vector $\mathbf{t}$ of target variables, and our goal is to predict $\mathbf{t}$ given a new value for $\mathbf{x}$. For regression problems, $\mathbf{t}$ will comprise continuous variables, whereas for classification problems $\mathbf{t}$ will represent class labels. The joint probability distribution $p(\mathbf{x}, \mathbf{t})$ provides a complete summary of the uncertainty associated with these variables. Determination of $p(\mathbf{x}, \mathbf{t})$ from a set of training data is an example of *inference* and is typically a very difficult problem whose solution forms the subject of much of this book. In a practical application, however, we must often make a specific prediction for the value of $\mathbf{t}$, or more generally take a specific action based on our understanding of the values $\mathbf{t}$ is likely to take, and this aspect is the subject of decision theory.

Consider, for example, a medical diagnosis problem in which we have taken an X-ray image of a patient, and we wish to determine whether the patient has cancer or not. In this case, the input vector $\mathbf{x}$ is the set of pixel intensities in the image, and output variable t will represent the presence of cancer, which we denote by the class $\mathcal{C}_1$, or the absence of cancer, which we denote by the class $\mathcal{C}_2$. We might, for instance, choose t to be a binary variable such that $t = 0$ corresponds to class $\mathcal{C}_1$ and $t = 1$ corresponds to class $\mathcal{C}_2$. We shall see later that this choice of label values is particularly convenient for probabilistic models. The general inference problem then involves determining the joint distribution $p(\mathbf{x}, \mathcal{C}_k)$, or equivalently $p(\mathbf{x}, t)$, which gives us the most complete probabilistic description of the situation. Although this can be a very useful and informative quantity, in the end we must decide either to give treatment to the patient or not, and we would like this choice to be optimal in some appropriate sense (Duda and Hart, 1973). This is the *decision* step, and it is the subject of decision theory to tell us how to make optimal decisions given the appropriate probabilities. We shall see that the decision stage is generally very simple, even trivial, once we have solved the inference problem.

Here we give an introduction to the key ideas of decision theory as required for the rest of the book. Further background, as well as more detailed accounts, can be found in Berger (1985) and Bather (2000).

Before giving a more detailed analysis, let us first consider informally how we might expect probabilities to play a role in making decisions. When we obtain the X-ray image $\mathbf{x}$ for a new patient, our goal is to decide which of the two classes to assign to the image. We are interested in the probabilities of the two classes given the image, which are given by $p(\mathcal{C}_k|\mathbf{x})$. Using Bayes' theorem, these probabilities can be expressed in the form

$$p(\mathcal{C}_k|\mathbf{x}) = \frac{p(\mathbf{x}|\mathcal{C}_k)p(\mathcal{C}_k)}{p(\mathbf{x})}. \tag{1.77}$$

Note that any of the quantities appearing in Bayes' theorem can be obtained from the joint distribution $p(\mathbf{x}, \mathcal{C}_k)$ by either marginalizing or conditioning with respect to the appropriate variables. We can now interpret $p(\mathcal{C}_k)$ as the prior probability for the class $\mathcal{C}_k$, and $p(\mathcal{C}_k|\mathbf{x})$ as the corresponding posterior probability. Thus $p(\mathcal{C}_1)$ represents the probability that a person has cancer, before we take the X-ray measurement. Similarly, $p(\mathcal{C}_1|\mathbf{x})$ is the corresponding probability, revised using Bayes' theorem in light of the information contained in the X-ray. If our aim is to minimize the chance of assigning $\mathbf{x}$ to the wrong class, then intuitively we would choose the class having the higher posterior probability. We now show that this intuition is correct, and we also discuss more general criteria for making decisions.

1.5.1 Minimizing the misclassification rate

Suppose that our goal is simply to make as few misclassifications as possible. We need a rule that assigns each value of $\mathbf{x}$ to one of the available classes. Such a rule will divide the input space into regions $\mathcal{R}_k$ called *decision regions*, one for each class, such that all points in $\mathcal{R}_k$ are assigned to class $\mathcal{C}_k$. The boundaries between decision regions are called *decision boundaries* or *decision surfaces*. Note that each decision region need not be contiguous but could comprise some number of disjoint regions. We shall encounter examples of decision boundaries and decision regions in later chapters. In order to find the optimal decision rule, consider first of all the case of two classes, as in the cancer problem for instance. A mistake occurs when an input vector belonging to class $\mathcal{C}_1$ is assigned to class $\mathcal{C}_2$ or vice versa. The probability of this occurring is given by

$$\begin{aligned} p(\text{mistake}) &= p(\mathbf{x} \in \mathcal{R}_1, \mathcal{C}_2) + p(\mathbf{x} \in \mathcal{R}_2, \mathcal{C}_1) \\ &= \int_{\mathcal{R}_1} p(\mathbf{x}, \mathcal{C}_2)\,\mathrm{d}\mathbf{x} + \int_{\mathcal{R}_2} p(\mathbf{x}, \mathcal{C}_1)\,\mathrm{d}\mathbf{x}. \end{aligned} \tag{1.78}$$

We are free to choose the decision rule that assigns each point $\mathbf{x}$ to one of the two classes. Clearly to minimize $p(\text{mistake})$ we should arrange that each $\mathbf{x}$ is assigned to whichever class has the smaller value of the integrand in (1.78). Thus, if $p(\mathbf{x}, \mathcal{C}_1) > p(\mathbf{x}, \mathcal{C}_2)$ for a given value of $\mathbf{x}$, then we should assign that $\mathbf{x}$ to class $\mathcal{C}_1$. From the product rule of probability we have $p(\mathbf{x}, \mathcal{C}_k) = p(\mathcal{C}_k|\mathbf{x})p(\mathbf{x})$. Because the factor $p(\mathbf{x})$ is common to both terms, we can restate this result as saying that the minimum

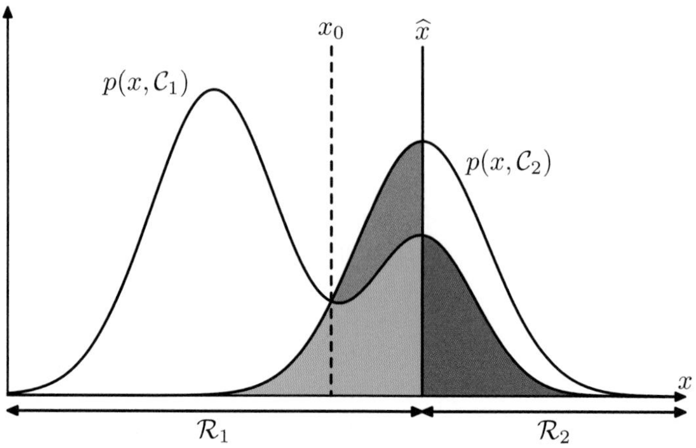

Figure 1.24 Schematic illustration of the joint probabilities $p(x, \mathcal{C}_k)$ for each of two classes plotted against x, together with the decision boundary $x = \widehat{x}$. Values of $x \geqslant \widehat{x}$ are classified as class $\mathcal{C}_2$ and hence belong to decision region $\mathcal{R}_2$, whereas points $x < \widehat{x}$ are classified as $\mathcal{C}_1$ and belong to $\mathcal{R}_1$. Errors arise from the blue, green, and red regions, so that for $x < \widehat{x}$ the errors are due to points from class $\mathcal{C}_2$ being misclassified as $\mathcal{C}_1$ (represented by the sum of the red and green regions), and conversely for points in the region $x \geqslant \widehat{x}$ the errors are due to points from class $\mathcal{C}_1$ being misclassified as $\mathcal{C}_2$ (represented by the blue region). As we vary the location $\widehat{x}$ of the decision boundary, the combined areas of the blue and green regions remains constant, whereas the size of the red region varies. The optimal choice for $\widehat{x}$ is where the curves for $p(x, \mathcal{C}_1)$ and $p(x, \mathcal{C}_2)$ cross, corresponding to $\widehat{x} = x_0$, because in this case the red region disappears. This is equivalent to the minimum misclassification rate decision rule, which assigns each value of x to the class having the higher posterior probability $p(\mathcal{C}_k|x)$.

probability of making a mistake is obtained if each value of $\mathbf{x}$ is assigned to the class for which the posterior probability $p(\mathcal{C}_k|\mathbf{x})$ is largest. This result is illustrated for two classes, and a single input variable x, in Figure 1.24.

For the more general case of K classes, it is slightly easier to maximize the probability of being correct, which is given by

$$\begin{aligned} p(\text{correct}) &= \sum_{k=1}^{K} p(\mathbf{x} \in \mathcal{R}_k, \mathcal{C}_k) \\ &= \sum_{k=1}^{K} \int_{\mathcal{R}_k} p(\mathbf{x}, \mathcal{C}_k) \, \mathrm{d}\mathbf{x} \end{aligned} \tag{1.79}$$

which is maximized when the regions $\mathcal{R}_k$ are chosen such that each $\mathbf{x}$ is assigned to the class for which $p(\mathbf{x}, \mathcal{C}_k)$ is largest. Again, using the product rule $p(\mathbf{x}, \mathcal{C}_k) = p(\mathcal{C}_k|\mathbf{x})p(\mathbf{x})$, and noting that the factor of $p(\mathbf{x})$ is common to all terms, we see that each $\mathbf{x}$ should be assigned to the class having the largest posterior probability $p(\mathcal{C}_k|\mathbf{x})$.

Figure 1.25 An example of a loss matrix with elements L_{kj} for the cancer treatment problem. The rows correspond to the true class, whereas the columns correspond to the assignment of class made by our decision criterion.

$$\begin{array}{c} \\ \text{cancer} \\ \text{normal} \end{array} \begin{array}{c} \begin{array}{cc} \text{cancer} & \text{normal} \end{array} \\ \left(\begin{array}{cc} 0 & 1000 \\ 1 & 0 \end{array} \right) \end{array}$$

1.5.2 Minimizing the expected loss

For many applications, our objective will be more complex than simply minimizing the number of misclassifications. Let us consider again the medical diagnosis problem. We note that, if a patient who does not have cancer is incorrectly diagnosed as having cancer, the consequences may be some patient distress plus the need for further investigations. Conversely, if a patient with cancer is diagnosed as healthy, the result may be premature death due to lack of treatment. Thus the consequences of these two types of mistake can be dramatically different. It would clearly be better to make fewer mistakes of the second kind, even if this was at the expense of making more mistakes of the first kind.

We can formalize such issues through the introduction of a *loss function*, also called a *cost function*, which is a single, overall measure of loss incurred in taking any of the available decisions or actions. Our goal is then to minimize the total loss incurred. Note that some authors consider instead a *utility function*, whose value they aim to maximize. These are equivalent concepts if we take the utility to be simply the negative of the loss, and throughout this text we shall use the loss function convention. Suppose that, for a new value of $\mathbf{x}$, the true class is $\mathcal{C}_k$ and that we assign $\mathbf{x}$ to class $\mathcal{C}_j$ (where j may or may not be equal to k). In so doing, we incur some level of loss that we denote by L_{kj}, which we can view as the k, j element of a *loss matrix*. For instance, in our cancer example, we might have a loss matrix of the form shown in Figure 1.25. This particular loss matrix says that there is no loss incurred if the correct decision is made, there is a loss of 1 if a healthy patient is diagnosed as having cancer, whereas there is a loss of 1000 if a patient having cancer is diagnosed as healthy.

The optimal solution is the one which minimizes the loss function. However, the loss function depends on the true class, which is unknown. For a given input vector $\mathbf{x}$, our uncertainty in the true class is expressed through the joint probability distribution $p(\mathbf{x}, \mathcal{C}_k)$ and so we seek instead to minimize the average loss, where the average is computed with respect to this distribution, which is given by

$$\mathbb{E}[L] = \sum_k \sum_j \int_{\mathcal{R}_j} L_{kj} p(\mathbf{x}, \mathcal{C}_k) \, \mathrm{d}\mathbf{x}. \tag{1.80}$$

Each $\mathbf{x}$ can be assigned independently to one of the decision regions $\mathcal{R}_j$. Our goal is to choose the regions $\mathcal{R}_j$ in order to minimize the expected loss (1.80), which implies that for each $\mathbf{x}$ we should minimize $\sum_k L_{kj} p(\mathbf{x}, \mathcal{C}_k)$. As before, we can use the product rule $p(\mathbf{x}, \mathcal{C}_k) = p(\mathcal{C}_k|\mathbf{x})p(\mathbf{x})$ to eliminate the common factor of $p(\mathbf{x})$. Thus the decision rule that minimizes the expected loss is the one that assigns each

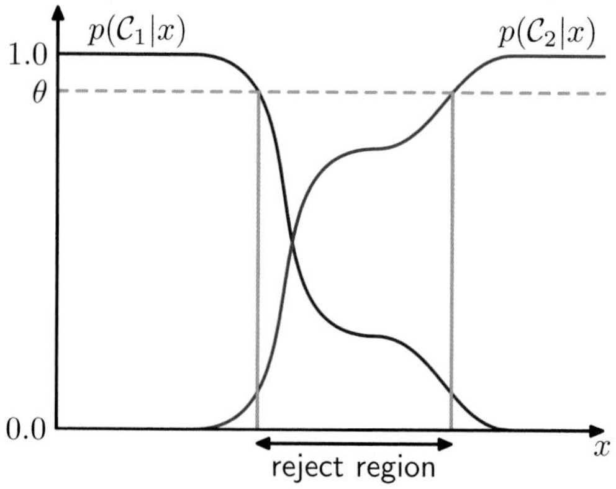

Figure 1.26 Illustration of the reject option. Inputs x such that the larger of the two posterior probabilities is less than or equal to some threshold θ will be rejected.

new $\mathbf{x}$ to the class j for which the quantity

$$\sum_k L_{kj} p(\mathcal{C}_k|\mathbf{x}) \tag{1.81}$$

is a minimum. This is clearly trivial to do, once we know the posterior class probabilities $p(\mathcal{C}_k|\mathbf{x})$.

1.5.3 The reject option

We have seen that classification errors arise from the regions of input space where the largest of the posterior probabilities $p(\mathcal{C}_k|\mathbf{x})$ is significantly less than unity, or equivalently where the joint distributions $p(\mathbf{x}, \mathcal{C}_k)$ have comparable values. These are the regions where we are relatively uncertain about class membership. In some applications, it will be appropriate to avoid making decisions on the difficult cases in anticipation of a lower error rate on those examples for which a classification decision is made. This is known as the *reject option*. For example, in our hypothetical medical illustration, it may be appropriate to use an automatic system to classify those X-ray images for which there is little doubt as to the correct class, while leaving a human expert to classify the more ambiguous cases. We can achieve this by introducing a threshold θ and rejecting those inputs $\mathbf{x}$ for which the largest of the posterior probabilities $p(\mathcal{C}_k|\mathbf{x})$ is less than or equal to θ. This is illustrated for the case of two classes, and a single continuous input variable x, in Figure 1.26. Note that setting $\theta = 1$ will ensure that all examples are rejected, whereas if there are K classes then setting $\theta < 1/K$ will ensure that no examples are rejected. Thus the fraction of examples that get rejected is controlled by the value of θ.

We can easily extend the reject criterion to minimize the expected loss, when a loss matrix is given, taking account of the loss incurred when a reject decision is made.

Exercise 1.24

1.5.4 Inference and decision

We have broken the classification problem down into two separate stages, the *inference stage* in which we use training data to learn a model for $p(\mathcal{C}_k|\mathbf{x})$, and the subsequent *decision* stage in which we use these posterior probabilities to make optimal class assignments. An alternative possibility would be to solve both problems together and simply learn a function that maps inputs $\mathbf{x}$ directly into decisions. Such a function is called a *discriminant function*.

In fact, we can identify three distinct approaches to solving decision problems, all of which have been used in practical applications. These are given, in decreasing order of complexity, by:

(a) First solve the inference problem of determining the class-conditional densities $p(\mathbf{x}|\mathcal{C}_k)$ for each class $\mathcal{C}_k$ individually. Also separately infer the prior class probabilities $p(\mathcal{C}_k)$. Then use Bayes' theorem in the form

$$p(\mathcal{C}_k|\mathbf{x}) = \frac{p(\mathbf{x}|\mathcal{C}_k)p(\mathcal{C}_k)}{p(\mathbf{x})} \tag{1.82}$$

to find the posterior class probabilities $p(\mathcal{C}_k|\mathbf{x})$. As usual, the denominator in Bayes' theorem can be found in terms of the quantities appearing in the numerator, because

$$p(\mathbf{x}) = \sum_k p(\mathbf{x}|\mathcal{C}_k)p(\mathcal{C}_k). \tag{1.83}$$

Equivalently, we can model the joint distribution $p(\mathbf{x}, \mathcal{C}_k)$ directly and then normalize to obtain the posterior probabilities. Having found the posterior probabilities, we use decision theory to determine class membership for each new input $\mathbf{x}$. Approaches that explicitly or implicitly model the distribution of inputs as well as outputs are known as *generative models*, because by sampling from them it is possible to generate synthetic data points in the input space.

(b) First solve the inference problem of determining the posterior class probabilities $p(\mathcal{C}_k|\mathbf{x})$, and then subsequently use decision theory to assign each new $\mathbf{x}$ to one of the classes. Approaches that model the posterior probabilities directly are called *discriminative models*.

(c) Find a function $f(\mathbf{x})$, called a discriminant function, which maps each input $\mathbf{x}$ directly onto a class label. For instance, in the case of two-class problems, $f(\cdot)$ might be binary valued and such that $f = 0$ represents class $\mathcal{C}_1$ and $f = 1$ represents class $\mathcal{C}_2$. In this case, probabilities play no role.

Let us consider the relative merits of these three alternatives. Approach (a) is the most demanding because it involves finding the joint distribution over both $\mathbf{x}$ and $\mathcal{C}_k$. For many applications, $\mathbf{x}$ will have high dimensionality, and consequently we may need a large training set in order to be able to determine the class-conditional densities to reasonable accuracy. Note that the class priors $p(\mathcal{C}_k)$ can often be estimated simply from the fractions of the training set data points in each of the classes. One advantage of approach (a), however, is that it also allows the marginal density of data $p(\mathbf{x})$ to be determined from (1.83). This can be useful for detecting new data points that have low probability under the model and for which the predictions may

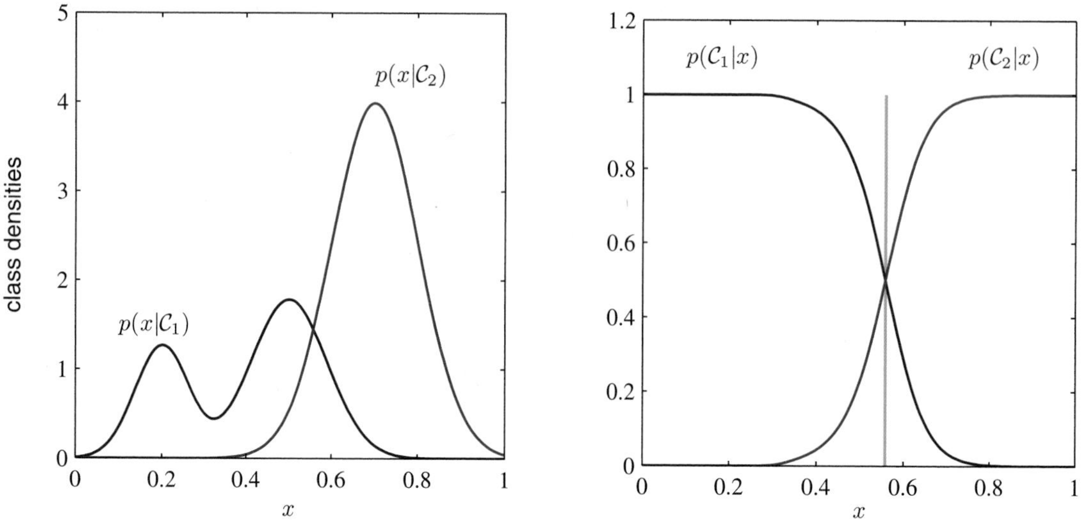

Figure 1.27 Example of the class-conditional densities for two classes having a single input variable x (left plot) together with the corresponding posterior probabilities (right plot). Note that the left-hand mode of the class-conditional density $p(\mathbf{x}|\mathcal{C}_1)$, shown in blue on the left plot, has no effect on the posterior probabilities. The vertical green line in the right plot shows the decision boundary in x that gives the minimum misclassification rate, assuming the prior class probabilities, $p(\mathcal{C}_1)$ and $p(\mathcal{C}_2)$, are equal.

be of low accuracy, which is known as *outlier detection* or *novelty detection* (Bishop, 1994; Tarassenko, 1995).

However, if we only wish to make classification decisions, then it can be wasteful of computational resources, and excessively demanding of data, to find the joint distribution $p(\mathbf{x}, \mathcal{C}_k)$ when in fact we only really need the posterior probabilities $p(\mathcal{C}_k|\mathbf{x})$, which can be obtained directly through approach (b). Indeed, the class-conditional densities may contain a lot of structure that has little effect on the posterior probabilities, as illustrated in Figure 1.27.There has been much interest in exploring the relative merits of generative and discriminative approaches to machine learning, and in finding ways to combine them (Jebara, 2004; Lasserre *et al.*, 2006).

An even simpler approach is (c) in which we use the training data to find a discriminant function $f(\mathbf{x})$ that maps each $\mathbf{x}$ directly onto a class label, thereby combining the inference and decision stages into a single learning problem. In the example of Figure 1.27, this would correspond to finding the value of x shown by the vertical green line, because this is the decision boundary giving the minimum probability of misclassification.

With option (c), however, we no longer have access to the posterior probabilities $p(\mathcal{C}_k|\mathbf{x})$. There are many powerful reasons for wanting to compute the posterior probabilities, even if we subsequently use them to make decisions. These include:

Minimizing risk. Consider a problem in which the elements of the loss matrix are subjected to revision from time to time (such as might occur in a financial application). If we know the posterior probabilities, we can trivially revise the minimum risk decision criterion by modifying (1.81) appropriately. If we have only a discriminant function, then any change to the loss matrix would require that we return to the training data and solve the classification problem afresh.

Reject option. Posterior probabilities allow us to determine a rejection criterion that will minimize the misclassification rate, or more generally the expected loss, for a given fraction of rejected data points.

Compensating for class priors. Consider our medical X-ray problem again, and suppose that we have collected a large number of X-ray images from the general population for use as training data in order to build an automated screening system. Because cancer is rare amongst the general population, we might find that, say, only 1 in every 1,000 examples corresponds to the presence of cancer. If we used such a data set to train an adaptive model, we could run into severe difficulties due to the small proportion of the cancer class. For instance, a classifier that assigned every point to the normal class would already achieve 99.9% accuracy and it would be difficult to avoid this trivial solution. Also, even a large data set will contain very few examples of X-ray images corresponding to cancer, and so the learning algorithm will not be exposed to a broad range of examples of such images and hence is not likely to generalize well. A balanced data set in which we have selected equal numbers of examples from each of the classes would allow us to find a more accurate model. However, we then have to compensate for the effects of our modifications to the training data. Suppose we have used such a modified data set and found models for the posterior probabilities. From Bayes' theorem (1.82), we see that the posterior probabilities are proportional to the prior probabilities, which we can interpret as the fractions of points in each class. We can therefore simply take the posterior probabilities obtained from our artificially balanced data set and first divide by the class fractions in that data set and then multiply by the class fractions in the population to which we wish to apply the model. Finally, we need to normalize to ensure that the new posterior probabilities sum to one. Note that this procedure cannot be applied if we have learned a discriminant function directly instead of determining posterior probabilities.

Combining models. For complex applications, we may wish to break the problem into a number of smaller subproblems each of which can be tackled by a separate module. For example, in our hypothetical medical diagnosis problem, we may have information available from, say, blood tests as well as X-ray images. Rather than combine all of this heterogeneous information into one huge input space, it may be more effective to build one system to interpret the X-ray images and a different one to interpret the blood data. As long as each of the two models gives posterior probabilities for the classes, we can combine the outputs systematically using the rules of probability. One simple way to do this is to assume that, for each class separately, the distributions of inputs for the X-ray images, denoted by $\mathbf{x}_{\mathrm{I}}$, and the blood data, denoted by $\mathbf{x}_{\mathrm{B}}$, are independent, so that

$$p(\mathbf{x}_\mathrm{I}, \mathbf{x}_\mathrm{B}|\mathcal{C}_k) = p(\mathbf{x}_\mathrm{I}|\mathcal{C}_k)p(\mathbf{x}_\mathrm{B}|\mathcal{C}_k). \tag{1.84}$$

Section 8.2

This is an example of *conditional independence* property, because the independence holds when the distribution is conditioned on the class $\mathcal{C}_k$. The posterior probability, given both the X-ray and blood data, is then given by

$$\begin{aligned} p(\mathcal{C}_k|\mathbf{x}_\mathrm{I}, \mathbf{x}_\mathrm{B}) &\propto p(\mathbf{x}_\mathrm{I}, \mathbf{x}_\mathrm{B}|\mathcal{C}_k)p(\mathcal{C}_k) \\ &\propto p(\mathbf{x}_\mathrm{I}|\mathcal{C}_k)p(\mathbf{x}_\mathrm{B}|\mathcal{C}_k)p(\mathcal{C}_k) \\ &\propto \frac{p(\mathcal{C}_k|\mathbf{x}_\mathrm{I})p(\mathcal{C}_k|\mathbf{x}_\mathrm{B})}{p(\mathcal{C}_k)} \end{aligned} \tag{1.85}$$

Section 8.2.2

Thus we need the class prior probabilities $p(\mathcal{C}_k)$, which we can easily estimate from the fractions of data points in each class, and then we need to normalize the resulting posterior probabilities so they sum to one. The particular conditional independence assumption (1.84) is an example of the *naive Bayes model*. Note that the joint marginal distribution $p(\mathbf{x}_\mathrm{I}, \mathbf{x}_\mathrm{B})$ will typically not factorize under this model. We shall see in later chapters how to construct models for combining data that do not require the conditional independence assumption (1.84).

1.5.5 Loss functions for regression

Section 1.1

So far, we have discussed decision theory in the context of classification problems. We now turn to the case of regression problems, such as the curve fitting example discussed earlier. The decision stage consists of choosing a specific estimate $y(\mathbf{x})$ of the value of t for each input $\mathbf{x}$. Suppose that in doing so, we incur a loss $L(t, y(\mathbf{x}))$. The average, or expected, loss is then given by

$$\mathbb{E}[L] = \iint L(t, y(\mathbf{x}))p(\mathbf{x}, t)\,\mathrm{d}\mathbf{x}\,\mathrm{d}t. \tag{1.86}$$

A common choice of loss function in regression problems is the squared loss given by $L(t, y(\mathbf{x})) = \{y(\mathbf{x}) - t\}^2$. In this case, the expected loss can be written

$$\mathbb{E}[L] = \iint \{y(\mathbf{x}) - t\}^2 p(\mathbf{x}, t)\,\mathrm{d}\mathbf{x}\,\mathrm{d}t. \tag{1.87}$$

Appendix D

Our goal is to choose $y(\mathbf{x})$ so as to minimize $\mathbb{E}[L]$. If we assume a completely flexible function $y(\mathbf{x})$, we can do this formally using the calculus of variations to give

$$\frac{\delta \mathbb{E}[L]}{\delta y(\mathbf{x})} = 2\int \{y(\mathbf{x}) - t\}p(\mathbf{x}, t)\,\mathrm{d}t = 0. \tag{1.88}$$

Solving for $y(\mathbf{x})$, and using the sum and product rules of probability, we obtain

$$y(\mathbf{x}) = \frac{\displaystyle\int tp(\mathbf{x}, t)\,\mathrm{d}t}{p(\mathbf{x})} = \int tp(t|\mathbf{x})\,\mathrm{d}t = \mathbb{E}_t[t|\mathbf{x}] \tag{1.89}$$

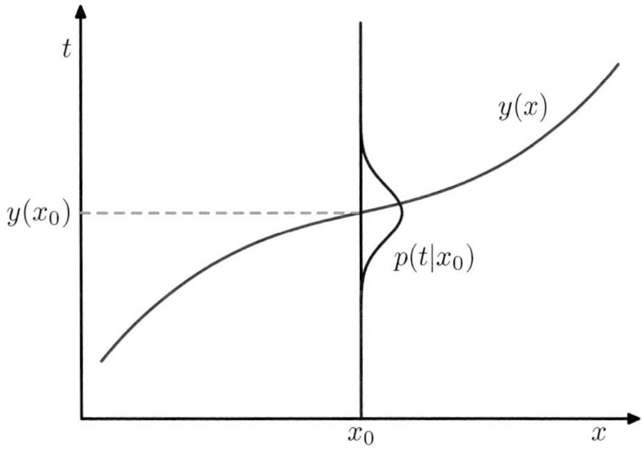

Figure 1.28 The regression function $y(x)$, which minimizes the expected squared loss, is given by the mean of the conditional distribution $p(t|x)$.

which is the conditional average of t conditioned on $\mathbf{x}$ and is known as the *regression function*. This result is illustrated in Figure 1.28. It can readily be extended to multiple target variables represented by the vector $\mathbf{t}$, in which case the optimal solution is the conditional average $\mathbf{y}(\mathbf{x}) = \mathbb{E}_t[\mathbf{t}|\mathbf{x}]$.

Exercise 1.25

We can also derive this result in a slightly different way, which will also shed light on the nature of the regression problem. Armed with the knowledge that the optimal solution is the conditional expectation, we can expand the square term as follows

$$\begin{aligned}\{y(\mathbf{x}) - t\}^2 &= \{y(\mathbf{x}) - \mathbb{E}[t|\mathbf{x}] + \mathbb{E}[t|\mathbf{x}] - t\}^2 \\ &= \{y(\mathbf{x}) - \mathbb{E}[t|\mathbf{x}]\}^2 + 2\{y(\mathbf{x}) - \mathbb{E}[t|\mathbf{x}]\}\{\mathbb{E}[t|\mathbf{x}] - t\} + \{\mathbb{E}[t|\mathbf{x}] - t\}^2\end{aligned}$$

where, to keep the notation uncluttered, we use $\mathbb{E}[t|\mathbf{x}]$ to denote $\mathbb{E}_t[t|\mathbf{x}]$. Substituting into the loss function and performing the integral over t, we see that the cross-term vanishes and we obtain an expression for the loss function in the form

$$\mathbb{E}[L] = \int \{y(\mathbf{x}) - \mathbb{E}[t|\mathbf{x}]\}^2\, p(\mathbf{x})\,\mathrm{d}\mathbf{x} + \int \mathrm{var}\,[t|\mathbf{x}]\, p(\mathbf{x})\,\mathrm{d}\mathbf{x}. \tag{1.90}$$

The function $y(\mathbf{x})$ we seek to determine enters only in the first term, which will be minimized when $y(\mathbf{x})$ is equal to $\mathbb{E}[t|\mathbf{x}]$, in which case this term will vanish. This is simply the result that we derived previously and that shows that the optimal least squares predictor is given by the conditional mean. The second term is the variance of the distribution of t, averaged over $\mathbf{x}$. It represents the intrinsic variability of the target data and can be regarded as noise. Because it is independent of $y(\mathbf{x})$, it represents the irreducible minimum value of the loss function.

As with the classification problem, we can either determine the appropriate probabilities and then use these to make optimal decisions, or we can build models that make decisions directly. Indeed, we can identify three distinct approaches to solving regression problems given, in order of decreasing complexity, by:

(a) First solve the inference problem of determining the joint density $p(\mathbf{x}, t)$. Then normalize to find the conditional density $p(t|\mathbf{x})$, and finally calculate the conditional mean given by (1.89).

(b) First solve the inference problem of determining the conditional density $p(t|\mathbf{x})$, and then subsequently calculate the conditional mean given by (1.89).

(c) Find a regression function $y(\mathbf{x})$ directly from the training data.

The relative merits of these three approaches follow the same lines as for classification problems above.

The squared loss is not the only possible choice of loss function for regression. Indeed, there are situations in which squared loss can lead to very poor results and where we need to develop more sophisticated approaches. An important example concerns situations in which the conditional distribution $p(t|\mathbf{x})$ is multimodal, as often arises in the solution of inverse problems. Here we consider briefly one simple generalization of the squared loss, called the *Minkowski* loss, whose expectation is given by

Section 5.6

$$\mathbb{E}[L_q] = \iint |y(\mathbf{x}) - t|^q p(\mathbf{x}, t) \, \mathrm{d}\mathbf{x} \, \mathrm{d}t \tag{1.91}$$

which reduces to the expected squared loss for $q = 2$. The function $|y - t|^q$ is plotted against $y - t$ for various values of q in Figure 1.29. The minimum of $\mathbb{E}[L_q]$ is given by the conditional mean for $q = 2$, the conditional median for $q = 1$, and the conditional mode for $q \to 0$.

Exercise 1.27

1.6. Information Theory

In this chapter, we have discussed a variety of concepts from probability theory and decision theory that will form the foundations for much of the subsequent discussion in this book. We close this chapter by introducing some additional concepts from the field of information theory, which will also prove useful in our development of pattern recognition and machine learning techniques. Again, we shall focus only on the key concepts, and we refer the reader elsewhere for more detailed discussions (Viterbi and Omura, 1979; Cover and Thomas, 1991; MacKay, 2003) .

We begin by considering a discrete random variable x and we ask how much information is received when we observe a specific value for this variable. The amount of information can be viewed as the 'degree of surprise' on learning the value of x. If we are told that a highly improbable event has just occurred, we will have received more information than if we were told that some very likely event has just occurred, and if we knew that the event was certain to happen we would receive no information. Our measure of information content will therefore depend on the probability distribution $p(x)$, and we therefore look for a quantity $h(x)$ that is a monotonic function of the probability $p(x)$ and that expresses the information content. The form of $h(\cdot)$ can be found by noting that if we have two events x and y that are unrelated, then the information gain from observing both of them should be the sum of the information gained from each of them separately, so that $h(x, y) = h(x) + h(y)$. Two unrelated events will be statistically independent and so $p(x, y) = p(x)p(y)$. From these two relationships, it is easily shown that $h(x)$ must be given by the logarithm of $p(x)$ and so we have

Exercise 1.28

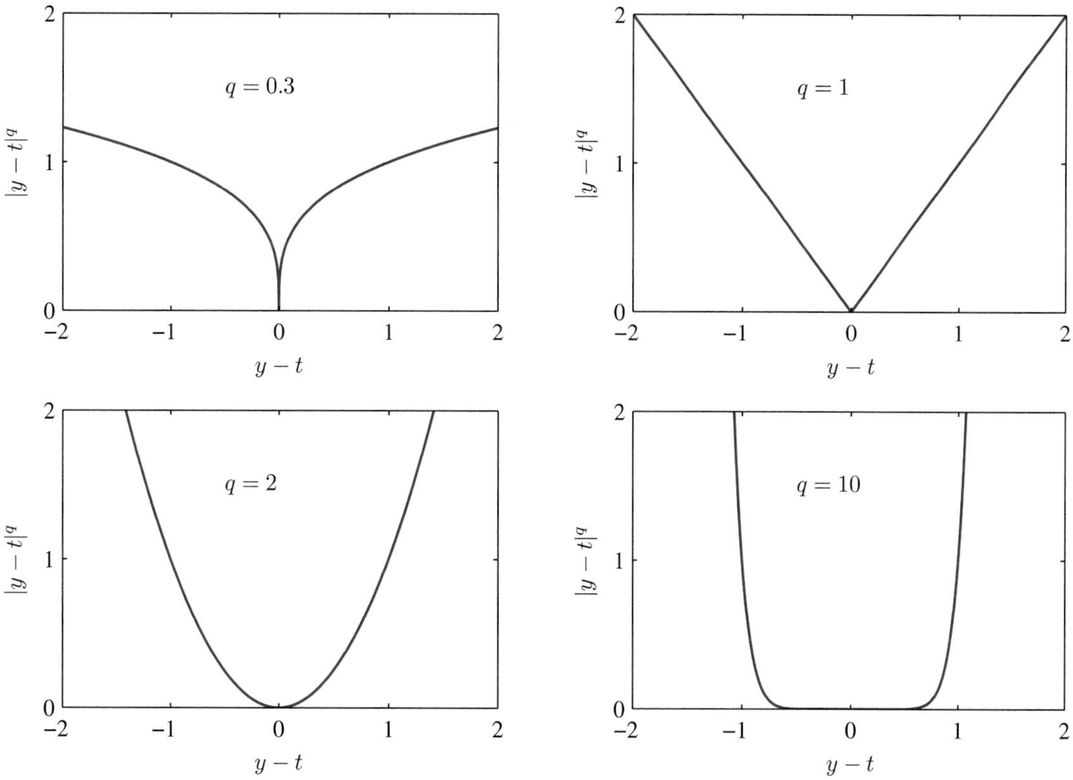

Figure 1.29 Plots of the quantity $L_q = |y - t|^q$ for various values of q.

$$h(x) = -\log_2 p(x) \tag{1.92}$$

where the negative sign ensures that information is positive or zero. Note that low probability events x correspond to high information content. The choice of basis for the logarithm is arbitrary, and for the moment we shall adopt the convention prevalent in information theory of using logarithms to the base of 2. In this case, as we shall see shortly, the units of $h(x)$ are bits ('binary digits').

Now suppose that a sender wishes to transmit the value of a random variable to a receiver. The average amount of information that they transmit in the process is obtained by taking the expectation of (1.92) with respect to the distribution $p(x)$ and is given by

$$\mathrm{H}[x] = -\sum_x p(x) \log_2 p(x). \tag{1.93}$$

This important quantity is called the *entropy* of the random variable x. Note that $\lim_{p \to 0} p \ln p = 0$ and so we shall take $p(x) \ln p(x) = 0$ whenever we encounter a value for x such that $p(x) = 0$.

So far we have given a rather heuristic motivation for the definition of information (1.92) and the corresponding entropy (1.93). We now show that these definitions indeed possess useful properties. Consider a random variable x having 8 possible states, each of which is equally likely. In order to communicate the value of x to a receiver, we would need to transmit a message of length 3 bits. Notice that the entropy of this variable is given by

$$\mathrm{H}[x] = -8 \times \frac{1}{8} \log_2 \frac{1}{8} = 3 \text{ bits.}$$

Now consider an example (Cover and Thomas, 1991) of a variable having 8 possible states $\{a, b, c, d, e, f, g, h\}$ for which the respective probabilities are given by $(\frac{1}{2}, \frac{1}{4}, \frac{1}{8}, \frac{1}{16}, \frac{1}{64}, \frac{1}{64}, \frac{1}{64}, \frac{1}{64})$. The entropy in this case is given by

$$\mathrm{H}[x] = -\frac{1}{2} \log_2 \frac{1}{2} - \frac{1}{4} \log_2 \frac{1}{4} - \frac{1}{8} \log_2 \frac{1}{8} - \frac{1}{16} \log_2 \frac{1}{16} - \frac{4}{64} \log_2 \frac{1}{64} = 2 \text{ bits.}$$

We see that the nonuniform distribution has a smaller entropy than the uniform one, and we shall gain some insight into this shortly when we discuss the interpretation of entropy in terms of disorder. For the moment, let us consider how we would transmit the identity of the variable's state to a receiver. We could do this, as before, using a 3-bit number. However, we can take advantage of the nonuniform distribution by using shorter codes for the more probable events, at the expense of longer codes for the less probable events, in the hope of getting a shorter average code length. This can be done by representing the states $\{a, b, c, d, e, f, g, h\}$ using, for instance, the following set of code strings: 0, 10, 110, 1110, 111100, 111101, 111110, 111111. The average length of the code that has to be transmitted is then

$$\text{average code length} = \frac{1}{2} \times 1 + \frac{1}{4} \times 2 + \frac{1}{8} \times 3 + \frac{1}{16} \times 4 + 4 \times \frac{1}{64} \times 6 = 2 \text{ bits}$$

which again is the same as the entropy of the random variable. Note that shorter code strings cannot be used because it must be possible to disambiguate a concatenation of such strings into its component parts. For instance, 11001110 decodes uniquely into the state sequence c, a, d.

This relation between entropy and shortest coding length is a general one. The *noiseless coding theorem* (Shannon, 1948) states that the entropy is a lower bound on the number of bits needed to transmit the state of a random variable.

From now on, we shall switch to the use of natural logarithms in defining entropy, as this will provide a more convenient link with ideas elsewhere in this book. In this case, the entropy is measured in units of 'nats' instead of bits, which differ simply by a factor of $\ln 2$.

We have introduced the concept of entropy in terms of the average amount of information needed to specify the state of a random variable. In fact, the concept of entropy has much earlier origins in physics where it was introduced in the context of equilibrium thermodynamics and later given a deeper interpretation as a measure of disorder through developments in statistical mechanics. We can understand this alternative view of entropy by considering a set of N identical objects that are to be divided amongst a set of bins, such that there are n_i objects in the ith bin. Consider the number of different ways of allocating the objects to the bins. There are N ways to choose the first object, $(N-1)$ ways to choose the second object, and so on, leading to a total of $N!$ ways to allocate all N objects to the bins, where $N!$ (pronounced 'factorial N') denotes the product $N \times (N-1) \times \cdots \times 2 \times 1$. However, we don't wish to distinguish between rearrangements of objects within each bin. In the ith bin there are $n_i!$ ways of reordering the objects, and so the total number of ways of allocating the N objects to the bins is given by

$$W = \frac{N!}{\prod_i n_i!} \tag{1.94}$$

which is called the *multiplicity*. The entropy is then defined as the logarithm of the multiplicity scaled by an appropriate constant

$$\mathrm{H} = \frac{1}{N} \ln W = \frac{1}{N} \ln N! - \frac{1}{N} \sum_i \ln n_i!. \tag{1.95}$$

We now consider the limit $N \to \infty$, in which the fractions n_i/N are held fixed, and apply Stirling's approximation

$$\ln N! \simeq N \ln N - N \tag{1.96}$$

which gives

$$\mathrm{H} = - \lim_{N\to\infty} \sum_i \left(\frac{n_i}{N}\right) \ln \left(\frac{n_i}{N}\right) = - \sum_i p_i \ln p_i \tag{1.97}$$

where we have used $\sum_i n_i = N$. Here $p_i = \lim_{N\to\infty}(n_i/N)$ is the probability of an object being assigned to the ith bin. In physics terminology, the specific arrangements of objects in the bins is called a *microstate*, and the overall distribution of occupation numbers, expressed through the ratios n_i/N, is called a *macrostate*. The multiplicity W is also known as the *weight* of the macrostate.

We can interpret the bins as the states x_i of a discrete random variable X, where $p(X = x_i) = p_i$. The entropy of the random variable X is then

$$\mathrm{H}[p] = - \sum_i p(x_i) \ln p(x_i). \tag{1.98}$$

Distributions $p(x_i)$ that are sharply peaked around a few values will have a relatively low entropy, whereas those that are spread more evenly across many values will have higher entropy, as illustrated in Figure 1.30. Because $0 \leqslant p_i \leqslant 1$, the entropy is nonnegative, and it will equal its minimum value of 0 when one of the $p_i = 1$ and all other $p_{j \neq i} = 0$. The maximum entropy configuration can be found by maximizing H using a Lagrange multiplier to enforce the normalization constraint on the probabilities. Thus we maximize

Appendix E

$$\widetilde{\mathrm{H}} = - \sum_i p(x_i) \ln p(x_i) + \lambda \left(\sum_i p(x_i) - 1 \right) \tag{1.99}$$

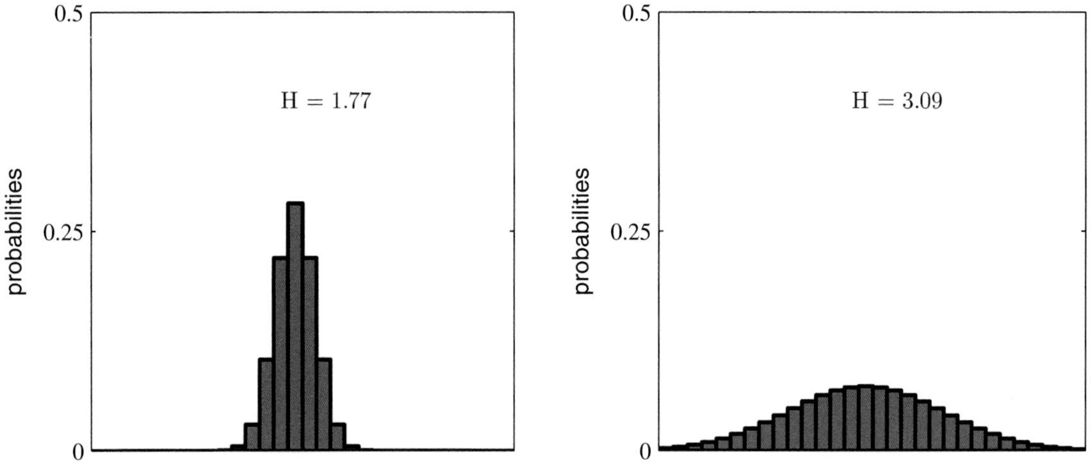

Figure 1.30 Histograms of two probability distributions over 30 bins illustrating the higher value of the entropy H for the broader distribution. The largest entropy would arise from a uniform distribution that would give $\mathrm{H} = -\ln(1/30) = 3.40$.

from which we find that all of the $p(x_i)$ are equal and are given by $p(x_i) = 1/M$ where M is the total number of states x_i. The corresponding value of the entropy is then $\mathrm{H} = \ln M$. This result can also be derived from Jensen's inequality (to be discussed shortly). To verify that the stationary point is indeed a maximum, we can evaluate the second derivative of the entropy, which gives

Exercise 1.29

$$\frac{\partial^2 \widetilde{\mathrm{H}}}{\partial p(x_i) \partial p(x_j)} = -I_{ij} \frac{1}{p_i} \tag{1.100}$$

where I_{ij} are the elements of the identity matrix.

We can extend the definition of entropy to include distributions $p(x)$ over continuous variables x as follows. First divide x into bins of width Δ. Then, assuming $p(x)$ is continuous, the *mean value theorem* (Weisstein, 1999) tells us that, for each such bin, there must exist a value x_i such that

$$\int_{i\Delta}^{(i+1)\Delta} p(x)\,\mathrm{d}x = p(x_i)\Delta. \tag{1.101}$$

We can now quantize the continuous variable x by assigning any value x to the value x_i whenever x falls in the ith bin. The probability of observing the value x_i is then $p(x_i)\Delta$. This gives a discrete distribution for which the entropy takes the form

$$\mathrm{H}_\Delta = -\sum_i p(x_i)\Delta \ln \left(p(x_i)\Delta\right) = -\sum_i p(x_i)\Delta \ln p(x_i) - \ln \Delta \tag{1.102}$$

where we have used $\sum_i p(x_i)\Delta = 1$, which follows from (1.101). We now omit the second term $-\ln \Delta$ on the right-hand side of (1.102) and then consider the limit

$\Delta \to 0$. The first term on the right-hand side of (1.102) will approach the integral of $p(x) \ln p(x)$ in this limit so that

$$\lim_{\Delta \to 0} \left\{ -\sum_i p(x_i)\Delta \ln p(x_i) \right\} = -\int p(x) \ln p(x)\, \mathrm{d}x \tag{1.103}$$

where the quantity on the right-hand side is called the *differential entropy*. We see that the discrete and continuous forms of the entropy differ by a quantity $\ln \Delta$, which diverges in the limit $\Delta \to 0$. This reflects the fact that to specify a continuous variable very precisely requires a large number of bits. For a density defined over multiple continuous variables, denoted collectively by the vector $\mathbf{x}$, the differential entropy is given by

$$\mathrm{H}[\mathbf{x}] = -\int p(\mathbf{x}) \ln p(\mathbf{x})\, \mathrm{d}\mathbf{x}. \tag{1.104}$$

In the case of discrete distributions, we saw that the maximum entropy configuration corresponded to an equal distribution of probabilities across the possible states of the variable. Let us now consider the maximum entropy configuration for a continuous variable. In order for this maximum to be well defined, it will be necessary to constrain the first and second moments of $p(x)$ as well as preserving the normalization constraint. We therefore maximize the differential entropy with the

Ludwig Boltzmann
1844–1906

Ludwig Eduard Boltzmann was an Austrian physicist who created the field of statistical mechanics. Prior to Boltzmann, the concept of entropy was already known from classical thermodynamics where it quantifies the fact that when we take energy from a system, not all of that energy is typically available to do useful work. Boltzmann showed that the thermodynamic entropy S, a macroscopic quantity, could be related to the statistical properties at the microscopic level. This is expressed through the famous equation $S = k \ln W$ in which W represents the number of possible microstates in a macrostate, and $k \simeq 1.38 \times 10^{-23}$ (in units of Joules per Kelvin) is known as Boltzmann's constant. Boltzmann's ideas were disputed by many scientists of the day. One difficulty they saw arose from the second law of thermodynamics, which states that the entropy of a closed system tends to increase with time. By contrast, at the microscopic level the classical Newtonian equations of physics are reversible, and so they found it difficult to see how the latter could explain the former. They didn't fully appreciate Boltzmann's arguments, which were statistical in nature and which concluded not that entropy could never decrease over time but simply that with overwhelming probability it would generally increase. Boltzmann even had a long-running dispute with the editor of the leading German physics journal who refused to let him refer to atoms and molecules as anything other than convenient theoretical constructs. The continued attacks on his work led to bouts of depression, and eventually he committed suicide. Shortly after Boltzmann's death, new experiments by Perrin on colloidal suspensions verified his theories and confirmed the value of the Boltzmann constant. The equation $S = k \ln W$ is carved on Boltzmann's tombstone.

three constraints

$$\int_{-\infty}^{\infty} p(x)\,\mathrm{d}x = 1 \tag{1.105}$$

$$\int_{-\infty}^{\infty} xp(x)\,\mathrm{d}x = \mu \tag{1.106}$$

$$\int_{-\infty}^{\infty} (x-\mu)^2 p(x)\,\mathrm{d}x = \sigma^2. \tag{1.107}$$

Appendix E

The constrained maximization can be performed using Lagrange multipliers so that we maximize the following functional with respect to $p(x)$

$$-\int_{-\infty}^{\infty} p(x)\ln p(x)\,\mathrm{d}x + \lambda_1\left(\int_{-\infty}^{\infty} p(x)\,\mathrm{d}x - 1\right)$$
$$+\lambda_2\left(\int_{-\infty}^{\infty} xp(x)\,\mathrm{d}x - \mu\right) + \lambda_3\left(\int_{-\infty}^{\infty} (x-\mu)^2 p(x)\,\mathrm{d}x - \sigma^2\right).$$

Appendix D

Using the calculus of variations, we set the derivative of this functional to zero giving

$$p(x) = \exp\left\{-1+\lambda_1+\lambda_2 x+\lambda_3(x-\mu)^2\right\}. \tag{1.108}$$

Exercise 1.34

The Lagrange multipliers can be found by back substitution of this result into the three constraint equations, leading finally to the result

$$p(x) = \frac{1}{(2\pi\sigma^2)^{1/2}}\exp\left\{-\frac{(x-\mu)^2}{2\sigma^2}\right\} \tag{1.109}$$

and so the distribution that maximizes the differential entropy is the Gaussian. Note that we did not constrain the distribution to be nonnegative when we maximized the entropy. However, because the resulting distribution is indeed nonnegative, we see with hindsight that such a constraint is not necessary.

Exercise 1.35

If we evaluate the differential entropy of the Gaussian, we obtain

$$\mathrm{H}[x] = \frac{1}{2}\left\{1+\ln(2\pi\sigma^2)\right\}. \tag{1.110}$$

Thus we see again that the entropy increases as the distribution becomes broader, i.e., as σ^2 increases. This result also shows that the differential entropy, unlike the discrete entropy, can be negative, because $\mathrm{H}(x) < 0$ in (1.110) for $\sigma^2 < 1/(2\pi e)$.

Suppose we have a joint distribution $p(\mathbf{x},\mathbf{y})$ from which we draw pairs of values of $\mathbf{x}$ and $\mathbf{y}$. If a value of $\mathbf{x}$ is already known, then the additional information needed to specify the corresponding value of $\mathbf{y}$ is given by $-\ln p(\mathbf{y}|\mathbf{x})$. Thus the average additional information needed to specify $\mathbf{y}$ can be written as

$$\mathrm{H}[\mathbf{y}|\mathbf{x}] = -\iint p(\mathbf{y},\mathbf{x})\ln p(\mathbf{y}|\mathbf{x})\,\mathrm{d}\mathbf{y}\,\mathrm{d}\mathbf{x} \tag{1.111}$$

which is called the *conditional entropy* of $\mathbf{y}$ given $\mathbf{x}$. It is easily seen, using the product rule, that the conditional entropy satisfies the relation

Exercise 1.37

$$\mathrm{H}[\mathbf{x}, \mathbf{y}] = \mathrm{H}[\mathbf{y}|\mathbf{x}] + \mathrm{H}[\mathbf{x}] \tag{1.112}$$

where $\mathrm{H}[\mathbf{x}, \mathbf{y}]$ is the differential entropy of $p(\mathbf{x}, \mathbf{y})$ and $\mathrm{H}[\mathbf{x}]$ is the differential entropy of the marginal distribution $p(\mathbf{x})$. Thus the information needed to describe $\mathbf{x}$ and $\mathbf{y}$ is given by the sum of the information needed to describe $\mathbf{x}$ alone plus the additional information required to specify $\mathbf{y}$ given $\mathbf{x}$.

1.6.1 Relative entropy and mutual information

So far in this section, we have introduced a number of concepts from information theory, including the key notion of entropy. We now start to relate these ideas to pattern recognition. Consider some unknown distribution $p(\mathbf{x})$, and suppose that we have modelled this using an approximating distribution $q(\mathbf{x})$. If we use $q(\mathbf{x})$ to construct a coding scheme for the purpose of transmitting values of $\mathbf{x}$ to a receiver, then the average *additional* amount of information (in nats) required to specify the value of $\mathbf{x}$ (assuming we choose an efficient coding scheme) as a result of using $q(\mathbf{x})$ instead of the true distribution $p(\mathbf{x})$ is given by

$$\begin{aligned} \mathrm{KL}(p\|q) &= -\int p(\mathbf{x}) \ln q(\mathbf{x})\, \mathrm{d}\mathbf{x} - \left(-\int p(\mathbf{x}) \ln p(\mathbf{x})\, \mathrm{d}\mathbf{x}\right) \\ &= -\int p(\mathbf{x}) \ln \left\{ \frac{q(\mathbf{x})}{p(\mathbf{x})} \right\} \mathrm{d}\mathbf{x}. \end{aligned} \tag{1.113}$$

This is known as the *relative entropy* or *Kullback-Leibler divergence*, or *KL divergence* (Kullback and Leibler, 1951), between the distributions $p(\mathbf{x})$ and $q(\mathbf{x})$. Note that it is not a symmetrical quantity, that is to say $\mathrm{KL}(p\|q) \not\equiv \mathrm{KL}(q\|p)$.

We now show that the Kullback-Leibler divergence satisfies $\mathrm{KL}(p\|q) \geqslant 0$ with equality if, and only if, $p(\mathbf{x}) = q(\mathbf{x})$. To do this we first introduce the concept of *convex* functions. A function $f(x)$ is said to be convex if it has the property that every chord lies on or above the function, as shown in Figure 1.31. Any value of x in the interval from $x = a$ to $x = b$ can be written in the form $\lambda a + (1 - \lambda) b$ where $0 \leqslant \lambda \leqslant 1$. The corresponding point on the chord is given by $\lambda f(a) + (1 - \lambda) f(b)$,

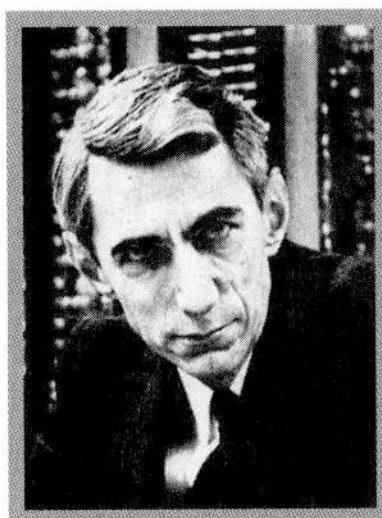

Claude Shannon
1916–2001

After graduating from Michigan and MIT, Shannon joined the AT&T Bell Telephone laboratories in 1941. His paper 'A Mathematical Theory of Communication' published in the *Bell System Technical Journal* in 1948 laid the foundations for modern information theory. This paper introduced the word 'bit', and his concept that information could be sent as a stream of 1s and 0s paved the way for the communications revolution. It is said that von Neumann recommended to Shannon that he use the term entropy, not only because of its similarity to the quantity used in physics, but also because "nobody knows what entropy really is, so in any discussion you will always have an advantage".

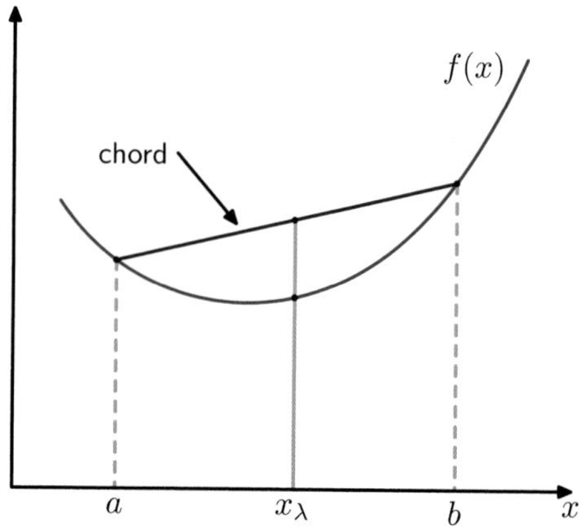

Figure 1.31 A convex function $f(x)$ is one for which every chord (shown in blue) lies on or above the function (shown in red).

and the corresponding value of the function is $f(\lambda a + (1-\lambda)b)$. Convexity then implies

$$f(\lambda a + (1-\lambda)b) \leqslant \lambda f(a) + (1-\lambda)f(b). \tag{1.114}$$

Exercise 1.36

This is equivalent to the requirement that the second derivative of the function be everywhere positive. Examples of convex functions are $x \ln x$ (for $x > 0$) and x^2. A function is called *strictly convex* if the equality is satisfied only for $\lambda = 0$ and $\lambda = 1$. If a function has the opposite property, namely that every chord lies on or below the function, it is called *concave*, with a corresponding definition for *strictly concave*. If a function $f(x)$ is convex, then $-f(x)$ will be concave.

Exercise 1.38

Using the technique of proof by induction, we can show from (1.114) that a convex function $f(x)$ satisfies

$$f\left(\sum_{i=1}^{M} \lambda_i x_i\right) \leqslant \sum_{i=1}^{M} \lambda_i f(x_i) \tag{1.115}$$

where $\lambda_i \geqslant 0$ and $\sum_i \lambda_i = 1$, for any set of points $\{x_i\}$. The result (1.115) is known as *Jensen's inequality*. If we interpret the λ_i as the probability distribution over a discrete variable x taking the values $\{x_i\}$, then (1.115) can be written

$$f\left(\mathbb{E}[x]\right) \leqslant \mathbb{E}[f(x)] \tag{1.116}$$

where $\mathbb{E}[\cdot]$ denotes the expectation. For continuous variables, Jensen's inequality takes the form

$$f\left(\int \mathbf{x} p(\mathbf{x})\,\mathrm{d}\mathbf{x}\right) \leqslant \int f(\mathbf{x}) p(\mathbf{x})\,\mathrm{d}\mathbf{x}. \tag{1.117}$$

We can apply Jensen's inequality in the form (1.117) to the Kullback-Leibler divergence (1.113) to give

$$\mathrm{KL}(p\|q) = -\int p(\mathbf{x}) \ln \left\{ \frac{q(\mathbf{x})}{p(\mathbf{x})} \right\} \mathrm{d}\mathbf{x} \geqslant -\ln \int q(\mathbf{x})\,\mathrm{d}\mathbf{x} = 0 \tag{1.118}$$

where we have used the fact that $-\ln x$ is a convex function, together with the normalization condition $\int q(\mathbf{x})\,\mathrm{d}\mathbf{x} = 1$. In fact, $-\ln x$ is a strictly convex function, so the equality will hold if, and only if, $q(\mathbf{x}) = p(\mathbf{x})$ for all $\mathbf{x}$. Thus we can interpret the Kullback-Leibler divergence as a measure of the dissimilarity of the two distributions $p(\mathbf{x})$ and $q(\mathbf{x})$.

We see that there is an intimate relationship between data compression and density estimation (i.e., the problem of modelling an unknown probability distribution) because the most efficient compression is achieved when we know the true distribution. If we use a distribution that is different from the true one, then we must necessarily have a less efficient coding, and on average the additional information that must be transmitted is (at least) equal to the Kullback-Leibler divergence between the two distributions.

Suppose that data is being generated from an unknown distribution $p(\mathbf{x})$ that we wish to model. We can try to approximate this distribution using some parametric distribution $q(\mathbf{x}|\boldsymbol{\theta})$, governed by a set of adjustable parameters $\boldsymbol{\theta}$, for example a multivariate Gaussian. One way to determine $\boldsymbol{\theta}$ is to minimize the Kullback-Leibler divergence between $p(\mathbf{x})$ and $q(\mathbf{x}|\boldsymbol{\theta})$ with respect to $\boldsymbol{\theta}$. We cannot do this directly because we don't know $p(\mathbf{x})$. Suppose, however, that we have observed a finite set of training points $\mathbf{x}_n$, for $n = 1, \ldots, N$, drawn from $p(\mathbf{x})$. Then the expectation with respect to $p(\mathbf{x})$ can be approximated by a finite sum over these points, using (1.35), so that

$$\mathrm{KL}(p\|q) \simeq \frac{1}{N}\sum_{n=1}^{N}\left\{-\ln q(\mathbf{x}_n|\boldsymbol{\theta}) + \ln p(\mathbf{x}_n)\right\}. \tag{1.119}$$

The second term on the right-hand side of (1.119) is independent of $\boldsymbol{\theta}$, and the first term is the negative log likelihood function for $\boldsymbol{\theta}$ under the distribution $q(\mathbf{x}|\boldsymbol{\theta})$ evaluated using the training set. Thus we see that minimizing this Kullback-Leibler divergence is equivalent to maximizing the likelihood function.

Now consider the joint distribution between two sets of variables $\mathbf{x}$ and $\mathbf{y}$ given by $p(\mathbf{x}, \mathbf{y})$. If the sets of variables are independent, then their joint distribution will factorize into the product of their marginals $p(\mathbf{x}, \mathbf{y}) = p(\mathbf{x})p(\mathbf{y})$. If the variables are not independent, we can gain some idea of whether they are 'close' to being independent by considering the Kullback-Leibler divergence between the joint distribution and the product of the marginals, given by

$$\begin{aligned}\mathrm{I}[\mathbf{x}, \mathbf{y}] &\equiv \mathrm{KL}(p(\mathbf{x}, \mathbf{y})\|p(\mathbf{x})p(\mathbf{y})) \\ &= -\iint p(\mathbf{x}, \mathbf{y}) \ln\left(\frac{p(\mathbf{x})p(\mathbf{y})}{p(\mathbf{x}, \mathbf{y})}\right) \mathrm{d}\mathbf{x}\,\mathrm{d}\mathbf{y}\end{aligned} \tag{1.120}$$

which is called the *mutual information* between the variables $\mathbf{x}$ and $\mathbf{y}$. From the properties of the Kullback-Leibler divergence, we see that $I(\mathbf{x}, \mathbf{y}) \geqslant 0$ with equality if, and only if, $\mathbf{x}$ and $\mathbf{y}$ are independent. Using the sum and product rules of probability, we see that the mutual information is related to the conditional entropy through

Exercise 1.41

$$\mathrm{I}[\mathbf{x}, \mathbf{y}] = \mathrm{H}[\mathbf{x}] - \mathrm{H}[\mathbf{x}|\mathbf{y}] = \mathrm{H}[\mathbf{y}] - \mathrm{H}[\mathbf{y}|\mathbf{x}]. \tag{1.121}$$

Thus we can view the mutual information as the reduction in the uncertainty about $\mathbf{x}$ by virtue of being told the value of $\mathbf{y}$ (or vice versa). From a Bayesian perspective, we can view $p(\mathbf{x})$ as the prior distribution for $\mathbf{x}$ and $p(\mathbf{x}|\mathbf{y})$ as the posterior distribution after we have observed new data $\mathbf{y}$. The mutual information therefore represents the reduction in uncertainty about $\mathbf{x}$ as a consequence of the new observation $\mathbf{y}$.

Exercises

1.1 ($\star$) **www** Consider the sum-of-squares error function given by (1.2) in which the function $y(x, \mathbf{w})$ is given by the polynomial (1.1). Show that the coefficients $\mathbf{w} = \{w_i\}$ that minimize this error function are given by the solution to the following set of linear equations

$$\sum_{j=0}^{M} A_{ij} w_j = T_i \tag{1.122}$$

where

$$A_{ij} = \sum_{n=1}^{N} (x_n)^{i+j}, \qquad T_i = \sum_{n=1}^{N} (x_n)^i t_n. \tag{1.123}$$

Here a suffix i or j denotes the index of a component, whereas $(x)^i$ denotes x raised to the power of i.

1.2 ($\star$) Write down the set of coupled linear equations, analogous to (1.122), satisfied by the coefficients w_i which minimize the regularized sum-of-squares error function given by (1.4).

1.3 ($\star\star$) Suppose that we have three coloured boxes r (red), b (blue), and g (green). Box r contains 3 apples, 4 oranges, and 3 limes, box b contains 1 apple, 1 orange, and 0 limes, and box g contains 3 apples, 3 oranges, and 4 limes. If a box is chosen at random with probabilities $p(r) = 0.2$, $p(b) = 0.2$, $p(g) = 0.6$, and a piece of fruit is removed from the box (with equal probability of selecting any of the items in the box), then what is the probability of selecting an apple? If we observe that the selected fruit is in fact an orange, what is the probability that it came from the green box?

1.4 ($\star\star$) **www** Consider a probability density $p_x(x)$ defined over a continuous variable x, and suppose that we make a nonlinear change of variable using $x = g(y)$, so that the density transforms according to (1.27). By differentiating (1.27), show that the location $\widehat{y}$ of the maximum of the density in y is not in general related to the location $\widehat{x}$ of the maximum of the density over x by the simple functional relation $\widehat{x} = g(\widehat{y})$ as a consequence of the Jacobian factor. This shows that the maximum of a probability density (in contrast to a simple function) is dependent on the choice of variable. Verify that, in the case of a linear transformation, the location of the maximum transforms in the same way as the variable itself.

1.5 ($\star$) Using the definition (1.38) show that $\mathrm{var}[f(x)]$ satisfies (1.39).

1.6 ($\star$) Show that if two variables x and y are independent, then their covariance is zero.

1.7 ($\star\star$) **www** In this exercise, we prove the normalization condition (1.48) for the univariate Gaussian. To do this consider, the integral

$$I = \int_{-\infty}^{\infty} \exp\left(-\frac{1}{2\sigma^2}x^2\right) \mathrm{d}x \tag{1.124}$$

which we can evaluate by first writing its square in the form

$$I^2 = \int_{-\infty}^{\infty}\int_{-\infty}^{\infty} \exp\left(-\frac{1}{2\sigma^2}x^2 - \frac{1}{2\sigma^2}y^2\right) \mathrm{d}x\,\mathrm{d}y. \tag{1.125}$$

Now make the transformation from Cartesian coordinates (x, y) to polar coordinates (r, θ) and then substitute $u = r^2$. Show that, by performing the integrals over θ and u, and then taking the square root of both sides, we obtain

$$I = \left(2\pi\sigma^2\right)^{1/2}. \tag{1.126}$$

Finally, use this result to show that the Gaussian distribution $\mathcal{N}(x|\mu, \sigma^2)$ is normalized.

1.8 ($\star\star$) **www** By using a change of variables, verify that the univariate Gaussian distribution given by (1.46) satisfies (1.49). Next, by differentiating both sides of the normalization condition

$$\int_{-\infty}^{\infty} \mathcal{N}\left(x|\mu, \sigma^2\right) \mathrm{d}x = 1 \tag{1.127}$$

with respect to σ^2, verify that the Gaussian satisfies (1.50). Finally, show that (1.51) holds.

1.9 ($\star$) **www** Show that the mode (i.e. the maximum) of the Gaussian distribution (1.46) is given by μ. Similarly, show that the mode of the multivariate Gaussian (1.52) is given by $\boldsymbol{\mu}$.

1.10 ($\star$) **www** Suppose that the two variables x and z are statistically independent. Show that the mean and variance of their sum satisfies

$$\mathbb{E}[x+z] = \mathbb{E}[x] + \mathbb{E}[z] \tag{1.128}$$

$$\mathrm{var}[x+z] = \mathrm{var}[x] + \mathrm{var}[z]. \tag{1.129}$$

1.11 ($\star$) By setting the derivatives of the log likelihood function (1.54) with respect to μ and σ^2 equal to zero, verify the results (1.55) and (1.56).

1.12 ($\star\star$) **www** Using the results (1.49) and (1.50), show that

$$\mathbb{E}[x_n x_m] = \mu^2 + I_{nm}\sigma^2 \tag{1.130}$$

where x_n and x_m denote data points sampled from a Gaussian distribution with mean μ and variance σ^2, and I_{nm} satisfies $I_{nm} = 1$ if $n = m$ and $I_{nm} = 0$ otherwise. Hence prove the results (1.57) and (1.58).

1.13 ($\star$) Suppose that the variance of a Gaussian is estimated using the result (1.56) but with the maximum likelihood estimate $\mu_{\rm ML}$ replaced with the true value μ of the mean. Show that this estimator has the property that its expectation is given by the true variance σ^2.

1.14 ($\star\star$) Show that an arbitrary square matrix with elements w_{ij} can be written in the form $w_{ij} = w_{ij}^{\rm S} + w_{ij}^{\rm A}$ where $w_{ij}^{\rm S}$ and $w_{ij}^{\rm A}$ are symmetric and anti-symmetric matrices, respectively, satisfying $w_{ij}^{\rm S} = w_{ji}^{\rm S}$ and $w_{ij}^{\rm A} = -w_{ji}^{\rm A}$ for all i and j. Now consider the second order term in a higher order polynomial in D dimensions, given by

$$\sum_{i=1}^{D}\sum_{j=1}^{D} w_{ij} x_i x_j. \tag{1.131}$$

Show that

$$\sum_{i=1}^{D}\sum_{j=1}^{D} w_{ij} x_i x_j = \sum_{i=1}^{D}\sum_{j=1}^{D} w_{ij}^{\rm S} x_i x_j \tag{1.132}$$

so that the contribution from the anti-symmetric matrix vanishes. We therefore see that, without loss of generality, the matrix of coefficients w_{ij} can be chosen to be symmetric, and so not all of the D^2 elements of this matrix can be chosen independently. Show that the number of independent parameters in the matrix $w_{ij}^{\rm S}$ is given by $D(D+1)/2$.

1.15 ($\star\star\star$) **www** In this exercise and the next, we explore how the number of independent parameters in a polynomial grows with the order M of the polynomial and with the dimensionality D of the input space. We start by writing down the $M^{\rm th}$ order term for a polynomial in D dimensions in the form

$$\sum_{i_1=1}^{D}\sum_{i_2=1}^{D}\cdots\sum_{i_M=1}^{D} w_{i_1 i_2 \cdots i_M} x_{i_1} x_{i_2} \cdots x_{i_M}. \tag{1.133}$$

The coefficients $w_{i_1 i_2 \cdots i_M}$ comprise D^M elements, but the number of independent parameters is significantly fewer due to the many interchange symmetries of the factor $x_{i_1} x_{i_2} \cdots x_{i_M}$. Begin by showing that the redundancy in the coefficients can be removed by rewriting this $M^{\rm th}$ order term in the form

$$\sum_{i_1=1}^{D}\sum_{i_2=1}^{i_1}\cdots\sum_{i_M=1}^{i_{M-1}} \widetilde{w}_{i_1 i_2 \cdots i_M} x_{i_1} x_{i_2} \cdots x_{i_M}. \tag{1.134}$$

Note that the precise relationship between the $\widetilde{w}$ coefficients and w coefficients need not be made explicit. Use this result to show that the number of *independent* parameters $n(D, M)$, which appear at order M, satisfies the following recursion relation

$$n(D, M) = \sum_{i=1}^{D} n(i, M-1). \tag{1.135}$$

Next use proof by induction to show that the following result holds

$$\sum_{i=1}^{D} \frac{(i+M-2)!}{(i-1)!\,(M-1)!} = \frac{(D+M-1)!}{(D-1)!\,M!} \tag{1.136}$$

which can be done by first proving the result for $D = 1$ and arbitrary M by making use of the result $0! = 1$, then assuming it is correct for dimension D and verifying that it is correct for dimension $D + 1$. Finally, use the two previous results, together with proof by induction, to show

$$n(D, M) = \frac{(D+M-1)!}{(D-1)!\,M!}. \tag{1.137}$$

To do this, first show that the result is true for $M = 2$, and any value of $D \geqslant 1$, by comparison with the result of Exercise 1.14. Then make use of (1.135), together with (1.136), to show that, if the result holds at order $M - 1$, then it will also hold at order M

1.16 ($\star\star\star$) In Exercise 1.15, we proved the result (1.135) for the number of independent parameters in the Mth order term of a D-dimensional polynomial. We now find an expression for the total number $N(D, M)$ of independent parameters in all of the terms up to and including the Mth order. First show that $N(D, M)$ satisfies

$$N(D, M) = \sum_{m=0}^{M} n(D, m) \tag{1.138}$$

where $n(D, m)$ is the number of independent parameters in the term of order m. Now make use of the result (1.137), together with proof by induction, to show that

$$N(D, M) = \frac{(D+M)!}{D!\,M!}. \tag{1.139}$$

This can be done by first proving that the result holds for $M = 0$ and arbitrary $D \geqslant 1$, then assuming that it holds at order M, and hence showing that it holds at order $M + 1$. Finally, make use of Stirling's approximation in the form

$$n! \simeq n^n e^{-n} \tag{1.140}$$

for large n to show that, for $D \gg M$, the quantity $N(D, M)$ grows like D^M, and for $M \gg D$ it grows like M^D. Consider a cubic ($M = 3$) polynomial in D dimensions, and evaluate numerically the total number of independent parameters for (i) $D = 10$ and (ii) $D = 100$, which correspond to typical small-scale and medium-scale machine learning applications.

1.17 ($\star\star$) **www** The gamma function is defined by

$$\Gamma(x) \equiv \int_0^\infty u^{x-1} e^{-u} \, \mathrm{d}u. \tag{1.141}$$

Using integration by parts, prove the relation $\Gamma(x+1) = x\Gamma(x)$. Show also that $\Gamma(1) = 1$ and hence that $\Gamma(x+1) = x!$ when x is an integer.

1.18 ($\star\star$) **www** We can use the result (1.126) to derive an expression for the surface area S_D, and the volume V_D, of a sphere of unit radius in D dimensions. To do this, consider the following result, which is obtained by transforming from Cartesian to polar coordinates

$$\prod_{i=1}^{D} \int_{-\infty}^{\infty} e^{-x_i^2} \, \mathrm{d}x_i = S_D \int_0^\infty e^{-r^2} r^{D-1} \, \mathrm{d}r. \tag{1.142}$$

Using the definition (1.141) of the Gamma function, together with (1.126), evaluate both sides of this equation, and hence show that

$$S_D = \frac{2\pi^{D/2}}{\Gamma(D/2)}. \tag{1.143}$$

Next, by integrating with respect to radius from 0 to 1, show that the volume of the unit sphere in D dimensions is given by

$$V_D = \frac{S_D}{D}. \tag{1.144}$$

Finally, use the results $\Gamma(1) = 1$ and $\Gamma(3/2) = \sqrt{\pi}/2$ to show that (1.143) and (1.144) reduce to the usual expressions for $D = 2$ and $D = 3$.

1.19 ($\star\star$) Consider a sphere of radius a in D-dimensions together with the concentric hypercube of side $2a$, so that the sphere touches the hypercube at the centres of each of its sides. By using the results of Exercise 1.18, show that the ratio of the volume of the sphere to the volume of the cube is given by

$$\frac{\text{volume of sphere}}{\text{volume of cube}} = \frac{\pi^{D/2}}{D 2^{D-1} \Gamma(D/2)}. \tag{1.145}$$

Now make use of Stirling's formula in the form

$$\Gamma(x+1) \simeq (2\pi)^{1/2} e^{-x} x^{x+1/2} \tag{1.146}$$

which is valid for $x \gg 1$, to show that, as $D \to \infty$, the ratio (1.145) goes to zero. Show also that the ratio of the distance from the centre of the hypercube to one of the corners, divided by the perpendicular distance to one of the sides, is $\sqrt{D}$, which therefore goes to ∞ as $D \to \infty$. From these results we see that, in a space of high dimensionality, most of the volume of a cube is concentrated in the large number of corners, which themselves become very long 'spikes'!

1.20 ($\star\star$) **www** In this exercise, we explore the behaviour of the Gaussian distribution in high-dimensional spaces. Consider a Gaussian distribution in D dimensions given by

$$p(\mathbf{x}) = \frac{1}{(2\pi\sigma^2)^{D/2}} \exp\left(-\frac{\|\mathbf{x}\|^2}{2\sigma^2}\right). \tag{1.147}$$

We wish to find the density with respect to radius in polar coordinates in which the direction variables have been integrated out. To do this, show that the integral of the probability density over a thin shell of radius r and thickness ϵ, where $\epsilon \ll 1$, is given by $p(r)\epsilon$ where

$$p(r) = \frac{S_D r^{D-1}}{(2\pi\sigma^2)^{D/2}} \exp\left(-\frac{r^2}{2\sigma^2}\right) \tag{1.148}$$

where S_D is the surface area of a unit sphere in D dimensions. Show that the function $p(r)$ has a single stationary point located, for large D, at $\widehat{r} \simeq \sqrt{D}\sigma$. By considering $p(\widehat{r} + \epsilon)$ where $\epsilon \ll \widehat{r}$, show that for large D,

$$p(\widehat{r} + \epsilon) = p(\widehat{r}) \exp\left(-\frac{\epsilon^2}{\sigma^2}\right) \tag{1.149}$$

which shows that $\widehat{r}$ is a maximum of the radial probability density and also that $p(r)$ decays exponentially away from its maximum at $\widehat{r}$ with length scale σ. We have already seen that $\sigma \ll \widehat{r}$ for large D, and so we see that most of the probability mass is concentrated in a thin shell at large radius. Finally, show that the probability density $p(\mathbf{x})$ is larger at the origin than at the radius $\widehat{r}$ by a factor of $\exp(D/2)$. We therefore see that most of the probability mass in a high-dimensional Gaussian distribution is located at a different radius from the region of high probability density. This property of distributions in spaces of high dimensionality will have important consequences when we consider Bayesian inference of model parameters in later chapters.

1.21 ($\star\star$) Consider two nonnegative numbers a and b, and show that, if $a \leqslant b$, then $a \leqslant (ab)^{1/2}$. Use this result to show that, if the decision regions of a two-class classification problem are chosen to minimize the probability of misclassification, this probability will satisfy

$$p(\text{mistake}) \leqslant \int \left\{p(\mathbf{x}, \mathcal{C}_1)p(\mathbf{x}, \mathcal{C}_2)\right\}^{1/2} \mathrm{d}\mathbf{x}. \tag{1.150}$$

1.22 ($\star$) **www** Given a loss matrix with elements L_{kj}, the expected risk is minimized if, for each $\mathbf{x}$, we choose the class that minimizes (1.81). Verify that, when the loss matrix is given by $L_{kj} = 1 - I_{kj}$, where I_{kj} are the elements of the identity matrix, this reduces to the criterion of choosing the class having the largest posterior probability. What is the interpretation of this form of loss matrix?

1.23 ($\star$) Derive the criterion for minimizing the expected loss when there is a general loss matrix and general prior probabilities for the classes.

1.24 ($\star\star$) **www** Consider a classification problem in which the loss incurred when an input vector from class $\mathcal{C}_k$ is classified as belonging to class $\mathcal{C}_j$ is given by the loss matrix L_{kj}, and for which the loss incurred in selecting the reject option is λ. Find the decision criterion that will give the minimum expected loss. Verify that this reduces to the reject criterion discussed in Section 1.5.3 when the loss matrix is given by $L_{kj} = 1 - I_{kj}$. What is the relationship between λ and the rejection threshold θ?

1.25 ($\star$) **www** Consider the generalization of the squared loss function (1.87) for a single target variable t to the case of multiple target variables described by the vector $\mathbf{t}$ given by

$$\mathbb{E}[L(\mathbf{t}, \mathbf{y}(\mathbf{x}))] = \iint \|\mathbf{y}(\mathbf{x}) - \mathbf{t}\|^2 p(\mathbf{x}, \mathbf{t})\, \mathrm{d}\mathbf{x}\, \mathrm{d}\mathbf{t}. \tag{1.151}$$

Using the calculus of variations, show that the function $\mathbf{y}(\mathbf{x})$ for which this expected loss is minimized is given by $\mathbf{y}(\mathbf{x}) = \mathbb{E}_{\mathbf{t}}[\mathbf{t}|\mathbf{x}]$. Show that this result reduces to (1.89) for the case of a single target variable t.

1.26 ($\star$) By expansion of the square in (1.151), derive a result analogous to (1.90) and hence show that the function $\mathbf{y}(\mathbf{x})$ that minimizes the expected squared loss for the case of a vector $\mathbf{t}$ of target variables is again given by the conditional expectation of $\mathbf{t}$.

1.27 ($\star\star$) **www** Consider the expected loss for regression problems under the L_q loss function given by (1.91). Write down the condition that $y(\mathbf{x})$ must satisfy in order to minimize $\mathbb{E}[L_q]$. Show that, for $q = 1$, this solution represents the conditional median, i.e., the function $y(\mathbf{x})$ such that the probability mass for $t < y(\mathbf{x})$ is the same as for $t \geqslant y(\mathbf{x})$. Also show that the minimum expected L_q loss for $q \to 0$ is given by the conditional mode, i.e., by the function $y(\mathbf{x})$ equal to the value of t that maximizes $p(t|\mathbf{x})$ for each $\mathbf{x}$.

1.28 ($\star$) In Section 1.6, we introduced the idea of entropy $h(x)$ as the information gained on observing the value of a random variable x having distribution $p(x)$. We saw that, for independent variables x and y for which $p(x, y) = p(x)p(y)$, the entropy functions are additive, so that $h(x, y) = h(x) + h(y)$. In this exercise, we derive the relation between h and p in the form of a function $h(p)$. First show that $h(p^2) = 2h(p)$, and hence by induction that $h(p^n) = nh(p)$ where n is a positive integer. Hence show that $h(p^{n/m}) = (n/m)h(p)$ where m is also a positive integer. This implies that $h(p^x) = xh(p)$ where x is a positive rational number, and hence by continuity when it is a positive real number. Finally, show that this implies $h(p)$ must take the form $h(p) \propto \ln p$.

1.29 ($\star$) **www** Consider an M-state discrete random variable x, and use Jensen's inequality in the form (1.115) to show that the entropy of its distribution $p(x)$ satisfies $\mathrm{H}[x] \leqslant \ln M$.

1.30 ($\star\star$) Evaluate the Kullback-Leibler divergence (1.113) between two Gaussians $p(x) = \mathcal{N}(x|\mu, \sigma^2)$ and $q(x) = \mathcal{N}(x|m, s^2)$.

Table 1.3 The joint distribution $p(x, y)$ for two binary variables x and y used in Exercise 1.39.

		$y=0$	$y=1$
x	0	1/3	1/3
	1	0	1/3

1.31 ($\star\star$) **www** Consider two variables $\mathbf{x}$ and $\mathbf{y}$ having joint distribution $p(\mathbf{x}, \mathbf{y})$. Show that the differential entropy of this pair of variables satisfies

$$\mathrm{H}[\mathbf{x}, \mathbf{y}] \leqslant \mathrm{H}[\mathbf{x}] + \mathrm{H}[\mathbf{y}] \tag{1.152}$$

with equality if, and only if, $\mathbf{x}$ and $\mathbf{y}$ are statistically independent.

1.32 ($\star$) Consider a vector $\mathbf{x}$ of continuous variables with distribution $p(\mathbf{x})$ and corresponding entropy $\mathrm{H}[\mathbf{x}]$. Suppose that we make a nonsingular linear transformation of $\mathbf{x}$ to obtain a new variable $\mathbf{y} = \mathbf{A}\mathbf{x}$. Show that the corresponding entropy is given by $\mathrm{H}[\mathbf{y}] = \mathrm{H}[\mathbf{x}] + \ln|\det(\mathbf{A})|$ where $|\det(\mathbf{A})|$ denotes the absolute value of the determinant of $\mathbf{A}$.

1.33 ($\star\star$) Suppose that the conditional entropy $\mathrm{H}[y|x]$ between two discrete random variables x and y is zero. Show that, for all values of x such that $p(x) > 0$, the variable y must be a function of x, in other words for each x there is only one value of y such that $p(y|x) \neq 0$.

1.34 ($\star\star$) **www** Use the calculus of variations to show that the stationary point of the functional preceding (1.108) is given by (1.108). Then use the constraints (1.105), (1.106), and (1.107) to eliminate the Lagrange multipliers and hence show that the maximum entropy solution is given by the Gaussian (1.109).

1.35 ($\star$) **www** Use the results (1.106) and (1.107) to show that the entropy of the univariate Gaussian (1.109) is given by (1.110).

1.36 ($\star$) A strictly convex function is defined as one for which every chord lies above the function. Show that this is equivalent to the condition that the second derivative of the function be positive.

1.37 ($\star$) Using the definition (1.111) together with the product rule of probability, prove the result (1.112).

1.38 ($\star\star$) **www** Using proof by induction, show that the inequality (1.114) for convex functions implies the result (1.115).

1.39 ($\star\star\star$) Consider two binary variables x and y having the joint distribution given in Table 1.3.

Evaluate the following quantities

(a) $\mathrm{H}[x]$ **(c)** $\mathrm{H}[y|x]$ **(e)** $\mathrm{H}[x, y]$
(b) $\mathrm{H}[y]$ **(d)** $\mathrm{H}[x|y]$ **(f)** $\mathrm{I}[x, y]$.

Draw a diagram to show the relationship between these various quantities.

1.40 ($\star$) By applying Jensen's inequality (1.115) with $f(x) = \ln x$, show that the arithmetic mean of a set of real numbers is never less than their geometrical mean.

1.41 ($\star$) **www** Using the sum and product rules of probability, show that the mutual information $I(\mathbf{x}, \mathbf{y})$ satisfies the relation (1.121).

2 Probability Distributions

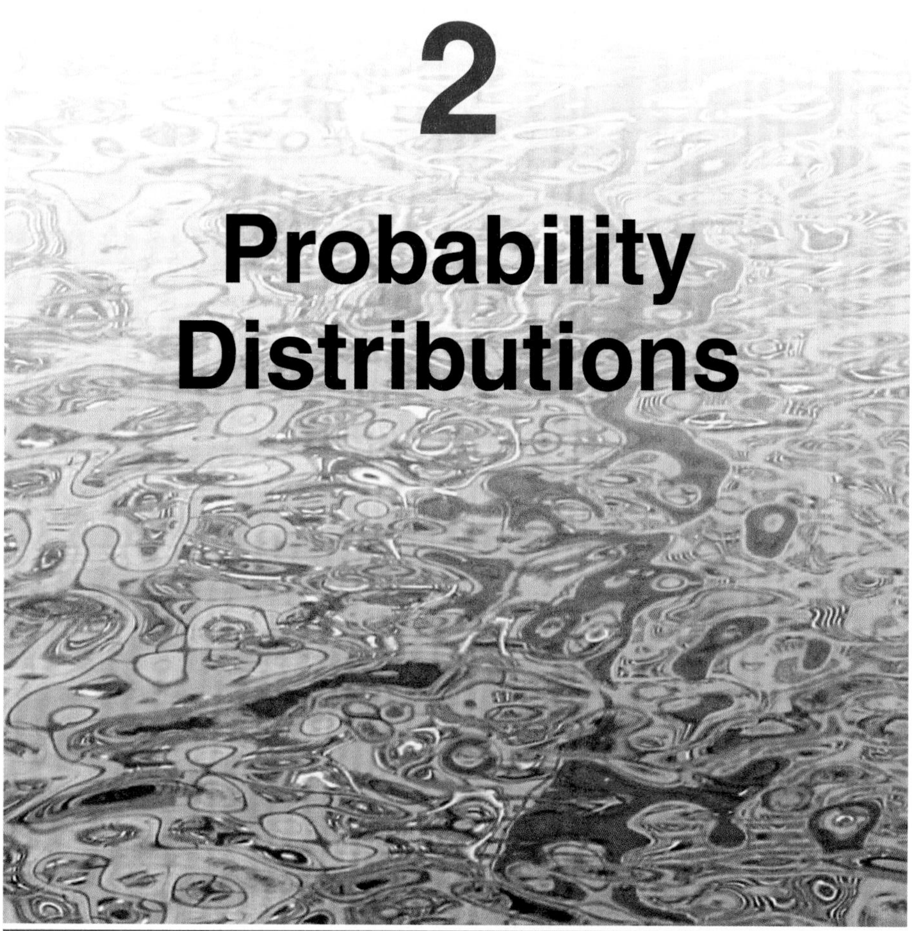

In Chapter 1, we emphasized the central role played by probability theory in the solution of pattern recognition problems. We turn now to an exploration of some particular examples of probability distributions and their properties. As well as being of great interest in their own right, these distributions can form building blocks for more complex models and will be used extensively throughout the book. The distributions introduced in this chapter will also serve another important purpose, namely to provide us with the opportunity to discuss some key statistical concepts, such as Bayesian inference, in the context of simple models before we encounter them in more complex situations in later chapters.

One role for the distributions discussed in this chapter is to model the probability distribution $p(\mathbf{x})$ of a random variable $\mathbf{x}$, given a finite set $\mathbf{x}_1, \ldots, \mathbf{x}_N$ of observations. This problem is known as *density estimation*. For the purposes of this chapter, we shall assume that the data points are independent and identically distributed. It should be emphasized that the problem of density estimation is fundamentally ill-posed, because there are infinitely many probability distributions that could have given rise to the observed finite data set. Indeed, any distribution $p(\mathbf{x})$ that is nonzero at each of the data points $\mathbf{x}_1, \ldots, \mathbf{x}_N$ is a potential candidate. The issue of choosing an appropriate distribution relates to the problem of model selection that has already been encountered in the context of polynomial curve fitting in Chapter 1 and that is a central issue in pattern recognition.

We begin by considering the binomial and multinomial distributions for discrete random variables and the Gaussian distribution for continuous random variables. These are specific examples of *parametric* distributions, so-called because they are governed by a small number of adaptive parameters, such as the mean and variance in the case of a Gaussian for example. To apply such models to the problem of density estimation, we need a procedure for determining suitable values for the parameters, given an observed data set. In a frequentist treatment, we choose specific values for the parameters by optimizing some criterion, such as the likelihood function. By contrast, in a Bayesian treatment we introduce prior distributions over the parameters and then use Bayes' theorem to compute the corresponding posterior distribution given the observed data.

We shall see that an important role is played by *conjugate* priors, that lead to posterior distributions having the same functional form as the prior, and that therefore lead to a greatly simplified Bayesian analysis. For example, the conjugate prior for the parameters of the multinomial distribution is called the *Dirichlet* distribution, while the conjugate prior for the mean of a Gaussian is another Gaussian. All of these distributions are examples of the *exponential family* of distributions, which possess a number of important properties, and which will be discussed in some detail.

One limitation of the parametric approach is that it assumes a specific functional form for the distribution, which may turn out to be inappropriate for a particular application. An alternative approach is given by *nonparametric* density estimation methods in which the form of the distribution typically depends on the size of the data set. Such models still contain parameters, but these control the model complexity rather than the form of the distribution. We end this chapter by considering three nonparametric methods based respectively on histograms, nearest-neighbours, and kernels.

2.1. Binary Variables

We begin by considering a single binary random variable $x \in \{0, 1\}$. For example, x might describe the outcome of flipping a coin, with $x = 1$ representing 'heads', and $x = 0$ representing 'tails'. We can imagine that this is a damaged coin so that the probability of landing heads is not necessarily the same as that of landing tails. The probability of $x = 1$ will be denoted by the parameter μ so that

$$p(x = 1|\mu) = \mu \tag{2.1}$$

where $0 \leqslant \mu \leqslant 1$, from which it follows that $p(x=0|\mu) = 1-\mu$. The probability distribution over x can therefore be written in the form

$$\mathrm{Bern}(x|\mu) = \mu^x (1-\mu)^{1-x} \tag{2.2}$$

Exercise 2.1

which is known as the *Bernoulli* distribution. It is easily verified that this distribution is normalized and that it has mean and variance given by

$$\mathbb{E}[x] = \mu \tag{2.3}$$
$$\mathrm{var}[x] = \mu(1-\mu). \tag{2.4}$$

Now suppose we have a data set $\mathcal{D} = \{x_1, \ldots, x_N\}$ of observed values of x. We can construct the likelihood function, which is a function of μ, on the assumption that the observations are drawn independently from $p(x|\mu)$, so that

$$p(\mathcal{D}|\mu) = \prod_{n=1}^{N} p(x_n|\mu) = \prod_{n=1}^{N} \mu^{x_n}(1-\mu)^{1-x_n}. \tag{2.5}$$

In a frequentist setting, we can estimate a value for μ by maximizing the likelihood function, or equivalently by maximizing the logarithm of the likelihood. In the case of the Bernoulli distribution, the log likelihood function is given by

$$\ln p(\mathcal{D}|\mu) = \sum_{n=1}^{N} \ln p(x_n|\mu) = \sum_{n=1}^{N} \{x_n \ln \mu + (1-x_n)\ln(1-\mu)\}. \tag{2.6}$$

Section 2.4

At this point, it is worth noting that the log likelihood function depends on the N observations x_n only through their sum $\sum_n x_n$. This sum provides an example of a *sufficient statistic* for the data under this distribution, and we shall study the important role of sufficient statistics in some detail. If we set the derivative of $\ln p(\mathcal{D}|\mu)$ with respect to μ equal to zero, we obtain the maximum likelihood estimator

$$\mu_{\mathrm{ML}} = \frac{1}{N}\sum_{n=1}^{N} x_n \tag{2.7}$$

Jacob Bernoulli
1654–1705

Jacob Bernoulli, also known as Jacques or James Bernoulli, was a Swiss mathematician and was the first of many in the Bernoulli family to pursue a career in science and mathematics. Although compelled to study philosophy and theology against his will by his parents, he travelled extensively after graduating in order to meet with many of the leading scientists of his time, including Boyle and Hooke in England. When he returned to Switzerland, he taught mechanics and became Professor of Mathematics at Basel in 1687. Unfortunately, rivalry between Jacob and his younger brother Johann turned an initially productive collaboration into a bitter and public dispute. Jacob's most significant contributions to mathematics appeared in *The Art of Conjecture* published in 1713, eight years after his death, which deals with topics in probability theory including what has become known as the Bernoulli distribution.

Figure 2.1 Histogram plot of the binomial distribution (2.9) as a function of m for $N = 10$ and $\mu = 0.25$.

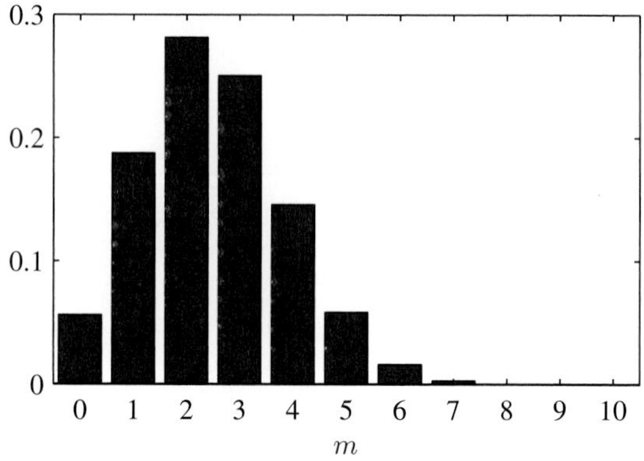

which is also known as the *sample mean*. If we denote the number of observations of $x = 1$ (heads) within this data set by m, then we can write (2.7) in the form

$$\mu_{\rm ML} = \frac{m}{N} \tag{2.8}$$

so that the probability of landing heads is given, in this maximum likelihood framework, by the fraction of observations of heads in the data set.

Now suppose we flip a coin, say, 3 times and happen to observe 3 heads. Then $N = m = 3$ and $\mu_{\rm ML} = 1$. In this case, the maximum likelihood result would predict that all future observations should give heads. Common sense tells us that this is unreasonable, and in fact this is an extreme example of the over-fitting associated with maximum likelihood. We shall see shortly how to arrive at more sensible conclusions through the introduction of a prior distribution over μ.

We can also work out the distribution of the number m of observations of $x = 1$, given that the data set has size N. This is called the *binomial* distribution, and from (2.5) we see that it is proportional to $\mu^m(1-\mu)^{N-m}$. In order to obtain the normalization coefficient we note that out of N coin flips, we have to add up all of the possible ways of obtaining m heads, so that the binomial distribution can be written

$$\mathrm{Bin}(m|N,\mu) = \binom{N}{m}\mu^m(1-\mu)^{N-m} \tag{2.9}$$

where

$$\binom{N}{m} \equiv \frac{N!}{(N-m)!m!} \tag{2.10}$$

Exercise 2.3

is the number of ways of choosing m objects out of a total of N identical objects. Figure 2.1 shows a plot of the binomial distribution for $N = 10$ and $\mu = 0.25$.

The mean and variance of the binomial distribution can be found by using the result of Exercise 1.10, which shows that for independent events the mean of the sum is the sum of the means, and the variance of the sum is the sum of the variances. Because $m = x_1 + \ldots + x_N$, and for each observation the mean and variance are given by (2.3) and (2.4), respectively, we have

$$\mathbb{E}[m] \equiv \sum_{m=0}^{N} m\mathrm{Bin}(m|N,\mu) = N\mu \tag{2.11}$$

$$\mathrm{var}[m] \equiv \sum_{m=0}^{N} (m - \mathbb{E}[m])^2 \,\mathrm{Bin}(m|N,\mu) = N\mu(1-\mu). \tag{2.12}$$

Exercise 2.4

These results can also be proved directly using calculus.

2.1.1 The beta distribution

We have seen in (2.8) that the maximum likelihood setting for the parameter μ in the Bernoulli distribution, and hence in the binomial distribution, is given by the fraction of the observations in the data set having $x = 1$. As we have already noted, this can give severely over-fitted results for small data sets. In order to develop a Bayesian treatment for this problem, we need to introduce a prior distribution $p(\mu)$ over the parameter μ. Here we consider a form of prior distribution that has a simple interpretation as well as some useful analytical properties. To motivate this prior, we note that the likelihood function takes the form of a product of factors of the form $\mu^x(1-\mu)^{1-x}$. If we choose a prior to be proportional to powers of μ and $(1-\mu)$, then the posterior distribution, which is proportional to the product of the prior and the likelihood function, will have the same functional form as the prior. This property is called *conjugacy* and we will see several examples of it later in this chapter. We therefore choose a prior, called the *beta* distribution, given by

$$\mathrm{Beta}(\mu|a,b) = \frac{\Gamma(a+b)}{\Gamma(a)\Gamma(b)}\mu^{a-1}(1-\mu)^{b-1} \tag{2.13}$$

Exercise 2.5

where $\Gamma(x)$ is the gamma function defined by (1.141), and the coefficient in (2.13) ensures that the beta distribution is normalized, so that

$$\int_0^1 \mathrm{Beta}(\mu|a,b)\,\mathrm{d}\mu = 1. \tag{2.14}$$

Exercise 2.6

The mean and variance of the beta distribution are given by

$$\mathbb{E}[\mu] = \frac{a}{a+b} \tag{2.15}$$

$$\mathrm{var}[\mu] = \frac{ab}{(a+b)^2(a+b+1)}. \tag{2.16}$$

The parameters a and b are often called *hyperparameters* because they control the distribution of the parameter μ. Figure 2.2 shows plots of the beta distribution for various values of the hyperparameters.

The posterior distribution of μ is now obtained by multiplying the beta prior (2.13) by the binomial likelihood function (2.9) and normalizing. Keeping only the factors that depend on μ, we see that this posterior distribution has the form

$$p(\mu|m,l,a,b) \propto \mu^{m+a-1}(1-\mu)^{l+b-1} \tag{2.17}$$

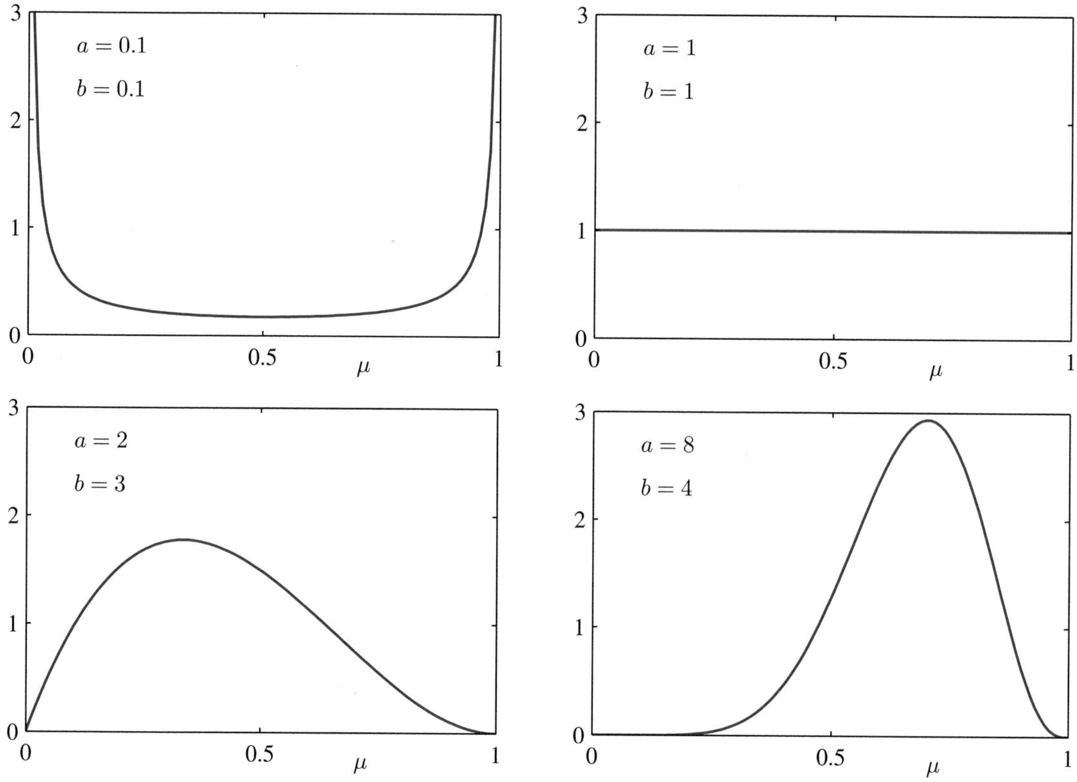

Figure 2.2 Plots of the beta distribution $\mathrm{Beta}(\mu|a, b)$ given by (2.13) as a function of μ for various values of the hyperparameters a and b.

where $l = N - m$, and therefore corresponds to the number of 'tails' in the coin example. We see that (2.17) has the same functional dependence on μ as the prior distribution, reflecting the conjugacy properties of the prior with respect to the likelihood function. Indeed, it is simply another beta distribution, and its normalization coefficient can therefore be obtained by comparison with (2.13) to give

$$p(\mu|m, l, a, b) = \frac{\Gamma(m + a + l + b)}{\Gamma(m + a)\Gamma(l + b)} \mu^{m+a-1}(1 - \mu)^{l+b-1}. \tag{2.18}$$

We see that the effect of observing a data set of m observations of $x = 1$ and l observations of $x = 0$ has been to increase the value of a by m, and the value of b by l, in going from the prior distribution to the posterior distribution. This allows us to provide a simple interpretation of the hyperparameters a and b in the prior as an *effective number of observations* of $x = 1$ and $x = 0$, respectively. Note that a and b need not be integers. Furthermore, the posterior distribution can act as the prior if we subsequently observe additional data. To see this, we can imagine taking observations one at a time and after each observation updating the current posterior

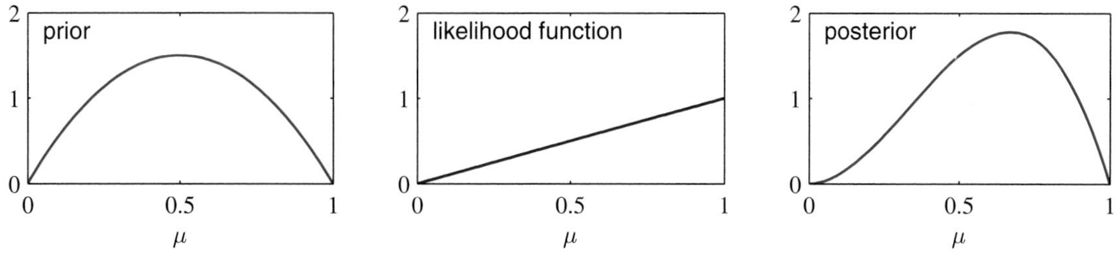

Figure 2.3 Illustration of one step of sequential Bayesian inference. The prior is given by a beta distribution with parameters $a = 2$, $b = 2$, and the likelihood function, given by (2.9) with $N = m = 1$, corresponds to a single observation of $x = 1$, so that the posterior is given by a beta distribution with parameters $a = 3$, $b = 2$.

distribution by multiplying by the likelihood function for the new observation and then normalizing to obtain the new, revised posterior distribution. At each stage, the posterior is a beta distribution with some total number of (prior and actual) observed values for $x = 1$ and $x = 0$ given by the parameters a and b. Incorporation of an additional observation of $x = 1$ simply corresponds to incrementing the value of a by 1, whereas for an observation of $x = 0$ we increment b by 1. Figure 2.3 illustrates one step in this process.

We see that this *sequential* approach to learning arises naturally when we adopt a Bayesian viewpoint. It is independent of the choice of prior and of the likelihood function and depends only on the assumption of i.i.d. data. Sequential methods make use of observations one at a time, or in small batches, and then discard them before the next observations are used. They can be used, for example, in real-time learning scenarios where a steady stream of data is arriving, and predictions must be made before all of the data is seen. Because they do not require the whole data set to be stored or loaded into memory, sequential methods are also useful for large data sets. Maximum likelihood methods can also be cast into a sequential framework.

Section 2.3.5

If our goal is to predict, as best we can, the outcome of the next trial, then we must evaluate the predictive distribution of x, given the observed data set $\mathcal{D}$. From the sum and product rules of probability, this takes the form

$$p(x = 1|\mathcal{D}) = \int_0^1 p(x = 1|\mu)p(\mu|\mathcal{D})\,\mathrm{d}\mu = \int_0^1 \mu p(\mu|\mathcal{D})\,\mathrm{d}\mu = \mathbb{E}[\mu|\mathcal{D}]. \tag{2.19}$$

Using the result (2.18) for the posterior distribution $p(\mu|\mathcal{D})$, together with the result (2.15) for the mean of the beta distribution, we obtain

$$p(x = 1|\mathcal{D}) = \frac{m + a}{m + a + l + b} \tag{2.20}$$

which has a simple interpretation as the total fraction of observations (both real observations and fictitious prior observations) that correspond to $x = 1$. Note that in the limit of an infinitely large data set $m, l \to \infty$ the result (2.20) reduces to the maximum likelihood result (2.8). As we shall see, it is a very general property that the Bayesian and maximum likelihood results will agree in the limit of an infinitely large data set. For a finite data set, the posterior mean for μ always lies between the prior mean and the maximum likelihood estimate for μ corresponding to the relative frequencies of events given by (2.7).

Exercise 2.7

From Figure 2.2, we see that as the number of observations increases, so the posterior distribution becomes more sharply peaked. This can also be seen from the result (2.16) for the variance of the beta distribution, in which we see that the variance goes to zero for $a \to \infty$ or $b \to \infty$. In fact, we might wonder whether it is a general property of Bayesian learning that, as we observe more and more data, the uncertainty represented by the posterior distribution will steadily decrease.

To address this, we can take a frequentist view of Bayesian learning and show that, on average, such a property does indeed hold. Consider a general Bayesian inference problem for a parameter $\boldsymbol{\theta}$ for which we have observed a data set $\mathcal{D}$, described by the joint distribution $p(\boldsymbol{\theta}, \mathcal{D})$. The following result

Exercise 2.8

$$\mathbb{E}_{\boldsymbol{\theta}}[\boldsymbol{\theta}] = \mathbb{E}_{\mathcal{D}}\left[\mathbb{E}_{\boldsymbol{\theta}}[\boldsymbol{\theta}|\mathcal{D}]\right] \tag{2.21}$$

where

$$\mathbb{E}_{\boldsymbol{\theta}}[\boldsymbol{\theta}] \equiv \int p(\boldsymbol{\theta})\boldsymbol{\theta}\,\mathrm{d}\boldsymbol{\theta} \tag{2.22}$$

$$\mathbb{E}_{\mathcal{D}}[\mathbb{E}_{\boldsymbol{\theta}}[\boldsymbol{\theta}|\mathcal{D}]] \equiv \int\left\{\int \boldsymbol{\theta}p(\boldsymbol{\theta}|\mathcal{D})\,\mathrm{d}\boldsymbol{\theta}\right\}p(\mathcal{D})\,\mathrm{d}\mathcal{D} \tag{2.23}$$

says that the posterior mean of $\boldsymbol{\theta}$, averaged over the distribution generating the data, is equal to the prior mean of $\boldsymbol{\theta}$. Similarly, we can show that

$$\mathrm{var}_{\boldsymbol{\theta}}[\boldsymbol{\theta}] = \mathbb{E}_{\mathcal{D}}\left[\mathrm{var}_{\boldsymbol{\theta}}[\boldsymbol{\theta}|\mathcal{D}]\right] + \mathrm{var}_{\mathcal{D}}\left[\mathbb{E}_{\boldsymbol{\theta}}[\boldsymbol{\theta}|\mathcal{D}]\right]. \tag{2.24}$$

The term on the left-hand side of (2.24) is the prior variance of $\boldsymbol{\theta}$. On the right-hand side, the first term is the average posterior variance of $\boldsymbol{\theta}$, and the second term measures the variance in the posterior mean of $\boldsymbol{\theta}$. Because this variance is a positive quantity, this result shows that, on average, the posterior variance of $\boldsymbol{\theta}$ is smaller than the prior variance. The reduction in variance is greater if the variance in the posterior mean is greater. Note, however, that this result only holds on average, and that for a particular observed data set it is possible for the posterior variance to be larger than the prior variance.

2.2. Multinomial Variables

Binary variables can be used to describe quantities that can take one of two possible values. Often, however, we encounter discrete variables that can take on one of K possible mutually exclusive states. Although there are various alternative ways to express such variables, we shall see shortly that a particularly convenient representation is the 1-of-K scheme in which the variable is represented by a K-dimensional vector $\mathbf{x}$ in which one of the elements x_k equals 1, and all remaining elements equal

0. So, for instance if we have a variable that can take $K = 6$ states and a particular observation of the variable happens to correspond to the state where $x_3 = 1$, then $\mathbf{x}$ will be represented by

$$\mathbf{x} = (0, 0, 1, 0, 0, 0)^{\mathrm{T}}. \tag{2.25}$$

Note that such vectors satisfy $\sum_{k=1}^{K} x_k = 1$. If we denote the probability of $x_k = 1$ by the parameter μ_k, then the distribution of $\mathbf{x}$ is given

$$p(\mathbf{x}|\boldsymbol{\mu}) = \prod_{k=1}^{K} \mu_k^{x_k} \tag{2.26}$$

where $\boldsymbol{\mu} = (\mu_1, \ldots, \mu_K)^{\mathrm{T}}$, and the parameters μ_k are constrained to satisfy $\mu_k \geqslant 0$ and $\sum_k \mu_k = 1$, because they represent probabilities. The distribution (2.26) can be regarded as a generalization of the Bernoulli distribution to more than two outcomes. It is easily seen that the distribution is normalized

$$\sum_{\mathbf{x}} p(\mathbf{x}|\boldsymbol{\mu}) = \sum_{k=1}^{K} \mu_k = 1 \tag{2.27}$$

and that

$$\mathbb{E}[\mathbf{x}|\boldsymbol{\mu}] = \sum_{\mathbf{x}} p(\mathbf{x}|\boldsymbol{\mu})\mathbf{x} = (\mu_1, \ldots, \mu_K)^{\mathrm{T}} = \boldsymbol{\mu}. \tag{2.28}$$

Now consider a data set $\mathcal{D}$ of N independent observations $\mathbf{x}_1, \ldots, \mathbf{x}_N$. The corresponding likelihood function takes the form

$$p(\mathcal{D}|\boldsymbol{\mu}) = \prod_{n=1}^{N} \prod_{k=1}^{K} \mu_k^{x_{nk}} = \prod_{k=1}^{K} \mu_k^{(\sum_n x_{nk})} = \prod_{k=1}^{K} \mu_k^{m_k}. \tag{2.29}$$

We see that the likelihood function depends on the N data points only through the K quantities

$$m_k = \sum_n x_{nk} \tag{2.30}$$

which represent the number of observations of $x_k = 1$. These are called the *sufficient statistics* for this distribution.

Section 2.4

In order to find the maximum likelihood solution for $\boldsymbol{\mu}$, we need to maximize $\ln p(\mathcal{D}|\boldsymbol{\mu})$ with respect to μ_k taking account of the constraint that the μ_k must sum to one. This can be achieved using a Lagrange multiplier λ and maximizing

Appendix E

$$\sum_{k=1}^{K} m_k \ln \mu_k + \lambda \left(\sum_{k=1}^{K} \mu_k - 1 \right). \tag{2.31}$$

Setting the derivative of (2.31) with respect to μ_k to zero, we obtain

$$\mu_k = -m_k/\lambda. \tag{2.32}$$

We can solve for the Lagrange multiplier λ by substituting (2.32) into the constraint $\sum_k \mu_k = 1$ to give $\lambda = -N$. Thus we obtain the maximum likelihood solution in the form

$$\mu_k^{\mathrm{ML}} = \frac{m_k}{N} \tag{2.33}$$

which is the fraction of the N observations for which $x_k = 1$.

We can consider the joint distribution of the quantities $m_1, \ldots, m_K$, conditioned on the parameters $\boldsymbol{\mu}$ and on the total number N of observations. From (2.29) this takes the form

$$\mathrm{Mult}(m_1, m_2, \ldots, m_K|\boldsymbol{\mu}, N) = \binom{N}{m_1 m_2 \ldots m_K} \prod_{k=1}^{K} \mu_k^{m_k} \tag{2.34}$$

which is known as the *multinomial* distribution. The normalization coefficient is the number of ways of partitioning N objects into K groups of size $m_1, \ldots, m_K$ and is given by

$$\binom{N}{m_1 m_2 \ldots m_K} = \frac{N!}{m_1! m_2! \ldots m_K!}. \tag{2.35}$$

Note that the variables m_k are subject to the constraint

$$\sum_{k=1}^{K} m_k = N. \tag{2.36}$$

2.2.1 The Dirichlet distribution

We now introduce a family of prior distributions for the parameters $\{\mu_k\}$ of the multinomial distribution (2.34). By inspection of the form of the multinomial distribution, we see that the conjugate prior is given by

$$p(\boldsymbol{\mu}|\boldsymbol{\alpha}) \propto \prod_{k=1}^{K} \mu_k^{\alpha_k - 1} \tag{2.37}$$

where $0 \leqslant \mu_k \leqslant 1$ and $\sum_k \mu_k = 1$. Here $\alpha_1, \ldots, \alpha_K$ are the parameters of the distribution, and $\boldsymbol{\alpha}$ denotes $(\alpha_1, \ldots, \alpha_K)^{\mathrm{T}}$. Note that, because of the summation constraint, the distribution over the space of the $\{\mu_k\}$ is confined to a *simplex* of dimensionality $K - 1$, as illustrated for $K = 3$ in Figure 2.4.

Exercise 2.9

The normalized form for this distribution is by

$$\mathrm{Dir}(\boldsymbol{\mu}|\boldsymbol{\alpha}) = \frac{\Gamma(\alpha_0)}{\Gamma(\alpha_1) \cdots \Gamma(\alpha_K)} \prod_{k=1}^{K} \mu_k^{\alpha_k - 1} \tag{2.38}$$

which is called the *Dirichlet* distribution. Here $\Gamma(x)$ is the gamma function defined by (1.141) while

$$\alpha_0 = \sum_{k=1}^{K} \alpha_k. \tag{2.39}$$

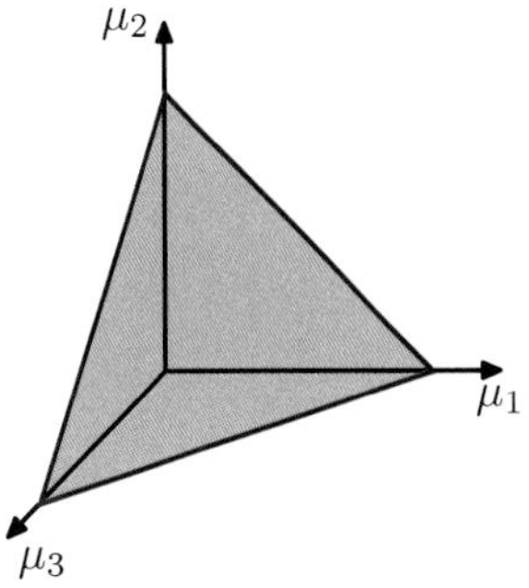

Figure 2.4 The Dirichlet distribution over three variables μ_1, μ_2, μ_3 is confined to a simplex (a bounded linear manifold) of the form shown, as a consequence of the constraints $0 \leqslant \mu_k \leqslant 1$ and $\sum_k \mu_k = 1$.

Plots of the Dirichlet distribution over the simplex, for various settings of the parameters α_k, are shown in Figure 2.5.

Multiplying the prior (2.38) by the likelihood function (2.34), we obtain the posterior distribution for the parameters $\{\mu_k\}$ in the form

$$p(\boldsymbol{\mu}|\mathcal{D}, \boldsymbol{\alpha}) \propto p(\mathcal{D}|\boldsymbol{\mu})p(\boldsymbol{\mu}|\boldsymbol{\alpha}) \propto \prod_{k=1}^{K} \mu_k^{\alpha_k + m_k - 1}. \tag{2.40}$$

We see that the posterior distribution again takes the form of a Dirichlet distribution, confirming that the Dirichlet is indeed a conjugate prior for the multinomial. This allows us to determine the normalization coefficient by comparison with (2.38) so that

$$\begin{aligned} p(\boldsymbol{\mu}|\mathcal{D}, \boldsymbol{\alpha}) &= \mathrm{Dir}(\boldsymbol{\mu}|\boldsymbol{\alpha} + \mathbf{m}) \\ &= \frac{\Gamma(\alpha_0 + N)}{\Gamma(\alpha_1 + m_1) \cdots \Gamma(\alpha_K + m_K)} \prod_{k=1}^{K} \mu_k^{\alpha_k + m_k - 1} \end{aligned} \tag{2.41}$$

where we have denoted $\mathbf{m} = (m_1, \ldots, m_K)^{\mathrm{T}}$. As for the case of the binomial distribution with its beta prior, we can interpret the parameters α_k of the Dirichlet prior as an effective number of observations of $x_k = 1$.

Note that two-state quantities can either be represented as binary variables and

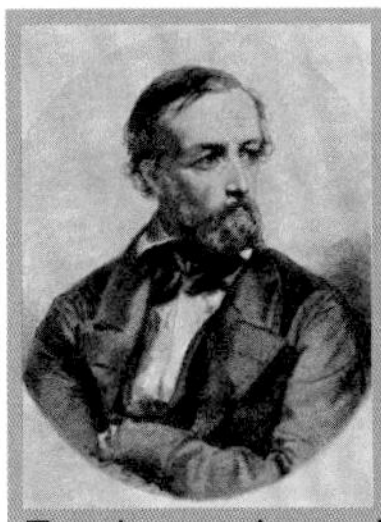

Lejeune Dirichlet
1805–1859

Johann Peter Gustav Lejeune Dirichlet was a modest and reserved mathematician who made contributions in number theory, mechanics, and astronomy, and who gave the first rigorous analysis of Fourier series. His family originated from Richelet in Belgium, and the name Lejeune Dirichlet comes from 'le jeune de Richelet' (the young person from Richelet). Dirichlet's first paper, which was published in 1825, brought him instant fame. It concerned Fermat's last theorem, which claims that there are no positive integer solutions to $x^n + y^n = z^n$ for $n > 2$. Dirichlet gave a partial proof for the case $n = 5$, which was sent to Legendre for review and who in turn completed the proof. Later, Dirichlet gave a complete proof for $n = 14$, although a full proof of Fermat's last theorem for arbitrary n had to wait until the work of Andrew Wiles in the closing years of the 20th century.

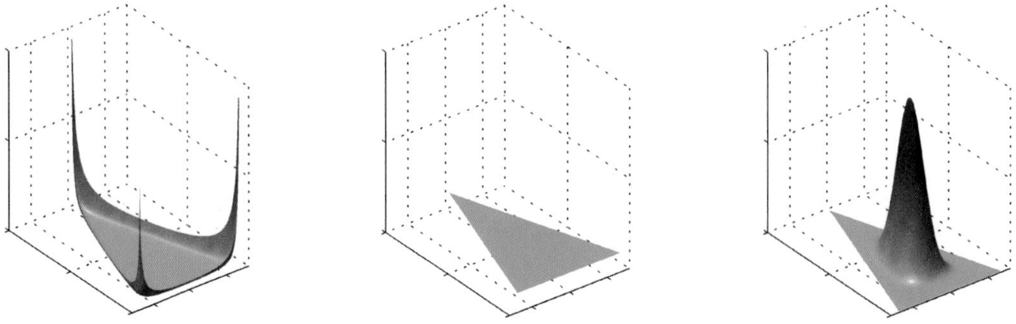

Figure 2.5 Plots of the Dirichlet distribution over three variables, where the two horizontal axes are coordinates in the plane of the simplex and the vertical axis corresponds to the value of the density. Here $\{\alpha_k\} = 0.1$ on the left plot, $\{\alpha_k\} = 1$ in the centre plot, and $\{\alpha_k\} = 10$ in the right plot.

modelled using the binomial distribution (2.9) or as 1-of-2 variables and modelled using the multinomial distribution (2.34) with $K = 2$.

2.3. The Gaussian Distribution

The Gaussian, also known as the normal distribution, is a widely used model for the distribution of continuous variables. In the case of a single variable x, the Gaussian distribution can be written in the form

$$\mathcal{N}(x|\mu, \sigma^2) = \frac{1}{(2\pi\sigma^2)^{1/2}} \exp\left\{-\frac{1}{2\sigma^2}(x-\mu)^2\right\} \tag{2.42}$$

where μ is the mean and σ^2 is the variance. For a D-dimensional vector $\mathbf{x}$, the multivariate Gaussian distribution takes the form

$$\mathcal{N}(\mathbf{x}|\boldsymbol{\mu}, \boldsymbol{\Sigma}) = \frac{1}{(2\pi)^{D/2}} \frac{1}{|\boldsymbol{\Sigma}|^{1/2}} \exp\left\{-\frac{1}{2}(\mathbf{x}-\boldsymbol{\mu})^{\mathrm{T}}\boldsymbol{\Sigma}^{-1}(\mathbf{x}-\boldsymbol{\mu})\right\} \tag{2.43}$$

where $\boldsymbol{\mu}$ is a D-dimensional mean vector, $\boldsymbol{\Sigma}$ is a $D \times D$ covariance matrix, and $|\boldsymbol{\Sigma}|$ denotes the determinant of $\boldsymbol{\Sigma}$.

The Gaussian distribution arises in many different contexts and can be motivated from a variety of different perspectives. For example, we have already seen that for a single real variable, the distribution that maximizes the entropy is the Gaussian. This property applies also to the multivariate Gaussian.

Section 1.6

Exercise 2.14

Another situation in which the Gaussian distribution arises is when we consider the sum of multiple random variables. The *central limit theorem* (due to Laplace) tells us that, subject to certain mild conditions, the sum of a set of random variables, which is of course itself a random variable, has a distribution that becomes increasingly Gaussian as the number of terms in the sum increases (Walker, 1969). We can

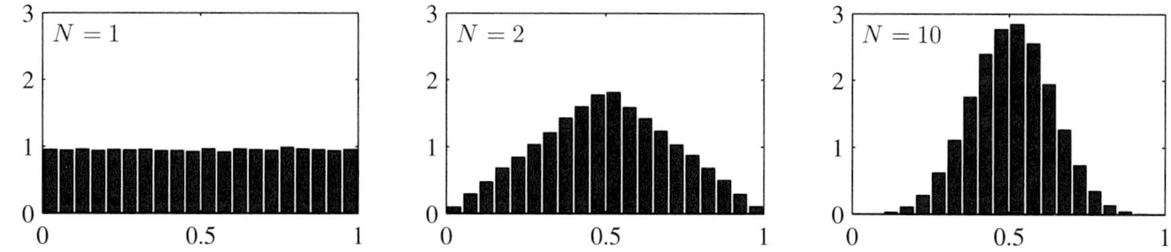

Figure 2.6 Histogram plots of the mean of N uniformly distributed numbers for various values of N. We observe that as N increases, the distribution tends towards a Gaussian.

illustrate this by considering N variables $x_1, \ldots, x_N$ each of which has a uniform distribution over the interval $[0, 1]$ and then considering the distribution of the mean $(x_1 + \cdots + x_N)/N$. For large N, this distribution tends to a Gaussian, as illustrated in Figure 2.6. In practice, the convergence to a Gaussian as N increases can be very rapid. One consequence of this result is that the binomial distribution (2.9), which is a distribution over m defined by the sum of N observations of the random binary variable x, will tend to a Gaussian as $N \to \infty$ (see Figure 2.1 for the case of $N = 10$).

Appendix C

The Gaussian distribution has many important analytical properties, and we shall consider several of these in detail. As a result, this section will be rather more technically involved than some of the earlier sections, and will require familiarity with various matrix identities. However, we strongly encourage the reader to become proficient in manipulating Gaussian distributions using the techniques presented here as this will prove invaluable in understanding the more complex models presented in later chapters.

We begin by considering the geometrical form of the Gaussian distribution. The

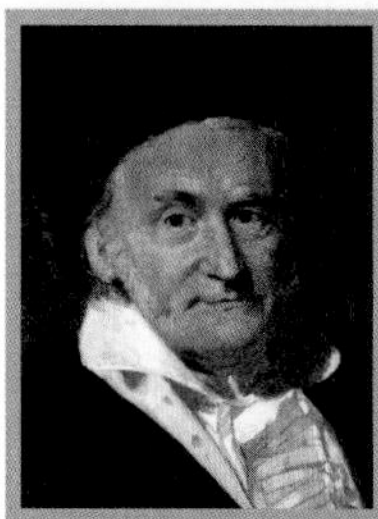

Carl Friedrich Gauss

1777–1855

It is said that when Gauss went to elementary school at age 7, his teacher Büttner, trying to keep the class occupied, asked the pupils to sum the integers from 1 to 100. To the teacher's amazement, Gauss arrived at the answer in a matter of moments by noting that the sum can be represented as 50 pairs ($1+100$, $2+99$, etc.) each of which added to 101, giving the answer 5,050. It is now believed that the problem which was actually set was of the same form but somewhat harder in that the sequence had a larger starting value and a larger increment. Gauss was a German mathematician and scientist with a reputation for being a hard-working perfectionist. One of his many contributions was to show that least squares can be derived under the assumption of normally distributed errors. He also created an early formulation of non-Euclidean geometry (a self-consistent geometrical theory that violates the axioms of Euclid) but was reluctant to discuss it openly for fear that his reputation might suffer if it were seen that he believed in such a geometry. At one point, Gauss was asked to conduct a geodetic survey of the state of Hanover, which led to his formulation of the normal distribution, now also known as the Gaussian. After his death, a study of his diaries revealed that he had discovered several important mathematical results years or even decades before they were published by others.

functional dependence of the Gaussian on $\mathbf{x}$ is through the quadratic form

$$\Delta^2 = (\mathbf{x} - \boldsymbol{\mu})^{\mathrm{T}} \boldsymbol{\Sigma}^{-1} (\mathbf{x} - \boldsymbol{\mu}) \tag{2.44}$$

which appears in the exponent. The quantity Δ is called the *Mahalanobis distance* from $\boldsymbol{\mu}$ to $\mathbf{x}$ and reduces to the Euclidean distance when $\boldsymbol{\Sigma}$ is the identity matrix. The Gaussian distribution will be constant on surfaces in $\mathbf{x}$-space for which this quadratic form is constant.

First of all, we note that the matrix $\boldsymbol{\Sigma}$ can be taken to be symmetric, without loss of generality, because any antisymmetric component would disappear from the exponent. Now consider the eigenvector equation for the covariance matrix

Exercise 2.17

$$\boldsymbol{\Sigma}\mathbf{u}_i = \lambda_i \mathbf{u}_i \tag{2.45}$$

where $i = 1, \ldots, D$. Because $\boldsymbol{\Sigma}$ is a real, symmetric matrix its eigenvalues will be real, and its eigenvectors can be chosen to form an orthonormal set, so that

Exercise 2.18

$$\mathbf{u}_i^{\mathrm{T}} \mathbf{u}_j = I_{ij} \tag{2.46}$$

where I_{ij} is the i, j element of the identity matrix and satisfies

$$I_{ij} = \begin{cases} 1, & \text{if } i = j \\ 0, & \text{otherwise.} \end{cases} \tag{2.47}$$

The covariance matrix $\boldsymbol{\Sigma}$ can be expressed as an expansion in terms of its eigenvectors in the form

Exercise 2.19

$$\boldsymbol{\Sigma} = \sum_{i=1}^{D} \lambda_i \mathbf{u}_i \mathbf{u}_i^{\mathrm{T}} \tag{2.48}$$

and similarly the inverse covariance matrix $\boldsymbol{\Sigma}^{-1}$ can be expressed as

$$\boldsymbol{\Sigma}^{-1} = \sum_{i=1}^{D} \frac{1}{\lambda_i} \mathbf{u}_i \mathbf{u}_i^{\mathrm{T}}. \tag{2.49}$$

Substituting (2.49) into (2.44), the quadratic form becomes

$$\Delta^2 = \sum_{i=1}^{D} \frac{y_i^2}{\lambda_i} \tag{2.50}$$

where we have defined

$$y_i = \mathbf{u}_i^{\mathrm{T}} (\mathbf{x} - \boldsymbol{\mu}). \tag{2.51}$$

We can interpret $\{y_i\}$ as a new coordinate system defined by the orthonormal vectors $\mathbf{u}_i$ that are shifted and rotated with respect to the original x_i coordinates. Forming the vector $\mathbf{y} = (y_1, \ldots, y_D)^{\mathrm{T}}$, we have

$$\mathbf{y} = \mathbf{U}(\mathbf{x} - \boldsymbol{\mu}) \tag{2.52}$$

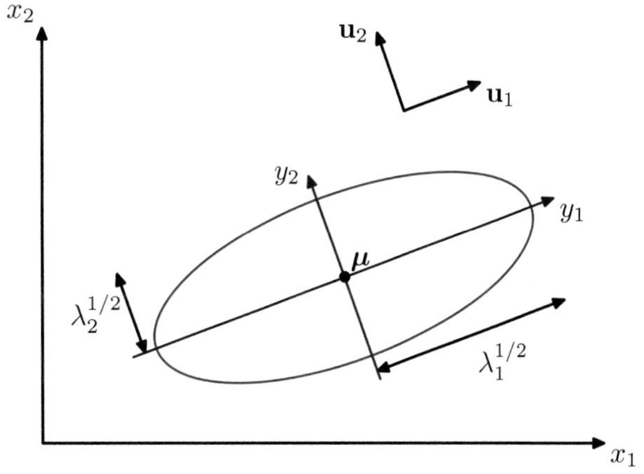

Figure 2.7 The red curve shows the elliptical surface of constant probability density for a Gaussian in a two-dimensional space $\mathbf{x} = (x_1, x_2)$ on which the density is $\exp(-1/2)$ of its value at $\mathbf{x} = \boldsymbol{\mu}$. The axes of the ellipse are defined by the eigenvectors $\mathbf{u}_i$ of the covariance matrix, with corresponding eigenvalues λ_i.

Appendix C

where $\mathbf{U}$ is a matrix whose rows are given by $\mathbf{u}_i^{\mathrm{T}}$. From (2.46) it follows that $\mathbf{U}$ is an *orthogonal* matrix, i.e., it satisfies $\mathbf{U}\mathbf{U}^{\mathrm{T}} = \mathbf{I}$, and hence also $\mathbf{U}^{\mathrm{T}}\mathbf{U} = \mathbf{I}$, where $\mathbf{I}$ is the identity matrix.

The quadratic form, and hence the Gaussian density, will be constant on surfaces for which (2.50) is constant. If all of the eigenvalues λ_i are positive, then these surfaces represent ellipsoids, with their centres at $\boldsymbol{\mu}$ and their axes oriented along $\mathbf{u}_i$, and with scaling factors in the directions of the axes given by $\lambda_i^{1/2}$, as illustrated in Figure 2.7.

For the Gaussian distribution to be well defined, it is necessary for all of the eigenvalues λ_i of the covariance matrix to be strictly positive, otherwise the distribution cannot be properly normalized. A matrix whose eigenvalues are strictly positive is said to be *positive definite*. In Chapter 12, we will encounter Gaussian distributions for which one or more of the eigenvalues are zero, in which case the distribution is singular and is confined to a subspace of lower dimensionality. If all of the eigenvalues are nonnegative, then the covariance matrix is said to be *positive semidefinite*.

Now consider the form of the Gaussian distribution in the new coordinate system defined by the y_i. In going from the $\mathbf{x}$ to the $\mathbf{y}$ coordinate system, we have a Jacobian matrix $\mathbf{J}$ with elements given by

$$J_{ij} = \frac{\partial x_i}{\partial y_j} = U_{ji} \tag{2.53}$$

where U_{ji} are the elements of the matrix $\mathbf{U}^{\mathrm{T}}$. Using the orthonormality property of the matrix $\mathbf{U}$, we see that the square of the determinant of the Jacobian matrix is

$$|\mathbf{J}|^2 = \left|\mathbf{U}^{\mathrm{T}}\right|^2 = \left|\mathbf{U}^{\mathrm{T}}\right| |\mathbf{U}| = \left|\mathbf{U}^{\mathrm{T}}\mathbf{U}\right| = |\mathbf{I}| = 1 \tag{2.54}$$

and hence $|\mathbf{J}| = 1$. Also, the determinant $|\boldsymbol{\Sigma}|$ of the covariance matrix can be written as the product of its eigenvalues, and hence

$$|\boldsymbol{\Sigma}|^{1/2} = \prod_{j=1}^{D} \lambda_j^{1/2}. \tag{2.55}$$

Thus in the y_j coordinate system, the Gaussian distribution takes the form

$$p(\mathbf{y}) = p(\mathbf{x})|\mathbf{J}| = \prod_{j=1}^{D} \frac{1}{(2\pi\lambda_j)^{1/2}} \exp\left\{-\frac{y_j^2}{2\lambda_j}\right\} \tag{2.56}$$

which is the product of D independent univariate Gaussian distributions. The eigenvectors therefore define a new set of shifted and rotated coordinates with respect to which the joint probability distribution factorizes into a product of independent distributions. The integral of the distribution in the $\mathbf{y}$ coordinate system is then

$$\int p(\mathbf{y})\,\mathrm{d}\mathbf{y} = \prod_{j=1}^{D} \int_{-\infty}^{\infty} \frac{1}{(2\pi\lambda_j)^{1/2}} \exp\left\{-\frac{y_j^2}{2\lambda_j}\right\} \mathrm{d}y_j = 1 \tag{2.57}$$

where we have used the result (1.48) for the normalization of the univariate Gaussian. This confirms that the multivariate Gaussian (2.43) is indeed normalized.

We now look at the moments of the Gaussian distribution and thereby provide an interpretation of the parameters $\boldsymbol{\mu}$ and $\boldsymbol{\Sigma}$. The expectation of $\mathbf{x}$ under the Gaussian distribution is given by

$$\begin{aligned}\mathbb{E}[\mathbf{x}] &= \frac{1}{(2\pi)^{D/2}} \frac{1}{|\boldsymbol{\Sigma}|^{1/2}} \int \exp\left\{-\frac{1}{2}(\mathbf{x}-\boldsymbol{\mu})^{\mathrm{T}}\boldsymbol{\Sigma}^{-1}(\mathbf{x}-\boldsymbol{\mu})\right\} \mathbf{x}\,\mathrm{d}\mathbf{x} \\ &= \frac{1}{(2\pi)^{D/2}} \frac{1}{|\boldsymbol{\Sigma}|^{1/2}} \int \exp\left\{-\frac{1}{2}\mathbf{z}^{\mathrm{T}}\boldsymbol{\Sigma}^{-1}\mathbf{z}\right\} (\mathbf{z}+\boldsymbol{\mu})\,\mathrm{d}\mathbf{z}\end{aligned} \tag{2.58}$$

where we have changed variables using $\mathbf{z} = \mathbf{x} - \boldsymbol{\mu}$. We now note that the exponent is an even function of the components of $\mathbf{z}$ and, because the integrals over these are taken over the range $(-\infty, \infty)$, the term in $\mathbf{z}$ in the factor $(\mathbf{z} + \boldsymbol{\mu})$ will vanish by symmetry. Thus

$$\mathbb{E}[\mathbf{x}] = \boldsymbol{\mu} \tag{2.59}$$

and so we refer to $\boldsymbol{\mu}$ as the mean of the Gaussian distribution.

We now consider second order moments of the Gaussian. In the univariate case, we considered the second order moment given by $\mathbb{E}[x^2]$. For the multivariate Gaussian, there are D^2 second order moments given by $\mathbb{E}[x_i x_j]$, which we can group together to form the matrix $\mathbb{E}[\mathbf{x}\mathbf{x}^{\mathrm{T}}]$. This matrix can be written as

$$\begin{aligned}\mathbb{E}[\mathbf{x}\mathbf{x}^{\mathrm{T}}] &= \frac{1}{(2\pi)^{D/2}} \frac{1}{|\boldsymbol{\Sigma}|^{1/2}} \int \exp\left\{-\frac{1}{2}(\mathbf{x}-\boldsymbol{\mu})^{\mathrm{T}}\boldsymbol{\Sigma}^{-1}(\mathbf{x}-\boldsymbol{\mu})\right\} \mathbf{x}\mathbf{x}^{\mathrm{T}}\,\mathrm{d}\mathbf{x} \\ &= \frac{1}{(2\pi)^{D/2}} \frac{1}{|\boldsymbol{\Sigma}|^{1/2}} \int \exp\left\{-\frac{1}{2}\mathbf{z}^{\mathrm{T}}\boldsymbol{\Sigma}^{-1}\mathbf{z}\right\} (\mathbf{z}+\boldsymbol{\mu})(\mathbf{z}+\boldsymbol{\mu})^{\mathrm{T}}\,\mathrm{d}\mathbf{z}\end{aligned}$$

where again we have changed variables using $\mathbf{z} = \mathbf{x} - \boldsymbol{\mu}$. Note that the cross-terms involving $\boldsymbol{\mu}\mathbf{z}^{\mathrm{T}}$ and $\mathbf{z}\boldsymbol{\mu}^{\mathrm{T}}$ will again vanish by symmetry. The term $\boldsymbol{\mu}\boldsymbol{\mu}^{\mathrm{T}}$ is constant and can be taken outside the integral, which itself is unity because the Gaussian distribution is normalized. Consider the term involving $\mathbf{z}\mathbf{z}^{\mathrm{T}}$. Again, we can make use of the eigenvector expansion of the covariance matrix given by (2.45), together with the completeness of the set of eigenvectors, to write

$$\mathbf{z} = \sum_{j=1}^{D} y_j \mathbf{u}_j \tag{2.60}$$

where $y_j = \mathbf{u}_j^{\mathrm{T}}\mathbf{z}$, which gives

$$\begin{aligned}
&\frac{1}{(2\pi)^{D/2}} \frac{1}{|\boldsymbol{\Sigma}|^{1/2}} \int \exp\left\{-\frac{1}{2}\mathbf{z}^{\mathrm{T}}\boldsymbol{\Sigma}^{-1}\mathbf{z}\right\} \mathbf{z}\mathbf{z}^{\mathrm{T}}\,\mathrm{d}\mathbf{z} \\
&= \frac{1}{(2\pi)^{D/2}} \frac{1}{|\boldsymbol{\Sigma}|^{1/2}} \sum_{i=1}^{D}\sum_{j=1}^{D} \mathbf{u}_i\mathbf{u}_j^{\mathrm{T}} \int \exp\left\{-\sum_{k=1}^{D}\frac{y_k^2}{2\lambda_k}\right\} y_i y_j\,\mathrm{d}\mathbf{y} \\
&= \sum_{i=1}^{D} \mathbf{u}_i\mathbf{u}_i^{\mathrm{T}}\lambda_i = \boldsymbol{\Sigma}
\end{aligned} \tag{2.61}$$

where we have made use of the eigenvector equation (2.45), together with the fact that the integral on the right-hand side of the middle line vanishes by symmetry unless $i = j$, and in the final line we have made use of the results (1.50) and (2.55), together with (2.48). Thus we have

$$\mathbb{E}[\mathbf{x}\mathbf{x}^{\mathrm{T}}] = \boldsymbol{\mu}\boldsymbol{\mu}^{\mathrm{T}} + \boldsymbol{\Sigma}. \tag{2.62}$$

For single random variables, we subtracted the mean before taking second moments in order to define a variance. Similarly, in the multivariate case it is again convenient to subtract off the mean, giving rise to the *covariance* of a random vector $\mathbf{x}$ defined by

$$\operatorname{cov}[\mathbf{x}] = \mathbb{E}\left[(\mathbf{x} - \mathbb{E}[\mathbf{x}])(\mathbf{x} - \mathbb{E}[\mathbf{x}])^{\mathrm{T}}\right]. \tag{2.63}$$

For the specific case of a Gaussian distribution, we can make use of $\mathbb{E}[\mathbf{x}] = \boldsymbol{\mu}$, together with the result (2.62), to give

$$\operatorname{cov}[\mathbf{x}] = \boldsymbol{\Sigma}. \tag{2.64}$$

Because the parameter matrix $\boldsymbol{\Sigma}$ governs the covariance of $\mathbf{x}$ under the Gaussian distribution, it is called the covariance matrix.

Although the Gaussian distribution (2.43) is widely used as a density model, it suffers from some significant limitations. Consider the number of free parameters in the distribution. A general symmetric covariance matrix $\boldsymbol{\Sigma}$ will have $D(D+1)/2$ independent parameters, and there are another D independent parameters in $\boldsymbol{\mu}$, giving $D(D+3)/2$ parameters in total. For large D, the total number of parameters

Exercise 2.21

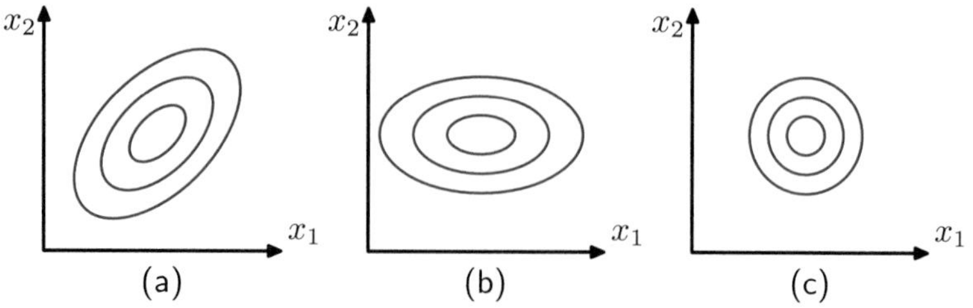

Figure 2.8 Contours of constant probability density for a Gaussian distribution in two dimensions in which the covariance matrix is (a) of general form, (b) diagonal, in which the elliptical contours are aligned with the coordinate axes, and (c) proportional to the identity matrix, in which the contours are concentric circles.

therefore grows quadratically with D, and the computational task of manipulating and inverting large matrices can become prohibitive. One way to address this problem is to use restricted forms of the covariance matrix. If we consider covariance matrices that are *diagonal*, so that $\mathbf{\Sigma} = \text{diag}(\sigma_i^2)$, we then have a total of $2D$ independent parameters in the density model. The corresponding contours of constant density are given by axis-aligned ellipsoids. We could further restrict the covariance matrix to be proportional to the identity matrix, $\mathbf{\Sigma} = \sigma^2\mathbf{I}$, known as an *isotropic* covariance, giving $D + 1$ independent parameters in the model and spherical surfaces of constant density. The three possibilities of general, diagonal, and isotropic covariance matrices are illustrated in Figure 2.8. Unfortunately, whereas such approaches limit the number of degrees of freedom in the distribution and make inversion of the covariance matrix a much faster operation, they also greatly restrict the form of the probability density and limit its ability to capture interesting correlations in the data.

A further limitation of the Gaussian distribution is that it is intrinsically unimodal (i.e., has a single maximum) and so is unable to provide a good approximation to multimodal distributions. Thus the Gaussian distribution can be both too flexible, in the sense of having too many parameters, while also being too limited in the range of distributions that it can adequately represent. We will see later that the introduction of *latent* variables, also called *hidden* variables or *unobserved* variables, allows both of these problems to be addressed. In particular, a rich family of multimodal distributions is obtained by introducing discrete latent variables leading to mixtures of Gaussians, as discussed in Section 2.3.9. Similarly, the introduction of continuous latent variables, as described in Chapter 12, leads to models in which the number of free parameters can be controlled independently of the dimensionality D of the data space while still allowing the model to capture the dominant correlations in the data set. Indeed, these two approaches can be combined and further extended to derive a very rich set of hierarchical models that can be adapted to a broad range of practical applications. For instance, the Gaussian version of the *Markov random field*, which is widely used as a probabilistic model of images, is a Gaussian distribution over the joint space of pixel intensities but rendered tractable through the imposition of considerable structure reflecting the spatial organization of the pixels. Similarly, the *linear dynamical system*, used to model time series data for applications such as tracking, is also a joint Gaussian distribution over a potentially large number of observed and latent variables and again is tractable due to the structure imposed on the distribution. A powerful framework for expressing the form and properties of

Section 8.3

Section 13.3

such complex distributions is that of probabilistic graphical models, which will form the subject of Chapter 8.

2.3.1 Conditional Gaussian distributions

An important property of the multivariate Gaussian distribution is that if two sets of variables are jointly Gaussian, then the conditional distribution of one set conditioned on the other is again Gaussian. Similarly, the marginal distribution of either set is also Gaussian.

Consider first the case of conditional distributions. Suppose $\mathbf{x}$ is a D-dimensional vector with Gaussian distribution $\mathcal{N}(\mathbf{x}|\boldsymbol{\mu}, \boldsymbol{\Sigma})$ and that we partition $\mathbf{x}$ into two disjoint subsets $\mathbf{x}_a$ and $\mathbf{x}_b$. Without loss of generality, we can take $\mathbf{x}_a$ to form the first M components of $\mathbf{x}$, with $\mathbf{x}_b$ comprising the remaining $D - M$ components, so that

$$\mathbf{x} = \begin{pmatrix} \mathbf{x}_a \\ \mathbf{x}_b \end{pmatrix}. \tag{2.65}$$

We also define corresponding partitions of the mean vector $\boldsymbol{\mu}$ given by

$$\boldsymbol{\mu} = \begin{pmatrix} \boldsymbol{\mu}_a \\ \boldsymbol{\mu}_b \end{pmatrix} \tag{2.66}$$

and of the covariance matrix $\boldsymbol{\Sigma}$ given by

$$\boldsymbol{\Sigma} = \begin{pmatrix} \boldsymbol{\Sigma}_{aa} & \boldsymbol{\Sigma}_{ab} \\ \boldsymbol{\Sigma}_{ba} & \boldsymbol{\Sigma}_{bb} \end{pmatrix}. \tag{2.67}$$

Note that the symmetry $\boldsymbol{\Sigma}^{\mathrm{T}} = \boldsymbol{\Sigma}$ of the covariance matrix implies that $\boldsymbol{\Sigma}_{aa}$ and $\boldsymbol{\Sigma}_{bb}$ are symmetric, while $\boldsymbol{\Sigma}_{ba} = \boldsymbol{\Sigma}_{ab}^{\mathrm{T}}$.

In many situations, it will be convenient to work with the inverse of the covariance matrix

$$\boldsymbol{\Lambda} \equiv \boldsymbol{\Sigma}^{-1} \tag{2.68}$$

which is known as the *precision matrix*. In fact, we shall see that some properties of Gaussian distributions are most naturally expressed in terms of the covariance, whereas others take a simpler form when viewed in terms of the precision. We therefore also introduce the partitioned form of the precision matrix

$$\boldsymbol{\Lambda} = \begin{pmatrix} \boldsymbol{\Lambda}_{aa} & \boldsymbol{\Lambda}_{ab} \\ \boldsymbol{\Lambda}_{ba} & \boldsymbol{\Lambda}_{bb} \end{pmatrix} \tag{2.69}$$

corresponding to the partitioning (2.65) of the vector $\mathbf{x}$. Because the inverse of a symmetric matrix is also symmetric, we see that $\boldsymbol{\Lambda}_{aa}$ and $\boldsymbol{\Lambda}_{bb}$ are symmetric, while $\boldsymbol{\Lambda}_{ab}^{\mathrm{T}} = \boldsymbol{\Lambda}_{ba}$. It should be stressed at this point that, for instance, $\boldsymbol{\Lambda}_{aa}$ is not simply given by the inverse of $\boldsymbol{\Sigma}_{aa}$. In fact, we shall shortly examine the relation between the inverse of a partitioned matrix and the inverses of its partitions.

Exercise 2.22

Let us begin by finding an expression for the conditional distribution $p(\mathbf{x}_a|\mathbf{x}_b)$. From the product rule of probability, we see that this conditional distribution can be evaluated from the joint distribution $p(\mathbf{x}) = p(\mathbf{x}_a, \mathbf{x}_b)$ simply by fixing $\mathbf{x}_b$ to the observed value and normalizing the resulting expression to obtain a valid probability distribution over $\mathbf{x}_a$. Instead of performing this normalization explicitly, we can obtain the solution more efficiently by considering the quadratic form in the exponent of the Gaussian distribution given by (2.44) and then reinstating the normalization coefficient at the end of the calculation. If we make use of the partitioning (2.65), (2.66), and (2.69), we obtain

$$\begin{aligned} -\frac{1}{2}(\mathbf{x}-\boldsymbol{\mu})^{\mathrm{T}}\boldsymbol{\Sigma}^{-1}(\mathbf{x}-\boldsymbol{\mu}) = \\ -\frac{1}{2}(\mathbf{x}_a-\boldsymbol{\mu}_a)^{\mathrm{T}}\boldsymbol{\Lambda}_{aa}(\mathbf{x}_a-\boldsymbol{\mu}_a) - \frac{1}{2}(\mathbf{x}_a-\boldsymbol{\mu}_a)^{\mathrm{T}}\boldsymbol{\Lambda}_{ab}(\mathbf{x}_b-\boldsymbol{\mu}_b) \\ -\frac{1}{2}(\mathbf{x}_b-\boldsymbol{\mu}_b)^{\mathrm{T}}\boldsymbol{\Lambda}_{ba}(\mathbf{x}_a-\boldsymbol{\mu}_a) - \frac{1}{2}(\mathbf{x}_b-\boldsymbol{\mu}_b)^{\mathrm{T}}\boldsymbol{\Lambda}_{bb}(\mathbf{x}_b-\boldsymbol{\mu}_b). \end{aligned} \quad (2.70)$$

We see that as a function of $\mathbf{x}_a$, this is again a quadratic form, and hence the corresponding conditional distribution $p(\mathbf{x}_a|\mathbf{x}_b)$ will be Gaussian. Because this distribution is completely characterized by its mean and its covariance, our goal will be to identify expressions for the mean and covariance of $p(\mathbf{x}_a|\mathbf{x}_b)$ by inspection of (2.70).

This is an example of a rather common operation associated with Gaussian distributions, sometimes called 'completing the square', in which we are given a quadratic form defining the exponent terms in a Gaussian distribution, and we need to determine the corresponding mean and covariance. Such problems can be solved straightforwardly by noting that the exponent in a general Gaussian distribution $\mathcal{N}(\mathbf{x}|\boldsymbol{\mu}, \boldsymbol{\Sigma})$ can be written

$$-\frac{1}{2}(\mathbf{x}-\boldsymbol{\mu})^{\mathrm{T}}\boldsymbol{\Sigma}^{-1}(\mathbf{x}-\boldsymbol{\mu}) = -\frac{1}{2}\mathbf{x}^{\mathrm{T}}\boldsymbol{\Sigma}^{-1}\mathbf{x} + \mathbf{x}^{\mathrm{T}}\boldsymbol{\Sigma}^{-1}\boldsymbol{\mu} + \mathrm{const} \quad (2.71)$$

where 'const' denotes terms which are independent of $\mathbf{x}$, and we have made use of the symmetry of $\boldsymbol{\Sigma}$. Thus if we take our general quadratic form and express it in the form given by the right-hand side of (2.71), then we can immediately equate the matrix of coefficients entering the second order term in $\mathbf{x}$ to the inverse covariance matrix $\boldsymbol{\Sigma}^{-1}$ and the coefficient of the linear term in $\mathbf{x}$ to $\boldsymbol{\Sigma}^{-1}\boldsymbol{\mu}$, from which we can obtain $\boldsymbol{\mu}$.

Now let us apply this procedure to the conditional Gaussian distribution $p(\mathbf{x}_a|\mathbf{x}_b)$ for which the quadratic form in the exponent is given by (2.70). We will denote the mean and covariance of this distribution by $\boldsymbol{\mu}_{a|b}$ and $\boldsymbol{\Sigma}_{a|b}$, respectively. Consider the functional dependence of (2.70) on $\mathbf{x}_a$ in which $\mathbf{x}_b$ is regarded as a constant. If we pick out all terms that are second order in $\mathbf{x}_a$, we have

$$-\frac{1}{2}\mathbf{x}_a^{\mathrm{T}}\boldsymbol{\Lambda}_{aa}\mathbf{x}_a \quad (2.72)$$

from which we can immediately conclude that the covariance (inverse precision) of $p(\mathbf{x}_a|\mathbf{x}_b)$ is given by

$$\boldsymbol{\Sigma}_{a|b} = \boldsymbol{\Lambda}_{aa}^{-1}. \quad (2.73)$$

Now consider all of the terms in (2.70) that are linear in $\mathbf{x}_a$

$$\mathbf{x}_a^{\mathrm{T}} \left\{ \boldsymbol{\Lambda}_{aa}\boldsymbol{\mu}_a - \boldsymbol{\Lambda}_{ab}(\mathbf{x}_b - \boldsymbol{\mu}_b) \right\} \tag{2.74}$$

where we have used $\boldsymbol{\Lambda}_{ba}^{\mathrm{T}} = \boldsymbol{\Lambda}_{ab}$. From our discussion of the general form (2.71), the coefficient of $\mathbf{x}_a$ in this expression must equal $\boldsymbol{\Sigma}_{a|b}^{-1}\boldsymbol{\mu}_{a|b}$ and hence

$$\begin{aligned} \boldsymbol{\mu}_{a|b} &= \boldsymbol{\Sigma}_{a|b} \left\{ \boldsymbol{\Lambda}_{aa}\boldsymbol{\mu}_a - \boldsymbol{\Lambda}_{ab}(\mathbf{x}_b - \boldsymbol{\mu}_b) \right\} \\ &= \boldsymbol{\mu}_a - \boldsymbol{\Lambda}_{aa}^{-1}\boldsymbol{\Lambda}_{ab}(\mathbf{x}_b - \boldsymbol{\mu}_b) \end{aligned} \tag{2.75}$$

where we have made use of (2.73).

The results (2.73) and (2.75) are expressed in terms of the partitioned precision matrix of the original joint distribution $p(\mathbf{x}_a, \mathbf{x}_b)$. We can also express these results in terms of the corresponding partitioned covariance matrix. To do this, we make use of the following identity for the inverse of a partitioned matrix

Exercise 2.24

$$\begin{pmatrix} \mathbf{A} & \mathbf{B} \\ \mathbf{C} & \mathbf{D} \end{pmatrix}^{-1} = \begin{pmatrix} \mathbf{M} & -\mathbf{M}\mathbf{B}\mathbf{D}^{-1} \\ -\mathbf{D}^{-1}\mathbf{C}\mathbf{M} & \mathbf{D}^{-1} + \mathbf{D}^{-1}\mathbf{C}\mathbf{M}\mathbf{B}\mathbf{D}^{-1} \end{pmatrix} \tag{2.76}$$

where we have defined

$$\mathbf{M} = (\mathbf{A} - \mathbf{B}\mathbf{D}^{-1}\mathbf{C})^{-1}. \tag{2.77}$$

The quantity $\mathbf{M}^{-1}$ is known as the *Schur complement* of the matrix on the left-hand side of (2.76) with respect to the submatrix $\mathbf{D}$. Using the definition

$$\begin{pmatrix} \boldsymbol{\Sigma}_{aa} & \boldsymbol{\Sigma}_{ab} \\ \boldsymbol{\Sigma}_{ba} & \boldsymbol{\Sigma}_{bb} \end{pmatrix}^{-1} = \begin{pmatrix} \boldsymbol{\Lambda}_{aa} & \boldsymbol{\Lambda}_{ab} \\ \boldsymbol{\Lambda}_{ba} & \boldsymbol{\Lambda}_{bb} \end{pmatrix} \tag{2.78}$$

and making use of (2.76), we have

$$\boldsymbol{\Lambda}_{aa} = (\boldsymbol{\Sigma}_{aa} - \boldsymbol{\Sigma}_{ab}\boldsymbol{\Sigma}_{bb}^{-1}\boldsymbol{\Sigma}_{ba})^{-1} \tag{2.79}$$

$$\boldsymbol{\Lambda}_{ab} = -(\boldsymbol{\Sigma}_{aa} - \boldsymbol{\Sigma}_{ab}\boldsymbol{\Sigma}_{bb}^{-1}\boldsymbol{\Sigma}_{ba})^{-1}\boldsymbol{\Sigma}_{ab}\boldsymbol{\Sigma}_{bb}^{-1}. \tag{2.80}$$

From these we obtain the following expressions for the mean and covariance of the conditional distribution $p(\mathbf{x}_a|\mathbf{x}_b)$

$$\boldsymbol{\mu}_{a|b} = \boldsymbol{\mu}_a + \boldsymbol{\Sigma}_{ab}\boldsymbol{\Sigma}_{bb}^{-1}(\mathbf{x}_b - \boldsymbol{\mu}_b) \tag{2.81}$$

$$\boldsymbol{\Sigma}_{a|b} = \boldsymbol{\Sigma}_{aa} - \boldsymbol{\Sigma}_{ab}\boldsymbol{\Sigma}_{bb}^{-1}\boldsymbol{\Sigma}_{ba}. \tag{2.82}$$

Comparing (2.73) and (2.82), we see that the conditional distribution $p(\mathbf{x}_a|\mathbf{x}_b)$ takes a simpler form when expressed in terms of the partitioned precision matrix than when it is expressed in terms of the partitioned covariance matrix. Note that the mean of the conditional distribution $p(\mathbf{x}_a|\mathbf{x}_b)$, given by (2.81), is a linear function of $\mathbf{x}_b$ and that the covariance, given by (2.82), is independent of $\mathbf{x}_b$. This represents an example of a *linear-Gaussian* model.

Section 8.1.4

2.3.2 Marginal Gaussian distributions

We have seen that if a joint distribution $p(\mathbf{x}_a, \mathbf{x}_b)$ is Gaussian, then the conditional distribution $p(\mathbf{x}_a|\mathbf{x}_b)$ will again be Gaussian. Now we turn to a discussion of the marginal distribution given by

$$p(\mathbf{x}_a) = \int p(\mathbf{x}_a, \mathbf{x}_b) \, \mathrm{d}\mathbf{x}_b \tag{2.83}$$

which, as we shall see, is also Gaussian. Once again, our strategy for evaluating this distribution efficiently will be to focus on the quadratic form in the exponent of the joint distribution and thereby to identify the mean and covariance of the marginal distribution $p(\mathbf{x}_a)$.

The quadratic form for the joint distribution can be expressed, using the partitioned precision matrix, in the form (2.70). Because our goal is to integrate out $\mathbf{x}_b$, this is most easily achieved by first considering the terms involving $\mathbf{x}_b$ and then completing the square in order to facilitate integration. Picking out just those terms that involve $\mathbf{x}_b$, we have

$$-\frac{1}{2}\mathbf{x}_b^{\mathrm{T}}\boldsymbol{\Lambda}_{bb}\mathbf{x}_b + \mathbf{x}_b^T\mathbf{m} = -\frac{1}{2}(\mathbf{x}_b - \boldsymbol{\Lambda}_{bb}^{-1}\mathbf{m})^{\mathrm{T}}\boldsymbol{\Lambda}_{bb}(\mathbf{x}_b - \boldsymbol{\Lambda}_{bb}^{-1}\mathbf{m}) + \frac{1}{2}\mathbf{m}^{\mathrm{T}}\boldsymbol{\Lambda}_{bb}^{-1}\mathbf{m} \tag{2.84}$$

where we have defined

$$\mathbf{m} = \boldsymbol{\Lambda}_{bb}\boldsymbol{\mu}_b - \boldsymbol{\Lambda}_{ba}(\mathbf{x}_a - \boldsymbol{\mu}_a). \tag{2.85}$$

We see that the dependence on $\mathbf{x}_b$ has been cast into the standard quadratic form of a Gaussian distribution corresponding to the first term on the right-hand side of (2.84), plus a term that does not depend on $\mathbf{x}_b$ (but that does depend on $\mathbf{x}_a$). Thus, when we take the exponential of this quadratic form, we see that the integration over $\mathbf{x}_b$ required by (2.83) will take the form

$$\int \exp\left\{-\frac{1}{2}(\mathbf{x}_b - \boldsymbol{\Lambda}_{bb}^{-1}\mathbf{m})^{\mathrm{T}}\boldsymbol{\Lambda}_{bb}(\mathbf{x}_b - \boldsymbol{\Lambda}_{bb}^{-1}\mathbf{m})\right\} \mathrm{d}\mathbf{x}_b. \tag{2.86}$$

This integration is easily performed by noting that it is the integral over an unnormalized Gaussian, and so the result will be the reciprocal of the normalization coefficient. We know from the form of the normalized Gaussian given by (2.43), that this coefficient is independent of the mean and depends only on the determinant of the covariance matrix. Thus, by completing the square with respect to $\mathbf{x}_b$, we can integrate out $\mathbf{x}_b$ and the only term remaining from the contributions on the left-hand side of (2.84) that depends on $\mathbf{x}_a$ is the last term on the right-hand side of (2.84) in which $\mathbf{m}$ is given by (2.85). Combining this term with the remaining terms from

(2.70) that depend on $\mathbf{x}_a$, we obtain

$$\begin{aligned}
&\frac{1}{2}\left[\boldsymbol{\Lambda}_{bb}\boldsymbol{\mu}_b - \boldsymbol{\Lambda}_{ba}(\mathbf{x}_a - \boldsymbol{\mu}_a)\right]^{\mathrm{T}} \boldsymbol{\Lambda}_{bb}^{-1} \left[\boldsymbol{\Lambda}_{bb}\boldsymbol{\mu}_b - \boldsymbol{\Lambda}_{ba}(\mathbf{x}_a - \boldsymbol{\mu}_a)\right] \\
&\qquad -\frac{1}{2}\mathbf{x}_a^{\mathrm{T}}\boldsymbol{\Lambda}_{aa}\mathbf{x}_a + \mathbf{x}_a^{\mathrm{T}}(\boldsymbol{\Lambda}_{aa}\boldsymbol{\mu}_a + \boldsymbol{\Lambda}_{ab}\boldsymbol{\mu}_b) + \text{const} \\
&= -\frac{1}{2}\mathbf{x}_a^{\mathrm{T}}(\boldsymbol{\Lambda}_{aa} - \boldsymbol{\Lambda}_{ab}\boldsymbol{\Lambda}_{bb}^{-1}\boldsymbol{\Lambda}_{ba})\mathbf{x}_a \\
&\qquad +\mathbf{x}_a^{\mathrm{T}}(\boldsymbol{\Lambda}_{aa} - \boldsymbol{\Lambda}_{ab}\boldsymbol{\Lambda}_{bb}^{-1}\boldsymbol{\Lambda}_{ba})\boldsymbol{\mu}_a + \text{const}
\end{aligned} \tag{2.87}$$

where 'const' denotes quantities independent of $\mathbf{x}_a$. Again, by comparison with (2.71), we see that the covariance of the marginal distribution $p(\mathbf{x}_a)$ is given by

$$\boldsymbol{\Sigma}_a = (\boldsymbol{\Lambda}_{aa} - \boldsymbol{\Lambda}_{ab}\boldsymbol{\Lambda}_{bb}^{-1}\boldsymbol{\Lambda}_{ba})^{-1}. \tag{2.88}$$

Similarly, the mean is given by

$$\boldsymbol{\Sigma}_a(\boldsymbol{\Lambda}_{aa} - \boldsymbol{\Lambda}_{ab}\boldsymbol{\Lambda}_{bb}^{-1}\boldsymbol{\Lambda}_{ba})\boldsymbol{\mu}_a = \boldsymbol{\mu}_a \tag{2.89}$$

where we have used (2.88). The covariance (2.88) is expressed in terms of the partitioned precision matrix given by (2.69). We can rewrite this in terms of the corresponding partitioning of the covariance matrix given by (2.67), as we did for the conditional distribution. These partitioned matrices are related by

$$\begin{pmatrix} \boldsymbol{\Lambda}_{aa} & \boldsymbol{\Lambda}_{ab} \\ \boldsymbol{\Lambda}_{ba} & \boldsymbol{\Lambda}_{bb} \end{pmatrix}^{-1} = \begin{pmatrix} \boldsymbol{\Sigma}_{aa} & \boldsymbol{\Sigma}_{ab} \\ \boldsymbol{\Sigma}_{ba} & \boldsymbol{\Sigma}_{bb} \end{pmatrix} \tag{2.90}$$

Making use of (2.76), we then have

$$\left(\boldsymbol{\Lambda}_{aa} - \boldsymbol{\Lambda}_{ab}\boldsymbol{\Lambda}_{bb}^{-1}\boldsymbol{\Lambda}_{ba}\right)^{-1} = \boldsymbol{\Sigma}_{aa}. \tag{2.91}$$

Thus we obtain the intuitively satisfying result that the marginal distribution $p(\mathbf{x}_a)$ has mean and covariance given by

$$\mathbb{E}[\mathbf{x}_a] = \boldsymbol{\mu}_a \tag{2.92}$$

$$\operatorname{cov}[\mathbf{x}_a] = \boldsymbol{\Sigma}_{aa}. \tag{2.93}$$

We see that for a marginal distribution, the mean and covariance are most simply expressed in terms of the partitioned covariance matrix, in contrast to the conditional distribution for which the partitioned precision matrix gives rise to simpler expressions.

Our results for the marginal and conditional distributions of a partitioned Gaussian are summarized below.

Partitioned Gaussians

Given a joint Gaussian distribution $\mathcal{N}(\mathbf{x}|\boldsymbol{\mu}, \boldsymbol{\Sigma})$ with $\boldsymbol{\Lambda} \equiv \boldsymbol{\Sigma}^{-1}$ and

$$\mathbf{x} = \begin{pmatrix} \mathbf{x}_a \\ \mathbf{x}_b \end{pmatrix}, \quad \boldsymbol{\mu} = \begin{pmatrix} \boldsymbol{\mu}_a \\ \boldsymbol{\mu}_b \end{pmatrix} \tag{2.94}$$

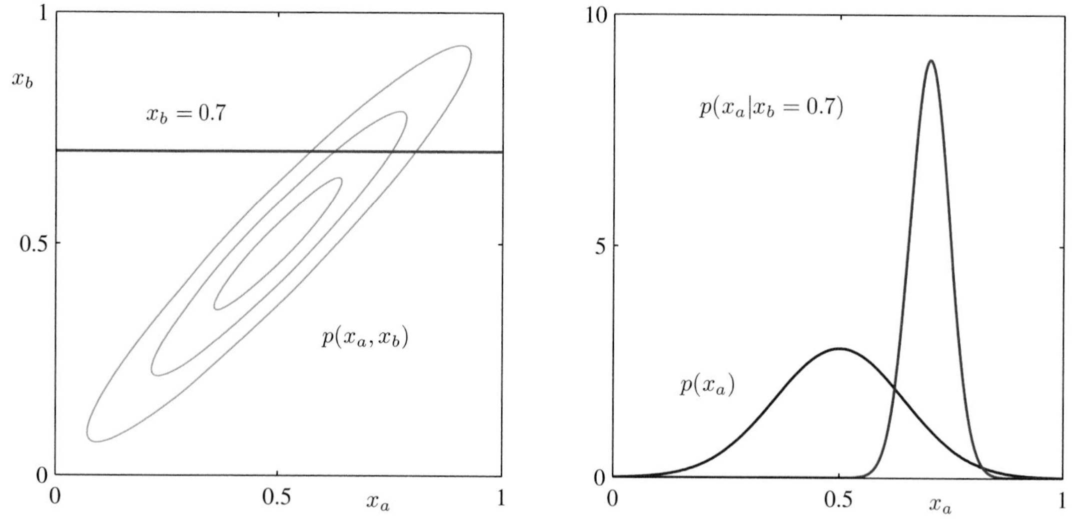

Figure 2.9 The plot on the left shows the contours of a Gaussian distribution $p(x_a, x_b)$ over two variables, and the plot on the right shows the marginal distribution $p(x_a)$ (blue curve) and the conditional distribution $p(x_a|x_b)$ for $x_b = 0.7$ (red curve).

$$\mathbf{\Sigma} = \begin{pmatrix} \mathbf{\Sigma}_{aa} & \mathbf{\Sigma}_{ab} \\ \mathbf{\Sigma}_{ba} & \mathbf{\Sigma}_{bb} \end{pmatrix}, \qquad \mathbf{\Lambda} = \begin{pmatrix} \mathbf{\Lambda}_{aa} & \mathbf{\Lambda}_{ab} \\ \mathbf{\Lambda}_{ba} & \mathbf{\Lambda}_{bb} \end{pmatrix}. \tag{2.95}$$

Conditional distribution:

$$p(\mathbf{x}_a|\mathbf{x}_b) = \mathcal{N}(\mathbf{x}|\boldsymbol{\mu}_{a|b}, \mathbf{\Lambda}_{aa}^{-1}) \tag{2.96}$$

$$\boldsymbol{\mu}_{a|b} = \boldsymbol{\mu}_a - \mathbf{\Lambda}_{aa}^{-1}\mathbf{\Lambda}_{ab}(\mathbf{x}_b - \boldsymbol{\mu}_b). \tag{2.97}$$

Marginal distribution:

$$p(\mathbf{x}_a) = \mathcal{N}(\mathbf{x}_a|\boldsymbol{\mu}_a, \mathbf{\Sigma}_{aa}). \tag{2.98}$$

We illustrate the idea of conditional and marginal distributions associated with a multivariate Gaussian using an example involving two variables in Figure 2.9.

2.3.3 Bayes' theorem for Gaussian variables

In Sections 2.3.1 and 2.3.2, we considered a Gaussian $p(\mathbf{x})$ in which we partitioned the vector $\mathbf{x}$ into two subvectors $\mathbf{x} = (\mathbf{x}_a, \mathbf{x}_b)$ and then found expressions for the conditional distribution $p(\mathbf{x}_a|\mathbf{x}_b)$ and the marginal distribution $p(\mathbf{x}_a)$. We noted that the mean of the conditional distribution $p(\mathbf{x}_a|\mathbf{x}_b)$ was a linear function of $\mathbf{x}_b$. Here we shall suppose that we are given a Gaussian marginal distribution $p(\mathbf{x})$ and a Gaussian conditional distribution $p(\mathbf{y}|\mathbf{x})$ in which $p(\mathbf{y}|\mathbf{x})$ has a mean that is a linear function of $\mathbf{x}$, and a covariance which is independent of $\mathbf{x}$. This is an example of a *linear Gaussian model* (Roweis and Ghahramani, 1999), which we shall study in greater generality in Section 8.1.4. We wish to find the marginal distribution $p(\mathbf{y})$ and the conditional distribution $p(\mathbf{x}|\mathbf{y})$. This is a problem that will arise frequently in subsequent chapters, and it will prove convenient to derive the general results here.

We shall take the marginal and conditional distributions to be

$$p(\mathbf{x}) = \mathcal{N}\left(\mathbf{x}|\boldsymbol{\mu}, \boldsymbol{\Lambda}^{-1}\right) \tag{2.99}$$

$$p(\mathbf{y}|\mathbf{x}) = \mathcal{N}\left(\mathbf{y}|\mathbf{A}\mathbf{x}+\mathbf{b}, \mathbf{L}^{-1}\right) \tag{2.100}$$

where $\boldsymbol{\mu}$, $\mathbf{A}$, and $\mathbf{b}$ are parameters governing the means, and $\boldsymbol{\Lambda}$ and $\mathbf{L}$ are precision matrices. If $\mathbf{x}$ has dimensionality M and $\mathbf{y}$ has dimensionality D, then the matrix $\mathbf{A}$ has size $D \times M$.

First we find an expression for the joint distribution over $\mathbf{x}$ and $\mathbf{y}$. To do this, we define

$$\mathbf{z} = \begin{pmatrix} \mathbf{x} \\ \mathbf{y} \end{pmatrix} \tag{2.101}$$

and then consider the log of the joint distribution

$$\begin{aligned} \ln p(\mathbf{z}) &= \ln p(\mathbf{x}) + \ln p(\mathbf{y}|\mathbf{x}) \\ &= -\frac{1}{2}(\mathbf{x}-\boldsymbol{\mu})^{\mathrm{T}}\boldsymbol{\Lambda}(\mathbf{x}-\boldsymbol{\mu}) \\ &\quad -\frac{1}{2}(\mathbf{y}-\mathbf{A}\mathbf{x}-\mathbf{b})^{\mathrm{T}}\mathbf{L}(\mathbf{y}-\mathbf{A}\mathbf{x}-\mathbf{b}) + \text{const} \end{aligned} \tag{2.102}$$

where 'const' denotes terms independent of $\mathbf{x}$ and $\mathbf{y}$. As before, we see that this is a quadratic function of the components of $\mathbf{z}$, and hence $p(\mathbf{z})$ is a Gaussian distribution. To find the precision of this Gaussian, we consider the second order terms in (2.102), which can be written as

$$\begin{aligned} &-\frac{1}{2}\mathbf{x}^{\mathrm{T}}(\boldsymbol{\Lambda}+\mathbf{A}^{\mathrm{T}}\mathbf{L}\mathbf{A})\mathbf{x} - \frac{1}{2}\mathbf{y}^{\mathrm{T}}\mathbf{L}\mathbf{y} + \frac{1}{2}\mathbf{y}^{\mathrm{T}}\mathbf{L}\mathbf{A}\mathbf{x} + \frac{1}{2}\mathbf{x}^{\mathrm{T}}\mathbf{A}^{\mathrm{T}}\mathbf{L}\mathbf{y} \\ &= -\frac{1}{2}\begin{pmatrix} \mathbf{x} \\ \mathbf{y} \end{pmatrix}^{\mathrm{T}} \begin{pmatrix} \boldsymbol{\Lambda}+\mathbf{A}^{\mathrm{T}}\mathbf{L}\mathbf{A} & -\mathbf{A}^{\mathrm{T}}\mathbf{L} \\ -\mathbf{L}\mathbf{A} & \mathbf{L} \end{pmatrix} \begin{pmatrix} \mathbf{x} \\ \mathbf{y} \end{pmatrix} = -\frac{1}{2}\mathbf{z}^{\mathrm{T}}\mathbf{R}\mathbf{z} \end{aligned} \tag{2.103}$$

and so the Gaussian distribution over $\mathbf{z}$ has precision (inverse covariance) matrix given by

$$\mathbf{R} = \begin{pmatrix} \boldsymbol{\Lambda}+\mathbf{A}^{\mathrm{T}}\mathbf{L}\mathbf{A} & -\mathbf{A}^{\mathrm{T}}\mathbf{L} \\ -\mathbf{L}\mathbf{A} & \mathbf{L} \end{pmatrix}. \tag{2.104}$$

Exercise 2.29

The covariance matrix is found by taking the inverse of the precision, which can be done using the matrix inversion formula (2.76) to give

$$\operatorname{cov}[\mathbf{z}] = \mathbf{R}^{-1} = \begin{pmatrix} \boldsymbol{\Lambda}^{-1} & \boldsymbol{\Lambda}^{-1}\mathbf{A}^{\mathrm{T}} \\ \mathbf{A}\boldsymbol{\Lambda}^{-1} & \mathbf{L}^{-1}+\mathbf{A}\boldsymbol{\Lambda}^{-1}\mathbf{A}^{\mathrm{T}} \end{pmatrix}. \tag{2.105}$$

Similarly, we can find the mean of the Gaussian distribution over $\mathbf{z}$ by identifying the linear terms in (2.102), which are given by

$$\mathbf{x}^{\mathrm{T}}\boldsymbol{\Lambda}\boldsymbol{\mu} - \mathbf{x}^{\mathrm{T}}\mathbf{A}^{\mathrm{T}}\mathbf{L}\mathbf{b} + \mathbf{y}^{\mathrm{T}}\mathbf{L}\mathbf{b} = \begin{pmatrix}\mathbf{x}\\ \mathbf{y}\end{pmatrix}^{\mathrm{T}} \begin{pmatrix}\boldsymbol{\Lambda}\boldsymbol{\mu} - \mathbf{A}^{\mathrm{T}}\mathbf{L}\mathbf{b}\\ \mathbf{L}\mathbf{b}\end{pmatrix}. \tag{2.106}$$

Using our earlier result (2.71) obtained by completing the square over the quadratic form of a multivariate Gaussian, we find that the mean of $\mathbf{z}$ is given by

$$\mathbb{E}[\mathbf{z}] = \mathbf{R}^{-1}\begin{pmatrix}\boldsymbol{\Lambda}\boldsymbol{\mu} - \mathbf{A}^{\mathrm{T}}\mathbf{L}\mathbf{b}\\ \mathbf{L}\mathbf{b}\end{pmatrix}. \tag{2.107}$$

Exercise 2.30

Making use of (2.105), we then obtain

$$\mathbb{E}[\mathbf{z}] = \begin{pmatrix}\boldsymbol{\mu}\\ \mathbf{A}\boldsymbol{\mu} + \mathbf{b}\end{pmatrix}. \tag{2.108}$$

Section 2.3

Next we find an expression for the marginal distribution $p(\mathbf{y})$ in which we have marginalized over $\mathbf{x}$. Recall that the marginal distribution over a subset of the components of a Gaussian random vector takes a particularly simple form when expressed in terms of the partitioned covariance matrix. Specifically, its mean and covariance are given by (2.92) and (2.93), respectively. Making use of (2.105) and (2.108) we see that the mean and covariance of the marginal distribution $p(\mathbf{y})$ are given by

$$\mathbb{E}[\mathbf{y}] = \mathbf{A}\boldsymbol{\mu} + \mathbf{b} \tag{2.109}$$

$$\mathrm{cov}[\mathbf{y}] = \mathbf{L}^{-1} + \mathbf{A}\boldsymbol{\Lambda}^{-1}\mathbf{A}^{\mathrm{T}}. \tag{2.110}$$

A special case of this result is when $\mathbf{A} = \mathbf{I}$, in which case it reduces to the convolution of two Gaussians, for which we see that the mean of the convolution is the sum of the mean of the two Gaussians, and the covariance of the convolution is the sum of their covariances.

Section 2.3

Finally, we seek an expression for the conditional $p(\mathbf{x}|\mathbf{y})$. Recall that the results for the conditional distribution are most easily expressed in terms of the partitioned precision matrix, using (2.73) and (2.75). Applying these results to (2.105) and (2.108) we see that the conditional distribution $p(\mathbf{x}|\mathbf{y})$ has mean and covariance given by

$$\mathbb{E}[\mathbf{x}|\mathbf{y}] = (\boldsymbol{\Lambda} + \mathbf{A}^{\mathrm{T}}\mathbf{L}\mathbf{A})^{-1}\left\{\mathbf{A}^{\mathrm{T}}\mathbf{L}(\mathbf{y} - \mathbf{b}) + \boldsymbol{\Lambda}\boldsymbol{\mu}\right\} \tag{2.111}$$

$$\mathrm{cov}[\mathbf{x}|\mathbf{y}] = (\boldsymbol{\Lambda} + \mathbf{A}^{\mathrm{T}}\mathbf{L}\mathbf{A})^{-1}. \tag{2.112}$$

The evaluation of this conditional can be seen as an example of Bayes' theorem. We can interpret the distribution $p(\mathbf{x})$ as a prior distribution over $\mathbf{x}$. If the variable $\mathbf{y}$ is observed, then the conditional distribution $p(\mathbf{x}|\mathbf{y})$ represents the corresponding posterior distribution over $\mathbf{x}$. Having found the marginal and conditional distributions, we effectively expressed the joint distribution $p(\mathbf{z}) = p(\mathbf{x})p(\mathbf{y}|\mathbf{x})$ in the form $p(\mathbf{x}|\mathbf{y})p(\mathbf{y})$. These results are summarized below.

Marginal and Conditional Gaussians

Given a marginal Gaussian distribution for $\mathbf{x}$ and a conditional Gaussian distribution for $\mathbf{y}$ given $\mathbf{x}$ in the form

$$p(\mathbf{x}) = \mathcal{N}(\mathbf{x}|\boldsymbol{\mu}, \boldsymbol{\Lambda}^{-1}) \tag{2.113}$$
$$p(\mathbf{y}|\mathbf{x}) = \mathcal{N}(\mathbf{y}|\mathbf{A}\mathbf{x} + \mathbf{b}, \mathbf{L}^{-1}) \tag{2.114}$$

the marginal distribution of $\mathbf{y}$ and the conditional distribution of $\mathbf{x}$ given $\mathbf{y}$ are given by

$$p(\mathbf{y}) = \mathcal{N}(\mathbf{y}|\mathbf{A}\boldsymbol{\mu} + \mathbf{b}, \mathbf{L}^{-1} + \mathbf{A}\boldsymbol{\Lambda}^{-1}\mathbf{A}^{\mathrm{T}}) \tag{2.115}$$
$$p(\mathbf{x}|\mathbf{y}) = \mathcal{N}(\mathbf{x}|\boldsymbol{\Sigma}\{\mathbf{A}^{\mathrm{T}}\mathbf{L}(\mathbf{y} - \mathbf{b}) + \boldsymbol{\Lambda}\boldsymbol{\mu}\}, \boldsymbol{\Sigma}) \tag{2.116}$$

where

$$\boldsymbol{\Sigma} = (\boldsymbol{\Lambda} + \mathbf{A}^{\mathrm{T}}\mathbf{L}\mathbf{A})^{-1}. \tag{2.117}$$

2.3.4 Maximum likelihood for the Gaussian

Given a data set $\mathbf{X} = (\mathbf{x}_1, \ldots, \mathbf{x}_N)^{\mathrm{T}}$ in which the observations $\{\mathbf{x}_n\}$ are assumed to be drawn independently from a multivariate Gaussian distribution, we can estimate the parameters of the distribution by maximum likelihood. The log likelihood function is given by

$$\ln p(\mathbf{X}|\boldsymbol{\mu}, \boldsymbol{\Sigma}) = -\frac{ND}{2}\ln(2\pi) - \frac{N}{2}\ln|\boldsymbol{\Sigma}| - \frac{1}{2}\sum_{n=1}^{N}(\mathbf{x}_n - \boldsymbol{\mu})^{\mathrm{T}}\boldsymbol{\Sigma}^{-1}(\mathbf{x}_n - \boldsymbol{\mu}). \tag{2.118}$$

By simple rearrangement, we see that the likelihood function depends on the data set only through the two quantities

$$\sum_{n=1}^{N}\mathbf{x}_n, \qquad \sum_{n=1}^{N}\mathbf{x}_n\mathbf{x}_n^{\mathrm{T}}. \tag{2.119}$$

Appendix C

These are known as the *sufficient statistics* for the Gaussian distribution. Using (C.19), the derivative of the log likelihood with respect to $\boldsymbol{\mu}$ is given by

$$\frac{\partial}{\partial\boldsymbol{\mu}}\ln p(\mathbf{X}|\boldsymbol{\mu}, \boldsymbol{\Sigma}) = \sum_{n=1}^{N}\boldsymbol{\Sigma}^{-1}(\mathbf{x}_n - \boldsymbol{\mu}) \tag{2.120}$$

and setting this derivative to zero, we obtain the solution for the maximum likelihood estimate of the mean given by

$$\boldsymbol{\mu}_{\mathrm{ML}} = \frac{1}{N}\sum_{n=1}^{N}\mathbf{x}_n \tag{2.121}$$

which is the mean of the observed set of data points. The maximization of (2.118) with respect to $\boldsymbol{\Sigma}$ is rather more involved. The simplest approach is to ignore the symmetry constraint and show that the resulting solution is symmetric as required. Alternative derivations of this result, which impose the symmetry and positive definiteness constraints explicitly, can be found in Magnus and Neudecker (1999). The result is as expected and takes the form

Exercise 2.34

$$\boldsymbol{\Sigma}_{\mathrm{ML}} = \frac{1}{N}\sum_{n=1}^{N}(\mathbf{x}_n - \boldsymbol{\mu}_{\mathrm{ML}})(\mathbf{x}_n - \boldsymbol{\mu}_{\mathrm{ML}})^{\mathrm{T}} \tag{2.122}$$

which involves $\boldsymbol{\mu}_{\mathrm{ML}}$ because this is the result of a joint maximization with respect to $\boldsymbol{\mu}$ and $\boldsymbol{\Sigma}$. Note that the solution (2.121) for $\boldsymbol{\mu}_{\mathrm{ML}}$ does not depend on $\boldsymbol{\Sigma}_{\mathrm{ML}}$, and so we can first evaluate $\boldsymbol{\mu}_{\mathrm{ML}}$ and then use this to evaluate $\boldsymbol{\Sigma}_{\mathrm{ML}}$.

If we evaluate the expectations of the maximum likelihood solutions under the true distribution, we obtain the following results

Exercise 2.35

$$\mathbb{E}[\boldsymbol{\mu}_{\mathrm{ML}}] = \boldsymbol{\mu} \tag{2.123}$$

$$\mathbb{E}[\boldsymbol{\Sigma}_{\mathrm{ML}}] = \frac{N-1}{N}\boldsymbol{\Sigma}. \tag{2.124}$$

We see that the expectation of the maximum likelihood estimate for the mean is equal to the true mean. However, the maximum likelihood estimate for the covariance has an expectation that is less than the true value, and hence it is biased. We can correct this bias by defining a different estimator $\widetilde{\boldsymbol{\Sigma}}$ given by

$$\widetilde{\boldsymbol{\Sigma}} = \frac{1}{N-1}\sum_{n=1}^{N}(\mathbf{x}_n - \boldsymbol{\mu}_{\mathrm{ML}})(\mathbf{x}_n - \boldsymbol{\mu}_{\mathrm{ML}})^{\mathrm{T}}. \tag{2.125}$$

Clearly from (2.122) and (2.124), the expectation of $\widetilde{\boldsymbol{\Sigma}}$ is equal to $\boldsymbol{\Sigma}$.

2.3.5 Sequential estimation

Our discussion of the maximum likelihood solution for the parameters of a Gaussian distribution provides a convenient opportunity to give a more general discussion of the topic of sequential estimation for maximum likelihood. Sequential methods allow data points to be processed one at a time and then discarded and are important for on-line applications, and also where large data sets are involved so that batch processing of all data points at once is infeasible.

Consider the result (2.121) for the maximum likelihood estimator of the mean $\boldsymbol{\mu}_{\mathrm{ML}}$, which we will denote by $\boldsymbol{\mu}_{\mathrm{ML}}^{(N)}$ when it is based on N observations. If we

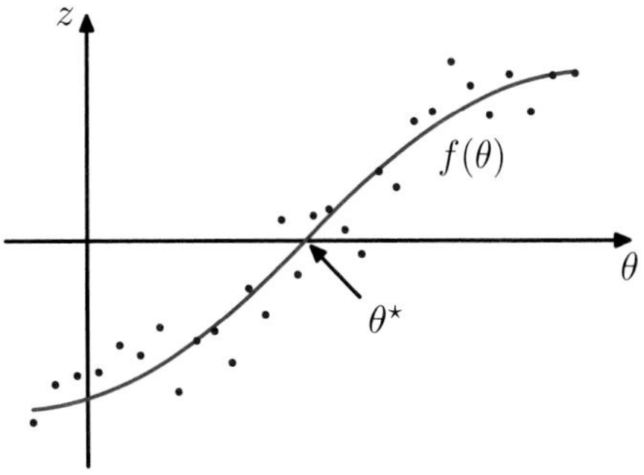

Figure 2.10 A schematic illustration of two correlated random variables z and θ, together with the regression function $f(\theta)$ given by the conditional expectation $\mathbb{E}[z|\theta]$. The Robbins-Monro algorithm provides a general sequential procedure for finding the root $\theta^\star$ of such functions.

dissect out the contribution from the final data point $\mathbf{x}_N$, we obtain

$$\begin{aligned}\boldsymbol{\mu}_{\mathrm{ML}}^{(N)} &= \frac{1}{N}\sum_{n=1}^{N}\mathbf{x}_n \\ &= \frac{1}{N}\mathbf{x}_N + \frac{1}{N}\sum_{n=1}^{N-1}\mathbf{x}_n \\ &= \frac{1}{N}\mathbf{x}_N + \frac{N-1}{N}\boldsymbol{\mu}_{\mathrm{ML}}^{(N-1)} \\ &= \boldsymbol{\mu}_{\mathrm{ML}}^{(N-1)} + \frac{1}{N}(\mathbf{x}_N - \boldsymbol{\mu}_{\mathrm{ML}}^{(N-1)}).\end{aligned} \tag{2.126}$$

This result has a nice interpretation, as follows. After observing $N-1$ data points we have estimated $\boldsymbol{\mu}$ by $\boldsymbol{\mu}_{\mathrm{ML}}^{(N-1)}$. We now observe data point $\mathbf{x}_N$, and we obtain our revised estimate $\boldsymbol{\mu}_{\mathrm{ML}}^{(N)}$ by moving the old estimate a small amount, proportional to $1/N$, in the direction of the 'error signal' $(\mathbf{x}_N - \boldsymbol{\mu}_{\mathrm{ML}}^{(N-1)})$. Note that, as N increases, so the contribution from successive data points gets smaller.

The result (2.126) will clearly give the same answer as the batch result (2.121) because the two formulae are equivalent. However, we will not always be able to derive a sequential algorithm by this route, and so we seek a more general formulation of sequential learning, which leads us to the *Robbins-Monro* algorithm. Consider a pair of random variables θ and z governed by a joint distribution $p(z, \theta)$. The conditional expectation of z given θ defines a deterministic function $f(\theta)$ that is given by

$$f(\theta) \equiv \mathbb{E}[z|\theta] = \int zp(z|\theta)\,\mathrm{d}z \tag{2.127}$$

and is illustrated schematically in Figure 2.10. Functions defined in this way are called *regression functions*.

Our goal is to find the root $\theta^\star$ at which $f(\theta^\star) = 0$. If we had a large data set of observations of z and θ, then we could model the regression function directly and then obtain an estimate of its root. Suppose, however, that we observe values of z one at a time and we wish to find a corresponding sequential estimation scheme for $\theta^\star$. The following general procedure for solving such problems was given by

Robbins and Monro (1951). We shall assume that the conditional variance of z is finite so that

$$\mathbb{E}\left[(z-f)^2 \mid \theta\right] < \infty \tag{2.128}$$

and we shall also, without loss of generality, consider the case where $f(\theta) > 0$ for $\theta > \theta^\star$ and $f(\theta) < 0$ for $\theta < \theta^\star$, as is the case in Figure 2.10. The Robbins-Monro procedure then defines a sequence of successive estimates of the root $\theta^\star$ given by

$$\theta^{(N)} = \theta^{(N-1)} - a_{N-1} z(\theta^{(N-1)}) \tag{2.129}$$

where $z(\theta^{(N)})$ is an observed value of z when θ takes the value $\theta^{(N)}$. The coefficients $\{a_N\}$ represent a sequence of positive numbers that satisfy the conditions

$$\lim_{N\to\infty} a_N = 0 \tag{2.130}$$

$$\sum_{N=1}^{\infty} a_N = \infty \tag{2.131}$$

$$\sum_{N=1}^{\infty} a_N^2 < \infty. \tag{2.132}$$

It can then be shown (Robbins and Monro, 1951; Fukunaga, 1990) that the sequence of estimates given by (2.129) does indeed converge to the root with probability one. Note that the first condition (2.130) ensures that the successive corrections decrease in magnitude so that the process can converge to a limiting value. The second condition (2.131) is required to ensure that the algorithm does not converge short of the root, and the third condition (2.132) is needed to ensure that the accumulated noise has finite variance and hence does not spoil convergence.

Now let us consider how a general maximum likelihood problem can be solved sequentially using the Robbins-Monro algorithm. By definition, the maximum likelihood solution θ_{ML} is a stationary point of the negative log likelihood function and hence satisfies

$$\frac{\partial}{\partial\theta}\left\{-\frac{1}{N}\sum_{n=1}^{N}\ln p(x_n|\theta)\right\}\Bigg|_{\theta_{\mathrm{ML}}} = 0. \tag{2.133}$$

Exchanging the derivative and the summation, and taking the limit $N \to \infty$ we have

$$-\lim_{N\to\infty}\frac{1}{N}\sum_{n=1}^{N}\frac{\partial}{\partial\theta}\ln p(x_n|\theta) = \mathbb{E}_x\left[-\frac{\partial}{\partial\theta}\ln p(x|\theta)\right] \tag{2.134}$$

and so we see that finding the maximum likelihood solution corresponds to finding the root of a regression function. We can therefore apply the Robbins-Monro procedure, which now takes the form

$$\theta^{(N)} = \theta^{(N-1)} - a_{N-1}\frac{\partial}{\partial\theta^{(N-1)}}\left[-\ln p(x_N|\theta^{(N-1)})\right]. \tag{2.135}$$

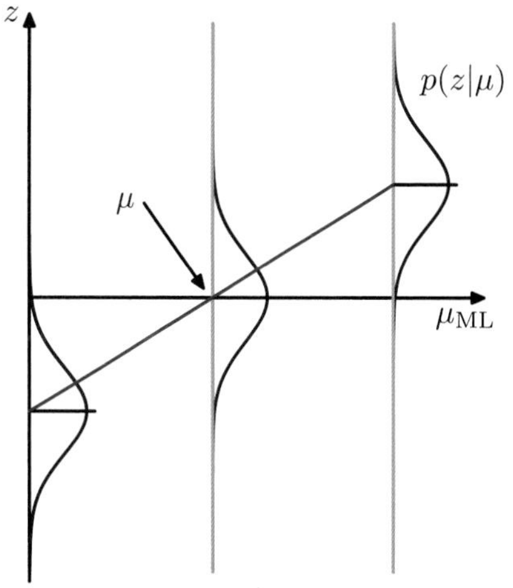

Figure 2.11 In the case of a Gaussian distribution, with θ corresponding to μ_{ML}, the regression function illustrated in Figure 2.10 takes the form of a straight line, as shown in red. In this case, the random variable z corresponds to the derivative of the negative log likelihood function and is given by $-(x - \mu_{\mathrm{ML}})/\sigma^2$, and its expectation that defines the regression function is a straight line given by $-(\mu - \mu_{\mathrm{ML}})/\sigma^2$. The root of the regression function corresponds to the true mean μ.

As a specific example, we consider once again the sequential estimation of the mean of a Gaussian distribution, in which case the parameter $\theta^{(N)}$ is the estimate $\mu_{\mathrm{ML}}^{(N)}$ of the mean of the Gaussian, and the random variable z is given by

$$z = \frac{\partial}{\partial \mu_{\mathrm{ML}}} \left[-\ln p(x|\mu_{\mathrm{ML}}, \sigma^2) \right] = -\frac{1}{\sigma^2}(x - \mu_{\mathrm{ML}}). \tag{2.136}$$

Thus the distribution of z is Gaussian with mean $-(\mu - \mu_{\mathrm{ML}})/\sigma^2$, as illustrated in Figure 2.11. Substituting (2.136) into (2.135), we obtain the univariate form of (2.126), provided we choose the coefficients a_N to have the form $a_N = \sigma^2/N$. Note that although we have focussed on the case of a single variable, the same technique, together with the same restrictions (2.130)–(2.132) on the coefficients a_N, apply equally to the multivariate case (Blum, 1965).

2.3.6 Bayesian inference for the Gaussian

The maximum likelihood framework gave point estimates for the parameters $\boldsymbol{\mu}$ and $\boldsymbol{\Sigma}$. Now we develop a Bayesian treatment by introducing prior distributions over these parameters. Let us begin with a simple example in which we consider a single Gaussian random variable x. We shall suppose that the variance σ^2 is known, and we consider the task of inferring the mean μ given a set of N observations $\mathbf{x} = \{x_1, \ldots, x_N\}$. The likelihood function, that is the probability of the observed data given μ, viewed as a function of μ, is given by

$$p(\mathbf{x}|\mu) = \prod_{n=1}^{N} p(x_n|\mu) = \frac{1}{(2\pi\sigma^2)^{N/2}} \exp\left\{ -\frac{1}{2\sigma^2} \sum_{n=1}^{N} (x_n - \mu)^2 \right\}. \tag{2.137}$$

Again we emphasize that the likelihood function $p(\mathbf{x}|\mu)$ is not a probability distribution over μ and is not normalized.

We see that the likelihood function takes the form of the exponential of a quadratic form in μ. Thus if we choose a prior $p(\mu)$ given by a Gaussian, it will be a conjugate distribution for this likelihood function because the corresponding posterior will be a product of two exponentials of quadratic functions of μ and hence will also be Gaussian. We therefore take our prior distribution to be

$$p(\mu) = \mathcal{N}\left(\mu|\mu_0, \sigma_0^2\right) \tag{2.138}$$

and the posterior distribution is given by

$$p(\mu|\mathbf{X}) \propto p(\mathbf{X}|\mu)p(\mu). \tag{2.139}$$

Exercise 2.38

Simple manipulation involving completing the square in the exponent shows that the posterior distribution is given by

$$p(\mu|\mathbf{X}) = \mathcal{N}\left(\mu|\mu_N, \sigma_N^2\right) \tag{2.140}$$

where

$$\mu_N = \frac{\sigma^2}{N\sigma_0^2 + \sigma^2}\mu_0 + \frac{N\sigma_0^2}{N\sigma_0^2 + \sigma^2}\mu_{\mathrm{ML}} \tag{2.141}$$

$$\frac{1}{\sigma_N^2} = \frac{1}{\sigma_0^2} + \frac{N}{\sigma^2} \tag{2.142}$$

in which μ_{ML} is the maximum likelihood solution for μ given by the sample mean

$$\mu_{\mathrm{ML}} = \frac{1}{N}\sum_{n=1}^{N} x_n. \tag{2.143}$$

It is worth spending a moment studying the form of the posterior mean and variance. First of all, we note that the mean of the posterior distribution given by (2.141) is a compromise between the prior mean μ_0 and the maximum likelihood solution μ_{ML}. If the number of observed data points $N = 0$, then (2.141) reduces to the prior mean as expected. For $N \to \infty$, the posterior mean is given by the maximum likelihood solution. Similarly, consider the result (2.142) for the variance of the posterior distribution. We see that this is most naturally expressed in terms of the inverse variance, which is called the precision. Furthermore, the precisions are additive, so that the precision of the posterior is given by the precision of the prior plus one contribution of the data precision from each of the observed data points. As we increase the number of observed data points, the precision steadily increases, corresponding to a posterior distribution with steadily decreasing variance. With no observed data points, we have the prior variance, whereas if the number of data points $N \to \infty$, the variance σ_N^2 goes to zero and the posterior distribution becomes infinitely peaked around the maximum likelihood solution. We therefore see that the maximum likelihood result of a point estimate for μ given by (2.143) is recovered precisely from the Bayesian formalism in the limit of an infinite number of observations. Note also that for finite N, if we take the limit $\sigma_0^2 \to \infty$ in which the prior has infinite variance then the posterior mean (2.141) reduces to the maximum likelihood result, while from (2.142) the posterior variance is given by $\sigma_N^2 = \sigma^2/N$.

Figure 2.12 Illustration of Bayesian inference for the mean μ of a Gaussian distribution, in which the variance is assumed to be known. The curves show the prior distribution over μ (the curve labelled $N = 0$), which in this case is itself Gaussian, along with the posterior distribution given by (2.140) for increasing numbers N of data points. The data points are generated from a Gaussian of mean 0.8 and variance 0.1, and the prior is chosen to have mean 0. In both the prior and the likelihood function, the variance is set to the true value.

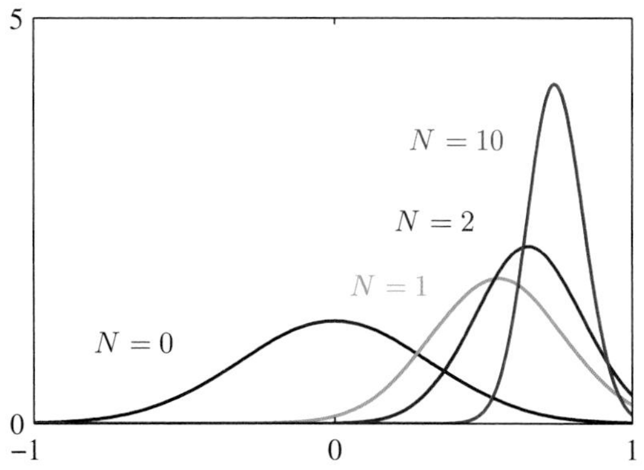

We illustrate our analysis of Bayesian inference for the mean of a Gaussian distribution in Figure 2.12. The generalization of this result to the case of a D-dimensional Gaussian random variable $\mathbf{x}$ with known covariance and unknown mean is straightforward.

Exercise 2.40

We have already seen how the maximum likelihood expression for the mean of a Gaussian can be re-cast as a sequential update formula in which the mean after observing N data points was expressed in terms of the mean after observing $N - 1$ data points together with the contribution from data point x_N. In fact, the Bayesian paradigm leads very naturally to a sequential view of the inference problem. To see this in the context of the inference of the mean of a Gaussian, we write the posterior distribution with the contribution from the final data point x_N separated out so that

Section 2.3.5

$$p(\mu|\mathbf{x}) \propto \left[p(\mu) \prod_{n=1}^{N-1} p(x_n|\mu) \right] p(x_N|\mu). \tag{2.144}$$

The term in square brackets is (up to a normalization coefficient) just the posterior distribution after observing $N - 1$ data points. We see that this can be viewed as a prior distribution, which is combined using Bayes' theorem with the likelihood function associated with data point x_N to arrive at the posterior distribution after observing N data points. This sequential view of Bayesian inference is very general and applies to any problem in which the observed data are assumed to be independent and identically distributed.

So far, we have assumed that the variance of the Gaussian distribution over the data is known and our goal is to infer the mean. Now let us suppose that the mean is known and we wish to infer the variance. Again, our calculations will be greatly simplified if we choose a conjugate form for the prior distribution. It turns out to be most convenient to work with the precision $\lambda \equiv 1/\sigma^2$. The likelihood function for λ takes the form

$$p(\mathbf{x}|\lambda) = \prod_{n=1}^{N} \mathcal{N}(x_n|\mu, \lambda^{-1}) \propto \lambda^{N/2} \exp\left\{ -\frac{\lambda}{2} \sum_{n=1}^{N} (x_n - \mu)^2 \right\}. \tag{2.145}$$

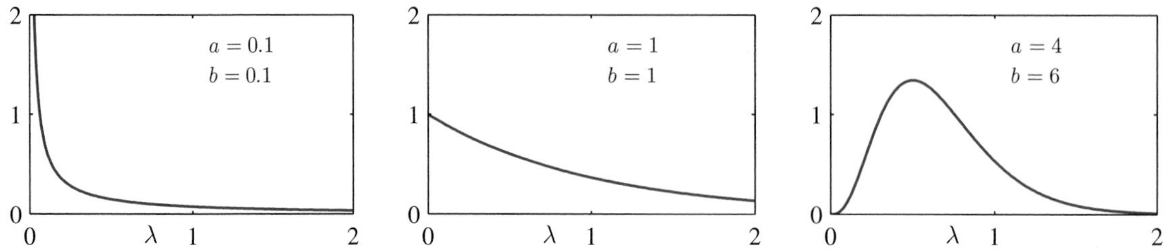

Figure 2.13 Plot of the gamma distribution $\mathrm{Gam}(\lambda|a,b)$ defined by (2.146) for various values of the parameters a and b.

The corresponding conjugate prior should therefore be proportional to the product of a power of λ and the exponential of a linear function of λ. This corresponds to the *gamma* distribution which is defined by

$$\mathrm{Gam}(\lambda|a,b) = \frac{1}{\Gamma(a)} b^a \lambda^{a-1} \exp(-b\lambda). \tag{2.146}$$

Here $\Gamma(a)$ is the gamma function that is defined by (1.141) and that ensures that (2.146) is correctly normalized. The gamma distribution has a finite integral if $a > 0$, and the distribution itself is finite if $a \geqslant 1$. It is plotted, for various values of a and b, in Figure 2.13. The mean and variance of the gamma distribution are given by

Exercise 2.41

Exercise 2.42

$$\mathbb{E}[\lambda] = \frac{a}{b} \tag{2.147}$$

$$\mathrm{var}[\lambda] = \frac{a}{b^2}. \tag{2.148}$$

Consider a prior distribution $\mathrm{Gam}(\lambda|a_0,b_0)$. If we multiply by the likelihood function (2.145), then we obtain a posterior distribution

$$p(\lambda|\mathbf{x}) \propto \lambda^{a_0-1}\lambda^{N/2} \exp\left\{-b_0\lambda - \frac{\lambda}{2}\sum_{n=1}^{N}(x_n-\mu)^2\right\} \tag{2.149}$$

which we recognize as a gamma distribution of the form $\mathrm{Gam}(\lambda|a_N,b_N)$ where

$$a_N = a_0 + \frac{N}{2} \tag{2.150}$$

$$b_N = b_0 + \frac{1}{2}\sum_{n=1}^{N}(x_n-\mu)^2 = b_0 + \frac{N}{2}\sigma^2_{\mathrm{ML}} \tag{2.151}$$

where σ^2_{ML} is the maximum likelihood estimator of the variance. Note that in (2.149) there is no need to keep track of the normalization constants in the prior and the likelihood function because, if required, the correct coefficient can be found at the end using the normalized form (2.146) for the gamma distribution.

From (2.150), we see that the effect of observing N data points is to increase the value of the coefficient a by $N/2$. Thus we can interpret the parameter a_0 in the prior in terms of $2a_0$ 'effective' prior observations. Similarly, from (2.151) we see that the N data points contribute $N\sigma^2_{\mathrm{ML}}/2$ to the parameter b, where σ^2_{ML} is the variance, and so we can interpret the parameter b_0 in the prior as arising from the $2a_0$ 'effective' prior observations having variance $2b_0/(2a_0) = b_0/a_0$. Recall that we made an analogous interpretation for the Dirichlet prior. These distributions are examples of the exponential family, and we shall see that the interpretation of a conjugate prior in terms of effective fictitious data points is a general one for the exponential family of distributions.

Section 2.2

Instead of working with the precision, we can consider the variance itself. The conjugate prior in this case is called the *inverse gamma* distribution, although we shall not discuss this further because we will find it more convenient to work with the precision.

Now suppose that both the mean and the precision are unknown. To find a conjugate prior, we consider the dependence of the likelihood function on μ and λ

$$
\begin{aligned}
p(\mathbf{x}|\mu,\lambda) &= \prod_{n=1}^{N}\left(\frac{\lambda}{2\pi}\right)^{1/2}\exp\left\{-\frac{\lambda}{2}(x_n-\mu)^2\right\} \\
&\propto \left[\lambda^{1/2}\exp\left(-\frac{\lambda\mu^2}{2}\right)\right]^N \exp\left\{\lambda\mu\sum_{n=1}^{N}x_n - \frac{\lambda}{2}\sum_{n=1}^{N}x_n^2\right\}. \qquad (2.152)
\end{aligned}
$$

We now wish to identify a prior distribution $p(\mu,\lambda)$ that has the same functional dependence on μ and λ as the likelihood function and that should therefore take the form

$$
\begin{aligned}
p(\mu,\lambda) &\propto \left[\lambda^{1/2}\exp\left(-\frac{\lambda\mu^2}{2}\right)\right]^{\beta}\exp\{c\lambda\mu - d\lambda\} \\
&= \exp\left\{-\frac{\beta\lambda}{2}(\mu-c/\beta)^2\right\}\lambda^{\beta/2}\exp\left\{-\left(d-\frac{c^2}{2\beta}\right)\lambda\right\} \qquad (2.153)
\end{aligned}
$$

where c, d, and β are constants. Since we can always write $p(\mu,\lambda) = p(\mu|\lambda)p(\lambda)$, we can find $p(\mu|\lambda)$ and $p(\lambda)$ by inspection. In particular, we see that $p(\mu|\lambda)$ is a Gaussian whose precision is a linear function of λ and that $p(\lambda)$ is a gamma distribution, so that the normalized prior takes the form

$$
p(\mu,\lambda) = \mathcal{N}(\mu|\mu_0,(\beta\lambda)^{-1})\mathrm{Gam}(\lambda|a,b) \qquad (2.154)
$$

where we have defined new constants given by $\mu_0 = c/\beta$, $a = 1+\beta/2$, $b = d-c^2/2\beta$. The distribution (2.154) is called the *normal-gamma* or *Gaussian-gamma* distribution and is plotted in Figure 2.14. Note that this is not simply the product of an independent Gaussian prior over μ and a gamma prior over λ, because the precision of μ is a linear function of λ. Even if we chose a prior in which μ and λ were independent, the posterior distribution would exhibit a coupling between the precision of μ and the value of λ.

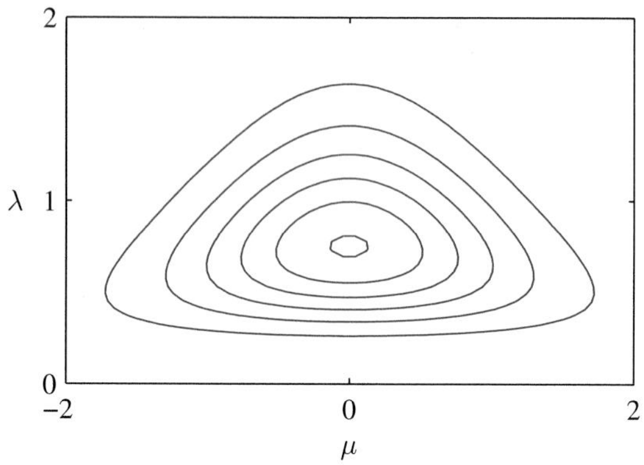

Figure 2.14 Contour plot of the normal-gamma distribution (2.154) for parameter values $\mu_0 = 0$, $\beta = 2$, $a = 5$ and $b = 6$.

In the case of the multivariate Gaussian distribution $\mathcal{N}\left(\mathbf{x}|\boldsymbol{\mu}, \boldsymbol{\Lambda}^{-1}\right)$ for a D-dimensional variable $\mathbf{x}$, the conjugate prior distribution for the mean $\boldsymbol{\mu}$, assuming the precision is known, is again a Gaussian. For known mean and unknown precision matrix $\boldsymbol{\Lambda}$, the conjugate prior is the *Wishart* distribution given by

Exercise 2.45

$$\mathcal{W}(\boldsymbol{\Lambda}|\mathbf{W}, \nu) = B|\boldsymbol{\Lambda}|^{(\nu-D-1)/2} \exp\left(-\frac{1}{2}\mathrm{Tr}(\mathbf{W}^{-1}\boldsymbol{\Lambda})\right) \tag{2.155}$$

where ν is called the number of *degrees of freedom* of the distribution, $\mathbf{W}$ is a $D \times D$ scale matrix, and $\mathrm{Tr}(\cdot)$ denotes the trace. The normalization constant B is given by

$$B(\mathbf{W}, \nu) = |\mathbf{W}|^{-\nu/2} \left(2^{\nu D/2}\,\pi^{D(D-1)/4} \prod_{i=1}^{D} \Gamma\left(\frac{\nu+1-i}{2}\right)\right)^{-1}. \tag{2.156}$$

Again, it is also possible to define a conjugate prior over the covariance matrix itself, rather than over the precision matrix, which leads to the *inverse Wishart* distribution, although we shall not discuss this further. If both the mean and the precision are unknown, then, following a similar line of reasoning to the univariate case, the conjugate prior is given by

$$p(\boldsymbol{\mu}, \boldsymbol{\Lambda}|\boldsymbol{\mu}_0, \beta, \mathbf{W}, \nu) = \mathcal{N}(\boldsymbol{\mu}|\boldsymbol{\mu}_0, (\beta\boldsymbol{\Lambda})^{-1})\,\mathcal{W}(\boldsymbol{\Lambda}|\mathbf{W}, \nu) \tag{2.157}$$

which is known as the *normal-Wishart* or *Gaussian-Wishart* distribution.

2.3.7 Student's t-distribution

We have seen that the conjugate prior for the precision of a Gaussian is given by a gamma distribution. If we have a univariate Gaussian $\mathcal{N}(x|\mu, \tau^{-1})$ together with a Gamma prior $\mathrm{Gam}(\tau|a, b)$ and we integrate out the precision, we obtain the marginal distribution of x in the form

Section 2.3.6

Exercise 2.46

Figure 2.15 Plot of Student's t-distribution (2.159) for $\mu = 0$ and $\lambda = 1$ for various values of ν. The limit $\nu \to \infty$ corresponds to a Gaussian distribution with mean μ and precision λ.

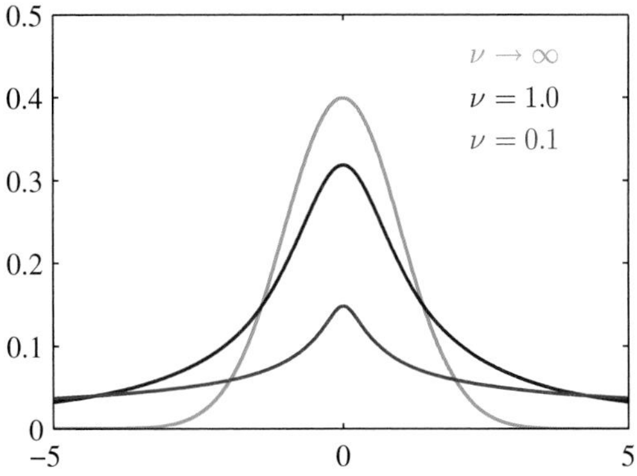

$$
\begin{aligned}
p(x|\mu, a, b) &= \int_0^\infty \mathcal{N}(x|\mu, \tau^{-1})\mathrm{Gam}(\tau|a, b)\,\mathrm{d}\tau \qquad (2.158)\\
&= \int_0^\infty \frac{b^a e^{(-b\tau)}\tau^{a-1}}{\Gamma(a)} \left(\frac{\tau}{2\pi}\right)^{1/2} \exp\left\{-\frac{\tau}{2}(x-\mu)^2\right\}\,\mathrm{d}\tau\\
&= \frac{b^a}{\Gamma(a)}\left(\frac{1}{2\pi}\right)^{1/2}\left[b + \frac{(x-\mu)^2}{2}\right]^{-a-1/2}\Gamma(a+1/2)
\end{aligned}
$$

where we have made the change of variable $z = \tau[b + (x-\mu)^2/2]$. By convention we define new parameters given by $\nu = 2a$ and $\lambda = a/b$, in terms of which the distribution $p(x|\mu, a, b)$ takes the form

$$
\mathrm{St}(x|\mu, \lambda, \nu) = \frac{\Gamma(\nu/2 + 1/2)}{\Gamma(\nu/2)}\left(\frac{\lambda}{\pi\nu}\right)^{1/2}\left[1 + \frac{\lambda(x-\mu)^2}{\nu}\right]^{-\nu/2-1/2} \qquad (2.159)
$$

which is known as *Student's t-distribution*. The parameter λ is sometimes called the *precision* of the t-distribution, even though it is not in general equal to the inverse of the variance. The parameter ν is called the *degrees of freedom*, and its effect is illustrated in Figure 2.15. For the particular case of $\nu = 1$, the t-distribution reduces to the *Cauchy* distribution, while in the limit $\nu \to \infty$ the t-distribution $\mathrm{St}(x|\mu, \lambda, \nu)$ becomes a Gaussian $\mathcal{N}(x|\mu, \lambda^{-1})$ with mean μ and precision λ.

Exercise 2.47

From (2.158), we see that Student's t-distribution is obtained by adding up an infinite number of Gaussian distributions having the same mean but different precisions. This can be interpreted as an infinite mixture of Gaussians (Gaussian mixtures will be discussed in detail in Section 2.3.9). The result is a distribution that in general has longer 'tails' than a Gaussian, as was seen in Figure 2.15. This gives the t-distribution an important property called *robustness*, which means that it is much less sensitive than the Gaussian to the presence of a few data points which are *outliers*. The robustness of the t-distribution is illustrated in Figure 2.16, which compares the maximum likelihood solutions for a Gaussian and a t-distribution. Note that the maximum likelihood solution for the t-distribution can be found using the expectation-maximization (EM) algorithm. Here we see that the effect of a small number of

Exercise 12.24

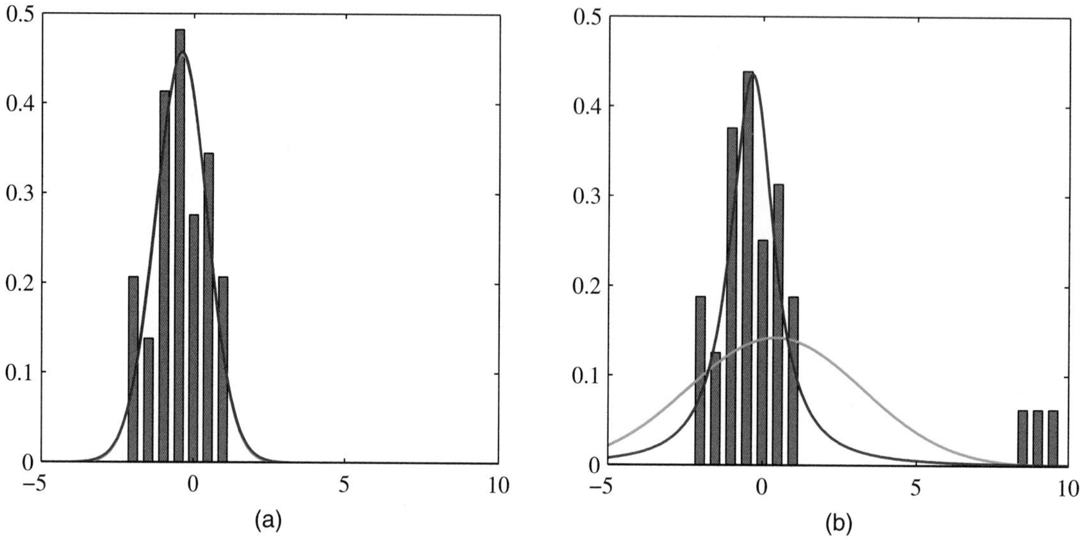

Figure 2.16 Illustration of the robustness of Student's t-distribution compared to a Gaussian. (a) Histogram distribution of 30 data points drawn from a Gaussian distribution, together with the maximum likelihood fit obtained from a t-distribution (red curve) and a Gaussian (green curve, largely hidden by the red curve). Because the t-distribution contains the Gaussian as a special case it gives almost the same solution as the Gaussian. (b) The same data set but with three additional outlying data points showing how the Gaussian (green curve) is strongly distorted by the outliers, whereas the t-distribution (red curve) is relatively unaffected.

outliers is much less significant for the t-distribution than for the Gaussian. Outliers can arise in practical applications either because the process that generates the data corresponds to a distribution having a heavy tail or simply through mislabelled data. Robustness is also an important property for regression problems. Unsurprisingly, the least squares approach to regression does not exhibit robustness, because it corresponds to maximum likelihood under a (conditional) Gaussian distribution. By basing a regression model on a heavy-tailed distribution such as a t-distribution, we obtain a more robust model.

If we go back to (2.158) and substitute the alternative parameters $\nu = 2a$, $\lambda = a/b$, and $\eta = \tau b/a$, we see that the t-distribution can be written in the form

$$\mathrm{St}(x|\mu, \lambda, \nu) = \int_0^\infty \mathcal{N}\left(x|\mu, (\eta\lambda)^{-1}\right) \mathrm{Gam}(\eta|\nu/2, \nu/2)\,\mathrm{d}\eta. \tag{2.160}$$

We can then generalize this to a multivariate Gaussian $\mathcal{N}(\mathbf{x}|\boldsymbol{\mu}, \boldsymbol{\Lambda})$ to obtain the corresponding multivariate Student's t-distribution in the form

$$\mathrm{St}(\mathbf{x}|\boldsymbol{\mu}, \boldsymbol{\Lambda}, \nu) = \int_0^\infty \mathcal{N}(\mathbf{x}|\boldsymbol{\mu}, (\eta\boldsymbol{\Lambda})^{-1})\mathrm{Gam}(\eta|\nu/2, \nu/2)\,\mathrm{d}\eta. \tag{2.161}$$

Exercise 2.48

Using the same technique as for the univariate case, we can evaluate this integral to give

$$\mathrm{St}(\mathbf{x}|\boldsymbol{\mu},\boldsymbol{\Lambda},\nu) = \frac{\Gamma(D/2+\nu/2)}{\Gamma(\nu/2)} \frac{|\boldsymbol{\Lambda}|^{1/2}}{(\pi\nu)^{D/2}} \left[1+\frac{\Delta^2}{\nu}\right]^{-D/2-\nu/2} \tag{2.162}$$

where D is the dimensionality of $\mathbf{x}$, and Δ^2 is the squared Mahalanobis distance defined by

$$\Delta^2 = (\mathbf{x}-\boldsymbol{\mu})^{\mathrm{T}}\boldsymbol{\Lambda}(\mathbf{x}-\boldsymbol{\mu}). \tag{2.163}$$

Exercise 2.49

This is the multivariate form of Student's t-distribution and satisfies the following properties

$$\mathbb{E}[\mathbf{x}] = \boldsymbol{\mu}, \qquad \text{if} \quad \nu > 1 \tag{2.164}$$

$$\mathrm{cov}[\mathbf{x}] = \frac{\nu}{(\nu-2)}\boldsymbol{\Lambda}^{-1}, \qquad \text{if} \quad \nu > 2 \tag{2.165}$$

$$\mathrm{mode}[\mathbf{x}] = \boldsymbol{\mu} \tag{2.166}$$

with corresponding results for the univariate case.

2.3.8 Periodic variables

Although Gaussian distributions are of great practical significance, both in their own right and as building blocks for more complex probabilistic models, there are situations in which they are inappropriate as density models for continuous variables. One important case, which arises in practical applications, is that of periodic variables.

An example of a periodic variable would be the wind direction at a particular geographical location. We might, for instance, measure values of wind direction on a number of days and wish to summarize this using a parametric distribution. Another example is calendar time, where we may be interested in modelling quantities that are believed to be periodic over 24 hours or over an annual cycle. Such quantities can conveniently be represented using an angular (polar) coordinate $0 \leqslant \theta < 2\pi$.

We might be tempted to treat periodic variables by choosing some direction as the origin and then applying a conventional distribution such as the Gaussian. Such an approach, however, would give results that were strongly dependent on the arbitrary choice of origin. Suppose, for instance, that we have two observations at $\theta_1 = 1^\circ$ and $\theta_2 = 359^\circ$, and we model them using a standard univariate Gaussian distribution. If we choose the origin at 0°, then the sample mean of this data set will be 180° with standard deviation 179°, whereas if we choose the origin at 180°, then the mean will be 0° and the standard deviation will be 1°. We clearly need to develop a special approach for the treatment of periodic variables.

Let us consider the problem of evaluating the mean of a set of observations $\mathcal{D} = \{\theta_1, \ldots, \theta_N\}$ of a periodic variable. From now on, we shall assume that θ is measured in radians. We have already seen that the simple average $(\theta_1 + \cdots + \theta_N)/N$ will be strongly coordinate dependent. To find an invariant measure of the mean, we note that the observations can be viewed as points on the unit circle and can therefore be described instead by two-dimensional unit vectors $\mathbf{x}_1, \ldots, \mathbf{x}_N$ where $\|\mathbf{x}_n\| = 1$ for $n = 1, \ldots, N$, as illustrated in Figure 2.17. We can average the vectors $\{\mathbf{x}_n\}$

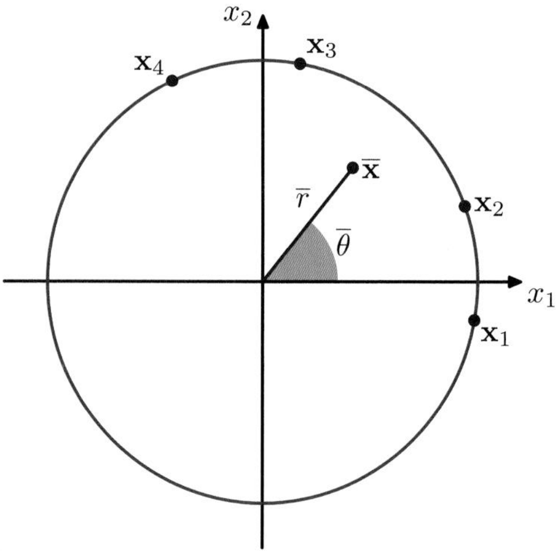

Figure 2.17 Illustration of the representation of values θ_n of a periodic variable as two-dimensional vectors $\mathbf{x}_n$ living on the unit circle. Also shown is the average $\overline{\mathbf{x}}$ of those vectors.

instead to give

$$\overline{\mathbf{x}} = \frac{1}{N} \sum_{n=1}^{N} \mathbf{x}_n \tag{2.167}$$

and then find the corresponding angle $\overline{\theta}$ of this average. Clearly, this definition will ensure that the location of the mean is independent of the origin of the angular coordinate. Note that $\overline{\mathbf{x}}$ will typically lie inside the unit circle. The Cartesian coordinates of the observations are given by $\mathbf{x}_n = (\cos\theta_n, \sin\theta_n)$, and we can write the Cartesian coordinates of the sample mean in the form $\overline{\mathbf{x}} = (\overline{r}\cos\overline{\theta}, \overline{r}\sin\overline{\theta})$. Substituting into (2.167) and equating the x_1 and x_2 components then gives

$$\overline{x}_1 = \overline{r}\cos\overline{\theta} = \frac{1}{N}\sum_{n=1}^{N}\cos\theta_n, \qquad \overline{x}_2 = \overline{r}\sin\overline{\theta} = \frac{1}{N}\sum_{n=1}^{N}\sin\theta_n. \tag{2.168}$$

Taking the ratio, and using the identity $\tan\theta = \sin\theta/\cos\theta$, we can solve for $\overline{\theta}$ to give

$$\overline{\theta} = \tan^{-1}\left\{ \frac{\sum_n \sin\theta_n}{\sum_n \cos\theta_n} \right\}. \tag{2.169}$$

Shortly, we shall see how this result arises naturally as the maximum likelihood estimator for an appropriately defined distribution over a periodic variable.

We now consider a periodic generalization of the Gaussian called the *von Mises* distribution. Here we shall limit our attention to univariate distributions, although periodic distributions can also be found over hyperspheres of arbitrary dimension. For an extensive discussion of periodic distributions, see Mardia and Jupp (2000).

By convention, we will consider distributions $p(\theta)$ that have period 2π. Any probability density $p(\theta)$ defined over θ must not only be nonnegative and integrate

Figure 2.18 The von Mises distribution can be derived by considering a two-dimensional Gaussian of the form (2.173), whose density contours are shown in blue and conditioning on the unit circle shown in red.

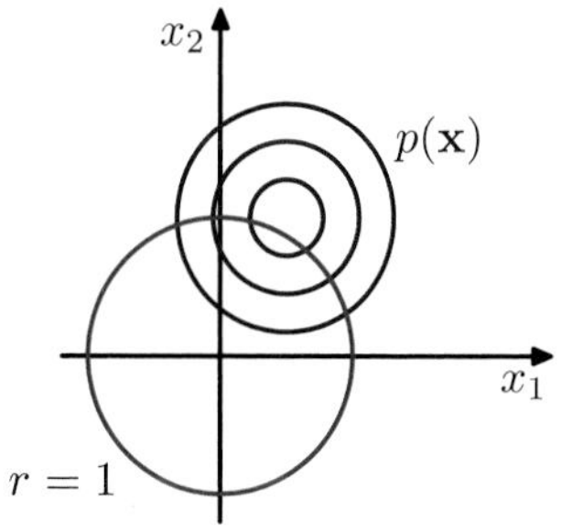

to one, but it must also be periodic. Thus $p(\theta)$ must satisfy the three conditions

$$p(\theta) \geqslant 0 \tag{2.170}$$

$$\int_0^{2\pi} p(\theta)\,\mathrm{d}\theta = 1 \tag{2.171}$$

$$p(\theta + 2\pi) = p(\theta). \tag{2.172}$$

From (2.172), it follows that $p(\theta + M2\pi) = p(\theta)$ for any integer M.

We can easily obtain a Gaussian-like distribution that satisfies these three properties as follows. Consider a Gaussian distribution over two variables $\mathbf{x} = (x_1, x_2)$ having mean $\boldsymbol{\mu} = (\mu_1, \mu_2)$ and a covariance matrix $\boldsymbol{\Sigma} = \sigma^2\mathbf{I}$ where $\mathbf{I}$ is the 2×2 identity matrix, so that

$$p(x_1, x_2) = \frac{1}{2\pi\sigma^2} \exp\left\{ -\frac{(x_1 - \mu_1)^2 + (x_2 - \mu_2)^2}{2\sigma^2} \right\}. \tag{2.173}$$

The contours of constant $p(\mathbf{x})$ are circles, as illustrated in Figure 2.18. Now suppose we consider the value of this distribution along a circle of fixed radius. Then by construction this distribution will be periodic, although it will not be normalized. We can determine the form of this distribution by transforming from Cartesian coordinates (x_1, x_2) to polar coordinates (r, θ) so that

$$x_1 = r\cos\theta, \qquad x_2 = r\sin\theta. \tag{2.174}$$

We also map the mean $\boldsymbol{\mu}$ into polar coordinates by writing

$$\mu_1 = r_0\cos\theta_0, \qquad \mu_2 = r_0\sin\theta_0. \tag{2.175}$$

Next we substitute these transformations into the two-dimensional Gaussian distribution (2.173), and then condition on the unit circle $r = 1$, noting that we are interested only in the dependence on θ. Focussing on the exponent in the Gaussian distribution we have

$$\begin{aligned}
&-\frac{1}{2\sigma^2}\left\{(r\cos\theta - r_0\cos\theta_0)^2 + (r\sin\theta - r_0\sin\theta_0)^2\right\} \\
&= -\frac{1}{2\sigma^2}\left\{1 + r_0^2 - 2r_0\cos\theta\cos\theta_0 - 2r_0\sin\theta\sin\theta_0\right\} \\
&= \frac{r_0}{\sigma^2}\cos(\theta - \theta_0) + \text{const}
\end{aligned} \tag{2.176}$$

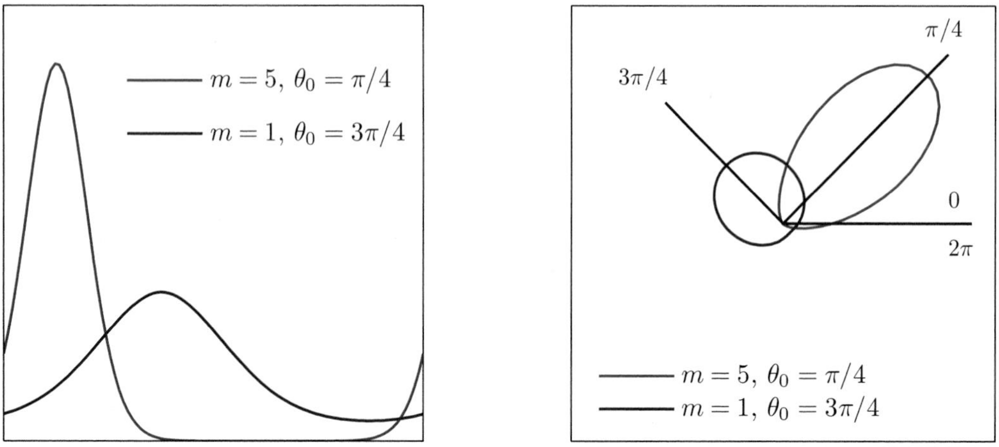

Figure 2.19 The von Mises distribution plotted for two different parameter values, shown as a Cartesian plot on the left and as the corresponding polar plot on the right.

Exercise 2.51

where 'const' denotes terms independent of θ, and we have made use of the following trigonometrical identities

$$\begin{aligned} \cos^2 A + \sin^2 A &= 1 && (2.177) \\ \cos A \cos B + \sin A \sin B &= \cos(A - B). && (2.178) \end{aligned}$$

If we now define $m = r_0/\sigma^2$, we obtain our final expression for the distribution of $p(\theta)$ along the unit circle $r = 1$ in the form

$$p(\theta|\theta_0, m) = \frac{1}{2\pi I_0(m)} \exp\{m \cos(\theta - \theta_0)\} \tag{2.179}$$

which is called the *von Mises* distribution, or the *circular normal*. Here the parameter θ_0 corresponds to the mean of the distribution, while m, which is known as the *concentration* parameter, is analogous to the inverse variance (precision) for the Gaussian. The normalization coefficient in (2.179) is expressed in terms of $I_0(m)$, which is the zeroth-order modified Bessel function of the first kind (Abramowitz and Stegun, 1965) and is defined by

$$I_0(m) = \frac{1}{2\pi} \int_0^{2\pi} \exp\{m \cos\theta\} \,\mathrm{d}\theta. \tag{2.180}$$

Exercise 2.52

For large m, the distribution becomes approximately Gaussian. The von Mises distribution is plotted in Figure 2.19, and the function $I_0(m)$ is plotted in Figure 2.20.

Now consider the maximum likelihood estimators for the parameters θ_0 and m for the von Mises distribution. The log likelihood function is given by

$$\ln p(\mathcal{D}|\theta_0, m) = -N \ln(2\pi) - N \ln I_0(m) + m \sum_{n=1}^{N} \cos(\theta_n - \theta_0). \tag{2.181}$$

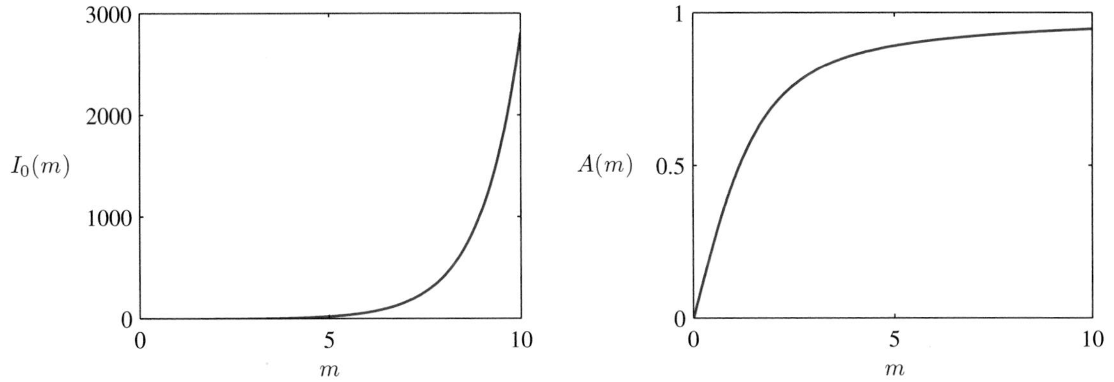

Figure 2.20 Plot of the Bessel function $I_0(m)$ defined by (2.180), together with the function $A(m)$ defined by (2.186).

Setting the derivative with respect to θ_0 equal to zero gives

$$\sum_{n=1}^{N} \sin(\theta_n - \theta_0) = 0. \tag{2.182}$$

To solve for θ_0, we make use of the trigonometric identity

$$\sin(A - B) = \cos B \sin A - \cos A \sin B \tag{2.183}$$

Exercise 2.53

from which we obtain

$$\theta_0^{\mathrm{ML}} = \tan^{-1}\left\{\frac{\sum_n \sin\theta_n}{\sum_n \cos\theta_n}\right\} \tag{2.184}$$

which we recognize as the result (2.169) obtained earlier for the mean of the observations viewed in a two-dimensional Cartesian space.

Similarly, maximizing (2.181) with respect to m, and making use of $I_0'(m) = I_1(m)$ (Abramowitz and Stegun, 1965), we have

$$A(m_{\mathrm{ML}}) = \frac{1}{N}\sum_{n=1}^{N} \cos(\theta_n - \theta_0^{\mathrm{ML}}) \tag{2.185}$$

where we have substituted for the maximum likelihood solution for θ_0^{ML} (recalling that we are performing a joint optimization over θ and m), and we have defined

$$A(m) = \frac{I_1(m)}{I_0(m)}. \tag{2.186}$$

The function $A(m)$ is plotted in Figure 2.20. Making use of the trigonometric identity (2.178), we can write (2.185) in the form

$$A(m_{\mathrm{ML}}) = \left(\frac{1}{N}\sum_{n=1}^{N}\cos\theta_n\right)\cos\theta_0^{\mathrm{ML}} + \left(\frac{1}{N}\sum_{n=1}^{N}\sin\theta_n\right)\sin\theta_0^{\mathrm{ML}}. \tag{2.187}$$

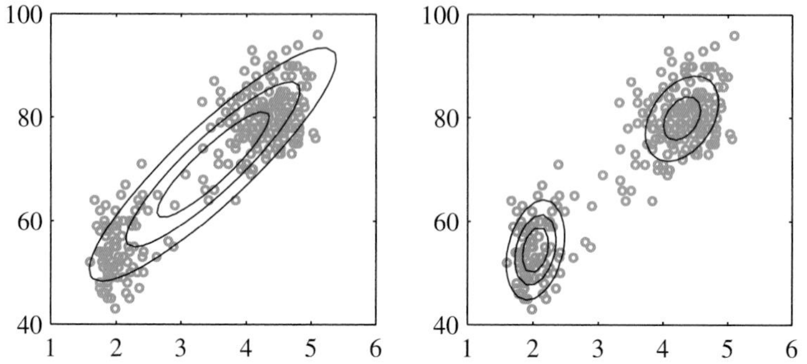

Figure 2.21 Plots of the 'old faithful' data in which the blue curves show contours of constant probability density. On the left is a single Gaussian distribution which has been fitted to the data using maximum likelihood. Note that this distribution fails to capture the two clumps in the data and indeed places much of its probability mass in the central region between the clumps where the data are relatively sparse. On the right the distribution is given by a linear combination of two Gaussians which has been fitted to the data by maximum likelihood using techniques discussed Chapter 9, and which gives a better representation of the data.

The right-hand side of (2.187) is easily evaluated, and the function $A(m)$ can be inverted numerically.

For completeness, we mention briefly some alternative techniques for the construction of periodic distributions. The simplest approach is to use a histogram of observations in which the angular coordinate is divided into fixed bins. This has the virtue of simplicity and flexibility but also suffers from significant limitations, as we shall see when we discuss histogram methods in more detail in Section 2.5. Another approach starts, like the von Mises distribution, from a Gaussian distribution over a Euclidean space but now marginalizes onto the unit circle rather than conditioning (Mardia and Jupp, 2000). However, this leads to more complex forms of distribution and will not be discussed further. Finally, any valid distribution over the real axis (such as a Gaussian) can be turned into a periodic distribution by mapping successive intervals of width 2π onto the periodic variable $(0, 2\pi)$, which corresponds to 'wrapping' the real axis around unit circle. Again, the resulting distribution is more complex to handle than the von Mises distribution.

One limitation of the von Mises distribution is that it is unimodal. By forming *mixtures* of von Mises distributions, we obtain a flexible framework for modelling periodic variables that can handle multimodality. For an example of a machine learning application that makes use of von Mises distributions, see Lawrence *et al.* (2002), and for extensions to modelling conditional densities for regression problems, see Bishop and Nabney (1996).

2.3.9 Mixtures of Gaussians

While the Gaussian distribution has some important analytical properties, it suffers from significant limitations when it comes to modelling real data sets. Consider the example shown in Figure 2.21.This is known as the 'Old Faithful' data set, and comprises 272 measurements of the eruption of the Old Faithful geyser at Yellowstone National Park in the USA. Each measurement comprises the duration of

Appendix A

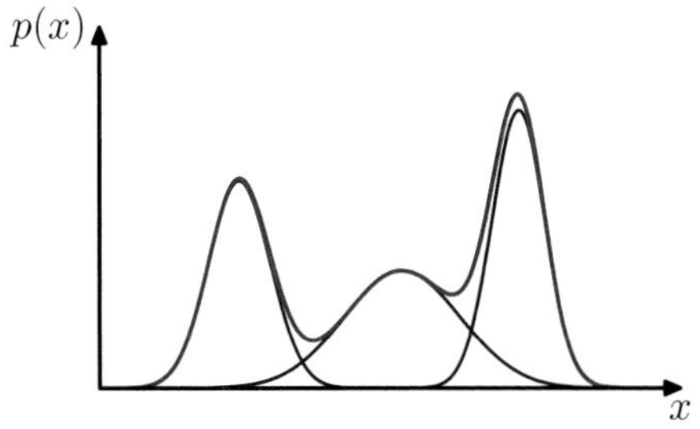

Figure 2.22 Example of a Gaussian mixture distribution in one dimension showing three Gaussians (each scaled by a coefficient) in blue and their sum in red.

the eruption in minutes (horizontal axis) and the time in minutes to the next eruption (vertical axis). We see that the data set forms two dominant clumps, and that a simple Gaussian distribution is unable to capture this structure, whereas a linear superposition of two Gaussians gives a better characterization of the data set.

Such superpositions, formed by taking linear combinations of more basic distributions such as Gaussians, can be formulated as probabilistic models known as *mixture distributions* (McLachlan and Basford, 1988; McLachlan and Peel, 2000). In Figure 2.22 we see that a linear combination of Gaussians can give rise to very complex densities. By using a sufficient number of Gaussians, and by adjusting their means and covariances as well as the coefficients in the linear combination, almost any continuous density can be approximated to arbitrary accuracy.

We therefore consider a superposition of K Gaussian densities of the form

$$p(\mathbf{x}) = \sum_{k=1}^{K} \pi_k \mathcal{N}(\mathbf{x}|\boldsymbol{\mu}_k, \boldsymbol{\Sigma}_k) \tag{2.188}$$

which is called a *mixture of Gaussians*. Each Gaussian density $\mathcal{N}(\mathbf{x}|\boldsymbol{\mu}_k, \boldsymbol{\Sigma}_k)$ is called a *component* of the mixture and has its own mean $\boldsymbol{\mu}_k$ and covariance $\boldsymbol{\Sigma}_k$. Contour and surface plots for a Gaussian mixture having 3 components are shown in Figure 2.23.

In this section we shall consider Gaussian components to illustrate the framework of mixture models. More generally, mixture models can comprise linear combinations of other distributions. For instance, in Section 9.3.3 we shall consider mixtures of Bernoulli distributions as an example of a mixture model for discrete variables.

Section 9.3.3

The parameters π_k in (2.188) are called *mixing coefficients*. If we integrate both sides of (2.188) with respect to $\mathbf{x}$, and note that both $p(\mathbf{x})$ and the individual Gaussian components are normalized, we obtain

$$\sum_{k=1}^{K} \pi_k = 1. \tag{2.189}$$

Also, given that $\mathcal{N}(\mathbf{x}|\boldsymbol{\mu}_k, \boldsymbol{\Sigma}_k) \geqslant 0$, a sufficient condition for the requirement $p(\mathbf{x}) \geqslant 0$ is that $\pi_k \geqslant 0$ for all k. Combining this with the condition (2.189) we obtain

$$0 \leqslant \pi_k \leqslant 1. \tag{2.190}$$

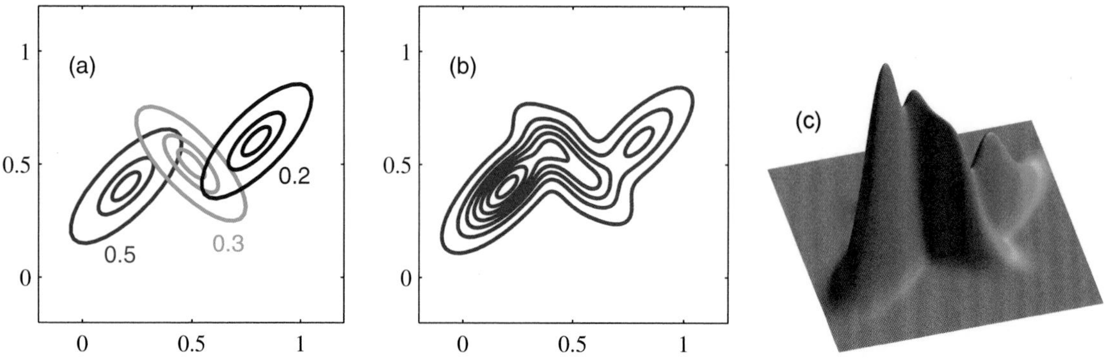

Figure 2.23 Illustration of a mixture of 3 Gaussians in a two-dimensional space. (a) Contours of constant density for each of the mixture components, in which the 3 components are denoted red, blue and green, and the values of the mixing coefficients are shown below each component. (b) Contours of the marginal probability density $p(\mathbf{x})$ of the mixture distribution. (c) A surface plot of the distribution $p(\mathbf{x})$.

We therefore see that the mixing coefficients satisfy the requirements to be probabilities.

From the sum and product rules, the marginal density is given by

$$p(\mathbf{x}) = \sum_{k=1}^{K} p(k)p(\mathbf{x}|k) \tag{2.191}$$

which is equivalent to (2.188) in which we can view $\pi_k = p(k)$ as the prior probability of picking the kth component, and the density $\mathcal{N}(\mathbf{x}|\boldsymbol{\mu}_k, \boldsymbol{\Sigma}_k) = p(\mathbf{x}|k)$ as the probability of $\mathbf{x}$ conditioned on k. As we shall see in later chapters, an important role is played by the posterior probabilities $p(k|\mathbf{x})$, which are also known as *responsibilities*. From Bayes' theorem these are given by

$$\begin{aligned} \gamma_k(\mathbf{x}) &\equiv p(k|\mathbf{x}) \\ &= \frac{p(k)p(\mathbf{x}|k)}{\sum_l p(l)p(\mathbf{x}|l)} \\ &= \frac{\pi_k \mathcal{N}(\mathbf{x}|\boldsymbol{\mu}_k, \boldsymbol{\Sigma}_k)}{\sum_l \pi_l \mathcal{N}(\mathbf{x}|\boldsymbol{\mu}_l, \boldsymbol{\Sigma}_l)}. \end{aligned} \tag{2.192}$$

We shall discuss the probabilistic interpretation of the mixture distribution in greater detail in Chapter 9.

The form of the Gaussian mixture distribution is governed by the parameters $\boldsymbol{\pi}$, $\boldsymbol{\mu}$ and $\boldsymbol{\Sigma}$, where we have used the notation $\boldsymbol{\pi} \equiv \{\pi_1, \ldots, \pi_K\}$, $\boldsymbol{\mu} \equiv \{\boldsymbol{\mu}_1, \ldots, \boldsymbol{\mu}_K\}$ and $\boldsymbol{\Sigma} \equiv \{\boldsymbol{\Sigma}_1, \ldots \boldsymbol{\Sigma}_K\}$. One way to set the values of these parameters is to use maximum likelihood. From (2.188) the log of the likelihood function is given by

$$\ln p(\mathbf{X}|\boldsymbol{\pi}, \boldsymbol{\mu}, \boldsymbol{\Sigma}) = \sum_{n=1}^{N} \ln \left\{ \sum_{k=1}^{K} \pi_k \mathcal{N}(\mathbf{x}_n|\boldsymbol{\mu}_k, \boldsymbol{\Sigma}_k) \right\} \tag{2.193}$$

where $\mathbf{X} = \{\mathbf{x}_1, \ldots, \mathbf{x}_N\}$. We immediately see that the situation is now much more complex than with a single Gaussian, due to the presence of the summation over k inside the logarithm. As a result, the maximum likelihood solution for the parameters no longer has a closed-form analytical solution. One approach to maximizing the likelihood function is to use iterative numerical optimization techniques (Fletcher, 1987; Nocedal and Wright, 1999; Bishop and Nabney, 2008). Alternatively we can employ a powerful framework called *expectation maximization*, which will be discussed at length in Chapter 9.

2.4. The Exponential Family

The probability distributions that we have studied so far in this chapter (with the exception of the Gaussian mixture) are specific examples of a broad class of distributions called the *exponential family* (Duda and Hart, 1973; Bernardo and Smith, 1994). Members of the exponential family have many important properties in common, and it is illuminating to discuss these properties in some generality.

The exponential family of distributions over $\mathbf{x}$, given parameters $\boldsymbol{\eta}$, is defined to be the set of distributions of the form

$$p(\mathbf{x}|\boldsymbol{\eta}) = h(\mathbf{x})g(\boldsymbol{\eta}) \exp\left\{\boldsymbol{\eta}^{\mathrm{T}}\mathbf{u}(\mathbf{x})\right\} \tag{2.194}$$

where $\mathbf{x}$ may be scalar or vector, and may be discrete or continuous. Here $\boldsymbol{\eta}$ are called the *natural parameters* of the distribution, and $\mathbf{u}(\mathbf{x})$ is some function of $\mathbf{x}$. The function $g(\boldsymbol{\eta})$ can be interpreted as the coefficient that ensures that the distribution is normalized and therefore satisfies

$$g(\boldsymbol{\eta}) \int h(\mathbf{x}) \exp\left\{\boldsymbol{\eta}^{\mathrm{T}}\mathbf{u}(\mathbf{x})\right\} \mathrm{d}\mathbf{x} = 1 \tag{2.195}$$

where the integration is replaced by summation if $\mathbf{x}$ is a discrete variable.

We begin by taking some examples of the distributions introduced earlier in the chapter and showing that they are indeed members of the exponential family. Consider first the Bernoulli distribution

$$p(x|\mu) = \mathrm{Bern}(x|\mu) = \mu^x(1-\mu)^{1-x}. \tag{2.196}$$

Expressing the right-hand side as the exponential of the logarithm, we have

$$\begin{aligned} p(x|\mu) &= \exp\left\{x \ln \mu + (1-x)\ln(1-\mu)\right\} \\ &= (1-\mu)\exp\left\{\ln\left(\frac{\mu}{1-\mu}\right)x\right\}. \end{aligned} \tag{2.197}$$

Comparison with (2.194) allows us to identify

$$\eta = \ln\left(\frac{\mu}{1-\mu}\right) \tag{2.198}$$

which we can solve for μ to give $\mu = \sigma(\eta)$, where

$$\sigma(\eta) = \frac{1}{1 + \exp(-\eta)} \tag{2.199}$$

is called the *logistic sigmoid* function. Thus we can write the Bernoulli distribution using the standard representation (2.194) in the form

$$p(x|\eta) = \sigma(-\eta) \exp(\eta x) \tag{2.200}$$

where we have used $1 - \sigma(\eta) = \sigma(-\eta)$, which is easily proved from (2.199). Comparison with (2.194) shows that

$$u(x) = x \tag{2.201}$$
$$h(x) = 1 \tag{2.202}$$
$$g(\eta) = \sigma(-\eta). \tag{2.203}$$

Next consider the multinomial distribution that, for a single observation $\mathbf{x}$, takes the form

$$p(\mathbf{x}|\boldsymbol{\mu}) = \prod_{k=1}^{M} \mu_k^{x_k} = \exp\left\{\sum_{k=1}^{M} x_k \ln \mu_k\right\} \tag{2.204}$$

where $\mathbf{x} = (x_1, \ldots, x_M)^{\mathrm{T}}$. Again, we can write this in the standard representation (2.194) so that

$$p(\mathbf{x}|\boldsymbol{\eta}) = \exp(\boldsymbol{\eta}^{\mathrm{T}}\mathbf{x}) \tag{2.205}$$

where $\eta_k = \ln \mu_k$, and we have defined $\boldsymbol{\eta} = (\eta_1, \ldots, \eta_M)^{\mathrm{T}}$. Again, comparing with (2.194) we have

$$\mathbf{u}(\mathbf{x}) = \mathbf{x} \tag{2.206}$$
$$h(\mathbf{x}) = 1 \tag{2.207}$$
$$g(\boldsymbol{\eta}) = 1. \tag{2.208}$$

Note that the parameters η_k are not independent because the parameters μ_k are subject to the constraint

$$\sum_{k=1}^{M} \mu_k = 1 \tag{2.209}$$

so that, given any $M - 1$ of the parameters μ_k, the value of the remaining parameter is fixed. In some circumstances, it will be convenient to remove this constraint by expressing the distribution in terms of only $M - 1$ parameters. This can be achieved by using the relationship (2.209) to eliminate μ_M by expressing it in terms of the remaining $\{\mu_k\}$ where $k = 1, \ldots, M - 1$, thereby leaving $M - 1$ parameters. Note that these remaining parameters are still subject to the constraints

$$0 \leqslant \mu_k \leqslant 1, \qquad \sum_{k=1}^{M-1} \mu_k \leqslant 1. \tag{2.210}$$

Making use of the constraint (2.209), the multinomial distribution in this representation then becomes

$$\begin{aligned}
&\exp\left\{\sum_{k=1}^{M} x_k \ln \mu_k\right\} \\
&= \exp\left\{\sum_{k=1}^{M-1} x_k \ln \mu_k + \left(1 - \sum_{k=1}^{M-1} x_k\right) \ln\left(1 - \sum_{k=1}^{M-1} \mu_k\right)\right\} \\
&= \exp\left\{\sum_{k=1}^{M-1} x_k \ln\left(\frac{\mu_k}{1 - \sum_{j=1}^{M-1} \mu_j}\right) + \ln\left(1 - \sum_{k=1}^{M-1} \mu_k\right)\right\}. \qquad (2.211)
\end{aligned}$$

We now identify

$$\ln\left(\frac{\mu_k}{1 - \sum_j \mu_j}\right) = \eta_k \qquad (2.212)$$

which we can solve for μ_k by first summing both sides over k and then rearranging and back-substituting to give

$$\mu_k = \frac{\exp(\eta_k)}{1 + \sum_j \exp(\eta_j)}. \qquad (2.213)$$

This is called the *softmax* function, or the *normalized exponential*. In this representation, the multinomial distribution therefore takes the form

$$p(\mathbf{x}|\boldsymbol{\eta}) = \left(1 + \sum_{k=1}^{M-1} \exp(\eta_k)\right)^{-1} \exp(\boldsymbol{\eta}^{\mathrm{T}}\mathbf{x}). \qquad (2.214)$$

This is the standard form of the exponential family, with parameter vector $\boldsymbol{\eta} = (\eta_1, \ldots, \eta_{M-1} 0)^{\mathrm{T}}$ in which

$$\mathbf{u}(\mathbf{x}) = \mathbf{x} \qquad (2.215)$$

$$h(\mathbf{x}) = 1 \qquad (2.216)$$

$$g(\boldsymbol{\eta}) = \left(1 + \sum_{k=1}^{M-1} \exp(\eta_k)\right)^{-1}. \qquad (2.217)$$

Finally, let us consider the Gaussian distribution. For the univariate Gaussian, we have

$$p(x|\mu, \sigma^2) = \frac{1}{(2\pi\sigma^2)^{1/2}} \exp\left\{-\frac{1}{2\sigma^2}(x-\mu)^2\right\} \qquad (2.218)$$

$$= \frac{1}{(2\pi\sigma^2)^{1/2}} \exp\left\{-\frac{1}{2\sigma^2}x^2 + \frac{\mu}{\sigma^2}x - \frac{1}{2\sigma^2}\mu^2\right\} \qquad (2.219)$$

which, after some simple rearrangement, can be cast in the standard exponential family form (2.194) with

Exercise 2.57

$$\boldsymbol{\eta} = \begin{pmatrix} \mu/\sigma^2 \\ -1/2\sigma^2 \end{pmatrix} \tag{2.220}$$

$$\mathbf{u}(x) = \begin{pmatrix} x \\ x^2 \end{pmatrix} \tag{2.221}$$

$$h(x) = (2\pi)^{-1/2} \tag{2.222}$$

$$g(\boldsymbol{\eta}) = (-2\eta_2)^{1/2} \exp\left(\frac{\eta_1^2}{4\eta_2}\right). \tag{2.223}$$

2.4.1 Maximum likelihood and sufficient statistics

Let us now consider the problem of estimating the parameter vector $\boldsymbol{\eta}$ in the general exponential family distribution (2.194) using the technique of maximum likelihood. Taking the gradient of both sides of (2.195) with respect to $\boldsymbol{\eta}$, we have

$$\nabla g(\boldsymbol{\eta}) \int h(\mathbf{x}) \exp\left\{\boldsymbol{\eta}^{\mathrm{T}}\mathbf{u}(\mathbf{x})\right\} \mathrm{d}\mathbf{x} + g(\boldsymbol{\eta}) \int h(\mathbf{x}) \exp\left\{\boldsymbol{\eta}^{\mathrm{T}}\mathbf{u}(\mathbf{x})\right\} \mathbf{u}(\mathbf{x})\, \mathrm{d}\mathbf{x} = 0. \tag{2.224}$$

Rearranging, and making use again of (2.195) then gives

$$-\frac{1}{g(\boldsymbol{\eta})}\nabla g(\boldsymbol{\eta}) = g(\boldsymbol{\eta}) \int h(\mathbf{x}) \exp\left\{\boldsymbol{\eta}^{\mathrm{T}}\mathbf{u}(\mathbf{x})\right\} \mathbf{u}(\mathbf{x})\, \mathrm{d}\mathbf{x} = \mathbb{E}[\mathbf{u}(\mathbf{x})]. \tag{2.225}$$

We therefore obtain the result

$$-\nabla \ln g(\boldsymbol{\eta}) = \mathbb{E}[\mathbf{u}(\mathbf{x})]. \tag{2.226}$$

Exercise 2.58

Note that the covariance of $\mathbf{u}(\mathbf{x})$ can be expressed in terms of the second derivatives of $g(\boldsymbol{\eta})$, and similarly for higher order moments. Thus, provided we can normalize a distribution from the exponential family, we can always find its moments by simple differentiation.

Now consider a set of independent identically distributed data denoted by $\mathbf{X} = \{\mathbf{x}_1, \ldots, \mathbf{x}_N\}$, for which the likelihood function is given by

$$p(\mathbf{X}|\boldsymbol{\eta}) = \left(\prod_{n=1}^{N} h(\mathbf{x}_n)\right) g(\boldsymbol{\eta})^N \exp\left\{\boldsymbol{\eta}^{\mathrm{T}} \sum_{n=1}^{N} \mathbf{u}(\mathbf{x}_n)\right\}. \tag{2.227}$$

Setting the gradient of $\ln p(\mathbf{X}|\boldsymbol{\eta})$ with respect to $\boldsymbol{\eta}$ to zero, we get the following condition to be satisfied by the maximum likelihood estimator $\boldsymbol{\eta}_{\mathrm{ML}}$

$$-\nabla \ln g(\boldsymbol{\eta}_{\mathrm{ML}}) = \frac{1}{N}\sum_{n=1}^{N} \mathbf{u}(\mathbf{x}_n) \tag{2.228}$$

which can in principle be solved to obtain $\boldsymbol{\eta}_{\mathrm{ML}}$. We see that the solution for the maximum likelihood estimator depends on the data only through $\sum_n \mathbf{u}(\mathbf{x}_n)$, which is therefore called the *sufficient statistic* of the distribution (2.194). We do not need to store the entire data set itself but only the value of the sufficient statistic. For the Bernoulli distribution, for example, the function $\mathbf{u}(x)$ is given just by x and so we need only keep the sum of the data points $\{x_n\}$, whereas for the Gaussian $\mathbf{u}(x) = (x, x^2)^{\mathrm{T}}$, and so we should keep both the sum of $\{x_n\}$ and the sum of $\{x_n^2\}$.

If we consider the limit $N \to \infty$, then the right-hand side of (2.228) becomes $\mathbb{E}[\mathbf{u}(\mathbf{x})]$, and so by comparing with (2.226) we see that in this limit $\boldsymbol{\eta}_{\mathrm{ML}}$ will equal the true value $\boldsymbol{\eta}$.

In fact, this sufficiency property holds also for Bayesian inference, although we shall defer discussion of this until Chapter 8 when we have equipped ourselves with the tools of graphical models and can thereby gain a deeper insight into these important concepts.

2.4.2 Conjugate priors

We have already encountered the concept of a conjugate prior several times, for example in the context of the Bernoulli distribution (for which the conjugate prior is the beta distribution) or the Gaussian (where the conjugate prior for the mean is a Gaussian, and the conjugate prior for the precision is the Wishart distribution). In general, for a given probability distribution $p(\mathbf{x}|\boldsymbol{\eta})$, we can seek a prior $p(\boldsymbol{\eta})$ that is conjugate to the likelihood function, so that the posterior distribution has the same functional form as the prior. For any member of the exponential family (2.194), there exists a conjugate prior that can be written in the form

$$p(\boldsymbol{\eta}|\boldsymbol{\chi}, \nu) = f(\boldsymbol{\chi}, \nu) g(\boldsymbol{\eta})^{\nu} \exp\left\{\nu \boldsymbol{\eta}^{\mathrm{T}} \boldsymbol{\chi}\right\} \tag{2.229}$$

where $f(\boldsymbol{\chi}, \nu)$ is a normalization coefficient, and $g(\boldsymbol{\eta})$ is the same function as appears in (2.194). To see that this is indeed conjugate, let us multiply the prior (2.229) by the likelihood function (2.227) to obtain the posterior distribution, up to a normalization coefficient, in the form

$$p(\boldsymbol{\eta}|\mathbf{X}, \boldsymbol{\chi}, \nu) \propto g(\boldsymbol{\eta})^{\nu+N} \exp\left\{\boldsymbol{\eta}^{\mathrm{T}}\left(\sum_{n=1}^{N} \mathbf{u}(\mathbf{x}_n) + \nu\boldsymbol{\chi}\right)\right\}. \tag{2.230}$$

This again takes the same functional form as the prior (2.229), confirming conjugacy. Furthermore, we see that the parameter ν can be interpreted as an effective number of pseudo-observations in the prior, each of which has a value for the sufficient statistic $\mathbf{u}(\mathbf{x})$ given by $\boldsymbol{\chi}$.

2.4.3 Noninformative priors

In some applications of probabilistic inference, we may have prior knowledge that can be conveniently expressed through the prior distribution. For example, if the prior assigns zero probability to some value of variable, then the posterior distribution will necessarily also assign zero probability to that value, irrespective of any subsequent observations of data. In many cases, however, we may have little idea of what form the distribution should take. We may then seek a form of prior distribution, called a *noninformative prior*, which is intended to have as little influence on the posterior distribution as possible (Jeffreys, 1946; Box and Tiao, 1973; Bernardo and Smith, 1994). This is sometimes referred to as 'letting the data speak for themselves'.

If we have a distribution $p(x|\lambda)$ governed by a parameter λ, we might be tempted to propose a prior distribution $p(\lambda) = \text{const}$ as a suitable prior. If λ is a discrete variable with K states, this simply amounts to setting the prior probability of each state to $1/K$. In the case of continuous parameters, however, there are two potential difficulties with this approach. The first is that, if the domain of λ is unbounded, this prior distribution cannot be correctly normalized because the integral over λ diverges. Such priors are called *improper*. In practice, improper priors can often be used provided the corresponding posterior distribution is *proper*, i.e., that it can be correctly normalized. For instance, if we put a uniform prior distribution over the mean of a Gaussian, then the posterior distribution for the mean, once we have observed at least one data point, will be proper.

A second difficulty arises from the transformation behaviour of a probability density under a nonlinear change of variables, given by (1.27). If a function $h(\lambda)$ is constant, and we change variables to $\lambda = \eta^2$, then $\widehat{h}(\eta) = h(\eta^2)$ will also be constant. However, if we choose the density $p_\lambda(\lambda)$ to be constant, then the density of η will be given, from (1.27), by

$$p_\eta(\eta) = p_\lambda(\lambda)\left|\frac{\mathrm{d}\lambda}{\mathrm{d}\eta}\right| = p_\lambda(\eta^2)2\eta \propto \eta \tag{2.231}$$

and so the density over η will not be constant. This issue does not arise when we use maximum likelihood, because the likelihood function $p(x|\lambda)$ is a simple function of λ and so we are free to use any convenient parameterization. If, however, we are to choose a prior distribution that is constant, we must take care to use an appropriate representation for the parameters.

Here we consider two simple examples of noninformative priors (Berger, 1985). First of all, if a density takes the form

$$p(x|\mu) = f(x - \mu) \tag{2.232}$$

then the parameter μ is known as a *location parameter*. This family of densities exhibits *translation invariance* because if we shift x by a constant to give $\widehat{x} = x + c$, then

$$p(\widehat{x}|\widehat{\mu}) = f(\widehat{x} - \widehat{\mu}) \tag{2.233}$$

where we have defined $\widehat{\mu} = \mu + c$. Thus the density takes the same form in the new variable as in the original one, and so the density is independent of the choice of origin. We would like to choose a prior distribution that reflects this translation invariance property, and so we choose a prior that assigns equal probability mass to an interval $A \leqslant \mu \leqslant B$ as to the shifted interval $A - c \leqslant \mu \leqslant B - c$. This implies

$$\int_A^B p(\mu)\,\mathrm{d}\mu = \int_{A-c}^{B-c} p(\mu)\,\mathrm{d}\mu = \int_A^B p(\mu - c)\,\mathrm{d}\mu \tag{2.234}$$

and because this must hold for all choices of A and B, we have

$$p(\mu - c) = p(\mu) \tag{2.235}$$

which implies that $p(\mu)$ is constant. An example of a location parameter would be the mean μ of a Gaussian distribution. As we have seen, the conjugate prior distribution for μ in this case is a Gaussian $p(\mu|\mu_0, \sigma_0^2) = \mathcal{N}(\mu|\mu_0, \sigma_0^2)$, and we obtain a noninformative prior by taking the limit $\sigma_0^2 \to \infty$. Indeed, from (2.141) and (2.142) we see that this gives a posterior distribution over μ in which the contributions from the prior vanish.

As a second example, consider a density of the form

$$p(x|\sigma) = \frac{1}{\sigma} f\left(\frac{x}{\sigma}\right) \tag{2.236}$$

where $\sigma > 0$. Note that this will be a normalized density provided $f(x)$ is correctly normalized. The parameter σ is known as a *scale parameter*, and the density exhibits *scale invariance* because if we scale x by a constant to give $\widehat{x} = cx$, then

Exercise 2.59

$$p(\widehat{x}|\widehat{\sigma}) = \frac{1}{\widehat{\sigma}} f\left(\frac{\widehat{x}}{\widehat{\sigma}}\right) \tag{2.237}$$

where we have defined $\widehat{\sigma} = c\sigma$. This transformation corresponds to a change of scale, for example from meters to kilometers if x is a length, and we would like to choose a prior distribution that reflects this scale invariance. If we consider an interval $A \leqslant \sigma \leqslant B$, and a scaled interval $A/c \leqslant \sigma \leqslant B/c$, then the prior should assign equal probability mass to these two intervals. Thus we have

$$\int_A^B p(\sigma)\,\mathrm{d}\sigma = \int_{A/c}^{B/c} p(\sigma)\,\mathrm{d}\sigma = \int_A^B p\left(\frac{1}{c}\sigma\right)\frac{1}{c}\,\mathrm{d}\sigma \tag{2.238}$$

and because this must hold for choices of A and B, we have

$$p(\sigma) = p\left(\frac{1}{c}\sigma\right)\frac{1}{c} \tag{2.239}$$

and hence $p(\sigma) \propto 1/\sigma$. Note that again this is an improper prior because the integral of the distribution over $0 \leqslant \sigma \leqslant \infty$ is divergent. It is sometimes also convenient to think of the prior distribution for a scale parameter in terms of the density of the log of the parameter. Using the transformation rule (1.27) for densities we see that $p(\ln \sigma) = \text{const}$. Thus, for this prior there is the same probability mass in the range $1 \leqslant \sigma \leqslant 10$ as in the range $10 \leqslant \sigma \leqslant 100$ and in $100 \leqslant \sigma \leqslant 1000$.

An example of a scale parameter would be the standard deviation σ of a Gaussian distribution, after we have taken account of the location parameter μ, because

$$\mathcal{N}(x|\mu, \sigma^2) \propto \sigma^{-1} \exp\left\{-(\widetilde{x}/\sigma)^2\right\} \tag{2.240}$$

where $\widetilde{x} = x - \mu$. As discussed earlier, it is often more convenient to work in terms of the precision $\lambda = 1/\sigma^2$ rather than σ itself. Using the transformation rule for densities, we see that a distribution $p(\sigma) \propto 1/\sigma$ corresponds to a distribution over λ of the form $p(\lambda) \propto 1/\lambda$. We have seen that the conjugate prior for λ was the gamma distribution $\mathrm{Gam}(\lambda|a_0, b_0)$ given by (2.146). The noninformative prior is obtained as the special case $a_0 = b_0 = 0$. Again, if we examine the results (2.150) and (2.151) for the posterior distribution of λ, we see that for $a_0 = b_0 = 0$, the posterior depends only on terms arising from the data and not from the prior.

Section 2.3

2.5. Nonparametric Methods

Throughout this chapter, we have focussed on the use of probability distributions having specific functional forms governed by a small number of parameters whose values are to be determined from a data set. This is called the *parametric* approach to density modelling. An important limitation of this approach is that the chosen density might be a poor model of the distribution that generates the data, which can result in poor predictive performance. For instance, if the process that generates the data is multimodal, then this aspect of the distribution can never be captured by a Gaussian, which is necessarily unimodal.

In this final section, we consider some *nonparametric* approaches to density estimation that make few assumptions about the form of the distribution. Here we shall focus mainly on simple frequentist methods. The reader should be aware, however, that nonparametric Bayesian methods are attracting increasing interest (Walker *et al.*, 1999; Neal, 2000; Müller and Quintana, 2004; Teh *et al.*, 2006).

Let us start with a discussion of histogram methods for density estimation, which we have already encountered in the context of marginal and conditional distributions in Figure 1.11 and in the context of the central limit theorem in Figure 2.6. Here we explore the properties of histogram density models in more detail, focussing on the case of a single continuous variable x. Standard histograms simply partition x into distinct bins of width Δ_i and then count the number n_i of observations of x falling in bin i. In order to turn this count into a normalized probability density, we simply divide by the total number N of observations and by the width Δ_i of the bins to obtain probability values for each bin given by

$$p_i = \frac{n_i}{N\Delta_i} \tag{2.241}$$

for which it is easily seen that $\int p(x)\,\mathrm{d}x = 1$. This gives a model for the density $p(x)$ that is constant over the width of each bin, and often the bins are chosen to have the same width $\Delta_i = \Delta$.

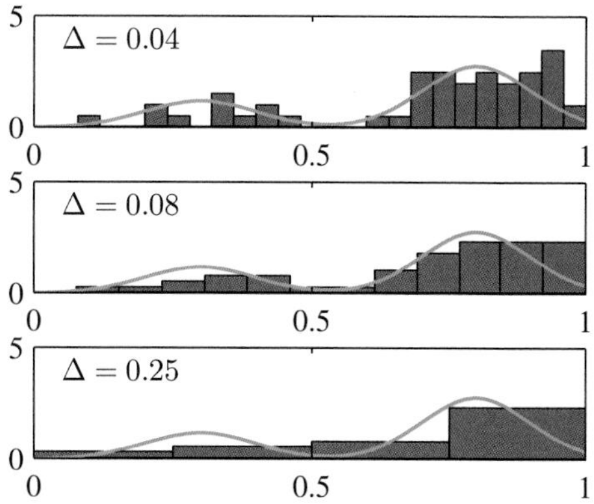

Figure 2.24 An illustration of the histogram approach to density estimation, in which a data set of 50 data points is generated from the distribution shown by the green curve. Histogram density estimates, based on (2.241), with a common bin width Δ are shown for various values of Δ.

In Figure 2.24, we show an example of histogram density estimation. Here the data is drawn from the distribution, corresponding to the green curve, which is formed from a mixture of two Gaussians. Also shown are three examples of histogram density estimates corresponding to three different choices for the bin width Δ. We see that when Δ is very small (top figure), the resulting density model is very spiky, with a lot of structure that is not present in the underlying distribution that generated the data set. Conversely, if Δ is too large (bottom figure) then the result is a model that is too smooth and that consequently fails to capture the bimodal property of the green curve. The best results are obtained for some intermediate value of Δ (middle figure). In principle, a histogram density model is also dependent on the choice of edge location for the bins, though this is typically much less significant than the value of Δ.

Note that the histogram method has the property (unlike the methods to be discussed shortly) that, once the histogram has been computed, the data set itself can be discarded, which can be advantageous if the data set is large. Also, the histogram approach is easily applied if the data points are arriving sequentially.

In practice, the histogram technique can be useful for obtaining a quick visualization of data in one or two dimensions but is unsuited to most density estimation applications. One obvious problem is that the estimated density has discontinuities that are due to the bin edges rather than any property of the underlying distribution that generated the data. Another major limitation of the histogram approach is its scaling with dimensionality. If we divide each variable in a D-dimensional space into M bins, then the total number of bins will be M^D. This exponential scaling with D is an example of the curse of dimensionality. In a space of high dimensionality, the quantity of data needed to provide meaningful estimates of local probability density would be prohibitive.

Section 1.4

The histogram approach to density estimation does, however, teach us two important lessons. First, to estimate the probability density at a particular location, we should consider the data points that lie within some local neighbourhood of that point. Note that the concept of locality requires that we assume some form of distance measure, and here we have been assuming Euclidean distance. For histograms, this neighbourhood property was defined by the bins, and there is a natural 'smoothing' parameter describing the spatial extent of the local region, in this case the bin width. Second, the value of the smoothing parameter should be neither too large nor too small in order to obtain good results. This is reminiscent of the choice of model complexity in polynomial curve fitting discussed in Chapter 1 where the degree M of the polynomial, or alternatively the value α of the regularization parameter, was optimal for some intermediate value, neither too large nor too small. Armed with these insights, we turn now to a discussion of two widely used nonparametric techniques for density estimation, kernel estimators and nearest neighbours, which have better scaling with dimensionality than the simple histogram model.

2.5.1 Kernel density estimators

Let us suppose that observations are being drawn from some unknown probability density $p(\mathbf{x})$ in some D-dimensional space, which we shall take to be Euclidean, and we wish to estimate the value of $p(\mathbf{x})$. From our earlier discussion of locality, let us consider some small region $\mathcal{R}$ containing $\mathbf{x}$. The probability mass associated with this region is given by

$$P = \int_{\mathcal{R}} p(\mathbf{x})\,\mathrm{d}\mathbf{x}. \tag{2.242}$$

Now suppose that we have collected a data set comprising N observations drawn from $p(\mathbf{x})$. Because each data point has a probability P of falling within $\mathcal{R}$, the total number K of points that lie inside $\mathcal{R}$ will be distributed according to the binomial distribution

Section 2.1

$$\mathrm{Bin}(K|N,P) = \frac{N!}{K!(N-K)!}P^K(1-P)^{N-K}. \tag{2.243}$$

Using (2.11), we see that the mean fraction of points falling inside the region is $\mathbb{E}[K/N] = P$, and similarly using (2.12) we see that the variance around this mean is $\mathrm{var}[K/N] = P(1-P)/N$. For large N, this distribution will be sharply peaked around the mean and so

$$K \simeq NP. \tag{2.244}$$

If, however, we also assume that the region $\mathcal{R}$ is sufficiently small that the probability density $p(\mathbf{x})$ is roughly constant over the region, then we have

$$P \simeq p(\mathbf{x})V \tag{2.245}$$

where V is the volume of $\mathcal{R}$. Combining (2.244) and (2.245), we obtain our density estimate in the form

$$p(\mathbf{x}) = \frac{K}{NV}. \tag{2.246}$$

Note that the validity of (2.246) depends on two contradictory assumptions, namely that the region $\mathcal{R}$ be sufficiently small that the density is approximately constant over the region and yet sufficiently large (in relation to the value of that density) that the number K of points falling inside the region is sufficient for the binomial distribution to be sharply peaked.

We can exploit the result (2.246) in two different ways. Either we can fix K and determine the value of V from the data, which gives rise to the K-nearest-neighbour technique discussed shortly, or we can fix V and determine K from the data, giving rise to the kernel approach. It can be shown that both the K-nearest-neighbour density estimator and the kernel density estimator converge to the true probability density in the limit $N \rightarrow \infty$ provided V shrinks suitably with N, and K grows with N (Duda and Hart, 1973).

We begin by discussing the kernel method in detail, and to start with we take the region $\mathcal{R}$ to be a small hypercube centred on the point $\mathbf{x}$ at which we wish to determine the probability density. In order to count the number K of points falling within this region, it is convenient to define the following function

$$k(\mathbf{u}) = \begin{cases} 1, & |u_i| \leqslant 1/2, \qquad i = 1, \ldots, D, \\ 0, & \text{otherwise} \end{cases} \tag{2.247}$$

which represents a unit cube centred on the origin. The function $k(\mathbf{u})$ is an example of a *kernel function*, and in this context is also called a *Parzen window*. From (2.247), the quantity $k((\mathbf{x} - \mathbf{x}_n)/h)$ will be one if the data point $\mathbf{x}_n$ lies inside a cube of side h centred on $\mathbf{x}$, and zero otherwise. The total number of data points lying inside this cube will therefore be

$$K = \sum_{n=1}^{N} k\left(\frac{\mathbf{x} - \mathbf{x}_n}{h}\right). \tag{2.248}$$

Substituting this expression into (2.246) then gives the following result for the estimated density at $\mathbf{x}$

$$p(\mathbf{x}) = \frac{1}{N} \sum_{n=1}^{N} \frac{1}{h^D} k\left(\frac{\mathbf{x} - \mathbf{x}_n}{h}\right) \tag{2.249}$$

where we have used $V = h^D$ for the volume of a hypercube of side h in D dimensions. Using the symmetry of the function $k(\mathbf{u})$, we can now re-interpret this equation, not as a single cube centred on $\mathbf{x}$ but as the sum over N cubes centred on the N data points $\mathbf{x}_n$.

As it stands, the kernel density estimator (2.249) will suffer from one of the same problems that the histogram method suffered from, namely the presence of artificial discontinuities, in this case at the boundaries of the cubes. We can obtain a smoother density model if we choose a smoother kernel function, and a common choice is the Gaussian, which gives rise to the following kernel density model

$$p(\mathbf{x}) = \frac{1}{N} \sum_{n=1}^{N} \frac{1}{(2\pi h^2)^{D/2}} \exp\left\{-\frac{\|\mathbf{x} - \mathbf{x}_n\|^2}{2h^2}\right\} \tag{2.250}$$

where h represents the standard deviation of the Gaussian components. Thus our density model is obtained by placing a Gaussian over each data point and then adding up the contributions over the whole data set, and then dividing by N so that the density is correctly normalized. In Figure 2.25, we apply the model (2.250) to the data

Figure 2.25 Illustration of the kernel density model (2.250) applied to the same data set used to demonstrate the histogram approach in Figure 2.24. We see that h acts as a smoothing parameter and that if it is set too small (top panel), the result is a very noisy density model, whereas if it is set too large (bottom panel), then the bimodal nature of the underlying distribution from which the data is generated (shown by the green curve) is washed out. The best density model is obtained for some intermediate value of h (middle panel).

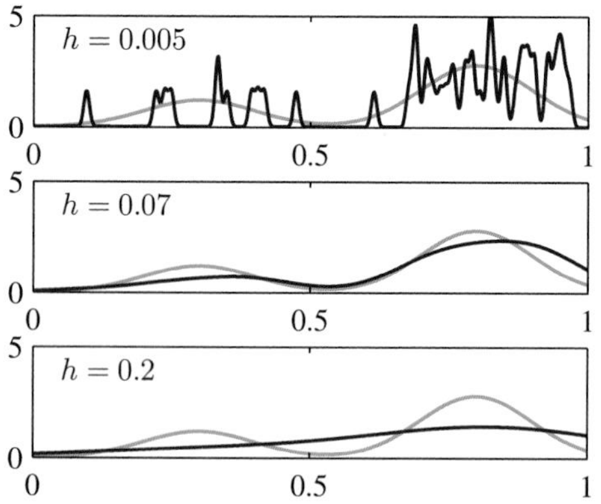

set used earlier to demonstrate the histogram technique. We see that, as expected, the parameter h plays the role of a smoothing parameter, and there is a trade-off between sensitivity to noise at small h and over-smoothing at large h. Again, the optimization of h is a problem in model complexity, analogous to the choice of bin width in histogram density estimation, or the degree of the polynomial used in curve fitting.

We can choose any other kernel function $k(\mathbf{u})$ in (2.249) subject to the conditions

$$k(\mathbf{u}) \geqslant 0, \tag{2.251}$$

$$\int k(\mathbf{u})\,\mathrm{d}\mathbf{u} = 1 \tag{2.252}$$

which ensure that the resulting probability distribution is nonnegative everywhere and integrates to one. The class of density model given by (2.249) is called a kernel density estimator, or *Parzen* estimator. It has a great merit that there is no computation involved in the 'training' phase because this simply requires storage of the training set. However, this is also one of its great weaknesses because the computational cost of evaluating the density grows linearly with the size of the data set.

2.5.2 Nearest-neighbour methods

One of the difficulties with the kernel approach to density estimation is that the parameter h governing the kernel width is fixed for all kernels. In regions of high data density, a large value of h may lead to over-smoothing and a washing out of structure that might otherwise be extracted from the data. However, reducing h may lead to noisy estimates elsewhere in data space where the density is smaller. Thus the optimal choice for h may be dependent on location within the data space. This issue is addressed by nearest-neighbour methods for density estimation.

We therefore return to our general result (2.246) for local density estimation, and instead of fixing V and determining the value of K from the data, we consider a fixed value of K and use the data to find an appropriate value for V. To do this, we consider a small sphere centred on the point $\mathbf{x}$ at which we wish to estimate the

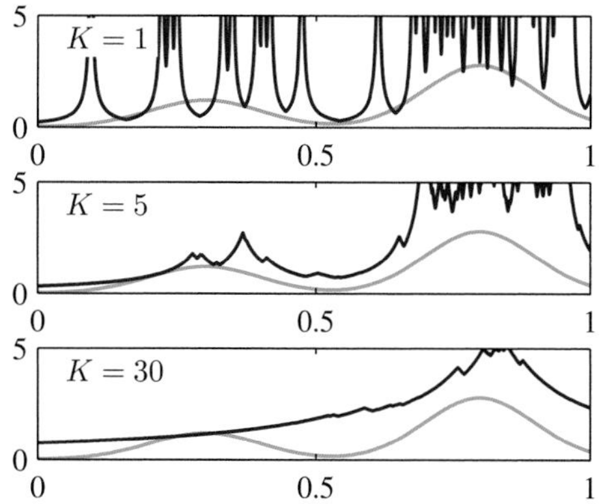

Figure 2.26 Illustration of K-nearest-neighbour density estimation using the same data set as in Figures 2.25 and 2.24. We see that the parameter K governs the degree of smoothing, so that a small value of K leads to a very noisy density model (top panel), whereas a large value (bottom panel) smoothes out the bimodal nature of the true distribution (shown by the green curve) from which the data set was generated.

density $p(\mathbf{x})$, and we allow the radius of the sphere to grow until it contains precisely K data points. The estimate of the density $p(\mathbf{x})$ is then given by (2.246) with V set to the volume of the resulting sphere. This technique is known as K *nearest neighbours* and is illustrated in Figure 2.26, for various choices of the parameter K, using the same data set as used in Figure 2.24 and Figure 2.25. We see that the value of K now governs the degree of smoothing and that again there is an optimum choice for K that is neither too large nor too small. Note that the model produced by K nearest neighbours is not a true density model because the integral over all space diverges.

Exercise 2.61

We close this chapter by showing how the K-nearest-neighbour technique for density estimation can be extended to the problem of classification. To do this, we apply the K-nearest-neighbour density estimation technique to each class separately and then make use of Bayes' theorem. Let us suppose that we have a data set comprising N_k points in class $\mathcal{C}_k$ with N points in total, so that $\sum_k N_k = N$. If we wish to classify a new point $\mathbf{x}$, we draw a sphere centred on $\mathbf{x}$ containing precisely K points irrespective of their class. Suppose this sphere has volume V and contains K_k points from class $\mathcal{C}_k$. Then (2.246) provides an estimate of the density associated with each class

$$p(\mathbf{x}|\mathcal{C}_k) = \frac{K_k}{N_k V}. \tag{2.253}$$

Similarly, the unconditional density is given by

$$p(\mathbf{x}) = \frac{K}{NV} \tag{2.254}$$

while the class priors are given by

$$p(\mathcal{C}_k) = \frac{N_k}{N}. \tag{2.255}$$

We can now combine (2.253), (2.254), and (2.255) using Bayes' theorem to obtain the posterior probability of class membership

$$p(\mathcal{C}_k|\mathbf{x}) = \frac{p(\mathbf{x}|\mathcal{C}_k)p(\mathcal{C}_k)}{p(\mathbf{x})} = \frac{K_k}{K}. \tag{2.256}$$

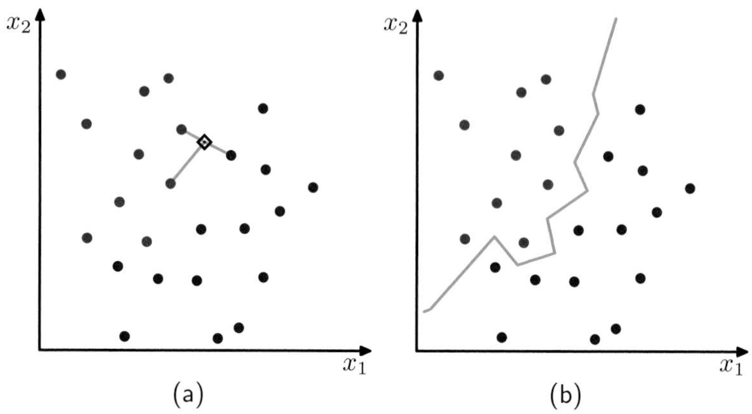

Figure 2.27 (a) In the K-nearest-neighbour classifier, a new point, shown by the black diamond, is classified according to the majority class membership of the K closest training data points, in this case $K = 3$. (b) In the nearest-neighbour ($K = 1$) approach to classification, the resulting decision boundary is composed of hyperplanes that form perpendicular bisectors of pairs of points from different classes.

If we wish to minimize the probability of misclassification, this is done by assigning the test point $\mathbf{x}$ to the class having the largest posterior probability, corresponding to the largest value of K_k/K. Thus to classify a new point, we identify the K nearest points from the training data set and then assign the new point to the class having the largest number of representatives amongst this set. Ties can be broken at random. The particular case of $K = 1$ is called the *nearest-neighbour* rule, because a test point is simply assigned to the same class as the nearest point from the training set. These concepts are illustrated in Figure 2.27.

In Figure 2.28, we show the results of applying the K-nearest-neighbour algorithm to the oil flow data, introduced in Chapter 1, for various values of K. As expected, we see that K controls the degree of smoothing, so that small K produces many small regions of each class, whereas large K leads to fewer larger regions.

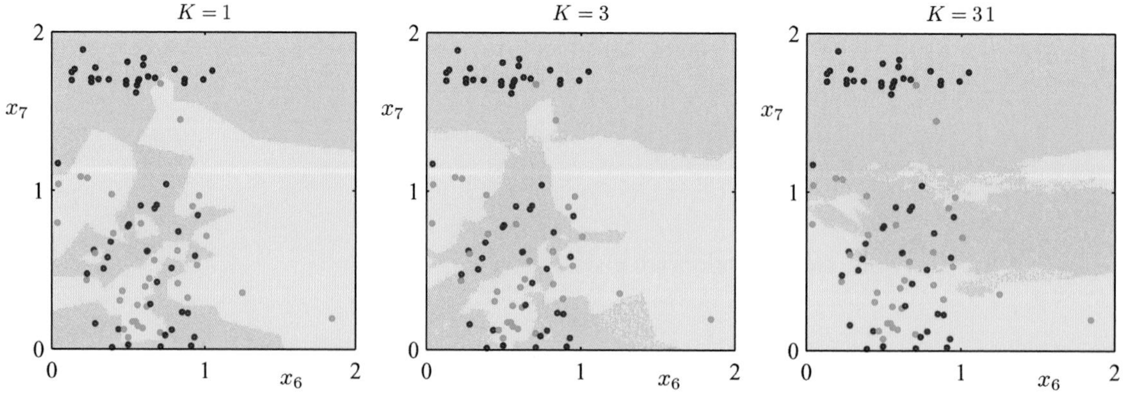

Figure 2.28 Plot of 200 data points from the oil data set showing values of x_6 plotted against x_7, where the red, green, and blue points correspond to the 'laminar', 'annular', and 'homogeneous' classes, respectively. Also shown are the classifications of the input space given by the K-nearest-neighbour algorithm for various values of K.

An interesting property of the nearest-neighbour ($K = 1$) classifier is that, in the limit $N \to \infty$, the error rate is never more than twice the minimum achievable error rate of an optimal classifier, i.e., one that uses the true class distributions (Cover and Hart, 1967) .

As discussed so far, both the K-nearest-neighbour method, and the kernel density estimator, require the entire training data set to be stored, leading to expensive computation if the data set is large. This effect can be offset, at the expense of some additional one-off computation, by constructing tree-based search structures to allow (approximate) near neighbours to be found efficiently without doing an exhaustive search of the data set. Nevertheless, these nonparametric methods are still severely limited. On the other hand, we have seen that simple parametric models are very restricted in terms of the forms of distribution that they can represent. We therefore need to find density models that are very flexible and yet for which the complexity of the models can be controlled independently of the size of the training set, and we shall see in subsequent chapters how to achieve this.

Exercises

2.1 ($\star$) **www** Verify that the Bernoulli distribution (2.2) satisfies the following properties

$$\sum_{x=0}^{1} p(x|\mu) = 1 \tag{2.257}$$

$$\mathbb{E}[x] = \mu \tag{2.258}$$

$$\mathrm{var}[x] = \mu(1-\mu). \tag{2.259}$$

Show that the entropy $\mathrm{H}[x]$ of a Bernoulli distributed random binary variable x is given by

$$\mathrm{H}[x] = -\mu \ln \mu - (1-\mu)\ln(1-\mu). \tag{2.260}$$

2.2 ($\star\star$) The form of the Bernoulli distribution given by (2.2) is not symmetric between the two values of x. In some situations, it will be more convenient to use an equivalent formulation for which $x \in \{-1, 1\}$, in which case the distribution can be written

$$p(x|\mu) = \left(\frac{1-\mu}{2}\right)^{(1-x)/2} \left(\frac{1+\mu}{2}\right)^{(1+x)/2} \tag{2.261}$$

where $\mu \in [-1, 1]$. Show that the distribution (2.261) is normalized, and evaluate its mean, variance, and entropy.

2.3 ($\star\star$) **www** In this exercise, we prove that the binomial distribution (2.9) is normalized. First use the definition (2.10) of the number of combinations of m identical objects chosen from a total of N to show that

$$\binom{N}{m} + \binom{N}{m-1} = \binom{N+1}{m}. \tag{2.262}$$

Use this result to prove by induction the following result

$$(1+x)^N = \sum_{m=0}^{N} \binom{N}{m} x^m \tag{2.263}$$

which is known as the *binomial theorem*, and which is valid for all real values of x. Finally, show that the binomial distribution is normalized, so that

$$\sum_{m=0}^{N} \binom{N}{m} \mu^m (1-\mu)^{N-m} = 1 \tag{2.264}$$

which can be done by first pulling out a factor $(1-\mu)^N$ out of the summation and then making use of the binomial theorem.

2.4 ($\star\star$) Show that the mean of the binomial distribution is given by (2.11). To do this, differentiate both sides of the normalization condition (2.264) with respect to μ and then rearrange to obtain an expression for the mean of n. Similarly, by differentiating (2.264) twice with respect to μ and making use of the result (2.11) for the mean of the binomial distribution prove the result (2.12) for the variance of the binomial.

2.5 ($\star\star$) **www** In this exercise, we prove that the beta distribution, given by (2.13), is correctly normalized, so that (2.14) holds. This is equivalent to showing that

$$\int_0^1 \mu^{a-1}(1-\mu)^{b-1}\,\mathrm{d}\mu = \frac{\Gamma(a)\Gamma(b)}{\Gamma(a+b)}. \tag{2.265}$$

From the definition (1.141) of the gamma function, we have

$$\Gamma(a)\Gamma(b) = \int_0^\infty \exp(-x)x^{a-1}\,\mathrm{d}x \int_0^\infty \exp(-y)y^{b-1}\,\mathrm{d}y. \tag{2.266}$$

Use this expression to prove (2.265) as follows. First bring the integral over y inside the integrand of the integral over x, next make the change of variable $t = y + x$ where x is fixed, then interchange the order of the x and t integrations, and finally make the change of variable $x = t\mu$ where t is fixed.

2.6 ($\star$) Make use of the result (2.265) to show that the mean, variance, and mode of the beta distribution (2.13) are given respectively by

$$\mathbb{E}[\mu] = \frac{a}{a+b} \tag{2.267}$$

$$\mathrm{var}[\mu] = \frac{ab}{(a+b)^2(a+b+1)} \tag{2.268}$$

$$\mathrm{mode}[\mu] = \frac{a-1}{a+b-2}. \tag{2.269}$$

2.7 ($\star\star$) Consider a binomial random variable x given by (2.9), with prior distribution for μ given by the beta distribution (2.13), and suppose we have observed m occurrences of $x = 1$ and l occurrences of $x = 0$. Show that the posterior mean value of μ lies between the prior mean and the maximum likelihood estimate for μ. To do this, show that the posterior mean can be written as λ times the prior mean plus $(1 - \lambda)$ times the maximum likelihood estimate, where $0 \leqslant \lambda \leqslant 1$. This illustrates the concept of the posterior distribution being a compromise between the prior distribution and the maximum likelihood solution.

2.8 ($\star$) Consider two variables x and y with joint distribution $p(x, y)$. Prove the following two results

$$\mathbb{E}[x] = \mathbb{E}_y\left[\mathbb{E}_x[x|y]\right] \tag{2.270}$$

$$\mathrm{var}[x] = \mathbb{E}_y\left[\mathrm{var}_x[x|y]\right] + \mathrm{var}_y\left[\mathbb{E}_x[x|y]\right]. \tag{2.271}$$

Here $\mathbb{E}_x[x|y]$ denotes the expectation of x under the conditional distribution $p(x|y)$, with a similar notation for the conditional variance.

2.9 ($\star\star\star$) **www** . In this exercise, we prove the normalization of the Dirichlet distribution (2.38) using induction. We have already shown in Exercise 2.5 that the beta distribution, which is a special case of the Dirichlet for $M = 2$, is normalized. We now assume that the Dirichlet distribution is normalized for $M - 1$ variables and prove that it is normalized for M variables. To do this, consider the Dirichlet distribution over M variables, and take account of the constraint $\sum_{k=1}^{M} \mu_k = 1$ by eliminating μ_M, so that the Dirichlet is written

$$p_M(\mu_1, \ldots, \mu_{M-1}) = C_M \prod_{k=1}^{M-1} \mu_k^{\alpha_k - 1} \left(1 - \sum_{j=1}^{M-1} \mu_j\right)^{\alpha_M - 1} \tag{2.272}$$

and our goal is to find an expression for C_M. To do this, integrate over μ_{M-1}, taking care over the limits of integration, and then make a change of variable so that this integral has limits 0 and 1. By assuming the correct result for C_{M-1} and making use of (2.265), derive the expression for C_M.

2.10 ($\star\star$) Using the property $\Gamma(x + 1) = x\Gamma(x)$ of the gamma function, derive the following results for the mean, variance, and covariance of the Dirichlet distribution given by (2.38)

$$\mathbb{E}[\mu_j] = \frac{\alpha_j}{\alpha_0} \tag{2.273}$$

$$\mathrm{var}[\mu_j] = \frac{\alpha_j(\alpha_0 - \alpha_j)}{\alpha_0^2(\alpha_0 + 1)} \tag{2.274}$$

$$\mathrm{cov}[\mu_j \mu_l] = -\frac{\alpha_j \alpha_l}{\alpha_0^2(\alpha_0 + 1)}, \qquad j \neq l \tag{2.275}$$

where α_0 is defined by (2.39).

2.11 ($\star$) **www** By expressing the expectation of $\ln \mu_j$ under the Dirichlet distribution (2.38) as a derivative with respect to α_j, show that

$$\mathbb{E}[\ln \mu_j] = \psi(\alpha_j) - \psi(\alpha_0) \tag{2.276}$$

where α_0 is given by (2.39) and

$$\psi(a) \equiv \frac{d}{da} \ln \Gamma(a) \tag{2.277}$$

is the *digamma* function.

2.12 ($\star$) The uniform distribution for a continuous variable x is defined by

$$\mathrm{U}(x|a, b) = \frac{1}{b-a}, \qquad a \leqslant x \leqslant b. \tag{2.278}$$

Verify that this distribution is normalized, and find expressions for its mean and variance.

2.13 ($\star\star$) Evaluate the Kullback-Leibler divergence (1.113) between two Gaussians $p(\mathbf{x}) = \mathcal{N}(\mathbf{x}|\boldsymbol{\mu}, \boldsymbol{\Sigma})$ and $q(\mathbf{x}) = \mathcal{N}(\mathbf{x}|\mathbf{m}, \mathbf{L})$.

2.14 ($\star\star$) **www** This exercise demonstrates that the multivariate distribution with maximum entropy, for a given covariance, is a Gaussian. The entropy of a distribution $p(\mathbf{x})$ is given by

$$\mathrm{H}[\mathbf{x}] = -\int p(\mathbf{x}) \ln p(\mathbf{x}) \,\mathrm{d}\mathbf{x}. \tag{2.279}$$

We wish to maximize $\mathrm{H}[\mathbf{x}]$ over all distributions $p(\mathbf{x})$ subject to the constraints that $p(\mathbf{x})$ be normalized and that it have a specific mean and covariance, so that

$$\int p(\mathbf{x}) \,\mathrm{d}\mathbf{x} = 1 \tag{2.280}$$

$$\int p(\mathbf{x})\mathbf{x} \,\mathrm{d}\mathbf{x} = \boldsymbol{\mu} \tag{2.281}$$

$$\int p(\mathbf{x})(\mathbf{x}-\boldsymbol{\mu})(\mathbf{x}-\boldsymbol{\mu})^{\mathrm{T}} \,\mathrm{d}\mathbf{x} = \boldsymbol{\Sigma}. \tag{2.282}$$

By performing a variational maximization of (2.279) and using Lagrange multipliers to enforce the constraints (2.280), (2.281), and (2.282), show that the maximum likelihood distribution is given by the Gaussian (2.43).

2.15 ($\star\star$) Show that the entropy of the multivariate Gaussian $\mathcal{N}(\mathbf{x}|\boldsymbol{\mu}, \boldsymbol{\Sigma})$ is given by

$$\mathrm{H}[\mathbf{x}] = \frac{1}{2} \ln |\boldsymbol{\Sigma}| + \frac{D}{2}(1 + \ln(2\pi)) \tag{2.283}$$

where D is the dimensionality of $\mathbf{x}$.

2.16 ($\star\star\star$) **www** Consider two random variables x_1 and x_2 having Gaussian distributions with means μ_1, μ_2 and precisions τ_1, τ_2 respectively. Derive an expression for the differential entropy of the variable $x = x_1 + x_2$. To do this, first find the distribution of x by using the relation

$$p(x) = \int_{-\infty}^{\infty} p(x|x_2)p(x_2)\,\mathrm{d}x_2 \tag{2.284}$$

and completing the square in the exponent. Then observe that this represents the convolution of two Gaussian distributions, which itself will be Gaussian, and finally make use of the result (1.110) for the entropy of the univariate Gaussian.

2.17 ($\star$) **www** Consider the multivariate Gaussian distribution given by (2.43). By writing the precision matrix (inverse covariance matrix) $\boldsymbol{\Sigma}^{-1}$ as the sum of a symmetric and an anti-symmetric matrix, show that the anti-symmetric term does not appear in the exponent of the Gaussian, and hence that the precision matrix may be taken to be symmetric without loss of generality. Because the inverse of a symmetric matrix is also symmetric (see Exercise 2.22), it follows that the covariance matrix may also be chosen to be symmetric without loss of generality.

2.18 ($\star\star\star$) Consider a real, symmetric matrix $\boldsymbol{\Sigma}$ whose eigenvalue equation is given by (2.45). By taking the complex conjugate of this equation and subtracting the original equation, and then forming the inner product with eigenvector $\mathbf{u}_i$, show that the eigenvalues λ_i are real. Similarly, use the symmetry property of $\boldsymbol{\Sigma}$ to show that two eigenvectors $\mathbf{u}_i$ and $\mathbf{u}_j$ will be orthogonal provided $\lambda_j \neq \lambda_i$. Finally, show that without loss of generality, the set of eigenvectors can be chosen to be orthonormal, so that they satisfy (2.46), even if some of the eigenvalues are zero.

2.19 ($\star\star$) Show that a real, symmetric matrix $\boldsymbol{\Sigma}$ having the eigenvector equation (2.45) can be expressed as an expansion in the eigenvectors, with coefficients given by the eigenvalues, of the form (2.48). Similarly, show that the inverse matrix $\boldsymbol{\Sigma}^{-1}$ has a representation of the form (2.49).

2.20 ($\star\star$) **www** A positive definite matrix $\boldsymbol{\Sigma}$ can be defined as one for which the quadratic form

$$\mathbf{a}^{\mathrm{T}}\boldsymbol{\Sigma}\mathbf{a} \tag{2.285}$$

is positive for any real value of the vector $\mathbf{a}$. Show that a necessary and sufficient condition for $\boldsymbol{\Sigma}$ to be positive definite is that all of the eigenvalues λ_i of $\boldsymbol{\Sigma}$, defined by (2.45), are positive.

2.21 ($\star$) Show that a real, symmetric matrix of size $D \times D$ has $D(D+1)/2$ independent parameters.

2.22 ($\star$) **www** Show that the inverse of a symmetric matrix is itself symmetric.

2.23 ($\star\star$) By diagonalizing the coordinate system using the eigenvector expansion (2.48), show that the volume contained within the hyperellipsoid corresponding to a constant

Mahalanobis distance Δ is given by

$$V_D|\boldsymbol{\Sigma}|^{1/2}\Delta^D \tag{2.286}$$

where V_D is the volume of the unit sphere in D dimensions, and the Mahalanobis distance is defined by (2.44).

2.24 ($\star\star$) **www** Prove the identity (2.76) by multiplying both sides by the matrix

$$\begin{pmatrix} \mathbf{A} & \mathbf{B} \\ \mathbf{C} & \mathbf{D} \end{pmatrix} \tag{2.287}$$

and making use of the definition (2.77).

2.25 ($\star\star$) In Sections 2.3.1 and 2.3.2, we considered the conditional and marginal distributions for a multivariate Gaussian. More generally, we can consider a partitioning of the components of $\mathbf{x}$ into three groups $\mathbf{x}_a$, $\mathbf{x}_b$, and $\mathbf{x}_c$, with a corresponding partitioning of the mean vector $\boldsymbol{\mu}$ and of the covariance matrix $\boldsymbol{\Sigma}$ in the form

$$\boldsymbol{\mu} = \begin{pmatrix} \boldsymbol{\mu}_a \\ \boldsymbol{\mu}_b \\ \boldsymbol{\mu}_c \end{pmatrix}, \qquad \boldsymbol{\Sigma} = \begin{pmatrix} \boldsymbol{\Sigma}_{aa} & \boldsymbol{\Sigma}_{ab} & \boldsymbol{\Sigma}_{ac} \\ \boldsymbol{\Sigma}_{ba} & \boldsymbol{\Sigma}_{bb} & \boldsymbol{\Sigma}_{bc} \\ \boldsymbol{\Sigma}_{ca} & \boldsymbol{\Sigma}_{cb} & \boldsymbol{\Sigma}_{cc} \end{pmatrix}. \tag{2.288}$$

By making use of the results of Section 2.3, find an expression for the conditional distribution $p(\mathbf{x}_a|\mathbf{x}_b)$ in which $\mathbf{x}_c$ has been marginalized out.

2.26 ($\star\star$) A very useful result from linear algebra is the *Woodbury* matrix inversion formula given by

$$(\mathbf{A} + \mathbf{BCD})^{-1} = \mathbf{A}^{-1} - \mathbf{A}^{-1}\mathbf{B}(\mathbf{C}^{-1} + \mathbf{D}\mathbf{A}^{-1}\mathbf{B})^{-1}\mathbf{D}\mathbf{A}^{-1}. \tag{2.289}$$

By multiplying both sides by $(\mathbf{A} + \mathbf{BCD})$ prove the correctness of this result.

2.27 ($\star$) Let $\mathbf{x}$ and $\mathbf{z}$ be two independent random vectors, so that $p(\mathbf{x}, \mathbf{z}) = p(\mathbf{x})p(\mathbf{z})$. Show that the mean of their sum $\mathbf{y} = \mathbf{x}+\mathbf{z}$ is given by the sum of the means of each of the variable separately. Similarly, show that the covariance matrix of $\mathbf{y}$ is given by the sum of the covariance matrices of $\mathbf{x}$ and $\mathbf{z}$. Confirm that this result agrees with that of Exercise 1.10.

2.28 ($\star\star\star$) **www** Consider a joint distribution over the variable

$$\mathbf{z} = \begin{pmatrix} \mathbf{x} \\ \mathbf{y} \end{pmatrix} \tag{2.290}$$

whose mean and covariance are given by (2.108) and (2.105) respectively. By making use of the results (2.92) and (2.93) show that the marginal distribution $p(\mathbf{x})$ is given (2.99). Similarly, by making use of the results (2.81) and (2.82) show that the conditional distribution $p(\mathbf{y}|\mathbf{x})$ is given by (2.100).

2.29 ($\star\star$) Using the partitioned matrix inversion formula (2.76), show that the inverse of the precision matrix (2.104) is given by the covariance matrix (2.105).

2.30 ($\star$) By starting from (2.107) and making use of the result (2.105), verify the result (2.108).

2.31 ($\star\star$) Consider two multidimensional random vectors $\mathbf{x}$ and $\mathbf{z}$ having Gaussian distributions $p(\mathbf{x}) = \mathcal{N}(\mathbf{x}|\boldsymbol{\mu}_{\mathbf{x}}, \boldsymbol{\Sigma}_{\mathbf{x}})$ and $p(\mathbf{z}) = \mathcal{N}(\mathbf{z}|\boldsymbol{\mu}_{\mathbf{z}}, \boldsymbol{\Sigma}_{\mathbf{z}})$ respectively, together with their sum $\mathbf{y} = \mathbf{x}+\mathbf{z}$. Use the results (2.109) and (2.110) to find an expression for the marginal distribution $p(\mathbf{y})$ by considering the linear-Gaussian model comprising the product of the marginal distribution $p(\mathbf{x})$ and the conditional distribution $p(\mathbf{y}|\mathbf{x})$.

2.32 ($\star\star\star$) **www** This exercise and the next provide practice at manipulating the quadratic forms that arise in linear-Gaussian models, as well as giving an independent check of results derived in the main text. Consider a joint distribution $p(\mathbf{x}, \mathbf{y})$ defined by the marginal and conditional distributions given by (2.99) and (2.100). By examining the quadratic form in the exponent of the joint distribution, and using the technique of 'completing the square' discussed in Section 2.3, find expressions for the mean and covariance of the marginal distribution $p(\mathbf{y})$ in which the variable $\mathbf{x}$ has been integrated out. To do this, make use of the Woodbury matrix inversion formula (2.289). Verify that these results agree with (2.109) and (2.110) obtained using the results of Chapter 2.

2.33 ($\star\star\star$) Consider the same joint distribution as in Exercise 2.32, but now use the technique of completing the square to find expressions for the mean and covariance of the conditional distribution $p(\mathbf{x}|\mathbf{y})$. Again, verify that these agree with the corresponding expressions (2.111) and (2.112).

2.34 ($\star\star$) **www** To find the maximum likelihood solution for the covariance matrix of a multivariate Gaussian, we need to maximize the log likelihood function (2.118) with respect to $\boldsymbol{\Sigma}$, noting that the covariance matrix must be symmetric and positive definite. Here we proceed by ignoring these constraints and doing a straightforward maximization. Using the results (C.21), (C.26), and (C.28) from Appendix C, show that the covariance matrix $\boldsymbol{\Sigma}$ that maximizes the log likelihood function (2.118) is given by the sample covariance (2.122). We note that the final result is necessarily symmetric and positive definite (provided the sample covariance is nonsingular).

2.35 ($\star\star$) Use the result (2.59) to prove (2.62). Now, using the results (2.59), and (2.62), show that

$$\mathbb{E}[\mathbf{x}_n \mathbf{x}_m^{\mathrm{T}}] = \boldsymbol{\mu}\boldsymbol{\mu}^{\mathrm{T}} + I_{nm}\boldsymbol{\Sigma} \tag{2.291}$$

where $\mathbf{x}_n$ denotes a data point sampled from a Gaussian distribution with mean $\boldsymbol{\mu}$ and covariance $\boldsymbol{\Sigma}$, and I_{nm} denotes the (n, m) element of the identity matrix. Hence prove the result (2.124).

2.36 ($\star\star$) **www** Using an analogous procedure to that used to obtain (2.126), derive an expression for the sequential estimation of the variance of a univariate Gaussian distribution, by starting with the maximum likelihood expression

$$\sigma^2_{\mathrm{ML}} = \frac{1}{N}\sum_{n=1}^{N}(x_n - \mu)^2. \tag{2.292}$$

Verify that substituting the expression for a Gaussian distribution into the Robbins-Monro sequential estimation formula (2.135) gives a result of the same form, and hence obtain an expression for the corresponding coefficients a_N.

2.37 ($\star\star$) Using an analogous procedure to that used to obtain (2.126), derive an expression for the sequential estimation of the covariance of a multivariate Gaussian distribution, by starting with the maximum likelihood expression (2.122). Verify that substituting the expression for a Gaussian distribution into the Robbins-Monro sequential estimation formula (2.135) gives a result of the same form, and hence obtain an expression for the corresponding coefficients a_N.

2.38 ($\star$) Use the technique of completing the square for the quadratic form in the exponent to derive the results (2.141) and (2.142).

2.39 ($\star\star$) Starting from the results (2.141) and (2.142) for the posterior distribution of the mean of a Gaussian random variable, dissect out the contributions from the first $N-1$ data points and hence obtain expressions for the sequential update of μ_N and σ^2_N. Now derive the same results starting from the posterior distribution $p(\mu|x_1,\ldots,x_{N-1}) = \mathcal{N}(\mu|\mu_{N-1},\sigma^2_{N-1})$ and multiplying by the likelihood function $p(x_N|\mu) = \mathcal{N}(x_N|\mu,\sigma^2)$ and then completing the square and normalizing to obtain the posterior distribution after N observations.

2.40 ($\star\star$) **www** Consider a D-dimensional Gaussian random variable $\mathbf{x}$ with distribution $\mathcal{N}(\mathbf{x}|\boldsymbol{\mu},\boldsymbol{\Sigma})$ in which the covariance $\boldsymbol{\Sigma}$ is known and for which we wish to infer the mean $\boldsymbol{\mu}$ from a set of observations $\mathbf{X} = \{\mathbf{x}_1,\ldots,\mathbf{x}_N\}$. Given a prior distribution $p(\boldsymbol{\mu}) = \mathcal{N}(\boldsymbol{\mu}|\boldsymbol{\mu}_0,\boldsymbol{\Sigma}_0)$, find the corresponding posterior distribution $p(\boldsymbol{\mu}|\mathbf{X})$.

2.41 ($\star$) Use the definition of the gamma function (1.141) to show that the gamma distribution (2.146) is normalized.

2.42 ($\star\star$) Evaluate the mean, variance, and mode of the gamma distribution (2.146).

2.43 ($\star$) The following distribution

$$p(x|\sigma^2, q) = \frac{q}{2(2\sigma^2)^{1/q}\Gamma(1/q)}\exp\left(-\frac{|x|^q}{2\sigma^2}\right) \tag{2.293}$$

is a generalization of the univariate Gaussian distribution. Show that this distribution is normalized so that

$$\int_{-\infty}^{\infty} p(x|\sigma^2, q)\,\mathrm{d}x = 1 \tag{2.294}$$

and that it reduces to the Gaussian when $q = 2$. Consider a regression model in which the target variable is given by $t = y(\mathbf{x},\mathbf{w}) + \epsilon$ and ϵ is a random noise variable drawn from the distribution (2.293). Show that the log likelihood function over $\mathbf{w}$ and σ^2, for an observed data set of input vectors $\mathbf{X} = \{\mathbf{x}_1, \ldots, \mathbf{x}_N\}$ and corresponding target variables $\mathbf{t} = (t_1, \ldots, t_N)^{\mathrm{T}}$, is given by

$$\ln p(\mathbf{t}|\mathbf{X}, \mathbf{w}, \sigma^2) = -\frac{1}{2\sigma^2} \sum_{n=1}^{N} |y(\mathbf{x}_n, \mathbf{w}) - t_n|^q - \frac{N}{q} \ln(2\sigma^2) + \text{const} \tag{2.295}$$

where 'const' denotes terms independent of both $\mathbf{w}$ and σ^2. Note that, as a function of $\mathbf{w}$, this is the L_q error function considered in Section 1.5.5.

2.44 ($\star\star$) Consider a univariate Gaussian distribution $\mathcal{N}(x|\mu, \tau^{-1})$ having conjugate Gaussian-gamma prior given by (2.154), and a data set $\mathbf{x} = \{x_1, \ldots, x_N\}$ of i.i.d. observations. Show that the posterior distribution is also a Gaussian-gamma distribution of the same functional form as the prior, and write down expressions for the parameters of this posterior distribution.

2.45 ($\star$) Verify that the Wishart distribution defined by (2.155) is indeed a conjugate prior for the precision matrix of a multivariate Gaussian.

2.46 ($\star$) **www** Verify that evaluating the integral in (2.158) leads to the result (2.159).

2.47 ($\star$) **www** Show that in the limit $\nu \to \infty$, the t-distribution (2.159) becomes a Gaussian. Hint: ignore the normalization coefficient, and simply look at the dependence on x.

2.48 ($\star$) By following analogous steps to those used to derive the univariate Student's t-distribution (2.159), verify the result (2.162) for the multivariate form of the Student's t-distribution, by marginalizing over the variable η in (2.161). Using the definition (2.161), show by exchanging integration variables that the multivariate t-distribution is correctly normalized.

2.49 ($\star\star$) By using the definition (2.161) of the multivariate Student's t-distribution as a convolution of a Gaussian with a gamma distribution, verify the properties (2.164), (2.165), and (2.166) for the multivariate t-distribution defined by (2.162).

2.50 ($\star$) Show that in the limit $\nu \to \infty$, the multivariate Student's t-distribution (2.162) reduces to a Gaussian with mean $\boldsymbol{\mu}$ and precision $\boldsymbol{\Lambda}$.

2.51 ($\star$) **www** The various trigonometric identities used in the discussion of periodic variables in this chapter can be proven easily from the relation

$$\exp(iA) = \cos A + i \sin A \tag{2.296}$$

in which i is the square root of minus one. By considering the identity

$$\exp(iA)\exp(-iA) = 1 \tag{2.297}$$

prove the result (2.177). Similarly, using the identity

$$\cos(A - B) = \Re \exp\{i(A - B)\} \tag{2.298}$$

where $\Re$ denotes the real part, prove (2.178). Finally, by using $\sin(A - B) = \Im \exp\{i(A - B)\}$, where $\Im$ denotes the imaginary part, prove the result (2.183).

2.52 ($\star\star$) For large m, the von Mises distribution (2.179) becomes sharply peaked around the mode θ_0. By defining $\xi = m^{1/2}(\theta - \theta_0)$ and making the Taylor expansion of the cosine function given by

$$\cos \alpha = 1 - \frac{\alpha^2}{2} + O(\alpha^4) \tag{2.299}$$

show that as $m \to \infty$, the von Mises distribution tends to a Gaussian.

2.53 ($\star$) Using the trigonometric identity (2.183), show that solution of (2.182) for θ_0 is given by (2.184).

2.54 ($\star$) By computing first and second derivatives of the von Mises distribution (2.179), and using $I_0(m) > 0$ for $m > 0$, show that the maximum of the distribution occurs when $\theta = \theta_0$ and that the minimum occurs when $\theta = \theta_0 + \pi \pmod{2\pi}$.

2.55 ($\star$) By making use of the result (2.168), together with (2.184) and the trigonometric identity (2.178), show that the maximum likelihood solution m_{ML} for the concentration of the von Mises distribution satisfies $A(m_{\text{ML}}) = \bar{r}$ where $\bar{r}$ is the radius of the mean of the observations viewed as unit vectors in the two-dimensional Euclidean plane, as illustrated in Figure 2.17.

2.56 ($\star\star$) **www** Express the beta distribution (2.13), the gamma distribution (2.146), and the von Mises distribution (2.179) as members of the exponential family (2.194) and thereby identify their natural parameters.

2.57 ($\star$) Verify that the multivariate Gaussian distribution can be cast in exponential family form (2.194) and derive expressions for $\boldsymbol{\eta}$, $\mathbf{u}(\mathbf{x})$, $h(\mathbf{x})$ and $g(\boldsymbol{\eta})$ analogous to (2.220)–(2.223).

2.58 ($\star$) The result (2.226) showed that the negative gradient of $\ln g(\boldsymbol{\eta})$ for the exponential family is given by the expectation of $\mathbf{u}(\mathbf{x})$. By taking the second derivatives of (2.195), show that

$$-\nabla\nabla \ln g(\boldsymbol{\eta}) = \mathbb{E}[\mathbf{u}(\mathbf{x})\mathbf{u}(\mathbf{x})^{\mathrm{T}}] - \mathbb{E}[\mathbf{u}(\mathbf{x})]\mathbb{E}[\mathbf{u}(\mathbf{x})^{\mathrm{T}}] = \text{cov}[\mathbf{u}(\mathbf{x})]. \tag{2.300}$$

2.59 ($\star$) By changing variables using $y = x/\sigma$, show that the density (2.236) will be correctly normalized, provided $f(x)$ is correctly normalized.

2.60 ($\star\star$) **www** Consider a histogram-like density model in which the space $\mathbf{x}$ is divided into fixed regions for which the density $p(\mathbf{x})$ takes the constant value h_i over the ith region, and that the volume of region i is denoted Δ_i. Suppose we have a set of N observations of $\mathbf{x}$ such that n_i of these observations fall in region i. Using a Lagrange multiplier to enforce the normalization constraint on the density, derive an expression for the maximum likelihood estimator for the $\{h_i\}$.

2.61 ($\star$) Show that the K-nearest-neighbour density model defines an improper distribution whose integral over all space is divergent.

3 Linear Models for Regression

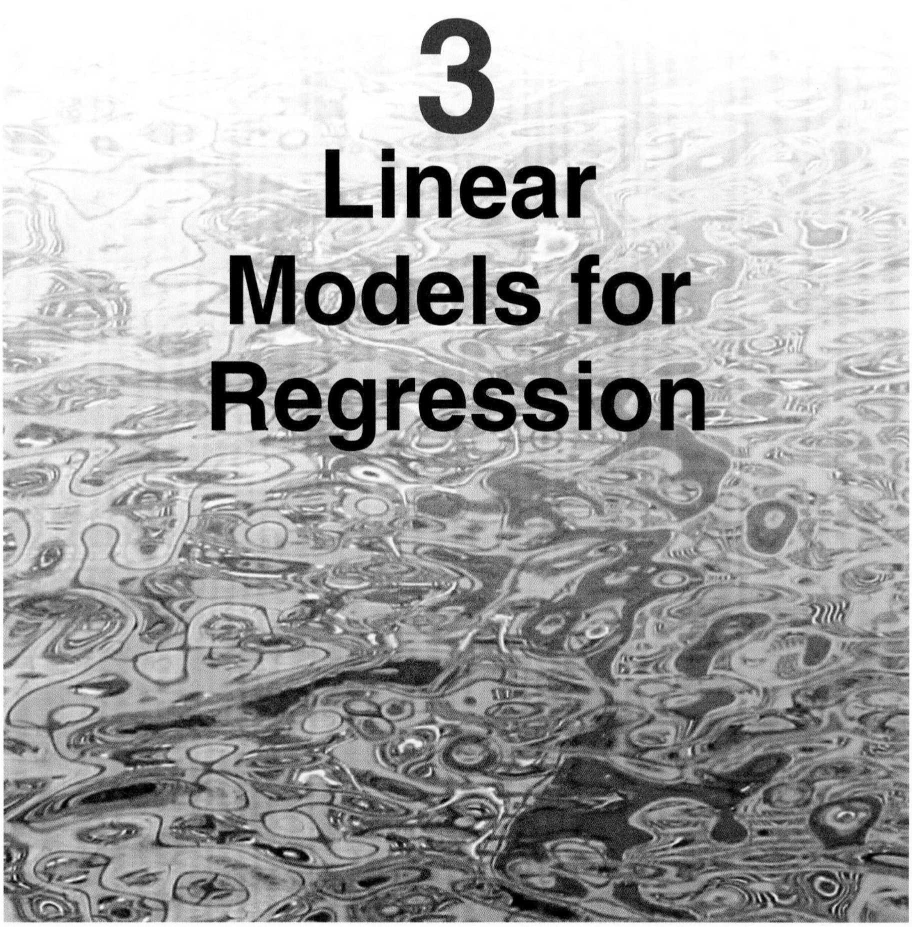

The focus so far in this book has been on unsupervised learning, including topics such as density estimation and data clustering. We turn now to a discussion of supervised learning, starting with regression. The goal of regression is to predict the value of one or more continuous *target* variables t given the value of a D-dimensional vector $\mathbf{x}$ of *input* variables. We have already encountered an example of a regression problem when we considered polynomial curve fitting in Chapter 1. The polynomial is a specific example of a broad class of functions called linear regression models, which share the property of being linear functions of the adjustable parameters, and which will form the focus of this chapter. The simplest form of linear regression models are also linear functions of the input variables. However, we can obtain a much more useful class of functions by taking linear combinations of a fixed set of nonlinear functions of the input variables, known as *basis functions*. Such models are linear functions of the parameters, which gives them simple analytical properties, and yet can be nonlinear with respect to the input variables.

Given a training data set comprising N observations $\{\mathbf{x}_n\}$, where $n = 1, \ldots, N$, together with corresponding target values $\{t_n\}$, the goal is to predict the value of t for a new value of $\mathbf{x}$. In the simplest approach, this can be done by directly constructing an appropriate function $y(\mathbf{x})$ whose values for new inputs $\mathbf{x}$ constitute the predictions for the corresponding values of t. More generally, from a probabilistic perspective, we aim to model the predictive distribution $p(t|\mathbf{x})$ because this expresses our uncertainty about the value of t for each value of $\mathbf{x}$. From this conditional distribution we can make predictions of t, for any new value of $\mathbf{x}$, in such a way as to minimize the expected value of a suitably chosen loss function. As discussed in Section 1.5.5, a common choice of loss function for real-valued variables is the squared loss, for which the optimal solution is given by the conditional expectation of t.

Although linear models have significant limitations as practical techniques for pattern recognition, particularly for problems involving input spaces of high dimensionality, they have nice analytical properties and form the foundation for more sophisticated models to be discussed in later chapters.

3.1. Linear Basis Function Models

The simplest linear model for regression is one that involves a linear combination of the input variables

$$y(\mathbf{x}, \mathbf{w}) = w_0 + w_1 x_1 + \ldots + w_D x_D \tag{3.1}$$

where $\mathbf{x} = (x_1, \ldots, x_D)^{\mathrm{T}}$. This is often simply known as *linear regression*. The key property of this model is that it is a linear function of the parameters $w_0, \ldots, w_D$. It is also, however, a linear function of the input variables x_i, and this imposes significant limitations on the model. We therefore extend the class of models by considering linear combinations of fixed nonlinear functions of the input variables, of the form

$$y(\mathbf{x}, \mathbf{w}) = w_0 + \sum_{j=1}^{M-1} w_j \phi_j(\mathbf{x}) \tag{3.2}$$

where $\phi_j(\mathbf{x})$ are known as *basis functions*. By denoting the maximum value of the index j by $M - 1$, the total number of parameters in this model will be M.

The parameter w_0 allows for any fixed offset in the data and is sometimes called a *bias* parameter (not to be confused with 'bias' in a statistical sense). It is often convenient to define an additional dummy 'basis function' $\phi_0(\mathbf{x}) = 1$ so that

$$y(\mathbf{x}, \mathbf{w}) = \sum_{j=0}^{M-1} w_j \phi_j(\mathbf{x}) = \mathbf{w}^{\mathrm{T}} \boldsymbol{\phi}(\mathbf{x}) \tag{3.3}$$

where $\mathbf{w} = (w_0, \ldots, w_{M-1})^{\mathrm{T}}$ and $\boldsymbol{\phi} = (\phi_0, \ldots, \phi_{M-1})^{\mathrm{T}}$. In many practical applications of pattern recognition, we will apply some form of fixed pre-processing, or feature extraction, to the original data variables. If the original variables comprise the vector $\mathbf{x}$, then the features can be expressed in terms of the basis functions $\{\phi_j(\mathbf{x})\}$.

By using nonlinear basis functions, we allow the function $y(\mathbf{x}, \mathbf{w})$ to be a nonlinear function of the input vector $\mathbf{x}$. Functions of the form (3.2) are called linear models, however, because this function is linear in $\mathbf{w}$. It is this linearity in the parameters that will greatly simplify the analysis of this class of models. However, it also leads to some significant limitations, as we discuss in Section 3.6.

The example of polynomial regression considered in Chapter 1 is a particular example of this model in which there is a single input variable x, and the basis functions take the form of powers of x so that $\phi_j(x) = x^j$. One limitation of polynomial basis functions is that they are global functions of the input variable, so that changes in one region of input space affect all other regions. This can be resolved by dividing the input space up into regions and fitting a different polynomial in each region, leading to *spline functions* (Hastie *et al.*, 2001).

There are many other possible choices for the basis functions, for example

$$\phi_j(x) = \exp\left\{-\frac{(x-\mu_j)^2}{2s^2}\right\} \tag{3.4}$$

where the μ_j govern the locations of the basis functions in input space, and the parameter s governs their spatial scale. These are usually referred to as 'Gaussian' basis functions, although it should be noted that they are not required to have a probabilistic interpretation, and in particular the normalization coefficient is unimportant because these basis functions will be multiplied by adaptive parameters w_j.

Another possibility is the sigmoidal basis function of the form

$$\phi_j(x) = \sigma\left(\frac{x-\mu_j}{s}\right) \tag{3.5}$$

where $\sigma(a)$ is the logistic sigmoid function defined by

$$\sigma(a) = \frac{1}{1+\exp(-a)}. \tag{3.6}$$

Equivalently, we can use the '$\tanh$' function because this is related to the logistic sigmoid by $\tanh(a) = 2\sigma(2a) - 1$, and so a general linear combination of logistic sigmoid functions is equivalent to a general linear combination of '$\tanh$' functions. These various choices of basis function are illustrated in Figure 3.1.

Yet another possible choice of basis function is the Fourier basis, which leads to an expansion in sinusoidal functions. Each basis function represents a specific frequency and has infinite spatial extent. By contrast, basis functions that are localized to finite regions of input space necessarily comprise a spectrum of different spatial frequencies. In many signal processing applications, it is of interest to consider basis functions that are localized in both space and frequency, leading to a class of functions known as *wavelets*. These are also defined to be mutually orthogonal, to simplify their application. Wavelets are most applicable when the input values live

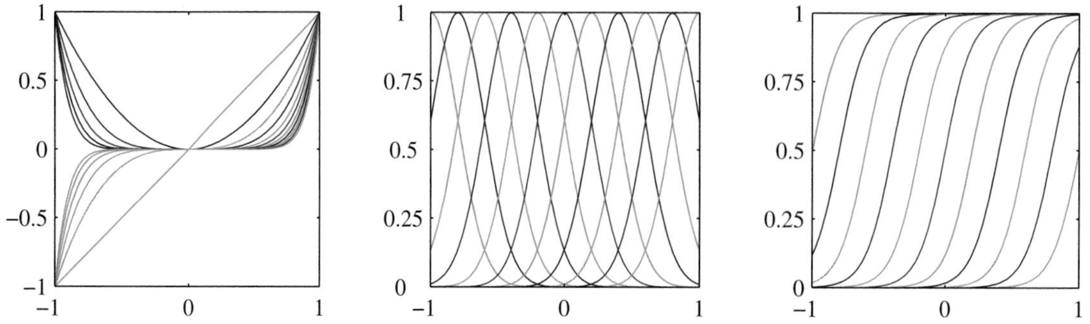

Figure 3.1 Examples of basis functions, showing polynomials on the left, Gaussians of the form (3.4) in the centre, and sigmoidal of the form (3.5) on the right.

on a regular lattice, such as the successive time points in a temporal sequence, or the pixels in an image. Useful texts on wavelets include Ogden (1997), Mallat (1999), and Vidakovic (1999).

Most of the discussion in this chapter, however, is independent of the particular choice of basis function set, and so for most of our discussion we shall not specify the particular form of the basis functions, except for the purposes of numerical illustration. Indeed, much of our discussion will be equally applicable to the situation in which the vector $\boldsymbol{\phi}(\mathbf{x})$ of basis functions is simply the identity $\boldsymbol{\phi}(\mathbf{x}) = \mathbf{x}$. Furthermore, in order to keep the notation simple, we shall focus on the case of a single target variable t. However, in Section 3.1.5, we consider briefly the modifications needed to deal with multiple target variables.

3.1.1 Maximum likelihood and least squares

In Chapter 1, we fitted polynomial functions to data sets by minimizing a sum-of-squares error function. We also showed that this error function could be motivated as the maximum likelihood solution under an assumed Gaussian noise model. Let us return to this discussion and consider the least squares approach, and its relation to maximum likelihood, in more detail.

As before, we assume that the target variable t is given by a deterministic function $y(\mathbf{x}, \mathbf{w})$ with additive Gaussian noise so that

$$t = y(\mathbf{x}, \mathbf{w}) + \epsilon \tag{3.7}$$

where ϵ is a zero mean Gaussian random variable with precision (inverse variance) β. Thus we can write

$$p(t|\mathbf{x}, \mathbf{w}, \beta) = \mathcal{N}(t|y(\mathbf{x}, \mathbf{w}), \beta^{-1}). \tag{3.8}$$

Section 1.5.5

Recall that, if we assume a squared loss function, then the optimal prediction, for a new value of $\mathbf{x}$, will be given by the conditional mean of the target variable. In the case of a Gaussian conditional distribution of the form (3.8), the conditional mean will be simply

$$\mathbb{E}[t|\mathbf{x}] = \int tp(t|\mathbf{x}) \, \mathrm{d}t = y(\mathbf{x}, \mathbf{w}). \tag{3.9}$$

Note that the Gaussian noise assumption implies that the conditional distribution of t given $\mathbf{x}$ is unimodal, which may be inappropriate for some applications. An extension to mixtures of conditional Gaussian distributions, which permit multimodal conditional distributions, will be discussed in Section 14.5.1.

Now consider a data set of inputs $\mathbf{X} = \{\mathbf{x}_1, \ldots, \mathbf{x}_N\}$ with corresponding target values $t_1, \ldots, t_N$. We group the target variables $\{t_n\}$ into a column vector that we denote by t where the typeface is chosen to distinguish it from a single observation of a multivariate target, which would be denoted $\mathbf{t}$. Making the assumption that these data points are drawn independently from the distribution (3.8), we obtain the following expression for the likelihood function, which is a function of the adjustable parameters $\mathbf{w}$ and β, in the form

$$p(\mathsf{t}|\mathbf{X}, \mathbf{w}, \beta) = \prod_{n=1}^{N} \mathcal{N}(t_n|\mathbf{w}^{\mathrm{T}}\boldsymbol{\phi}(\mathbf{x}_n), \beta^{-1}) \tag{3.10}$$

where we have used (3.3). Note that in supervised learning problems such as regression (and classification), we are not seeking to model the distribution of the input variables. Thus $\mathbf{x}$ will always appear in the set of conditioning variables, and so from now on we will drop the explicit $\mathbf{x}$ from expressions such as $p(\mathsf{t}|\mathbf{x}, \mathbf{w}, \beta)$ in order to keep the notation uncluttered. Taking the logarithm of the likelihood function, and making use of the standard form (1.46) for the univariate Gaussian, we have

$$\begin{aligned} \ln p(\mathsf{t}|\mathbf{w}, \beta) &= \sum_{n=1}^{N} \ln \mathcal{N}(t_n|\mathbf{w}^{\mathrm{T}}\boldsymbol{\phi}(\mathbf{x}_n), \beta^{-1}) \\ &= \frac{N}{2}\ln\beta - \frac{N}{2}\ln(2\pi) - \beta E_D(\mathbf{w}) \end{aligned} \tag{3.11}$$

where the sum-of-squares error function is defined by

$$E_D(\mathbf{w}) = \frac{1}{2}\sum_{n=1}^{N}\{t_n - \mathbf{w}^{\mathrm{T}}\boldsymbol{\phi}(\mathbf{x}_n)\}^2. \tag{3.12}$$

Having written down the likelihood function, we can use maximum likelihood to determine $\mathbf{w}$ and β. Consider first the maximization with respect to $\mathbf{w}$. As observed already in Section 1.2.5, we see that maximization of the likelihood function under a conditional Gaussian noise distribution for a linear model is equivalent to minimizing a sum-of-squares error function given by $E_D(\mathbf{w})$. The gradient of the log likelihood function (3.11) takes the form

$$\nabla \ln p(\mathsf{t}|\mathbf{w}, \beta) = \beta \sum_{n=1}^{N} \left\{t_n - \mathbf{w}^{\mathrm{T}}\boldsymbol{\phi}(\mathbf{x}_n)\right\} \boldsymbol{\phi}(\mathbf{x}_n)^{\mathrm{T}}. \tag{3.13}$$

Setting this gradient to zero gives

$$0 = \sum_{n=1}^{N} t_n \boldsymbol{\phi}(\mathbf{x}_n)^{\mathrm{T}} - \mathbf{w}^{\mathrm{T}} \left(\sum_{n=1}^{N} \boldsymbol{\phi}(\mathbf{x}_n) \boldsymbol{\phi}(\mathbf{x}_n)^{\mathrm{T}} \right). \tag{3.14}$$

Solving for $\mathbf{w}$ we obtain

$$\mathbf{w}_{\mathrm{ML}} = \left(\boldsymbol{\Phi}^{\mathrm{T}} \boldsymbol{\Phi} \right)^{-1} \boldsymbol{\Phi}^{\mathrm{T}} \mathbf{t} \tag{3.15}$$

which are known as the *normal equations* for the least squares problem. Here $\boldsymbol{\Phi}$ is an $N \times M$ matrix, called the *design matrix*, whose elements are given by $\Phi_{nj} = \phi_j(\mathbf{x}_n)$, so that

$$\boldsymbol{\Phi} = \begin{pmatrix} \phi_0(\mathbf{x}_1) & \phi_1(\mathbf{x}_1) & \cdots & \phi_{M-1}(\mathbf{x}_1) \\ \phi_0(\mathbf{x}_2) & \phi_1(\mathbf{x}_2) & \cdots & \phi_{M-1}(\mathbf{x}_2) \\ \vdots & \vdots & \ddots & \vdots \\ \phi_0(\mathbf{x}_N) & \phi_1(\mathbf{x}_N) & \cdots & \phi_{M-1}(\mathbf{x}_N) \end{pmatrix}. \tag{3.16}$$

The quantity

$$\boldsymbol{\Phi}^{\dagger} \equiv \left(\boldsymbol{\Phi}^{\mathrm{T}} \boldsymbol{\Phi} \right)^{-1} \boldsymbol{\Phi}^{\mathrm{T}} \tag{3.17}$$

is known as the *Moore-Penrose pseudo-inverse* of the matrix $\boldsymbol{\Phi}$ (Rao and Mitra, 1971; Golub and Van Loan, 1996). It can be regarded as a generalization of the notion of matrix inverse to nonsquare matrices. Indeed, if $\boldsymbol{\Phi}$ is square and invertible, then using the property $(\mathbf{AB})^{-1} = \mathbf{B}^{-1}\mathbf{A}^{-1}$ we see that $\boldsymbol{\Phi}^{\dagger} \equiv \boldsymbol{\Phi}^{-1}$.

At this point, we can gain some insight into the role of the bias parameter w_0. If we make the bias parameter explicit, then the error function (3.12) becomes

$$E_D(\mathbf{w}) = \frac{1}{2} \sum_{n=1}^{N} \{ t_n - w_0 - \sum_{j=1}^{M-1} w_j \phi_j(\mathbf{x}_n) \}^2. \tag{3.18}$$

Setting the derivative with respect to w_0 equal to zero, and solving for w_0, we obtain

$$w_0 = \overline{t} - \sum_{j=1}^{M-1} w_j \overline{\phi_j} \tag{3.19}$$

where we have defined

$$\overline{t} = \frac{1}{N} \sum_{n=1}^{N} t_n, \qquad \overline{\phi_j} = \frac{1}{N} \sum_{n=1}^{N} \phi_j(\mathbf{x}_n). \tag{3.20}$$

Thus the bias w_0 compensates for the difference between the averages (over the training set) of the target values and the weighted sum of the averages of the basis function values.

We can also maximize the log likelihood function (3.11) with respect to the noise precision parameter β, giving

$$\frac{1}{\beta_{\mathrm{ML}}} = \frac{1}{N} \sum_{n=1}^{N} \{ t_n - \mathbf{w}_{\mathrm{ML}}^{\mathrm{T}} \boldsymbol{\phi}(\mathbf{x}_n) \}^2 \tag{3.21}$$

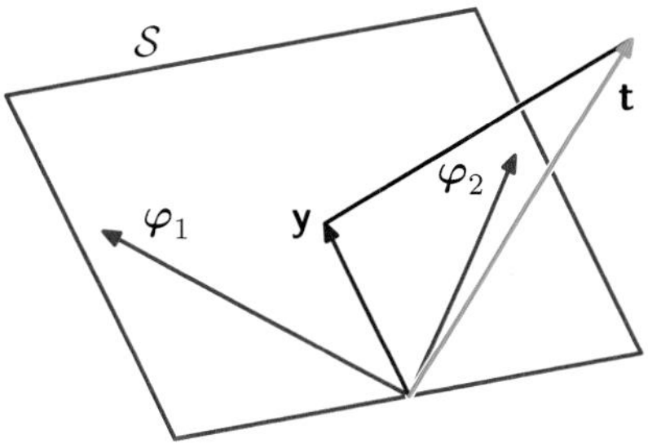

Figure 3.2 Geometrical interpretation of the least-squares solution, in an N-dimensional space whose axes are the values of $t_1, \ldots, t_N$. The least-squares regression function is obtained by finding the orthogonal projection of the data vector $\mathbf{t}$ onto the subspace spanned by the basis functions $\phi_j(\mathbf{x})$ in which each basis function is viewed as a vector $\boldsymbol{\varphi}_j$ of length N with elements $\phi_j(\mathbf{x}_n)$.

and so we see that the inverse of the noise precision is given by the residual variance of the target values around the regression function.

3.1.2 Geometry of least squares

At this point, it is instructive to consider the geometrical interpretation of the least-squares solution. To do this we consider an N-dimensional space whose axes are given by the t_n, so that $\mathbf{t} = (t_1, \ldots, t_N)^{\mathrm{T}}$ is a vector in this space. Each basis function $\phi_j(\mathbf{x}_n)$, evaluated at the N data points, can also be represented as a vector in the same space, denoted by $\boldsymbol{\varphi}_j$, as illustrated in Figure 3.2. Note that $\boldsymbol{\varphi}_j$ corresponds to the jth column of $\boldsymbol{\Phi}$, whereas $\boldsymbol{\phi}(\mathbf{x}_n)$ corresponds to the nth row of $\boldsymbol{\Phi}$. If the number M of basis functions is smaller than the number N of data points, then the M vectors $\boldsymbol{\varphi}_j$ will span a linear subspace $\mathcal{S}$ of dimensionality M. We define $\mathbf{y}$ to be an N-dimensional vector whose n^{th} element is given by $y(\mathbf{x}_n, \mathbf{w})$, where $n = 1, \ldots, N$. Because $\mathbf{y}$ is an arbitrary linear combination of the vectors $\boldsymbol{\varphi}_j$, it can live anywhere in the M-dimensional subspace. The sum-of-squares error (3.12) is then equal (up to a factor of $1/2$) to the squared Euclidean distance between $\mathbf{y}$ and $\mathbf{t}$. Thus the least-squares solution for $\mathbf{w}$ corresponds to that choice of $\mathbf{y}$ that lies in subspace $\mathcal{S}$ and that is closest to $\mathbf{t}$. Intuitively, from Figure 3.2, we anticipate that this solution corresponds to the orthogonal projection of $\mathbf{t}$ onto the subspace $\mathcal{S}$. This is indeed the case, as can easily be verified by noting that the solution for $\mathbf{y}$ is given by $\boldsymbol{\Phi}\mathbf{w}_{\mathrm{ML}}$, and then confirming that this takes the form of an orthogonal projection.

Exercise 3.2

In practice, a direct solution of the normal equations can lead to numerical difficulties when $\boldsymbol{\Phi}^{\mathrm{T}}\boldsymbol{\Phi}$ is close to singular. In particular, when two or more of the basis vectors $\boldsymbol{\varphi}_j$ are co-linear, or nearly so, the resulting parameter values can have large magnitudes. Such near degeneracies will not be uncommon when dealing with real data sets. The resulting numerical difficulties can be addressed using the technique of *singular value decomposition*, or *SVD* (Press *et al.*, 1992; Bishop and Nabney, 2008). Note that the addition of a regularization term ensures that the matrix is non-singular, even in the presence of degeneracies.

3.1.3 Sequential learning

Batch techniques, such as the maximum likelihood solution (3.15), which involve processing the entire training set in one go, can be computationally costly for large data sets. As we have discussed in Chapter 1, if the data set is sufficiently large, it may be worthwhile to use *sequential* algorithms, also known as *on-line* algorithms, in which the data points are considered one at a time, and the model parameters updated after each such presentation. Sequential learning is also appropriate for real-time applications in which the data observations are arriving in a continuous stream, and predictions must be made before all of the data points are seen.

We can obtain a sequential learning algorithm by applying the technique of *stochastic gradient descent*, also known as *sequential gradient descent*, as follows. If the error function comprises a sum over data points $E = \sum_n E_n$, then after presentation of pattern n, the stochastic gradient descent algorithm updates the parameter vector $\mathbf{w}$ using

$$\mathbf{w}^{(\tau+1)} = \mathbf{w}^{(\tau)} - \eta \nabla E_n \tag{3.22}$$

where τ denotes the iteration number, and η is a learning rate parameter. We shall discuss the choice of value for η shortly. The value of $\mathbf{w}$ is initialized to some starting vector $\mathbf{w}^{(0)}$. For the case of the sum-of-squares error function (3.12), this gives

$$\mathbf{w}^{(\tau+1)} = \mathbf{w}^{(\tau)} + \eta(t_n - \mathbf{w}^{(\tau)\mathrm{T}}\boldsymbol{\phi}_n)\boldsymbol{\phi}_n \tag{3.23}$$

where $\boldsymbol{\phi}_n = \boldsymbol{\phi}(\mathbf{x}_n)$. This is known as *least-mean-squares* or the *LMS algorithm*. The value of η needs to be chosen with care to ensure that the algorithm converges (Bishop and Nabney, 2008).

3.1.4 Regularized least squares

In Section 1.1, we introduced the idea of adding a regularization term to an error function in order to control over-fitting, so that the total error function to be minimized takes the form

$$E_D(\mathbf{w}) + \lambda E_W(\mathbf{w}) \tag{3.24}$$

where λ is the regularization coefficient that controls the relative importance of the data-dependent error $E_D(\mathbf{w})$ and the regularization term $E_W(\mathbf{w})$. One of the simplest forms of regularizer is given by the sum-of-squares of the weight vector elements

$$E_W(\mathbf{w}) = \frac{1}{2}\mathbf{w}^{\mathrm{T}}\mathbf{w}. \tag{3.25}$$

If we also consider the sum-of-squares error function given by

$$E_D(\mathbf{w}) = \frac{1}{2}\sum_{n=1}^{N}\{t_n - \mathbf{w}^{\mathrm{T}}\boldsymbol{\phi}(\mathbf{x}_n)\}^2 \tag{3.26}$$

then the total error function becomes

$$\frac{1}{2}\sum_{n=1}^{N}\{t_n - \mathbf{w}^{\mathrm{T}}\boldsymbol{\phi}(\mathbf{x}_n)\}^2 + \frac{\lambda}{2}\mathbf{w}^{\mathrm{T}}\mathbf{w}. \tag{3.27}$$

This particular choice of regularizer is known in the machine learning literature as *weight decay* because in sequential learning algorithms, it encourages weight values to decay towards zero, unless supported by the data. In statistics, it provides an example of a *parameter shrinkage* method because it shrinks parameter values towards

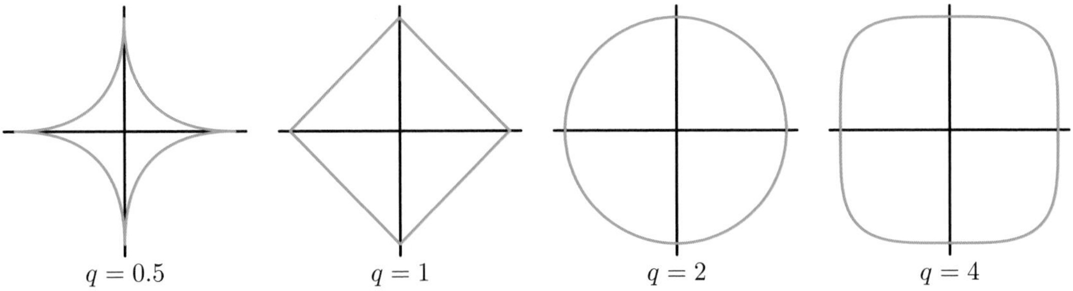

Figure 3.3 Contours of the regularization term in (3.29) for various values of the parameter q.

zero. It has the advantage that the error function remains a quadratic function of $\mathbf{w}$, and so its exact minimizer can be found in closed form. Specifically, setting the gradient of (3.27) with respect to $\mathbf{w}$ to zero, and solving for $\mathbf{w}$ as before, we obtain

$$\mathbf{w} = \left(\lambda\mathbf{I} + \boldsymbol{\Phi}^{\mathrm{T}}\boldsymbol{\Phi}\right)^{-1}\boldsymbol{\Phi}^{\mathrm{T}}\mathbf{t}. \tag{3.28}$$

This represents a simple extension of the least-squares solution (3.15).

A more general regularizer is sometimes used, for which the regularized error takes the form

$$\frac{1}{2}\sum_{n=1}^{N}\{t_n - \mathbf{w}^{\mathrm{T}}\boldsymbol{\phi}(\mathbf{x}_n)\}^2 + \frac{\lambda}{2}\sum_{j=1}^{M}|w_j|^q \tag{3.29}$$

where $q = 2$ corresponds to the quadratic regularizer (3.27). Figure 3.3 shows contours of the regularization function for different values of q.

The case of $q = 1$ is known as the *lasso* in the statistics literature (Tibshirani, 1996). It has the property that if λ is sufficiently large, some of the coefficients w_j are driven to zero, leading to a *sparse* model in which the corresponding basis functions play no role. To see this, we first note that minimizing (3.29) is equivalent to minimizing the unregularized sum-of-squares error (3.12) subject to the constraint

Exercise 3.5

$$\sum_{j=1}^{M}|w_j|^q \leqslant \eta \tag{3.30}$$

for an appropriate value of the parameter η, where the two approaches can be related using Lagrange multipliers. The origin of the sparsity can be seen from Figure 3.4, which shows the minimum of the error function, subject to the constraint (3.30). As λ is increased, so an increasing number of parameters are driven to zero.

Appendix E

Regularization allows complex models to be trained on data sets of limited size without severe over-fitting, essentially by limiting the effective model complexity. However, the problem of determining the optimal model complexity is then shifted from one of finding the appropriate number of basis functions to one of determining a suitable value of the regularization coefficient λ. We shall return to the issue of model complexity later in this chapter.

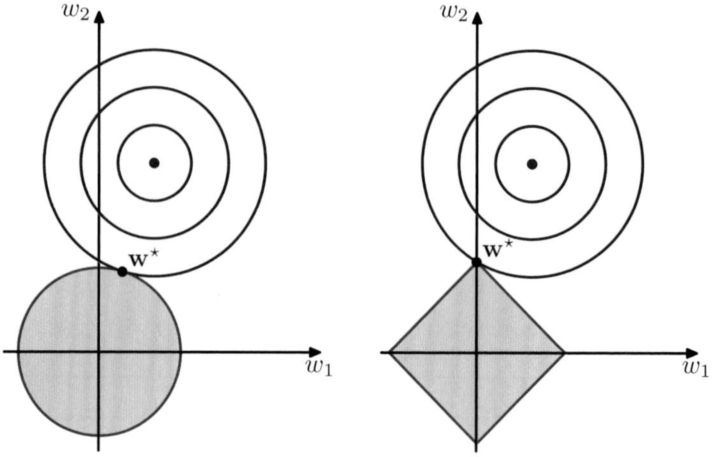

Figure 3.4 Plot of the contours of the unregularized error function (blue) along with the constraint region (3.30) for the quadratic regularizer $q = 2$ on the left and the lasso regularizer $q = 1$ on the right, in which the optimum value for the parameter vector $\mathbf{w}$ is denoted by $\mathbf{w}^\star$. The lasso gives a sparse solution in which $w_1^\star = 0$.

For the remainder of this chapter we shall focus on the quadratic regularizer (3.27) both for its practical importance and its analytical tractability.

3.1.5 Multiple outputs

So far, we have considered the case of a single target variable t. In some applications, we may wish to predict $K > 1$ target variables, which we denote collectively by the target vector $\mathbf{t}$. This could be done by introducing a different set of basis functions for each component of $\mathbf{t}$, leading to multiple, independent regression problems. However, a more interesting, and more common, approach is to use the same set of basis functions to model all of the components of the target vector so that

$$\mathbf{y}(\mathbf{x}, \mathbf{w}) = \mathbf{W}^{\mathrm{T}}\boldsymbol{\phi}(\mathbf{x}) \tag{3.31}$$

where $\mathbf{y}$ is a K-dimensional column vector, $\mathbf{W}$ is an $M \times K$ matrix of parameters, and $\boldsymbol{\phi}(\mathbf{x})$ is an M-dimensional column vector with elements $\phi_j(\mathbf{x})$, with $\phi_0(\mathbf{x}) = 1$ as before. Suppose we take the conditional distribution of the target vector to be an isotropic Gaussian of the form

$$p(\mathbf{t}|\mathbf{x}, \mathbf{W}, \beta) = \mathcal{N}(\mathbf{t}|\mathbf{W}^{\mathrm{T}}\boldsymbol{\phi}(\mathbf{x}), \beta^{-1}\mathbf{I}). \tag{3.32}$$

If we have a set of observations $\mathbf{t}_1, \ldots, \mathbf{t}_N$, we can combine these into a matrix $\mathbf{T}$ of size $N \times K$ such that the nth row is given by $\mathbf{t}_n^{\mathrm{T}}$. Similarly, we can combine the input vectors $\mathbf{x}_1, \ldots, \mathbf{x}_N$ into a matrix $\mathbf{X}$. The log likelihood function is then given by

$$\begin{aligned}\ln p(\mathbf{T}|\mathbf{X}, \mathbf{W}, \beta) &= \sum_{n=1}^{N} \ln \mathcal{N}(\mathbf{t}_n|\mathbf{W}^{\mathrm{T}}\boldsymbol{\phi}(\mathbf{x}_n), \beta^{-1}\mathbf{I}) \\ &= \frac{NK}{2}\ln\left(\frac{\beta}{2\pi}\right) - \frac{\beta}{2}\sum_{n=1}^{N}\left\|\mathbf{t}_n - \mathbf{W}^{\mathrm{T}}\boldsymbol{\phi}(\mathbf{x}_n)\right\|^2. \end{aligned} \tag{3.33}$$

As before, we can maximize this function with respect to $\mathbf{W}$, giving

$$\mathbf{W}_{\mathrm{ML}} = \left(\mathbf{\Phi}^{\mathrm{T}}\mathbf{\Phi}\right)^{-1}\mathbf{\Phi}^{\mathrm{T}}\mathbf{T}. \tag{3.34}$$

If we examine this result for each target variable t_k, we have

$$\mathbf{w}_k = \left(\mathbf{\Phi}^{\mathrm{T}}\mathbf{\Phi}\right)^{-1}\mathbf{\Phi}^{\mathrm{T}}\mathbf{t}_k = \mathbf{\Phi}^{\dagger}\mathbf{t}_k \tag{3.35}$$

where $\mathbf{t}_k$ is an N-dimensional column vector with components t_{nk} for $n = 1, \ldots N$. Thus the solution to the regression problem decouples between the different target variables, and we need only compute a single pseudo-inverse matrix $\mathbf{\Phi}^{\dagger}$, which is shared by all of the vectors $\mathbf{w}_k$.

Exercise 3.6

The extension to general Gaussian noise distributions having arbitrary covariance matrices is straightforward. Again, this leads to a decoupling into K independent regression problems. This result is unsurprising because the parameters $\mathbf{W}$ define only the mean of the Gaussian noise distribution, and we know from Section 2.3.4 that the maximum likelihood solution for the mean of a multivariate Gaussian is independent of the covariance. From now on, we shall therefore consider a single target variable t for simplicity.

3.2. The Bias-Variance Decomposition

So far in our discussion of linear models for regression, we have assumed that the form and number of basis functions are both fixed. As we have seen in Chapter 1, the use of maximum likelihood, or equivalently least squares, can lead to severe over-fitting if complex models are trained using data sets of limited size. However, limiting the number of basis functions in order to avoid over-fitting has the side effect of limiting the flexibility of the model to capture interesting and important trends in the data. Although the introduction of regularization terms can control over-fitting for models with many parameters, this raises the question of how to determine a suitable value for the regularization coefficient λ. Seeking the solution that minimizes the regularized error function with respect to both the weight vector $\mathbf{w}$ and the regularization coefficient λ is clearly not the right approach since this leads to the unregularized solution with $\lambda = 0$.

As we have seen in earlier chapters, the phenomenon of over-fitting is really an unfortunate property of maximum likelihood and does not arise when we marginalize over parameters in a Bayesian setting. In this chapter, we shall consider the Bayesian view of model complexity in some depth. Before doing so, however, it is instructive to consider a frequentist viewpoint of the model complexity issue, known as the *bias-variance* trade-off. Although we shall introduce this concept in the context of linear basis function models, where it is easy to illustrate the ideas using simple examples, the discussion has more general applicability.

In Section 1.5.5, when we discussed decision theory for regression problems, we considered various loss functions each of which leads to a corresponding optimal prediction once we are given the conditional distribution $p(t|\mathbf{x})$. A popular choice is the squared loss function, for which the optimal prediction is given by the conditional expectation, which we denote by $h(\mathbf{x})$ and which is given by

$$h(\mathbf{x}) = \mathbb{E}[t|\mathbf{x}] = \int tp(t|\mathbf{x}) \, \mathrm{d}t. \tag{3.36}$$

At this point, it is worth distinguishing between the squared loss function arising from decision theory and the sum-of-squares error function that arose in the maximum likelihood estimation of model parameters. We might use more sophisticated techniques than least squares, for example regularization or a fully Bayesian approach, to determine the conditional distribution $p(t|\mathbf{x})$. These can all be combined with the squared loss function for the purpose of making predictions.

We showed in Section 1.5.5 that the expected squared loss can be written in the form

$$\mathbb{E}[L] = \int \{y(\mathbf{x}) - h(\mathbf{x})\}^2 \, p(\mathbf{x}) \, \mathrm{d}\mathbf{x} + \iint \{h(\mathbf{x}) - t\}^2 p(\mathbf{x}, t) \, \mathrm{d}\mathbf{x} \, \mathrm{d}t. \tag{3.37}$$

Recall that the second term, which is independent of $y(\mathbf{x})$, arises from the intrinsic noise on the data and represents the minimum achievable value of the expected loss. The first term depends on our choice for the function $y(\mathbf{x})$, and we will seek a solution for $y(\mathbf{x})$ which makes this term a minimum. Because it is nonnegative, the smallest that we can hope to make this term is zero. If we had an unlimited supply of data (and unlimited computational resources), we could in principle find the regression function $h(\mathbf{x})$ to any desired degree of accuracy, and this would represent the optimal choice for $y(\mathbf{x})$. However, in practice we have a data set $\mathcal{D}$ containing only a finite number N of data points, and consequently we do not know the regression function $h(\mathbf{x})$ exactly.

If we model the $h(\mathbf{x})$ using a parametric function $y(\mathbf{x}, \mathbf{w})$ governed by a parameter vector $\mathbf{w}$, then from a Bayesian perspective the uncertainty in our model is expressed through a posterior distribution over $\mathbf{w}$. A frequentist treatment, however, involves making a point estimate of $\mathbf{w}$ based on the data set $\mathcal{D}$, and tries instead to interpret the uncertainty of this estimate through the following thought experiment. Suppose we had a large number of data sets each of size N and each drawn independently from the distribution $p(t, \mathbf{x})$. For any given data set $\mathcal{D}$, we can run our learning algorithm and obtain a prediction function $y(\mathbf{x}; \mathcal{D})$. Different data sets from the ensemble will give different functions and consequently different values of the squared loss. The performance of a particular learning algorithm is then assessed by taking the average over this ensemble of data sets.

Consider the integrand of the first term in (3.37), which for a particular data set $\mathcal{D}$ takes the form

$$\{y(\mathbf{x}; \mathcal{D}) - h(\mathbf{x})\}^2. \tag{3.38}$$

Because this quantity will be dependent on the particular data set $\mathcal{D}$, we take its average over the ensemble of data sets. If we add and subtract the quantity $\mathbb{E}_{\mathcal{D}}[y(\mathbf{x}; \mathcal{D})]$

inside the braces, and then expand, we obtain

$$\begin{aligned}
&\{y(\mathbf{x};\mathcal{D}) - \mathbb{E}_{\mathcal{D}}[y(\mathbf{x};\mathcal{D})] + \mathbb{E}_{\mathcal{D}}[y(\mathbf{x};\mathcal{D})] - h(\mathbf{x})\}^2 \\
&= \{y(\mathbf{x};\mathcal{D}) - \mathbb{E}_{\mathcal{D}}[y(\mathbf{x};\mathcal{D})]\}^2 + \{\mathbb{E}_{\mathcal{D}}[y(\mathbf{x};\mathcal{D})] - h(\mathbf{x})\}^2 \\
&\quad +2\{y(\mathbf{x};\mathcal{D}) - \mathbb{E}_{\mathcal{D}}[y(\mathbf{x};\mathcal{D})]\}\{\mathbb{E}_{\mathcal{D}}[y(\mathbf{x};\mathcal{D})] - h(\mathbf{x})\}.
\end{aligned} \tag{3.39}$$

We now take the expectation of this expression with respect to $\mathcal{D}$ and note that the final term will vanish, giving

$$\begin{aligned}
&\mathbb{E}_{\mathcal{D}}\left[\{y(\mathbf{x};\mathcal{D}) - h(\mathbf{x})\}^2\right] \\
&= \underbrace{\{\mathbb{E}_{\mathcal{D}}[y(\mathbf{x};\mathcal{D})] - h(\mathbf{x})\}^2}_{(\text{bias})^2} + \underbrace{\mathbb{E}_{\mathcal{D}}\left[\{y(\mathbf{x};\mathcal{D}) - \mathbb{E}_{\mathcal{D}}[y(\mathbf{x};\mathcal{D})]\}^2\right]}_{\text{variance}}.
\end{aligned} \tag{3.40}$$

We see that the expected squared difference between $y(\mathbf{x};\mathcal{D})$ and the regression function $h(\mathbf{x})$ can be expressed as the sum of two terms. The first term, called the squared *bias*, represents the extent to which the average prediction over all data sets differs from the desired regression function. The second term, called the *variance*, measures the extent to which the solutions for individual data sets vary around their average, and hence this measures the extent to which the function $y(\mathbf{x};\mathcal{D})$ is sensitive to the particular choice of data set. We shall provide some intuition to support these definitions shortly when we consider a simple example.

So far, we have considered a single input value $\mathbf{x}$. If we substitute this expansion back into (3.37), we obtain the following decomposition of the expected squared loss

$$\text{expected loss} = (\text{bias})^2 + \text{variance} + \text{noise} \tag{3.41}$$

where

$$(\text{bias})^2 = \int \{\mathbb{E}_{\mathcal{D}}[y(\mathbf{x};\mathcal{D})] - h(\mathbf{x})\}^2 p(\mathbf{x})\,\mathrm{d}\mathbf{x} \tag{3.42}$$

$$\text{variance} = \int \mathbb{E}_{\mathcal{D}}\left[\{y(\mathbf{x};\mathcal{D}) - \mathbb{E}_{\mathcal{D}}[y(\mathbf{x};\mathcal{D})]\}^2\right] p(\mathbf{x})\,\mathrm{d}\mathbf{x} \tag{3.43}$$

$$\text{noise} = \iint \{h(\mathbf{x}) - t\}^2 p(\mathbf{x},t)\,\mathrm{d}\mathbf{x}\,\mathrm{d}t \tag{3.44}$$

and the bias and variance terms now refer to integrated quantities.

Our goal is to minimize the expected loss, which we have decomposed into the sum of a (squared) bias, a variance, and a constant noise term. As we shall see, there is a trade-off between bias and variance, with very flexible models having low bias and high variance, and relatively rigid models having high bias and low variance. The model with the optimal predictive capability is the one that leads to the best balance between bias and variance. This is illustrated by considering the sinusoidal data set from Chapter 1. Here we generate 100 data sets, each containing $N = 25$ data points, independently from the sinusoidal curve $h(x) = \sin(2\pi x)$. The data sets are indexed by $l = 1, \ldots, L$, where $L = 100$, and for each data set $\mathcal{D}^{(l)}$ we

Appendix A

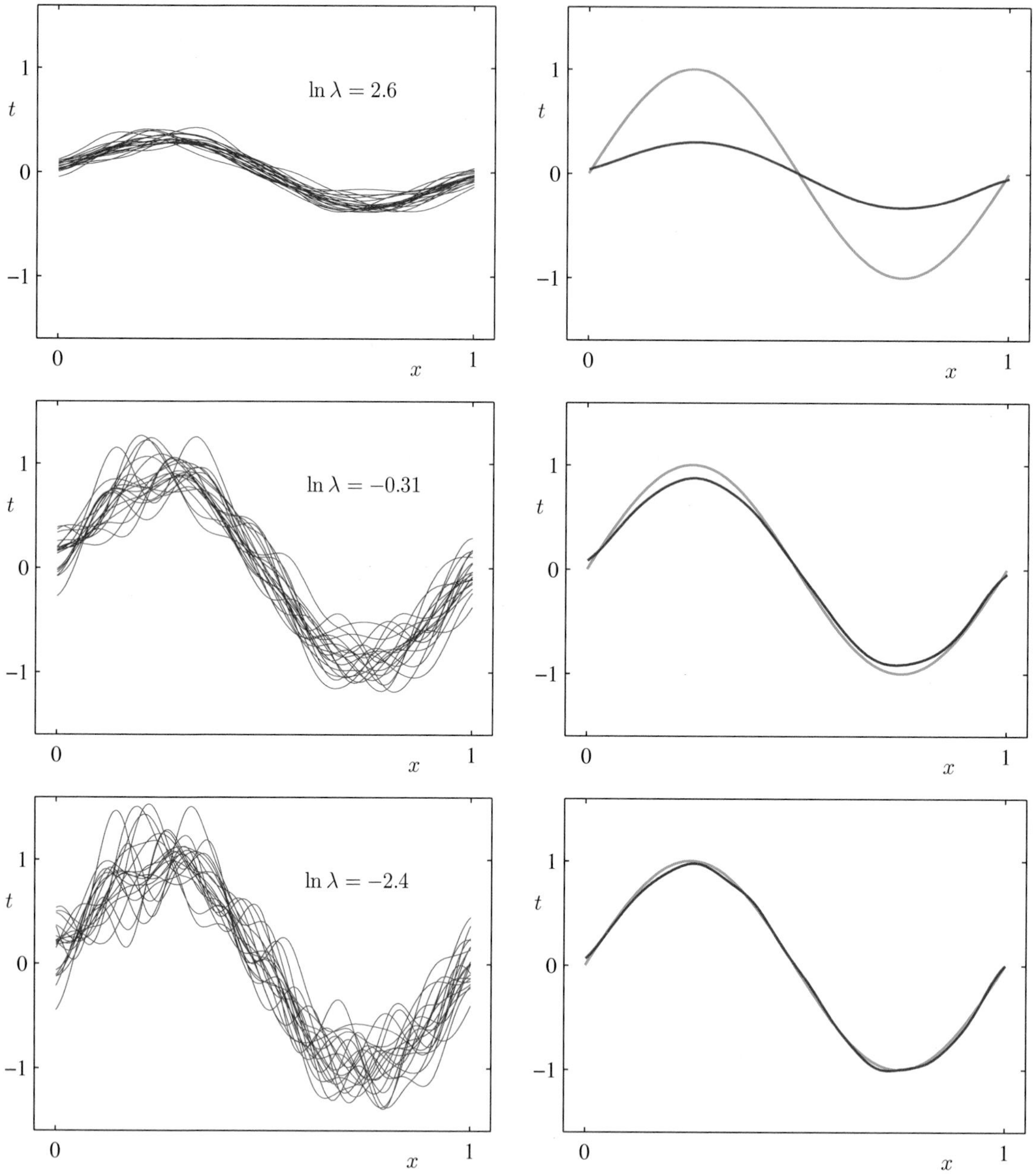

Figure 3.5 Illustration of the dependence of bias and variance on model complexity, governed by a regularization parameter λ, using the sinusoidal data set from Chapter 1. There are $L = 100$ data sets, each having $N = 25$ data points, and there are 24 Gaussian basis functions in the model so that the total number of parameters is $M = 25$ including the bias parameter. The left column shows the result of fitting the model to the data sets for various values of $\ln \lambda$ (for clarity, only 20 of the 100 fits are shown). The right column shows the corresponding average of the 100 fits (red) along with the sinusoidal function from which the data sets were generated (green).

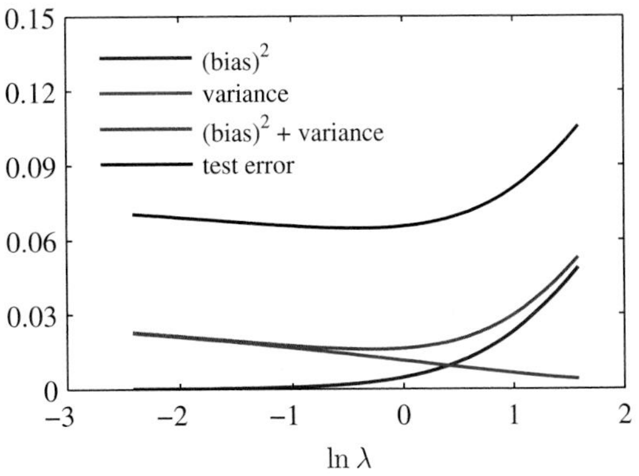

Figure 3.6 Plot of squared bias and variance, together with their sum, corresponding to the results shown in Figure 3.5. Also shown is the average test set error for a test data set size of 1000 points. The minimum value of $(\text{bias})^2 + \text{variance}$ occurs around $\ln \lambda = -0.31$, which is close to the value that gives the minimum error on the test data.

fit a model with 24 Gaussian basis functions by minimizing the regularized error function (3.27) to give a prediction function $y^{(l)}(x)$ as shown in Figure 3.5. The top row corresponds to a large value of the regularization coefficient λ that gives low variance (because the red curves in the left plot look similar) but high bias (because the two curves in the right plot are very different). Conversely on the bottom row, for which λ is small, there is large variance (shown by the high variability between the red curves in the left plot) but low bias (shown by the good fit between the average model fit and the original sinusoidal function). Note that the result of averaging many solutions for the complex model with $M = 25$ is a very good fit to the regression function, which suggests that averaging may be a beneficial procedure. Indeed, a weighted averaging of multiple solutions lies at the heart of a Bayesian approach, although the averaging is with respect to the posterior distribution of parameters, not with respect to multiple data sets.

We can also examine the bias-variance trade-off quantitatively for this example. The average prediction is estimated from

$$\overline{y}(x) = \frac{1}{L} \sum_{l=1}^{L} y^{(l)}(x) \tag{3.45}$$

and the integrated squared bias and integrated variance are then given by

$$(\text{bias})^2 = \frac{1}{N} \sum_{n=1}^{N} \{\overline{y}(x_n) - h(x_n)\}^2 \tag{3.46}$$

$$\text{variance} = \frac{1}{N} \sum_{n=1}^{N} \frac{1}{L} \sum_{l=1}^{L} \left\{ y^{(l)}(x_n) - \overline{y}(x_n) \right\}^2 \tag{3.47}$$

where the integral over x weighted by the distribution $p(x)$ is approximated by a finite sum over data points drawn from that distribution. These quantities, along with their sum, are plotted as a function of $\ln \lambda$ in Figure 3.6. We see that small values of λ allow the model to become finely tuned to the noise on each individual data set leading to large variance. Conversely, a large value of λ pulls the weight parameters towards zero leading to large bias.

Although the bias-variance decomposition may provide some interesting insights into the model complexity issue from a frequentist perspective, it is of limited practical value, because the bias-variance decomposition is based on averages with respect to ensembles of data sets, whereas in practice we have only the single observed data set. If we had a large number of independent training sets of a given size, we would be better off combining them into a single large training set, which of course would reduce the level of over-fitting for a given model complexity.

Given these limitations, we turn in the next section to a Bayesian treatment of linear basis function models, which not only provides powerful insights into the issues of over-fitting but which also leads to practical techniques for addressing the question of model complexity.

3.3. Bayesian Linear Regression

In our discussion of maximum likelihood for setting the parameters of a linear regression model, we have seen that the effective model complexity, governed by the number of basis functions, needs to be controlled according to the size of the data set. Adding a regularization term to the log likelihood function means the effective model complexity can then be controlled by the value of the regularization coefficient, although the choice of the number and form of the basis functions is of course still important in determining the overall behaviour of the model.

This leaves the issue of deciding the appropriate model complexity for the particular problem, which cannot be decided simply by maximizing the likelihood function, because this always leads to excessively complex models and over-fitting. Independent hold-out data can be used to determine model complexity, as discussed in Section 1.3, but this can be both computationally expensive and wasteful of valuable data. We therefore turn to a Bayesian treatment of linear regression, which will avoid the over-fitting problem of maximum likelihood, and which will also lead to automatic methods of determining model complexity using the training data alone. Again, for simplicity we will focus on the case of a single target variable t. Extension to multiple target variables is straightforward and follows the discussion of Section 3.1.5.

3.3.1 Parameter distribution

We begin our discussion of the Bayesian treatment of linear regression by introducing a prior probability distribution over the model parameters $\mathbf{w}$. For the moment, we shall treat the noise precision parameter β as a known constant. First note that the likelihood function $p(\mathbf{t}|\mathbf{w})$ defined by (3.10) is the exponential of a quadratic function of $\mathbf{w}$. The corresponding conjugate prior is therefore given by a Gaussian distribution of the form

$$p(\mathbf{w}) = \mathcal{N}(\mathbf{w}|\mathbf{m}_0, \mathbf{S}_0) \tag{3.48}$$

having mean $\mathbf{m}_0$ and covariance $\mathbf{S}_0$.

Next we compute the posterior distribution, which is proportional to the product of the likelihood function and the prior. Due to the choice of a conjugate Gaussian prior distribution, the posterior will also be Gaussian. We can evaluate this distribution by the usual procedure of completing the square in the exponential, and then finding the normalization coefficient using the standard result for a normalized Gaussian. However, we have already done the necessary work in deriving the general result (2.116), which allows us to write down the posterior distribution directly in the form

Exercise 3.7

$$p(\mathbf{w}|\mathbf{t}) = \mathcal{N}(\mathbf{w}|\mathbf{m}_N, \mathbf{S}_N) \tag{3.49}$$

where

$$\mathbf{m}_N = \mathbf{S}_N \left(\mathbf{S}_0^{-1}\mathbf{m}_0 + \beta\boldsymbol{\Phi}^{\mathrm{T}}\mathbf{t}\right) \tag{3.50}$$

$$\mathbf{S}_N^{-1} = \mathbf{S}_0^{-1} + \beta\boldsymbol{\Phi}^{\mathrm{T}}\boldsymbol{\Phi}. \tag{3.51}$$

Note that because the posterior distribution is Gaussian, its mode coincides with its mean. Thus the maximum posterior weight vector is simply given by $\mathbf{w}_{\mathrm{MAP}} = \mathbf{m}_N$. If we consider an infinitely broad prior $\mathbf{S}_0 = \alpha^{-1}\mathbf{I}$ with $\alpha \to 0$, the mean $\mathbf{m}_N$ of the posterior distribution reduces to the maximum likelihood value $\mathbf{w}_{\mathrm{ML}}$ given by (3.15). Similarly, if $N = 0$, then the posterior distribution reverts to the prior. Furthermore, if data points arrive sequentially, then the posterior distribution at any stage acts as the prior distribution for the subsequent data point, such that the new posterior distribution is again given by (3.49).

Exercise 3.8

For the remainder of this chapter, we shall consider a particular form of Gaussian prior in order to simplify the treatment. Specifically, we consider a zero-mean isotropic Gaussian governed by a single precision parameter α so that

$$p(\mathbf{w}|\alpha) = \mathcal{N}(\mathbf{w}|\mathbf{0}, \alpha^{-1}\mathbf{I}) \tag{3.52}$$

and the corresponding posterior distribution over $\mathbf{w}$ is then given by (3.49) with

$$\mathbf{m}_N = \beta\mathbf{S}_N\boldsymbol{\Phi}^{\mathrm{T}}\mathbf{t} \tag{3.53}$$

$$\mathbf{S}_N^{-1} = \alpha\mathbf{I} + \beta\boldsymbol{\Phi}^{\mathrm{T}}\boldsymbol{\Phi}. \tag{3.54}$$

The log of the posterior distribution is given by the sum of the log likelihood and the log of the prior and, as a function of $\mathbf{w}$, takes the form

$$\ln p(\mathbf{w}|\mathbf{t}) = -\frac{\beta}{2}\sum_{n=1}^{N}\{t_n - \mathbf{w}^{\mathrm{T}}\boldsymbol{\phi}(\mathbf{x}_n)\}^2 - \frac{\alpha}{2}\mathbf{w}^{\mathrm{T}}\mathbf{w} + \mathrm{const}. \tag{3.55}$$

Maximization of this posterior distribution with respect to $\mathbf{w}$ is therefore equivalent to the minimization of the sum-of-squares error function with the addition of a quadratic regularization term, corresponding to (3.27) with $\lambda = \alpha/\beta$.

We can illustrate Bayesian learning in a linear basis function model, as well as the sequential update of a posterior distribution, using a simple example involving straight-line fitting. Consider a single input variable x, a single target variable t and a linear model of the form $y(x, \mathbf{w}) = w_0 + w_1 x$. Because this has just two adaptive parameters, we can plot the prior and posterior distributions directly in parameter space. We generate synthetic data from the function $f(x, \mathbf{a}) = a_0 + a_1 x$ with parameter values $a_0 = -0.3$ and $a_1 = 0.5$ by first choosing values of x_n from the uniform distribution $\mathrm{U}(x|-1, 1)$, then evaluating $f(x_n, \mathbf{a})$, and finally adding Gaussian noise with standard deviation of 0.2 to obtain the target values t_n. Our goal is to recover the values of a_0 and a_1 from such data, and we will explore the dependence on the size of the data set. We assume here that the noise variance is known and hence we set the precision parameter to its true value $\beta = (1/0.2)^2 = 25$. Similarly, we fix the parameter α to 2.0. We shall shortly discuss strategies for determining α and β from the training data. Figure 3.7 shows the results of Bayesian learning in this model as the size of the data set is increased and demonstrates the sequential nature of Bayesian learning in which the current posterior distribution forms the prior when a new data point is observed. It is worth taking time to study this figure in detail as it illustrates several important aspects of Bayesian inference. The first row of this figure corresponds to the situation before any data points are observed and shows a plot of the prior distribution in $\mathbf{w}$ space together with six samples of the function $y(x, \mathbf{w})$ in which the values of $\mathbf{w}$ are drawn from the prior. In the second row, we see the situation after observing a single data point. The location (x, t) of the data point is shown by a blue circle in the right-hand column. In the left-hand column is a plot of the likelihood function $p(t|x, \mathbf{w})$ for this data point as a function of $\mathbf{w}$. Note that the likelihood function provides a soft constraint that the line must pass close to the data point, where close is determined by the noise precision β. For comparison, the true parameter values $a_0 = -0.3$ and $a_1 = 0.5$ used to generate the data set are shown by a white cross in the plots in the left column of Figure 3.7. When we multiply this likelihood function by the prior from the top row, and normalize, we obtain the posterior distribution shown in the middle plot on the second row. Samples of the regression function $y(x, \mathbf{w})$ obtained by drawing samples of $\mathbf{w}$ from this posterior distribution are shown in the right-hand plot. Note that these sample lines all pass close to the data point. The third row of this figure shows the effect of observing a second data point, again shown by a blue circle in the plot in the right-hand column. The corresponding likelihood function for this second data point alone is shown in the left plot. When we multiply this likelihood function by the posterior distribution from the second row, we obtain the posterior distribution shown in the middle plot of the third row. Note that this is exactly the same posterior distribution as would be obtained by combining the original prior with the likelihood function for the two data points. This posterior has now been influenced by two data points, and because two points are sufficient to define a line this already gives a relatively compact posterior distribution. Samples from this posterior distribution give rise to the functions shown in red in the third column, and we see that these functions pass close to both of the data points. The fourth row shows the effect of observing a total of 20 data points. The left-hand plot shows the likelihood function for the 20th data point alone, and the middle plot shows the resulting posterior distribution that has now absorbed information from all 20 observations. Note how the posterior is much sharper than in the third row. In the limit of an infinite number of data points, the

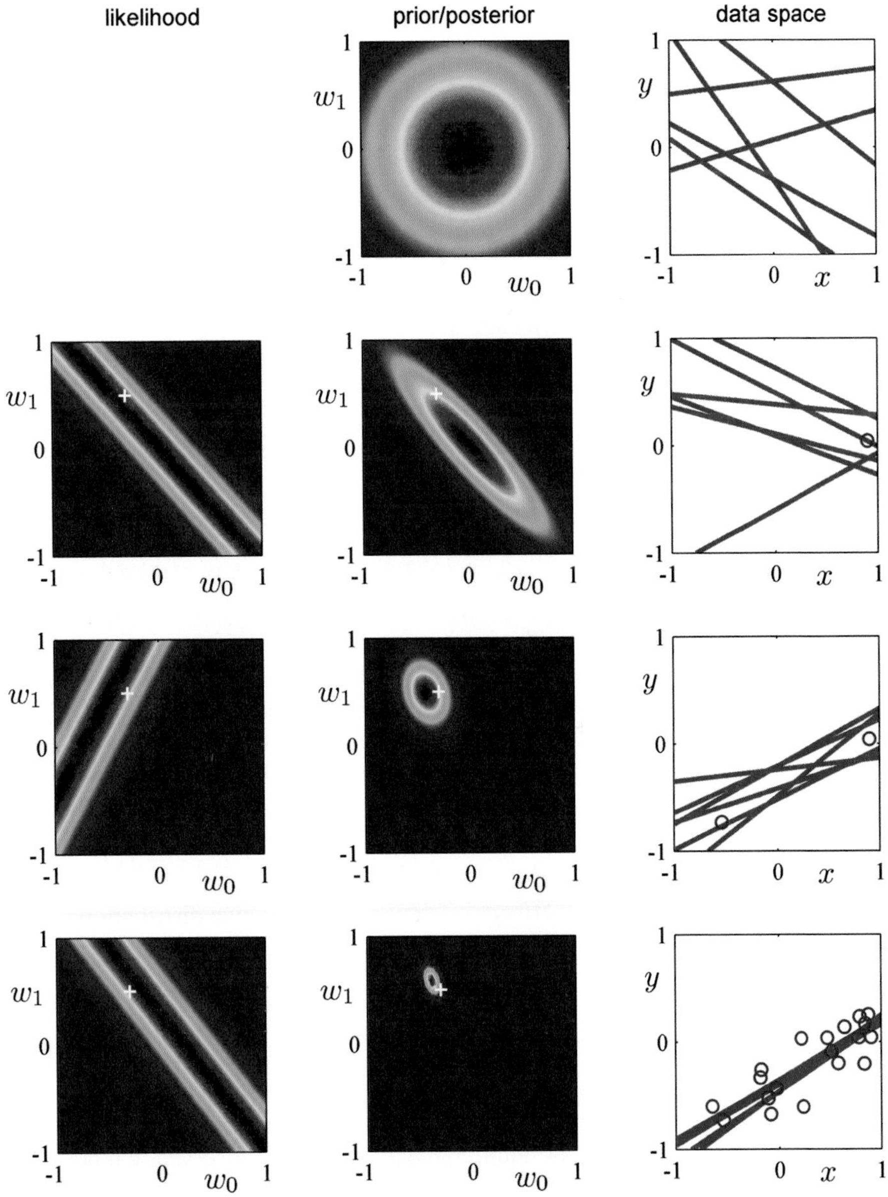

Figure 3.7 Illustration of sequential Bayesian learning for a simple linear model of the form $y(x, \mathbf{w}) = w_0 + w_1 x$. A detailed description of this figure is given in the text.

posterior distribution would become a delta function centred on the true parameter values, shown by the white cross.

Other forms of prior over the parameters can be considered. For instance, we can generalize the Gaussian prior to give

$$p(\mathbf{w}|\alpha) = \left[\frac{q}{2}\left(\frac{\alpha}{2}\right)^{1/q}\frac{1}{\Gamma(1/q)}\right]^M \exp\left(-\frac{\alpha}{2}\sum_{j=0}^{M-1}|w_j|^q\right) \tag{3.56}$$

in which $q = 2$ corresponds to the Gaussian distribution, and only in this case is the prior conjugate to the likelihood function (3.10). Finding the maximum of the posterior distribution over $\mathbf{w}$ corresponds to minimization of the regularized error function (3.29). In the case of the Gaussian prior, the mode of the posterior distribution was equal to the mean, although this will no longer hold if $q \neq 2$.

3.3.2 Predictive distribution

In practice, we are not usually interested in the value of $\mathbf{w}$ itself but rather in making predictions of t for new values of $\mathbf{x}$. This requires that we evaluate the *predictive distribution* defined by

$$p(t|\mathbf{t}, \alpha, \beta) = \int p(t|\mathbf{w}, \beta)p(\mathbf{w}|\mathbf{t}, \alpha, \beta)\,\mathrm{d}\mathbf{w} \tag{3.57}$$

in which $\mathbf{t}$ is the vector of target values from the training set, and we have omitted the corresponding input vectors from the right-hand side of the conditioning statements to simplify the notation. The conditional distribution $p(t|\mathbf{x}, \mathbf{w}, \beta)$ of the target variable is given by (3.8), and the posterior weight distribution is given by (3.49). We see that (3.57) involves the convolution of two Gaussian distributions, and so making use of the result (2.115) from Section 2.3.3, we see that the predictive distribution takes the form

Exercise 3.10

$$p(t|\mathbf{x}, \mathbf{t}, \alpha, \beta) = \mathcal{N}(t|\mathbf{m}_N^{\mathrm{T}}\boldsymbol{\phi}(\mathbf{x}), \sigma_N^2(\mathbf{x})) \tag{3.58}$$

where the variance $\sigma_N^2(\mathbf{x})$ of the predictive distribution is given by

$$\sigma_N^2(\mathbf{x}) = \frac{1}{\beta} + \boldsymbol{\phi}(\mathbf{x})^{\mathrm{T}}\mathbf{S}_N\boldsymbol{\phi}(\mathbf{x}). \tag{3.59}$$

The first term in (3.59) represents the noise on the data whereas the second term reflects the uncertainty associated with the parameters $\mathbf{w}$. Because the noise process and the distribution of $\mathbf{w}$ are independent Gaussians, their variances are additive. Note that, as additional data points are observed, the posterior distribution becomes narrower. As a consequence it can be shown (Qazaz *et al.*, 1997) that $\sigma_{N+1}^2(\mathbf{x}) \leqslant \sigma_N^2(\mathbf{x})$. In the limit $N \to \infty$, the second term in (3.59) goes to zero, and the variance of the predictive distribution arises solely from the additive noise governed by the parameter β.

Exercise 3.11

As an illustration of the predictive distribution for Bayesian linear regression models, let us return to the synthetic sinusoidal data set of Section 1.1. In Figure 3.8,

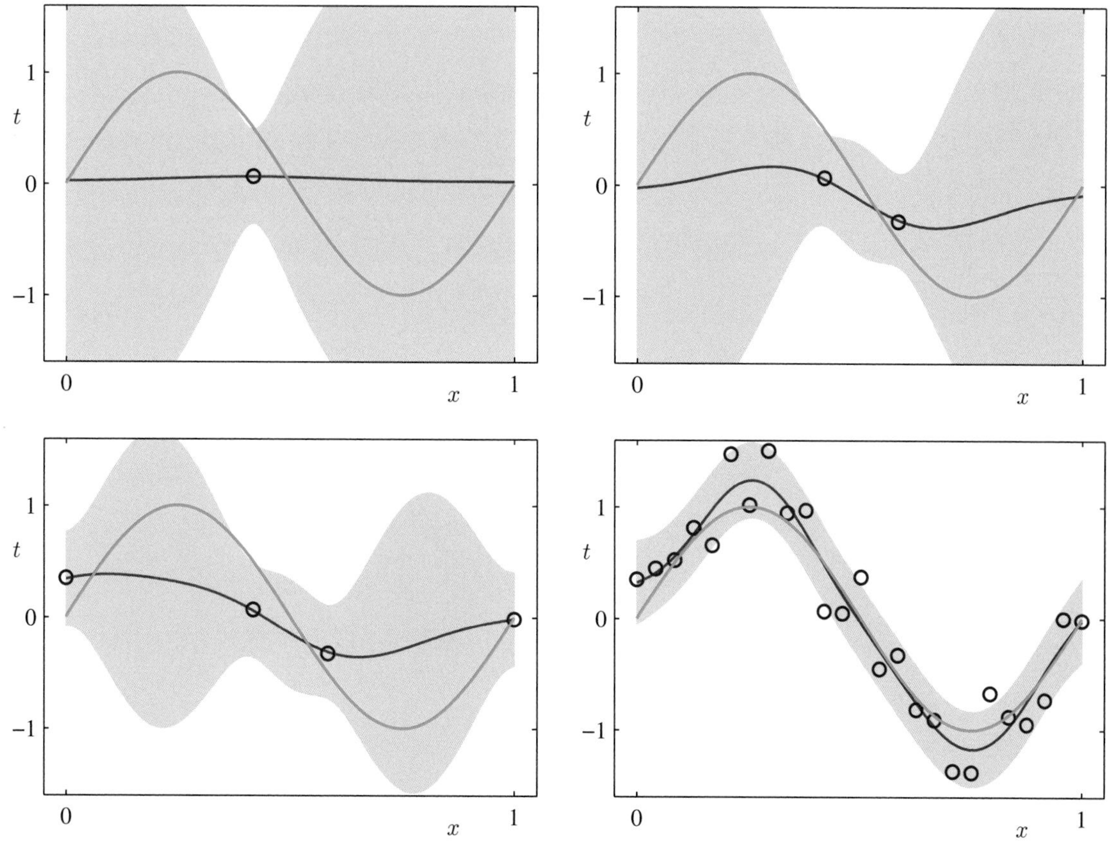

Figure 3.8 Examples of the predictive distribution (3.58) for a model consisting of 9 Gaussian basis functions of the form (3.4) using the synthetic sinusoidal data set of Section 1.1. See the text for a detailed discussion.

we fit a model comprising a linear combination of Gaussian basis functions to data sets of various sizes and then look at the corresponding posterior distributions. Here the green curves correspond to the function $\sin(2\pi x)$ from which the data points were generated (with the addition of Gaussian noise). Data sets of size $N = 1$, $N = 2$, $N = 4$, and $N = 25$ are shown in the four plots by the blue circles. For each plot, the red curve shows the mean of the corresponding Gaussian predictive distribution, and the red shaded region spans one standard deviation either side of the mean. Note that the predictive uncertainty depends on x and is smallest in the neighbourhood of the data points. Also note that the level of uncertainty decreases as more data points are observed.

The plots in Figure 3.8 only show the point-wise predictive variance as a function of x. In order to gain insight into the covariance between the predictions at different values of x, we can draw samples from the posterior distribution over $\mathbf{w}$, and then plot the corresponding functions $y(x, \mathbf{w})$, as shown in Figure 3.9.

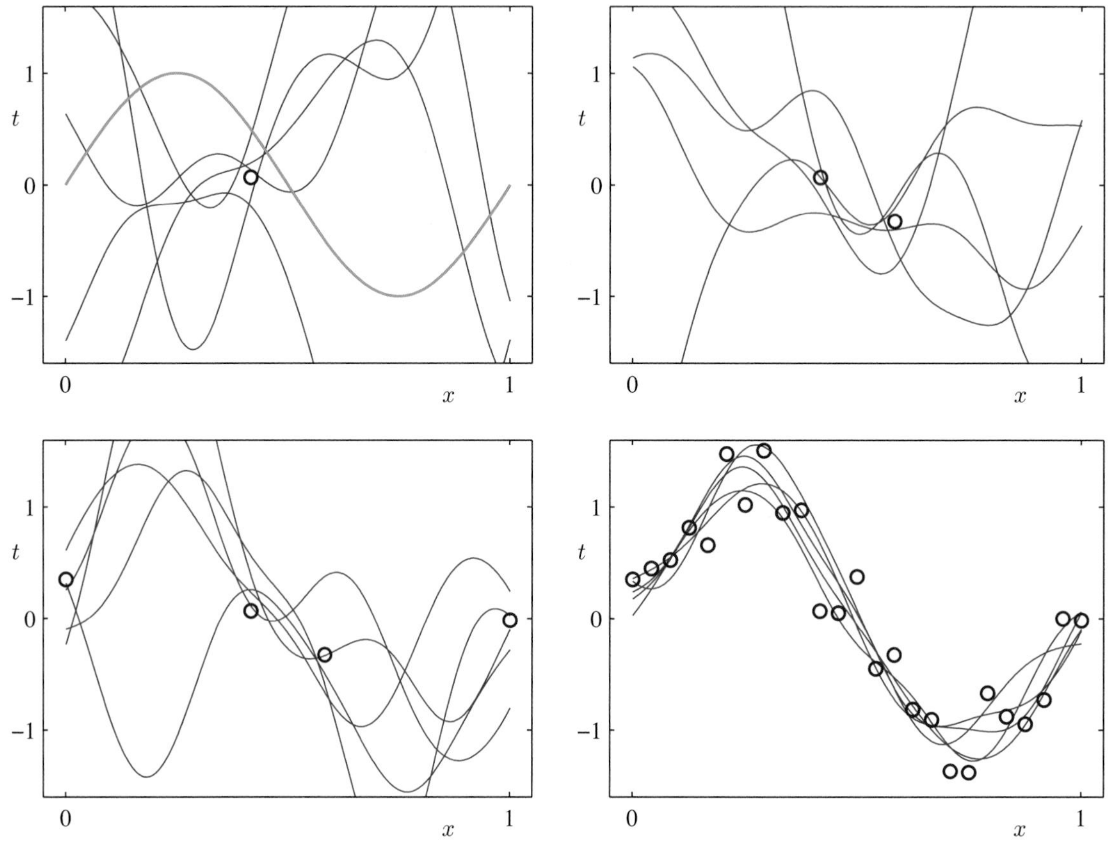

Figure 3.9 Plots of the function $y(x, \mathbf{w})$ using samples from the posterior distributions over $\mathbf{w}$ corresponding to the plots in Figure 3.8.

If we used localized basis functions such as Gaussians, then in regions away from the basis function centres, the contribution from the second term in the predictive variance (3.59) will go to zero, leaving only the noise contribution β^{-1}. Thus, the model becomes very confident in its predictions when extrapolating outside the region occupied by the basis functions, which is generally an undesirable behaviour. This problem can be avoided by adopting an alternative Bayesian approach to regression known as a Gaussian process.

Section 6.4

Note that, if both $\mathbf{w}$ and β are treated as unknown, then we can introduce a conjugate prior distribution $p(\mathbf{w}, \beta)$ that, from the discussion in Section 2.3.6, will be given by a Gaussian-gamma distribution (Denison *et al.*, 2002). In this case, the predictive distribution is a Student's t-distribution.

Exercise 3.12
Exercise 3.13

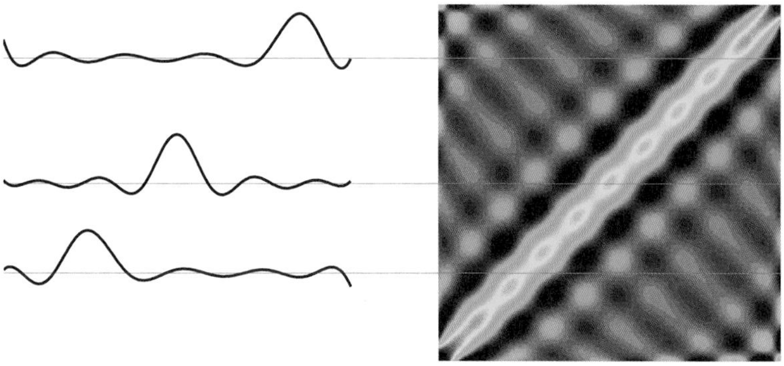

Figure 3.10 The equivalent kernel $k(x, x')$ for the Gaussian basis functions in Figure 3.1, shown as a plot of x versus x', together with three slices through this matrix corresponding to three different values of x. The data set used to generate this kernel comprised 200 values of x equally spaced over the interval $(-1, 1)$.

3.3.3 Equivalent kernel

The posterior mean solution (3.53) for the linear basis function model has an interesting interpretation that will set the stage for kernel methods, including Gaussian processes. If we substitute (3.53) into the expression (3.3), we see that the predictive mean can be written in the form

Chapter 6

$$y(\mathbf{x}, \mathbf{m}_N) = \mathbf{m}_N^{\mathrm{T}} \boldsymbol{\phi}(\mathbf{x}) = \beta \boldsymbol{\phi}(\mathbf{x})^{\mathrm{T}} \mathbf{S}_N \boldsymbol{\Phi}^{\mathrm{T}} \mathbf{t} = \sum_{n=1}^{N} \beta \boldsymbol{\phi}(\mathbf{x})^{\mathrm{T}} \mathbf{S}_N \boldsymbol{\phi}(\mathbf{x}_n) t_n \tag{3.60}$$

where $\mathbf{S}_N$ is defined by (3.51). Thus the mean of the predictive distribution at a point $\mathbf{x}$ is given by a linear combination of the training set target variables t_n, so that we can write

$$y(\mathbf{x}, \mathbf{m}_N) = \sum_{n=1}^{N} k(\mathbf{x}, \mathbf{x}_n) t_n \tag{3.61}$$

where the function

$$k(\mathbf{x}, \mathbf{x}') = \beta \boldsymbol{\phi}(\mathbf{x})^{\mathrm{T}} \mathbf{S}_N \boldsymbol{\phi}(\mathbf{x}') \tag{3.62}$$

is known as the *smoother matrix* or the *equivalent kernel*. Regression functions, such as this, which make predictions by taking linear combinations of the training set target values are known as *linear smoothers*. Note that the equivalent kernel depends on the input values $\mathbf{x}_n$ from the data set because these appear in the definition of $\mathbf{S}_N$. The equivalent kernel is illustrated for the case of Gaussian basis functions in Figure 3.10 in which the kernel functions $k(x, x')$ have been plotted as a function of x' for three different values of x. We see that they are localized around x, and so the mean of the predictive distribution at x, given by $y(x, \mathbf{m}_N)$, is obtained by forming a weighted combination of the target values in which data points close to x are given higher weight than points further removed from x. Intuitively, it seems reasonable that we should weight local evidence more strongly than distant evidence. Note that this localization property holds not only for the localized Gaussian basis functions but also for the nonlocal polynomial and sigmoidal basis functions, as illustrated in Figure 3.11.

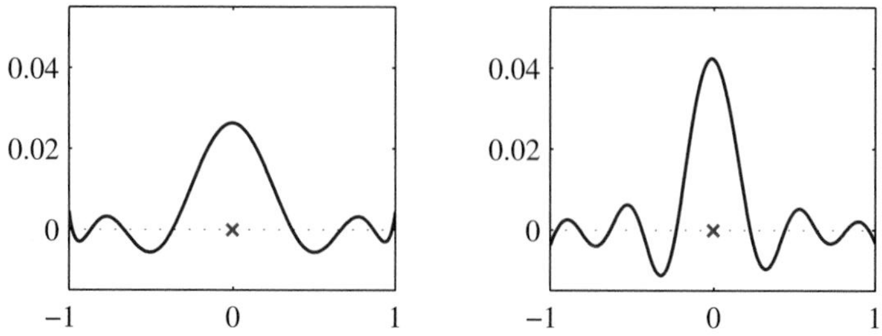

Figure 3.11 Examples of equivalent kernels $k(x, x')$ for $x = 0$ plotted as a function of x', corresponding (left) to the polynomial basis functions and (right) to the sigmoidal basis functions shown in Figure 3.1. Note that these are localized functions of x' even though the corresponding basis functions are nonlocal.

Further insight into the role of the equivalent kernel can be obtained by considering the covariance between $y(\mathbf{x})$ and $y(\mathbf{x}')$, which is given by

$$\begin{aligned} \text{cov}[y(\mathbf{x}), y(\mathbf{x}')] &= \text{cov}[\boldsymbol{\phi}(\mathbf{x})^{\mathrm{T}}\mathbf{w}, \mathbf{w}^{\mathrm{T}}\boldsymbol{\phi}(\mathbf{x}')] \\ &= \boldsymbol{\phi}(\mathbf{x})^{\mathrm{T}}\mathbf{S}_N\boldsymbol{\phi}(\mathbf{x}') = \beta^{-1}k(\mathbf{x}, \mathbf{x}') \end{aligned} \tag{3.63}$$

where we have made use of (3.49) and (3.62). From the form of the equivalent kernel, we see that the predictive mean at nearby points will be highly correlated, whereas for more distant pairs of points the correlation will be smaller.

The predictive distribution shown in Figure 3.8 allows us to visualize the pointwise uncertainty in the predictions, governed by (3.59). However, by drawing samples from the posterior distribution over $\mathbf{w}$, and plotting the corresponding model functions $y(\mathbf{x}, \mathbf{w})$ as in Figure 3.9, we are visualizing the joint uncertainty in the posterior distribution between the y values at two (or more) x values, as governed by the equivalent kernel.

The formulation of linear regression in terms of a kernel function suggests an alternative approach to regression as follows. Instead of introducing a set of basis functions, which implicitly determines an equivalent kernel, we can instead define a localized kernel directly and use this to make predictions for new input vectors $\mathbf{x}$, given the observed training set. This leads to a practical framework for regression (and classification) called *Gaussian processes*, which will be discussed in detail in Section 6.4.

We have seen that the equivalent kernel defines the weights by which the training set target values are combined in order to make a prediction at a new value of $\mathbf{x}$, and it can be shown that these weights sum to one, in other words

$$\sum_{n=1}^{N} k(\mathbf{x}, \mathbf{x}_n) = 1 \tag{3.64}$$

Exercise 3.14

for all values of $\mathbf{x}$. This intuitively pleasing result can easily be proven informally by noting that the summation is equivalent to considering the predictive mean $\widehat{y}(\mathbf{x})$ for a set of target data in which $t_n = 1$ for all n. Provided the basis functions are linearly independent, that there are more data points than basis functions, and that one of the basis functions is constant (corresponding to the bias parameter), then it is clear that we can fit the training data exactly and hence that the predictive mean will be simply $\widehat{y}(\mathbf{x}) = 1$, from which we obtain (3.64). Note that the kernel function can be negative as well as positive, so although it satisfies a summation constraint, the corresponding predictions are not necessarily convex combinations of the training set target variables.

Chapter 6

Finally, we note that the equivalent kernel (3.62) satisfies an important property shared by kernel functions in general, namely that it can be expressed in the form of an inner product with respect to a vector $\psi(\mathbf{x})$ of nonlinear functions, so that

$$k(\mathbf{x}, \mathbf{z}) = \psi(\mathbf{x})^{\mathrm{T}}\psi(\mathbf{z}) \tag{3.65}$$

where $\psi(\mathbf{x}) = \beta^{1/2}\mathbf{S}_N^{1/2}\phi(\mathbf{x})$.

3.4. Bayesian Model Comparison

In Chapter 1, we highlighted the problem of over-fitting as well as the use of cross-validation as a technique for setting the values of regularization parameters or for choosing between alternative models. Here we consider the problem of model selection from a Bayesian perspective. In this section, our discussion will be very general, and then in Section 3.5 we shall see how these ideas can be applied to the determination of regularization parameters in linear regression.

As we shall see, the over-fitting associated with maximum likelihood can be avoided by marginalizing (summing or integrating) over the model parameters instead of making point estimates of their values. Models can then be compared directly on the training data, without the need for a validation set. This allows all available data to be used for training and avoids the multiple training runs for each model associated with cross-validation. It also allows multiple complexity parameters to be determined simultaneously as part of the training process. For example, in Chapter 7 we shall introduce the *relevance vector machine*, which is a Bayesian model having one complexity parameter for every training data point.

The Bayesian view of model comparison simply involves the use of probabilities to represent uncertainty in the choice of model, along with a consistent application of the sum and product rules of probability. Suppose we wish to compare a set of L models $\{\mathcal{M}_i\}$ where $i = 1, \ldots, L$. Here a model refers to a probability distribution over the observed data $\mathcal{D}$. In the case of the polynomial curve-fitting problem, the distribution is defined over the set of target values $\mathbf{t}$, while the set of input values $\mathbf{X}$ is assumed to be known. Other types of model define a joint distributions over $\mathbf{X}$ and $\mathbf{t}$. We shall suppose that the data is generated from one of these models but we are uncertain which one. Our uncertainty is expressed through a prior probability distribution $p(\mathcal{M}_i)$. Given a training set $\mathcal{D}$, we then wish to evaluate the posterior distribution

Section 1.5.4

$$p(\mathcal{M}_i|\mathcal{D}) \propto p(\mathcal{M}_i)p(\mathcal{D}|\mathcal{M}_i). \tag{3.66}$$

The prior allows us to express a preference for different models. Let us simply assume that all models are given equal prior probability. The interesting term is the *model evidence* $p(\mathcal{D}|\mathcal{M}_i)$ which expresses the preference shown by the data for different models, and we shall examine this term in more detail shortly. The model evidence is sometimes also called the *marginal likelihood* because it can be viewed as a likelihood function over the space of models, in which the parameters have been marginalized out. The ratio of model evidences $p(\mathcal{D}|\mathcal{M}_i)/p(\mathcal{D}|\mathcal{M}_j)$ for two models is known as a *Bayes factor* (Kass and Raftery, 1995).

Once we know the posterior distribution over models, the predictive distribution is given, from the sum and product rules, by

$$p(t|\mathbf{x}, \mathcal{D}) = \sum_{i=1}^{L} p(t|\mathbf{x}, \mathcal{M}_i, \mathcal{D})p(\mathcal{M}_i|\mathcal{D}). \tag{3.67}$$

This is an example of a *mixture distribution* in which the overall predictive distribution is obtained by averaging the predictive distributions $p(t|\mathbf{x}, \mathcal{M}_i, \mathcal{D})$ of individual models, weighted by the posterior probabilities $p(\mathcal{M}_i|\mathcal{D})$ of those models. For instance, if we have two models that are a-posteriori equally likely and one predicts a narrow distribution around $t = a$ while the other predicts a narrow distribution around $t = b$, the overall predictive distribution will be a bimodal distribution with modes at $t = a$ and $t = b$, not a single model at $t = (a + b)/2$.

A simple approximation to model averaging is to use the single most probable model alone to make predictions. This is known as *model selection*.

For a model governed by a set of parameters $\mathbf{w}$, the model evidence is given, from the sum and product rules of probability, by

$$p(\mathcal{D}|\mathcal{M}_i) = \int p(\mathcal{D}|\mathbf{w}, \mathcal{M}_i)p(\mathbf{w}|\mathcal{M}_i)\,\mathrm{d}\mathbf{w}. \tag{3.68}$$

Chapter 11

From a sampling perspective, the marginal likelihood can be viewed as the probability of generating the data set $\mathcal{D}$ from a model whose parameters are sampled at random from the prior. It is also interesting to note that the evidence is precisely the normalizing term that appears in the denominator in Bayes' theorem when evaluating the posterior distribution over parameters because

$$p(\mathbf{w}|\mathcal{D}, \mathcal{M}_i) = \frac{p(\mathcal{D}|\mathbf{w}, \mathcal{M}_i)p(\mathbf{w}|\mathcal{M}_i)}{p(\mathcal{D}|\mathcal{M}_i)}. \tag{3.69}$$

We can obtain some insight into the model evidence by making a simple approximation to the integral over parameters. Consider first the case of a model having a single parameter w. The posterior distribution over parameters is proportional to $p(\mathcal{D}|w)p(w)$, where we omit the dependence on the model $\mathcal{M}_i$ to keep the notation uncluttered. If we assume that the posterior distribution is sharply peaked around the most probable value w_{MAP}, with width $\Delta w_{\text{posterior}}$, then we can approximate the integral by the value of the integrand at its maximum times the width of the peak. If we further assume that the prior is flat with width Δw_{prior} so that $p(w) = 1/\Delta w_{\text{prior}}$, then we have

$$p(\mathcal{D}) = \int p(\mathcal{D}|w)p(w)\,\mathrm{d}w \simeq p(\mathcal{D}|w_{\text{MAP}})\frac{\Delta w_{\text{posterior}}}{\Delta w_{\text{prior}}} \tag{3.70}$$

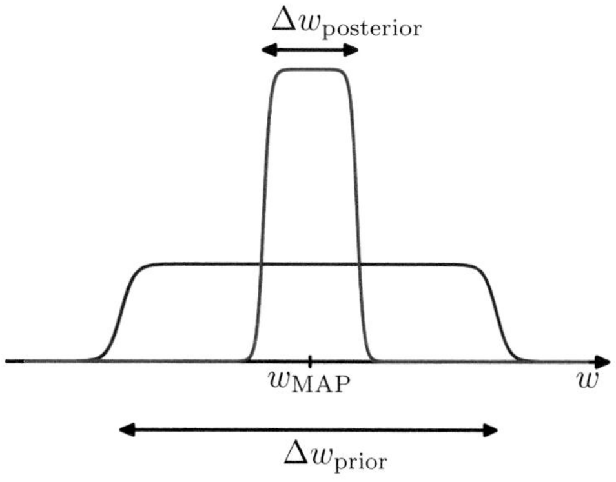

Figure 3.12 We can obtain a rough approximation to the model evidence if we assume that the posterior distribution over parameters is sharply peaked around its mode w_{MAP}.

and so taking logs we obtain

$$\ln p(\mathcal{D}) \simeq \ln p(\mathcal{D}|w_{\text{MAP}}) + \ln \left(\frac{\Delta w_{\text{posterior}}}{\Delta w_{\text{prior}}} \right). \tag{3.71}$$

This approximation is illustrated in Figure 3.12. The first term represents the fit to the data given by the most probable parameter values, and for a flat prior this would correspond to the log likelihood. The second term penalizes the model according to its complexity. Because $\Delta w_{\text{posterior}} < \Delta w_{\text{prior}}$ this term is negative, and it increases in magnitude as the ratio $\Delta w_{\text{posterior}}/\Delta w_{\text{prior}}$ gets smaller. Thus, if parameters are finely tuned to the data in the posterior distribution, then the penalty term is large.

For a model having a set of M parameters, we can make a similar approximation for each parameter in turn. Assuming that all parameters have the same ratio of $\Delta w_{\text{posterior}}/\Delta w_{\text{prior}}$, we obtain

$$\ln p(\mathcal{D}) \simeq \ln p(\mathcal{D}|\mathbf{w}_{\text{MAP}}) + M \ln \left(\frac{\Delta w_{\text{posterior}}}{\Delta w_{\text{prior}}} \right). \tag{3.72}$$

Thus, in this very simple approximation, the size of the complexity penalty increases linearly with the number M of adaptive parameters in the model. As we increase the complexity of the model, the first term will typically increase, because a more complex model is better able to fit the data, whereas the second term will decrease due to the dependence on M. The optimal model complexity, as determined by the maximum evidence, will be given by a trade-off between these two competing terms. We shall later develop a more refined version of this approximation, based on a Gaussian approximation to the posterior distribution.

Section 4.4.1

We can gain further insight into Bayesian model comparison and understand how the marginal likelihood can favour models of intermediate complexity by considering Figure 3.13. Here the horizontal axis is a one-dimensional representation of the space of possible data sets, so that each point on this axis corresponds to a specific data set. We now consider three models $\mathcal{M}_1$, $\mathcal{M}_2$ and $\mathcal{M}_3$ of successively increasing complexity. Imagine running these models generatively to produce example data sets, and then looking at the distribution of data sets that result. Any given

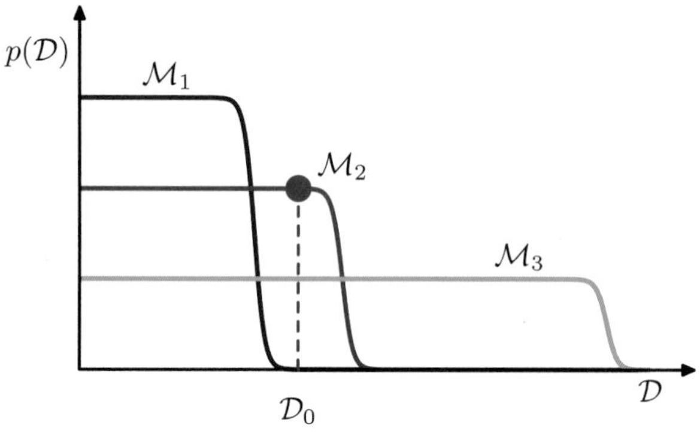

Figure 3.13 Schematic illustration of the distribution of data sets for three models of different complexity, in which $\mathcal{M}_1$ is the simplest and $\mathcal{M}_3$ is the most complex. Note that the distributions are normalized. In this example, for the particular observed data set $\mathcal{D}_0$, the model $\mathcal{M}_2$ with intermediate complexity has the largest evidence.

model can generate a variety of different data sets since the parameters are governed by a prior probability distribution, and for any choice of the parameters there may be random noise on the target variables. To generate a particular data set from a specific model, we first choose the values of the parameters from their prior distribution $p(\mathbf{w})$, and then for these parameter values we sample the data from $p(\mathcal{D}|\mathbf{w})$. A simple model (for example, based on a first order polynomial) has little variability and so will generate data sets that are fairly similar to each other. Its distribution $p(\mathcal{D})$ is therefore confined to a relatively small region of the horizontal axis. By contrast, a complex model (such as a ninth order polynomial) can generate a great variety of different data sets, and so its distribution $p(\mathcal{D})$ is spread over a large region of the space of data sets. Because the distributions $p(\mathcal{D}|\mathcal{M}_i)$ are normalized, we see that the particular data set $\mathcal{D}_0$ can have the highest value of the evidence for the model of intermediate complexity. Essentially, the simpler model cannot fit the data well, whereas the more complex model spreads its predictive probability over too broad a range of data sets and so assigns relatively small probability to any one of them.

Implicit in the Bayesian model comparison framework is the assumption that the true distribution from which the data are generated is contained within the set of models under consideration. Provided this is so, we can show that Bayesian model comparison will on average favour the correct model. To see this, consider two models $\mathcal{M}_1$ and $\mathcal{M}_2$ in which the truth corresponds to $\mathcal{M}_1$. For a given finite data set, it is possible for the Bayes factor to be larger for the incorrect model. However, if we average the Bayes factor over the distribution of data sets, we obtain the expected Bayes factor in the form

$$\int p(\mathcal{D}|\mathcal{M}_1) \ln \frac{p(\mathcal{D}|\mathcal{M}_1)}{p(\mathcal{D}|\mathcal{M}_2)} \, \mathrm{d}\mathcal{D} \tag{3.73}$$

where the average has been taken with respect to the true distribution of the data. *Section 1.6.1* This quantity is an example of the *Kullback-Leibler* divergence and satisfies the property of always being positive unless the two distributions are equal in which case it is zero. Thus on average the Bayes factor will always favour the correct model.

We have seen that the Bayesian framework avoids the problem of over-fitting and allows models to be compared on the basis of the training data alone. However, a Bayesian approach, like any approach to pattern recognition, needs to make assumptions about the form of the model, and if these are invalid then the results can be misleading. In particular, we see from Figure 3.12 that the model evidence can be sensitive to many aspects of the prior, such as the behaviour in the tails. Indeed, the evidence is not defined if the prior is improper, as can be seen by noting that an improper prior has an arbitrary scaling factor (in other words, the normalization coefficient is not defined because the distribution cannot be normalized). If we consider a proper prior and then take a suitable limit in order to obtain an improper prior (for example, a Gaussian prior in which we take the limit of infinite variance) then the evidence will go to zero, as can be seen from (3.70) and Figure 3.12. It may, however, be possible to consider the evidence ratio between two models first and then take a limit to obtain a meaningful answer.

In a practical application, therefore, it will be wise to keep aside an independent test set of data on which to evaluate the overall performance of the final system.

3.5. The Evidence Approximation

In a fully Bayesian treatment of the linear basis function model, we would introduce prior distributions over the hyperparameters α and β and make predictions by marginalizing with respect to these hyperparameters as well as with respect to the parameters $\mathbf{w}$. However, although we can integrate analytically either over $\mathbf{w}$ or over the hyperparameters, the complete marginalization over all of these variables is analytically intractable. Here we discuss an approximation in which we set the hyperparameters to specific values determined by maximizing the *marginal likelihood function* obtained by first integrating over the parameters $\mathbf{w}$. This framework is known in the statistics literature as *empirical Bayes* (Bernardo and Smith, 1994; Gelman *et al.*, 2004), or *type 2 maximum likelihood* (Berger, 1985), or *generalized maximum likelihood* (Wahba, 1975), and in the machine learning literature is also called the *evidence approximation* (Gull, 1989; MacKay, 1992a).

If we introduce hyperpriors over α and β, the predictive distribution is obtained by marginalizing over $\mathbf{w}$, α and β so that

$$p(t|\mathbf{t}) = \iiint p(t|\mathbf{w},\beta)p(\mathbf{w}|\mathbf{t},\alpha,\beta)p(\alpha,\beta|\mathbf{t})\,\mathrm{d}\mathbf{w}\,\mathrm{d}\alpha\,\mathrm{d}\beta \tag{3.74}$$

where $p(t|\mathbf{w},\beta)$ is given by (3.8) and $p(\mathbf{w}|\mathbf{t},\alpha,\beta)$ is given by (3.49) with $\mathbf{m}_N$ and $\mathbf{S}_N$ defined by (3.53) and (3.54) respectively. Here we have omitted the dependence on the input variable $\mathbf{x}$ to keep the notation uncluttered. If the posterior distribution $p(\alpha,\beta|\mathbf{t})$ is sharply peaked around values $\widehat{\alpha}$ and $\widehat{\beta}$, then the predictive distribution is obtained simply by marginalizing over $\mathbf{w}$ in which α and β are fixed to the values $\widehat{\alpha}$ and $\widehat{\beta}$, so that

$$p(t|\mathbf{t}) \simeq p(t|\mathbf{t},\widehat{\alpha},\widehat{\beta}) = \int p(t|\mathbf{w},\widehat{\beta})p(\mathbf{w}|\mathbf{t},\widehat{\alpha},\widehat{\beta})\,\mathrm{d}\mathbf{w}. \tag{3.75}$$

From Bayes' theorem, the posterior distribution for α and β is given by

$$p(\alpha, \beta|\mathbf{t}) \propto p(\mathbf{t}|\alpha, \beta)p(\alpha, \beta). \tag{3.76}$$

If the prior is relatively flat, then in the evidence framework the values of $\widehat{\alpha}$ and $\widehat{\beta}$ are obtained by maximizing the marginal likelihood function $p(\mathbf{t}|\alpha, \beta)$. We shall proceed by evaluating the marginal likelihood for the linear basis function model and then finding its maxima. This will allow us to determine values for these hyperparameters from the training data alone, without recourse to cross-validation. Recall that the ratio α/β is analogous to a regularization parameter.

As an aside it is worth noting that, if we define conjugate (Gamma) prior distributions over α and β, then the marginalization over these hyperparameters in (3.74) can be performed analytically to give a Student's t-distribution over $\mathbf{w}$ (see Section 2.3.7). Although the resulting integral over $\mathbf{w}$ is no longer analytically tractable, it might be thought that approximating this integral, for example using the Laplace approximation discussed in Section 4.4, which is based on a local Gaussian approximation centred on the mode of the posterior distribution, might provide a practical alternative to the evidence framework (Buntine and Weigend, 1991). However, the integrand as a function of $\mathbf{w}$ typically has a strongly skewed mode so that the Laplace approximation fails to capture the bulk of the probability mass, leading to poorer results than those obtained by maximizing the evidence (MacKay, 1999).

Returning to the evidence framework, we note that there are two approaches that we can take to the maximization of the log evidence. We can evaluate the evidence function analytically and then set its derivative equal to zero to obtain re-estimation equations for α and β, which we shall do in Section 3.5.2. Alternatively we use a technique called the expectation maximization (EM) algorithm, which will be discussed in Section 9.3.4 where we shall also show that these two approaches converge to the same solution.

3.5.1 Evaluation of the evidence function

The marginal likelihood function $p(\mathbf{t}|\alpha, \beta)$ is obtained by integrating over the weight parameters $\mathbf{w}$, so that

$$p(\mathbf{t}|\alpha, \beta) = \int p(\mathbf{t}|\mathbf{w}, \beta)p(\mathbf{w}|\alpha)\,\mathrm{d}\mathbf{w}. \tag{3.77}$$

Exercise 3.16

One way to evaluate this integral is to make use once again of the result (2.115) for the conditional distribution in a linear-Gaussian model. Here we shall evaluate the integral instead by completing the square in the exponent and making use of the standard form for the normalization coefficient of a Gaussian.

Exercise 3.17

From (3.11), (3.12), and (3.52), we can write the evidence function in the form

$$p(\mathbf{t}|\alpha, \beta) = \left(\frac{\beta}{2\pi}\right)^{N/2} \left(\frac{\alpha}{2\pi}\right)^{M/2} \int \exp\left\{-E(\mathbf{w})\right\} \mathrm{d}\mathbf{w} \tag{3.78}$$

where M is the dimensionality of $\mathbf{w}$, and we have defined

$$\begin{aligned} E(\mathbf{w}) &= \beta E_D(\mathbf{w}) + \alpha E_W(\mathbf{w}) \\ &= \frac{\beta}{2} \|\mathbf{t} - \boldsymbol{\Phi}\mathbf{w}\|^2 + \frac{\alpha}{2}\mathbf{w}^{\mathrm{T}}\mathbf{w}. \end{aligned} \tag{3.79}$$

Exercise 3.18

We recognize (3.79) as being equal, up to a constant of proportionality, to the regularized sum-of-squares error function (3.27). We now complete the square over $\mathbf{w}$ giving

$$E(\mathbf{w}) = E(\mathbf{m}_N) + \frac{1}{2}(\mathbf{w} - \mathbf{m}_N)^{\mathrm{T}}\mathbf{A}(\mathbf{w} - \mathbf{m}_N) \tag{3.80}$$

where we have introduced

$$\mathbf{A} = \alpha\mathbf{I} + \beta\boldsymbol{\Phi}^{\mathrm{T}}\boldsymbol{\Phi} \tag{3.81}$$

together with

$$E(\mathbf{m}_N) = \frac{\beta}{2} \|\mathbf{t} - \boldsymbol{\Phi}\mathbf{m}_N\|^2 + \frac{\alpha}{2}\mathbf{m}_N^{\mathrm{T}}\mathbf{m}_N. \tag{3.82}$$

Note that $\mathbf{A}$ corresponds to the matrix of second derivatives of the error function

$$\mathbf{A} = \nabla\nabla E(\mathbf{w}) \tag{3.83}$$

and is known as the *Hessian matrix*. Here we have also defined $\mathbf{m}_N$ given by

$$\mathbf{m}_N = \beta\mathbf{A}^{-1}\boldsymbol{\Phi}^{\mathrm{T}}\mathbf{t}. \tag{3.84}$$

Using (3.54), we see that $\mathbf{A} = \mathbf{S}_N^{-1}$, and hence (3.84) is equivalent to the previous definition (3.53), and therefore represents the mean of the posterior distribution.

Exercise 3.19

The integral over $\mathbf{w}$ can now be evaluated simply by appealing to the standard result for the normalization coefficient of a multivariate Gaussian, giving

$$\begin{aligned} &\int \exp\{-E(\mathbf{w})\}\,\mathrm{d}\mathbf{w} \\ &= \exp\{-E(\mathbf{m}_N)\}\int \exp\left\{-\frac{1}{2}(\mathbf{w} - \mathbf{m}_N)^{\mathrm{T}}\mathbf{A}(\mathbf{w} - \mathbf{m}_N)\right\}\,\mathrm{d}\mathbf{w} \\ &= \exp\{-E(\mathbf{m}_N)\}(2\pi)^{M/2}|\mathbf{A}|^{-1/2}. \end{aligned} \tag{3.85}$$

Using (3.78) we can then write the log of the marginal likelihood in the form

$$\ln p(\mathbf{t}|\alpha, \beta) = \frac{M}{2}\ln\alpha + \frac{N}{2}\ln\beta - E(\mathbf{m}_N) - \frac{1}{2}\ln|\mathbf{A}| - \frac{N}{2}\ln(2\pi) \tag{3.86}$$

which is the required expression for the evidence function.

Returning to the polynomial regression problem, we can plot the model evidence against the order of the polynomial, as shown in Figure 3.14. Here we have assumed a prior of the form (1.65) with the parameter α fixed at $\alpha = 5 \times 10^{-3}$. The form of this plot is very instructive. Referring back to Figure 1.4, we see that the $M = 0$ polynomial has very poor fit to the data and consequently gives a relatively low value

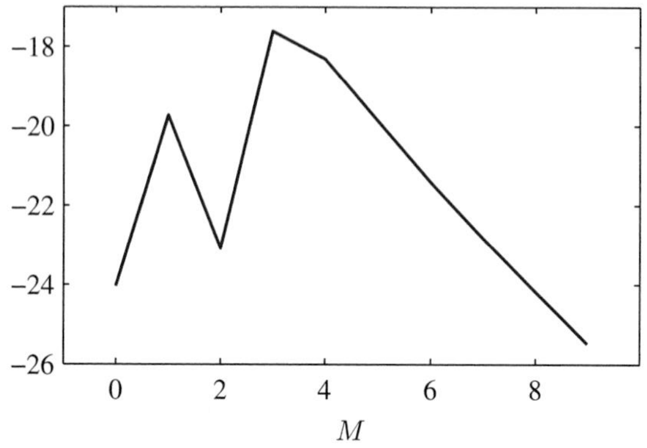

Figure 3.14 Plot of the model log evidence versus the order M, for the polynomial regression model, showing that the evidence favours the model with $M = 3$.

for the evidence. Going to the $M = 1$ polynomial greatly improves the data fit, and hence the evidence is significantly higher. However, in going to $M = 2$, the data fit is improved only very marginally, due to the fact that the underlying sinusoidal function from which the data is generated is an odd function and so has no even terms in a polynomial expansion. Indeed, Figure 1.5 shows that the residual data error is reduced only slightly in going from $M = 1$ to $M = 2$. Because this richer model suffers a greater complexity penalty, the evidence actually falls in going from $M = 1$ to $M = 2$. When we go to $M = 3$ we obtain a significant further improvement in data fit, as seen in Figure 1.4, and so the evidence is increased again, giving the highest overall evidence for any of the polynomials. Further increases in the value of M produce only small improvements in the fit to the data but suffer increasing complexity penalty, leading overall to a decrease in the evidence values. Looking again at Figure 1.5, we see that the generalization error is roughly constant between $M = 3$ and $M = 8$, and it would be difficult to choose between these models on the basis of this plot alone. The evidence values, however, show a clear preference for $M = 3$, since this is the simplest model which gives a good explanation for the observed data.

3.5.2 Maximizing the evidence function

Let us first consider the maximization of $p(\mathbf{t}|\alpha, \beta)$ with respect to α. This can be done by first defining the following eigenvector equation

$$\left(\beta\boldsymbol{\Phi}^{\mathrm{T}}\boldsymbol{\Phi}\right)\mathbf{u}_i = \lambda_i\mathbf{u}_i. \tag{3.87}$$

From (3.81), it then follows that $\mathbf{A}$ has eigenvalues $\alpha + \lambda_i$. Now consider the derivative of the term involving $\ln|\mathbf{A}|$ in (3.86) with respect to α. We have

$$\frac{d}{d\alpha}\ln|\mathbf{A}| = \frac{d}{d\alpha}\ln\prod_i(\lambda_i + \alpha) = \frac{d}{d\alpha}\sum_i\ln(\lambda_i + \alpha) = \sum_i\frac{1}{\lambda_i + \alpha}. \tag{3.88}$$

Thus the stationary points of (3.86) with respect to α satisfy

$$0 = \frac{M}{2\alpha} - \frac{1}{2}\mathbf{m}_N^{\mathrm{T}}\mathbf{m}_N - \frac{1}{2}\sum_i\frac{1}{\lambda_i + \alpha}. \tag{3.89}$$

where we neglect derivatives of $\mathbf{m}_N^{\mathrm{T}}\mathbf{m}_N$ with respect to α. Multiplying through by 2α and rearranging, we obtain

$$\alpha \mathbf{m}_N^{\mathrm{T}}\mathbf{m}_N = M - \alpha \sum_i \frac{1}{\lambda_i + \alpha} = \gamma. \tag{3.90}$$

Since there are M terms in the sum over i, the quantity γ can be written

$$\gamma = \sum_i \frac{\lambda_i}{\alpha + \lambda_i}. \tag{3.91}$$

Exercise 3.20

The interpretation of the quantity γ will be discussed shortly. From (3.90) we see that the value of α that maximizes the marginal likelihood satisfies

$$\alpha = \frac{\gamma}{\mathbf{m}_N^{\mathrm{T}}\mathbf{m}_N}. \tag{3.92}$$

Note that this is an implicit solution for α not only because γ depends on α, but also because the mode $\mathbf{m}_N$ of the posterior distribution itself depends on the choice of α. We therefore adopt an iterative procedure in which we make an initial choice for α and use this to find $\mathbf{m}_N$, which is given by (3.53), and also to evaluate γ, which is given by (3.91). These values are then used to re-estimate α using (3.92), and the process repeated until convergence. Note that because the matrix $\boldsymbol{\Phi}^{\mathrm{T}}\boldsymbol{\Phi}$ is fixed, we can compute its eigenvalues once at the start and then simply multiply these by β to obtain the λ_i.

It should be emphasized that the value of α has been determined purely by looking at the training data. In contrast to maximum likelihood methods, no independent data set is required in order to optimize the model complexity.

We can similarly maximize the log marginal likelihood (3.86) with respect to β. To do this, we note that the eigenvalues λ_i defined by (3.87) are proportional to β, and hence $d\lambda_i/d\beta = \lambda_i/\beta$ giving

$$\frac{d}{d\beta} \ln |\mathbf{A}| = \frac{d}{d\beta} \sum_i \ln(\lambda_i + \alpha) = \frac{1}{\beta} \sum_i \frac{\lambda_i}{\lambda_i + \alpha} = \frac{\gamma}{\beta}. \tag{3.93}$$

The stationary point of the marginal likelihood therefore satisfies

$$0 = \frac{N}{2\beta} - \frac{1}{2} \sum_{n=1}^{N} \left\{ t_n - \mathbf{m}_N^{\mathrm{T}} \boldsymbol{\phi}(\mathbf{x}_n) \right\}^2 - \frac{\gamma}{2\beta} \tag{3.94}$$

Exercise 3.22

and rearranging we obtain

$$\frac{1}{\beta} = \frac{1}{N - \gamma} \sum_{n=1}^{N} \left\{ t_n - \mathbf{m}_N^{\mathrm{T}} \boldsymbol{\phi}(\mathbf{x}_n) \right\}^2. \tag{3.95}$$

Again, this is an implicit solution for β and can be solved by choosing an initial value for β and then using this to calculate $\mathbf{m}_N$ and γ and then re-estimate β using (3.95), repeating until convergence. If both α and β are to be determined from the data, then their values can be re-estimated together after each update of γ.

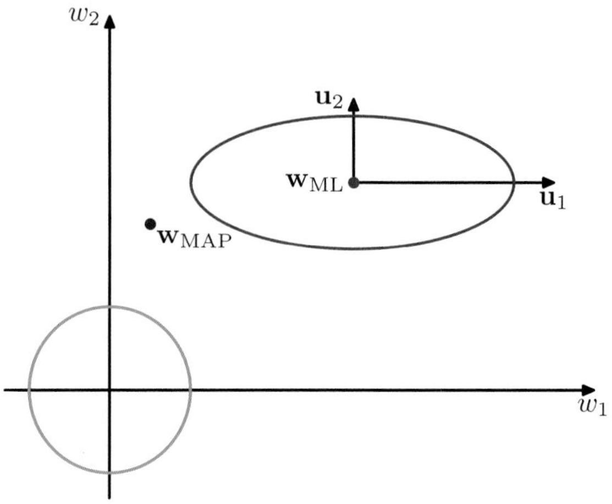

Figure 3.15 Contours of the likelihood function (red) and the prior (green) in which the axes in parameter space have been rotated to align with the eigenvectors $\mathbf{u}_i$ of the Hessian. For $\alpha = 0$, the mode of the posterior is given by the maximum likelihood solution $\mathbf{w}_{\rm ML}$, whereas for nonzero α the mode is at $\mathbf{w}_{\rm MAP} = \mathbf{m}_N$. In the direction w_1 the eigenvalue λ_1, defined by (3.87), is small compared with α and so the quantity $\lambda_1/(\lambda_1 + \alpha)$ is close to zero, and the corresponding MAP value of w_1 is also close to zero. By contrast, in the direction w_2 the eigenvalue λ_2 is large compared with α and so the quantity $\lambda_2/(\lambda_2+\alpha)$ is close to unity, and the MAP value of w_2 is close to its maximum likelihood value.

3.5.3 Effective number of parameters

The result (3.92) has an elegant interpretation (MacKay, 1992a), which provides insight into the Bayesian solution for α. To see this, consider the contours of the likelihood function and the prior as illustrated in Figure 3.15. Here we have implicitly transformed to a rotated set of axes in parameter space aligned with the eigenvectors $\mathbf{u}_i$ defined in (3.87). Contours of the likelihood function are then axis-aligned ellipses. The eigenvalues λ_i measure the curvature of the likelihood function, and so in Figure 3.15 the eigenvalue λ_1 is small compared with λ_2 (because a smaller curvature corresponds to a greater elongation of the contours of the likelihood function). Because $\beta\mathbf{\Phi}^{\rm T}\mathbf{\Phi}$ is a positive definite matrix, it will have positive eigenvalues, and so the ratio $\lambda_i/(\lambda_i + \alpha)$ will lie between 0 and 1. Consequently, the quantity γ defined by (3.91) will lie in the range $0 \leqslant \gamma \leqslant M$. For directions in which $\lambda_i \gg \alpha$, the corresponding parameter w_i will be close to its maximum likelihood value, and the ratio $\lambda_i/(\lambda_i + \alpha)$ will be close to 1. Such parameters are called *well determined* because their values are tightly constrained by the data. Conversely, for directions in which $\lambda_i \ll \alpha$, the corresponding parameters w_i will be close to zero, as will the ratios $\lambda_i/(\lambda_i+\alpha)$. These are directions in which the likelihood function is relatively insensitive to the parameter value and so the parameter has been set to a small value by the prior. The quantity γ defined by (3.91) therefore measures the effective total number of well determined parameters.

We can obtain some insight into the result (3.95) for re-estimating β by comparing it with the corresponding maximum likelihood result given by (3.21). Both of these formulae express the variance (the inverse precision) as an average of the squared differences between the targets and the model predictions. However, they differ in that the number of data points N in the denominator of the maximum likelihood result is replaced by $N - \gamma$ in the Bayesian result. We recall from (1.56) that the maximum likelihood estimate of the variance for a Gaussian distribution over a single variable x is given by

$$\sigma_{\mathrm{ML}}^2 = \frac{1}{N}\sum_{n=1}^{N}(x_n - \mu_{\mathrm{ML}})^2 \tag{3.96}$$

and that this estimate is biased because the maximum likelihood solution μ_{ML} for the mean has fitted some of the noise on the data. In effect, this has used up one degree of freedom in the model. The corresponding unbiased estimate is given by (1.59) and takes the form

$$\sigma_{\mathrm{MAP}}^2 = \frac{1}{N-1}\sum_{n=1}^{N}(x_n - \mu_{\mathrm{ML}})^2. \tag{3.97}$$

The factor of $N-1$ in the denominator of the Bayesian result takes account of the fact that one degree of freedom has been used in fitting the mean and removes the bias of maximum likelihood. Now consider the corresponding results for the linear regression model. The mean of the target distribution is now given by the function $\mathbf{w}^{\mathrm{T}}\boldsymbol{\phi}(\mathbf{x})$, which contains M parameters. However, not all of these parameters are tuned to the data. The effective number of parameters that are determined by the data is γ, with the remaining $M-\gamma$ parameters set to small values by the prior. This is reflected in the Bayesian result for the variance that has a factor $N-\gamma$ in the denominator, thereby correcting for the bias of the maximum likelihood result.

We can illustrate the evidence framework for setting hyperparameters using the sinusoidal synthetic data set from Section 1.1, together with the Gaussian basis function model comprising 9 basis functions, so that the total number of parameters in the model is given by $M = 10$ including the bias. Here, for simplicity of illustration, we have set β to its true value of 11.1 and then used the evidence framework to determine α, as shown in Figure 3.16.

We can also see how the parameter α controls the magnitude of the parameters $\{w_i\}$, by plotting the individual parameters versus the effective number γ of parameters, as shown in Figure 3.17.

If we consider the limit $N \gg M$ in which the number of data points is large in relation to the number of parameters, then from (3.87) all of the parameters will be well determined by the data because $\boldsymbol{\Phi}^{\mathrm{T}}\boldsymbol{\Phi}$ involves an implicit sum over data points, and so the eigenvalues λ_i increase with the size of the data set. In this case, $\gamma = M$, and the re-estimation equations for α and β become

$$\alpha = \frac{M}{2E_W(\mathbf{m}_N)} \tag{3.98}$$

$$\beta = \frac{N}{2E_D(\mathbf{m}_N)} \tag{3.99}$$

where E_W and E_D are defined by (3.25) and (3.26), respectively. These results can be used as an easy-to-compute approximation to the full evidence re-estimation

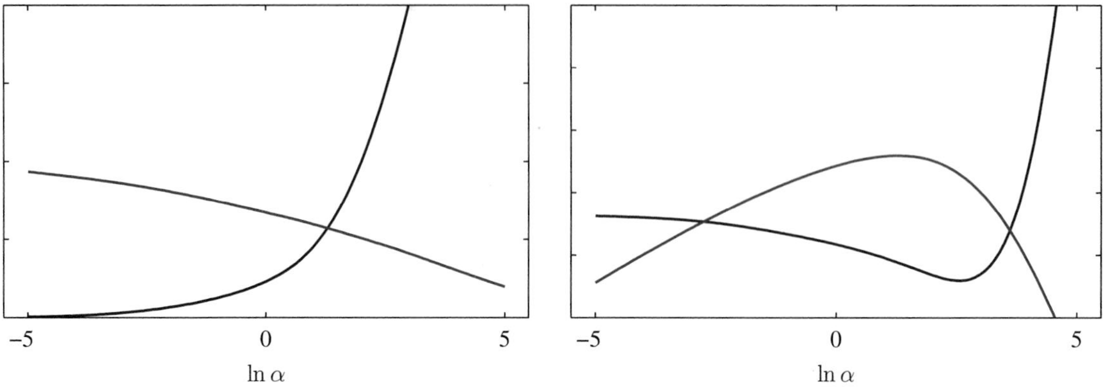

Figure 3.16 The left plot shows γ (red curve) and $2\alpha E_W(\mathbf{m}_N)$ (blue curve) versus $\ln \alpha$ for the sinusoidal synthetic data set. It is the intersection of these two curves that defines the optimum value for α given by the evidence procedure. The right plot shows the corresponding graph of log evidence $\ln p(\mathbf{t}|\alpha, \beta)$ versus $\ln \alpha$ (red curve) showing that the peak coincides with the crossing point of the curves in the left plot. Also shown is the test set error (blue curve) showing that the evidence maximum occurs close to the point of best generalization.

formulae, because they do not require evaluation of the eigenvalue spectrum of the Hessian.

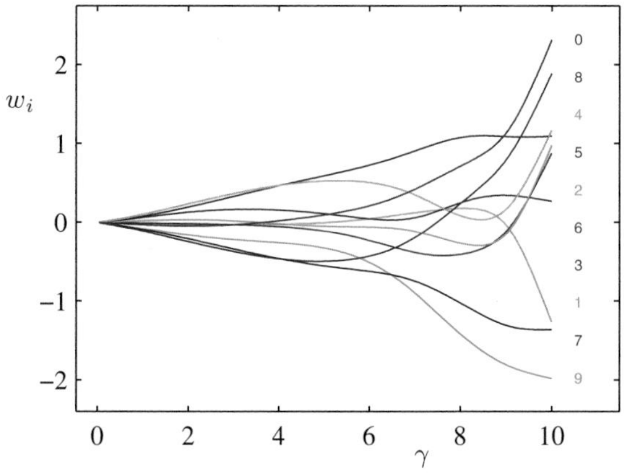

Figure 3.17 Plot of the 10 parameters w_i from the Gaussian basis function model versus the effective number of parameters γ, in which the hyperparameter α is varied in the range $0 \leqslant \alpha \leqslant \infty$ causing γ to vary in the range $0 \leqslant \gamma \leqslant M$.

3.6. Limitations of Fixed Basis Functions

Throughout this chapter, we have focussed on models comprising a linear combination of fixed, nonlinear basis functions. We have seen that the assumption of linearity in the parameters led to a range of useful properties including closed-form solutions to the least-squares problem, as well as a tractable Bayesian treatment. Furthermore, for a suitable choice of basis functions, we can model arbitrary nonlinearities in the mapping from input variables to targets. In the next chapter, we shall study an analogous class of models for classification.

It might appear, therefore, that such linear models constitute a general purpose framework for solving problems in pattern recognition. Unfortunately, there are some significant shortcomings with linear models, which will cause us to turn in later chapters to more complex models such as support vector machines and neural networks.

The difficulty stems from the assumption that the basis functions $\phi_j(\mathbf{x})$ are fixed before the training data set is observed and is a manifestation of the curse of dimensionality discussed in Section 1.4. As a consequence, the number of basis functions needs to grow rapidly, often exponentially, with the dimensionality D of the input space.

Fortunately, there are two properties of real data sets that we can exploit to help alleviate this problem. First of all, the data vectors $\{\mathbf{x}_n\}$ typically lie close to a nonlinear manifold whose intrinsic dimensionality is smaller than that of the input space as a result of strong correlations between the input variables. We will see an example of this when we consider images of handwritten digits in Chapter 12. If we are using localized basis functions, we can arrange that they are scattered in input space only in regions containing data. This approach is used in radial basis function networks and also in support vector and relevance vector machines. Neural network models, which use adaptive basis functions having sigmoidal nonlinearities, can adapt the parameters so that the regions of input space over which the basis functions vary corresponds to the data manifold. The second property is that target variables may have significant dependence on only a small number of possible directions within the data manifold. Neural networks can exploit this property by choosing the directions in input space to which the basis functions respond.

Exercises

3.1 ($\star$) **www** Show that the 'tanh' function and the logistic sigmoid function (3.6) are related by

$$\tanh(a) = 2\sigma(2a) - 1. \tag{3.100}$$

Hence show that a general linear combination of logistic sigmoid functions of the form

$$y(x, \mathbf{w}) = w_0 + \sum_{j=1}^{M} w_j \sigma\left(\frac{x - \mu_j}{s}\right) \tag{3.101}$$

is equivalent to a linear combination of 'tanh' functions of the form

$$y(x, \mathbf{u}) = u_0 + \sum_{j=1}^{M} u_j \tanh\left(\frac{x - \mu_j}{2s}\right) \tag{3.102}$$

and find expressions to relate the new parameters $\{u_0, \ldots, u_M\}$ to the original parameters $\{w_0, \ldots, w_M\}$.

3.2 ($\star\star$) Show that the matrix

$$\boldsymbol{\Phi}(\boldsymbol{\Phi}^{\mathrm{T}}\boldsymbol{\Phi})^{-1}\boldsymbol{\Phi}^{\mathrm{T}} \tag{3.103}$$

takes any vector $\mathbf{v}$ and projects it onto the space spanned by the columns of $\boldsymbol{\Phi}$. Use this result to show that the least-squares solution (3.15) corresponds to an orthogonal projection of the vector $\mathbf{t}$ onto the manifold $\mathcal{S}$ as shown in Figure 3.2.

3.3 ($\star$) Consider a data set in which each data point t_n is associated with a weighting factor $r_n > 0$, so that the sum-of-squares error function becomes

$$E_D(\mathbf{w}) = \frac{1}{2}\sum_{n=1}^{N} r_n \left\{t_n - \mathbf{w}^{\mathrm{T}}\boldsymbol{\phi}(\mathbf{x}_n)\right\}^2. \tag{3.104}$$

Find an expression for the solution $\mathbf{w}^\star$ that minimizes this error function. Give two alternative interpretations of the weighted sum-of-squares error function in terms of (i) data dependent noise variance and (ii) replicated data points.

3.4 ($\star$) **www** Consider a linear model of the form

$$y(\mathbf{x}, \mathbf{w}) = w_0 + \sum_{i=1}^{D} w_i x_i \tag{3.105}$$

together with a sum-of-squares error function of the form

$$E_D(\mathbf{w}) = \frac{1}{2}\sum_{n=1}^{N} \left\{y(\mathbf{x}_n, \mathbf{w}) - t_n\right\}^2. \tag{3.106}$$

Now suppose that Gaussian noise ϵ_i with zero mean and variance σ^2 is added independently to each of the input variables x_i. By making use of $\mathbb{E}[\epsilon_i] = 0$ and $\mathbb{E}[\epsilon_i\epsilon_j] = \delta_{ij}\sigma^2$, show that minimizing E_D averaged over the noise distribution is equivalent to minimizing the sum-of-squares error for noise-free input variables with the addition of a weight-decay regularization term, in which the bias parameter w_0 is omitted from the regularizer.

3.5 ($\star$) **www** Using the technique of Lagrange multipliers, discussed in Appendix E, show that minimization of the regularized error function (3.29) is equivalent to minimizing the unregularized sum-of-squares error (3.12) subject to the constraint (3.30). Discuss the relationship between the parameters η and λ.

3.6 ($\star$) **www** Consider a linear basis function regression model for a multivariate target variable $\mathbf{t}$ having a Gaussian distribution of the form

$$p(\mathbf{t}|\mathbf{W}, \boldsymbol{\Sigma}) = \mathcal{N}(\mathbf{t}|\mathbf{y}(\mathbf{x}, \mathbf{W}), \boldsymbol{\Sigma}) \tag{3.107}$$

where

$$\mathbf{y}(\mathbf{x}, \mathbf{W}) = \mathbf{W}^{\mathrm{T}}\boldsymbol{\phi}(\mathbf{x}) \tag{3.108}$$

together with a training data set comprising input basis vectors $\phi(\mathbf{x}_n)$ and corresponding target vectors $\mathbf{t}_n$, with $n = 1, \ldots, N$. Show that the maximum likelihood solution $\mathbf{W}_{\mathrm{ML}}$ for the parameter matrix $\mathbf{W}$ has the property that each column is given by an expression of the form (3.15), which was the solution for an isotropic noise distribution. Note that this is independent of the covariance matrix $\boldsymbol{\Sigma}$. Show that the maximum likelihood solution for $\boldsymbol{\Sigma}$ is given by

$$\boldsymbol{\Sigma} = \frac{1}{N} \sum_{n=1}^{N} \left(\mathbf{t}_n - \mathbf{W}_{\mathrm{ML}}^{\mathrm{T}} \phi(\mathbf{x}_n)\right) \left(\mathbf{t}_n - \mathbf{W}_{\mathrm{ML}}^{\mathrm{T}} \phi(\mathbf{x}_n)\right)^{\mathrm{T}}. \tag{3.109}$$

3.7 $(\star)$ By using the technique of completing the square, verify the result (3.49) for the posterior distribution of the parameters $\mathbf{w}$ in the linear basis function model in which $\mathbf{m}_N$ and $\mathbf{S}_N$ are defined by (3.50) and (3.51) respectively.

3.8 $(\star\star)$ **www** Consider the linear basis function model in Section 3.1, and suppose that we have already observed N data points, so that the posterior distribution over $\mathbf{w}$ is given by (3.49). This posterior can be regarded as the prior for the next observation. By considering an additional data point $(\mathbf{x}_{N+1}, t_{N+1})$, and by completing the square in the exponential, show that the resulting posterior distribution is again given by (3.49) but with $\mathbf{S}_N$ replaced by $\mathbf{S}_{N+1}$ and $\mathbf{m}_N$ replaced by $\mathbf{m}_{N+1}$.

3.9 $(\star\star)$ Repeat the previous exercise but instead of completing the square by hand, make use of the general result for linear-Gaussian models given by (2.116).

3.10 $(\star\star)$ **www** By making use of the result (2.115) to evaluate the integral in (3.57), verify that the predictive distribution for the Bayesian linear regression model is given by (3.58) in which the input-dependent variance is given by (3.59).

3.11 $(\star\star)$ We have seen that, as the size of a data set increases, the uncertainty associated with the posterior distribution over model parameters decreases. Make use of the matrix identity (Appendix C)

$$\left(\mathbf{M} + \mathbf{v}\mathbf{v}^{\mathrm{T}}\right)^{-1} = \mathbf{M}^{-1} - \frac{(\mathbf{M}^{-1}\mathbf{v})\left(\mathbf{v}^{\mathrm{T}}\mathbf{M}^{-1}\right)}{1 + \mathbf{v}^{\mathrm{T}}\mathbf{M}^{-1}\mathbf{v}} \tag{3.110}$$

to show that the uncertainty $\sigma_N^2(\mathbf{x})$ associated with the linear regression function given by (3.59) satisfies

$$\sigma_{N+1}^2(\mathbf{x}) \leqslant \sigma_N^2(\mathbf{x}). \tag{3.111}$$

3.12 $(\star\star)$ We saw in Section 2.3.6 that the conjugate prior for a Gaussian distribution with unknown mean and unknown precision (inverse variance) is a normal-gamma distribution. This property also holds for the case of the conditional Gaussian distribution $p(t|\mathbf{x}, \mathbf{w}, \beta)$ of the linear regression model. If we consider the likelihood function (3.10), then the conjugate prior for $\mathbf{w}$ and β is given by

$$p(\mathbf{w}, \beta) = \mathcal{N}(\mathbf{w}|\mathbf{m}_0, \beta^{-1}\mathbf{S}_0)\mathrm{Gam}(\beta|a_0, b_0). \tag{3.112}$$

Show that the corresponding posterior distribution takes the same functional form, so that

$$p(\mathbf{w}, \beta|\mathbf{t}) = \mathcal{N}(\mathbf{w}|\mathbf{m}_N, \beta^{-1}\mathbf{S}_N)\mathrm{Gam}(\beta|a_N, b_N) \tag{3.113}$$

and find expressions for the posterior parameters $\mathbf{m}_N$, $\mathbf{S}_N$, a_N, and b_N.

3.13 ($\star\star$) Show that the predictive distribution $p(t|\mathbf{x}, \mathbf{t})$ for the model discussed in Exercise 3.12 is given by a Student's t-distribution of the form

$$p(t|\mathbf{x}, \mathbf{t}) = \mathrm{St}(t|\mu, \lambda, \nu) \tag{3.114}$$

and obtain expressions for μ, λ and ν.

3.14 ($\star\star$) In this exercise, we explore in more detail the properties of the equivalent kernel defined by (3.62), where $\mathbf{S}_N$ is defined by (3.54). Suppose that the basis functions $\phi_j(\mathbf{x})$ are linearly independent and that the number N of data points is greater than the number M of basis functions. Furthermore, let one of the basis functions be constant, say $\phi_0(\mathbf{x}) = 1$. By taking suitable linear combinations of these basis functions, we can construct a new basis set $\psi_j(\mathbf{x})$ spanning the same space but that are orthonormal, so that

$$\sum_{n=1}^{N} \psi_j(\mathbf{x}_n)\psi_k(\mathbf{x}_n) = I_{jk} \tag{3.115}$$

where I_{jk} is defined to be 1 if $j = k$ and 0 otherwise, and we take $\psi_0(\mathbf{x}) = 1$. Show that for $\alpha = 0$, the equivalent kernel can be written as $k(\mathbf{x}, \mathbf{x}') = \boldsymbol{\psi}(\mathbf{x})^{\mathrm{T}}\boldsymbol{\psi}(\mathbf{x}')$ where $\boldsymbol{\psi} = (\psi_0, \ldots, \psi_{M-1})^{\mathrm{T}}$. Use this result to show that the kernel satisfies the summation constraint

$$\sum_{n=1}^{N} k(\mathbf{x}, \mathbf{x}_n) = 1. \tag{3.116}$$

3.15 ($\star$) **www** Consider a linear basis function model for regression in which the parameters α and β are set using the evidence framework. Show that the function $E(\mathbf{m}_N)$ defined by (3.82) satisfies the relation $2E(\mathbf{m}_N) = N$.

3.16 ($\star\star$) Derive the result (3.86) for the log evidence function $p(\mathbf{t}|\alpha, \beta)$ of the linear regression model by making use of (2.115) to evaluate the integral (3.77) directly.

3.17 ($\star$) Show that the evidence function for the Bayesian linear regression model can be written in the form (3.78) in which $E(\mathbf{w})$ is defined by (3.79).

3.18 ($\star\star$) **www** By completing the square over $\mathbf{w}$, show that the error function (3.79) in Bayesian linear regression can be written in the form (3.80).

3.19 ($\star\star$) Show that the integration over $\mathbf{w}$ in the Bayesian linear regression model gives the result (3.85). Hence show that the log marginal likelihood is given by (3.86).

3.20 (⋆⋆) **www** Verify all of the steps needed to show that maximization of the log marginal likelihood function (3.86) with respect to α leads to the re-estimation equation (3.92).

3.21 (⋆⋆) An alternative way to derive the result (3.92) for the optimal value of α in the evidence framework is to make use of the identity

$$\frac{d}{d\alpha}\ln|\mathbf{A}| = \mathrm{Tr}\left(\mathbf{A}^{-1}\frac{d}{d\alpha}\mathbf{A}\right). \tag{3.117}$$

Prove this identity by considering the eigenvalue expansion of a real, symmetric matrix $\mathbf{A}$, and making use of the standard results for the determinant and trace of $\mathbf{A}$ expressed in terms of its eigenvalues (Appendix C). Then make use of (3.117) to derive (3.92) starting from (3.86).

3.22 (⋆⋆) Verify all of the steps needed to show that maximization of the log marginal likelihood function (3.86) with respect to β leads to the re-estimation equation (3.95).

3.23 (⋆⋆) **www** Show that the marginal probability of the data, in other words the model evidence, for the model described in Exercise 3.12 is given by

$$p(\mathbf{t}) = \frac{1}{(2\pi)^{N/2}}\frac{b_0^{a_0}}{b_N^{a_N}}\frac{\Gamma(a_N)}{\Gamma(a_0)}\frac{|\mathbf{S}_N|^{1/2}}{|\mathbf{S}_0|^{1/2}} \tag{3.118}$$

by first marginalizing with respect to $\mathbf{w}$ and then with respect to β.

3.24 (⋆⋆) Repeat the previous exercise but now use Bayes' theorem in the form

$$p(\mathbf{t}) = \frac{p(\mathbf{t}|\mathbf{w},\beta)p(\mathbf{w},\beta)}{p(\mathbf{w},\beta|\mathbf{t})} \tag{3.119}$$

and then substitute for the prior and posterior distributions and the likelihood function in order to derive the result (3.118).

4 Linear Models for Classification

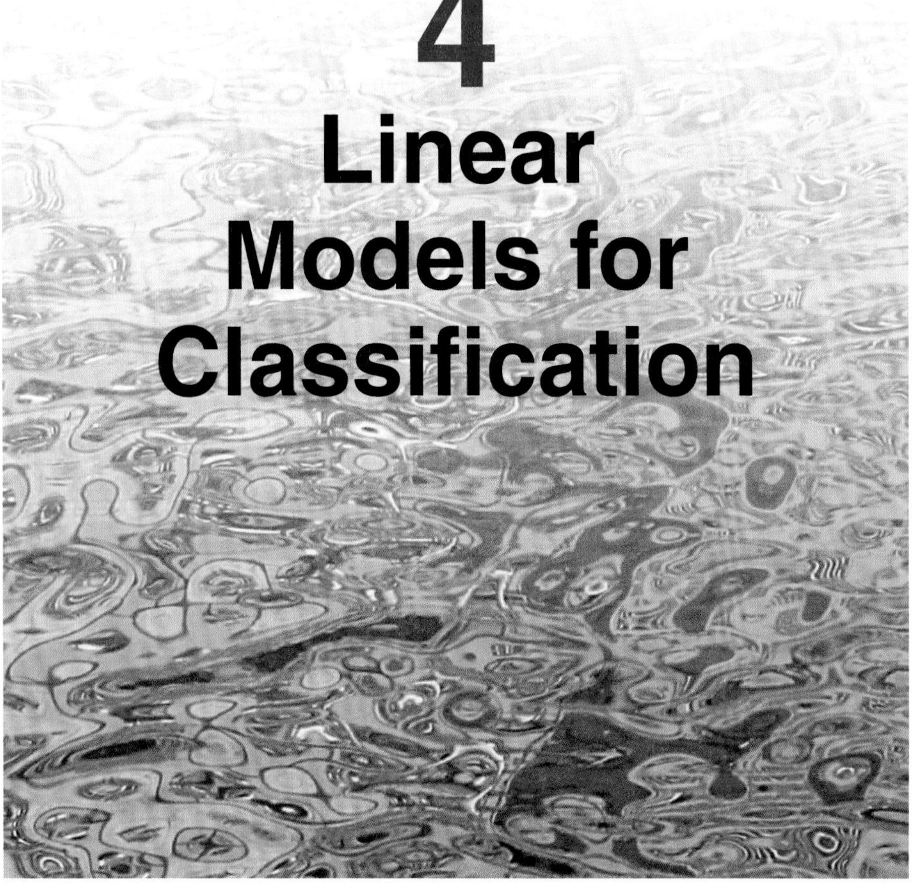

In the previous chapter, we explored a class of regression models having particularly simple analytical and computational properties. We now discuss an analogous class of models for solving classification problems. The goal in classification is to take an input vector $\mathbf{x}$ and to assign it to one of K discrete classes $\mathcal{C}_k$ where $k = 1, \ldots, K$. In the most common scenario, the classes are taken to be disjoint, so that each input is assigned to one and only one class. The input space is thereby divided into *decision regions* whose boundaries are called *decision boundaries* or *decision surfaces*. In this chapter, we consider linear models for classification, by which we mean that the decision surfaces are linear functions of the input vector $\mathbf{x}$ and hence are defined by $(D-1)$-dimensional hyperplanes within the D-dimensional input space. Data sets whose classes can be separated exactly by linear decision surfaces are said to be *linearly separable*.

For regression problems, the target variable $\mathbf{t}$ was simply the vector of real numbers whose values we wish to predict. In the case of classification, there are various ways of using target values to represent class labels. For probabilistic models, the most convenient, in the case of two-class problems, is the binary representation in which there is a single target variable $t \in \{0,1\}$ such that $t=1$ represents class $\mathcal{C}_1$ and $t=0$ represents class $\mathcal{C}_2$. We can interpret the value of t as the probability that the class is $\mathcal{C}_1$, with the values of probability taking only the extreme values of 0 and 1. For $K > 2$ classes, it is convenient to use a 1-of-K coding scheme in which $\mathbf{t}$ is a vector of length K such that if the class is $\mathcal{C}_j$, then all elements t_k of $\mathbf{t}$ are zero except element t_j, which takes the value 1. For instance, if we have $K=5$ classes, then a pattern from class 2 would be given the target vector

$$\mathbf{t} = (0,1,0,0,0)^{\mathrm{T}}. \tag{4.1}$$

Again, we can interpret the value of t_k as the probability that the class is $\mathcal{C}_k$. For nonprobabilistic models, alternative choices of target variable representation will sometimes prove convenient.

In Chapter 1, we identified three distinct approaches to the classification problem. The simplest involves constructing a *discriminant function* that directly assigns each vector $\mathbf{x}$ to a specific class. A more powerful approach, however, models the conditional probability distribution $p(\mathcal{C}_k|\mathbf{x})$ in an inference stage, and then subsequently uses this distribution to make optimal decisions. By separating inference and decision, we gain numerous benefits, as discussed in Section 1.5.4. There are two different approaches to determining the conditional probabilities $p(\mathcal{C}_k|\mathbf{x})$. One technique is to model them directly, for example by representing them as parametric models and then optimizing the parameters using a training set. Alternatively, we can adopt a generative approach in which we model the class-conditional densities given by $p(\mathbf{x}|\mathcal{C}_k)$, together with the prior probabilities $p(\mathcal{C}_k)$ for the classes, and then we compute the required posterior probabilities using Bayes' theorem

$$p(\mathcal{C}_k|\mathbf{x}) = \frac{p(\mathbf{x}|\mathcal{C}_k)p(\mathcal{C}_k)}{p(\mathbf{x})}. \tag{4.2}$$

We shall discuss examples of all three approaches in this chapter.

In the linear regression models considered in Chapter 3, the model prediction $y(\mathbf{x}, \mathbf{w})$ was given by a linear function of the parameters $\mathbf{w}$. In the simplest case, the model is also linear in the input variables and therefore takes the form $y(\mathbf{x}) = \mathbf{w}^{\mathrm{T}}\mathbf{x} + w_0$, so that y is a real number. For classification problems, however, we wish to predict discrete class labels, or more generally posterior probabilities that lie in the range $(0,1)$. To achieve this, we consider a generalization of this model in which we transform the linear function of $\mathbf{w}$ using a nonlinear function $f(\,\cdot\,)$ so that

$$y(\mathbf{x}) = f\left(\mathbf{w}^{\mathrm{T}}\mathbf{x} + w_0\right). \tag{4.3}$$

In the machine learning literature $f(\,\cdot\,)$ is known as an *activation function*, whereas its inverse is called a *link function* in the statistics literature. The decision surfaces correspond to $y(\mathbf{x}) = \text{constant}$, so that $\mathbf{w}^{\mathrm{T}}\mathbf{x} + w_0 = \text{constant}$ and hence the decision surfaces are linear functions of $\mathbf{x}$, even if the function $f(\cdot)$ is nonlinear. For this reason, the class of models described by (4.3) are called *generalized linear models*

(McCullagh and Nelder, 1989). Note, however, that in contrast to the models used for regression, they are no longer linear in the parameters due to the presence of the nonlinear function $f(\cdot)$. This will lead to more complex analytical and computational properties than for linear regression models. Nevertheless, these models are still relatively simple compared to the more general nonlinear models that will be studied in subsequent chapters.

The algorithms discussed in this chapter will be equally applicable if we first make a fixed nonlinear transformation of the input variables using a vector of basis functions $\boldsymbol{\phi}(\mathbf{x})$ as we did for regression models in Chapter 3. We begin by considering classification directly in the original input space $\mathbf{x}$, while in Section 4.3 we shall find it convenient to switch to a notation involving basis functions for consistency with later chapters.

4.1. Discriminant Functions

A discriminant is a function that takes an input vector $\mathbf{x}$ and assigns it to one of K classes, denoted $\mathcal{C}_k$. In this chapter, we shall restrict attention to *linear discriminants*, namely those for which the decision surfaces are hyperplanes. To simplify the discussion, we consider first the case of two classes and then investigate the extension to $K > 2$ classes.

4.1.1 Two classes

The simplest representation of a linear discriminant function is obtained by taking a linear function of the input vector so that

$$y(\mathbf{x}) = \mathbf{w}^{\mathrm{T}}\mathbf{x} + w_0 \tag{4.4}$$

where $\mathbf{w}$ is called a *weight vector*, and w_0 is a *bias* (not to be confused with bias in the statistical sense). The negative of the bias is sometimes called a *threshold*. An input vector $\mathbf{x}$ is assigned to class $\mathcal{C}_1$ if $y(\mathbf{x}) \geqslant 0$ and to class $\mathcal{C}_2$ otherwise. The corresponding decision boundary is therefore defined by the relation $y(\mathbf{x}) = 0$, which corresponds to a $(D-1)$-dimensional hyperplane within the D-dimensional input space. Consider two points $\mathbf{x}_\mathrm{A}$ and $\mathbf{x}_\mathrm{B}$ both of which lie on the decision surface. Because $y(\mathbf{x}_\mathrm{A}) = y(\mathbf{x}_\mathrm{B}) = 0$, we have $\mathbf{w}^{\mathrm{T}}(\mathbf{x}_\mathrm{A} - \mathbf{x}_\mathrm{B}) = 0$ and hence the vector $\mathbf{w}$ is orthogonal to every vector lying within the decision surface, and so $\mathbf{w}$ determines the orientation of the decision surface. Similarly, if $\mathbf{x}$ is a point on the decision surface, then $y(\mathbf{x}) = 0$, and so the normal distance from the origin to the decision surface is given by

$$\frac{\mathbf{w}^{\mathrm{T}}\mathbf{x}}{\|\mathbf{w}\|} = -\frac{w_0}{\|\mathbf{w}\|}. \tag{4.5}$$

We therefore see that the bias parameter w_0 determines the location of the decision surface. These properties are illustrated for the case of $D = 2$ in Figure 4.1.

Furthermore, we note that the value of $y(\mathbf{x})$ gives a signed measure of the perpendicular distance r of the point $\mathbf{x}$ from the decision surface. To see this, consider

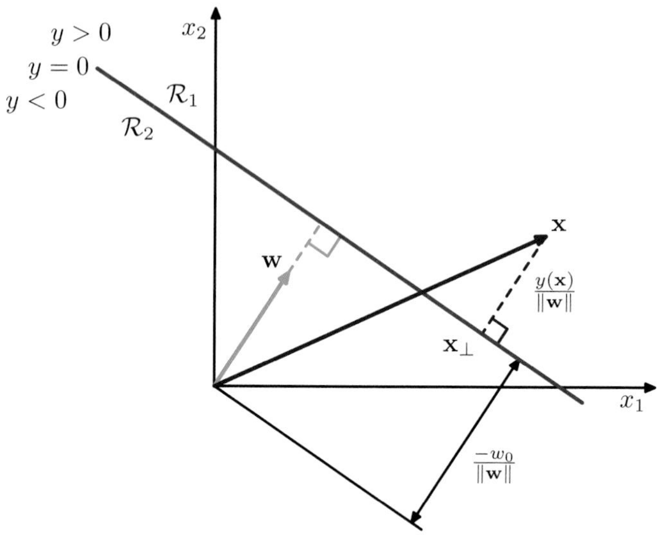

Figure 4.1 Illustration of the geometry of a linear discriminant function in two dimensions. The decision surface, shown in red, is perpendicular to $\mathbf{w}$, and its displacement from the origin is controlled by the bias parameter w_0. Also, the signed orthogonal distance of a general point $\mathbf{x}$ from the decision surface is given by $y(\mathbf{x})/\|\mathbf{w}\|$.

an arbitrary point $\mathbf{x}$ and let $\mathbf{x}_\perp$ be its orthogonal projection onto the decision surface, so that

$$\mathbf{x} = \mathbf{x}_\perp + r\frac{\mathbf{w}}{\|\mathbf{w}\|}. \tag{4.6}$$

Multiplying both sides of this result by $\mathbf{w}^{\mathrm{T}}$ and adding w_0, and making use of $y(\mathbf{x}) = \mathbf{w}^{\mathrm{T}}\mathbf{x} + w_0$ and $y(\mathbf{x}_\perp) = \mathbf{w}^{\mathrm{T}}\mathbf{x}_\perp + w_0 = 0$, we have

$$r = \frac{y(\mathbf{x})}{\|\mathbf{w}\|}. \tag{4.7}$$

This result is illustrated in Figure 4.1.

As with the linear regression models in Chapter 3, it is sometimes convenient to use a more compact notation in which we introduce an additional dummy 'input' value $x_0 = 1$ and then define $\widetilde{\mathbf{w}} = (w_0, \mathbf{w})$ and $\widetilde{\mathbf{x}} = (x_0, \mathbf{x})$ so that

$$y(\mathbf{x}) = \widetilde{\mathbf{w}}^{\mathrm{T}}\widetilde{\mathbf{x}}. \tag{4.8}$$

In this case, the decision surfaces are D-dimensional hyperplanes passing through the origin of the $D+1$-dimensional expanded input space.

4.1.2 Multiple classes

Now consider the extension of linear discriminants to $K > 2$ classes. We might be tempted to build a K-class discriminant by combining a number of two-class discriminant functions. However, this leads to some serious difficulties (Duda and Hart, 1973) as we now show.

Consider the use of $K-1$ classifiers each of which solves a two-class problem of separating points in a particular class $\mathcal{C}_k$ from points not in that class. This is known as a *one-versus-the-rest* classifier. The left-hand example in Figure 4.2 shows an

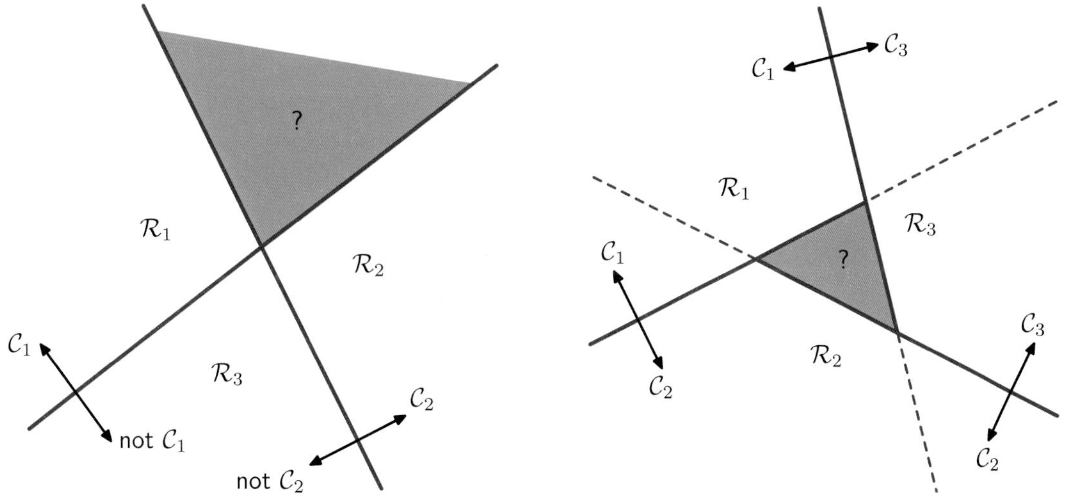

Figure 4.2 Attempting to construct a K class discriminant from a set of two class discriminants leads to ambiguous regions, shown in green. On the left is an example involving the use of two discriminants designed to distinguish points in class $\mathcal{C}_k$ from points not in class $\mathcal{C}_k$. On the right is an example involving three discriminant functions each of which is used to separate a pair of classes $\mathcal{C}_k$ and $\mathcal{C}_j$.

example involving three classes where this approach leads to regions of input space that are ambiguously classified.

An alternative is to introduce $K(K-1)/2$ binary discriminant functions, one for every possible pair of classes. This is known as a *one-versus-one* classifier. Each point is then classified according to a majority vote amongst the discriminant functions. However, this too runs into the problem of ambiguous regions, as illustrated in the right-hand diagram of Figure 4.2.

We can avoid these difficulties by considering a single K-class discriminant comprising K linear functions of the form

$$y_k(\mathbf{x}) = \mathbf{w}_k^{\mathrm{T}}\mathbf{x} + w_{k0} \tag{4.9}$$

and then assigning a point $\mathbf{x}$ to class $\mathcal{C}_k$ if $y_k(\mathbf{x}) > y_j(\mathbf{x})$ for all $j \neq k$. The decision boundary between class $\mathcal{C}_k$ and class $\mathcal{C}_j$ is therefore given by $y_k(\mathbf{x}) = y_j(\mathbf{x})$ and hence corresponds to a $(D-1)$-dimensional hyperplane defined by

$$(\mathbf{w}_k - \mathbf{w}_j)^{\mathrm{T}}\mathbf{x} + (w_{k0} - w_{j0}) = 0. \tag{4.10}$$

This has the same form as the decision boundary for the two-class case discussed in Section 4.1.1, and so analogous geometrical properties apply.

The decision regions of such a discriminant are always singly connected and convex. To see this, consider two points $\mathbf{x}_{\mathrm{A}}$ and $\mathbf{x}_{\mathrm{B}}$ both of which lie inside decision region $\mathcal{R}_k$, as illustrated in Figure 4.3. Any point $\widehat{\mathbf{x}}$ that lies on the line connecting $\mathbf{x}_{\mathrm{A}}$ and $\mathbf{x}_{\mathrm{B}}$ can be expressed in the form

$$\widehat{\mathbf{x}} = \lambda\mathbf{x}_{\mathrm{A}} + (1-\lambda)\mathbf{x}_{\mathrm{B}} \tag{4.11}$$

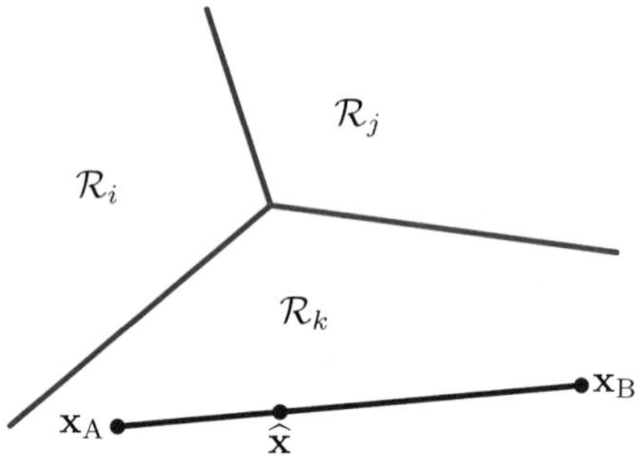

Figure 4.3 Illustration of the decision regions for a multiclass linear discriminant, with the decision boundaries shown in red. If two points $\mathbf{x}_A$ and $\mathbf{x}_B$ both lie inside the same decision region $\mathcal{R}_k$, then any point $\widehat{\mathbf{x}}$ that lies on the line connecting these two points must also lie in $\mathcal{R}_k$, and hence the decision region must be singly connected and convex.

where $0 \leqslant \lambda \leqslant 1$. From the linearity of the discriminant functions, it follows that

$$y_k(\widehat{\mathbf{x}}) = \lambda y_k(\mathbf{x}_A) + (1 - \lambda) y_k(\mathbf{x}_B). \tag{4.12}$$

Because both $\mathbf{x}_A$ and $\mathbf{x}_B$ lie inside $\mathcal{R}_k$, it follows that $y_k(\mathbf{x}_A) > y_j(\mathbf{x}_A)$, and $y_k(\mathbf{x}_B) > y_j(\mathbf{x}_B)$, for all $j \neq k$, and hence $y_k(\widehat{\mathbf{x}}) > y_j(\widehat{\mathbf{x}})$, and so $\widehat{\mathbf{x}}$ also lies inside $\mathcal{R}_k$. Thus $\mathcal{R}_k$ is singly connected and convex.

Note that for two classes, we can either employ the formalism discussed here, based on two discriminant functions $y_1(\mathbf{x})$ and $y_2(\mathbf{x})$, or else use the simpler but equivalent formulation described in Section 4.1.1 based on a single discriminant function $y(\mathbf{x})$.

We now explore three approaches to learning the parameters of linear discriminant functions, based on least squares, Fisher's linear discriminant, and the perceptron algorithm.

4.1.3 Least squares for classification

In Chapter 3, we considered models that were linear functions of the parameters, and we saw that the minimization of a sum-of-squares error function led to a simple closed-form solution for the parameter values. It is therefore tempting to see if we can apply the same formalism to classification problems. Consider a general classification problem with K classes, with a 1-of-K binary coding scheme for the target vector $\mathbf{t}$. One justification for using least squares in such a context is that it approximates the conditional expectation $\mathbb{E}[\mathbf{t}|\mathbf{x}]$ of the target values given the input vector. For the binary coding scheme, this conditional expectation is given by the vector of posterior class probabilities. Unfortunately, however, these probabilities are typically approximated rather poorly, indeed the approximations can have values outside the range $(0, 1)$, due to the limited flexibility of a linear model as we shall see shortly.

Each class $\mathcal{C}_k$ is described by its own linear model so that

$$y_k(\mathbf{x}) = \mathbf{w}_k^{\mathrm{T}} \mathbf{x} + w_{k0} \tag{4.13}$$

where $k = 1, \ldots, K$. We can conveniently group these together using vector notation so that

$$\mathbf{y}(\mathbf{x}) = \widetilde{\mathbf{W}}^{\mathrm{T}} \widetilde{\mathbf{x}} \tag{4.14}$$

where $\widetilde{\mathbf{W}}$ is a matrix whose k^{th} column comprises the $D+1$-dimensional vector $\widetilde{\mathbf{w}}_k = (w_{k0}, \mathbf{w}_k^{\mathrm{T}})^{\mathrm{T}}$ and $\widetilde{\mathbf{x}}$ is the corresponding augmented input vector $(1, \mathbf{x}^{\mathrm{T}})^{\mathrm{T}}$ with a dummy input $x_0 = 1$. This representation was discussed in detail in Section 3.1. A new input $\mathbf{x}$ is then assigned to the class for which the output $y_k = \widetilde{\mathbf{w}}_k^{\mathrm{T}}\widetilde{\mathbf{x}}$ is largest.

We now determine the parameter matrix $\widetilde{\mathbf{W}}$ by minimizing a sum-of-squares error function, as we did for regression in Chapter 3. Consider a training data set $\{\mathbf{x}_n, \mathbf{t}_n\}$ where $n = 1, \ldots, N$, and define a matrix $\mathbf{T}$ whose n^{th} row is the vector $\mathbf{t}_n^{\mathrm{T}}$, together with a matrix $\widetilde{\mathbf{X}}$ whose n^{th} row is $\widetilde{\mathbf{x}}_n^{\mathrm{T}}$. The sum-of-squares error function can then be written as

$$E_D(\widetilde{\mathbf{W}}) = \frac{1}{2}\mathrm{Tr}\left\{(\widetilde{\mathbf{X}}\widetilde{\mathbf{W}} - \mathbf{T})^{\mathrm{T}}(\widetilde{\mathbf{X}}\widetilde{\mathbf{W}} - \mathbf{T})\right\}. \tag{4.15}$$

Setting the derivative with respect to $\widetilde{\mathbf{W}}$ to zero, and rearranging, we then obtain the solution for $\widetilde{\mathbf{W}}$ in the form

$$\widetilde{\mathbf{W}} = (\widetilde{\mathbf{X}}^{\mathrm{T}}\widetilde{\mathbf{X}})^{-1}\widetilde{\mathbf{X}}^{\mathrm{T}}\mathbf{T} = \widetilde{\mathbf{X}}^{\dagger}\mathbf{T} \tag{4.16}$$

where $\widetilde{\mathbf{X}}^{\dagger}$ is the pseudo-inverse of the matrix $\widetilde{\mathbf{X}}$, as discussed in Section 3.1.1. We then obtain the discriminant function in the form

$$\mathbf{y}(\mathbf{x}) = \widetilde{\mathbf{W}}^{\mathrm{T}}\widetilde{\mathbf{x}} = \mathbf{T}^{\mathrm{T}}\left(\widetilde{\mathbf{X}}^{\dagger}\right)^{\mathrm{T}}\widetilde{\mathbf{x}}. \tag{4.17}$$

An interesting property of least-squares solutions with multiple target variables is that if every target vector in the training set satisfies some linear constraint

$$\mathbf{a}^{\mathrm{T}}\mathbf{t}_n + b = 0 \tag{4.18}$$

for some constants $\mathbf{a}$ and b, then the model prediction for any value of $\mathbf{x}$ will satisfy the same constraint so that

Exercise 4.2

$$\mathbf{a}^{\mathrm{T}}\mathbf{y}(\mathbf{x}) + b = 0. \tag{4.19}$$

Thus if we use a 1-of-K coding scheme for K classes, then the predictions made by the model will have the property that the elements of $\mathbf{y}(\mathbf{x})$ will sum to 1 for any value of $\mathbf{x}$. However, this summation constraint alone is not sufficient to allow the model outputs to be interpreted as probabilities because they are not constrained to lie within the interval $(0, 1)$.

The least-squares approach gives an exact closed-form solution for the discriminant function parameters. However, even as a discriminant function (where we use it to make decisions directly and dispense with any probabilistic interpretation) it suffers from some severe problems. We have already seen that least-squares solutions lack robustness to outliers, and this applies equally to the classification application, as illustrated in Figure 4.4. Here we see that the additional data points in the right-hand figure produce a significant change in the location of the decision boundary, even though these points would be correctly classified by the original decision boundary in the left-hand figure. The sum-of-squares error function penalizes predictions that are 'too correct' in that they lie a long way on the correct side of the decision

Section 2.3.7

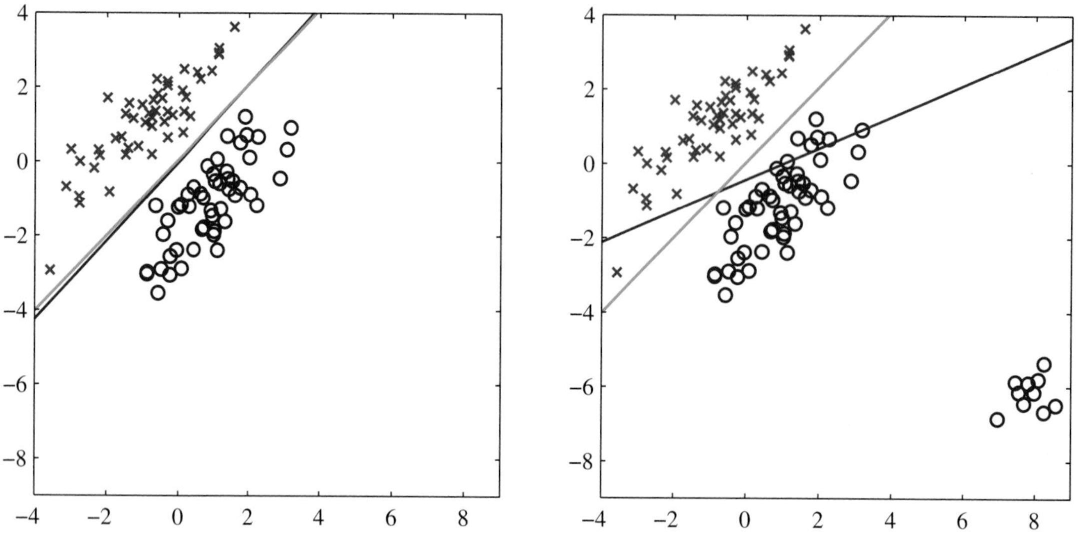

Figure 4.4 The left plot shows data from two classes, denoted by red crosses and blue circles, together with the decision boundary found by least squares (magenta curve) and also by the logistic regression model (green curve), which is discussed later in Section 4.3.2 The right-hand plot shows the corresponding results obtained when extra data points are added at the bottom left of the diagram, showing that least squares is highly sensitive to outliers, unlike logistic regression.

boundary. In Section 7.1.2, we shall consider several alternative error functions for classification and we shall see that they do not suffer from this difficulty.

However, problems with least squares can be more severe than simply lack of robustness, as illustrated in Figure 4.5.This shows a synthetic data set drawn from three classes in a two-dimensional input space (x_1, x_2), having the property that linear decision boundaries can give excellent separation between the classes. Indeed, the technique of logistic regression, described later in this chapter, gives a satisfactory solution as seen in the right-hand plot. However, the least-squares solution gives poor results, with only a small region of the input space assigned to the green class.

The failure of least squares should not surprise us when we recall that it corresponds to maximum likelihood under the assumption of a Gaussian conditional distribution, whereas binary target vectors clearly have a distribution that is far from Gaussian. By adopting more appropriate probabilistic models, we shall obtain classification techniques with much better properties than least squares. For the moment, however, we continue to explore alternative nonprobabilistic methods for setting the parameters in the linear classification models.

4.1.4 Fisher's linear discriminant

One way to view a linear classification model is in terms of dimensionality reduction. Consider first the case of two classes, and suppose we take the D-

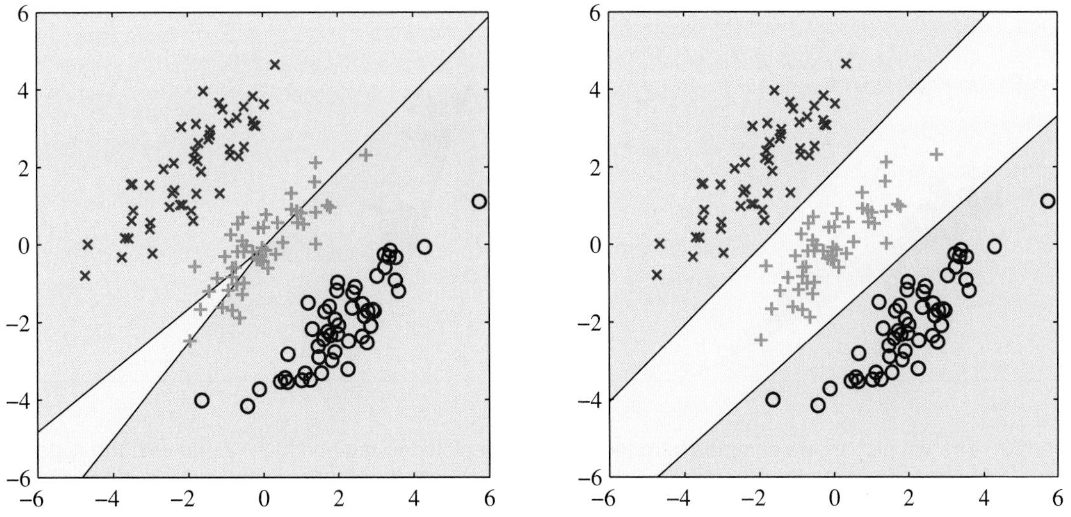

Figure 4.5 Example of a synthetic data set comprising three classes, with training data points denoted in red ($\times$), green ($+$), and blue ($\circ$). Lines denote the decision boundaries, and the background colours denote the respective classes of the decision regions. On the left is the result of using a least-squares discriminant. We see that the region of input space assigned to the green class is too small and so most of the points from this class are misclassified. On the right is the result of using logistic regressions as described in Section 4.3.2 showing correct classification of the training data.

dimensional input vector $\mathbf{x}$ and project it down to one dimension using

$$y = \mathbf{w}^{\mathrm{T}}\mathbf{x}. \tag{4.20}$$

If we place a threshold on y and classify $y \geqslant -w_0$ as class $\mathcal{C}_1$, and otherwise class $\mathcal{C}_2$, then we obtain our standard linear classifier discussed in the previous section. In general, the projection onto one dimension leads to a considerable loss of information, and classes that are well separated in the original D-dimensional space may become strongly overlapping in one dimension. However, by adjusting the components of the weight vector $\mathbf{w}$, we can select a projection that maximizes the class separation. To begin with, consider a two-class problem in which there are N_1 points of class $\mathcal{C}_1$ and N_2 points of class $\mathcal{C}_2$, so that the mean vectors of the two classes are given by

$$\mathbf{m}_1 = \frac{1}{N_1}\sum_{n \in \mathcal{C}_1} \mathbf{x}_n, \qquad \mathbf{m}_2 = \frac{1}{N_2}\sum_{n \in \mathcal{C}_2} \mathbf{x}_n. \tag{4.21}$$

The simplest measure of the separation of the classes, when projected onto $\mathbf{w}$, is the separation of the projected class means. This suggests that we might choose $\mathbf{w}$ so as to maximize

$$m_2 - m_1 = \mathbf{w}^{\mathrm{T}}(\mathbf{m}_2 - \mathbf{m}_1) \tag{4.22}$$

where

$$m_k = \mathbf{w}^{\mathrm{T}}\mathbf{m}_k \tag{4.23}$$

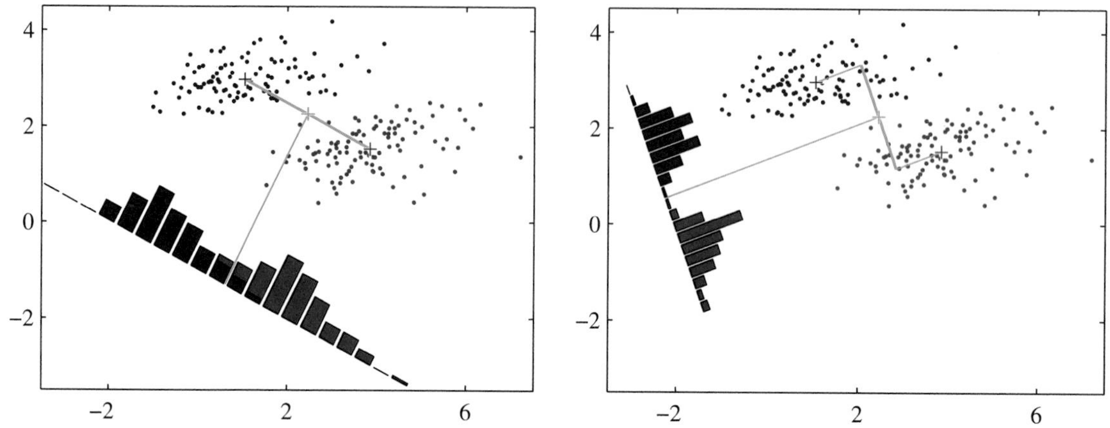

Figure 4.6 The left plot shows samples from two classes (depicted in red and blue) along with the histograms resulting from projection onto the line joining the class means. Note that there is considerable class overlap in the projected space. The right plot shows the corresponding projection based on the Fisher linear discriminant, showing the greatly improved class separation.

is the mean of the projected data from class $\mathcal{C}_k$. However, this expression can be made arbitrarily large simply by increasing the magnitude of $\mathbf{w}$. To solve this problem, we could constrain $\mathbf{w}$ to have unit length, so that $\sum_i w_i^2 = 1$. Using a Lagrange multiplier to perform the constrained maximization, we then find that $\mathbf{w} \propto (\mathbf{m}_2 - \mathbf{m}_1)$. There is still a problem with this approach, however, as illustrated in Figure 4.6. This shows two classes that are well separated in the original two-dimensional space (x_1, x_2) but that have considerable overlap when projected onto the line joining their means. This difficulty arises from the strongly nondiagonal covariances of the class distributions. The idea proposed by Fisher is to maximize a function that will give a large separation between the projected class means while also giving a small variance within each class, thereby minimizing the class overlap.

Appendix E

Exercise 4.4

The projection formula (4.20) transforms the set of labelled data points in $\mathbf{x}$ into a labelled set in the one-dimensional space y. The within-class variance of the transformed data from class $\mathcal{C}_k$ is therefore given by

$$s_k^2 = \sum_{n \in \mathcal{C}_k} (y_n - m_k)^2 \tag{4.24}$$

where $y_n = \mathbf{w}^{\mathrm{T}}\mathbf{x}_n$. We can define the total within-class variance for the whole data set to be simply $s_1^2 + s_2^2$. The Fisher criterion is defined to be the ratio of the between-class variance to the within-class variance and is given by

$$J(\mathbf{w}) = \frac{(m_2 - m_1)^2}{s_1^2 + s_2^2}. \tag{4.25}$$

We can make the dependence on $\mathbf{w}$ explicit by using (4.20), (4.23), and (4.24) to rewrite the Fisher criterion in the form

Exercise 4.5

$$J(\mathbf{w}) = \frac{\mathbf{w}^{\mathrm{T}}\mathbf{S}_{\mathrm{B}}\mathbf{w}}{\mathbf{w}^{\mathrm{T}}\mathbf{S}_{\mathrm{W}}\mathbf{w}} \tag{4.26}$$

where $\mathbf{S}_{\mathrm{B}}$ is the *between-class* covariance matrix and is given by

$$\mathbf{S}_{\mathrm{B}} = (\mathbf{m}_2 - \mathbf{m}_1)(\mathbf{m}_2 - \mathbf{m}_1)^{\mathrm{T}} \tag{4.27}$$

and $\mathbf{S}_{\mathrm{W}}$ is the total *within-class* covariance matrix, given by

$$\mathbf{S}_{\mathrm{W}} = \sum_{n \in \mathcal{C}_1} (\mathbf{x}_n - \mathbf{m}_1)(\mathbf{x}_n - \mathbf{m}_1)^{\mathrm{T}} + \sum_{n \in \mathcal{C}_2} (\mathbf{x}_n - \mathbf{m}_2)(\mathbf{x}_n - \mathbf{m}_2)^{\mathrm{T}}. \tag{4.28}$$

Differentiating (4.26) with respect to $\mathbf{w}$, we find that $J(\mathbf{w})$ is maximized when

$$(\mathbf{w}^{\mathrm{T}}\mathbf{S}_{\mathrm{B}}\mathbf{w})\mathbf{S}_{\mathrm{W}}\mathbf{w} = (\mathbf{w}^{\mathrm{T}}\mathbf{S}_{\mathrm{W}}\mathbf{w})\mathbf{S}_{\mathrm{B}}\mathbf{w}. \tag{4.29}$$

From (4.27), we see that $\mathbf{S}_{\mathrm{B}}\mathbf{w}$ is always in the direction of $(\mathbf{m}_2 - \mathbf{m}_1)$. Furthermore, we do not care about the magnitude of $\mathbf{w}$, only its direction, and so we can drop the scalar factors $(\mathbf{w}^{\mathrm{T}}\mathbf{S}_{\mathrm{B}}\mathbf{w})$ and $(\mathbf{w}^{\mathrm{T}}\mathbf{S}_{\mathrm{W}}\mathbf{w})$. Multiplying both sides of (4.29) by $\mathbf{S}_{\mathrm{W}}^{-1}$ we then obtain

$$\mathbf{w} \propto \mathbf{S}_{\mathrm{W}}^{-1}(\mathbf{m}_2 - \mathbf{m}_1). \tag{4.30}$$

Note that if the within-class covariance is isotropic, so that $\mathbf{S}_{\mathrm{W}}$ is proportional to the unit matrix, we find that $\mathbf{w}$ is proportional to the difference of the class means, as discussed above.

The result (4.30) is known as *Fisher's linear discriminant*, although strictly it is not a discriminant but rather a specific choice of direction for projection of the data down to one dimension. However, the projected data can subsequently be used to construct a discriminant, by choosing a threshold y_0 so that we classify a new point as belonging to $\mathcal{C}_1$ if $y(\mathbf{x}) \geqslant y_0$ and classify it as belonging to $\mathcal{C}_2$ otherwise. For example, we can model the class-conditional densities $p(y|\mathcal{C}_k)$ using Gaussian distributions and then use the techniques of Section 1.2.4 to find the parameters of the Gaussian distributions by maximum likelihood. Having found Gaussian approximations to the projected classes, the formalism of Section 1.5.1 then gives an expression for the optimal threshold. Some justification for the Gaussian assumption comes from the central limit theorem by noting that $y = \mathbf{w}^{\mathrm{T}}\mathbf{x}$ is the sum of a set of random variables.

4.1.5 Relation to least squares

The least-squares approach to the determination of a linear discriminant was based on the goal of making the model predictions as close as possible to a set of target values. By contrast, the Fisher criterion was derived by requiring maximum class separation in the output space. It is interesting to see the relationship between these two approaches. In particular, we shall show that, for the two-class problem, the Fisher criterion can be obtained as a special case of least squares.

So far we have considered 1-of-K coding for the target values. If, however, we adopt a slightly different target coding scheme, then the least-squares solution for the weights becomes equivalent to the Fisher solution (Duda and Hart, 1973). In particular, we shall take the targets for class $\mathcal{C}_1$ to be N/N_1, where N_1 is the number of patterns in class $\mathcal{C}_1$, and N is the total number of patterns. This target value approximates the reciprocal of the prior probability for class $\mathcal{C}_1$. For class $\mathcal{C}_2$, we shall take the targets to be $-N/N_2$, where N_2 is the number of patterns in class $\mathcal{C}_2$.

The sum-of-squares error function can be written

$$E = \frac{1}{2}\sum_{n=1}^{N}\left(\mathbf{w}^{\mathrm{T}}\mathbf{x}_n + w_0 - t_n\right)^2. \tag{4.31}$$

Setting the derivatives of E with respect to w_0 and $\mathbf{w}$ to zero, we obtain respectively

$$\sum_{n=1}^{N}\left(\mathbf{w}^{\mathrm{T}}\mathbf{x}_n + w_0 - t_n\right) = 0 \tag{4.32}$$

$$\sum_{n=1}^{N}\left(\mathbf{w}^{\mathrm{T}}\mathbf{x}_n + w_0 - t_n\right)\mathbf{x}_n = 0. \tag{4.33}$$

From (4.32), and making use of our choice of target coding scheme for the t_n, we obtain an expression for the bias in the form

$$w_0 = -\mathbf{w}^{\mathrm{T}}\mathbf{m} \tag{4.34}$$

where we have used

$$\sum_{n=1}^{N} t_n = N_1\frac{N}{N_1} - N_2\frac{N}{N_2} = 0 \tag{4.35}$$

and where $\mathbf{m}$ is the mean of the total data set and is given by

$$\mathbf{m} = \frac{1}{N}\sum_{n=1}^{N}\mathbf{x}_n = \frac{1}{N}(N_1\mathbf{m}_1 + N_2\mathbf{m}_2). \tag{4.36}$$

Exercise 4.6

After some straightforward algebra, and again making use of the choice of t_n, the second equation (4.33) becomes

$$\left(\mathbf{S}_{\mathrm{W}} + \frac{N_1 N_2}{N}\mathbf{S}_{\mathrm{B}}\right)\mathbf{w} = N(\mathbf{m}_1 - \mathbf{m}_2) \tag{4.37}$$

where $\mathbf{S}_{\mathrm{W}}$ is defined by (4.28), $\mathbf{S}_{\mathrm{B}}$ is defined by (4.27), and we have substituted for the bias using (4.34). Using (4.27), we note that $\mathbf{S}_{\mathrm{B}}\mathbf{w}$ is always in the direction of $(\mathbf{m}_2 - \mathbf{m}_1)$. Thus we can write

$$\mathbf{w} \propto \mathbf{S}_{\mathrm{W}}^{-1}(\mathbf{m}_2 - \mathbf{m}_1) \tag{4.38}$$

where we have ignored irrelevant scale factors. Thus the weight vector coincides with that found from the Fisher criterion. In addition, we have also found an expression for the bias value w_0 given by (4.34). This tells us that a new vector $\mathbf{x}$ should be classified as belonging to class $\mathcal{C}_1$ if $y(\mathbf{x}) = \mathbf{w}^{\mathrm{T}}(\mathbf{x}-\mathbf{m}) > 0$ and class $\mathcal{C}_2$ otherwise.

4.1.6 Fisher's discriminant for multiple classes

We now consider the generalization of the Fisher discriminant to $K > 2$ classes, and we shall assume that the dimensionality D of the input space is greater than the number K of classes. Next, we introduce $D' > 1$ linear 'features' $y_k = \mathbf{w}_k^{\mathrm{T}}\mathbf{x}$, where $k = 1, \ldots, D'$. These feature values can conveniently be grouped together to form a vector $\mathbf{y}$. Similarly, the weight vectors $\{\mathbf{w}_k\}$ can be considered to be the columns of a matrix $\mathbf{W}$, so that

$$\mathbf{y} = \mathbf{W}^{\mathrm{T}}\mathbf{x}. \tag{4.39}$$

Note that again we are not including any bias parameters in the definition of $\mathbf{y}$. The generalization of the within-class covariance matrix to the case of K classes follows from (4.28) to give

$$\mathbf{S}_{\mathrm{W}} = \sum_{k=1}^{K} \mathbf{S}_k \tag{4.40}$$

where

$$\mathbf{S}_k = \sum_{n \in \mathcal{C}_k} (\mathbf{x}_n - \mathbf{m}_k)(\mathbf{x}_n - \mathbf{m}_k)^{\mathrm{T}} \tag{4.41}$$

$$\mathbf{m}_k = \frac{1}{N_k} \sum_{n \in \mathcal{C}_k} \mathbf{x}_n \tag{4.42}$$

and N_k is the number of patterns in class $\mathcal{C}_k$. In order to find a generalization of the between-class covariance matrix, we follow Duda and Hart (1973) and consider first the total covariance matrix

$$\mathbf{S}_{\mathrm{T}} = \sum_{n=1}^{N} (\mathbf{x}_n - \mathbf{m})(\mathbf{x}_n - \mathbf{m})^{\mathrm{T}} \tag{4.43}$$

where $\mathbf{m}$ is the mean of the total data set

$$\mathbf{m} = \frac{1}{N} \sum_{n=1}^{N} \mathbf{x}_n = \frac{1}{N} \sum_{k=1}^{K} N_k \mathbf{m}_k \tag{4.44}$$

and $N = \sum_k N_k$ is the total number of data points. The total covariance matrix can be decomposed into the sum of the within-class covariance matrix, given by (4.40) and (4.41), plus an additional matrix $\mathbf{S}_{\mathrm{B}}$, which we identify as a measure of the between-class covariance

$$\mathbf{S}_{\mathrm{T}} = \mathbf{S}_{\mathrm{W}} + \mathbf{S}_{\mathrm{B}} \tag{4.45}$$

where

$$\mathbf{S}_{\mathrm{B}} = \sum_{k=1}^{K} N_k (\mathbf{m}_k - \mathbf{m})(\mathbf{m}_k - \mathbf{m})^{\mathrm{T}}. \tag{4.46}$$

These covariance matrices have been defined in the original $\mathbf{x}$-space. We can now define similar matrices in the projected D'-dimensional $\mathbf{y}$-space

$$\mathbf{s}_{\mathrm{W}} = \sum_{k=1}^{K} \sum_{n \in \mathcal{C}_k} (\mathbf{y}_n - \boldsymbol{\mu}_k)(\mathbf{y}_n - \boldsymbol{\mu}_k)^{\mathrm{T}} \tag{4.47}$$

and

$$\mathbf{s}_{\mathrm{B}} = \sum_{k=1}^{K} N_k (\boldsymbol{\mu}_k - \boldsymbol{\mu})(\boldsymbol{\mu}_k - \boldsymbol{\mu})^{\mathrm{T}} \tag{4.48}$$

where

$$\boldsymbol{\mu}_k = \frac{1}{N_k} \sum_{n \in \mathcal{C}_k} \mathbf{y}_n, \qquad \boldsymbol{\mu} = \frac{1}{N} \sum_{k=1}^{K} N_k \boldsymbol{\mu}_k. \tag{4.49}$$

Again we wish to construct a scalar that is large when the between-class covariance is large and when the within-class covariance is small. There are now many possible choices of criterion (Fukunaga, 1990). One example is given by

$$J(\mathbf{W}) = \mathrm{Tr}\left\{ \mathbf{s}_{\mathrm{W}}^{-1} \mathbf{s}_{\mathrm{B}} \right\}. \tag{4.50}$$

This criterion can then be rewritten as an explicit function of the projection matrix $\mathbf{W}$ in the form

$$J(\mathbf{w}) = \mathrm{Tr}\left\{ (\mathbf{W}\mathbf{S}_{\mathrm{W}}\mathbf{W}^{\mathrm{T}})^{-1} (\mathbf{W}\mathbf{S}_{\mathrm{B}}\mathbf{W}^{\mathrm{T}}) \right\}. \tag{4.51}$$

Maximization of such criteria is straightforward, though somewhat involved, and is discussed at length in Fukunaga (1990). The weight values are determined by those eigenvectors of $\mathbf{S}_{\mathrm{W}}^{-1}\mathbf{S}_{\mathrm{B}}$ that correspond to the D' largest eigenvalues.

There is one important result that is common to all such criteria, which is worth emphasizing. We first note from (4.46) that $\mathbf{S}_{\mathrm{B}}$ is composed of the sum of K matrices, each of which is an outer product of two vectors and therefore of rank 1. In addition, only $(K-1)$ of these matrices are independent as a result of the constraint (4.44). Thus, $\mathbf{S}_{\mathrm{B}}$ has rank at most equal to $(K-1)$ and so there are at most $(K-1)$ nonzero eigenvalues. This shows that the projection onto the $(K-1)$-dimensional subspace spanned by the eigenvectors of $\mathbf{S}_{\mathrm{B}}$ does not alter the value of $J(\mathbf{w})$, and so we are therefore unable to find more than $(K-1)$ linear 'features' by this means (Fukunaga, 1990).

4.1.7 The perceptron algorithm

Another example of a linear discriminant model is the perceptron of Rosenblatt (1962), which occupies an important place in the history of pattern recognition algorithms. It corresponds to a two-class model in which the input vector $\mathbf{x}$ is first transformed using a fixed nonlinear transformation to give a feature vector $\boldsymbol{\phi}(\mathbf{x})$, and this is then used to construct a generalized linear model of the form

$$y(\mathbf{x}) = f\left(\mathbf{w}^{\mathrm{T}}\boldsymbol{\phi}(\mathbf{x})\right) \tag{4.52}$$

where the nonlinear activation function $f(\cdot)$ is given by a step function of the form

$$f(a) = \begin{cases} +1, & a \geqslant 0 \\ -1, & a < 0. \end{cases} \tag{4.53}$$

The vector $\boldsymbol{\phi}(\mathbf{x})$ will typically include a bias component $\phi_0(\mathbf{x}) = 1$. In earlier discussions of two-class classification problems, we have focussed on a target coding scheme in which $t \in \{0, 1\}$, which is appropriate in the context of probabilistic models. For the perceptron, however, it is more convenient to use target values $t = +1$ for class $\mathcal{C}_1$ and $t = -1$ for class $\mathcal{C}_2$, which matches the choice of activation function.

The algorithm used to determine the parameters $\mathbf{w}$ of the perceptron can most easily be motivated by error function minimization. A natural choice of error function would be the total number of misclassified patterns. However, this does not lead to a simple learning algorithm because the error is a piecewise constant function of $\mathbf{w}$, with discontinuities wherever a change in $\mathbf{w}$ causes the decision boundary to move across one of the data points. Methods based on changing $\mathbf{w}$ using the gradient of the error function cannot then be applied, because the gradient is zero almost everywhere.

We therefore consider an alternative error function known as the *perceptron criterion*. To derive this, we note that we are seeking a weight vector $\mathbf{w}$ such that patterns $\mathbf{x}_n$ in class $\mathcal{C}_1$ will have $\mathbf{w}^{\mathrm{T}}\boldsymbol{\phi}(\mathbf{x}_n) > 0$, whereas patterns $\mathbf{x}_n$ in class $\mathcal{C}_2$ have $\mathbf{w}^{\mathrm{T}}\boldsymbol{\phi}(\mathbf{x}_n) < 0$. Using the $t \in \{-1, +1\}$ target coding scheme it follows that we would like all patterns to satisfy $\mathbf{w}^{\mathrm{T}}\boldsymbol{\phi}(\mathbf{x}_n)t_n > 0$. The perceptron criterion associates zero error with any pattern that is correctly classified, whereas for a misclassified pattern $\mathbf{x}_n$ it tries to minimize the quantity $-\mathbf{w}^{\mathrm{T}}\boldsymbol{\phi}(\mathbf{x}_n)t_n$. The perceptron criterion is therefore given by

$$E_{\mathrm{P}}(\mathbf{w}) = -\sum_{n \in \mathcal{M}} \mathbf{w}^{\mathrm{T}}\boldsymbol{\phi}_n t_n \tag{4.54}$$

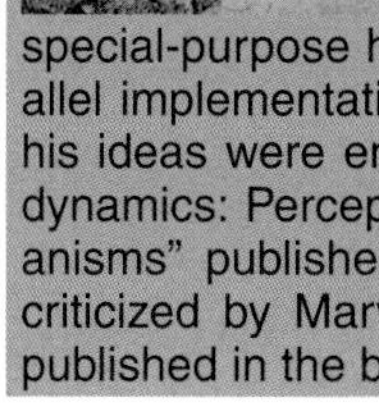

Frank Rosenblatt
1928–1969

Rosenblatt's perceptron played an important role in the history of machine learning. Initially, Rosenblatt simulated the perceptron on an IBM 704 computer at Cornell in 1957, but by the early 1960s he had built special-purpose hardware that provided a direct, parallel implementation of perceptron learning. Many of his ideas were encapsulated in "Principles of Neurodynamics: Perceptrons and the Theory of Brain Mechanisms" published in 1962. Rosenblatt's work was criticized by Marvin Minksy, whose objections were published in the book "Perceptrons", co-authored with Seymour Papert. This book was widely misinterpreted at the time as showing that neural networks were fatally flawed and could only learn solutions for linearly separable problems. In fact, it only proved such limitations in the case of single-layer networks such as the perceptron and merely conjectured (incorrectly) that they applied to more general network models. Unfortunately, however, this book contributed to the substantial decline in research funding for neural computing, a situation that was not reversed until the mid-1980s. Today, there are many hundreds, if not thousands, of applications of neural networks in widespread use, with examples in areas such as handwriting recognition and information retrieval being used routinely by millions of people.

where $\boldsymbol{\phi}_n = \boldsymbol{\phi}(\mathbf{x}_n)$ and $\mathcal{M}$ denotes the set of all misclassified patterns. The contribution to the error associated with a particular misclassified pattern is a linear function of $\mathbf{w}$ in regions of $\mathbf{w}$ space where the pattern is misclassified and zero in regions where it is correctly classified. The total error function is therefore piecewise linear.

Section 3.1.3

We now apply the stochastic gradient descent algorithm to this error function. The change in the weight vector $\mathbf{w}$ is then given by

$$\mathbf{w}^{(\tau+1)} = \mathbf{w}^{(\tau)} - \eta \nabla E_{\mathrm{P}}(\mathbf{w}) = \mathbf{w}^{(\tau)} + \eta \boldsymbol{\phi}_n t_n \tag{4.55}$$

where η is the learning rate parameter and τ is an integer that indexes the steps of the algorithm. Because the perceptron function $y(\mathbf{x}, \mathbf{w})$ is unchanged if we multiply $\mathbf{w}$ by a constant, we can set the learning rate parameter η equal to 1 without loss of generality. Note that, as the weight vector evolves during training, the set of patterns that are misclassified will change.

The perceptron learning algorithm has a simple interpretation, as follows. We cycle through the training patterns in turn, and for each pattern $\mathbf{x}_n$ we evaluate the perceptron function (4.52). If the pattern is correctly classified, then the weight vector remains unchanged, whereas if it is incorrectly classified, then for class $\mathcal{C}_1$ we add the vector $\boldsymbol{\phi}(\mathbf{x}_n)$ onto the current estimate of weight vector $\mathbf{w}$ while for class $\mathcal{C}_2$ we subtract the vector $\boldsymbol{\phi}(\mathbf{x}_n)$ from $\mathbf{w}$. The perceptron learning algorithm is illustrated in Figure 4.7.

If we consider the effect of a single update in the perceptron learning algorithm, we see that the contribution to the error from a misclassified pattern will be reduced because from (4.55) we have

$$-\mathbf{w}^{(\tau+1)\mathrm{T}} \boldsymbol{\phi}_n t_n = -\mathbf{w}^{(\tau)\mathrm{T}} \boldsymbol{\phi}_n t_n - (\boldsymbol{\phi}_n t_n)^{\mathrm{T}} \boldsymbol{\phi}_n t_n < -\mathbf{w}^{(\tau)\mathrm{T}} \boldsymbol{\phi}_n t_n \tag{4.56}$$

where we have set $\eta = 1$, and made use of $\|\boldsymbol{\phi}_n t_n\|^2 > 0$. Of course, this does not imply that the contribution to the error function from the other misclassified patterns will have been reduced. Furthermore, the change in weight vector may have caused some previously correctly classified patterns to become misclassified. Thus the perceptron learning rule is not guaranteed to reduce the total error function at each stage.

However, the *perceptron convergence theorem* states that if there exists an exact solution (in other words, if the training data set is linearly separable), then the perceptron learning algorithm is guaranteed to find an exact solution in a finite number of steps. Proofs of this theorem can be found for example in Rosenblatt (1962), Block (1962), Nilsson (1965), Minsky and Papert (1969), Hertz *et al.* (1991), and Bishop (1995a). Note, however, that the number of steps required to achieve convergence could still be substantial, and in practice, until convergence is achieved, we will not be able to distinguish between a nonseparable problem and one that is simply slow to converge.

Even when the data set is linearly separable, there may be many solutions, and which one is found will depend on the initialization of the parameters and on the order of presentation of the data points. Furthermore, for data sets that are not linearly separable, the perceptron learning algorithm will never converge.

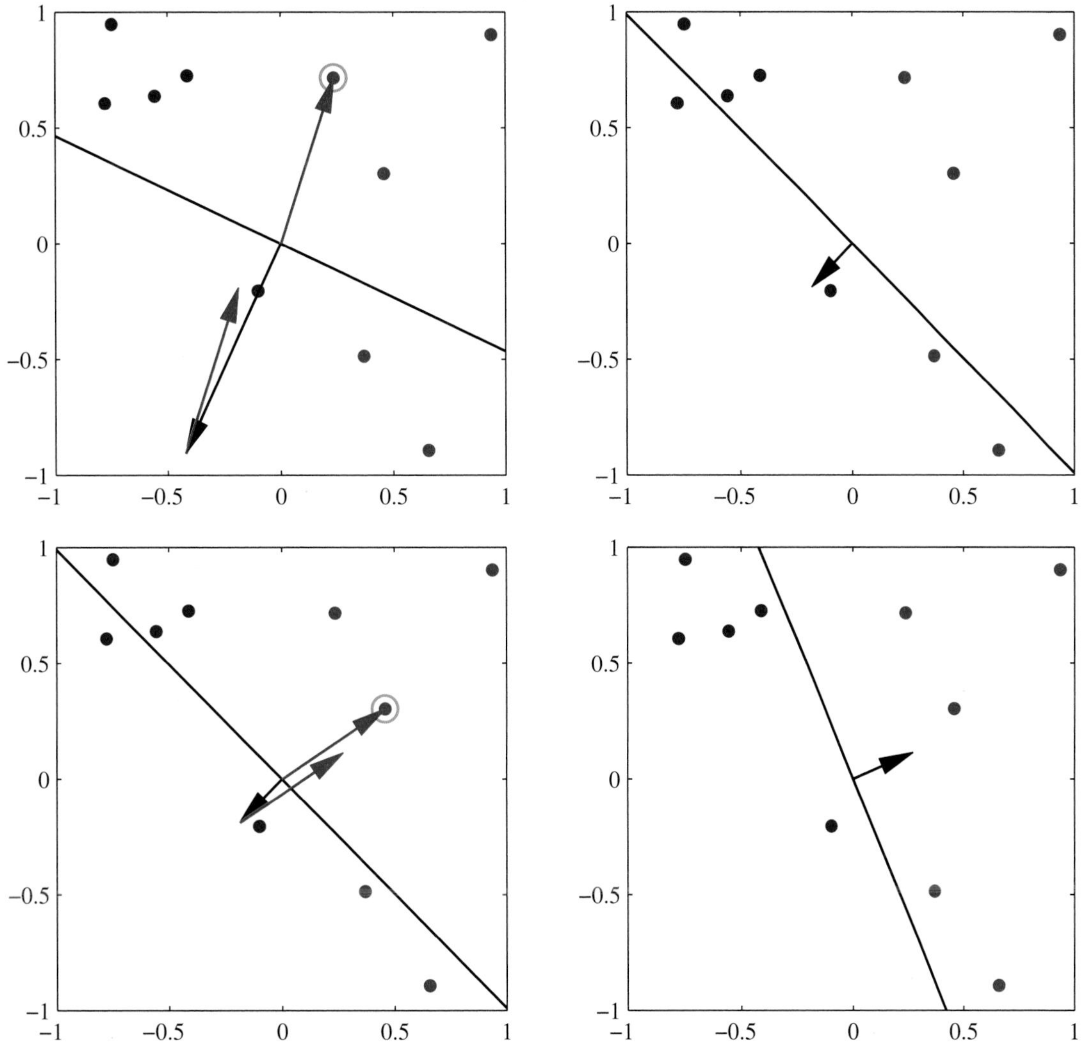

Figure 4.7 Illustration of the convergence of the perceptron learning algorithm, showing data points from two classes (red and blue) in a two-dimensional feature space (ϕ_1, ϕ_2). The top left plot shows the initial parameter vector $\mathbf{w}$ shown as a black arrow together with the corresponding decision boundary (black line), in which the arrow points towards the decision region which classified as belonging to the red class. The data point circled in green is misclassified and so its feature vector is added to the current weight vector, giving the new decision boundary shown in the top right plot. The bottom left plot shows the next misclassified point to be considered, indicated by the green circle, and its feature vector is again added to the weight vector giving the decision boundary shown in the bottom right plot for which all data points are correctly classified.

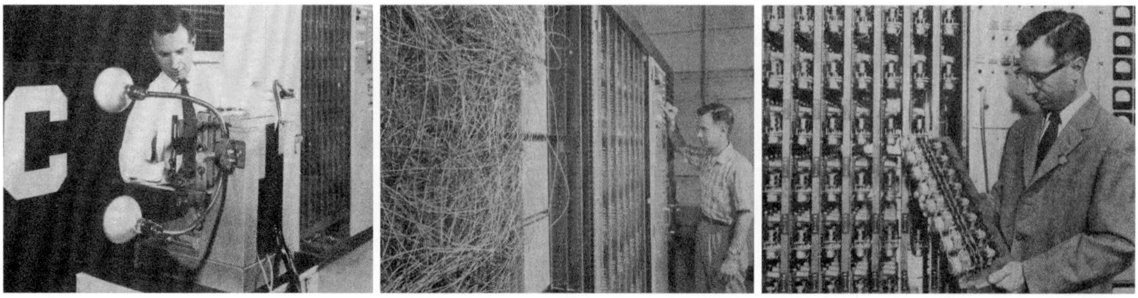

Figure 4.8 Illustration of the Mark 1 perceptron hardware. The photograph on the left shows how the inputs were obtained using a simple camera system in which an input scene, in this case a printed character, was illuminated by powerful lights, and an image focussed onto a 20×20 array of cadmium sulphide photocells, giving a primitive 400 pixel image. The perceptron also had a patch board, shown in the middle photograph, which allowed different configurations of input features to be tried. Often these were wired up at random to demonstrate the ability of the perceptron to learn without the need for precise wiring, in contrast to a modern digital computer. The photograph on the right shows one of the racks of adaptive weights. Each weight was implemented using a rotary variable resistor, also called a potentiometer, driven by an electric motor thereby allowing the value of the weight to be adjusted automatically by the learning algorithm.

Aside from difficulties with the learning algorithm, the perceptron does not provide probabilistic outputs, nor does it generalize readily to $K > 2$ classes. The most important limitation, however, arises from the fact that (in common with all of the models discussed in this chapter and the previous one) it is based on linear combinations of fixed basis functions. More detailed discussions of the limitations of perceptrons can be found in Minsky and Papert (1969) and Bishop (1995a).

Analogue hardware implementations of the perceptron were built by Rosenblatt, based on motor-driven variable resistors to implement the adaptive parameters w_j. These are illustrated in Figure 4.8. The inputs were obtained from a simple camera system based on an array of photo-sensors, while the basis functions ϕ could be chosen in a variety of ways, for example based on simple fixed functions of randomly chosen subsets of pixels from the input image. Typical applications involved learning to discriminate simple shapes or characters.

At the same time that the perceptron was being developed, a closely related system called the *adaline*, which is short for 'adaptive linear element', was being explored by Widrow and co-workers. The functional form of the model was the same as for the perceptron, but a different approach to training was adopted (Widrow and Hoff, 1960; Widrow and Lehr, 1990).

4.2. Probabilistic Generative Models

We turn next to a probabilistic view of classification and show how models with linear decision boundaries arise from simple assumptions about the distribution of the data. In Section 1.5.4, we discussed the distinction between the discriminative and the generative approaches to classification. Here we shall adopt a generative

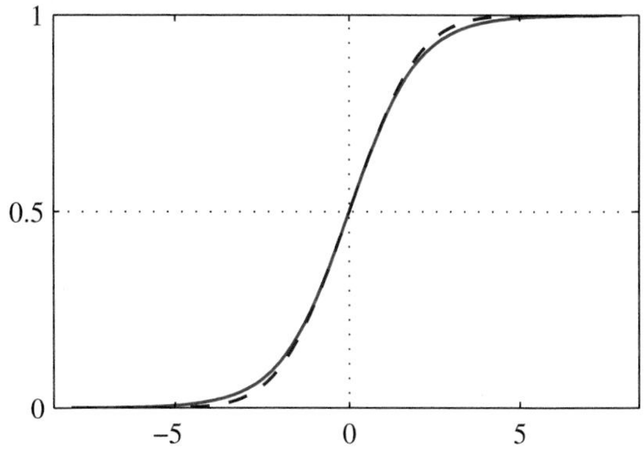

Figure 4.9 Plot of the logistic sigmoid function $\sigma(a)$ defined by (4.59), shown in red, together with the scaled inverse probit function $\Phi(\lambda a)$, for $\lambda^2 = \pi/8$, shown in dashed blue, where $\Phi(a)$ is defined by (4.114). The scaling factor $\pi/8$ is chosen so that the derivatives of the two curves are equal for $a = 0$.

approach in which we model the class-conditional densities $p(\mathbf{x}|\mathcal{C}_k)$, as well as the class priors $p(\mathcal{C}_k)$, and then use these to compute posterior probabilities $p(\mathcal{C}_k|\mathbf{x})$ through Bayes' theorem.

Consider first of all the case of two classes. The posterior probability for class $\mathcal{C}_1$ can be written as

$$\begin{aligned} p(\mathcal{C}_1|\mathbf{x}) &= \frac{p(\mathbf{x}|\mathcal{C}_1)p(\mathcal{C}_1)}{p(\mathbf{x}|\mathcal{C}_1)p(\mathcal{C}_1) + p(\mathbf{x}|\mathcal{C}_2)p(\mathcal{C}_2)} \\ &= \frac{1}{1+\exp(-a)} = \sigma(a) \end{aligned} \tag{4.57}$$

where we have defined

$$a = \ln \frac{p(\mathbf{x}|\mathcal{C}_1)p(\mathcal{C}_1)}{p(\mathbf{x}|\mathcal{C}_2)p(\mathcal{C}_2)} \tag{4.58}$$

and $\sigma(a)$ is the *logistic sigmoid* function defined by

$$\sigma(a) = \frac{1}{1+\exp(-a)} \tag{4.59}$$

which is plotted in Figure 4.9. The term 'sigmoid' means S-shaped. This type of function is sometimes also called a 'squashing function' because it maps the whole real axis into a finite interval. The logistic sigmoid has been encountered already in earlier chapters and plays an important role in many classification algorithms. It satisfies the following symmetry property

$$\sigma(-a) = 1 - \sigma(a) \tag{4.60}$$

as is easily verified. The inverse of the logistic sigmoid is given by

$$a = \ln\left(\frac{\sigma}{1-\sigma}\right) \tag{4.61}$$

and is known as the *logit* function. It represents the log of the ratio of probabilities $\ln\left[p(\mathcal{C}_1|\mathbf{x})/p(\mathcal{C}_2|\mathbf{x})\right]$ for the two classes, also known as the *log odds*.

Note that in (4.57) we have simply rewritten the posterior probabilities in an equivalent form, and so the appearance of the logistic sigmoid may seem rather vacuous. However, it will have significance provided $a(\mathbf{x})$ takes a simple functional form. We shall shortly consider situations in which $a(\mathbf{x})$ is a linear function of $\mathbf{x}$, in which case the posterior probability is governed by a generalized linear model.

For the case of $K > 2$ classes, we have

$$\begin{aligned} p(\mathcal{C}_k|\mathbf{x}) &= \frac{p(\mathbf{x}|\mathcal{C}_k)p(\mathcal{C}_k)}{\sum_j p(\mathbf{x}|\mathcal{C}_j)p(\mathcal{C}_j)} \\ &= \frac{\exp(a_k)}{\sum_j \exp(a_j)} \end{aligned} \tag{4.62}$$

which is known as the *normalized exponential* and can be regarded as a multiclass generalization of the logistic sigmoid. Here the quantities a_k are defined by

$$a_k = \ln\left(p(\mathbf{x}|\mathcal{C}_k)p(\mathcal{C}_k)\right). \tag{4.63}$$

The normalized exponential is also known as the *softmax function*, as it represents a smoothed version of the 'max' function because, if $a_k \gg a_j$ for all $j \neq k$, then $p(\mathcal{C}_k|\mathbf{x}) \simeq 1$, and $p(\mathcal{C}_j|\mathbf{x}) \simeq 0$.

We now investigate the consequences of choosing specific forms for the class-conditional densities, looking first at continuous input variables $\mathbf{x}$ and then discussing briefly the case of discrete inputs.

4.2.1 Continuous inputs

Let us assume that the class-conditional densities are Gaussian and then explore the resulting form for the posterior probabilities. To start with, we shall assume that all classes share the same covariance matrix. Thus the density for class $\mathcal{C}_k$ is given by

$$p(\mathbf{x}|\mathcal{C}_k) = \frac{1}{(2\pi)^{D/2}}\frac{1}{|\mathbf{\Sigma}|^{1/2}}\exp\left\{-\frac{1}{2}(\mathbf{x}-\boldsymbol{\mu}_k)^{\mathrm{T}}\mathbf{\Sigma}^{-1}(\mathbf{x}-\boldsymbol{\mu}_k)\right\}. \tag{4.64}$$

Consider first the case of two classes. From (4.57) and (4.58), we have

$$p(\mathcal{C}_1|\mathbf{x}) = \sigma(\mathbf{w}^{\mathrm{T}}\mathbf{x} + w_0) \tag{4.65}$$

where we have defined

$$\mathbf{w} = \mathbf{\Sigma}^{-1}(\boldsymbol{\mu}_1 - \boldsymbol{\mu}_2) \tag{4.66}$$

$$w_0 = -\frac{1}{2}\boldsymbol{\mu}_1^{\mathrm{T}}\mathbf{\Sigma}^{-1}\boldsymbol{\mu}_1 + \frac{1}{2}\boldsymbol{\mu}_2^{\mathrm{T}}\mathbf{\Sigma}^{-1}\boldsymbol{\mu}_2 + \ln\frac{p(\mathcal{C}_1)}{p(\mathcal{C}_2)}. \tag{4.67}$$

We see that the quadratic terms in $\mathbf{x}$ from the exponents of the Gaussian densities have cancelled (due to the assumption of common covariance matrices) leading to a linear function of $\mathbf{x}$ in the argument of the logistic sigmoid. This result is illustrated for the case of a two-dimensional input space $\mathbf{x}$ in Figure 4.10. The resulting

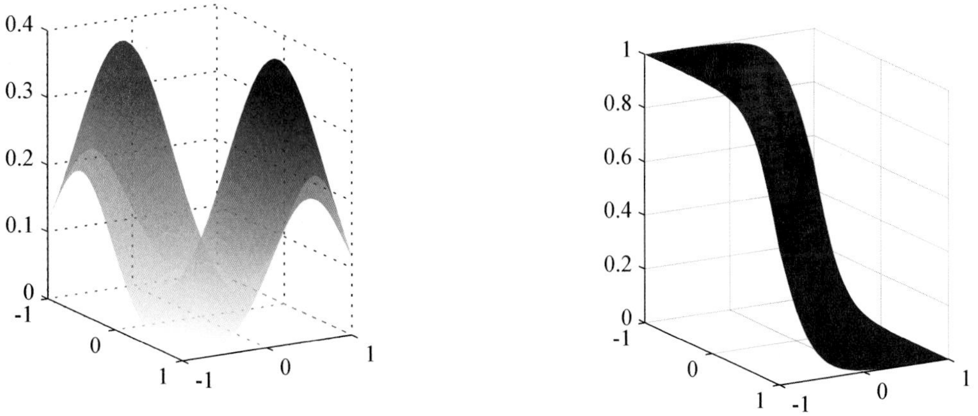

Figure 4.10 The left-hand plot shows the class-conditional densities for two classes, denoted red and blue. On the right is the corresponding posterior probability $p(\mathcal{C}_1|\mathbf{x})$, which is given by a logistic sigmoid of a linear function of $\mathbf{x}$. The surface in the right-hand plot is coloured using a proportion of red ink given by $p(\mathcal{C}_1|\mathbf{x})$ and a proportion of blue ink given by $p(\mathcal{C}_2|\mathbf{x}) = 1 - p(\mathcal{C}_1|\mathbf{x})$.

decision boundaries correspond to surfaces along which the posterior probabilities $p(\mathcal{C}_k|\mathbf{x})$ are constant and so will be given by linear functions of $\mathbf{x}$, and therefore the decision boundaries are linear in input space. The prior probabilities $p(\mathcal{C}_k)$ enter only through the bias parameter w_0 so that changes in the priors have the effect of making parallel shifts of the decision boundary and more generally of the parallel contours of constant posterior probability.

For the general case of K classes we have, from (4.62) and (4.63),

$$a_k(\mathbf{x}) = \mathbf{w}_k^{\mathrm{T}}\mathbf{x} + w_{k0} \tag{4.68}$$

where we have defined

$$\mathbf{w}_k = \boldsymbol{\Sigma}^{-1}\boldsymbol{\mu}_k \tag{4.69}$$

$$w_{k0} = -\frac{1}{2}\boldsymbol{\mu}_k^{\mathrm{T}}\boldsymbol{\Sigma}^{-1}\boldsymbol{\mu}_k + \ln p(\mathcal{C}_k). \tag{4.70}$$

We see that the $a_k(\mathbf{x})$ are again linear functions of $\mathbf{x}$ as a consequence of the cancellation of the quadratic terms due to the shared covariances. The resulting decision boundaries, corresponding to the minimum misclassification rate, will occur when two of the posterior probabilities (the two largest) are equal, and so will be defined by linear functions of $\mathbf{x}$, and so again we have a generalized linear model.

If we relax the assumption of a shared covariance matrix and allow each class-conditional density $p(\mathbf{x}|\mathcal{C}_k)$ to have its own covariance matrix $\boldsymbol{\Sigma}_k$, then the earlier cancellations will no longer occur, and we will obtain quadratic functions of $\mathbf{x}$, giving rise to a *quadratic discriminant*. The linear and quadratic decision boundaries are illustrated in Figure 4.11.

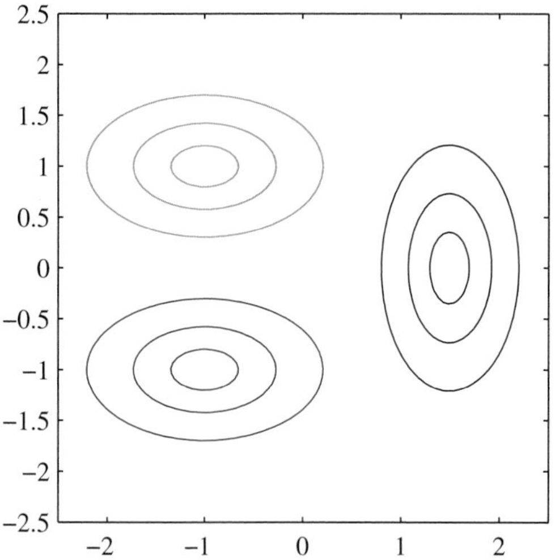

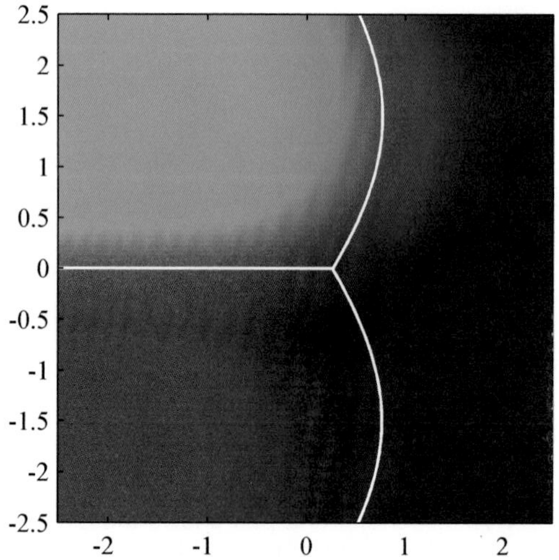

Figure 4.11 The left-hand plot shows the class-conditional densities for three classes each having a Gaussian distribution, coloured red, green, and blue, in which the red and green classes have the same covariance matrix. The right-hand plot shows the corresponding posterior probabilities, in which the RGB colour vector represents the posterior probabilities for the respective three classes. The decision boundaries are also shown. Notice that the boundary between the red and green classes, which have the same covariance matrix, is linear, whereas those between the other pairs of classes are quadratic.

4.2.2 Maximum likelihood solution

Once we have specified a parametric functional form for the class-conditional densities $p(\mathbf{x}|\mathcal{C}_k)$, we can then determine the values of the parameters, together with the prior class probabilities $p(\mathcal{C}_k)$, using maximum likelihood. This requires a data set comprising observations of $\mathbf{x}$ along with their corresponding class labels.

Consider first the case of two classes, each having a Gaussian class-conditional density with a shared covariance matrix, and suppose we have a data set $\{\mathbf{x}_n, t_n\}$ where $n = 1, \ldots, N$. Here $t_n = 1$ denotes class $\mathcal{C}_1$ and $t_n = 0$ denotes class $\mathcal{C}_2$. We denote the prior class probability $p(\mathcal{C}_1) = \pi$, so that $p(\mathcal{C}_2) = 1 - \pi$. For a data point $\mathbf{x}_n$ from class $\mathcal{C}_1$, we have $t_n = 1$ and hence

$$p(\mathbf{x}_n, \mathcal{C}_1) = p(\mathcal{C}_1)p(\mathbf{x}_n|\mathcal{C}_1) = \pi\mathcal{N}(\mathbf{x}_n|\boldsymbol{\mu}_1, \boldsymbol{\Sigma}).$$

Similarly for class $\mathcal{C}_2$, we have $t_n = 0$ and hence

$$p(\mathbf{x}_n, \mathcal{C}_2) = p(\mathcal{C}_2)p(\mathbf{x}_n|\mathcal{C}_2) = (1 - \pi)\mathcal{N}(\mathbf{x}_n|\boldsymbol{\mu}_2, \boldsymbol{\Sigma}).$$

Thus the likelihood function is given by

$$p(\mathsf{t}, \mathbf{X}|\pi, \boldsymbol{\mu}_1, \boldsymbol{\mu}_2, \boldsymbol{\Sigma}) = \prod_{n=1}^{N} \left[\pi\mathcal{N}(\mathbf{x}_n|\boldsymbol{\mu}_1, \boldsymbol{\Sigma})\right]^{t_n} \left[(1 - \pi)\mathcal{N}(\mathbf{x}_n|\boldsymbol{\mu}_2, \boldsymbol{\Sigma})\right]^{1-t_n} \tag{4.71}$$

where $\mathsf{t} = (t_1, \ldots, t_N)^{\mathrm{T}}$. As usual, it is convenient to maximize the log of the likelihood function. Consider first the maximization with respect to π. The terms in the log likelihood function that depend on π are

$$\sum_{n=1}^{N} \{t_n \ln \pi + (1 - t_n) \ln(1 - \pi)\}. \tag{4.72}$$

Setting the derivative with respect to π equal to zero and rearranging, we obtain

$$\pi = \frac{1}{N} \sum_{n=1}^{N} t_n = \frac{N_1}{N} = \frac{N_1}{N_1 + N_2} \tag{4.73}$$

where N_1 denotes the total number of data points in class $\mathcal{C}_1$, and N_2 denotes the total number of data points in class $\mathcal{C}_2$. Thus the maximum likelihood estimate for π is simply the fraction of points in class $\mathcal{C}_1$ as expected. This result is easily generalized to the multiclass case where again the maximum likelihood estimate of the prior probability associated with class $\mathcal{C}_k$ is given by the fraction of the training set points assigned to that class.

Exercise 4.9

Now consider the maximization with respect to $\boldsymbol{\mu}_1$. Again we can pick out of the log likelihood function those terms that depend on $\boldsymbol{\mu}_1$ giving

$$\sum_{n=1}^{N} t_n \ln \mathcal{N}(\mathbf{x}_n | \boldsymbol{\mu}_1, \boldsymbol{\Sigma}) = -\frac{1}{2} \sum_{n=1}^{N} t_n (\mathbf{x}_n - \boldsymbol{\mu}_1)^{\mathrm{T}} \boldsymbol{\Sigma}^{-1} (\mathbf{x}_n - \boldsymbol{\mu}_1) + \text{const}. \tag{4.74}$$

Setting the derivative with respect to $\boldsymbol{\mu}_1$ to zero and rearranging, we obtain

$$\boldsymbol{\mu}_1 = \frac{1}{N_1} \sum_{n=1}^{N} t_n \mathbf{x}_n \tag{4.75}$$

which is simply the mean of all the input vectors $\mathbf{x}_n$ assigned to class $\mathcal{C}_1$. By a similar argument, the corresponding result for $\boldsymbol{\mu}_2$ is given by

$$\boldsymbol{\mu}_2 = \frac{1}{N_2} \sum_{n=1}^{N} (1 - t_n) \mathbf{x}_n \tag{4.76}$$

which again is the mean of all the input vectors $\mathbf{x}_n$ assigned to class $\mathcal{C}_2$.

Finally, consider the maximum likelihood solution for the shared covariance matrix $\boldsymbol{\Sigma}$. Picking out the terms in the log likelihood function that depend on $\boldsymbol{\Sigma}$, we have

$$\begin{aligned}
& -\frac{1}{2} \sum_{n=1}^{N} t_n \ln |\boldsymbol{\Sigma}| - \frac{1}{2} \sum_{n=1}^{N} t_n (\mathbf{x}_n - \boldsymbol{\mu}_1)^{\mathrm{T}} \boldsymbol{\Sigma}^{-1} (\mathbf{x}_n - \boldsymbol{\mu}_1) \\
& -\frac{1}{2} \sum_{n=1}^{N} (1 - t_n) \ln |\boldsymbol{\Sigma}| - \frac{1}{2} \sum_{n=1}^{N} (1 - t_n) (\mathbf{x}_n - \boldsymbol{\mu}_2)^{\mathrm{T}} \boldsymbol{\Sigma}^{-1} (\mathbf{x}_n - \boldsymbol{\mu}_2) \\
& = -\frac{N}{2} \ln |\boldsymbol{\Sigma}| - \frac{N}{2} \mathrm{Tr} \left\{ \boldsymbol{\Sigma}^{-1} \mathbf{S} \right\}
\end{aligned} \tag{4.77}$$

where we have defined

$$\mathbf{S} = \frac{N_1}{N}\mathbf{S}_1 + \frac{N_2}{N}\mathbf{S}_2 \tag{4.78}$$

$$\mathbf{S}_1 = \frac{1}{N_1}\sum_{n\in\mathcal{C}_1}(\mathbf{x}_n - \boldsymbol{\mu}_1)(\mathbf{x}_n - \boldsymbol{\mu}_1)^{\mathrm{T}} \tag{4.79}$$

$$\mathbf{S}_2 = \frac{1}{N_2}\sum_{n\in\mathcal{C}_2}(\mathbf{x}_n - \boldsymbol{\mu}_2)(\mathbf{x}_n - \boldsymbol{\mu}_2)^{\mathrm{T}}. \tag{4.80}$$

Using the standard result for the maximum likelihood solution for a Gaussian distribution, we see that $\boldsymbol{\Sigma} = \mathbf{S}$, which represents a weighted average of the covariance matrices associated with each of the two classes separately.

This result is easily extended to the K class problem to obtain the corresponding maximum likelihood solutions for the parameters in which each class-conditional density is Gaussian with a shared covariance matrix. Note that the approach of fitting Gaussian distributions to the classes is not robust to outliers, because the maximum likelihood estimation of a Gaussian is not robust.

Exercise 4.10

Section 2.3.7

4.2.3 Discrete features

Let us now consider the case of discrete feature values x_i. For simplicity, we begin by looking at binary feature values $x_i \in \{0, 1\}$ and discuss the extension to more general discrete features shortly. If there are D inputs, then a general distribution would correspond to a table of 2^D numbers for each class, containing $2^D - 1$ independent variables (due to the summation constraint). Because this grows exponentially with the number of features, we might seek a more restricted representation. Here we will make the *naive Bayes* assumption in which the feature values are treated as independent, conditioned on the class $\mathcal{C}_k$. Thus we have class-conditional distributions of the form

Section 8.2.2

$$p(\mathbf{x}|\mathcal{C}_k) = \prod_{i=1}^{D} \mu_{ki}^{x_i}(1 - \mu_{ki})^{1-x_i} \tag{4.81}$$

which contain D independent parameters for each class. Substituting into (4.63) then gives

$$a_k(\mathbf{x}) = \sum_{i=1}^{D}\{x_i \ln \mu_{ki} + (1 - x_i)\ln(1 - \mu_{ki})\} + \ln p(\mathcal{C}_k) \tag{4.82}$$

which again are linear functions of the input values x_i. For the case of $K = 2$ classes, we can alternatively consider the logistic sigmoid formulation given by (4.57). Analogous results are obtained for discrete variables each of which can take $M > 2$ states.

Exercise 4.11

4.2.4 Exponential family

As we have seen, for both Gaussian distributed and discrete inputs, the posterior class probabilities are given by generalized linear models with logistic sigmoid ($K =$

2 classes) or softmax ($K \geqslant 2$ classes) activation functions. These are particular cases of a more general result obtained by assuming that the class-conditional densities $p(\mathbf{x}|\mathcal{C}_k)$ are members of the exponential family of distributions.

Using the form (2.194) for members of the exponential family, we see that the distribution of $\mathbf{x}$ can be written in the form

$$p(\mathbf{x}|\boldsymbol{\lambda}_k) = h(\mathbf{x})g(\boldsymbol{\lambda}_k)\exp\left\{\boldsymbol{\lambda}_k^{\mathrm{T}}\mathbf{u}(\mathbf{x})\right\}. \tag{4.83}$$

We now restrict attention to the subclass of such distributions for which $\mathbf{u}(\mathbf{x}) = \mathbf{x}$. Then we make use of (2.236) to introduce a scaling parameter s, so that we obtain the restricted set of exponential family class-conditional densities of the form

$$p(\mathbf{x}|\boldsymbol{\lambda}_k, s) = \frac{1}{s}h\left(\frac{1}{s}\mathbf{x}\right)g(\boldsymbol{\lambda}_k)\exp\left\{\frac{1}{s}\boldsymbol{\lambda}_k^{\mathrm{T}}\mathbf{x}\right\}. \tag{4.84}$$

Note that we are allowing each class to have its own parameter vector $\boldsymbol{\lambda}_k$ but we are assuming that the classes share the same scale parameter s.

For the two-class problem, we substitute this expression for the class-conditional densities into (4.58) and we see that the posterior class probability is again given by a logistic sigmoid acting on a linear function $a(\mathbf{x})$ which is given by

$$a(\mathbf{x}) = \frac{1}{s}(\boldsymbol{\lambda}_1 - \boldsymbol{\lambda}_2)^{\mathrm{T}}\mathbf{x} + \ln g(\boldsymbol{\lambda}_1) - \ln g(\boldsymbol{\lambda}_2) + \ln p(\mathcal{C}_1) - \ln p(\mathcal{C}_2). \tag{4.85}$$

Similarly, for the K-class problem, we substitute the class-conditional density expression into (4.63) to give

$$a_k(\mathbf{x}) = \frac{1}{s}\boldsymbol{\lambda}_k^{\mathrm{T}}\mathbf{x} + \ln g(\boldsymbol{\lambda}_k) + \ln p(\mathcal{C}_k) \tag{4.86}$$

and so again is a linear function of $\mathbf{x}$.

4.3. Probabilistic Discriminative Models

For the two-class classification problem, we have seen that the posterior probability of class $\mathcal{C}_1$ can be written as a logistic sigmoid acting on a linear function of $\mathbf{x}$, for a wide choice of class-conditional distributions $p(\mathbf{x}|\mathcal{C}_k)$. Similarly, for the multiclass case, the posterior probability of class $\mathcal{C}_k$ is given by a softmax transformation of a linear function of $\mathbf{x}$. For specific choices of the class-conditional densities $p(\mathbf{x}|\mathcal{C}_k)$, we have used maximum likelihood to determine the parameters of the densities as well as the class priors $p(\mathcal{C}_k)$ and then used Bayes' theorem to find the posterior class probabilities.

However, an alternative approach is to use the functional form of the generalized linear model explicitly and to determine its parameters directly by using maximum likelihood. We shall see that there is an efficient algorithm finding such solutions known as *iterative reweighted least squares*, or *IRLS*.

The indirect approach to finding the parameters of a generalized linear model, by fitting class-conditional densities and class priors separately and then applying

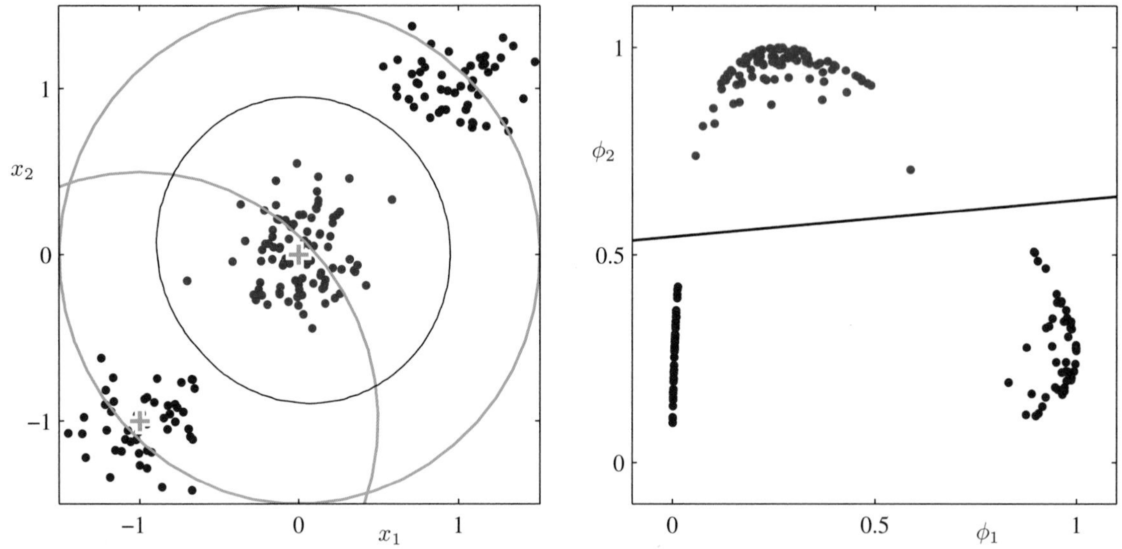

Figure 4.12 Illustration of the role of nonlinear basis functions in linear classification models. The left plot shows the original input space (x_1, x_2) together with data points from two classes labelled red and blue. Two 'Gaussian' basis functions $\phi_1(\mathbf{x})$ and $\phi_2(\mathbf{x})$ are defined in this space with centres shown by the green crosses and with contours shown by the green circles. The right-hand plot shows the corresponding feature space (ϕ_1, ϕ_2) together with the linear decision boundary obtained given by a logistic regression model of the form discussed in Section 4.3.2. This corresponds to a nonlinear decision boundary in the original input space, shown by the black curve in the left-hand plot.

Bayes' theorem, represents an example of *generative* modelling, because we could take such a model and generate synthetic data by drawing values of $\mathbf{x}$ from the marginal distribution $p(\mathbf{x})$. In the direct approach, we are maximizing a likelihood function defined through the conditional distribution $p(\mathcal{C}_k|\mathbf{x})$, which represents a form of *discriminative* training. One advantage of the discriminative approach is that there will typically be fewer adaptive parameters to be determined, as we shall see shortly. It may also lead to improved predictive performance, particularly when the class-conditional density assumptions give a poor approximation to the true distributions.

4.3.1 Fixed basis functions

So far in this chapter, we have considered classification models that work directly with the original input vector $\mathbf{x}$. However, all of the algorithms are equally applicable if we first make a fixed nonlinear transformation of the inputs using a vector of basis functions $\boldsymbol{\phi}(\mathbf{x})$. The resulting decision boundaries will be linear in the feature space $\boldsymbol{\phi}$, and these correspond to nonlinear decision boundaries in the original $\mathbf{x}$ space, as illustrated in Figure 4.12. Classes that are linearly separable in the feature space $\boldsymbol{\phi}(\mathbf{x})$ need not be linearly separable in the original observation space $\mathbf{x}$. Note that as in our discussion of linear models for regression, one of the basis functions is typically set to a constant, say $\phi_0(\mathbf{x}) = 1$, so that the corresponding parameter w_0 plays the role of a bias. For the remainder of this chapter, we shall include a fixed basis function transformation $\boldsymbol{\phi}(\mathbf{x})$, as this will highlight some useful similarities to the regression models discussed in Chapter 3.

For many problems of practical interest, there is significant overlap between the class-conditional densities $p(\mathbf{x}|\mathcal{C}_k)$. This corresponds to posterior probabilities $p(\mathcal{C}_k|\mathbf{x})$, which, for at least some values of $\mathbf{x}$, are not 0 or 1. In such cases, the optimal solution is obtained by modelling the posterior probabilities accurately and then applying standard decision theory, as discussed in Chapter 1. Note that nonlinear transformations $\boldsymbol{\phi}(\mathbf{x})$ cannot remove such class overlap. Indeed, they can increase the level of overlap, or create overlap where none existed in the original observation space. However, suitable choices of nonlinearity can make the process of modelling the posterior probabilities easier.

Section 3.6

Such fixed basis function models have important limitations, and these will be resolved in later chapters by allowing the basis functions themselves to adapt to the data. Notwithstanding these limitations, models with fixed nonlinear basis functions play an important role in applications, and a discussion of such models will introduce many of the key concepts needed for an understanding of their more complex counterparts.

4.3.2 Logistic regression

We begin our treatment of generalized linear models by considering the problem of two-class classification. In our discussion of generative approaches in Section 4.2, we saw that under rather general assumptions, the posterior probability of class $\mathcal{C}_1$ can be written as a logistic sigmoid acting on a linear function of the feature vector $\boldsymbol{\phi}$ so that

$$p(\mathcal{C}_1|\boldsymbol{\phi}) = y(\boldsymbol{\phi}) = \sigma\left(\mathbf{w}^{\mathrm{T}}\boldsymbol{\phi}\right) \tag{4.87}$$

with $p(\mathcal{C}_2|\boldsymbol{\phi}) = 1 - p(\mathcal{C}_1|\boldsymbol{\phi})$. Here $\sigma(\cdot)$ is the *logistic sigmoid* function defined by (4.59). In the terminology of statistics, this model is known as *logistic regression*, although it should be emphasized that this is a model for classification rather than regression.

For an M-dimensional feature space $\boldsymbol{\phi}$, this model has M adjustable parameters. By contrast, if we had fitted Gaussian class conditional densities using maximum likelihood, we would have used $2M$ parameters for the means and $M(M+1)/2$ parameters for the (shared) covariance matrix. Together with the class prior $p(\mathcal{C}_1)$, this gives a total of $M(M+5)/2+1$ parameters, which grows quadratically with M, in contrast to the linear dependence on M of the number of parameters in logistic regression. For large values of M, there is a clear advantage in working with the logistic regression model directly.

We now use maximum likelihood to determine the parameters of the logistic regression model. To do this, we shall make use of the derivative of the logistic sigmoid function, which can conveniently be expressed in terms of the sigmoid function itself

Exercise 4.12

$$\frac{d\sigma}{da} = \sigma(1-\sigma). \tag{4.88}$$

For a data set $\{\boldsymbol{\phi}_n, t_n\}$, where $t_n \in \{0, 1\}$ and $\boldsymbol{\phi}_n = \boldsymbol{\phi}(\mathbf{x}_n)$, with $n = 1, \ldots, N$, the likelihood function can be written

$$p(\mathbf{t}|\mathbf{w}) = \prod_{n=1}^{N} y_n^{t_n} \left\{1 - y_n\right\}^{1-t_n} \tag{4.89}$$

where $\mathbf{t} = (t_1, \ldots, t_N)^{\mathrm{T}}$ and $y_n = p(\mathcal{C}_1|\boldsymbol{\phi}_n)$. As usual, we can define an error function by taking the negative logarithm of the likelihood, which gives the *cross-entropy* error function in the form

$$E(\mathbf{w}) = -\ln p(\mathbf{t}|\mathbf{w}) = -\sum_{n=1}^{N} \left\{t_n \ln y_n + (1 - t_n) \ln(1 - y_n)\right\} \tag{4.90}$$

where $y_n = \sigma(a_n)$ and $a_n = \mathbf{w}^{\mathrm{T}}\boldsymbol{\phi}_n$. Taking the gradient of the error function with respect to $\mathbf{w}$, we obtain

Exercise 4.13

$$\nabla E(\mathbf{w}) = \sum_{n=1}^{N} (y_n - t_n)\boldsymbol{\phi}_n \tag{4.91}$$

where we have made use of (4.88). We see that the factor involving the derivative of the logistic sigmoid has cancelled, leading to a simplified form for the gradient of the log likelihood. In particular, the contribution to the gradient from data point n is given by the 'error' $y_n - t_n$ between the target value and the prediction of the model, times the basis function vector $\boldsymbol{\phi}_n$. Furthermore, comparison with (3.13) shows that this takes precisely the same form as the gradient of the sum-of-squares error function for the linear regression model.

Section 3.1.1

If desired, we could make use of the result (4.91) to give a sequential algorithm in which patterns are presented one at a time, in which each of the weight vectors is updated using (3.22) in which ∇E_n is the nth term in (4.91).

It is worth noting that maximum likelihood can exhibit severe over-fitting for data sets that are linearly separable. This arises because the maximum likelihood solution occurs when the hyperplane corresponding to $\sigma = 0.5$, equivalent to $\mathbf{w}^{\mathrm{T}}\boldsymbol{\phi} = 0$, separates the two classes and the magnitude of $\mathbf{w}$ goes to infinity. In this case, the logistic sigmoid function becomes infinitely steep in feature space, corresponding to a Heaviside step function, so that every training point from each class k is assigned a posterior probability $p(\mathcal{C}_k|\mathbf{x}) = 1$. Furthermore, there is typically a continuum of such solutions because any separating hyperplane will give rise to the same posterior probabilities at the training data points, as will be seen later in Figure 10.13. Maximum likelihood provides no way to favour one such solution over another, and which solution is found in practice will depend on the choice of optimization algorithm and on the parameter initialization. Note that the problem will arise even if the number of data points is large compared with the number of parameters in the model, so long as the training data set is linearly separable. The singularity can be avoided by inclusion of a prior and finding a MAP solution for $\mathbf{w}$, or equivalently by adding a regularization term to the error function.

Exercise 4.14

4.3.3 Iterative reweighted least squares

In the case of the linear regression models discussed in Chapter 3, the maximum likelihood solution, on the assumption of a Gaussian noise model, leads to a closed-form solution. This was a consequence of the quadratic dependence of the log likelihood function on the parameter vector $\mathbf{w}$. For logistic regression, there is no longer a closed-form solution, due to the nonlinearity of the logistic sigmoid function. However, the departure from a quadratic form is not substantial. To be precise, the error function is convex, as we shall see shortly, and hence has a unique minimum. Furthermore, the error function can be minimized by an efficient iterative technique based on the *Newton-Raphson* iterative optimization scheme, which uses a local quadratic approximation to the log likelihood function. The Newton-Raphson update, for minimizing a function $E(\mathbf{w})$, takes the form (Fletcher, 1987; Bishop and Nabney, 2008)

$$\mathbf{w}^{(\text{new})} = \mathbf{w}^{(\text{old})} - \mathbf{H}^{-1}\nabla E(\mathbf{w}). \tag{4.92}$$

where $\mathbf{H}$ is the Hessian matrix whose elements comprise the second derivatives of $E(\mathbf{w})$ with respect to the components of $\mathbf{w}$.

Let us first of all apply the Newton-Raphson method to the linear regression model (3.3) with the sum-of-squares error function (3.12). The gradient and Hessian of this error function are given by

$$\nabla E(\mathbf{w}) = \sum_{n=1}^{N}(\mathbf{w}^{\mathrm{T}}\boldsymbol{\phi}_n - t_n)\boldsymbol{\phi}_n = \boldsymbol{\Phi}^{\mathrm{T}}\boldsymbol{\Phi}\mathbf{w} - \boldsymbol{\Phi}^{\mathrm{T}}\mathbf{t} \tag{4.93}$$

$$\mathbf{H} = \nabla\nabla E(\mathbf{w}) = \sum_{n=1}^{N}\boldsymbol{\phi}_n\boldsymbol{\phi}_n^{\mathrm{T}} = \boldsymbol{\Phi}^{\mathrm{T}}\boldsymbol{\Phi} \tag{4.94}$$

Section 3.1.1

where $\boldsymbol{\Phi}$ is the $N \times M$ design matrix, whose n^{th} row is given by $\boldsymbol{\phi}_n^{\mathrm{T}}$. The Newton-Raphson update then takes the form

$$\begin{aligned}\mathbf{w}^{(\text{new})} &= \mathbf{w}^{(\text{old})} - (\boldsymbol{\Phi}^{\mathrm{T}}\boldsymbol{\Phi})^{-1}\left\{\boldsymbol{\Phi}^{\mathrm{T}}\boldsymbol{\Phi}\mathbf{w}^{(\text{old})} - \boldsymbol{\Phi}^{\mathrm{T}}\mathbf{t}\right\} \\ &= (\boldsymbol{\Phi}^{\mathrm{T}}\boldsymbol{\Phi})^{-1}\boldsymbol{\Phi}^{\mathrm{T}}\mathbf{t}\end{aligned} \tag{4.95}$$

which we recognize as the standard least-squares solution. Note that the error function in this case is quadratic and hence the Newton-Raphson formula gives the exact solution in one step.

Now let us apply the Newton-Raphson update to the cross-entropy error function (4.90) for the logistic regression model. From (4.91) we see that the gradient and Hessian of this error function are given by

$$\nabla E(\mathbf{w}) = \sum_{n=1}^{N}(y_n - t_n)\boldsymbol{\phi}_n = \boldsymbol{\Phi}^{\mathrm{T}}(\mathbf{y} - \mathbf{t}) \tag{4.96}$$

$$\mathbf{H} = \nabla\nabla E(\mathbf{w}) = \sum_{n=1}^{N} y_n(1 - y_n)\boldsymbol{\phi}_n\boldsymbol{\phi}_n^{\mathrm{T}} = \boldsymbol{\Phi}^{\mathrm{T}}\mathbf{R}\boldsymbol{\Phi} \tag{4.97}$$

where we have made use of (4.88). Also, we have introduced the $N \times N$ diagonal matrix $\mathbf{R}$ with elements

$$R_{nn} = y_n(1 - y_n). \tag{4.98}$$

We see that the Hessian is no longer constant but depends on $\mathbf{w}$ through the weighting matrix $\mathbf{R}$, corresponding to the fact that the error function is no longer quadratic. Using the property $0 < y_n < 1$, which follows from the form of the logistic sigmoid function, we see that $\mathbf{u}^{\mathrm{T}}\mathbf{H}\mathbf{u} > 0$ for an arbitrary vector $\mathbf{u}$, and so the Hessian matrix $\mathbf{H}$ is positive definite. It follows that the error function is a convex function of $\mathbf{w}$ and hence has a unique minimum.

Exercise 4.15

The Newton-Raphson update formula for the logistic regression model then becomes

$$\begin{aligned}
\mathbf{w}^{(\text{new})} &= \mathbf{w}^{(\text{old})} - (\boldsymbol{\Phi}^{\mathrm{T}}\mathbf{R}\boldsymbol{\Phi})^{-1}\boldsymbol{\Phi}^{\mathrm{T}}(\mathbf{y} - \mathbf{t}) \\
&= (\boldsymbol{\Phi}^{\mathrm{T}}\mathbf{R}\boldsymbol{\Phi})^{-1}\left\{\boldsymbol{\Phi}^{\mathrm{T}}\mathbf{R}\boldsymbol{\Phi}\mathbf{w}^{(\text{old})} - \boldsymbol{\Phi}^{\mathrm{T}}(\mathbf{y} - \mathbf{t})\right\} \\
&= (\boldsymbol{\Phi}^{\mathrm{T}}\mathbf{R}\boldsymbol{\Phi})^{-1}\boldsymbol{\Phi}^{\mathrm{T}}\mathbf{R}\mathbf{z}
\end{aligned} \tag{4.99}$$

where $\mathbf{z}$ is an N-dimensional vector with elements

$$\mathbf{z} = \boldsymbol{\Phi}\mathbf{w}^{(\text{old})} - \mathbf{R}^{-1}(\mathbf{y} - \mathbf{t}). \tag{4.100}$$

We see that the update formula (4.99) takes the form of a set of normal equations for a weighted least-squares problem. Because the weighing matrix $\mathbf{R}$ is not constant but depends on the parameter vector $\mathbf{w}$, we must apply the normal equations iteratively, each time using the new weight vector $\mathbf{w}$ to compute a revised weighing matrix $\mathbf{R}$. For this reason, the algorithm is known as *iterative reweighted least squares*, or *IRLS* (Rubin, 1983). As in the weighted least-squares problem, the elements of the diagonal weighting matrix $\mathbf{R}$ can be interpreted as variances because the mean and variance of t in the logistic regression model are given by

$$\mathbb{E}[t] = \sigma(\mathbf{x}) = y \tag{4.101}$$

$$\mathrm{var}[t] = \mathbb{E}[t^2] - \mathbb{E}[t]^2 = \sigma(\mathbf{x}) - \sigma(\mathbf{x})^2 = y(1 - y) \tag{4.102}$$

where we have used the property $t^2 = t$ for $t \in \{0, 1\}$. In fact, we can interpret IRLS as the solution to a linearized problem in the space of the variable $a = \mathbf{w}^{\mathrm{T}}\boldsymbol{\phi}$. The quantity z_n, which corresponds to the nth element of $\mathbf{z}$, can then be given a simple interpretation as an effective target value in this space obtained by making a local linear approximation to the logistic sigmoid function around the current operating point $\mathbf{w}^{(\text{old})}$

$$\begin{aligned}
a_n(\mathbf{w}) &\simeq a_n(\mathbf{w}^{(\text{old})}) + \left.\frac{\mathrm{d}a_n}{\mathrm{d}y_n}\right|_{\mathbf{w}^{(\text{old})}} (t_n - y_n) \\
&= \boldsymbol{\phi}_n^{\mathrm{T}}\mathbf{w}^{(\text{old})} - \frac{(y_n - t_n)}{y_n(1 - y_n)} = z_n.
\end{aligned} \tag{4.103}$$

4.3.4 Multiclass logistic regression

Section 4.2

In our discussion of generative models for multiclass classification, we have seen that for a large class of distributions, the posterior probabilities are given by a softmax transformation of linear functions of the feature variables, so that

$$p(\mathcal{C}_k|\boldsymbol{\phi}) = y_k(\boldsymbol{\phi}) = \frac{\exp(a_k)}{\sum_j \exp(a_j)} \tag{4.104}$$

where the 'activations' a_k are given by

$$a_k = \mathbf{w}_k^{\mathrm{T}}\boldsymbol{\phi}. \tag{4.105}$$

There we used maximum likelihood to determine separately the class-conditional densities and the class priors and then found the corresponding posterior probabilities using Bayes' theorem, thereby implicitly determining the parameters $\{\mathbf{w}_k\}$. Here we consider the use of maximum likelihood to determine the parameters $\{\mathbf{w}_k\}$ of this model directly. To do this, we will require the derivatives of y_k with respect to all of the activations a_j. These are given by

Exercise 4.17

$$\frac{\partial y_k}{\partial a_j} = y_k(I_{kj} - y_j) \tag{4.106}$$

where I_{kj} are the elements of the identity matrix.

Next we write down the likelihood function. This is most easily done using the 1-of-K coding scheme in which the target vector $\mathbf{t}_n$ for a feature vector $\boldsymbol{\phi}_n$ belonging to class $\mathcal{C}_k$ is a binary vector with all elements zero except for element k, which equals one. The likelihood function is then given by

$$p(\mathbf{T}|\mathbf{w}_1, \ldots, \mathbf{w}_K) = \prod_{n=1}^{N}\prod_{k=1}^{K} p(\mathcal{C}_k|\boldsymbol{\phi}_n)^{t_{nk}} = \prod_{n=1}^{N}\prod_{k=1}^{K} y_{nk}^{t_{nk}} \tag{4.107}$$

where $y_{nk} = y_k(\boldsymbol{\phi}_n)$, and $\mathbf{T}$ is an $N \times K$ matrix of target variables with elements t_{nk}. Taking the negative logarithm then gives

$$E(\mathbf{w}_1, \ldots, \mathbf{w}_K) = -\ln p(\mathbf{T}|\mathbf{w}_1, \ldots, \mathbf{w}_K) = -\sum_{n=1}^{N}\sum_{k=1}^{K} t_{nk} \ln y_{nk} \tag{4.108}$$

which is known as the *cross-entropy* error function for the multiclass classification problem.

We now take the gradient of the error function with respect to one of the parameter vectors $\mathbf{w}_j$. Making use of the result (4.106) for the derivatives of the softmax function, we obtain

Exercise 4.18

$$\nabla_{\mathbf{w}_j} E(\mathbf{w}_1, \ldots, \mathbf{w}_K) = \sum_{n=1}^{N} (y_{nj} - t_{nj})\,\boldsymbol{\phi}_n \tag{4.109}$$

where we have made use of $\sum_k t_{nk} = 1$. Once again, we see the same form arising for the gradient as was found for the sum-of-squares error function with the linear model and the cross-entropy error for the logistic regression model, namely the product of the error $(y_{nj} - t_{nj})$ times the basis function ϕ_n. Again, we could use this to formulate a sequential algorithm in which patterns are presented one at a time, in which each of the weight vectors is updated using (3.22).

We have seen that the derivative of the log likelihood function for a linear regression model with respect to the parameter vector $\mathbf{w}$ for a data point n took the form of the 'error' $y_n - t_n$ times the feature vector ϕ_n. Similarly, for the combination of logistic sigmoid activation function and cross-entropy error function (4.90), and for the softmax activation function with the multiclass cross-entropy error function (4.108), we again obtain this same simple form. This is an example of a more general result, as we shall see in Section 4.3.6.

To find a batch algorithm, we again appeal to the Newton-Raphson update to obtain the corresponding IRLS algorithm for the multiclass problem. This requires evaluation of the Hessian matrix that comprises blocks of size $M \times M$ in which block j, k is given by

$$\nabla_{\mathbf{w}_k} \nabla_{\mathbf{w}_j} E(\mathbf{w}_1, \ldots, \mathbf{w}_K) = \sum_{n=1}^{N} y_{nk}(I_{kj} - y_{nj}) \phi_n \phi_n^{\mathrm{T}}. \tag{4.110}$$

Exercise 4.20

As with the two-class problem, the Hessian matrix for the multiclass logistic regression model is positive definite and so the error function again has a unique minimum. Practical details of IRLS for the multiclass case can be found in Bishop and Nabney (2008).

4.3.5 Probit regression

We have seen that, for a broad range of class-conditional distributions, described by the exponential family, the resulting posterior class probabilities are given by a logistic (or softmax) transformation acting on a linear function of the feature variables. However, not all choices of class-conditional density give rise to such a simple form for the posterior probabilities (for instance, if the class-conditional densities are modelled using Gaussian mixtures). This suggests that it might be worth exploring other types of discriminative probabilistic model. For the purposes of this chapter, however, we shall return to the two-class case, and again remain within the framework of generalized linear models so that

$$p(t = 1|a) = f(a) \tag{4.111}$$

where $a = \mathbf{w}^{\mathrm{T}}\phi$, and $f(\cdot)$ is the activation function.

One way to motivate an alternative choice for the link function is to consider a noisy threshold model, as follows. For each input ϕ_n, we evaluate $a_n = \mathbf{w}^{\mathrm{T}}\phi_n$ and then we set the target value according to

$$\begin{cases} t_n = 1 & \text{if } a_n \geqslant \theta \\ t_n = 0 & \text{otherwise.} \end{cases} \tag{4.112}$$

Figure 4.13 Schematic example of a probability density $p(\theta)$ shown by the blue curve, given in this example by a mixture of two Gaussians, along with its cumulative distribution function $f(a)$, shown by the red curve. Note that the value of the blue curve at any point, such as that indicated by the vertical green line, corresponds to the slope of the red curve at the same point. Conversely, the value of the red curve at this point corresponds to the area under the blue curve indicated by the shaded green region. In the stochastic threshold model, the class label takes the value $t = 1$ if the value of $a = \mathbf{w}^{\mathrm{T}}\boldsymbol{\phi}$ exceeds a threshold, otherwise it takes the value $t = 0$. This is equivalent to an activation function given by the cumulative distribution function $f(a)$.

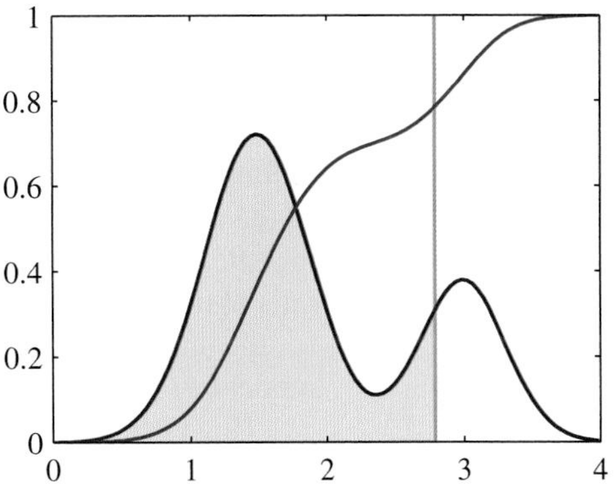

If the value of θ is drawn from a probability density $p(\theta)$, then the corresponding activation function will be given by the cumulative distribution function

$$f(a) = \int_{-\infty}^{a} p(\theta)\,\mathrm{d}\theta \tag{4.113}$$

as illustrated in Figure 4.13.

As a specific example, suppose that the density $p(\theta)$ is given by a zero mean, unit variance Gaussian. The corresponding cumulative distribution function is given by

$$\Phi(a) = \int_{-\infty}^{a} \mathcal{N}(\theta|0,1)\,\mathrm{d}\theta \tag{4.114}$$

which is known as the *inverse probit* function. It has a sigmoidal shape and is compared with the logistic sigmoid function in Figure 4.9. Note that the use of a more general Gaussian distribution does not change the model because this is equivalent to a re-scaling of the linear coefficients $\mathbf{w}$. Many numerical packages provide for the evaluation of a closely related function defined by

$$\mathrm{erf}(a) = \frac{2}{\sqrt{\pi}} \int_{0}^{a} \exp(-\theta^2)\,\mathrm{d}\theta \tag{4.115}$$

Exercise 4.21

and known as the *erf function* or *error function* (not to be confused with the error function of a machine learning model). It is related to the inverse probit function by

$$\Phi(a) = \frac{1}{2}\left\{1 + \mathrm{erf}\left(\frac{a}{\sqrt{2}}\right)\right\}. \tag{4.116}$$

The generalized linear model based on an inverse probit activation function is known as *probit regression*.

We can determine the parameters of this model using maximum likelihood, by a straightforward extension of the ideas discussed earlier. In practice, the results found using probit regression tend to be similar to those of logistic regression. We shall, however, find another use for the probit model when we discuss Bayesian treatments of logistic regression in Section 4.5.

One issue that can occur in practical applications is that of *outliers*, which can arise for instance through errors in measuring the input vector $\mathbf{x}$ or through mislabelling of the target value t. Because such points can lie a long way to the wrong side of the ideal decision boundary, they can seriously distort the classifier. Note that the logistic and probit regression models behave differently in this respect because the tails of the logistic sigmoid decay asymptotically like $\exp(-x)$ for $x \to \infty$, whereas for the inverse probit activation function they decay like $\exp(-x^2)$, and so the probit model can be significantly more sensitive to outliers.

However, both the logistic and the probit models assume the data is correctly labelled. The effect of mislabelling is easily incorporated into a probabilistic model by introducing a probability ϵ that the target value t has been flipped to the wrong value (Opper and Winther, 2000a), leading to a target value distribution for data point $\mathbf{x}$ of the form

$$\begin{aligned} p(t|\mathbf{x}) &= (1-\epsilon)\sigma(\mathbf{x}) + \epsilon(1-\sigma(\mathbf{x})) \\ &= \epsilon + (1-2\epsilon)\sigma(\mathbf{x}) \end{aligned} \tag{4.117}$$

where $\sigma(\mathbf{x})$ is the activation function with input vector $\mathbf{x}$. Here ϵ may be set in advance, or it may be treated as a hyperparameter whose value is inferred from the data.

4.3.6 Canonical link functions

For the linear regression model with a Gaussian noise distribution, the error function, corresponding to the negative log likelihood, is given by (3.12). If we take the derivative with respect to the parameter vector $\mathbf{w}$ of the contribution to the error function from a data point n, this takes the form of the 'error' $y_n - t_n$ times the feature vector $\boldsymbol{\phi}_n$, where $y_n = \mathbf{w}^{\mathrm{T}}\boldsymbol{\phi}_n$. Similarly, for the combination of the logistic sigmoid activation function and the cross-entropy error function (4.90), and for the softmax activation function with the multiclass cross-entropy error function (4.108), we again obtain this same simple form. We now show that this is a general result of assuming a conditional distribution for the target variable from the exponential family, along with a corresponding choice for the activation function known as the *canonical link function*.

We again make use of the restricted form (4.84) of exponential family distributions. Note that here we are applying the assumption of exponential family distribution to the target variable t, in contrast to Section 4.2.4 where we applied it to the input vector $\mathbf{x}$. We therefore consider conditional distributions of the target variable of the form

$$p(t|\eta, s) = \frac{1}{s} h\left(\frac{t}{s}\right) g(\eta) \exp\left\{\frac{\eta t}{s}\right\}. \tag{4.118}$$

Using the same line of argument as led to the derivation of the result (2.226), we see that the conditional mean of t, which we denote by y, is given by

$$y \equiv \mathbb{E}[t|\eta] = -s \frac{d}{d\eta} \ln g(\eta). \tag{4.119}$$

Thus y and η must related, and we denote this relation through $\eta = \psi(y)$.

Following Nelder and Wedderburn (1972), we define a *generalized linear model* to be one for which y is a nonlinear function of a linear combination of the input (or feature) variables so that

$$y = f(\mathbf{w}^{\mathrm{T}}\boldsymbol{\phi}) \tag{4.120}$$

where $f(\cdot)$ is known as the *activation function* in the machine learning literature, and $f^{-1}(\cdot)$ is known as the *link function* in statistics.

Now consider the log likelihood function for this model, which, as a function of η, is given by

$$\ln p(\mathbf{t}|\eta, s) = \sum_{n=1}^{N} \ln p(t_n|\eta, s) = \sum_{n=1}^{N} \left\{ \ln g(\eta_n) + \frac{\eta_n t_n}{s} \right\} + \text{const} \tag{4.121}$$

where we are assuming that all observations share a common scale parameter (which corresponds to the noise variance for a Gaussian distribution for instance) and so s is independent of n. The derivative of the log likelihood with respect to the model parameters $\mathbf{w}$ is then given by

$$\begin{aligned} \nabla_{\mathbf{w}} \ln p(\mathbf{t}|\eta, s) &= \sum_{n=1}^{N} \left\{ \frac{d}{d\eta_n} \ln g(\eta_n) + \frac{t_n}{s} \right\} \frac{d\eta_n}{dy_n} \frac{dy_n}{da_n} \nabla a_n \\ &= \sum_{n=1}^{N} \frac{1}{s} \{t_n - y_n\} \psi'(y_n) f'(a_n) \boldsymbol{\phi}_n \end{aligned} \tag{4.122}$$

where $a_n = \mathbf{w}^{\mathrm{T}}\boldsymbol{\phi}_n$, and we have used $y_n = f(a_n)$ together with the result (4.119) for $\mathbb{E}[t|\eta]$. We now see that there is a considerable simplification if we choose a particular form for the link function $f^{-1}(y)$ given by

$$f^{-1}(y) = \psi(y) \tag{4.123}$$

which gives $f(\psi(y)) = y$ and hence $f'(\psi)\psi'(y) = 1$. Also, because $a = f^{-1}(y)$, we have $a = \psi$ and hence $f'(a)\psi'(y) = 1$. In this case, the gradient of the error function reduces to

$$\nabla E(\mathbf{w}) = \frac{1}{s} \sum_{n=1}^{N} \{y_n - t_n\} \boldsymbol{\phi}_n. \tag{4.124}$$

For the Gaussian $s = \beta^{-1}$, whereas for the logistic model $s = 1$.

4.4. The Laplace Approximation

In Section 4.5 we shall discuss the Bayesian treatment of logistic regression. As we shall see, this is more complex than the Bayesian treatment of linear regression models, discussed in Sections 3.3 and 3.5. In particular, we cannot integrate exactly over the parameter vector $\mathbf{w}$ since the posterior distribution is no longer Gaussian. It is therefore necessary to introduce some form of approximation. Later in the book we shall consider a range of techniques based on analytical approximations and numerical sampling.

Chapter 10
Chapter 11

Here we introduce a simple, but widely used, framework called the Laplace approximation, that aims to find a Gaussian approximation to a probability density defined over a set of continuous variables. Consider first the case of a single continuous variable z, and suppose the distribution $p(z)$ is defined by

$$p(z) = \frac{1}{Z} f(z) \tag{4.125}$$

where $Z = \int f(z)\,\mathrm{d}z$ is the normalization coefficient. We shall suppose that the value of Z is unknown. In the Laplace method the goal is to find a Gaussian approximation $q(z)$ which is centred on a mode of the distribution $p(z)$. The first step is to find a mode of $p(z)$, in other words a point z_0 such that $p'(z_0) = 0$, or equivalently

$$\left. \frac{df(z)}{dz} \right|_{z=z_0} = 0. \tag{4.126}$$

A Gaussian distribution has the property that its logarithm is a quadratic function of the variables. We therefore consider a Taylor expansion of $\ln f(z)$ centred on the mode z_0 so that

$$\ln f(z) \simeq \ln f(z_0) - \frac{1}{2} A (z - z_0)^2 \tag{4.127}$$

where

$$A = - \left. \frac{d^2}{dz^2} \ln f(z) \right|_{z=z_0}. \tag{4.128}$$

Note that the first-order term in the Taylor expansion does not appear since z_0 is a local maximum of the distribution. Taking the exponential we obtain

$$f(z) \simeq f(z_0) \exp\left\{ -\frac{A}{2}(z - z_0)^2 \right\}. \tag{4.129}$$

We can then obtain a normalized distribution $q(z)$ by making use of the standard result for the normalization of a Gaussian, so that

$$q(z) = \left(\frac{A}{2\pi} \right)^{1/2} \exp\left\{ -\frac{A}{2}(z - z_0)^2 \right\}. \tag{4.130}$$

The Laplace approximation is illustrated in Figure 4.14. Note that the Gaussian approximation will only be well defined if its precision $A > 0$, in other words the stationary point z_0 must be a local maximum, so that the second derivative of $f(z)$ at the point z_0 is negative.

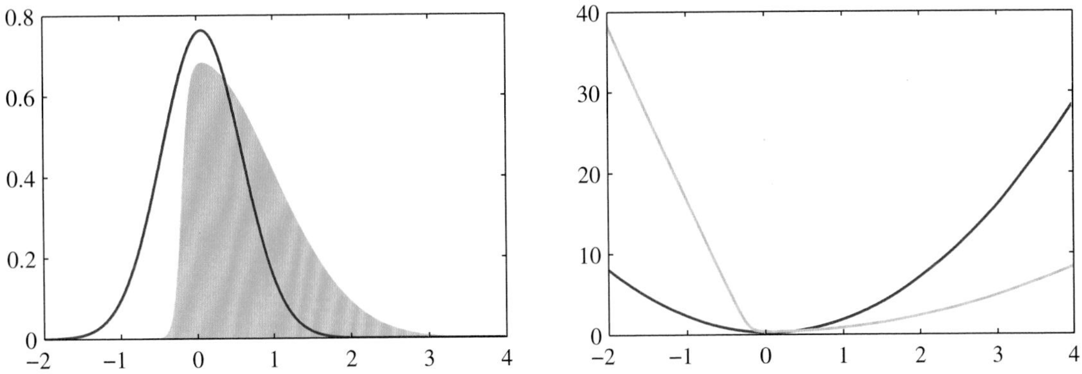

Figure 4.14 Illustration of the Laplace approximation applied to the distribution $p(z) \propto \exp(-z^2/2)\sigma(20z+4)$ where $\sigma(z)$ is the logistic sigmoid function defined by $\sigma(z) = (1+e^{-z})^{-1}$. The left plot shows the normalized distribution $p(z)$ in yellow, together with the Laplace approximation centred on the mode z_0 of $p(z)$ in red. The right plot shows the negative logarithms of the corresponding curves.

We can extend the Laplace method to approximate a distribution $p(\mathbf{z}) = f(\mathbf{z})/Z$ defined over an M-dimensional space $\mathbf{z}$. At a stationary point $\mathbf{z}_0$ the gradient $\nabla f(\mathbf{z})$ will vanish. Expanding around this stationary point we have

$$\ln f(\mathbf{z}) \simeq \ln f(\mathbf{z}_0) - \frac{1}{2}(\mathbf{z}-\mathbf{z}_0)^{\mathrm{T}}\mathbf{A}(\mathbf{z}-\mathbf{z}_0) \tag{4.131}$$

where the $M \times M$ Hessian matrix $\mathbf{A}$ is defined by

$$\mathbf{A} = -\left.\nabla\nabla \ln f(\mathbf{z})\right|_{\mathbf{z}=\mathbf{z}_0} \tag{4.132}$$

and ∇ is the gradient operator. Taking the exponential of both sides we obtain

$$f(\mathbf{z}) \simeq f(\mathbf{z}_0)\exp\left\{-\frac{1}{2}(\mathbf{z}-\mathbf{z}_0)^{\mathrm{T}}\mathbf{A}(\mathbf{z}-\mathbf{z}_0)\right\}. \tag{4.133}$$

The distribution $q(\mathbf{z})$ is proportional to $f(\mathbf{z})$ and the appropriate normalization coefficient can be found by inspection, using the standard result (2.43) for a normalized multivariate Gaussian, giving

$$q(\mathbf{z}) = \frac{|\mathbf{A}|^{1/2}}{(2\pi)^{M/2}}\exp\left\{-\frac{1}{2}(\mathbf{z}-\mathbf{z}_0)^{\mathrm{T}}\mathbf{A}(\mathbf{z}-\mathbf{z}_0)\right\} = \mathcal{N}(\mathbf{z}|\mathbf{z}_0, \mathbf{A}^{-1}) \tag{4.134}$$

where $|\mathbf{A}|$ denotes the determinant of $\mathbf{A}$. This Gaussian distribution will be well defined provided its precision matrix, given by $\mathbf{A}$, is positive definite, which implies that the stationary point $\mathbf{z}_0$ must be a local maximum, not a minimum or a saddle point.

In order to apply the Laplace approximation we first need to find the mode $\mathbf{z}_0$, and then evaluate the Hessian matrix at that mode. In practice a mode will typically be found by running some form of numerical optimization algorithm (Bishop and Nabney, 2008). Many of the distributions encountered in practice will be multimodal and so there will be different Laplace approximations according to which mode is being considered. Note that the normalization constant Z of the true distribution does not need to be known in order to apply the Laplace method. As a result of the central limit theorem, the posterior distribution for a model is expected to become increasingly better approximated by a Gaussian as the number of observed data points is increased, and so we would expect the Laplace approximation to be most useful in situations where the number of data points is relatively large.

One major weakness of the Laplace approximation is that, since it is based on a Gaussian distribution, it is only directly applicable to real variables. In other cases it may be possible to apply the Laplace approximation to a transformation of the variable. For instance if $0 \leqslant \tau < \infty$ then we can consider a Laplace approximation of $\ln \tau$. The most serious limitation of the Laplace framework, however, is that it is based purely on the aspects of the true distribution at a specific value of the variable, and so can fail to capture important global properties. In Chapter 10 we shall consider alternative approaches which adopt a more global perspective.

4.4.1 Model comparison and BIC

As well as approximating the distribution $p(\mathbf{z})$ we can also obtain an approximation to the normalization constant Z. Using the approximation (4.133) we have

$$\begin{aligned} Z &= \int f(\mathbf{z})\,\mathrm{d}\mathbf{z} \\ &\simeq f(\mathbf{z}_0)\int \exp\left\{-\frac{1}{2}(\mathbf{z}-\mathbf{z}_0)^{\mathrm{T}}\mathbf{A}(\mathbf{z}-\mathbf{z}_0)\right\}\mathrm{d}\mathbf{z} \\ &= f(\mathbf{z}_0)\frac{(2\pi)^{M/2}}{|\mathbf{A}|^{1/2}} \end{aligned} \tag{4.135}$$

where we have noted that the integrand is Gaussian and made use of the standard result (2.43) for a normalized Gaussian distribution. We can use the result (4.135) to obtain an approximation to the model evidence which, as discussed in Section 3.4, plays a central role in Bayesian model comparison.

Consider a data set $\mathcal{D}$ and a set of models $\{\mathcal{M}_i\}$ having parameters $\{\boldsymbol{\theta}_i\}$. For each model we define a likelihood function $p(\mathcal{D}|\boldsymbol{\theta}_i, \mathcal{M}_i)$. If we introduce a prior $p(\boldsymbol{\theta}_i|\mathcal{M}_i)$ over the parameters, then we are interested in computing the model evidence $p(\mathcal{D}|\mathcal{M}_i)$ for the various models. From now on we omit the conditioning on $\mathcal{M}_i$ to keep the notation uncluttered. From Bayes' theorem the model evidence is given by

$$p(\mathcal{D}) = \int p(\mathcal{D}|\boldsymbol{\theta})p(\boldsymbol{\theta})\,\mathrm{d}\boldsymbol{\theta}. \tag{4.136}$$

Exercise 4.22

Identifying $f(\boldsymbol{\theta}) = p(\mathcal{D}|\boldsymbol{\theta})p(\boldsymbol{\theta})$ and $Z = p(\mathcal{D})$, and applying the result (4.135), we obtain

$$\ln p(\mathcal{D}) \simeq \ln p(\mathcal{D}|\boldsymbol{\theta}_{\mathrm{MAP}}) + \underbrace{\ln p(\boldsymbol{\theta}_{\mathrm{MAP}}) + \frac{M}{2}\ln(2\pi) - \frac{1}{2}\ln|\mathbf{A}|}_{\text{Occam factor}} \tag{4.137}$$

where $\boldsymbol{\theta}_{\text{MAP}}$ is the value of $\boldsymbol{\theta}$ at the mode of the posterior distribution, and $\mathbf{A}$ is the *Hessian* matrix of second derivatives of the negative log posterior

$$\mathbf{A} = -\nabla\nabla \ln p(\mathcal{D}|\boldsymbol{\theta}_{\text{MAP}})p(\boldsymbol{\theta}_{\text{MAP}}) = -\nabla\nabla \ln p(\boldsymbol{\theta}_{\text{MAP}}|\mathcal{D}). \tag{4.138}$$

The first term on the right hand side of (4.137) represents the log likelihood evaluated using the optimized parameters, while the remaining three terms comprise the 'Occam factor' which penalizes model complexity.

Exercise 4.23

If we assume that the Gaussian prior distribution over parameters is broad, and that the Hessian has full rank, then we can approximate (4.137) very roughly using

$$\ln p(\mathcal{D}) \simeq \ln p(\mathcal{D}|\boldsymbol{\theta}_{\text{MAP}}) - \frac{1}{2}M \ln N \tag{4.139}$$

where N is the number of data points, M is the number of parameters in $\boldsymbol{\theta}$ and we have omitted additive constants. This is known as the *Bayesian Information Criterion* (BIC) or the *Schwarz criterion* (Schwarz, 1978). Note that, compared to AIC given by (1.73), this penalizes model complexity more heavily.

Section 3.5.3

Complexity measures such as AIC and BIC have the virtue of being easy to evaluate, but can also give misleading results. In particular, the assumption that the Hessian matrix has full rank is often not valid since many of the parameters are not 'well-determined'. We can use the result (4.137) to obtain a more accurate estimate of the model evidence starting from the Laplace approximation, as we illustrate in the context of neural networks in Section 5.7.

4.5. Bayesian Logistic Regression

We now turn to a Bayesian treatment of logistic regression. Exact Bayesian inference for logistic regression is intractable. In particular, evaluation of the posterior distribution would require normalization of the product of a prior distribution and a likelihood function that itself comprises a product of logistic sigmoid functions, one for every data point. Evaluation of the predictive distribution is similarly intractable. Here we consider the application of the Laplace approximation to the problem of Bayesian logistic regression (Spiegelhalter and Lauritzen, 1990; MacKay, 1992b).

4.5.1 Laplace approximation

Recall from Section 4.4 that the Laplace approximation is obtained by finding the mode of the posterior distribution and then fitting a Gaussian centred at that mode. This requires evaluation of the second derivatives of the log posterior, which is equivalent to finding the Hessian matrix.

Because we seek a Gaussian representation for the posterior distribution, it is natural to begin with a Gaussian prior, which we write in the general form

$$p(\mathbf{w}) = \mathcal{N}(\mathbf{w}|\mathbf{m}_0, \mathbf{S}_0) \tag{4.140}$$

where $\mathbf{m}_0$ and $\mathbf{S}_0$ are fixed hyperparameters. The posterior distribution over $\mathbf{w}$ is given by

$$p(\mathbf{w}|\mathbf{t}) \propto p(\mathbf{w})p(\mathbf{t}|\mathbf{w}) \tag{4.141}$$

where $\mathbf{t} = (t_1, \ldots, t_N)^{\mathrm{T}}$. Taking the log of both sides, and substituting for the prior distribution using (4.140), and for the likelihood function using (4.89), we obtain

$$\begin{aligned} \ln p(\mathbf{w}|\mathbf{t}) &= -\frac{1}{2}(\mathbf{w} - \mathbf{m}_0)^{\mathrm{T}}\mathbf{S}_0^{-1}(\mathbf{w} - \mathbf{m}_0) \\ &\quad + \sum_{n=1}^{N} \{t_n \ln y_n + (1 - t_n)\ln(1 - y_n)\} + \text{const} \end{aligned} \tag{4.142}$$

where $y_n = \sigma(\mathbf{w}^{\mathrm{T}}\boldsymbol{\phi}_n)$. To obtain a Gaussian approximation to the posterior distribution, we first maximize the posterior distribution to give the MAP (maximum posterior) solution $\mathbf{w}_{\mathrm{MAP}}$, which defines the mean of the Gaussian. The covariance is then given by the inverse of the matrix of second derivatives of the negative log likelihood, which takes the form

$$\mathbf{S}_N^{-1} = -\nabla\nabla \ln p(\mathbf{w}|\mathbf{t}) = \mathbf{S}_0^{-1} + \sum_{n=1}^{N} y_n(1 - y_n)\boldsymbol{\phi}_n\boldsymbol{\phi}_n^{\mathrm{T}}. \tag{4.143}$$

The Gaussian approximation to the posterior distribution therefore takes the form

$$q(\mathbf{w}) = \mathcal{N}(\mathbf{w}|\mathbf{w}_{\mathrm{MAP}}, \mathbf{S}_N). \tag{4.144}$$

Having obtained a Gaussian approximation to the posterior distribution, there remains the task of marginalizing with respect to this distribution in order to make predictions.

4.5.2 Predictive distribution

The predictive distribution for class $\mathcal{C}_1$, given a new feature vector $\boldsymbol{\phi}(\mathbf{x})$, is obtained by marginalizing with respect to the posterior distribution $p(\mathbf{w}|\mathbf{t})$, which is itself approximated by a Gaussian distribution $q(\mathbf{w})$ so that

$$p(\mathcal{C}_1|\boldsymbol{\phi}, \mathbf{t}) = \int p(\mathcal{C}_1|\boldsymbol{\phi}, \mathbf{w})p(\mathbf{w}|\mathbf{t})\,\mathrm{d}\mathbf{w} \simeq \int \sigma(\mathbf{w}^{\mathrm{T}}\boldsymbol{\phi})q(\mathbf{w})\,\mathrm{d}\mathbf{w} \tag{4.145}$$

with the corresponding probability for class $\mathcal{C}_2$ given by $p(\mathcal{C}_2|\boldsymbol{\phi}, \mathbf{t}) = 1 - p(\mathcal{C}_1|\boldsymbol{\phi}, \mathbf{t})$. To evaluate the predictive distribution, we first note that the function $\sigma(\mathbf{w}^{\mathrm{T}}\boldsymbol{\phi})$ depends on $\mathbf{w}$ only through its projection onto $\boldsymbol{\phi}$. Denoting $a = \mathbf{w}^{\mathrm{T}}\boldsymbol{\phi}$, we have

$$\sigma(\mathbf{w}^{\mathrm{T}}\boldsymbol{\phi}) = \int \delta(a - \mathbf{w}^{\mathrm{T}}\boldsymbol{\phi})\sigma(a)\,\mathrm{d}a \tag{4.146}$$

where $\delta(\cdot)$ is the Dirac delta function. From this we obtain

$$\int \sigma(\mathbf{w}^{\mathrm{T}}\boldsymbol{\phi})q(\mathbf{w})\,\mathrm{d}\mathbf{w} = \int \sigma(a)p(a)\,\mathrm{d}a \tag{4.147}$$

where

$$p(a) = \int \delta(a - \mathbf{w}^{\mathrm{T}}\boldsymbol{\phi})q(\mathbf{w})\,\mathrm{d}\mathbf{w}. \tag{4.148}$$

We can evaluate $p(a)$ by noting that the delta function imposes a linear constraint on $\mathbf{w}$ and so forms a marginal distribution from the joint distribution $q(\mathbf{w})$ by integrating out all directions orthogonal to $\boldsymbol{\phi}$. Because $q(\mathbf{w})$ is Gaussian, we know from Section 2.3.2 that the marginal distribution will also be Gaussian. We can evaluate the mean and covariance of this distribution by taking moments, and interchanging the order of integration over a and $\mathbf{w}$, so that

$$\mu_a = \mathbb{E}[a] = \int p(a)a\,\mathrm{d}a = \int q(\mathbf{w})\mathbf{w}^{\mathrm{T}}\boldsymbol{\phi}\,\mathrm{d}\mathbf{w} = \mathbf{w}_{\mathrm{MAP}}^{\mathrm{T}}\boldsymbol{\phi} \tag{4.149}$$

where we have used the result (4.144) for the variational posterior distribution $q(\mathbf{w})$. Similarly

$$\begin{aligned}\sigma_a^2 &= \mathrm{var}[a] = \int p(a)\left\{a^2 - \mathbb{E}[a]^2\right\}\mathrm{d}a \\ &= \int q(\mathbf{w})\left\{(\mathbf{w}^{\mathrm{T}}\boldsymbol{\phi})^2 - (\mathbf{m}_N^{\mathrm{T}}\boldsymbol{\phi})^2\right\}\mathrm{d}\mathbf{w} = \boldsymbol{\phi}^{\mathrm{T}}\mathbf{S}_N\boldsymbol{\phi}. \end{aligned} \tag{4.150}$$

Note that the distribution of a takes the same form as the predictive distribution (3.58) for the linear regression model, with the noise variance set to zero. Thus our variational approximation to the predictive distribution becomes

$$p(\mathcal{C}_1|\mathbf{t}) = \int \sigma(a)p(a)\,\mathrm{d}a = \int \sigma(a)\mathcal{N}(a|\mu_a, \sigma_a^2)\,\mathrm{d}a. \tag{4.151}$$

This result can also be derived directly by making use of the results for the marginal of a Gaussian distribution given in Section 2.3.2.

Exercise 4.24

The integral over a represents the convolution of a Gaussian with a logistic sigmoid, and cannot be evaluated analytically. We can, however, obtain a good approximation (Spiegelhalter and Lauritzen, 1990; MacKay, 1992b; Barber and Bishop, 1998a) by making use of the close similarity between the logistic sigmoid function $\sigma(a)$ defined by (4.59) and the inverse probit function $\Phi(a)$ defined by (4.114). In order to obtain the best approximation to the logistic function we need to re-scale the horizontal axis, so that we approximate $\sigma(a)$ by $\Phi(\lambda a)$. We can find a suitable value of λ by requiring that the two functions have the same slope at the origin, which gives $\lambda^2 = \pi/8$. The similarity of the logistic sigmoid and the inverse probit function, for this choice of λ, is illustrated in Figure 4.9.

Exercise 4.25

The advantage of using an inverse probit function is that its convolution with a Gaussian can be expressed analytically in terms of another inverse probit function. Specifically we can show that

Exercise 4.26

$$\int \Phi(\lambda a)\mathcal{N}(a|\mu, \sigma^2)\,\mathrm{d}a = \Phi\left(\frac{\mu}{(\lambda^{-2} + \sigma^2)^{1/2}}\right). \tag{4.152}$$

We now apply the approximation $\sigma(a) \simeq \Phi(\lambda a)$ to the inverse probit functions appearing on both sides of this equation, leading to the following approximation for the convolution of a logistic sigmoid with a Gaussian

$$\int \sigma(a)\mathcal{N}(a|\mu, \sigma^2)\,\mathrm{d}a \simeq \sigma\left(\kappa(\sigma^2)\mu\right) \tag{4.153}$$

where we have defined

$$\kappa(\sigma^2) = (1 + \pi\sigma^2/8)^{-1/2}. \tag{4.154}$$

Applying this result to (4.151) we obtain the approximate predictive distribution in the form

$$p(\mathcal{C}_1|\boldsymbol{\phi}, \mathbf{t}) = \sigma\left(\kappa(\sigma_a^2)\mu_a\right) \tag{4.155}$$

where μ_a and σ_a^2 are defined by (4.149) and (4.150), respectively, and $\kappa(\sigma_a^2)$ is defined by (4.154).

Note that the decision boundary corresponding to $p(\mathcal{C}_1|\boldsymbol{\phi}, \mathbf{t}) = 0.5$ is given by $\mu_a = 0$, which is the same as the decision boundary obtained by using the MAP value for $\mathbf{w}$. Thus if the decision criterion is based on minimizing misclassification rate, with equal prior probabilities, then the marginalization over $\mathbf{w}$ has no effect. However, for more complex decision criteria it will play an important role. Marginalization of the logistic sigmoid model under a Gaussian approximation to the posterior distribution will be illustrated in the context of variational inference in Figure 10.13.

Exercises

4.1 ($\star\star$) Given a set of data points $\{\mathbf{x}_n\}$, we can define the *convex hull* to be the set of all points $\mathbf{x}$ given by

$$\mathbf{x} = \sum_n \alpha_n \mathbf{x}_n \tag{4.156}$$

where $\alpha_n \geqslant 0$ and $\sum_n \alpha_n = 1$. Consider a second set of points $\{\mathbf{y}_n\}$ together with their corresponding convex hull. By definition, the two sets of points will be linearly separable if there exists a vector $\widehat{\mathbf{w}}$ and a scalar w_0 such that $\widehat{\mathbf{w}}^{\mathrm{T}}\mathbf{x}_n + w_0 > 0$ for all $\mathbf{x}_n$, and $\widehat{\mathbf{w}}^{\mathrm{T}}\mathbf{y}_n + w_0 < 0$ for all $\mathbf{y}_n$. Show that if their convex hulls intersect, the two sets of points cannot be linearly separable, and conversely that if they are linearly separable, their convex hulls do not intersect.

4.2 ($\star\star$) **www** Consider the minimization of a sum-of-squares error function (4.15), and suppose that all of the target vectors in the training set satisfy a linear constraint

$$\mathbf{a}^{\mathrm{T}}\mathbf{t}_n + b = 0 \tag{4.157}$$

where $\mathbf{t}_n$ corresponds to the nth row of the matrix $\mathbf{T}$ in (4.15). Show that as a consequence of this constraint, the elements of the model prediction $\mathbf{y}(\mathbf{x})$ given by the least-squares solution (4.17) also satisfy this constraint, so that

$$\mathbf{a}^{\mathrm{T}}\mathbf{y}(\mathbf{x}) + b = 0. \tag{4.158}$$

To do so, assume that one of the basis functions $\phi_0(\mathbf{x}) = 1$ so that the corresponding parameter w_0 plays the role of a bias.

4.3 ($\star\star$) Extend the result of Exercise 4.2 to show that if multiple linear constraints are satisfied simultaneously by the target vectors, then the same constraints will also be satisfied by the least-squares prediction of a linear model.

4.4 ($\star$) **www** Show that maximization of the class separation criterion given by (4.22) with respect to $\mathbf{w}$, using a Lagrange multiplier to enforce the constraint $\mathbf{w}^{\mathrm{T}}\mathbf{w} = 1$, leads to the result that $\mathbf{w} \propto (\mathbf{m}_2 - \mathbf{m}_1)$.

4.5 ($\star$) By making use of (4.20), (4.23), and (4.24), show that the Fisher criterion (4.25) can be written in the form (4.26).

4.6 ($\star$) Using the definitions of the between-class and within-class covariance matrices given by (4.27) and (4.28), respectively, together with (4.34) and (4.36) and the choice of target values described in Section 4.1.5, show that the expression (4.33) that minimizes the sum-of-squares error function can be written in the form (4.37).

4.7 ($\star$) **www** Show that the logistic sigmoid function (4.59) satisfies the property $\sigma(-a) = 1 - \sigma(a)$ and that its inverse is given by $\sigma^{-1}(y) = \ln\{y/(1-y)\}$.

4.8 ($\star$) Using (4.57) and (4.58), derive the result (4.65) for the posterior class probability in the two-class generative model with Gaussian densities, and verify the results (4.66) and (4.67) for the parameters $\mathbf{w}$ and w_0.

4.9 ($\star$) **www** Consider a generative classification model for K classes defined by prior class probabilities $p(\mathcal{C}_k) = \pi_k$ and general class-conditional densities $p(\boldsymbol{\phi}|\mathcal{C}_k)$ where $\boldsymbol{\phi}$ is the input feature vector. Suppose we are given a training data set $\{\boldsymbol{\phi}_n, \mathbf{t}_n\}$ where $n = 1, \ldots, N$, and $\mathbf{t}_n$ is a binary target vector of length K that uses the 1-of-K coding scheme, so that it has components $t_{nj} = I_{jk}$ if pattern n is from class $\mathcal{C}_k$. Assuming that the data points are drawn independently from this model, show that the maximum-likelihood solution for the prior probabilities is given by

$$\pi_k = \frac{N_k}{N} \tag{4.159}$$

where N_k is the number of data points assigned to class $\mathcal{C}_k$.

4.10 ($\star\star$) Consider the classification model of Exercise 4.9 and now suppose that the class-conditional densities are given by Gaussian distributions with a shared covariance matrix, so that

$$p(\boldsymbol{\phi}|\mathcal{C}_k) = \mathcal{N}(\boldsymbol{\phi}|\boldsymbol{\mu}_k, \boldsymbol{\Sigma}). \tag{4.160}$$

Show that the maximum likelihood solution for the mean of the Gaussian distribution for class $\mathcal{C}_k$ is given by

$$\boldsymbol{\mu}_k = \frac{1}{N_k}\sum_{n=1}^{N} t_{nk}\boldsymbol{\phi}_n \tag{4.161}$$

which represents the mean of those feature vectors assigned to class $\mathcal{C}_k$. Similarly, show that the maximum likelihood solution for the shared covariance matrix is given by

$$\boldsymbol{\Sigma} = \sum_{k=1}^{K} \frac{N_k}{N} \mathbf{S}_k \tag{4.162}$$

where

$$\mathbf{S}_k = \frac{1}{N_k} \sum_{n=1}^{N} t_{nk} (\boldsymbol{\phi}_n - \boldsymbol{\mu}_k)(\boldsymbol{\phi}_n - \boldsymbol{\mu}_k)^{\mathrm{T}}. \tag{4.163}$$

Thus $\boldsymbol{\Sigma}$ is given by a weighted average of the covariances of the data associated with each class, in which the weighting coefficients are given by the prior probabilities of the classes.

4.11 ($\star\star$) Consider a classification problem with K classes for which the feature vector $\boldsymbol{\phi}$ has M components each of which can take L discrete states. Let the values of the components be represented by a 1-of-L binary coding scheme. Further suppose that, conditioned on the class $\mathcal{C}_k$, the M components of $\boldsymbol{\phi}$ are independent, so that the class-conditional density factorizes with respect to the feature vector components. Show that the quantities a_k given by (4.63), which appear in the argument to the softmax function describing the posterior class probabilities, are linear functions of the components of $\boldsymbol{\phi}$. Note that this represents an example of the naive Bayes model which is discussed in Section 8.2.2.

4.12 ($\star$) **www** Verify the relation (4.88) for the derivative of the logistic sigmoid function defined by (4.59).

4.13 ($\star$) **www** By making use of the result (4.88) for the derivative of the logistic sigmoid, show that the derivative of the error function (4.90) for the logistic regression model is given by (4.91).

4.14 ($\star$) Show that for a linearly separable data set, the maximum likelihood solution for the logistic regression model is obtained by finding a vector $\mathbf{w}$ whose decision boundary $\mathbf{w}^{\mathrm{T}}\boldsymbol{\phi}(\mathbf{x}) = 0$ separates the classes and then taking the magnitude of $\mathbf{w}$ to infinity.

4.15 ($\star\star$) Show that the Hessian matrix $\mathbf{H}$ for the logistic regression model, given by (4.97), is positive definite. Here $\mathbf{R}$ is a diagonal matrix with elements $y_n(1 - y_n)$, and y_n is the output of the logistic regression model for input vector $\mathbf{x}_n$. Hence show that the error function is a convex function of $\mathbf{w}$ and that it has a unique minimum.

4.16 ($\star$) Consider a binary classification problem in which each observation $\mathbf{x}_n$ is known to belong to one of two classes, corresponding to $t_n = 0$ and $t_n = 1$, and suppose that the procedure for collecting training data is imperfect, so that training points are sometimes mislabelled. For every data point $\mathbf{x}_n$, instead of having a value t for the class label, we have instead a value π_n representing the probability that $t_n = 1$. Given a probabilistic model $p(t = 1|\boldsymbol{\phi})$, write down the log likelihood function appropriate to such a data set.

4.17 ($\star$) **www** Show that the derivatives of the softmax activation function (4.104), where the a_k are defined by (4.105), are given by (4.106).

4.18 ($\star$) Using the result (4.106) for the derivatives of the softmax activation function, show that the gradients of the cross-entropy error (4.108) are given by (4.109).

4.19 ($\star$) **www** Write down expressions for the gradient of the log likelihood, as well as the corresponding Hessian matrix, for the probit regression model defined in Section 4.3.5. These are the quantities that would be required to train such a model using IRLS.

4.20 ($\star\star$) Show that the Hessian matrix for the multiclass logistic regression problem, defined by (4.110), is positive semidefinite. Note that the full Hessian matrix for this problem is of size $MK \times MK$, where M is the number of parameters and K is the number of classes. To prove the positive semidefinite property, consider the product $\mathbf{u}^{\mathrm{T}}\mathbf{H}\mathbf{u}$ where $\mathbf{u}$ is an arbitrary vector of length MK, and then apply Jensen's inequality.

4.21 ($\star$) Show that the inverse probit function (4.114) and the erf function (4.115) are related by (4.116).

4.22 ($\star$) Using the result (4.135), derive the expression (4.137) for the log model evidence under the Laplace approximation.

4.23 ($\star\star$) **www** In this exercise, we derive the BIC result (4.139) starting from the Laplace approximation to the model evidence given by (4.137). Show that if the prior over parameters is Gaussian of the form $p(\boldsymbol{\theta}) = \mathcal{N}(\boldsymbol{\theta}|\mathbf{m}, \mathbf{V}_0)$, the log model evidence under the Laplace approximation takes the form

$$\ln p(\mathcal{D}) \simeq \ln p(\mathcal{D}|\boldsymbol{\theta}_{\mathrm{MAP}}) - \frac{1}{2}(\boldsymbol{\theta}_{\mathrm{MAP}} - \mathbf{m})^{\mathrm{T}}\mathbf{V}_0^{-1}(\boldsymbol{\theta}_{\mathrm{MAP}} - \mathbf{m}) - \frac{1}{2}\ln|\mathbf{H}| + \mathrm{const}$$

where $\mathbf{H}$ is the matrix of second derivatives of the negative log likelihood $\ln p(\mathcal{D}|\boldsymbol{\theta})$ evaluated at $\boldsymbol{\theta}_{\mathrm{MAP}}$. Now assume that the prior is broad so that $\mathbf{V}_0^{-1}$ is small and the second term on the right-hand side above can be neglected. Furthermore, consider the case of independent, identically distributed data so that $\mathbf{H}$ is the sum of terms one for each data point. Show that the log model evidence can then be written approximately in the form of the BIC expression (4.139).

4.24 ($\star\star$) Use the results from Section 2.3.2 to derive the result (4.151) for the marginalization of the logistic regression model with respect to a Gaussian posterior distribution over the parameters $\mathbf{w}$.

4.25 ($\star\star$) Suppose we wish to approximate the logistic sigmoid $\sigma(a)$ defined by (4.59) by a scaled inverse probit function $\Phi(\lambda a)$, where $\Phi(a)$ is defined by (4.114). Show that if λ is chosen so that the derivatives of the two functions are equal at $a = 0$, then $\lambda^2 = \pi/8$.

4.26 ($\star\star$) In this exercise, we prove the relation (4.152) for the convolution of an inverse probit function with a Gaussian distribution. To do this, show that the derivative of the left-hand side with respect to μ is equal to the derivative of the right-hand side, and then integrate both sides with respect to μ and then show that the constant of integration vanishes. Note that before differentiating the left-hand side, it is convenient first to introduce a change of variable given by $a = \mu + \sigma z$ so that the integral over a is replaced by an integral over z. When we differentiate the left-hand side of the relation (4.152), we will then obtain a Gaussian integral over z that can be evaluated analytically.

5 Neural Networks

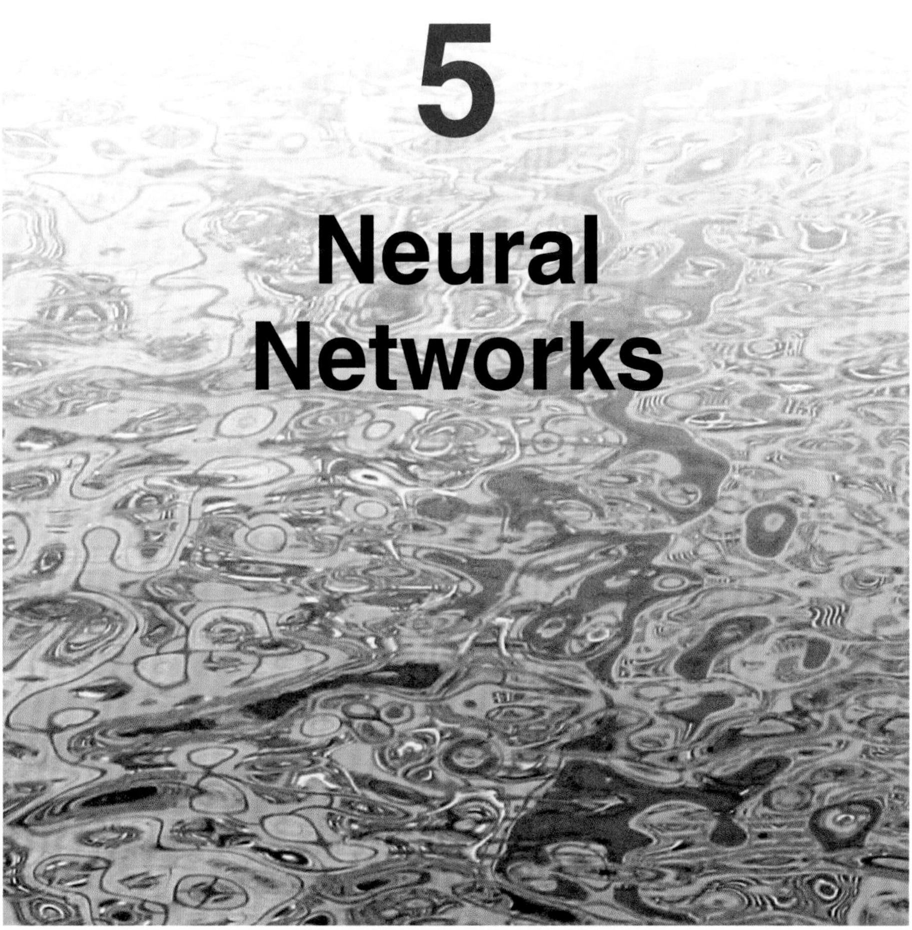

In Chapters 3 and 4 we considered models for regression and classification that comprised linear combinations of fixed basis functions. We saw that such models have useful analytical and computational properties but that their practical applicability was limited by the curse of dimensionality. In order to apply such models to large-scale problems, it is necessary to adapt the basis functions to the data.

Support vector machines (SVMs), discussed in Chapter 7, address this by first defining basis functions that are centred on the training data points and then selecting a subset of these during training. One advantage of SVMs is that, although the training involves nonlinear optimization, the objective function is convex, and so the solution of the optimization problem is relatively straightforward. The number of basis functions in the resulting models is generally much smaller than the number of training points, although it is often still relatively large and typically increases with the size of the training set. The relevance vector machine, discussed in Section 7.2, also chooses a subset from a fixed set of basis functions and typically results in much sparser models. Unlike the SVM it also produces probabilistic outputs, although this is at the expense of a nonconvex optimization during training.

An alternative approach is to fix the number of basis functions in advance but allow them to be adaptive, in other words to use parametric forms for the basis functions in which the parameter values are adapted during training. The most successful model of this type in the context of pattern recognition is the feed-forward neural network, also known as the *multilayer perceptron*, discussed in this chapter. In fact, 'multilayer perceptron' is really a misnomer, because the model comprises multiple layers of logistic regression models (with continuous nonlinearities) rather than multiple perceptrons (with discontinuous nonlinearities). For many applications, the resulting model can be significantly more compact, and hence faster to evaluate, than a support vector machine having the same generalization performance. The price to be paid for this compactness, as with the relevance vector machine, is that the likelihood function, which forms the basis for network training, is no longer a convex function of the model parameters. In practice, however, it is often worth investing substantial computational resources during the training phase in order to obtain a compact model that is fast at processing new data.

The term 'neural network' has its origins in attempts to find mathematical representations of information processing in biological systems (McCulloch and Pitts, 1943; Widrow and Hoff, 1960; Rosenblatt, 1962; Rumelhart *et al.*, 1986). Indeed, it has been used very broadly to cover a wide range of different models, many of which have been the subject of exaggerated claims regarding their biological plausibility. From the perspective of practical applications of pattern recognition, however, biological realism would impose entirely unnecessary constraints. Our focus in this chapter is therefore on neural networks as efficient models for statistical pattern recognition. In particular, we shall restrict our attention to the specific class of neural networks that have proven to be of greatest practical value, namely the multilayer perceptron.

We begin by considering the functional form of the network model, including the specific parameterization of the basis functions, and we then discuss the problem of determining the network parameters within a maximum likelihood framework, which involves the solution of a nonlinear optimization problem. This requires the evaluation of derivatives of the log likelihood function with respect to the network parameters, and we shall see how these can be obtained efficiently using the technique of *error backpropagation*. We shall also show how the backpropagation framework can be extended to allow other derivatives to be evaluated, such as the Jacobian and Hessian matrices. Next we discuss various approaches to regularization of neural network training and the relationships between them. We also consider some extensions to the neural network model, and in particular we describe a general framework for modelling conditional probability distributions known as *mixture density networks*. Finally, we discuss the use of Bayesian treatments of neural networks. Additional background on neural network models can be found in Bishop (1995a).

5.1. Feed-forward Network Functions

The linear models for regression and classification discussed in Chapters 3 and 4, respectively, are based on linear combinations of fixed nonlinear basis functions $\phi_j(\mathbf{x})$ and take the form

$$y(\mathbf{x}, \mathbf{w}) = f\left(\sum_{j=1}^{M} w_j \phi_j(\mathbf{x})\right) \tag{5.1}$$

where $f(\cdot)$ is a nonlinear activation function in the case of classification and is the identity in the case of regression. Our goal is to extend this model by making the basis functions $\phi_j(\mathbf{x})$ depend on parameters and then to allow these parameters to be adjusted, along with the coefficients $\{w_j\}$, during training. There are, of course, many ways to construct parametric nonlinear basis functions. Neural networks use basis functions that follow the same form as (5.1), so that each basis function is itself a nonlinear function of a linear combination of the inputs, where the coefficients in the linear combination are adaptive parameters.

This leads to the basic neural network model, which can be described as a series of functional transformations. First we construct M linear combinations of the input variables $x_1, \ldots, x_D$ in the form

$$a_j = \sum_{i=1}^{D} w_{ji}^{(1)} x_i + w_{j0}^{(1)} \tag{5.2}$$

where $j = 1, \ldots, M$, and the superscript (1) indicates that the corresponding parameters are in the first 'layer' of the network. We shall refer to the parameters $w_{ji}^{(1)}$ as *weights* and the parameters $w_{j0}^{(1)}$ as *biases*, following the nomenclature of Chapter 3. The quantities a_j are known as *activations*. Each of them is then transformed using a differentiable, nonlinear *activation function* $h(\cdot)$ to give

$$z_j = h(a_j). \tag{5.3}$$

These quantities correspond to the outputs of the basis functions in (5.1) that, in the context of neural networks, are called *hidden units*. The nonlinear functions $h(\cdot)$ are generally chosen to be sigmoidal functions such as the logistic sigmoid or the '$\tanh$' function. Following (5.1), these values are again linearly combined to give *output unit activations*

Exercise 5.1

$$a_k = \sum_{j=1}^{M} w_{kj}^{(2)} z_j + w_{k0}^{(2)} \tag{5.4}$$

where $k = 1, \ldots, K$, and K is the total number of outputs. This transformation corresponds to the second layer of the network, and again the $w_{k0}^{(2)}$ are bias parameters. Finally, the output unit activations are transformed using an appropriate activation function to give a set of network outputs y_k. The choice of activation function is determined by the nature of the data and the assumed distribution of target variables

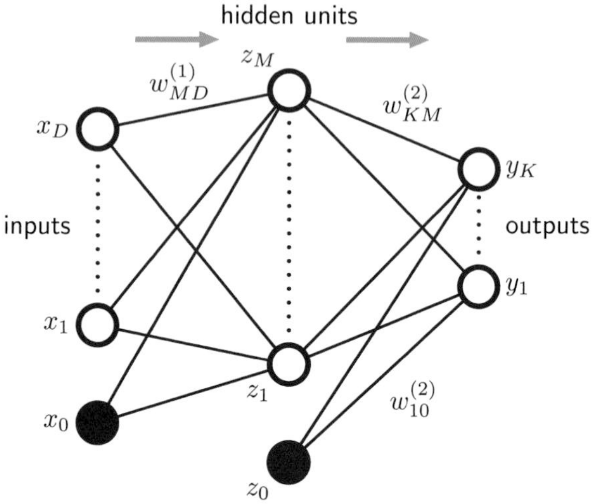

Figure 5.1 Network diagram for the two-layer neural network corresponding to (5.7). The input, hidden, and output variables are represented by nodes, and the weight parameters are represented by links between the nodes, in which the bias parameters are denoted by links coming from additional input and hidden variables x_0 and z_0. Arrows denote the direction of information flow through the network during forward propagation.

and follows the same considerations as for linear models discussed in Chapters 3 and 4. Thus for standard regression problems, the activation function is the identity so that $y_k = a_k$. Similarly, for multiple binary classification problems, each output unit activation is transformed using a logistic sigmoid function so that

$$y_k = \sigma(a_k) \tag{5.5}$$

where

$$\sigma(a) = \frac{1}{1 + \exp(-a)}. \tag{5.6}$$

Finally, for multiclass problems, a softmax activation function of the form (4.62) is used. The choice of output unit activation function is discussed in detail in Section 5.2.

We can combine these various stages to give the overall network function that, for sigmoidal output unit activation functions, takes the form

$$y_k(\mathbf{x}, \mathbf{w}) = \sigma\left(\sum_{j=1}^{M} w_{kj}^{(2)} h\left(\sum_{i=1}^{D} w_{ji}^{(1)} x_i + w_{j0}^{(1)}\right) + w_{k0}^{(2)}\right) \tag{5.7}$$

where the set of all weight and bias parameters have been grouped together into a vector $\mathbf{w}$. Thus the neural network model is simply a nonlinear function from a set of input variables $\{x_i\}$ to a set of output variables $\{y_k\}$ controlled by a vector $\mathbf{w}$ of adjustable parameters.

This function can be represented in the form of a network diagram as shown in Figure 5.1. The process of evaluating (5.7) can then be interpreted as a *forward propagation* of information through the network. It should be emphasized that these diagrams do not represent probabilistic graphical models of the kind to be considered in Chapter 8 because the internal nodes represent deterministic variables rather than stochastic ones. For this reason, we have adopted a slightly different graphical notation for the two kinds of model. We shall see later how to give a probabilistic interpretation to a neural network.

As discussed in Section 3.1, the bias parameters in (5.2) can be absorbed into the set of weight parameters by defining an additional input variable x_0 whose value is clamped at $x_0 = 1$, so that (5.2) takes the form

$$a_j = \sum_{i=0}^{D} w_{ji}^{(1)} x_i. \tag{5.8}$$

We can similarly absorb the second-layer biases into the second-layer weights, so that the overall network function becomes

$$y_k(\mathbf{x}, \mathbf{w}) = \sigma\left(\sum_{j=0}^{M} w_{kj}^{(2)} h\left(\sum_{i=0}^{D} w_{ji}^{(1)} x_i\right)\right). \tag{5.9}$$

As can be seen from Figure 5.1, the neural network model comprises two stages of processing, each of which resembles the perceptron model of Section 4.1.7, and for this reason the neural network is also known as the *multilayer perceptron*, or MLP. A key difference compared to the perceptron, however, is that the neural network uses continuous sigmoidal nonlinearities in the hidden units, whereas the perceptron uses step-function nonlinearities. This means that the neural network function is differentiable with respect to the network parameters, and this property will play a central role in network training.

If the activation functions of all the hidden units in a network are taken to be linear, then for any such network we can always find an equivalent network without hidden units. This follows from the fact that the composition of successive linear transformations is itself a linear transformation. However, if the number of hidden units is smaller than either the number of input or output units, then the transformations that the network can generate are not the most general possible linear transformations from inputs to outputs because information is lost in the dimensionality reduction at the hidden units. In Section 12.4.2, we show that networks of linear units give rise to principal component analysis. In general, however, there is little interest in multilayer networks of linear units.

The network architecture shown in Figure 5.1 is the most commonly used one in practice. However, it is easily generalized, for instance by considering additional layers of processing each consisting of a weighted linear combination of the form (5.4) followed by an element-wise transformation using a nonlinear activation function. Note that there is some confusion in the literature regarding the terminology for counting the number of layers in such networks. Thus the network in Figure 5.1 may be described as a 3-layer network (which counts the number of layers of units, and treats the inputs as units) or sometimes as a single-hidden-layer network (which counts the number of layers of hidden units). We recommend a terminology in which Figure 5.1 is called a two-layer network, because it is the number of layers of adaptive weights that is important for determining the network properties.

Another generalization of the network architecture is to include *skip-layer* connections, each of which is associated with a corresponding adaptive parameter. For

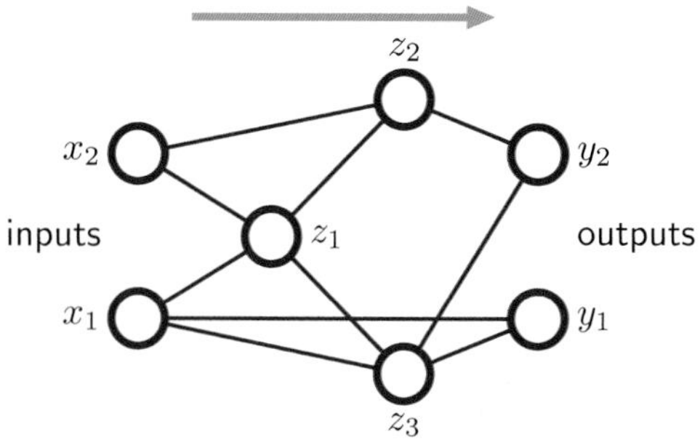

Figure 5.2 Example of a neural network having a general feed-forward topology. Note that each hidden and output unit has an associated bias parameter (omitted for clarity).

instance, in a two-layer network these would go directly from inputs to outputs. In principle, a network with sigmoidal hidden units can always mimic skip layer connections (for bounded input values) by using a sufficiently small first-layer weight that, over its operating range, the hidden unit is effectively linear, and then compensating with a large weight value from the hidden unit to the output. In practice, however, it may be advantageous to include skip-layer connections explicitly.

Furthermore, the network can be sparse, with not all possible connections within a layer being present. We shall see an example of a sparse network architecture when we consider convolutional neural networks in Section 5.5.6.

Because there is a direct correspondence between a network diagram and its mathematical function, we can develop more general network mappings by considering more complex network diagrams. However, these must be restricted to a *feed-forward* architecture, in other words to one having no closed directed cycles, to ensure that the outputs are deterministic functions of the inputs. This is illustrated with a simple example in Figure 5.2. Each (hidden or output) unit in such a network computes a function given by

$$z_k = h\left(\sum_j w_{kj} z_j\right) \tag{5.10}$$

where the sum runs over all units that send connections to unit k (and a bias parameter is included in the summation). For a given set of values applied to the inputs of the network, successive application of (5.10) allows the activations of all units in the network to be evaluated including those of the output units.

The approximation properties of feed-forward networks have been widely studied (Funahashi, 1989; Cybenko, 1989; Hornik *et al.*, 1989; Stinchecombe and White, 1989; Cotter, 1990; Ito, 1991; Hornik, 1991; Kreinovich, 1991; Ripley, 1996) and found to be very general. Neural networks are therefore said to be *universal approximators*. For example, a two-layer network with linear outputs can uniformly approximate any continuous function on a compact input domain to arbitrary accuracy provided the network has a sufficiently large number of hidden units. This result holds for a wide range of hidden unit activation functions, but excluding polynomials. Although such theorems are reassuring, the key problem is how to find suitable parameter values given a set of training data, and in later sections of this chapter we

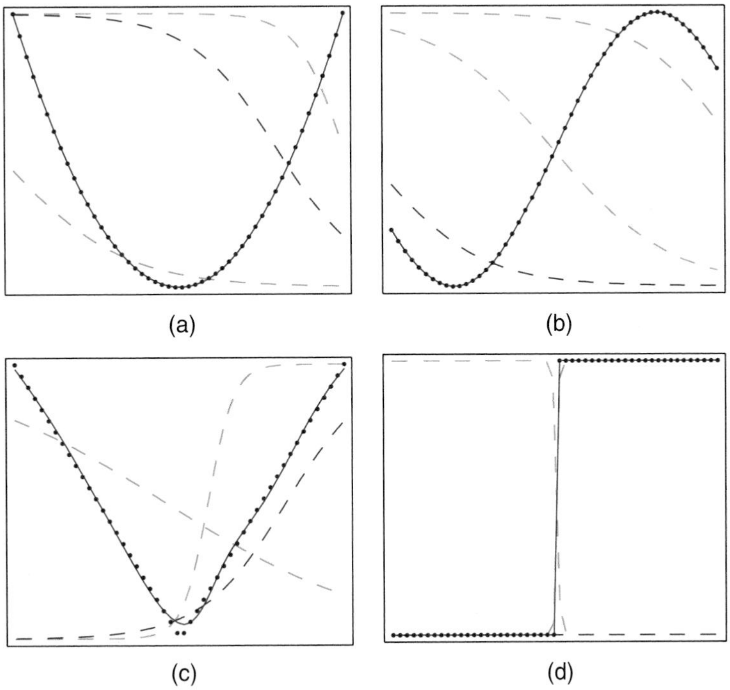

Figure 5.3 Illustration of the capability of a multilayer perceptron to approximate four different functions comprising (a) $f(x) = x^2$, (b) $f(x) = \sin(x)$, (c), $f(x) = |x|$, and (d) $f(x) = H(x)$ where $H(x)$ is the Heaviside step function. In each case, $N = 50$ data points, shown as blue dots, have been sampled uniformly in x over the interval $(-1, 1)$ and the corresponding values of $f(x)$ evaluated. These data points are then used to train a two-layer network having 3 hidden units with '$\tanh$' activation functions and linear output units. The resulting network functions are shown by the red curves, and the outputs of the three hidden units are shown by the three dashed curves.

will show that there exist effective solutions to this problem based on both maximum likelihood and Bayesian approaches.

The capability of a two-layer network to model a broad range of functions is illustrated in Figure 5.3. This figure also shows how individual hidden units work collaboratively to approximate the final function. The role of hidden units in a simple classification problem is illustrated in Figure 5.4 using the synthetic classification data set described in Appendix A.

5.1.1 Weight-space symmetries

One property of feed-forward networks, which will play a role when we consider Bayesian model comparison, is that multiple distinct choices for the weight vector $\mathbf{w}$ can all give rise to the same mapping function from inputs to outputs (Chen *et al.*, 1993). Consider a two-layer network of the form shown in Figure 5.1 with M hidden units having '$\tanh$' activation functions and full connectivity in both layers. If we change the sign of all of the weights and the bias feeding into a particular hidden unit, then, for a given input pattern, the sign of the activation of the hidden unit will be reversed, because '$\tanh$' is an odd function, so that $\tanh(-a) = -\tanh(a)$. This transformation can be exactly compensated by changing the sign of all of the weights leading out of that hidden unit. Thus, by changing the signs of a particular group of weights (and a bias), the input–output mapping function represented by the network is unchanged, and so we have found two different weight vectors that give rise to the same mapping function. For M hidden units, there will be M such 'sign-flip'

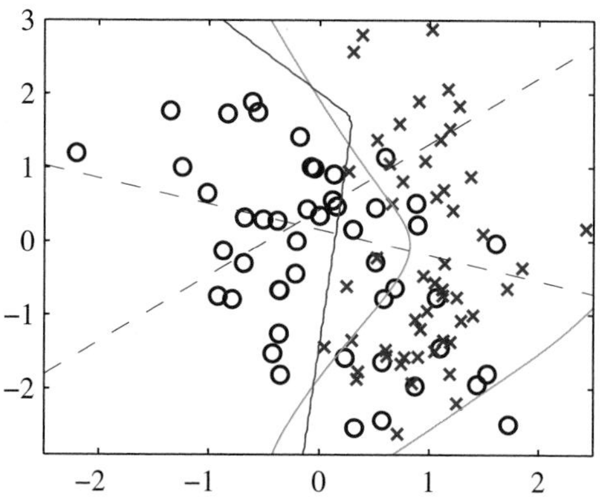

Figure 5.4 Example of the solution of a simple two-class classification problem involving synthetic data using a neural network having two inputs, two hidden units with 'tanh' activation functions, and a single output having a logistic sigmoid activation function. The dashed blue lines show the $z = 0.5$ contours for each of the hidden units, and the red line shows the $y = 0.5$ decision surface for the network. For comparison, the green line denotes the optimal decision boundary computed from the distributions used to generate the data.

symmetries, and thus any given weight vector will be one of a set 2^M equivalent weight vectors .

Similarly, imagine that we interchange the values of all of the weights (and the bias) leading both into and out of a particular hidden unit with the corresponding values of the weights (and bias) associated with a different hidden unit. Again, this clearly leaves the network input–output mapping function unchanged, but it corresponds to a different choice of weight vector. For M hidden units, any given weight vector will belong to a set of $M!$ equivalent weight vectors associated with this interchange symmetry, corresponding to the $M!$ different orderings of the hidden units. The network will therefore have an overall weight-space symmetry factor of $M!2^M$. For networks with more than two layers of weights, the total level of symmetry will be given by the product of such factors, one for each layer of hidden units.

It turns out that these factors account for all of the symmetries in weight space (except for possible accidental symmetries due to specific choices for the weight values). Furthermore, the existence of these symmetries is not a particular property of the 'tanh' function but applies to a wide range of activation functions (Kůrková and Kainen, 1994). In many cases, these symmetries in weight space are of little practical consequence, although in Section 5.7 we shall encounter a situation in which we need to take them into account.

5.2. Network Training

So far, we have viewed neural networks as a general class of parametric nonlinear functions from a vector $\mathbf{x}$ of input variables to a vector $\mathbf{y}$ of output variables. A simple approach to the problem of determining the network parameters is to make an analogy with the discussion of polynomial curve fitting in Section 1.1, and therefore to minimize a sum-of-squares error function. Given a training set comprising a set of input vectors $\{\mathbf{x}_n\}$, where $n = 1, \ldots, N$, together with a corresponding set of target vectors $\{\mathbf{t}_n\}$, we minimize the error function

$$E(\mathbf{w}) = \frac{1}{2}\sum_{n=1}^{N} \|\mathbf{y}(\mathbf{x}_n, \mathbf{w}) - \mathbf{t}_n\|^2. \tag{5.11}$$

However, we can provide a much more general view of network training by first giving a probabilistic interpretation to the network outputs. We have already seen many advantages of using probabilistic predictions in Section 1.5.4. Here it will also provide us with a clearer motivation both for the choice of output unit nonlinearity and the choice of error function.

We start by discussing regression problems, and for the moment we consider a single target variable t that can take any real value. Following the discussions in Section 1.2.5 and 3.1, we assume that t has a Gaussian distribution with an $\mathbf{x}$-dependent mean, which is given by the output of the neural network, so that

$$p(t|\mathbf{x}, \mathbf{w}) = \mathcal{N}\left(t|y(\mathbf{x}, \mathbf{w}), \beta^{-1}\right) \tag{5.12}$$

where β is the precision (inverse variance) of the Gaussian noise. Of course this is a somewhat restrictive assumption, and in Section 5.6 we shall see how to extend this approach to allow for more general conditional distributions. For the conditional distribution given by (5.12), it is sufficient to take the output unit activation function to be the identity, because such a network can approximate any continuous function from $\mathbf{x}$ to y. Given a data set of N independent, identically distributed observations $\mathbf{X} = \{\mathbf{x}_1, \ldots, \mathbf{x}_N\}$, along with corresponding target values $\mathsf{t} = \{t_1, \ldots, t_N\}$, we can construct the corresponding likelihood function

$$p(\mathsf{t}|\mathbf{X}, \mathbf{w}, \beta) = \prod_{n=1}^{N} p(t_n|\mathbf{x}_n, \mathbf{w}, \beta).$$

Taking the negative logarithm, we obtain the error function

$$\frac{\beta}{2}\sum_{n=1}^{N}\{y(\mathbf{x}_n, \mathbf{w}) - t_n\}^2 - \frac{N}{2}\ln\beta + \frac{N}{2}\ln(2\pi) \tag{5.13}$$

which can be used to learn the parameters $\mathbf{w}$ and β. In Section 5.7, we shall discuss the Bayesian treatment of neural networks, while here we consider a maximum likelihood approach. Note that in the neural networks literature, it is usual to consider the minimization of an error function rather than the maximization of the (log) likelihood, and so here we shall follow this convention. Consider first the determination of $\mathbf{w}$. Maximizing the likelihood function is equivalent to minimizing the sum-of-squares error function given by

$$E(\mathbf{w}) = \frac{1}{2}\sum_{n=1}^{N}\{y(\mathbf{x}_n, \mathbf{w}) - t_n\}^2 \tag{5.14}$$

where we have discarded additive and multiplicative constants. The value of $\mathbf{w}$ found by minimizing $E(\mathbf{w})$ will be denoted $\mathbf{w}_{\mathrm{ML}}$ because it corresponds to the maximum likelihood solution. In practice, the nonlinearity of the network function $y(\mathbf{x}_n, \mathbf{w})$ causes the error $E(\mathbf{w})$ to be nonconvex, and so in practice local maxima of the likelihood may be found, corresponding to local minima of the error function, as discussed in Section 5.2.1.

Having found $\mathbf{w}_{\mathrm{ML}}$, the value of β can be found by minimizing the negative log likelihood to give

$$\frac{1}{\beta_{\mathrm{ML}}} = \frac{1}{N}\sum_{n=1}^{N}\{y(\mathbf{x}_n, \mathbf{w}_{\mathrm{ML}}) - t_n\}^2. \tag{5.15}$$

Note that this can be evaluated once the iterative optimization required to find $\mathbf{w}_{\mathrm{ML}}$ is completed. If we have multiple target variables, and we assume that they are independent conditional on $\mathbf{x}$ and $\mathbf{w}$ with shared noise precision β, then the conditional distribution of the target values is given by

$$p(\mathbf{t}|\mathbf{x}, \mathbf{w}) = \mathcal{N}\left(\mathbf{t}|\mathbf{y}(\mathbf{x}, \mathbf{w}), \beta^{-1}\mathbf{I}\right). \tag{5.16}$$

Exercise 5.2

Following the same argument as for a single target variable, we see that the maximum likelihood weights are determined by minimizing the sum-of-squares error function (5.11). The noise precision is then given by

$$\frac{1}{\beta_{\mathrm{ML}}} = \frac{1}{NK}\sum_{n=1}^{N}\|\mathbf{y}(\mathbf{x}_n, \mathbf{w}_{\mathrm{ML}}) - \mathbf{t}_n\|^2 \tag{5.17}$$

Exercise 5.3

where K is the number of target variables. The assumption of independence can be dropped at the expense of a slightly more complex optimization problem.

Recall from Section 4.3.6 that there is a natural pairing of the error function (given by the negative log likelihood) and the output unit activation function. In the regression case, we can view the network as having an output activation function that is the identity, so that $y_k = a_k$. The corresponding sum-of-squares error function has the property

$$\frac{\partial E}{\partial a_k} = y_k - t_k \tag{5.18}$$

which we shall make use of when discussing error backpropagation in Section 5.3.

Now consider the case of binary classification in which we have a single target variable t such that $t = 1$ denotes class $\mathcal{C}_1$ and $t = 0$ denotes class $\mathcal{C}_2$. Following the discussion of canonical link functions in Section 4.3.6, we consider a network having a single output whose activation function is a logistic sigmoid

$$y = \sigma(a) \equiv \frac{1}{1 + \exp(-a)} \tag{5.19}$$

so that $0 \leqslant y(\mathbf{x}, \mathbf{w}) \leqslant 1$. We can interpret $y(\mathbf{x}, \mathbf{w})$ as the conditional probability $p(\mathcal{C}_1|\mathbf{x})$, with $p(\mathcal{C}_2|\mathbf{x})$ given by $1 - y(\mathbf{x}, \mathbf{w})$. The conditional distribution of targets given inputs is then a Bernoulli distribution of the form

$$p(t|\mathbf{x}, \mathbf{w}) = y(\mathbf{x}, \mathbf{w})^t \left\{1 - y(\mathbf{x}, \mathbf{w})\right\}^{1-t}. \tag{5.20}$$

If we consider a training set of independent observations, then the error function, which is given by the negative log likelihood, is then a *cross-entropy* error function of the form

$$E(\mathbf{w}) = -\sum_{n=1}^{N} \{t_n \ln y_n + (1 - t_n) \ln(1 - y_n)\} \tag{5.21}$$

where y_n denotes $y(\mathbf{x}_n, \mathbf{w})$. Note that there is no analogue of the noise precision β because the target values are assumed to be correctly labelled. However, the model is easily extended to allow for labelling errors. Simard *et al.* (2003) found that using the cross-entropy error function instead of the sum-of-squares for a classification problem leads to faster training as well as improved generalization.

Exercise 5.4

If we have K separate binary classifications to perform, then we can use a network having K outputs each of which has a logistic sigmoid activation function. Associated with each output is a binary class label $t_k \in \{0, 1\}$, where $k = 1, \ldots, K$. If we assume that the class labels are independent, given the input vector, then the conditional distribution of the targets is

$$p(\mathbf{t}|\mathbf{x}, \mathbf{w}) = \prod_{k=1}^{K} y_k(\mathbf{x}, \mathbf{w})^{t_k} \left[1 - y_k(\mathbf{x}, \mathbf{w})\right]^{1-t_k}. \tag{5.22}$$

Taking the negative logarithm of the corresponding likelihood function then gives the following error function

Exercise 5.5

$$E(\mathbf{w}) = -\sum_{n=1}^{N} \sum_{k=1}^{K} \{t_{nk} \ln y_{nk} + (1 - t_{nk}) \ln(1 - y_{nk})\} \tag{5.23}$$

where y_{nk} denotes $y_k(\mathbf{x}_n, \mathbf{w})$. Again, the derivative of the error function with respect to the activation for a particular output unit takes the form (5.18) just as in the regression case.

Exercise 5.6

It is interesting to contrast the neural network solution to this problem with the corresponding approach based on a linear classification model of the kind discussed in Chapter 4. Suppose that we are using a standard two-layer network of the kind shown in Figure 5.1. We see that the weight parameters in the first layer of the network are shared between the various outputs, whereas in the linear model each classification problem is solved independently. The first layer of the network can be viewed as performing a nonlinear feature extraction, and the sharing of features between the different outputs can save on computation and can also lead to improved generalization.

Finally, we consider the standard multiclass classification problem in which each input is assigned to one of K mutually exclusive classes. The binary target variables $t_k \in \{0, 1\}$ have a 1-of-K coding scheme indicating the class, and the network outputs are interpreted as $y_k(\mathbf{x}, \mathbf{w}) = p(t_k = 1|\mathbf{x})$, leading to the following error function

$$E(\mathbf{w}) = -\sum_{n=1}^{N} \sum_{k=1}^{K} t_{nk} \ln y_k(\mathbf{x}_n, \mathbf{w}). \tag{5.24}$$

Figure 5.5 Geometrical view of the error function $E(\mathbf{w})$ as a surface sitting over weight space. Point $\mathbf{w}_A$ is a local minimum and $\mathbf{w}_B$ is the global minimum. At any point $\mathbf{w}_C$, the local gradient of the error surface is given by the vector ∇E.

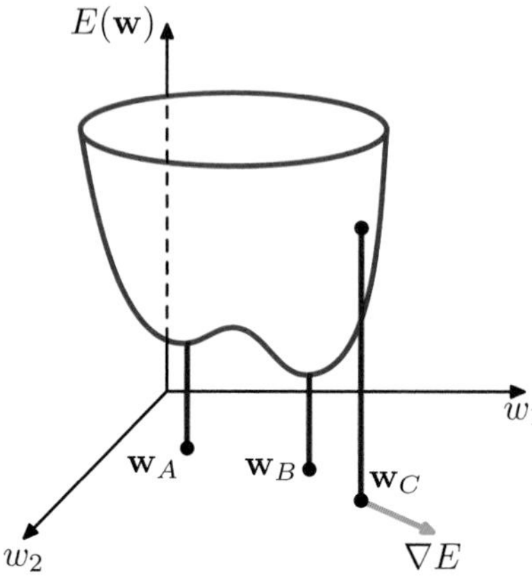

Following the discussion of Section 4.3.4, we see that the output unit activation function, which corresponds to the canonical link, is given by the softmax function

$$y_k(\mathbf{x}, \mathbf{w}) = \frac{\exp(a_k(\mathbf{x}, \mathbf{w}))}{\sum_j \exp(a_j(\mathbf{x}, \mathbf{w}))} \tag{5.25}$$

which satisfies $0 \leqslant y_k \leqslant 1$ and $\sum_k y_k = 1$. Note that the $y_k(\mathbf{x}, \mathbf{w})$ are unchanged if a constant is added to all of the $a_k(\mathbf{x}, \mathbf{w})$, causing the error function to be constant for some directions in weight space. This degeneracy is removed if an appropriate regularization term (Section 5.5) is added to the error function.

Exercise 5.7

Once again, the derivative of the error function with respect to the activation for a particular output unit takes the familiar form (5.18).

In summary, there is a natural choice of both output unit activation function and matching error function, according to the type of problem being solved. For regression we use linear outputs and a sum-of-squares error, for (multiple independent) binary classifications we use logistic sigmoid outputs and a cross-entropy error function, and for multiclass classification we use softmax outputs with the corresponding multiclass cross-entropy error function. For classification problems involving two classes, we can use a single logistic sigmoid output, or alternatively we can use a network with two outputs having a softmax output activation function.

5.2.1 Parameter optimization

We turn next to the task of finding a weight vector $\mathbf{w}$ which minimizes the chosen function $E(\mathbf{w})$. At this point, it is useful to have a geometrical picture of the error function, which we can view as a surface sitting over weight space as shown in Figure 5.5. First note that if we make a small step in weight space from $\mathbf{w}$ to $\mathbf{w}+\delta\mathbf{w}$ then the change in the error function is $\delta E \simeq \delta\mathbf{w}^{\mathrm{T}}\nabla E(\mathbf{w})$, where the vector $\nabla E(\mathbf{w})$ points in the direction of greatest rate of increase of the error function. Because the error $E(\mathbf{w})$ is a smooth continuous function of $\mathbf{w}$, its smallest value will occur at a point in weight space such that the gradient of the error function vanishes, so that

$$\nabla E(\mathbf{w}) = 0 \tag{5.26}$$

as otherwise we could make a small step in the direction of $-\nabla E(\mathbf{w})$ and thereby further reduce the error. Points at which the gradient vanishes are called stationary points, and may be further classified into minima, maxima, and saddle points.

Our goal is to find a vector $\mathbf{w}$ such that $E(\mathbf{w})$ takes its smallest value. However, the error function typically has a highly nonlinear dependence on the weights and bias parameters, and so there will be many points in weight space at which the gradient vanishes (or is numerically very small). Indeed, from the discussion in Section 5.1.1 we see that for any point $\mathbf{w}$ that is a local minimum, there will be other points in weight space that are equivalent minima. For instance, in a two-layer network of the kind shown in Figure 5.1, with M hidden units, each point in weight space is a member of a family of $M!2^M$ equivalent points.

Section 5.1.1

Furthermore, there will typically be multiple inequivalent stationary points and in particular multiple inequivalent minima. A minimum that corresponds to the smallest value of the error function for any weight vector is said to be a *global minimum*. Any other minima corresponding to higher values of the error function are said to be *local minima*. For a successful application of neural networks, it may not be necessary to find the global minimum (and in general it will not be known whether the global minimum has been found) but it may be necessary to compare several local minima in order to find a sufficiently good solution.

Because there is clearly no hope of finding an analytical solution to the equation $\nabla E(\mathbf{w}) = 0$ we resort to iterative numerical procedures. The optimization of continuous nonlinear functions is a widely studied problem and there exists an extensive literature on how to solve it efficiently. Most techniques involve choosing some initial value $\mathbf{w}^{(0)}$ for the weight vector and then moving through weight space in a succession of steps of the form

$$\mathbf{w}^{(\tau+1)} = \mathbf{w}^{(\tau)} + \Delta\mathbf{w}^{(\tau)} \tag{5.27}$$

where τ labels the iteration step. Different algorithms involve different choices for the weight vector update $\Delta\mathbf{w}^{(\tau)}$. Many algorithms make use of gradient information and therefore require that, after each update, the value of $\nabla E(\mathbf{w})$ is evaluated at the new weight vector $\mathbf{w}^{(\tau+1)}$. In order to understand the importance of gradient information, it is useful to consider a local approximation to the error function based on a Taylor expansion.

5.2.2 Local quadratic approximation

Insight into the optimization problem, and into the various techniques for solving it, can be obtained by considering a local quadratic approximation to the error function.

Consider the Taylor expansion of $E(\mathbf{w})$ around some point $\widehat{\mathbf{w}}$ in weight space

$$E(\mathbf{w}) \simeq E(\widehat{\mathbf{w}}) + (\mathbf{w} - \widehat{\mathbf{w}})^{\mathrm{T}}\mathbf{b} + \frac{1}{2}(\mathbf{w} - \widehat{\mathbf{w}})^{\mathrm{T}}\mathbf{H}(\mathbf{w} - \widehat{\mathbf{w}}) \tag{5.28}$$

where cubic and higher terms have been omitted. Here $\mathbf{b}$ is defined to be the gradient of E evaluated at $\widehat{\mathbf{w}}$

$$\mathbf{b} \equiv \nabla E|_{\mathbf{w}=\widehat{\mathbf{w}}} \tag{5.29}$$

and the Hessian matrix $\mathbf{H} = \nabla\nabla E$ has elements

$$(\mathbf{H})_{ij} \equiv \left.\frac{\partial E}{\partial w_i \partial w_j}\right|_{\mathbf{w}=\widehat{\mathbf{w}}}. \tag{5.30}$$

From (5.28), the corresponding local approximation to the gradient is given by

$$\nabla E \simeq \mathbf{b} + \mathbf{H}(\mathbf{w} - \widehat{\mathbf{w}}). \tag{5.31}$$

For points $\mathbf{w}$ that are sufficiently close to $\widehat{\mathbf{w}}$, these expressions will give reasonable approximations for the error and its gradient.

Consider the particular case of a local quadratic approximation around a point $\mathbf{w}^\star$ that is a minimum of the error function. In this case there is no linear term, because $\nabla E = 0$ at $\mathbf{w}^\star$, and (5.28) becomes

$$E(\mathbf{w}) \simeq E(\mathbf{w}^\star) + \frac{1}{2}(\mathbf{w} - \mathbf{w}^\star)^{\mathrm{T}}\mathbf{H}(\mathbf{w} - \mathbf{w}^\star) \tag{5.32}$$

where the Hessian $\mathbf{H}$ is evaluated at $\mathbf{w}^\star$. In order to interpret this geometrically, consider the eigenvalue equation for the Hessian matrix

$$\mathbf{H}\mathbf{u}_i = \lambda_i \mathbf{u}_i \tag{5.33}$$

where the eigenvectors $\mathbf{u}_i$ form a complete orthonormal set (Appendix C) so that

$$\mathbf{u}_i^{\mathrm{T}}\mathbf{u}_j = \delta_{ij}. \tag{5.34}$$

We now expand $(\mathbf{w} - \mathbf{w}^\star)$ as a linear combination of the eigenvectors in the form

$$\mathbf{w} - \mathbf{w}^\star = \sum_i \alpha_i \mathbf{u}_i. \tag{5.35}$$

This can be regarded as a transformation of the coordinate system in which the origin is translated to the point $\mathbf{w}^\star$, and the axes are rotated to align with the eigenvectors (through the orthogonal matrix whose columns are the $\mathbf{u}_i$), and is discussed in more detail in Appendix C. Substituting (5.35) into (5.32), and using (5.33) and (5.34), allows the error function to be written in the form

$$E(\mathbf{w}) = E(\mathbf{w}^\star) + \frac{1}{2}\sum_i \lambda_i \alpha_i^2. \tag{5.36}$$

A matrix $\mathbf{H}$ is said to be *positive definite* if, and only if,

$$\mathbf{v}^{\mathrm{T}}\mathbf{H}\mathbf{v} > 0 \qquad \text{for all } \mathbf{v}. \tag{5.37}$$

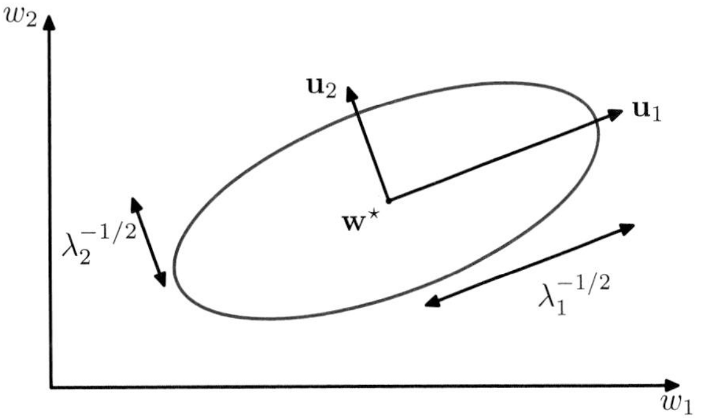

Figure 5.6 In the neighbourhood of a minimum $\mathbf{w}^\star$, the error function can be approximated by a quadratic. Contours of constant error are then ellipses whose axes are aligned with the eigenvectors $\mathbf{u}_i$ of the Hessian matrix, with lengths that are inversely proportional to the square roots of the corresponding eigenvectors λ_i.

Because the eigenvectors $\{\mathbf{u}_i\}$ form a complete set, an arbitrary vector $\mathbf{v}$ can be written in the form

$$\mathbf{v} = \sum_i c_i \mathbf{u}_i. \tag{5.38}$$

From (5.33) and (5.34), we then have

$$\mathbf{v}^{\mathrm{T}}\mathbf{H}\mathbf{v} = \sum_i c_i^2 \lambda_i \tag{5.39}$$

Exercise 5.10

Exercise 5.11

and so $\mathbf{H}$ will be positive definite if, and only if, all of its eigenvalues are positive. In the new coordinate system, whose basis vectors are given by the eigenvectors $\{\mathbf{u}_i\}$, the contours of constant E are ellipses centred on the origin, as illustrated in Figure 5.6. For a one-dimensional weight space, a stationary point $w^\star$ will be a minimum if

$$\left.\frac{\partial^2 E}{\partial w^2}\right|_{w^\star} > 0. \tag{5.40}$$

Exercise 5.12

The corresponding result in D-dimensions is that the Hessian matrix, evaluated at $\mathbf{w}^\star$, should be positive definite.

5.2.3 Use of gradient information

As we shall see in Section 5.3, it is possible to evaluate the gradient of an error function efficiently by means of the backpropagation procedure. The use of this gradient information can lead to significant improvements in the speed with which the minima of the error function can be located. We can see why this is so, as follows.

Exercise 5.13

In the quadratic approximation to the error function, given in (5.28), the error surface is specified by the quantities $\mathbf{b}$ and $\mathbf{H}$, which contain a total of $W(W+3)/2$ independent elements (because the matrix $\mathbf{H}$ is symmetric), where W is the dimensionality of $\mathbf{w}$ (i.e., the total number of adaptive parameters in the network). The location of the minimum of this quadratic approximation therefore depends on $O(W^2)$ parameters, and we should not expect to be able to locate the minimum until we have gathered $O(W^2)$ independent pieces of information. If we do not make use of gradient information, we would expect to have to perform $O(W^2)$ function evaluations, each of which would require $O(W)$ steps. Thus, the computational effort needed to find the minimum using such an approach would be $O(W^3)$.

Now compare this with an algorithm that makes use of the gradient information. Because each evaluation of ∇E brings W items of information, we might hope to find the minimum of the function in $O(W)$ gradient evaluations. As we shall see, by using error backpropagation, each such evaluation takes only $O(W)$ steps and so the minimum can now be found in $O(W^2)$ steps. For this reason, the use of gradient information forms the basis of practical algorithms for training neural networks.

5.2.4 Gradient descent optimization

The simplest approach to using gradient information is to choose the weight update in (5.27) to comprise a small step in the direction of the negative gradient, so that

$$\mathbf{w}^{(\tau+1)} = \mathbf{w}^{(\tau)} - \eta \nabla E(\mathbf{w}^{(\tau)}) \tag{5.41}$$

where the parameter $\eta > 0$ is known as the *learning rate*. After each such update, the gradient is re-evaluated for the new weight vector and the process repeated. Note that the error function is defined with respect to a training set, and so each step requires that the entire training set be processed in order to evaluate ∇E. Techniques that use the whole data set at once are called *batch* methods. At each step the weight vector is moved in the direction of the greatest rate of decrease of the error function, and so this approach is known as *gradient descent* or *steepest descent*. Although such an approach might intuitively seem reasonable, in fact it turns out to be a poor algorithm, for reasons discussed in Bishop and Nabney (2008).

For batch optimization, there are more efficient methods, such as *conjugate gradients* and *quasi-Newton* methods, which are much more robust and much faster than simple gradient descent (Gill *et al.*, 1981; Fletcher, 1987; Nocedal and Wright, 1999). Unlike gradient descent, these algorithms have the property that the error function always decreases at each iteration unless the weight vector has arrived at a local or global minimum.

In order to find a sufficiently good minimum, it may be necessary to run a gradient-based algorithm multiple times, each time using a different randomly chosen starting point, and comparing the resulting performance on an independent validation set.

There is, however, an on-line version of gradient descent that has proved useful in practice for training neural networks on large data sets (Le Cun *et al.*, 1989). Error functions based on maximum likelihood for a set of independent observations comprise a sum of terms, one for each data point

$$E(\mathbf{w}) = \sum_{n=1}^{N} E_n(\mathbf{w}). \tag{5.42}$$

On-line gradient descent, also known as *sequential gradient descent* or *stochastic gradient descent*, makes an update to the weight vector based on one data point at a time, so that

$$\mathbf{w}^{(\tau+1)} = \mathbf{w}^{(\tau)} - \eta \nabla E_n(\mathbf{w}^{(\tau)}). \tag{5.43}$$

This update is repeated by cycling through the data either in sequence or by selecting points at random with replacement. There are of course intermediate scenarios in which the updates are based on batches of data points.

One advantage of on-line methods compared to batch methods is that the former handle redundancy in the data much more efficiently. To see, this consider an extreme example in which we take a data set and double its size by duplicating every data point. Note that this simply multiplies the error function by a factor of 2 and so is equivalent to using the original error function. Batch methods will require double the computational effort to evaluate the batch error function gradient, whereas online methods will be unaffected. Another property of on-line gradient descent is the possibility of escaping from local minima, since a stationary point with respect to the error function for the whole data set will generally not be a stationary point for each data point individually.

Nonlinear optimization algorithms, and their practical application to neural network training, are discussed in detail in Bishop and Nabney (2008).

5.3. Error Backpropagation

Our goal in this section is to find an efficient technique for evaluating the gradient of an error function $E(\mathbf{w})$ for a feed-forward neural network. We shall see that this can be achieved using a local message passing scheme in which information is sent alternately forwards and backwards through the network and is known as *error backpropagation*, or sometimes simply as *backprop*.

It should be noted that the term backpropagation is used in the neural computing literature to mean a variety of different things. For instance, the multilayer perceptron architecture is sometimes called a backpropagation network. The term backpropagation is also used to describe the training of a multilayer perceptron using gradient descent applied to a sum-of-squares error function. In order to clarify the terminology, it is useful to consider the nature of the training process more carefully. Most training algorithms involve an iterative procedure for minimization of an error function, with adjustments to the weights being made in a sequence of steps. At each such step, we can distinguish between two distinct stages. In the first stage, the derivatives of the error function with respect to the weights must be evaluated. As we shall see, the important contribution of the backpropagation technique is in providing a computationally efficient method for evaluating such derivatives. Because it is at this stage that errors are propagated backwards through the network, we shall use the term backpropagation specifically to describe the evaluation of derivatives. In the second stage, the derivatives are then used to compute the adjustments to be made to the weights. The simplest such technique, and the one originally considered by Rumelhart *et al.* (1986), involves gradient descent. It is important to recognize that the two stages are distinct. Thus, the first stage, namely the propagation of errors backwards through the network in order to evaluate derivatives, can be applied to many other kinds of network and not just the multilayer perceptron. It can also be applied to error functions other that just the simple sum-of-squares, and to the evaluation of other derivatives such as the Jacobian and Hessian matrices, as we shall see later in this chapter. Similarly, the second stage of weight adjustment using the calculated derivatives can be tackled using a variety of optimization schemes, many of which are substantially more powerful than simple gradient descent.

5.3.1 Evaluation of error-function derivatives

We now derive the backpropagation algorithm for a general network having arbitrary feed-forward topology, arbitrary differentiable nonlinear activation functions, and a broad class of error function. The resulting formulae will then be illustrated using a simple layered network structure having a single layer of sigmoidal hidden units together with a sum-of-squares error.

Many error functions of practical interest, for instance those defined by maximum likelihood for a set of i.i.d. data, comprise a sum of terms, one for each data point in the training set, so that

$$E(\mathbf{w}) = \sum_{n=1}^{N} E_n(\mathbf{w}). \tag{5.44}$$

Here we shall consider the problem of evaluating $\nabla E_n(\mathbf{w})$ for one such term in the error function. This may be used directly for sequential optimization, or the results can be accumulated over the training set in the case of batch methods.

Consider first a simple linear model in which the outputs y_k are linear combinations of the input variables x_i so that

$$y_k = \sum_i w_{ki} x_i \tag{5.45}$$

together with an error function that, for a particular input pattern n, takes the form

$$E_n = \frac{1}{2} \sum_k (y_{nk} - t_{nk})^2 \tag{5.46}$$

where $y_{nk} = y_k(\mathbf{x}_n, \mathbf{w})$. The gradient of this error function with respect to a weight w_{ji} is given by

$$\frac{\partial E_n}{\partial w_{ji}} = (y_{nj} - t_{nj}) x_{ni} \tag{5.47}$$

which can be interpreted as a 'local' computation involving the product of an 'error signal' $y_{nj} - t_{nj}$ associated with the output end of the link w_{ji} and the variable x_{ni} associated with the input end of the link. In Section 4.3.2, we saw how a similar formula arises with the logistic sigmoid activation function together with the cross entropy error function, and similarly for the softmax activation function together with its matching cross-entropy error function. We shall now see how this simple result extends to the more complex setting of multilayer feed-forward networks.

In a general feed-forward network, each unit computes a weighted sum of its inputs of the form

$$a_j = \sum_i w_{ji} z_i \tag{5.48}$$

where z_i is the activation of a unit, or input, that sends a connection to unit j, and w_{ji} is the weight associated with that connection. In Section 5.1, we saw that biases can be included in this sum by introducing an extra unit, or input, with activation fixed at $+1$. We therefore do not need to deal with biases explicitly. The sum in (5.48) is transformed by a nonlinear activation function $h(\cdot)$ to give the activation z_j of unit j in the form

$$z_j = h(a_j). \tag{5.49}$$

Note that one or more of the variables z_i in the sum in (5.48) could be an input, and similarly, the unit j in (5.49) could be an output.

For each pattern in the training set, we shall suppose that we have supplied the corresponding input vector to the network and calculated the activations of all of the hidden and output units in the network by successive application of (5.48) and (5.49). This process is often called *forward propagation* because it can be regarded as a forward flow of information through the network.

Now consider the evaluation of the derivative of E_n with respect to a weight w_{ji}. The outputs of the various units will depend on the particular input pattern n. However, in order to keep the notation uncluttered, we shall omit the subscript n from the network variables. First we note that E_n depends on the weight w_{ji} only via the summed input a_j to unit j. We can therefore apply the chain rule for partial derivatives to give

$$\frac{\partial E_n}{\partial w_{ji}} = \frac{\partial E_n}{\partial a_j}\frac{\partial a_j}{\partial w_{ji}}. \tag{5.50}$$

We now introduce a useful notation

$$\delta_j \equiv \frac{\partial E_n}{\partial a_j} \tag{5.51}$$

where the δ's are often referred to as *errors* for reasons we shall see shortly. Using (5.48), we can write

$$\frac{\partial a_j}{\partial w_{ji}} = z_i. \tag{5.52}$$

Substituting (5.51) and (5.52) into (5.50), we then obtain

$$\frac{\partial E_n}{\partial w_{ji}} = \delta_j z_i. \tag{5.53}$$

Equation (5.53) tells us that the required derivative is obtained simply by multiplying the value of δ for the unit at the output end of the weight by the value of z for the unit at the input end of the weight (where $z = 1$ in the case of a bias). Note that this takes the same form as for the simple linear model considered at the start of this section. Thus, in order to evaluate the derivatives, we need only to calculate the value of δ_j for each hidden and output unit in the network, and then apply (5.53).

As we have seen already, for the output units, we have

$$\delta_k = y_k - t_k \tag{5.54}$$

Figure 5.7 Illustration of the calculation of δ_j for hidden unit j by backpropagation of the δ's from those units k to which unit j sends connections. The blue arrow denotes the direction of information flow during forward propagation, and the red arrows indicate the backward propagation of error information.

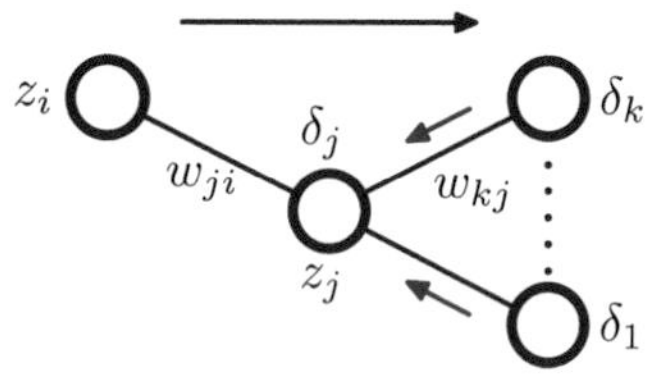

provided we are using the canonical link as the output-unit activation function. To evaluate the δ's for hidden units, we again make use of the chain rule for partial derivatives,

$$\delta_j \equiv \frac{\partial E_n}{\partial a_j} = \sum_k \frac{\partial E_n}{\partial a_k}\frac{\partial a_k}{\partial a_j} \tag{5.55}$$

where the sum runs over all units k to which unit j sends connections. The arrangement of units and weights is illustrated in Figure 5.7. Note that the units labelled k could include other hidden units and/or output units. In writing down (5.55), we are making use of the fact that variations in a_j give rise to variations in the error function only through variations in the variables a_k. If we now substitute the definition of δ given by (5.51) into (5.55), and make use of (5.48) and (5.49), we obtain the following *backpropagation* formula

$$\delta_j = h'(a_j)\sum_k w_{kj}\delta_k \tag{5.56}$$

which tells us that the value of δ for a particular hidden unit can be obtained by propagating the δ's backwards from units higher up in the network, as illustrated in Figure 5.7. Note that the summation in (5.56) is taken over the first index on w_{kj} (corresponding to backward propagation of information through the network), whereas in the forward propagation equation (5.10) it is taken over the second index. Because we already know the values of the δ's for the output units, it follows that by recursively applying (5.56) we can evaluate the δ's for all of the hidden units in a feed-forward network, regardless of its topology.

The backpropagation procedure can therefore be summarized as follows.

Error Backpropagation

1. Apply an input vector $\mathbf{x}_n$ to the network and forward propagate through the network using (5.48) and (5.49) to find the activations of all the hidden and output units.
2. Evaluate the δ_k for all the output units using (5.54).
3. Backpropagate the δ's using (5.56) to obtain δ_j for each hidden unit in the network.
4. Use (5.53) to evaluate the required derivatives.

For batch methods, the derivative of the total error E can then be obtained by repeating the above steps for each pattern in the training set and then summing over all patterns:

$$\frac{\partial E}{\partial w_{ji}} = \sum_n \frac{\partial E_n}{\partial w_{ji}}. \tag{5.57}$$

In the above derivation we have implicitly assumed that each hidden or output unit in the network has the same activation function $h(\cdot)$. The derivation is easily generalized, however, to allow different units to have individual activation functions, simply by keeping track of which form of $h(\cdot)$ goes with which unit.

5.3.2 A simple example

The above derivation of the backpropagation procedure allowed for general forms for the error function, the activation functions, and the network topology. In order to illustrate the application of this algorithm, we shall consider a particular example. This is chosen both for its simplicity and for its practical importance, because many applications of neural networks reported in the literature make use of this type of network. Specifically, we shall consider a two-layer network of the form illustrated in Figure 5.1, together with a sum-of-squares error, in which the output units have linear activation functions, so that $y_k = a_k$, while the hidden units have sigmoidal activation functions given by

$$h(a) \equiv \tanh(a) \tag{5.58}$$

where

$$\tanh(a) = \frac{e^a - e^{-a}}{e^a + e^{-a}}. \tag{5.59}$$

A useful feature of this function is that its derivative can be expressed in a particularly simple form:

$$h'(a) = 1 - h(a)^2. \tag{5.60}$$

We also consider a standard sum-of-squares error function, so that for pattern n the error is given by

$$E_n = \frac{1}{2}\sum_{k=1}^{K}(y_k - t_k)^2 \tag{5.61}$$

where y_k is the activation of output unit k, and t_k is the corresponding target, for a particular input pattern $\mathbf{x}_n$.

For each pattern in the training set in turn, we first perform a forward propagation using

$$a_j = \sum_{i=0}^{D} w_{ji}^{(1)} x_i \tag{5.62}$$

$$z_j = \tanh(a_j) \tag{5.63}$$

$$y_k = \sum_{j=0}^{M} w_{kj}^{(2)} z_j. \tag{5.64}$$

Next we compute the δ's for each output unit using

$$\delta_k = y_k - t_k. \tag{5.65}$$

Then we backpropagate these to obtain δs for the hidden units using

$$\delta_j = (1 - z_j^2) \sum_{k=1}^{K} w_{kj} \delta_k. \tag{5.66}$$

Finally, the derivatives with respect to the first-layer and second-layer weights are given by

$$\frac{\partial E_n}{\partial w_{ji}^{(1)}} = \delta_j x_i, \qquad \frac{\partial E_n}{\partial w_{kj}^{(2)}} = \delta_k z_j. \tag{5.67}$$

5.3.3 Efficiency of backpropagation

One of the most important aspects of backpropagation is its computational efficiency. To understand this, let us examine how the number of computer operations required to evaluate the derivatives of the error function scales with the total number W of weights and biases in the network. A single evaluation of the error function (for a given input pattern) would require $O(W)$ operations, for sufficiently large W. This follows from the fact that, except for a network with very sparse connections, the number of weights is typically much greater than the number of units, and so the bulk of the computational effort in forward propagation is concerned with evaluating the sums in (5.48), with the evaluation of the activation functions representing a small overhead. Each term in the sum in (5.48) requires one multiplication and one addition, leading to an overall computational cost that is $O(W)$.

An alternative approach to backpropagation for computing the derivatives of the error function is to use finite differences. This can be done by perturbing each weight in turn, and approximating the derivatives by the expression

$$\frac{\partial E_n}{\partial w_{ji}} = \frac{E_n(w_{ji} + \epsilon) - E_n(w_{ji})}{\epsilon} + O(\epsilon) \tag{5.68}$$

where $\epsilon \ll 1$. In a software simulation, the accuracy of the approximation to the derivatives can be improved by making ϵ smaller, until numerical roundoff problems arise. The accuracy of the finite differences method can be improved significantly by using symmetrical *central differences* of the form

$$\frac{\partial E_n}{\partial w_{ji}} = \frac{E_n(w_{ji} + \epsilon) - E_n(w_{ji} - \epsilon)}{2\epsilon} + O(\epsilon^2). \tag{5.69}$$

Exercise 5.14

In this case, the $O(\epsilon)$ corrections cancel, as can be verified by Taylor expansion on the right-hand side of (5.69), and so the residual corrections are $O(\epsilon^2)$. The number of computational steps is, however, roughly doubled compared with (5.68).

The main problem with numerical differentiation is that the highly desirable $O(W)$ scaling has been lost. Each forward propagation requires $O(W)$ steps, and

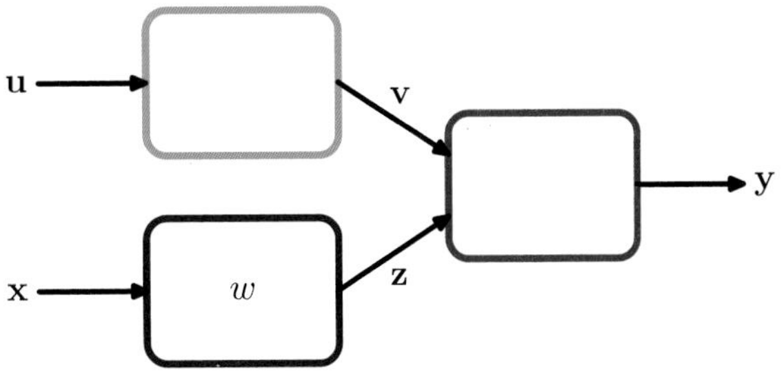

Figure 5.8 Illustration of a modular pattern recognition system in which the Jacobian matrix can be used to backpropagate error signals from the outputs through to earlier modules in the system.

there are W weights in the network each of which must be perturbed individually, so that the overall scaling is $O(W^2)$.

However, numerical differentiation plays an important role in practice, because a comparison of the derivatives calculated by backpropagation with those obtained using central differences provides a powerful check on the correctness of any software implementation of the backpropagation algorithm. When training networks in practice, derivatives should be evaluated using backpropagation, because this gives the greatest accuracy and numerical efficiency. However, the results should be compared with numerical differentiation using (5.69) for some test cases in order to check the correctness of the implementation.

5.3.4 The Jacobian matrix

We have seen how the derivatives of an error function with respect to the weights can be obtained by the propagation of errors backwards through the network. The technique of backpropagation can also be applied to the calculation of other derivatives. Here we consider the evaluation of the *Jacobian* matrix, whose elements are given by the derivatives of the network outputs with respect to the inputs

$$J_{ki} \equiv \frac{\partial y_k}{\partial x_i} \tag{5.70}$$

where each such derivative is evaluated with all other inputs held fixed. Jacobian matrices play a useful role in systems built from a number of distinct modules, as illustrated in Figure 5.8. Each module can comprise a fixed or adaptive function, which can be linear or nonlinear, so long as it is differentiable. Suppose we wish to minimize an error function E with respect to the parameter w in Figure 5.8. The derivative of the error function is given by

$$\frac{\partial E}{\partial w} = \sum_{k,j} \frac{\partial E}{\partial y_k} \frac{\partial y_k}{\partial z_j} \frac{\partial z_j}{\partial w} \tag{5.71}$$

in which the Jacobian matrix for the red module in Figure 5.8 appears in the middle term.

Because the Jacobian matrix provides a measure of the local sensitivity of the outputs to changes in each of the input variables, it also allows any known errors Δx_i

associated with the inputs to be propagated through the trained network in order to estimate their contribution Δy_k to the errors at the outputs, through the relation

$$\Delta y_k \simeq \sum_i \frac{\partial y_k}{\partial x_i} \Delta x_i \tag{5.72}$$

which is valid provided the $|\Delta x_i|$ are small. In general, the network mapping represented by a trained neural network will be nonlinear, and so the elements of the Jacobian matrix will not be constants but will depend on the particular input vector used. Thus (5.72) is valid only for small perturbations of the inputs, and the Jacobian itself must be re-evaluated for each new input vector.

The Jacobian matrix can be evaluated using a backpropagation procedure that is similar to the one derived earlier for evaluating the derivatives of an error function with respect to the weights. We start by writing the element J_{ki} in the form

$$\begin{aligned} J_{ki} = \frac{\partial y_k}{\partial x_i} &= \sum_j \frac{\partial y_k}{\partial a_j} \frac{\partial a_j}{\partial x_i} \\ &= \sum_j w_{ji} \frac{\partial y_k}{\partial a_j} \end{aligned} \tag{5.73}$$

where we have made use of (5.48). The sum in (5.73) runs over all units j to which the input unit i sends connections (for example, over all units in the first hidden layer in the layered topology considered earlier). We now write down a recursive backpropagation formula to determine the derivatives $\partial y_k / \partial a_j$

$$\begin{aligned} \frac{\partial y_k}{\partial a_j} &= \sum_l \frac{\partial y_k}{\partial a_l} \frac{\partial a_l}{\partial a_j} \\ &= h'(a_j) \sum_l w_{lj} \frac{\partial y_k}{\partial a_l} \end{aligned} \tag{5.74}$$

where the sum runs over all units l to which unit j sends connections (corresponding to the first index of w_{lj}). Again, we have made use of (5.48) and (5.49). This backpropagation starts at the output units for which the required derivatives can be found directly from the functional form of the output-unit activation function. For instance, if we have individual sigmoidal activation functions at each output unit, then

$$\frac{\partial y_k}{\partial a_l} = \delta_{kl} \sigma'(a_l) \tag{5.75}$$

whereas for softmax outputs we have

$$\frac{\partial y_k}{\partial a_l} = \delta_{kl} y_k - y_k y_l. \tag{5.76}$$

We can summarize the procedure for evaluating the Jacobian matrix as follows. Apply the input vector corresponding to the point in input space at which the Jacobian matrix is to be found, and forward propagate in the usual way to obtain the activations of all of the hidden and output units in the network. Next, for each row k of the Jacobian matrix, corresponding to the output unit k, backpropagate using the recursive relation (5.74), starting with (5.75) or (5.76), for all of the hidden units in the network. Finally, use (5.73) to do the backpropagation to the inputs. The Jacobian can also be evaluated using an alternative *forward* propagation formalism, which can be derived in an analogous way to the backpropagation approach given here.

Exercise 5.15

Again, the implementation of such algorithms can be checked by using numerical differentiation in the form

$$\frac{\partial y_k}{\partial x_i} = \frac{y_k(x_i + \epsilon) - y_k(x_i - \epsilon)}{2\epsilon} + O(\epsilon^2) \tag{5.77}$$

which involves $2D$ forward propagations for a network having D inputs.

5.4. The Hessian Matrix

We have shown how the technique of backpropagation can be used to obtain the first derivatives of an error function with respect to the weights in the network. Backpropagation can also be used to evaluate the second derivatives of the error, given by

$$\frac{\partial^2 E}{\partial w_{ji} \partial w_{lk}}. \tag{5.78}$$

Note that it is sometimes convenient to consider all of the weight and bias parameters as elements w_i of a single vector, denoted $\mathbf{w}$, in which case the second derivatives form the elements H_{ij} of the *Hessian* matrix $\mathbf{H}$, where $i, j \in \{1, \ldots, W\}$ and W is the total number of weights and biases. The Hessian plays an important role in many aspects of neural computing, including the following:

1. Several nonlinear optimization algorithms used for training neural networks are based on considerations of the second-order properties of the error surface, which are controlled by the Hessian matrix (Bishop and Nabney, 2008).

2. The Hessian forms the basis of a fast procedure for re-training a feed-forward network following a small change in the training data (Bishop, 1991).

3. The inverse of the Hessian has been used to identify the least significant weights in a network as part of network 'pruning' algorithms (Le Cun *et al.*, 1990).

4. The Hessian plays a central role in the Laplace approximation for a Bayesian neural network (see Section 5.7). Its inverse is used to determine the predictive distribution for a trained network, its eigenvalues determine the values of hyperparameters, and its determinant is used to evaluate the model evidence.

Various approximation schemes have been used to evaluate the Hessian matrix for a neural network. However, the Hessian can also be calculated exactly using an extension of the backpropagation technique.

An important consideration for many applications of the Hessian is the efficiency with which it can be evaluated. If there are W parameters (weights and biases) in the network, then the Hessian matrix has dimensions $W \times W$ and so the computational effort needed to evaluate the Hessian will scale like $O(W^2)$ for each pattern in the data set. As we shall see, there are efficient methods for evaluating the Hessian whose scaling is indeed $O(W^2)$.

5.4.1 Diagonal approximation

Some of the applications for the Hessian matrix discussed above require the inverse of the Hessian, rather than the Hessian itself. For this reason, there has been some interest in using a diagonal approximation to the Hessian, in other words one that simply replaces the off-diagonal elements with zeros, because its inverse is trivial to evaluate. Again, we shall consider an error function that consists of a sum of terms, one for each pattern in the data set, so that $E = \sum_n E_n$. The Hessian can then be obtained by considering one pattern at a time, and then summing the results over all patterns. From (5.48), the diagonal elements of the Hessian, for pattern n, can be written

$$\frac{\partial^2 E_n}{\partial w_{ji}^2} = \frac{\partial^2 E_n}{\partial a_j^2} z_i^2. \tag{5.79}$$

Using (5.48) and (5.49), the second derivatives on the right-hand side of (5.79) can be found recursively using the chain rule of differential calculus to give a backpropagation equation of the form

$$\frac{\partial^2 E_n}{\partial a_j^2} = h'(a_j)^2 \sum_k \sum_{k'} w_{kj} w_{k'j} \frac{\partial^2 E_n}{\partial a_k \partial a_{k'}} + h''(a_j) \sum_k w_{kj} \frac{\partial E_n}{\partial a_k}. \tag{5.80}$$

If we now neglect off-diagonal elements in the second-derivative terms, we obtain (Becker and Le Cun, 1989; Le Cun *et al.*, 1990)

$$\frac{\partial^2 E_n}{\partial a_j^2} = h'(a_j)^2 \sum_k w_{kj}^2 \frac{\partial^2 E_n}{\partial a_k^2} + h''(a_j) \sum_k w_{kj} \frac{\partial E_n}{\partial a_k}. \tag{5.81}$$

Note that the number of computational steps required to evaluate this approximation is $O(W)$, where W is the total number of weight and bias parameters in the network, compared with $O(W^2)$ for the full Hessian.

Ricotti *et al.* (1988) also used the diagonal approximation to the Hessian, but they retained all terms in the evaluation of $\partial^2 E_n / \partial a_j^2$ and so obtained exact expressions for the diagonal terms. Note that this no longer has $O(W)$ scaling. The major problem with diagonal approximations, however, is that in practice the Hessian is typically found to be strongly nondiagonal, and so these approximations, which are driven mainly by computational convenience, must be treated with care.

5.4.2 Outer product approximation

When neural networks are applied to regression problems, it is common to use a sum-of-squares error function of the form

$$E = \frac{1}{2}\sum_{n=1}^{N}(y_n - t_n)^2 \tag{5.82}$$

where we have considered the case of a single output in order to keep the notation simple (the extension to several outputs is straightforward). We can then write the Hessian matrix in the form

Exercise 5.16

$$\mathbf{H} = \nabla\nabla E = \sum_{n=1}^{N}\nabla y_n(\nabla y_n)^{\mathrm{T}} + \sum_{n=1}^{N}(y_n - t_n)\nabla\nabla y_n. \tag{5.83}$$

If the network has been trained on the data set, and its outputs y_n happen to be very close to the target values t_n, then the second term in (5.83) will be small and can be neglected. More generally, however, it may be appropriate to neglect this term by the following argument. Recall from Section 1.5.5 that the optimal function that minimizes a sum-of-squares loss is the conditional average of the target data. The quantity $(y_n - t_n)$ is then a random variable with zero mean. If we assume that its value is uncorrelated with the value of the second derivative term on the right-hand side of (5.83), then the whole term will average to zero in the summation over n.

Exercise 5.17

By neglecting the second term in (5.83), we arrive at the *Levenberg–Marquardt* approximation or *outer product* approximation (because the Hessian matrix is built up from a sum of outer products of vectors), given by

$$\mathbf{H} \simeq \sum_{n=1}^{N}\mathbf{b}_n\mathbf{b}_n^{\mathrm{T}} \tag{5.84}$$

where $\mathbf{b}_n \equiv \nabla a_n = \nabla y_n$ because the activation function for the output units is simply the identity. Evaluation of the outer product approximation for the Hessian is straightforward as it only involves first derivatives of the error function, which can be evaluated efficiently in $O(W)$ steps using standard backpropagation. The elements of the matrix can then be found in $O(W^2)$ steps by simple multiplication. It is important to emphasize that this approximation is only likely to be valid for a network that has been trained appropriately, and that for a general network mapping the second derivative terms on the right-hand side of (5.83) will typically not be negligible.

In the case of the cross-entropy error function for a network with logistic sigmoid output-unit activation functions, the corresponding approximation is given by

Exercise 5.19

$$\mathbf{H} \simeq \sum_{n=1}^{N}y_n(1 - y_n)\mathbf{b}_n\mathbf{b}_n^{\mathrm{T}}. \tag{5.85}$$

An analogous result can be obtained for multiclass networks having softmax output-unit activation functions.

Exercise 5.20

5.4.3 Inverse Hessian

We can use the outer-product approximation to develop a computationally efficient procedure for approximating the inverse of the Hessian (Hassibi and Stork, 1993). First we write the outer-product approximation in matrix notation as

$$\mathbf{H}_N = \sum_{n=1}^{N} \mathbf{b}_n \mathbf{b}_n^{\mathrm{T}} \tag{5.86}$$

where $\mathbf{b}_n \equiv \nabla_{\mathbf{w}} a_n$ is the contribution to the gradient of the output unit activation arising from data point n. We now derive a sequential procedure for building up the Hessian by including data points one at a time. Suppose we have already obtained the inverse Hessian using the first L data points. By separating off the contribution from data point $L+1$, we obtain

$$\mathbf{H}_{L+1} = \mathbf{H}_L + \mathbf{b}_{L+1}\mathbf{b}_{L+1}^{\mathrm{T}}. \tag{5.87}$$

In order to evaluate the inverse of the Hessian, we now consider the matrix identity

$$\left(\mathbf{M} + \mathbf{v}\mathbf{v}^{\mathrm{T}}\right)^{-1} = \mathbf{M}^{-1} - \frac{(\mathbf{M}^{-1}\mathbf{v})\left(\mathbf{v}^{\mathrm{T}}\mathbf{M}^{-1}\right)}{1 + \mathbf{v}^{\mathrm{T}}\mathbf{M}^{-1}\mathbf{v}} \tag{5.88}$$

which is simply a special case of the Woodbury identity (C.7). If we now identify $\mathbf{H}_L$ with $\mathbf{M}$ and $\mathbf{b}_{L+1}$ with $\mathbf{v}$, we obtain

$$\mathbf{H}_{L+1}^{-1} = \mathbf{H}_L^{-1} - \frac{\mathbf{H}_L^{-1}\mathbf{b}_{L+1}\mathbf{b}_{L+1}^{\mathrm{T}}\mathbf{H}_L^{-1}}{1 + \mathbf{b}_{L+1}^{\mathrm{T}}\mathbf{H}_L^{-1}\mathbf{b}_{L+1}}. \tag{5.89}$$

In this way, data points are sequentially absorbed until $L+1 = N$ and the whole data set has been processed. This result therefore represents a procedure for evaluating the inverse of the Hessian using a single pass through the data set. The initial matrix $\mathbf{H}_0$ is chosen to be $\alpha\mathbf{I}$, where α is a small quantity, so that the algorithm actually finds the inverse of $\mathbf{H} + \alpha\mathbf{I}$. The results are not particularly sensitive to the precise value of α. Extension of this algorithm to networks having more than one output is straightforward.

Exercise 5.21

We note here that the Hessian matrix can sometimes be calculated indirectly as part of the network training algorithm. In particular, quasi-Newton nonlinear optimization algorithms gradually build up an approximation to the inverse of the Hessian during training. Such algorithms are discussed in detail in Bishop and Nabney (2008).

5.4.4 Finite differences

As in the case of the first derivatives of the error function, we can find the second derivatives by using finite differences, with accuracy limited by numerical precision. If we perturb each possible pair of weights in turn, we obtain

$$\begin{aligned}\frac{\partial^2 E}{\partial w_{ji}\partial w_{lk}} &= \frac{1}{4\epsilon^2}\left\{E(w_{ji}+\epsilon, w_{lk}+\epsilon) - E(w_{ji}+\epsilon, w_{lk}-\epsilon)\right. \\ &\quad \left. - E(w_{ji}-\epsilon, w_{lk}+\epsilon) + E(w_{ji}-\epsilon, w_{lk}-\epsilon)\right\} + O(\epsilon^2). \end{aligned} \tag{5.90}$$

Again, by using a symmetrical central differences formulation, we ensure that the residual errors are $O(\epsilon^2)$ rather than $O(\epsilon)$. Because there are W^2 elements in the Hessian matrix, and because the evaluation of each element requires four forward propagations each needing $O(W)$ operations (per pattern), we see that this approach will require $O(W^3)$ operations to evaluate the complete Hessian. It therefore has poor scaling properties, although in practice it is very useful as a check on the software implementation of backpropagation methods.

A more efficient version of numerical differentiation can be found by applying central differences to the first derivatives of the error function, which are themselves calculated using backpropagation. This gives

$$\frac{\partial^2 E}{\partial w_{ji} \partial w_{lk}} = \frac{1}{2\epsilon} \left\{ \frac{\partial E}{\partial w_{ji}}(w_{lk} + \epsilon) - \frac{\partial E}{\partial w_{ji}}(w_{lk} - \epsilon) \right\} + O(\epsilon^2). \tag{5.91}$$

Because there are now only W weights to be perturbed, and because the gradients can be evaluated in $O(W)$ steps, we see that this method gives the Hessian in $O(W^2)$ operations.

5.4.5 Exact evaluation of the Hessian

So far, we have considered various approximation schemes for evaluating the Hessian matrix or its inverse. The Hessian can also be evaluated exactly, for a network of arbitrary feed-forward topology, using extension of the technique of backpropagation used to evaluate first derivatives, which shares many of its desirable features including computational efficiency (Bishop, 1991; Bishop, 1992). It can be applied to any differentiable error function that can be expressed as a function of the network outputs and to networks having arbitrary differentiable activation functions. The number of computational steps needed to evaluate the Hessian scales like $O(W^2)$. Similar algorithms have also been considered by Buntine and Weigend (1993).

Exercise 5.22

Here we consider the specific case of a network having two layers of weights, for which the required equations are easily derived. We shall use indices i and i' to denote inputs, indices j and j' to denoted hidden units, and indices k and k' to denote outputs. We first define

$$\delta_k = \frac{\partial E_n}{\partial a_k}, \qquad M_{kk'} \equiv \frac{\partial^2 E_n}{\partial a_k \partial a_{k'}} \tag{5.92}$$

where E_n is the contribution to the error from data point n. The Hessian matrix for this network can then be considered in three separate blocks as follows.

1. Both weights in the second layer:

$$\frac{\partial^2 E_n}{\partial w_{kj}^{(2)} \partial w_{k'j'}^{(2)}} = z_j z_{j'} M_{kk'}. \tag{5.93}$$

2. Both weights in the first layer:

$$\frac{\partial^2 E_n}{\partial w_{ji}^{(1)} \partial w_{j'i'}^{(1)}} = x_i x_{i'} h''(a_{j'}) I_{jj'} \sum_k w_{kj'}^{(2)} \delta_k + x_i x_{i'} h'(a_{j'}) h'(a_j) \sum_k \sum_{k'} w_{k'j'}^{(2)} w_{kj}^{(2)} M_{kk'}. \tag{5.94}$$

3. One weight in each layer:

$$\frac{\partial^2 E_n}{\partial w_{ji}^{(1)} \partial w_{kj'}^{(2)}} = x_i h'(a_j) \left\{ \delta_k I_{jj'} + z_{j'} \sum_{k'} w_{k'j}^{(2)} M_{kk'} \right\}. \tag{5.95}$$

Here $I_{jj'}$ is the j, j' element of the identity matrix. If one or both of the weights is a bias term, then the corresponding expressions are obtained simply by setting the appropriate activation(s) to 1. Inclusion of skip-layer connections is straightforward.

Exercise 5.23

5.4.6 Fast multiplication by the Hessian

For many applications of the Hessian, the quantity of interest is not the Hessian matrix $\mathbf{H}$ itself but the product of $\mathbf{H}$ with some vector $\mathbf{v}$. We have seen that the evaluation of the Hessian takes $O(W^2)$ operations, and it also requires storage that is $O(W^2)$. The vector $\mathbf{v}^{\mathrm{T}}\mathbf{H}$ that we wish to calculate, however, has only W elements, so instead of computing the Hessian as an intermediate step, we can instead try to find an efficient approach to evaluating $\mathbf{v}^{\mathrm{T}}\mathbf{H}$ directly in a way that requires only $O(W)$ operations.

To do this, we first note that

$$\mathbf{v}^{\mathrm{T}}\mathbf{H} = \mathbf{v}^{\mathrm{T}}\nabla(\nabla E) \tag{5.96}$$

where ∇ denotes the gradient operator in weight space. We can then write down the standard forward-propagation and backpropagation equations for the evaluation of ∇E and apply (5.96) to these equations to give a set of forward-propagation and backpropagation equations for the evaluation of $\mathbf{v}^{\mathrm{T}}\mathbf{H}$ (Møller, 1993; Pearlmutter, 1994). This corresponds to acting on the original forward-propagation and backpropagation equations with a differential operator $\mathbf{v}^{\mathrm{T}}\nabla$. Pearlmutter (1994) used the notation $\mathcal{R}\{\cdot\}$ to denote the operator $\mathbf{v}^{\mathrm{T}}\nabla$, and we shall follow this convention. The analysis is straightforward and makes use of the usual rules of differential calculus, together with the result

$$\mathcal{R}\{\mathbf{w}\} = \mathbf{v}. \tag{5.97}$$

The technique is best illustrated with a simple example, and again we choose a two-layer network of the form shown in Figure 5.1, with linear output units and a sum-of-squares error function. As before, we consider the contribution to the error function from one pattern in the data set. The required vector is then obtained as usual by summing over the contributions from each of the patterns separately. For the two-layer network, the forward-propagation equations are given by

$$a_j = \sum_i w_{ji} x_i \tag{5.98}$$

$$z_j = h(a_j) \tag{5.99}$$

$$y_k = \sum_j w_{kj} z_j. \tag{5.100}$$

We now act on these equations using the $\mathcal{R}\{\cdot\}$ operator to obtain a set of forward propagation equations in the form

$$\mathcal{R}\{a_j\} = \sum_i v_{ji} x_i \tag{5.101}$$

$$\mathcal{R}\{z_j\} = h'(a_j)\mathcal{R}\{a_j\} \tag{5.102}$$

$$\mathcal{R}\{y_k\} = \sum_j w_{kj}\mathcal{R}\{z_j\} + \sum_j v_{kj} z_j \tag{5.103}$$

where v_{ji} is the element of the vector $\mathbf{v}$ that corresponds to the weight w_{ji}. Quantities of the form $\mathcal{R}\{z_j\}$, $\mathcal{R}\{a_j\}$ and $\mathcal{R}\{y_k\}$ are to be regarded as new variables whose values are found using the above equations.

Because we are considering a sum-of-squares error function, we have the following standard backpropagation expressions:

$$\delta_k = y_k - t_k \tag{5.104}$$

$$\delta_j = h'(a_j)\sum_k w_{kj}\delta_k. \tag{5.105}$$

Again, we act on these equations with the $\mathcal{R}\{\cdot\}$ operator to obtain a set of backpropagation equations in the form

$$\mathcal{R}\{\delta_k\} = \mathcal{R}\{y_k\} \tag{5.106}$$

$$\mathcal{R}\{\delta_j\} = h''(a_j)\mathcal{R}\{a_j\}\sum_k w_{kj}\delta_k + h'(a_j)\sum_k v_{kj}\delta_k + h'(a_j)\sum_k w_{kj}\mathcal{R}\{\delta_k\}. \tag{5.107}$$

Finally, we have the usual equations for the first derivatives of the error

$$\frac{\partial E}{\partial w_{kj}} = \delta_k z_j \tag{5.108}$$

$$\frac{\partial E}{\partial w_{ji}} = \delta_j x_i \tag{5.109}$$

and acting on these with the $\mathcal{R}\{\cdot\}$ operator, we obtain expressions for the elements of the vector $\mathbf{v}^{\mathrm{T}}\mathbf{H}$

$$\mathcal{R}\left\{\frac{\partial E}{\partial w_{kj}}\right\} = \mathcal{R}\{\delta_k\}z_j + \delta_k\mathcal{R}\{z_j\} \tag{5.110}$$

$$\mathcal{R}\left\{\frac{\partial E}{\partial w_{ji}}\right\} = x_i\mathcal{R}\{\delta_j\}. \tag{5.111}$$

The implementation of this algorithm involves the introduction of additional variables $\mathcal{R}\{a_j\}$, $\mathcal{R}\{z_j\}$ and $\mathcal{R}\{\delta_j\}$ for the hidden units and $\mathcal{R}\{\delta_k\}$ and $\mathcal{R}\{y_k\}$ for the output units. For each input pattern, the values of these quantities can be found using the above results, and the elements of $\mathbf{v}^{\mathrm{T}}\mathbf{H}$ are then given by (5.110) and (5.111). An elegant aspect of this technique is that the equations for evaluating $\mathbf{v}^{\mathrm{T}}\mathbf{H}$ mirror closely those for standard forward and backward propagation, and so the extension of existing software to compute this product is typically straightforward.

If desired, the technique can be used to evaluate the full Hessian matrix by choosing the vector $\mathbf{v}$ to be given successively by a series of unit vectors of the form $(0, 0, \ldots, 1, \ldots, 0)$ each of which picks out one column of the Hessian. This leads to a formalism that is analytically equivalent to the backpropagation procedure of Bishop (1992), as described in Section 5.4.5, though with some loss of efficiency due to redundant calculations.

5.5. Regularization in Neural Networks

The number of input and output units in a neural network is generally determined by the dimensionality of the data set, whereas the number M of hidden units is a free parameter that can be adjusted to give the best predictive performance. Note that M controls the number of parameters (weights and biases) in the network, and so we might expect that in a maximum likelihood setting there will be an optimum value of M that gives the best generalization performance, corresponding to the optimum balance between under-fitting and over-fitting. Figure 5.9 shows an example of the effect of different values of M for the sinusoidal regression problem.

The generalization error, however, is not a simple function of M due to the presence of local minima in the error function, as illustrated in Figure 5.10. Here we see the effect of choosing multiple random initializations for the weight vector for a range of values of M. The overall best validation set performance in this case occurred for a particular solution having $M = 8$. In practice, one approach to choosing M is in fact to plot a graph of the kind shown in Figure 5.10 and then to choose the specific solution having the smallest validation set error.

There are, however, other ways to control the complexity of a neural network model in order to avoid over-fitting. From our discussion of polynomial curve fitting in Chapter 1, we see that an alternative approach is to choose a relatively large value for M and then to control complexity by the addition of a regularization term to the error function. The simplest regularizer is the quadratic, giving a regularized error

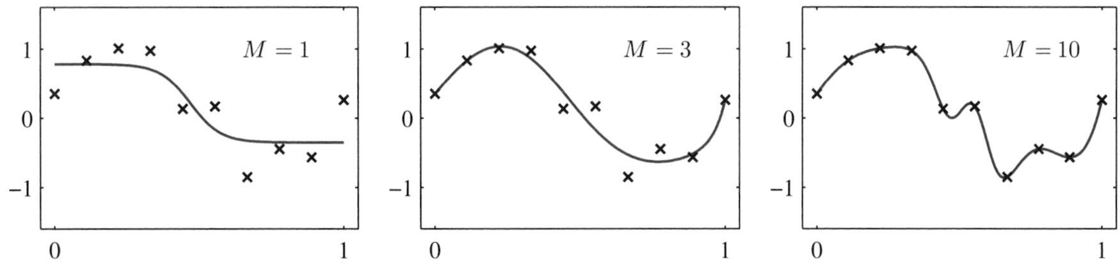

Figure 5.9 Examples of two-layer networks trained on 10 data points drawn from the sinusoidal data set. The graphs show the result of fitting networks having $M = 1$, 3 and 10 hidden units, respectively, by minimizing a sum-of-squares error function using a scaled conjugate-gradient algorithm.

of the form

$$\widetilde{E}(\mathbf{w}) = E(\mathbf{w}) + \frac{\lambda}{2}\mathbf{w}^{\mathrm{T}}\mathbf{w}. \tag{5.112}$$

This regularizer is also known as *weight decay* and has been discussed at length in Chapter 3. The effective model complexity is then determined by the choice of the regularization coefficient λ. As we have seen previously, this regularizer can be interpreted as the negative logarithm of a zero-mean Gaussian prior distribution over the weight vector $\mathbf{w}$.

5.5.1 Consistent Gaussian priors

One of the limitations of simple weight decay in the form (5.112) is that is inconsistent with certain scaling properties of network mappings. To illustrate this, consider a multilayer perceptron network having two layers of weights and linear output units, which performs a mapping from a set of input variables $\{x_i\}$ to a set of output variables $\{y_k\}$. The activations of the hidden units in the first hidden layer

Figure 5.10 Plot of the sum-of-squares test-set error for the polynomial data set versus the number of hidden units in the network, with 30 random starts for each network size, showing the effect of local minima. For each new start, the weight vector was initialized by sampling from an isotropic Gaussian distribution having a mean of zero and a variance of 10.

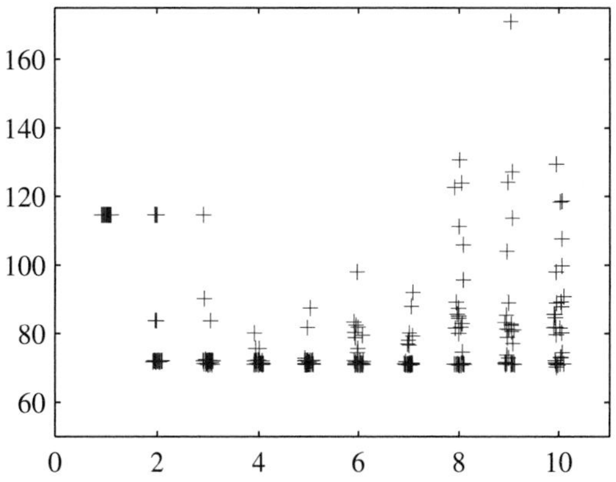

take the form

$$z_j = h\left(\sum_i w_{ji}x_i + w_{j0}\right) \tag{5.113}$$

while the activations of the output units are given by

$$y_k = \sum_j w_{kj}z_j + w_{k0}. \tag{5.114}$$

Suppose we perform a linear transformation of the input data of the form

$$x_i \to \widetilde{x}_i = ax_i + b. \tag{5.115}$$

Then we can arrange for the mapping performed by the network to be unchanged by making a corresponding linear transformation of the weights and biases from the inputs to the units in the hidden layer of the form

Exercise 5.24

$$w_{ji} \to \widetilde{w}_{ji} = \frac{1}{a}w_{ji} \tag{5.116}$$

$$w_{j0} \to \widetilde{w}_{j0} = w_{j0} - \frac{b}{a}\sum_i w_{ji}. \tag{5.117}$$

Similarly, a linear transformation of the output variables of the network of the form

$$y_k \to \widetilde{y}_k = cy_k + d \tag{5.118}$$

can be achieved by making a transformation of the second-layer weights and biases using

$$w_{kj} \to \widetilde{w}_{kj} = cw_{kj} \tag{5.119}$$

$$w_{k0} \to \widetilde{w}_{k0} = cw_{k0} + d. \tag{5.120}$$

If we train one network using the original data and one network using data for which the input and/or target variables are transformed by one of the above linear transformations, then consistency requires that we should obtain equivalent networks that differ only by the linear transformation of the weights as given. Any regularizer should be consistent with this property, otherwise it arbitrarily favours one solution over another, equivalent one. Clearly, simple weight decay (5.112), that treats all weights and biases on an equal footing, does not satisfy this property.

We therefore look for a regularizer which is invariant under the linear transformations (5.116), (5.117), (5.119) and (5.120). These require that the regularizer should be invariant to re-scaling of the weights and to shifts of the biases. Such a regularizer is given by

$$\frac{\lambda_1}{2}\sum_{w\in\mathcal{W}_1} w^2 + \frac{\lambda_2}{2}\sum_{w\in\mathcal{W}_2} w^2 \tag{5.121}$$

where $\mathcal{W}_1$ denotes the set of weights in the first layer, $\mathcal{W}_2$ denotes the set of weights in the second layer, and biases are excluded from the summations. This regularizer will remain unchanged under the weight transformations provided the regularization parameters are re-scaled using $\lambda_1 \to a^{1/2}\lambda_1$ and $\lambda_2 \to c^{-1/2}\lambda_2$.

The regularizer (5.121) corresponds to a prior of the form

$$p(\mathbf{w}|\alpha_1, \alpha_2) \propto \exp\left(-\frac{\alpha_1}{2}\sum_{w\in\mathcal{W}_1} w^2 - \frac{\alpha_2}{2}\sum_{w\in\mathcal{W}_2} w^2\right). \tag{5.122}$$

Note that priors of this form are *improper* (they cannot be normalized) because the bias parameters are unconstrained. The use of improper priors can lead to difficulties in selecting regularization coefficients and in model comparison within the Bayesian framework, because the corresponding evidence is zero. It is therefore common to include separate priors for the biases (which then break shift invariance) having their own hyperparameters. We can illustrate the effect of the resulting four hyperparameters by drawing samples from the prior and plotting the corresponding network functions, as shown in Figure 5.11.

More generally, we can consider priors in which the weights are divided into any number of groups $\mathcal{W}_k$ so that

$$p(\mathbf{w}) \propto \exp\left(-\frac{1}{2}\sum_k \alpha_k \|\mathbf{w}\|_k^2\right) \tag{5.123}$$

where

$$\|\mathbf{w}\|_k^2 = \sum_{j\in\mathcal{W}_k} w_j^2. \tag{5.124}$$

As a special case of this prior, if we choose the groups to correspond to the sets of weights associated with each of the input units, and we optimize the marginal likelihood with respect to the corresponding parameters α_k, we obtain *automatic relevance determination* as discussed in Section 7.2.2.

5.5.2 Early stopping

An alternative to regularization as a way of controlling the effective complexity of a network is the procedure of *early stopping*. The training of nonlinear network models corresponds to an iterative reduction of the error function defined with respect to a set of training data. For many of the optimization algorithms used for network training, such as conjugate gradients, the error is a nonincreasing function of the iteration index. However, the error measured with respect to independent data, generally called a validation set, often shows a decrease at first, followed by an increase as the network starts to over-fit. Training can therefore be stopped at the point of smallest error with respect to the validation data set, as indicated in Figure 5.12, in order to obtain a network having good generalization performance.

The behaviour of the network in this case is sometimes explained qualitatively in terms of the effective number of degrees of freedom in the network, in which this number starts out small and then grows during the training process, corresponding to a steady increase in the effective complexity of the model. Halting training before

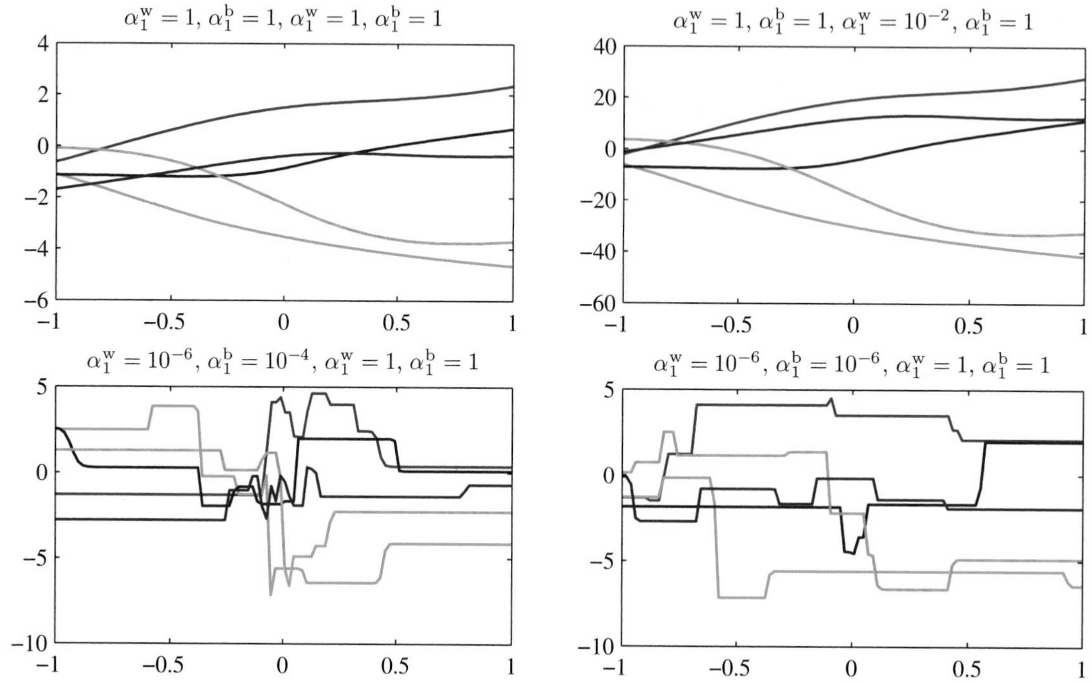

Figure 5.11 Illustration of the effect of the hyperparameters governing the prior distribution over weights and biases in a two-layer network having a single input, a single linear output, and 12 hidden units having 'tanh' activation functions. The priors are governed by four hyperparameters α_1^{b}, α_1^{w}, α_2^{b}, and α_2^{w}, which represent the precisions of the Gaussian distributions of the first-layer biases, first-layer weights, second-layer biases, and second-layer weights, respectively. We see that the parameter α_2^{w} governs the vertical scale of functions (note the different vertical axis ranges on the top two diagrams), α_1^{w} governs the horizontal scale of variations in the function values, and α_1^{b} governs the horizontal range over which variations occur. The parameter α_2^{b}, whose effect is not illustrated here, governs the range of vertical offsets of the functions.

a minimum of the training error has been reached then represents a way of limiting the effective network complexity.

In the case of a quadratic error function, we can verify this insight, and show that early stopping should exhibit similar behaviour to regularization using a simple weight-decay term. This can be understood from Figure 5.13, in which the axes in weight space have been rotated to be parallel to the eigenvectors of the Hessian matrix. If, in the absence of weight decay, the weight vector starts at the origin and proceeds during training along a path that follows the local negative gradient vector, then the weight vector will move initially parallel to the w_2 axis through a point corresponding roughly to $\widetilde{\mathbf{w}}$ and then move towards the minimum of the error function $\mathbf{w}_{\mathrm{ML}}$. This follows from the shape of the error surface and the widely differing eigenvalues of the Hessian. Stopping at a point near $\widetilde{\mathbf{w}}$ is therefore similar to weight decay. The relationship between early stopping and weight decay can be made quantitative, thereby showing that the quantity $\tau\eta$ (where τ is the iteration index, and η is the learning rate parameter) plays the role of the reciprocal of the regularization

Exercise 5.25

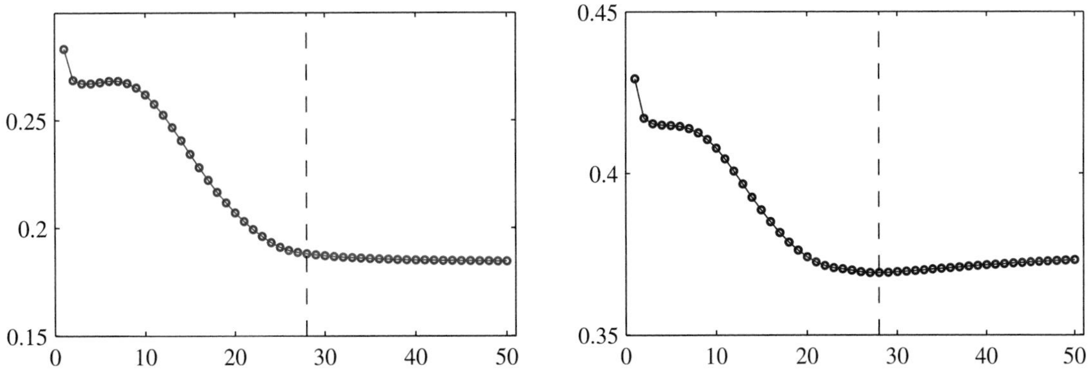

Figure 5.12 An illustration of the behaviour of training set error (left) and validation set error (right) during a typical training session, as a function of the iteration step, for the sinusoidal data set. The goal of achieving the best generalization performance suggests that training should be stopped at the point shown by the vertical dashed lines, corresponding to the minimum of the validation set error.

parameter λ. The effective number of parameters in the network therefore grows during the course of training.

5.5.3 Invariances

In many applications of pattern recognition, it is known that predictions should be unchanged, or *invariant*, under one or more transformations of the input variables. For example, in the classification of objects in two-dimensional images, such as handwritten digits, a particular object should be assigned the same classification irrespective of its position within the image (*translation invariance*) or of its size (*scale invariance*). Such transformations produce significant changes in the raw data, expressed in terms of the intensities at each of the pixels in the image, and yet should give rise to the same output from the classification system. Similarly in speech recognition, small levels of nonlinear warping along the time axis, which preserve temporal ordering, should not change the interpretation of the signal.

If sufficiently large numbers of training patterns are available, then an adaptive model such as a neural network can learn the invariance, at least approximately. This involves including within the training set a sufficiently large number of examples of the effects of the various transformations. Thus, for translation invariance in an image, the training set should include examples of objects at many different positions.

This approach may be impractical, however, if the number of training examples is limited, or if there are several invariants (because the number of combinations of transformations grows exponentially with the number of such transformations). We therefore seek alternative approaches for encouraging an adaptive model to exhibit the required invariances. These can broadly be divided into four categories:

1. The training set is augmented using replicas of the training patterns, transformed according to the desired invariances. For instance, in our digit recognition example, we could make multiple copies of each example in which the

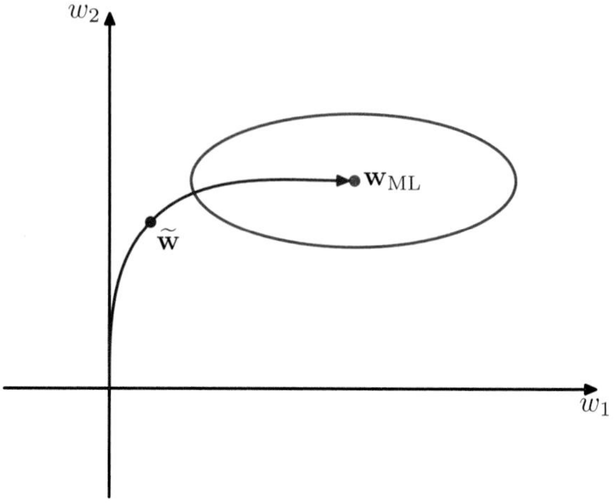

Figure 5.13 A schematic illustration of why early stopping can give similar results to weight decay in the case of a quadratic error function. The ellipse shows a contour of constant error, and $\mathbf{w}_{\mathrm{ML}}$ denotes the minimum of the error function. If the weight vector starts at the origin and moves according to the local negative gradient direction, then it will follow the path shown by the curve. By stopping training early, a weight vector $\widetilde{\mathbf{w}}$ is found that is qualitatively similar to that obtained with a simple weight-decay regularizer and training to the minimum of the regularized error, as can be seen by comparing with Figure 3.15.

digit is shifted to a different position in each image.

2. A regularization term is added to the error function that penalizes changes in the model output when the input is transformed. This leads to the technique of *tangent propagation*, discussed in Section 5.5.4.

3. Invariance is built into the pre-processing by extracting features that are invariant under the required transformations. Any subsequent regression or classification system that uses such features as inputs will necessarily also respect these invariances.

4. The final option is to build the invariance properties into the structure of a neural network (or into the definition of a kernel function in the case of techniques such as the relevance vector machine). One way to achieve this is through the use of local receptive fields and shared weights, as discussed in the context of convolutional neural networks in Section 5.5.6.

Approach 1 is often relatively easy to implement and can be used to encourage complex invariances such as those illustrated in Figure 5.14. For sequential training algorithms, this can be done by transforming each input pattern before it is presented to the model so that, if the patterns are being recycled, a different transformation (drawn from an appropriate distribution) is added each time. For batch methods, a similar effect can be achieved by replicating each data point a number of times and transforming each copy independently. The use of such augmented data can lead to significant improvements in generalization (Simard *et al.*, 2003), although it can also be computationally costly.

Approach 2 leaves the data set unchanged but modifies the error function through the addition of a regularizer. In Section 5.5.5, we shall show that this approach is closely related to approach 1.

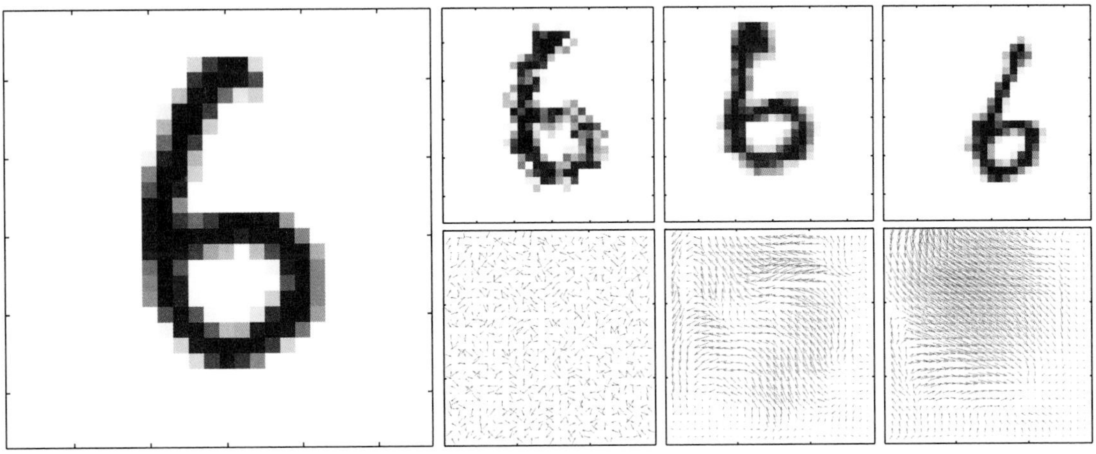

Figure 5.14 Illustration of the synthetic warping of a handwritten digit. The original image is shown on the left. On the right, the top row shows three examples of warped digits, with the corresponding displacement fields shown on the bottom row. These displacement fields are generated by sampling random displacements $\Delta x, \Delta y \in (0, 1)$ at each pixel and then smoothing by convolution with Gaussians of width 0.01, 30 and 60 respectively.

One advantage of approach 3 is that it can correctly extrapolate well beyond the range of transformations included in the training set. However, it can be difficult to find hand-crafted features with the required invariances that do not also discard information that can be useful for discrimination.

5.5.4 Tangent propagation

We can use regularization to encourage models to be invariant to transformations of the input through the technique of *tangent propagation* (Simard *et al.*, 1992). Consider the effect of a transformation on a particular input vector $\mathbf{x}_n$. Provided the transformation is continuous (such as translation or rotation, but not mirror reflection for instance), then the transformed pattern will sweep out a manifold $\mathcal{M}$ within the D-dimensional input space. This is illustrated in Figure 5.15, for the case of $D = 2$ for simplicity. Suppose the transformation is governed by a single parameter ξ

Figure 5.15 Illustration of a two-dimensional input space showing the effect of a continuous transformation on a particular input vector $\mathbf{x}_n$. A one-dimensional transformation, parameterized by the continuous variable ξ, applied to $\mathbf{x}_n$ causes it to sweep out a one-dimensional manifold $\mathcal{M}$. Locally, the effect of the transformation can be approximated by the tangent vector $\boldsymbol{\tau}_n$.

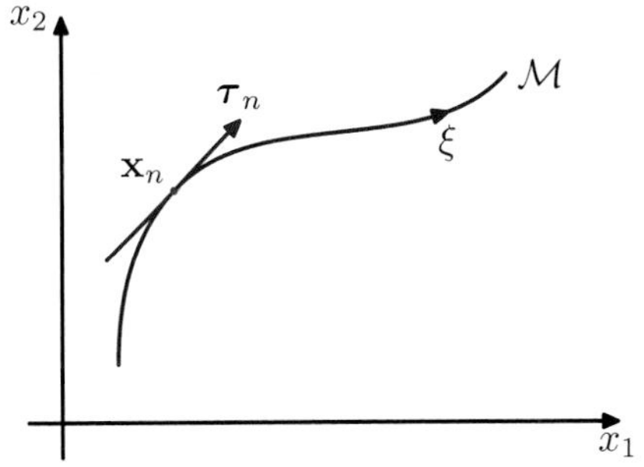

(which might be rotation angle for instance). Then the subspace $\mathcal{M}$ swept out by $\mathbf{x}_n$ will be one-dimensional, and will be parameterized by ξ. Let the vector that results from acting on $\mathbf{x}_n$ by this transformation be denoted by $\mathbf{s}(\mathbf{x}_n, \xi)$, which is defined so that $\mathbf{s}(\mathbf{x}, 0) = \mathbf{x}$. Then the tangent to the curve $\mathcal{M}$ is given by the directional derivative $\boldsymbol{\tau} = \partial\mathbf{s}/\partial\xi$, and the tangent vector at the point $\mathbf{x}_n$ is given by

$$\boldsymbol{\tau}_n = \left.\frac{\partial\mathbf{s}(\mathbf{x}_n, \xi)}{\partial\xi}\right|_{\xi=0}. \tag{5.125}$$

Under a transformation of the input vector, the network output vector will, in general, change. The derivative of output k with respect to ξ is given by

$$\left.\frac{\partial y_k}{\partial\xi}\right|_{\xi=0} = \sum_{i=1}^{D} \left.\frac{\partial y_k}{\partial x_i}\frac{\partial x_i}{\partial\xi}\right|_{\xi=0} = \sum_{i=1}^{D} J_{ki}\tau_i \tag{5.126}$$

where J_{ki} is the (k, i) element of the Jacobian matrix $\mathbf{J}$, as discussed in Section 5.3.4. The result (5.126) can be used to modify the standard error function, so as to encourage local invariance in the neighbourhood of the data points, by the addition to the original error function E of a regularization function Ω to give a total error function of the form

$$\widetilde{E} = E + \lambda\Omega \tag{5.127}$$

where λ is a regularization coefficient and

$$\Omega = \frac{1}{2}\sum_n\sum_k\left(\left.\frac{\partial y_{nk}}{\partial\xi}\right|_{\xi=0}\right)^2 = \frac{1}{2}\sum_n\sum_k\left(\sum_{i=1}^{D} J_{nki}\tau_{ni}\right)^2. \tag{5.128}$$

The regularization function will be zero when the network mapping function is invariant under the transformation in the neighbourhood of each pattern vector, and the value of the parameter λ determines the balance between fitting the training data and learning the invariance property.

In a practical implementation, the tangent vector $\boldsymbol{\tau}_n$ can be approximated using finite differences, by subtracting the original vector $\mathbf{x}_n$ from the corresponding vector after transformation using a small value of ξ, and then dividing by ξ. This is illustrated in Figure 5.16.

Exercise 5.26

The regularization function depends on the network weights through the Jacobian $\mathbf{J}$. A backpropagation formalism for computing the derivatives of the regularizer with respect to the network weights is easily obtained by extension of the techniques introduced in Section 5.3.

If the transformation is governed by L parameters (e.g., $L = 3$ for the case of translations combined with in-plane rotations in a two-dimensional image), then the manifold $\mathcal{M}$ will have dimensionality L, and the corresponding regularizer is given by the sum of terms of the form (5.128), one for each transformation. If several transformations are considered at the same time, and the network mapping is made invariant to each separately, then it will be (locally) invariant to combinations of the transformations (Simard *et al.*, 1992).

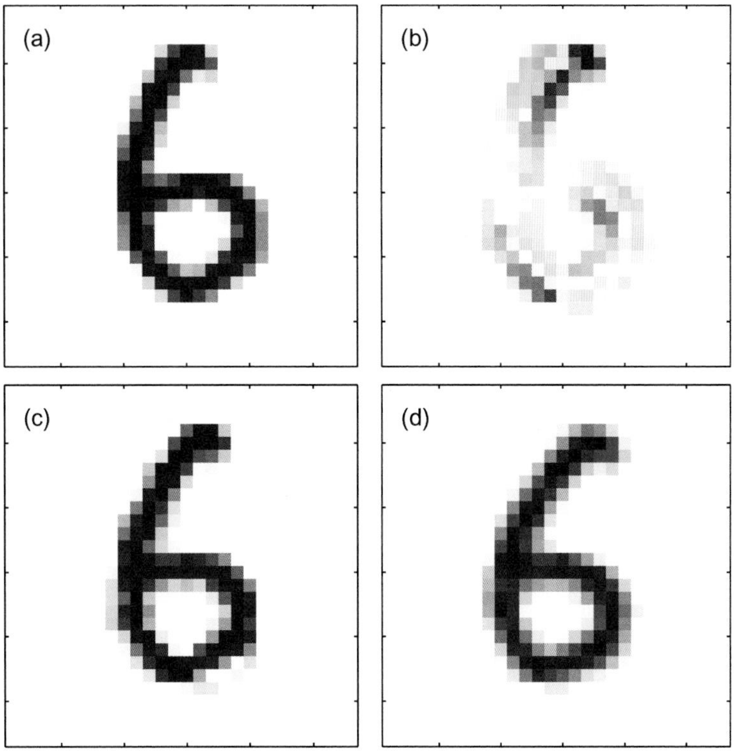

Figure 5.16 Illustration showing (a) the original image $\mathbf{x}$ of a hand-written digit, (b) the tangent vector $\boldsymbol{\tau}$ corresponding to an infinitesimal clockwise rotation, where blue and yellow correspond to positive and negative values, respectively, (c) the result of adding a small contribution from the tangent vector to the original image giving $\mathbf{x} + \epsilon\boldsymbol{\tau}$ with $\epsilon = 15$ degrees, and (d) the true image rotated for comparison.

A related technique, called *tangent distance*, can be used to build invariance properties into distance-based methods such as nearest-neighbour classifiers (Simard *et al.*, 1993).

5.5.5 Training with transformed data

We have seen that one way to encourage invariance of a model to a set of transformations is to expand the training set using transformed versions of the original input patterns. Here we show that this approach is closely related to the technique of tangent propagation (Bishop, 1995b; Leen, 1995).

As in Section 5.5.4, we shall consider a transformation governed by a single parameter ξ and described by the function $\mathbf{s}(\mathbf{x}, \xi)$, with $\mathbf{s}(\mathbf{x}, 0) = \mathbf{x}$. We shall also consider a sum-of-squares error function. The error function for untransformed inputs can be written (in the infinite data set limit) in the form

$$E = \frac{1}{2} \iint \{y(\mathbf{x}) - t\}^2 p(t|\mathbf{x}) p(\mathbf{x}) \, \mathrm{d}\mathbf{x} \, \mathrm{d}t \tag{5.129}$$

as discussed in Section 1.5.5. Here we have considered a network having a single output, in order to keep the notation uncluttered. If we now consider an infinite number of copies of each data point, each of which is perturbed by the transformation in which the parameter ξ is drawn from a distribution $p(\xi)$, then the error function defined over this expanded data set can be written as

$$\widetilde{E} = \frac{1}{2} \iiint \{y(\mathbf{s}(\mathbf{x}, \xi)) - t\}^2 p(t|\mathbf{x}) p(\mathbf{x}) p(\xi) \, \mathrm{d}\mathbf{x} \, \mathrm{d}t \, \mathrm{d}\xi. \tag{5.130}$$

We now assume that the distribution $p(\xi)$ has zero mean with small variance, so that we are only considering small transformations of the original input vectors. We can then expand the transformation function as a Taylor series in powers of ξ to give

$$\begin{aligned} \mathbf{s}(\mathbf{x}, \xi) &= \mathbf{s}(\mathbf{x}, 0) + \xi \left. \frac{\partial}{\partial \xi} \mathbf{s}(\mathbf{x}, \xi) \right|_{\xi=0} + \frac{\xi^2}{2} \left. \frac{\partial^2}{\partial \xi^2} \mathbf{s}(\mathbf{x}, \xi) \right|_{\xi=0} + O(\xi^3) \\ &= \mathbf{x} + \xi \boldsymbol{\tau} + \frac{1}{2} \xi^2 \boldsymbol{\tau}' + O(\xi^3) \end{aligned}$$

where $\boldsymbol{\tau}'$ denotes the second derivative of $\mathbf{s}(\mathbf{x}, \xi)$ with respect to ξ evaluated at $\xi = 0$. This allows us to expand the model function to give

$$y(\mathbf{s}(\mathbf{x}, \xi)) = y(\mathbf{x}) + \xi \boldsymbol{\tau}^{\mathrm{T}} \nabla y(\mathbf{x}) + \frac{\xi^2}{2} \left[(\boldsymbol{\tau}')^{\mathrm{T}} \nabla y(\mathbf{x}) + \boldsymbol{\tau}^{\mathrm{T}} \nabla \nabla y(\mathbf{x}) \boldsymbol{\tau} \right] + O(\xi^3).$$

Substituting into the mean error function (5.130) and expanding, we then have

$$\begin{aligned} \widetilde{E} &= \frac{1}{2} \iint \{y(\mathbf{x}) - t\}^2 p(t|\mathbf{x}) p(\mathbf{x}) \, \mathrm{d}\mathbf{x} \, \mathrm{d}t \\ &+ \mathbb{E}[\xi] \iint \{y(\mathbf{x}) - t\} \boldsymbol{\tau}^{\mathrm{T}} \nabla y(\mathbf{x}) p(t|\mathbf{x}) p(\mathbf{x}) \, \mathrm{d}\mathbf{x} \, \mathrm{d}t \\ &+ \mathbb{E}[\xi^2] \frac{1}{2} \iint \left[\{y(\mathbf{x}) - t\} \left\{ (\boldsymbol{\tau}')^{\mathrm{T}} \nabla y(\mathbf{x}) + \boldsymbol{\tau}^{\mathrm{T}} \nabla \nabla y(\mathbf{x}) \boldsymbol{\tau} \right\} \right. \\ &\left. + \left(\boldsymbol{\tau}^{\mathrm{T}} \nabla y(\mathbf{x}) \right)^2 \right] p(t|\mathbf{x}) p(\mathbf{x}) \, \mathrm{d}\mathbf{x} \, \mathrm{d}t. \end{aligned}$$

Because the distribution of transformations has zero mean we have $\mathbb{E}[\xi] = 0$. Also, we shall denote $\mathbb{E}[\xi^2]$ by λ. Omitting terms of $O(\xi^3)$, the average error function then becomes

$$\widetilde{E} = E + \lambda \Omega \tag{5.131}$$

where E is the original sum-of-squares error, and the regularization term Ω takes the form

$$\begin{aligned} \Omega &= \frac{1}{2} \int \left[\{y(\mathbf{x}) - \mathbb{E}[t|\mathbf{x}]\} \left\{ (\boldsymbol{\tau}')^{\mathrm{T}} \nabla y(\mathbf{x}) + \boldsymbol{\tau}^{\mathrm{T}} \nabla \nabla y(\mathbf{x}) \boldsymbol{\tau} \right\} \right. \\ &\left. + \left(\boldsymbol{\tau}^{T} \nabla y(\mathbf{x}) \right)^2 \right] p(\mathbf{x}) \, \mathrm{d}\mathbf{x} \end{aligned} \tag{5.132}$$

in which we have performed the integration over t.

We can further simplify this regularization term as follows. In Section 1.5.5 we saw that the function that minimizes the sum-of-squares error is given by the conditional average $\mathbb{E}[t|\mathbf{x}]$ of the target values t. From (5.131) we see that the regularized error will equal the unregularized sum-of-squares plus terms which are $O(\xi)$, and so the network function that minimizes the total error will have the form

$$y(\mathbf{x}) = \mathbb{E}[t|\mathbf{x}] + O(\xi). \tag{5.133}$$

Thus, to leading order in ξ, the first term in the regularizer vanishes and we are left with

$$\Omega = \frac{1}{2} \int \left(\boldsymbol{\tau}^{\mathrm{T}} \nabla y(\mathbf{x})\right)^2 p(\mathbf{x})\,\mathrm{d}\mathbf{x} \tag{5.134}$$

which is equivalent to the tangent propagation regularizer (5.128).

If we consider the special case in which the transformation of the inputs simply consists of the addition of random noise, so that $\mathbf{x} \to \mathbf{x} + \boldsymbol{\xi}$, then the regularizer takes the form

Exercise 5.27

$$\Omega = \frac{1}{2} \int \|\nabla y(\mathbf{x})\|^2 p(\mathbf{x})\,\mathrm{d}\mathbf{x} \tag{5.135}$$

which is known as *Tikhonov* regularization (Tikhonov and Arsenin, 1977; Bishop, 1995b). Derivatives of this regularizer with respect to the network weights can be found using an extended backpropagation algorithm (Bishop, 1993). We see that, for small noise amplitudes, Tikhonov regularization is related to the addition of random noise to the inputs, which has been shown to improve generalization in appropriate circumstances (Sietsma and Dow, 1991).

5.5.6 Convolutional networks

Another approach to creating models that are invariant to certain transformation of the inputs is to build the invariance properties into the structure of a neural network. This is the basis for the *convolutional neural network* (Le Cun *et al.*, 1989; LeCun *et al.*, 1998), which has been widely applied to image data.

Consider the specific task of recognizing handwritten digits. Each input image comprises a set of pixel intensity values, and the desired output is a posterior probability distribution over the ten digit classes. We know that the identity of the digit is invariant under translations and scaling as well as (small) rotations. Furthermore, the network must also exhibit invariance to more subtle transformations such as elastic deformations of the kind illustrated in Figure 5.14. One simple approach would be to treat the image as the input to a fully connected network, such as the kind shown in Figure 5.1. Given a sufficiently large training set, such a network could in principle yield a good solution to this problem and would learn the appropriate invariances by example.

However, this approach ignores a key property of images, which is that nearby pixels are more strongly correlated than more distant pixels. Many of the modern approaches to computer vision exploit this property by extracting *local* features that depend only on small subregions of the image. Information from such features can then be merged in later stages of processing in order to detect higher-order features

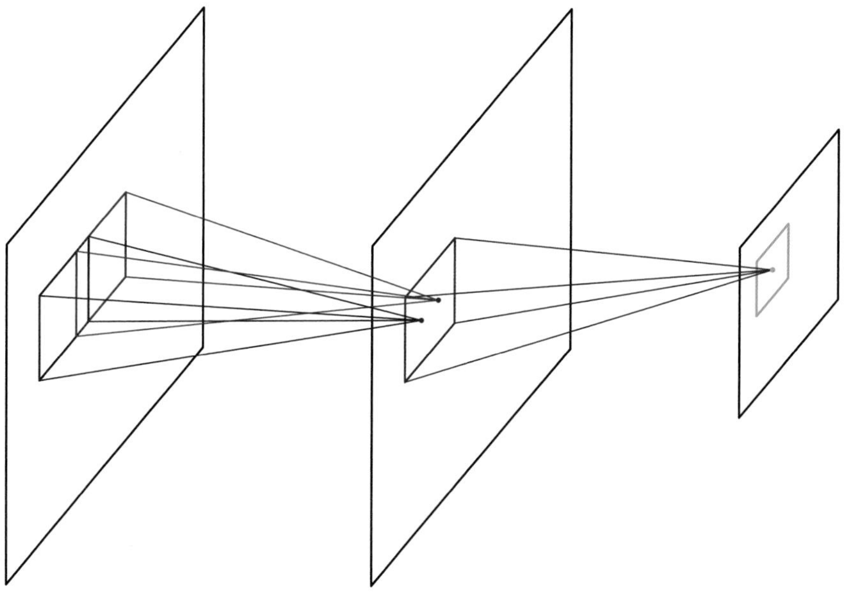

Figure 5.17 Diagram illustrating part of a convolutional neural network, showing a layer of convolutional units followed by a layer of subsampling units. Several successive pairs of such layers may be used.

and ultimately to yield information about the image as whole. Also, local features that are useful in one region of the image are likely to be useful in other regions of the image, for instance if the object of interest is translated.

These notions are incorporated into convolutional neural networks through three mechanisms: (i) local receptive fields, (ii) weight sharing, and (iii) subsampling. The structure of a convolutional network is illustrated in Figure 5.17. In the convolutional layer the units are organized into planes, each of which is called a *feature map*. Units in a feature map each take inputs only from a small subregion of the image, and all of the units in a feature map are constrained to share the same weight values. For instance, a feature map might consist of 100 units arranged in a 10×10 grid, with each unit taking inputs from a 5×5 pixel patch of the image. The whole feature map therefore has 25 adjustable weight parameters plus one adjustable bias parameter. Input values from a patch are linearly combined using the weights and the bias, and the result transformed by a sigmoidal nonlinearity using (5.1). If we think of the units as feature detectors, then all of the units in a feature map detect the same pattern but at different locations in the input image. Due to the weight sharing, the evaluation of the activations of these units is equivalent to a convolution of the image pixel intensities with a 'kernel' comprising the weight parameters. If the input image is shifted, the activations of the feature map will be shifted by the same amount but will otherwise be unchanged. This provides the basis for the (approximate) invariance of the network outputs to translations and distortions of the input image. Because we will typically need to detect multiple features in order to build an effective model, there will generally be multiple feature maps in the convolutional layer, each having its own set of weight and bias parameters.

The outputs of the convolutional units form the inputs to the subsampling layer of the network. For each feature map in the convolutional layer, there is a plane of units in the subsampling layer and each unit takes inputs from a small receptive field in the corresponding feature map of the convolutional layer. These units perform subsampling. For instance, each subsampling unit might take inputs from a 2×2 unit region in the corresponding feature map and would compute the average of those inputs, multiplied by an adaptive weight with the addition of an adaptive bias parameter, and then transformed using a sigmoidal nonlinear activation function. The receptive fields are chosen to be contiguous and nonoverlapping so that there are half the number of rows and columns in the subsampling layer compared with the convolutional layer. In this way, the response of a unit in the subsampling layer will be relatively insensitive to small shifts of the image in the corresponding regions of the input space.

In a practical architecture, there may be several pairs of convolutional and subsampling layers. At each stage there is a larger degree of invariance to input transformations compared to the previous layer. There may be several feature maps in a given convolutional layer for each plane of units in the previous subsampling layer, so that the gradual reduction in spatial resolution is then compensated by an increasing number of features. The final layer of the network would typically be a fully connected, fully adaptive layer, with a softmax output nonlinearity in the case of multiclass classification.

The whole network can be trained by error minimization using backpropagation to evaluate the gradient of the error function. This involves a slight modification of the usual backpropagation algorithm to ensure that the shared-weight constraints are satisfied. Due to the use of local receptive fields, the number of weights in the network is smaller than if the network were fully connected. Furthermore, the number of independent parameters to be learned from the data is much smaller still, due to the substantial numbers of constraints on the weights.

Exercise 5.28

5.5.7 Soft weight sharing

One way to reduce the effective complexity of a network with a large number of weights is to constrain weights within certain groups to be equal. This is the technique of weight sharing that was discussed in Section 5.5.6 as a way of building translation invariance into networks used for image interpretation. It is only applicable, however, to particular problems in which the form of the constraints can be specified in advance. Here we consider a form of *soft weight sharing* (Nowlan and Hinton, 1992) in which the hard constraint of equal weights is replaced by a form of regularization in which groups of weights are encouraged to have similar values. Furthermore, the division of weights into groups, the mean weight value for each group, and the spread of values within the groups are all determined as part of the learning process.

Recall that the simple weight decay regularizer, given in (5.112), can be viewed as the negative log of a Gaussian prior distribution over the weights. We can encourage the weight values to form several groups, rather than just one group, by considering instead a probability distribution that is a *mixture* of Gaussians. The centres and variances of the Gaussian components, as well as the mixing coefficients, will be considered as adjustable parameters to be determined as part of the learning process. Thus, we have a probability density of the form

Section 2.3.9

$$p(\mathbf{w}) = \prod_i p(w_i) \tag{5.136}$$

where

$$p(w_i) = \sum_{j=1}^{M} \pi_j \mathcal{N}(w_i|\mu_j, \sigma_j^2) \tag{5.137}$$

and π_j are the mixing coefficients. Taking the negative logarithm then leads to a regularization function of the form

$$\Omega(\mathbf{w}) = -\sum_i \ln \left(\sum_{j=1}^{M} \pi_j \mathcal{N}(w_i|\mu_j, \sigma_j^2) \right). \tag{5.138}$$

The total error function is then given by

$$\widetilde{E}(\mathbf{w}) = E(\mathbf{w}) + \lambda \Omega(\mathbf{w}) \tag{5.139}$$

where λ is the regularization coefficient. This error is minimized both with respect to the weights w_i and with respect to the parameters $\{\pi_j, \mu_j, \sigma_j\}$ of the mixture model. If the weights were constant, then the parameters of the mixture model could be determined by using the EM algorithm discussed in Chapter 9. However, the distribution of weights is itself evolving during the learning process, and so to avoid numerical instability, a joint optimization is performed simultaneously over the weights and the mixture-model parameters. This can be done using a standard optimization algorithm such as conjugate gradients or quasi-Newton methods.

In order to minimize the total error function, it is necessary to be able to evaluate its derivatives with respect to the various adjustable parameters. To do this it is convenient to regard the $\{\pi_j\}$ as *prior* probabilities and to introduce the corresponding posterior probabilities which, following (2.192), are given by Bayes' theorem in the form

$$\gamma_j(w) = \frac{\pi_j \mathcal{N}(w|\mu_j, \sigma_j^2)}{\sum_k \pi_k \mathcal{N}(w|\mu_k, \sigma_k^2)}. \tag{5.140}$$

The derivatives of the total error function with respect to the weights are then given by

Exercise 5.29

$$\frac{\partial \widetilde{E}}{\partial w_i} = \frac{\partial E}{\partial w_i} + \lambda \sum_j \gamma_j(w_i) \frac{(w_i - \mu_j)}{\sigma_j^2}. \tag{5.141}$$

The effect of the regularization term is therefore to pull each weight towards the centre of the jth Gaussian, with a force proportional to the posterior probability of that Gaussian for the given weight. This is precisely the kind of effect that we are seeking.

Derivatives of the error with respect to the centres of the Gaussians are also easily computed to give

Exercise 5.30

$$\frac{\partial \widetilde{E}}{\partial \mu_j} = \lambda \sum_i \gamma_j(w_i) \frac{(\mu_j - w_i)}{\sigma_j^2} \tag{5.142}$$

which has a simple intuitive interpretation, because it pushes μ_j towards an average of the weight values, weighted by the posterior probabilities that the respective weight parameters were generated by component j. Similarly, the derivatives with respect to the variances are given by

Exercise 5.31

$$\frac{\partial \widetilde{E}}{\partial \sigma_j} = \lambda \sum_i \gamma_j(w_i) \left(\frac{1}{\sigma_j} - \frac{(w_i - \mu_j)^2}{\sigma_j^3} \right) \tag{5.143}$$

which drives σ_j towards the weighted average of the squared deviations of the weights around the corresponding centre μ_j, where the weighting coefficients are again given by the posterior probability that each weight is generated by component j. Note that in a practical implementation, new variables ξ_j defined by

$$\sigma_j^2 = \exp(\xi_j) \tag{5.144}$$

are introduced, and the minimization is performed with respect to the ξ_j. This ensures that the parameters σ_j remain positive. It also has the effect of discouraging pathological solutions in which one or more of the σ_j goes to zero, corresponding to a Gaussian component collapsing onto one of the weight parameter values. Such solutions are discussed in more detail in the context of Gaussian mixture models in Section 9.2.1.

For the derivatives with respect to the mixing coefficients π_j, we need to take account of the constraints

$$\sum_j \pi_j = 1, \qquad 0 \leqslant \pi_i \leqslant 1 \tag{5.145}$$

which follow from the interpretation of the π_j as prior probabilities. This can be done by expressing the mixing coefficients in terms of a set of auxiliary variables $\{\eta_j\}$ using the *softmax* function given by

$$\pi_j = \frac{\exp(\eta_j)}{\sum_{k=1}^{M} \exp(\eta_k)}. \tag{5.146}$$

The derivatives of the regularized error function with respect to the $\{\eta_j\}$ then take the form

Exercise 5.32

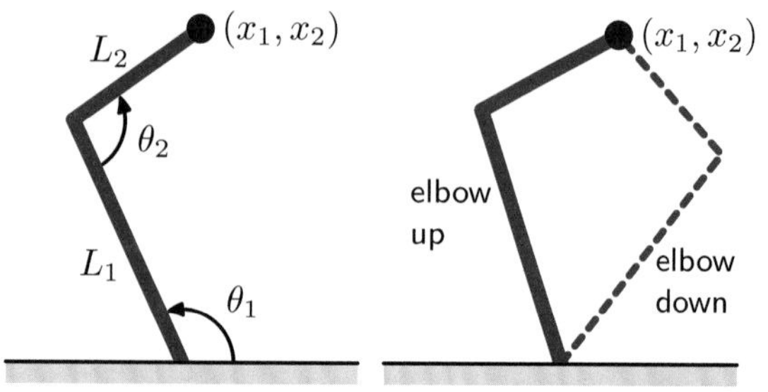

Figure 5.18 The left figure shows a two-link robot arm, in which the Cartesian coordinates (x_1, x_2) of the end effector are determined uniquely by the two joint angles θ_1 and θ_2 and the (fixed) lengths L_1 and L_2 of the arms. This is know as the *forward kinematics* of the arm. In practice, we have to find the joint angles that will give rise to a desired end effector position and, as shown in the right figure, this *inverse kinematics* has two solutions corresponding to 'elbow up' and 'elbow down'.

$$\frac{\partial \widetilde{E}}{\partial \eta_j} = \lambda \sum_i \{\pi_j - \gamma_j(w_i)\}. \tag{5.147}$$

We see that π_j is therefore driven towards the average posterior probability for component j.

5.6. Mixture Density Networks

The goal of supervised learning is to model a conditional distribution $p(\mathbf{t}|\mathbf{x})$, which for many simple regression problems is chosen to be Gaussian. However, practical machine learning problems can often have significantly non-Gaussian distributions. These can arise, for example, with *inverse problems* in which the distribution can be multimodal, in which case the Gaussian assumption can lead to very poor predictions.

Exercise 5.33

As a simple example of an inverse problem, consider the kinematics of a robot arm, as illustrated in Figure 5.18. The *forward problem* involves finding the end effector position given the joint angles and has a unique solution. However, in practice we wish to move the end effector of the robot to a specific position, and to do this we must set appropriate joint angles. We therefore need to solve the inverse problem, which has two solutions as seen in Figure 5.18.

Forward problems often corresponds to causality in a physical system and generally have a unique solution. For instance, a specific pattern of symptoms in the human body may be caused by the presence of a particular disease. In pattern recognition, however, we typically have to solve an inverse problem, such as trying to predict the presence of a disease given a set of symptoms. If the forward problem involves a many-to-one mapping, then the inverse problem will have multiple solutions. For instance, several different diseases may result in the same symptoms.

In the robotics example, the kinematics is defined by geometrical equations, and the multimodality is readily apparent. However, in many machine learning problems the presence of multimodality, particularly in problems involving spaces of high dimensionality, can be less obvious. For tutorial purposes, however, we shall consider a simple toy problem for which we can easily visualize the multimodality. Data for this problem is generated by sampling a variable x uniformly over the interval $(0, 1)$, to give a set of values $\{x_n\}$, and the corresponding target values t_n are obtained

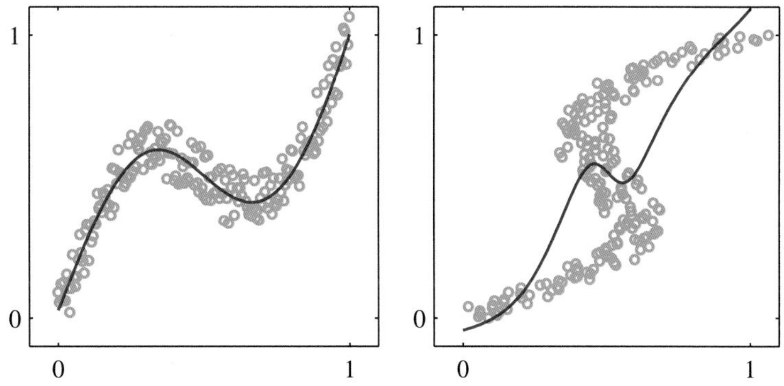

Figure 5.19 On the left is the data set for a simple 'forward problem' in which the red curve shows the result of fitting a two-layer neural network by minimizing the sum-of-squares error function. The corresponding inverse problem, shown on the right, is obtained by exchanging the roles of x and t. Here the same network trained again by minimizing the sum-of-squares error function gives a very poor fit to the data due to the multimodality of the data set.

by computing the function $x_n + 0.3\sin(2\pi x_n)$ and then adding uniform noise over the interval $(-0.1, 0.1)$. The inverse problem is then obtained by keeping the same data points but exchanging the roles of x and t. Figure 5.19 shows the data sets for the forward and inverse problems, along with the results of fitting two-layer neural networks having 6 hidden units and a single linear output unit by minimizing a sum-of-squares error function. Least squares corresponds to maximum likelihood under a Gaussian assumption. We see that this leads to a very poor model for the highly non-Gaussian inverse problem.

We therefore seek a general framework for modelling conditional probability distributions. This can be achieved by using a mixture model for $p(\mathbf{t}|\mathbf{x})$ in which both the mixing coefficients as well as the component densities are flexible functions of the input vector $\mathbf{x}$, giving rise to the *mixture density network*. For any given value of $\mathbf{x}$, the mixture model provides a general formalism for modelling an arbitrary conditional density function $p(\mathbf{t}|\mathbf{x})$. Provided we consider a sufficiently flexible network, we then have a framework for approximating arbitrary conditional distributions.

Here we shall develop the model explicitly for Gaussian components, so that

$$p(\mathbf{t}|\mathbf{x}) = \sum_{k=1}^{K} \pi_k(\mathbf{x}) \mathcal{N}\left(\mathbf{t}|\boldsymbol{\mu}_k(\mathbf{x}), \sigma_k^2(\mathbf{x})\right). \tag{5.148}$$

This is an example of a *heteroscedastic* model since the noise variance on the data is a function of the input vector $\mathbf{x}$. Instead of Gaussians, we can use other distributions for the components, such as Bernoulli distributions if the target variables are binary rather than continuous. We have also specialized to the case of isotropic covariances for the components, although the mixture density network can readily be extended to allow for general covariance matrices by representing the covariances using a Cholesky factorization (Williams, 1996). Even with isotropic components, the conditional distribution $p(\mathbf{t}|\mathbf{x})$ does not assume factorization with respect to the components of $\mathbf{t}$ (in contrast to the standard sum-of-squares regression model) as a consequence of the mixture distribution.

We now take the various parameters of the mixture model, namely the mixing coefficients $\pi_k(\mathbf{x})$, the means $\boldsymbol{\mu}_k(\mathbf{x})$, and the variances $\sigma_k^2(\mathbf{x})$, to be governed by

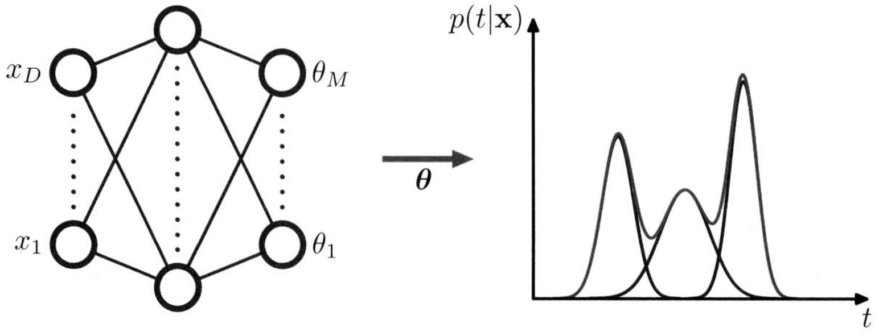

Figure 5.20 The *mixture density network* can represent general conditional probability densities $p(\mathbf{t}|\mathbf{x})$ by considering a parametric mixture model for the distribution of $\mathbf{t}$ whose parameters are determined by the outputs of a neural network that takes $\mathbf{x}$ as its input vector.

the outputs of a conventional neural network that takes $\mathbf{x}$ as its input. The structure of this mixture density network is illustrated in Figure 5.20. The mixture density network is closely related to the mixture of experts discussed in Section 14.5.3. The principle difference is that in the mixture density network the same function is used to predict the parameters of all of the component densities as well as the mixing coefficients, and so the nonlinear hidden units are shared amongst the input-dependent functions.

The neural network in Figure 5.20 can, for example, be a two-layer network having sigmoidal ('tanh') hidden units. If there are K components in the mixture model (5.148), and if $\mathbf{t}$ has L components, then the network will have K output unit activations denoted by a_k^{π} that determine the mixing coefficients $\pi_k(\mathbf{x})$, K outputs denoted by a_k^{σ} that determine the kernel widths $\sigma_k(\mathbf{x})$, and $K \times L$ outputs denoted by a_{kj}^{μ} that determine the components $\mu_{kj}(\mathbf{x})$ of the kernel centres $\boldsymbol{\mu}_k(\mathbf{x})$. The total number of network outputs is given by $(L+2)K$, as compared with the usual L outputs for a network, which simply predicts the conditional means of the target variables.

The mixing coefficients must satisfy the constraints

$$\sum_{k=1}^{K} \pi_k(\mathbf{x}) = 1, \qquad 0 \leqslant \pi_k(\mathbf{x}) \leqslant 1 \tag{5.149}$$

which can be achieved using a set of softmax outputs

$$\pi_k(\mathbf{x}) = \frac{\exp(a_k^{\pi})}{\sum_{l=1}^{K} \exp(a_l^{\pi})}. \tag{5.150}$$

Similarly, the variances must satisfy $\sigma_k^2(\mathbf{x}) \geqslant 0$ and so can be represented in terms of the exponentials of the corresponding network activations using

$$\sigma_k(\mathbf{x}) = \exp(a_k^{\sigma}). \tag{5.151}$$

Finally, because the means $\boldsymbol{\mu}_k(\mathbf{x})$ have real components, they can be represented directly by the network output activations

$$\mu_{kj}(\mathbf{x}) = a^{\mu}_{kj}. \tag{5.152}$$

The adaptive parameters of the mixture density network comprise the vector $\mathbf{w}$ of weights and biases in the neural network, that can be set by maximum likelihood, or equivalently by minimizing an error function defined to be the negative logarithm of the likelihood. For independent data, this error function takes the form

$$E(\mathbf{w}) = -\sum_{n=1}^{N} \ln \left\{ \sum_{k=1}^{K} \pi_k(\mathbf{x}_n, \mathbf{w}) \mathcal{N}\left(\mathbf{t}_n | \boldsymbol{\mu}_k(\mathbf{x}_n, \mathbf{w}), \sigma_k^2(\mathbf{x}_n, \mathbf{w})\right) \right\} \tag{5.153}$$

where we have made the dependencies on $\mathbf{w}$ explicit.

In order to minimize the error function, we need to calculate the derivatives of the error $E(\mathbf{w})$ with respect to the components of $\mathbf{w}$. These can be evaluated by using the standard backpropagation procedure, provided we obtain suitable expressions for the derivatives of the error with respect to the output-unit activations. These represent error signals δ for each pattern and for each output unit, and can be backpropagated to the hidden units and the error function derivatives evaluated in the usual way. Because the error function (5.153) is composed of a sum of terms, one for each training data point, we can consider the derivatives for a particular pattern n and then find the derivatives of E by summing over all patterns.

Because we are dealing with mixture distributions, it is convenient to view the mixing coefficients $\pi_k(\mathbf{x})$ as $\mathbf{x}$-dependent prior probabilities and to introduce the corresponding posterior probabilities given by

$$\gamma_{nk} = \gamma_k(\mathbf{t}_n|\mathbf{x}_n) = \frac{\pi_k \mathcal{N}_{nk}}{\sum_{l=1}^{K} \pi_l \mathcal{N}_{nl}} \tag{5.154}$$

where $\mathcal{N}_{nk}$ denotes $\mathcal{N}\left(\mathbf{t}_n|\boldsymbol{\mu}_k(\mathbf{x}_n), \sigma_k^2(\mathbf{x}_n)\right)$.

Exercise 5.34

The derivatives with respect to the network output activations governing the mixing coefficients are given by

$$\frac{\partial E_n}{\partial a^{\pi}_k} = \pi_k - \gamma_{nk}. \tag{5.155}$$

Exercise 5.35

Similarly, the derivatives with respect to the output activations controlling the component means are given by

$$\frac{\partial E_n}{\partial a^{\mu}_{kl}} = \gamma_{nk} \left\{ \frac{\mu_{kl} - t_{nl}}{\sigma_k^2} \right\}. \tag{5.156}$$

Exercise 5.36

Finally, the derivatives with respect to the output activations controlling the component variances are given by

$$\frac{\partial E_n}{\partial a^{\sigma}_k} = \gamma_{nk} \left\{ L - \frac{\|\mathbf{t}_n - \boldsymbol{\mu}_k\|^2}{\sigma_k^2} \right\}. \tag{5.157}$$

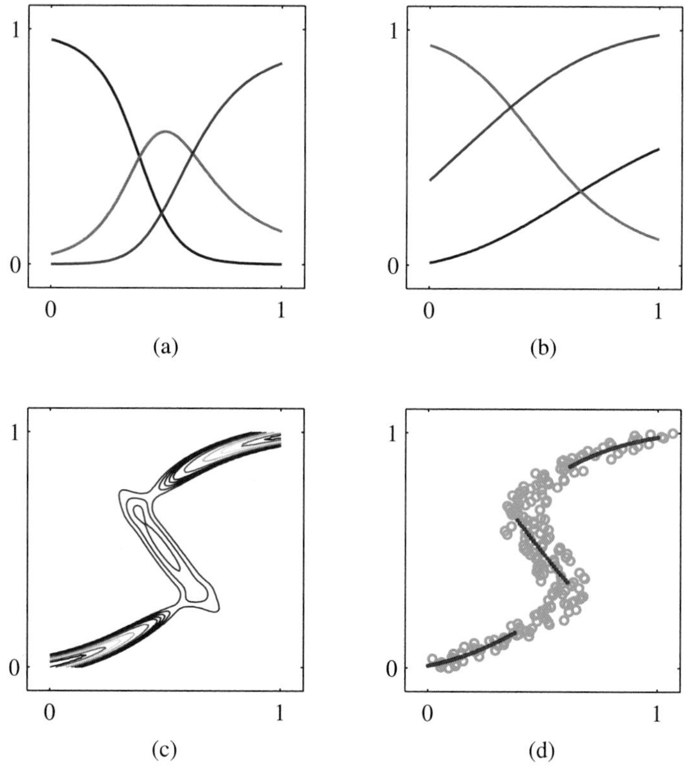

Figure 5.21 (a) Plot of the mixing coefficients $\pi_k(x)$ as a function of x for the three kernel functions in a mixture density network trained on the data shown in Figure 5.19. The model has three Gaussian components, and uses a two-layer multi-layer perceptron with five 'tanh' sigmoidal units in the hidden layer, and nine outputs (corresponding to the 3 means and 3 variances of the Gaussian components and the 3 mixing coefficients). At both small and large values of x, where the conditional probability density of the target data is unimodal, only one of the kernels has a high value for its prior probability, while at intermediate values of x, where the conditional density is trimodal, the three mixing coefficients have comparable values. (b) Plots of the means $\mu_k(x)$ using the same colour coding as for the mixing coefficients. (c) Plot of the contours of the corresponding conditional probability density of the target data for the same mixture density network. (d) Plot of the approximate conditional mode, shown by the red points, of the conditional density.

We illustrate the use of a mixture density network by returning to the toy example of an inverse problem shown in Figure 5.19. Plots of the mixing coefficients $\pi_k(x)$, the means $\mu_k(x)$, and the conditional density contours corresponding to $p(t|x)$, are shown in Figure 5.21. The outputs of the neural network, and hence the parameters in the mixture model, are necessarily continuous single-valued functions of the input variables. However, we see from Figure 5.21(c) that the model is able to produce a conditional density that is unimodal for some values of x and trimodal for other values by modulating the amplitudes of the mixing components $\pi_k(\mathbf{x})$.

Once a mixture density network has been trained, it can predict the conditional density function of the target data for any given value of the input vector. This conditional density represents a complete description of the generator of the data, so far as the problem of predicting the value of the output vector is concerned. From this density function we can calculate more specific quantities that may be of interest in different applications. One of the simplest of these is the mean, corresponding to the conditional average of the target data, and is given by

$$\mathbb{E}[\mathbf{t}|\mathbf{x}] = \int \mathbf{t}p(\mathbf{t}|\mathbf{x})\,\mathrm{d}\mathbf{t} = \sum_{k=1}^{K} \pi_k(\mathbf{x})\boldsymbol{\mu}_k(\mathbf{x}) \tag{5.158}$$

where we have used (5.148). Because a standard network trained by least squares is approximating the conditional mean, we see that a mixture density network can reproduce the conventional least-squares result as a special case. Of course, as we have already noted, for a multimodal distribution the conditional mean is of limited value.

We can similarly evaluate the variance of the density function about the conditional average, to give

Exercise 5.37

$$s^2(\mathbf{x}) = \mathbb{E}\left[\|\mathbf{t} - \mathbb{E}[\mathbf{t}|\mathbf{x}]\|^2 |\mathbf{x}\right] \tag{5.159}$$

$$= \sum_{k=1}^{K} \pi_k(\mathbf{x}) \left\{ \sigma_k^2(\mathbf{x}) + \left\| \boldsymbol{\mu}_k(\mathbf{x}) - \sum_{l=1}^{K} \pi_l(\mathbf{x}) \boldsymbol{\mu}_l(\mathbf{x}) \right\|^2 \right\} \tag{5.160}$$

where we have used (5.148) and (5.158). This is more general than the corresponding least-squares result because the variance is a function of $\mathbf{x}$.

We have seen that for multimodal distributions, the conditional mean can give a poor representation of the data. For instance, in controlling the simple robot arm shown in Figure 5.18, we need to pick one of the two possible joint angle settings in order to achieve the desired end-effector location, whereas the average of the two solutions is not itself a solution. In such cases, the conditional mode may be of more value. Because the conditional mode for the mixture density network does not have a simple analytical solution, this would require numerical iteration. A simple alternative is to take the mean of the most probable component (i.e., the one with the largest mixing coefficient) at each value of $\mathbf{x}$. This is shown for the toy data set in Figure 5.21(d).

5.7. Bayesian Neural Networks

So far, our discussion of neural networks has focussed on the use of maximum likelihood to determine the network parameters (weights and biases). Regularized maximum likelihood can be interpreted as a MAP (maximum posterior) approach in which the regularizer can be viewed as the logarithm of a prior parameter distribution. However, in a Bayesian treatment we need to marginalize over the distribution of parameters in order to make predictions.

In Section 3.3, we developed a Bayesian solution for a simple linear regression model under the assumption of Gaussian noise. We saw that the posterior distribution, which is Gaussian, could be evaluated exactly and that the predictive distribution could also be found in closed form. In the case of a multilayered network, the highly nonlinear dependence of the network function on the parameter values means that an exact Bayesian treatment can no longer be found. In fact, the log of the posterior distribution will be nonconvex, corresponding to the multiple local minima in the error function.

The technique of variational inference, to be discussed in Chapter 10, has been applied to Bayesian neural networks using a factorized Gaussian approximation to the posterior distribution (Hinton and van Camp, 1993) and also using a full-covariance Gaussian (Barber and Bishop, 1998a; Barber and Bishop, 1998b). The most complete treatment, however, has been based on the Laplace approximation (MacKay, 1992c; MacKay, 1992b) and forms the basis for the discussion given here. We will approximate the posterior distribution by a Gaussian, centred at a mode of the true posterior. Furthermore, we shall assume that the covariance of this Gaussian is small so that the network function is approximately linear with respect to the parameters over the region of parameter space for which the posterior probability is significantly nonzero. With these two approximations, we will obtain models that are analogous to the linear regression and classification models discussed in earlier chapters and so we can exploit the results obtained there. We can then make use of the evidence framework to provide point estimates for the hyperparameters and to compare alternative models (for example, networks having different numbers of hidden units). To start with, we shall discuss the regression case and then later consider the modifications needed for solving classification tasks.

5.7.1 Posterior parameter distribution

Consider the problem of predicting a single continuous target variable t from a vector $\mathbf{x}$ of inputs (the extension to multiple targets is straightforward). We shall suppose that the conditional distribution $p(t|\mathbf{x})$ is Gaussian, with an $\mathbf{x}$-dependent mean given by the output of a neural network model $y(\mathbf{x}, \mathbf{w})$, and with precision (inverse variance) β

$$p(t|\mathbf{x}, \mathbf{w}, \beta) = \mathcal{N}(t|y(\mathbf{x}, \mathbf{w}), \beta^{-1}). \tag{5.161}$$

Similarly, we shall choose a prior distribution over the weights $\mathbf{w}$ that is Gaussian of the form

$$p(\mathbf{w}|\alpha) = \mathcal{N}(\mathbf{w}|\mathbf{0}, \alpha^{-1}\mathbf{I}). \tag{5.162}$$

For an i.i.d. data set of N observations $\mathbf{x}_1, \ldots, \mathbf{x}_N$, with a corresponding set of target values $\mathcal{D} = \{t_1, \ldots, t_N\}$, the likelihood function is given by

$$p(\mathcal{D}|\mathbf{w}, \beta) = \prod_{n=1}^{N} \mathcal{N}(t_n|y(\mathbf{x}_n, \mathbf{w}), \beta^{-1}) \tag{5.163}$$

and so the resulting posterior distribution is then

$$p(\mathbf{w}|\mathcal{D}, \alpha, \beta) \propto p(\mathbf{w}|\alpha)p(\mathcal{D}|\mathbf{w}, \beta). \tag{5.164}$$

which, as a consequence of the nonlinear dependence of $y(\mathbf{x}, \mathbf{w})$ on $\mathbf{w}$, will be non-Gaussian.

We can find a Gaussian approximation to the posterior distribution by using the Laplace approximation. To do this, we must first find a (local) maximum of the posterior, and this must be done using iterative numerical optimization. As usual, it is convenient to maximize the logarithm of the posterior, which can be written in the form

$$\ln p(\mathbf{w}|\mathcal{D}) = -\frac{\alpha}{2}\mathbf{w}^{\mathrm{T}}\mathbf{w} - \frac{\beta}{2}\sum_{n=1}^{N}\{y(\mathbf{x}_n, \mathbf{w}) - t_n\}^2 + \text{const} \tag{5.165}$$

which corresponds to a regularized sum-of-squares error function. Assuming for the moment that α and β are fixed, we can find a maximum of the posterior, which we denote $\mathbf{w}_{\mathrm{MAP}}$, by standard nonlinear optimization algorithms such as conjugate gradients, using error backpropagation to evaluate the required derivatives.

Having found a mode $\mathbf{w}_{\mathrm{MAP}}$, we can then build a local Gaussian approximation by evaluating the matrix of second derivatives of the negative log posterior distribution. From (5.165), this is given by

$$\mathbf{A} = -\nabla\nabla \ln p(\mathbf{w}|\mathcal{D}, \alpha, \beta) = \alpha\mathbf{I} + \beta\mathbf{H} \tag{5.166}$$

where $\mathbf{H}$ is the Hessian matrix comprising the second derivatives of the sum-of-squares error function with respect to the components of $\mathbf{w}$. Algorithms for computing and approximating the Hessian were discussed in Section 5.4. The corresponding Gaussian approximation to the posterior is then given from (4.134) by

$$q(\mathbf{w}|\mathcal{D}) = \mathcal{N}(\mathbf{w}|\mathbf{w}_{\mathrm{MAP}}, \mathbf{A}^{-1}). \tag{5.167}$$

Similarly, the predictive distribution is obtained by marginalizing with respect to this posterior distribution

$$p(t|\mathbf{x}, \mathcal{D}) = \int p(t|\mathbf{x}, \mathbf{w})q(\mathbf{w}|\mathcal{D})\,\mathrm{d}\mathbf{w}. \tag{5.168}$$

However, even with the Gaussian approximation to the posterior, this integration is still analytically intractable due to the nonlinearity of the network function $y(\mathbf{x}, \mathbf{w})$ as a function of $\mathbf{w}$. To make progress, we now assume that the posterior distribution has small variance compared with the characteristic scales of $\mathbf{w}$ over which $y(\mathbf{x}, \mathbf{w})$ is varying. This allows us to make a Taylor series expansion of the network function around $\mathbf{w}_{\mathrm{MAP}}$ and retain only the linear terms

$$y(\mathbf{x}, \mathbf{w}) \simeq y(\mathbf{x}, \mathbf{w}_{\mathrm{MAP}}) + \mathbf{g}^{\mathrm{T}}(\mathbf{w} - \mathbf{w}_{\mathrm{MAP}}) \tag{5.169}$$

where we have defined

$$\mathbf{g} = \nabla_{\mathbf{w}} y(\mathbf{x}, \mathbf{w})\big|_{\mathbf{w}=\mathbf{w}_{\mathrm{MAP}}}. \tag{5.170}$$

With this approximation, we now have a linear-Gaussian model with a Gaussian distribution for $p(\mathbf{w})$ and a Gaussian for $p(t|\mathbf{w})$ whose mean is a linear function of $\mathbf{w}$ of the form

$$p(t|\mathbf{x}, \mathbf{w}, \beta) \simeq \mathcal{N}\left(t|y(\mathbf{x}, \mathbf{w}_{\mathrm{MAP}}) + \mathbf{g}^{\mathrm{T}}(\mathbf{w} - \mathbf{w}_{\mathrm{MAP}}), \beta^{-1}\right). \tag{5.171}$$

Exercise 5.38

We can therefore make use of the general result (2.115) for the marginal $p(t)$ to give

$$p(t|\mathbf{x}, \mathcal{D}, \alpha, \beta) = \mathcal{N}\left(t|y(\mathbf{x}, \mathbf{w}_{\mathrm{MAP}}), \sigma^2(\mathbf{x})\right) \tag{5.172}$$

where the input-dependent variance is given by

$$\sigma^2(\mathbf{x}) = \beta^{-1} + \mathbf{g}^{\mathrm{T}}\mathbf{A}^{-1}\mathbf{g}. \tag{5.173}$$

We see that the predictive distribution $p(t|\mathbf{x}, \mathcal{D})$ is a Gaussian whose mean is given by the network function $y(\mathbf{x}, \mathbf{w}_{\mathrm{MAP}})$ with the parameter set to their MAP value. The variance has two terms, the first of which arises from the intrinsic noise on the target variable, whereas the second is an $\mathbf{x}$-dependent term that expresses the uncertainty in the interpolant due to the uncertainty in the model parameters $\mathbf{w}$. This should be compared with the corresponding predictive distribution for the linear regression model, given by (3.58) and (3.59).

5.7.2 Hyperparameter optimization

So far, we have assumed that the hyperparameters α and β are fixed and known. We can make use of the evidence framework, discussed in Section 3.5, together with the Gaussian approximation to the posterior obtained using the Laplace approximation, to obtain a practical procedure for choosing the values of such hyperparameters.

The marginal likelihood, or evidence, for the hyperparameters is obtained by integrating over the network weights

$$p(\mathcal{D}|\alpha, \beta) = \int p(\mathcal{D}|\mathbf{w}, \beta)p(\mathbf{w}|\alpha)\,\mathrm{d}\mathbf{w}. \tag{5.174}$$

Exercise 5.39

This is easily evaluated by making use of the Laplace approximation result (4.135). Taking logarithms then gives

$$\ln p(\mathcal{D}|\alpha, \beta) \simeq -E(\mathbf{w}_{\mathrm{MAP}}) - \frac{1}{2}\ln|\mathbf{A}| + \frac{W}{2}\ln\alpha + \frac{N}{2}\ln\beta - \frac{N}{2}\ln(2\pi) \tag{5.175}$$

where W is the total number of parameters in $\mathbf{w}$, and the regularized error function is defined by

$$E(\mathbf{w}_{\mathrm{MAP}}) = \frac{\beta}{2}\sum_{n=1}^{N}\{y(\mathbf{x}_n, \mathbf{w}_{\mathrm{MAP}}) - t_n\}^2 + \frac{\alpha}{2}\mathbf{w}_{\mathrm{MAP}}^{\mathrm{T}}\mathbf{w}_{\mathrm{MAP}}. \tag{5.176}$$

We see that this takes the same form as the corresponding result (3.86) for the linear regression model.

In the evidence framework, we make point estimates for α and β by maximizing $\ln p(\mathcal{D}|\alpha, \beta)$. Consider first the maximization with respect to α, which can be done by analogy with the linear regression case discussed in Section 3.5.2. We first define the eigenvalue equation

$$\beta\mathbf{H}\mathbf{u}_i = \lambda_i\mathbf{u}_i \tag{5.177}$$

where $\mathbf{H}$ is the Hessian matrix comprising the second derivatives of the sum-of-squares error function, evaluated at $\mathbf{w} = \mathbf{w}_{\mathrm{MAP}}$. By analogy with (3.92), we obtain

$$\alpha = \frac{\gamma}{\mathbf{w}_{\mathrm{MAP}}^{\mathrm{T}}\mathbf{w}_{\mathrm{MAP}}} \tag{5.178}$$

Section 3.5.3

where γ represents the effective number of parameters and is defined by

$$\gamma = \sum_{i=1}^{W} \frac{\lambda_i}{\alpha + \lambda_i}. \tag{5.179}$$

Note that this result was exact for the linear regression case. For the nonlinear neural network, however, it ignores the fact that changes in α will cause changes in the Hessian $\mathbf{H}$, which in turn will change the eigenvalues. We have therefore implicitly ignored terms involving the derivatives of λ_i with respect to α.

Similarly, from (3.95) we see that maximizing the evidence with respect to β gives the re-estimation formula

$$\frac{1}{\beta} = \frac{1}{N-\gamma} \sum_{n=1}^{N} \{y(\mathbf{x}_n, \mathbf{w}_{\mathrm{MAP}}) - t_n\}^2. \tag{5.180}$$

As with the linear model, we need to alternate between re-estimation of the hyperparameters α and β and updating of the posterior distribution. The situation with a neural network model is more complex, however, due to the multimodality of the posterior distribution. As a consequence, the solution for $\mathbf{w}_{\mathrm{MAP}}$ found by maximizing the log posterior will depend on the initialization of $\mathbf{w}$. Solutions that differ only as a consequence of the interchange and sign reversal symmetries in the hidden units are identical so far as predictions are concerned, and it is irrelevant which of the equivalent solutions is found. However, there may be inequivalent solutions as well, and these will generally yield different values for the optimized hyperparameters.

Section 5.1.1

In order to compare different models, for example neural networks having different numbers of hidden units, we need to evaluate the model evidence $p(\mathcal{D})$. This can be approximated by taking (5.175) and substituting the values of α and β obtained from the iterative optimization of these hyperparameters. A more careful evaluation is obtained by marginalizing over α and β, again by making a Gaussian approximation (MacKay, 1992c; Bishop, 1995a). In either case, it is necessary to evaluate the determinant $|\mathbf{A}|$ of the Hessian matrix. This can be problematic in practice because the determinant, unlike the trace, is sensitive to the small eigenvalues that are often difficult to determine accurately.

The Laplace approximation is based on a local quadratic expansion around a mode of the posterior distribution over weights. We have seen in Section 5.1.1 that any given mode in a two-layer network is a member of a set of $M!2^M$ equivalent modes that differ by interchange and sign-change symmetries, where M is the number of hidden units. When comparing networks having different numbers of hidden units, this can be taken into account by multiplying the evidence by a factor of $M!2^M$.

5.7.3 Bayesian neural networks for classification

So far, we have used the Laplace approximation to develop a Bayesian treatment of neural network regression models. We now discuss the modifications to this framework that arise when it is applied to classification. Here we shall consider a network having a single logistic sigmoid output corresponding to a two-class classification problem. The extension to networks with multiclass softmax outputs is straightforward. We shall build extensively on the analogous results for linear classification models discussed in Section 4.5, and so we encourage the reader to familiarize themselves with that material before studying this section.

Exercise 5.40

The log likelihood function for this model is given by

$$\ln p(\mathcal{D}|\mathbf{w}) = \sum_{n=1}^{N} \{t_n \ln y_n + (1 - t_n) \ln(1 - y_n)\} \tag{5.181}$$

where $t_n \in \{0, 1\}$ are the target values, and $y_n \equiv y(\mathbf{x}_n, \mathbf{w})$. Note that there is no hyperparameter β, because the data points are assumed to be correctly labelled. As before, the prior is taken to be an isotropic Gaussian of the form (5.162).

The first stage in applying the Laplace framework to this model is to initialize the hyperparameter α, and then to determine the parameter vector $\mathbf{w}$ by maximizing the log posterior distribution. This is equivalent to minimizing the regularized error function

$$E(\mathbf{w}) = -\ln p(\mathcal{D}|\mathbf{w}) + \frac{\alpha}{2}\mathbf{w}^{\mathrm{T}}\mathbf{w} \tag{5.182}$$

and can be achieved using error backpropagation combined with standard optimization algorithms, as discussed in Section 5.3.

Having found a solution $\mathbf{w}_{\mathrm{MAP}}$ for the weight vector, the next step is to evaluate the Hessian matrix $\mathbf{H}$ comprising the second derivatives of the negative log likelihood function. This can be done, for instance, using the exact method of Section 5.4.5, or using the outer product approximation given by (5.85). The second derivatives of the negative log posterior can again be written in the form (5.166), and the Gaussian approximation to the posterior is then given by (5.167).

To optimize the hyperparameter α, we again maximize the marginal likelihood, which is easily shown to take the form

Exercise 5.41

$$\ln p(\mathcal{D}|\alpha) \simeq -E(\mathbf{w}_{\mathrm{MAP}}) - \frac{1}{2}\ln|\mathbf{A}| + \frac{W}{2}\ln\alpha \tag{5.183}$$

where the regularized error function is defined by

$$E(\mathbf{w}_{\mathrm{MAP}}) = -\sum_{n=1}^{N} \{t_n \ln y_n + (1 - t_n) \ln(1 - y_n)\} + \frac{\alpha}{2}\mathbf{w}_{\mathrm{MAP}}^{\mathrm{T}}\mathbf{w}_{\mathrm{MAP}} \tag{5.184}$$

in which $y_n \equiv y(\mathbf{x}_n, \mathbf{w}_{\mathrm{MAP}})$. Maximizing this evidence function with respect to α again leads to the re-estimation equation given by (5.178).

The use of the evidence procedure to determine α is illustrated in Figure 5.22 for the synthetic two-dimensional data discussed in Appendix A.

Finally, we need the predictive distribution, which is defined by (5.168). Again, this integration is intractable due to the nonlinearity of the network function. The

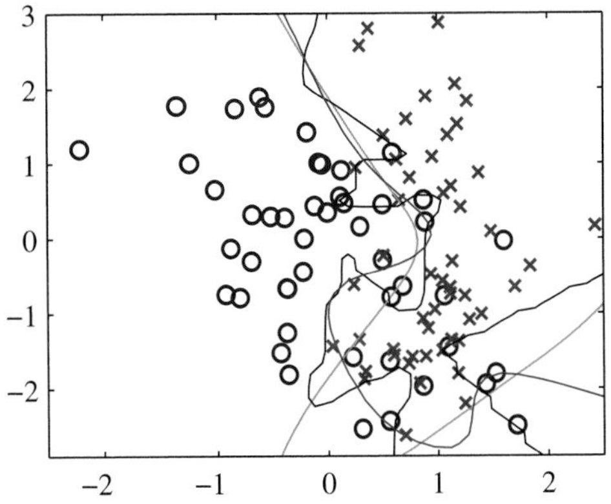

Figure 5.22 Illustration of the evidence framework applied to a synthetic two-class data set. The green curve shows the optimal decision boundary, the black curve shows the result of fitting a two-layer network with 8 hidden units by maximum likelihood, and the red curve shows the result of including a regularizer in which α is optimized using the evidence procedure, starting from the initial value $\alpha = 0$. Note that the evidence procedure greatly reduces the over-fitting of the network.

simplest approximation is to assume that the posterior distribution is very narrow and hence make the approximation

$$p(t|\mathbf{x}, \mathcal{D}) \simeq p(t|\mathbf{x}, \mathbf{w}_{\text{MAP}}). \tag{5.185}$$

We can improve on this, however, by taking account of the variance of the posterior distribution. In this case, a linear approximation for the network outputs, as was used in the case of regression, would be inappropriate due to the logistic sigmoid output-unit activation function that constrains the output to lie in the range $(0, 1)$. Instead, we make a linear approximation for the output unit activation in the form

$$a(\mathbf{x}, \mathbf{w}) \simeq a_{\text{MAP}}(\mathbf{x}) + \mathbf{b}^{\mathrm{T}}(\mathbf{w} - \mathbf{w}_{\text{MAP}}) \tag{5.186}$$

where $a_{\text{MAP}}(\mathbf{x}) = a(\mathbf{x}, \mathbf{w}_{\text{MAP}})$, and the vector $\mathbf{b} \equiv \nabla a(\mathbf{x}, \mathbf{w}_{\text{MAP}})$ can be found by backpropagation.

Because we now have a Gaussian approximation for the posterior distribution over $\mathbf{w}$, and a model for a that is a linear function of $\mathbf{w}$, we can now appeal to the results of Section 4.5.2. The distribution of output unit activation values, induced by the distribution over network weights, is given by

$$p(a|\mathbf{x}, \mathcal{D}) = \int \delta\left(a - a_{\text{MAP}}(\mathbf{x}) - \mathbf{b}^{\mathrm{T}}(\mathbf{x})(\mathbf{w} - \mathbf{w}_{\text{MAP}})\right) q(\mathbf{w}|\mathcal{D})\, \mathrm{d}\mathbf{w} \tag{5.187}$$

where $q(\mathbf{w}|\mathcal{D})$ is the Gaussian approximation to the posterior distribution given by (5.167). From Section 4.5.2, we see that this distribution is Gaussian with mean $a_{\text{MAP}} \equiv a(\mathbf{x}, \mathbf{w}_{\text{MAP}})$, and variance

$$\sigma_a^2(\mathbf{x}) = \mathbf{b}^{\mathrm{T}}(\mathbf{x})\mathbf{A}^{-1}\mathbf{b}(\mathbf{x}). \tag{5.188}$$

Finally, to obtain the predictive distribution, we must marginalize over a using

$$p(t = 1|\mathbf{x}, \mathcal{D}) = \int \sigma(a) p(a|\mathbf{x}, \mathcal{D})\, \mathrm{d}a. \tag{5.189}$$

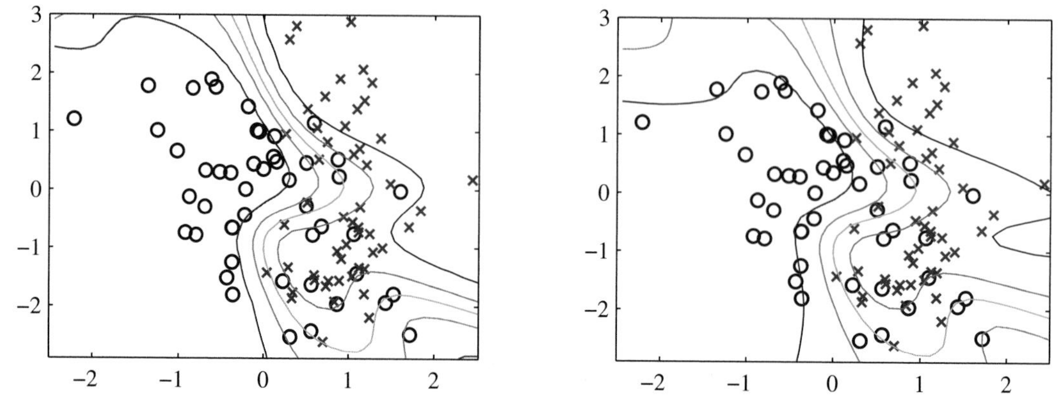

Figure 5.23 An illustration of the Laplace approximation for a Bayesian neural network having 8 hidden units with 'tanh' activation functions and a single logistic-sigmoid output unit. The weight parameters were found using scaled conjugate gradients, and the hyperparameter α was optimized using the evidence framework. On the left is the result of using the simple approximation (5.185) based on a point estimate $\mathbf{w}_{\mathrm{MAP}}$ of the parameters, in which the green curve shows the $y = 0.5$ decision boundary, and the other contours correspond to output probabilities of $y = 0.1$, 0.3, 0.7, and 0.9. On the right is the corresponding result obtained using (5.190). Note that the effect of marginalization is to spread out the contours and to make the predictions less confident, so that at each input point $\mathbf{x}$, the posterior probabilities are shifted towards 0.5, while the $y = 0.5$ contour itself is unaffected.

The convolution of a Gaussian with a logistic sigmoid is intractable. We therefore apply the approximation (4.153) to (5.189) giving

$$p(t = 1|\mathbf{x}, \mathcal{D}) = \sigma\left(\kappa(\sigma_a^2)a_{\mathrm{MAP}}\right) \tag{5.190}$$

where $\kappa(\cdot)$ is defined by (4.154). Recall that both σ_a^2 and $\mathbf{b}$ are functions of $\mathbf{x}$.

Figure 5.23 shows an example of this framework applied to the synthetic classification data set described in Appendix A.

Exercises

5.1 ($\star\star$) Consider a two-layer network function of the form (5.7) in which the hidden-unit nonlinear activation functions $h(\cdot)$ are given by logistic sigmoid functions of the form

$$\sigma(a) = \{1 + \exp(-a)\}^{-1}. \tag{5.191}$$

Show that there exists an equivalent network, which computes exactly the same function, but with hidden unit activation functions given by $\tanh(a)$ where the $\tanh$ function is defined by (5.59). Hint: first find the relation between $\sigma(a)$ and $\tanh(a)$, and then show that the parameters of the two networks differ by linear transformations.

5.2 ($\star$) **www** Show that maximizing the likelihood function under the conditional distribution (5.16) for a multioutput neural network is equivalent to minimizing the sum-of-squares error function (5.11).

5.3 ($\star\star$) Consider a regression problem involving multiple target variables in which it is assumed that the distribution of the targets, conditioned on the input vector $\mathbf{x}$, is a Gaussian of the form

$$p(\mathbf{t}|\mathbf{x}, \mathbf{w}) = \mathcal{N}(\mathbf{t}|\mathbf{y}(\mathbf{x}, \mathbf{w}), \boldsymbol{\Sigma}) \tag{5.192}$$

where $\mathbf{y}(\mathbf{x}, \mathbf{w})$ is the output of a neural network with input vector $\mathbf{x}$ and weight vector $\mathbf{w}$, and $\boldsymbol{\Sigma}$ is the covariance of the assumed Gaussian noise on the targets. Given a set of independent observations of $\mathbf{x}$ and $\mathbf{t}$, write down the error function that must be minimized in order to find the maximum likelihood solution for $\mathbf{w}$, if we assume that $\boldsymbol{\Sigma}$ is fixed and known. Now assume that $\boldsymbol{\Sigma}$ is also to be determined from the data, and write down an expression for the maximum likelihood solution for $\boldsymbol{\Sigma}$. Note that the optimizations of $\mathbf{w}$ and $\boldsymbol{\Sigma}$ are now coupled, in contrast to the case of independent target variables discussed in Section 5.2.

5.4 ($\star\star$) Consider a binary classification problem in which the target values are $t \in \{0, 1\}$, with a network output $y(\mathbf{x}, \mathbf{w})$ that represents $p(t = 1|\mathbf{x})$, and suppose that there is a probability ϵ that the class label on a training data point has been incorrectly set. Assuming independent and identically distributed data, write down the error function corresponding to the negative log likelihood. Verify that the error function (5.21) is obtained when $\epsilon = 0$. Note that this error function makes the model robust to incorrectly labelled data, in contrast to the usual error function.

5.5 ($\star$) **www** Show that maximizing likelihood for a multiclass neural network model in which the network outputs have the interpretation $y_k(\mathbf{x}, \mathbf{w}) = p(t_k = 1|\mathbf{x})$ is equivalent to the minimization of the cross-entropy error function (5.24).

5.6 ($\star$) **www** Show the derivative of the error function (5.21) with respect to the activation a_k for an output unit having a logistic sigmoid activation function satisfies (5.18).

5.7 ($\star$) Show the derivative of the error function (5.24) with respect to the activation a_k for output units having a softmax activation function satisfies (5.18).

5.8 ($\star$) We saw in (4.88) that the derivative of the logistic sigmoid activation function can be expressed in terms of the function value itself. Derive the corresponding result for the '$\tanh$' activation function defined by (5.59).

5.9 ($\star$) **www** The error function (5.21) for binary classification problems was derived for a network having a logistic-sigmoid output activation function, so that $0 \leqslant y(\mathbf{x}, \mathbf{w}) \leqslant 1$, and data having target values $t \in \{0, 1\}$. Derive the corresponding error function if we consider a network having an output $-1 \leqslant y(\mathbf{x}, \mathbf{w}) \leqslant 1$ and target values $t = 1$ for class $\mathcal{C}_1$ and $t = -1$ for class $\mathcal{C}_2$. What would be the appropriate choice of output unit activation function?

5.10 ($\star$) **www** Consider a Hessian matrix $\mathbf{H}$ with eigenvector equation (5.33). By setting the vector $\mathbf{v}$ in (5.39) equal to each of the eigenvectors $\mathbf{u}_i$ in turn, show that $\mathbf{H}$ is positive definite if, and only if, all of its eigenvalues are positive.

5.11 ($\star\star$) **www** Consider a quadratic error function defined by (5.32), in which the Hessian matrix $\mathbf{H}$ has an eigenvalue equation given by (5.33). Show that the contours of constant error are ellipses whose axes are aligned with the eigenvectors $\mathbf{u}_i$, with lengths that are inversely proportional to the square root of the corresponding eigenvalues λ_i.

5.12 ($\star\star$) **www** By considering the local Taylor expansion (5.32) of an error function about a stationary point $\mathbf{w}^\star$, show that the necessary and sufficient condition for the stationary point to be a local minimum of the error function is that the Hessian matrix $\mathbf{H}$, defined by (5.30) with $\widehat{\mathbf{w}} = \mathbf{w}^\star$, be positive definite.

5.13 ($\star$) Show that as a consequence of the symmetry of the Hessian matrix $\mathbf{H}$, the number of independent elements in the quadratic error function (5.28) is given by $W(W+3)/2$.

5.14 ($\star$) By making a Taylor expansion, verify that the terms that are $O(\epsilon)$ cancel on the right-hand side of (5.69).

5.15 ($\star\star$) In Section 5.3.4, we derived a procedure for evaluating the Jacobian matrix of a neural network using a backpropagation procedure. Derive an alternative formalism for finding the Jacobian based on *forward propagation* equations.

5.16 ($\star$) The outer product approximation to the Hessian matrix for a neural network using a sum-of-squares error function is given by (5.84). Extend this result to the case of multiple outputs.

5.17 ($\star$) Consider a squared loss function of the form

$$E = \frac{1}{2}\iint \{y(\mathbf{x}, \mathbf{w}) - t\}^2 \, p(\mathbf{x}, t)\, \mathrm{d}\mathbf{x}\, \mathrm{d}t \tag{5.193}$$

where $y(\mathbf{x}, \mathbf{w})$ is a parametric function such as a neural network. The result (1.89) shows that the function $y(\mathbf{x}, \mathbf{w})$ that minimizes this error is given by the conditional expectation of t given $\mathbf{x}$. Use this result to show that the second derivative of E with respect to two elements w_r and w_s of the vector $\mathbf{w}$, is given by

$$\frac{\partial^2 E}{\partial w_r \partial w_s} = \int \frac{\partial y}{\partial w_r}\frac{\partial y}{\partial w_s} p(\mathbf{x})\, \mathrm{d}\mathbf{x}. \tag{5.194}$$

Note that, for a finite sample from $p(\mathbf{x})$, we obtain (5.84).

5.18 ($\star$) Consider a two-layer network of the form shown in Figure 5.1 with the addition of extra parameters corresponding to skip-layer connections that go directly from the inputs to the outputs. By extending the discussion of Section 5.3.2, write down the equations for the derivatives of the error function with respect to these additional parameters.

5.19 ($\star$) **www** Derive the expression (5.85) for the outer product approximation to the Hessian matrix for a network having a single output with a logistic sigmoid output-unit activation function and a cross-entropy error function, corresponding to the result (5.84) for the sum-of-squares error function.

5.20 ($\star$) Derive an expression for the outer product approximation to the Hessian matrix for a network having K outputs with a softmax output-unit activation function and a cross-entropy error function, corresponding to the result (5.84) for the sum-of-squares error function.

5.21 ($\star\star\star$) Extend the expression (5.86) for the outer product approximation of the Hessian matrix to the case of $K > 1$ output units. Hence, derive a form that allows (5.87) to be used to incorporate sequentially contributions from individual outputs as well as individual patterns. This, together with the identity (5.88), will allow the use of (5.89) for finding the inverse of the Hessian by sequentially incorporating contributions from individual outputs and patterns.

5.22 ($\star\star$) Derive the results (5.93), (5.94), and (5.95) for the elements of the Hessian matrix of a two-layer feed-forward network by application of the chain rule of calculus.

5.23 ($\star\star$) Extend the results of Section 5.4.5 for the exact Hessian of a two-layer network to include skip-layer connections that go directly from inputs to outputs.

5.24 ($\star$) Verify that the network function defined by (5.113) and (5.114) is invariant under the transformation (5.115) applied to the inputs, provided the weights and biases are simultaneously transformed using (5.116) and (5.117). Similarly, show that the network outputs can be transformed according (5.118) by applying the transformation (5.119) and (5.120) to the second-layer weights and biases.

5.25 ($\star\star\star$) **www** Consider a quadratic error function of the form

$$E = E_0 + \frac{1}{2}(\mathbf{w} - \mathbf{w}^\star)^{\mathrm{T}}\mathbf{H}(\mathbf{w} - \mathbf{w}^\star) \tag{5.195}$$

where $\mathbf{w}^\star$ represents the minimum, and the Hessian matrix $\mathbf{H}$ is positive definite and constant. Suppose the initial weight vector $\mathbf{w}^{(0)}$ is chosen to be at the origin and is updated using simple gradient descent

$$\mathbf{w}^{(\tau)} = \mathbf{w}^{(\tau-1)} - \rho\nabla E \tag{5.196}$$

where τ denotes the step number, and ρ is the learning rate (which is assumed to be small). Show that, after τ steps, the components of the weight vector parallel to the eigenvectors of $\mathbf{H}$ can be written

$$w_j^{(\tau)} = \{1 - (1 - \rho\eta_j)^\tau\}\, w_j^\star \tag{5.197}$$

where $w_j = \mathbf{w}^{\mathrm{T}}\mathbf{u}_j$, and $\mathbf{u}_j$ and η_j are the eigenvectors and eigenvalues, respectively, of $\mathbf{H}$ so that

$$\mathbf{H}\mathbf{u}_j = \eta_j\mathbf{u}_j. \tag{5.198}$$

Show that as $\tau \to \infty$, this gives $\mathbf{w}^{(\tau)} \to \mathbf{w}^\star$ as expected, provided $|1 - \rho\eta_j| < 1$. Now suppose that training is halted after a finite number τ of steps. Show that the components of the weight vector parallel to the eigenvectors of the Hessian satisfy

$$w_j^{(\tau)} \simeq w_j^\star \quad \text{when} \quad \eta_j \gg (\rho\tau)^{-1} \tag{5.199}$$

$$|w_j^{(\tau)}| \ll |w_j^\star| \quad \text{when} \quad \eta_j \ll (\rho\tau)^{-1}. \tag{5.200}$$

Compare this result with the discussion in Section 3.5.3 of regularization with simple weight decay, and hence show that $(\rho\tau)^{-1}$ is analogous to the regularization parameter λ. The above results also show that the effective number of parameters in the network, as defined by (3.91), grows as the training progresses.

5.26 ($\star\star$) Consider a multilayer perceptron with arbitrary feed-forward topology, which is to be trained by minimizing the *tangent propagation* error function (5.127) in which the regularizing function is given by (5.128). Show that the regularization term Ω can be written as a sum over patterns of terms of the form

$$\Omega_n = \frac{1}{2}\sum_k (\mathcal{G}y_k)^2 \tag{5.201}$$

where $\mathcal{G}$ is a differential operator defined by

$$\mathcal{G} \equiv \sum_i \tau_i \frac{\partial}{\partial x_i}. \tag{5.202}$$

By acting on the forward propagation equations

$$z_j = h(a_j), \qquad a_j = \sum_i w_{ji} z_i \tag{5.203}$$

with the operator $\mathcal{G}$, show that Ω_n can be evaluated by forward propagation using the following equations:

$$\alpha_j = h'(a_j)\beta_j, \qquad \beta_j = \sum_i w_{ji}\alpha_i. \tag{5.204}$$

where we have defined the new variables

$$\alpha_j \equiv \mathcal{G}z_j, \qquad \beta_j \equiv \mathcal{G}a_j. \tag{5.205}$$

Now show that the derivatives of Ω_n with respect to a weight w_{rs} in the network can be written in the form

$$\frac{\partial \Omega_n}{\partial w_{rs}} = \sum_k \alpha_k \left\{\phi_{kr} z_s + \delta_{kr}\alpha_s\right\} \tag{5.206}$$

where we have defined

$$\delta_{kr} \equiv \frac{\partial y_k}{\partial a_r}, \qquad \phi_{kr} \equiv \mathcal{G}\delta_{kr}. \tag{5.207}$$

Write down the backpropagation equations for δ_{kr}, and hence derive a set of backpropagation equations for the evaluation of the ϕ_{kr}.

5.27 ($\star\star$) **www** Consider the framework for training with transformed data in the special case in which the transformation consists simply of the addition of random noise $\mathbf{x} \rightarrow \mathbf{x} + \boldsymbol{\xi}$ where $\boldsymbol{\xi}$ has a Gaussian distribution with zero mean and unit covariance. By following an argument analogous to that of Section 5.5.5, show that the resulting regularizer reduces to the Tikhonov form (5.135).

5.28 ($\star$) **www** Consider a neural network, such as the convolutional network discussed in Section 5.5.6, in which multiple weights are constrained to have the same value. Discuss how the standard backpropagation algorithm must be modified in order to ensure that such constraints are satisfied when evaluating the derivatives of an error function with respect to the adjustable parameters in the network.

5.29 ($\star$) **www** Verify the result (5.141).

5.30 ($\star$) Verify the result (5.142).

5.31 ($\star$) Verify the result (5.143).

5.32 ($\star\star$) Show that the derivatives of the mixing coefficients $\{\pi_k\}$, defined by (5.146), with respect to the auxiliary parameters $\{\eta_j\}$ are given by

$$\frac{\partial \pi_k}{\partial \eta_j} = \delta_{jk}\pi_j - \pi_j \pi_k. \tag{5.208}$$

Hence, by making use of the constraint $\sum_k \gamma_k(w_i) = 1$ for all i, derive the result (5.147).

5.33 ($\star$) Write down a pair of equations that express the Cartesian coordinates (x_1, x_2) for the robot arm shown in Figure 5.18 in terms of the joint angles θ_1 and θ_2 and the lengths L_1 and L_2 of the links. Assume the origin of the coordinate system is given by the attachment point of the lower arm. These equations define the 'forward kinematics' of the robot arm.

5.34 ($\star$) **www** Derive the result (5.155) for the derivative of the error function with respect to the network output activations controlling the mixing coefficients in the mixture density network.

5.35 ($\star$) Derive the result (5.156) for the derivative of the error function with respect to the network output activations controlling the component means in the mixture density network.

5.36 ($\star$) Derive the result (5.157) for the derivative of the error function with respect to the network output activations controlling the component variances in the mixture density network.

5.37 ($\star$) Verify the results (5.158) and (5.160) for the conditional mean and variance of the mixture density network model.

5.38 ($\star$) Using the general result (2.115), derive the predictive distribution (5.172) for the Laplace approximation to the Bayesian neural network model.

5.39 ($\star$) **www** Make use of the Laplace approximation result (4.135) to show that the evidence function for the hyperparameters α and β in the Bayesian neural network model can be approximated by (5.175).

5.40 ($\star$) **www** Outline the modifications needed to the framework for Bayesian neural networks, discussed in Section 5.7.3, to handle multiclass problems using networks having softmax output-unit activation functions.

5.41 ($\star\star$) By following analogous steps to those given in Sections 5.7.1 and 5.7.2 for regression networks, derive the result (5.183) for the marginal likelihood in the case of a network having a cross-entropy error function and logistic-sigmoid output-unit activation function.

6 Kernel Methods

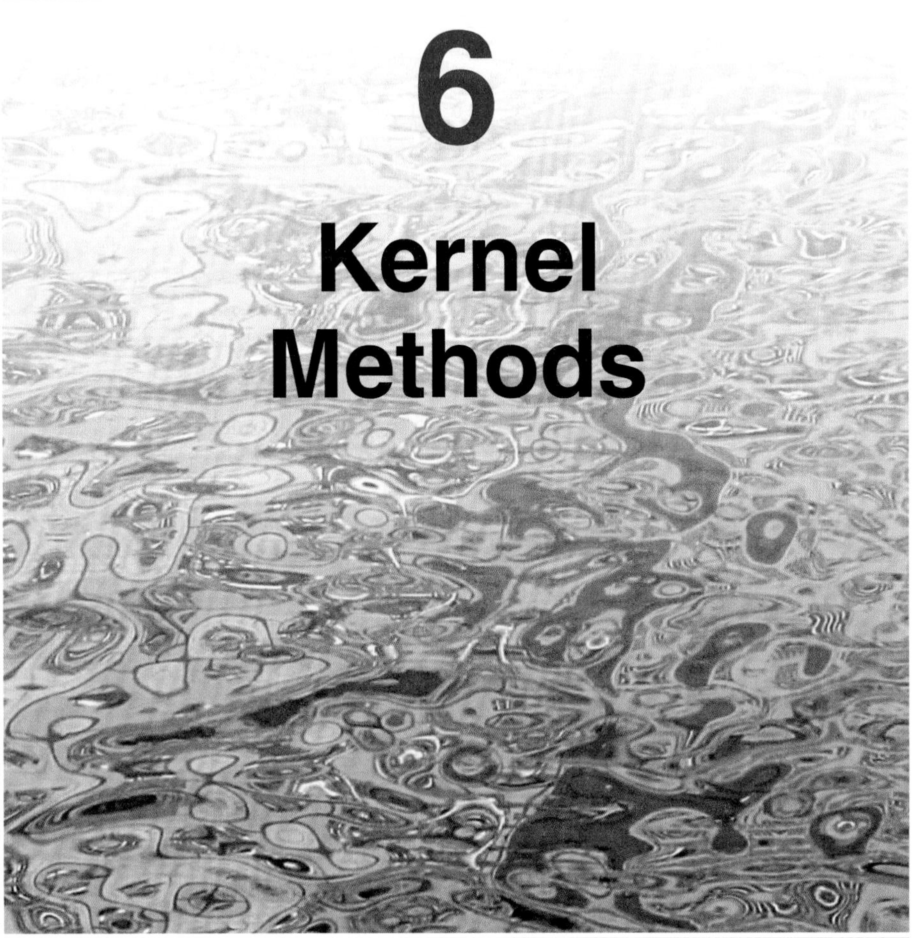

In Chapters 3 and 4, we considered linear parametric models for regression and classification in which the form of the mapping $y(\mathbf{x}, \mathbf{w})$ from input $\mathbf{x}$ to output y is governed by a vector $\mathbf{w}$ of adaptive parameters. During the learning phase, a set of training data is used either to obtain a point estimate of the parameter vector or to determine a posterior distribution over this vector. The training data is then discarded, and predictions for new inputs are based purely on the learned parameter vector $\mathbf{w}$. This approach is also used in nonlinear parametric models such as neural networks.

Chapter 5

However, there is a class of pattern recognition techniques, in which the training data points, or a subset of them, are kept and used also during the prediction phase. For instance, the Parzen probability density model comprised a linear combination of 'kernel' functions each one centred on one of the training data points. Similarly, in Section 2.5.2 we introduced a simple technique for classification called nearest neighbours, which involved assigning to each new test vector the same label as the

Section 2.5.1

closest example from the training set. These are examples of *memory-based* methods that involve storing the entire training set in order to make predictions for future data points. They typically require a metric to be defined that measures the similarity of any two vectors in input space, and are generally fast to 'train' but slow at making predictions for test data points.

Many linear parametric models can be re-cast into an equivalent 'dual representation' in which the predictions are also based on linear combinations of a *kernel function* evaluated at the training data points. As we shall see, for models which are based on a fixed nonlinear *feature space* mapping $\phi(\mathbf{x})$, the kernel function is given by the relation

$$k(\mathbf{x}, \mathbf{x}') = \phi(\mathbf{x})^{\mathrm{T}} \phi(\mathbf{x}'). \tag{6.1}$$

From this definition, we see that the kernel is a symmetric function of its arguments so that $k(\mathbf{x}, \mathbf{x}') = k(\mathbf{x}', \mathbf{x})$. The kernel concept was introduced into the field of pattern recognition by Aizerman *et al.* (1964) in the context of the method of potential functions, so-called because of an analogy with electrostatics. Although neglected for many years, it was re-introduced into machine learning in the context of large-margin classifiers by Boser *et al.* (1992) giving rise to the technique of *support vector machines*. Since then, there has been considerable interest in this topic, both in terms of theory and applications. One of the most significant developments has been the extension of kernels to handle symbolic objects, thereby greatly expanding the range of problems that can be addressed.

Chapter 7

The simplest example of a kernel function is obtained by considering the identity mapping for the feature space in (6.1) so that $\phi(\mathbf{x}) = \mathbf{x}$, in which case $k(\mathbf{x}, \mathbf{x}') = \mathbf{x}^{\mathrm{T}}\mathbf{x}'$. We shall refer to this as the linear kernel.

The concept of a kernel formulated as an inner product in a feature space allows us to build interesting extensions of many well-known algorithms by making use of the *kernel trick*, also known as *kernel substitution*. The general idea is that, if we have an algorithm formulated in such a way that the input vector $\mathbf{x}$ enters only in the form of scalar products, then we can replace that scalar product with some other choice of kernel. For instance, the technique of kernel substitution can be applied to principal component analysis in order to develop a nonlinear variant of PCA (Schölkopf *et al.*, 1998). Other examples of kernel substitution include nearest-neighbour classifiers and the kernel Fisher discriminant (Mika *et al.*, 1999; Roth and Steinhage, 2000; Baudat and Anouar, 2000).

Section 12.3

There are numerous forms of kernel functions in common use, and we shall encounter several examples in this chapter. Many have the property of being a function only of the difference between the arguments, so that $k(\mathbf{x}, \mathbf{x}') = k(\mathbf{x} - \mathbf{x}')$, which are known as *stationary* kernels because they are invariant to translations in input space. A further specialization involves *homogeneous* kernels, also known as *radial basis functions*, which depend only on the magnitude of the distance (typically Euclidean) between the arguments so that $k(\mathbf{x}, \mathbf{x}') = k(\|\mathbf{x} - \mathbf{x}'\|)$.

Section 6.3

For recent textbooks on kernel methods, see Schölkopf and Smola (2002), Herbrich (2002), and Shawe-Taylor and Cristianini (2004).

6.1. Dual Representations

Many linear models for regression and classification can be reformulated in terms of a dual representation in which the kernel function arises naturally. This concept will play an important role when we consider support vector machines in the next chapter. Here we consider a linear regression model whose parameters are determined by minimizing a regularized sum-of-squares error function given by

$$J(\mathbf{w}) = \frac{1}{2}\sum_{n=1}^{N}\left\{\mathbf{w}^{\mathrm{T}}\boldsymbol{\phi}(\mathbf{x}_n) - t_n\right\}^2 + \frac{\lambda}{2}\mathbf{w}^{\mathrm{T}}\mathbf{w} \tag{6.2}$$

where $\lambda \geqslant 0$. If we set the gradient of $J(\mathbf{w})$ with respect to $\mathbf{w}$ equal to zero, we see that the solution for $\mathbf{w}$ takes the form of a linear combination of the vectors $\boldsymbol{\phi}(\mathbf{x}_n)$, with coefficients that are functions of $\mathbf{w}$, of the form

$$\mathbf{w} = -\frac{1}{\lambda}\sum_{n=1}^{N}\left\{\mathbf{w}^{\mathrm{T}}\boldsymbol{\phi}(\mathbf{x}_n) - t_n\right\}\boldsymbol{\phi}(\mathbf{x}_n) = \sum_{n=1}^{N} a_n\boldsymbol{\phi}(\mathbf{x}_n) = \boldsymbol{\Phi}^{\mathrm{T}}\mathbf{a} \tag{6.3}$$

where $\boldsymbol{\Phi}$ is the design matrix, whose n^{th} row is given by $\boldsymbol{\phi}(\mathbf{x}_n)^{\mathrm{T}}$. Here the vector $\mathbf{a} = (a_1, \ldots, a_N)^{\mathrm{T}}$, and we have defined

$$a_n = -\frac{1}{\lambda}\left\{\mathbf{w}^{\mathrm{T}}\boldsymbol{\phi}(\mathbf{x}_n) - t_n\right\}. \tag{6.4}$$

Instead of working with the parameter vector $\mathbf{w}$, we can now reformulate the least-squares algorithm in terms of the parameter vector $\mathbf{a}$, giving rise to a *dual representation*. If we substitute $\mathbf{w} = \boldsymbol{\Phi}^{\mathrm{T}}\mathbf{a}$ into $J(\mathbf{w})$, we obtain

$$J(\mathbf{a}) = \frac{1}{2}\mathbf{a}^{\mathrm{T}}\boldsymbol{\Phi}\boldsymbol{\Phi}^{\mathrm{T}}\boldsymbol{\Phi}\boldsymbol{\Phi}^{\mathrm{T}}\mathbf{a} - \mathbf{a}^{\mathrm{T}}\boldsymbol{\Phi}\boldsymbol{\Phi}^{\mathrm{T}}\mathbf{t} + \frac{1}{2}\mathbf{t}^{\mathrm{T}}\mathbf{t} + \frac{\lambda}{2}\mathbf{a}^{\mathrm{T}}\boldsymbol{\Phi}\boldsymbol{\Phi}^{\mathrm{T}}\mathbf{a} \tag{6.5}$$

where $\mathbf{t} = (t_1, \ldots, t_N)^{\mathrm{T}}$. We now define the *Gram* matrix $\mathbf{K} = \boldsymbol{\Phi}\boldsymbol{\Phi}^{\mathrm{T}}$, which is an $N \times N$ symmetric matrix with elements

$$K_{nm} = \boldsymbol{\phi}(\mathbf{x}_n)^{\mathrm{T}}\boldsymbol{\phi}(\mathbf{x}_m) = k(\mathbf{x}_n, \mathbf{x}_m) \tag{6.6}$$

where we have introduced the *kernel function* $k(\mathbf{x}, \mathbf{x}')$ defined by (6.1). In terms of the Gram matrix, the sum-of-squares error function can be written as

$$J(\mathbf{a}) = \frac{1}{2}\mathbf{a}^{\mathrm{T}}\mathbf{K}\mathbf{K}\mathbf{a} - \mathbf{a}^{\mathrm{T}}\mathbf{K}\mathbf{t} + \frac{1}{2}\mathbf{t}^{\mathrm{T}}\mathbf{t} + \frac{\lambda}{2}\mathbf{a}^{\mathrm{T}}\mathbf{K}\mathbf{a}. \tag{6.7}$$

Using (6.3) to eliminate $\mathbf{w}$ from (6.4) and solving for $\mathbf{a}$ we obtain

$$\mathbf{a} = (\mathbf{K} + \lambda\mathbf{I}_N)^{-1}\mathbf{t}. \tag{6.8}$$

If we substitute this back into the linear regression model, we obtain the following prediction for a new input $\mathbf{x}$

$$y(\mathbf{x}) = \mathbf{w}^{\mathrm{T}}\boldsymbol{\phi}(\mathbf{x}) = \mathbf{a}^{\mathrm{T}}\boldsymbol{\Phi}\boldsymbol{\phi}(\mathbf{x}) = \mathbf{k}(\mathbf{x})^{\mathrm{T}}\left(\mathbf{K} + \lambda\mathbf{I}_N\right)^{-1}\mathbf{t} \tag{6.9}$$

where we have defined the vector $\mathbf{k}(\mathbf{x})$ with elements $k_n(\mathbf{x}) = k(\mathbf{x}_n, \mathbf{x})$. Thus we see that the dual formulation allows the solution to the least-squares problem to be expressed entirely in terms of the kernel function $k(\mathbf{x}, \mathbf{x}')$. This is known as a dual formulation because, by noting that the solution for $\mathbf{a}$ can be expressed as a linear combination of the elements of $\boldsymbol{\phi}(\mathbf{x})$, we recover the original formulation in terms of the parameter vector $\mathbf{w}$. Note that the prediction at $\mathbf{x}$ is given by a linear combination of the target values from the training set. In fact, we have already obtained this result, using a slightly different notation, in Section 3.3.3.

Exercise 6.1

In the dual formulation, we determine the parameter vector $\mathbf{a}$ by inverting an $N \times N$ matrix, whereas in the original parameter space formulation we had to invert an $M \times M$ matrix in order to determine $\mathbf{w}$. Because N is typically much larger than M, the dual formulation does not seem to be particularly useful. However, the advantage of the dual formulation, as we shall see, is that it is expressed entirely in terms of the kernel function $k(\mathbf{x}, \mathbf{x}')$. We can therefore work directly in terms of kernels and avoid the explicit introduction of the feature vector $\boldsymbol{\phi}(\mathbf{x})$, which allows us implicitly to use feature spaces of high, even infinite, dimensionality.

The existence of a dual representation based on the Gram matrix is a property of many linear models, including the perceptron. In Section 6.4, we will develop a duality between probabilistic linear models for regression and the technique of Gaussian processes. Duality will also play an important role when we discuss support vector machines in Chapter 7.

Exercise 6.2

6.2. Constructing Kernels

In order to exploit kernel substitution, we need to be able to construct valid kernel functions. One approach is to choose a feature space mapping $\boldsymbol{\phi}(\mathbf{x})$ and then use this to find the corresponding kernel, as is illustrated in Figure 6.1. Here the kernel function is defined for a one-dimensional input space by

$$k(x, x') = \boldsymbol{\phi}(x)^{\mathrm{T}}\boldsymbol{\phi}(x') = \sum_{i=1}^{M} \phi_i(x)\phi_i(x') \tag{6.10}$$

where $\phi_i(x)$ are the basis functions.

An alternative approach is to construct kernel functions directly. In this case, we must ensure that the function we choose is a valid kernel, in other words that it corresponds to a scalar product in some (perhaps infinite dimensional) feature space. As a simple example, consider a kernel function given by

$$k(\mathbf{x}, \mathbf{z}) = \left(\mathbf{x}^{\mathrm{T}}\mathbf{z}\right)^2. \tag{6.11}$$

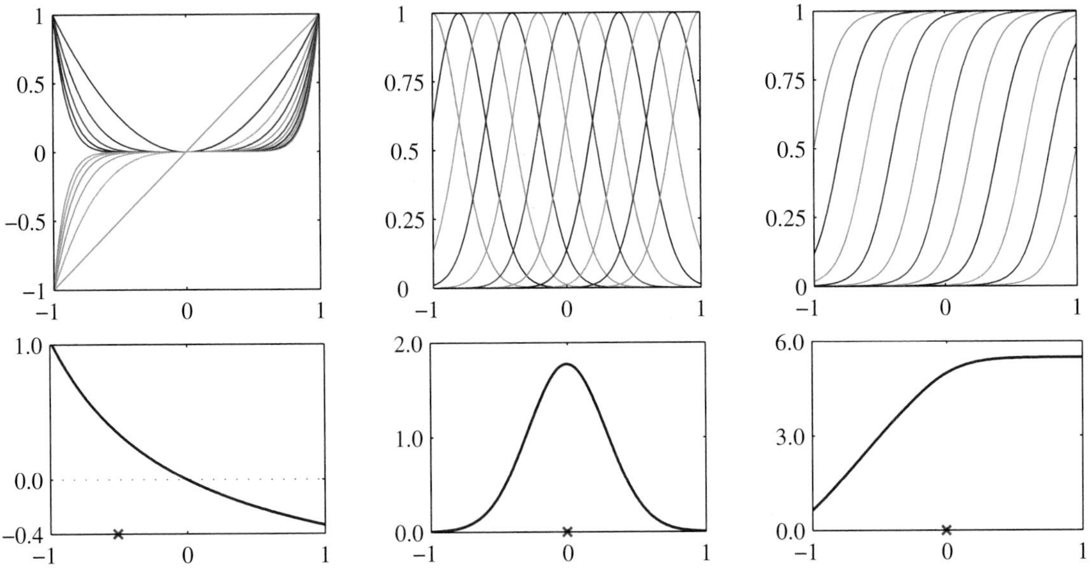

Figure 6.1 Illustration of the construction of kernel functions starting from a corresponding set of basis functions. In each column the lower plot shows the kernel function $k(x, x')$ defined by (6.10) plotted as a function of x, where x' is given by the red cross ($\times$), while the upper plot shows the corresponding basis functions given by polynomials (left column), 'Gaussians' (centre column), and logistic sigmoids (right column).

If we take the particular case of a two-dimensional input space $\mathbf{x} = (x_1, x_2)$ we can expand out the terms and thereby identify the corresponding nonlinear feature mapping

$$\begin{aligned} k(\mathbf{x}, \mathbf{z}) &= \left(\mathbf{x}^{\mathrm{T}}\mathbf{z}\right)^2 = (x_1 z_1 + x_2 z_2)^2 \\ &= x_1^2 z_1^2 + 2x_1 z_1 x_2 z_2 + x_2^2 z_2^2 \\ &= (x_1^2, \sqrt{2}x_1 x_2, x_2^2)(z_1^2, \sqrt{2}z_1 z_2, z_2^2)^{\mathrm{T}} \\ &= \boldsymbol{\phi}(\mathbf{x})^{\mathrm{T}}\boldsymbol{\phi}(\mathbf{z}). \end{aligned} \tag{6.12}$$

We see that the feature mapping takes the form $\boldsymbol{\phi}(\mathbf{x}) = (x_1^2, \sqrt{2}x_1 x_2, x_2^2)^{\mathrm{T}}$ and therefore comprises all possible second order terms, with a specific weighting between them.

More generally, however, we need a simple way to test whether a function constitutes a valid kernel without having to construct the function $\boldsymbol{\phi}(\mathbf{x})$ explicitly. A necessary and sufficient condition for a function $k(\mathbf{x}, \mathbf{x}')$ to be a valid kernel (Shawe-Taylor and Cristianini, 2004) is that the Gram matrix $\mathbf{K}$, whose elements are given by $k(\mathbf{x}_n, \mathbf{x}_m)$, should be positive semidefinite for all possible choices of the set $\{\mathbf{x}_n\}$. Note that a positive semidefinite matrix is not the same thing as a matrix whose elements are nonnegative.

Appendix C

One powerful technique for constructing new kernels is to build them out of simpler kernels as building blocks. This can be done using the following properties:

Techniques for Constructing New Kernels.

Given valid kernels $k_1(\mathbf{x}, \mathbf{x}')$ and $k_2(\mathbf{x}, \mathbf{x}')$, the following new kernels will also be valid:

$$
\begin{aligned}
k(\mathbf{x}, \mathbf{x}') &= ck_1(\mathbf{x}, \mathbf{x}') && (6.13)\\
k(\mathbf{x}, \mathbf{x}') &= f(\mathbf{x})k_1(\mathbf{x}, \mathbf{x}')f(\mathbf{x}') && (6.14)\\
k(\mathbf{x}, \mathbf{x}') &= q\left(k_1(\mathbf{x}, \mathbf{x}')\right) && (6.15)\\
k(\mathbf{x}, \mathbf{x}') &= \exp\left(k_1(\mathbf{x}, \mathbf{x}')\right) && (6.16)\\
k(\mathbf{x}, \mathbf{x}') &= k_1(\mathbf{x}, \mathbf{x}') + k_2(\mathbf{x}, \mathbf{x}') && (6.17)\\
k(\mathbf{x}, \mathbf{x}') &= k_1(\mathbf{x}, \mathbf{x}')k_2(\mathbf{x}, \mathbf{x}') && (6.18)\\
k(\mathbf{x}, \mathbf{x}') &= k_3\left(\boldsymbol{\phi}(\mathbf{x}), \boldsymbol{\phi}(\mathbf{x}')\right) && (6.19)\\
k(\mathbf{x}, \mathbf{x}') &= \mathbf{x}^{\mathrm{T}}\mathbf{A}\mathbf{x}' && (6.20)\\
k(\mathbf{x}, \mathbf{x}') &= k_a(\mathbf{x}_a, \mathbf{x}_a') + k_b(\mathbf{x}_b, \mathbf{x}_b') && (6.21)\\
k(\mathbf{x}, \mathbf{x}') &= k_a(\mathbf{x}_a, \mathbf{x}_a')k_b(\mathbf{x}_b, \mathbf{x}_b') && (6.22)
\end{aligned}
$$

where $c > 0$ is a constant, $f(\cdot)$ is any function, $q(\cdot)$ is a polynomial with nonnegative coefficients, $\boldsymbol{\phi}(\mathbf{x})$ is a function from $\mathbf{x}$ to $\mathbb{R}^M$, $k_3(\cdot, \cdot)$ is a valid kernel in $\mathbb{R}^M$, $\mathbf{A}$ is a symmetric positive semidefinite matrix, $\mathbf{x}_a$ and $\mathbf{x}_b$ are variables (not necessarily disjoint) with $\mathbf{x} = (\mathbf{x}_a, \mathbf{x}_b)$, and k_a and k_b are valid kernel functions over their respective spaces.

Equipped with these properties, we can now embark on the construction of more complex kernels appropriate to specific applications. We require that the kernel $k(\mathbf{x}, \mathbf{x}')$ be symmetric and positive semidefinite and that it expresses the appropriate form of similarity between $\mathbf{x}$ and $\mathbf{x}'$ according to the intended application. Here we consider a few common examples of kernel functions. For a more extensive discussion of 'kernel engineering', see Shawe-Taylor and Cristianini (2004).

We saw that the simple polynomial kernel $k(\mathbf{x}, \mathbf{x}') = \left(\mathbf{x}^{\mathrm{T}}\mathbf{x}'\right)^2$ contains only terms of degree two. If we consider the slightly generalized kernel $k(\mathbf{x}, \mathbf{x}') = \left(\mathbf{x}^{\mathrm{T}}\mathbf{x}' + c\right)^2$ with $c > 0$, then the corresponding feature mapping $\boldsymbol{\phi}(\mathbf{x})$ contains constant and linear terms as well as terms of order two. Similarly, $k(\mathbf{x}, \mathbf{x}') = \left(\mathbf{x}^{\mathrm{T}}\mathbf{x}'\right)^M$ contains all monomials of order M. For instance, if $\mathbf{x}$ and $\mathbf{x}'$ are two images, then the kernel represents a particular weighted sum of all possible products of M pixels in the first image with M pixels in the second image. This can similarly be generalized to include all terms up to degree M by considering $k(\mathbf{x}, \mathbf{x}') = \left(\mathbf{x}^{\mathrm{T}}\mathbf{x}' + c\right)^M$ with $c > 0$. Using the results (6.17) and (6.18) for combining kernels we see that these will all be valid kernel functions.

Another commonly used kernel takes the form

$$k(\mathbf{x}, \mathbf{x}') = \exp\left(-\|\mathbf{x} - \mathbf{x}'\|^2/2\sigma^2\right) \tag{6.23}$$

and is often called a 'Gaussian' kernel. Note, however, that in this context it is not interpreted as a probability density, and hence the normalization coefficient is omitted. We can see that this is a valid kernel by expanding the square

$$\|\mathbf{x} - \mathbf{x}'\|^2 = \mathbf{x}^{\mathrm{T}}\mathbf{x} + (\mathbf{x}')^{\mathrm{T}}\mathbf{x}' - 2\mathbf{x}^{\mathrm{T}}\mathbf{x}' \tag{6.24}$$

to give

$$k(\mathbf{x}, \mathbf{x}') = \exp\left(-\mathbf{x}^{\mathrm{T}}\mathbf{x}/2\sigma^2\right) \exp\left(\mathbf{x}^{\mathrm{T}}\mathbf{x}'/\sigma^2\right) \exp\left(-(\mathbf{x}')^{\mathrm{T}}\mathbf{x}'/2\sigma^2\right) \tag{6.25}$$

and then making use of (6.14) and (6.16), together with the validity of the linear kernel $k(\mathbf{x}, \mathbf{x}') = \mathbf{x}^{\mathrm{T}}\mathbf{x}'$. Note that the feature vector that corresponds to the Gaussian kernel has infinite dimensionality.

Exercise 6.11

The Gaussian kernel is not restricted to the use of Euclidean distance. If we use kernel substitution in (6.24) to replace $\mathbf{x}^{\mathrm{T}}\mathbf{x}'$ with a nonlinear kernel $\kappa(\mathbf{x}, \mathbf{x}')$, we obtain

$$k(\mathbf{x}, \mathbf{x}') = \exp\left\{-\frac{1}{2\sigma^2}\left(\kappa(\mathbf{x}, \mathbf{x}) + \kappa(\mathbf{x}', \mathbf{x}') - 2\kappa(\mathbf{x}, \mathbf{x}')\right)\right\}. \tag{6.26}$$

An important contribution to arise from the kernel viewpoint has been the extension to inputs that are symbolic, rather than simply vectors of real numbers. Kernel functions can be defined over objects as diverse as graphs, sets, strings, and text documents. Consider, for instance, a fixed set and define a nonvectorial space consisting of all possible subsets of this set. If A_1 and A_2 are two such subsets then one simple choice of kernel would be

$$k(A_1, A_2) = 2^{|A_1 \cap A_2|} \tag{6.27}$$

where $A_1 \cap A_2$ denotes the intersection of sets A_1 and A_2, and $|A|$ denotes the number of elements in A. This is a valid kernel function because it can be shown to correspond to an inner product in a feature space.

Exercise 6.12

One powerful approach to the construction of kernels starts from a probabilistic generative model (Haussler, 1999), which allows us to apply generative models in a discriminative setting. Generative models can deal naturally with missing data and in the case of hidden Markov models can handle sequences of varying length. By contrast, discriminative models generally give better performance on discriminative tasks than generative models. It is therefore of some interest to combine these two approaches (Lasserre *et al.*, 2006). One way to combine them is to use a generative model to define a kernel, and then use this kernel in a discriminative approach.

Given a generative model $p(\mathbf{x})$ we can define a kernel by

$$k(\mathbf{x}, \mathbf{x}') = p(\mathbf{x})p(\mathbf{x}'). \tag{6.28}$$

This is clearly a valid kernel function because we can interpret it as an inner product in the one-dimensional feature space defined by the mapping $p(\mathbf{x})$. It says that two inputs $\mathbf{x}$ and $\mathbf{x}'$ are similar if they both have high probabilities. We can use (6.13) and (6.17) to extend this class of kernels by considering sums over products of different probability distributions, with positive weighting coefficients $p(i)$, of the form

$$k(\mathbf{x}, \mathbf{x}') = \sum_i p(\mathbf{x}|i)p(\mathbf{x}'|i)p(i). \tag{6.29}$$

This is equivalent, up to an overall multiplicative constant, to a mixture distribution in which the components factorize, with the index i playing the role of a 'latent' variable. Two inputs $\mathbf{x}$ and $\mathbf{x}'$ will give a large value for the kernel function, and hence appear similar, if they have significant probability under a range of different components. Taking the limit of an infinite sum, we can also consider kernels of the form

Section 9.2

$$k(\mathbf{x}, \mathbf{x}') = \int p(\mathbf{x}|\mathbf{z})p(\mathbf{x}'|\mathbf{z})p(\mathbf{z})\,\mathrm{d}\mathbf{z} \tag{6.30}$$

where $\mathbf{z}$ is a continuous latent variable.

Now suppose that our data consists of ordered sequences of length L so that an observation is given by $\mathbf{X} = \{\mathbf{x}_1, \ldots, \mathbf{x}_L\}$. A popular generative model for sequences is the hidden Markov model, which expresses the distribution $p(\mathbf{X})$ as a marginalization over a corresponding sequence of hidden states $\mathbf{Z} = \{\mathbf{z}_1, \ldots, \mathbf{z}_L\}$. We can use this approach to define a kernel function measuring the similarity of two sequences $\mathbf{X}$ and $\mathbf{X}'$ by extending the mixture representation (6.29) to give

Section 13.2

$$k(\mathbf{X}, \mathbf{X}') = \sum_{\mathbf{Z}} p(\mathbf{X}|\mathbf{Z})p(\mathbf{X}'|\mathbf{Z})p(\mathbf{Z}) \tag{6.31}$$

so that both observed sequences are generated by the same hidden sequence $\mathbf{Z}$. This model can easily be extended to allow sequences of differing length to be compared.

An alternative technique for using generative models to define kernel functions is known as the *Fisher kernel* (Jaakkola and Haussler, 1999). Consider a parametric generative model $p(\mathbf{x}|\boldsymbol{\theta})$ where $\boldsymbol{\theta}$ denotes the vector of parameters. The goal is to find a kernel that measures the similarity of two input vectors $\mathbf{x}$ and $\mathbf{x}'$ induced by the generative model. Jaakkola and Haussler (1999) consider the gradient with respect to $\boldsymbol{\theta}$, which defines a vector in a 'feature' space having the same dimensionality as $\boldsymbol{\theta}$. In particular, they consider the *Fisher score*

$$\mathbf{g}(\boldsymbol{\theta}, \mathbf{x}) = \nabla_{\boldsymbol{\theta}} \ln p(\mathbf{x}|\boldsymbol{\theta}) \tag{6.32}$$

from which the Fisher kernel is defined by

$$k(\mathbf{x}, \mathbf{x}') = \mathbf{g}(\boldsymbol{\theta}, \mathbf{x})^{\mathrm{T}}\mathbf{F}^{-1}\mathbf{g}(\boldsymbol{\theta}, \mathbf{x}'). \tag{6.33}$$

Here $\mathbf{F}$ is the *Fisher information matrix*, given by

$$\mathbf{F} = \mathbb{E}_{\mathbf{x}}\left[\mathbf{g}(\boldsymbol{\theta}, \mathbf{x})\mathbf{g}(\boldsymbol{\theta}, \mathbf{x})^{\mathrm{T}}\right] \tag{6.34}$$

where the expectation is with respect to $\mathbf{x}$ under the distribution $p(\mathbf{x}|\boldsymbol{\theta})$. This can be motivated from the perspective of *information geometry* (Amari, 1998), which considers the differential geometry of the space of model parameters. Here we simply note that the presence of the Fisher information matrix causes this kernel to be invariant under a nonlinear re-parameterization of the density model $\boldsymbol{\theta} \to \boldsymbol{\psi}(\boldsymbol{\theta})$.

Exercise 6.13

In practice, it is often infeasible to evaluate the Fisher information matrix. One approach is simply to replace the expectation in the definition of the Fisher information with the sample average, giving

$$\mathbf{F} \simeq \frac{1}{N}\sum_{n=1}^{N} \mathbf{g}(\boldsymbol{\theta}, \mathbf{x}_n)\mathbf{g}(\boldsymbol{\theta}, \mathbf{x}_n)^{\mathrm{T}}. \tag{6.35}$$

This is the covariance matrix of the Fisher scores, and so the Fisher kernel corresponds to a whitening of these scores. More simply, we can just omit the Fisher information matrix altogether and use the noninvariant kernel

Section 12.1.3

$$k(\mathbf{x}, \mathbf{x}') = \mathbf{g}(\boldsymbol{\theta}, \mathbf{x})^{\mathrm{T}} \mathbf{g}(\boldsymbol{\theta}, \mathbf{x}'). \tag{6.36}$$

An application of Fisher kernels to document retrieval is given by Hofmann (2000).

A final example of a kernel function is the sigmoidal kernel given by

$$k(\mathbf{x}, \mathbf{x}') = \tanh\left(a\mathbf{x}^{\mathrm{T}}\mathbf{x}' + b\right) \tag{6.37}$$

whose Gram matrix in general is not positive semidefinite. This form of kernel has, however, been used in practice (Vapnik, 1995), possibly because it gives kernel expansions such as the support vector machine a superficial resemblance to neural network models. As we shall see, in the limit of an infinite number of basis functions, a Bayesian neural network with an appropriate prior reduces to a Gaussian process, thereby providing a deeper link between neural networks and kernel methods.

Section 6.4.7

6.3. Radial Basis Function Networks

In Chapter 3, we discussed regression models based on linear combinations of fixed basis functions, although we did not discuss in detail what form those basis functions might take. One choice that has been widely used is that of *radial basis functions*, which have the property that each basis function depends only on the radial distance (typically Euclidean) from a centre $\boldsymbol{\mu}_j$, so that $\phi_j(\mathbf{x}) = h(\|\mathbf{x} - \boldsymbol{\mu}_j\|)$.

Historically, radial basis functions were introduced for the purpose of exact function interpolation (Powell, 1987). Given a set of input vectors $\{\mathbf{x}_1, \ldots, \mathbf{x}_N\}$ along with corresponding target values $\{t_1, \ldots, t_N\}$, the goal is to find a smooth function $f(\mathbf{x})$ that fits every target value exactly, so that $f(\mathbf{x}_n) = t_n$ for $n = 1, \ldots, N$. This is achieved by expressing $f(\mathbf{x})$ as a linear combination of radial basis functions, one centred on every data point

$$f(\mathbf{x}) = \sum_{n=1}^{N} w_n h(\|\mathbf{x} - \mathbf{x}_n\|). \tag{6.38}$$

The values of the coefficients $\{w_n\}$ are found by least squares, and because there are the same number of coefficients as there are constraints, the result is a function that fits every target value exactly. In pattern recognition applications, however, the target values are generally noisy, and exact interpolation is undesirable because this corresponds to an over-fitted solution.

Expansions in radial basis functions also arise from regularization theory (Poggio and Girosi, 1990; Bishop, 1995a). For a sum-of-squares error function with a regularizer defined in terms of a differential operator, the optimal solution is given by an expansion in the *Green's functions* of the operator (which are analogous to the eigenvectors of a discrete matrix), again with one basis function centred on each data point. If the differential operator is isotropic then the Green's functions depend only on the radial distance from the corresponding data point. Due to the presence of the regularizer, the solution no longer interpolates the training data exactly.

Another motivation for radial basis functions comes from a consideration of the interpolation problem when the input (rather than the target) variables are noisy (Webb, 1994; Bishop, 1995a). If the noise on the input variable $\mathbf{x}$ is described by a variable $\boldsymbol{\xi}$ having a distribution $\nu(\boldsymbol{\xi})$, then the sum-of-squares error function becomes

$$E = \frac{1}{2}\sum_{n=1}^{N}\int \{y(\mathbf{x}_n + \boldsymbol{\xi}) - t_n\}^2 \nu(\boldsymbol{\xi})\,\mathrm{d}\boldsymbol{\xi}. \tag{6.39}$$

Appendix D
Exercise 6.17

Using the calculus of variations, we can optimize with respect to the function $y(\mathbf{x})$ to give

$$y(\mathbf{x}) = \sum_{n=1}^{N} t_n h(\mathbf{x} - \mathbf{x}_n) \tag{6.40}$$

where the basis functions are given by

$$h(\mathbf{x} - \mathbf{x}_n) = \frac{\nu(\mathbf{x} - \mathbf{x}_n)}{\displaystyle\sum_{n=1}^{N} \nu(\mathbf{x} - \mathbf{x}_n)}. \tag{6.41}$$

We see that there is one basis function centred on every data point. This is known as the *Nadaraya-Watson* model and will be derived again from a different perspective in Section 6.3.1. If the noise distribution $\nu(\boldsymbol{\xi})$ is isotropic, so that it is a function only of $\|\boldsymbol{\xi}\|$, then the basis functions will be radial.

Note that the basis functions (6.41) are normalized, so that $\sum_n h(\mathbf{x} - \mathbf{x}_n) = 1$ for any value of $\mathbf{x}$. The effect of such normalization is shown in Figure 6.2. Normalization is sometimes used in practice as it avoids having regions of input space where all of the basis functions take small values, which would necessarily lead to predictions in such regions that are either small or controlled purely by the bias parameter.

Another situation in which expansions in normalized radial basis functions arise is in the application of kernel density estimation to the problem of regression, as we shall discuss in Section 6.3.1.

Because there is one basis function associated with every data point, the corresponding model can be computationally costly to evaluate when making predictions for new data points. Models have therefore been proposed (Broomhead and Lowe, 1988; Moody and Darken, 1989; Poggio and Girosi, 1990), which retain the expansion in radial basis functions but where the number M of basis functions is smaller than the number N of data points. Typically, the number of basis functions, and the locations $\boldsymbol{\mu}_i$ of their centres, are determined based on the input data $\{\mathbf{x}_n\}$ alone. The basis functions are then kept fixed and the coefficients $\{w_i\}$ are determined by least squares by solving the usual set of linear equations, as discussed in Section 3.1.1.

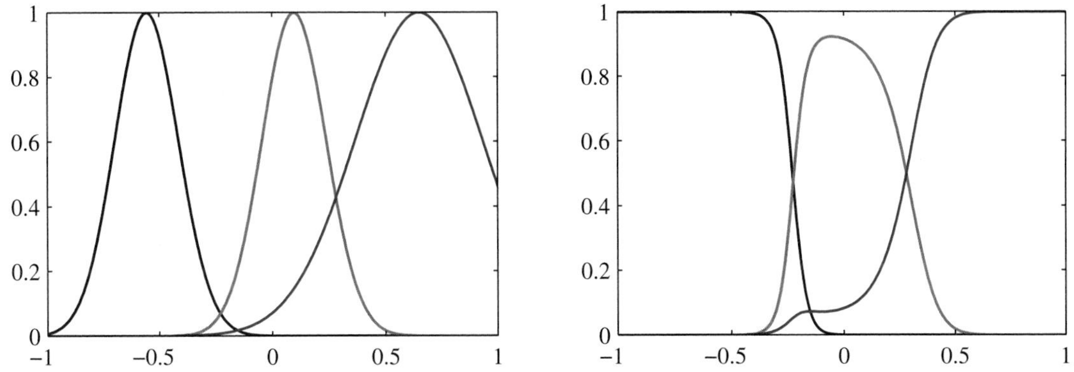

Figure 6.2 Plot of a set of Gaussian basis functions on the left, together with the corresponding normalized basis functions on the right.

One of the simplest ways of choosing basis function centres is to use a randomly chosen subset of the data points. A more systematic approach is called *orthogonal least squares* (Chen *et al.*, 1991). This is a sequential selection process in which at each step the next data point to be chosen as a basis function centre corresponds to the one that gives the greatest reduction in the sum-of-squares error. Values for the expansion coefficients are determined as part of the algorithm. Clustering algorithms such as K-means have also been used, which give a set of basis function centres that no longer coincide with training data points.

Section 9.1

6.3.1 Nadaraya-Watson model

In Section 3.3.3, we saw that the prediction of a linear regression model for a new input $\mathbf{x}$ takes the form of a linear combination of the training set target values with coefficients given by the 'equivalent kernel' (3.62) where the equivalent kernel satisfies the summation constraint (3.64).

We can motivate the kernel regression model (3.61) from a different perspective, starting with kernel density estimation. Suppose we have a training set $\{\mathbf{x}_n, t_n\}$ and we use a Parzen density estimator to model the joint distribution $p(\mathbf{x}, t)$, so that

Section 2.5.1

$$p(\mathbf{x}, t) = \frac{1}{N} \sum_{n=1}^{N} f(\mathbf{x} - \mathbf{x}_n, t - t_n) \tag{6.42}$$

where $f(\mathbf{x}, t)$ is the component density function, and there is one such component centred on each data point. We now find an expression for the regression function $y(\mathbf{x})$, corresponding to the conditional average of the target variable conditioned on the input variable, which is given by

$$\begin{aligned}
y(\mathbf{x}) &= \mathbb{E}[t|\mathbf{x}] = \int_{-\infty}^{\infty} tp(t|\mathbf{x})\,\mathrm{d}t \\
&= \frac{\int tp(\mathbf{x},t)\,\mathrm{d}t}{\int p(\mathbf{x},t)\,\mathrm{d}t} \\
&= \frac{\sum_n \int tf(\mathbf{x}-\mathbf{x}_n, t-t_n)\,\mathrm{d}t}{\sum_m \int f(\mathbf{x}-\mathbf{x}_m, t-t_m)\,\mathrm{d}t}. \qquad (6.43)
\end{aligned}$$

We now assume for simplicity that the component density functions have zero mean so that

$$\int_{-\infty}^{\infty} f(\mathbf{x},t)t\,\mathrm{d}t = 0 \qquad (6.44)$$

for all values of $\mathbf{x}$. Using a simple change of variable, we then obtain

$$\begin{aligned}
y(\mathbf{x}) &= \frac{\sum_n g(\mathbf{x}-\mathbf{x}_n)t_n}{\sum_m g(\mathbf{x}-\mathbf{x}_m)} \\
&= \sum_n k(\mathbf{x},\mathbf{x}_n)t_n \qquad (6.45)
\end{aligned}$$

where $n, m = 1, \ldots, N$ and the kernel function $k(\mathbf{x}, \mathbf{x}_n)$ is given by

$$k(\mathbf{x},\mathbf{x}_n) = \frac{g(\mathbf{x}-\mathbf{x}_n)}{\sum_m g(\mathbf{x}-\mathbf{x}_m)} \qquad (6.46)$$

and we have defined

$$g(\mathbf{x}) = \int_{-\infty}^{\infty} f(\mathbf{x},t)\,\mathrm{d}t. \qquad (6.47)$$

The result (6.45) is known as the *Nadaraya-Watson* model, or *kernel regression* (Nadaraya, 1964; Watson, 1964). For a localized kernel function, it has the property of giving more weight to the data points $\mathbf{x}_n$ that are close to $\mathbf{x}$. Note that the kernel (6.46) satisfies the summation constraint

$$\sum_{n=1}^{N} k(\mathbf{x},\mathbf{x}_n) = 1.$$

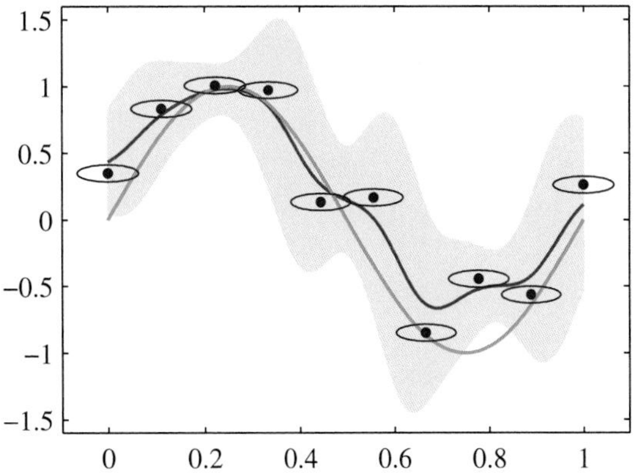

Figure 6.3 Illustration of the Nadaraya-Watson kernel regression model using isotropic Gaussian kernels, for the sinusoidal data set. The original sine function is shown by the green curve, the data points are shown in blue, and each is the centre of an isotropic Gaussian kernel. The resulting regression function, given by the conditional mean, is shown by the red line, along with the two-standard-deviation region for the conditional distribution $p(t|x)$ shown by the red shading. The blue ellipse around each data point shows one standard deviation contour for the corresponding kernel. These appear noncircular due to the different scales on the horizontal and vertical axes.

In fact, this model defines not only a conditional expectation but also a full conditional distribution given by

$$p(t|\mathbf{x}) = \frac{p(t, \mathbf{x})}{\int p(t, \mathbf{x})\,\mathrm{d}t} = \frac{\sum_n f(\mathbf{x} - \mathbf{x}_n, t - t_n)}{\sum_m \int f(\mathbf{x} - \mathbf{x}_m, t - t_m)\,\mathrm{d}t} \tag{6.48}$$

from which other expectations can be evaluated.

As an illustration we consider the case of a single input variable x in which $f(x, t)$ is given by a zero-mean isotropic Gaussian over the variable $\mathbf{z} = (x, t)$ with variance σ^2. The corresponding conditional distribution (6.48) is given by a Gaussian mixture, and is shown, together with the conditional mean, for the sinusoidal synthetic data set in Figure 6.3.

Exercise 6.18

An obvious extension of this model is to allow for more flexible forms of Gaussian components, for instance having different variance parameters for the input and target variables. More generally, we could model the joint distribution $p(t, \mathbf{x})$ using a Gaussian mixture model, trained using techniques discussed in Chapter 9 (Ghahramani and Jordan, 1994), and then find the corresponding conditional distribution $p(t|\mathbf{x})$. In this latter case we no longer have a representation in terms of kernel functions evaluated at the training set data points. However, the number of components in the mixture model can be smaller than the number of training set points, resulting in a model that is faster to evaluate for test data points. We have thereby accepted an increased computational cost during the training phase in order to have a model that is faster at making predictions.

6.4. Gaussian Processes

In Section 6.1, we introduced kernels by applying the concept of duality to a non-probabilistic model for regression. Here we extend the role of kernels to probabilistic discriminative models, leading to the framework of Gaussian processes. We shall thereby see how kernels arise naturally in a Bayesian setting.

In Chapter 3, we considered linear regression models of the form $y(\mathbf{x}, \mathbf{w}) = \mathbf{w}^{\mathrm{T}}\boldsymbol{\phi}(\mathbf{x})$ in which $\mathbf{w}$ is a vector of parameters and $\boldsymbol{\phi}(\mathbf{x})$ is a vector of fixed nonlinear basis functions that depend on the input vector $\mathbf{x}$. We showed that a prior distribution over $\mathbf{w}$ induced a corresponding prior distribution over functions $y(\mathbf{x}, \mathbf{w})$. Given a training data set, we then evaluated the posterior distribution over $\mathbf{w}$ and thereby obtained the corresponding posterior distribution over regression functions, which in turn (with the addition of noise) implies a predictive distribution $p(t|\mathbf{x})$ for new input vectors $\mathbf{x}$.

In the Gaussian process viewpoint, we dispense with the parametric model and instead define a prior probability distribution over functions directly. At first sight, it might seem difficult to work with a distribution over the uncountably infinite space of functions. However, as we shall see, for a finite training set we only need to consider the values of the function at the discrete set of input values $\mathbf{x}_n$ corresponding to the training set and test set data points, and so in practice we can work in a finite space.

Models equivalent to Gaussian processes have been widely studied in many different fields. For instance, in the geostatistics literature Gaussian process regression is known as *kriging* (Cressie, 1993). Similarly, ARMA (autoregressive moving average) models, Kalman filters, and radial basis function networks can all be viewed as forms of Gaussian process models. Reviews of Gaussian processes from a machine learning perspective can be found in MacKay (1998), Williams (1999), and MacKay (2003), and a comparison of Gaussian process models with alternative approaches is given in Rasmussen (1996). See also Rasmussen and Williams (2006) for a recent textbook on Gaussian processes.

6.4.1 Linear regression revisited

In order to motivate the Gaussian process viewpoint, let us return to the linear regression example and re-derive the predictive distribution by working in terms of distributions over functions $y(\mathbf{x}, \mathbf{w})$. This will provide a specific example of a Gaussian process.

Consider a model defined in terms of a linear combination of M fixed basis functions given by the elements of the vector $\boldsymbol{\phi}(\mathbf{x})$ so that

$$y(\mathbf{x}) = \mathbf{w}^{\mathrm{T}}\boldsymbol{\phi}(\mathbf{x}) \tag{6.49}$$

where $\mathbf{x}$ is the input vector and $\mathbf{w}$ is the M-dimensional weight vector. Now consider a prior distribution over $\mathbf{w}$ given by an isotropic Gaussian of the form

$$p(\mathbf{w}) = \mathcal{N}(\mathbf{w}|\mathbf{0}, \alpha^{-1}\mathbf{I}) \tag{6.50}$$

governed by the hyperparameter α, which represents the precision (inverse variance) of the distribution. For any given value of $\mathbf{w}$, the definition (6.49) defines a particular function of $\mathbf{x}$. The probability distribution over $\mathbf{w}$ defined by (6.50) therefore induces a probability distribution over functions $y(\mathbf{x})$. In practice, we wish to evaluate this function at specific values of $\mathbf{x}$, for example at the training data points

$\mathbf{x}_1, \ldots, \mathbf{x}_N$. We are therefore interested in the joint distribution of the function values $y(\mathbf{x}_1), \ldots, y(\mathbf{x}_N)$, which we denote by the vector $\mathbf{y}$ with elements $y_n = y(\mathbf{x}_n)$ for $n = 1, \ldots, N$. From (6.49), this vector is given by

$$\mathbf{y} = \boldsymbol{\Phi}\mathbf{w} \tag{6.51}$$

where $\boldsymbol{\Phi}$ is the design matrix with elements $\Phi_{nk} = \phi_k(\mathbf{x}_n)$. We can find the probability distribution of $\mathbf{y}$ as follows. First of all we note that $\mathbf{y}$ is a linear combination of Gaussian distributed variables given by the elements of $\mathbf{w}$ and hence is itself Gaussian. We therefore need only to find its mean and covariance, which are given from (6.50) by

Exercise 2.31

$$\begin{aligned} \mathbb{E}[\mathbf{y}] &= \boldsymbol{\Phi}\mathbb{E}[\mathbf{w}] = \mathbf{0} & (6.52)\\ \operatorname{cov}[\mathbf{y}] &= \mathbb{E}\left[\mathbf{y}\mathbf{y}^{\mathrm{T}}\right] = \boldsymbol{\Phi}\mathbb{E}\left[\mathbf{w}\mathbf{w}^{\mathrm{T}}\right]\boldsymbol{\Phi}^{\mathrm{T}} = \frac{1}{\alpha}\boldsymbol{\Phi}\boldsymbol{\Phi}^{\mathrm{T}} = \mathbf{K} & (6.53) \end{aligned}$$

where $\mathbf{K}$ is the Gram matrix with elements

$$K_{nm} = k(\mathbf{x}_n, \mathbf{x}_m) = \frac{1}{\alpha}\boldsymbol{\phi}(\mathbf{x}_n)^{\mathrm{T}}\boldsymbol{\phi}(\mathbf{x}_m) \tag{6.54}$$

and $k(\mathbf{x}, \mathbf{x}')$ is the kernel function.

This model provides us with a particular example of a Gaussian process. In general, a Gaussian process is defined as a probability distribution over functions $y(\mathbf{x})$ such that the set of values of $y(\mathbf{x})$ evaluated at an arbitrary set of points $\mathbf{x}_1, \ldots, \mathbf{x}_N$ jointly have a Gaussian distribution. In cases where the input vector $\mathbf{x}$ is two dimensional, this may also be known as a *Gaussian random field.* More generally, a *stochastic process* $y(\mathbf{x})$ is specified by giving the joint probability distribution for any finite set of values $y(\mathbf{x}_1), \ldots, y(\mathbf{x}_N)$ in a consistent manner.

A key point about Gaussian stochastic processes is that the joint distribution over N variables $y_1, \ldots, y_N$ is specified completely by the second-order statistics, namely the mean and the covariance. In most applications, we will not have any prior knowledge about the mean of $y(\mathbf{x})$ and so by symmetry we take it to be zero. This is equivalent to choosing the mean of the prior over weight values $p(\mathbf{w}|\alpha)$ to be zero in the basis function viewpoint. The specification of the Gaussian process is then completed by giving the covariance of $y(\mathbf{x})$ evaluated at any two values of $\mathbf{x}$, which is given by the kernel function

$$\mathbb{E}\left[y(\mathbf{x}_n)y(\mathbf{x}_m)\right] = k(\mathbf{x}_n, \mathbf{x}_m). \tag{6.55}$$

For the specific case of a Gaussian process defined by the linear regression model (6.49) with a weight prior (6.50), the kernel function is given by (6.54).

We can also define the kernel function directly, rather than indirectly through a choice of basis function. Figure 6.4 shows samples of functions drawn from Gaussian processes for two different choices of kernel function. The first of these is a 'Gaussian' kernel of the form (6.23), and the second is the exponential kernel given by

$$k(x, x') = \exp\left(-\theta\,|x - x'|\right) \tag{6.56}$$

which corresponds to the *Ornstein-Uhlenbeck process* originally introduced by Uhlenbeck and Ornstein (1930) to describe Brownian motion.

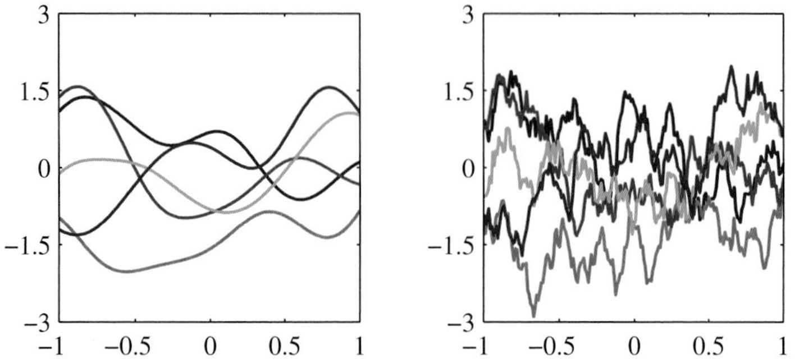

Figure 6.4 Samples from Gaussian processes for a 'Gaussian' kernel (left) and an exponential kernel (right).

6.4.2 Gaussian processes for regression

In order to apply Gaussian process models to the problem of regression, we need to take account of the noise on the observed target values, which are given by

$$t_n = y_n + \epsilon_n \tag{6.57}$$

where $y_n = y(\mathbf{x}_n)$, and ϵ_n is a random noise variable whose value is chosen independently for each observation n. Here we shall consider noise processes that have a Gaussian distribution, so that

$$p(t_n|y_n) = \mathcal{N}(t_n|y_n, \beta^{-1}) \tag{6.58}$$

where β is a hyperparameter representing the precision of the noise. Because the noise is independent for each data point, the joint distribution of the target values $\mathbf{t} = (t_1, \ldots, t_N)^{\mathrm{T}}$ conditioned on the values of $\mathbf{y} = (y_1, \ldots, y_N)^{\mathrm{T}}$ is given by an isotropic Gaussian of the form

$$p(\mathbf{t}|\mathbf{y}) = \mathcal{N}(\mathbf{t}|\mathbf{y}, \beta^{-1}\mathbf{I}_N) \tag{6.59}$$

where $\mathbf{I}_N$ denotes the $N \times N$ unit matrix. From the definition of a Gaussian process, the marginal distribution $p(\mathbf{y})$ is given by a Gaussian whose mean is zero and whose covariance is defined by a Gram matrix $\mathbf{K}$ so that

$$p(\mathbf{y}) = \mathcal{N}(\mathbf{y}|\mathbf{0}, \mathbf{K}). \tag{6.60}$$

The kernel function that determines $\mathbf{K}$ is typically chosen to express the property that, for points $\mathbf{x}_n$ and $\mathbf{x}_m$ that are similar, the corresponding values $y(\mathbf{x}_n)$ and $y(\mathbf{x}_m)$ will be more strongly correlated than for dissimilar points. Here the notion of similarity will depend on the application.

In order to find the marginal distribution $p(\mathbf{t})$, conditioned on the input values $\mathbf{x}_1, \ldots, \mathbf{x}_N$, we need to integrate over $\mathbf{y}$. This can be done by making use of the results from Section 2.3.3 for the linear-Gaussian model. Using (2.115), we see that the marginal distribution of $\mathbf{t}$ is given by

$$p(\mathbf{t}) = \int p(\mathbf{t}|\mathbf{y})p(\mathbf{y})\,\mathrm{d}\mathbf{y} = \mathcal{N}(\mathbf{t}|\mathbf{0}, \mathbf{C}) \tag{6.61}$$

where the covariance matrix $\mathbf{C}$ has elements

$$C(\mathbf{x}_n, \mathbf{x}_m) = k(\mathbf{x}_n, \mathbf{x}_m) + \beta^{-1}\delta_{nm}. \tag{6.62}$$

This result reflects the fact that the two Gaussian sources of randomness, namely that associated with $y(\mathbf{x})$ and that associated with ϵ, are independent and so their covariances simply add.

One widely used kernel function for Gaussian process regression is given by the exponential of a quadratic form, with the addition of constant and linear terms to give

$$k(\mathbf{x}_n, \mathbf{x}_m) = \theta_0 \exp\left\{-\frac{\theta_1}{2}\|\mathbf{x}_n - \mathbf{x}_m\|^2\right\} + \theta_2 + \theta_3 \mathbf{x}_n^{\mathrm{T}}\mathbf{x}_m. \tag{6.63}$$

Note that the term involving θ_3 corresponds to a parametric model that is a linear function of the input variables. Samples from this prior are plotted for various values of the parameters $\theta_0, \ldots, \theta_3$ in Figure 6.5, and Figure 6.6 shows a set of points sampled from the joint distribution (6.60) along with the corresponding values defined by (6.61).

So far, we have used the Gaussian process viewpoint to build a model of the joint distribution over sets of data points. Our goal in regression, however, is to make predictions of the target variables for new inputs, given a set of training data. Let us suppose that $\mathbf{t}_N = (t_1, \ldots, t_N)^{\mathrm{T}}$, corresponding to input values $\mathbf{x}_1, \ldots, \mathbf{x}_N$, comprise the observed training set, and our goal is to predict the target variable t_{N+1} for a new input vector $\mathbf{x}_{N+1}$. This requires that we evaluate the predictive distribution $p(t_{N+1}|\mathbf{t}_N)$. Note that this distribution is conditioned also on the variables $\mathbf{x}_1, \ldots, \mathbf{x}_N$ and $\mathbf{x}_{N+1}$. However, to keep the notation simple we will not show these conditioning variables explicitly.

To find the conditional distribution $p(t_{N+1}|\mathbf{t})$, we begin by writing down the joint distribution $p(\mathbf{t}_{N+1})$, where $\mathbf{t}_{N+1}$ denotes the vector $(t_1, \ldots, t_N, t_{N+1})^{\mathrm{T}}$. We then apply the results from Section 2.3.1 to obtain the required conditional distribution, as illustrated in Figure 6.7.

From (6.61), the joint distribution over $t_1, \ldots, t_{N+1}$ will be given by

$$p(\mathbf{t}_{N+1}) = \mathcal{N}(\mathbf{t}_{N+1}|\mathbf{0}, \mathbf{C}_{N+1}) \tag{6.64}$$

where $\mathbf{C}_{N+1}$ is an $(N+1) \times (N+1)$ covariance matrix with elements given by (6.62). Because this joint distribution is Gaussian, we can apply the results from Section 2.3.1 to find the conditional Gaussian distribution. To do this, we partition the covariance matrix as follows

$$\mathbf{C}_{N+1} = \begin{pmatrix} \mathbf{C}_N & \mathbf{k} \\ \mathbf{k}^{\mathrm{T}} & c \end{pmatrix} \tag{6.65}$$

where $\mathbf{C}_N$ is the $N \times N$ covariance matrix with elements given by (6.62) for $n, m = 1, \ldots, N$, the vector $\mathbf{k}$ has elements $k(\mathbf{x}_n, \mathbf{x}_{N+1})$ for $n = 1, \ldots, N$, and the scalar

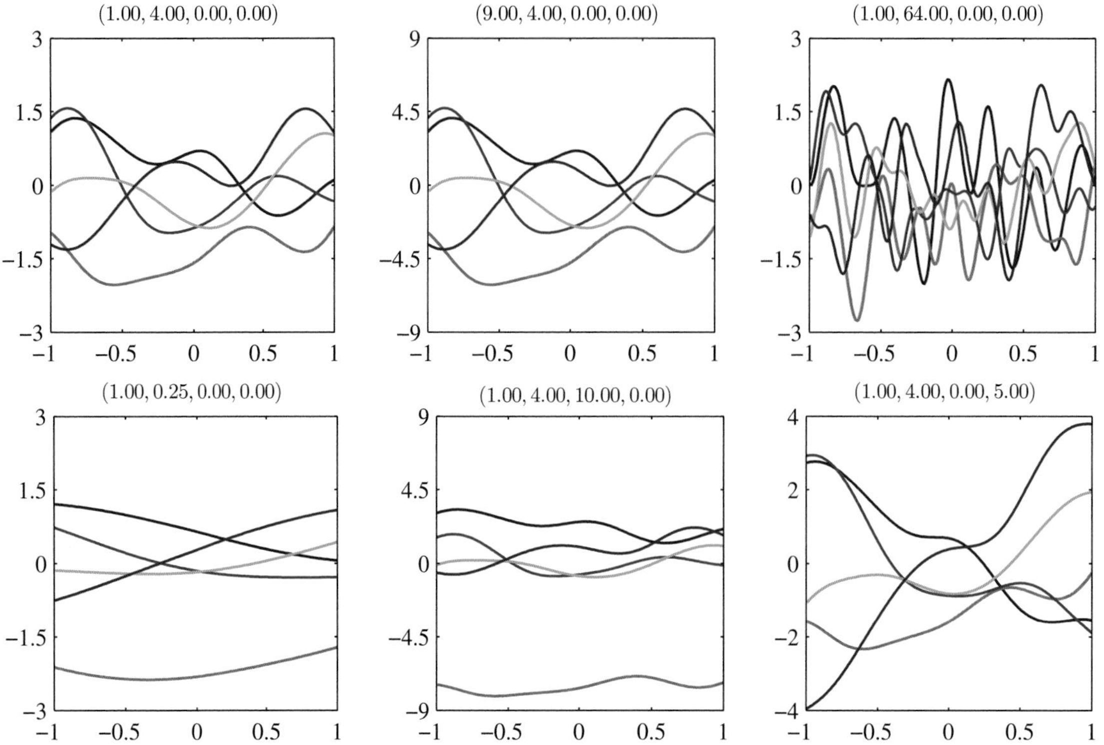

Figure 6.5 Samples from a Gaussian process prior defined by the covariance function (6.63). The title above each plot denotes $(\theta_0, \theta_1, \theta_2, \theta_3)$.

$c = k(\mathbf{x}_{N+1}, \mathbf{x}_{N+1}) + \beta^{-1}$. Using the results (2.81) and (2.82), we see that the conditional distribution $p(t_{N+1}|\mathbf{t})$ is a Gaussian distribution with mean and covariance given by

$$m(\mathbf{x}_{N+1}) = \mathbf{k}^{\mathrm{T}}\mathbf{C}_N^{-1}\mathbf{t} \tag{6.66}$$

$$\sigma^2(\mathbf{x}_{N+1}) = c - \mathbf{k}^{\mathrm{T}}\mathbf{C}_N^{-1}\mathbf{k}. \tag{6.67}$$

These are the key results that define Gaussian process regression. Because the vector $\mathbf{k}$ is a function of the test point input value $\mathbf{x}_{N+1}$, we see that the predictive distribution is a Gaussian whose mean and variance both depend on $\mathbf{x}_{N+1}$. An example of Gaussian process regression is shown in Figure 6.8.

The only restriction on the kernel function is that the covariance matrix given by (6.62) must be positive definite. If λ_i is an eigenvalue of $\mathbf{K}$, then the corresponding eigenvalue of $\mathbf{C}$ will be $\lambda_i + \beta^{-1}$. It is therefore sufficient that the kernel matrix $k(\mathbf{x}_n, \mathbf{x}_m)$ be positive semidefinite for any pair of points $\mathbf{x}_n$ and $\mathbf{x}_m$, so that $\lambda_i \geqslant 0$, because any eigenvalue λ_i that is zero will still give rise to a positive eigenvalue for $\mathbf{C}$ because $\beta > 0$. This is the same restriction on the kernel function discussed earlier, and so we can again exploit all of the techniques in Section 6.2 to construct

Figure 6.6 Illustration of the sampling of data points $\{t_n\}$ from a Gaussian process. The blue curve shows a sample function from the Gaussian process prior over functions, and the red points show the values of y_n obtained by evaluating the function at a set of input values $\{x_n\}$. The corresponding values of $\{t_n\}$, shown in green, are obtained by adding independent Gaussian noise to each of the $\{y_n\}$.

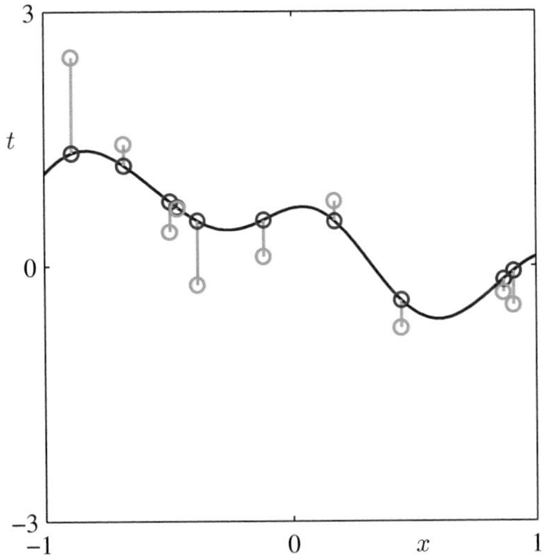

suitable kernels.

Note that the mean (6.66) of the predictive distribution can be written, as a function of $\mathbf{x}_{N+1}$, in the form

$$m(\mathbf{x}_{N+1}) = \sum_{n=1}^{N} a_n k(\mathbf{x}_n, \mathbf{x}_{N+1}) \tag{6.68}$$

where a_n is the nth component of $\mathbf{C}_N^{-1}\mathbf{t}$. Thus, if the kernel function $k(\mathbf{x}_n, \mathbf{x}_m)$ depends only on the distance $\|\mathbf{x}_n - \mathbf{x}_m\|$, then we obtain an expansion in radial basis functions.

The results (6.66) and (6.67) define the predictive distribution for Gaussian process regression with an arbitrary kernel function $k(\mathbf{x}_n, \mathbf{x}_m)$. In the particular case in which the kernel function $k(\mathbf{x}, \mathbf{x}')$ is defined in terms of a finite set of basis functions, we can derive the results obtained previously in Section 3.3.2 for linear regression starting from the Gaussian process viewpoint.

Exercise 6.21

For such models, we can therefore obtain the predictive distribution either by taking a parameter space viewpoint and using the linear regression result or by taking a function space viewpoint and using the Gaussian process result.

The central computational operation in using Gaussian processes will involve the inversion of a matrix of size $N \times N$, for which standard methods require $O(N^3)$ computations. By contrast, in the basis function model we have to invert a matrix $\mathbf{S}_N$ of size $M \times M$, which has $O(M^3)$ computational complexity. Note that for both viewpoints, the matrix inversion must be performed once for the given training set. For each new test point, both methods require a vector-matrix multiply, which has cost $O(N^2)$ in the Gaussian process case and $O(M^2)$ for the linear basis function model. If the number M of basis functions is smaller than the number N of data points, it will be computationally more efficient to work in the basis function

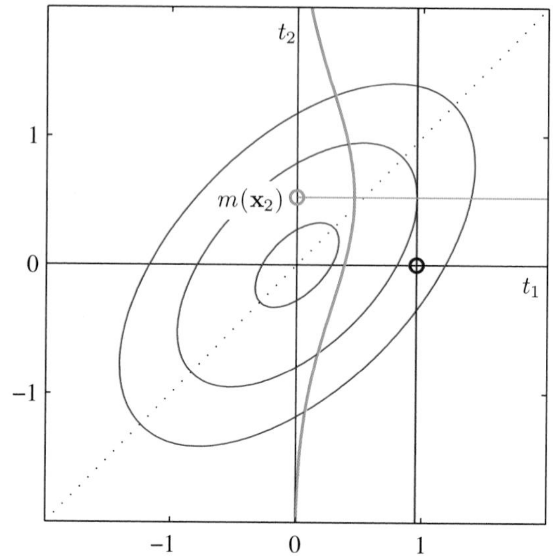

Figure 6.7 Illustration of the mechanism of Gaussian process regression for the case of one training point and one test point, in which the red ellipses show contours of the joint distribution $p(t_1, t_2)$. Here t_1 is the training data point, and conditioning on the value of t_1, corresponding to the vertical blue line, we obtain $p(t_2|t_1)$ shown as a function of t_2 by the green curve.

framework. However, an advantage of a Gaussian processes viewpoint is that we can consider covariance functions that can only be expressed in terms of an infinite number of basis functions.

For large training data sets, however, the direct application of Gaussian process methods can become infeasible, and so a range of approximation schemes have been developed that have better scaling with training set size than the exact approach (Gibbs, 1997; Tresp, 2001; Smola and Bartlett, 2001; Williams and Seeger, 2001; Csató and Opper, 2002; Seeger *et al.*, 2003).

Exercise 6.23

We have introduced Gaussian process regression for the case of a single target variable. The extension of this formalism to multiple target variables, known as co-kriging (Cressie, 1993), is straightforward. Various other extensions of Gaus-

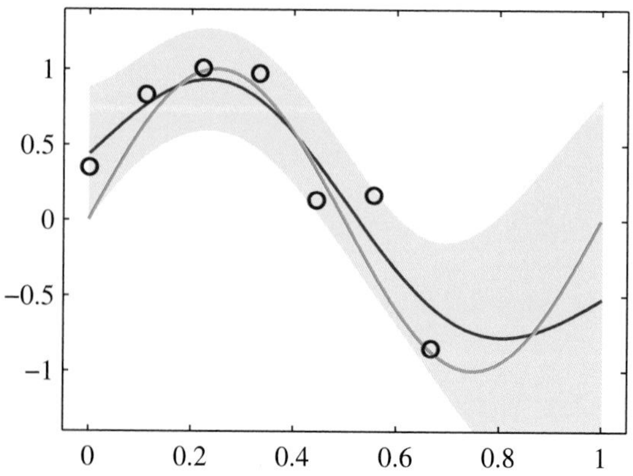

Figure 6.8 Illustration of Gaussian process regression applied to the sinusoidal data set in Figure A.6 in which the three right-most data points have been omitted. The green curve shows the sinusoidal function from which the data points, shown in blue, are obtained by sampling and addition of Gaussian noise. The red line shows the mean of the Gaussian process predictive distribution, and the shaded region corresponds to plus and minus two standard deviations. Notice how the uncertainty increases in the region to the right of the data points.

sian process regression have also been considered, for purposes such as modelling the distribution over low-dimensional manifolds for unsupervised learning (Bishop *et al.*, 1998a) and the solution of stochastic differential equations (Graepel, 2003).

6.4.3 Learning the hyperparameters

The predictions of a Gaussian process model will depend, in part, on the choice of covariance function. In practice, rather than fixing the covariance function, we may prefer to use a parametric family of functions and then infer the parameter values from the data. These parameters govern such things as the length scale of the correlations and the precision of the noise and correspond to the hyperparameters in a standard parametric model.

Techniques for learning the hyperparameters are based on the evaluation of the likelihood function $p(\mathbf{t}|\boldsymbol{\theta})$ where $\boldsymbol{\theta}$ denotes the hyperparameters of the Gaussian process model. The simplest approach is to make a point estimate of $\boldsymbol{\theta}$ by maximizing the log likelihood function. Because $\boldsymbol{\theta}$ represents a set of hyperparameters for the regression problem, this can be viewed as analogous to the type 2 maximum likelihood procedure for linear regression models. Maximization of the log likelihood can be done using efficient gradient-based optimization algorithms such as conjugate gradients (Fletcher, 1987; Nocedal and Wright, 1999; Bishop and Nabney, 2008).

Section 3.5

The log likelihood function for a Gaussian process regression model is easily evaluated using the standard form for a multivariate Gaussian distribution, giving

$$\ln p(\mathbf{t}|\boldsymbol{\theta}) = -\frac{1}{2}\ln|\mathbf{C}_N| - \frac{1}{2}\mathbf{t}^{\mathrm{T}}\mathbf{C}_N^{-1}\mathbf{t} - \frac{N}{2}\ln(2\pi). \tag{6.69}$$

For nonlinear optimization, we also need the gradient of the log likelihood function with respect to the parameter vector $\boldsymbol{\theta}$. We shall assume that evaluation of the derivatives of $\mathbf{C}_N$ is straightforward, as would be the case for the covariance functions considered in this chapter. Making use of the result (C.21) for the derivative of $\mathbf{C}_N^{-1}$, together with the result (C.22) for the derivative of $\ln|\mathbf{C}_N|$, we obtain

$$\frac{\partial}{\partial\theta_i}\ln p(\mathbf{t}|\boldsymbol{\theta}) = -\frac{1}{2}\mathrm{Tr}\left(\mathbf{C}_N^{-1}\frac{\partial\mathbf{C}_N}{\partial\theta_i}\right) + \frac{1}{2}\mathbf{t}^{\mathrm{T}}\mathbf{C}_N^{-1}\frac{\partial\mathbf{C}_N}{\partial\theta_i}\mathbf{C}_N^{-1}\mathbf{t}. \tag{6.70}$$

Because $\ln p(\mathbf{t}|\boldsymbol{\theta})$ will in general be a nonconvex function, it can have multiple maxima.

It is straightforward to introduce a prior over $\boldsymbol{\theta}$ and to maximize the log posterior using gradient-based methods. In a fully Bayesian treatment, we need to evaluate marginals over $\boldsymbol{\theta}$ weighted by the product of the prior $p(\boldsymbol{\theta})$ and the likelihood function $p(\mathbf{t}|\boldsymbol{\theta})$. In general, however, exact marginalization will be intractable, and we must resort to approximations.

The Gaussian process regression model gives a predictive distribution whose mean and variance are functions of the input vector $\mathbf{x}$. However, we have assumed that the contribution to the predictive variance arising from the additive noise, governed by the parameter β, is a constant. For some problems, known as *heteroscedastic*, the noise variance itself will also depend on $\mathbf{x}$. To model this, we can extend the

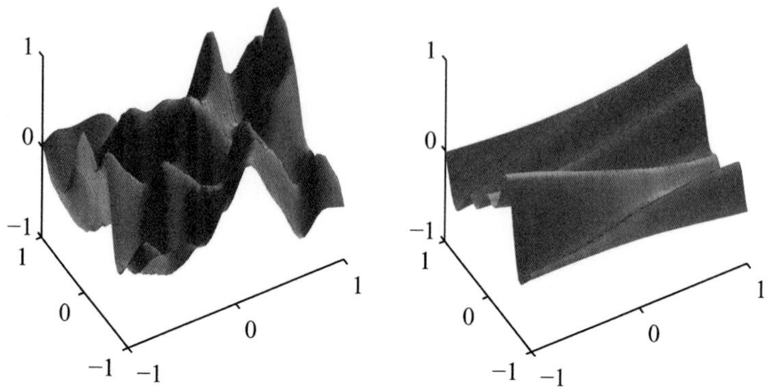

Figure 6.9 Samples from the ARD prior for Gaussian processes, in which the kernel function is given by (6.71). The left plot corresponds to $\eta_1 = \eta_2 = 1$, and the right plot corresponds to $\eta_1 = 1$, $\eta_2 = 0.01$.

Gaussian process framework by introducing a second Gaussian process to represent the dependence of β on the input $\mathbf{x}$ (Goldberg *et al.*, 1998). Because β is a variance, and hence nonnegative, we use the Gaussian process to model $\ln \beta(\mathbf{x})$.

6.4.4 Automatic relevance determination

In the previous section, we saw how maximum likelihood could be used to determine a value for the correlation length-scale parameter in a Gaussian process. This technique can usefully be extended by incorporating a separate parameter for each input variable (Rasmussen and Williams, 2006). The result, as we shall see, is that the optimization of these parameters by maximum likelihood allows the relative importance of different inputs to be inferred from the data. This represents an example in the Gaussian process context of *automatic relevance determination*, or *ARD*, which was originally formulated in the framework of neural networks (MacKay, 1994; Neal, 1996). The mechanism by which appropriate inputs are preferred is discussed in Section 7.2.2.

Consider a Gaussian process with a two-dimensional input space $\mathbf{x} = (x_1, x_2)$, having a kernel function of the form

$$k(\mathbf{x}, \mathbf{x}') = \theta_0 \exp\left\{-\frac{1}{2}\sum_{i=1}^{2}\eta_i(x_i - x_i')^2\right\}. \tag{6.71}$$

Samples from the resulting prior over functions $y(\mathbf{x})$ are shown for two different settings of the precision parameters η_i in Figure 6.9. We see that, as a particular parameter η_i becomes small, the function becomes relatively insensitive to the corresponding input variable x_i. By adapting these parameters to a data set using maximum likelihood, it becomes possible to detect input variables that have little effect on the predictive distribution, because the corresponding values of η_i will be small. This can be useful in practice because it allows such inputs to be discarded. ARD is illustrated using a simple synthetic data set having three inputs x_1, x_2 and x_3 (Nabney, 2002) in Figure 6.10. The target variable t, is generated by sampling 100 values of x_1 from a Gaussian, evaluating the function $\sin(2\pi x_1)$, and then adding

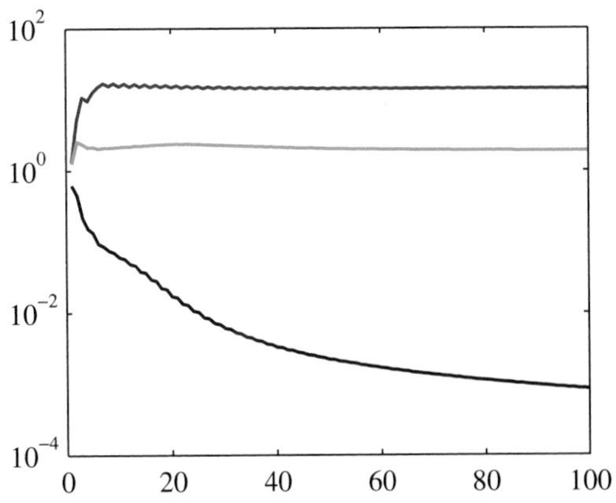

Figure 6.10 Illustration of automatic relevance determination in a Gaussian process for a synthetic problem having three inputs x_1, x_2, and x_3, for which the curves show the corresponding values of the hyperparameters η_1 (red), η_2 (green), and η_3 (blue) as a function of the number of iterations when optimizing the marginal likelihood. Details are given in the text. Note the logarithmic scale on the vertical axis.

Gaussian noise. Values of x_2 are given by copying the corresponding values of x_1 and adding noise, and values of x_3 are sampled from an independent Gaussian distribution. Thus x_1 is a good predictor of t, x_2 is a more noisy predictor of t, and x_3 has only chance correlations with t. The marginal likelihood for a Gaussian process with ARD parameters η_1, η_2, η_3 is optimized using the scaled conjugate gradients algorithm. We see from Figure 6.10 that η_1 converges to a relatively large value, η_2 converges to a much smaller value, and η_3 becomes very small indicating that x_3 is irrelevant for predicting t.

The ARD framework is easily incorporated into the exponential-quadratic kernel (6.63) to give the following form of kernel function, which has been found useful for applications of Gaussian processes to a range of regression problems

$$k(\mathbf{x}_n, \mathbf{x}_m) = \theta_0 \exp\left\{ -\frac{1}{2} \sum_{i=1}^{D} \eta_i (x_{ni} - x_{mi})^2 \right\} + \theta_2 + \theta_3 \sum_{i=1}^{D} x_{ni} x_{mi} \tag{6.72}$$

where D is the dimensionality of the input space.

6.4.5 Gaussian processes for classification

In a probabilistic approach to classification, our goal is to model the posterior probabilities of the target variable for a new input vector, given a set of training data. These probabilities must lie in the interval $(0, 1)$, whereas a Gaussian process model makes predictions that lie on the entire real axis. However, we can easily adapt Gaussian processes to classification problems by transforming the output of the Gaussian process using an appropriate nonlinear activation function.

Consider first the two-class problem with a target variable $t \in \{0, 1\}$. If we define a Gaussian process over a function $a(\mathbf{x})$ and then transform the function using a logistic sigmoid $y = \sigma(a)$, given by (4.59), then we will obtain a non-Gaussian stochastic process over functions $y(\mathbf{x})$ where $y \in (0, 1)$. This is illustrated for the case of a one-dimensional input space in Figure 6.11 in which the probability distri-

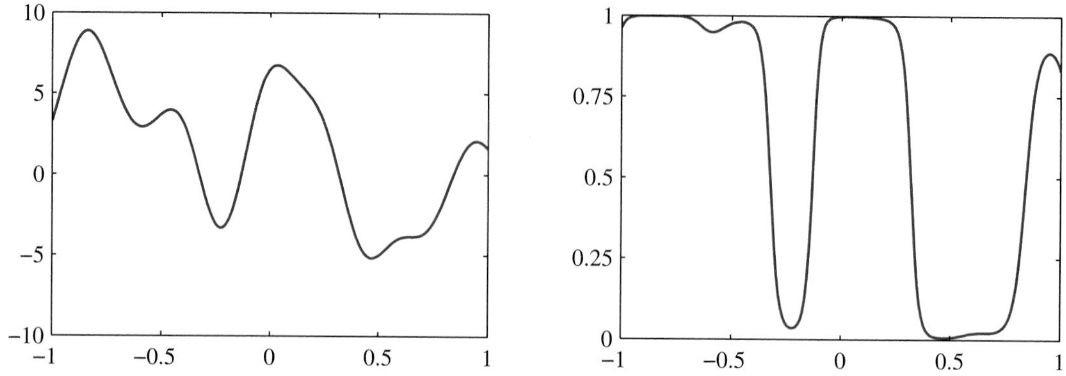

Figure 6.11 The left plot shows a sample from a Gaussian process prior over functions $a(\mathbf{x})$, and the right plot shows the result of transforming this sample using a logistic sigmoid function.

bution over the target variable t is then given by the Bernoulli distribution

$$p(t|a) = \sigma(a)^t(1 - \sigma(a))^{1-t}. \tag{6.73}$$

As usual, we denote the training set inputs by $\mathbf{x}_1, \ldots, \mathbf{x}_N$ with corresponding observed target variables $\mathbf{t}_N = (t_1, \ldots, t_N)^{\mathrm{T}}$. We also consider a single test point $\mathbf{x}_{N+1}$ with target value t_{N+1}. Our goal is to determine the predictive distribution $p(t_{N+1}|\mathbf{t})$, where we have left the conditioning on the input variables implicit. To do this we introduce a Gaussian process prior over the vector $\mathbf{a}_{N+1}$, which has components $a(\mathbf{x}_1), \ldots, a(\mathbf{x}_{N+1})$. This in turn defines a non-Gaussian process over $\mathbf{t}_{N+1}$, and by conditioning on the training data $\mathbf{t}_N$ we obtain the required predictive distribution. The Gaussian process prior for $\mathbf{a}_{N+1}$ takes the form

$$p(\mathbf{a}_{N+1}) = \mathcal{N}(\mathbf{a}_{N+1}|\mathbf{0}, \mathbf{C}_{N+1}). \tag{6.74}$$

Unlike the regression case, the covariance matrix no longer includes a noise term because we assume that all of the training data points are correctly labelled. However, for numerical reasons it is convenient to introduce a noise-like term governed by a parameter ν that ensures that the covariance matrix is positive definite. Thus the covariance matrix $\mathbf{C}_{N+1}$ has elements given by

$$C(\mathbf{x}_n, \mathbf{x}_m) = k(\mathbf{x}_n, \mathbf{x}_m) + \nu\delta_{nm} \tag{6.75}$$

where $k(\mathbf{x}_n, \mathbf{x}_m)$ is any positive semidefinite kernel function of the kind considered in Section 6.2, and the value of ν is typically fixed in advance. We shall assume that the kernel function $k(\mathbf{x}, \mathbf{x}')$ is governed by a vector $\boldsymbol{\theta}$ of parameters, and we shall later discuss how $\boldsymbol{\theta}$ may be learned from the training data.

For two-class problems, it is sufficient to predict $p(t_{N+1} = 1|\mathbf{t}_N)$ because the value of $p(t_{N+1} = 0|\mathbf{t}_N)$ is then given by $1 - p(t_{N+1} = 1|\mathbf{t}_N)$. The required predictive distribution is given by

$$p(t_{N+1} = 1|\mathbf{t}_N) = \int p(t_{N+1} = 1|a_{N+1})p(a_{N+1}|\mathbf{t}_N)\,\mathrm{d}a_{N+1} \tag{6.76}$$

where $p(t_{N+1} = 1|a_{N+1}) = \sigma(a_{N+1})$.

This integral is analytically intractable, and so may be approximated using sampling methods (Neal, 1997). Alternatively, we can consider techniques based on an analytical approximation. In Section 4.5.2, we derived the approximate formula (4.153) for the convolution of a logistic sigmoid with a Gaussian distribution. We can use this result to evaluate the integral in (6.76) provided we have a Gaussian approximation to the posterior distribution $p(a_{N+1}|\mathbf{t}_N)$. The usual justification for a Gaussian approximation to a posterior distribution is that the true posterior will tend to a Gaussian as the number of data points increases as a consequence of the central limit theorem. In the case of Gaussian processes, the number of variables grows with the number of data points, and so this argument does not apply directly. However, if we consider increasing the number of data points falling in a fixed region of $\mathbf{x}$ space, then the corresponding uncertainty in the function $a(\mathbf{x})$ will decrease, again leading asymptotically to a Gaussian (Williams and Barber, 1998).

Section 2.3

Three different approaches to obtaining a Gaussian approximation have been considered. One technique is based on *variational inference* (Gibbs and MacKay, 2000) and makes use of the local variational bound (10.144) on the logistic sigmoid. This allows the product of sigmoid functions to be approximated by a product of Gaussians thereby allowing the marginalization over $\mathbf{a}_N$ to be performed analytically. The approach also yields a lower bound on the likelihood function $p(\mathbf{t}_N|\boldsymbol{\theta})$. The variational framework for Gaussian process classification can also be extended to multiclass ($K > 2$) problems by using a Gaussian approximation to the softmax function (Gibbs, 1997).

Section 10.1

A second approach uses *expectation propagation* (Opper and Winther, 2000b; Minka, 2001b; Seeger, 2003). Because the true posterior distribution is unimodal, as we shall see shortly, the expectation propagation approach can give good results.

Section 10.7

6.4.6 Laplace approximation

The third approach to Gaussian process classification is based on the Laplace approximation, which we now consider in detail. In order to evaluate the predictive distribution (6.76), we seek a Gaussian approximation to the posterior distribution over a_{N+1}, which, using Bayes' theorem, is given by

Section 4.4

$$\begin{aligned}
p(a_{N+1}|\mathbf{t}_N) &= \int p(a_{N+1}, \mathbf{a}_N|\mathbf{t}_N)\,\mathrm{d}\mathbf{a}_N \\
&= \frac{1}{p(\mathbf{t}_N)}\int p(a_{N+1}, \mathbf{a}_N)p(\mathbf{t}_N|a_{N+1}, \mathbf{a}_N)\,\mathrm{d}\mathbf{a}_N \\
&= \frac{1}{p(\mathbf{t}_N)}\int p(a_{N+1}|\mathbf{a}_N)p(\mathbf{a}_N)p(\mathbf{t}_N|\mathbf{a}_N)\,\mathrm{d}\mathbf{a}_N \\
&= \int p(a_{N+1}|\mathbf{a}_N)p(\mathbf{a}_N|\mathbf{t}_N)\,\mathrm{d}\mathbf{a}_N
\end{aligned} \tag{6.77}$$

where we have used $p(\mathbf{t}_N|a_{N+1}, \mathbf{a}_N) = p(\mathbf{t}_N|\mathbf{a}_N)$. The conditional distribution $p(a_{N+1}|\mathbf{a}_N)$ is obtained by invoking the results (6.66) and (6.67) for Gaussian process regression, to give

$$p(a_{N+1}|\mathbf{a}_N) = \mathcal{N}(a_{N+1}|\mathbf{k}^{\mathrm{T}}\mathbf{C}_N^{-1}\mathbf{a}_N, c - \mathbf{k}^{\mathrm{T}}\mathbf{C}_N^{-1}\mathbf{k}). \tag{6.78}$$

We can therefore evaluate the integral in (6.77) by finding a Laplace approximation for the posterior distribution $p(\mathbf{a}_N|\mathbf{t}_N)$, and then using the standard result for the convolution of two Gaussian distributions.

The prior $p(\mathbf{a}_N)$ is given by a zero-mean Gaussian process with covariance matrix $\mathbf{C}_N$, and the data term (assuming independence of the data points) is given by

$$p(\mathbf{t}_N|\mathbf{a}_N) = \prod_{n=1}^{N} \sigma(a_n)^{t_n}(1 - \sigma(a_n))^{1-t_n} = \prod_{n=1}^{N} e^{a_n t_n}\sigma(-a_n). \tag{6.79}$$

We then obtain the Laplace approximation by Taylor expanding the logarithm of $p(\mathbf{a}_N|\mathbf{t}_N)$, which up to an additive normalization constant is given by the quantity

$$\begin{aligned}\Psi(\mathbf{a}_N) &= \ln p(\mathbf{a}_N) + \ln p(\mathbf{t}_N|\mathbf{a}_N) \\ &= -\frac{1}{2}\mathbf{a}_N^{\mathrm{T}}\mathbf{C}_N^{-1}\mathbf{a}_N - \frac{N}{2}\ln(2\pi) - \frac{1}{2}\ln|\mathbf{C}_N| + \mathbf{t}_N^{\mathrm{T}}\mathbf{a}_N \\ &\quad - \sum_{n=1}^{N}\ln(1 + e^{a_n}).\end{aligned} \tag{6.80}$$

First we need to find the mode of the posterior distribution, and this requires that we evaluate the gradient of $\Psi(\mathbf{a}_N)$, which is given by

$$\nabla\Psi(\mathbf{a}_N) = \mathbf{t}_N - \boldsymbol{\sigma}_N - \mathbf{C}_N^{-1}\mathbf{a}_N \tag{6.81}$$

where $\boldsymbol{\sigma}_N$ is a vector with elements $\sigma(a_n)$. We cannot simply find the mode by setting this gradient to zero, because $\boldsymbol{\sigma}_N$ depends nonlinearly on $\mathbf{a}_N$, and so we resort to an iterative scheme based on the Newton-Raphson method, which gives rise to an iterative reweighted least squares (IRLS) algorithm. This requires the second derivatives of $\Psi(\mathbf{a}_N)$, which we also require for the Laplace approximation anyway, and which are given by

Section 4.3.3

$$\nabla\nabla\Psi(\mathbf{a}_N) = -\mathbf{W}_N - \mathbf{C}_N^{-1} \tag{6.82}$$

where $\mathbf{W}_N$ is a diagonal matrix with elements $\sigma(a_n)(1 - \sigma(a_n))$, and we have used the result (4.88) for the derivative of the logistic sigmoid function. Note that these diagonal elements lie in the range $(0, 1/4)$, and hence $\mathbf{W}_N$ is a positive definite matrix. Because $\mathbf{C}_N$ (and hence its inverse) is positive definite by construction, and because the sum of two positive definite matrices is also positive definite, we see that the Hessian matrix $\mathbf{A} = -\nabla\nabla\Psi(\mathbf{a}_N)$ is positive definite and so the posterior distribution $p(\mathbf{a}_N|\mathbf{t}_N)$ is log convex and therefore has a single mode that is the global

Exercise 6.24

maximum. The posterior distribution is not Gaussian, however, because the Hessian is a function of $\mathbf{a}_N$.

Using the Newton-Raphson formula (4.92), the iterative update equation for $\mathbf{a}_N$ is given by

Exercise 6.25

$$\mathbf{a}_N^{\text{new}} = \mathbf{C}_N(\mathbf{I} + \mathbf{W}_N\mathbf{C}_N)^{-1}\left\{\mathbf{t}_N - \boldsymbol{\sigma}_N + \mathbf{W}_N\mathbf{a}_N\right\}. \tag{6.83}$$

These equations are iterated until they converge to the mode which we denote by $\mathbf{a}_N^\star$. At the mode, the gradient $\nabla\Psi(\mathbf{a}_N)$ will vanish, and hence $\mathbf{a}_N^\star$ will satisfy

$$\mathbf{a}_N^\star = \mathbf{C}_N(\mathbf{t}_N - \boldsymbol{\sigma}_N). \tag{6.84}$$

Once we have found the mode $\mathbf{a}_N^\star$ of the posterior, we can evaluate the Hessian matrix given by

$$\mathbf{H} = -\nabla\nabla\Psi(\mathbf{a}_N) = \mathbf{W}_N + \mathbf{C}_N^{-1} \tag{6.85}$$

where the elements of $\mathbf{W}_N$ are evaluated using $\mathbf{a}_N^\star$. This defines our Gaussian approximation to the posterior distribution $p(\mathbf{a}_N|\mathbf{t}_N)$ given by

$$q(\mathbf{a}_N) = \mathcal{N}(\mathbf{a}_N|\mathbf{a}_N^\star, \mathbf{H}^{-1}). \tag{6.86}$$

We can now combine this with (6.78) and hence evaluate the integral (6.77). Because this corresponds to a linear-Gaussian model, we can use the general result (2.115) to give

Exercise 6.26

$$\mathbb{E}[a_{N+1}|\mathbf{t}_N] = \mathbf{k}^{\mathrm{T}}(\mathbf{t}_N - \boldsymbol{\sigma}_N) \tag{6.87}$$

$$\operatorname{var}[a_{N+1}|\mathbf{t}_N] = c - \mathbf{k}^{\mathrm{T}}(\mathbf{W}_N^{-1} + \mathbf{C}_N)^{-1}\mathbf{k}. \tag{6.88}$$

Now that we have a Gaussian distribution for $p(a_{N+1}|\mathbf{t}_N)$, we can approximate the integral (6.76) using the result (4.153). As with the Bayesian logistic regression model of Section 4.5, if we are only interested in the decision boundary corresponding to $p(t_{N+1}|\mathbf{t}_N) = 0.5$, then we need only consider the mean and we can ignore the effect of the variance.

We also need to determine the parameters $\boldsymbol{\theta}$ of the covariance function. One approach is to maximize the likelihood function given by $p(\mathbf{t}_N|\boldsymbol{\theta})$ for which we need expressions for the log likelihood and its gradient. If desired, suitable regularization terms can also be added, leading to a penalized maximum likelihood solution. The likelihood function is defined by

$$p(\mathbf{t}_N|\boldsymbol{\theta}) = \int p(\mathbf{t}_N|\mathbf{a}_N)p(\mathbf{a}_N|\boldsymbol{\theta})\,\mathrm{d}\mathbf{a}_N. \tag{6.89}$$

This integral is analytically intractable, so again we make use of the Laplace approximation. Using the result (4.135), we obtain the following approximation for the log of the likelihood function

$$\ln p(\mathbf{t}_N|\boldsymbol{\theta}) = \Psi(\mathbf{a}_N^\star) - \frac{1}{2}\ln|\mathbf{W}_N + \mathbf{C}_N^{-1}| + \frac{N}{2}\ln(2\pi) \tag{6.90}$$

where $\Psi(\mathbf{a}_N^\star) = \ln p(\mathbf{a}_N^\star|\boldsymbol{\theta}) + \ln p(\mathbf{t}_N|\mathbf{a}_N^\star)$. We also need to evaluate the gradient of $\ln p(\mathbf{t}_N|\boldsymbol{\theta})$ with respect to the parameter vector $\boldsymbol{\theta}$. Note that changes in $\boldsymbol{\theta}$ will cause changes in $\mathbf{a}_N^\star$, leading to additional terms in the gradient. Thus, when we differentiate (6.90) with respect to $\boldsymbol{\theta}$, we obtain two sets of terms, the first arising from the dependence of the covariance matrix $\mathbf{C}_N$ on $\boldsymbol{\theta}$, and the rest arising from dependence of $\mathbf{a}_N^\star$ on $\boldsymbol{\theta}$.

The terms arising from the explicit dependence on $\boldsymbol{\theta}$ can be found by using (6.80) together with the results (C.21) and (C.22), and are given by

$$\begin{aligned}\frac{\partial \ln p(\mathbf{t}_N|\boldsymbol{\theta})}{\partial \theta_j} &= \frac{1}{2}\mathbf{a}_N^{\star\mathrm{T}}\mathbf{C}_N^{-1}\frac{\partial \mathbf{C}_N}{\partial \theta_j}\mathbf{C}_N^{-1}\mathbf{a}_N^\star \\ &\quad -\frac{1}{2}\mathrm{Tr}\left[(\mathbf{I}+\mathbf{C}_N\mathbf{W}_N)^{-1}\mathbf{W}_N\frac{\partial \mathbf{C}_N}{\partial \theta_j}\right]. \qquad (6.91)\end{aligned}$$

To compute the terms arising from the dependence of $\mathbf{a}_N^\star$ on $\boldsymbol{\theta}$, we note that the Laplace approximation has been constructed such that $\Psi(\mathbf{a}_N)$ has zero gradient at $\mathbf{a}_N = \mathbf{a}_N^\star$, and so $\Psi(\mathbf{a}_N^\star)$ gives no contribution to the gradient as a result of its dependence on $\mathbf{a}_N^\star$. This leaves the following contribution to the derivative with respect to a component θ_j of $\boldsymbol{\theta}$

$$\begin{aligned}&-\frac{1}{2}\sum_{n=1}^{N}\frac{\partial \ln|\mathbf{W}_N+\mathbf{C}_N^{-1}|}{\partial a_n^\star}\frac{\partial a_n^\star}{\partial \theta_j} \\ &\quad = -\frac{1}{2}\sum_{n=1}^{N}\left[(\mathbf{I}+\mathbf{C}_N\mathbf{W}_N)^{-1}\mathbf{C}_N\right]_{nn}\sigma_n^\star(1-\sigma_n^\star)(1-2\sigma_n^\star)\frac{\partial a_n^\star}{\partial \theta_j} \qquad (6.92)\end{aligned}$$

where $\sigma_n^\star = \sigma(a_n^\star)$, and again we have used the result (C.22) together with the definition of $\mathbf{W}_N$. We can evaluate the derivative of $a_N^\star$ with respect to θ_j by differentiating the relation (6.84) with respect to θ_j to give

$$\frac{\partial a_n^\star}{\partial \theta_j} = \frac{\partial \mathbf{C}_N}{\partial \theta_j}(\mathbf{t}_N - \boldsymbol{\sigma}_N) - \mathbf{C}_N\mathbf{W}_N\frac{\partial a_n^\star}{\partial \theta_j}. \qquad (6.93)$$

Rearranging then gives

$$\frac{\partial a_n^\star}{\partial \theta_j} = (\mathbf{I}+\mathbf{W}_N\mathbf{C}_N)^{-1}\frac{\partial \mathbf{C}_N}{\partial \theta_j}(\mathbf{t}_N - \boldsymbol{\sigma}_N). \qquad (6.94)$$

Combining (6.91), (6.92), and (6.94), we can evaluate the gradient of the log likelihood function, which can be used with standard nonlinear optimization algorithms in order to determine a value for $\boldsymbol{\theta}$.

Appendix A

We can illustrate the application of the Laplace approximation for Gaussian processes using the synthetic two-class data set shown in Figure 6.12. Extension of the Laplace approximation to Gaussian processes involving $K > 2$ classes, using the softmax activation function, is straightforward (Williams and Barber, 1998).

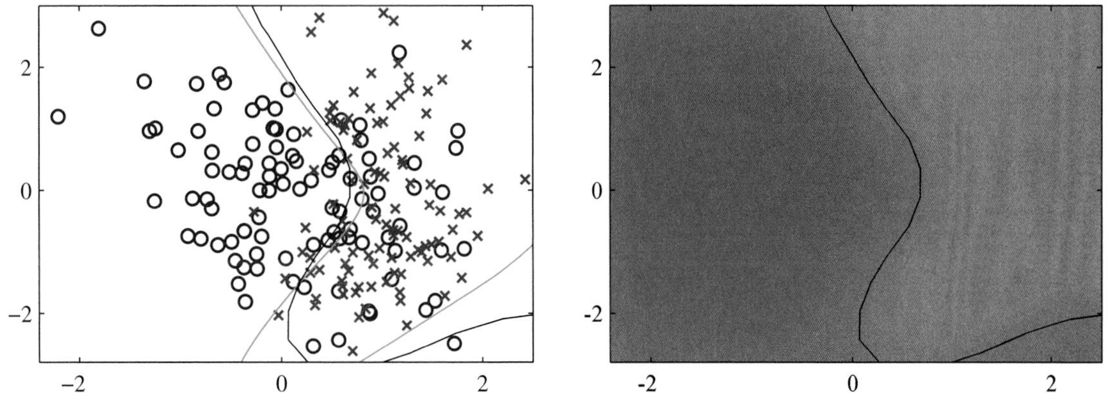

Figure 6.12 Illustration of the use of a Gaussian process for classification, showing the data on the left together with the optimal decision boundary from the true distribution in green, and the decision boundary from the Gaussian process classifier in black. On the right is the predicted posterior probability for the blue and red classes together with the Gaussian process decision boundary.

6.4.7 Connection to neural networks

We have seen that the range of functions which can be represented by a neural network is governed by the number M of hidden units, and that, for sufficiently large M, a two-layer network can approximate any given function with arbitrary accuracy. In the framework of maximum likelihood, the number of hidden units needs to be limited (to a level dependent on the size of the training set) in order to avoid over-fitting. However, from a Bayesian perspective it makes little sense to limit the number of parameters in the network according to the size of the training set.

In a Bayesian neural network, the prior distribution over the parameter vector $\mathbf{w}$, in conjunction with the network function $f(\mathbf{x}, \mathbf{w})$, produces a prior distribution over functions from $y(\mathbf{x})$ where $\mathbf{y}$ is the vector of network outputs. Neal (1996) has shown that, for a broad class of prior distributions over $\mathbf{w}$, the distribution of functions generated by a neural network will tend to a Gaussian process in the limit $M \to \infty$. It should be noted, however, that in this limit the output variables of the neural network become independent. One of the great merits of neural networks is that the outputs share the hidden units and so they can 'borrow statistical strength' from each other, that is, the weights associated with each hidden unit are influenced by all of the output variables not just by one of them. This property is therefore lost in the Gaussian process limit.

We have seen that a Gaussian process is determined by its covariance (kernel) function. Williams (1998) has given explicit forms for the covariance in the case of two specific choices for the hidden unit activation function (probit and Gaussian). These kernel functions $k(\mathbf{x}, \mathbf{x}')$ are nonstationary, i.e. they cannot be expressed as a function of the difference $\mathbf{x} - \mathbf{x}'$, as a consequence of the Gaussian weight prior being centred on zero which breaks translation invariance in weight space.

By working directly with the covariance function we have implicitly marginalized over the distribution of weights. If the weight prior is governed by hyperparameters, then their values will determine the length scales of the distribution over functions, as can be understood by studying the examples in Figure 5.11 for the case of a finite number of hidden units. Note that we cannot marginalize out the hyperparameters analytically, and must instead resort to techniques of the kind discussed in Section 6.4.

Exercises

6.1 ($\star\star$) **www** Consider the dual formulation of the least squares linear regression problem given in Section 6.1. Show that the solution for the components a_n of the vector $\mathbf{a}$ can be expressed as a linear combination of the elements of the vector $\boldsymbol{\phi}(\mathbf{x}_n)$. Denoting these coefficients by the vector $\mathbf{w}$, show that the dual of the dual formulation is given by the original representation in terms of the parameter vector $\mathbf{w}$.

6.2 ($\star\star$) In this exercise, we develop a dual formulation of the perceptron learning algorithm. Using the perceptron learning rule (4.55), show that the learned weight vector $\mathbf{w}$ can be written as a linear combination of the vectors $t_n\boldsymbol{\phi}(\mathbf{x}_n)$ where $t_n \in \{-1, +1\}$. Denote the coefficients of this linear combination by α_n and derive a formulation of the perceptron learning algorithm, and the predictive function for the perceptron, in terms of the α_n. Show that the feature vector $\boldsymbol{\phi}(\mathbf{x})$ enters only in the form of the kernel function $k(\mathbf{x}, \mathbf{x}') = \boldsymbol{\phi}(\mathbf{x})^{\mathrm{T}}\boldsymbol{\phi}(\mathbf{x}')$.

6.3 ($\star$) The nearest-neighbour classifier (Section 2.5.2) assigns a new input vector $\mathbf{x}$ to the same class as that of the nearest input vector $\mathbf{x}_n$ from the training set, where in the simplest case, the distance is defined by the Euclidean metric $\|\mathbf{x} - \mathbf{x}_n\|^2$. By expressing this rule in terms of scalar products and then making use of kernel substitution, formulate the nearest-neighbour classifier for a general nonlinear kernel.

6.4 ($\star$) In Appendix C, we give an example of a matrix that has positive elements but that has a negative eigenvalue and hence that is not positive definite. Find an example of the converse property, namely a 2×2 matrix with positive eigenvalues yet that has at least one negative element.

6.5 ($\star$) **www** Verify the results (6.13) and (6.14) for constructing valid kernels.

6.6 ($\star$) Verify the results (6.15) and (6.16) for constructing valid kernels.

6.7 ($\star$) **www** Verify the results (6.17) and (6.18) for constructing valid kernels.

6.8 ($\star$) Verify the results (6.19) and (6.20) for constructing valid kernels.

6.9 ($\star$) Verify the results (6.21) and (6.22) for constructing valid kernels.

6.10 ($\star$) Show that an excellent choice of kernel for learning a function $f(\mathbf{x})$ is given by $k(\mathbf{x}, \mathbf{x}') = f(\mathbf{x})f(\mathbf{x}')$ by showing that a linear learning machine based on this kernel will always find a solution proportional to $f(\mathbf{x})$.

6.11 ($\star$) By making use of the expansion (6.25), and then expanding the middle factor as a power series, show that the Gaussian kernel (6.23) can be expressed as the inner product of an infinite-dimensional feature vector.

6.12 ($\star\star$) **www** Consider the space of all possible subsets A of a given fixed set D. Show that the kernel function (6.27) corresponds to an inner product in a feature space of dimensionality $2^{|D|}$ defined by the mapping $\boldsymbol{\phi}(A)$ where A is a subset of D and the element $\phi_U(A)$, indexed by the subset U, is given by

$$\phi_U(A) = \begin{cases} 1, & \text{if } U \subseteq A; \\ 0, & \text{otherwise.} \end{cases} \tag{6.95}$$

Here $U \subseteq A$ denotes that U is either a subset of A or is equal to A.

6.13 ($\star$) Show that the Fisher kernel, defined by (6.33), remains invariant if we make a nonlinear transformation of the parameter vector $\boldsymbol{\theta} \rightarrow \boldsymbol{\psi}(\boldsymbol{\theta})$, where the function $\boldsymbol{\psi}(\cdot)$ is invertible and differentiable.

6.14 ($\star$) **www** Write down the form of the Fisher kernel, defined by (6.33), for the case of a distribution $p(\mathbf{x}|\boldsymbol{\mu}) = \mathcal{N}(\mathbf{x}|\boldsymbol{\mu}, \mathbf{S})$ that is Gaussian with mean $\boldsymbol{\mu}$ and fixed covariance $\mathbf{S}$.

6.15 ($\star$) By considering the determinant of a 2×2 Gram matrix, show that a positive-definite kernel function $k(x, x')$ satisfies the Cauchy-Schwartz inequality

$$k(x_1, x_2)^2 \leqslant k(x_1, x_1)k(x_2, x_2). \tag{6.96}$$

6.16 ($\star\star$) Consider a parametric model governed by the parameter vector $\mathbf{w}$ together with a data set of input values $\mathbf{x}_1, \ldots, \mathbf{x}_N$ and a nonlinear feature mapping $\boldsymbol{\phi}(\mathbf{x})$. Suppose that the dependence of the error function on $\mathbf{w}$ takes the form

$$J(\mathbf{w}) = f(\mathbf{w}^{\mathrm{T}}\boldsymbol{\phi}(\mathbf{x}_1), \ldots, \mathbf{w}^{\mathrm{T}}\boldsymbol{\phi}(\mathbf{x}_N)) + g(\mathbf{w}^{\mathrm{T}}\mathbf{w}) \tag{6.97}$$

where $g(\cdot)$ is a monotonically increasing function. By writing $\mathbf{w}$ in the form

$$\mathbf{w} = \sum_{n=1}^{N} \alpha_n \boldsymbol{\phi}(\mathbf{x}_n) + \mathbf{w}_{\perp} \tag{6.98}$$

where $\mathbf{w}_{\perp}^{\mathrm{T}}\boldsymbol{\phi}(\mathbf{x}_n) = 0$ for all n, show that the value of $\mathbf{w}$ that minimizes $J(\mathbf{w})$ takes the form of a linear combination of the basis functions $\boldsymbol{\phi}(\mathbf{x}_n)$ for $n = 1, \ldots, N$.

6.17 ($\star\star$) **www** Consider the sum-of-squares error function (6.39) for data having noisy inputs, where $\nu(\boldsymbol{\xi})$ is the distribution of the noise. Use the calculus of variations to minimize this error function with respect to the function $y(\mathbf{x})$, and hence show that the optimal solution is given by an expansion of the form (6.40) in which the basis functions are given by (6.41).

6.18 ($\star$) Consider a Nadaraya-Watson model with one input variable x and one target variable t having Gaussian components with isotropic covariances, so that the covariance matrix is given by $\sigma^2\mathbf{I}$ where $\mathbf{I}$ is the unit matrix. Write down expressions for the conditional density $p(t|x)$ and for the conditional mean $\mathbb{E}[t|x]$ and variance $\mathrm{var}[t|x]$, in terms of the kernel function $k(x, x_n)$.

6.19 ($\star\star$) Another viewpoint on kernel regression comes from a consideration of regression problems in which the input variables as well as the target variables are corrupted with additive noise. Suppose each target value t_n is generated as usual by taking a function $y(\mathbf{z}_n)$ evaluated at a point $\mathbf{z}_n$, and adding Gaussian noise. The value of $\mathbf{z}_n$ is not directly observed, however, but only a noise corrupted version $\mathbf{x}_n = \mathbf{z}_n + \boldsymbol{\xi}_n$ where the random variable $\boldsymbol{\xi}$ is governed by some distribution $g(\boldsymbol{\xi})$. Consider a set of observations $\{\mathbf{x}_n, t_n\}$, where $n = 1, \ldots, N$, together with a corresponding sum-of-squares error function defined by averaging over the distribution of input noise to give

$$E = \frac{1}{2}\sum_{n=1}^{N}\int \{y(\mathbf{x}_n - \boldsymbol{\xi}_n) - t_n\}^2 g(\boldsymbol{\xi}_n)\,\mathrm{d}\boldsymbol{\xi}_n. \tag{6.99}$$

By minimizing E with respect to the function $y(\mathbf{z})$ using the calculus of variations (Appendix D), show that optimal solution for $y(\mathbf{x})$ is given by a Nadaraya-Watson kernel regression solution of the form (6.45) with a kernel of the form (6.46).

6.20 ($\star\star$) **www** Verify the results (6.66) and (6.67).

6.21 ($\star\star$) **www** Consider a Gaussian process regression model in which the kernel function is defined in terms of a fixed set of nonlinear basis functions. Show that the predictive distribution is identical to the result (3.58) obtained in Section 3.3.2 for the Bayesian linear regression model. To do this, note that both models have Gaussian predictive distributions, and so it is only necessary to show that the conditional mean and variance are the same. For the mean, make use of the matrix identity (C.6), and for the variance, make use of the matrix identity (C.7).

6.22 ($\star\star$) Consider a regression problem with N training set input vectors $\mathbf{x}_1, \ldots, \mathbf{x}_N$ and L test set input vectors $\mathbf{x}_{N+1}, \ldots, \mathbf{x}_{N+L}$, and suppose we define a Gaussian process prior over functions $t(\mathbf{x})$. Derive an expression for the joint predictive distribution for $t(\mathbf{x}_{N+1}), \ldots, t(\mathbf{x}_{N+L})$, given the values of $t(\mathbf{x}_1), \ldots, t(\mathbf{x}_N)$. Show the marginal of this distribution for one of the test observations t_j where $N + 1 \leqslant j \leqslant N + L$ is given by the usual Gaussian process regression result (6.66) and (6.67).

6.23 ($\star\star$) **www** Consider a Gaussian process regression model in which the target variable $\mathbf{t}$ has dimensionality D. Write down the conditional distribution of $\mathbf{t}_{N+1}$ for a test input vector $\mathbf{x}_{N+1}$, given a training set of input vectors $\mathbf{x}_1, \ldots, \mathbf{x}_N$ and corresponding target observations $\mathbf{t}_1, \ldots, \mathbf{t}_N$.

6.24 ($\star$) Show that a diagonal matrix $\mathbf{W}$ whose elements satisfy $0 < W_{ii} < 1$ is positive definite. Show that the sum of two positive definite matrices is itself positive definite.

6.25 ($\star$) **www** Using the Newton-Raphson formula (4.92), derive the iterative update formula (6.83) for finding the mode $\mathbf{a}_N^\star$ of the posterior distribution in the Gaussian process classification model.

6.26 ($\star$) Using the result (2.115), derive the expressions (6.87) and (6.88) for the mean and variance of the posterior distribution $p(a_{N+1}|\mathbf{t}_N)$ in the Gaussian process classification model.

6.27 ($\star\star\star$) Derive the result (6.90) for the log likelihood function in the Laplace approximation framework for Gaussian process classification. Similarly, derive the results (6.91), (6.92), and (6.94) for the terms in the gradient of the log likelihood.

7
Sparse Kernel Machines

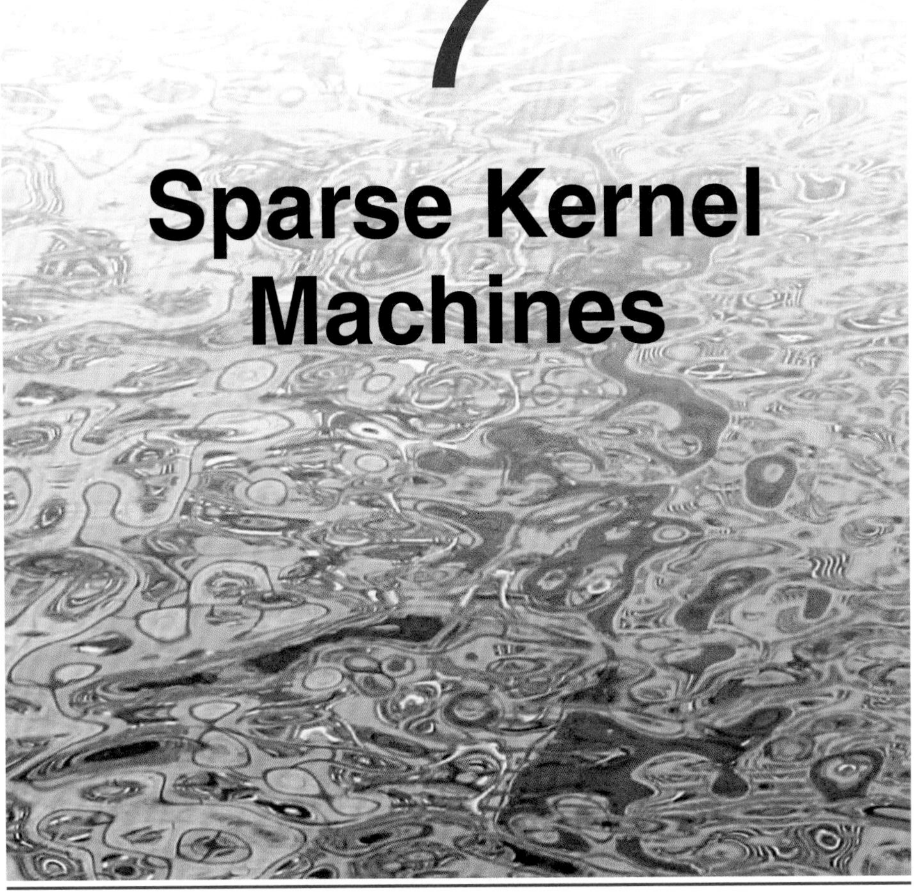

In the previous chapter, we explored a variety of learning algorithms based on nonlinear kernels. One of the significant limitations of many such algorithms is that the kernel function $k(\mathbf{x}_n, \mathbf{x}_m)$ must be evaluated for all possible pairs $\mathbf{x}_n$ and $\mathbf{x}_m$ of training points, which can be computationally infeasible during training and can lead to excessive computation times when making predictions for new data points. In this chapter we shall look at kernel-based algorithms that have *sparse* solutions, so that predictions for new inputs depend only on the kernel function evaluated at a subset of the training data points.

We begin by looking in some detail at the *support vector machine* (SVM), which became popular in some years ago for solving problems in classification, regression, and novelty detection. An important property of support vector machines is that the determination of the model parameters corresponds to a convex optimization problem, and so any local solution is also a global optimum. Because the discussion of support vector machines makes extensive use of Lagrange multipliers, the reader is encouraged to review the key concepts covered in Appendix E. Additional information on support vector machines can be found in Vapnik (1995), Burges (1998), Cristianini and Shawe-Taylor (2000), Müller *et al.* (2001), Schölkopf and Smola (2002), and Herbrich (2002).

The SVM is a decision machine and so does not provide posterior probabilities. We have already discussed some of the benefits of determining probabilities in Section 1.5.4. An alternative sparse kernel technique, known as the *relevance vector machine* (RVM), is based on a Bayesian formulation and provides posterior probabilistic outputs, as well as having typically much sparser solutions than the SVM.

Section 7.2

7.1. Maximum Margin Classifiers

We begin our discussion of support vector machines by returning to the two-class classification problem using linear models of the form

$$y(\mathbf{x}) = \mathbf{w}^{\mathrm{T}}\boldsymbol{\phi}(\mathbf{x}) + b \tag{7.1}$$

where $\boldsymbol{\phi}(\mathbf{x})$ denotes a fixed feature-space transformation, and we have made the bias parameter b explicit. Note that we shall shortly introduce a dual representation expressed in terms of kernel functions, which avoids having to work explicitly in feature space. The training data set comprises N input vectors $\mathbf{x}_1, \ldots, \mathbf{x}_N$, with corresponding target values $t_1, \ldots, t_N$ where $t_n \in \{-1, 1\}$, and new data points $\mathbf{x}$ are classified according to the sign of $y(\mathbf{x})$.

We shall assume for the moment that the training data set is linearly separable in feature space, so that by definition there exists at least one choice of the parameters $\mathbf{w}$ and b such that a function of the form (7.1) satisfies $y(\mathbf{x}_n) > 0$ for points having $t_n = +1$ and $y(\mathbf{x}_n) < 0$ for points having $t_n = -1$, so that $t_n y(\mathbf{x}_n) > 0$ for all training data points.

There may of course exist many such solutions that separate the classes exactly. In Section 4.1.7, we described the perceptron algorithm that is guaranteed to find a solution in a finite number of steps. The solution that it finds, however, will be dependent on the (arbitrary) initial values chosen for $\mathbf{w}$ and b as well as on the order in which the data points are presented. If there are multiple solutions all of which classify the training data set exactly, then we should try to find the one that will give the smallest generalization error. The support vector machine approaches this problem through the concept of the *margin*, which is defined to be the smallest distance between the decision boundary and any of the samples, as illustrated in Figure 7.1.

In support vector machines the decision boundary is chosen to be the one for which the margin is maximized. The maximum margin solution can be motivated using *computational learning theory*, also known as *statistical learning theory*. However, a simple insight into the origins of maximum margin has been given by Tong and Koller (2000) who consider a framework for classification based on a hybrid of generative and discriminative approaches. They first model the distribution over input vectors $\mathbf{x}$ for each class using a Parzen density estimator with Gaussian kernels

Section 7.1.5

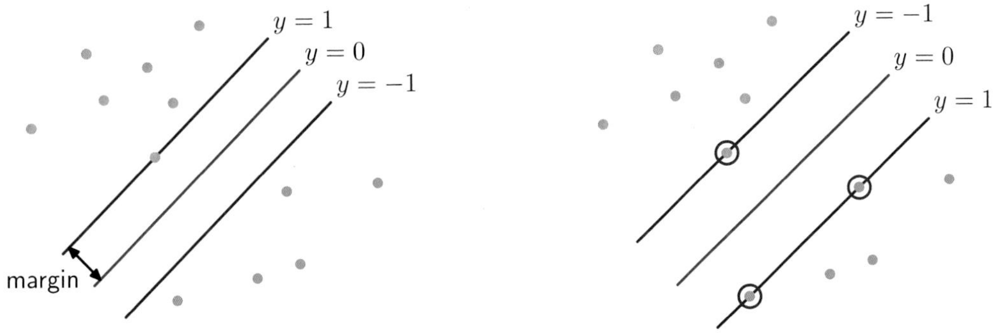

Figure 7.1 The margin is defined as the perpendicular distance between the decision boundary and the closest of the data points, as shown on the left figure. Maximizing the margin leads to a particular choice of decision boundary, as shown on the right. The location of this boundary is determined by a subset of the data points, known as support vectors, which are indicated by the circles.

having a common parameter σ^2. Together with the class priors, this defines an optimal misclassification-rate decision boundary. However, instead of using this optimal boundary, they determine the best hyperplane by minimizing the probability of error relative to the learned density model. In the limit $\sigma^2 \to 0$, the optimal hyperplane is shown to be the one having maximum margin. The intuition behind this result is that as σ^2 is reduced, the hyperplane is increasingly dominated by nearby data points relative to more distant ones. In the limit, the hyperplane becomes independent of data points that are not support vectors.

We shall see in Figure 10.13 that marginalization with respect to the prior distribution of the parameters in a Bayesian approach for a simple linearly separable data set leads to a decision boundary that lies in the middle of the region separating the data points. The large margin solution has similar behaviour.

Recall from Figure 4.1 that the perpendicular distance of a point $\mathbf{x}$ from a hyperplane defined by $y(\mathbf{x}) = 0$ where $y(\mathbf{x})$ takes the form (7.1) is given by $|y(\mathbf{x})|/\|\mathbf{w}\|$. Furthermore, we are only interested in solutions for which all data points are correctly classified, so that $t_n y(\mathbf{x}_n) > 0$ for all n. Thus the distance of a point $\mathbf{x}_n$ to the decision surface is given by

$$\frac{t_n y(\mathbf{x}_n)}{\|\mathbf{w}\|} = \frac{t_n(\mathbf{w}^{\mathrm{T}}\boldsymbol{\phi}(\mathbf{x}_n) + b)}{\|\mathbf{w}\|}. \tag{7.2}$$

The margin is given by the perpendicular distance to the closest point $\mathbf{x}_n$ from the data set, and we wish to optimize the parameters $\mathbf{w}$ and b in order to maximize this distance. Thus the maximum margin solution is found by solving

$$\arg\max_{\mathbf{w},b} \left\{ \frac{1}{\|\mathbf{w}\|} \min_n \left[t_n \left(\mathbf{w}^{\mathrm{T}}\boldsymbol{\phi}(\mathbf{x}_n) + b \right) \right] \right\} \tag{7.3}$$

where we have taken the factor $1/\|\mathbf{w}\|$ outside the optimization over n because $\mathbf{w}$

does not depend on n. Direct solution of this optimization problem would be very complex, and so we shall convert it into an equivalent problem that is much easier to solve. To do this we note that if we make the rescaling $\mathbf{w} \rightarrow \kappa\mathbf{w}$ and $b \rightarrow \kappa b$, then the distance from any point $\mathbf{x}_n$ to the decision surface, given by $t_n y(\mathbf{x}_n)/\|\mathbf{w}\|$, is unchanged. We can use this freedom to set

$$t_n \left(\mathbf{w}^{\mathrm{T}}\boldsymbol{\phi}(\mathbf{x}_n) + b\right) = 1 \tag{7.4}$$

for the point that is closest to the surface. In this case, all data points will satisfy the constraints

$$t_n \left(\mathbf{w}^{\mathrm{T}}\boldsymbol{\phi}(\mathbf{x}_n) + b\right) \geqslant 1, \qquad n = 1, \ldots, N. \tag{7.5}$$

This is known as the canonical representation of the decision hyperplane. In the case of data points for which the equality holds, the constraints are said to be *active*, whereas for the remainder they are said to be *inactive*. By definition, there will always be at least one active constraint, because there will always be a closest point, and once the margin has been maximized there will be at least two active constraints. The optimization problem then simply requires that we maximize $\|\mathbf{w}\|^{-1}$, which is equivalent to minimizing $\|\mathbf{w}\|^2$, and so we have to solve the optimization problem

$$\underset{\mathbf{w},b}{\arg\min}\ \frac{1}{2}\|\mathbf{w}\|^2 \tag{7.6}$$

subject to the constraints given by (7.5). The factor of $1/2$ in (7.6) is included for later convenience. This is an example of a *quadratic programming* problem in which we are trying to minimize a quadratic function subject to a set of linear inequality constraints. It appears that the bias parameter b has disappeared from the optimization. However, it is determined implicitly via the constraints, because these require that changes to $\|\mathbf{w}\|$ be compensated by changes to b. We shall see how this works shortly.

Appendix E

In order to solve this constrained optimization problem, we introduce Lagrange multipliers $a_n \geqslant 0$, with one multiplier a_n for each of the constraints in (7.5), giving the Lagrangian function

$$L(\mathbf{w}, b, \mathbf{a}) = \frac{1}{2}\|\mathbf{w}\|^2 - \sum_{n=1}^{N} a_n \left\{t_n(\mathbf{w}^{\mathrm{T}}\boldsymbol{\phi}(\mathbf{x}_n) + b) - 1\right\} \tag{7.7}$$

where $\mathbf{a} = (a_1, \ldots, a_N)^{\mathrm{T}}$. Note the minus sign in front of the Lagrange multiplier term, because we are minimizing with respect to $\mathbf{w}$ and b, and maximizing with respect to $\mathbf{a}$. Setting the derivatives of $L(\mathbf{w}, b, \mathbf{a})$ with respect to $\mathbf{w}$ and b equal to zero, we obtain the following two conditions

$$\mathbf{w} = \sum_{n=1}^{N} a_n t_n \boldsymbol{\phi}(\mathbf{x}_n) \tag{7.8}$$

$$0 = \sum_{n=1}^{N} a_n t_n. \tag{7.9}$$

Eliminating $\mathbf{w}$ and b from $L(\mathbf{w}, b, \mathbf{a})$ using these conditions then gives the *dual representation* of the maximum margin problem in which we maximize

$$\widetilde{L}(\mathbf{a}) = \sum_{n=1}^{N} a_n - \frac{1}{2}\sum_{n=1}^{N}\sum_{m=1}^{N} a_n a_m t_n t_m k(\mathbf{x}_n, \mathbf{x}_m) \tag{7.10}$$

with respect to $\mathbf{a}$ subject to the constraints

$$a_n \geqslant 0, \qquad n = 1, \ldots, N, \tag{7.11}$$

$$\sum_{n=1}^{N} a_n t_n = 0. \tag{7.12}$$

Here the kernel function is defined by $k(\mathbf{x}, \mathbf{x}') = \boldsymbol{\phi}(\mathbf{x})^{\mathrm{T}}\boldsymbol{\phi}(\mathbf{x}')$. Again, this takes the form of a quadratic programming problem in which we optimize a quadratic function of $\mathbf{a}$ subject to a set of inequality constraints. We shall discuss techniques for solving such quadratic programming problems in Section 7.1.1.

The solution to a quadratic programming problem in M variables in general has computational complexity that is $O(M^3)$. In going to the dual formulation we have turned the original optimization problem, which involved minimizing (7.6) over M variables, into the dual problem (7.10), which has N variables. For a fixed set of basis functions whose number M is smaller than the number N of data points, the move to the dual problem appears disadvantageous. However, it allows the model to be reformulated using kernels, and so the maximum margin classifier can be applied efficiently to feature spaces whose dimensionality exceeds the number of data points, including infinite feature spaces. The kernel formulation also makes clear the role of the constraint that the kernel function $k(\mathbf{x}, \mathbf{x}')$ be positive definite, because this ensures that the Lagrangian function $\widetilde{L}(\mathbf{a})$ is bounded below, giving rise to a well-defined optimization problem.

In order to classify new data points using the trained model, we evaluate the sign of $y(\mathbf{x})$ defined by (7.1). This can be expressed in terms of the parameters $\{a_n\}$ and the kernel function by substituting for $\mathbf{w}$ using (7.8) to give

$$y(\mathbf{x}) = \sum_{n=1}^{N} a_n t_n k(\mathbf{x}, \mathbf{x}_n) + b. \tag{7.13}$$

Joseph-Louis Lagrange
1736–1813

Although widely considered to be a French mathematician, Lagrange was born in Turin in Italy. By the age of nineteen, he had already made important contributions to mathematics and had been appointed as Professor at the Royal Artillery School in Turin. For many years, Euler worked hard to persuade Lagrange to move to Berlin, which he eventually did in 1766 where he succeeded Euler as Director of Mathematics at the Berlin Academy. Later he moved to Paris, narrowly escaping with his life during the French revolution thanks to the personal intervention of Lavoisier (the French chemist who discovered oxygen) who himself was later executed at the guillotine. Lagrange made key contributions to the calculus of variations and the foundations of dynamics.

In Appendix E, we show that a constrained optimization of this form satisfies the *Karush-Kuhn-Tucker* (KKT) conditions, which in this case require that the following three properties hold

$$\begin{aligned} a_n &\geqslant 0 & (7.14)\\ t_n y(\mathbf{x}_n) - 1 &\geqslant 0 & (7.15)\\ a_n \{t_n y(\mathbf{x}_n) - 1\} &= 0. & (7.16)\end{aligned}$$

Thus for every data point, either $a_n = 0$ or $t_n y(\mathbf{x}_n) = 1$. Any data point for which $a_n = 0$ will not appear in the sum in (7.13) and hence plays no role in making predictions for new data points. The remaining data points are called *support vectors*, and because they satisfy $t_n y(\mathbf{x}_n) = 1$, they correspond to points that lie on the maximum margin hyperplanes in feature space, as illustrated in Figure 7.1. This property is central to the practical applicability of support vector machines. Once the model is trained, a significant proportion of the data points can be discarded and only the support vectors retained.

Having solved the quadratic programming problem and found a value for $\mathbf{a}$, we can then determine the value of the threshold parameter b by noting that any support vector $\mathbf{x}_n$ satisfies $t_n y(\mathbf{x}_n) = 1$. Using (7.13) this gives

$$t_n \left(\sum_{m \in \mathcal{S}} a_m t_m k(\mathbf{x}_n, \mathbf{x}_m) + b \right) = 1 \tag{7.17}$$

where $\mathcal{S}$ denotes the set of indices of the support vectors. Although we can solve this equation for b using an arbitrarily chosen support vector $\mathbf{x}_n$, a numerically more stable solution is obtained by first multiplying through by t_n, making use of $t_n^2 = 1$, and then averaging these equations over all support vectors and solving for b to give

$$b = \frac{1}{N_{\mathcal{S}}} \sum_{n \in \mathcal{S}} \left(t_n - \sum_{m \in \mathcal{S}} a_m t_m k(\mathbf{x}_n, \mathbf{x}_m) \right) \tag{7.18}$$

where $N_{\mathcal{S}}$ is the total number of support vectors.

For later comparison with alternative models, we can express the maximum-margin classifier in terms of the minimization of an error function, with a simple quadratic regularizer, in the form

$$\sum_{n=1}^{N} E_{\infty}(y(\mathbf{x}_n)t_n - 1) + \lambda \|\mathbf{w}\|^2 \tag{7.19}$$

where $E_{\infty}(z)$ is a function that is zero if $z \geqslant 0$ and ∞ otherwise and ensures that the constraints (7.5) are satisfied. Note that as long as the regularization parameter satisfies $\lambda > 0$, its precise value plays no role.

Figure 7.2 shows an example of the classification resulting from training a support vector machine on a simple synthetic data set using a Gaussian kernel of the

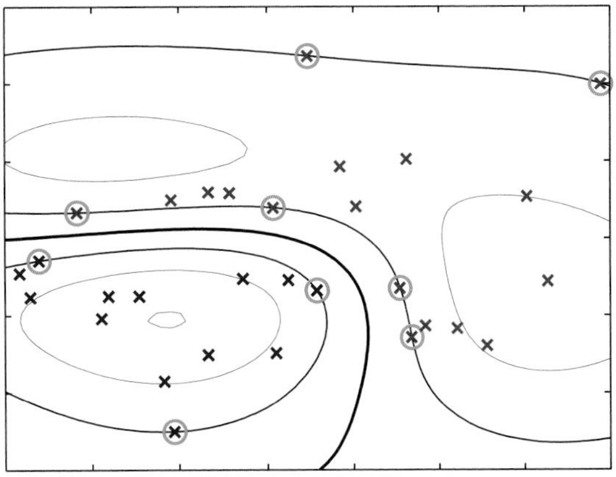

Figure 7.2 Example of synthetic data from two classes in two dimensions showing contours of constant $y(\mathbf{x})$ obtained from a support vector machine having a Gaussian kernel function. Also shown are the decision boundary, the margin boundaries, and the support vectors.

form (6.23). Although the data set is not linearly separable in the two-dimensional data space $\mathbf{x}$, it is linearly separable in the nonlinear feature space defined implicitly by the nonlinear kernel function. Thus the training data points are perfectly separated in the original data space.

This example also provides a geometrical insight into the origin of sparsity in the SVM. The maximum margin hyperplane is defined by the location of the support vectors. Other data points can be moved around freely (so long as they remain outside the margin region) without changing the decision boundary, and so the solution will be independent of such data points.

7.1.1 Overlapping class distributions

So far, we have assumed that the training data points are linearly separable in the feature space $\phi(\mathbf{x})$. The resulting support vector machine will give exact separation of the training data in the original input space $\mathbf{x}$, although the corresponding decision boundary will be nonlinear. In practice, however, the class-conditional distributions may overlap, in which case exact separation of the training data can lead to poor generalization.

We therefore need a way to modify the support vector machine so as to allow some of the training points to be misclassified. From (7.19) we see that in the case of separable classes, we implicitly used an error function that gave infinite error if a data point was misclassified and zero error if it was classified correctly, and then optimized the model parameters to maximize the margin. We now modify this approach so that data points are allowed to be on the 'wrong side' of the margin boundary, but with a penalty that increases with the distance from that boundary. For the subsequent optimization problem, it is convenient to make this penalty a linear function of this distance. To do this, we introduce *slack variables*, $\xi_n \geqslant 0$ where $n = 1, \ldots, N$, with one slack variable for each training data point (Bennett, 1992; Cortes and Vapnik, 1995). These are defined by $\xi_n = 0$ for data points that are on or inside the correct margin boundary and $\xi_n = |t_n - y(\mathbf{x}_n)|$ for other points. Thus a data point that is on the decision boundary $y(\mathbf{x}_n) = 0$ will have $\xi_n = 1$, and points

Figure 7.3 Illustration of the slack variables $\xi_n \geqslant 0$. Data points with circles around them are support vectors.

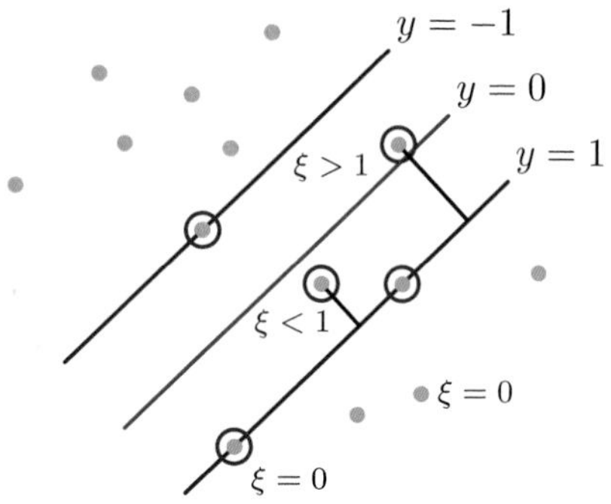

with $\xi_n > 1$ will be misclassified. The exact classification constraints (7.5) are then replaced with

$$t_n y(\mathbf{x}_n) \geqslant 1 - \xi_n, \qquad n = 1, \ldots, N \tag{7.20}$$

in which the slack variables are constrained to satisfy $\xi_n \geqslant 0$. Data points for which $\xi_n = 0$ are correctly classified and are either on the margin or on the correct side of the margin. Points for which $0 < \xi_n \leqslant 1$ lie inside the margin, but on the correct side of the decision boundary, and those data points for which $\xi_n > 1$ lie on the wrong side of the decision boundary and are misclassified, as illustrated in Figure 7.3. This is sometimes described as relaxing the hard margin constraint to give a *soft margin* and allows some of the training set data points to be misclassified. Note that while slack variables allow for overlapping class distributions, this framework is still sensitive to outliers because the penalty for misclassification increases linearly with ξ.

Our goal is now to maximize the margin while softly penalizing points that lie on the wrong side of the margin boundary. We therefore minimize

$$C \sum_{n=1}^{N} \xi_n + \frac{1}{2}\|\mathbf{w}\|^2 \tag{7.21}$$

where the parameter $C > 0$ controls the trade-off between the slack variable penalty and the margin. Because any point that is misclassified has $\xi_n > 1$, it follows that $\sum_n \xi_n$ is an upper bound on the number of misclassified points. The parameter C is therefore analogous to (the inverse of) a regularization coefficient because it controls the trade-off between minimizing training errors and controlling model complexity. In the limit $C \to \infty$, we will recover the earlier support vector machine for separable data.

We now wish to minimize (7.21) subject to the constraints (7.20) together with $\xi_n \geqslant 0$. The corresponding Lagrangian is given by

$$L(\mathbf{w}, b, \boldsymbol{\xi}, \mathbf{a}, \boldsymbol{\mu}) = \frac{1}{2}\|\mathbf{w}\|^2 + C\sum_{n=1}^{N} \xi_n - \sum_{n=1}^{N} a_n \left\{ t_n y(\mathbf{x}_n) - 1 + \xi_n \right\} - \sum_{n=1}^{N} \mu_n \xi_n \tag{7.22}$$

Appendix E

where $\{a_n \geqslant 0\}$ and $\{\mu_n \geqslant 0\}$ are Lagrange multipliers. The corresponding set of KKT conditions are given by

$$\begin{aligned} a_n &\geqslant 0 && (7.23)\\ t_n y(\mathbf{x}_n) - 1 + \xi_n &\geqslant 0 && (7.24)\\ a_n\left(t_n y(\mathbf{x}_n) - 1 + \xi_n\right) &= 0 && (7.25)\\ \mu_n &\geqslant 0 && (7.26)\\ \xi_n &\geqslant 0 && (7.27)\\ \mu_n \xi_n &= 0 && (7.28) \end{aligned}$$

where $n = 1, \ldots, N$.

We now optimize out $\mathbf{w}$, b, and $\{\xi_n\}$ making use of the definition (7.1) of $y(\mathbf{x})$ to give

$$\frac{\partial L}{\partial \mathbf{w}} = 0 \quad \Rightarrow \quad \mathbf{w} = \sum_{n=1}^{N} a_n t_n \boldsymbol{\phi}(\mathbf{x}_n) \tag{7.29}$$

$$\frac{\partial L}{\partial b} = 0 \quad \Rightarrow \quad \sum_{n=1}^{N} a_n t_n = 0 \tag{7.30}$$

$$\frac{\partial L}{\partial \xi_n} = 0 \quad \Rightarrow \quad a_n = C - \mu_n. \tag{7.31}$$

Using these results to eliminate $\mathbf{w}$, b, and $\{\xi_n\}$ from the Lagrangian, we obtain the dual Lagrangian in the form

$$\widetilde{L}(\mathbf{a}) = \sum_{n=1}^{N} a_n - \frac{1}{2}\sum_{n=1}^{N}\sum_{m=1}^{N} a_n a_m t_n t_m k(\mathbf{x}_n, \mathbf{x}_m) \tag{7.32}$$

which is identical to the separable case, except that the constraints are somewhat different. To see what these constraints are, we note that $a_n \geqslant 0$ is required because these are Lagrange multipliers. Furthermore, (7.31) together with $\mu_n \geqslant 0$ implies $a_n \leqslant C$. We therefore have to maximize (7.32) with respect to the dual variables $\{a_n\}$ subject to

$$0 \leqslant a_n \leqslant C \tag{7.33}$$

$$\sum_{n=1}^{N} a_n t_n = 0 \tag{7.34}$$

for $n = 1, \ldots, N$, where (7.33) are known as *box constraints*. This again represents a quadratic programming problem. If we substitute (7.29) into (7.1), we see that predictions for new data points are again made by using (7.13).

We can now interpret the resulting solution. As before, a subset of the data points may have $a_n = 0$, in which case they do not contribute to the predictive model (7.13). The remaining data points constitute the support vectors. These have $a_n > 0$ and hence from (7.25) must satisfy

$$t_n y(\mathbf{x}_n) = 1 - \xi_n. \tag{7.35}$$

If $a_n < C$, then (7.31) implies that $\mu_n > 0$, which from (7.28) requires $\xi_n = 0$ and hence such points lie on the margin. Points with $a_n = C$ can lie inside the margin and can either be correctly classified if $\xi_n \leqslant 1$ or misclassified if $\xi_n > 1$.

To determine the parameter b in (7.1), we note that those support vectors for which $0 < a_n < C$ have $\xi_n = 0$ so that $t_n y(\mathbf{x}_n) = 1$ and hence will satisfy

$$t_n \left(\sum_{m \in \mathcal{S}} a_m t_m k(\mathbf{x}_n, \mathbf{x}_m) + b \right) = 1. \tag{7.36}$$

Again, a numerically stable solution is obtained by averaging to give

$$b = \frac{1}{N_{\mathcal{M}}} \sum_{n \in \mathcal{M}} \left(t_n - \sum_{m \in \mathcal{S}} a_m t_m k(\mathbf{x}_n, \mathbf{x}_m) \right) \tag{7.37}$$

where $\mathcal{M}$ denotes the set of indices of data points having $0 < a_n < C$.

An alternative, equivalent formulation of the support vector machine, known as the ν-*SVM*, has been proposed by Schölkopf *et al.* (2000). This involves maximizing

$$\widetilde{L}(\mathbf{a}) = -\frac{1}{2} \sum_{n=1}^{N} \sum_{m=1}^{N} a_n a_m t_n t_m k(\mathbf{x}_n, \mathbf{x}_m) \tag{7.38}$$

subject to the constraints

$$0 \leqslant a_n \leqslant 1/N \tag{7.39}$$

$$\sum_{n=1}^{N} a_n t_n = 0 \tag{7.40}$$

$$\sum_{n=1}^{N} a_n \geqslant \nu. \tag{7.41}$$

This approach has the advantage that the parameter ν, which replaces C, can be interpreted as both an upper bound on the fraction of *margin errors* (points for which $\xi_n > 0$ and hence which lie on the wrong side of the margin boundary and which may or may not be misclassified) and a lower bound on the fraction of support vectors. An example of the ν-SVM applied to a synthetic data set is shown in Figure 7.4. Here Gaussian kernels of the form $\exp\left(-\gamma \|\mathbf{x} - \mathbf{x}'\|^2\right)$ have been used, with $\gamma = 0.45$.

Although predictions for new inputs are made using only the support vectors, the training phase (i.e., the determination of the parameters $\mathbf{a}$ and b) makes use of the whole data set, and so it is important to have efficient algorithms for solving

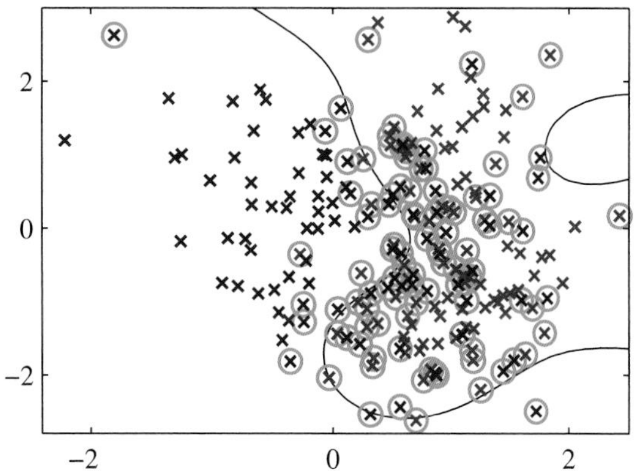

Figure 7.4 Illustration of the ν-SVM applied to a nonseparable data set in two dimensions. The support vectors are indicated by circles.

the quadratic programming problem. We first note that the objective function $\widetilde{L}(\mathbf{a})$ given by (7.10) or (7.32) is quadratic and so any local optimum will also be a global optimum provided the constraints define a convex region (which they do as a consequence of being linear). Direct solution of the quadratic programming problem using traditional techniques is often infeasible due to the demanding computation and memory requirements, and so more practical approaches need to be found. The technique of *chunking* (Vapnik, 1982) exploits the fact that the value of the Lagrangian is unchanged if we remove the rows and columns of the kernel matrix corresponding to Lagrange multipliers that have value zero. This allows the full quadratic programming problem to be broken down into a series of smaller ones, whose goal is eventually to identify all of the nonzero Lagrange multipliers and discard the others. Chunking can be implemented using *protected conjugate gradients* (Burges, 1998). Although chunking reduces the size of the matrix in the quadratic function from the number of data points squared to approximately the number of nonzero Lagrange multipliers squared, even this may be too big to fit in memory for large-scale applications. *Decomposition methods* (Osuna *et al.*, 1996) also solve a series of smaller quadratic programming problems but are designed so that each of these is of a fixed size, and so the technique can be applied to arbitrarily large data sets. However, it still involves numerical solution of quadratic programming subproblems and these can be problematic and expensive. One of the most popular approaches to training support vector machines is called *sequential minimal optimization*, or *SMO* (Platt, 1999). It takes the concept of chunking to the extreme limit and considers just two Lagrange multipliers at a time. In this case, the subproblem can be solved analytically, thereby avoiding numerical quadratic programming altogether. Heuristics are given for choosing the pair of Lagrange multipliers to be considered at each step. In practice, SMO is found to have a scaling with the number of data points that is somewhere between linear and quadratic depending on the particular application.

We have seen that kernel functions correspond to inner products in feature spaces that can have high, or even infinite, dimensionality. By working directly in terms of the kernel function, without introducing the feature space explicitly, it might therefore seem that support vector machines somehow manage to avoid the curse of di-

Section 1.4

mensionality. This is not the case, however, because there are constraints amongst the feature values that restrict the effective dimensionality of feature space. To see this consider a simple second-order polynomial kernel that we can expand in terms of its components

$$\begin{aligned} k(\mathbf{x}, \mathbf{z}) &= \left(1 + \mathbf{x}^{\mathrm{T}}\mathbf{z}\right)^2 = (1 + x_1 z_1 + x_2 z_2)^2 \\ &= 1 + 2x_1 z_1 + 2x_2 z_2 + x_1^2 z_1^2 + 2x_1 z_1 x_2 z_2 + x_2^2 z_2^2 \\ &= (1, \sqrt{2}x_1, \sqrt{2}x_2, x_1^2, \sqrt{2}x_1 x_2, x_2^2)(1, \sqrt{2}z_1, \sqrt{2}z_2, z_1^2, \sqrt{2}z_1 z_2, z_2^2)^{\mathrm{T}} \\ &= \boldsymbol{\phi}(\mathbf{x})^{\mathrm{T}}\boldsymbol{\phi}(\mathbf{z}). \end{aligned} \tag{7.42}$$

This kernel function therefore represents an inner product in a feature space having six dimensions, in which the mapping from input space to feature space is described by the vector function $\boldsymbol{\phi}(\mathbf{x})$. However, the coefficients weighting these different features are constrained to have specific forms. Thus any set of points in the original two-dimensional space $\mathbf{x}$ would be constrained to lie exactly on a two-dimensional nonlinear manifold embedded in the six-dimensional feature space.

We have already highlighted the fact that the support vector machine does not provide probabilistic outputs but instead makes classification decisions for new input vectors. Veropoulos *et al.* (1999) discuss modifications to the SVM to allow the trade-off between false positive and false negative errors to be controlled. However, if we wish to use the SVM as a module in a larger probabilistic system, then probabilistic predictions of the class label t for new inputs $\mathbf{x}$ are required.

To address this issue, Platt (2000) has proposed fitting a logistic sigmoid to the outputs of a previously trained support vector machine. Specifically, the required conditional probability is assumed to be of the form

$$p(t = 1|\mathbf{x}) = \sigma\left(Ay(\mathbf{x}) + B\right) \tag{7.43}$$

where $y(\mathbf{x})$ is defined by (7.1). Values for the parameters A and B are found by minimizing the cross-entropy error function defined by a training set consisting of pairs of values $y(\mathbf{x}_n)$ and t_n. The data used to fit the sigmoid needs to be independent of that used to train the original SVM in order to avoid severe over-fitting. This two-stage approach is equivalent to assuming that the output $y(\mathbf{x})$ of the support vector machine represents the log-odds of $\mathbf{x}$ belonging to class $t = 1$. Because the SVM training procedure is not specifically intended to encourage this, the SVM can give a poor approximation to the posterior probabilities (Tipping, 2001).

7.1.2 Relation to logistic regression

As with the separable case, we can re-cast the SVM for nonseparable distributions in terms of the minimization of a regularized error function. This will also allow us to highlight similarities, and differences, compared to the logistic regression model.

Section 4.3.2

We have seen that for data points that are on the correct side of the margin boundary, and which therefore satisfy $y_n t_n \geqslant 1$, we have $\xi_n = 0$, and for the

Figure 7.5 Plot of the 'hinge' error function used in support vector machines, shown in blue, along with the error function for logistic regression, rescaled by a factor of $1/\ln(2)$ so that it passes through the point $(0, 1)$, shown in red. Also shown are the misclassification error in black and the squared error in green.

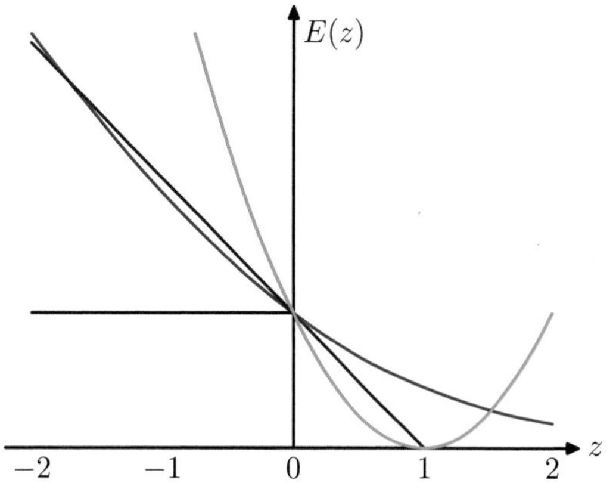

remaining points we have $\xi_n = 1 - y_n t_n$. Thus the objective function (7.21) can be written (up to an overall multiplicative constant) in the form

$$\sum_{n=1}^{N} E_{\mathrm{SV}}(y_n t_n) + \lambda \|\mathbf{w}\|^2 \tag{7.44}$$

where $\lambda = (2C)^{-1}$, and $E_{\mathrm{SV}}(\cdot)$ is the *hinge* error function defined by

$$E_{\mathrm{SV}}(y_n t_n) = [1 - y_n t_n]_+ \tag{7.45}$$

where $[\,\cdot\,]_+$ denotes the positive part. The hinge error function, so-called because of its shape, is plotted in Figure 7.5. It can be viewed as an approximation to the misclassification error, i.e., the error function that ideally we would like to minimize, which is also shown in Figure 7.5.

When we considered the logistic regression model in Section 4.3.2, we found it convenient to work with target variable $t \in \{0, 1\}$. For comparison with the support vector machine, we first reformulate maximum likelihood logistic regression using the target variable $t \in \{-1, 1\}$. To do this, we note that $p(t = 1|y) = \sigma(y)$ where $y(\mathbf{x})$ is given by (7.1), and $\sigma(y)$ is the logistic sigmoid function defined by (4.59). It follows that $p(t = -1|y) = 1 - \sigma(y) = \sigma(-y)$, where we have used the properties of the logistic sigmoid function, and so we can write

$$p(t|y) = \sigma(yt). \tag{7.46}$$

Exercise 7.6

From this we can construct an error function by taking the negative logarithm of the likelihood function that, with a quadratic regularizer, takes the form

$$\sum_{n=1}^{N} E_{\mathrm{LR}}(y_n t_n) + \lambda \|\mathbf{w}\|^2. \tag{7.47}$$

where

$$E_{\mathrm{LR}}(yt) = \ln\left(1 + \exp(-yt)\right). \tag{7.48}$$

For comparison with other error functions, we can divide by $\ln(2)$ so that the error function passes through the point $(0, 1)$. This rescaled error function is also plotted in Figure 7.5 and we see that it has a similar form to the support vector error function. The key difference is that the flat region in $E_{\mathrm{SV}}(yt)$ leads to sparse solutions.

Both the logistic error and the hinge loss can be viewed as continuous approximations to the misclassification error. Another continuous error function that has sometimes been used to solve classification problems is the squared error, which is again plotted in Figure 7.5. It has the property, however, of placing increasing emphasis on data points that are correctly classified but that are a long way from the decision boundary on the correct side. Such points will be strongly weighted at the expense of misclassified points, and so if the objective is to minimize the misclassification rate, then a monotonically decreasing error function would be a better choice.

7.1.3 Multiclass SVMs

The support vector machine is fundamentally a two-class classifier. In practice, however, we often have to tackle problems involving $K > 2$ classes. Various methods have therefore been proposed for combining multiple two-class SVMs in order to build a multiclass classifier.

One commonly used approach (Vapnik, 1998) is to construct K separate SVMs, in which the kth model $y_k(\mathbf{x})$ is trained using the data from class $\mathcal{C}_k$ as the positive examples and the data from the remaining $K - 1$ classes as the negative examples. This is known as the *one-versus-the-rest* approach. However, in Figure 4.2 we saw that using the decisions of the individual classifiers can lead to inconsistent results in which an input is assigned to multiple classes simultaneously. This problem is sometimes addressed by making predictions for new inputs $\mathbf{x}$ using

$$y(\mathbf{x}) = \max_k y_k(\mathbf{x}). \tag{7.49}$$

Unfortunately, this heuristic approach suffers from the problem that the different classifiers were trained on different tasks, and there is no guarantee that the real-valued quantities $y_k(\mathbf{x})$ for different classifiers will have appropriate scales.

Another problem with the one-versus-the-rest approach is that the training sets are imbalanced. For instance, if we have ten classes each with equal numbers of training data points, then the individual classifiers are trained on data sets comprising 90% negative examples and only 10% positive examples, and the symmetry of the original problem is lost. A variant of the one-versus-the-rest scheme was proposed by Lee *et al.* (2001) who modify the target values so that the positive class has target $+1$ and the negative class has target $-1/(K - 1)$.

Weston and Watkins (1999) define a single objective function for training all K SVMs simultaneously, based on maximizing the margin from each to remaining classes. However, this can result in much slower training because, instead of solving K separate optimization problems each over N data points with an overall cost of $O(KN^2)$, a single optimization problem of size $(K - 1)N$ must be solved giving an overall cost of $O(K^2N^2)$.

Another approach is to train $K(K-1)/2$ different 2-class SVMs on all possible pairs of classes, and then to classify test points according to which class has the highest number of 'votes', an approach that is sometimes called *one-versus-one*. Again, we saw in Figure 4.2 that this can lead to ambiguities in the resulting classification. Also, for large K this approach requires significantly more training time than the one-versus-the-rest approach. Similarly, to evaluate test points, significantly more computation is required.

The latter problem can be alleviated by organizing the pairwise classifiers into a directed acyclic graph (not to be confused with a probabilistic graphical model) leading to the *DAGSVM* (Platt *et al.*, 2000). For K classes, the DAGSVM has a total of $K(K-1)/2$ classifiers, and to classify a new test point only $K-1$ pairwise classifiers need to be evaluated, with the particular classifiers used depending on which path through the graph is traversed.

A different approach to multiclass classification, based on error-correcting output codes, was developed by Dietterich and Bakiri (1995) and applied to support vector machines by Allwein *et al.* (2000). This can be viewed as a generalization of the voting scheme of the one-versus-one approach in which more general partitions of the classes are used to train the individual classifiers. The K classes themselves are represented as particular sets of responses from the two-class classifiers chosen, and together with a suitable decoding scheme, this gives robustness to errors and to ambiguity in the outputs of the individual classifiers. Although the application of SVMs to multiclass classification problems remains an open issue, in practice the one-versus-the-rest approach is the most widely used in spite of its ad-hoc formulation and its practical limitations.

There are also *single-class* support vector machines, which solve an unsupervised learning problem related to probability density estimation. Instead of modelling the density of data, however, these methods aim to find a smooth boundary enclosing a region of high density. The boundary is chosen to represent a quantile of the density, that is, the probability that a data point drawn from the distribution will land inside that region is given by a fixed number between 0 and 1 that is specified in advance. This is a more restricted problem than estimating the full density but may be sufficient in specific applications. Two approaches to this problem using support vector machines have been proposed. The algorithm of Schölkopf *et al.* (2001) tries to find a hyperplane that separates all but a fixed fraction ν of the training data from the origin while at the same time maximizing the distance (margin) of the hyperplane from the origin, while Tax and Duin (1999) look for the smallest sphere in feature space that contains all but a fraction ν of the data points. For kernels $k(\mathbf{x}, \mathbf{x}')$ that are functions only of $\mathbf{x} - \mathbf{x}'$, the two algorithms are equivalent.

7.1.4 SVMs for regression

Section 3.1.4

We now extend support vector machines to regression problems while at the same time preserving the property of sparseness. In simple linear regression, we

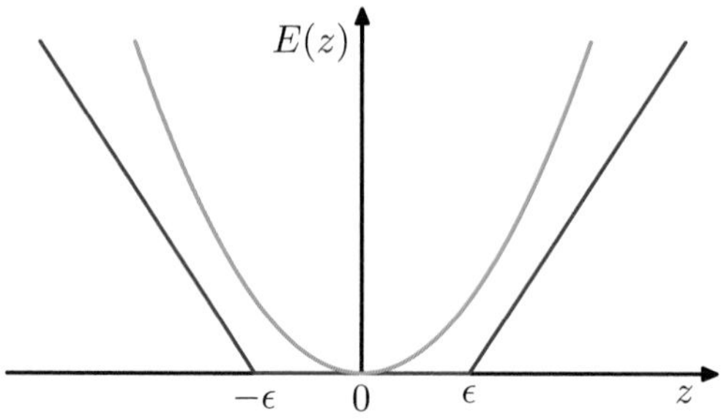

Figure 7.6 Plot of an ϵ-insensitive error function (in red) in which the error increases linearly with distance beyond the insensitive region. Also shown for comparison is the quadratic error function (in green).

minimize a regularized error function given by

$$\frac{1}{2}\sum_{n=1}^{N}\{y_n - t_n\}^2 + \frac{\lambda}{2}\|\mathbf{w}\|^2. \tag{7.50}$$

To obtain sparse solutions, the quadratic error function is replaced by an *ϵ-insensitive error function* (Vapnik, 1995), which gives zero error if the absolute difference between the prediction $y(\mathbf{x})$ and the target t is less than ϵ where $\epsilon > 0$. A simple example of an ϵ-insensitive error function, having a linear cost associated with errors outside the insensitive region, is given by

$$E_\epsilon(y(\mathbf{x}) - t) = \begin{cases} 0, & \text{if } |y(\mathbf{x}) - t| < \epsilon; \\ |y(\mathbf{x}) - t| - \epsilon, & \text{otherwise} \end{cases} \tag{7.51}$$

and is illustrated in Figure 7.6.

We therefore minimize a regularized error function given by

$$C\sum_{n=1}^{N} E_\epsilon(y(\mathbf{x}_n) - t_n) + \frac{1}{2}\|\mathbf{w}\|^2 \tag{7.52}$$

where $y(\mathbf{x})$ is given by (7.1). By convention the (inverse) regularization parameter, denoted C, appears in front of the error term.

As before, we can re-express the optimization problem by introducing slack variables. For each data point $\mathbf{x}_n$, we now need two slack variables $\xi_n \geqslant 0$ and $\widehat{\xi}_n \geqslant 0$, where $\xi_n > 0$ corresponds to a point for which $t_n > y(\mathbf{x}_n) + \epsilon$, and $\widehat{\xi}_n > 0$ corresponds to a point for which $t_n < y(\mathbf{x}_n) - \epsilon$, as illustrated in Figure 7.7.

The condition for a target point to lie inside the ϵ-tube is that $y_n - \epsilon \leqslant t_n \leqslant y_n + \epsilon$, where $y_n = y(\mathbf{x}_n)$. Introducing the slack variables allows points to lie outside the tube provided the slack variables are nonzero, and the corresponding conditions are

$$t_n \leqslant y(\mathbf{x}_n) + \epsilon + \xi_n \tag{7.53}$$

$$t_n \geqslant y(\mathbf{x}_n) - \epsilon - \widehat{\xi}_n. \tag{7.54}$$

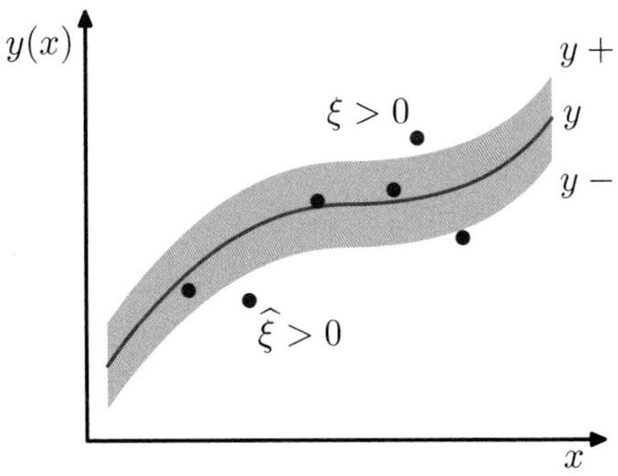

Figure 7.7 Illustration of SVM regression, showing the regression curve together with the ϵ-insensitive 'tube'. Also shown are examples of the slack variables ξ and $\widehat{\xi}$. Points above the ϵ-tube have $\xi > 0$ and $\widehat{\xi} = 0$, points below the ϵ-tube have $\xi = 0$ and $\widehat{\xi} > 0$, and points inside the ϵ-tube have $\xi = \widehat{\xi} = 0$.

The error function for support vector regression can then be written as

$$C\sum_{n=1}^{N}(\xi_n + \widehat{\xi}_n) + \frac{1}{2}\|\mathbf{w}\|^2 \tag{7.55}$$

which must be minimized subject to the constraints $\xi_n \geqslant 0$ and $\widehat{\xi}_n \geqslant 0$ as well as (7.53) and (7.54). This can be achieved by introducing Lagrange multipliers $a_n \geqslant 0$, $\widehat{a}_n \geqslant 0$, $\mu_n \geqslant 0$, and $\widehat{\mu}_n \geqslant 0$ and optimizing the Lagrangian

$$\begin{aligned} L &= C\sum_{n=1}^{N}(\xi_n + \widehat{\xi}_n) + \frac{1}{2}\|\mathbf{w}\|^2 - \sum_{n=1}^{N}(\mu_n\xi_n + \widehat{\mu}_n\widehat{\xi}_n) \\ &\quad - \sum_{n=1}^{N} a_n(\epsilon + \xi_n + y_n - t_n) - \sum_{n=1}^{N}\widehat{a}_n(\epsilon + \widehat{\xi}_n - y_n + t_n). \end{aligned} \tag{7.56}$$

We now substitute for $y(\mathbf{x})$ using (7.1) and then set the derivatives of the Lagrangian with respect to $\mathbf{w}$, b, ξ_n, and $\widehat{\xi}_n$ to zero, giving

$$\frac{\partial L}{\partial \mathbf{w}} = 0 \quad \Rightarrow \quad \mathbf{w} = \sum_{n=1}^{N}(a_n - \widehat{a}_n)\boldsymbol{\phi}(\mathbf{x}_n) \tag{7.57}$$

$$\frac{\partial L}{\partial b} = 0 \quad \Rightarrow \quad \sum_{n=1}^{N}(a_n - \widehat{a}_n) = 0 \tag{7.58}$$

$$\frac{\partial L}{\partial \xi_n} = 0 \quad \Rightarrow \quad a_n + \mu_n = C \tag{7.59}$$

$$\frac{\partial L}{\partial \widehat{\xi}_n} = 0 \quad \Rightarrow \quad \widehat{a}_n + \widehat{\mu}_n = C. \tag{7.60}$$

Exercise 7.7

Using these results to eliminate the corresponding variables from the Lagrangian, we see that the dual problem involves maximizing

$$\widetilde{L}(\mathbf{a}, \widehat{\mathbf{a}}) = -\frac{1}{2}\sum_{n=1}^{N}\sum_{m=1}^{N}(a_n - \widehat{a}_n)(a_m - \widehat{a}_m)k(\mathbf{x}_n, \mathbf{x}_m) - \epsilon\sum_{n=1}^{N}(a_n + \widehat{a}_n) + \sum_{n=1}^{N}(a_n - \widehat{a}_n)t_n \tag{7.61}$$

with respect to $\{a_n\}$ and $\{\widehat{a}_n\}$, where we have introduced the kernel $k(\mathbf{x}, \mathbf{x}') = \boldsymbol{\phi}(\mathbf{x})^{\mathrm{T}}\boldsymbol{\phi}(\mathbf{x}')$. Again, this is a constrained maximization, and to find the constraints we note that $a_n \geqslant 0$ and $\widehat{a}_n \geqslant 0$ are both required because these are Lagrange multipliers. Also $\mu_n \geqslant 0$ and $\widehat{\mu}_n \geqslant 0$ together with (7.59) and (7.60), require $a_n \leqslant C$ and $\widehat{a}_n \leqslant C$, and so again we have the box constraints

$$0 \leqslant a_n \leqslant C \tag{7.62}$$

$$0 \leqslant \widehat{a}_n \leqslant C \tag{7.63}$$

together with the condition (7.58).

Substituting (7.57) into (7.1), we see that predictions for new inputs can be made using

$$y(\mathbf{x}) = \sum_{n=1}^{N}(a_n - \widehat{a}_n)k(\mathbf{x}, \mathbf{x}_n) + b \tag{7.64}$$

which is again expressed in terms of the kernel function.

The corresponding Karush-Kuhn-Tucker (KKT) conditions, which state that at the solution the product of the dual variables and the constraints must vanish, are given by

$$a_n(\epsilon + \xi_n + y_n - t_n) = 0 \tag{7.65}$$

$$\widehat{a}_n(\epsilon + \widehat{\xi}_n - y_n + t_n) = 0 \tag{7.66}$$

$$(C - a_n)\xi_n = 0 \tag{7.67}$$

$$(C - \widehat{a}_n)\widehat{\xi}_n = 0. \tag{7.68}$$

From these we can obtain several useful results. First of all, we note that a coefficient a_n can only be nonzero if $\epsilon + \xi_n + y_n - t_n = 0$, which implies that the data point either lies on the upper boundary of the ϵ-tube ($\xi_n = 0$) or lies above the upper boundary ($\xi_n > 0$). Similarly, a nonzero value for $\widehat{a}_n$ implies $\epsilon + \widehat{\xi}_n - y_n + t_n = 0$, and such points must lie either on or below the lower boundary of the ϵ-tube.

Furthermore, the two constraints $\epsilon + \xi_n + y_n - t_n = 0$ and $\epsilon + \widehat{\xi}_n - y_n + t_n = 0$ are incompatible, as is easily seen by adding them together and noting that ξ_n and $\widehat{\xi}_n$ are nonnegative while ϵ is strictly positive, and so for every data point $\mathbf{x}_n$, either a_n or $\widehat{a}_n$ (or both) must be zero.

The support vectors are those data points that contribute to predictions given by (7.64), in other words those for which either $a_n \neq 0$ or $\widehat{a}_n \neq 0$. These are points that lie on the boundary of the ϵ-tube or outside the tube. All points within the tube have

$a_n = \widehat{a}_n = 0$. We again have a sparse solution, and the only terms that have to be evaluated in the predictive model (7.64) are those that involve the support vectors.

The parameter b can be found by considering a data point for which $0 < a_n < C$, which from (7.67) must have $\xi_n = 0$, and from (7.65) must therefore satisfy $\epsilon + y_n - t_n = 0$. Using (7.1) and solving for b, we obtain

$$\begin{aligned} b &= t_n - \epsilon - \mathbf{w}^{\mathrm{T}}\boldsymbol{\phi}(\mathbf{x}_n) \\ &= t_n - \epsilon - \sum_{m=1}^{N}(a_m - \widehat{a}_m)k(\mathbf{x}_n, \mathbf{x}_m) \end{aligned} \tag{7.69}$$

where we have used (7.57). We can obtain an analogous result by considering a point for which $0 < \widehat{a}_n < C$. In practice, it is better to average over all such estimates of b.

As with the classification case, there is an alternative formulation of the SVM for regression in which the parameter governing complexity has a more intuitive interpretation (Schölkopf *et al.*, 2000). In particular, instead of fixing the width ϵ of the insensitive region, we fix instead a parameter ν that bounds the fraction of points lying outside the tube. This involves maximizing

$$\begin{aligned} \widetilde{L}(\mathbf{a}, \widehat{\mathbf{a}}) &= -\frac{1}{2}\sum_{n=1}^{N}\sum_{m=1}^{N}(a_n - \widehat{a}_n)(a_m - \widehat{a}_m)k(\mathbf{x}_n, \mathbf{x}_m) \\ &\quad + \sum_{n=1}^{N}(a_n - \widehat{a}_n)t_n \end{aligned} \tag{7.70}$$

subject to the constraints

$$0 \leqslant a_n \leqslant C/N \tag{7.71}$$

$$0 \leqslant \widehat{a}_n \leqslant C/N \tag{7.72}$$

$$\sum_{n=1}^{N}(a_n - \widehat{a}_n) = 0 \tag{7.73}$$

$$\sum_{n=1}^{N}(a_n + \widehat{a}_n) \leqslant \nu C. \tag{7.74}$$

It can be shown that there are at most νN data points falling outside the insensitive tube, while at least νN data points are support vectors and so lie either on the tube or outside it.

Appendix A

The use of a support vector machine to solve a regression problem is illustrated using the sinusoidal data set in Figure 7.8. Here the parameters ν and C have been chosen by hand. In practice, their values would typically be determined by cross-validation.

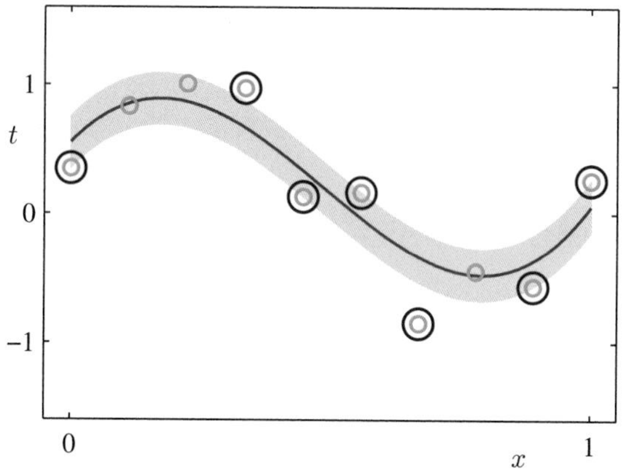

Figure 7.8 Illustration of the ν-SVM for regression applied to the sinusoidal synthetic data set using Gaussian kernels. The predicted regression curve is shown by the red line, and the ϵ-insensitive tube corresponds to the shaded region. Also, the data points are shown in green, and those with support vectors are indicated by blue circles.

7.1.5 Computational learning theory

Historically, support vector machines have largely been motivated and analysed using a theoretical framework known as *computational learning theory*, also sometimes called *statistical learning theory* (Anthony and Biggs, 1992; Kearns and Vazirani, 1994; Vapnik, 1995; Vapnik, 1998). This has its origins with Valiant (1984) who formulated the *probably approximately correct*, or PAC, learning framework. The goal of the PAC framework is to understand how large a data set needs to be in order to give good generalization. It also gives bounds for the computational cost of learning, although we do not consider these here.

Suppose that a data set $\mathcal{D}$ of size N is drawn from some joint distribution $p(\mathbf{x}, \mathbf{t})$ where $\mathbf{x}$ is the input variable and $\mathbf{t}$ represents the class label, and that we restrict attention to 'noise free' situations in which the class labels are determined by some (unknown) deterministic function $\mathbf{t} = \mathbf{g}(\mathbf{x})$. In PAC learning we say that a function $\mathbf{f}(\mathbf{x}; \mathcal{D})$, drawn from a space $\mathcal{F}$ of such functions on the basis of the training set $\mathcal{D}$, has good generalization if its expected error rate is below some pre-specified threshold ϵ, so that

$$\mathbb{E}_{\mathbf{x},\mathbf{t}} \left[I \left(\mathbf{f}(\mathbf{x}; \mathcal{D}) \neq \mathbf{t} \right) \right] < \epsilon \tag{7.75}$$

where $I(\cdot)$ is the indicator function, and the expectation is with respect to the distribution $p(\mathbf{x}, \mathbf{t})$. The quantity on the left-hand side is a random variable, because it depends on the training set $\mathcal{D}$, and the PAC framework requires that (7.75) holds, with probability greater than $1 - \delta$, for a data set $\mathcal{D}$ drawn randomly from $p(\mathbf{x}, \mathbf{t})$. Here δ is another pre-specified parameter, and the terminology 'probably approximately correct' comes from the requirement that with high probability (greater than $1 - \delta$), the error rate be small (less than ϵ). For a given choice of model space $\mathcal{F}$, and for given parameters ϵ and δ, PAC learning aims to provide bounds on the minimum size N of data set needed to meet this criterion. A key quantity in PAC learning is the *Vapnik-Chervonenkis dimension*, or VC dimension, which provides a measure of the complexity of a space of functions, and which allows the PAC framework to be extended to spaces containing an infinite number of functions.

The bounds derived within the PAC framework are often described as worst-case, because they apply to *any* choice for the distribution $p(\mathbf{x}, \mathbf{t})$, so long as both the training and the test examples are drawn (independently) from the same distribution, and for *any* choice for the function $\mathbf{f}(\mathbf{x})$ so long as it belongs to $\mathcal{F}$. In real-world applications of machine learning, we deal with distributions that have significant regularity, for example in which large regions of input space carry the same class label. As a consequence of the lack of any assumptions about the form of the distribution, the PAC bounds are very conservative, in other words they strongly over-estimate the size of data sets required to achieve a given generalization performance. For this reason, PAC bounds have found few, if any, practical applications.

One attempt to improve the tightness of the PAC bounds is the *PAC-Bayesian* framework (McAllester, 2003), which considers a distribution over the space $\mathcal{F}$ of functions, somewhat analogous to the prior in a Bayesian treatment. This still considers any possible choice for $p(\mathbf{x}, \mathbf{t})$, and so although the bounds are tighter, they are still very conservative.

7.2. Relevance Vector Machines

Support vector machines have been used in a variety of classification and regression applications. Nevertheless, they suffer from a number of limitations, several of which have been highlighted already in this chapter. In particular, the outputs of an SVM represent decisions rather than posterior probabilities. Also, the SVM was originally formulated for two classes, and the extension to $K > 2$ classes is problematic. There is a complexity parameter C, or ν (as well as a parameter ϵ in the case of regression), that must be found using a hold-out method such as cross-validation. Finally, predictions are expressed as linear combinations of kernel functions that are centred on training data points and that are required to be positive definite.

The *relevance vector machine* or RVM (Tipping, 2001) is a Bayesian sparse kernel technique for regression and classification that shares many of the characteristics of the SVM whilst avoiding its principal limitations. Additionally, it typically leads to much sparser models resulting in correspondingly faster performance on test data whilst maintaining comparable generalization error.

In contrast to the SVM we shall find it more convenient to introduce the regression form of the RVM first and then consider the extension to classification tasks.

7.2.1 RVM for regression

The relevance vector machine for regression is a linear model of the form studied in Chapter 3 but with a modified prior that results in sparse solutions. The model defines a conditional distribution for a real-valued target variable t, given an input vector $\mathbf{x}$, which takes the form

$$p(t|\mathbf{x}, \mathbf{w}, \beta) = \mathcal{N}(t|y(\mathbf{x}), \beta^{-1}) \tag{7.76}$$

where $\beta = \sigma^{-2}$ is the noise precision (inverse noise variance), and the mean is given by a linear model of the form

$$y(\mathbf{x}) = \sum_{i=1}^{M} w_i \phi_i(\mathbf{x}) = \mathbf{w}^{\mathrm{T}} \boldsymbol{\phi}(\mathbf{x}) \tag{7.77}$$

with fixed nonlinear basis functions $\phi_i(\mathbf{x})$, which will typically include a constant term so that the corresponding weight parameter represents a 'bias'.

The relevance vector machine is a specific instance of this model, which is intended to mirror the structure of the support vector machine. In particular, the basis functions are given by kernels, with one kernel associated with each of the data points from the training set. The general expression (7.77) then takes the SVM-like form

$$y(\mathbf{x}) = \sum_{n=1}^{N} w_n k(\mathbf{x}, \mathbf{x}_n) + b \tag{7.78}$$

where b is a bias parameter. The number of parameters in this case is $M = N + 1$, and $y(\mathbf{x})$ has the same form as the predictive model (7.64) for the SVM, except that the coefficients a_n are here denoted w_n. It should be emphasized that the subsequent analysis is valid for arbitrary choices of basis function, and for generality we shall work with the form (7.77). In contrast to the SVM, there is no restriction to positive-definite kernels, nor are the basis functions tied in either number or location to the training data points.

Suppose we are given a set of N observations of the input vector $\mathbf{x}$, which we denote collectively by a data matrix $\mathbf{X}$ whose n^{th} row is $\mathbf{x}_n^{\mathrm{T}}$ with $n = 1, \ldots, N$. The corresponding target values are given by $\mathbf{t} = (t_1, \ldots, t_N)^{\mathrm{T}}$. Thus, the likelihood function is given by

$$p(\mathbf{t}|\mathbf{X}, \mathbf{w}, \beta) = \prod_{n=1}^{N} p(t_n|\mathbf{x}_n, \mathbf{w}, \beta). \tag{7.79}$$

Next we introduce a prior distribution over the parameter vector $\mathbf{w}$ and as in Chapter 3, we shall consider a zero-mean Gaussian prior. However, the key difference in the RVM is that we introduce a separate hyperparameter α_i for each of the weight parameters w_i instead of a single shared hyperparameter. Thus the weight prior takes the form

$$p(\mathbf{w}|\boldsymbol{\alpha}) = \prod_{i=1}^{M} \mathcal{N}(w_i|0, \alpha_i^{-1}) \tag{7.80}$$

where α_i represents the precision of the corresponding parameter w_i, and $\boldsymbol{\alpha}$ denotes $(\alpha_1, \ldots, \alpha_M)^{\mathrm{T}}$. We shall see that, when we maximize the evidence with respect to these hyperparameters, a significant proportion of them go to infinity, and the corresponding weight parameters have posterior distributions that are concentrated at zero. The basis functions associated with these parameters therefore play no role in the predictions made by the model and so are effectively pruned out, resulting in a sparse model.

Using the result (3.49) for linear regression models, we see that the posterior distribution for the weights is again Gaussian and takes the form

$$p(\mathbf{w}|\mathbf{t}, \mathbf{X}, \boldsymbol{\alpha}, \beta) = \mathcal{N}(\mathbf{w}|\mathbf{m}, \boldsymbol{\Sigma}) \tag{7.81}$$

where the mean and covariance are given by

$$\mathbf{m} = \beta \boldsymbol{\Sigma} \boldsymbol{\Phi}^{\mathrm{T}} \mathbf{t} \tag{7.82}$$

$$\boldsymbol{\Sigma} = \left(\mathbf{A} + \beta \boldsymbol{\Phi}^{\mathrm{T}} \boldsymbol{\Phi}\right)^{-1} \tag{7.83}$$

where $\boldsymbol{\Phi}$ is the $N \times M$ design matrix with elements $\Phi_{ni} = \phi_i(\mathbf{x}_n)$, and $\mathbf{A} = \mathrm{diag}(\alpha_i)$. Note that in the specific case of the model (7.78), we have $\boldsymbol{\Phi} = \mathbf{K}$, where $\mathbf{K}$ is the symmetric $(N+1) \times (N+1)$ kernel matrix with elements $k(\mathbf{x}_n, \mathbf{x}_m)$.

Section 3.5

The values of $\boldsymbol{\alpha}$ and β are determined using type-2 maximum likelihood, also known as the *evidence approximation*, in which we maximize the marginal likelihood function obtained by integrating out the weight parameters

$$p(\mathbf{t}|\mathbf{X}, \boldsymbol{\alpha}, \beta) = \int p(\mathbf{t}|\mathbf{X}, \mathbf{w}, \beta) p(\mathbf{w}|\boldsymbol{\alpha}) \, \mathrm{d}\mathbf{w}. \tag{7.84}$$

Exercise 7.10

Because this represents the convolution of two Gaussians, it is readily evaluated to give the log marginal likelihood in the form

$$\begin{aligned} \ln p(\mathbf{t}|\mathbf{X}, \boldsymbol{\alpha}, \beta) &= \ln \mathcal{N}(\mathbf{t}|\mathbf{0}, \mathbf{C}) \\ &= -\frac{1}{2}\left\{N \ln(2\pi) + \ln|\mathbf{C}| + \mathbf{t}^{\mathrm{T}} \mathbf{C}^{-1} \mathbf{t}\right\} \end{aligned} \tag{7.85}$$

where $\mathbf{t} = (t_1, \ldots, t_N)^{\mathrm{T}}$, and we have defined the $N \times N$ matrix $\mathbf{C}$ given by

$$\mathbf{C} = \beta^{-1}\mathbf{I} + \boldsymbol{\Phi} \mathbf{A}^{-1} \boldsymbol{\Phi}^{\mathrm{T}}. \tag{7.86}$$

Our goal is now to maximize (7.85) with respect to the hyperparameters $\boldsymbol{\alpha}$ and β. This requires only a small modification to the results obtained in Section 3.5 for the evidence approximation in the linear regression model. Again, we can identify two approaches. In the first, we simply set the required derivatives of the marginal likelihood to zero and obtain the following re-estimation equations

Exercise 7.12

$$\alpha_i^{\mathrm{new}} = \frac{\gamma_i}{m_i^2} \tag{7.87}$$

$$(\beta^{\mathrm{new}})^{-1} = \frac{\|\mathbf{t} - \boldsymbol{\Phi}\mathbf{m}\|^2}{N - \sum_i \gamma_i} \tag{7.88}$$

where m_i is the ith component of the posterior mean $\mathbf{m}$ defined by (7.82). The quantity γ_i measures how well the corresponding parameter w_i is determined by the data and is defined by

Section 3.5.3

$$\gamma_i = 1 - \alpha_i \Sigma_{ii} \tag{7.89}$$

in which Σ_{ii} is the ith diagonal component of the posterior covariance $\boldsymbol{\Sigma}$ given by (7.83). Learning therefore proceeds by choosing initial values for $\boldsymbol{\alpha}$ and β, evaluating the mean and covariance of the posterior using (7.82) and (7.83), respectively, and then alternately re-estimating the hyperparameters, using (7.87) and (7.88), and re-estimating the posterior mean and covariance, using (7.82) and (7.83), until a suitable convergence criterion is satisfied.

The second approach is to use the EM algorithm, and is discussed in Section 9.3.4. These two approaches to finding the values of the hyperparameters that maximize the evidence are formally equivalent. Numerically, however, it is found that the direct optimization approach corresponding to (7.87) and (7.88) gives somewhat faster convergence (Tipping, 2001).

Exercise 9.23

As a result of the optimization, we find that a proportion of the hyperparameters $\{\alpha_i\}$ are driven to large (in principle infinite) values, and so the weight parameters w_i corresponding to these hyperparameters have posterior distributions with mean and variance both zero. Thus those parameters, and the corresponding basis functions $\phi_i(\mathbf{x})$, are removed from the model and play no role in making predictions for new inputs. In the case of models of the form (7.78), the inputs $\mathbf{x}_n$ corresponding to the remaining nonzero weights are called *relevance vectors*, because they are identified through the mechanism of automatic relevance determination, and are analogous to the support vectors of an SVM. It is worth emphasizing, however, that this mechanism for achieving sparsity in probabilistic models through automatic relevance determination is quite general and can be applied to any model expressed as an adaptive linear combination of basis functions.

Section 7.2.2

Having found values $\boldsymbol{\alpha}^\star$ and $\beta^\star$ for the hyperparameters that maximize the marginal likelihood, we can evaluate the predictive distribution over t for a new input $\mathbf{x}$. Using (7.76) and (7.81), this is given by

Exercise 7.14

$$\begin{aligned} p(t|\mathbf{x}, \mathbf{X}, \mathbf{t}, \boldsymbol{\alpha}^\star, \beta^\star) &= \int p(t|\mathbf{x}, \mathbf{w}, \beta^\star) p(\mathbf{w}|\mathbf{X}, \mathbf{t}, \boldsymbol{\alpha}^\star, \beta^\star)\,\mathrm{d}\mathbf{w} \\ &= \mathcal{N}\left(t|\mathbf{m}^{\mathrm{T}}\boldsymbol{\phi}(\mathbf{x}), \sigma^2(\mathbf{x})\right). \end{aligned} \tag{7.90}$$

Thus the predictive mean is given by (7.76) with $\mathbf{w}$ set equal to the posterior mean $\mathbf{m}$, and the variance of the predictive distribution is given by

$$\sigma^2(\mathbf{x}) = (\beta^\star)^{-1} + \boldsymbol{\phi}(\mathbf{x})^{\mathrm{T}} \boldsymbol{\Sigma} \boldsymbol{\phi}(\mathbf{x}) \tag{7.91}$$

where $\boldsymbol{\Sigma}$ is given by (7.83) in which $\boldsymbol{\alpha}$ and β are set to their optimized values $\boldsymbol{\alpha}^\star$ and $\beta^\star$. This is just the familiar result (3.59) obtained in the context of linear regression. Recall that for localized basis functions, the predictive variance for linear regression models becomes small in regions of input space where there are no basis functions. In the case of an RVM with the basis functions centred on data points, the model will therefore become increasingly certain of its predictions when extrapolating outside the domain of the data (Rasmussen and Quiñonero-Candela, 2005), which of course is undesirable. The predictive distribution in Gaussian process regression does not

Section 6.4.2

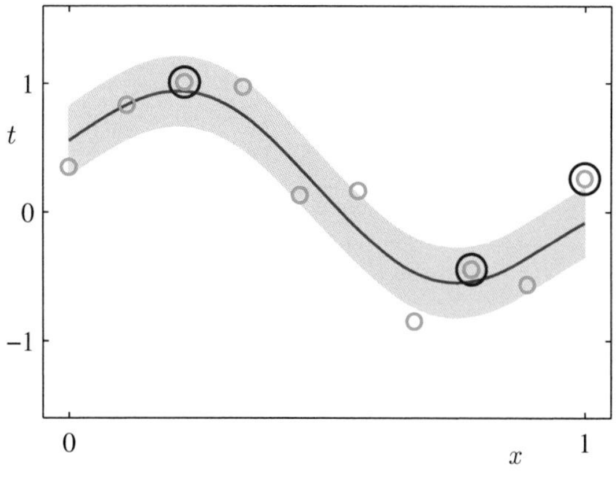

Figure 7.9 Illustration of RVM regression using the same data set, and the same Gaussian kernel functions, as used in Figure 7.8 for the ν-SVM regression model. The mean of the predictive distribution for the RVM is shown by the red line, and the one standard-deviation predictive distribution is shown by the shaded region. Also, the data points are shown in green, and the relevance vectors are indicated by blue circles. Note that there are only 3 relevance vectors compared to 7 support vectors for the ν-SVM in Figure 7.8.

suffer from this problem. However, the computational cost of making predictions with a Gaussian processes is typically much higher than with an RVM.

Figure 7.9 shows an example of the RVM applied to the sinusoidal regression data set. Here the noise precision parameter β is also determined through evidence maximization. We see that the number of relevance vectors in the RVM is significantly smaller than the number of support vectors used by the SVM. For a wide range of regression and classification tasks, the RVM is found to give models that are typically an order of magnitude more compact than the corresponding support vector machine, resulting in a significant improvement in the speed of processing on test data. Remarkably, this greater sparsity is achieved with little or no reduction in generalization error compared with the corresponding SVM.

The principal disadvantage of the RVM compared to the SVM is that training involves optimizing a nonconvex function, and training times can be longer than for a comparable SVM. For a model with M basis functions, the RVM requires inversion of a matrix of size $M \times M$, which in general requires $O(M^3)$ computation. In the specific case of the SVM-like model (7.78), we have $M = N+1$. As we have noted, there are techniques for training SVMs whose cost is roughly quadratic in N. Of course, in the case of the RVM we always have the option of starting with a smaller number of basis functions than $N + 1$. More significantly, in the relevance vector machine the parameters governing complexity and noise variance are determined automatically from a single training run, whereas in the support vector machine the parameters C and ϵ (or ν) are generally found using cross-validation, which involves multiple training runs. Furthermore, in the next section we shall derive an alternative procedure for training the relevance vector machine that improves training speed significantly.

7.2.2 Analysis of sparsity

We have noted earlier that the mechanism of *automatic relevance determination* causes a subset of parameters to be driven to zero. We now examine in more detail

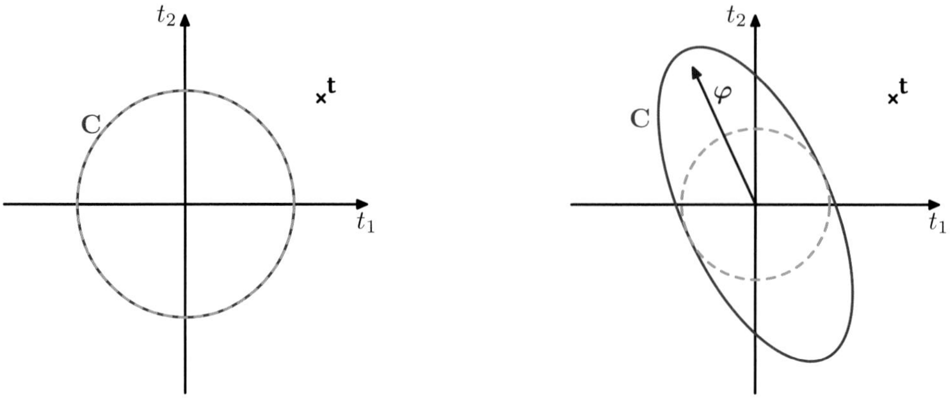

Figure 7.10 Illustration of the mechanism for sparsity in a Bayesian linear regression model, showing a training set vector of target values given by $\mathbf{t} = (t_1, t_2)^{\mathrm{T}}$, indicated by the cross, for a model with one basis vector $\boldsymbol{\varphi} = (\phi(\mathbf{x}_1), \phi(\mathbf{x}_2))^{\mathrm{T}}$, which is poorly aligned with the target data vector $\mathbf{t}$. On the left we see a model having only isotropic noise, so that $\mathbf{C} = \beta^{-1}\mathbf{I}$, corresponding to $\alpha = \infty$, with β set to its most probable value. On the right we see the same model but with a finite value of α. In each case the red ellipse corresponds to unit Mahalanobis distance, with $|\mathbf{C}|$ taking the same value for both plots, while the dashed green circle shows the contribution arising from the noise term β^{-1}. We see that any finite value of α reduces the probability of the observed data, and so for the most probable solution the basis vector is removed.

the mechanism of sparsity in the context of the relevance vector machine. In the process, we will arrive at a significantly faster procedure for optimizing the hyperparameters compared to the direct techniques given above.

Before proceeding with a mathematical analysis, we first give some informal insight into the origin of sparsity in Bayesian linear models. Consider a data set comprising $N = 2$ observations t_1 and t_2, together with a model having a single basis function $\phi(\mathbf{x})$, with hyperparameter α, along with isotropic noise having precision β. From (7.85), the marginal likelihood is given by $p(\mathbf{t}|\alpha, \beta) = \mathcal{N}(\mathbf{t}|\mathbf{0}, \mathbf{C})$ in which the covariance matrix takes the form

$$\mathbf{C} = \frac{1}{\beta}\mathbf{I} + \frac{1}{\alpha}\boldsymbol{\varphi}\boldsymbol{\varphi}^{\mathrm{T}} \tag{7.92}$$

where $\boldsymbol{\varphi}$ denotes the N-dimensional vector $(\phi(\mathbf{x}_1), \phi(\mathbf{x}_2))^{\mathrm{T}}$, and similarly $\mathbf{t} = (t_1, t_2)^{\mathrm{T}}$. Notice that this is just a zero-mean Gaussian process model over $\mathbf{t}$ with covariance $\mathbf{C}$. Given a particular observation for $\mathbf{t}$, our goal is to find $\alpha^\star$ and $\beta^\star$ by maximizing the marginal likelihood. We see from Figure 7.10 that, if there is a poor alignment between the direction of $\boldsymbol{\varphi}$ and that of the training data vector $\mathbf{t}$, then the corresponding hyperparameter α will be driven to ∞, and the basis vector will be pruned from the model. This arises because any finite value for α will always assign a lower probability to the data, thereby decreasing the value of the density at $\mathbf{t}$, provided that β is set to its optimal value. We see that any finite value for α would cause the distribution to be elongated in a direction away from the data, thereby increasing the probability mass in regions away from the observed data and hence reducing the value of the density at the target data vector itself. For the more general case of M

basis vectors $\boldsymbol{\varphi}_1, \ldots, \boldsymbol{\varphi}_M$ a similar intuition holds, namely that if a particular basis vector is poorly aligned with the data vector $\mathbf{t}$, then it is likely to be pruned from the model.

We now investigate the mechanism for sparsity from a more mathematical perspective, for a general case involving M basis functions. To motivate this analysis we first note that, in the result (7.87) for re-estimating the parameter α_i, the terms on the right-hand side are themselves also functions of α_i. These results therefore represent implicit solutions, and iteration would be required even to determine a single α_i with all other α_j for $j \neq i$ fixed.

This suggests a different approach to solving the optimization problem for the RVM, in which we make explicit all of the dependence of the marginal likelihood (7.85) on a particular α_i and then determine its stationary points explicitly (Faul and Tipping, 2002; Tipping and Faul, 2003). To do this, we first pull out the contribution from α_i in the matrix $\mathbf{C}$ defined by (7.86) to give

$$\begin{aligned} \mathbf{C} &= \beta^{-1}\mathbf{I} + \sum_{j \neq i} \alpha_j^{-1} \boldsymbol{\varphi}_j \boldsymbol{\varphi}_j^{\mathrm{T}} + \alpha_i^{-1} \boldsymbol{\varphi}_i \boldsymbol{\varphi}_i^{\mathrm{T}} \\ &= \mathbf{C}_{-i} + \alpha_i^{-1} \boldsymbol{\varphi}_i \boldsymbol{\varphi}_i^{\mathrm{T}} \end{aligned} \tag{7.93}$$

where $\boldsymbol{\varphi}_i$ denotes the ith column of $\boldsymbol{\Phi}$, in other words the N-dimensional vector with elements $(\phi_i(\mathbf{x}_1), \ldots, \phi_i(\mathbf{x}_N))$, in contrast to $\boldsymbol{\phi}_n$, which denotes the nth row of $\boldsymbol{\Phi}$. The matrix $\mathbf{C}_{-i}$ represents the matrix $\mathbf{C}$ with the contribution from basis function i removed. Using the matrix identities (C.7) and (C.15), the determinant and inverse of $\mathbf{C}$ can then be written

$$|\mathbf{C}| = |\mathbf{C}_{-i}| \left(1 + \alpha_i^{-1} \boldsymbol{\varphi}_i^{\mathrm{T}} \mathbf{C}_{-i}^{-1} \boldsymbol{\varphi}_i\right) \tag{7.94}$$

$$\mathbf{C}^{-1} = \mathbf{C}_{-i}^{-1} - \frac{\mathbf{C}_{-i}^{-1} \boldsymbol{\varphi}_i \boldsymbol{\varphi}_i^{\mathrm{T}} \mathbf{C}_{-i}^{-1}}{\alpha_i + \boldsymbol{\varphi}_i^{\mathrm{T}} \mathbf{C}_{-i}^{-1} \boldsymbol{\varphi}_i}. \tag{7.95}$$

Exercise 7.15

Using these results, we can then write the log marginal likelihood function (7.85) in the form

$$L(\boldsymbol{\alpha}) = L(\boldsymbol{\alpha}_{-i}) + \lambda(\alpha_i) \tag{7.96}$$

where $L(\boldsymbol{\alpha}_{-i})$ is simply the log marginal likelihood with basis function $\boldsymbol{\varphi}_i$ omitted, and the quantity $\lambda(\alpha_i)$ is defined by

$$\lambda(\alpha_i) = \frac{1}{2}\left[\ln \alpha_i - \ln(\alpha_i + s_i) + \frac{q_i^2}{\alpha_i + s_i}\right] \tag{7.97}$$

and contains all of the dependence on α_i. Here we have introduced the two quantities

$$s_i = \boldsymbol{\varphi}_i^{\mathrm{T}} \mathbf{C}_{-i}^{-1} \boldsymbol{\varphi}_i \tag{7.98}$$

$$q_i = \boldsymbol{\varphi}_i^{\mathrm{T}} \mathbf{C}_{-i}^{-1} \mathbf{t}. \tag{7.99}$$

Here s_i is called the *sparsity* and q_i is known as the *quality* of $\boldsymbol{\varphi}_i$, and as we shall see, a large value of s_i relative to the value of q_i means that the basis function $\boldsymbol{\varphi}_i$

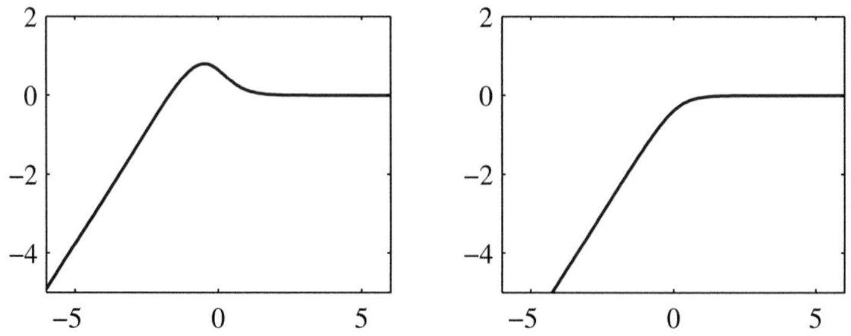

Figure 7.11 Plots of the log marginal likelihood $\lambda(\alpha_i)$ versus $\ln \alpha_i$ showing on the left, the single maximum at a finite α_i for $q_i^2 = 4$ and $s_i = 1$ (so that $q_i^2 > s_i$) and on the right, the maximum at $\alpha_i = \infty$ for $q_i^2 = 1$ and $s_i = 2$ (so that $q_i^2 < s_i$).

is more likely to be pruned from the model. The 'sparsity' measures the extent to which basis function $\boldsymbol{\varphi}_i$ overlaps with the other basis vectors in the model, and the 'quality' represents a measure of the alignment of the basis vector $\boldsymbol{\varphi}_i$ with the error between the training set values $\mathbf{t} = (t_1, \ldots, t_N)^{\mathrm{T}}$ and the vector $\mathbf{y}_{-i}$ of predictions that would result from the model with the vector $\boldsymbol{\varphi}_i$ excluded (Tipping and Faul, 2003).

The stationary points of the marginal likelihood with respect to α_i occur when the derivative

$$\frac{\mathrm{d}\lambda(\alpha_i)}{\mathrm{d}\alpha_i} = \frac{\alpha_i^{-1} s_i^2 - (q_i^2 - s_i)}{2(\alpha_i + s_i)^2} \tag{7.100}$$

is equal to zero. There are two possible forms for the solution. Recalling that $\alpha_i \geqslant 0$, we see that if $q_i^2 < s_i$, then $\alpha_i \to \infty$ provides a solution. Conversely, if $q_i^2 > s_i$, we can solve for α_i to obtain

$$\alpha_i = \frac{s_i^2}{q_i^2 - s_i}. \tag{7.101}$$

These two solutions are illustrated in Figure 7.11. We see that the relative size of the quality and sparsity terms determines whether a particular basis vector will be pruned from the model or not. A more complete analysis (Faul and Tipping, 2002), based on the second derivatives of the marginal likelihood, confirms these solutions are indeed the unique maxima of $\lambda(\alpha_i)$.

Exercise 7.16

Note that this approach has yielded a closed-form solution for α_i, for given values of the other hyperparameters. As well as providing insight into the origin of sparsity in the RVM, this analysis also leads to a practical algorithm for optimizing the hyperparameters that has significant speed advantages. This uses a fixed set of candidate basis vectors, and then cycles through them in turn to decide whether each vector should be included in the model or not. The resulting sequential sparse Bayesian learning algorithm is described below.

Sequential Sparse Bayesian Learning Algorithm

1. If solving a regression problem, initialize β.
2. Initialize using one basis function $\boldsymbol{\varphi}_1$, with hyperparameter α_1 set using (7.101), with the remaining hyperparameters α_j for $j \neq 1$ initialized to infinity, so that only $\boldsymbol{\varphi}_1$ is included in the model.

3. Evaluate $\boldsymbol{\Sigma}$ and $\mathbf{m}$, along with q_i and s_i for all basis functions.
4. Select a candidate basis function $\boldsymbol{\varphi}_i$.
5. If $q_i^2 > s_i$, and $\alpha_i < \infty$, so that the basis vector $\boldsymbol{\varphi}_i$ is already included in the model, then update α_i using (7.101).
6. If $q_i^2 > s_i$, and $\alpha_i = \infty$, then add $\boldsymbol{\varphi}_i$ to the model, and evaluate hyperparameter α_i using (7.101).
7. If $q_i^2 \leqslant s_i$, and $\alpha_i < \infty$ then remove basis function $\boldsymbol{\varphi}_i$ from the model, and set $\alpha_i = \infty$.
8. If solving a regression problem, update β.
9. If converged terminate, otherwise go to 3.

Note that if $q_i^2 \leqslant s_i$ and $\alpha_i = \infty$, then the basis function $\boldsymbol{\varphi}_i$ is already excluded from the model and no action is required.

In practice, it is convenient to evaluate the quantities

$$Q_i = \boldsymbol{\varphi}_i^{\mathrm{T}} \mathbf{C}^{-1} \mathbf{t} \tag{7.102}$$

$$S_i = \boldsymbol{\varphi}_i^{\mathrm{T}} \mathbf{C}^{-1} \boldsymbol{\varphi}_i. \tag{7.103}$$

The quality and sparseness variables can then be expressed in the form

$$q_i = \frac{\alpha_i Q_i}{\alpha_i - S_i} \tag{7.104}$$

$$s_i = \frac{\alpha_i S_i}{\alpha_i - S_i}. \tag{7.105}$$

Exercise 7.17

Note that when $\alpha_i = \infty$, we have $q_i = Q_i$ and $s_i = S_i$. Using (C.7), we can write

$$Q_i = \beta \boldsymbol{\varphi}_i^{\mathrm{T}} \mathbf{t} - \beta^2 \boldsymbol{\varphi}_i^{\mathrm{T}} \boldsymbol{\Phi} \boldsymbol{\Sigma} \boldsymbol{\Phi}^{\mathrm{T}} \mathbf{t} \tag{7.106}$$

$$S_i = \beta \boldsymbol{\varphi}_i^{\mathrm{T}} \boldsymbol{\varphi}_i - \beta^2 \boldsymbol{\varphi}_i^{\mathrm{T}} \boldsymbol{\Phi} \boldsymbol{\Sigma} \boldsymbol{\Phi}^{\mathrm{T}} \boldsymbol{\varphi}_i \tag{7.107}$$

where $\boldsymbol{\Phi}$ and $\boldsymbol{\Sigma}$ involve only those basis vectors that correspond to finite hyperparameters α_i. At each stage the required computations therefore scale like $O(M^3)$, where M is the number of active basis vectors in the model and is typically much smaller than the number N of training patterns.

7.2.3 RVM for classification

We can extend the relevance vector machine framework to classification problems by applying the ARD prior over weights to a probabilistic linear classification model of the kind studied in Chapter 4. To start with, we consider two-class problems with a binary target variable $t \in \{0, 1\}$. The model now takes the form of a linear combination of basis functions transformed by a logistic sigmoid function

$$y(\mathbf{x}, \mathbf{w}) = \sigma\left(\mathbf{w}^{\mathrm{T}} \boldsymbol{\phi}(\mathbf{x})\right) \tag{7.108}$$

where $\sigma(\cdot)$ is the logistic sigmoid function defined by (4.59). If we introduce a Gaussian prior over the weight vector $\mathbf{w}$, then we obtain the model that has been considered already in Chapter 4. The difference here is that in the RVM, this model uses the ARD prior (7.80) in which there is a separate precision hyperparameter associated with each weight parameter.

Section 4.4

In contrast to the regression model, we can no longer integrate analytically over the parameter vector $\mathbf{w}$. Here we follow Tipping (2001) and use the Laplace approximation, which was applied to the closely related problem of Bayesian logistic regression in Section 4.5.1.

We begin by initializing the hyperparameter vector $\boldsymbol{\alpha}$. For this given value of $\boldsymbol{\alpha}$, we then build a Gaussian approximation to the posterior distribution and thereby obtain an approximation to the marginal likelihood. Maximization of this approximate marginal likelihood then leads to a re-estimated value for $\boldsymbol{\alpha}$, and the process is repeated until convergence.

Let us consider the Laplace approximation for this model in more detail. For a fixed value of $\boldsymbol{\alpha}$, the mode of the posterior distribution over $\mathbf{w}$ is obtained by maximizing

$$\begin{aligned}\ln p(\mathbf{w}|\mathbf{t}, \boldsymbol{\alpha}) &= \ln\{p(\mathbf{t}|\mathbf{w})p(\mathbf{w}|\boldsymbol{\alpha})\} - \ln p(\mathbf{t}|\boldsymbol{\alpha}) \\ &= \sum_{n=1}^{N}\{t_n \ln y_n + (1-t_n)\ln(1-y_n)\} - \frac{1}{2}\mathbf{w}^{\mathrm{T}}\mathbf{A}\mathbf{w} + \text{const} \qquad (7.109)\end{aligned}$$

Exercise 7.18

where $\mathbf{A} = \text{diag}(\alpha_i)$. This can be done using iterative reweighted least squares (IRLS) as discussed in Section 4.3.3. For this, we need the gradient vector and Hessian matrix of the log posterior distribution, which from (7.109) are given by

$$\nabla \ln p(\mathbf{w}|\mathbf{t}, \boldsymbol{\alpha}) = \boldsymbol{\Phi}^{\mathrm{T}}(\mathbf{t}-\mathbf{y}) - \mathbf{A}\mathbf{w} \qquad (7.110)$$

$$\nabla\nabla \ln p(\mathbf{w}|\mathbf{t}, \boldsymbol{\alpha}) = -\left(\boldsymbol{\Phi}^{\mathrm{T}}\mathbf{B}\boldsymbol{\Phi} + \mathbf{A}\right) \qquad (7.111)$$

where $\mathbf{B}$ is an $N \times N$ diagonal matrix with elements $b_n = y_n(1-y_n)$, the vector $\mathbf{y} = (y_1, \ldots, y_N)^{\mathrm{T}}$, and $\boldsymbol{\Phi}$ is the design matrix with elements $\Phi_{ni} = \phi_i(\mathbf{x}_n)$. Here we have used the property (4.88) for the derivative of the logistic sigmoid function. At convergence of the IRLS algorithm, the negative Hessian represents the inverse covariance matrix for the Gaussian approximation to the posterior distribution.

The mode of the resulting approximation to the posterior distribution, corresponding to the mean of the Gaussian approximation, is obtained setting (7.110) to zero, giving the mean and covariance of the Laplace approximation in the form

$$\mathbf{w}^{\star} = \mathbf{A}^{-1}\boldsymbol{\Phi}^{\mathrm{T}}(\mathbf{t}-\mathbf{y}) \qquad (7.112)$$

$$\boldsymbol{\Sigma} = \left(\boldsymbol{\Phi}^{\mathrm{T}}\mathbf{B}\boldsymbol{\Phi} + \mathbf{A}\right)^{-1}. \qquad (7.113)$$

We can now use this Laplace approximation to evaluate the marginal likelihood. Using the general result (4.135) for an integral evaluated using the Laplace approximation, we have

$$p(\mathbf{t}|\boldsymbol{\alpha}) = \int p(\mathbf{t}|\mathbf{w})p(\mathbf{w}|\boldsymbol{\alpha})\,\mathrm{d}\mathbf{w}$$
$$\simeq p(\mathbf{t}|\mathbf{w}^\star)p(\mathbf{w}^\star|\boldsymbol{\alpha})(2\pi)^{M/2}|\boldsymbol{\Sigma}|^{1/2}. \tag{7.114}$$

Exercise 7.19

If we substitute for $p(\mathbf{t}|\mathbf{w}^\star)$ and $p(\mathbf{w}^\star|\boldsymbol{\alpha})$ and then set the derivative of the marginal likelihood with respect to α_i equal to zero, we obtain

$$-\frac{1}{2}(w_i^\star)^2 + \frac{1}{2\alpha_i} - \frac{1}{2}\Sigma_{ii} = 0. \tag{7.115}$$

Defining $\gamma_i = 1 - \alpha_i\Sigma_{ii}$ and rearranging then gives

$$\alpha_i^{\text{new}} = \frac{\gamma_i}{(w_i^\star)^2} \tag{7.116}$$

which is identical to the re-estimation formula (7.87) obtained for the regression RVM.

If we define

$$\widehat{\mathbf{t}} = \boldsymbol{\Phi}\mathbf{w}^\star + \mathbf{B}^{-1}(\mathbf{t} - \mathbf{y}) \tag{7.117}$$

we can write the approximate log marginal likelihood in the form

$$\ln p(\mathbf{t}|\boldsymbol{\alpha}) = -\frac{1}{2}\left\{N\ln(2\pi) + \ln|\mathbf{C}| + (\widehat{\mathbf{t}})^{\mathrm{T}}\mathbf{C}^{-1}\widehat{\mathbf{t}}\right\} \tag{7.118}$$

where

$$\mathbf{C} = \mathbf{B} + \boldsymbol{\Phi}\mathbf{A}\boldsymbol{\Phi}^{\mathrm{T}}. \tag{7.119}$$

This takes the same form as (7.85) in the regression case, and so we can apply the same analysis of sparsity and obtain the same fast learning algorithm in which we fully optimize a single hyperparameter α_i at each step.

Appendix A

Figure 7.12 shows the relevance vector machine applied to a synthetic classification data set. We see that the relevance vectors tend not to lie in the region of the decision boundary, in contrast to the support vector machine. This is consistent with our earlier discussion of sparsity in the RVM, because a basis function $\phi_i(\mathbf{x})$ centred on a data point near the boundary will have a vector $\boldsymbol{\varphi}_i$ that is poorly aligned with the training data vector $\mathbf{t}$.

Section 13.3

One of the potential advantages of the relevance vector machine compared with the SVM is that it makes probabilistic predictions. For example, this allows the RVM to be used to help construct an emission density in a nonlinear extension of the linear dynamical system for tracking faces in video sequences (Williams *et al.*, 2005).

So far, we have considered the RVM for binary classification problems. For $K > 2$ classes, we again make use of the probabilistic approach in Section 4.3.4 in which there are K linear models of the form

$$a_k = \mathbf{w}_k^{\mathrm{T}}\mathbf{x} \tag{7.120}$$

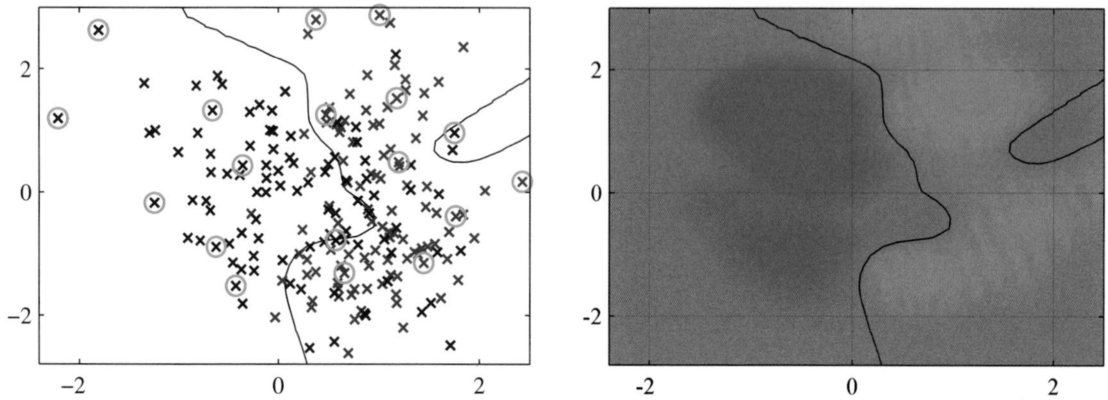

Figure 7.12 Example of the relevance vector machine applied to a synthetic data set, in which the left-hand plot shows the decision boundary and the data points, with the relevance vectors indicated by circles. Comparison with the results shown in Figure 7.4 for the corresponding support vector machine shows that the RVM gives a much sparser model. The right-hand plot shows the posterior probability given by the RVM output in which the proportion of red (blue) ink indicates the probability of that point belonging to the red (blue) class.

which are combined using a softmax function to give outputs

$$y_k(\mathbf{x}) = \frac{\exp(a_k)}{\sum_j \exp(a_j)}. \tag{7.121}$$

The log likelihood function is then given by

$$\ln p(\mathbf{T}|\mathbf{w}_1, \ldots, \mathbf{w}_K) = \prod_{n=1}^{N} \prod_{k=1}^{K} y_{nk}^{t_{nk}} \tag{7.122}$$

where the target values t_{nk} have a 1-of-K coding for each data point n, and $\mathbf{T}$ is a matrix with elements t_{nk}. Again, the Laplace approximation can be used to optimize the hyperparameters (Tipping, 2001), in which the model and its Hessian are found using IRLS. This gives a more principled approach to multiclass classification than the pairwise method used in the support vector machine and also provides probabilistic predictions for new data points. The principal disadvantage is that the Hessian matrix has size $MK \times MK$, where M is the number of active basis functions, which gives an additional factor of K^3 in the computational cost of training compared with the two-class RVM.

The principal disadvantage of the relevance vector machine is the relatively long training times compared with the SVM. This is offset, however, by the avoidance of cross-validation runs to set the model complexity parameters. Furthermore, because it yields sparser models, the computation time on test points, which is usually the more important consideration in practice, is typically much less.

Exercises

7.1 ($\star\star$) **www** Suppose we have a data set of input vectors $\{\mathbf{x}_n\}$ with corresponding target values $t_n \in \{-1, 1\}$, and suppose that we model the density of input vectors within each class separately using a Parzen kernel density estimator (see Section 2.5.1) with a kernel $k(\mathbf{x}, \mathbf{x}')$. Write down the minimum misclassification-rate decision rule assuming the two classes have equal prior probability. Show also that, if the kernel is chosen to be $k(\mathbf{x}, \mathbf{x}') = \mathbf{x}^{\mathrm{T}}\mathbf{x}'$, then the classification rule reduces to simply assigning a new input vector to the class having the closest mean. Finally, show that, if the kernel takes the form $k(\mathbf{x}, \mathbf{x}') = \boldsymbol{\phi}(\mathbf{x})^{\mathrm{T}}\boldsymbol{\phi}(\mathbf{x}')$, that the classification is based on the closest mean in the feature space $\boldsymbol{\phi}(\mathbf{x})$.

7.2 ($\star$) Show that, if the 1 on the right-hand side of the constraint (7.5) is replaced by some arbitrary constant $\gamma > 0$, the solution for the maximum margin hyperplane is unchanged.

7.3 ($\star\star$) Show that, irrespective of the dimensionality of the data space, a data set consisting of just two data points, one from each class, is sufficient to determine the location of the maximum-margin hyperplane.

7.4 ($\star\star$) **www** Show that the value ρ of the margin for the maximum-margin hyperplane is given by

$$\frac{1}{\rho^2} = \sum_{n=1}^{N} a_n \tag{7.123}$$

where $\{a_n\}$ are given by maximizing (7.10) subject to the constraints (7.11) and (7.12).

7.5 ($\star\star$) Show that the values of ρ and $\{a_n\}$ in the previous exercise also satisfy

$$\frac{1}{\rho^2} = 2\widetilde{L}(\mathbf{a}) \tag{7.124}$$

where $\widetilde{L}(\mathbf{a})$ is defined by (7.10). Similarly, show that

$$\frac{1}{\rho^2} = \|\mathbf{w}\|^2. \tag{7.125}$$

7.6 ($\star$) Consider the logistic regression model with a target variable $t \in \{-1, 1\}$. If we define $p(t = 1|y) = \sigma(y)$ where $y(\mathbf{x})$ is given by (7.1), show that the negative log likelihood, with the addition of a quadratic regularization term, takes the form (7.47).

7.7 ($\star$) Consider the Lagrangian (7.56) for the regression support vector machine. By setting the derivatives of the Lagrangian with respect to $\mathbf{w}$, b, ξ_n, and $\widehat{\xi}_n$ to zero and then back substituting to eliminate the corresponding variables, show that the dual Lagrangian is given by (7.61).

7.8 ($\star$) **www** For the regression support vector machine considered in Section 7.1.4, show that all training data points for which $\xi_n > 0$ will have $a_n = C$, and similarly all points for which $\widehat{\xi}_n > 0$ will have $\widehat{a}_n = C$.

7.9 ($\star$) Verify the results (7.82) and (7.83) for the mean and covariance of the posterior distribution over weights in the regression RVM.

7.10 ($\star\star$) **www** Derive the result (7.85) for the marginal likelihood function in the regression RVM, by performing the Gaussian integral over $\mathbf{w}$ in (7.84) using the technique of completing the square in the exponential.

7.11 ($\star\star$) Repeat the above exercise, but this time make use of the general result (2.115).

7.12 ($\star\star$) **www** Show that direct maximization of the log marginal likelihood (7.85) for the regression relevance vector machine leads to the re-estimation equations (7.87) and (7.88) where γ_i is defined by (7.89).

7.13 ($\star\star$) In the evidence framework for RVM regression, we obtained the re-estimation formulae (7.87) and (7.88) by maximizing the marginal likelihood given by (7.85). Extend this approach by inclusion of hyperpriors given by gamma distributions of the form (B.26) and obtain the corresponding re-estimation formulae for $\boldsymbol{\alpha}$ and β by maximizing the corresponding posterior probability $p(\mathbf{t}, \boldsymbol{\alpha}, \beta|\mathbf{X})$ with respect to $\boldsymbol{\alpha}$ and β.

7.14 ($\star\star$) Derive the result (7.90) for the predictive distribution in the relevance vector machine for regression. Show that the predictive variance is given by (7.91).

7.15 ($\star\star$) **www** Using the results (7.94) and (7.95), show that the marginal likelihood (7.85) can be written in the form (7.96), where $\lambda(\alpha_n)$ is defined by (7.97) and the sparsity and quality factors are defined by (7.98) and (7.99), respectively.

7.16 ($\star$) By taking the second derivative of the log marginal likelihood (7.97) for the regression RVM with respect to the hyperparameter α_i, show that the stationary point given by (7.101) is a maximum of the marginal likelihood.

7.17 ($\star\star$) Using (7.83) and (7.86), together with the matrix identity (C.7), show that the quantities S_n and Q_n defined by (7.102) and (7.103) can be written in the form (7.106) and (7.107).

7.18 ($\star$) **www** Show that the gradient vector and Hessian matrix of the log posterior distribution (7.109) for the classification relevance vector machine are given by (7.110) and (7.111).

7.19 ($\star\star$) Verify that maximization of the approximate marginal likelihood function (7.114) for the classification relevance vector machine leads to the result (7.116) for re-estimation of the hyperparameters.

8 Graphical Models

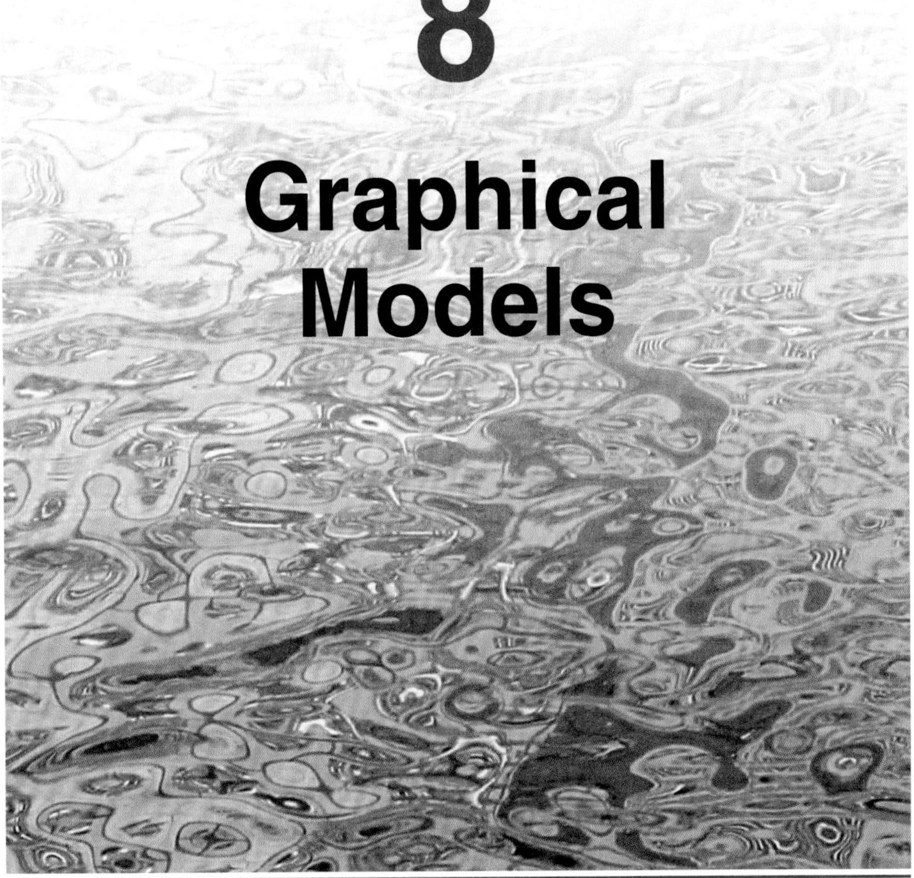

Probabilities play a central role in modern pattern recognition. We have seen in Chapter 1 that probability theory can be expressed in terms of two simple equations corresponding to the sum rule and the product rule. All of the probabilistic inference and learning manipulations discussed in this book, no matter how complex, amount to repeated application of these two equations. We could therefore proceed to formulate and solve complicated probabilistic models purely by algebraic manipulation. However, we shall find it highly advantageous to augment the analysis using diagrammatic representations of probability distributions, called *probabilistic graphical models*. These offer several useful properties:

1. They provide a simple way to visualize the structure of a probabilistic model and can be used to design and motivate new models.

2. Insights into the properties of the model, including conditional independence properties, can be obtained by inspection of the graph.

3. Complex computations, required to perform inference and learning in sophisticated models, can be expressed in terms of graphical manipulations, in which underlying mathematical expressions are carried along implicitly.

A graph comprises *nodes* (also called *vertices*) connected by *links* (also known as *edges* or *arcs*). In a probabilistic graphical model, each node represents a random variable (or group of random variables), and the links express probabilistic relationships between these variables. The graph then captures the way in which the joint distribution over all of the random variables can be decomposed into a product of factors each depending only on a subset of the variables. We shall begin by discussing *Bayesian networks*, also known as *directed graphical models*, in which the links of the graphs have a particular directionality indicated by arrows. The other major class of graphical models are *Markov random fields*, also known as *undirected graphical models*, in which the links do not carry arrows and have no directional significance. Directed graphs are useful for expressing causal relationships between random variables, whereas undirected graphs are better suited to expressing soft constraints between random variables. For the purposes of solving inference problems, it is often convenient to convert both directed and undirected graphs into a different representation called a *factor graph*.

In this chapter, we shall focus on the key aspects of graphical models as needed for applications in pattern recognition and machine learning. More general treatments of graphical models can be found in the books by Whittaker (1990), Lauritzen (1996), Jensen (1996), Castillo *et al.* (1997), Jordan (1999), Cowell *et al.* (1999), and Jordan (2007).

8.1. Bayesian Networks

In order to motivate the use of directed graphs to describe probability distributions, consider first an arbitrary joint distribution $p(a, b, c)$ over three variables a, b, and c. Note that at this stage, we do not need to specify anything further about these variables, such as whether they are discrete or continuous. Indeed, one of the powerful aspects of graphical models is that a specific graph can make probabilistic statements for a broad class of distributions. By application of the product rule of probability (1.11), we can write the joint distribution in the form

$$p(a, b, c) = p(c|a, b)p(a, b). \tag{8.1}$$

A second application of the product rule, this time to the second term on the right-hand side of (8.1), gives

$$p(a, b, c) = p(c|a, b)p(b|a)p(a). \tag{8.2}$$

Note that this decomposition holds for any choice of the joint distribution. We now represent the right-hand side of (8.2) in terms of a simple graphical model as follows. First we introduce a node for each of the random variables a, b, and c and associate each node with the corresponding conditional distribution on the right-hand side of

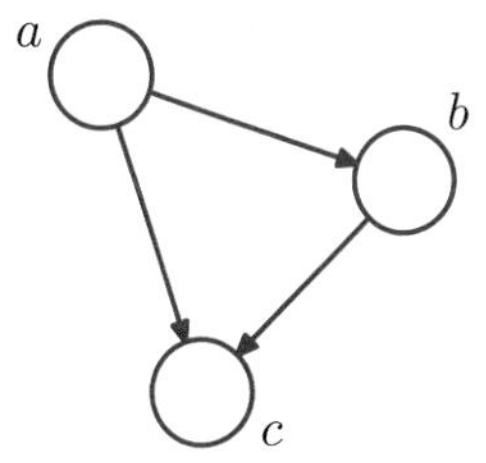

Figure 8.1 A directed graphical model representing the joint probability distribution over three variables a, b, and c, corresponding to the decomposition on the right-hand side of (8.2).

(8.2). Then, for each conditional distribution we add directed links (arrows) to the graph from the nodes corresponding to the variables on which the distribution is conditioned. Thus for the factor $p(c|a, b)$, there will be links from nodes a and b to node c, whereas for the factor $p(a)$ there will be no incoming links. The result is the graph shown in Figure 8.1. If there is a link going from a node a to a node b, then we say that node a is the *parent* of node b, and we say that node b is the *child* of node a. Note that we shall not make any formal distinction between a node and the variable to which it corresponds but will simply use the same symbol to refer to both.

An interesting point to note about (8.2) is that the left-hand side is symmetrical with respect to the three variables a, b, and c, whereas the right-hand side is not. Indeed, in making the decomposition in (8.2), we have implicitly chosen a particular ordering, namely a, b, c, and had we chosen a different ordering we would have obtained a different decomposition and hence a different graphical representation. We shall return to this point later.

For the moment let us extend the example of Figure 8.1 by considering the joint distribution over K variables given by $p(x_1, \ldots, x_K)$. By repeated application of the product rule of probability, this joint distribution can be written as a product of conditional distributions, one for each of the variables

$$p(x_1, \ldots, x_K) = p(x_K|x_1, \ldots, x_{K-1}) \ldots p(x_2|x_1)p(x_1). \tag{8.3}$$

For a given choice of K, we can again represent this as a directed graph having K nodes, one for each conditional distribution on the right-hand side of (8.3), with each node having incoming links from all lower numbered nodes. We say that this graph is *fully connected* because there is a link between every pair of nodes.

So far, we have worked with completely general joint distributions, so that the decompositions, and their representations as fully connected graphs, will be applicable to any choice of distribution. As we shall see shortly, it is the *absence* of links in the graph that conveys interesting information about the properties of the class of distributions that the graph represents. Consider the graph shown in Figure 8.2. This is not a fully connected graph because, for instance, there is no link from x_1 to x_2 or from x_3 to x_7.

We shall now go from this graph to the corresponding representation of the joint probability distribution written in terms of the product of a set of conditional distributions, one for each node in the graph. Each such conditional distribution will be conditioned only on the parents of the corresponding node in the graph. For instance, x_5 will be conditioned on x_1 and x_3. The joint distribution of all 7 variables

Figure 8.2 Example of a directed acyclic graph describing the joint distribution over variables $x_1, \ldots, x_7$. The corresponding decomposition of the joint distribution is given by (8.4).

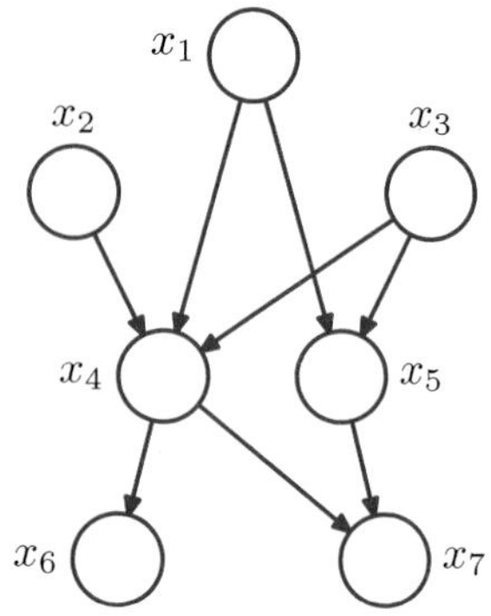

is therefore given by

$$p(x_1)p(x_2)p(x_3)p(x_4|x_1, x_2, x_3)p(x_5|x_1, x_3)p(x_6|x_4)p(x_7|x_4, x_5). \tag{8.4}$$

The reader should take a moment to study carefully the correspondence between (8.4) and Figure 8.2.

We can now state in general terms the relationship between a given directed graph and the corresponding distribution over the variables. The joint distribution defined by a graph is given by the product, over all of the nodes of the graph, of a conditional distribution for each node conditioned on the variables corresponding to the parents of that node in the graph. Thus, for a graph with K nodes, the joint distribution is given by

$$p(\mathbf{x}) = \prod_{k=1}^{K} p(x_k|\mathrm{pa}_k) \tag{8.5}$$

where pa_k denotes the set of parents of x_k, and $\mathbf{x} = \{x_1, \ldots, x_K\}$. This key equation expresses the *factorization* properties of the joint distribution for a directed graphical model. Although we have considered each node to correspond to a single variable, we can equally well associate sets of variables and vector-valued variables with the nodes of a graph. It is easy to show that the representation on the right-hand side of (8.5) is always correctly normalized provided the individual conditional distributions are normalized.

Exercise 8.1

The directed graphs that we are considering are subject to an important restriction namely that there must be no *directed cycles*, in other words there are no closed paths within the graph such that we can move from node to node along links following the direction of the arrows and end up back at the starting node. Such graphs are also called *directed acyclic graphs*, or *DAGs*. This is equivalent to the statement that there exists an ordering of the nodes such that there are no links that go from any node to any lower numbered node.

Exercise 8.2

8.1.1 Example: Polynomial regression

As an illustration of the use of directed graphs to describe probability distributions, we consider the Bayesian polynomial regression model introduced in Sec-

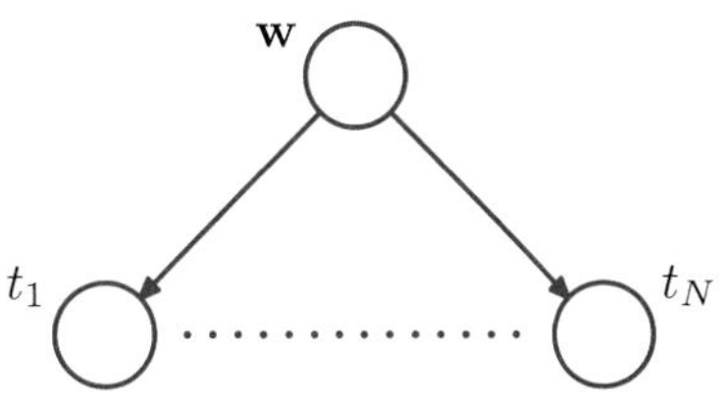

Figure 8.3 Directed graphical model representing the joint distribution (8.6) corresponding to the Bayesian polynomial regression model introduced in Section 1.2.6.

tion 1.2.6. The random variables in this model are the vector of polynomial coefficients $\mathbf{w}$ and the observed data $\mathbf{t} = (t_1, \ldots, t_N)^{\mathrm{T}}$. In addition, this model contains the input data $\mathbf{x} = (x_1, \ldots, x_N)^{\mathrm{T}}$, the noise variance σ^2, and the hyperparameter α representing the precision of the Gaussian prior over $\mathbf{w}$, all of which are parameters of the model rather than random variables. Focussing just on the random variables for the moment, we see that the joint distribution is given by the product of the prior $p(\mathbf{w})$ and N conditional distributions $p(t_n|\mathbf{w})$ for $n = 1, \ldots, N$ so that

$$p(\mathbf{t}, \mathbf{w}) = p(\mathbf{w}) \prod_{n=1}^{N} p(t_n|\mathbf{w}). \tag{8.6}$$

This joint distribution can be represented by a graphical model shown in Figure 8.3.

When we start to deal with more complex models later in the book, we shall find it inconvenient to have to write out multiple nodes of the form $t_1, \ldots, t_N$ explicitly as in Figure 8.3. We therefore introduce a graphical notation that allows such multiple nodes to be expressed more compactly, in which we draw a single representative node t_n and then surround this with a box, called a *plate*, labelled with N indicating that there are N nodes of this kind. Re-writing the graph of Figure 8.3 in this way, we obtain the graph shown in Figure 8.4.

We shall sometimes find it helpful to make the parameters of a model, as well as its stochastic variables, explicit. In this case, (8.6) becomes

$$p(\mathbf{t}, \mathbf{w}|\mathbf{x}, \alpha, \sigma^2) = p(\mathbf{w}|\alpha) \prod_{n=1}^{N} p(t_n|\mathbf{w}, x_n, \sigma^2).$$

Correspondingly, we can make $\mathbf{x}$ and α explicit in the graphical representation. To do this, we shall adopt the convention that random variables will be denoted by open circles, and deterministic parameters will be denoted by smaller solid circles. If we take the graph of Figure 8.4 and include the deterministic parameters, we obtain the graph shown in Figure 8.5.

When we apply a graphical model to a problem in machine learning or pattern recognition, we will typically set some of the random variables to specific observed

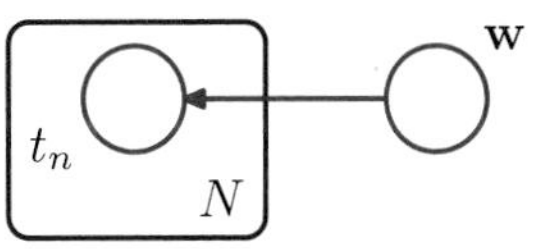

Figure 8.4 An alternative, more compact, representation of the graph shown in Figure 8.3 in which we have introduced a *plate* (the box labelled N) that represents N nodes of which only a single example t_n is shown explicitly.

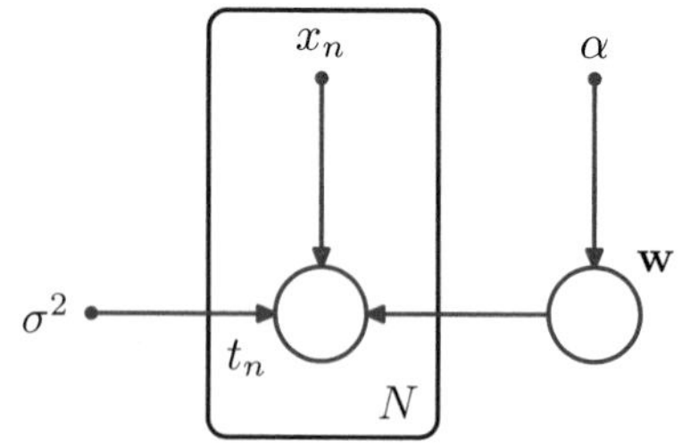

Figure 8.5 This shows the same model as in Figure 8.4 but with the deterministic parameters shown explicitly by the smaller solid nodes.

values, for example the variables $\{t_n\}$ from the training set in the case of polynomial curve fitting. In a graphical model, we will denote such *observed variables* by shading the corresponding nodes. Thus the graph corresponding to Figure 8.5 in which the variables $\{t_n\}$ are observed is shown in Figure 8.6. Note that the value of $\mathbf{w}$ is not observed, and so $\mathbf{w}$ is an example of a *latent* variable, also known as a *hidden* variable. Such variables play a crucial role in many probabilistic models and will form the focus of Chapters 9 and 12.

Having observed the values $\{t_n\}$ we can, if desired, evaluate the posterior distribution of the polynomial coefficients $\mathbf{w}$ as discussed in Section 1.2.5. For the moment, we note that this involves a straightforward application of Bayes' theorem

$$p(\mathbf{w}|\mathbf{t}) \propto p(\mathbf{w}) \prod_{n=1}^{N} p(t_n|\mathbf{w}) \tag{8.7}$$

where again we have omitted the deterministic parameters in order to keep the notation uncluttered.

In general, model parameters such as $\mathbf{w}$ are of little direct interest in themselves, because our ultimate goal is to make predictions for new input values. Suppose we are given a new input value $\widehat{x}$ and we wish to find the corresponding probability distribution for $\widehat{t}$ conditioned on the observed data. The graphical model that describes this problem is shown in Figure 8.7, and the corresponding joint distribution of all of the random variables in this model, conditioned on the deterministic parameters, is then given by

$$p(\widehat{t}, \mathbf{t}, \mathbf{w}|\widehat{x}, \mathbf{x}, \alpha, \sigma^2) = \left[\prod_{n=1}^{N} p(t_n|x_n, \mathbf{w}, \sigma^2)\right] p(\mathbf{w}|\alpha)p(\widehat{t}|\widehat{x}, \mathbf{w}, \sigma^2). \tag{8.8}$$

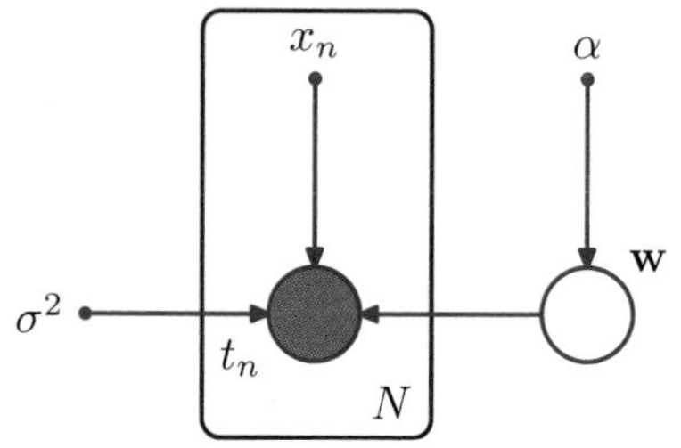

Figure 8.6 As in Figure 8.5 but with the nodes $\{t_n\}$ shaded to indicate that the corresponding random variables have been set to their observed (training set) values.

Figure 8.7 The polynomial regression model, corresponding to Figure 8.6, showing also a new input value $\widehat{x}$ together with the corresponding model prediction $\widehat{t}$.

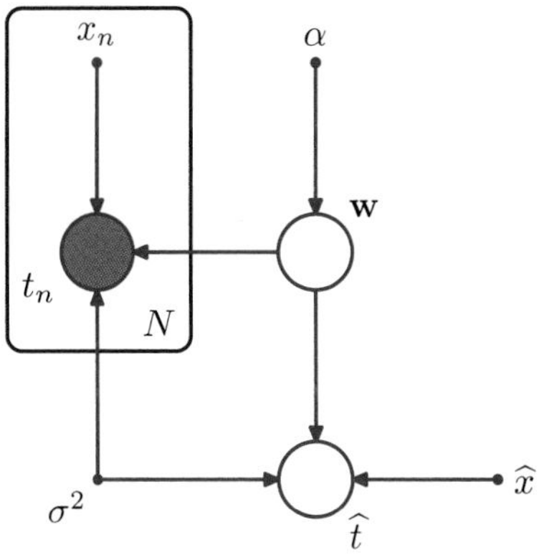

The required predictive distribution for $\widehat{t}$ is then obtained, from the sum rule of probability, by integrating out the model parameters $\mathbf{w}$ so that

$$p(\widehat{t}|\widehat{x}, \mathbf{x}, \mathbf{t}, \alpha, \sigma^2) \propto \int p(\widehat{t}, \mathbf{t}, \mathbf{w}|\widehat{x}, \mathbf{x}, \alpha, \sigma^2)\, \mathrm{d}\mathbf{w}$$

where we are implicitly setting the random variables in $\mathbf{t}$ to the specific values observed in the data set. The details of this calculation were discussed in Chapter 3.

8.1.2 Generative models

There are many situations in which we wish to draw samples from a given probability distribution. Although we shall devote the whole of Chapter 11 to a detailed discussion of sampling methods, it is instructive to outline here one technique, called *ancestral sampling*, which is particularly relevant to graphical models. Consider a joint distribution $p(x_1, \ldots, x_K)$ over K variables that factorizes according to (8.5) corresponding to a directed acyclic graph. We shall suppose that the variables have been ordered such that there are no links from any node to any lower numbered node, in other words each node has a higher number than any of its parents. Our goal is to draw a sample $\widehat{x}_1, \ldots, \widehat{x}_K$ from the joint distribution.

To do this, we start with the lowest-numbered node and draw a sample from the distribution $p(x_1)$, which we call $\widehat{x}_1$. We then work through each of the nodes in order, so that for node n we draw a sample from the conditional distribution $p(x_n|\mathrm{pa}_n)$ in which the parent variables have been set to their sampled values. Note that at each stage, these parent values will always be available because they correspond to lower-numbered nodes that have already been sampled. Techniques for sampling from specific distributions will be discussed in detail in Chapter 11. Once we have sampled from the final variable x_K, we will have achieved our objective of obtaining a sample from the joint distribution. To obtain a sample from some marginal distribution corresponding to a subset of the variables, we simply take the sampled values for the required nodes and ignore the sampled values for the remaining nodes. For example, to draw a sample from the distribution $p(x_2, x_4)$, we simply sample from the full joint distribution and then retain the values $\widehat{x}_2, \widehat{x}_4$ and discard the remaining values $\{\widehat{x}_{j\neq 2,4}\}$.

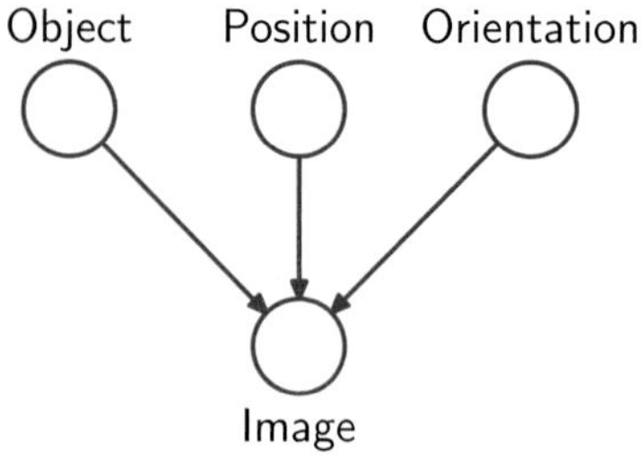

Figure 8.8 A graphical model representing the process by which images of objects are created, in which the identity of an object (a discrete variable) and the position and orientation of that object (continuous variables) have independent prior probabilities. The image (a vector of pixel intensities) has a probability distribution that is dependent on the identity of the object as well as on its position and orientation.

For practical applications of probabilistic models, it will typically be the higher-numbered variables corresponding to terminal nodes of the graph that represent the observations, with lower-numbered nodes corresponding to latent variables. The primary role of the latent variables is to allow a complicated distribution over the observed variables to be represented in terms of a model constructed from simpler (typically exponential family) conditional distributions.

We can interpret such models as expressing the processes by which the observed data arose. For instance, consider an object recognition task in which each observed data point corresponds to an image (comprising a vector of pixel intensities) of one of the objects. In this case, the latent variables might have an interpretation as the position and orientation of the object. Given a particular observed image, our goal is to find the posterior distribution over objects, in which we integrate over all possible positions and orientations. We can represent this problem using a graphical model of the form shown in Figure 8.8.

The graphical model captures the *causal* process (Pearl, 1988) by which the observed data was generated. For this reason, such models are often called *generative* models. By contrast, the polynomial regression model described by Figure 8.5 is not generative because there is no probability distribution associated with the input variable x, and so it is not possible to generate synthetic data points from this model. We could make it generative by introducing a suitable prior distribution $p(x)$, at the expense of a more complex model.

The hidden variables in a probabilistic model need not, however, have any explicit physical interpretation but may be introduced simply to allow a more complex joint distribution to be constructed from simpler components. In either case, the technique of ancestral sampling applied to a generative model mimics the creation of the observed data and would therefore give rise to 'fantasy' data whose probability distribution (if the model were a perfect representation of reality) would be the same as that of the observed data. In practice, producing synthetic observations from a generative model can prove informative in understanding the form of the probability distribution represented by that model.

8.1.3 Discrete variables

Section 2.4

We have discussed the importance of probability distributions that are members of the exponential family, and we have seen that this family includes many well-known distributions as particular cases. Although such distributions are relatively simple, they form useful building blocks for constructing more complex probability

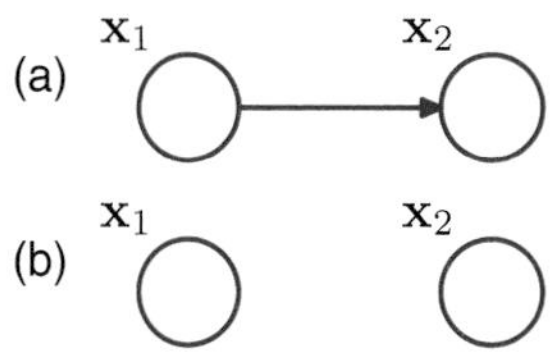

Figure 8.9 (a) This fully-connected graph describes a general distribution over two K-state discrete variables having a total of $K^2 - 1$ parameters. (b) By dropping the link between the nodes, the number of parameters is reduced to $2(K-1)$.

distributions, and the framework of graphical models is very useful in expressing the way in which these building blocks are linked together.

Such models have particularly nice properties if we choose the relationship between each parent-child pair in a directed graph to be conjugate, and we shall explore several examples of this shortly. Two cases are particularly worthy of note, namely when the parent and child node each correspond to discrete variables and when they each correspond to Gaussian variables, because in these two cases the relationship can be extended hierarchically to construct arbitrarily complex directed acyclic graphs. We begin by examining the discrete case.

The probability distribution $p(\mathbf{x}|\boldsymbol{\mu})$ for a single discrete variable $\mathbf{x}$ having K possible states (using the 1-of-K representation) is given by

$$p(\mathbf{x}|\boldsymbol{\mu}) = \prod_{k=1}^{K} \mu_k^{x_k} \tag{8.9}$$

and is governed by the parameters $\boldsymbol{\mu} = (\mu_1, \ldots, \mu_K)^{\mathrm{T}}$. Due to the constraint $\sum_k \mu_k = 1$, only $K-1$ values for μ_k need to be specified in order to define the distribution.

Now suppose that we have two discrete variables, $\mathbf{x}_1$ and $\mathbf{x}_2$, each of which has K states, and we wish to model their joint distribution. We denote the probability of observing both $x_{1k} = 1$ and $x_{2l} = 1$ by the parameter μ_{kl}, where x_{1k} denotes the kth component of $\mathbf{x}_1$, and similarly for x_{2l}. The joint distribution can be written

$$p(\mathbf{x}_1, \mathbf{x}_2|\boldsymbol{\mu}) = \prod_{k=1}^{K}\prod_{l=1}^{K} \mu_{kl}^{x_{1k}x_{2l}}.$$

Because the parameters μ_{kl} are subject to the constraint $\sum_k \sum_l \mu_{kl} = 1$, this distribution is governed by $K^2 - 1$ parameters. It is easily seen that the total number of parameters that must be specified for an arbitrary joint distribution over M variables is $K^M - 1$ and therefore grows exponentially with the number M of variables.

Using the product rule, we can factor the joint distribution $p(\mathbf{x}_1, \mathbf{x}_2)$ in the form $p(\mathbf{x}_2|\mathbf{x}_1)p(\mathbf{x}_1)$, which corresponds to a two-node graph with a link going from the $\mathbf{x}_1$ node to the $\mathbf{x}_2$ node as shown in Figure 8.9(a).The marginal distribution $p(\mathbf{x}_1)$ is governed by $K-1$ parameters, as before. Similarly, the conditional distribution $p(\mathbf{x}_2|\mathbf{x}_1)$ requires the specification of $K-1$ parameters for each of the K possible values of $\mathbf{x}_1$. The total number of parameters that must be specified in the joint distribution is therefore $(K-1) + K(K-1) = K^2 - 1$ as before.

Now suppose that the variables $\mathbf{x}_1$ and $\mathbf{x}_2$ were independent, corresponding to the graphical model shown in Figure 8.9(b). Each variable is then described by

Figure 8.10 This chain of M discrete nodes, each having K states, requires the specification of $K-1+(M-1)K(K-1)$ parameters, which grows linearly with the length M of the chain. In contrast, a fully connected graph of M nodes would have K^M-1 parameters, which grows exponentially with M.

a separate multinomial distribution, and the total number of parameters would be $2(K-1)$. For a distribution over M independent discrete variables, each having K states, the total number of parameters would be $M(K-1)$, which therefore grows linearly with the number of variables. From a graphical perspective, we have reduced the number of parameters by dropping links in the graph, at the expense of having a restricted class of distributions.

More generally, if we have M discrete variables $\mathbf{x}_1, \ldots, \mathbf{x}_M$, we can model the joint distribution using a directed graph with one variable corresponding to each node. The conditional distribution at each node is given by a set of nonnegative parameters subject to the usual normalization constraint. If the graph is fully connected then we have a completely general distribution having $K^M - 1$ parameters, whereas if there are no links in the graph the joint distribution factorizes into the product of the marginals, and the total number of parameters is $M(K-1)$. Graphs having intermediate levels of connectivity allow for more general distributions than the fully factorized one while requiring fewer parameters than the general joint distribution. As an illustration, consider the chain of nodes shown in Figure 8.10. The marginal distribution $p(\mathbf{x}_1)$ requires $K-1$ parameters, whereas each of the $M-1$ conditional distributions $p(\mathbf{x}_i|\mathbf{x}_{i-1})$, for $i = 2, \ldots, M$, requires $K(K-1)$ parameters. This gives a total parameter count of $K-1+(M-1)K(K-1)$, which is quadratic in K and which grows linearly (rather than exponentially) with the length M of the chain.

An alternative way to reduce the number of independent parameters in a model is by *sharing* parameters (also known as *tying* of parameters). For instance, in the chain example of Figure 8.10, we can arrange that all of the conditional distributions $p(\mathbf{x}_i|\mathbf{x}_{i-1})$, for $i = 2, \ldots, M$, are governed by the same set of $K(K-1)$ parameters. Together with the $K-1$ parameters governing the distribution of $\mathbf{x}_1$, this gives a total of $K^2 - 1$ parameters that must be specified in order to define the joint distribution.

We can turn a graph over discrete variables into a Bayesian model by introducing Dirichlet priors for the parameters. From a graphical point of view, each node then acquires an additional parent representing the Dirichlet distribution over the parameters associated with the corresponding discrete node. This is illustrated for the chain model in Figure 8.11. The corresponding model in which we tie the parameters governing the conditional distributions $p(\mathbf{x}_i|\mathbf{x}_{i-1})$, for $i = 2, \ldots, M$, is shown in Figure 8.12.

Another way of controlling the exponential growth in the number of parameters in models of discrete variables is to use parameterized models for the conditional distributions instead of complete tables of conditional probability values. To illustrate this idea, consider the graph in Figure 8.13 in which all of the nodes represent binary variables. Each of the parent variables x_i is governed by a single parame-

Figure 8.11 An extension of the model of Figure 8.10 to include Dirichlet priors over the parameters governing the discrete distributions.

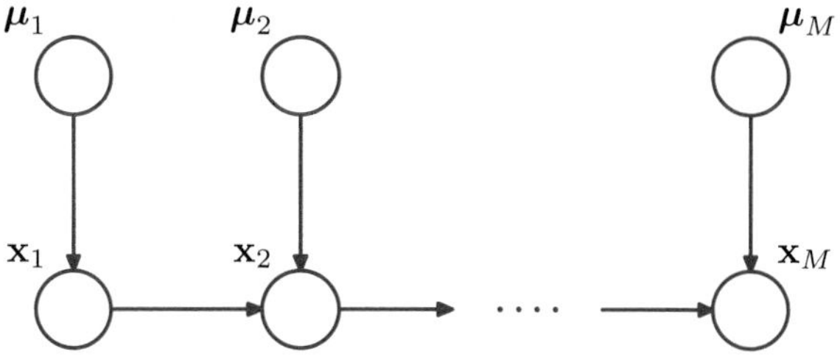

Figure 8.12 As in Figure 8.11 but with a single set of parameters $\boldsymbol{\mu}$ shared amongst all of the conditional distributions $p(\mathbf{x}_i|\mathbf{x}_{i-1})$.

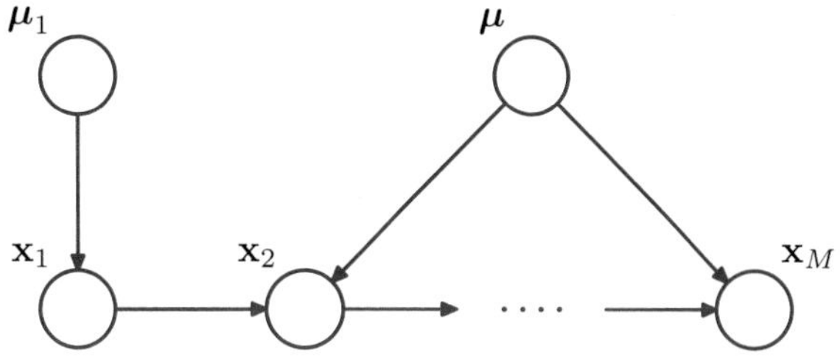

ter μ_i representing the probability $p(x_i = 1)$, giving M parameters in total for the parent nodes. The conditional distribution $p(y|x_1, \ldots, x_M)$, however, would require 2^M parameters representing the probability $p(y = 1)$ for each of the 2^M possible settings of the parent variables. Thus in general the number of parameters required to specify this conditional distribution will grow exponentially with M. We can obtain a more parsimonious form for the conditional distribution by using a logistic sigmoid function acting on a linear combination of the parent variables, giving

Section 2.4

$$p(y = 1|x_1, \ldots, x_M) = \sigma\left(w_0 + \sum_{i=1}^{M} w_i x_i\right) = \sigma(\mathbf{w}^{\mathrm{T}}\mathbf{x}) \tag{8.10}$$

where $\sigma(a) = (1+\exp(-a))^{-1}$ is the logistic sigmoid, $\mathbf{x} = (x_0, x_1, \ldots, x_M)^{\mathrm{T}}$ is an $(M + 1)$-dimensional vector of parent states augmented with an additional variable x_0 whose value is clamped to 1, and $\mathbf{w} = (w_0, w_1, \ldots, w_M)^{\mathrm{T}}$ is a vector of $M + 1$ parameters. This is a more restricted form of conditional distribution than the general case but is now governed by a number of parameters that grows linearly with M. In this sense, it is analogous to the choice of a restrictive form of covariance matrix (for example, a diagonal matrix) in a multivariate Gaussian distribution. The motivation for the logistic sigmoid representation was discussed in Section 4.2.

Figure 8.13 A graph comprising M parents $x_1, \ldots, x_M$ and a single child y, used to illustrate the idea of parameterized conditional distributions for discrete variables.

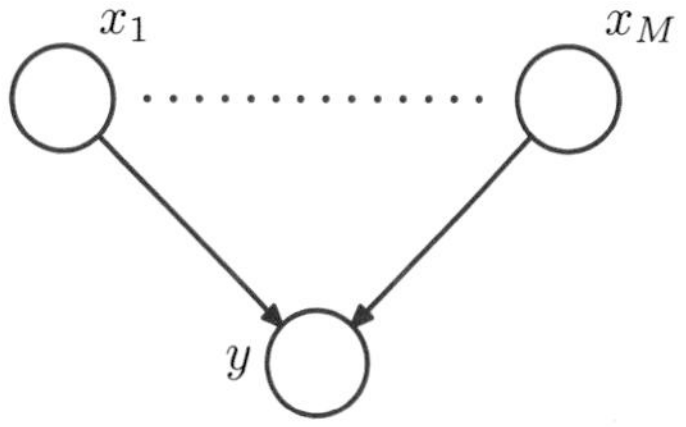

8.1.4 Linear-Gaussian models

In the previous section, we saw how to construct joint probability distributions over a set of discrete variables by expressing the variables as nodes in a directed acyclic graph. Here we show how a multivariate Gaussian can be expressed as a directed graph corresponding to a linear-Gaussian model over the component variables. This allows us to impose interesting structure on the distribution, with the general Gaussian and the diagonal covariance Gaussian representing opposite extremes. Several widely used techniques are examples of linear-Gaussian models, such as probabilistic principal component analysis, factor analysis, and linear dynamical systems (Roweis and Ghahramani, 1999). We shall make extensive use of the results of this section in later chapters when we consider some of these techniques in detail.

Consider an arbitrary directed acyclic graph over D variables in which node i represents a single continuous random variable x_i having a Gaussian distribution. The mean of this distribution is taken to be a linear combination of the states of its parent nodes pa_i of node i

$$p(x_i|\mathrm{pa}_i) = \mathcal{N}\left(x_i \middle| \sum_{j\in\mathrm{pa}_i} w_{ij}x_j + b_i, v_i\right) \tag{8.11}$$

where w_{ij} and b_i are parameters governing the mean, and v_i is the variance of the conditional distribution for x_i. The log of the joint distribution is then the log of the product of these conditionals over all nodes in the graph and hence takes the form

$$\begin{aligned}\ln p(\mathbf{x}) &= \sum_{i=1}^{D} \ln p(x_i|\mathrm{pa}_i) &(8.12)\\ &= -\sum_{i=1}^{D} \frac{1}{2v_i}\left(x_i - \sum_{j\in\mathrm{pa}_i} w_{ij}x_j - b_i\right)^2 + \text{const} &(8.13)\end{aligned}$$

where $\mathbf{x} = (x_1, \ldots, x_D)^{\mathrm{T}}$ and 'const' denotes terms independent of $\mathbf{x}$. We see that this is a quadratic function of the components of $\mathbf{x}$, and hence the joint distribution $p(\mathbf{x})$ is a multivariate Gaussian.

We can determine the mean and covariance of the joint distribution recursively as follows. Each variable x_i has (conditional on the states of its parents) a Gaussian distribution of the form (8.11) and so

$$x_i = \sum_{j\in\mathrm{pa}_i} w_{ij}x_j + b_i + \sqrt{v_i}\epsilon_i \tag{8.14}$$

where ϵ_i is a zero mean, unit variance Gaussian random variable satisfying $\mathbb{E}[\epsilon_i] = 0$ and $\mathbb{E}[\epsilon_i\epsilon_j] = I_{ij}$, where I_{ij} is the i,j element of the identity matrix. Taking the expectation of (8.14), we have

$$\mathbb{E}[x_i] = \sum_{j\in\mathrm{pa}_i} w_{ij}\mathbb{E}[x_j] + b_i. \tag{8.15}$$

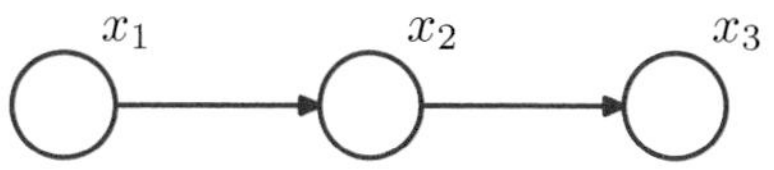

Figure 8.14 A directed graph over three Gaussian variables, with one missing link.

Thus we can find the components of $\mathbb{E}[\mathbf{x}] = (\mathbb{E}[x_1], \ldots, \mathbb{E}[x_D])^{\mathrm{T}}$ by starting at the lowest numbered node and working recursively through the graph (here we again assume that the nodes are numbered such that each node has a higher number than its parents). Similarly, we can use (8.14) and (8.15) to obtain the i, j element of the covariance matrix for $p(\mathbf{x})$ in the form of a recursion relation

$$\begin{aligned} \mathrm{cov}[x_i, x_j] &= \mathbb{E}\left[(x_i - \mathbb{E}[x_i])(x_j - \mathbb{E}[x_j])\right] \\ &= \mathbb{E}\left[(x_i - \mathbb{E}[x_i])\left\{\sum_{k \in \mathrm{pa}_j} w_{jk}(x_k - \mathbb{E}[x_k]) + \sqrt{v_j}\epsilon_j\right\}\right] \\ &= \sum_{k \in \mathrm{pa}_j} w_{jk}\mathrm{cov}[x_i, x_k] + I_{ij}v_j \end{aligned} \tag{8.16}$$

and so the covariance can similarly be evaluated recursively starting from the lowest numbered node.

Let us consider two extreme cases. First of all, suppose that there are no links in the graph, which therefore comprises D isolated nodes. In this case, there are no parameters w_{ij} and so there are just D parameters b_i and D parameters v_i. From the recursion relations (8.15) and (8.16), we see that the mean of $p(\mathbf{x})$ is given by $(b_1, \ldots, b_D)^{\mathrm{T}}$ and the covariance matrix is diagonal of the form $\mathrm{diag}(v_1, \ldots, v_D)$. The joint distribution has a total of $2D$ parameters and represents a set of D independent univariate Gaussian distributions.

Now consider a fully connected graph in which each node has all lower numbered nodes as parents. The matrix w_{ij} then has $i - 1$ entries on the ith row and hence is a lower triangular matrix (with no entries on the leading diagonal). Then the total number of parameters w_{ij} is obtained by taking the number D^2 of elements in a $D \times D$ matrix, subtracting D to account for the absence of elements on the leading diagonal, and then dividing by 2 because the matrix has elements only below the diagonal, giving a total of $D(D-1)/2$. The total number of independent parameters $\{w_{ij}\}$ and $\{v_i\}$ in the covariance matrix is therefore $D(D+1)/2$ corresponding to a general symmetric covariance matrix.

Section 2.3

Graphs having some intermediate level of complexity correspond to joint Gaussian distributions with partially constrained covariance matrices. Consider for example the graph shown in Figure 8.14, which has a link missing between variables x_1 and x_3. Using the recursion relations (8.15) and (8.16), we see that the mean and covariance of the joint distribution are given by

Exercise 8.7

$$\boldsymbol{\mu} = (b_1, b_2 + w_{21}b_1, b_3 + w_{32}b_2 + w_{32}w_{21}b_1)^{\mathrm{T}} \tag{8.17}$$

$$\boldsymbol{\Sigma} = \begin{pmatrix} v_1 & w_{21}v_1 & w_{32}w_{21}v_1 \\ w_{21}v_1 & v_2 + w_{21}^2 v_1 & w_{32}(v_2 + w_{21}^2 v_1) \\ w_{32}w_{21}v_1 & w_{32}(v_2 + w_{21}^2 v_1) & v_3 + w_{32}^2(v_2 + w_{21}^2 v_1) \end{pmatrix}. \tag{8.18}$$

We can readily extend the linear-Gaussian graphical model to the case in which the nodes of the graph represent multivariate Gaussian variables. In this case, we can write the conditional distribution for node i in the form

$$p(\mathbf{x}_i|\mathrm{pa}_i) = \mathcal{N}\left(\mathbf{x}_i \middle| \sum_{j\in\mathrm{pa}_i} \mathbf{W}_{ij}\mathbf{x}_j + \mathbf{b}_i, \boldsymbol{\Sigma}_i\right) \tag{8.19}$$

where now $\mathbf{W}_{ij}$ is a matrix (which is nonsquare if $\mathbf{x}_i$ and $\mathbf{x}_j$ have different dimensionalities). Again it is easy to verify that the joint distribution over all variables is Gaussian.

Section 2.3.6

Note that we have already encountered a specific example of the linear-Gaussian relationship when we saw that the conjugate prior for the mean $\boldsymbol{\mu}$ of a Gaussian variable $\mathbf{x}$ is itself a Gaussian distribution over $\boldsymbol{\mu}$. The joint distribution over $\mathbf{x}$ and $\boldsymbol{\mu}$ is therefore Gaussian. This corresponds to a simple two-node graph in which the node representing $\boldsymbol{\mu}$ is the parent of the node representing $\mathbf{x}$. The mean of the distribution over $\boldsymbol{\mu}$ is a parameter controlling a prior, and so it can be viewed as a hyperparameter. Because the value of this hyperparameter may itself be unknown, we can again treat it from a Bayesian perspective by introducing a prior over the hyperparameter, sometimes called a *hyperprior*, which is again given by a Gaussian distribution. This type of construction can be extended in principle to any level and is an illustration of a *hierarchical Bayesian model*, of which we shall encounter further examples in later chapters.

8.2. Conditional Independence

An important concept for probability distributions over multiple variables is that of *conditional independence* (Dawid, 1980). Consider three variables a, b, and c, and suppose that the conditional distribution of a, given b and c, is such that it does not depend on the value of b, so that

$$p(a|b, c) = p(a|c). \tag{8.20}$$

We say that a is conditionally independent of b given c. This can be expressed in a slightly different way if we consider the joint distribution of a and b conditioned on c, which we can write in the form

$$\begin{aligned} p(a, b|c) &= p(a|b, c)p(b|c) \\ &= p(a|c)p(b|c). \end{aligned} \tag{8.21}$$

where we have used the product rule of probability together with (8.20). Thus we see that, conditioned on c, the joint distribution of a and b factorizes into the product of the marginal distribution of a and the marginal distribution of b (again both conditioned on c). This says that the variables a and b are statistically independent, given c. Note that our definition of conditional independence will require that (8.20),

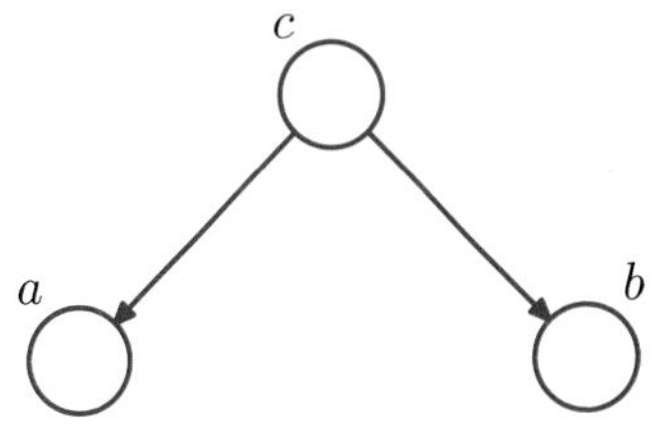

Figure 8.15 The first of three examples of graphs over three variables a, b, and c used to discuss conditional independence properties of directed graphical models.

or equivalently (8.21), must hold for every possible value of c, and not just for some values. We shall sometimes use a shorthand notation for conditional independence (Dawid, 1979) in which

$$a \perp\!\!\!\perp b \mid c \tag{8.22}$$

denotes that a is conditionally independent of b given c and is equivalent to (8.20).

Conditional independence properties play an important role in using probabilistic models for pattern recognition by simplifying both the structure of a model and the computations needed to perform inference and learning under that model. We shall see examples of this shortly.

If we are given an expression for the joint distribution over a set of variables in terms of a product of conditional distributions (i.e., the mathematical representation underlying a directed graph), then we could in principle test whether any potential conditional independence property holds by repeated application of the sum and product rules of probability. In practice, such an approach would be very time consuming. An important and elegant feature of graphical models is that conditional independence properties of the joint distribution can be read directly from the graph without having to perform any analytical manipulations. The general framework for achieving this is called *d-separation*, where the 'd' stands for 'directed' (Pearl, 1988). Here we shall motivate the concept of d-separation and give a general statement of the d-separation criterion. A formal proof can be found in Lauritzen (1996).

8.2.1 Three example graphs

We begin our discussion of the conditional independence properties of directed graphs by considering three simple examples each involving graphs having just three nodes. Together, these will motivate and illustrate the key concepts of d-separation. The first of the three examples is shown in Figure 8.15, and the joint distribution corresponding to this graph is easily written down using the general result (8.5) to give

$$p(a, b, c) = p(a|c)p(b|c)p(c). \tag{8.23}$$

If none of the variables are observed, then we can investigate whether a and b are independent by marginalizing both sides of (8.23) with respect to c to give

$$p(a, b) = \sum_{c} p(a|c)p(b|c)p(c). \tag{8.24}$$

In general, this does not factorize into the product $p(a)p(b)$, and so

$$a \not\perp\!\!\!\perp b \mid \emptyset \tag{8.25}$$

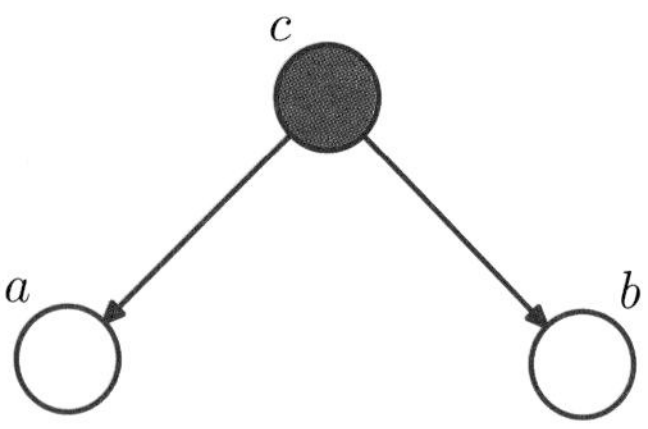

Figure 8.16 As in Figure 8.15 but where we have conditioned on the value of variable c.

where $\emptyset$ denotes the empty set, and the symbol $\not\!\perp\!\!\!\perp$ means that the conditional independence property does not hold in general. Of course, it may hold for a particular distribution by virtue of the specific numerical values associated with the various conditional probabilities, but it does not follow in general from the structure of the graph.

Now suppose we condition on the variable c, as represented by the graph of Figure 8.16. From (8.23), we can easily write down the conditional distribution of a and b, given c, in the form

$$\begin{aligned} p(a, b|c) &= \frac{p(a, b, c)}{p(c)} \\ &= p(a|c)p(b|c) \end{aligned}$$

and so we obtain the conditional independence property

$$a \perp\!\!\!\perp b \mid c.$$

We can provide a simple graphical interpretation of this result by considering the path from node a to node b via c. The node c is said to be *tail-to-tail* with respect to this path because the node is connected to the tails of the two arrows, and the presence of such a path connecting nodes a and b causes these nodes to be dependent. However, when we condition on node c, as in Figure 8.16, the conditioned node 'blocks' the path from a to b and causes a and b to become (conditionally) independent.

We can similarly consider the graph shown in Figure 8.17. The joint distribution corresponding to this graph is again obtained from our general formula (8.5) to give

$$p(a, b, c) = p(a)p(c|a)p(b|c). \tag{8.26}$$

First of all, suppose that none of the variables are observed. Again, we can test to see if a and b are independent by marginalizing over c to give

$$p(a, b) = p(a) \sum_c p(c|a)p(b|c) = p(a)p(b|a).$$

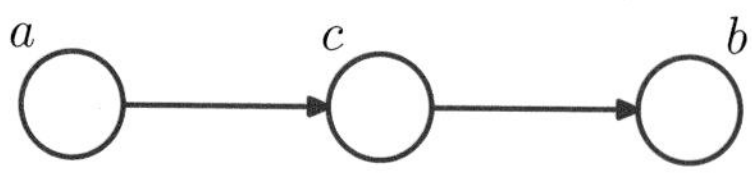

Figure 8.17 The second of our three examples of 3-node graphs used to motivate the conditional independence framework for directed graphical models.

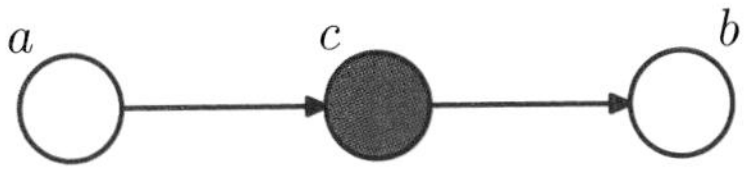

Figure 8.18 As in Figure 8.17 but now conditioning on node c.

which in general does not factorize into $p(a)p(b)$, and so

$$a \not\!\perp\!\!\!\perp b \mid \emptyset \tag{8.27}$$

as before.

Now suppose we condition on node c, as shown in Figure 8.18. Using Bayes' theorem, together with (8.26), we obtain

$$\begin{aligned} p(a, b|c) &= \frac{p(a, b, c)}{p(c)} \\ &= \frac{p(a)p(c|a)p(b|c)}{p(c)} \\ &= p(a|c)p(b|c) \end{aligned}$$

and so again we obtain the conditional independence property

$$a \perp\!\!\!\perp b \mid c.$$

As before, we can interpret these results graphically. The node c is said to be *head-to-tail* with respect to the path from node a to node b. Such a path connects nodes a and b and renders them dependent. If we now observe c, as in Figure 8.18, then this observation 'blocks' the path from a to b and so we obtain the conditional independence property $a \perp\!\!\!\perp b \mid c$.

Finally, we consider the third of our 3-node examples, shown by the graph in Figure 8.19. As we shall see, this has a more subtle behaviour than the two previous graphs.

The joint distribution can again be written down using our general result (8.5) to give

$$p(a, b, c) = p(a)p(b)p(c|a, b). \tag{8.28}$$

Consider first the case where none of the variables are observed. Marginalizing both sides of (8.28) over c we obtain

$$p(a, b) = p(a)p(b)$$

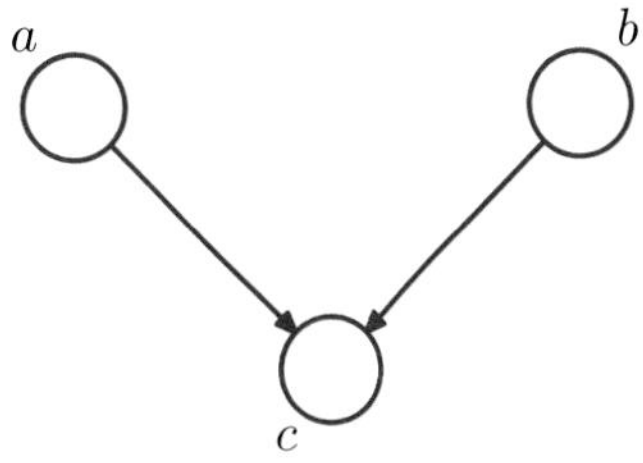

Figure 8.19 The last of our three examples of 3-node graphs used to explore conditional independence properties in graphical models. This graph has rather different properties from the two previous examples.

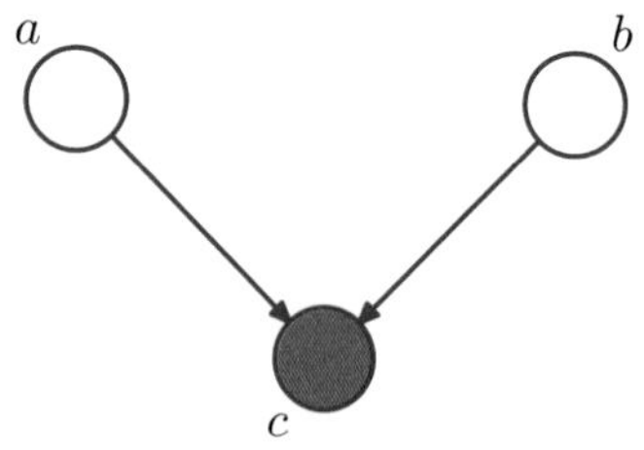

Figure 8.20 As in Figure 8.19 but conditioning on the value of node c. In this graph, the act of conditioning induces a dependence between a and b.

and so a and b are independent with no variables observed, in contrast to the two previous examples. We can write this result as

$$a \perp\!\!\!\perp b \mid \emptyset. \tag{8.29}$$

Now suppose we condition on c, as indicated in Figure 8.20. The conditional distribution of a and b is then given by

$$\begin{aligned} p(a, b|c) &= \frac{p(a, b, c)}{p(c)} \\ &= \frac{p(a)p(b)p(c|a, b)}{p(c)} \end{aligned}$$

which in general does not factorize into the product $p(a|c)p(b|c)$, and so

$$a \not\perp\!\!\!\perp b \mid c.$$

Thus our third example has the opposite behaviour from the first two. Graphically, we say that node c is *head-to-head* with respect to the path from a to b because it connects to the heads of the two arrows. When node c is unobserved, it 'blocks' the path, and the variables a and b are independent. However, conditioning on c 'unblocks' the path and renders a and b dependent.

There is one more subtlety associated with this third example that we need to consider. First we introduce some more terminology. We say that node y is a *descendant* of node x if there is a path from x to y in which each step of the path follows the directions of the arrows. Then it can be shown that a head-to-head path will become unblocked if either the node, *or any of its descendants*, is observed.

Exercise 8.10

In summary, a tail-to-tail node or a head-to-tail node leaves a path unblocked unless it is observed in which case it blocks the path. By contrast, a head-to-head node blocks a path if it is unobserved, but once the node, and/or at least one of its descendants, is observed the path becomes unblocked.

It is worth spending a moment to understand further the unusual behaviour of the graph of Figure 8.20. Consider a particular instance of such a graph corresponding to a problem with three binary random variables relating to the fuel system on a car, as shown in Figure 8.21. The variables are called B, representing the state of a battery that is either charged ($B = 1$) or flat ($B = 0$), F representing the state of the fuel tank that is either full of fuel ($F = 1$) or empty ($F = 0$), and G, which is the state of an electric fuel gauge and which indicates either full ($G = 1$) or empty

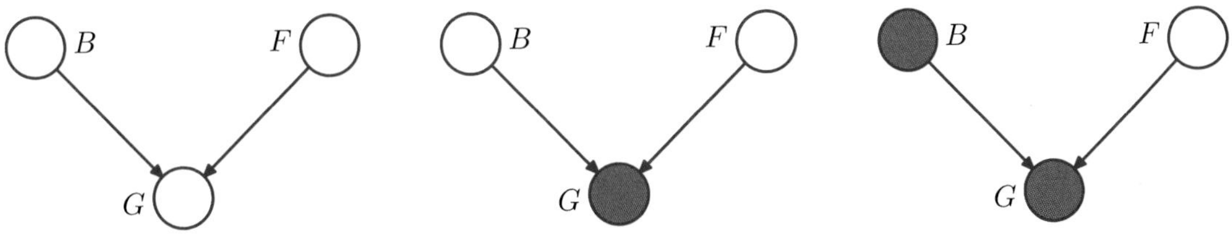

Figure 8.21 An example of a 3-node graph used to illustrate the phenomenon of 'explaining away'. The three nodes represent the state of the battery (B), the state of the fuel tank (F) and the reading on the electric fuel gauge (G). See the text for details.

($G = 0$). The battery is either charged or flat, and independently the fuel tank is either full or empty, with prior probabilities

$$\begin{aligned} p(B=1) &= 0.9 \\ p(F=1) &= 0.9. \end{aligned}$$

Given the state of the fuel tank and the battery, the fuel gauge reads full with probabilities given by

$$\begin{aligned} p(G=1|B=1,F=1) &= 0.8 \\ p(G=1|B=1,F=0) &= 0.2 \\ p(G=1|B=0,F=1) &= 0.2 \\ p(G=1|B=0,F=0) &= 0.1 \end{aligned}$$

so this is a rather unreliable fuel gauge! All remaining probabilities are determined by the requirement that probabilities sum to one, and so we have a complete specification of the probabilistic model.

Before we observe any data, the prior probability of the fuel tank being empty is $p(F = 0) = 0.1$. Now suppose that we observe the fuel gauge and discover that it reads empty, i.e., $G = 0$, corresponding to the middle graph in Figure 8.21. We can use Bayes' theorem to evaluate the posterior probability of the fuel tank being empty. First we evaluate the denominator for Bayes' theorem given by

$$p(G=0) = \sum_{B\in\{0,1\}} \sum_{F\in\{0,1\}} p(G=0|B,F)p(B)p(F) = 0.315 \tag{8.30}$$

and similarly we evaluate

$$p(G=0|F=0) = \sum_{B\in\{0,1\}} p(G=0|B,F=0)p(B) = 0.81 \tag{8.31}$$

and using these results we have

$$p(F=0|G=0) = \frac{p(G=0|F=0)p(F=0)}{p(G=0)} \simeq 0.257 \tag{8.32}$$

and so $p(F = 0|G = 0) > p(F = 0)$. Thus observing that the gauge reads empty makes it more likely that the tank is indeed empty, as we would intuitively expect. Next suppose that we also check the state of the battery and find that it is flat, i.e., $B = 0$. We have now observed the states of both the fuel gauge and the battery, as shown by the right-hand graph in Figure 8.21. The posterior probability that the fuel tank is empty given the observations of both the fuel gauge and the battery state is then given by

$$p(F = 0|G = 0, B = 0) = \frac{p(G = 0|B = 0, F = 0)p(F = 0)}{\sum_{F\in\{0,1\}} p(G = 0|B = 0, F)p(F)} \simeq 0.111 \quad (8.33)$$

where the prior probability $p(B = 0)$ has cancelled between numerator and denominator. Thus the probability that the tank is empty has *decreased* (from 0.257 to 0.111) as a result of the observation of the state of the battery. This accords with our intuition that finding out that the battery is flat *explains away* the observation that the fuel gauge reads empty. We see that the state of the fuel tank and that of the battery have indeed become dependent on each other as a result of observing the reading on the fuel gauge. In fact, this would also be the case if, instead of observing the fuel gauge directly, we observed the state of some descendant of G. Note that the probability $p(F = 0|G = 0, B = 0) \simeq 0.111$ is greater than the prior probability $p(F = 0) = 0.1$ because the observation that the fuel gauge reads zero still provides some evidence in favour of an empty fuel tank.

8.2.2 D-separation

We now give a general statement of the d-separation property (Pearl, 1988) for directed graphs. Consider a general directed graph in which A, B, and C are arbitrary nonintersecting sets of nodes (whose union may be smaller than the complete set of nodes in the graph). We wish to ascertain whether a particular conditional independence statement $A \perp\!\!\!\perp B \mid C$ is implied by a given directed acyclic graph. To do so, we consider all possible paths from any node in A to any node in B. Any such path is said to be *blocked* if it includes a node such that either

(a) the arrows on the path meet either head-to-tail or tail-to-tail at the node, and the node is in the set C, or

(b) the arrows meet head-to-head at the node, and neither the node, nor any of its descendants, is in the set C.

If all paths are blocked, then A is said to be d-separated from B by C, and the joint distribution over all of the variables in the graph will satisfy $A \perp\!\!\!\perp B \mid C$.

The concept of d-separation is illustrated in Figure 8.22. In graph (a), the path from a to b is not blocked by node f because it is a tail-to-tail node for this path and is not observed, nor is it blocked by node e because, although the latter is a head-to-head node, it has a descendant c in the conditioning set. Thus the conditional independence statement $a \perp\!\!\!\perp b \mid c$ does *not* follow from this graph. In graph (b), the path from a to b is blocked by node f because this is a tail-to-tail node that is observed, and so the conditional independence property $a \perp\!\!\!\perp b \mid f$ will be satisfied

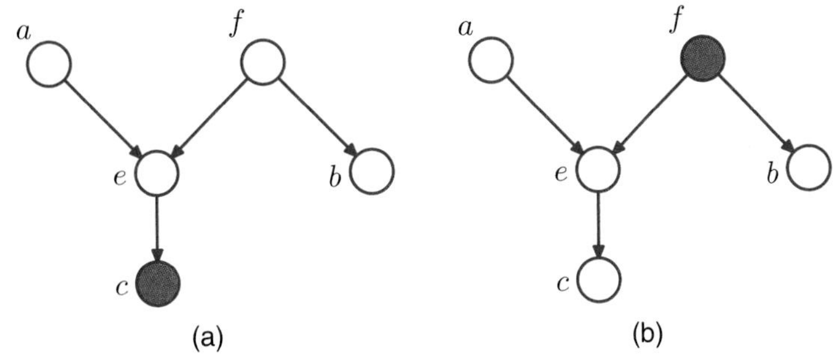

Figure 8.22 Illustration of the concept of d-separation. See the text for details.

by any distribution that factorizes according to this graph. Note that this path is also blocked by node e because e is a head-to-head node and neither it nor its descendant are in the conditioning set.

For the purposes of d-separation, parameters such as α and σ^2 in Figure 8.5, indicated by small filled circles, behave in the same way as observed nodes. However, there are no marginal distributions associated with such nodes. Consequently parameter nodes never themselves have parents and so all paths through these nodes will always be tail-to-tail and hence blocked. Consequently they play no role in d-separation.

Another example of conditional independence and d-separation is provided by the concept of i.i.d. (independent identically distributed) data introduced in Section 1.2.4. Consider the problem of finding the posterior distribution for the mean of a univariate Gaussian distribution. This can be represented by the directed graph shown in Figure 8.23 in which the joint distribution is defined by a prior $p(\mu)$ together with a set of conditional distributions $p(x_n|\mu)$ for $n = 1, \ldots, N$. In practice, we observe $\mathcal{D} = \{x_1, \ldots, x_N\}$ and our goal is to infer μ. Suppose, for a moment, that we condition on μ and consider the joint distribution of the observations. Using d-separation, we note that there is a unique path from any x_i to any other $x_{j \neq i}$ and that this path is tail-to-tail with respect to the observed node μ. Every such path is blocked and so the observations $D = \{x_1, \ldots, x_N\}$ are independent given μ, so that

Section 2.3

$$p(\mathcal{D}|\mu) = \prod_{n=1}^{N} p(x_n|\mu). \tag{8.34}$$

Figure 8.23 (a) Directed graph corresponding to the problem of inferring the mean μ of a univariate Gaussian distribution from observations $x_1, \ldots, x_N$. (b) The same graph drawn using the plate notation.

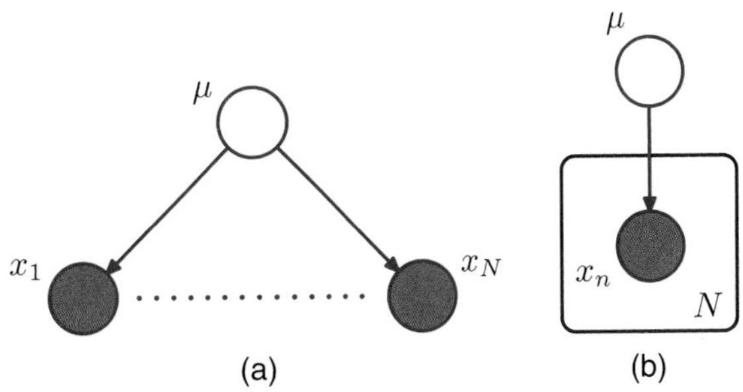

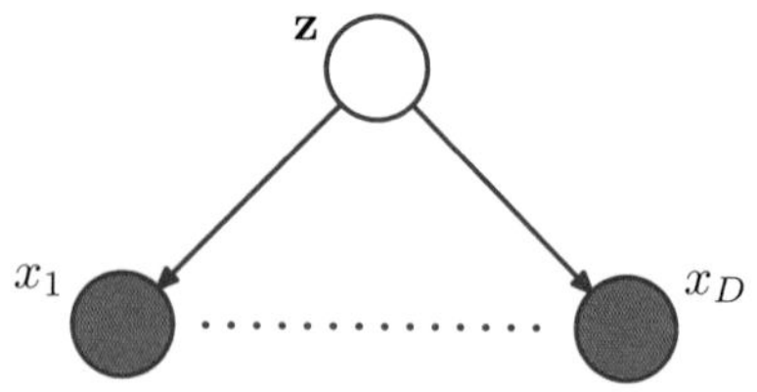

Figure 8.24 A graphical representation of the 'naive Bayes' model for classification. Conditioned on the class label z, the components of the observed vector $\mathbf{x} = (x_1, \ldots, x_D)^{\mathrm{T}}$ are assumed to be independent.

However, if we integrate over μ, the observations are in general no longer independent

$$p(\mathcal{D}) = \int_{-\infty}^{\infty} p(\mathcal{D}|\mu)p(\mu)\,\mathrm{d}\mu \neq \prod_{n=1}^{N} p(x_n). \tag{8.35}$$

Here μ is a latent variable, because its value is not observed.

Another example of a model representing i.i.d. data is the graph in Figure 8.7 corresponding to Bayesian polynomial regression. Here the stochastic nodes correspond to $\{t_n\}$, $\mathbf{w}$ and $\widehat{t}$. We see that the node for $\mathbf{w}$ is tail-to-tail with respect to the path from $\widehat{t}$ to any one of the nodes t_n and so we have the following conditional independence property

$$\widehat{t} \perp\!\!\!\perp t_n \mid \mathbf{w}. \tag{8.36}$$

Thus, conditioned on the polynomial coefficients $\mathbf{w}$, the predictive distribution for $\widehat{t}$ is independent of the training data $\{t_1, \ldots, t_N\}$. We can therefore first use the training data to determine the posterior distribution over the coefficients $\mathbf{w}$ and then we can discard the training data and use the posterior distribution for $\mathbf{w}$ to make predictions of $\widehat{t}$ for new input observations $\widehat{x}$. *Section 3.3*

A related graphical structure arises in an approach to classification called the *naive Bayes* model, in which we use conditional independence assumptions to simplify the model structure. Suppose our observed variable consists of a D-dimensional vector $\mathbf{x} = (x_1, \ldots, x_D)^{\mathrm{T}}$, and we wish to assign observed values of $\mathbf{x}$ to one of K classes. Using the 1-of-K encoding scheme, we can represent these classes by a K-dimensional binary vector $\mathbf{z}$. We can then define a generative model by introducing a multinomial prior $p(\mathbf{z}|\boldsymbol{\mu})$ over the class labels, where the kth component μ_k of $\boldsymbol{\mu}$ is the prior probability of class $\mathcal{C}_k$, together with a conditional distribution $p(\mathbf{x}|\mathbf{z})$ for the observed vector $\mathbf{x}$. The key assumption of the naive Bayes model is that, conditioned on the class $\mathbf{z}$, the distributions of the input variables $x_1, \ldots, x_D$ are independent. The graphical representation of this model is shown in Figure 8.24. We see that observation of $\mathbf{z}$ blocks the path between x_i and x_j for $j \neq i$ (because such paths are tail-to-tail at the node $\mathbf{z}$) and so x_i and x_j are conditionally independent given $\mathbf{z}$. If, however, we marginalize out $\mathbf{z}$ (so that $\mathbf{z}$ is unobserved) the tail-to-tail path from x_i to x_j is no longer blocked. This tells us that in general the marginal density $p(\mathbf{x})$ will not factorize with respect to the components of $\mathbf{x}$. We encountered a simple application of the naive Bayes model in the context of fusing data from different sources for medical diagnosis in Section 1.5.

If we are given a labelled training set, comprising inputs $\{\mathbf{x}_1, \ldots, \mathbf{x}_N\}$ together with their class labels, then we can fit the naive Bayes model to the training data using maximum likelihood assuming that the data are drawn independently from the model. The solution is obtained by fitting the model for each class separately using the correspondingly labelled data. As an example, suppose that the probability density within each class is chosen to be Gaussian. In this case, the naive Bayes assumption then implies that the covariance matrix for each Gaussian is diagonal, and the contours of constant density within each class will be axis-aligned ellipsoids. The marginal density, however, is given by a superposition of diagonal Gaussians (with weighting coefficients given by the class priors) and so will no longer factorize with respect to its components.

The naive Bayes assumption is helpful when the dimensionality D of the input space is high, making density estimation in the full D-dimensional space more challenging. It is also useful if the input vector contains both discrete and continuous variables, since each can be represented separately using appropriate models (e.g., Bernoulli distributions for binary observations or Gaussians for real-valued variables). The conditional independence assumption of this model is clearly a strong one that may lead to rather poor representations of the class-conditional densities. Nevertheless, even if this assumption is not precisely satisfied, the model may still give good classification performance in practice because the decision boundaries can be insensitive to some of the details in the class-conditional densities, as illustrated in Figure 1.27.

We have seen that a particular directed graph represents a specific decomposition of a joint probability distribution into a product of conditional probabilities. The graph also expresses a set of conditional independence statements obtained through the d-separation criterion, and the d-separation theorem is really an expression of the equivalence of these two properties. In order to make this clear, it is helpful to think of a directed graph as a filter. Suppose we consider a particular joint probability distribution $p(\mathbf{x})$ over the variables $\mathbf{x}$ corresponding to the (nonobserved) nodes of the graph. The filter will allow this distribution to pass through if, and only if, it can be expressed in terms of the factorization (8.5) implied by the graph. If we present to the filter the set of all possible distributions $p(\mathbf{x})$ over the set of variables $\mathbf{x}$, then the subset of distributions that are passed by the filter will be denoted $\mathcal{DF}$, for *directed factorization*. This is illustrated in Figure 8.25. Alternatively, we can use the graph as a different kind of filter by first listing all of the conditional independence properties obtained by applying the d-separation criterion to the graph, and then allowing a distribution to pass only if it satisfies all of these properties. If we present all possible distributions $p(\mathbf{x})$ to this second kind of filter, then the d-separation theorem tells us that the set of distributions that will be allowed through is precisely the set $\mathcal{DF}$.

It should be emphasized that the conditional independence properties obtained from d-separation apply to any probabilistic model described by that particular directed graph. This will be true, for instance, whether the variables are discrete or continuous or a combination of these. Again, we see that a particular graph is describing a whole family of probability distributions.

At one extreme we have a fully connected graph that exhibits no conditional independence properties at all, and which can represent any possible joint probability distribution over the given variables. The set $\mathcal{DF}$ will contain all possible distribu-

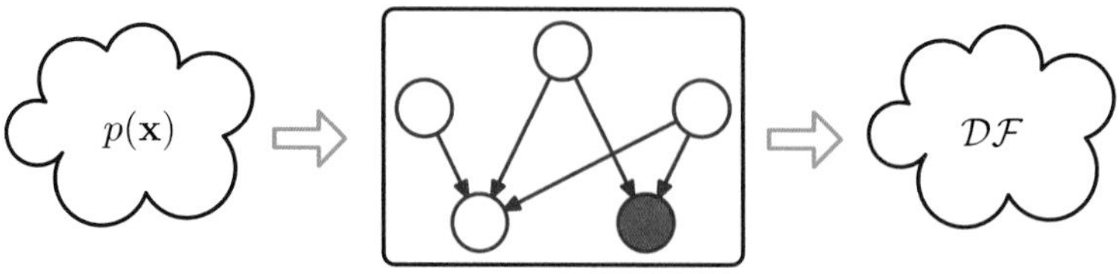

Figure 8.25 We can view a graphical model (in this case a directed graph) as a filter in which a probability distribution $p(\mathbf{x})$ is allowed through the filter if, and only if, it satisfies the directed factorization property (8.5). The set of all possible probability distributions $p(\mathbf{x})$ that pass through the filter is denoted $\mathcal{DF}$. We can alternatively use the graph to filter distributions according to whether they respect all of the conditional independencies implied by the d-separation properties of the graph. The d-separation theorem says that it is the same set of distributions $\mathcal{DF}$ that will be allowed through this second kind of filter.

tions $p(\mathbf{x})$. At the other extreme, we have the fully disconnected graph, i.e., one having no links at all. This corresponds to joint distributions which factorize into the product of the marginal distributions over the variables comprising the nodes of the graph.

Note that for any given graph, the set of distributions $\mathcal{DF}$ will include any distributions that have additional independence properties beyond those described by the graph. For instance, a fully factorized distribution will always be passed through the filter implied by any graph over the corresponding set of variables.

We end our discussion of conditional independence properties by exploring the concept of a *Markov blanket* or *Markov boundary*. Consider a joint distribution $p(\mathbf{x}_1, \ldots, \mathbf{x}_D)$ represented by a directed graph having D nodes, and consider the conditional distribution of a particular node with variables $\mathbf{x}_i$ conditioned on all of the remaining variables $\mathbf{x}_{j \neq i}$. Using the factorization property (8.5), we can express this conditional distribution in the form

$$\begin{aligned} p(\mathbf{x}_i|\mathbf{x}_{\{j\neq i\}}) &= \frac{p(\mathbf{x}_1, \ldots, \mathbf{x}_D)}{\displaystyle\int p(\mathbf{x}_1, \ldots, \mathbf{x}_D)\,\mathrm{d}\mathbf{x}_i} \\ &= \frac{\displaystyle\prod_k p(\mathbf{x}_k|\mathrm{pa}_k)}{\displaystyle\int \prod_k p(\mathbf{x}_k|\mathrm{pa}_k)\,\mathrm{d}\mathbf{x}_i} \end{aligned}$$

in which the integral is replaced by a summation in the case of discrete variables. We now observe that any factor $p(\mathbf{x}_k|\mathrm{pa}_k)$ that does not have any functional dependence on $\mathbf{x}_i$ can be taken outside the integral over $\mathbf{x}_i$, and will therefore cancel between numerator and denominator. The only factors that remain will be the conditional distribution $p(\mathbf{x}_i|\mathrm{pa}_i)$ for node $\mathbf{x}_i$ itself, together with the conditional distributions for any nodes $\mathbf{x}_k$ such that node $\mathbf{x}_i$ is in the conditioning set of $p(\mathbf{x}_k|\mathrm{pa}_k)$, in other words for which $\mathbf{x}_i$ is a parent of $\mathbf{x}_k$. The conditional $p(\mathbf{x}_i|\mathrm{pa}_i)$ will depend on the parents of node $\mathbf{x}_i$, whereas the conditionals $p(\mathbf{x}_k|\mathrm{pa}_k)$ will depend on the children

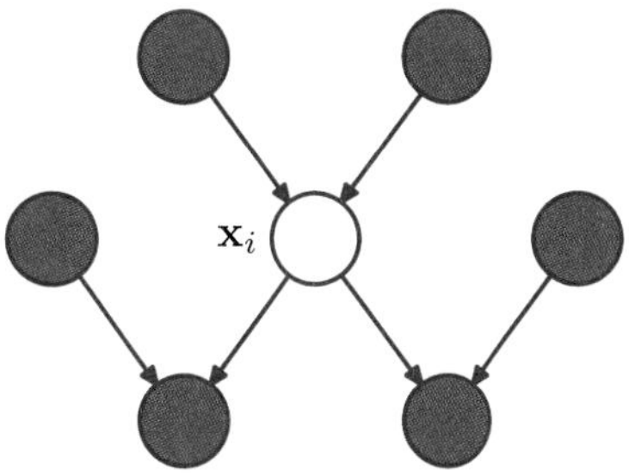

Figure 8.26 The Markov blanket of a node $\mathbf{x}_i$ comprises the set of parents, children and co-parents of the node. It has the property that the conditional distribution of $\mathbf{x}_i$, conditioned on all the remaining variables in the graph, is dependent only on the variables in the Markov blanket.

of $\mathbf{x}_i$ as well as on the *co-parents*, in other words variables corresponding to parents of node $\mathbf{x}_k$ other than node $\mathbf{x}_i$. The set of nodes comprising the parents, the children and the co-parents is called the Markov blanket and is illustrated in Figure 8.26. We can think of the Markov blanket of a node $\mathbf{x}_i$ as being the minimal set of nodes that isolates $\mathbf{x}_i$ from the rest of the graph. Note that it is not sufficient to include only the parents and children of node $\mathbf{x}_i$ because the phenomenon of explaining away means that observations of the child nodes will not block paths to the co-parents. We must therefore observe the co-parent nodes also.

8.3. Markov Random Fields

We have seen that directed graphical models specify a factorization of the joint distribution over a set of variables into a product of local conditional distributions. They also define a set of conditional independence properties that must be satisfied by any distribution that factorizes according to the graph. We turn now to the second major class of graphical models that are described by undirected graphs and that again specify both a factorization and a set of conditional independence relations.

A *Markov random field*, also known as a *Markov network* or an *undirected graphical model* (Kindermann and Snell, 1980), has a set of nodes each of which corresponds to a variable or group of variables, as well as a set of links each of which connects a pair of nodes. The links are undirected, that is they do not carry arrows. In the case of undirected graphs, it is convenient to begin with a discussion of conditional independence properties.

8.3.1 Conditional independence properties

Section 8.2

In the case of directed graphs, we saw that it was possible to test whether a particular conditional independence property holds by applying a graphical test called d-separation. This involved testing whether or not the paths connecting two sets of nodes were 'blocked'. The definition of blocked, however, was somewhat subtle due to the presence of paths having head-to-head nodes. We might ask whether it is possible to define an alternative graphical semantics for probability distributions such that conditional independence is determined by simple graph separation. This is indeed the case and corresponds to undirected graphical models. By removing the

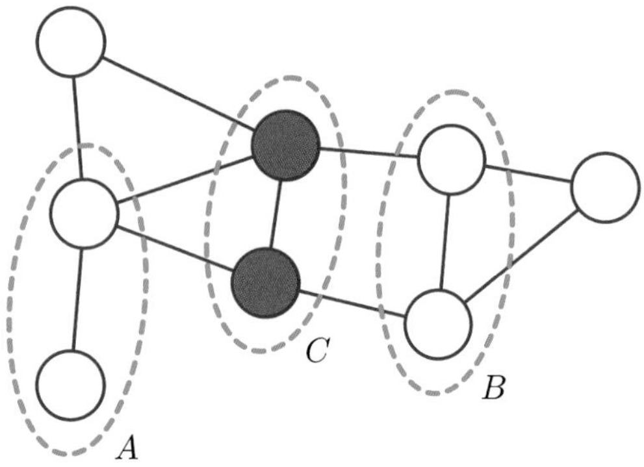

Figure 8.27 An example of an undirected graph in which every path from any node in set A to any node in set B passes through at least one node in set C. Consequently the conditional independence property $A \perp\!\!\!\perp B \mid C$ holds for any probability distribution described by this graph.

directionality from the links of the graph, the asymmetry between parent and child nodes is removed, and so the subtleties associated with head-to-head nodes no longer arise.

Suppose that in an undirected graph we identify three sets of nodes, denoted A, B, and C, and that we consider the conditional independence property

$$A \perp\!\!\!\perp B \mid C. \tag{8.37}$$

To test whether this property is satisfied by a probability distribution defined by a graph we consider all possible paths that connect nodes in set A to nodes in set B. If all such paths pass through one or more nodes in set C, then all such paths are 'blocked' and so the conditional independence property holds. However, if there is at least one such path that is not blocked, then the property does not necessarily hold, or more precisely there will exist at least some distributions corresponding to the graph that do not satisfy this conditional independence relation. This is illustrated with an example in Figure 8.27. Note that this is exactly the same as the d-separation criterion except that there is no 'explaining away' phenomenon. Testing for conditional independence in undirected graphs is therefore simpler than in directed graphs.

An alternative way to view the conditional independence test is to imagine removing all nodes in set C from the graph together with any links that connect to those nodes. We then ask if there exists a path that connects any node in A to any node in B. If there are no such paths, then the conditional independence property must hold.

The Markov blanket for an undirected graph takes a particularly simple form, because a node will be conditionally independent of all other nodes conditioned only on the neighbouring nodes, as illustrated in Figure 8.28.

8.3.2 Factorization properties

We now seek a factorization rule for undirected graphs that will correspond to the above conditional independence test. Again, this will involve expressing the joint distribution $p(\mathbf{x})$ as a product of functions defined over sets of variables that are local to the graph. We therefore need to decide what is the appropriate notion of locality in this case.

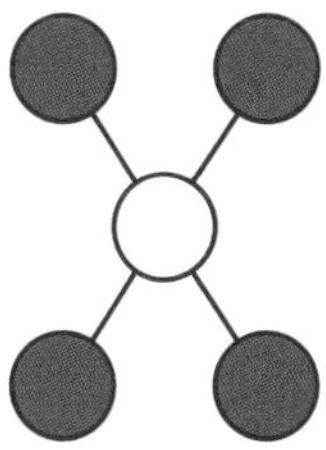

Figure 8.28 For an undirected graph, the Markov blanket of a node x_i consists of the set of neighbouring nodes. It has the property that the conditional distribution of x_i, conditioned on all the remaining variables in the graph, is dependent only on the variables in the Markov blanket.

If we consider two nodes x_i and x_j that are not connected by a link, then these variables must be conditionally independent given all other nodes in the graph. This follows from the fact that there is no direct path between the two nodes, and all other paths pass through nodes that are observed, and hence those paths are blocked. This conditional independence property can be expressed as

$$p(x_i, x_j|\mathbf{x}_{\setminus\{i,j\}}) = p(x_i|\mathbf{x}_{\setminus\{i,j\}})p(x_j|\mathbf{x}_{\setminus\{i,j\}}) \tag{8.38}$$

where $\mathbf{x}_{\setminus\{i,j\}}$ denotes the set $\mathbf{x}$ of all variables with x_i and x_j removed. The factorization of the joint distribution must therefore be such that x_i and x_j do not appear in the same factor in order for the conditional independence property to hold for all possible distributions belonging to the graph.

This leads us to consider a graphical concept called a *clique*, which is defined as a subset of the nodes in a graph such that there exists a link between all pairs of nodes in the subset. In other words, the set of nodes in a clique is fully connected. Furthermore, a *maximal clique* is a clique such that it is not possible to include any other nodes from the graph in the set without it ceasing to be a clique. These concepts are illustrated by the undirected graph over four variables shown in Figure 8.29. This graph has five cliques of two nodes given by $\{x_1, x_2\}$, $\{x_2, x_3\}$, $\{x_3, x_4\}$, $\{x_4, x_2\}$, and $\{x_1, x_3\}$, as well as two maximal cliques given by $\{x_1, x_2, x_3\}$ and $\{x_2, x_3, x_4\}$. The set $\{x_1, x_2, x_3, x_4\}$ is not a clique because of the missing link from x_1 to x_4.

We can therefore define the factors in the decomposition of the joint distribution to be functions of the variables in the cliques. In fact, we can consider functions of the maximal cliques, without loss of generality, because other cliques must be subsets of maximal cliques. Thus, if $\{x_1, x_2, x_3\}$ is a maximal clique and we define an arbitrary function over this clique, then including another factor defined over a subset of these variables would be redundant.

Let us denote a clique by C and the set of variables in that clique by $\mathbf{x}_C$. Then

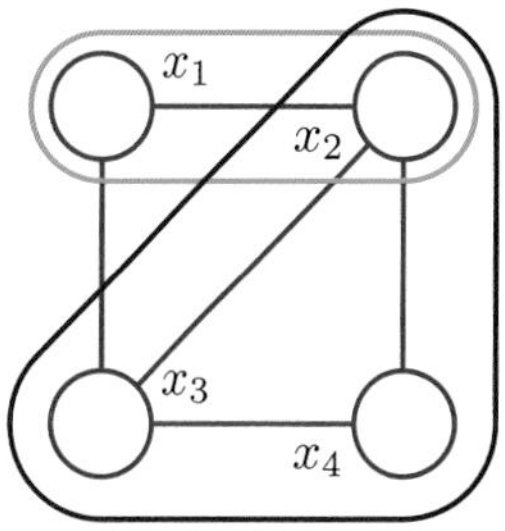

Figure 8.29 A four-node undirected graph showing a clique (outlined in green) and a maximal clique (outlined in blue).

the joint distribution is written as a product of *potential functions* $\psi_C(\mathbf{x}_C)$ over the maximal cliques of the graph

$$p(\mathbf{x}) = \frac{1}{Z}\prod_C \psi_C(\mathbf{x}_C). \tag{8.39}$$

Here the quantity Z, sometimes called the *partition function*, is a normalization constant and is given by

$$Z = \sum_{\mathbf{x}}\prod_C \psi_C(\mathbf{x}_C) \tag{8.40}$$

which ensures that the distribution $p(\mathbf{x})$ given by (8.39) is correctly normalized. By considering only potential functions which satisfy $\psi_C(\mathbf{x}_C) \geqslant 0$ we ensure that $p(\mathbf{x}) \geqslant 0$. In (8.40) we have assumed that $\mathbf{x}$ comprises discrete variables, but the framework is equally applicable to continuous variables, or a combination of the two, in which the summation is replaced by the appropriate combination of summation and integration.

Note that we do not restrict the choice of potential functions to those that have a specific probabilistic interpretation as marginal or conditional distributions. This is in contrast to directed graphs in which each factor represents the conditional distribution of the corresponding variable, conditioned on the state of its parents. However, in special cases, for instance where the undirected graph is constructed by starting with a directed graph, the potential functions may indeed have such an interpretation, as we shall see shortly.

One consequence of the generality of the potential functions $\psi_C(\mathbf{x}_C)$ is that their product will in general not be correctly normalized. We therefore have to introduce an explicit normalization factor given by (8.40). Recall that for directed graphs, the joint distribution was automatically normalized as a consequence of the normalization of each of the conditional distributions in the factorization.

The presence of this normalization constant is one of the major limitations of undirected graphs. If we have a model with M discrete nodes each having K states, then the evaluation of the normalization term involves summing over K^M states and so (in the worst case) is exponential in the size of the model. The partition function is needed for parameter learning because it will be a function of any parameters that govern the potential functions $\psi_C(\mathbf{x}_C)$. However, for evaluation of local conditional distributions, the partition function is not needed because a conditional is the ratio of two marginals, and the partition function cancels between numerator and denominator when evaluating this ratio. Similarly, for evaluating local marginal probabilities we can work with the unnormalized joint distribution and then normalize the marginals explicitly at the end. Provided the marginals only involves a small number of variables, the evaluation of their normalization coefficient will be feasible.

So far, we have discussed the notion of conditional independence based on simple graph separation and we have proposed a factorization of the joint distribution that is intended to correspond to this conditional independence structure. However, we have not made any formal connection between conditional independence and factorization for undirected graphs. To do so we need to restrict attention to potential functions $\psi_C(\mathbf{x}_C)$ that are strictly positive (i.e., never zero or negative for any choice of $\mathbf{x}_C$). Given this restriction, we can make a precise relationship between factorization and conditional independence.

To do this we again return to the concept of a graphical model as a filter, corresponding to Figure 8.25. Consider the set of all possible distributions defined over a fixed set of variables corresponding to the nodes of a particular undirected graph. We can define $\mathcal{UI}$ to be the set of such distributions that are consistent with the set of conditional independence statements that can be read from the graph using graph separation. Similarly, we can define $\mathcal{UF}$ to be the set of such distributions that can be expressed as a factorization of the form (8.39) with respect to the maximal cliques of the graph. The *Hammersley-Clifford* theorem (Clifford, 1990) states that the sets $\mathcal{UI}$ and $\mathcal{UF}$ are identical.

Because we are restricted to potential functions which are strictly positive it is convenient to express them as exponentials, so that

$$\psi_C(\mathbf{x}_C) = \exp\{-E(\mathbf{x}_C)\} \tag{8.41}$$

where $E(\mathbf{x}_C)$ is called an *energy function*, and the exponential representation is called the *Boltzmann distribution*. The joint distribution is defined as the product of potentials, and so the total energy is obtained by adding the energies of each of the maximal cliques.

In contrast to the factors in the joint distribution for a directed graph, the potentials in an undirected graph do not have a specific probabilistic interpretation. Although this gives greater flexibility in choosing the potential functions, because there is no normalization constraint, it does raise the question of how to motivate a choice of potential function for a particular application. This can be done by viewing the potential function as expressing which configurations of the local variables are preferred to others. Global configurations that have a relatively high probability are those that find a good balance in satisfying the (possibly conflicting) influences of the clique potentials. We turn now to a specific example to illustrate the use of undirected graphs.

8.3.3 Illustration: Image de-noising

We can illustrate the application of undirected graphs using an example of noise removal from a binary image (Besag, 1974; Geman and Geman, 1984; Besag, 1986). Although a very simple example, this is typical of more sophisticated applications. Let the observed noisy image be described by an array of binary pixel values $y_i \in \{-1, +1\}$, where the index $i = 1, \ldots, D$ runs over all pixels. We shall suppose that the image is obtained by taking an unknown noise-free image, described by binary pixel values $x_i \in \{-1, +1\}$ and randomly flipping the sign of pixels with some small probability. An example binary image, together with a noise corrupted image obtained by flipping the sign of the pixels with probability 10%, is shown in Figure 8.30. Given the noisy image, our goal is to recover the original noise-free image.

Because the noise level is small, we know that there will be a strong correlation between x_i and y_i. We also know that neighbouring pixels x_i and x_j in an image are strongly correlated. This prior knowledge can be captured using the Markov

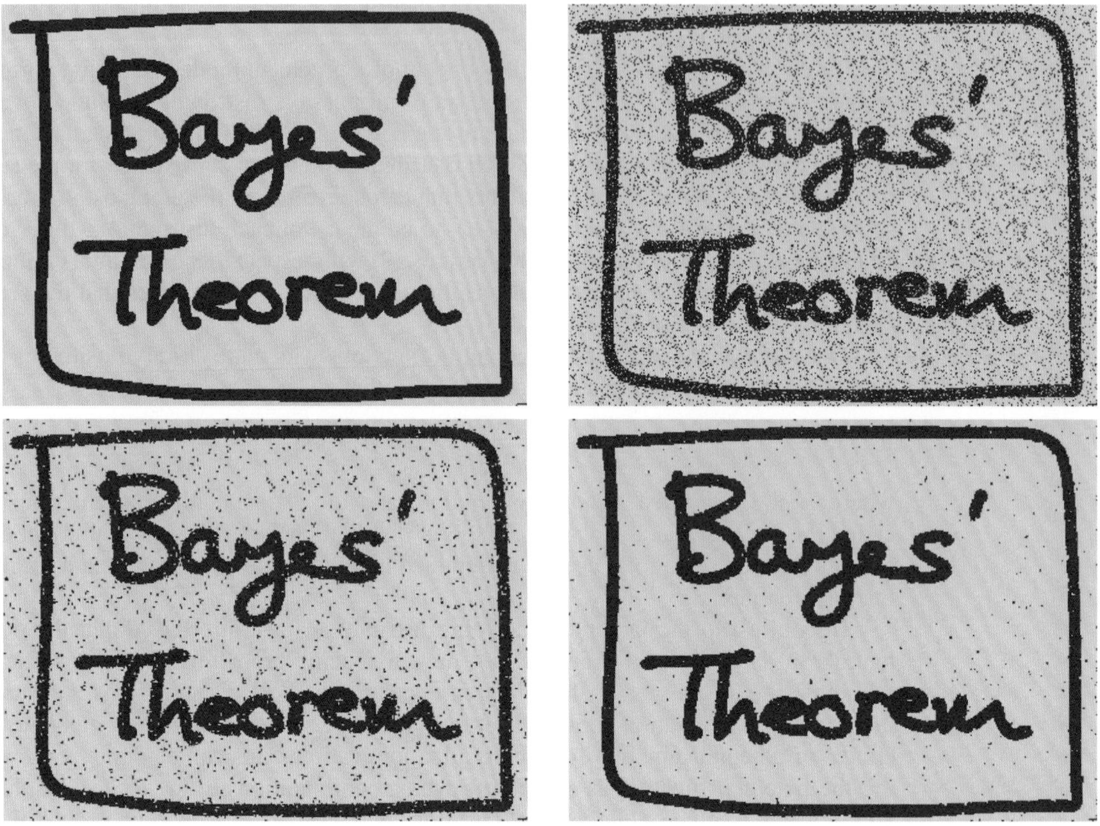

Figure 8.30 Illustration of image de-noising using a Markov random field. The top row shows the original binary image on the left and the corrupted image after randomly changing 10% of the pixels on the right. The bottom row shows the restored images obtained using iterated conditional models (ICM) on the left and using the graph-cut algorithm on the right. ICM produces an image where 96% of the pixels agree with the original image, whereas the corresponding number for graph-cut is 99%.

random field model whose undirected graph is shown in Figure 8.31. This graph has two types of cliques, each of which contains two variables. The cliques of the form $\{x_i, y_i\}$ have an associated energy function that expresses the correlation between these variables. We choose a very simple energy function for these cliques of the form $-\eta x_i y_i$ where η is a positive constant. This has the desired effect of giving a lower energy (thus encouraging a higher probability) when x_i and y_i have the same sign and a higher energy when they have the opposite sign.

The remaining cliques comprise pairs of variables $\{x_i, x_j\}$ where i and j are indices of neighbouring pixels. Again, we want the energy to be lower when the pixels have the same sign than when they have the opposite sign, and so we choose an energy given by $-\beta x_i x_j$ where β is a positive constant.

Because a potential function is an arbitrary, nonnegative function over a maximal clique, we can multiply it by any nonnegative functions of subsets of the clique, or

Figure 8.31 An undirected graphical model representing a Markov random field for image de-noising, in which x_i is a binary variable denoting the state of pixel i in the unknown noise-free image, and y_i denotes the corresponding value of pixel i in the observed noisy image.

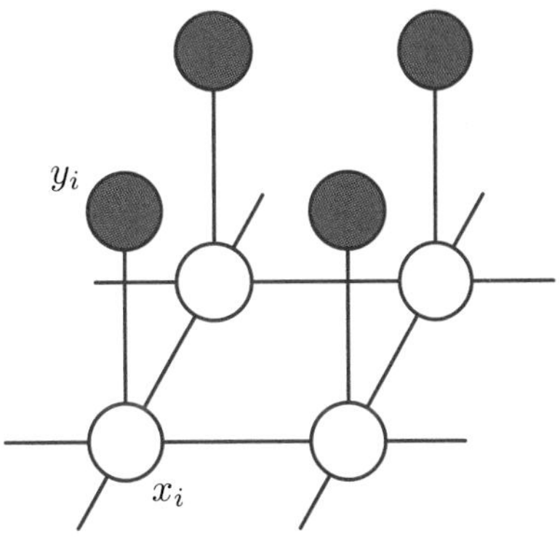

equivalently we can add the corresponding energies. In this example, this allows us to add an extra term hx_i for each pixel i in the noise-free image. Such a term has the effect of biasing the model towards pixel values that have one particular sign in preference to the other.

The complete energy function for the model then takes the form

$$E(\mathbf{x}, \mathbf{y}) = h \sum_i x_i - \beta \sum_{\{i,j\}} x_i x_j - \eta \sum_i x_i y_i \tag{8.42}$$

which defines a joint distribution over $\mathbf{x}$ and $\mathbf{y}$ given by

$$p(\mathbf{x}, \mathbf{y}) = \frac{1}{Z} \exp\{-E(\mathbf{x}, \mathbf{y})\}. \tag{8.43}$$

We now fix the elements of $\mathbf{y}$ to the observed values given by the pixels of the noisy image, which implicitly defines a conditional distribution $p(\mathbf{x}|\mathbf{y})$ over noise-free images. This is an example of the *Ising model*, which has been widely studied in statistical physics. For the purposes of image restoration, we wish to find an image $\mathbf{x}$ having a high probability (ideally the maximum probability). To do this we shall use a simple iterative technique called *iterated conditional modes*, or *ICM* (Kittler and Föglein, 1984), which is simply an application of coordinate-wise gradient ascent. The idea is first to initialize the variables $\{x_i\}$, which we do by simply setting $x_i = y_i$ for all i. Then we take one node x_j at a time and we evaluate the total energy for the two possible states $x_j = +1$ and $x_j = -1$, keeping all other node variables fixed, and set x_j to whichever state has the lower energy. This will either leave the probability unchanged, if x_j is unchanged, or will increase it. Because only one variable is changed, this is a simple local computation that can be performed efficiently. We then repeat the update for another site, and so on, until some suitable stopping criterion is satisfied. The nodes may be updated in a systematic way, for instance by repeatedly raster scanning through the image, or by choosing nodes at random.

Exercise 8.13

If we have a sequence of updates in which every site is visited at least once, and in which no changes to the variables are made, then by definition the algorithm

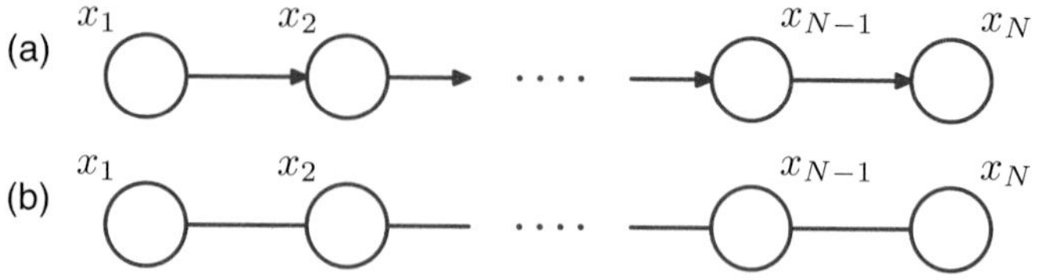

Figure 8.32 (a) Example of a directed graph. (b) The equivalent undirected graph.

will have converged to a local maximum of the probability. This need not, however, correspond to the global maximum.

For the purposes of this simple illustration, we have fixed the parameters to be $\beta = 1.0$, $\eta = 2.1$ and $h = 0$. Note that leaving $h = 0$ simply means that the prior probabilities of the two states of x_i are equal. Starting with the observed noisy image as the initial configuration, we run ICM until convergence, leading to the de-noised image shown in the lower left panel of Figure 8.30. Note that if we set $\beta = 0$, which effectively removes the links between neighbouring pixels, then the global most probable solution is given by $x_i = y_i$ for all i, corresponding to the observed noisy image.

Exercise 8.14

Section 8.4

Later we shall discuss a more effective algorithm for finding high probability solutions called the max-sum algorithm, which typically leads to better solutions, although this is still not guaranteed to find the global maximum of the posterior distribution. However, for certain classes of model, including the one given by (8.42), there exist efficient algorithms based on *graph cuts* that are guaranteed to find the global maximum (Greig *et al.*, 1989; Boykov *et al.*, 2001; Kolmogorov and Zabih, 2004). The lower right panel of Figure 8.30 shows the result of applying a graph-cut algorithm to the de-noising problem.

8.3.4 Relation to directed graphs

We have introduced two graphical frameworks for representing probability distributions, corresponding to directed and undirected graphs, and it is instructive to discuss the relation between these. Consider first the problem of taking a model that is specified using a directed graph and trying to convert it to an undirected graph. In some cases this is straightforward, as in the simple example in Figure 8.32.Here the joint distribution for the directed graph is given as a product of conditionals in the form

$$p(\mathbf{x}) = p(x_1)p(x_2|x_1)p(x_3|x_2)\cdots p(x_N|x_{N-1}). \tag{8.44}$$

Now let us convert this to an undirected graph representation, as shown in Figure 8.32. In the undirected graph, the maximal cliques are simply the pairs of neighbouring nodes, and so from (8.39) we wish to write the joint distribution in the form

$$p(\mathbf{x}) = \frac{1}{Z}\psi_{1,2}(x_1, x_2)\psi_{2,3}(x_2, x_3)\cdots\psi_{N-1,N}(x_{N-1}, x_N). \tag{8.45}$$

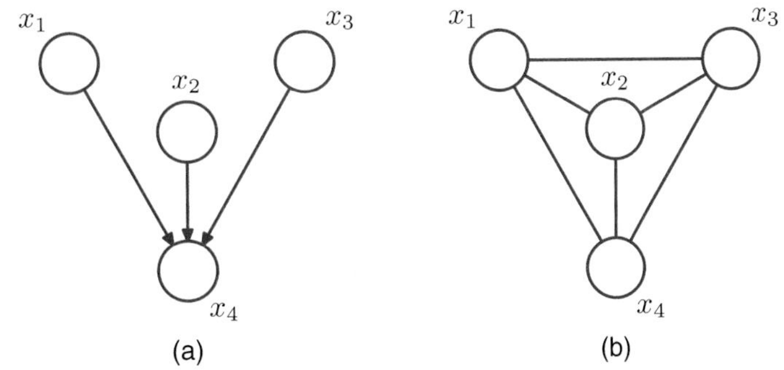

Figure 8.33 Example of a simple directed graph (a) and the corresponding moral graph (b).

This is easily done by identifying

$$\begin{aligned}
\psi_{1,2}(x_1, x_2) &= p(x_1)p(x_2|x_1) \\
\psi_{2,3}(x_2, x_3) &= p(x_3|x_2) \\
&\vdots \\
\psi_{N-1,N}(x_{N-1}, x_N) &= p(x_N|x_{N-1})
\end{aligned}$$

where we have absorbed the marginal $p(x_1)$ for the first node into the first potential function. Note that in this case, the partition function $Z = 1$.

Let us consider how to generalize this construction, so that we can convert any distribution specified by a factorization over a directed graph into one specified by a factorization over an undirected graph. This can be achieved if the clique potentials of the undirected graph are given by the conditional distributions of the directed graph. In order for this to be valid, we must ensure that the set of variables that appears in each of the conditional distributions is a member of at least one clique of the undirected graph. For nodes on the directed graph having just one parent, this is achieved simply by replacing the directed link with an undirected link. However, for nodes in the directed graph having more than one parent, this is not sufficient. These are nodes that have 'head-to-head' paths encountered in our discussion of conditional independence. Consider a simple directed graph over 4 nodes shown in Figure 8.33. The joint distribution for the directed graph takes the form

$$p(\mathbf{x}) = p(x_1)p(x_2)p(x_3)p(x_4|x_1, x_2, x_3). \tag{8.46}$$

We see that the factor $p(x_4|x_1, x_2, x_3)$ involves the four variables x_1, x_2, x_3, and x_4, and so these must all belong to a single clique if this conditional distribution is to be absorbed into a clique potential. To ensure this, we add extra links between all pairs of parents of the node x_4. Anachronistically, this process of 'marrying the parents' has become known as *moralization*, and the resulting undirected graph, after dropping the arrows, is called the *moral graph*. It is important to observe that the moral graph in this example is fully connected and so exhibits no conditional independence properties, in contrast to the original directed graph.

Thus in general to convert a directed graph into an undirected graph, we first add additional undirected links between all pairs of parents for each node in the graph and then drop the arrows on the original links to give the moral graph. Then we initialize all of the clique potentials of the moral graph to 1. We then take each conditional distribution factor in the original directed graph and multiply it into one of the clique potentials. There will always exist at least one maximal clique that contains all of the variables in the factor as a result of the moralization step. Note that in all cases the partition function is given by $Z = 1$.

Section 8.4

The process of converting a directed graph into an undirected graph plays an important role in exact inference techniques such as the *junction tree algorithm*. Converting from an undirected to a directed representation is much less common and in general presents problems due to the normalization constraints.

We saw that in going from a directed to an undirected representation we had to discard some conditional independence properties from the graph. Of course, we could always trivially convert any distribution over a directed graph into one over an undirected graph by simply using a fully connected undirected graph. This would, however, discard all conditional independence properties and so would be vacuous. The process of moralization adds the fewest extra links and so retains the maximum number of independence properties.

We have seen that the procedure for determining the conditional independence properties is different between directed and undirected graphs. It turns out that the two types of graph can express different conditional independence properties, and it is worth exploring this issue in more detail. To do so, we return to the view of a specific (directed or undirected) graph as a filter, so that the set of all possible distributions over the given variables could be reduced to a subset that respects the conditional independencies implied by the graph. A graph is said to be a *D map* (for 'dependency map') of a distribution if every conditional independence statement satisfied by the distribution is reflected in the graph. Thus a completely disconnected graph (no links) will be a trivial D map for any distribution.

Section 8.2

Alternatively, we can consider a specific distribution and ask which graphs have the appropriate conditional independence properties. If every conditional independence statement implied by a graph is satisfied by a specific distribution, then the graph is said to be an *I map* (for 'independence map') of that distribution. Clearly a fully connected graph will be a trivial I map for any distribution.

If it is the case that every conditional independence property of the distribution is reflected in the graph, and vice versa, then the graph is said to be a *perfect map* for

Figure 8.34 Venn diagram illustrating the set of all distributions P over a given set of variables, together with the set of distributions D that can be represented as a perfect map using a directed graph, and the set U that can be represented as a perfect map using an undirected graph.

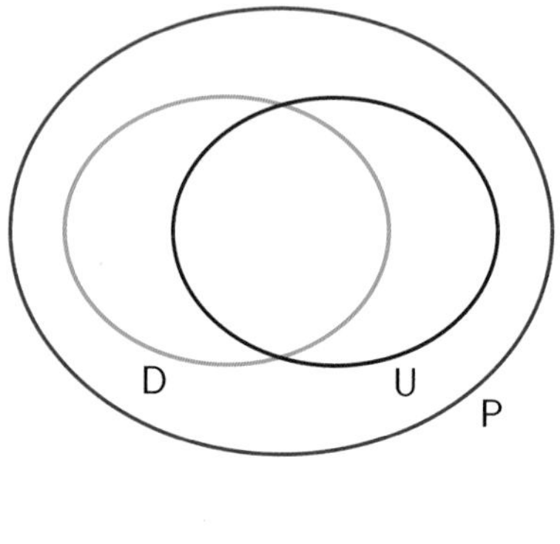

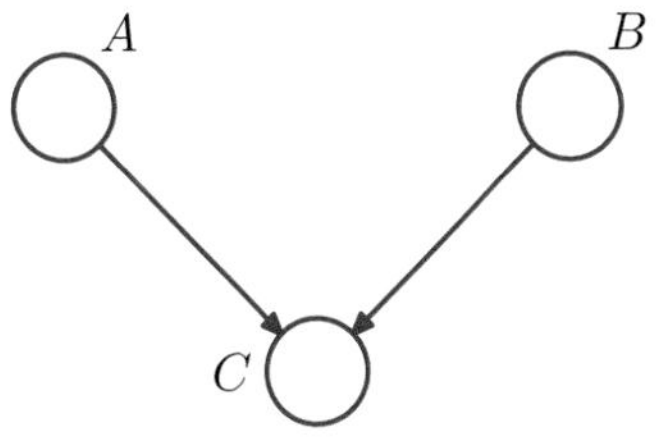

Figure 8.35 A directed graph whose conditional independence properties cannot be expressed using an undirected graph over the same three variables.

that distribution. A perfect map is therefore both an I map and a D map.

Consider the set of distributions such that for each distribution there exists a directed graph that is a perfect map. This set is distinct from the set of distributions such that for each distribution there exists an undirected graph that is a perfect map. In addition there are distributions for which neither directed nor undirected graphs offer a perfect map. This is illustrated as a Venn diagram in Figure 8.34.

Figure 8.35 shows an example of a directed graph that is a perfect map for a distribution satisfying the conditional independence properties $A \perp\!\!\!\perp B \mid \emptyset$ and $A \not\perp\!\!\!\perp B \mid C$. There is no corresponding undirected graph over the same three variables that is a perfect map.

Conversely, consider the undirected graph over four variables shown in Figure 8.36. This graph exhibits the properties $A \not\perp\!\!\!\perp B \mid \emptyset$, $C \perp\!\!\!\perp D \mid A \cup B$ and $A \perp\!\!\!\perp B \mid C \cup D$. There is no directed graph over four variables that implies the same set of conditional independence properties.

The graphical framework can be extended in a consistent way to graphs that include both directed and undirected links. These are called *chain graphs* (Lauritzen and Wermuth, 1989; Frydenberg, 1990), and contain the directed and undirected graphs considered so far as special cases. Although such graphs can represent a broader class of distributions than either directed or undirected alone, there remain distributions for which even a chain graph cannot provide a perfect map. Chain graphs are not discussed further in this book.

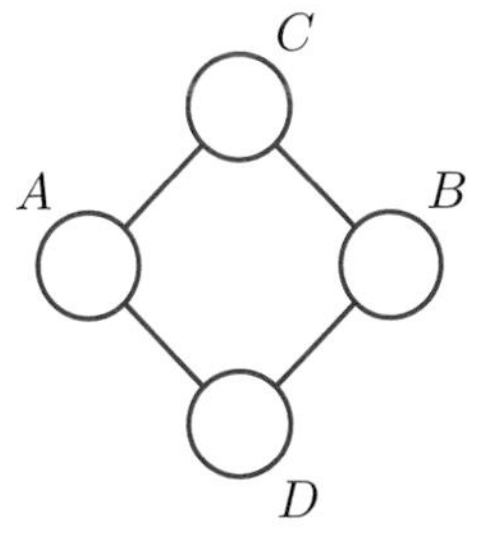

Figure 8.36 An undirected graph whose conditional independence properties cannot be expressed in terms of a directed graph over the same variables.

8.4. Inference in Graphical Models

We turn now to the problem of inference in graphical models, in which some of the nodes in a graph are clamped to observed values, and we wish to compute the posterior distributions of one or more subsets of other nodes. As we shall see, we can exploit the graphical structure both to find efficient algorithms for inference, and

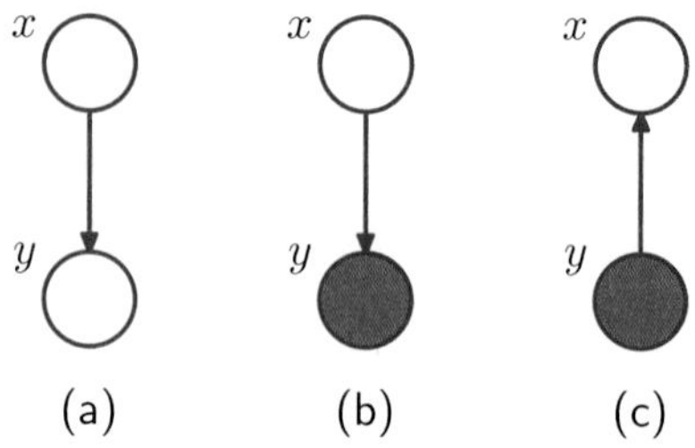

Figure 8.37 A graphical representation of Bayes' theorem. See the text for details.

to make the structure of those algorithms transparent. Specifically, we shall see that many algorithms can be expressed in terms of the propagation of local *messages* around the graph. In this section, we shall focus primarily on techniques for exact inference, and in Chapter 10 we shall consider a number of approximate inference algorithms.

To start with, let us consider the graphical interpretation of Bayes' theorem. Suppose we decompose the joint distribution $p(x, y)$ over two variables x and y into a product of factors in the form $p(x, y) = p(x)p(y|x)$. This can be represented by the directed graph shown in Figure 8.37(a). Now suppose we observe the value of y, as indicated by the shaded node in Figure 8.37(b). We can view the marginal distribution $p(x)$ as a prior over the latent variable x, and our goal is to infer the corresponding posterior distribution over x. Using the sum and product rules of probability we can evaluate

$$p(y) = \sum_{x'} p(y|x')p(x') \tag{8.47}$$

which can then be used in Bayes' theorem to calculate

$$p(x|y) = \frac{p(y|x)p(x)}{p(y)}. \tag{8.48}$$

Thus the joint distribution is now expressed in terms of $p(y)$ and $p(x|y)$. From a graphical perspective, the joint distribution $p(x, y)$ is now represented by the graph shown in Figure 8.37(c), in which the direction of the arrow is reversed. This is the simplest example of an inference problem for a graphical model.

8.4.1 Inference on a chain

Now consider a more complex problem involving the chain of nodes of the form shown in Figure 8.32. This example will lay the foundation for a discussion of exact inference in more general graphs later in this section.

Specifically, we shall consider the undirected graph in Figure 8.32(b). We have already seen that the directed chain can be transformed into an equivalent undirected chain. Because the directed graph does not have any nodes with more than one parent, this does not require the addition of any extra links, and the directed and undirected versions of this graph express exactly the same set of conditional independence statements.

The joint distribution for this graph takes the form

$$p(\mathbf{x}) = \frac{1}{Z}\psi_{1,2}(x_1, x_2)\psi_{2,3}(x_2, x_3) \cdots \psi_{N-1,N}(x_{N-1}, x_N). \tag{8.49}$$

We shall consider the specific case in which the N nodes represent discrete variables each having K states, in which case each potential function $\psi_{n-1,n}(x_{n-1}, x_n)$ comprises an $K \times K$ table, and so the joint distribution has $(N-1)K^2$ parameters.

Let us consider the inference problem of finding the marginal distribution $p(x_n)$ for a specific node x_n that is part way along the chain. Note that, for the moment, there are no observed nodes. By definition, the required marginal is obtained by summing the joint distribution over all variables except x_n, so that

$$p(x_n) = \sum_{x_1} \cdots \sum_{x_{n-1}} \sum_{x_{n+1}} \cdots \sum_{x_N} p(\mathbf{x}). \tag{8.50}$$

In a naive implementation, we would first evaluate the joint distribution and then perform the summations explicitly. The joint distribution can be represented as a set of numbers, one for each possible value for $\mathbf{x}$. Because there are N variables each with K states, there are K^N values for $\mathbf{x}$ and so evaluation and storage of the joint distribution, as well as marginalization to obtain $p(x_n)$, all involve storage and computation that scale exponentially with the length N of the chain.

We can, however, obtain a much more efficient algorithm by exploiting the conditional independence properties of the graphical model. If we substitute the factorized expression (8.49) for the joint distribution into (8.50), then we can rearrange the order of the summations and the multiplications to allow the required marginal to be evaluated much more efficiently. Consider for instance the summation over x_N. The potential $\psi_{N-1,N}(x_{N-1}, x_N)$ is the only one that depends on x_N, and so we can perform the summation

$$\sum_{x_N} \psi_{N-1,N}(x_{N-1}, x_N) \tag{8.51}$$

first to give a function of x_{N-1}. We can then use this to perform the summation over x_{N-1}, which will involve only this new function together with the potential $\psi_{N-2,N-1}(x_{N-2}, x_{N-1})$, because this is the only other place that x_{N-1} appears. Similarly, the summation over x_1 involves only the potential $\psi_{1,2}(x_1, x_2)$ and so can be performed separately to give a function of x_2, and so on. Because each summation effectively removes a variable from the distribution, this can be viewed as the removal of a node from the graph.

If we group the potentials and summations together in this way, we can express the desired marginal in the form

$$
\begin{aligned}
p(x_n) &= \frac{1}{Z} \\
&\underbrace{\left[\sum_{x_{n-1}} \psi_{n-1,n}(x_{n-1}, x_n) \cdots \left[\sum_{x_2} \psi_{2,3}(x_2, x_3) \left[\sum_{x_1} \psi_{1,2}(x_1, x_2)\right]\right] \cdots \right]}_{\mu_\alpha(x_n)} \\
&\underbrace{\left[\sum_{x_{n+1}} \psi_{n,n+1}(x_n, x_{n+1}) \cdots \left[\sum_{x_N} \psi_{N-1,N}(x_{N-1}, x_N)\right] \cdots \right]}_{\mu_\beta(x_n)}. \qquad (8.52)
\end{aligned}
$$

The reader is encouraged to study this re-ordering carefully as the underlying idea forms the basis for the later discussion of the general sum-product algorithm. Here the key concept that we are exploiting is that multiplication is distributive over addition, so that

$$ab + ac = a(b + c) \qquad (8.53)$$

in which the left-hand side involves three arithmetic operations whereas the right-hand side reduces this to two operations.

Let us work out the computational cost of evaluating the required marginal using this re-ordered expression. We have to perform $N - 1$ summations each of which is over K states and each of which involves a function of two variables. For instance, the summation over x_1 involves only the function $\psi_{1,2}(x_1, x_2)$, which is a table of $K \times K$ numbers. We have to sum this table over x_1 for each value of x_2 and so this has $O(K^2)$ cost. The resulting vector of K numbers is multiplied by the matrix of numbers $\psi_{2,3}(x_2, x_3)$ and so is again $O(K^2)$. Because there are $N - 1$ summations and multiplications of this kind, the total cost of evaluating the marginal $p(x_n)$ is $O(NK^2)$. This is linear in the length of the chain, in contrast to the exponential cost of a naive approach. We have therefore been able to exploit the many conditional independence properties of this simple graph in order to obtain an efficient calculation. If the graph had been fully connected, there would have been no conditional independence properties, and we would have been forced to work directly with the full joint distribution.

We now give a powerful interpretation of this calculation in terms of the passing of local *messages* around on the graph. From (8.52) we see that the expression for the marginal $p(x_n)$ decomposes into the product of two factors times the normalization constant

$$p(x_n) = \frac{1}{Z}\mu_\alpha(x_n)\mu_\beta(x_n). \qquad (8.54)$$

We shall interpret $\mu_\alpha(x_n)$ as a message passed forwards along the chain from node x_{n-1} to node x_n. Similarly, $\mu_\beta(x_n)$ can be viewed as a message passed backwards

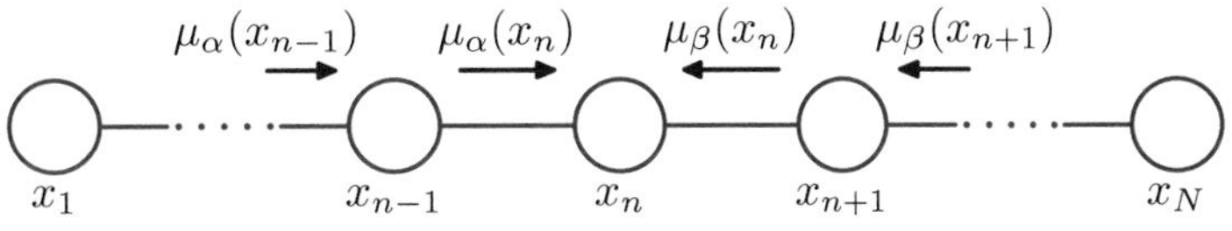

Figure 8.38 The marginal distribution $p(x_n)$ for a node x_n along the chain is obtained by multiplying the two messages $\mu_\alpha(x_n)$ and $\mu_\beta(x_n)$, and then normalizing. These messages can themselves be evaluated recursively by passing messages from both ends of the chain towards node x_n.

along the chain to node x_n from node x_{n+1}. Note that each of the messages comprises a set of K values, one for each choice of x_n, and so the product of two messages should be interpreted as the point-wise multiplication of the elements of the two messages to give another set of K values.

The message $\mu_\alpha(x_n)$ can be evaluated recursively because

$$\begin{aligned}\mu_\alpha(x_n) &= \sum_{x_{n-1}} \psi_{n-1,n}(x_{n-1}, x_n) \left[\sum_{x_{n-2}} \cdots \right] \\ &= \sum_{x_{n-1}} \psi_{n-1,n}(x_{n-1}, x_n) \mu_\alpha(x_{n-1}). \end{aligned} \tag{8.55}$$

We therefore first evaluate

$$\mu_\alpha(x_2) = \sum_{x_1} \psi_{1,2}(x_1, x_2) \tag{8.56}$$

and then apply (8.55) repeatedly until we reach the desired node. Note carefully the structure of the message passing equation. The outgoing message $\mu_\alpha(x_n)$ in (8.55) is obtained by multiplying the incoming message $\mu_\alpha(x_{n-1})$ by the local potential involving the node variable and the outgoing variable and then summing over the node variable.

Similarly, the message $\mu_\beta(x_n)$ can be evaluated recursively by starting with node x_N and using

$$\begin{aligned}\mu_\beta(x_n) &= \sum_{x_{n+1}} \psi_{n,n+1}(x_n, x_{n+1}) \left[\sum_{x_{n+2}} \cdots \right] \\ &= \sum_{x_{n+1}} \psi_{n,n+1}(x_n, x_{n+1}) \mu_\beta(x_{n+1}). \end{aligned} \tag{8.57}$$

This recursive message passing is illustrated in Figure 8.38. The normalization constant Z is easily evaluated by summing the right-hand side of (8.54) over all states of x_n, an operation that requires only $O(K)$ computation.

Graphs of the form shown in Figure 8.38 are called *Markov chains*, and the corresponding message passing equations represent an example of the *Chapman-Kolmogorov* equations for Markov processes (Papoulis, 1984).

Now suppose we wish to evaluate the marginals $p(x_n)$ for every node $n \in \{1, \ldots, N\}$ in the chain. Simply applying the above procedure separately for each node will have computational cost that is $O(N^2K^2)$. However, such an approach would be very wasteful of computation. For instance, to find $p(x_1)$ we need to propagate a message $\mu_\beta(\cdot)$ from node x_N back to node x_2. Similarly, to evaluate $p(x_2)$ we need to propagate a messages $\mu_\beta(\cdot)$ from node x_N back to node x_3. This will involve much duplicated computation because most of the messages will be identical in the two cases.

Suppose instead we first launch a message $\mu_\beta(x_{N-1})$ starting from node x_N and propagate corresponding messages all the way back to node x_1, and suppose we similarly launch a message $\mu_\alpha(x_2)$ starting from node x_1 and propagate the corresponding messages all the way forward to node x_N. Provided we store all of the intermediate messages along the way, then any node can evaluate its marginal simply by applying (8.54). The computational cost is only twice that for finding the marginal of a single node, rather than N times as much. Observe that a message has passed once in each direction across each link in the graph. Note also that the normalization constant Z need be evaluated only once, using any convenient node.

If some of the nodes in the graph are observed, then the corresponding variables are simply clamped to their observed values and there is no summation. To see this, note that the effect of clamping a variable x_n to an observed value $\widehat{x}_n$ can be expressed by multiplying the joint distribution by (one or more copies of) an additional function $I(x_n, \widehat{x}_n)$, which takes the value 1 when $x_n = \widehat{x}_n$ and the value 0 otherwise. One such function can then be absorbed into each of the potentials that contain x_n. Summations over x_n then contain only one term in which $x_n = \widehat{x}_n$.

Now suppose we wish to calculate the joint distribution $p(x_{n-1}, x_n)$ for two neighbouring nodes on the chain. This is similar to the evaluation of the marginal for a single node, except that there are now two variables that are not summed out.

Exercise 8.15

A few moments thought will show that the required joint distribution can be written in the form

$$p(x_{n-1}, x_n) = \frac{1}{Z}\mu_\alpha(x_{n-1})\psi_{n-1,n}(x_{n-1}, x_n)\mu_\beta(x_n). \tag{8.58}$$

Thus we can obtain the joint distributions over all of the sets of variables in each of the potentials directly once we have completed the message passing required to obtain the marginals.

This is a useful result because in practice we may wish to use parametric forms for the clique potentials, or equivalently for the conditional distributions if we started from a directed graph. In order to learn the parameters of these potentials in situations where not all of the variables are observed, we can employ the *EM algorithm*, and it turns out that the local joint distributions of the cliques, conditioned on any observed data, is precisely what is needed in the E step. We shall consider some examples of this in detail in Chapter 13.

Chapter 9

8.4.2 Trees

We have seen that exact inference on a graph comprising a chain of nodes can be performed efficiently in time that is linear in the number of nodes, using an algorithm

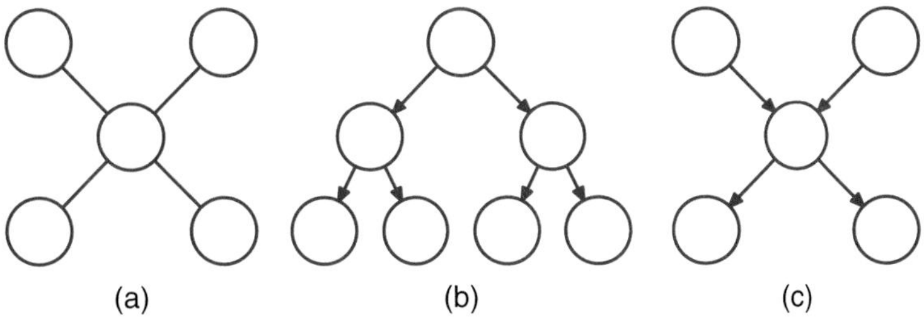

Figure 8.39 Examples of tree-structured graphs, showing (a) an undirected tree, (b) a directed tree, and (c) a directed polytree.

that can be interpreted in terms of messages passed along the chain. More generally, inference can be performed efficiently using local message passing on a broader class of graphs called *trees*. In particular, we shall shortly generalize the message passing formalism derived above for chains to give the *sum-product* algorithm, which provides an efficient framework for exact inference in tree-structured graphs.

In the case of an undirected graph, a tree is defined as a graph in which there is one, and only one, path between any pair of nodes. Such graphs therefore do not have loops. In the case of directed graphs, a tree is defined such that there is a single node, called the *root*, which has no parents, and all other nodes have one parent. If we convert a directed tree into an undirected graph, we see that the moralization step will not add any links as all nodes have at most one parent, and as a consequence the corresponding moralized graph will be an undirected tree. Examples of undirected and directed trees are shown in Figure 8.39(a) and 8.39(b). Note that a distribution represented as a directed tree can easily be converted into one represented by an undirected tree, and vice versa.

Exercise 8.18

If there are nodes in a directed graph that have more than one parent, but there is still only one path (ignoring the direction of the arrows) between any two nodes, then the graph is a called a *polytree*, as illustrated in Figure 8.39(c). Such a graph will have more than one node with the property of having no parents, and furthermore, the corresponding moralized undirected graph will have loops.

8.4.3 Factor graphs

The sum-product algorithm that we derive in the next section is applicable to undirected and directed trees and to polytrees. It can be cast in a particularly simple and general form if we first introduce a new graphical construction called a *factor graph* (Frey, 1998; Kschischnang *et al.*, 2001).

Both directed and undirected graphs allow a global function of several variables to be expressed as a product of factors over subsets of those variables. Factor graphs make this decomposition explicit by introducing additional nodes for the factors themselves in addition to the nodes representing the variables. They also allow us to be more explicit about the details of the factorization, as we shall see.

Let us write the joint distribution over a set of variables in the form of a product of factors

$$p(\mathbf{x}) = \prod_s f_s(\mathbf{x}_s) \tag{8.59}$$

where $\mathbf{x}_s$ denotes a subset of the variables. For convenience, we shall denote the

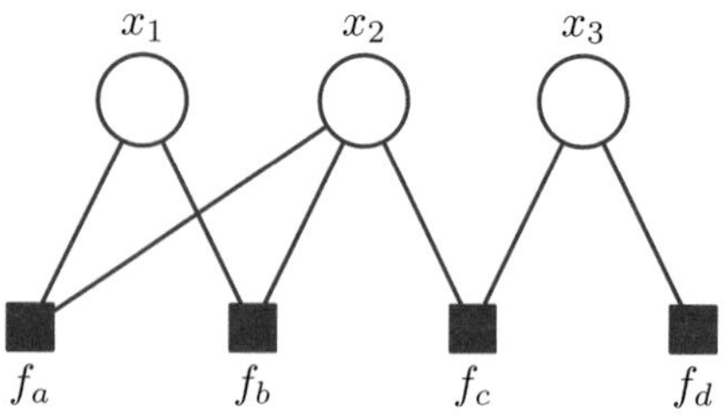

Figure 8.40 Example of a factor graph, which corresponds to the factorization (8.60).

individual variables by x_i, however, as in earlier discussions, these can comprise groups of variables (such as vectors or matrices). Each factor f_s is a function of a corresponding set of variables $\mathbf{x}_s$.

Directed graphs, whose factorization is defined by (8.5), represent special cases of (8.59) in which the factors $f_s(\mathbf{x}_s)$ are local conditional distributions. Similarly, undirected graphs, given by (8.39), are a special case in which the factors are potential functions over the maximal cliques (the normalizing coefficient $1/Z$ can be viewed as a factor defined over the empty set of variables).

In a factor graph, there is a node (depicted as usual by a circle) for every variable in the distribution, as was the case for directed and undirected graphs. There are also additional nodes (depicted by small squares) for each factor $f_s(\mathbf{x}_s)$ in the joint distribution. Finally, there are undirected links connecting each factor node to all of the variables nodes on which that factor depends. Consider, for example, a distribution that is expressed in terms of the factorization

$$p(\mathbf{x}) = f_a(x_1, x_2)f_b(x_1, x_2)f_c(x_2, x_3)f_d(x_3). \tag{8.60}$$

This can be expressed by the factor graph shown in Figure 8.40. Note that there are two factors $f_a(x_1, x_2)$ and $f_b(x_1, x_2)$ that are defined over the same set of variables. In an undirected graph, the product of two such factors would simply be lumped together into the same clique potential. Similarly, $f_c(x_2, x_3)$ and $f_d(x_3)$ could be combined into a single potential over x_2 and x_3. The factor graph, however, keeps such factors explicit and so is able to convey more detailed information about the underlying factorization.

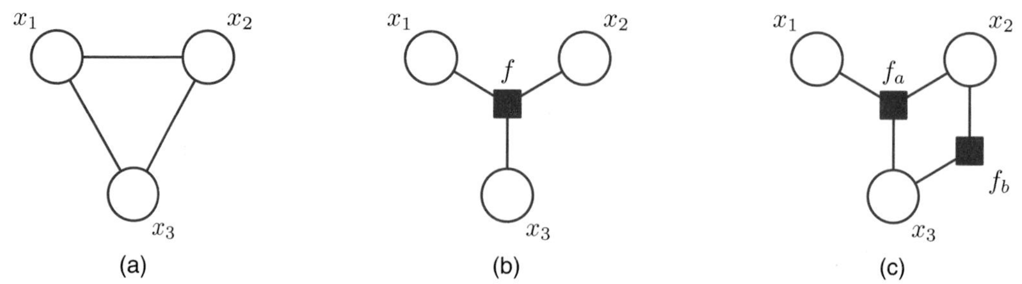

Figure 8.41 (a) An undirected graph with a single clique potential $\psi(x_1, x_2, x_3)$. (b) A factor graph with factor $f(x_1, x_2, x_3) = \psi(x_1, x_2, x_3)$ representing the same distribution as the undirected graph. (c) A different factor graph representing the same distribution, whose factors satisfy $f_a(x_1, x_2, x_3)f_b(x_2, x_3) = \psi(x_1, x_2, x_3)$.

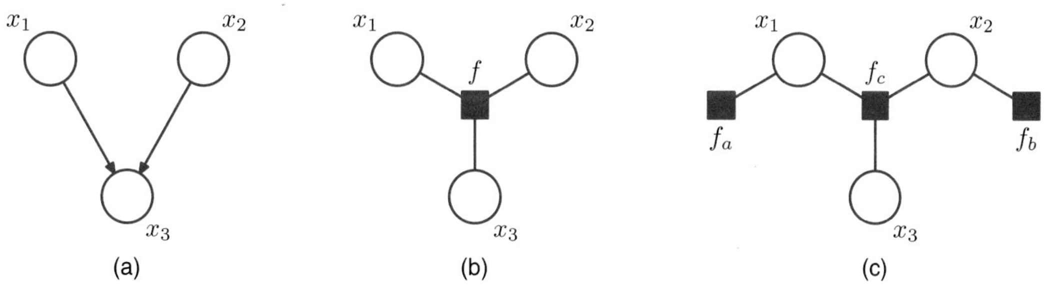

Figure 8.42 (a) A directed graph with the factorization $p(x_1)p(x_2)p(x_3|x_1, x_2)$. (b) A factor graph representing the same distribution as the directed graph, whose factor satisfies $f(x_1, x_2, x_3) = p(x_1)p(x_2)p(x_3|x_1, x_2)$. (c) A different factor graph representing the same distribution with factors $f_a(x_1) = p(x_1)$, $f_b(x_2) = p(x_2)$ and $f_c(x_1, x_2, x_3) = p(x_3|x_1, x_2)$.

Factor graphs are said to be *bipartite* because they consist of two distinct kinds of nodes, and all links go between nodes of opposite type. In general, factor graphs can therefore always be drawn as two rows of nodes (variable nodes at the top and factor nodes at the bottom) with links between the rows, as shown in the example in Figure 8.40. In some situations, however, other ways of laying out the graph may be more intuitive, for example when the factor graph is derived from a directed or undirected graph, as we shall see.

If we are given a distribution that is expressed in terms of an undirected graph, then we can readily convert it to a factor graph. To do this, we create variable nodes corresponding to the nodes in the original undirected graph, and then create additional factor nodes corresponding to the maximal cliques $\mathbf{x}_s$. The factors $f_s(\mathbf{x}_s)$ are then set equal to the clique potentials. Note that there may be several different factor graphs that correspond to the same undirected graph. These concepts are illustrated in Figure 8.41.

Similarly, to convert a directed graph to a factor graph, we simply create variable nodes in the factor graph corresponding to the nodes of the directed graph, and then create factor nodes corresponding to the conditional distributions, and then finally add the appropriate links. Again, there can be multiple factor graphs all of which correspond to the same directed graph. The conversion of a directed graph to a factor graph is illustrated in Figure 8.42.

We have already noted the importance of tree-structured graphs for performing efficient inference. If we take a directed or undirected tree and convert it into a factor graph, then the result will again be a tree (in other words, the factor graph will have no loops, and there will be one and only one path connecting any two nodes). In the case of a directed polytree, conversion to an undirected graph results in loops due to the moralization step, whereas conversion to a factor graph again results in a tree, as illustrated in Figure 8.43. In fact, local cycles in a directed graph due to links connecting parents of a node can be removed on conversion to a factor graph by defining the appropriate factor function, as shown in Figure 8.44.

We have seen that multiple different factor graphs can represent the same directed or undirected graph. This allows factor graphs to be more specific about the

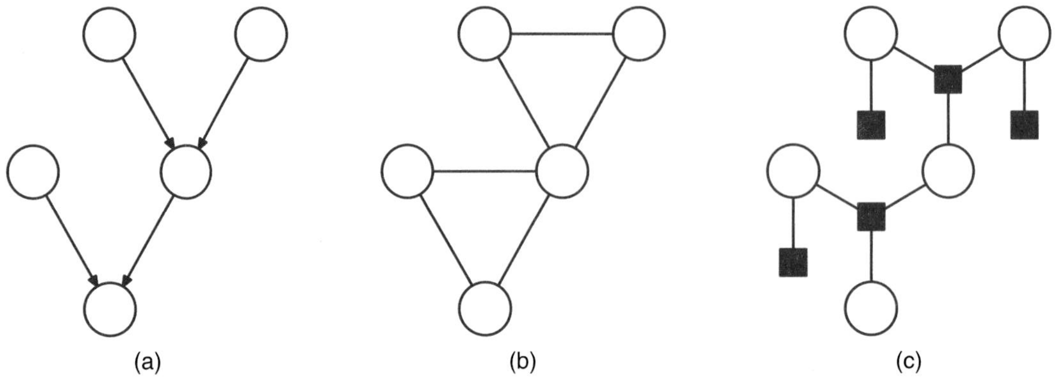

Figure 8.43 (a) A directed polytree. (b) The result of converting the polytree into an undirected graph showing the creation of loops. (c) The result of converting the polytree into a factor graph, which retains the tree structure.

precise form of the factorization. Figure 8.45 shows an example of a fully connected undirected graph along with two different factor graphs. In (b), the joint distribution is given by a general form $p(\mathbf{x}) = f(x_1, x_2, x_3)$, whereas in (c), it is given by the more specific factorization $p(\mathbf{x}) = f_a(x_1, x_2) f_b(x_1, x_3) f_c(x_2, x_3)$. It should be emphasized that the factorization in (c) does not correspond to any conditional independence properties.

8.4.4 The sum-product algorithm

We shall now make use of the factor graph framework to derive a powerful class of efficient, exact inference algorithms that are applicable to tree-structured graphs. Here we shall focus on the problem of evaluating local marginals over nodes or subsets of nodes, which will lead us to the *sum-product* algorithm. Later we shall modify the technique to allow the most probable state to be found, giving rise to the *max-sum* algorithm.

Also we shall suppose that all of the variables in the model are discrete, and so marginalization corresponds to performing sums. The framework, however, is equally applicable to linear-Gaussian models in which case marginalization involves integration, and we shall consider an example of this in detail when we discuss linear dynamical systems.

Section 13.3

Figure 8.44 (a) A fragment of a directed graph having a local cycle. (b) Conversion to a fragment of a factor graph having a tree structure, in which $f(x_1, x_2, x_3) = p(x_1)p(x_2|x_1)p(x_3|x_1, x_2)$.

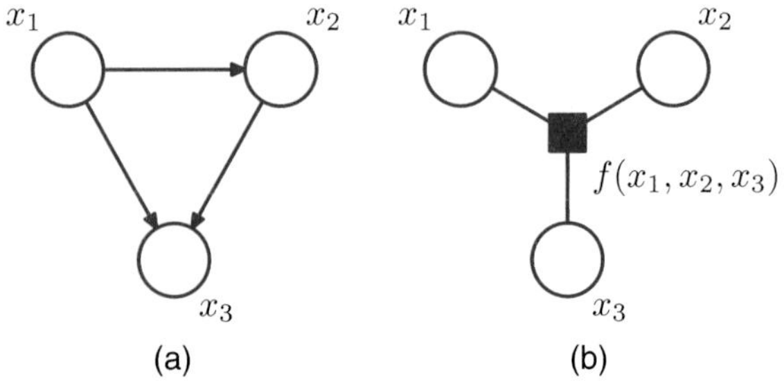

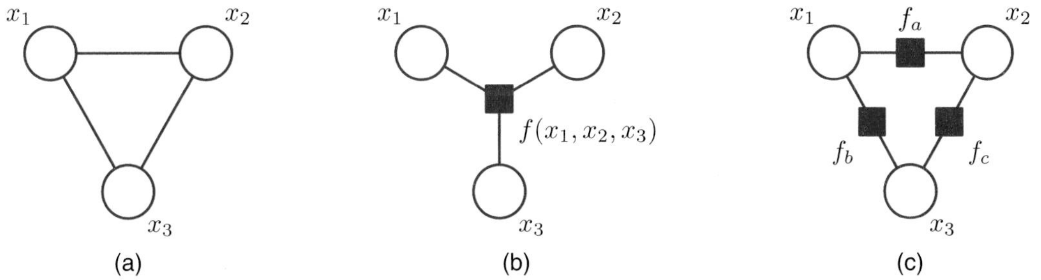

Figure 8.45 (a) A fully connected undirected graph. (b) and (c) Two factor graphs each of which corresponds to the undirected graph in (a).

There is an algorithm for exact inference on directed graphs without loops known as *belief propagation* (Pearl, 1988; Lauritzen and Spiegelhalter, 1988), and is equivalent to a special case of the sum-product algorithm. Here we shall consider only the sum-product algorithm because it is simpler to derive and to apply, as well as being more general.

We shall assume that the original graph is an undirected tree or a directed tree or polytree, so that the corresponding factor graph has a tree structure. We first convert the original graph into a factor graph so that we can deal with both directed and undirected models using the same framework. Our goal is to exploit the structure of the graph to achieve two things: (i) to obtain an efficient, exact inference algorithm for finding marginals; (ii) in situations where several marginals are required to allow computations to be shared efficiently.

We begin by considering the problem of finding the marginal $p(x)$ for particular variable node x. For the moment, we shall suppose that all of the variables are hidden. Later we shall see how to modify the algorithm to incorporate evidence corresponding to observed variables. By definition, the marginal is obtained by summing the joint distribution over all variables except x so that

$$p(x) = \sum_{\mathbf{x} \setminus x} p(\mathbf{x}) \tag{8.61}$$

where $\mathbf{x} \setminus x$ denotes the set of variables in $\mathbf{x}$ with variable x omitted. The idea is to substitute for $p(\mathbf{x})$ using the factor graph expression (8.59) and then interchange summations and products in order to obtain an efficient algorithm. Consider the fragment of graph shown in Figure 8.46 in which we see that the tree structure of the graph allows us to partition the factors in the joint distribution into groups, with one group associated with each of the factor nodes that is a neighbour of the variable node x. We see that the joint distribution can be written as a product of the form

$$p(\mathbf{x}) = \prod_{s \in \mathrm{ne}(x)} F_s(x, X_s) \tag{8.62}$$

$\mathrm{ne}(x)$ denotes the set of factor nodes that are neighbours of x, and X_s denotes the set of all variables in the subtree connected to the variable node x via the factor node

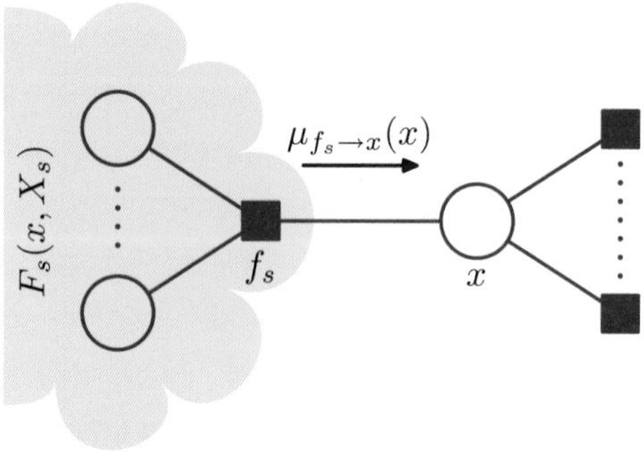

Figure 8.46 A fragment of a factor graph illustrating the evaluation of the marginal $p(x)$.

f_s, and $F_s(x, X_s)$ represents the product of all the factors in the group associated with factor f_s.

Substituting (8.62) into (8.61) and interchanging the sums and products, we obtain

$$\begin{aligned} p(x) &= \prod_{s \in \mathrm{ne}(x)} \left[\sum_{X_s} F_s(x, X_s) \right] \\ &= \prod_{s \in \mathrm{ne}(x)} \mu_{f_s \to x}(x). \end{aligned} \tag{8.63}$$

Here we have introduced a set of functions $\mu_{f_s \to x}(x)$, defined by

$$\mu_{f_s \to x}(x) \equiv \sum_{X_s} F_s(x, X_s) \tag{8.64}$$

which can be viewed as *messages* from the factor nodes f_s to the variable node x. We see that the required marginal $p(x)$ is given by the product of all the incoming messages arriving at node x.

In order to evaluate these messages, we again turn to Figure 8.46 and note that each factor $F_s(x, X_s)$ is described by a factor (sub-)graph and so can itself be factorized. In particular, we can write

$$F_s(x, X_s) = f_s(x, x_1, \ldots, x_M) G_1(x_1, X_{s1}) \ldots G_M(x_M, X_{sM}) \tag{8.65}$$

where, for convenience, we have denoted the variables associated with factor f_s, in addition to x, by $x_1, \ldots, x_M$. This factorization is illustrated in Figure 8.47. Note that the set of variables $\{x, x_1, \ldots, x_M\}$ is the set of variables on which the factor f_s depends, and so it can also be denoted $\mathbf{x}_s$, using the notation of (8.59).

Substituting (8.65) into (8.64) we obtain

$$\begin{aligned} \mu_{f_s \to x}(x) &= \sum_{x_1} \ldots \sum_{x_M} f_s(x, x_1, \ldots, x_M) \prod_{m \in \mathrm{ne}(f_s) \setminus x} \left[\sum_{X_{sm}} G_m(x_m, X_{sm}) \right] \\ &= \sum_{x_1} \ldots \sum_{x_M} f_s(x, x_1, \ldots, x_M) \prod_{m \in \mathrm{ne}(f_s) \setminus x} \mu_{x_m \to f_s}(x_m) \end{aligned} \tag{8.66}$$

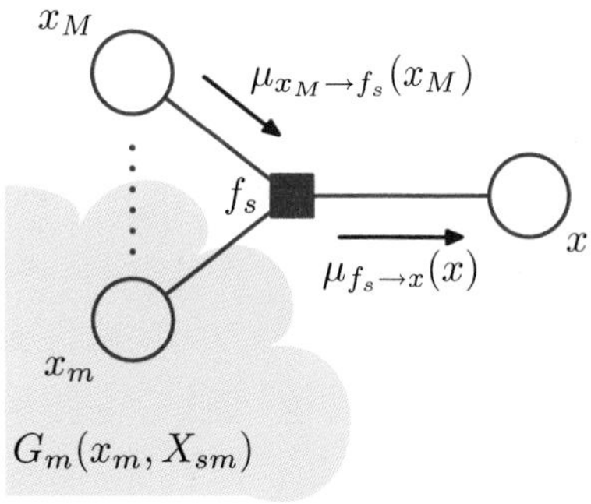

Figure 8.47 Illustration of the factorization of the subgraph associated with factor node f_s.

where $\mathrm{ne}(f_s)$ denotes the set of variable nodes that are neighbours of the factor node f_s, and $\mathrm{ne}(f_s) \setminus x$ denotes the same set but with node x removed. Here we have defined the following messages from variable nodes to factor nodes

$$\mu_{x_m \to f_s}(x_m) \equiv \sum_{X_{sm}} G_m(x_m, X_{sm}). \tag{8.67}$$

We have therefore introduced two distinct kinds of message, those that go from factor nodes to variable nodes denoted $\mu_{f \to x}(x)$, and those that go from variable nodes to factor nodes denoted $\mu_{x \to f}(x)$. In each case, we see that messages passed along a link are always a function of the variable associated with the variable node that link connects to.

The result (8.66) says that to evaluate the message sent by a factor node to a variable node along the link connecting them, take the product of the incoming messages along all other links coming into the factor node, multiply by the factor associated with that node, and then marginalize over all of the variables associated with the incoming messages. This is illustrated in Figure 8.47. It is important to note that a factor node can send a message to a variable node once it has received incoming messages from all other neighbouring variable nodes.

Finally, we derive an expression for evaluating the messages from variable nodes to factor nodes, again by making use of the (sub-)graph factorization. From Figure 8.48, we see that term $G_m(x_m, X_{sm})$ associated with node x_m is given by a product of terms $F_l(x_m, X_{ml})$ each associated with one of the factor nodes f_l that is linked to node x_m (excluding node f_s), so that

$$G_m(x_m, X_{sm}) = \prod_{l \in \mathrm{ne}(x_m) \setminus f_s} F_l(x_m, X_{ml}) \tag{8.68}$$

where the product is taken over all neighbours of node x_m except for node f_s. Note that each of the factors $F_l(x_m, X_{ml})$ represents a subtree of the original graph of precisely the same kind as introduced in (8.62). Substituting (8.68) into (8.67), we

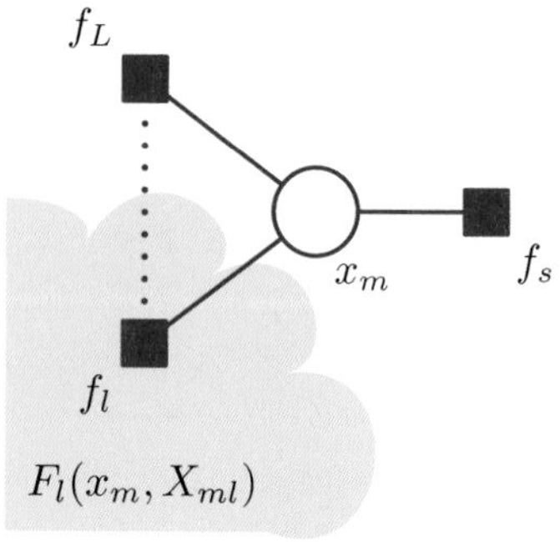

Figure 8.48 Illustration of the evaluation of the message sent by a variable node to an adjacent factor node.

then obtain

$$\begin{aligned}\mu_{x_m \to f_s}(x_m) &= \prod_{l \in \mathrm{ne}(x_m) \setminus f_s} \left[\sum_{X_{ml}} F_l(x_m, X_{ml}) \right] \\ &= \prod_{l \in \mathrm{ne}(x_m) \setminus f_s} \mu_{f_l \to x_m}(x_m) \end{aligned} \tag{8.69}$$

where we have used the definition (8.64) of the messages passed from factor nodes to variable nodes. Thus to evaluate the message sent by a variable node to an adjacent factor node along the connecting link, we simply take the product of the incoming messages along all of the other links. Note that any variable node that has only two neighbours performs no computation but simply passes messages through unchanged. Also, we note that a variable node can send a message to a factor node once it has received incoming messages from all other neighbouring factor nodes.

Recall that our goal is to calculate the marginal for variable node x, and that this marginal is given by the product of incoming messages along all of the links arriving at that node. Each of these messages can be computed recursively in terms of other messages. In order to start this recursion, we can view the node x as the root of the tree and begin at the leaf nodes. From the definition (8.69), we see that if a leaf node is a variable node, then the message that it sends along its one and only link is given by

$$\mu_{x \to f}(x) = 1 \tag{8.70}$$

as illustrated in Figure 8.49(a). Similarly, if the leaf node is a factor node, we see from (8.66) that the message sent should take the form

$$\mu_{f \to x}(x) = f(x) \tag{8.71}$$

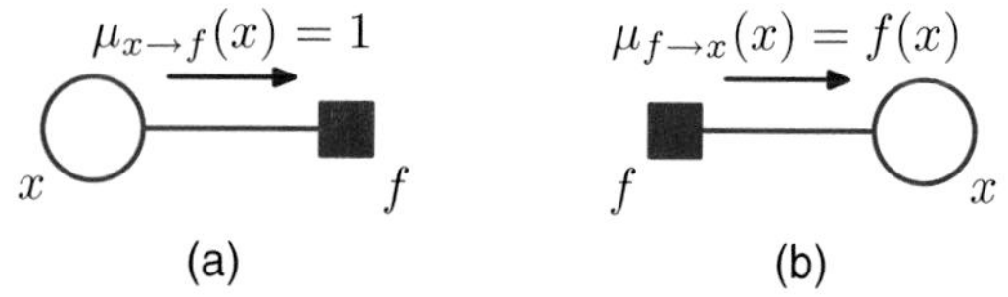

Figure 8.49 The sum-product algorithm begins with messages sent by the leaf nodes, which depend on whether the leaf node is (a) a variable node, or (b) a factor node.

as illustrated in Figure 8.49(b).

At this point, it is worth pausing to summarize the particular version of the sum-product algorithm obtained so far for evaluating the marginal $p(x)$. We start by viewing the variable node x as the root of the factor graph and initiating messages at the leaves of the graph using (8.70) and (8.71). The message passing steps (8.66) and (8.69) are then applied recursively until messages have been propagated along every link, and the root node has received messages from all of its neighbours. Each node can send a message towards the root once it has received messages from all of its other neighbours. Once the root node has received messages from all of its neighbours, the required marginal can be evaluated using (8.63). We shall illustrate this process shortly.

To see that each node will always receive enough messages to be able to send out a message, we can use a simple inductive argument as follows. Clearly, for a graph comprising a variable root node connected directly to several factor leaf nodes, the algorithm trivially involves sending messages of the form (8.71) directly from the leaves to the root. Now imagine building up a general graph by adding nodes one at a time, and suppose that for some particular graph we have a valid algorithm. When one more (variable or factor) node is added, it can be connected only by a single link because the overall graph must remain a tree, and so the new node will be a leaf node. It therefore sends a message to the node to which it is linked, which in turn will therefore receive all the messages it requires in order to send its own message towards the root, and so again we have a valid algorithm, thereby completing the proof.

Now suppose we wish to find the marginals for every variable node in the graph. This could be done by simply running the above algorithm afresh for each such node. However, this would be very wasteful as many of the required computations would be repeated. We can obtain a much more efficient procedure by 'overlaying' these multiple message passing algorithms to obtain the general sum-product algorithm as follows. Arbitrarily pick any (variable or factor) node and designate it as the root. Propagate messages from the leaves to the root as before. At this point, the root node will have received messages from all of its neighbours. It can therefore send out messages to all of its neighbours. These in turn will then have received messages from all of their neighbours and so can send out messages along the links going away from the root, and so on. In this way, messages are passed outwards from the root all the way to the leaves. By now, a message will have passed in both directions across every link in the graph, and every node will have received a message from all of its neighbours. Again a simple inductive argument can be used to verify the validity of this message passing protocol. Because every variable node will have received messages from all of its neighbours, we can readily calculate the marginal distribution for every variable in the graph. The number of messages that have to be computed is given by twice the number of links in the graph and so involves only twice the computation involved in finding a single marginal. By comparison, if we had run the sum-product algorithm separately for each node, the amount of computation would grow quadratically with the size of the graph. Note that this algorithm is in fact independent of which node was designated as the root,

Exercise 8.20

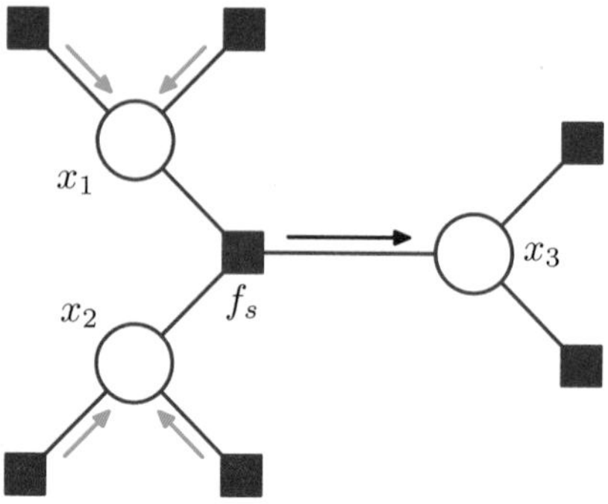

Figure 8.50 The sum-product algorithm can be viewed purely in terms of messages sent out by factor nodes to other factor nodes. In this example, the outgoing message shown by the blue arrow is obtained by taking the product of all the incoming messages shown by green arrows, multiplying by the factor f_s, and marginalizing over the variables x_1 and x_2.

and indeed the notion of one node having a special status was introduced only as a convenient way to explain the message passing protocol.

Next suppose we wish to find the marginal distributions $p(\mathbf{x}_s)$ associated with the sets of variables belonging to each of the factors. By a similar argument to that used above, it is easy to see that the marginal associated with a factor is given by the product of messages arriving at the factor node and the local factor at that node

Exercise 8.21

$$p(\mathbf{x}_s) = f_s(\mathbf{x}_s) \prod_{i \in \mathrm{ne}(f_s)} \mu_{x_i \to f_s}(x_i) \tag{8.72}$$

in complete analogy with the marginals at the variable nodes. If the factors are parameterized functions and we wish to learn the values of the parameters using the EM algorithm, then these marginals are precisely the quantities we will need to calculate in the E step, as we shall see in detail when we discuss the hidden Markov model in Chapter 13.

The message sent by a variable node to a factor node, as we have seen, is simply the product of the incoming messages on other links. We can if we wish view the sum-product algorithm in a slightly different form by eliminating messages from variable nodes to factor nodes and simply considering messages that are sent out by factor nodes. This is most easily seen by considering the example in Figure 8.50.

So far, we have rather neglected the issue of normalization. If the factor graph was derived from a directed graph, then the joint distribution is already correctly normalized, and so the marginals obtained by the sum-product algorithm will similarly be normalized correctly. However, if we started from an undirected graph, then in general there will be an unknown normalization coefficient $1/Z$. As with the simple chain example of Figure 8.38, this is easily handled by working with an unnormalized version $\widetilde{p}(\mathbf{x})$ of the joint distribution, where $p(\mathbf{x}) = \widetilde{p}(\mathbf{x})/Z$. We first run the sum-product algorithm to find the corresponding unnormalized marginals $\widetilde{p}(x_i)$. The coefficient $1/Z$ is then easily obtained by normalizing any one of these marginals, and this is computationally efficient because the normalization is done over a single variable rather than over the entire set of variables as would be required to normalize $\widetilde{p}(\mathbf{x})$ directly.

At this point, it may be helpful to consider a simple example to illustrate the operation of the sum-product algorithm. Figure 8.51 shows a simple 4-node factor

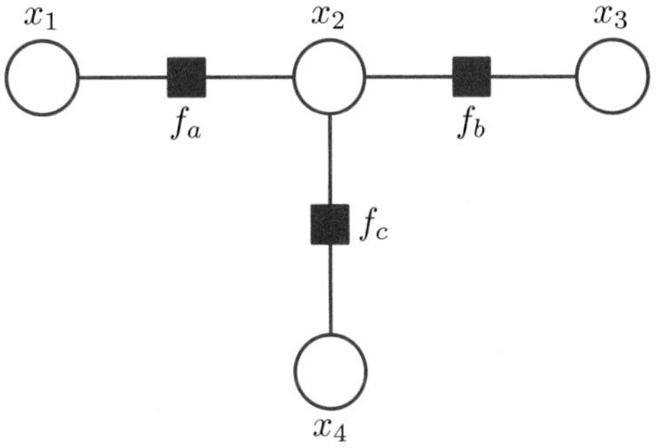

Figure 8.51 A simple factor graph used to illustrate the sum-product algorithm.

graph whose unnormalized joint distribution is given by

$$\widetilde{p}(\mathbf{x}) = f_a(x_1, x_2) f_b(x_2, x_3) f_c(x_2, x_4). \tag{8.73}$$

In order to apply the sum-product algorithm to this graph, let us designate node x_3 as the root, in which case there are two leaf nodes x_1 and x_4. Starting with the leaf nodes, we then have the following sequence of six messages

$$\mu_{x_1 \to f_a}(x_1) = 1 \tag{8.74}$$
$$\mu_{f_a \to x_2}(x_2) = \sum_{x_1} f_a(x_1, x_2) \tag{8.75}$$
$$\mu_{x_4 \to f_c}(x_4) = 1 \tag{8.76}$$
$$\mu_{f_c \to x_2}(x_2) = \sum_{x_4} f_c(x_2, x_4) \tag{8.77}$$
$$\mu_{x_2 \to f_b}(x_2) = \mu_{f_a \to x_2}(x_2) \mu_{f_c \to x_2}(x_2) \tag{8.78}$$
$$\mu_{f_b \to x_3}(x_3) = \sum_{x_2} f_b(x_2, x_3) \mu_{x_2 \to f_b}(x_2). \tag{8.79}$$

The direction of flow of these messages is illustrated in Figure 8.52. Once this message propagation is complete, we can then propagate messages from the root node out to the leaf nodes, and these are given by

$$\mu_{x_3 \to f_b}(x_3) = 1 \tag{8.80}$$
$$\mu_{f_b \to x_2}(x_2) = \sum_{x_3} f_b(x_2, x_3) \tag{8.81}$$
$$\mu_{x_2 \to f_a}(x_2) = \mu_{f_b \to x_2}(x_2) \mu_{f_c \to x_2}(x_2) \tag{8.82}$$
$$\mu_{f_a \to x_1}(x_1) = \sum_{x_2} f_a(x_1, x_2) \mu_{x_2 \to f_a}(x_2) \tag{8.83}$$
$$\mu_{x_2 \to f_c}(x_2) = \mu_{f_a \to x_2}(x_2) \mu_{f_b \to x_2}(x_2) \tag{8.84}$$
$$\mu_{f_c \to x_4}(x_4) = \sum_{x_2} f_c(x_2, x_4) \mu_{x_2 \to f_c}(x_2). \tag{8.85}$$

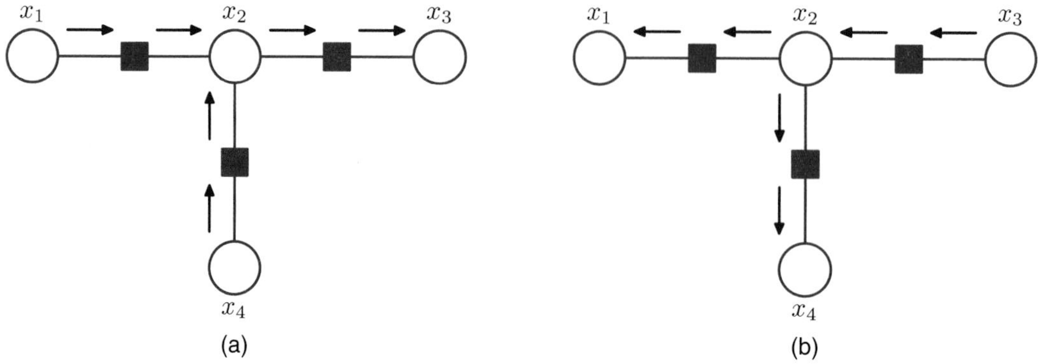

Figure 8.52 Flow of messages for the sum-product algorithm applied to the example graph in Figure 8.51. (a) From the leaf nodes x_1 and x_4 towards the root node x_3. (b) From the root node towards the leaf nodes.

One message has now passed in each direction across each link, and we can now evaluate the marginals. As a simple check, let us verify that the marginal $p(x_2)$ is given by the correct expression. Using (8.63) and substituting for the messages using the above results, we have

$$\begin{aligned}\widetilde{p}(x_2) &= \mu_{f_a \to x_2}(x_2)\mu_{f_b \to x_2}(x_2)\mu_{f_c \to x_2}(x_2) \\ &= \left[\sum_{x_1} f_a(x_1, x_2)\right]\left[\sum_{x_3} f_b(x_2, x_3)\right]\left[\sum_{x_4} f_c(x_2, x_4)\right] \\ &= \sum_{x_1}\sum_{x_3}\sum_{x_4} f_a(x_1, x_2) f_b(x_2, x_3) f_c(x_2, x_4) \\ &= \sum_{x_1}\sum_{x_3}\sum_{x_4} \widetilde{p}(\mathbf{x}) \end{aligned} \tag{8.86}$$

as required.

So far, we have assumed that all of the variables in the graph are hidden. In most practical applications, a subset of the variables will be observed, and we wish to calculate posterior distributions conditioned on these observations. Observed nodes are easily handled within the sum-product algorithm as follows. Suppose we partition $\mathbf{x}$ into hidden variables $\mathbf{h}$ and observed variables $\mathbf{v}$, and that the observed value of $\mathbf{v}$ is denoted $\widehat{\mathbf{v}}$. Then we simply multiply the joint distribution $p(\mathbf{x})$ by $\prod_i I(v_i, \widehat{v}_i)$, where $I(v, \widehat{v}) = 1$ if $v = \widehat{v}$ and $I(v, \widehat{v}) = 0$ otherwise. This product corresponds to $p(\mathbf{h}, \mathbf{v} = \widehat{\mathbf{v}})$ and hence is an unnormalized version of $p(\mathbf{h}|\mathbf{v} = \widehat{\mathbf{v}})$. By running the sum-product algorithm, we can efficiently calculate the posterior marginals $p(h_i|\mathbf{v} = \widehat{\mathbf{v}})$ up to a normalization coefficient whose value can be found efficiently using a local computation. Any summations over variables in $\mathbf{v}$ then collapse into a single term.

We have assumed throughout this section that we are dealing with discrete variables. However, there is nothing specific to discrete variables either in the graphical framework or in the probabilistic construction of the sum-product algorithm. For

Table 8.1 Example of a joint distribution over two binary variables for which the maximum of the joint distribution occurs for different variable values compared to the maxima of the two marginals.

	$x = 0$	$x = 1$
$y = 0$	0.3	0.4
$y = 1$	0.3	0.0

continuous variables the summations are simply replaced by integrations. We shall give an example of the sum-product algorithm applied to a graph of linear-Gaussian variables when we consider linear dynamical systems.

Section 13.3

8.4.5 The max-sum algorithm

The sum-product algorithm allows us to take a joint distribution $p(\mathbf{x})$ expressed as a factor graph and efficiently find marginals over the component variables. Two other common tasks are to find a setting of the variables that has the largest probability and to find the value of that probability. These can be addressed through a closely related algorithm called *max-sum*, which can be viewed as an application of *dynamic programming* in the context of graphical models (Cormen *et al.*, 2001).

A simple approach to finding latent variable values having high probability would be to run the sum-product algorithm to obtain the marginals $p(x_i)$ for every variable, and then, for each marginal in turn, to find the value $x_i^\star$ that maximizes that marginal. However, this would give the set of values that are *individually* the most probable. In practice, we typically wish to find the set of values that *jointly* have the largest probability, in other words the vector $\mathbf{x}^{\max}$ that maximizes the joint distribution, so that

$$\mathbf{x}^{\max} = \arg\max_{\mathbf{x}} p(\mathbf{x}) \tag{8.87}$$

for which the corresponding value of the joint probability will be given by

$$p(\mathbf{x}^{\max}) = \max_{\mathbf{x}} p(\mathbf{x}). \tag{8.88}$$

In general, $\mathbf{x}^{\max}$ is not the same as the set of $x_i^\star$ values, as we can easily show using a simple example. Consider the joint distribution $p(x, y)$ over two binary variables $x, y \in \{0, 1\}$ given in Table 8.1. The joint distribution is maximized by setting $x = 1$ and $y = 0$, corresponding the value 0.4. However, the marginal for $p(x)$, obtained by summing over both values of y, is given by $p(x = 0) = 0.6$ and $p(x = 1) = 0.4$, and similarly the marginal for y is given by $p(y = 0) = 0.7$ and $p(y = 1) = 0.3$, and so the marginals are maximized by $x = 0$ and $y = 0$, which corresponds to a value of 0.3 for the joint distribution. In fact, it is not difficult to construct examples for which the set of individually most probable values has probability zero under the joint distribution.

Exercise 8.27

We therefore seek an efficient algorithm for finding the value of $\mathbf{x}$ that maximizes the joint distribution $p(\mathbf{x})$ and that will allow us to obtain the value of the joint distribution at its maximum. To address the second of these problems, we shall simply write out the max operator in terms of its components

$$\max_{\mathbf{x}} p(\mathbf{x}) = \max_{x_1} \ldots \max_{x_M} p(\mathbf{x}) \tag{8.89}$$

where M is the total number of variables, and then substitute for $p(\mathbf{x})$ using its expansion in terms of a product of factors. In deriving the sum-product algorithm, we made use of the distributive law (8.53) for multiplication. Here we make use of the analogous law for the max operator

$$\max(ab, ac) = a \max(b, c) \tag{8.90}$$

which holds if $a \geqslant 0$ (as will always be the case for the factors in a graphical model). This allows us to exchange products with maximizations.

Consider first the simple example of a chain of nodes described by (8.49). The evaluation of the probability maximum can be written as

$$\begin{aligned}\max_{\mathbf{x}} p(\mathbf{x}) &= \frac{1}{Z} \max_{x_1} \cdots \max_{x_N} [\psi_{1,2}(x_1, x_2) \cdots \psi_{N-1,N}(x_{N-1}, x_N)] \\ &= \frac{1}{Z} \max_{x_1} \left[\max_{x_2} \left[\psi_{1,2}(x_1, x_2) \left[\cdots \max_{x_N} \psi_{N-1,N}(x_{N-1}, x_N) \right] \cdots \right] \right].\end{aligned}$$

As with the calculation of marginals, we see that exchanging the max and product operators results in a much more efficient computation, and one that is easily interpreted in terms of messages passed from node x_N backwards along the chain to node x_1.

We can readily generalize this result to arbitrary tree-structured factor graphs by substituting the expression (8.59) for the factor graph expansion into (8.89) and again exchanging maximizations with products. The structure of this calculation is identical to that of the sum-product algorithm, and so we can simply translate those results into the present context. In particular, suppose that we designate a particular variable node as the 'root' of the graph. Then we start a set of messages propagating inwards from the leaves of the tree towards the root, with each node sending its message towards the root once it has received all incoming messages from its other neighbours. The final maximization is performed over the product of all messages arriving at the root node, and gives the maximum value for $p(\mathbf{x})$. This could be called the *max-product* algorithm and is identical to the sum-product algorithm except that summations are replaced by maximizations. Note that at this stage, messages have been sent from leaves to the root, but not in the other direction.

In practice, products of many small probabilities can lead to numerical underflow problems, and so it is convenient to work with the logarithm of the joint distribution. The logarithm is a monotonic function, so that if $a > b$ then $\ln a > \ln b$, and hence the max operator and the logarithm function can be interchanged, so that

$$\ln \left(\max_{\mathbf{x}} p(\mathbf{x}) \right) = \max_{\mathbf{x}} \ln p(\mathbf{x}). \tag{8.91}$$

The distributive property is preserved because

$$\max(a + b, a + c) = a + \max(b, c). \tag{8.92}$$

Thus taking the logarithm simply has the effect of replacing the products in the max-product algorithm with sums, and so we obtain the *max-sum* algorithm. From the results (8.66) and (8.69) derived earlier for the sum-product algorithm, we can readily write down the max-sum algorithm in terms of message passing simply by replacing 'sum' with 'max' and replacing products with sums of logarithms to give

$$\mu_{f\to x}(x) = \max_{x_1,\ldots,x_M}\left[\ln f(x,x_1,\ldots,x_M) + \sum_{m\in\mathrm{ne}(f_s)\backslash x}\mu_{x_m\to f}(x_m)\right] \tag{8.93}$$

$$\mu_{x\to f}(x) = \sum_{l\in\mathrm{ne}(x)\backslash f}\mu_{f_l\to x}(x). \tag{8.94}$$

The initial messages sent by the leaf nodes are obtained by analogy with (8.70) and (8.71) and are given by

$$\mu_{x\to f}(x) = 0 \tag{8.95}$$

$$\mu_{f\to x}(x) = \ln f(x) \tag{8.96}$$

while at the root node the maximum probability can then be computed, by analogy with (8.63), using

$$p^{\max} = \max_x\left[\sum_{s\in\mathrm{ne}(x)}\mu_{f_s\to x}(x)\right]. \tag{8.97}$$

So far, we have seen how to find the maximum of the joint distribution by propagating messages from the leaves to an arbitrarily chosen root node. The result will be the same irrespective of which node is chosen as the root. Now we turn to the second problem of finding the configuration of the variables for which the joint distribution attains this maximum value. So far, we have sent messages from the leaves to the root. The process of evaluating (8.97) will also give the value $x^{\max}$ for the most probable value of the root node variable, defined by

$$x^{\max} = \arg\max_x\left[\sum_{s\in\mathrm{ne}(x)}\mu_{f_s\to x}(x)\right]. \tag{8.98}$$

At this point, we might be tempted simply to continue with the message passing algorithm and send messages from the root back out to the leaves, using (8.93) and (8.94), then apply (8.98) to all of the remaining variable nodes. However, because we are now maximizing rather than summing, it is possible that there may be multiple configurations of $\mathbf{x}$ all of which give rise to the maximum value for $p(\mathbf{x})$. In such cases, this strategy can fail because it is possible for the individual variable values obtained by maximizing the product of messages at each node to belong to different maximizing configurations, giving an overall configuration that no longer corresponds to a maximum.

The problem can be resolved by adopting a rather different kind of message passing from the root node to the leaves. To see how this works, let us return once again to the simple chain example of N variables $x_1,\ldots,x_N$ each having K states,

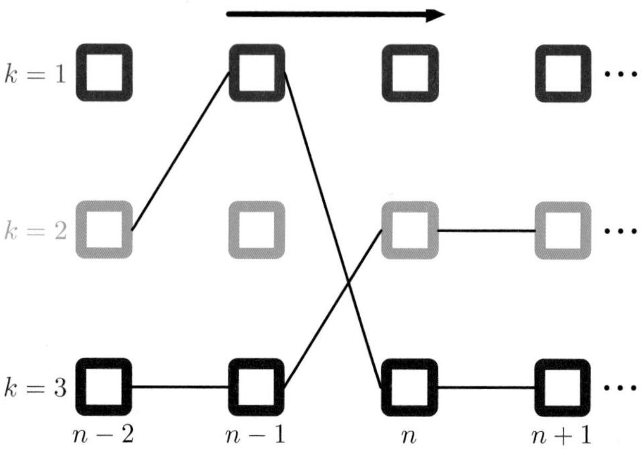

Figure 8.53 A lattice, or trellis, diagram showing explicitly the K possible states (one per row of the diagram) for each of the variables x_n in the chain model. In this illustration $K = 3$. The arrow shows the direction of message passing in the max-product algorithm. For every state k of each variable x_n (corresponding to column n of the diagram) the function $\phi(x_n)$ defines a unique state at the previous variable, indicated by the black lines. The two paths through the lattice correspond to configurations that give the global maximum of the joint probability distribution, and either of these can be found by tracing back along the black lines in the opposite direction to the arrow.

corresponding to the graph shown in Figure 8.38. Suppose we take node x_N to be the root node. Then in the first phase, we propagate messages from the leaf node x_1 to the root node using

$$\begin{aligned}\mu_{x_n \to f_{n,n+1}}(x_n) &= \mu_{f_{n-1,n} \to x_n}(x_n) \\ \mu_{f_{n-1,n} \to x_n}(x_n) &= \max_{x_{n-1}} \left[\ln f_{n-1,n}(x_{n-1}, x_n) + \mu_{x_{n-1} \to f_{n-1,n}}(x_n)\right]\end{aligned}$$

which follow from applying (8.94) and (8.93) to this particular graph. The initial message sent from the leaf node is simply

$$\mu_{x_1 \to f_{1,2}}(x_1) = 0. \tag{8.99}$$

The most probable value for x_N is then given by

$$x_N^{\max} = \arg\max_{x_N} \left[\mu_{f_{N-1,N} \to x_N}(x_N)\right]. \tag{8.100}$$

Now we need to determine the states of the previous variables that correspond to the same maximizing configuration. This can be done by keeping track of which values of the variables gave rise to the maximum state of each variable, in other words by storing quantities given by

$$\phi(x_n) = \arg\max_{x_{n-1}} \left[\ln f_{n-1,n}(x_{n-1}, x_n) + \mu_{x_{n-1} \to f_{n-1,n}}(x_n)\right]. \tag{8.101}$$

To understand better what is happening, it is helpful to represent the chain of variables in terms of a *lattice* or *trellis* diagram as shown in Figure 8.53. Note that this is not a probabilistic graphical model because the nodes represent individual states of variables, while each variable corresponds to a column of such states in the diagram. For each state of a given variable, there is a unique state of the previous variable that maximizes the probability (ties are broken either systematically or at random), corresponding to the function $\phi(x_n)$ given by (8.101), and this is indicated by the lines connecting the nodes. Once we know the most probable value of the final node x_N, we can then simply follow the link back to find the most probable state of node x_{N-1} and so on back to the initial node x_1. This corresponds to propagating a message back down the chain using

$$x_{n-1}^{\max} = \phi(x_n^{\max}) \tag{8.102}$$

and is known as *back-tracking*. Note that there could be several values of x_{n-1} all of which give the maximum value in (8.101). Provided we chose one of these values when we do the back-tracking, we are assured of a globally consistent maximizing configuration.

In Figure 8.53, we have indicated two paths, each of which we shall suppose corresponds to a global maximum of the joint probability distribution. If $k = 2$ and $k = 3$ each represent possible values of $x_N^{\max}$, then starting from either state and tracing back along the black lines, which corresponds to iterating (8.102), we obtain a valid global maximum configuration. Note that if we had run a forward pass of max-sum message passing followed by a backward pass and then applied (8.98) at each node separately, we could end up selecting some states from one path and some from the other path, giving an overall configuration that is not a global maximizer. We see that it is necessary instead to keep track of the maximizing states during the forward pass using the functions $\phi(x_n)$ and then use back-tracking to find a consistent solution.

The extension to a general tree-structured factor graph should now be clear. If a message is sent from a factor node f to a variable node x, a maximization is performed over all other variable nodes $x_1, \ldots, x_M$ that are neighbours of that factor node, using (8.93). When we perform this maximization, we keep a record of which values of the variables $x_1, \ldots, x_M$ gave rise to the maximum. Then in the back-tracking step, having found $x^{\max}$, we can then use these stored values to assign consistent maximizing states $x_1^{\max}, \ldots, x_M^{\max}$. The max-sum algorithm, with back-tracking, gives an exact maximizing configuration for the variables provided the factor graph is a tree. An important application of this technique is for finding the most probable sequence of hidden states in a hidden Markov model, in which case it is known as the *Viterbi* algorithm.

Section 13.2

As with the sum-product algorithm, the inclusion of evidence in the form of observed variables is straightforward. The observed variables are clamped to their observed values, and the maximization is performed over the remaining hidden variables. This can be shown formally by including identity functions for the observed variables into the factor functions, as we did for the sum-product algorithm.

It is interesting to compare max-sum with the iterated conditional modes (ICM) algorithm described on page 389. Each step in ICM is computationally simpler because the 'messages' that are passed from one node to the next comprise a single value consisting of the new state of the node for which the conditional distribution is maximized. The max-sum algorithm is more complex because the messages are functions of node variables x and hence comprise a set of K values for each possible state of x. Unlike max-sum, however, ICM is not guaranteed to find a global maximum even for tree-structured graphs.

8.4.6 Exact inference in general graphs

The sum-product and max-sum algorithms provide efficient and exact solutions to inference problems in tree-structured graphs. For many practical applications, however, we have to deal with graphs having loops.

The message passing framework can be generalized to arbitrary graph topologies, giving an exact inference procedure known as the *junction tree algorithm* (Lauritzen and Spiegelhalter, 1988; Jordan, 2007). Here we give a brief outline of the key steps involved. This is not intended to convey a detailed understanding of the algorithm, but rather to give a flavour of the various stages involved. If the starting point is a directed graph, it is first converted to an undirected graph by moralization, whereas if starting from an undirected graph this step is not required. Next the graph is *triangulated*, which involves finding chord-less cycles containing four or more nodes and adding extra links to eliminate such chord-less cycles. For instance, in the graph in Figure 8.36, the cycle A–C–B–D–A is chord-less and so a link should be added between A and B or alternatively between C and D. Note that the joint distribution for the resulting triangulated graph is still defined by a product of the same potential functions, but these are now considered to be functions over expanded sets of variables. Next the triangulated graph is used to construct a new tree-structured undirected graph called a *junction tree*, whose nodes correspond to the maximal cliques of the triangulated graph, and whose links connect pairs of cliques that have variables in common. The selection of which pairs of cliques to connect in this way is important and is done so as to give a *maximal spanning tree* defined as follows. Of all possible trees that link up the cliques, the one that is chosen is one for which the *weight* of the tree is largest, where the weight for a link is the number of nodes shared by the two cliques it connects, and the weight for the tree is the sum of the weights for the links. As a consequence of the triangulation step, the resulting tree satisfies the *running intersection property*, which means that if a variable is contained in two cliques, then it must also be contained in every clique on the path that connects them. This ensures that inference about variables will be consistent across the graph. Finally, a two-stage message passing algorithm, essentially equivalent to the sum-product algorithm, can now be applied to this junction tree in order to find marginals and conditionals. Although the junction tree algorithm sounds complicated, at its heart is the simple idea that we have used already of exploiting the factorization properties of the distribution to allow sums and products to be interchanged so that partial summations can be performed, thereby avoiding having to work directly with the joint distribution. The role of the junction tree is to provide a precise and efficient way to organize these computations. It is worth emphasizing that this is achieved using purely graphical operations!

The junction tree is exact for arbitrary graphs and is efficient in the sense that for a given graph there does not in general exist a computationally cheaper approach. Unfortunately, the algorithm must work with the joint distributions within each node (each of which corresponds to a clique of the triangulated graph) and so the computational cost of the algorithm is determined by the number of variables in the largest clique and will grow exponentially with this number in the case of discrete variables. An important concept is the *treewidth* of a graph (Bodlaender, 1993), which is defined in terms of the number of variables in the largest clique. In fact, it is defined to be as one less than the size of the largest clique, to ensure that a tree has a treewidth of 1. Because there in general there can be multiple different junction trees that can be constructed from a given starting graph, the treewidth is defined by the junction tree for which the largest clique has the fewest variables. If the treewidth of the original graph is high, the junction tree algorithm becomes impractical.

8.4.7 Loopy belief propagation

For many problems of practical interest, it will not be feasible to use exact inference, and so we need to exploit effective approximation methods. An important class of such approximations, that can broadly be called *variational* methods, will be discussed in detail in Chapter 10. Complementing these deterministic approaches is a wide range of *sampling* methods, also called *Monte Carlo* methods, that are based on stochastic numerical sampling from distributions and that will be discussed at length in Chapter 11.

Here we consider one simple approach to approximate inference in graphs with loops, which builds directly on the previous discussion of exact inference in trees. The idea is simply to apply the sum-product algorithm even though there is no guarantee that it will yield good results. This approach is known as *loopy belief propagation* (Frey and MacKay, 1998) and is possible because the message passing rules (8.66) and (8.69) for the sum-product algorithm are purely local. However, because the graph now has cycles, information can flow many times around the graph. For some models, the algorithm will converge, whereas for others it will not.

In order to apply this approach, we need to define a *message passing schedule*. Let us assume that one message is passed at a time on any given link and in any given direction. Each message sent from a node replaces any previous message sent in the same direction across the same link and will itself be a function only of the most recent messages received by that node at previous steps of the algorithm.

We have seen that a message can only be sent across a link from a node when all other messages have been received by that node across its other links. Because there are loops in the graph, this raises the problem of how to initiate the message passing algorithm. To resolve this, we suppose that an initial message given by the unit function has been passed across every link in each direction. Every node is then in a position to send a message.

There are now many possible ways to organize the message passing schedule. For example, the *flooding schedule* simultaneously passes a message across every link in both directions at each time step, whereas schedules that pass one message at a time are called *serial schedules*.

Following Kschischnang *et al.* (2001), we will say that a (variable or factor) node a has a message *pending* on its link to a node b if node a has received any message on any of its other links since the last time it send a message to b. Thus, when a node receives a message on one of its links, this creates pending messages on all of its other links. Only pending messages need to be transmitted because other messages would simply duplicate the previous message on the same link. For graphs that have a tree structure, any schedule that sends only pending messages will eventually terminate once a message has passed in each direction across every link. At this point, there are no pending messages, and the product of the received messages at every variable give the exact marginal. In graphs having loops, however, the algorithm may never terminate because there might always be pending messages, although in practice it is generally found to converge within a reasonable time for most applications. Once the algorithm has converged, or once it has been stopped if convergence is not observed, the (approximate) local marginals can be computed using the product of the most recently received incoming messages to each variable node or factor node on every link.

Exercise 8.29

In some applications, the loopy belief propagation algorithm can give poor results, whereas in other applications it has proven to be very effective. In particular, state-of-the-art algorithms for decoding certain kinds of error-correcting codes are equivalent to loopy belief propagation (Gallager, 1963; Berrou *et al.*, 1993; McEliece *et al.*, 1998; MacKay and Neal, 1999; Frey, 1998).

8.4.8 Learning the graph structure

In our discussion of inference in graphical models, we have assumed that the structure of the graph is known and fixed. However, there is also interest in going beyond the inference problem and learning the graph structure itself from data (Friedman and Koller, 2003). This requires that we define a space of possible structures as well as a measure that can be used to score each structure.

From a Bayesian viewpoint, we would ideally like to compute a posterior distribution over graph structures and to make predictions by averaging with respect to this distribution. If we have a prior $p(m)$ over graphs indexed by m, then the posterior distribution is given by

$$p(m|\mathcal{D}) \propto p(m)p(\mathcal{D}|m) \tag{8.103}$$

where $\mathcal{D}$ is the observed data set. The model evidence $p(\mathcal{D}|m)$ then provides the score for each model. However, evaluation of the evidence involves marginalization over the latent variables and presents a challenging computational problem for many models.

Exploring the space of structures can also be problematic. Because the number of different graph structures grows exponentially with the number of nodes, it is often necessary to resort to heuristics to find good candidates.

Exercises

8.1 ($\star$) **www** By marginalizing out the variables in order, show that the representation (8.5) for the joint distribution of a directed graph is correctly normalized, provided each of the conditional distributions is normalized.

8.2 ($\star$) **www** Show that the property of there being no directed cycles in a directed graph follows from the statement that there exists an ordered numbering of the nodes such that for each node there are no links going to a lower-numbered node.

Table 8.2 The joint distribution over three binary variables.

a	b	c	$p(a,b,c)$
0	0	0	0.192
0	0	1	0.144
0	1	0	0.048
0	1	1	0.216
1	0	0	0.192
1	0	1	0.064
1	1	0	0.048
1	1	1	0.096

8.3 ($\star\star$) Consider three binary variables $a, b, c \in \{0, 1\}$ having the joint distribution given in Table 8.2. Show by direct evaluation that this distribution has the property that a and b are marginally dependent, so that $p(a, b) \neq p(a)p(b)$, but that they become independent when conditioned on c, so that $p(a, b|c) = p(a|c)p(b|c)$ for both $c = 0$ and $c = 1$.

8.4 ($\star\star$) Evaluate the distributions $p(a)$, $p(b|c)$, and $p(c|a)$ corresponding to the joint distribution given in Table 8.2. Hence show by direct evaluation that $p(a, b, c) = p(a)p(c|a)p(b|c)$. Draw the corresponding directed graph.

8.5 ($\star$) **www** Draw a directed probabilistic graphical model corresponding to the relevance vector machine described by (7.79) and (7.80).

8.6 ($\star$) For the model shown in Figure 8.13, we have seen that the number of parameters required to specify the conditional distribution $p(y|x_1, \ldots, x_M)$, where $x_i \in \{0, 1\}$, could be reduced from 2^M to $M+1$ by making use of the logistic sigmoid representation (8.10). An alternative representation (Pearl, 1988) is given by

$$p(y = 1|x_1, \ldots, x_M) = 1 - (1 - \mu_0) \prod_{i=1}^{M} (1 - \mu_i)^{x_i} \tag{8.104}$$

where $0 \leqslant \mu_i \leqslant 1$ for $i = 1, \ldots, M$. The conditional distribution (8.104) is known as the *noisy-OR*. Show that this can be interpreted as a 'soft' (probabilistic) form of the logical OR function (i.e., the function that gives $y = 1$ whenever at least one of the $x_i = 1$). Discuss the interpretation of the μ_is.

8.7 ($\star\star$) Using the recursion relations (8.15) and (8.16), show that the mean and covariance of the joint distribution for the graph shown in Figure 8.14 are given by (8.17) and (8.18), respectively.

8.8 ($\star$) **www** Show that $a \perp\!\!\!\perp b, c \mid d$ implies $a \perp\!\!\!\perp b \mid d$.

8.9 ($\star$) **www** Using the d-separation criterion, show that the conditional distribution for a node x in a directed graph, conditioned on all of the nodes in the Markov blanket, is independent of the remaining variables in the graph.

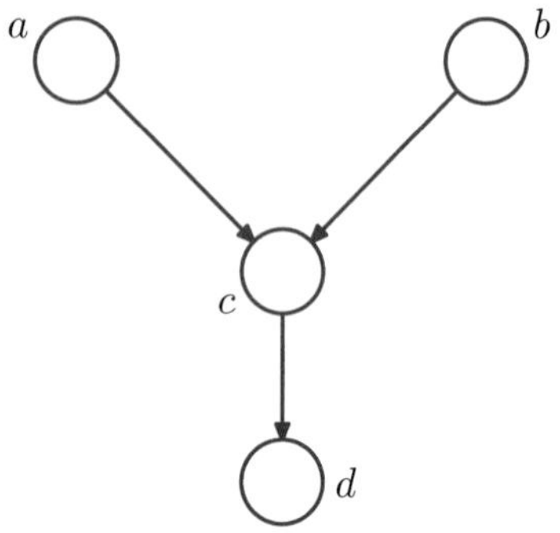

Figure 8.54 Example of a graphical model used to explore the conditional independence properties of the head-to-head path a–c–b when a descendant of c, namely the node d, is observed.

8.10 ($\star$) Consider the directed graph shown in Figure 8.54 in which none of the variables is observed. Show that $a \perp\!\!\!\perp b \mid \emptyset$. Suppose we now observe the variable d. Show that in general $a \not\perp\!\!\!\perp b \mid d$.

8.11 ($\star\star$) Consider the example of the car fuel system shown in Figure 8.21, and suppose that instead of observing the state of the fuel gauge G directly, the gauge is seen by the driver D who reports to us the reading on the gauge. This report is either that the gauge shows full $D = 1$ or that it shows empty $D = 0$. Our driver is a bit unreliable, as expressed through the following probabilities

$$p(D = 1|G = 1) = 0.9 \tag{8.105}$$
$$p(D = 0|G = 0) = 0.9. \tag{8.106}$$

Suppose that the driver tells us that the fuel gauge shows empty, in other words that we observe $D = 0$. Evaluate the probability that the tank is empty given only this observation. Similarly, evaluate the corresponding probability given also the observation that the battery is flat, and note that this second probability is lower. Discuss the intuition behind this result, and relate the result to Figure 8.54.

8.12 ($\star$) **www** Show that there are $2^{M(M-1)/2}$ distinct undirected graphs over a set of M distinct random variables. Draw the 8 possibilities for the case of $M = 3$.

8.13 ($\star$) Consider the use of iterated conditional modes (ICM) to minimize the energy function given by (8.42). Write down an expression for the difference in the values of the energy associated with the two states of a particular variable x_j, with all other variables held fixed, and show that it depends only on quantities that are local to x_j in the graph.

8.14 ($\star$) Consider a particular case of the energy function given by (8.42) in which the coefficients $\beta = h = 0$. Show that the most probable configuration of the latent variables is given by $x_i = y_i$ for all i.

8.15 ($\star\star$) **www** Show that the joint distribution $p(x_{n-1}, x_n)$ for two neighbouring nodes in the graph shown in Figure 8.38 is given by an expression of the form (8.58).

8.16 ($\star\star$) Consider the inference problem of evaluating $p(x_n|x_N)$ for the graph shown in Figure 8.38, for all nodes $n \in \{1, \ldots, N-1\}$. Show that the message passing algorithm discussed in Section 8.4.1 can be used to solve this efficiently, and discuss which messages are modified and in what way.

8.17 ($\star\star$) Consider a graph of the form shown in Figure 8.38 having $N = 5$ nodes, in which nodes x_3 and x_5 are observed. Use d-separation to show that $x_2 \perp\!\!\!\perp x_5 \mid x_3$. Show that if the message passing algorithm of Section 8.4.1 is applied to the evaluation of $p(x_2|x_3, x_5)$, the result will be independent of the value of x_5.

8.18 ($\star\star$) **www** Show that a distribution represented by a directed tree can trivially be written as an equivalent distribution over the corresponding undirected tree. Also show that a distribution expressed as an undirected tree can, by suitable normalization of the clique potentials, be written as a directed tree. Calculate the number of distinct directed trees that can be constructed from a given undirected tree.

8.19 ($\star\star$) Apply the sum-product algorithm derived in Section 8.4.4 to the chain-of-nodes model discussed in Section 8.4.1 and show that the results (8.54), (8.55), and (8.57) are recovered as a special case.

8.20 ($\star$) **www** Consider the message passing protocol for the sum-product algorithm on a tree-structured factor graph in which messages are first propagated from the leaves to an arbitrarily chosen root node and then from the root node out to the leaves. Use proof by induction to show that the messages can be passed in such an order that at every step, each node that must send a message has received all of the incoming messages necessary to construct its outgoing messages.

8.21 ($\star\star$) **www** Show that the marginal distributions $p(\mathbf{x}_s)$ over the sets of variables $\mathbf{x}_s$ associated with each of the factors $f_s(\mathbf{x}_s)$ in a factor graph can be found by first running the sum-product message passing algorithm and then evaluating the required marginals using (8.72).

8.22 ($\star$) Consider a tree-structured factor graph, in which a given subset of the variable nodes form a connected subgraph (i.e., any variable node of the subset is connected to at least one of the other variable nodes via a single factor node). Show how the sum-product algorithm can be used to compute the marginal distribution over that subset.

8.23 ($\star\star$) **www** In Section 8.4.4, we showed that the marginal distribution $p(x_i)$ for a variable node x_i in a factor graph is given by the product of the messages arriving at this node from neighbouring factor nodes in the form (8.63). Show that the marginal $p(x_i)$ can also be written as the product of the incoming message along any one of the links with the outgoing message along the same link.

8.24 ($\star\star$) Show that the marginal distribution for the variables $\mathbf{x}_s$ in a factor $f_s(\mathbf{x}_s)$ in a tree-structured factor graph, after running the sum-product message passing algorithm, can be written as the product of the message arriving at the factor node along all its links, times the local factor $f(\mathbf{x}_s)$, in the form (8.72).

8.25 ($\star\star$) In (8.86), we verified that the sum-product algorithm run on the graph in Figure 8.51 with node x_3 designated as the root node gives the correct marginal for x_2. Show that the correct marginals are obtained also for x_1 and x_3. Similarly, show that the use of the result (8.72) after running the sum-product algorithm on this graph gives the correct joint distribution for x_1, x_2.

8.26 ($\star$) Consider a tree-structured factor graph over discrete variables, and suppose we wish to evaluate the joint distribution $p(x_a, x_b)$ associated with two variables x_a and x_b that do not belong to a common factor. Define a procedure for using the sum-product algorithm to evaluate this joint distribution in which one of the variables is successively clamped to each of its allowed values.

8.27 ($\star\star$) Consider two discrete variables x and y each having three possible states, for example $x, y \in \{0, 1, 2\}$. Construct a joint distribution $p(x, y)$ over these variables having the property that the value $\widehat{x}$ that maximizes the marginal $p(x)$, along with the value $\widehat{y}$ that maximizes the marginal $p(y)$, together have probability zero under the joint distribution, so that $p(\widehat{x}, \widehat{y}) = 0$.

8.28 ($\star\star$) **www** The concept of a *pending* message in the sum-product algorithm for a factor graph was defined in Section 8.4.7. Show that if the graph has one or more cycles, there will always be at least one pending message irrespective of how long the algorithm runs.

8.29 ($\star\star$) **www** Show that if the sum-product algorithm is run on a factor graph with a tree structure (no loops), then after a finite number of messages have been sent, there will be no pending messages.

9 Mixture Models and EM

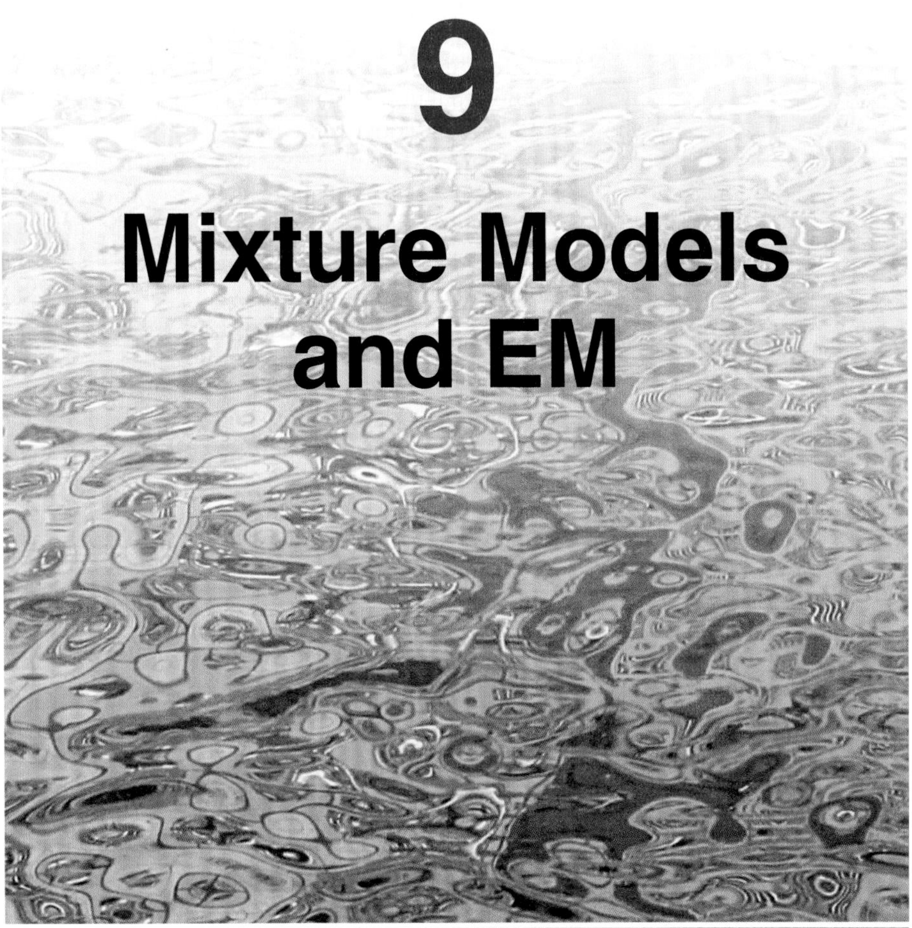

If we define a joint distribution over observed and latent variables, the corresponding distribution of the observed variables alone is obtained by marginalization. This allows relatively complex marginal distributions over observed variables to be expressed in terms of more tractable joint distributions over the expanded space of observed and latent variables. The introduction of latent variables thereby allows complicated distributions to be formed from simpler components. In this chapter, we shall see that mixture distributions, such as the Gaussian mixture discussed in Section 2.3.9, can be interpreted in terms of discrete latent variables. Continuous latent variables will form the subject of Chapter 12.

As well as providing a framework for building more complex probability distributions, mixture models can also be used to cluster data. We therefore begin our discussion of mixture distributions by considering the problem of finding clusters in a set of data points, which we approach first using a nonprobabilistic technique called the K-means algorithm (Lloyd, 1982). Then we introduce the latent variable

Section 9.1

view of mixture distributions in which the discrete latent variables can be interpreted as defining assignments of data points to specific components of the mixture. A general technique for finding maximum likelihood estimators in latent variable models is the expectation-maximization (EM) algorithm. We first of all use the Gaussian mixture distribution to motivate the EM algorithm in a fairly informal way, and then we give a more careful treatment based on the latent variable viewpoint. We shall see that the K-means algorithm corresponds to a particular nonprobabilistic limit of EM applied to mixtures of Gaussians. Finally, we discuss EM in some generality.

Section 9.2
Section 9.3
Section 9.4

Gaussian mixture models are widely used in data mining, pattern recognition, machine learning, and statistical analysis. In many applications, their parameters are determined by maximum likelihood, typically using the EM algorithm. However, as we shall see there are some significant limitations to the maximum likelihood approach, and in Chapter 10 we shall show that an elegant Bayesian treatment can be given using the framework of variational inference. This requires little additional computation compared with EM, and it resolves the principal difficulties of maximum likelihood while also allowing the number of components in the mixture to be inferred automatically from the data.

9.1. K-means Clustering

We begin by considering the problem of identifying groups, or clusters, of data points in a multidimensional space. Suppose we have a data set $\{\mathbf{x}_1, \ldots, \mathbf{x}_N\}$ consisting of N observations of a random D-dimensional Euclidean variable $\mathbf{x}$. Our goal is to partition the data set into some number K of clusters, where we shall suppose for the moment that the value of K is given. Intuitively, we might think of a cluster as comprising a group of data points whose inter-point distances are small compared with the distances to points outside of the cluster. We can formalize this notion by first introducing a set of D-dimensional vectors $\boldsymbol{\mu}_k$, where $k = 1, \ldots, K$, in which $\boldsymbol{\mu}_k$ is a prototype associated with the kth cluster. As we shall see shortly, we can think of the $\boldsymbol{\mu}_k$ as representing the centres of the clusters. Our goal is then to find an assignment of data points to clusters, as well as a set of vectors $\{\boldsymbol{\mu}_k\}$, such that the sum of the squares of the distances of each data point to its closest vector $\boldsymbol{\mu}_k$, is a minimum.

It is convenient at this point to define some notation to describe the assignment of data points to clusters. For each data point $\mathbf{x}_n$, we introduce a corresponding set of binary indicator variables $r_{nk} \in \{0, 1\}$, where $k = 1, \ldots, K$ describing which of the K clusters the data point $\mathbf{x}_n$ is assigned to, so that if data point $\mathbf{x}_n$ is assigned to cluster k then $r_{nk} = 1$, and $r_{nj} = 0$ for $j \neq k$. This is known as the 1-of-K coding scheme. We can then define an objective function, sometimes called a *distortion measure*, given by

$$J = \sum_{n=1}^{N} \sum_{k=1}^{K} r_{nk} \|\mathbf{x}_n - \boldsymbol{\mu}_k\|^2 \tag{9.1}$$

which represents the sum of the squares of the distances of each data point to its assigned vector $\boldsymbol{\mu}_k$. Our goal is to find values for the $\{r_{nk}\}$ and the $\{\boldsymbol{\mu}_k\}$ so as to minimize J. We can do this through an iterative procedure in which each iteration involves two successive steps corresponding to successive optimizations with respect to the r_{nk} and the $\boldsymbol{\mu}_k$. First we choose some initial values for the $\boldsymbol{\mu}_k$. Then in the first phase we minimize J with respect to the r_{nk}, keeping the $\boldsymbol{\mu}_k$ fixed. In the second phase we minimize J with respect to the $\boldsymbol{\mu}_k$, keeping r_{nk} fixed. This two-stage optimization is then repeated until convergence. We shall see that these two stages of updating r_{nk} and updating $\boldsymbol{\mu}_k$ correspond respectively to the E (expectation) and M (maximization) steps of the EM algorithm, and to emphasize this we shall use the terms E step and M step in the context of the K-means algorithm.

Section 9.4

Consider first the determination of the r_{nk}. Because J in (9.1) is a linear function of r_{nk}, this optimization can be performed easily to give a closed form solution. The terms involving different n are independent and so we can optimize for each n separately by choosing r_{nk} to be 1 for whichever value of k gives the minimum value of $\|\mathbf{x}_n - \boldsymbol{\mu}_k\|^2$. In other words, we simply assign the nth data point to the closest cluster centre. More formally, this can be expressed as

$$r_{nk} = \begin{cases} 1 & \text{if } k = \arg\min_j \|\mathbf{x}_n - \boldsymbol{\mu}_j\|^2 \\ 0 & \text{otherwise.} \end{cases} \tag{9.2}$$

Now consider the optimization of the $\boldsymbol{\mu}_k$ with the r_{nk} held fixed. The objective function J is a quadratic function of $\boldsymbol{\mu}_k$, and it can be minimized by setting its derivative with respect to $\boldsymbol{\mu}_k$ to zero giving

$$2\sum_{n=1}^{N} r_{nk}(\mathbf{x}_n - \boldsymbol{\mu}_k) = 0 \tag{9.3}$$

which we can easily solve for $\boldsymbol{\mu}_k$ to give

$$\boldsymbol{\mu}_k = \frac{\sum_n r_{nk}\mathbf{x}_n}{\sum_n r_{nk}}. \tag{9.4}$$

The denominator in this expression is equal to the number of points assigned to cluster k, and so this result has a simple interpretation, namely set $\boldsymbol{\mu}_k$ equal to the mean of all of the data points $\mathbf{x}_n$ assigned to cluster k. For this reason, the procedure is known as the K-*means* algorithm.

The two phases of re-assigning data points to clusters and re-computing the cluster means are repeated in turn until there is no further change in the assignments (or until some maximum number of iterations is exceeded). Because each phase reduces the value of the objective function J, convergence of the algorithm is assured. However, it may converge to a local rather than global minimum of J. The convergence properties of the K-means algorithm were studied by MacQueen (1967).

Exercise 9.1

Appendix A

The K-means algorithm is illustrated using the Old Faithful data set in Figure 9.1. For the purposes of this example, we have made a linear re-scaling of the data, known as *standardizing*, such that each of the variables has zero mean and unit standard deviation. For this example, we have chosen $K = 2$, and so in this

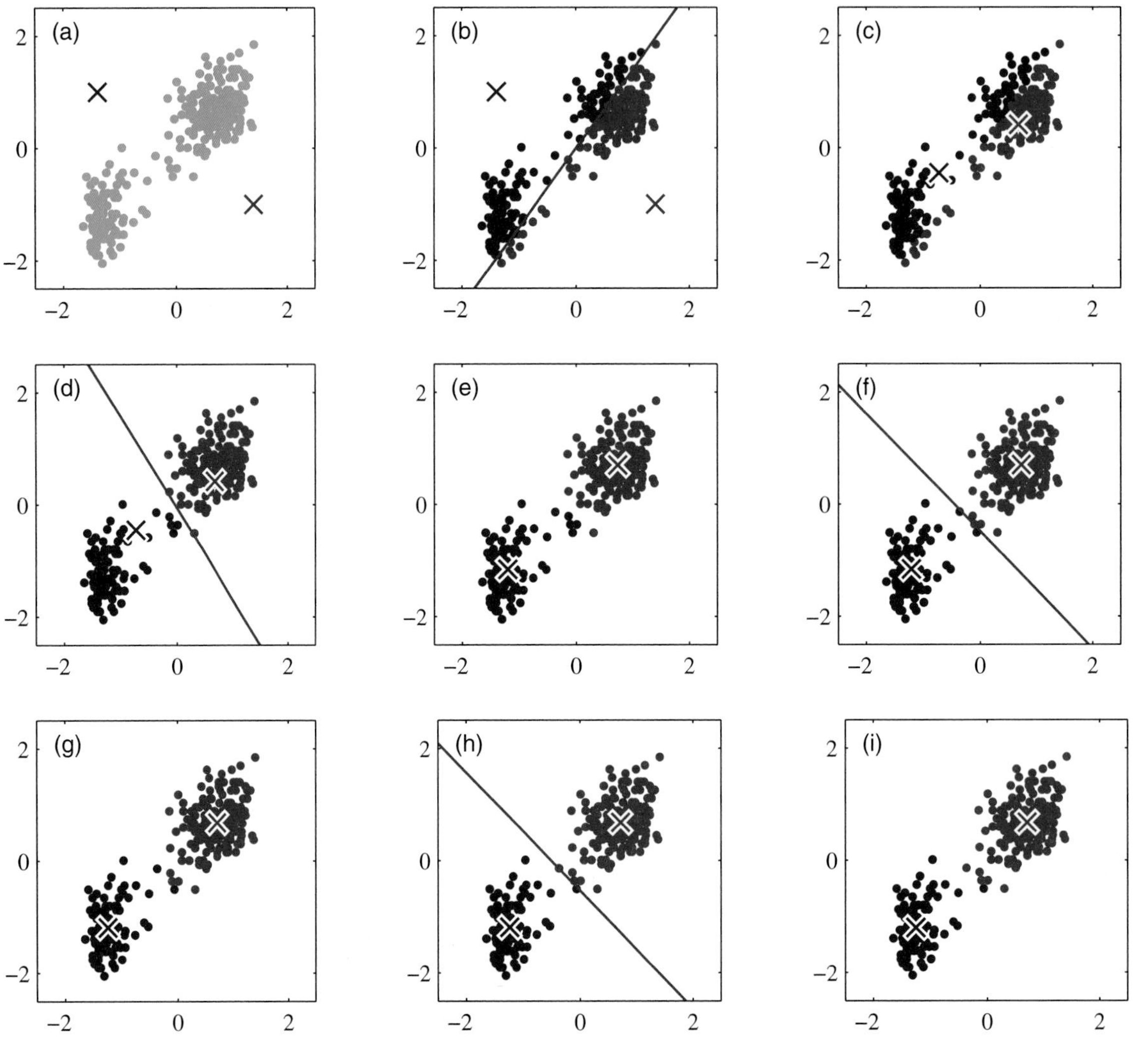

Figure 9.1 Illustration of the K-means algorithm using the re-scaled Old Faithful data set. (a) Green points denote the data set in a two-dimensional Euclidean space. The initial choices for centres $\boldsymbol{\mu}_1$ and $\boldsymbol{\mu}_2$ are shown by the red and blue crosses, respectively. (b) In the initial E step, each data point is assigned either to the red cluster or to the blue cluster, according to which cluster centre is nearer. This is equivalent to classifying the points according to which side of the perpendicular bisector of the two cluster centres, shown by the magenta line, they lie on. (c) In the subsequent M step, each cluster centre is re-computed to be the mean of the points assigned to the corresponding cluster. (d)–(i) show successive E and M steps through to final convergence of the algorithm.

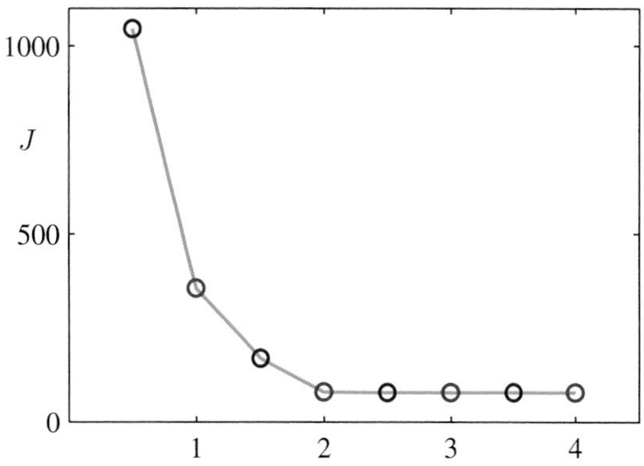

Figure 9.2 Plot of the cost function J given by (9.1) after each E step (blue points) and M step (red points) of the K-means algorithm for the example shown in Figure 9.1. The algorithm has converged after the third M step, and the final EM cycle produces no changes in either the assignments or the prototype vectors.

case, the assignment of each data point to the nearest cluster centre is equivalent to a classification of the data points according to which side they lie of the perpendicular bisector of the two cluster centres. A plot of the cost function J given by (9.1) for the Old Faithful example is shown in Figure 9.2.

Note that we have deliberately chosen poor initial values for the cluster centres so that the algorithm takes several steps before convergence. In practice, a better initialization procedure would be to choose the cluster centres $\boldsymbol{\mu}_k$ to be equal to a random subset of K data points. It is also worth noting that the K-means algorithm itself is often used to initialize the parameters in a Gaussian mixture model before applying the EM algorithm.

Section 9.2.2

A direct implementation of the K-means algorithm as discussed here can be relatively slow, because in each E step it is necessary to compute the Euclidean distance between every prototype vector and every data point. Various schemes have been proposed for speeding up the K-means algorithm, some of which are based on precomputing a data structure such as a tree such that nearby points are in the same subtree (Ramasubramanian and Paliwal, 1990; Moore, 2000). Other approaches make use of the triangle inequality for distances, thereby avoiding unnecessary distance calculations (Hodgson, 1998; Elkan, 2003).

So far, we have considered a batch version of K-means in which the whole data set is used together to update the prototype vectors. We can also derive an on-line stochastic algorithm (MacQueen, 1967) by applying the Robbins-Monro procedure to the problem of finding the roots of the regression function given by the derivatives of J in (9.1) with respect to $\boldsymbol{\mu}_k$. This leads to a sequential update in which, for each data point $\mathbf{x}_n$ in turn, we update the nearest prototype $\boldsymbol{\mu}_k$ using

Section 2.3.5

Exercise 9.2

$$\boldsymbol{\mu}_k^{\text{new}} = \boldsymbol{\mu}_k^{\text{old}} + \eta_n(\mathbf{x}_n - \boldsymbol{\mu}_k^{\text{old}}) \tag{9.5}$$

where η_n is the learning rate parameter, which is typically made to decrease monotonically as more data points are considered.

The K-means algorithm is based on the use of squared Euclidean distance as the measure of dissimilarity between a data point and a prototype vector. Not only does this limit the type of data variables that can be considered (it would be inappropriate for cases where some or all of the variables represent categorical labels for instance),

Section 2.3.7

but it can also make the determination of the cluster means nonrobust to outliers. We can generalize the K-means algorithm by introducing a more general dissimilarity measure $\mathcal{V}(\mathbf{x}, \mathbf{x}')$ between two vectors $\mathbf{x}$ and $\mathbf{x}'$ and then minimizing the following distortion measure

$$\widetilde{J} = \sum_{n=1}^{N} \sum_{k=1}^{K} r_{nk} \mathcal{V}(\mathbf{x}_n, \boldsymbol{\mu}_k) \tag{9.6}$$

which gives the K*-medoids* algorithm. The E step again involves, for given cluster prototypes $\boldsymbol{\mu}_k$, assigning each data point to the cluster for which the dissimilarity to the corresponding prototype is smallest. The computational cost of this is $O(KN)$, as is the case for the standard K-means algorithm. For a general choice of dissimilarity measure, the M step is potentially more complex than for K-means, and so it is common to restrict each cluster prototype to be equal to one of the data vectors assigned to that cluster, as this allows the algorithm to be implemented for any choice of dissimilarity measure $\mathcal{V}(\cdot, \cdot)$ so long as it can be readily evaluated. Thus the M step involves, for each cluster k, a discrete search over the N_k points assigned to that cluster, which requires $O(N_k^2)$ evaluations of $\mathcal{V}(\cdot, \cdot)$.

One notable feature of the K-means algorithm is that at each iteration, every data point is assigned uniquely to one, and only one, of the clusters. Whereas some data points will be much closer to a particular centre $\boldsymbol{\mu}_k$ than to any other centre, there may be other data points that lie roughly midway between cluster centres. In the latter case, it is not clear that the hard assignment to the nearest cluster is the most appropriate. We shall see in the next section that by adopting a probabilistic approach, we obtain 'soft' assignments of data points to clusters in a way that reflects the level of uncertainty over the most appropriate assignment. This probabilistic formulation brings with it numerous benefits.

9.1.1 Image segmentation and compression

As an illustration of the application of the K-means algorithm, we consider the related problems of image segmentation and image compression. The goal of segmentation is to partition an image into regions each of which has a reasonably homogeneous visual appearance or which corresponds to objects or parts of objects (Forsyth and Ponce, 2003). Each pixel in an image is a point in a 3-dimensional space comprising the intensities of the red, blue, and green channels, and our segmentation algorithm simply treats each pixel in the image as a separate data point. Note that strictly this space is not Euclidean because the channel intensities are bounded by the interval $[0, 1]$. Nevertheless, we can apply the K-means algorithm without difficulty. We illustrate the result of running K-means to convergence, for any particular value of K, by re-drawing the image replacing each pixel vector with the $\{R, G, B\}$ intensity triplet given by the centre $\boldsymbol{\mu}_k$ to which that pixel has been assigned. Results for various values of K are shown in Figure 9.3. We see that for a given value of K, the algorithm is representing the image using a palette of only K colours. It should be emphasized that this use of K-means is not a particularly sophisticated approach to image segmentation, not least because it takes no account of the spatial proximity of different pixels. The image segmentation problem is in general extremely difficult

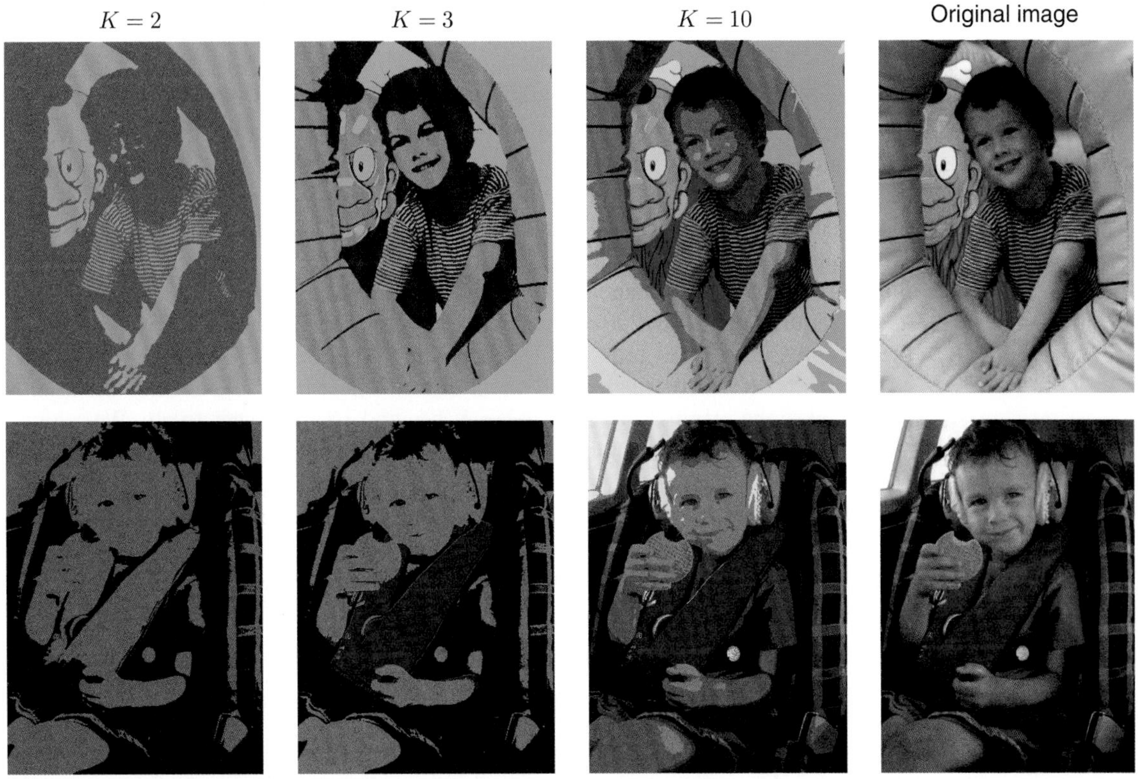

Figure 9.3 Two examples of the application of the K-means clustering algorithm to image segmentation showing the initial images together with their K-means segmentations obtained using various values of K. This also illustrates of the use of vector quantization for data compression, in which smaller values of K give higher compression at the expense of poorer image quality.

and remains the subject of active research and is introduced here simply to illustrate the behaviour of the K-means algorithm.

We can also use the result of a clustering algorithm to perform data compression. It is important to distinguish between *lossless data compression*, in which the goal is to be able to reconstruct the original data exactly from the compressed representation, and *lossy data compression*, in which we accept some errors in the reconstruction in return for higher levels of compression than can be achieved in the lossless case. We can apply the K-means algorithm to the problem of lossy data compression as follows. For each of the N data points, we store only the identity k of the cluster to which it is assigned. We also store the values of the K cluster centres $\boldsymbol{\mu}_k$, which typically requires significantly less data, provided we choose $K \ll N$. Each data point is then approximated by its nearest centre $\boldsymbol{\mu}_k$. New data points can similarly be compressed by first finding the nearest $\boldsymbol{\mu}_k$ and then storing the label k instead of the original data vector. This framework is often called *vector quantization*, and the vectors $\boldsymbol{\mu}_k$ are called *code-book vectors*.

The image segmentation problem discussed above also provides an illustration of the use of clustering for data compression. Suppose the original image has N pixels comprising $\{R, G, B\}$ values each of which is stored with 8 bits of precision. Then to transmit the whole image directly would cost $24N$ bits. Now suppose we first run K-means on the image data, and then instead of transmitting the original pixel intensity vectors we transmit the identity of the nearest vector $\boldsymbol{\mu}_k$. Because there are K such vectors, this requires $\log_2 K$ bits per pixel. We must also transmit the K code book vectors $\boldsymbol{\mu}_k$, which requires $24K$ bits, and so the total number of bits required to transmit the image is $24K + N \log_2 K$ (rounding up to the nearest integer). The original image shown in Figure 9.3 has $240 \times 180 = 43,200$ pixels and so requires $24 \times 43,200 = 1,036,800$ bits to transmit directly. By comparison, the compressed images require $43,248$ bits ($K = 2$), $86,472$ bits ($K = 3$), and $173,040$ bits ($K = 10$), respectively, to transmit. These represent compression ratios compared to the original image of 4.2%, 8.3%, and 16.7%, respectively. We see that there is a trade-off between degree of compression and image quality. Note that our aim in this example is to illustrate the K-means algorithm. If we had been aiming to produce a good image compressor, then it would be more fruitful to consider small blocks of adjacent pixels, for instance 5×5, and thereby exploit the correlations that exist in natural images between nearby pixels.

9.2. Mixtures of Gaussians

In Section 2.3.9 we motivated the Gaussian mixture model as a simple linear superposition of Gaussian components, aimed at providing a richer class of density models than the single Gaussian. We now turn to a formulation of Gaussian mixtures in terms of discrete *latent* variables. This will provide us with a deeper insight into this important distribution, and will also serve to motivate the expectation-maximization algorithm.

Recall from (2.188) that the Gaussian mixture distribution can be written as a linear superposition of Gaussians in the form

$$p(\mathbf{x}) = \sum_{k=1}^{K} \pi_k \mathcal{N}(\mathbf{x}|\boldsymbol{\mu}_k, \boldsymbol{\Sigma}_k). \tag{9.7}$$

Let us introduce a K-dimensional binary random variable $\mathbf{z}$ having a 1-of-K representation in which a particular element z_k is equal to 1 and all other elements are equal to 0. The values of z_k therefore satisfy $z_k \in \{0, 1\}$ and $\sum_k z_k = 1$, and we see that there are K possible states for the vector $\mathbf{z}$ according to which element is nonzero. We shall define the joint distribution $p(\mathbf{x}, \mathbf{z})$ in terms of a marginal distribution $p(\mathbf{z})$ and a conditional distribution $p(\mathbf{x}|\mathbf{z})$, corresponding to the graphical model in Figure 9.4. The marginal distribution over $\mathbf{z}$ is specified in terms of the mixing coefficients π_k, such that

$$p(z_k = 1) = \pi_k$$

Figure 9.4 Graphical representation of a mixture model, in which the joint distribution is expressed in the form $p(\mathbf{x}, \mathbf{z}) = p(\mathbf{z})p(\mathbf{x}|\mathbf{z})$.

where the parameters $\{\pi_k\}$ must satisfy

$$0 \leqslant \pi_k \leqslant 1 \tag{9.8}$$

together with

$$\sum_{k=1}^{K} \pi_k = 1 \tag{9.9}$$

in order to be valid probabilities. Because $\mathbf{z}$ uses a 1-of-K representation, we can also write this distribution in the form

$$p(\mathbf{z}) = \prod_{k=1}^{K} \pi_k^{z_k}. \tag{9.10}$$

Similarly, the conditional distribution of $\mathbf{x}$ given a particular value for $\mathbf{z}$ is a Gaussian

$$p(\mathbf{x}|z_k = 1) = \mathcal{N}(\mathbf{x}|\boldsymbol{\mu}_k, \boldsymbol{\Sigma}_k)$$

which can also be written in the form

$$p(\mathbf{x}|\mathbf{z}) = \prod_{k=1}^{K} \mathcal{N}(\mathbf{x}|\boldsymbol{\mu}_k, \boldsymbol{\Sigma}_k)^{z_k}. \tag{9.11}$$

Exercise 9.3

The joint distribution is given by $p(\mathbf{z})p(\mathbf{x}|\mathbf{z})$, and the marginal distribution of $\mathbf{x}$ is then obtained by summing the joint distribution over all possible states of $\mathbf{z}$ to give

$$p(\mathbf{x}) = \sum_{\mathbf{z}} p(\mathbf{z})p(\mathbf{x}|\mathbf{z}) = \sum_{k=1}^{K} \pi_k \mathcal{N}(\mathbf{x}|\boldsymbol{\mu}_k, \boldsymbol{\Sigma}_k) \tag{9.12}$$

where we have made use of (9.10) and (9.11). Thus the marginal distribution of $\mathbf{x}$ is a Gaussian mixture of the form (9.7). If we have several observations $\mathbf{x}_1, \ldots, \mathbf{x}_N$, then, because we have represented the marginal distribution in the form $p(\mathbf{x}) = \sum_{\mathbf{z}} p(\mathbf{x}, \mathbf{z})$, it follows that for every observed data point $\mathbf{x}_n$ there is a corresponding latent variable $\mathbf{z}_n$.

We have therefore found an equivalent formulation of the Gaussian mixture involving an explicit latent variable. It might seem that we have not gained much by doing so. However, we are now able to work with the joint distribution $p(\mathbf{x}, \mathbf{z})$

instead of the marginal distribution $p(\mathbf{x})$, and this will lead to significant simplifications, most notably through the introduction of the expectation-maximization (EM) algorithm.

Another quantity that will play an important role is the conditional probability of $\mathbf{z}$ given $\mathbf{x}$. We shall use $\gamma(z_k)$ to denote $p(z_k = 1|\mathbf{x})$, whose value can be found using Bayes' theorem

$$\begin{aligned}\gamma(z_k) \equiv p(z_k = 1|\mathbf{x}) &= \frac{p(z_k = 1)p(\mathbf{x}|z_k = 1)}{\sum_{j=1}^{K} p(z_j = 1)p(\mathbf{x}|z_j = 1)} \\ &= \frac{\pi_k \mathcal{N}(\mathbf{x}|\boldsymbol{\mu}_k, \boldsymbol{\Sigma}_k)}{\sum_{j=1}^{K} \pi_j \mathcal{N}(\mathbf{x}|\boldsymbol{\mu}_j, \boldsymbol{\Sigma}_j)}. \end{aligned} \tag{9.13}$$

We shall view π_k as the prior probability of $z_k = 1$, and the quantity $\gamma(z_k)$ as the corresponding posterior probability once we have observed $\mathbf{x}$. As we shall see later, $\gamma(z_k)$ can also be viewed as the *responsibility* that component k takes for 'explaining' the observation $\mathbf{x}$.

Section 8.1.2

We can use the technique of ancestral sampling to generate random samples distributed according to the Gaussian mixture model. To do this, we first generate a value for $\mathbf{z}$, which we denote $\widehat{\mathbf{z}}$, from the marginal distribution $p(\mathbf{z})$ and then generate a value for $\mathbf{x}$ from the conditional distribution $p(\mathbf{x}|\widehat{\mathbf{z}})$. Techniques for sampling from standard distributions are discussed in Chapter 11. We can depict samples from the joint distribution $p(\mathbf{x}, \mathbf{z})$ by plotting points at the corresponding values of $\mathbf{x}$ and then colouring them according to the value of $\mathbf{z}$, in other words according to which Gaussian component was responsible for generating them, as shown in Figure 9.5(a). Similarly samples from the marginal distribution $p(\mathbf{x})$ are obtained by taking the samples from the joint distribution and ignoring the values of $\mathbf{z}$. These are illustrated in Figure 9.5(b) by plotting the $\mathbf{x}$ values without any coloured labels.

We can also use this synthetic data set to illustrate the 'responsibilities' by evaluating, for every data point, the posterior probability for each component in the mixture distribution from which this data set was generated. In particular, we can represent the value of the responsibilities $\gamma(z_{nk})$ associated with data point $\mathbf{x}_n$ by plotting the corresponding point using proportions of red, blue, and green ink given by $\gamma(z_{nk})$ for $k = 1, 2, 3$, respectively, as shown in Figure 9.5(c). So, for instance, a data point for which $\gamma(z_{n1}) = 1$ will be coloured red, whereas one for which $\gamma(z_{n2}) = \gamma(z_{n3}) = 0.5$ will be coloured with equal proportions of blue and green ink and so will appear cyan. This should be compared with Figure 9.5(a) in which the data points were labelled using the true identity of the component from which they were generated.

9.2.1 Maximum likelihood

Suppose we have a data set of observations $\{\mathbf{x}_1, \ldots, \mathbf{x}_N\}$, and we wish to model this data using a mixture of Gaussians. We can represent this data set as an $N \times D$

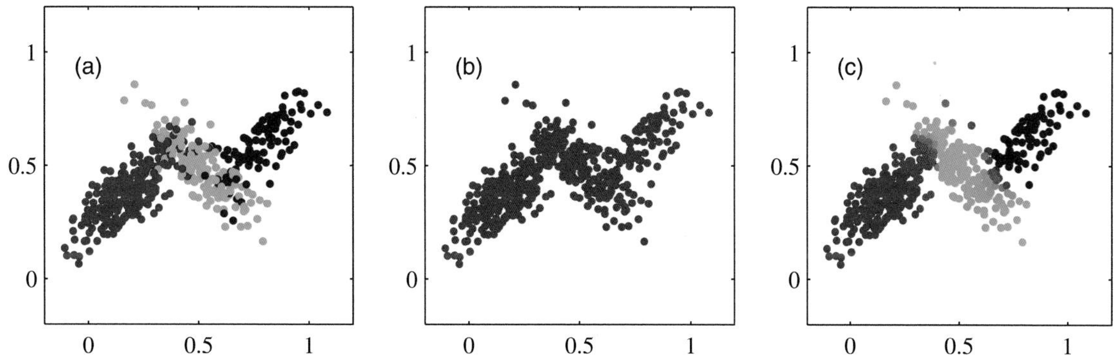

Figure 9.5 Example of 500 points drawn from the mixture of 3 Gaussians shown in Figure 2.23. (a) Samples from the joint distribution $p(\mathbf{z})p(\mathbf{x}|\mathbf{z})$ in which the three states of $\mathbf{z}$, corresponding to the three components of the mixture, are depicted in red, green, and blue, and (b) the corresponding samples from the marginal distribution $p(\mathbf{x})$, which is obtained by simply ignoring the values of $\mathbf{z}$ and just plotting the $\mathbf{x}$ values. The data set in (a) is said to be *complete*, whereas that in (b) is *incomplete*. (c) The same samples in which the colours represent the value of the responsibilities $\gamma(z_{nk})$ associated with data point $\mathbf{x}_n$, obtained by plotting the corresponding point using proportions of red, blue, and green ink given by $\gamma(z_{nk})$ for $k = 1, 2, 3$, respectively

matrix $\mathbf{X}$ in which the nth row is given by $\mathbf{x}_n^{\mathrm{T}}$. Similarly, the corresponding latent variables will be denoted by an $N \times K$ matrix $\mathbf{Z}$ with rows $\mathbf{z}_n^{\mathrm{T}}$. If we assume that the data points are drawn independently from the distribution, then we can express the Gaussian mixture model for this i.i.d. data set using the graphical representation shown in Figure 9.6. From (9.7) the log of the likelihood function is given by

$$\ln p(\mathbf{X}|\boldsymbol{\pi}, \boldsymbol{\mu}, \boldsymbol{\Sigma}) = \sum_{n=1}^{N} \ln \left\{ \sum_{k=1}^{K} \pi_k \mathcal{N}(\mathbf{x}_n|\boldsymbol{\mu}_k, \boldsymbol{\Sigma}_k) \right\}. \tag{9.14}$$

Before discussing how to maximize this function, it is worth emphasizing that there is a significant problem associated with the maximum likelihood framework applied to Gaussian mixture models, due to the presence of singularities. For simplicity, consider a Gaussian mixture whose components have covariance matrices given by $\boldsymbol{\Sigma}_k = \sigma_k^2 \mathbf{I}$, where $\mathbf{I}$ is the unit matrix, although the conclusions will hold for general covariance matrices. Suppose that one of the components of the mixture model, let us say the jth component, has its mean $\boldsymbol{\mu}_j$ exactly equal to one of the data

Figure 9.6 Graphical representation of a Gaussian mixture model for a set of N i.i.d. data points $\{\mathbf{x}_n\}$, with corresponding latent points $\{\mathbf{z}_n\}$, where $n = 1, \ldots, N$.

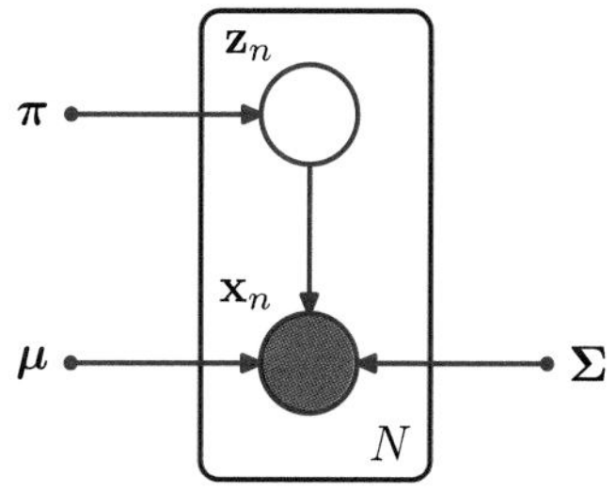

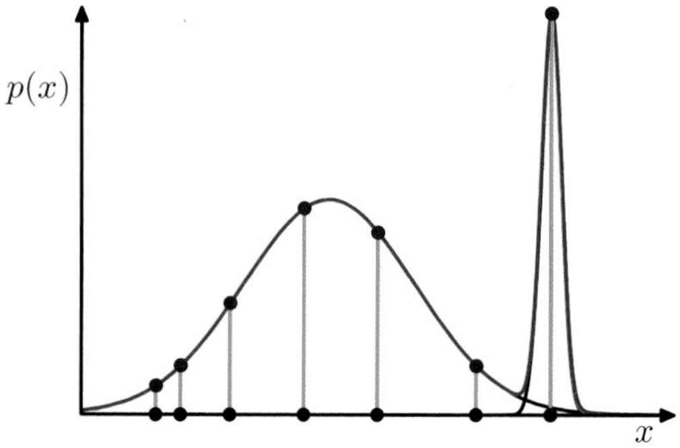

Figure 9.7 Illustration of how singularities in the likelihood function arise with mixtures of Gaussians. This should be compared with the case of a single Gaussian shown in Figure 1.14 for which no singularities arise.

points so that $\boldsymbol{\mu}_j = \mathbf{x}_n$ for some value of n. This data point will then contribute a term in the likelihood function of the form

$$\mathcal{N}(\mathbf{x}_n|\mathbf{x}_n, \sigma_j^2\mathbf{I}) = \frac{1}{(2\pi)^{1/2}}\frac{1}{\sigma_j}. \tag{9.15}$$

If we consider the limit $\sigma_j \to 0$, then we see that this term goes to infinity and so the log likelihood function will also go to infinity. Thus the maximization of the log likelihood function is not a well posed problem because such singularities will always be present and will occur whenever one of the Gaussian components 'collapses' onto a specific data point. Recall that this problem did not arise in the case of a single Gaussian distribution. To understand the difference, note that if a single Gaussian collapses onto a data point it will contribute multiplicative factors to the likelihood function arising from the other data points and these factors will go to zero exponentially fast, giving an overall likelihood that goes to zero rather than infinity. However, once we have (at least) two components in the mixture, one of the components can have a finite variance and therefore assign finite probability to all of the data points while the other component can shrink onto one specific data point and thereby contribute an ever increasing additive value to the log likelihood. This is illustrated in Figure 9.7. These singularities provide another example of the severe over-fitting that can occur in a maximum likelihood approach. We shall see that this difficulty does not occur if we adopt a Bayesian approach. For the moment, however, we simply note that in applying maximum likelihood to Gaussian mixture models we must take steps to avoid finding such pathological solutions and instead seek local maxima of the likelihood function that are well behaved. We can hope to avoid the singularities by using suitable heuristics, for instance by detecting when a Gaussian component is collapsing and resetting its mean to a randomly chosen value while also resetting its covariance to some large value, and then continuing with the optimization.

Section 10.1

A further issue in finding maximum likelihood solutions arises from the fact that for any given maximum likelihood solution, a K-component mixture will have a total of $K!$ equivalent solutions corresponding to the $K!$ ways of assigning K sets of parameters to K components. In other words, for any given (nondegenerate) point in the space of parameter values there will be a further $K! - 1$ additional points all of which give rise to exactly the same distribution. This problem is known as *identifiability* (Casella and Berger, 2002) and is an important issue when we wish to interpret the parameter values discovered by a model. Identifiability will also arise when we discuss models having continuous latent variables in Chapter 12. However, for the purposes of finding a good density model, it is irrelevant because any of the equivalent solutions is as good as any other.

Maximizing the log likelihood function (9.14) for a Gaussian mixture model turns out to be a more complex problem than for the case of a single Gaussian. The difficulty arises from the presence of the summation over k that appears inside the logarithm in (9.14), so that the logarithm function no longer acts directly on the Gaussian. If we set the derivatives of the log likelihood to zero, we will no longer obtain a closed form solution, as we shall see shortly.

One approach is to apply gradient-based optimization techniques (Fletcher, 1987; Nocedal and Wright, 1999; Bishop and Nabney, 2008). Although gradient-based techniques are feasible, and indeed will play an important role when we discuss mixture density networks in Chapter 5, we now consider an alternative approach known as the EM algorithm which has broad applicability and which will lay the foundations for a discussion of variational inference techniques in Chapter 10.

9.2.2 EM for Gaussian mixtures

An elegant and powerful method for finding maximum likelihood solutions for models with latent variables is called the *expectation-maximization* algorithm, or *EM* algorithm (Dempster *et al.*, 1977; McLachlan and Krishnan, 1997). Later we shall give a general treatment of EM, and we shall also show how EM can be generalized to obtain the variational inference framework. Initially, we shall motivate the EM algorithm by giving a relatively informal treatment in the context of the Gaussian mixture model. We emphasize, however, that EM has broad applicability, and indeed it will be encountered in the context of a variety of different models in this book.

Section 10.1

Let us begin by writing down the conditions that must be satisfied at a maximum of the likelihood function. Setting the derivatives of $\ln p(\mathbf{X}|\boldsymbol{\pi}, \boldsymbol{\mu}, \boldsymbol{\Sigma})$ in (9.14) with respect to the means $\boldsymbol{\mu}_k$ of the Gaussian components to zero, we obtain

$$0 = \sum_{n=1}^{N} \underbrace{\frac{\pi_k \mathcal{N}(\mathbf{x}_n|\boldsymbol{\mu}_k, \boldsymbol{\Sigma}_k)}{\sum_j \pi_j \mathcal{N}(\mathbf{x}_n|\boldsymbol{\mu}_j, \boldsymbol{\Sigma}_j)}}_{\gamma(z_{nk})} \boldsymbol{\Sigma}_k^{-1}(\mathbf{x}_n - \boldsymbol{\mu}_k) \tag{9.16}$$

where we have made use of the form (2.43) for the Gaussian distribution. Note that the posterior probabilities, or responsibilities, given by (9.13) appear naturally on the right-hand side. Multiplying by $\boldsymbol{\Sigma}_k$ (which we assume to be nonsingular) and rearranging we obtain

$$\boldsymbol{\mu}_k = \frac{1}{N_k} \sum_{n=1}^{N} \gamma(z_{nk}) \mathbf{x}_n \tag{9.17}$$

where we have defined

$$N_k = \sum_{n=1}^{N} \gamma(z_{nk}). \tag{9.18}$$

We can interpret N_k as the effective number of points assigned to cluster k. Note carefully the form of this solution. We see that the mean $\boldsymbol{\mu}_k$ for the kth Gaussian component is obtained by taking a weighted mean of all of the points in the data set, in which the weighting factor for data point $\mathbf{x}_n$ is given by the posterior probability $\gamma(z_{nk})$ that component k was responsible for generating $\mathbf{x}_n$.

If we set the derivative of $\ln p(\mathbf{X}|\boldsymbol{\pi}, \boldsymbol{\mu}, \boldsymbol{\Sigma})$ with respect to $\boldsymbol{\Sigma}_k$ to zero, and follow a similar line of reasoning, making use of the result for the maximum likelihood solution for the covariance matrix of a single Gaussian, we obtain

Section 2.3.4

$$\boldsymbol{\Sigma}_k = \frac{1}{N_k} \sum_{n=1}^{N} \gamma(z_{nk})(\mathbf{x}_n - \boldsymbol{\mu}_k)(\mathbf{x}_n - \boldsymbol{\mu}_k)^{\mathrm{T}} \tag{9.19}$$

which has the same form as the corresponding result for a single Gaussian fitted to the data set, but again with each data point weighted by the corresponding posterior probability and with the denominator given by the effective number of points associated with the corresponding component.

Finally, we maximize $\ln p(\mathbf{X}|\boldsymbol{\pi}, \boldsymbol{\mu}, \boldsymbol{\Sigma})$ with respect to the mixing coefficients π_k. Here we must take account of the constraint (9.9), which requires the mixing coefficients to sum to one. This can be achieved using a Lagrange multiplier and maximizing the following quantity

Appendix E

$$\ln p(\mathbf{X}|\boldsymbol{\pi}, \boldsymbol{\mu}, \boldsymbol{\Sigma}) + \lambda \left(\sum_{k=1}^{K} \pi_k - 1 \right) \tag{9.20}$$

which gives

$$0 = \sum_{n=1}^{N} \frac{\mathcal{N}(\mathbf{x}_n|\boldsymbol{\mu}_k, \boldsymbol{\Sigma}_k)}{\sum_j \pi_j \mathcal{N}(\mathbf{x}_n|\boldsymbol{\mu}_j, \boldsymbol{\Sigma}_j)} + \lambda \tag{9.21}$$

where again we see the appearance of the responsibilities. If we now multiply both sides by π_k and sum over k making use of the constraint (9.9), we find $\lambda = -N$. Using this to eliminate λ and rearranging we obtain

$$\pi_k = \frac{N_k}{N} \tag{9.22}$$

so that the mixing coefficient for the kth component is given by the average responsibility which that component takes for explaining the data points.

It is worth emphasizing that the results (9.17), (9.19), and (9.22) do not constitute a closed-form solution for the parameters of the mixture model because the responsibilities $\gamma(z_{nk})$ depend on those parameters in a complex way through (9.13). However, these results do suggest a simple iterative scheme for finding a solution to the maximum likelihood problem, which as we shall see turns out to be an instance of the EM algorithm for the particular case of the Gaussian mixture model. We first choose some initial values for the means, covariances, and mixing coefficients. Then we alternate between the following two updates that we shall call the E step

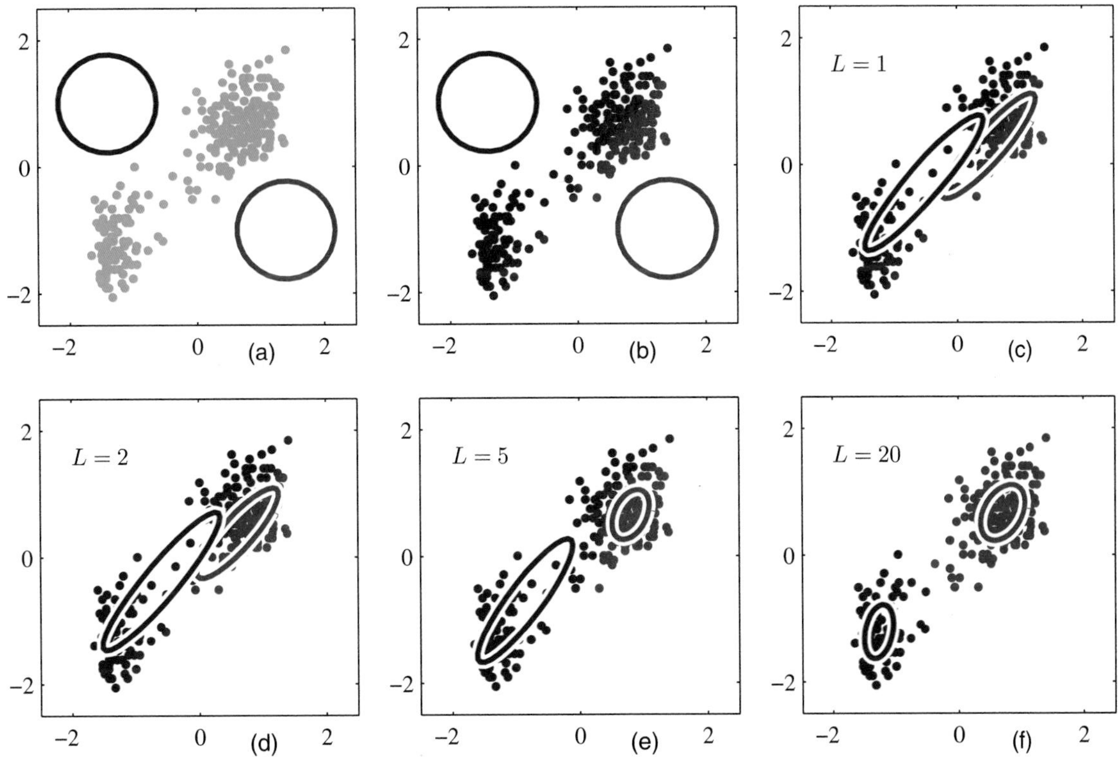

Figure 9.8 Illustration of the EM algorithm using the Old Faithful set as used for the illustration of the K-means algorithm in Figure 9.1. See the text for details.

and the M step, for reasons that will become apparent shortly. In the *expectation* step, or E step, we use the current values for the parameters to evaluate the posterior probabilities, or responsibilities, given by (9.13). We then use these probabilities in the *maximization* step, or M step, to re-estimate the means, covariances, and mixing coefficients using the results (9.17), (9.19), and (9.22). Note that in so doing we first evaluate the new means using (9.17) and then use these new values to find the covariances using (9.19), in keeping with the corresponding result for a single Gaussian distribution. We shall show that each update to the parameters resulting from an E step followed by an M step is guaranteed to increase the log likelihood function. In practice, the algorithm is deemed to have converged when the change in the log likelihood function, or alternatively in the parameters, falls below some threshold. We illustrate the EM algorithm for a mixture of two Gaussians applied to the rescaled Old Faithful data set in Figure 9.8. Here a mixture of two Gaussians is used, with centres initialized using the same values as for the K-means algorithm in Figure 9.1, and with precision matrices initialized to be proportional to the unit matrix. Plot (a) shows the data points in green, together with the initial configuration of the mixture model in which the one standard-deviation contours for the two

Section 9.4

Gaussian components are shown as blue and red circles. Plot (b) shows the result of the initial E step, in which each data point is depicted using a proportion of blue ink equal to the posterior probability of having been generated from the blue component, and a corresponding proportion of red ink given by the posterior probability of having been generated by the red component. Thus, points that have a significant probability for belonging to either cluster appear purple. The situation after the first M step is shown in plot (c), in which the mean of the blue Gaussian has moved to the mean of the data set, weighted by the probabilities of each data point belonging to the blue cluster, in other words it has moved to the centre of mass of the blue ink. Similarly, the covariance of the blue Gaussian is set equal to the covariance of the blue ink. Analogous results hold for the red component. Plots (d), (e), and (f) show the results after 2, 5, and 20 complete cycles of EM, respectively. In plot (f) the algorithm is close to convergence.

Note that the EM algorithm takes many more iterations to reach (approximate) convergence compared with the K-means algorithm, and that each cycle requires significantly more computation. It is therefore common to run the K-means algorithm in order to find a suitable initialization for a Gaussian mixture model that is subsequently adapted using EM. The covariance matrices can conveniently be initialized to the sample covariances of the clusters found by the K-means algorithm, and the mixing coefficients can be set to the fractions of data points assigned to the respective clusters. As with gradient-based approaches for maximizing the log likelihood, techniques must be employed to avoid singularities of the likelihood function in which a Gaussian component collapses onto a particular data point. It should be emphasized that there will generally be multiple local maxima of the log likelihood function, and that EM is not guaranteed to find the largest of these maxima. Because the EM algorithm for Gaussian mixtures plays such an important role, we summarize it below.

EM for Gaussian Mixtures

Given a Gaussian mixture model, the goal is to maximize the likelihood function with respect to the parameters (comprising the means and covariances of the components and the mixing coefficients).

1. Initialize the means $\boldsymbol{\mu}_k$, covariances $\boldsymbol{\Sigma}_k$ and mixing coefficients π_k, and evaluate the initial value of the log likelihood.

2. **E step**. Evaluate the responsibilities using the current parameter values

$$\gamma(z_{nk}) = \frac{\pi_k \mathcal{N}(\mathbf{x}_n|\boldsymbol{\mu}_k, \boldsymbol{\Sigma}_k)}{\sum_{j=1}^{K} \pi_j \mathcal{N}(\mathbf{x}_n|\boldsymbol{\mu}_j, \boldsymbol{\Sigma}_j)}. \tag{9.23}$$

3. **M step**. Re-estimate the parameters using the current responsibilities

$$\boldsymbol{\mu}_k^{\text{new}} = \frac{1}{N_k}\sum_{n=1}^{N}\gamma(z_{nk})\mathbf{x}_n \tag{9.24}$$

$$\boldsymbol{\Sigma}_k^{\text{new}} = \frac{1}{N_k}\sum_{n=1}^{N}\gamma(z_{nk})\left(\mathbf{x}_n - \boldsymbol{\mu}_k^{\text{new}}\right)\left(\mathbf{x}_n - \boldsymbol{\mu}_k^{\text{new}}\right)^{\mathrm{T}} \tag{9.25}$$

$$\pi_k^{\text{new}} = \frac{N_k}{N} \tag{9.26}$$

where

$$N_k = \sum_{n=1}^{N}\gamma(z_{nk}). \tag{9.27}$$

4. Evaluate the log likelihood

$$\ln p(\mathbf{X}|\boldsymbol{\mu}, \boldsymbol{\Sigma}, \boldsymbol{\pi}) = \sum_{n=1}^{N}\ln\left\{\sum_{k=1}^{K}\pi_k\mathcal{N}(\mathbf{x}_n|\boldsymbol{\mu}_k, \boldsymbol{\Sigma}_k)\right\} \tag{9.28}$$

and check for convergence of either the parameters or the log likelihood. If the convergence criterion is not satisfied return to step 2.

9.3. An Alternative View of EM

In this section, we present a complementary view of the EM algorithm that recognizes the key role played by latent variables. We discuss this approach first of all in an abstract setting, and then for illustration we consider once again the case of Gaussian mixtures.

The goal of the EM algorithm is to find maximum likelihood solutions for models having latent variables. We denote the set of all observed data by $\mathbf{X}$, in which the nth row represents $\mathbf{x}_n^{\mathrm{T}}$, and similarly we denote the set of all latent variables by $\mathbf{Z}$, with a corresponding row $\mathbf{z}_n^{\mathrm{T}}$. The set of all model parameters is denoted by $\boldsymbol{\theta}$, and so the log likelihood function is given by

$$\ln p(\mathbf{X}|\boldsymbol{\theta}) = \ln\left\{\sum_{\mathbf{Z}} p(\mathbf{X}, \mathbf{Z}|\boldsymbol{\theta})\right\}. \tag{9.29}$$

Note that our discussion will apply equally well to continuous latent variables simply by replacing the sum over $\mathbf{Z}$ with an integral.

A key observation is that the summation over the latent variables appears inside the logarithm. Even if the joint distribution $p(\mathbf{X}, \mathbf{Z}|\boldsymbol{\theta})$ belongs to the exponential family, the marginal distribution $p(\mathbf{X}|\boldsymbol{\theta})$ typically does not as a result of this summation. The presence of the sum prevents the logarithm from acting directly on the joint distribution, resulting in complicated expressions for the maximum likelihood solution.

Now suppose that, for each observation in $\mathbf{X}$, we were told the corresponding value of the latent variable $\mathbf{Z}$. We shall call $\{\mathbf{X}, \mathbf{Z}\}$ the *complete* data set, and we shall refer to the actual observed data $\mathbf{X}$ as *incomplete*, as illustrated in Figure 9.5. The likelihood function for the complete data set simply takes the form $\ln p(\mathbf{X}, \mathbf{Z}|\boldsymbol{\theta})$, and we shall suppose that maximization of this complete-data log likelihood function is straightforward.

In practice, however, we are not given the complete data set $\{\mathbf{X}, \mathbf{Z}\}$, but only the incomplete data $\mathbf{X}$. Our state of knowledge of the values of the latent variables in $\mathbf{Z}$ is given only by the posterior distribution $p(\mathbf{Z}|\mathbf{X}, \boldsymbol{\theta})$. Because we cannot use the complete-data log likelihood, we consider instead its expected value under the posterior distribution of the latent variable, which corresponds (as we shall see) to the E step of the EM algorithm. In the subsequent M step, we maximize this expectation. If the current estimate for the parameters is denoted $\boldsymbol{\theta}^{\text{old}}$, then a pair of successive E and M steps gives rise to a revised estimate $\boldsymbol{\theta}^{\text{new}}$. The algorithm is initialized by choosing some starting value for the parameters $\boldsymbol{\theta}_0$. The use of the expectation may seem somewhat arbitrary. However, we shall see the motivation for this choice when we give a deeper treatment of EM in Section 9.4.

In the E step, we use the current parameter values $\boldsymbol{\theta}^{\text{old}}$ to find the posterior distribution of the latent variables given by $p(\mathbf{Z}|\mathbf{X}, \boldsymbol{\theta}^{\text{old}})$. We then use this posterior distribution to find the expectation of the complete-data log likelihood evaluated for some general parameter value $\boldsymbol{\theta}$. This expectation, denoted $\mathcal{Q}(\boldsymbol{\theta}, \boldsymbol{\theta}^{\text{old}})$, is given by

$$\mathcal{Q}(\boldsymbol{\theta}, \boldsymbol{\theta}^{\text{old}}) = \sum_{\mathbf{Z}} p(\mathbf{Z}|\mathbf{X}, \boldsymbol{\theta}^{\text{old}}) \ln p(\mathbf{X}, \mathbf{Z}|\boldsymbol{\theta}). \tag{9.30}$$

In the M step, we determine the revised parameter estimate $\boldsymbol{\theta}^{\text{new}}$ by maximizing this function

$$\boldsymbol{\theta}^{\text{new}} = \arg\max_{\boldsymbol{\theta}} \mathcal{Q}(\boldsymbol{\theta}, \boldsymbol{\theta}^{\text{old}}). \tag{9.31}$$

Note that in the definition of $\mathcal{Q}(\boldsymbol{\theta}, \boldsymbol{\theta}^{\text{old}})$, the logarithm acts directly on the joint distribution $p(\mathbf{X}, \mathbf{Z}|\boldsymbol{\theta})$, and so the corresponding M-step maximization will, by supposition, be tractable.

The general EM algorithm is summarized below. It has the property, as we shall show later, that each cycle of EM will increase the incomplete-data log likelihood (unless it is already at a local maximum).

Section 9.4

The General EM Algorithm

Given a joint distribution $p(\mathbf{X}, \mathbf{Z}|\boldsymbol{\theta})$ over observed variables $\mathbf{X}$ and latent variables $\mathbf{Z}$, governed by parameters $\boldsymbol{\theta}$, the goal is to maximize the likelihood function $p(\mathbf{X}|\boldsymbol{\theta})$ with respect to $\boldsymbol{\theta}$.

1. Choose an initial setting for the parameters $\boldsymbol{\theta}^{\text{old}}$.

2. **E step** Evaluate $p(\mathbf{Z}|\mathbf{X}, \boldsymbol{\theta}^{\text{old}})$.

3. **M step** Evaluate $\boldsymbol{\theta}^{\text{new}}$ given by

$$\boldsymbol{\theta}^{\text{new}} = \underset{\boldsymbol{\theta}}{\arg\max}\, \mathcal{Q}(\boldsymbol{\theta}, \boldsymbol{\theta}^{\text{old}}) \tag{9.32}$$

where

$$\mathcal{Q}(\boldsymbol{\theta}, \boldsymbol{\theta}^{\text{old}}) = \sum_{\mathbf{Z}} p(\mathbf{Z}|\mathbf{X}, \boldsymbol{\theta}^{\text{old}}) \ln p(\mathbf{X}, \mathbf{Z}|\boldsymbol{\theta}). \tag{9.33}$$

4. Check for convergence of either the log likelihood or the parameter values. If the convergence criterion is not satisfied, then let

$$\boldsymbol{\theta}^{\text{old}} \leftarrow \boldsymbol{\theta}^{\text{new}} \tag{9.34}$$

and return to step 2.

Exercise 9.4

The EM algorithm can also be used to find MAP (maximum posterior) solutions for models in which a prior $p(\boldsymbol{\theta})$ is defined over the parameters. In this case the E step remains the same as in the maximum likelihood case, whereas in the M step the quantity to be maximized is given by $\mathcal{Q}(\boldsymbol{\theta}, \boldsymbol{\theta}^{\text{old}}) + \ln p(\boldsymbol{\theta})$. Suitable choices for the prior will remove the singularities of the kind illustrated in Figure 9.7.

Here we have considered the use of the EM algorithm to maximize a likelihood function when there are discrete latent variables. However, it can also be applied when the unobserved variables correspond to missing values in the data set. The distribution of the observed values is obtained by taking the joint distribution of all the variables and then marginalizing over the missing ones. EM can then be used to maximize the corresponding likelihood function. We shall show an example of the application of this technique in the context of principal component analysis in Figure 12.11. This will be a valid procedure if the data values are *missing at random*, meaning that the mechanism causing values to be missing does not depend on the unobserved values. In many situations this will not be the case, for instance if a sensor fails to return a value whenever the quantity it is measuring exceeds some threshold.

9.3.1 Gaussian mixtures revisited

We now consider the application of this latent variable view of EM to the specific case of a Gaussian mixture model. Recall that our goal is to maximize the log likelihood function (9.14), which is computed using the observed data set $\mathbf{X}$, and we saw that this was more difficult than for the case of a single Gaussian distribution due to the presence of the summation over k that occurs inside the logarithm. Suppose then that in addition to the observed data set $\mathbf{X}$, we were also given the values of the corresponding discrete variables $\mathbf{Z}$. Recall that Figure 9.5(a) shows a 'complete' data set (i.e., one that includes labels showing which component generated each data point) while Figure 9.5(b) shows the corresponding 'incomplete' data set. The graphical model for the complete data is shown in Figure 9.9.

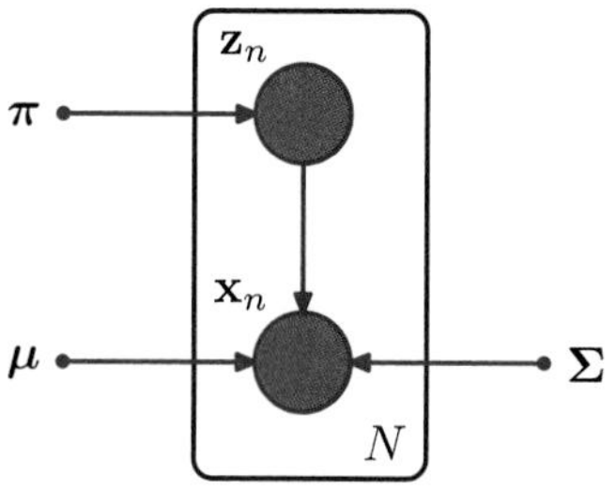

Figure 9.9 This shows the same graph as in Figure 9.6 except that we now suppose that the discrete variables $\mathbf{z}_n$ are observed, as well as the data variables $\mathbf{x}_n$.

Now consider the problem of maximizing the likelihood for the complete data set $\{\mathbf{X}, \mathbf{Z}\}$. From (9.10) and (9.11), this likelihood function takes the form

$$p(\mathbf{X}, \mathbf{Z}|\boldsymbol{\mu}, \boldsymbol{\Sigma}, \boldsymbol{\pi}) = \prod_{n=1}^{N} \prod_{k=1}^{K} \pi_k^{z_{nk}} \mathcal{N}(\mathbf{x}_n|\boldsymbol{\mu}_k, \boldsymbol{\Sigma}_k)^{z_{nk}} \tag{9.35}$$

where z_{nk} denotes the kth component of $\mathbf{z}_n$. Taking the logarithm, we obtain

$$\ln p(\mathbf{X}, \mathbf{Z}|\boldsymbol{\mu}, \boldsymbol{\Sigma}, \boldsymbol{\pi}) = \sum_{n=1}^{N} \sum_{k=1}^{K} z_{nk} \left\{ \ln \pi_k + \ln \mathcal{N}(\mathbf{x}_n|\boldsymbol{\mu}_k, \boldsymbol{\Sigma}_k) \right\}. \tag{9.36}$$

Comparison with the log likelihood function (9.14) for the incomplete data shows that the summation over k and the logarithm have been interchanged. The logarithm now acts directly on the Gaussian distribution, which itself is a member of the exponential family. Not surprisingly, this leads to a much simpler solution to the maximum likelihood problem, as we now show. Consider first the maximization with respect to the means and covariances. Because $\mathbf{z}_n$ is a K-dimensional vector with all elements equal to 0 except for a single element having the value 1, the complete-data log likelihood function is simply a sum of K independent contributions, one for each mixture component. Thus the maximization with respect to a mean or a covariance is exactly as for a single Gaussian, except that it involves only the subset of data points that are 'assigned' to that component. For the maximization with respect to the mixing coefficients, we note that these are coupled for different values of k by virtue of the summation constraint (9.9). Again, this can be enforced using a Lagrange multiplier as before, and leads to the result

$$\pi_k = \frac{1}{N} \sum_{n=1}^{N} z_{nk} \tag{9.37}$$

so that the mixing coefficients are equal to the fractions of data points assigned to the corresponding components.

Thus we see that the complete-data log likelihood function can be maximized trivially in closed form. In practice, however, we do not have values for the latent variables so, as discussed earlier, we consider the expectation, with respect to the posterior distribution of the latent variables, of the complete-data log likelihood.

Using (9.10) and (9.11) together with Bayes' theorem, we see that this posterior distribution takes the form

$$p(\mathbf{Z}|\mathbf{X}, \boldsymbol{\mu}, \boldsymbol{\Sigma}, \boldsymbol{\pi}) \propto \prod_{n=1}^{N} \prod_{k=1}^{K} [\pi_k \mathcal{N}(\mathbf{x}_n|\boldsymbol{\mu}_k, \boldsymbol{\Sigma}_k)]^{z_{nk}}. \tag{9.38}$$

Exercise 9.5
Section 8.2

and hence factorizes over n so that under the posterior distribution the $\{\mathbf{z}_n\}$ are independent. This is easily verified by inspection of the directed graph in Figure 9.6 and making use of the d-separation criterion. The expected value of the indicator variable z_{nk} under this posterior distribution is then given by

$$\begin{aligned}\mathbb{E}[z_{nk}] &= \frac{\displaystyle\sum_{\mathbf{z}_n} z_{nk} \prod_{k'} [\pi_{k'} \mathcal{N}(\mathbf{x}_n|\boldsymbol{\mu}_{k'}, \boldsymbol{\Sigma}_{k'})]^{z_{nk'}}}{\displaystyle\sum_{\mathbf{z}_n} \prod_{j} [\pi_j \mathcal{N}(\mathbf{x}_n|\boldsymbol{\mu}_j, \boldsymbol{\Sigma}_j)]^{z_{nj}}} \\ &= \frac{\pi_k \mathcal{N}(\mathbf{x}_n|\boldsymbol{\mu}_k, \boldsymbol{\Sigma}_k)}{\displaystyle\sum_{j=1}^{K} \pi_j \mathcal{N}(\mathbf{x}_n|\boldsymbol{\mu}_j, \boldsymbol{\Sigma}_j)} = \gamma(z_{nk}) \end{aligned} \tag{9.39}$$

which is just the responsibility of component k for data point $\mathbf{x}_n$. The expected value of the complete-data log likelihood function is therefore given by

$$\mathbb{E}_{\mathbf{Z}}[\ln p(\mathbf{X}, \mathbf{Z}|\boldsymbol{\mu}, \boldsymbol{\Sigma}, \boldsymbol{\pi})] = \sum_{n=1}^{N} \sum_{k=1}^{K} \gamma(z_{nk}) \{\ln \pi_k + \ln \mathcal{N}(\mathbf{x}_n|\boldsymbol{\mu}_k, \boldsymbol{\Sigma}_k)\}. \tag{9.40}$$

We can now proceed as follows. First we choose some initial values for the parameters $\boldsymbol{\mu}^{\text{old}}$, $\boldsymbol{\Sigma}^{\text{old}}$ and $\boldsymbol{\pi}^{\text{old}}$, and use these to evaluate the responsibilities (the E step). We then keep the responsibilities fixed and maximize (9.40) with respect to $\boldsymbol{\mu}_k$, $\boldsymbol{\Sigma}_k$ and π_k (the M step). This leads to closed form solutions for $\boldsymbol{\mu}^{\text{new}}$, $\boldsymbol{\Sigma}^{\text{new}}$ and $\boldsymbol{\pi}^{\text{new}}$ given by (9.17), (9.19), and (9.22) as before. This is precisely the EM algorithm for Gaussian mixtures as derived earlier. We shall gain more insight into the role of the expected complete-data log likelihood function when we give a proof of convergence of the EM algorithm in Section 9.4.

Exercise 9.8

9.3.2 Relation to K-means

Comparison of the K-means algorithm with the EM algorithm for Gaussian mixtures shows that there is a close similarity. Whereas the K-means algorithm performs a *hard* assignment of data points to clusters, in which each data point is associated uniquely with one cluster, the EM algorithm makes a *soft* assignment based on the posterior probabilities. In fact, we can derive the K-means algorithm as a particular limit of EM for Gaussian mixtures as follows.

Consider a Gaussian mixture model in which the covariance matrices of the mixture components are given by $\epsilon\mathbf{I}$, where ϵ is a variance parameter that is shared by all of the components, and $\mathbf{I}$ is the identity matrix, so that

$$p(\mathbf{x}|\boldsymbol{\mu}_k, \boldsymbol{\Sigma}_k) = \frac{1}{(2\pi\epsilon)^{M/2}} \exp\left\{-\frac{1}{2\epsilon}\|\mathbf{x}-\boldsymbol{\mu}_k\|^2\right\}. \tag{9.41}$$

We now consider the EM algorithm for a mixture of K Gaussians of this form in which we treat ϵ as a fixed constant, instead of a parameter to be re-estimated. From (9.13) the posterior probabilities, or responsibilities, for a particular data point $\mathbf{x}_n$, are given by

$$\gamma(z_{nk}) = \frac{\pi_k \exp\left\{-\|\mathbf{x}_n - \boldsymbol{\mu}_k\|^2/2\epsilon\right\}}{\sum_j \pi_j \exp\left\{-\|\mathbf{x}_n - \boldsymbol{\mu}_j\|^2/2\epsilon\right\}}. \tag{9.42}$$

If we consider the limit $\epsilon \to 0$, we see that in the denominator the term for which $\|\mathbf{x}_n - \boldsymbol{\mu}_j\|^2$ is smallest will go to zero most slowly, and hence the responsibilities $\gamma(z_{nk})$ for the data point $\mathbf{x}_n$ all go to zero except for term j, for which the responsibility $\gamma(z_{nj})$ will go to unity. Note that this holds independently of the values of the π_k so long as none of the π_k is zero. Thus, in this limit, we obtain a hard assignment of data points to clusters, just as in the K-means algorithm, so that $\gamma(z_{nk}) \to r_{nk}$ where r_{nk} is defined by (9.2). Each data point is thereby assigned to the cluster having the closest mean.

The EM re-estimation equation for the $\boldsymbol{\mu}_k$, given by (9.17), then reduces to the K-means result (9.4). Note that the re-estimation formula for the mixing coefficients (9.22) simply re-sets the value of π_k to be equal to the fraction of data points assigned to cluster k, although these parameters no longer play an active role in the algorithm.

Exercise 9.11

Finally, in the limit $\epsilon \to 0$ the expected complete-data log likelihood, given by (9.40), becomes

$$\mathbb{E}_{\mathbf{Z}}[\ln p(\mathbf{X}, \mathbf{Z}|\boldsymbol{\mu}, \boldsymbol{\Sigma}, \boldsymbol{\pi})] \to -\frac{1}{2}\sum_{n=1}^{N}\sum_{k=1}^{K} r_{nk}\|\mathbf{x}_n - \boldsymbol{\mu}_k\|^2 + \text{const}. \tag{9.43}$$

Thus we see that in this limit, maximizing the expected complete-data log likelihood is equivalent to minimizing the distortion measure J for the K-means algorithm given by (9.1).

Note that the K-means algorithm does not estimate the covariances of the clusters but only the cluster means. A hard-assignment version of the Gaussian mixture model with general covariance matrices, known as the *elliptical K-means* algorithm, has been considered by Sung and Poggio (1994).

9.3.3 Mixtures of Bernoulli distributions

So far in this chapter, we have focussed on distributions over continuous variables described by mixtures of Gaussians. As a further example of mixture modelling, and to illustrate the EM algorithm in a different context, we now discuss mixtures of discrete binary variables described by Bernoulli distributions. This model is also known as *latent class analysis* (Lazarsfeld and Henry, 1968; McLachlan and Peel, 2000). As well as being of practical importance in its own right, our discussion of Bernoulli mixtures will also lay the foundation for a consideration of hidden Markov models over discrete variables.

Section 13.2

Consider a set of D binary variables x_i, where $i = 1, \ldots, D$, each of which is governed by a Bernoulli distribution with parameter μ_i, so that

$$p(\mathbf{x}|\boldsymbol{\mu}) = \prod_{i=1}^{D} \mu_i^{x_i}(1-\mu_i)^{(1-x_i)} \tag{9.44}$$

where $\mathbf{x} = (x_1, \ldots, x_D)^{\mathrm{T}}$ and $\boldsymbol{\mu} = (\mu_1, \ldots, \mu_D)^{\mathrm{T}}$. We see that the individual variables x_i are independent, given $\boldsymbol{\mu}$. The mean and covariance of this distribution are easily seen to be

$$\mathbb{E}[\mathbf{x}] = \boldsymbol{\mu} \tag{9.45}$$

$$\mathrm{cov}[\mathbf{x}] = \mathrm{diag}\{\mu_i(1-\mu_i)\}. \tag{9.46}$$

Now let us consider a finite mixture of these distributions given by

$$p(\mathbf{x}|\boldsymbol{\mu}, \boldsymbol{\pi}) = \sum_{k=1}^{K} \pi_k p(\mathbf{x}|\boldsymbol{\mu}_k) \tag{9.47}$$

where $\boldsymbol{\mu} = \{\boldsymbol{\mu}_1, \ldots, \boldsymbol{\mu}_K\}$, $\boldsymbol{\pi} = \{\pi_1, \ldots, \pi_K\}$, and

$$p(\mathbf{x}|\boldsymbol{\mu}_k) = \prod_{i=1}^{D} \mu_{ki}^{x_i}(1-\mu_{ki})^{(1-x_i)}. \tag{9.48}$$

Exercise 9.12

The mean and covariance of this mixture distribution are given by

$$\mathbb{E}[\mathbf{x}] = \sum_{k=1}^{K} \pi_k \boldsymbol{\mu}_k \tag{9.49}$$

$$\mathrm{cov}[\mathbf{x}] = \sum_{k=1}^{K} \pi_k \left\{\boldsymbol{\Sigma}_k + \boldsymbol{\mu}_k\boldsymbol{\mu}_k^{\mathrm{T}}\right\} - \mathbb{E}[\mathbf{x}]\mathbb{E}[\mathbf{x}]^{\mathrm{T}} \tag{9.50}$$

where $\boldsymbol{\Sigma}_k = \mathrm{diag}\{\mu_{ki}(1-\mu_{ki})\}$. Because the covariance matrix $\mathrm{cov}[\mathbf{x}]$ is no longer diagonal, the mixture distribution can capture correlations between the variables, unlike a single Bernoulli distribution.

If we are given a data set $\mathbf{X} = \{\mathbf{x}_1, \ldots, \mathbf{x}_N\}$ then the log likelihood function for this model is given by

$$\ln p(\mathbf{X}|\boldsymbol{\mu}, \boldsymbol{\pi}) = \sum_{n=1}^{N} \ln \left\{ \sum_{k=1}^{K} \pi_k p(\mathbf{x}_n|\boldsymbol{\mu}_k) \right\}. \tag{9.51}$$

Again we see the appearance of the summation inside the logarithm, so that the maximum likelihood solution no longer has closed form.

We now derive the EM algorithm for maximizing the likelihood function for the mixture of Bernoulli distributions. To do this, we first introduce an explicit latent variable $\mathbf{z}$ associated with each instance of $\mathbf{x}$. As in the case of the Gaussian mixture, $\mathbf{z} = (z_1, \ldots, z_K)^{\mathrm{T}}$ is a binary K-dimensional variable having a single component equal to 1, with all other components equal to 0. We can then write the conditional distribution of $\mathbf{x}$, given the latent variable, as

$$p(\mathbf{x}|\mathbf{z}, \boldsymbol{\mu}) = \prod_{k=1}^{K} p(\mathbf{x}|\boldsymbol{\mu}_k)^{z_k} \tag{9.52}$$

while the prior distribution for the latent variables is the same as for the mixture of Gaussians model, so that

$$p(\mathbf{z}|\boldsymbol{\pi}) = \prod_{k=1}^{K} \pi_k^{z_k}. \tag{9.53}$$

Exercise 9.14

If we form the product of $p(\mathbf{x}|\mathbf{z}, \boldsymbol{\mu})$ and $p(\mathbf{z}|\boldsymbol{\pi})$ and then marginalize over $\mathbf{z}$, then we recover (9.47).

In order to derive the EM algorithm, we first write down the complete-data log likelihood function, which is given by

$$\ln p(\mathbf{X}, \mathbf{Z}|\boldsymbol{\mu}, \boldsymbol{\pi}) = \sum_{n=1}^{N} \sum_{k=1}^{K} z_{nk} \left\{ \ln \pi_k + \sum_{i=1}^{D} [x_{ni} \ln \mu_{ki} + (1 - x_{ni}) \ln(1 - \mu_{ki})] \right\} \tag{9.54}$$

where $\mathbf{X} = \{\mathbf{x}_n\}$ and $\mathbf{Z} = \{\mathbf{z}_n\}$. Next we take the expectation of the complete-data log likelihood with respect to the posterior distribution of the latent variables to give

$$\mathbb{E}_{\mathbf{Z}}[\ln p(\mathbf{X}, \mathbf{Z}|\boldsymbol{\mu}, \boldsymbol{\pi})] = \sum_{n=1}^{N} \sum_{k=1}^{K} \gamma(z_{nk}) \left\{ \ln \pi_k + \sum_{i=1}^{D} [x_{ni} \ln \mu_{ki} + (1 - x_{ni}) \ln(1 - \mu_{ki})] \right\} \tag{9.55}$$

where $\gamma(z_{nk}) = \mathbb{E}[z_{nk}]$ is the posterior probability, or responsibility, of component k given data point $\mathbf{x}_n$. In the E step, these responsibilities are evaluated using Bayes' theorem, which takes the form

$$\begin{aligned} \gamma(z_{nk}) = \mathbb{E}[z_{nk}] &= \frac{\sum_{\mathbf{z}_n} z_{nk} \prod_{k'} [\pi_{k'} p(\mathbf{x}_n|\boldsymbol{\mu}_{k'})]^{z_{nk'}}}{\sum_{\mathbf{z}_n} \prod_{j} [\pi_j p(\mathbf{x}_n|\boldsymbol{\mu}_j)]^{z_{nj}}} \\ &= \frac{\pi_k p(\mathbf{x}_n|\boldsymbol{\mu}_k)}{\sum_{j=1}^{K} \pi_j p(\mathbf{x}_n|\boldsymbol{\mu}_j)}. \end{aligned} \tag{9.56}$$

If we consider the sum over n in (9.55), we see that the responsibilities enter only through two terms, which can be written as

$$N_k = \sum_{n=1}^{N} \gamma(z_{nk}) \tag{9.57}$$

$$\overline{\mathbf{x}}_k = \frac{1}{N_k} \sum_{n=1}^{N} \gamma(z_{nk}) \mathbf{x}_n \tag{9.58}$$

where N_k is the effective number of data points associated with component k. In the M step, we maximize the expected complete-data log likelihood with respect to the parameters $\boldsymbol{\mu}_k$ and $\boldsymbol{\pi}$. If we set the derivative of (9.55) with respect to $\boldsymbol{\mu}_k$ equal to zero and rearrange the terms, we obtain

Exercise 9.15

$$\boldsymbol{\mu}_k = \overline{\mathbf{x}}_k. \tag{9.59}$$

We see that this sets the mean of component k equal to a weighted mean of the data, with weighting coefficients given by the responsibilities that component k takes for data points. For the maximization with respect to π_k, we need to introduce a Lagrange multiplier to enforce the constraint $\sum_k \pi_k = 1$. Following analogous steps to those used for the mixture of Gaussians, we then obtain

Exercise 9.16

$$\pi_k = \frac{N_k}{N} \tag{9.60}$$

which represents the intuitively reasonable result that the mixing coefficient for component k is given by the effective fraction of points in the data set explained by that component.

Note that in contrast to the mixture of Gaussians, there are no singularities in which the likelihood function goes to infinity. This can be seen by noting that the likelihood function is bounded above because $0 \leqslant p(\mathbf{x}_n|\boldsymbol{\mu}_k) \leqslant 1$. There exist singularities at which the likelihood function goes to zero, but these will not be found by EM provided it is not initialized to a pathological starting point, because the EM algorithm always increases the value of the likelihood function, until a local maximum is found. We illustrate the Bernoulli mixture model in Figure 9.10 by using it to model handwritten digits. Here the digit images have been turned into binary vectors by setting all elements whose values exceed 0.5 to 1 and setting the remaining elements to 0. We now fit a data set of $N = 600$ such digits, comprising the digits '2', '3', and '4', with a mixture of $K = 3$ Bernoulli distributions by running 10 iterations of the EM algorithm. The mixing coefficients were initialized to $\pi_k = 1/K$, and the parameters μ_{kj} were set to random values chosen uniformly in the range $(0.25, 0.75)$ and then normalized to satisfy the constraint that $\sum_j \mu_{kj} = 1$. We see that a mixture of 3 Bernoulli distributions is able to find the three clusters in the data set corresponding to the different digits.

Exercise 9.17

Section 9.4

The conjugate prior for the parameters of a Bernoulli distribution is given by the beta distribution, and we have seen that a beta prior is equivalent to introducing

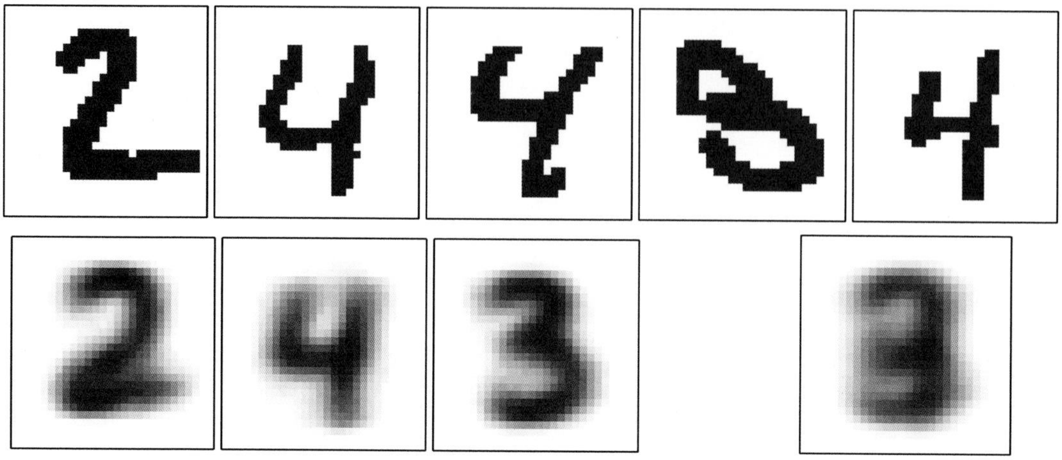

Figure 9.10 Illustration of the Bernoulli mixture model in which the top row shows examples from the digits data set after converting the pixel values from grey scale to binary using a threshold of 0.5. On the bottom row the first three images show the parameters μ_{ki} for each of the three components in the mixture model. As a comparison, we also fit the same data set using a single multivariate Bernoulli distribution, again using maximum likelihood. This amounts to simply averaging the counts in each pixel and is shown by the right-most image on the bottom row.

Section 2.1.1

Exercise 9.18

Exercise 9.19

additional effective observations of $\mathbf{x}$. We can similarly introduce priors into the Bernoulli mixture model, and use EM to maximize the posterior probability distributions.

It is straightforward to extend the analysis of Bernoulli mixtures to the case of multinomial binary variables having $M > 2$ states by making use of the discrete distribution (2.26). Again, we can introduce Dirichlet priors over the model parameters if desired.

9.3.4 EM for Bayesian linear regression

As a third example of the application of EM, we return to the evidence approximation for Bayesian linear regression. In Section 3.5.2, we obtained the re-estimation equations for the hyperparameters α and β by evaluation of the evidence and then setting the derivatives of the resulting expression to zero. We now turn to an alternative approach for finding α and β based on the EM algorithm. Recall that our goal is to maximize the evidence function $p(\mathbf{t}|\alpha, \beta)$ given by (3.77) with respect to α and β. Because the parameter vector $\mathbf{w}$ is marginalized out, we can regard it as a latent variable, and hence we can optimize this marginal likelihood function using EM. In the E step, we compute the posterior distribution of $\mathbf{w}$ given the current setting of the parameters α and β and then use this to find the expected complete-data log likelihood. In the M step, we maximize this quantity with respect to α and β. We have already derived the posterior distribution of $\mathbf{w}$ because this is given by (3.49). The complete-data log likelihood function is then given by

$$\ln p(\mathbf{t}, \mathbf{w}|\alpha, \beta) = \ln p(\mathbf{t}|\mathbf{w}, \beta) + \ln p(\mathbf{w}|\alpha) \tag{9.61}$$

where the likelihood $p(\mathbf{t}|\mathbf{w}, \beta)$ and the prior $p(\mathbf{w}|\alpha)$ are given by (3.10) and (3.52), respectively. Taking the expectation with respect to the posterior distribution of $\mathbf{w}$ then gives

$$\begin{aligned}\mathbb{E}\left[\ln p(\mathbf{t}, \mathbf{w}|\alpha, \beta)\right] &= \frac{M}{2}\ln\left(\frac{\alpha}{2\pi}\right) - \frac{\alpha}{2}\mathbb{E}\left[\mathbf{w}^{\mathrm{T}}\mathbf{w}\right] + \frac{N}{2}\ln\left(\frac{\beta}{2\pi}\right) \\ &\quad -\frac{\beta}{2}\sum_{n=1}^{N}\mathbb{E}\left[(t_n - \mathbf{w}^{\mathrm{T}}\boldsymbol{\phi}_n)^2\right]. \end{aligned} \tag{9.62}$$

Exercise 9.20

Setting the derivatives with respect to α to zero, we obtain the M step re-estimation equation

$$\alpha = \frac{M}{\mathbb{E}\left[\mathbf{w}^{\mathrm{T}}\mathbf{w}\right]} = \frac{M}{\mathbf{m}_N^{\mathrm{T}}\mathbf{m}_N + \mathrm{Tr}(\mathbf{S}_N)}. \tag{9.63}$$

Exercise 9.21

An analogous result holds for β.

Note that this re-estimation equation takes a slightly different form from the corresponding result (3.92) derived by direct evaluation of the evidence function. However, they each involve computation and inversion (or eigen decomposition) of an $M \times M$ matrix and hence will have comparable computational cost per iteration.

These two approaches to determining α should of course converge to the same result (assuming they find the same local maximum of the evidence function). This can be verified by first noting that the quantity γ is defined by

$$\gamma = M - \alpha\sum_{i=1}^{M}\frac{1}{\lambda_i + \alpha} = M - \alpha\mathrm{Tr}(\mathbf{S}_N). \tag{9.64}$$

At a stationary point of the evidence function, the re-estimation equation (3.92) will be self-consistently satisfied, and hence we can substitute for γ to give

$$\alpha\mathbf{m}_N^{\mathrm{T}}\mathbf{m}_N = \gamma = M - \alpha\mathrm{Tr}(\mathbf{S}_N) \tag{9.65}$$

and solving for α we obtain (9.63), which is precisely the EM re-estimation equation.

As a final example, we consider a closely related model, namely the relevance vector machine for regression discussed in Section 7.2.1. There we used direct maximization of the marginal likelihood to derive re-estimation equations for the hyperparameters $\boldsymbol{\alpha}$ and β. Here we consider an alternative approach in which we view the weight vector $\mathbf{w}$ as a latent variable and apply the EM algorithm. The E step involves finding the posterior distribution over the weights, and this is given by (7.81). In the M step we maximize the expected complete-data log likelihood, which is defined by

$$\mathbb{E}_{\mathbf{w}}\left[\ln\left\{p(\mathbf{t}|\mathbf{X}, \mathbf{w}, \beta)p(\mathbf{w}|\boldsymbol{\alpha})\right\}\right] \tag{9.66}$$

where the expectation is taken with respect to the posterior distribution computed using the 'old' parameter values. To compute the new parameter values we maximize with respect to $\boldsymbol{\alpha}$ and β to give

Exercise 9.22

$$\alpha_i^{\text{new}} = \frac{1}{m_i^2 + \Sigma_{ii}} \tag{9.67}$$

$$(\beta^{\text{new}})^{-1} = \frac{\|\mathbf{t} - \boldsymbol{\Phi}\mathbf{m}\|^2 + \beta^{-1}\sum_i \gamma_i}{N} \tag{9.68}$$

These re-estimation equations are formally equivalent to those obtained by direct maxmization.

Exercise 9.23

9.4. The EM Algorithm in General

The *expectation maximization* algorithm, or EM algorithm, is a general technique for finding maximum likelihood solutions for probabilistic models having latent variables (Dempster *et al.*, 1977; McLachlan and Krishnan, 1997). Here we give a very general treatment of the EM algorithm and in the process provide a proof that the EM algorithm derived heuristically in Sections 9.2 and 9.3 for Gaussian mixtures does indeed maximize the likelihood function (Csiszàr and Tusnàdy, 1984; Hathaway, 1986; Neal and Hinton, 1999). Our discussion will also form the basis for the derivation of the variational inference framework.

Section 10.1

Consider a probabilistic model in which we collectively denote all of the observed variables by $\mathbf{X}$ and all of the hidden variables by $\mathbf{Z}$. The joint distribution $p(\mathbf{X}, \mathbf{Z}|\boldsymbol{\theta})$ is governed by a set of parameters denoted $\boldsymbol{\theta}$. Our goal is to maximize the likelihood function that is given by

$$p(\mathbf{X}|\boldsymbol{\theta}) = \sum_{\mathbf{Z}} p(\mathbf{X}, \mathbf{Z}|\boldsymbol{\theta}). \tag{9.69}$$

Here we are assuming $\mathbf{Z}$ is discrete, although the discussion is identical if $\mathbf{Z}$ comprises continuous variables or a combination of discrete and continuous variables, with summation replaced by integration as appropriate.

We shall suppose that direct optimization of $p(\mathbf{X}|\boldsymbol{\theta})$ is difficult, but that optimization of the complete-data likelihood function $p(\mathbf{X}, \mathbf{Z}|\boldsymbol{\theta})$ is significantly easier. Next we introduce a distribution $q(\mathbf{Z})$ defined over the latent variables, and we observe that, for any choice of $q(\mathbf{Z})$, the following decomposition holds

$$\ln p(\mathbf{X}|\boldsymbol{\theta}) = \mathcal{L}(q, \boldsymbol{\theta}) + \mathrm{KL}(q\|p) \tag{9.70}$$

where we have defined

$$\mathcal{L}(q, \boldsymbol{\theta}) = \sum_{\mathbf{Z}} q(\mathbf{Z}) \ln \left\{ \frac{p(\mathbf{X}, \mathbf{Z}|\boldsymbol{\theta})}{q(\mathbf{Z})} \right\} \tag{9.71}$$

$$\mathrm{KL}(q\|p) = -\sum_{\mathbf{Z}} q(\mathbf{Z}) \ln \left\{ \frac{p(\mathbf{Z}|\mathbf{X}, \boldsymbol{\theta})}{q(\mathbf{Z})} \right\}. \tag{9.72}$$

Note that $\mathcal{L}(q, \boldsymbol{\theta})$ is a functional (see Appendix D for a discussion of functionals) of the distribution $q(\mathbf{Z})$, and a function of the parameters $\boldsymbol{\theta}$. It is worth studying

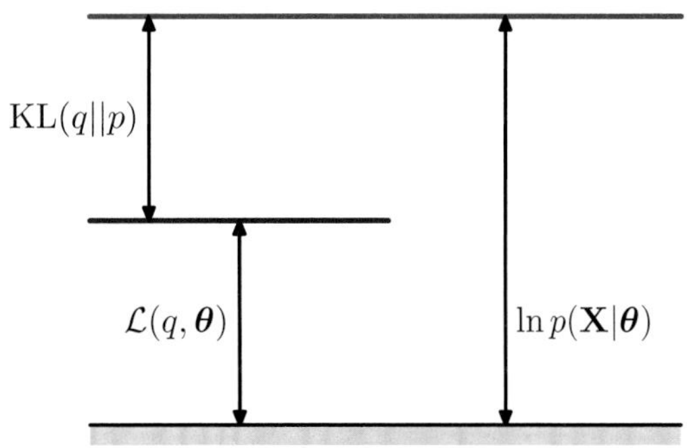

Figure 9.11 Illustration of the decomposition given by (9.70), which holds for any choice of distribution $q(\mathbf{Z})$. Because the Kullback-Leibler divergence satisfies $\mathrm{KL}(q\|p) \geqslant 0$, we see that the quantity $\mathcal{L}(q, \boldsymbol{\theta})$ is a lower bound on the log likelihood function $\ln p(\mathbf{X}|\boldsymbol{\theta})$.

carefully the forms of the expressions (9.71) and (9.72), and in particular noting that they differ in sign and also that $\mathcal{L}(q, \boldsymbol{\theta})$ contains the joint distribution of $\mathbf{X}$ and $\mathbf{Z}$ while $\mathrm{KL}(q\|p)$ contains the conditional distribution of $\mathbf{Z}$ given $\mathbf{X}$. To verify the decomposition (9.70), we first make use of the product rule of probability to give

Exercise 9.24

$$\ln p(\mathbf{X}, \mathbf{Z}|\boldsymbol{\theta}) = \ln p(\mathbf{Z}|\mathbf{X}, \boldsymbol{\theta}) + \ln p(\mathbf{X}|\boldsymbol{\theta}) \tag{9.73}$$

which we then substitute into the expression for $\mathcal{L}(q, \boldsymbol{\theta})$. This gives rise to two terms, one of which cancels $\mathrm{KL}(q\|p)$ while the other gives the required log likelihood $\ln p(\mathbf{X}|\boldsymbol{\theta})$ after noting that $q(\mathbf{Z})$ is a normalized distribution that sums to 1.

From (9.72), we see that $\mathrm{KL}(q\|p)$ is the Kullback-Leibler divergence between $q(\mathbf{Z})$ and the posterior distribution $p(\mathbf{Z}|\mathbf{X}, \boldsymbol{\theta})$. Recall that the Kullback-Leibler divergence satisfies $\mathrm{KL}(q\|p) \geqslant 0$, with equality if, and only if, $q(\mathbf{Z}) = p(\mathbf{Z}|\mathbf{X}, \boldsymbol{\theta})$. It therefore follows from (9.70) that $\mathcal{L}(q, \boldsymbol{\theta}) \leqslant \ln p(\mathbf{X}|\boldsymbol{\theta})$, in other words that $\mathcal{L}(q, \boldsymbol{\theta})$ is a lower bound on $\ln p(\mathbf{X}|\boldsymbol{\theta})$. The decomposition (9.70) is illustrated in Figure 9.11.

Section 1.6.1

The EM algorithm is a two-stage iterative optimization technique for finding maximum likelihood solutions. We can use the decomposition (9.70) to define the EM algorithm and to demonstrate that it does indeed maximize the log likelihood. Suppose that the current value of the parameter vector is $\boldsymbol{\theta}^{\mathrm{old}}$. In the E step, the lower bound $\mathcal{L}(q, \boldsymbol{\theta}^{\mathrm{old}})$ is maximized with respect to $q(\mathbf{Z})$ while holding $\boldsymbol{\theta}^{\mathrm{old}}$ fixed. The solution to this maximization problem is easily seen by noting that the value of $\ln p(\mathbf{X}|\boldsymbol{\theta}^{\mathrm{old}})$ does not depend on $q(\mathbf{Z})$ and so the largest value of $\mathcal{L}(q, \boldsymbol{\theta}^{\mathrm{old}})$ will occur when the Kullback-Leibler divergence vanishes, in other words when $q(\mathbf{Z})$ is equal to the posterior distribution $p(\mathbf{Z}|\mathbf{X}, \boldsymbol{\theta}^{\mathrm{old}})$. In this case, the lower bound will equal the log likelihood, as illustrated in Figure 9.12.

In the subsequent M step, the distribution $q(\mathbf{Z})$ is held fixed and the lower bound $\mathcal{L}(q, \boldsymbol{\theta})$ is maximized with respect to $\boldsymbol{\theta}$ to give some new value $\boldsymbol{\theta}^{\mathrm{new}}$. This will cause the lower bound $\mathcal{L}$ to increase (unless it is already at a maximum), which will necessarily cause the corresponding log likelihood function to increase. Because the distribution q is determined using the old parameter values rather than the new values and is held fixed during the M step, it will not equal the new posterior distribution $p(\mathbf{Z}|\mathbf{X}, \boldsymbol{\theta}^{\mathrm{new}})$, and hence there will be a nonzero KL divergence. The increase in the log likelihood function is therefore greater than the increase in the lower bound, as

Figure 9.12 Illustration of the E step of the EM algorithm. The q distribution is set equal to the posterior distribution for the current parameter values $\boldsymbol{\theta}^{\text{old}}$, causing the lower bound to move up to the same value as the log likelihood function, with the KL divergence vanishing.

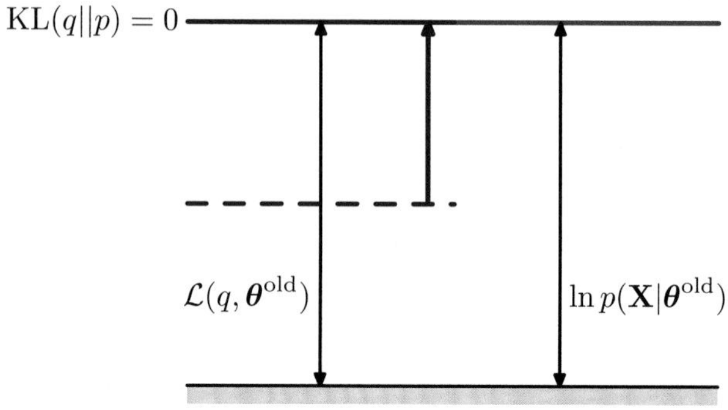

shown in Figure 9.13. If we substitute $q(\mathbf{Z}) = p(\mathbf{Z}|\mathbf{X}, \boldsymbol{\theta}^{\text{old}})$ into (9.71), we see that, after the E step, the lower bound takes the form

$$\begin{aligned}\mathcal{L}(q, \boldsymbol{\theta}) &= \sum_{\mathbf{Z}} p(\mathbf{Z}|\mathbf{X}, \boldsymbol{\theta}^{\text{old}}) \ln p(\mathbf{X}, \mathbf{Z}|\boldsymbol{\theta}) - \sum_{\mathbf{Z}} p(\mathbf{Z}|\mathbf{X}, \boldsymbol{\theta}^{\text{old}}) \ln p(\mathbf{Z}|\mathbf{X}, \boldsymbol{\theta}^{\text{old}}) \\ &= \mathcal{Q}(\boldsymbol{\theta}, \boldsymbol{\theta}^{\text{old}}) + \text{const} \end{aligned} \tag{9.74}$$

where the constant is simply the entropy of the q distribution and is therefore independent of $\boldsymbol{\theta}$. Thus in the M step, the quantity that is being maximized is the expectation of the complete-data log likelihood, as we saw earlier in the case of mixtures of Gaussians. Note that the variable $\boldsymbol{\theta}$ over which we are optimizing appears only inside the logarithm. If the joint distribution $p(\mathbf{Z}, \mathbf{X}|\boldsymbol{\theta})$ comprises a member of the exponential family, or a product of such members, then we see that the logarithm will cancel the exponential and lead to an M step that will be typically much simpler than the maximization of the corresponding incomplete-data log likelihood function $p(\mathbf{X}|\boldsymbol{\theta})$.

The operation of the EM algorithm can also be viewed in the space of parameters, as illustrated schematically in Figure 9.14. Here the red curve depicts the (in-

Figure 9.13 Illustration of the M step of the EM algorithm. The distribution $q(\mathbf{Z})$ is held fixed and the lower bound $\mathcal{L}(q, \boldsymbol{\theta})$ is maximized with respect to the parameter vector $\boldsymbol{\theta}$ to give a revised value $\boldsymbol{\theta}^{\text{new}}$. Because the KL divergence is nonnegative, this causes the log likelihood $\ln p(\mathbf{X}|\boldsymbol{\theta})$ to increase by at least as much as the lower bound does.

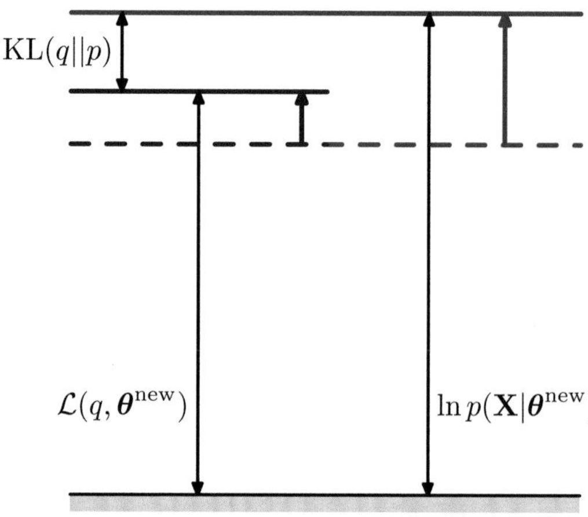

Figure 9.14 The EM algorithm involves alternately computing a lower bound on the log likelihood for the current parameter values and then maximizing this bound to obtain the new parameter values. See the text for a full discussion.

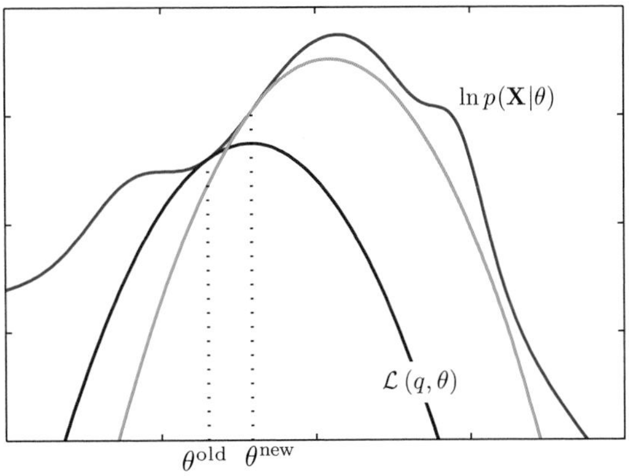

complete data) log likelihood function whose value we wish to maximize. We start with some initial parameter value $\boldsymbol{\theta}^{\text{old}}$, and in the first E step we evaluate the posterior distribution over latent variables, which gives rise to a lower bound $\mathcal{L}(q, \boldsymbol{\theta}^{(\text{old})})$ whose value equals the log likelihood at $\boldsymbol{\theta}^{(\text{old})}$, as shown by the blue curve. Note that the bound makes a tangential contact with the log likelihood at $\boldsymbol{\theta}^{(\text{old})}$, so that both curves have the same gradient. This bound is a convex function having a unique maximum (for mixture components from the exponential family). In the M step, the bound is maximized giving the value $\boldsymbol{\theta}^{(\text{new})}$, which gives a larger value of log likelihood than $\boldsymbol{\theta}^{(\text{old})}$. The subsequent E step then constructs a bound that is tangential at $\boldsymbol{\theta}^{(\text{new})}$ as shown by the green curve.

Exercise 9.25

For the particular case of an independent, identically distributed data set, $\mathbf{X}$ will comprise N data points $\{\mathbf{x}_n\}$ while $\mathbf{Z}$ will comprise N corresponding latent variables $\{\mathbf{z}_n\}$, where $n = 1, \ldots, N$. From the independence assumption, we have $p(\mathbf{X}, \mathbf{Z}) = \prod_n p(\mathbf{x}_n, \mathbf{z}_n)$ and, by marginalizing over the $\{\mathbf{z}_n\}$ we have $p(\mathbf{X}) = \prod_n p(\mathbf{x}_n)$. Using the sum and product rules, we see that the posterior probability that is evaluated in the E step takes the form

$$p(\mathbf{Z}|\mathbf{X}, \boldsymbol{\theta}) = \frac{p(\mathbf{X}, \mathbf{Z}|\boldsymbol{\theta})}{\sum_{\mathbf{Z}} p(\mathbf{X}, \mathbf{Z}|\boldsymbol{\theta})} = \frac{\prod_{n=1}^{N} p(\mathbf{x}_n, \mathbf{z}_n|\boldsymbol{\theta})}{\sum_{\mathbf{Z}} \prod_{n=1}^{N} p(\mathbf{x}_n, \mathbf{z}_n|\boldsymbol{\theta})} = \prod_{n=1}^{N} p(\mathbf{z}_n|\mathbf{x}_n, \boldsymbol{\theta}) \tag{9.75}$$

and so the posterior distribution also factorizes with respect to n. In the case of the Gaussian mixture model this simply says that the responsibility that each of the mixture components takes for a particular data point $\mathbf{x}_n$ depends only on the value of $\mathbf{x}_n$ and on the parameters $\boldsymbol{\theta}$ of the mixture components, not on the values of the other data points.

We have seen that both the E and the M steps of the EM algorithm are increasing the value of a well-defined bound on the log likelihood function and that the complete EM cycle will change the model parameters in such a way as to cause the log likelihood to increase (unless it is already at a maximum, in which case the parameters remain unchanged).

We can also use the EM algorithm to maximize the posterior distribution $p(\boldsymbol{\theta}|\mathbf{X})$ for models in which we have introduced a prior $p(\boldsymbol{\theta})$ over the parameters. To see this, we note that as a function of $\boldsymbol{\theta}$, we have $p(\boldsymbol{\theta}|\mathbf{X}) = p(\boldsymbol{\theta}, \mathbf{X})/p(\mathbf{X})$ and so

$$\ln p(\boldsymbol{\theta}|\mathbf{X}) = \ln p(\boldsymbol{\theta}, \mathbf{X}) - \ln p(\mathbf{X}). \tag{9.76}$$

Making use of the decomposition (9.70), we have

$$\begin{aligned} \ln p(\boldsymbol{\theta}|\mathbf{X}) &= \mathcal{L}(q, \boldsymbol{\theta}) + \mathrm{KL}(q\|p) + \ln p(\boldsymbol{\theta}) - \ln p(\mathbf{X}) \\ &\geqslant \mathcal{L}(q, \boldsymbol{\theta}) + \ln p(\boldsymbol{\theta}) - \ln p(\mathbf{X}). \end{aligned} \tag{9.77}$$

where $\ln p(\mathbf{X})$ is a constant. We can again optimize the right-hand side alternately with respect to q and $\boldsymbol{\theta}$. The optimization with respect to q gives rise to the same E-step equations as for the standard EM algorithm, because q only appears in $\mathcal{L}(q, \boldsymbol{\theta})$. The M-step equations are modified through the introduction of the prior term $\ln p(\boldsymbol{\theta})$, which typically requires only a small modification to the standard maximum likelihood M-step equations.

The EM algorithm breaks down the potentially difficult problem of maximizing the likelihood function into two stages, the E step and the M step, each of which will often prove simpler to implement. Nevertheless, for complex models it may be the case that either the E step or the M step, or indeed both, remain intractable. This leads to two possible extensions of the EM algorithm, as follows.

The *generalized EM*, or *GEM*, algorithm addresses the problem of an intractable M step. Instead of aiming to maximize $\mathcal{L}(q, \boldsymbol{\theta})$ with respect to $\boldsymbol{\theta}$, it seeks instead to change the parameters in such a way as to increase its value. Again, because $\mathcal{L}(q, \boldsymbol{\theta})$ is a lower bound on the log likelihood function, each complete EM cycle of the GEM algorithm is guaranteed to increase the value of the log likelihood (unless the parameters already correspond to a local maximum). One way to exploit the GEM approach would be to use one of the nonlinear optimization strategies, such as the conjugate gradients algorithm, during the M step. Another form of GEM algorithm, known as the *expectation conditional maximization*, or ECM, algorithm, involves making several constrained optimizations within each M step (Meng and Rubin, 1993). For instance, the parameters might be partitioned into groups, and the M step is broken down into multiple steps each of which involves optimizing one of the subset with the remainder held fixed.

We can similarly generalize the E step of the EM algorithm by performing a partial, rather than complete, optimization of $\mathcal{L}(q, \boldsymbol{\theta})$ with respect to $q(\mathbf{Z})$ (Neal and Hinton, 1999). As we have seen, for any given value of $\boldsymbol{\theta}$ there is a unique maximum of $\mathcal{L}(q, \boldsymbol{\theta})$ with respect to $q(\mathbf{Z})$ that corresponds to the posterior distribution $q_{\boldsymbol{\theta}}(\mathbf{Z}) = p(\mathbf{Z}|\mathbf{X}, \boldsymbol{\theta})$ and that for this choice of $q(\mathbf{Z})$ the bound $\mathcal{L}(q, \boldsymbol{\theta})$ is equal to the log likelihood function $\ln p(\mathbf{X}|\boldsymbol{\theta})$. It follows that any algorithm that converges to the global maximum of $\mathcal{L}(q, \boldsymbol{\theta})$ will find a value of $\boldsymbol{\theta}$ that is also a global maximum of the log likelihood $\ln p(\mathbf{X}|\boldsymbol{\theta})$. Provided $p(\mathbf{X}, \mathbf{Z}|\boldsymbol{\theta})$ is a continuous function of $\boldsymbol{\theta}$

then, by continuity, any local maximum of $\mathcal{L}(q, \boldsymbol{\theta})$ will also be a local maximum of $\ln p(\mathbf{X}|\boldsymbol{\theta})$.

Consider the case of N independent data points $\mathbf{x}_1, \ldots, \mathbf{x}_N$ with corresponding latent variables $\mathbf{z}_1, \ldots, \mathbf{z}_N$. The joint distribution $p(\mathbf{X}, \mathbf{Z}|\boldsymbol{\theta})$ factorizes over the data points, and this structure can be exploited in an incremental form of EM in which at each EM cycle only one data point is processed at a time. In the E step, instead of recomputing the responsibilities for all of the data points, we just re-evaluate the responsibilities for one data point. It might appear that the subsequent M step would require computation involving the responsibilities for all of the data points. However, if the mixture components are members of the exponential family, then the responsibilities enter only through simple sufficient statistics, and these can be updated efficiently. Consider, for instance, the case of a Gaussian mixture, and suppose we perform an update for data point m in which the corresponding old and new values of the responsibilities are denoted $\gamma^{\text{old}}(z_{mk})$ and $\gamma^{\text{new}}(z_{mk})$. In the M step, the required sufficient statistics can be updated incrementally. For instance, for the means the sufficient statistics are defined by (9.17) and (9.18) from which we obtain

Exercise 9.26

$$\boldsymbol{\mu}_k^{\text{new}} = \boldsymbol{\mu}_k^{\text{old}} + \left(\frac{\gamma^{\text{new}}(z_{mk}) - \gamma^{\text{old}}(z_{mk})}{N_k^{\text{new}}} \right) \left(\mathbf{x}_m - \boldsymbol{\mu}_k^{\text{old}} \right) \tag{9.78}$$

together with

$$N_k^{\text{new}} = N_k^{\text{old}} + \gamma^{\text{new}}(z_{mk}) - \gamma^{\text{old}}(z_{mk}). \tag{9.79}$$

The corresponding results for the covariances and the mixing coefficients are analogous.

Thus both the E step and the M step take fixed time that is independent of the total number of data points. Because the parameters are revised after each data point, rather than waiting until after the whole data set is processed, this incremental version can converge faster than the batch version. Each E or M step in this incremental algorithm is increasing the value of $\mathcal{L}(q, \boldsymbol{\theta})$ and, as we have shown above, if the algorithm converges to a local (or global) maximum of $\mathcal{L}(q, \boldsymbol{\theta})$, this will correspond to a local (or global) maximum of the log likelihood function $\ln p(\mathbf{X}|\boldsymbol{\theta})$.

Exercises

9.1 ($\star$) **www** Consider the K-means algorithm discussed in Section 9.1. Show that as a consequence of there being a finite number of possible assignments for the set of discrete indicator variables r_{nk}, and that for each such assignment there is a unique optimum for the $\{\boldsymbol{\mu}_k\}$, the K-means algorithm must converge after a finite number of iterations.

9.2 ($\star$) Apply the Robbins-Monro sequential estimation procedure described in Section 2.3.5 to the problem of finding the roots of the regression function given by the derivatives of J in (9.1) with respect to $\boldsymbol{\mu}_k$. Show that this leads to a stochastic K-means algorithm in which, for each data point $\mathbf{x}_n$, the nearest prototype $\boldsymbol{\mu}_k$ is updated using (9.5).

9.3 ($\star$) **www** Consider a Gaussian mixture model in which the marginal distribution $p(\mathbf{z})$ for the latent variable is given by (9.10), and the conditional distribution $p(\mathbf{x}|\mathbf{z})$ for the observed variable is given by (9.11). Show that the marginal distribution $p(\mathbf{x})$, obtained by summing $p(\mathbf{z})p(\mathbf{x}|\mathbf{z})$ over all possible values of $\mathbf{z}$, is a Gaussian mixture of the form (9.7).

9.4 ($\star$) Suppose we wish to use the EM algorithm to maximize the posterior distribution over parameters $p(\boldsymbol{\theta}|\mathbf{X})$ for a model containing latent variables, where $\mathbf{X}$ is the observed data set. Show that the E step remains the same as in the maximum likelihood case, whereas in the M step the quantity to be maximized is given by $\mathcal{Q}(\boldsymbol{\theta}, \boldsymbol{\theta}^{\text{old}}) + \ln p(\boldsymbol{\theta})$ where $\mathcal{Q}(\boldsymbol{\theta}, \boldsymbol{\theta}^{\text{old}})$ is defined by (9.30).

9.5 ($\star$) Consider the directed graph for a Gaussian mixture model shown in Figure 9.6. By making use of the d-separation criterion discussed in Section 8.2, show that the posterior distribution of the latent variables factorizes with respect to the different data points so that

$$p(\mathbf{Z}|\mathbf{X}, \boldsymbol{\mu}, \boldsymbol{\Sigma}, \boldsymbol{\pi}) = \prod_{n=1}^{N} p(\mathbf{z}_n|\mathbf{x}_n, \boldsymbol{\mu}, \boldsymbol{\Sigma}, \boldsymbol{\pi}). \tag{9.80}$$

9.6 ($\star\star$) Consider a special case of a Gaussian mixture model in which the covariance matrices $\boldsymbol{\Sigma}_k$ of the components are all constrained to have a common value $\boldsymbol{\Sigma}$. Derive the EM equations for maximizing the likelihood function under such a model.

9.7 ($\star$) **www** Verify that maximization of the complete-data log likelihood (9.36) for a Gaussian mixture model leads to the result that the means and covariances of each component are fitted independently to the corresponding group of data points, and the mixing coefficients are given by the fractions of points in each group.

9.8 ($\star$) **www** Show that if we maximize (9.40) with respect to $\boldsymbol{\mu}_k$ while keeping the responsibilities $\gamma(z_{nk})$ fixed, we obtain the closed form solution given by (9.17).

9.9 ($\star$) Show that if we maximize (9.40) with respect to $\boldsymbol{\Sigma}_k$ and π_k while keeping the responsibilities $\gamma(z_{nk})$ fixed, we obtain the closed form solutions given by (9.19) and (9.22).

9.10 ($\star\star$) Consider a density model given by a mixture distribution

$$p(\mathbf{x}) = \sum_{k=1}^{K} \pi_k p(\mathbf{x}|k) \tag{9.81}$$

and suppose that we partition the vector $\mathbf{x}$ into two parts so that $\mathbf{x} = (\mathbf{x}_a, \mathbf{x}_b)$. Show that the conditional density $p(\mathbf{x}_b|\mathbf{x}_a)$ is itself a mixture distribution and find expressions for the mixing coefficients and for the component densities.

9.11 ($\star$) In Section 9.3.2, we obtained a relationship between K means and EM for Gaussian mixtures by considering a mixture model in which all components have covariance $\epsilon\mathbf{I}$. Show that in the limit $\epsilon \rightarrow 0$, maximizing the expected complete-data log likelihood for this model, given by (9.40), is equivalent to minimizing the distortion measure J for the K-means algorithm given by (9.1).

9.12 ($\star$) **www** Consider a mixture distribution of the form

$$p(\mathbf{x}) = \sum_{k=1}^{K} \pi_k p(\mathbf{x}|k) \tag{9.82}$$

where the elements of $\mathbf{x}$ could be discrete or continuous or a combination of these. Denote the mean and covariance of $p(\mathbf{x}|k)$ by $\boldsymbol{\mu}_k$ and $\boldsymbol{\Sigma}_k$, respectively. Show that the mean and covariance of the mixture distribution are given by (9.49) and (9.50).

9.13 ($\star\star$) Using the re-estimation equations for the EM algorithm, show that a mixture of Bernoulli distributions, with its parameters set to values corresponding to a maximum of the likelihood function, has the property that

$$\mathbb{E}[\mathbf{x}] = \frac{1}{N}\sum_{n=1}^{N} \mathbf{x}_n \equiv \overline{\mathbf{x}}. \tag{9.83}$$

Hence show that if the parameters of this model are initialized such that all components have the same mean $\boldsymbol{\mu}_k = \widehat{\boldsymbol{\mu}}$ for $k = 1, \ldots, K$, then the EM algorithm will converge after one iteration, for any choice of the initial mixing coefficients, and that this solution has the property $\boldsymbol{\mu}_k = \overline{\mathbf{x}}$. Note that this represents a degenerate case of the mixture model in which all of the components are identical, and in practice we try to avoid such solutions by using an appropriate initialization.

9.14 ($\star$) Consider the joint distribution of latent and observed variables for the Bernoulli distribution obtained by forming the product of $p(\mathbf{x}|\mathbf{z}, \boldsymbol{\mu})$ given by (9.52) and $p(\mathbf{z}|\boldsymbol{\pi})$ given by (9.53). Show that if we marginalize this joint distribution with respect to $\mathbf{z}$, then we obtain (9.47).

9.15 ($\star$) **www** Show that if we maximize the expected complete-data log likelihood function (9.55) for a mixture of Bernoulli distributions with respect to $\boldsymbol{\mu}_k$, we obtain the M step equation (9.59).

9.16 ($\star$) Show that if we maximize the expected complete-data log likelihood function (9.55) for a mixture of Bernoulli distributions with respect to the mixing coefficients π_k, using a Lagrange multiplier to enforce the summation constraint, we obtain the M step equation (9.60).

9.17 ($\star$) **www** Show that as a consequence of the constraint $0 \leqslant p(\mathbf{x}_n|\boldsymbol{\mu}_k) \leqslant 1$ for the discrete variable $\mathbf{x}_n$, the incomplete-data log likelihood function for a mixture of Bernoulli distributions is bounded above, and hence that there are no singularities for which the likelihood goes to infinity.

9.18 ($\star\star$) Consider a Bernoulli mixture model as discussed in Section 9.3.3, together with a prior distribution $p(\boldsymbol{\mu}_k|a_k, b_k)$ over each of the parameter vectors $\boldsymbol{\mu}_k$ given by the beta distribution (2.13), and a Dirichlet prior $p(\boldsymbol{\pi}|\boldsymbol{\alpha})$ given by (2.38). Derive the EM algorithm for maximizing the posterior probability $p(\boldsymbol{\mu}, \boldsymbol{\pi}|\mathbf{X})$.

9.19 ($\star\star$) Consider a D-dimensional variable $\mathbf{x}$ each of whose components i is itself a multinomial variable of degree M so that $\mathbf{x}$ is a binary vector with components x_{ij} where $i = 1, \ldots, D$ and $j = 1, \ldots, M$, subject to the constraint that $\sum_j x_{ij} = 1$ for all i. Suppose that the distribution of these variables is described by a mixture of the discrete multinomial distributions considered in Section 2.2 so that

$$p(\mathbf{x}) = \sum_{k=1}^{K} \pi_k p(\mathbf{x}|\boldsymbol{\mu}_k) \tag{9.84}$$

where

$$p(\mathbf{x}|\boldsymbol{\mu}_k) = \prod_{i=1}^{D} \prod_{j=1}^{M} \mu_{kij}^{x_{ij}}. \tag{9.85}$$

The parameters μ_{kij} represent the probabilities $p(x_{ij} = 1|\boldsymbol{\mu}_k)$ and must satisfy $0 \leqslant \mu_{kij} \leqslant 1$ together with the constraint $\sum_j \mu_{kij} = 1$ for all values of k and i. Given an observed data set $\{\mathbf{x}_n\}$, where $n = 1, \ldots, N$, derive the E and M step equations of the EM algorithm for optimizing the mixing coefficients π_k and the component parameters μ_{kij} of this distribution by maximum likelihood.

9.20 ($\star$) **www** Show that maximization of the expected complete-data log likelihood function (9.62) for the Bayesian linear regression model leads to the M step re-estimation result (9.63) for α.

9.21 ($\star\star$) Using the evidence framework of Section 3.5, derive the M-step re-estimation equations for the parameter β in the Bayesian linear regression model, analogous to the result (9.63) for α.

9.22 ($\star\star$) By maximization of the expected complete-data log likelihood defined by (9.66), derive the M step equations (9.67) and (9.68) for re-estimating the hyperparameters of the relevance vector machine for regression.

9.23 ($\star\star$) **www** In Section 7.2.1 we used direct maximization of the marginal likelihood to derive the re-estimation equations (7.87) and (7.88) for finding values of the hyperparameters $\boldsymbol{\alpha}$ and β for the regression RVM. Similarly, in Section 9.3.4 we used the EM algorithm to maximize the same marginal likelihood, giving the re-estimation equations (9.67) and (9.68). Show that, at any stationary point, these two sets of re-estimation equations are formally equivalent.

9.24 ($\star$) Verify the relation (9.70) in which $\mathcal{L}(q, \boldsymbol{\theta})$ and $\mathrm{KL}(q\|p)$ are defined by (9.71) and (9.72), respectively.

9.25 ($\star$) **www** Show that the lower bound $\mathcal{L}(q, \boldsymbol{\theta})$ given by (9.71), with $q(\mathbf{Z}) = p(\mathbf{Z}|\mathbf{X}, \boldsymbol{\theta}^{(\text{old})})$, has the same gradient with respect to $\boldsymbol{\theta}$ as the log likelihood function $\ln p(\mathbf{X}|\boldsymbol{\theta})$ at the point $\boldsymbol{\theta} = \boldsymbol{\theta}^{(\text{old})}$.

9.26 ($\star$) **www** Consider the incremental form of the EM algorithm for a mixture of Gaussians, in which the responsibilities are recomputed only for a specific data point $\mathbf{x}_m$. Starting from the M-step formulae (9.17) and (9.18), derive the results (9.78) and (9.79) for updating the component means.

9.27 ($\star\star$) Derive M-step formulae for updating the covariance matrices and mixing coefficients in a Gaussian mixture model when the responsibilities are updated incrementally, analogous to the result (9.78) for updating the means.

10 Approximate Inference

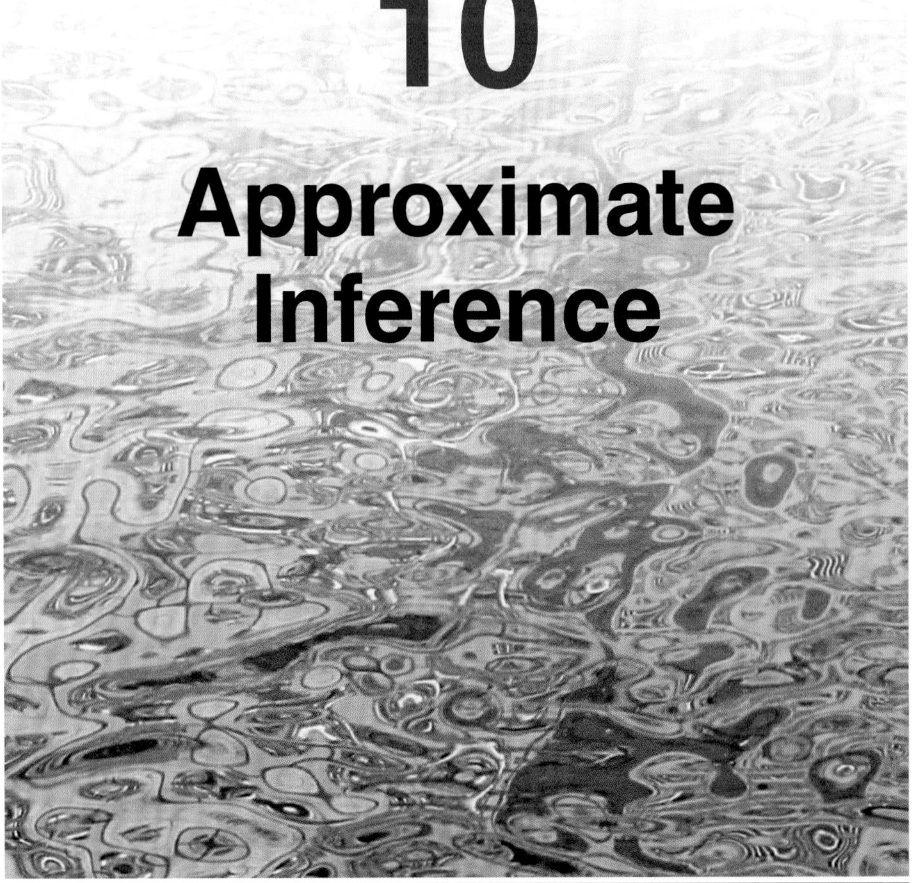

A central task in the application of probabilistic models is the evaluation of the posterior distribution $p(\mathbf{Z}|\mathbf{X})$ of the latent variables $\mathbf{Z}$ given the observed (visible) data variables $\mathbf{X}$, and the evaluation of expectations computed with respect to this distribution. The model might also contain some deterministic parameters, which we will leave implicit for the moment, or it may be a fully Bayesian model in which any unknown parameters are given prior distributions and are absorbed into the set of latent variables denoted by the vector $\mathbf{Z}$. For instance, in the EM algorithm we need to evaluate the expectation of the complete-data log likelihood with respect to the posterior distribution of the latent variables. For many models of practical interest, it will be infeasible to evaluate the posterior distribution or indeed to compute expectations with respect to this distribution. This could be because the dimensionality of the latent space is too high to work with directly or because the posterior distribution has a highly complex form for which expectations are not analytically tractable. In the case of continuous variables, the required integrations may not have closed-form analytical solutions, while the dimensionality of the space and the complexity of the integrand may prohibit numerical integration. For discrete variables, the marginalizations involve summing over all possible configurations of the hidden variables, and though this is always possible in principle, we often find in practice that there may be exponentially many hidden states so that exact calculation is prohibitively expensive.

In such situations, we need to resort to approximation schemes, and these fall broadly into two classes, according to whether they rely on stochastic or deterministic approximations. Stochastic techniques such as Markov chain Monte Carlo, described in Chapter 11, have enabled the widespread use of Bayesian methods across many domains. They generally have the property that given infinite computational resource, they can generate exact results, and the approximation arises from the use of a finite amount of processor time. In practice, sampling methods can be computationally demanding, often limiting their use to small-scale problems. Also, it can be difficult to know whether a sampling scheme is generating independent samples from the required distribution.

In this chapter, we introduce a range of deterministic approximation schemes, some of which scale well to large applications. These are based on analytical approximations to the posterior distribution, for example by assuming that it factorizes in a particular way or that it has a specific parametric form such as a Gaussian. As such, they can never generate exact results, and so their strengths and weaknesses are complementary to those of sampling methods.

In Section 4.4, we discussed the Laplace approximation, which is based on a local Gaussian approximation to a mode (i.e., a maximum) of the distribution. Here we turn to a family of approximation techniques called *variational inference* or *variational Bayes*, which use more global criteria and which have been widely applied. We conclude with a brief introduction to an alternative variational framework known as *expectation propagation*.

10.1. Variational Inference

Variational methods have their origins in the 18th century with the work of Euler, Lagrange, and others on the *calculus of variations*. Standard calculus is concerned with finding derivatives of functions. We can think of a function as a mapping that takes the value of a variable as the input and returns the value of the function as the output. The derivative of the function then describes how the output value varies as we make infinitesimal changes to the input value. Similarly, we can define a *functional* as a mapping that takes a function as the input and that returns the value of the functional as the output. An example would be the entropy $\mathrm{H}[p]$, which takes a probability distribution $p(x)$ as the input and returns the quantity

$$\mathrm{H}[p] = -\int p(x) \ln p(x) \, \mathrm{d}x \tag{10.1}$$

as the output. We can then introduce the concept of a *functional derivative*, which expresses how the value of the functional changes in response to infinitesimal changes to the input function (Feynman *et al.*, 1964). The rules for the calculus of variations mirror those of standard calculus and are discussed in Appendix D. Many problems can be expressed in terms of an optimization problem in which the quantity being optimized is a functional. The solution is obtained by exploring all possible input functions to find the one that maximizes, or minimizes, the functional. Variational methods have broad applicability and include such areas as finite element methods (Kapur, 1989) and maximum entropy (Schwarz, 1988).

Although there is nothing intrinsically approximate about variational methods, they do naturally lend themselves to finding approximate solutions. This is done by restricting the range of functions over which the optimization is performed, for instance by considering only quadratic functions or by considering functions composed of a linear combination of fixed basis functions in which only the coefficients of the linear combination can vary. In the case of applications to probabilistic inference, the restriction may for example take the form of factorization assumptions (Jordan *et al.*, 1999; Jaakkola, 2001).

Now let us consider in more detail how the concept of variational optimization can be applied to the inference problem. Suppose we have a fully Bayesian model in which all parameters are given prior distributions. The model may also have latent variables as well as parameters, and we shall denote the set of all latent variables and parameters by $\mathbf{Z}$. Similarly, we denote the set of all observed variables by $\mathbf{X}$. For example, we might have a set of N independent, identically distributed data, for which $\mathbf{X} = \{\mathbf{x}_1, \ldots, \mathbf{x}_N\}$ and $\mathbf{Z} = \{\mathbf{z}_1, \ldots, \mathbf{z}_N\}$. Our probabilistic model specifies the joint distribution $p(\mathbf{X}, \mathbf{Z})$, and our goal is to find an approximation for the posterior distribution $p(\mathbf{Z}|\mathbf{X})$ as well as for the model evidence $p(\mathbf{X})$. As in our discussion of EM, we can decompose the log marginal probability using

$$\ln p(\mathbf{X}) = \mathcal{L}(q) + \mathrm{KL}(q\|p) \tag{10.2}$$

where we have defined

$$\mathcal{L}(q) = \int q(\mathbf{Z}) \ln \left\{ \frac{p(\mathbf{X}, \mathbf{Z})}{q(\mathbf{Z})} \right\} \mathrm{d}\mathbf{Z} \tag{10.3}$$

$$\mathrm{KL}(q\|p) = -\int q(\mathbf{Z}) \ln \left\{ \frac{p(\mathbf{Z}|\mathbf{X})}{q(\mathbf{Z})} \right\} \mathrm{d}\mathbf{Z}. \tag{10.4}$$

This differs from our discussion of EM only in that the parameter vector $\boldsymbol{\theta}$ no longer appears, because the parameters are now stochastic variables and are absorbed into $\mathbf{Z}$. Since in this chapter we will mainly be interested in continuous variables we have used integrations rather than summations in formulating this decomposition. However, the analysis goes through unchanged if some or all of the variables are discrete simply by replacing the integrations with summations as required. As before, we can maximize the lower bound $\mathcal{L}(q)$ by optimization with respect to the distribution $q(\mathbf{Z})$, which is equivalent to minimizing the KL divergence. If we allow any possible choice for $q(\mathbf{Z})$, then the maximum of the lower bound occurs when the KL divergence vanishes, which occurs when $q(\mathbf{Z})$ equals the posterior distribution $p(\mathbf{Z}|\mathbf{X})$.

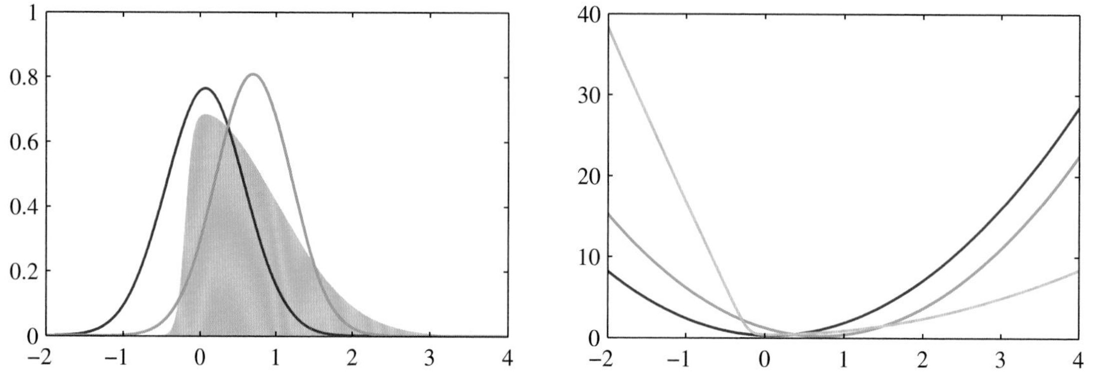

Figure 10.1 Illustration of the variational approximation for the example considered earlier in Figure 4.14. The left-hand plot shows the original distribution (yellow) along with the Laplace (red) and variational (green) approximations, and the right-hand plot shows the negative logarithms of the corresponding curves.

However, we shall suppose the model is such that working with the true posterior distribution is intractable.

We therefore consider instead a restricted family of distributions $q(\mathbf{Z})$ and then seek the member of this family for which the KL divergence is minimized. Our goal is to restrict the family sufficiently that they comprise only tractable distributions, while at the same time allowing the family to be sufficiently rich and flexible that it can provide a good approximation to the true posterior distribution. It is important to emphasize that the restriction is imposed purely to achieve tractability, and that subject to this requirement we should use as rich a family of approximating distributions as possible. In particular, there is no 'over-fitting' associated with highly flexible distributions. Using more flexible approximations simply allows us to approach the true posterior distribution more closely.

One way to restrict the family of approximating distributions is to use a parametric distribution $q(\mathbf{Z}|\boldsymbol{\omega})$ governed by a set of parameters $\boldsymbol{\omega}$. The lower bound $\mathcal{L}(q)$ then becomes a function of $\boldsymbol{\omega}$, and we can exploit standard nonlinear optimization techniques to determine the optimal values for the parameters. An example of this approach, in which the variational distribution is a Gaussian and we have optimized with respect to its mean and variance, is shown in Figure 10.1.

10.1.1 Factorized distributions

Here we consider an alternative way in which to restrict the family of distributions $q(\mathbf{Z})$. Suppose we partition the elements of $\mathbf{Z}$ into disjoint groups that we denote by $\mathbf{Z}_i$ where $i = 1, \ldots, M$. We then assume that the q distribution factorizes with respect to these groups, so that

$$q(\mathbf{Z}) = \prod_{i=1}^{M} q_i(\mathbf{Z}_i). \tag{10.5}$$

It should be emphasized that we are making no further assumptions about the distribution. In particular, we place no restriction on the functional forms of the individual factors $q_i(\mathbf{Z}_i)$. This factorized form of variational inference corresponds to an approximation framework developed in physics called *mean field theory* (Parisi, 1988).

Amongst all distributions $q(\mathbf{Z})$ having the form (10.5), we now seek that distribution for which the lower bound $\mathcal{L}(q)$ is largest. We therefore wish to make a free form (variational) optimization of $\mathcal{L}(q)$ with respect to all of the distributions $q_i(\mathbf{Z}_i)$, which we do by optimizing with respect to each of the factors in turn. To achieve this, we first substitute (10.5) into (10.3) and then dissect out the dependence on one of the factors $q_j(\mathbf{Z}_j)$. Denoting $q_j(\mathbf{Z}_j)$ by simply q_j to keep the notation uncluttered, we then obtain

$$\begin{aligned}\mathcal{L}(q) &= \int \prod_i q_i \left\{ \ln p(\mathbf{X}, \mathbf{Z}) - \sum_i \ln q_i \right\} \mathrm{d}\mathbf{Z} \\ &= \int q_j \left\{ \int \ln p(\mathbf{X}, \mathbf{Z}) \prod_{i \neq j} q_i \,\mathrm{d}\mathbf{Z}_i \right\} \mathrm{d}\mathbf{Z}_j - \int q_j \ln q_j \,\mathrm{d}\mathbf{Z}_j + \text{const} \\ &= \int q_j \ln \widetilde{p}(\mathbf{X}, \mathbf{Z}_j) \,\mathrm{d}\mathbf{Z}_j - \int q_j \ln q_j \,\mathrm{d}\mathbf{Z}_j + \text{const} \end{aligned} \tag{10.6}$$

where we have defined a new distribution $\widetilde{p}(\mathbf{X}, \mathbf{Z}_j)$ by the relation

$$\ln \widetilde{p}(\mathbf{X}, \mathbf{Z}_j) = \mathbb{E}_{i \neq j}[\ln p(\mathbf{X}, \mathbf{Z})] + \text{const}. \tag{10.7}$$

Here the notation $\mathbb{E}_{i \neq j}[\cdots]$ denotes an expectation with respect to the q distributions over all variables $\mathbf{z}_i$ for $i \neq j$, so that

$$\mathbb{E}_{i \neq j}[\ln p(\mathbf{X}, \mathbf{Z})] = \int \ln p(\mathbf{X}, \mathbf{Z}) \prod_{i \neq j} q_i \,\mathrm{d}\mathbf{Z}_i. \tag{10.8}$$

Now suppose we keep the $\{q_{i \neq j}\}$ fixed and maximize $\mathcal{L}(q)$ in (10.6) with respect to all possible forms for the distribution $q_j(\mathbf{Z}_j)$. This is easily done by recognizing that (10.6) is a negative Kullback-Leibler divergence between $q_j(\mathbf{Z}_j)$ and $\widetilde{p}(\mathbf{X}, \mathbf{Z}_j)$. Thus maximizing (10.6) is equivalent to minimizing the Kullback-Leibler

Leonhard Euler
1707–1783

Euler was a Swiss mathematician and physicist who worked in St. Petersburg and Berlin and who is widely considered to be one of the greatest mathematicians of all time. He is certainly the most prolific, and his collected works fill 75 volumes. Amongst his many contributions, he formulated the modern theory of the function, he developed (together with Lagrange) the calculus of variations, and he discovered the formula $e^{i\pi} = -1$, which relates four of the most important numbers in mathematics. During the last 17 years of his life, he was almost totally blind, and yet he produced nearly half of his results during this period.

divergence, and the minimum occurs when $q_j(\mathbf{Z}_j) = \widetilde{p}(\mathbf{X}, \mathbf{Z}_j)$. Thus we obtain a general expression for the optimal solution $q_j^\star(\mathbf{Z}_j)$ given by

$$\ln q_j^\star(\mathbf{Z}_j) = \mathbb{E}_{i \neq j}[\ln p(\mathbf{X}, \mathbf{Z})] + \text{const}. \tag{10.9}$$

It is worth taking a few moments to study the form of this solution as it provides the basis for applications of variational methods. It says that the log of the optimal solution for factor q_j is obtained simply by considering the log of the joint distribution over all hidden and visible variables and then taking the expectation with respect to all of the other factors $\{q_i\}$ for $i \neq j$.

The additive constant in (10.9) is set by normalizing the distribution $q_j^\star(\mathbf{Z}_j)$. Thus if we take the exponential of both sides and normalize, we have

$$q_j^\star(\mathbf{Z}_j) = \frac{\exp\left(\mathbb{E}_{i \neq j}[\ln p(\mathbf{X}, \mathbf{Z})]\right)}{\int \exp\left(\mathbb{E}_{i \neq j}[\ln p(\mathbf{X}, \mathbf{Z})]\right) \mathrm{d}\mathbf{Z}_j}.$$

In practice, we shall find it more convenient to work with the form (10.9) and then reinstate the normalization constant (where required) by inspection. This will become clear from subsequent examples.

The set of equations given by (10.9) for $j = 1, \ldots, M$ represent a set of consistency conditions for the maximum of the lower bound subject to the factorization constraint. However, they do not represent an explicit solution because the expression on the right-hand side of (10.9) for the optimum $q_j^\star(\mathbf{Z}_j)$ depends on expectations computed with respect to the other factors $q_i(\mathbf{Z}_i)$ for $i \neq j$. We will therefore seek a consistent solution by first initializing all of the factors $q_i(\mathbf{Z}_i)$ appropriately and then cycling through the factors and replacing each in turn with a revised estimate given by the right-hand side of (10.9) evaluated using the current estimates for all of the other factors. Convergence is guaranteed because bound is convex with respect to each of the factors $q_i(\mathbf{Z}_i)$ (Boyd and Vandenberghe, 2004).

10.1.2 Properties of factorized approximations

Our approach to variational inference is based on a factorized approximation to the true posterior distribution. Let us consider for a moment the problem of approximating a general distribution by a factorized distribution. To begin with, we discuss the problem of approximating a Gaussian distribution using a factorized Gaussian, which will provide useful insight into the types of inaccuracy introduced in using factorized approximations. Consider a Gaussian distribution $p(\mathbf{z}) = \mathcal{N}(\mathbf{z}|\boldsymbol{\mu}, \boldsymbol{\Lambda}^{-1})$ over two correlated variables $\mathbf{z} = (z_1, z_2)$ in which the mean and precision have elements

$$\boldsymbol{\mu} = \begin{pmatrix} \mu_1 \\ \mu_2 \end{pmatrix}, \qquad \boldsymbol{\Lambda} = \begin{pmatrix} \Lambda_{11} & \Lambda_{12} \\ \Lambda_{21} & \Lambda_{22} \end{pmatrix} \tag{10.10}$$

and $\Lambda_{21} = \Lambda_{12}$ due to the symmetry of the precision matrix. Now suppose we wish to approximate this distribution using a factorized Gaussian of the form $q(\mathbf{z}) = q_1(z_1)q_2(z_2)$. We first apply the general result (10.9) to find an expression for the optimal factor $q_1^\star(z_1)$. In doing so it is useful to note that on the right-hand side we only need to retain those terms that have some functional dependence on z_1 because all other terms can be absorbed into the normalization constant. Thus we have

$$\begin{aligned}\ln q_1^\star(z_1) &= \mathbb{E}_{z_2}[\ln p(\mathbf{z})] + \text{const} \\ &= \mathbb{E}_{z_2}\left[-\frac{1}{2}(z_1-\mu_1)^2\Lambda_{11} - (z_1-\mu_1)\Lambda_{12}(z_2-\mu_2)\right] + \text{const} \\ &= -\frac{1}{2}z_1^2\Lambda_{11} + z_1\mu_1\Lambda_{11} - z_1\Lambda_{12}\left(\mathbb{E}[z_2]-\mu_2\right) + \text{const.} \qquad (10.11)\end{aligned}$$

Next we observe that the right-hand side of this expression is a quadratic function of z_1, and so we can identify $q^\star(z_1)$ as a Gaussian distribution. It is worth emphasizing that we did not assume that $q(z_i)$ is Gaussian, but rather we derived this result by variational optimization of the KL divergence over all possible distributions $q(z_i)$. Note also that we do not need to consider the additive constant in (10.9) explicitly because it represents the normalization constant that can be found at the end by inspection if required. Using the technique of completing the square, we can identify the mean and precision of this Gaussian, giving

Section 2.3.1

$$q_1^\star(z_1) = \mathcal{N}(z_1|m_1, \Lambda_{11}^{-1}) \qquad (10.12)$$

where

$$m_1 = \mu_1 - \Lambda_{11}^{-1}\Lambda_{12}\left(\mathbb{E}[z_2]-\mu_2\right). \qquad (10.13)$$

By symmetry, $q_2^\star(z_2)$ is also Gaussian and can be written as

$$q_2^\star(z_2) = \mathcal{N}(z_2|m_2, \Lambda_{22}^{-1}) \qquad (10.14)$$

in which

$$m_2 = \mu_2 - \Lambda_{22}^{-1}\Lambda_{21}\left(\mathbb{E}[z_1]-\mu_1\right). \qquad (10.15)$$

Note that these solutions are coupled, so that $q^\star(z_1)$ depends on expectations computed with respect to $q^\star(z_2)$ and vice versa. In general, we address this by treating the variational solutions as re-estimation equations and cycling through the variables in turn updating them until some convergence criterion is satisfied. We shall see an example of this shortly. Here, however, we note that the problem is sufficiently simple that a closed form solution can be found. In particular, because $\mathbb{E}[z_1] = m_1$ and $\mathbb{E}[z_2] = m_2$, we see that the two equations are satisfied if we take $\mathbb{E}[z_1] = \mu_1$ and $\mathbb{E}[z_2] = \mu_2$, and it is easily shown that this is the only solution provided the distribution is nonsingular. This result is illustrated in Figure 10.2(a). We see that the mean is correctly captured but that the variance of $q(\mathbf{z})$ is controlled by the direction of smallest variance of $p(\mathbf{z})$, and that the variance along the orthogonal direction is significantly under-estimated. It is a general result that a factorized variational approximation tends to give approximations to the posterior distribution that are too compact.

Exercise 10.2

By way of comparison, suppose instead that we had been minimizing the reverse Kullback-Leibler divergence $\mathrm{KL}(p\|q)$. As we shall see, this form of KL divergence

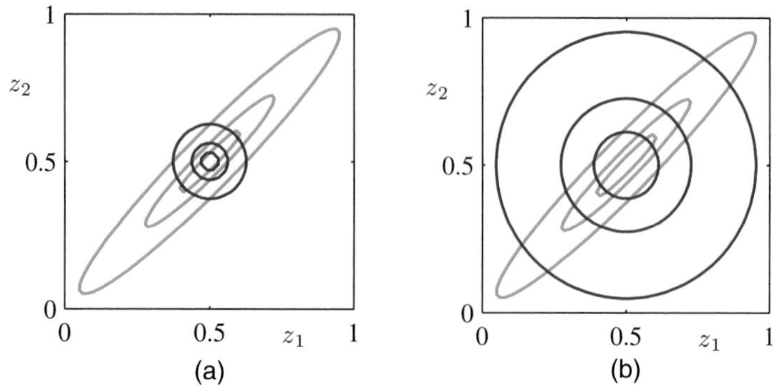

Figure 10.2 Comparison of the two alternative forms for the Kullback-Leibler divergence. The green contours corresponding to 1, 2, and 3 standard deviations for a correlated Gaussian distribution $p(\mathbf{z})$ over two variables z_1 and z_2, and the red contours represent the corresponding levels for an approximating distribution $q(\mathbf{z})$ over the same variables given by the product of two independent univariate Gaussian distributions whose parameters are obtained by minimization of (a) the Kullback-Leibler divergence $\mathrm{KL}(q\|p)$, and (b) the reverse Kullback-Leibler divergence $\mathrm{KL}(p\|q)$.

Section 10.7

is used in an alternative approximate inference framework called *expectation propagation*. We therefore consider the general problem of minimizing $\mathrm{KL}(p\|q)$ when $q(\mathbf{Z})$ is a factorized approximation of the form (10.5). The KL divergence can then be written in the form

$$\mathrm{KL}(p\|q) = -\int p(\mathbf{Z}) \left[\sum_{i=1}^{M} \ln q_i(\mathbf{Z}_i) \right] \mathrm{d}\mathbf{Z} + \mathrm{const} \tag{10.16}$$

Exercise 10.3

where the constant term is simply the entropy of $p(\mathbf{Z})$ and so does not depend on $q(\mathbf{Z})$. We can now optimize with respect to each of the factors $q_j(\mathbf{Z}_j)$, which is easily done using a Lagrange multiplier to give

$$q_j^\star(\mathbf{Z}_j) = \int p(\mathbf{Z}) \prod_{i \neq j} \mathrm{d}\mathbf{Z}_i = p(\mathbf{Z}_j). \tag{10.17}$$

In this case, we find that the optimal solution for $q_j(\mathbf{Z}_j)$ is just given by the corresponding marginal distribution of $p(\mathbf{Z})$. Note that this is a closed-form solution and so does not require iteration.

To apply this result to the illustrative example of a Gaussian distribution $p(\mathbf{z})$ over a vector $\mathbf{z}$ we can use (2.98), which gives the result shown in Figure 10.2(b). We see that once again the mean of the approximation is correct, but that it places significant probability mass in regions of variable space that have very low probability.

The difference between these two results can be understood by noting that there is a large positive contribution to the Kullback-Leibler divergence

$$\mathrm{KL}(q\|p) = -\int q(\mathbf{Z}) \ln \left\{ \frac{p(\mathbf{Z})}{q(\mathbf{Z})} \right\} \mathrm{d}\mathbf{Z} \tag{10.18}$$

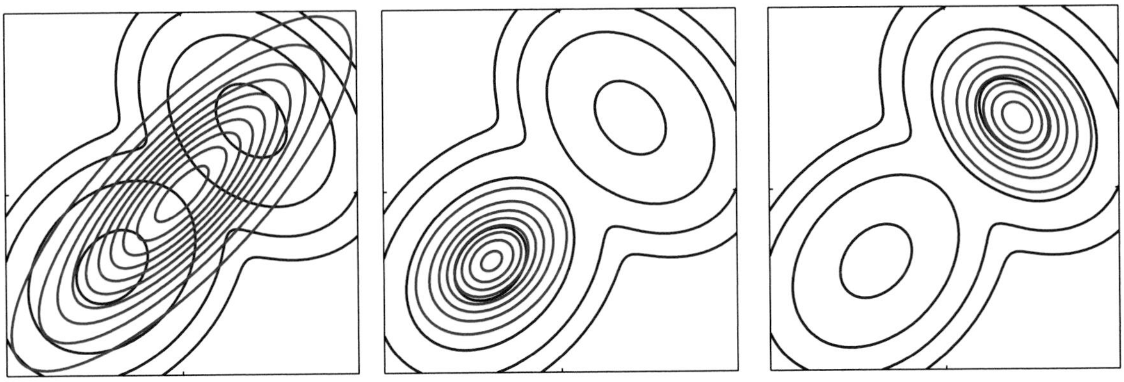

Figure 10.3 Another comparison of the two alternative forms for the Kullback-Leibler divergence. (a) The blue contours show a bimodal distribution $p(\mathbf{Z})$ given by a mixture of two Gaussians, and the red contours correspond to the single Gaussian distribution $q(\mathbf{Z})$ that best approximates $p(\mathbf{Z})$ in the sense of minimizing the Kullback-Leibler divergence $\mathrm{KL}(p\|q)$. (b) As in (a) but now the red contours correspond to a Gaussian distribution $q(\mathbf{Z})$ found by numerical minimization of the Kullback-Leibler divergence $\mathrm{KL}(q\|p)$. (c) As in (b) but showing a different local minimum of the Kullback-Leibler divergence.

from regions of $\mathbf{Z}$ space in which $p(\mathbf{Z})$ is near zero unless $q(\mathbf{Z})$ is also close to zero. Thus minimizing this form of KL divergence leads to distributions $q(\mathbf{Z})$ that avoid regions in which $p(\mathbf{Z})$ is small. Conversely, the Kullback-Leibler divergence $\mathrm{KL}(p\|q)$ is minimized by distributions $q(\mathbf{Z})$ that are nonzero in regions where $p(\mathbf{Z})$ is nonzero.

We can gain further insight into the different behaviour of the two KL divergences if we consider approximating a multimodal distribution by a unimodal one, as illustrated in Figure 10.3. In practical applications, the true posterior distribution will often be multimodal, with most of the posterior mass concentrated in some number of relatively small regions of parameter space. These multiple modes may arise through nonidentifiability in the latent space or through complex nonlinear dependence on the parameters. Both types of multimodality were encountered in Chapter 9 in the context of Gaussian mixtures, where they manifested themselves as multiple maxima in the likelihood function, and a variational treatment based on the minimization of $\mathrm{KL}(q\|p)$ will tend to find one of these modes. By contrast, if we were to minimize $\mathrm{KL}(p\|q)$, the resulting approximations would average across all of the modes and, in the context of the mixture model, would lead to poor predictive distributions (because the average of two good parameter values is typically itself not a good parameter value). It is possible to make use of $\mathrm{KL}(p\|q)$ to define a useful inference procedure, but this requires a rather different approach to the one discussed here, and will be considered in detail when we discuss expectation propagation.

Section 10.7

The two forms of Kullback-Leibler divergence are members of the *alpha family* of divergences (Ali and Silvey, 1966; Amari, 1985; Minka, 2005) defined by

$$\mathrm{D}_\alpha(p\|q) = \frac{4}{1-\alpha^2}\left(1 - \int p(x)^{(1+\alpha)/2} q(x)^{(1-\alpha)/2}\,\mathrm{d}x\right) \tag{10.19}$$

where $-\infty < \alpha < \infty$ is a continuous parameter. The Kullback-Leibler divergence $\mathrm{KL}(p\|q)$ corresponds to the limit $\alpha \to 1$, whereas $\mathrm{KL}(q\|p)$ corresponds to the limit $\alpha \to -1$. For all values of α we have $\mathrm{D}_\alpha(p\|q) \geqslant 0$, with equality if, and only if, $p(x) = q(x)$. Suppose $p(x)$ is a fixed distribution, and we minimize $\mathrm{D}_\alpha(p\|q)$ with respect to some set of distributions $q(x)$. Then for $\alpha \leqslant -1$ the divergence is *zero forcing*, so that any values of x for which $p(x) = 0$ will have $q(x) = 0$, and typically $q(x)$ will under-estimate the support of $p(x)$ and will tend to seek the mode with the largest mass. Conversely for $\alpha \geqslant 1$ the divergence is *zero-avoiding*, so that values of x for which $p(x) > 0$ will have $q(x) > 0$, and typically $q(x)$ will stretch to cover all of $p(x)$, and will over-estimate the support of $p(x)$. When $\alpha = 0$ we obtain a symmetric divergence that is linearly related to the *Hellinger distance* given by

Exercise 10.6

$$\mathrm{D}_\mathrm{H}(p\|q) = \int \left(p(x)^{1/2} - q(x)^{1/2}\right)^2 \mathrm{d}x. \tag{10.20}$$

The square root of the Hellinger distance is a valid distance metric.

10.1.3 Example: The univariate Gaussian

We now illustrate the factorized variational approximation using a Gaussian distribution over a single variable x (MacKay, 2003). Our goal is to infer the posterior distribution for the mean μ and precision τ, given a data set $\mathcal{D} = \{x_1, \ldots, x_N\}$ of observed values of x which are assumed to be drawn independently from the Gaussian. The likelihood function is given by

$$p(\mathcal{D}|\mu, \tau) = \left(\frac{\tau}{2\pi}\right)^{N/2} \exp\left\{-\frac{\tau}{2}\sum_{n=1}^{N}(x_n - \mu)^2\right\}. \tag{10.21}$$

We now introduce conjugate prior distributions for μ and τ given by

$$p(\mu|\tau) = \mathcal{N}\left(\mu|\mu_0, (\lambda_0\tau)^{-1}\right) \tag{10.22}$$

$$p(\tau) = \mathrm{Gam}(\tau|a_0, b_0) \tag{10.23}$$

where $\mathrm{Gam}(\tau|a_0, b_0)$ is the gamma distribution defined by (2.146). Together these distributions constitute a Gaussian-Gamma conjugate prior distribution.

Section 2.3.6

For this simple problem the posterior distribution can be found exactly, and again takes the form of a Gaussian-gamma distribution. However, for tutorial purposes we will consider a factorized variational approximation to the posterior distribution given by

Exercise 2.44

$$q(\mu, \tau) = q_\mu(\mu) q_\tau(\tau). \tag{10.24}$$

Note that the true posterior distribution does not factorize in this way. The optimum factors $q_\mu(\mu)$ and $q_\tau(\tau)$ can be obtained from the general result (10.9) as follows. For $q_\mu(\mu)$ we have

$$\begin{aligned}\ln q_\mu^\star(\mu) &= \mathbb{E}_\tau\left[\ln p(\mathcal{D}|\mu,\tau) + \ln p(\mu|\tau)\right] + \text{const} \\ &= -\frac{\mathbb{E}[\tau]}{2}\left\{\lambda_0(\mu-\mu_0)^2 + \sum_{n=1}^N (x_n-\mu)^2\right\} + \text{const}. \quad (10.25)\end{aligned}$$

Exercise 10.7

Completing the square over μ we see that $q_\mu(\mu)$ is a Gaussian $\mathcal{N}\left(\mu|\mu_N, \lambda_N^{-1}\right)$ with mean and precision given by

$$\mu_N = \frac{\lambda_0\mu_0 + N\overline{x}}{\lambda_0 + N} \quad (10.26)$$

$$\lambda_N = (\lambda_0 + N)\mathbb{E}[\tau]. \quad (10.27)$$

Note that for $N \to \infty$ this gives the maximum likelihood result in which $\mu_N = \overline{x}$ and the precision is infinite.

Similarly, the optimal solution for the factor $q_\tau(\tau)$ is given by

$$\begin{aligned}\ln q_\tau^\star(\tau) &= \mathbb{E}_\mu\left[\ln p(\mathcal{D}|\mu,\tau) + \ln p(\mu|\tau)\right] + \ln p(\tau) + \text{const} \\ &= (a_0 - 1)\ln\tau - b_0\tau + \frac{1}{2}\ln\tau + \frac{N}{2}\ln\tau \\ &\quad -\frac{\tau}{2}\mathbb{E}_\mu\left[\sum_{n=1}^N (x_n-\mu)^2 + \lambda_0(\mu-\mu_0)^2\right] + \text{const} \quad (10.28)\end{aligned}$$

and hence $q_\tau(\tau)$ is a gamma distribution $\text{Gam}(\tau|a_N, b_N)$ with parameters

$$a_N = a_0 + \frac{N+1}{2} \quad (10.29)$$

$$b_N = b_0 + \frac{1}{2}\mathbb{E}_\mu\left[\sum_{n=1}^N (x_n-\mu)^2 + \lambda_0(\mu-\mu_0)^2\right]. \quad (10.30)$$

Exercise 10.8

Again this exhibits the expected behaviour when $N \to \infty$.

It should be emphasized that we did not assume these specific functional forms for the optimal distributions $q_\mu(\mu)$ and $q_\tau(\tau)$. They arose naturally from the structure of the likelihood function and the corresponding conjugate priors.

Section 10.4.1

Thus we have expressions for the optimal distributions $q_\mu(\mu)$ and $q_\tau(\tau)$ each of which depends on moments evaluated with respect to the other distribution. One approach to finding a solution is therefore to make an initial guess for, say, the moment $\mathbb{E}[\tau]$ and use this to re-compute the distribution $q_\mu(\mu)$. Given this revised distribution we can then extract the required moments $\mathbb{E}[\mu]$ and $\mathbb{E}[\mu^2]$, and use these to recompute the distribution $q_\tau(\tau)$, and so on. Since the space of hidden variables for this example is only two dimensional, we can illustrate the variational approximation to the posterior distribution by plotting contours of both the true posterior and the factorized approximation, as illustrated in Figure 10.4.

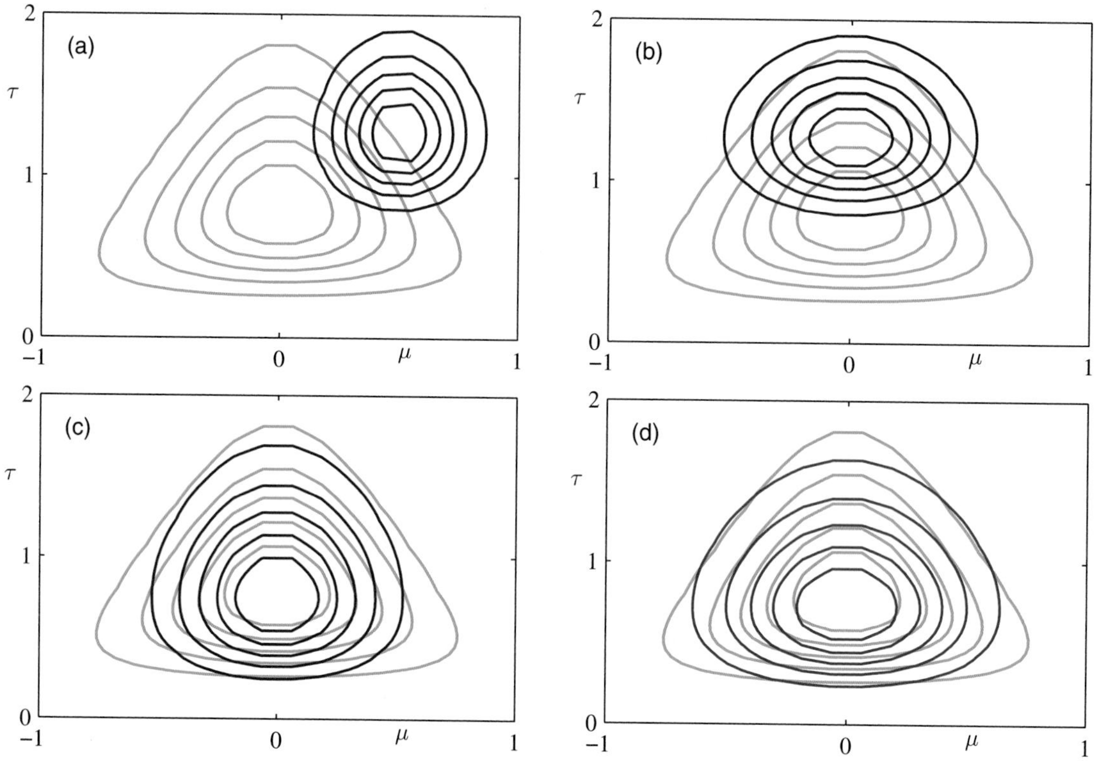

Figure 10.4 Illustration of variational inference for the mean μ and precision τ of a univariate Gaussian distribution. Contours of the true posterior distribution $p(\mu, \tau|D)$ are shown in green. (a) Contours of the initial factorized approximation $q_\mu(\mu)q_\tau(\tau)$ are shown in blue. (b) After re-estimating the factor $q_\mu(\mu)$. (c) After re-estimating the factor $q_\tau(\tau)$. (d) Contours of the optimal factorized approximation, to which the iterative scheme converges, are shown in red.

In general, we will need to use an iterative approach such as this in order to solve for the optimal factorized posterior distribution. For the very simple example we are considering here, however, we can find an explicit solution by solving the simultaneous equations for the optimal factors $q_\mu(\mu)$ and $q_\tau(\tau)$. Before doing this, we can simplify these expressions by considering broad, noninformative priors in which $\mu_0 = a_0 = b_0 = \lambda_0 = 0$. Although these parameter settings correspond to improper priors, we see that the posterior distribution is still well defined. Using the standard result $\mathbb{E}[\tau] = a_N/b_N$ for the mean of a gamma distribution, together with (10.29) and (10.30), we have

Appendix B

$$\frac{1}{\mathbb{E}[\tau]} = \mathbb{E}\left[\frac{1}{N+1}\sum_{n=1}^{N}(x_n - \mu)^2\right] = \frac{N}{N+1}\left(\overline{x^2} - 2\overline{x}\mathbb{E}[\mu] + \mathbb{E}[\mu^2]\right). \quad (10.31)$$

Then, using (10.26) and (10.27), we obtain the first and second order moments of

$q_\mu(\mu)$ in the form

$$\mathbb{E}[\mu] = \overline{x}, \qquad \mathbb{E}[\mu^2] = \overline{x}^2 + \frac{1}{N\mathbb{E}[\tau]}. \tag{10.32}$$

Exercise 10.9

We can now substitute these moments into (10.31) and then solve for $\mathbb{E}[\tau]$ to give

$$\frac{1}{\mathbb{E}[\tau]} = (\overline{x^2} - \overline{x}^2) = \frac{1}{N}\sum_{n=1}^{N}(x_n - \overline{x})^2. \tag{10.33}$$

For a comprehensive treatment of Bayesian inference for the Gaussian distribution, including a discussion of the advantages compared to maximum likelihood, see Minka (1998).

10.1.4 Model comparison

As well as performing inference over the hidden variables $\mathbf{Z}$, we may also wish to compare a set of candidate models, labelled by the index m, and having prior probabilities $p(m)$. Our goal is then to approximate the posterior probabilities $p(m|\mathbf{X})$, where $\mathbf{X}$ is the observed data. This is a slightly more complex situation than that considered so far because different models may have different structure and indeed different dimensionality for the hidden variables $\mathbf{Z}$. We cannot therefore simply consider a factorized approximation $q(\mathbf{Z})q(m)$, but must instead recognize that the posterior over $\mathbf{Z}$ must be conditioned on m, and so we must consider $q(\mathbf{Z}, m) = q(\mathbf{Z}|m)q(m)$. We can readily verify the following decomposition based on this variational distribution

Exercise 10.10

$$\ln p(\mathbf{X}) = \mathcal{L} - \sum_m \sum_{\mathbf{Z}} q(\mathbf{Z}|m)q(m) \ln \left\{ \frac{p(\mathbf{Z}, m|\mathbf{X})}{q(\mathbf{Z}|m)q(m)} \right\} \tag{10.34}$$

where the $\mathcal{L}$ is a lower bound on $\ln p(\mathbf{X})$ and is given by

$$\mathcal{L} = \sum_m \sum_{\mathbf{Z}} q(\mathbf{Z}|m)q(m) \ln \left\{ \frac{p(\mathbf{Z}, \mathbf{X}, m)}{q(\mathbf{Z}|m)q(m)} \right\}. \tag{10.35}$$

Here we are assuming discrete $\mathbf{Z}$, but the same analysis applies to continuous latent variables provided the summations are replaced with integrations. We can maximize $\mathcal{L}$ with respect to the distribution $q(m)$ using a Lagrange multiplier, with the result

Exercise 10.11

$$q(m) \propto p(m) \exp\{\mathcal{L}_m\} \tag{10.36}$$

where

$$\mathcal{L}_m = \sum_{\mathbf{Z}} q(\mathbf{Z}|m) \ln \left\{ \frac{p(\mathbf{Z}, \mathbf{X}|m)}{q(\mathbf{Z}|m)} \right\}.$$

However, if we maximize $\mathcal{L}$ with respect to the $q(\mathbf{Z}|m)$, we find that the solutions for different m are coupled, as we expect because they are conditioned on m. We proceed instead by first optimizing each of the $q(\mathbf{Z}|m)$ individually by optimization of (10.35), or equivalently by optimization of $\mathcal{L}_m$, and then subsequently determining the $q(m)$ using (10.36). After normalization the resulting values for $q(m)$ can be used for model selection or model averaging in the usual way.

10.2. Illustration: Variational Mixture of Gaussians

We now return to our discussion of the Gaussian mixture model and apply the variational inference machinery developed in the previous section. This will provide a good illustration of the application of variational methods and will also demonstrate how a Bayesian treatment elegantly resolves many of the difficulties associated with the maximum likelihood approach (Attias, 1999b). The reader is encouraged to work through this example in detail as it provides many insights into the practical application of variational methods. Many Bayesian models, corresponding to much more sophisticated distributions, can be solved by straightforward extensions and generalizations of this analysis.

Our starting point is the likelihood function for the Gaussian mixture model, illustrated by the graphical model in Figure 9.6. For each observation $\mathbf{x}_n$ we have a corresponding latent variable $\mathbf{z}_n$ comprising a 1-of-K binary vector with elements z_{nk} for $k = 1, \ldots, K$. As before we denote the observed data set by $\mathbf{X} = \{\mathbf{x}_1, \ldots, \mathbf{x}_N\}$, and similarly we denote the latent variables by $\mathbf{Z} = \{\mathbf{z}_1, \ldots, \mathbf{z}_N\}$. From (9.10) we can write down the conditional distribution of $\mathbf{Z}$, given the mixing coefficients $\boldsymbol{\pi}$, in the form

$$p(\mathbf{Z}|\boldsymbol{\pi}) = \prod_{n=1}^{N} \prod_{k=1}^{K} \pi_k^{z_{nk}}. \tag{10.37}$$

Similarly, from (9.11), we can write down the conditional distribution of the observed data vectors, given the latent variables and the component parameters

$$p(\mathbf{X}|\mathbf{Z}, \boldsymbol{\mu}, \boldsymbol{\Lambda}) = \prod_{n=1}^{N} \prod_{k=1}^{K} \mathcal{N}\left(\mathbf{x}_n|\boldsymbol{\mu}_k, \boldsymbol{\Lambda}_k^{-1}\right)^{z_{nk}} \tag{10.38}$$

where $\boldsymbol{\mu} = \{\boldsymbol{\mu}_k\}$ and $\boldsymbol{\Lambda} = \{\boldsymbol{\Lambda}_k\}$. Note that we are working in terms of precision matrices rather than covariance matrices as this somewhat simplifies the mathematics.

Section 10.4.1

Next we introduce priors over the parameters $\boldsymbol{\mu}$, $\boldsymbol{\Lambda}$ and $\boldsymbol{\pi}$. The analysis is considerably simplified if we use conjugate prior distributions. We therefore choose a Dirichlet distribution over the mixing coefficients $\boldsymbol{\pi}$

$$p(\boldsymbol{\pi}) = \mathrm{Dir}(\boldsymbol{\pi}|\boldsymbol{\alpha}_0) = C(\boldsymbol{\alpha}_0) \prod_{k=1}^{K} \pi_k^{\alpha_0 - 1} \tag{10.39}$$

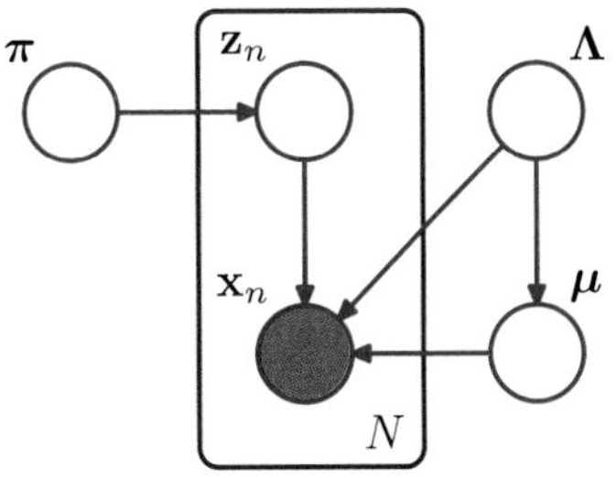

Figure 10.5 Directed acyclic graph representing the Bayesian mixture of Gaussians model, in which the box (plate) denotes a set of N i.i.d. observations. Here $\boldsymbol{\mu}$ denotes $\{\boldsymbol{\mu}_k\}$ and $\boldsymbol{\Lambda}$ denotes $\{\boldsymbol{\Lambda}_k\}$.

where by symmetry we have chosen the same parameter α_0 for each of the components, and $C(\boldsymbol{\alpha}_0)$ is the normalization constant for the Dirichlet distribution defined by (B.23). As we have seen, the parameter α_0 can be interpreted as the effective prior number of observations associated with each component of the mixture. If the value of α_0 is small, then the posterior distribution will be influenced primarily by the data rather than by the prior.

Section 2.2.1

Similarly, we introduce an independent Gaussian-Wishart prior governing the mean and precision of each Gaussian component, given by

$$\begin{aligned} p(\boldsymbol{\mu}, \boldsymbol{\Lambda}) &= p(\boldsymbol{\mu}|\boldsymbol{\Lambda})p(\boldsymbol{\Lambda}) \\ &= \prod_{k=1}^{K} \mathcal{N}\left(\boldsymbol{\mu}_k|\mathbf{m}_0, (\beta_0\boldsymbol{\Lambda}_k)^{-1}\right) \mathcal{W}(\boldsymbol{\Lambda}_k|\mathbf{W}_0, \nu_0) \end{aligned} \tag{10.40}$$

because this represents the conjugate prior distribution when both the mean and precision are unknown. Typically we would choose $\mathbf{m}_0 = \mathbf{0}$ by symmetry.

Section 2.3.6

The resulting model can be represented as a directed graph as shown in Figure 10.5. Note that there is a link from $\boldsymbol{\Lambda}$ to $\boldsymbol{\mu}$ since the variance of the distribution over $\boldsymbol{\mu}$ in (10.40) is a function of $\boldsymbol{\Lambda}$.

This example provides a nice illustration of the distinction between latent variables and parameters. Variables such as $\mathbf{z}_n$ that appear inside the plate are regarded as latent variables because the number of such variables grows with the size of the data set. By contrast, variables such as $\boldsymbol{\mu}$ that are outside the plate are fixed in number independently of the size of the data set, and so are regarded as parameters. From the perspective of graphical models, however, there is really no fundamental difference between them.

10.2.1 Variational distribution

In order to formulate a variational treatment of this model, we next write down the joint distribution of all of the random variables, which is given by

$$p(\mathbf{X}, \mathbf{Z}, \boldsymbol{\pi}, \boldsymbol{\mu}, \boldsymbol{\Lambda}) = p(\mathbf{X}|\mathbf{Z}, \boldsymbol{\mu}, \boldsymbol{\Lambda})p(\mathbf{Z}|\boldsymbol{\pi})p(\boldsymbol{\pi})p(\boldsymbol{\mu}|\boldsymbol{\Lambda})p(\boldsymbol{\Lambda}) \tag{10.41}$$

in which the various factors are defined above. The reader should take a moment to verify that this decomposition does indeed correspond to the probabilistic graphical model shown in Figure 10.5. Note that only the variables $\mathbf{X} = \{\mathbf{x}_1, \ldots, \mathbf{x}_N\}$ are observed.

We now consider a variational distribution which factorizes between the latent variables and the parameters so that

$$q(\mathbf{Z}, \boldsymbol{\pi}, \boldsymbol{\mu}, \boldsymbol{\Lambda}) = q(\mathbf{Z})q(\boldsymbol{\pi}, \boldsymbol{\mu}, \boldsymbol{\Lambda}). \tag{10.42}$$

It is remarkable that this is the *only* assumption that we need to make in order to obtain a tractable practical solution to our Bayesian mixture model. In particular, the functional form of the factors $q(\mathbf{Z})$ and $q(\boldsymbol{\pi}, \boldsymbol{\mu}, \boldsymbol{\Lambda})$ will be determined automatically by optimization of the variational distribution. Note that we are omitting the subscripts on the q distributions, much as we do with the p distributions in (10.41), and are relying on the arguments to distinguish the different distributions.

The corresponding sequential update equations for these factors can be easily derived by making use of the general result (10.9). Let us consider the derivation of the update equation for the factor $q(\mathbf{Z})$. The log of the optimized factor is given by

$$\ln q^{\star}(\mathbf{Z}) = \mathbb{E}_{\boldsymbol{\pi},\boldsymbol{\mu},\boldsymbol{\Lambda}}[\ln p(\mathbf{X}, \mathbf{Z}, \boldsymbol{\pi}, \boldsymbol{\mu}, \boldsymbol{\Lambda})] + \text{const}. \tag{10.43}$$

We now make use of the decomposition (10.41). Note that we are only interested in the functional dependence of the right-hand side on the variable $\mathbf{Z}$. Thus any terms that do not depend on $\mathbf{Z}$ can be absorbed into the additive normalization constant, giving

$$\ln q^{\star}(\mathbf{Z}) = \mathbb{E}_{\boldsymbol{\pi}}[\ln p(\mathbf{Z}|\boldsymbol{\pi})] + \mathbb{E}_{\boldsymbol{\mu},\boldsymbol{\Lambda}}[\ln p(\mathbf{X}|\mathbf{Z}, \boldsymbol{\mu}, \boldsymbol{\Lambda})] + \text{const}. \tag{10.44}$$

Substituting for the two conditional distributions on the right-hand side, and again absorbing any terms that are independent of $\mathbf{Z}$ into the additive constant, we have

$$\ln q^{\star}(\mathbf{Z}) = \sum_{n=1}^{N} \sum_{k=1}^{K} z_{nk} \ln \rho_{nk} + \text{const} \tag{10.45}$$

where we have defined

$$\begin{aligned}\ln \rho_{nk} \quad = \quad & \mathbb{E}[\ln \pi_k] + \frac{1}{2}\mathbb{E}\left[\ln |\boldsymbol{\Lambda}_k|\right] - \frac{D}{2}\ln(2\pi) \\ & -\frac{1}{2}\mathbb{E}_{\boldsymbol{\mu}_k,\boldsymbol{\Lambda}_k}\left[(\mathbf{x}_n - \boldsymbol{\mu}_k)^{\mathrm{T}}\boldsymbol{\Lambda}_k(\mathbf{x}_n - \boldsymbol{\mu}_k)\right]\end{aligned} \tag{10.46}$$

where D is the dimensionality of the data variable $\mathbf{x}$. Taking the exponential of both sides of (10.45) we obtain

$$q^{\star}(\mathbf{Z}) \propto \prod_{n=1}^{N} \prod_{k=1}^{K} \rho_{nk}^{z_{nk}}. \tag{10.47}$$

Exercise 10.12

Requiring that this distribution be normalized, and noting that for each value of n the quantities z_{nk} are binary and sum to 1 over all values of k, we obtain

$$q^{\star}(\mathbf{Z}) = \prod_{n=1}^{N} \prod_{k=1}^{K} r_{nk}^{z_{nk}} \tag{10.48}$$

where

$$r_{nk} = \frac{\rho_{nk}}{\displaystyle\sum_{j=1}^{K} \rho_{nj}}. \tag{10.49}$$

We see that the optimal solution for the factor $q(\mathbf{Z})$ takes the same functional form as the prior $p(\mathbf{Z}|\boldsymbol{\pi})$. Note that because ρ_{nk} is given by the exponential of a real quantity, the quantities r_{nk} will be nonnegative and will sum to one, as required.

For the discrete distribution $q^\star(\mathbf{Z})$ we have the standard result

$$\mathbb{E}[z_{nk}] = r_{nk} \tag{10.50}$$

from which we see that the quantities r_{nk} are playing the role of responsibilities. Note that the optimal solution for $q^\star(\mathbf{Z})$ depends on moments evaluated with respect to the distributions of other variables, and so again the variational update equations are coupled and must be solved iteratively.

At this point, we shall find it convenient to define three statistics of the observed data set evaluated with respect to the responsibilities, given by

$$N_k = \sum_{n=1}^{N} r_{nk} \tag{10.51}$$

$$\overline{\mathbf{x}}_k = \frac{1}{N_k} \sum_{n=1}^{N} r_{nk} \mathbf{x}_n \tag{10.52}$$

$$\mathbf{S}_k = \frac{1}{N_k} \sum_{n=1}^{N} r_{nk} (\mathbf{x}_n - \overline{\mathbf{x}}_k)(\mathbf{x}_n - \overline{\mathbf{x}}_k)^{\mathrm{T}}. \tag{10.53}$$

Note that these are analogous to quantities evaluated in the maximum likelihood EM algorithm for the Gaussian mixture model.

Now let us consider the factor $q(\boldsymbol{\pi}, \boldsymbol{\mu}, \boldsymbol{\Lambda})$ in the variational posterior distribution. Again using the general result (10.9) we have

$$\begin{aligned} \ln q^\star(\boldsymbol{\pi}, \boldsymbol{\mu}, \boldsymbol{\Lambda}) &= \ln p(\boldsymbol{\pi}) + \sum_{k=1}^{K} \ln p(\boldsymbol{\mu}_k, \boldsymbol{\Lambda}_k) + \mathbb{E}_{\mathbf{Z}} \left[\ln p(\mathbf{Z}|\boldsymbol{\pi})\right] \\ &\quad + \sum_{k=1}^{K} \sum_{n=1}^{N} \mathbb{E}[z_{nk}] \ln \mathcal{N}\left(\mathbf{x}_n|\boldsymbol{\mu}_k, \boldsymbol{\Lambda}_k^{-1}\right) + \text{const.} \end{aligned} \tag{10.54}$$

We observe that the right-hand side of this expression decomposes into a sum of terms involving only $\boldsymbol{\pi}$ together with terms only involving $\boldsymbol{\mu}$ and $\boldsymbol{\Lambda}$, which implies that the variational posterior $q(\boldsymbol{\pi}, \boldsymbol{\mu}, \boldsymbol{\Lambda})$ factorizes to give $q(\boldsymbol{\pi})q(\boldsymbol{\mu}, \boldsymbol{\Lambda})$. Furthermore, the terms involving $\boldsymbol{\mu}$ and $\boldsymbol{\Lambda}$ themselves comprise a sum over k of terms involving $\boldsymbol{\mu}_k$ and $\boldsymbol{\Lambda}_k$ leading to the further factorization

$$q(\boldsymbol{\pi}, \boldsymbol{\mu}, \boldsymbol{\Lambda}) = q(\boldsymbol{\pi}) \prod_{k=1}^{K} q(\boldsymbol{\mu}_k, \boldsymbol{\Lambda}_k). \tag{10.55}$$

Identifying the terms on the right-hand side of (10.54) that depend on $\boldsymbol{\pi}$, we have

$$\ln q^\star(\boldsymbol{\pi}) = (\alpha_0 - 1)\sum_{k=1}^{K} \ln \pi_k + \sum_{k=1}^{K}\sum_{n=1}^{N} r_{nk} \ln \pi_k + \text{const} \tag{10.56}$$

where we have used (10.50). Taking the exponential of both sides, we recognize $q^\star(\boldsymbol{\pi})$ as a Dirichlet distribution

$$q^\star(\boldsymbol{\pi}) = \text{Dir}(\boldsymbol{\pi}|\boldsymbol{\alpha}) \tag{10.57}$$

where $\boldsymbol{\alpha}$ has components α_k given by

$$\alpha_k = \alpha_0 + N_k. \tag{10.58}$$

Finally, the variational posterior distribution $q^\star(\boldsymbol{\mu}_k, \boldsymbol{\Lambda}_k)$ does not factorize into the product of the marginals, but we can always use the product rule to write it in the form $q^\star(\boldsymbol{\mu}_k, \boldsymbol{\Lambda}_k) = q^\star(\boldsymbol{\mu}_k|\boldsymbol{\Lambda}_k)q^\star(\boldsymbol{\Lambda}_k)$. The two factors can be found by inspecting (10.54) and reading off those terms that involve $\boldsymbol{\mu}_k$ and $\boldsymbol{\Lambda}_k$. The result, as expected, is a Gaussian-Wishart distribution and is given by

Exercise 10.13

$$q^\star(\boldsymbol{\mu}_k, \boldsymbol{\Lambda}_k) = \mathcal{N}\left(\boldsymbol{\mu}_k|\mathbf{m}_k, (\beta_k\boldsymbol{\Lambda}_k)^{-1}\right)\,\mathcal{W}(\boldsymbol{\Lambda}_k|\mathbf{W}_k, \nu_k) \tag{10.59}$$

where we have defined

$$\beta_k = \beta_0 + N_k \tag{10.60}$$

$$\mathbf{m}_k = \frac{1}{\beta_k}(\beta_0\mathbf{m}_0 + N_k\overline{\mathbf{x}}_k) \tag{10.61}$$

$$\mathbf{W}_k^{-1} = \mathbf{W}_0^{-1} + N_k\mathbf{S}_k + \frac{\beta_0 N_k}{\beta_0 + N_k}(\overline{\mathbf{x}}_k - \mathbf{m}_0)(\overline{\mathbf{x}}_k - \mathbf{m}_0)^{\mathrm{T}} \tag{10.62}$$

$$\nu_k = \nu_0 + N_k + 1. \tag{10.63}$$

These update equations are analogous to the M-step equations of the EM algorithm for the maximum likelihood solution of the mixture of Gaussians. We see that the computations that must be performed in order to update the variational posterior distribution over the model parameters involve evaluation of the same sums over the data set, as arose in the maximum likelihood treatment.

In order to perform this variational M step, we need the expectations $\mathbb{E}[z_{nk}] = r_{nk}$ representing the responsibilities. These are obtained by normalizing the ρ_{nk} that are given by (10.46). We see that this expression involves expectations with respect to the variational distributions of the parameters, and these are easily evaluated to give

Exercise 10.14

$$\mathbb{E}_{\boldsymbol{\mu}_k,\boldsymbol{\Lambda}_k}\left[(\mathbf{x}_n - \boldsymbol{\mu}_k)^{\mathrm{T}}\boldsymbol{\Lambda}_k(\mathbf{x}_n - \boldsymbol{\mu}_k)\right] = D\beta_k^{-1} + \nu_k(\mathbf{x}_n - \mathbf{m}_k)^{\mathrm{T}}\mathbf{W}_k(\mathbf{x}_n - \mathbf{m}_k) \tag{10.64}$$

$$\ln\widetilde{\Lambda}_k \equiv \mathbb{E}\left[\ln|\boldsymbol{\Lambda}_k|\right] = \sum_{i=1}^{D}\psi\left(\frac{\nu_k + 1 - i}{2}\right) + D\ln 2 + \ln|\mathbf{W}_k| \tag{10.65}$$

$$\ln\widetilde{\pi}_k \equiv \mathbb{E}\left[\ln \pi_k\right] = \psi(\alpha_k) - \psi(\widehat{\alpha}) \tag{10.66}$$

where we have introduced definitions of $\widetilde{\Lambda}_k$ and $\widetilde{\pi}_k$, and $\psi(\cdot)$ is the digamma function defined by (B.25), with $\widehat{\alpha} = \sum_k \alpha_k$. The results (10.65) and (10.66) follow from the standard properties of the Wishart and Dirichlet distributions.

Appendix B

If we substitute (10.64), (10.65), and (10.66) into (10.46) and make use of (10.49), we obtain the following result for the responsibilities

$$r_{nk} \propto \widetilde{\pi}_k \widetilde{\Lambda}_k^{1/2} \exp\left\{ -\frac{D}{2\beta_k} - \frac{\nu_k}{2}(\mathbf{x}_n - \mathbf{m}_k)^{\mathrm{T}} \mathbf{W}_k (\mathbf{x}_n - \mathbf{m}_k) \right\}. \tag{10.67}$$

Notice the similarity to the corresponding result for the responsibilities in maximum likelihood EM, which from (9.13) can be written in the form

$$r_{nk} \propto \pi_k |\boldsymbol{\Lambda}_k|^{1/2} \exp\left\{ -\frac{1}{2}(\mathbf{x}_n - \boldsymbol{\mu}_k)^{\mathrm{T}} \boldsymbol{\Lambda}_k (\mathbf{x}_n - \boldsymbol{\mu}_k) \right\} \tag{10.68}$$

where we have used the precision in place of the covariance to highlight the similarity to (10.67).

Thus the optimization of the variational posterior distribution involves cycling between two stages analogous to the E and M steps of the maximum likelihood EM algorithm. In the variational equivalent of the E step, we use the current distributions over the model parameters to evaluate the moments in (10.64), (10.65), and (10.66) and hence evaluate $\mathbb{E}[z_{nk}] = r_{nk}$. Then in the subsequent variational equivalent of the M step, we keep these responsibilities fixed and use them to re-compute the variational distribution over the parameters using (10.57) and (10.59). In each case, we see that the variational posterior distribution has the same functional form as the corresponding factor in the joint distribution (10.41). This is a general result and is a consequence of the choice of conjugate distributions.

Section 10.4.1

Figure 10.6 shows the results of applying this approach to the rescaled Old Faithful data set for a Gaussian mixture model having $K = 6$ components. We see that after convergence, there are only two components for which the expected values of the mixing coefficients are numerically distinguishable from their prior values. This effect can be understood qualitatively in terms of the automatic trade-off in a Bayesian model between fitting the data and the complexity of the model, in which the complexity penalty arises from components whose parameters are pushed away from their prior values. Components that take essentially no responsibility for explaining the data points have $r_{nk} \simeq 0$ and hence $N_k \simeq 0$. From (10.58), we see that $\alpha_k \simeq \alpha_0$ and from (10.60)–(10.63) we see that the other parameters revert to their prior values. In principle such components are fitted slightly to the data points, but for broad priors this effect is too small to be seen numerically. For the variational Gaussian mixture model the expected values of the mixing coefficients in the posterior distribution are given by

Section 3.4

Exercise 10.15

$$\mathbb{E}[\pi_k] = \frac{\alpha_0 + N_k}{K\alpha_0 + N}. \tag{10.69}$$

Consider a component for which $N_k \simeq 0$ and $\alpha_k \simeq \alpha_0$. If the prior is broad so that $\alpha_0 \to 0$, then $\mathbb{E}[\pi_k] \to 0$ and the component plays no role in the model, whereas if

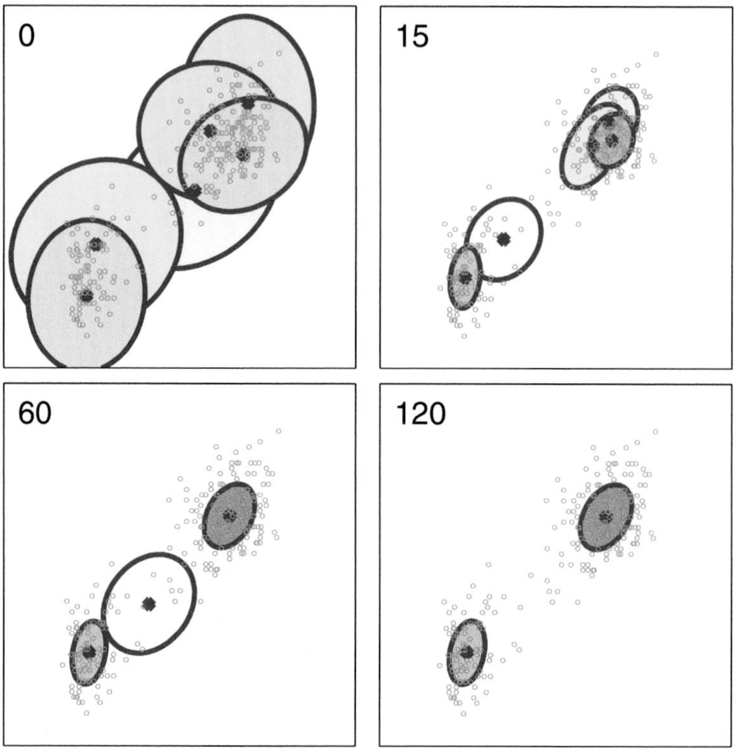

Figure 10.6 Variational Bayesian mixture of $K = 6$ Gaussians applied to the Old Faithful data set, in which the ellipses denote the one standard-deviation density contours for each of the components, and the density of red ink inside each ellipse corresponds to the mean value of the mixing coefficient for each component. The number in the top left of each diagram shows the number of iterations of variational inference. Components whose expected mixing coefficient are numerically indistinguishable from zero are not plotted.

the prior tightly constrains the mixing coefficients so that $\alpha_0 \to \infty$, then $\mathbb{E}[\pi_k] \to 1/K$.

In Figure 10.6, the prior over the mixing coefficients is a Dirichlet of the form (10.39). Recall from Figure 2.5 that for $\alpha_0 < 1$ the prior favours solutions in which some of the mixing coefficients are zero. Figure 10.6 was obtained using $\alpha_0 = 10^{-3}$, and resulted in two components having nonzero mixing coefficients. If instead we choose $\alpha_0 = 1$ we obtain three components with nonzero mixing coefficients, and for $\alpha = 10$ all six components have nonzero mixing coefficients.

As we have seen there is a close similarity between the variational solution for the Bayesian mixture of Gaussians and the EM algorithm for maximum likelihood. In fact if we consider the limit $N \to \infty$ then the Bayesian treatment converges to the maximum likelihood EM algorithm. For anything other than very small data sets, the dominant computational cost of the variational algorithm for Gaussian mixtures arises from the evaluation of the responsibilities, together with the evaluation and inversion of the weighted data covariance matrices. These computations mirror precisely those that arise in the maximum likelihood EM algorithm, and so there is little computational overhead in using this Bayesian approach as compared to the traditional maximum likelihood one. There are, however, some substantial advantages. First of all, the singularities that arise in maximum likelihood when a Gaussian component 'collapses' onto a specific data point are absent in the Bayesian treatment.

Indeed, these singularities are removed if we simply introduce a prior and then use a MAP estimate instead of maximum likelihood. Furthermore, there is no over-fitting if we choose a large number K of components in the mixture, as we saw in Figure 10.6. Finally, the variational treatment opens up the possibility of determining the optimal number of components in the mixture without resorting to techniques such as cross validation.

Section 10.2.4

10.2.2 Variational lower bound

We can also straightforwardly evaluate the lower bound (10.3) for this model. In practice, it is useful to be able to monitor the bound during the re-estimation in order to test for convergence. It can also provide a valuable check on both the mathematical expressions for the solutions and their software implementation, because at each step of the iterative re-estimation procedure the value of this bound should not decrease. We can take this a stage further to provide a deeper test of the correctness of both the mathematical derivation of the update equations and of their software implementation by using finite differences to check that each update does indeed give a (constrained) maximum of the bound (Svensén and Bishop, 2004).

For the variational mixture of Gaussians, the lower bound (10.3) is given by

$$\begin{aligned}\mathcal{L} &= \sum_{\mathbf{Z}} \iiint q(\mathbf{Z}, \boldsymbol{\pi}, \boldsymbol{\mu}, \boldsymbol{\Lambda}) \ln \left\{ \frac{p(\mathbf{X}, \mathbf{Z}, \boldsymbol{\pi}, \boldsymbol{\mu}, \boldsymbol{\Lambda})}{q(\mathbf{Z}, \boldsymbol{\pi}, \boldsymbol{\mu}, \boldsymbol{\Lambda})} \right\} \mathrm{d}\boldsymbol{\pi}\, \mathrm{d}\boldsymbol{\mu}\, \mathrm{d}\boldsymbol{\Lambda} \\ &= \mathbb{E}[\ln p(\mathbf{X}, \mathbf{Z}, \boldsymbol{\pi}, \boldsymbol{\mu}, \boldsymbol{\Lambda})] - \mathbb{E}[\ln q(\mathbf{Z}, \boldsymbol{\pi}, \boldsymbol{\mu}, \boldsymbol{\Lambda})] \\ &= \mathbb{E}[\ln p(\mathbf{X}|\mathbf{Z}, \boldsymbol{\mu}, \boldsymbol{\Lambda})] + \mathbb{E}[\ln p(\mathbf{Z}|\boldsymbol{\pi})] + \mathbb{E}[\ln p(\boldsymbol{\pi})] + \mathbb{E}[\ln p(\boldsymbol{\mu}, \boldsymbol{\Lambda})] \\ &\quad - \mathbb{E}[\ln q(\mathbf{Z})] - \mathbb{E}[\ln q(\boldsymbol{\pi})] - \mathbb{E}[\ln q(\boldsymbol{\mu}, \boldsymbol{\Lambda})] \qquad (10.70)\end{aligned}$$

where, to keep the notation uncluttered, we have omitted the $\star$ superscript on the q distributions, along with the subscripts on the expectation operators because each expectation is taken with respect to all of the random variables in its argument. The various terms in the bound are easily evaluated to give the following results

Exercise 10.16

$$\begin{aligned}\mathbb{E}[\ln p(\mathbf{X}|\mathbf{Z}, \boldsymbol{\mu}, \boldsymbol{\Lambda})] = \frac{1}{2} \sum_{k=1}^{K} N_k \Big\{ & \ln \widetilde{\Lambda}_k - D\beta_k^{-1} - \nu_k \mathrm{Tr}(\mathbf{S}_k \mathbf{W}_k) \\ & - \nu_k (\overline{\mathbf{x}}_k - \mathbf{m}_k)^{\mathrm{T}} \mathbf{W}_k (\overline{\mathbf{x}}_k - \mathbf{m}_k) - D \ln(2\pi) \Big\} \qquad (10.71)\end{aligned}$$

$$\mathbb{E}[\ln p(\mathbf{Z}|\boldsymbol{\pi})] = \sum_{n=1}^{N} \sum_{k=1}^{K} r_{nk} \ln \widetilde{\pi}_k \qquad (10.72)$$

$$\mathbb{E}[\ln p(\boldsymbol{\pi})] = \ln C(\boldsymbol{\alpha}_0) + (\alpha_0 - 1) \sum_{k=1}^{K} \ln \widetilde{\pi}_k \qquad (10.73)$$

$$\mathbb{E}[\ln p(\boldsymbol{\mu}, \boldsymbol{\Lambda})] = \frac{1}{2}\sum_{k=1}^{K}\left\{D\ln(\beta_0/2\pi) + \ln\widetilde{\Lambda}_k - \frac{D\beta_0}{\beta_k} - \beta_0\nu_k(\mathbf{m}_k - \mathbf{m}_0)^{\mathrm{T}}\mathbf{W}_k(\mathbf{m}_k - \mathbf{m}_0)\right\} + K\ln B(\mathbf{W}_0, \nu_0) + \frac{(\nu_0 - D - 1)}{2}\sum_{k=1}^{K}\ln\widetilde{\Lambda}_k - \frac{1}{2}\sum_{k=1}^{K}\nu_k\mathrm{Tr}(\mathbf{W}_0^{-1}\mathbf{W}_k) \tag{10.74}$$

$$\mathbb{E}[\ln q(\mathbf{Z})] = \sum_{n=1}^{N}\sum_{k=1}^{K} r_{nk}\ln r_{nk} \tag{10.75}$$

$$\mathbb{E}[\ln q(\boldsymbol{\pi})] = \sum_{k=1}^{K}(\alpha_k - 1)\ln\widetilde{\pi}_k + \ln C(\boldsymbol{\alpha}) \tag{10.76}$$

$$\mathbb{E}[\ln q(\boldsymbol{\mu}, \boldsymbol{\Lambda})] = \sum_{k=1}^{K}\left\{\frac{1}{2}\ln\widetilde{\Lambda}_k + \frac{D}{2}\ln\left(\frac{\beta_k}{2\pi}\right) - \frac{D}{2} - \mathrm{H}\left[q(\boldsymbol{\Lambda}_k)\right]\right\} \tag{10.77}$$

where D is the dimensionality of $\mathbf{x}$, $\mathrm{H}[q(\boldsymbol{\Lambda}_k)]$ is the entropy of the Wishart distribution given by (B.82), and the coefficients $C(\boldsymbol{\alpha})$ and $B(\mathbf{W}, \nu)$ are defined by (B.23) and (B.79), respectively. Note that the terms involving expectations of the logs of the q distributions simply represent the negative entropies of those distributions. Some simplifications and combination of terms can be performed when these expressions are summed to give the lower bound. However, we have kept the expressions separate for ease of understanding.

Finally, it is worth noting that the lower bound provides an alternative approach for deriving the variational re-estimation equations obtained in Section 10.2.1. To do this we use the fact that, since the model has conjugate priors, the functional form of the factors in the variational posterior distribution is known, namely discrete for $\mathbf{Z}$, Dirichlet for π, and Gaussian-Wishart for $(\boldsymbol{\mu}_k, \boldsymbol{\Lambda}_k)$. By taking general parametric forms for these distributions we can derive the form of the lower bound as a function of the parameters of the distributions. Maximizing the bound with respect to these parameters then gives the required re-estimation equations.

Exercise 10.18

10.2.3 Predictive density

In applications of the Bayesian mixture of Gaussians model we will often be interested in the predictive density for a new value $\widehat{\mathbf{x}}$ of the observed variable. Associated with this observation will be a corresponding latent variable $\widehat{\mathbf{z}}$, and the predictive density is then given by

$$p(\widehat{\mathbf{x}}|\mathbf{X}) = \sum_{\widehat{\mathbf{z}}}\iiint p(\widehat{\mathbf{x}}|\widehat{\mathbf{z}}, \boldsymbol{\mu}, \boldsymbol{\Lambda})p(\widehat{\mathbf{z}}|\boldsymbol{\pi})p(\boldsymbol{\pi}, \boldsymbol{\mu}, \boldsymbol{\Lambda}|\mathbf{X})\,\mathrm{d}\boldsymbol{\pi}\,\mathrm{d}\boldsymbol{\mu}\,\mathrm{d}\boldsymbol{\Lambda} \tag{10.78}$$

where $p(\boldsymbol{\pi}, \boldsymbol{\mu}, \boldsymbol{\Lambda}|\mathbf{X})$ is the (unknown) true posterior distribution of the parameters. Using (10.37) and (10.38) we can first perform the summation over $\widehat{\mathbf{z}}$ to give

$$p(\widehat{\mathbf{x}}|\mathbf{X}) = \sum_{k=1}^{K} \iiint \pi_k \mathcal{N}\left(\widehat{\mathbf{x}}|\boldsymbol{\mu}_k, \boldsymbol{\Lambda}_k^{-1}\right) p(\boldsymbol{\pi}, \boldsymbol{\mu}, \boldsymbol{\Lambda}|\mathbf{X})\, \mathrm{d}\boldsymbol{\pi}\, \mathrm{d}\boldsymbol{\mu}\, \mathrm{d}\boldsymbol{\Lambda}. \tag{10.79}$$

Because the remaining integrations are intractable, we approximate the predictive density by replacing the true posterior distribution $p(\boldsymbol{\pi}, \boldsymbol{\mu}, \boldsymbol{\Lambda}|\mathbf{X})$ with its variational approximation $q(\boldsymbol{\pi})q(\boldsymbol{\mu}, \boldsymbol{\Lambda})$ to give

$$p(\widehat{\mathbf{x}}|\mathbf{X}) \simeq \sum_{k=1}^{K} \iiint \pi_k \mathcal{N}\left(\widehat{\mathbf{x}}|\boldsymbol{\mu}_k, \boldsymbol{\Lambda}_k^{-1}\right) q(\boldsymbol{\pi})q(\boldsymbol{\mu}_k, \boldsymbol{\Lambda}_k)\, \mathrm{d}\boldsymbol{\pi}\, \mathrm{d}\boldsymbol{\mu}_k\, \mathrm{d}\boldsymbol{\Lambda}_k \tag{10.80}$$

where we have made use of the factorization (10.55) and in each term we have implicitly integrated out all variables $\{\boldsymbol{\mu}_j, \boldsymbol{\Lambda}_j\}$ for $j \neq k$. The remaining integrations can now be evaluated analytically giving a mixture of Student's t-distributions

Exercise 10.19

$$p(\widehat{\mathbf{x}}|\mathbf{X}) \simeq \frac{1}{\widehat{\alpha}} \sum_{k=1}^{K} \alpha_k \mathrm{St}(\widehat{\mathbf{x}}|\mathbf{m}_k, \mathbf{L}_k, \nu_k + 1 - D) \tag{10.81}$$

in which the kth component has mean $\mathbf{m}_k$, and the precision is given by

$$\mathbf{L}_k = \frac{(\nu_k + 1 - D)\beta_k}{(1 + \beta_k)} \mathbf{W}_k \tag{10.82}$$

in which ν_k is given by (10.63). When the size N of the data set is large the predictive distribution (10.81) reduces to a mixture of Gaussians.

Exercise 10.20

10.2.4 Determining the number of components

Section 10.1.4

We have seen that the variational lower bound can be used to determine a posterior distribution over the number K of components in the mixture model. There is, however, one subtlety that needs to be addressed. For any given setting of the parameters in a Gaussian mixture model (except for specific degenerate settings), there will exist other parameter settings for which the density over the observed variables will be identical. These parameter values differ only through a re-labelling of the components. For instance, consider a mixture of two Gaussians and a single observed variable x, in which the parameters have the values $\pi_1 = a$, $\pi_2 = b$, $\mu_1 = c$, $\mu_2 = d$, $\sigma_1 = e$, $\sigma_2 = f$. Then the parameter values $\pi_1 = b$, $\pi_2 = a$, $\mu_1 = d$, $\mu_2 = c$, $\sigma_1 = f$, $\sigma_2 = e$, in which the two components have been exchanged, will by symmetry give rise to the same value of $p(x)$. If we have a mixture model comprising K components, then each parameter setting will be a member of a family of $K!$ equivalent settings.

Exercise 10.21

In the context of maximum likelihood, this redundancy is irrelevant because the parameter optimization algorithm (for example EM) will, depending on the initialization of the parameters, find one specific solution, and the other equivalent solutions play no role. In a Bayesian setting, however, we marginalize over all possible

Figure 10.7 Plot of the variational lower bound $\mathcal{L}$ versus the number K of components in the Gaussian mixture model, for the Old Faithful data, showing a distinct peak at $K = 2$ components. For each value of K, the model is trained from 100 different random starts, and the results shown as '+' symbols plotted with small random horizontal perturbations so that they can be distinguished. Note that some solutions find suboptimal local maxima, but that this happens infrequently.

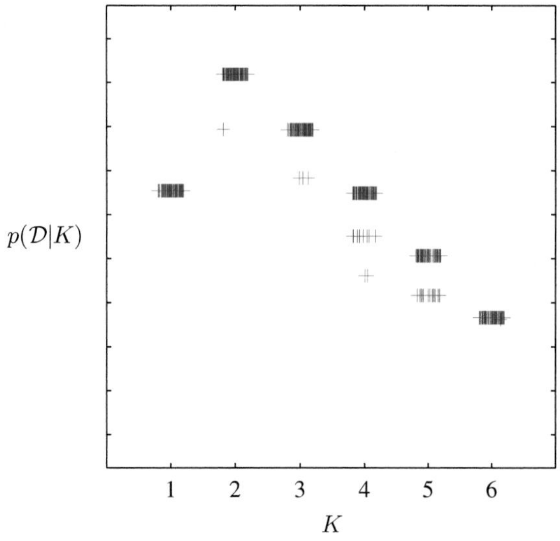

parameter values. We have seen in Figure 10.3 that if the true posterior distribution is multimodal, variational inference based on the minimization of $\mathrm{KL}(q\|p)$ will tend to approximate the distribution in the neighbourhood of one of the modes and ignore the others. Again, because equivalent modes have equivalent predictive densities, this is of no concern provided we are considering a model having a specific number K of components. If, however, we wish to compare different values of K, then we need to take account of this multimodality. A simple approximate solution is to add a term $\ln K!$ onto the lower bound when used for model comparison and averaging.

Exercise 10.22

Figure 10.7 shows a plot of the lower bound, including the multimodality factor, versus the number K of components for the Old Faithful data set. It is worth emphasizing once again that maximum likelihood would lead to values of the likelihood function that increase monotonically with K (assuming the singular solutions have been avoided, and discounting the effects of local maxima) and so cannot be used to determine an appropriate model complexity. By contrast, Bayesian inference automatically makes the trade-off between model complexity and fitting the data.

Section 3.4

This approach to the determination of K requires that a range of models having different K values be trained and compared. An alternative approach to determining a suitable value for K is to treat the mixing coefficients $\boldsymbol{\pi}$ as parameters and make point estimates of their values by maximizing the lower bound (Corduneanu and Bishop, 2001) with respect to $\boldsymbol{\pi}$ instead of maintaining a probability distribution over them as in the fully Bayesian approach. This leads to the re-estimation equation

Exercise 10.23

$$\pi_k = \frac{1}{N}\sum_{n=1}^{N} r_{nk} \tag{10.83}$$

and this maximization is interleaved with the variational updates for the q distribution over the remaining parameters. Components that provide insufficient contribution to explaining the data will have their mixing coefficients driven to zero during the optimization, and so they are effectively removed from the model through *automatic relevance determination*. This allows us to make a single training run in which we start with a relatively large initial value of K, and allow surplus components to be pruned out of the model. The origins of the sparsity when optimizing with respect to hyperparameters is discussed in detail in the context of the relevance vector machine.

Section 7.2.2

10.2.5 Induced factorizations

In deriving these variational update equations for the Gaussian mixture model, we assumed a particular factorization of the variational posterior distribution given by (10.42). However, the optimal solutions for the various factors exhibit additional factorizations. In particular, the solution for $q^\star(\boldsymbol{\mu}, \boldsymbol{\Lambda})$ is given by the product of an independent distribution $q^\star(\boldsymbol{\mu}_k, \boldsymbol{\Lambda}_k)$ over each of the components k of the mixture, whereas the variational posterior distribution $q^\star(\mathbf{Z})$ over the latent variables, given by (10.48), factorizes into an independent distribution $q^\star(\mathbf{z}_n)$ for each observation n (note that it does not further factorize with respect to k because, for each value of n, the z_{nk} are constrained to sum to one over k). These additional factorizations are a consequence of the interaction between the assumed factorization and the conditional independence properties of the true distribution, as characterized by the directed graph in Figure 10.5.

We shall refer to these additional factorizations as *induced factorizations* because they arise from an interaction between the factorization assumed in the variational posterior distribution and the conditional independence properties of the true joint distribution. In a numerical implementation of the variational approach it is important to take account of such additional factorizations. For instance, it would be very inefficient to maintain a full precision matrix for the Gaussian distribution over a set of variables if the optimal form for that distribution always had a diagonal precision matrix (corresponding to a factorization with respect to the individual variables described by that Gaussian).

Such induced factorizations can easily be detected using a simple graphical test based on d-separation as follows. We partition the latent variables into three disjoint groups $\mathbf{A}, \mathbf{B}, \mathbf{C}$ and then let us suppose that we are assuming a factorization between $\mathbf{C}$ and the remaining latent variables, so that

$$q(\mathbf{A}, \mathbf{B}, \mathbf{C}) = q(\mathbf{A}, \mathbf{B})q(\mathbf{C}). \tag{10.84}$$

Using the general result (10.9), together with the product rule for probabilities, we see that the optimal solution for $q(\mathbf{A}, \mathbf{B})$ is given by

$$\begin{aligned} \ln q^\star(\mathbf{A}, \mathbf{B}) &= \mathbb{E}_{\mathbf{C}}[\ln p(\mathbf{X}, \mathbf{A}, \mathbf{B}, \mathbf{C})] + \text{const} \\ &= \mathbb{E}_{\mathbf{C}}[\ln p(\mathbf{A}, \mathbf{B}|\mathbf{X}, \mathbf{C})] + \text{const}. \end{aligned} \tag{10.85}$$

We now ask whether this resulting solution will factorize between $\mathbf{A}$ and $\mathbf{B}$, in other words whether $q^\star(\mathbf{A}, \mathbf{B}) = q^\star(\mathbf{A})q^\star(\mathbf{B})$. This will happen if, and only if, $\ln p(\mathbf{A}, \mathbf{B}|\mathbf{X}, \mathbf{C}) = \ln p(\mathbf{A}|\mathbf{X}, \mathbf{C}) + \ln p(\mathbf{B}|\mathbf{X}, \mathbf{C})$, that is, if the conditional independence relation

$$\mathbf{A} \perp\!\!\!\perp \mathbf{B} \mid \mathbf{X}, \mathbf{C} \tag{10.86}$$

is satisfied. We can test to see if this relation does hold, for any choice of $\mathbf{A}$ and $\mathbf{B}$ by making use of the d-separation criterion.

To illustrate this, consider again the Bayesian mixture of Gaussians represented by the directed graph in Figure 10.5, in which we are assuming a variational factorization given by (10.42). We can see immediately that the variational posterior distribution over the parameters must factorize between $\boldsymbol{\pi}$ and the remaining parameters $\boldsymbol{\mu}$ and $\boldsymbol{\Lambda}$ because all paths connecting $\boldsymbol{\pi}$ to either $\boldsymbol{\mu}$ or $\boldsymbol{\Lambda}$ must pass through one of the nodes $\mathbf{z}_n$ all of which are in the conditioning set for our conditional independence test and all of which are head-to-tail with respect to such paths.

10.3. Variational Linear Regression

As a second illustration of variational inference, we return to the Bayesian linear regression model of Section 3.3. In the evidence framework, we approximated the integration over α and β by making point estimates obtained by maximizing the log marginal likelihood. A fully Bayesian approach would integrate over the hyperparameters as well as over the parameters. Although exact integration is intractable, we can use variational methods to find a tractable approximation. In order to simplify the discussion, we shall suppose that the noise precision parameter β is known, and is fixed to its true value, although the framework is easily extended to include the distribution over β. For the linear regression model, the variational treatment will turn out to be equivalent to the evidence framework. Nevertheless, it provides a good exercise in the use of variational methods and will also lay the foundation for variational treatment of Bayesian logistic regression in Section 10.6.

Exercise 10.26

Recall that the likelihood function for $\mathbf{w}$, and the prior over $\mathbf{w}$, are given by

$$p(\mathbf{t}|\mathbf{w}) = \prod_{n=1}^{N} \mathcal{N}(t_n|\mathbf{w}^{\mathrm{T}}\boldsymbol{\phi}_n, \beta^{-1}) \tag{10.87}$$

$$p(\mathbf{w}|\alpha) = \mathcal{N}(\mathbf{w}|\mathbf{0}, \alpha^{-1}\mathbf{I}) \tag{10.88}$$

where $\boldsymbol{\phi}_n = \boldsymbol{\phi}(\mathbf{x}_n)$. We now introduce a prior distribution over α. From our discussion in Section 2.3.6, we know that the conjugate prior for the precision of a Gaussian is given by a gamma distribution, and so we choose

$$p(\alpha) = \mathrm{Gam}(\alpha|a_0, b_0) \tag{10.89}$$

where $\mathrm{Gam}(\cdot|\cdot, \cdot)$ is defined by (B.26). Thus the joint distribution of all the variables is given by

$$p(\mathbf{t}, \mathbf{w}, \alpha) = p(\mathbf{t}|\mathbf{w})p(\mathbf{w}|\alpha)p(\alpha). \tag{10.90}$$

This can be represented as a directed graphical model as shown in Figure 10.8.

10.3.1 Variational distribution

Our first goal is to find an approximation to the posterior distribution $p(\mathbf{w}, \alpha|\mathbf{t})$. To do this, we employ the variational framework of Section 10.1, with a variational

Figure 10.8 Probabilistic graphical model representing the joint distribution (10.90) for the Bayesian linear regression model.

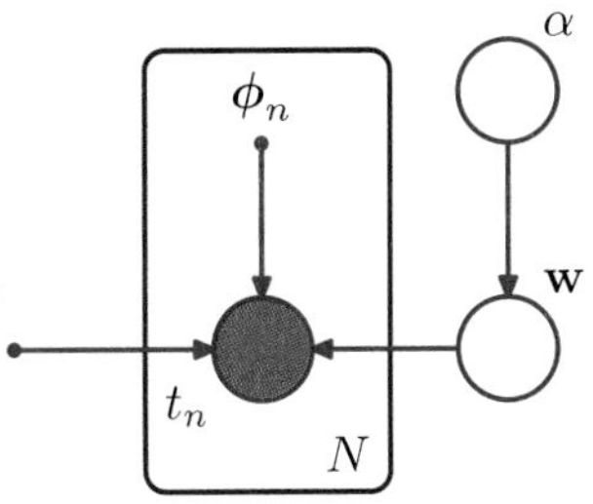

posterior distribution given by the factorized expression

$$q(\mathbf{w}, \alpha) = q(\mathbf{w})q(\alpha). \tag{10.91}$$

We can find re-estimation equations for the factors in this distribution by making use of the general result (10.9). Recall that for each factor, we take the log of the joint distribution over all variables and then average with respect to those variables not in that factor. Consider first the distribution over α. Keeping only terms that have a functional dependence on α, we have

$$\begin{aligned}\ln q^\star(\alpha) &= \ln p(\alpha) + \mathbb{E}_{\mathbf{w}}\left[\ln p(\mathbf{w}|\alpha)\right] + \text{const} \\ &= (a_0 - 1)\ln\alpha - b_0\alpha + \frac{M}{2}\ln\alpha - \frac{\alpha}{2}\mathbb{E}[\mathbf{w}^{\mathrm{T}}\mathbf{w}] + \text{const.}\end{aligned} \tag{10.92}$$

We recognize this as the log of a gamma distribution, and so identifying the coefficients of α and $\ln\alpha$ we obtain

$$q^\star(\alpha) = \text{Gam}(\alpha|a_N, b_N) \tag{10.93}$$

where

$$a_N = a_0 + \frac{M}{2} \tag{10.94}$$

$$b_N = b_0 + \frac{1}{2}\mathbb{E}[\mathbf{w}^{\mathrm{T}}\mathbf{w}]. \tag{10.95}$$

Similarly, we can find the variational re-estimation equation for the posterior distribution over $\mathbf{w}$. Again, using the general result (10.9), and keeping only those terms that have a functional dependence on $\mathbf{w}$, we have

$$\ln q^\star(\mathbf{w}) = \ln p(\mathbf{t}|\mathbf{w}) + \mathbb{E}_\alpha\left[\ln p(\mathbf{w}|\alpha)\right] + \text{const} \tag{10.96}$$

$$= -\frac{\beta}{2}\sum_{n=1}^{N}\{\mathbf{w}^{\mathrm{T}}\boldsymbol{\phi}_n - t_n\}^2 - \frac{1}{2}\mathbb{E}[\alpha]\mathbf{w}^{\mathrm{T}}\mathbf{w} + \text{const} \tag{10.97}$$

$$= -\frac{1}{2}\mathbf{w}^{\mathrm{T}}\left(\mathbb{E}[\alpha]\mathbf{I} + \beta\boldsymbol{\Phi}^{\mathrm{T}}\boldsymbol{\Phi}\right)\mathbf{w} + \beta\mathbf{w}^{\mathrm{T}}\boldsymbol{\Phi}^{\mathrm{T}}\mathbf{t} + \text{const.} \tag{10.98}$$

Because this is a quadratic form, the distribution $q^\star(\mathbf{w})$ is Gaussian, and so we can complete the square in the usual way to identify the mean and covariance, giving

$$q^\star(\mathbf{w}) = \mathcal{N}(\mathbf{w}|\mathbf{m}_N, \mathbf{S}_N) \tag{10.99}$$

where

$$\mathbf{m}_N = \beta \mathbf{S}_N \mathbf{\Phi}^{\mathrm{T}} \mathbf{t} \tag{10.100}$$

$$\mathbf{S}_N = \left(\mathbb{E}[\alpha]\mathbf{I} + \beta \mathbf{\Phi}^{\mathrm{T}}\mathbf{\Phi}\right)^{-1}. \tag{10.101}$$

Note the close similarity to the posterior distribution (3.52) obtained when α was treated as a fixed parameter. The difference is that here α is replaced by its expectation $\mathbb{E}[\alpha]$ under the variational distribution. Indeed, we have chosen to use the same notation for the covariance matrix $\mathbf{S}_N$ in both cases.

Using the standard results (B.27), (B.38), and (B.39), we can obtain the required moments as follows

$$\mathbb{E}[\alpha] = a_N/b_N \tag{10.102}$$

$$\mathbb{E}[\mathbf{w}\mathbf{w}^{\mathrm{T}}] = \mathbf{m}_N\mathbf{m}_N^{\mathrm{T}} + \mathbf{S}_N. \tag{10.103}$$

The evaluation of the variational posterior distribution begins by initializing the parameters of one of the distributions $q(\mathbf{w})$ or $q(\alpha)$, and then alternately re-estimates these factors in turn until a suitable convergence criterion is satisfied (usually specified in terms of the lower bound to be discussed shortly).

It is instructive to relate the variational solution to that found using the evidence framework in Section 3.5. To do this consider the case $a_0 = b_0 = 0$, corresponding to the limit of an infinitely broad prior over α. The mean of the variational posterior distribution $q(\alpha)$ is then given by

$$\mathbb{E}[\alpha] = \frac{a_N}{b_N} = \frac{M/2}{\mathbb{E}[\mathbf{w}^{\mathrm{T}}\mathbf{w}]/2} = \frac{M}{\mathbf{m}_N^{\mathrm{T}}\mathbf{m}_N + \mathrm{Tr}(\mathbf{S}_N)}. \tag{10.104}$$

Comparison with (9.63) shows that in the case of this particularly simple model, the variational approach gives precisely the same expression as that obtained by maximizing the evidence function using EM except that the point estimate for α is replaced by its expected value. Because the distribution $q(\mathbf{w})$ depends on $q(\alpha)$ only through the expectation $\mathbb{E}[\alpha]$, we see that the two approaches will give identical results for the case of an infinitely broad prior.

10.3.2 Predictive distribution

The predictive distribution over t, given a new input $\mathbf{x}$, is easily evaluated for this model using the Gaussian variational posterior for the parameters

$$\begin{aligned} p(t|\mathbf{x}, \mathbf{t}) &= \int p(t|\mathbf{x}, \mathbf{w})p(\mathbf{w}|\mathbf{t})\,\mathrm{d}\mathbf{w} \\ &\simeq \int p(t|\mathbf{x}, \mathbf{w})q(\mathbf{w})\,\mathrm{d}\mathbf{w} \\ &= \int \mathcal{N}(t|\mathbf{w}^{\mathrm{T}}\boldsymbol{\phi}(\mathbf{x}), \beta^{-1})\mathcal{N}(\mathbf{w}|\mathbf{m}_N, \mathbf{S}_N)\,\mathrm{d}\mathbf{w} \\ &= \mathcal{N}(t|\mathbf{m}_N^{\mathrm{T}}\boldsymbol{\phi}(\mathbf{x}), \sigma^2(\mathbf{x})) \end{aligned} \tag{10.105}$$

where we have evaluated the integral by making use of the result (2.115) for the linear-Gaussian model. Here the input-dependent variance is given by

$$\sigma^2(\mathbf{x}) = \frac{1}{\beta} + \boldsymbol{\phi}(\mathbf{x})^{\mathrm{T}} \mathbf{S}_N \boldsymbol{\phi}(\mathbf{x}). \tag{10.106}$$

Note that this takes the same form as the result (3.59) obtained with fixed α except that now the expected value $\mathbb{E}[\alpha]$ appears in the definition of $\mathbf{S}_N$.

10.3.3 Lower bound

Another quantity of importance is the lower bound $\mathcal{L}$ defined by

$$\begin{aligned} \mathcal{L}(q) &= \mathbb{E}[\ln p(\mathbf{w}, \alpha, \mathbf{t})] - \mathbb{E}[\ln q(\mathbf{w}, \alpha)] \\ &= \mathbb{E}_{\mathbf{w}}[\ln p(\mathbf{t}|\mathbf{w})] + \mathbb{E}_{\mathbf{w},\alpha}[\ln p(\mathbf{w}|\alpha)] + \mathbb{E}_{\alpha}[\ln p(\alpha)] \\ &\quad - \mathbb{E}_{\alpha}[\ln q(\mathbf{w})]_{\mathbf{w}} - \mathbb{E}[\ln q(\alpha)]. \end{aligned} \tag{10.107}$$

Exercise 10.27

Evaluation of the various terms is straightforward, making use of results obtained in previous chapters, and gives

$$\begin{aligned} \mathbb{E}[\ln p(\mathbf{t}|\mathbf{w})]_{\mathbf{w}} &= \frac{N}{2} \ln \left(\frac{\beta}{2\pi} \right) - \frac{\beta}{2} \mathbf{t}^{\mathrm{T}} \mathbf{t} + \beta \mathbf{m}_N^{\mathrm{T}} \boldsymbol{\Phi}^{\mathrm{T}} \mathbf{t} \\ &\quad - \frac{\beta}{2} \mathrm{Tr} \left[\boldsymbol{\Phi}^{\mathrm{T}} \boldsymbol{\Phi} (\mathbf{m}_N \mathbf{m}_N^{\mathrm{T}} + \mathbf{S}_N) \right] \end{aligned} \tag{10.108}$$

$$\begin{aligned} \mathbb{E}[\ln p(\mathbf{w}|\alpha)]_{\mathbf{w},\alpha} &= -\frac{M}{2} \ln(2\pi) + \frac{M}{2} (\psi(a_N) - \ln b_N) \\ &\quad - \frac{a_N}{2b_N} \left[\mathbf{m}_N^{\mathrm{T}} \mathbf{m}_N + \mathrm{Tr}(\mathbf{S}_N) \right] \end{aligned} \tag{10.109}$$

$$\begin{aligned} \mathbb{E}[\ln p(\alpha)]_{\alpha} &= a_0 \ln b_0 + (a_0 - 1) \left[\psi(a_N) - \ln b_N \right] \\ &\quad - b_0 \frac{a_N}{b_N} - \ln \Gamma(a_0) \end{aligned} \tag{10.110}$$

$$-\mathbb{E}[\ln q(\mathbf{w})]_{\mathbf{w}} = \frac{1}{2} \ln |\mathbf{S}_N| + \frac{M}{2} [1 + \ln(2\pi)] \tag{10.111}$$

$$-\mathbb{E}[\ln q(\alpha)]_{\alpha} = \ln \Gamma(a_N) - (a_N - 1)\psi(a_N) - \ln b_N + a_N. \tag{10.112}$$

Figure 10.9 shows a plot of the lower bound $\mathcal{L}(q)$ versus the degree of a polynomial model for a synthetic data set generated from a degree three polynomial. Here the prior parameters have been set to $a_0 = b_0 = 0$, corresponding to the noninformative prior $p(\alpha) \propto 1/\alpha$, which is uniform over $\ln \alpha$ as discussed in Section 2.3.6. As we saw in Section 10.1, the quantity $\mathcal{L}$ represents lower bound on the log marginal likelihood $\ln p(\mathbf{t}|M)$ for the model. If we assign equal prior probabilities $p(M)$ to the different values of M, then we can interpret $\mathcal{L}$ as an approximation to the posterior model probability $p(M|\mathbf{t})$. Thus the variational framework assigns the highest probability to the model with $M = 3$. This should be contrasted with the maximum likelihood result, which assigns ever smaller residual error to models of increasing complexity until the residual error is driven to zero, causing maximum likelihood to favour severely over-fitted models.

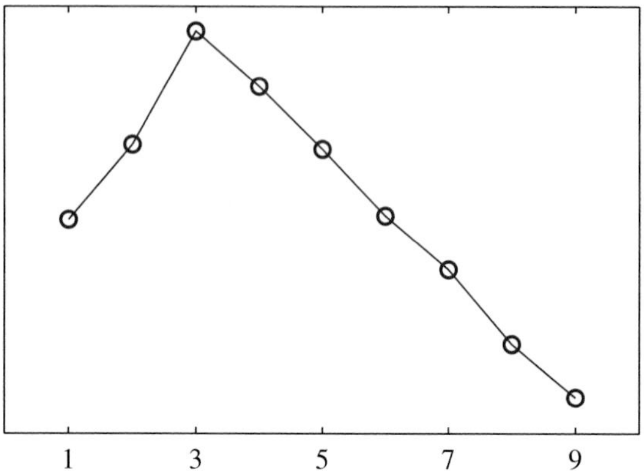

Figure 10.9 Plot of the lower bound $\mathcal{L}$ versus the order M of the polynomial, for a polynomial model, in which a set of 10 data points is generated from a polynomial with $M = 3$ sampled over the interval $(-5, 5)$ with additive Gaussian noise of variance 0.09. The value of the bound gives the log probability of the model, and we see that the value of the bound peaks at $M = 3$, corresponding to the true model from which the data set was generated.

10.4. Exponential Family Distributions

In Chapter 2, we discussed the important role played by the exponential family of distributions and their conjugate priors. For many of the models discussed in this book, the complete-data likelihood is drawn from the exponential family. However, in general this will not be the case for the marginal likelihood function for the observed data. For example, in a mixture of Gaussians, the joint distribution of observations $\mathbf{x}_n$ and corresponding hidden variables $\mathbf{z}_n$ is a member of the exponential family, whereas the marginal distribution of $\mathbf{x}_n$ is a mixture of Gaussians and hence is not.

Up to now we have grouped the variables in the model into observed variables and hidden variables. We now make a further distinction between latent variables, denoted $\mathbf{Z}$, and parameters, denoted $\boldsymbol{\theta}$, where parameters are *intensive* (fixed in number independent of the size of the data set), whereas latent variables are *extensive* (scale in number with the size of the data set). For example, in a Gaussian mixture model, the indicator variables z_{kn} (which specify which component k is responsible for generating data point $\mathbf{x}_n$) represent the latent variables, whereas the means $\boldsymbol{\mu}_k$, precisions $\boldsymbol{\Lambda}_k$ and mixing proportions π_k represent the parameters.

Consider the case of independent identically distributed data. We denote the data values by $\mathbf{X} = \{\mathbf{x}_n\}$, where $n = 1, \ldots N$, with corresponding latent variables $\mathbf{Z} = \{\mathbf{z}_n\}$. Now suppose that the joint distribution of observed and latent variables is a member of the exponential family, parameterized by natural parameters $\boldsymbol{\eta}$ so that

$$p(\mathbf{X}, \mathbf{Z}|\boldsymbol{\eta}) = \prod_{n=1}^{N} h(\mathbf{x}_n, \mathbf{z}_n) g(\boldsymbol{\eta}) \exp\left\{\boldsymbol{\eta}^{\mathrm{T}} \mathbf{u}(\mathbf{x}_n, \mathbf{z}_n)\right\}. \tag{10.113}$$

We shall also use a conjugate prior for $\boldsymbol{\eta}$, which can be written as

$$p(\boldsymbol{\eta}|\nu_0, \boldsymbol{\chi}_0) = f(\nu_0, \boldsymbol{\chi}_0) g(\boldsymbol{\eta})^{\nu_0} \exp\left\{\nu_o \boldsymbol{\eta}^{\mathrm{T}} \boldsymbol{\chi}_0\right\}. \tag{10.114}$$

Recall that the conjugate prior distribution can be interpreted as a prior number ν_0 of observations all having the value $\boldsymbol{\chi}_0$ for the $\mathbf{u}$ vector. Now consider a variational distribution that factorizes between the latent variables and the parameters, so that $q(\mathbf{Z}, \boldsymbol{\eta}) = q(\mathbf{Z})q(\boldsymbol{\eta})$. Using the general result (10.9), we can solve for the two factors as follows

$$\begin{aligned} \ln q^{\star}(\mathbf{Z}) &= \mathbb{E}_{\boldsymbol{\eta}}[\ln p(\mathbf{X}, \mathbf{Z}|\boldsymbol{\eta})] + \text{const} \\ &= \sum_{n=1}^{N} \left\{ \ln h(\mathbf{x}_n, \mathbf{z}_n) + \mathbb{E}[\boldsymbol{\eta}^{\mathrm{T}}]\mathbf{u}(\mathbf{x}_n, \mathbf{z}_n) \right\} + \text{const.} \end{aligned} \tag{10.115}$$

Section 10.2.5

Thus we see that this decomposes into a sum of independent terms, one for each value of n, and hence the solution for $q^{\star}(\mathbf{Z})$ will factorize over n so that $q^{\star}(\mathbf{Z}) = \prod_n q^{\star}(\mathbf{z}_n)$. This is an example of an induced factorization. Taking the exponential of both sides, we have

$$q^{\star}(\mathbf{z}_n) = h(\mathbf{x}_n, \mathbf{z}_n) g\left(\mathbb{E}[\boldsymbol{\eta}]\right) \exp\left\{ \mathbb{E}[\boldsymbol{\eta}^{\mathrm{T}}]\mathbf{u}(\mathbf{x}_n, \mathbf{z}_n) \right\} \tag{10.116}$$

where the normalization coefficient has been re-instated by comparison with the standard form for the exponential family.

Similarly, for the variational distribution over the parameters, we have

$$\ln q^{\star}(\boldsymbol{\eta}) = \ln p(\boldsymbol{\eta}|\nu_0, \boldsymbol{\chi}_0) + \mathbb{E}_{\mathbf{Z}}[\ln p(\mathbf{X}, \mathbf{Z}|\boldsymbol{\eta})] + \text{const} \tag{10.117}$$

$$= \nu_0 \ln g(\boldsymbol{\eta}) + \nu_0 \boldsymbol{\eta}^{\mathrm{T}} \boldsymbol{\chi}_0 + \sum_{n=1}^{N} \left\{ \ln g(\boldsymbol{\eta}) + \boldsymbol{\eta}^{\mathrm{T}} \mathbb{E}_{\mathbf{z}_n}[\mathbf{u}(\mathbf{x}_n, \mathbf{z}_n)] \right\} + \text{const.} \tag{10.118}$$

Again, taking the exponential of both sides, and re-instating the normalization coefficient by inspection, we have

$$q^{\star}(\boldsymbol{\eta}) = f(\nu_N, \boldsymbol{\chi}_N) g(\boldsymbol{\eta})^{\nu_N} \exp\left\{ \nu_N \boldsymbol{\eta}^{\mathrm{T}} \boldsymbol{\chi}_N \right\} \tag{10.119}$$

where we have defined

$$\nu_N = \nu_0 + N \tag{10.120}$$

$$\nu_N \boldsymbol{\chi}_N = \nu_0 \boldsymbol{\chi}_0 + \sum_{n=1}^{N} \mathbb{E}_{\mathbf{z}_n}[\mathbf{u}(\mathbf{x}_n, \mathbf{z}_n)]. \tag{10.121}$$

Note that the solutions for $q^{\star}(\mathbf{z}_n)$ and $q^{\star}(\boldsymbol{\eta})$ are coupled, and so we solve them iteratively in a two-stage procedure. In the variational E step, we evaluate the expected sufficient statistics $\mathbb{E}[\mathbf{u}(\mathbf{x}_n, \mathbf{z}_n)]$ using the current posterior distribution $q(\mathbf{z}_n)$ over the latent variables and use this to compute a revised posterior distribution $q(\boldsymbol{\eta})$ over the parameters. Then in the subsequent variational M step, we use this revised parameter posterior distribution to find the expected natural parameters $\mathbb{E}[\boldsymbol{\eta}^{\mathrm{T}}]$, which gives rise to a revised variational distribution over the latent variables.

10.4.1 Variational message passing

We have illustrated the application of variational methods by considering a specific model, the Bayesian mixture of Gaussians, in some detail. This model can be described by the directed graph shown in Figure 10.5. Here we consider more generally the use of variational methods for models described by directed graphs and derive a number of widely applicable results.

The joint distribution corresponding to a directed graph can be written using the decomposition

$$p(\mathbf{x}) = \prod_i p(\mathbf{x}_i|\mathrm{pa}_i) \tag{10.122}$$

where $\mathbf{x}_i$ denotes the variable(s) associated with node i, and pa_i denotes the parent set corresponding to node i. Note that $\mathbf{x}_i$ may be a latent variable or it may belong to the set of observed variables. Now consider a variational approximation in which the distribution $q(\mathbf{x})$ is assumed to factorize with respect to the $\mathbf{x}_i$ so that

$$q(\mathbf{x}) = \prod_i q_i(\mathbf{x}_i). \tag{10.123}$$

Note that for observed nodes, there is no factor $q(\mathbf{x}_i)$ in the variational distribution. We now substitute (10.122) into our general result (10.9) to give

$$\ln q_j^\star(\mathbf{x}_j) = \mathbb{E}_{i\neq j}\left[\sum_i \ln p(\mathbf{x}_i|\mathrm{pa}_i)\right] + \text{const.} \tag{10.124}$$

Any terms on the right-hand side that do not depend on $\mathbf{x}_j$ can be absorbed into the additive constant. In fact, the only terms that do depend on $\mathbf{x}_j$ are the conditional distribution for $\mathbf{x}_j$ given by $p(\mathbf{x}_j|\mathrm{pa}_j)$ together with any other conditional distributions that have $\mathbf{x}_j$ in the conditioning set. By definition, these conditional distributions correspond to the children of node j, and they therefore also depend on the *co-parents* of the child nodes, i.e., the other parents of the child nodes besides node $\mathbf{x}_j$ itself. We see that the set of all nodes on which $q_j^\star(\mathbf{x}_j)$ depends corresponds to the Markov blanket of node $\mathbf{x}_j$, as illustrated in Figure 8.26. Thus the update of the factors in the variational posterior distribution represents a local calculation on the graph. This makes possible the construction of general purpose software for variational inference in which the form of the model does not need to be specified in advance (Bishop *et al.*, 2003).

If we now specialize to the case of a model in which all of the conditional distributions have a conjugate-exponential structure, then the variational update procedure can be cast in terms of a local message passing algorithm (Winn and Bishop, 2005). In particular, the distribution associated with a particular node can be updated once that node has received messages from all of its parents and all of its children. This in turn requires that the children have already received messages from their coparents. The evaluation of the lower bound can also be simplified because many of the required quantities are already evaluated as part of the message passing scheme. This distributed message passing formulation has good scaling properties and is well suited to large networks.

10.5. Local Variational Methods

The variational framework discussed in Sections 10.1 and 10.2 can be considered a 'global' method in the sense that it directly seeks an approximation to the full posterior distribution over all random variables. An alternative 'local' approach involves finding bounds on functions over individual variables or groups of variables within a model. For instance, we might seek a bound on a conditional distribution $p(y|x)$, which is itself just one factor in a much larger probabilistic model specified by a directed graph. The purpose of introducing the bound of course is to simplify the resulting distribution. This local approximation can be applied to multiple variables in turn until a tractable approximation is obtained, and in Section 10.6.1 we shall give a practical example of this approach in the context of logistic regression. Here we focus on developing the bounds themselves.

We have already seen in our discussion of the Kullback-Leibler divergence that the convexity of the logarithm function played a key role in developing the lower bound in the global variational approach. We have defined a (strictly) convex function as one for which every chord lies above the function. Convexity also plays a central role in the local variational framework. Note that our discussion will apply equally to concave functions with 'min' and 'max' interchanged and with lower bounds replaced by upper bounds.

Section 1.6.1

Let us begin by considering a simple example, namely the function $f(x) = \exp(-x)$, which is a convex function of x, and which is shown in the left-hand plot of Figure 10.10. Our goal is to approximate $f(x)$ by a simpler function, in particular a linear function of x. From Figure 10.10, we see that this linear function will be a lower bound on $f(x)$ if it corresponds to a tangent. We can obtain the tangent line $y(x)$ at a specific value of x, say $x = \xi$, by making a first order Taylor expansion

$$y(x) = f(\xi) + f'(\xi)(x - \xi) \tag{10.125}$$

so that $y(x) \leqslant f(x)$ with equality when $x = \xi$. For our example function $f(x) =$

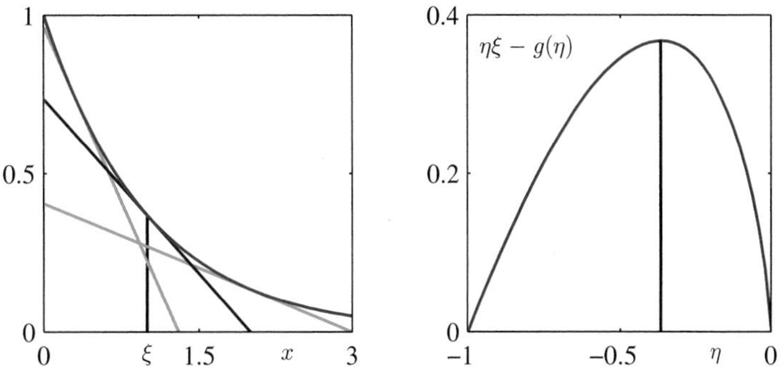

Figure 10.10 In the left-hand figure the red curve shows the function $\exp(-x)$, and the blue line shows the tangent at $x = \xi$ defined by (10.125) with $\xi = 1$. This line has slope $\eta = f'(\xi) = -\exp(-\xi)$. Note that any other tangent line, for example the ones shown in green, will have a smaller value of y at $x = \xi$. The right-hand figure shows the corresponding plot of the function $\eta\xi - g(\eta)$, where $g(\eta)$ is given by (10.131), versus η for $\xi = 1$, in which the maximum corresponds to $\eta = -\exp(-\xi) = -1/e$.

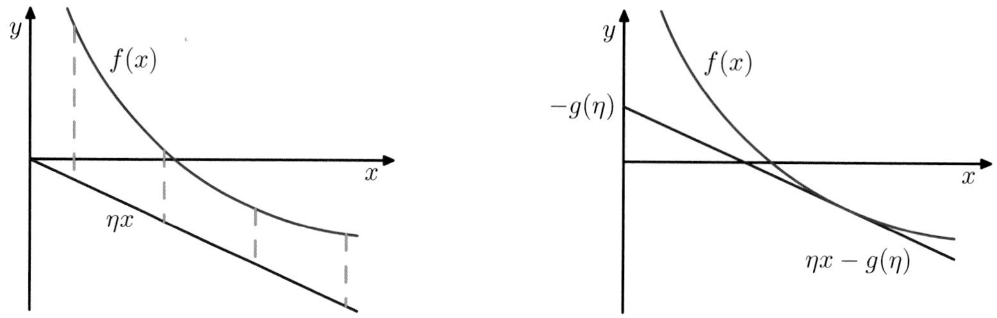

Figure 10.11 In the left-hand plot the red curve shows a convex function $f(x)$, and the blue line represents the linear function ηx, which is a lower bound on $f(x)$ because $f(x) > \eta x$ for all x. For the given value of slope η the contact point of the tangent line having the same slope is found by minimizing with respect to x the discrepancy (shown by the green dashed lines) given by $f(x) - \eta x$. This defines the dual function $g(\eta)$, which corresponds to the (negative of the) intercept of the tangent line having slope η.

$\exp(-x)$, we therefore obtain the tangent line in the form

$$y(x) = \exp(-\xi) - \exp(-\xi)(x - \xi) \tag{10.126}$$

which is a linear function parameterized by ξ. For consistency with subsequent discussion, let us define $\eta = -\exp(-\xi)$ so that

$$y(x, \eta) = \eta x - \eta + \eta \ln(-\eta). \tag{10.127}$$

Different values of η correspond to different tangent lines, and because all such lines are lower bounds on the function, we have $f(x) \geqslant y(x, \eta)$. Thus we can write the function in the form

$$f(x) = \max_{\eta} \{\eta x - \eta + \eta \ln(-\eta)\}. \tag{10.128}$$

We have succeeded in approximating the convex function $f(x)$ by a simpler, linear function $y(x, \eta)$. The price we have paid is that we have introduced a variational parameter η, and to obtain the tightest bound we must optimize with respect to η.

We can formulate this approach more generally using the framework of *convex duality* (Rockafellar, 1972; Jordan *et al.*, 1999). Consider the illustration of a convex function $f(x)$ shown in the left-hand plot in Figure 10.11. In this example, the function ηx is a lower bound on $f(x)$ but it is not the best lower bound that can be achieved by a linear function having slope η, because the tightest bound is given by the tangent line. Let us write the equation of the tangent line, having slope η as $\eta x - g(\eta)$ where the (negative) intercept $g(\eta)$ clearly depends on the slope η of the tangent. To determine the intercept, we note that the line must be moved vertically by an amount equal to the smallest vertical distance between the line and the function, as shown in Figure 10.11. Thus

$$\begin{aligned} g(\eta) &= -\min_{x} \{f(x) - \eta x\} \\ &= \max_{x} \{\eta x - f(x)\}. \end{aligned} \tag{10.129}$$

Now, instead of fixing η and varying x, we can consider a particular x and then adjust η until the tangent plane is tangent at that particular x. Because the y value of the tangent line at a particular x is maximized when that value coincides with its contact point, we have

$$f(x) = \max_{\eta} \left\{ \eta x - g(\eta) \right\}. \tag{10.130}$$

We see that the functions $f(x)$ and $g(\eta)$ play a dual role, and are related through (10.129) and (10.130).

Let us apply these duality relations to our simple example $f(x) = \exp(-x)$. From (10.129) we see that the maximizing value of x is given by $\xi = -\ln(-\eta)$, and back-substituting we obtain the conjugate function $g(\eta)$ in the form

$$g(\eta) = \eta - \eta \ln(-\eta) \tag{10.131}$$

as obtained previously. The function $\eta\xi - g(\eta)$ is shown, for $\xi = 1$ in the right-hand plot in Figure 10.10. As a check, we can substitute (10.131) into (10.130), which gives the maximizing value of $\eta = -\exp(-x)$, and back-substituting then recovers the original function $f(x) = \exp(-x)$.

For concave functions, we can follow a similar argument to obtain upper bounds, in which 'max' is replaced with 'min', so that

$$f(x) = \min_{\eta} \left\{ \eta x - g(\eta) \right\} \tag{10.132}$$

$$g(\eta) = \min_{x} \left\{ \eta x - f(x) \right\}. \tag{10.133}$$

If the function of interest is not convex (or concave), then we cannot directly apply the method above to obtain a bound. However, we can first seek invertible transformations either of the function or of its argument which change it into a convex form. We then calculate the conjugate function and then transform back to the original variables.

An important example, which arises frequently in pattern recognition, is the logistic sigmoid function defined by

$$\sigma(x) = \frac{1}{1 + e^{-x}}. \tag{10.134}$$

As it stands this function is neither convex nor concave. However, if we take the logarithm we obtain a function which is concave, as is easily verified by finding the second derivative. From (10.133) the corresponding conjugate function then takes the form

Exercise 10.30

$$g(\eta) = \min_{x} \left\{ \eta x - f(x) \right\} = -\eta \ln \eta - (1 - \eta) \ln(1 - \eta) \tag{10.135}$$

which we recognize as the binary entropy function for a variable whose probability of having the value 1 is η. Using (10.132), we then obtain an upper bound on the log sigmoid

Appendix B

$$\ln \sigma(x) \leqslant \eta x - g(\eta) \tag{10.136}$$

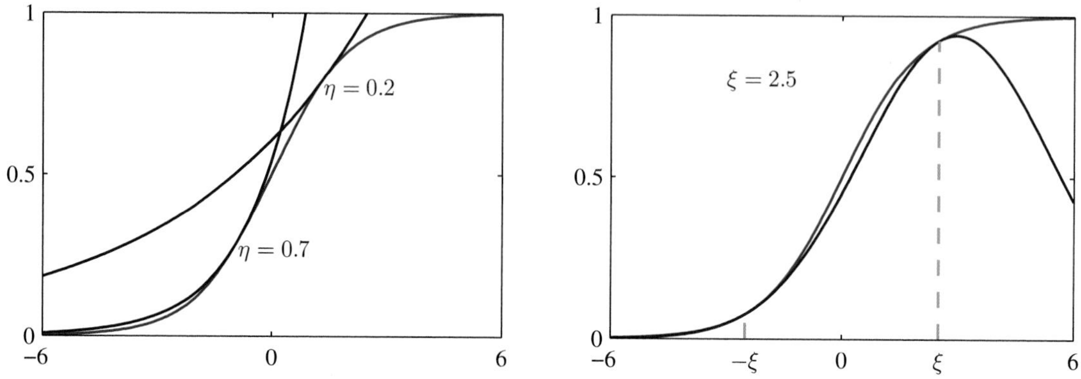

Figure 10.12 The left-hand plot shows the logistic sigmoid function $\sigma(x)$ defined by (10.134) in red, together with two examples of the exponential upper bound (10.137) shown in blue. The right-hand plot shows the logistic sigmoid again in red together with the Gaussian lower bound (10.144) shown in blue. Here the parameter $\xi = 2.5$, and the bound is exact at $x = \xi$ and $x = -\xi$, denoted by the dashed green lines.

and taking the exponential, we obtain an upper bound on the logistic sigmoid itself of the form

$$\sigma(x) \leqslant \exp(\eta x - g(\eta)) \tag{10.137}$$

which is plotted for two values of η on the left-hand plot in Figure 10.12.

We can also obtain a lower bound on the sigmoid having the functional form of a Gaussian. To do this, we follow Jaakkola and Jordan (2000) and make transformations both of the input variable and of the function itself. First we take the log of the logistic function and then decompose it so that

$$\begin{aligned} \ln \sigma(x) &= -\ln(1 + e^{-x}) = -\ln\left\{e^{-x/2}(e^{x/2} + e^{-x/2})\right\} \\ &= x/2 - \ln(e^{x/2} + e^{-x/2}). \end{aligned} \tag{10.138}$$

Exercise 10.31

We now note that the function $f(x) = -\ln(e^{x/2} + e^{-x/2})$ is a convex function of the variable x^2, as can again be verified by finding the second derivative. This leads to a lower bound on $f(x)$, which is a linear function of x^2 whose conjugate function is given by

$$g(\eta) = \max_{x^2}\left\{\eta x^2 - f\left(\sqrt{x^2}\right)\right\}. \tag{10.139}$$

The stationarity condition leads to

$$0 = \eta - \frac{dx}{dx^2}\frac{d}{dx}f(x) = \eta + \frac{1}{4x}\tanh\left(\frac{x}{2}\right). \tag{10.140}$$

If we denote this value of x, corresponding to the contact point of the tangent line for this particular value of η, by ξ, then we have

$$\eta = -\frac{1}{4\xi}\tanh\left(\frac{\xi}{2}\right) = -\frac{1}{2\xi}\left[\sigma(\xi) - \frac{1}{2}\right] = -\lambda(\xi) \tag{10.141}$$

where we have defined $\lambda = -\eta$ to maintain consistency with Jaakkola and Jordan (2000). Instead of thinking of λ as the variational parameter, we can let ξ play this role as this leads to simpler expressions for the conjugate function, which is then given by

$$g(\lambda(\xi)) = -\lambda(\xi)\xi^2 - f(\xi) = -\lambda(\xi)\xi^2 + \ln(e^{\xi/2} + e^{-\xi/2}). \tag{10.142}$$

Hence the bound on $f(x)$ can be written as

$$f(x) \geqslant -\lambda(\xi)x^2 - g(\lambda(\xi)) = -\lambda(\xi)x^2 + \lambda(\xi)\xi^2 - \ln(e^{\xi/2} + e^{-\xi/2}). \tag{10.143}$$

The bound on the sigmoid then becomes

$$\sigma(x) \geqslant \sigma(\xi) \exp\left\{(x-\xi)/2 - \lambda(\xi)(x^2 - \xi^2)\right\} \tag{10.144}$$

where $\lambda(\xi)$ is defined by (10.141). This bound is illustrated in the right-hand plot of Figure 10.12. We see that the bound has the form of the exponential of a quadratic function of x, which will prove useful when we seek Gaussian representations of posterior distributions defined through logistic sigmoid functions.

Section 4.5

The logistic sigmoid arises frequently in probabilistic models over binary variables because it is the function that transforms a log odds ratio into a posterior probability. The corresponding transformation for a multiclass distribution is given by the softmax function. Unfortunately, the lower bound derived here for the logistic sigmoid does not directly extend to the softmax. Gibbs (1997) proposes a method for constructing a Gaussian distribution that is conjectured to be a bound (although no rigorous proof is given), which may be used to apply local variational methods to multiclass problems.

Section 4.3

We shall see an example of the use of local variational bounds in Sections 10.6.1. For the moment, however, it is instructive to consider in general terms how these bounds can be used. Suppose we wish to evaluate an integral of the form

$$I = \int \sigma(a)p(a)\,\mathrm{d}a \tag{10.145}$$

where $\sigma(a)$ is the logistic sigmoid, and $p(a)$ is a Gaussian probability density. Such integrals arise in Bayesian models when, for instance, we wish to evaluate the predictive distribution, in which case $p(a)$ represents a posterior parameter distribution. Because the integral is intractable, we employ the variational bound (10.144), which we write in the form $\sigma(a) \geqslant f(a, \xi)$ where ξ is a variational parameter. The integral now becomes the product of two exponential-quadratic functions and so can be integrated analytically to give a bound on I

$$I \geqslant \int f(a, \xi)p(a)\,\mathrm{d}a = F(\xi). \tag{10.146}$$

We now have the freedom to choose the variational parameter ξ, which we do by finding the value $\xi^\star$ that maximizes the function $F(\xi)$. The resulting value $F(\xi^\star)$

represents the tightest bound within this family of bounds and can be used as an approximation to I. This optimized bound, however, will in general not be exact. Although the bound $\sigma(a) \geqslant f(a, \xi)$ on the logistic sigmoid can be optimized exactly, the required choice for ξ depends on the value of a, so that the bound is exact for one value of a only. Because the quantity $F(\xi)$ is obtained by integrating over all values of a, the value of $\xi^\star$ represents a compromise, weighted by the distribution $p(a)$.

10.6. Variational Logistic Regression

We now illustrate the use of local variational methods by returning to the Bayesian logistic regression model studied in Section 4.5. There we focussed on the use of the Laplace approximation, while here we consider a variational treatment based on the approach of Jaakkola and Jordan (2000). Like the Laplace method, this also leads to a Gaussian approximation to the posterior distribution. However, the greater flexibility of the variational approximation leads to improved accuracy compared to the Laplace method. Furthermore (unlike the Laplace method), the variational approach is optimizing a well defined objective function given by a rigourous bound on the model evidence. Logistic regression has also been treated by Dybowski and Roberts (2005) from a Bayesian perspective using Monte Carlo sampling techniques.

10.6.1 Variational posterior distribution

Here we shall make use of a variational approximation based on the local bounds introduced in Section 10.5. This allows the likelihood function for logistic regression, which is governed by the logistic sigmoid, to be approximated by the exponential of a quadratic form. It is therefore again convenient to choose a conjugate Gaussian prior of the form (4.140). For the moment, we shall treat the hyperparameters $\mathbf{m}_0$ and $\mathbf{S}_0$ as fixed constants. In Section 10.6.3, we shall demonstrate how the variational formalism can be extended to the case where there are unknown hyperparameters whose values are to be inferred from the data.

In the variational framework, we seek to maximize a lower bound on the marginal likelihood. For the Bayesian logistic regression model, the marginal likelihood takes the form

$$p(\mathbf{t}) = \int p(\mathbf{t}|\mathbf{w})p(\mathbf{w})\,\mathrm{d}\mathbf{w} = \int \left[\prod_{n=1}^{N} p(t_n|\mathbf{w})\right] p(\mathbf{w})\,\mathrm{d}\mathbf{w}. \tag{10.147}$$

We first note that the conditional distribution for t can be written as

$$\begin{aligned} p(t|\mathbf{w}) &= \sigma(a)^t \{1-\sigma(a)\}^{1-t} \\ &= \left(\frac{1}{1+e^{-a}}\right)^t \left(1 - \frac{1}{1+e^{-a}}\right)^{1-t} \\ &= e^{at}\frac{e^{-a}}{1+e^{-a}} = e^{at}\sigma(-a) \end{aligned} \tag{10.148}$$

where $a = \mathbf{w}^{\mathrm{T}}\boldsymbol{\phi}$. In order to obtain a lower bound on $p(\mathbf{t})$, we make use of the variational lower bound on the logistic sigmoid function given by (10.144), which we reproduce here for convenience

$$\sigma(z) \geqslant \sigma(\xi)\exp\left\{(z-\xi)/2 - \lambda(\xi)(z^2-\xi^2)\right\} \tag{10.149}$$

where

$$\lambda(\xi) = \frac{1}{2\xi}\left[\sigma(\xi) - \frac{1}{2}\right]. \tag{10.150}$$

We can therefore write

$$p(t|\mathbf{w}) = e^{at}\sigma(-a) \geqslant e^{at}\sigma(\xi)\exp\left\{-(a+\xi)/2 - \lambda(\xi)(a^2-\xi^2)\right\}. \tag{10.151}$$

Note that because this bound is applied to each of the terms in the likelihood function separately, there is a variational parameter ξ_n corresponding to each training set observation $(\boldsymbol{\phi}_n, t_n)$. Using $a = \mathbf{w}^{\mathrm{T}}\boldsymbol{\phi}$, and multiplying by the prior distribution, we obtain the following bound on the joint distribution of $\mathbf{t}$ and $\mathbf{w}$

$$p(\mathbf{t},\mathbf{w}) = p(\mathbf{t}|\mathbf{w})p(\mathbf{w}) \geqslant h(\mathbf{w},\boldsymbol{\xi})p(\mathbf{w}) \tag{10.152}$$

where $\boldsymbol{\xi}$ denotes the set $\{\xi_n\}$ of variational parameters, and

$$\begin{aligned} h(\mathbf{w},\boldsymbol{\xi}) &= \prod_{n=1}^{N}\sigma(\xi_n)\exp\left\{\mathbf{w}^{\mathrm{T}}\boldsymbol{\phi}_n t_n - (\mathbf{w}^{\mathrm{T}}\boldsymbol{\phi}_n + \xi_n)/2\right. \\ &\qquad \left. - \lambda(\xi_n)([\mathbf{w}^{\mathrm{T}}\boldsymbol{\phi}_n]^2 - \xi_n^2)\right\}. \end{aligned} \tag{10.153}$$

Evaluation of the exact posterior distribution would require normalization of the left-hand side of this inequality. Because this is intractable, we work instead with the right-hand side. Note that the function on the right-hand side cannot be interpreted as a probability density because it is not normalized. Once it is normalized to give a variational posterior distribution $q(\mathbf{w})$, however, it no longer represents a bound.

Because the logarithm function is monotonically increasing, the inequality $A \geqslant B$ implies $\ln A \geqslant \ln B$. This gives a lower bound on the log of the joint distribution of $\mathbf{t}$ and $\mathbf{w}$ of the form

$$\begin{aligned} \ln\{p(\mathbf{t}|\mathbf{w})p(\mathbf{w})\} &\geqslant \ln p(\mathbf{w}) + \sum_{n=1}^{N}\left\{\ln\sigma(\xi_n) + \mathbf{w}^{\mathrm{T}}\boldsymbol{\phi}_n t_n\right. \\ &\qquad \left. - (\mathbf{w}^{\mathrm{T}}\boldsymbol{\phi}_n + \xi_n)/2 - \lambda(\xi_n)([\mathbf{w}^{\mathrm{T}}\boldsymbol{\phi}_n]^2 - \xi_n^2)\right\}. \end{aligned} \tag{10.154}$$

Substituting for the prior $p(\mathbf{w})$, the right-hand side of this inequality becomes, as a function of $\mathbf{w}$

$$\begin{aligned} &-\frac{1}{2}(\mathbf{w}-\mathbf{m}_0)^{\mathrm{T}}\mathbf{S}_0^{-1}(\mathbf{w}-\mathbf{m}_0) \\ &+\sum_{n=1}^{N}\left\{\mathbf{w}^{\mathrm{T}}\boldsymbol{\phi}_n(t_n - 1/2) - \lambda(\xi_n)\mathbf{w}^{\mathrm{T}}(\boldsymbol{\phi}_n\boldsymbol{\phi}_n^{\mathrm{T}})\mathbf{w}\right\} + \text{const}. \end{aligned} \tag{10.155}$$

This is a quadratic function of $\mathbf{w}$, and so we can obtain the corresponding variational approximation to the posterior distribution by identifying the linear and quadratic terms in $\mathbf{w}$, giving a Gaussian variational posterior of the form

$$q(\mathbf{w}) = \mathcal{N}(\mathbf{w}|\mathbf{m}_N, \mathbf{S}_N) \tag{10.156}$$

where

$$\mathbf{m}_N = \mathbf{S}_N \left(\mathbf{S}_0^{-1}\mathbf{m}_0 + \sum_{n=1}^{N} (t_n - 1/2)\boldsymbol{\phi}_n \right) \tag{10.157}$$

$$\mathbf{S}_N^{-1} = \mathbf{S}_0^{-1} + 2\sum_{n=1}^{N} \lambda(\xi_n)\boldsymbol{\phi}_n\boldsymbol{\phi}_n^{\mathrm{T}}. \tag{10.158}$$

As with the Laplace framework, we have again obtained a Gaussian approximation to the posterior distribution. However, the additional flexibility provided by the variational parameters $\{\xi_n\}$ leads to improved accuracy in the approximation (Jaakkola and Jordan, 2000).

Here we have considered a batch learning context in which all of the training data is available at once. However, Bayesian methods are intrinsically well suited to sequential learning in which the data points are processed one at a time and then discarded. The formulation of this variational approach for the sequential case is straightforward.

Exercise 10.32

Note that the bound given by (10.149) applies only to the two-class problem and so this approach does not directly generalize to classification problems with $K > 2$ classes. An alternative bound for the multiclass case has been explored by Gibbs (1997).

10.6.2 Optimizing the variational parameters

We now have a normalized Gaussian approximation to the posterior distribution, which we shall use shortly to evaluate the predictive distribution for new data points. First, however, we need to determine the variational parameters $\{\xi_n\}$ by maximizing the lower bound on the marginal likelihood.

To do this, we substitute the inequality (10.152) back into the marginal likelihood to give

$$\ln p(\mathbf{t}) = \ln \int p(\mathbf{t}|\mathbf{w})p(\mathbf{w})\,\mathrm{d}\mathbf{w} \geqslant \ln \int h(\mathbf{w}, \boldsymbol{\xi})p(\mathbf{w})\,\mathrm{d}\mathbf{w} = \mathcal{L}(\boldsymbol{\xi}). \tag{10.159}$$

As with the optimization of the hyperparameter α in the linear regression model of Section 3.5, there are two approaches to determining the ξ_n. In the first approach, we recognize that the function $\mathcal{L}(\boldsymbol{\xi})$ is defined by an integration over $\mathbf{w}$ and so we can view $\mathbf{w}$ as a latent variable and invoke the EM algorithm. In the second approach, we integrate over $\mathbf{w}$ analytically and then perform a direct maximization over $\boldsymbol{\xi}$. Let us begin by considering the EM approach.

The EM algorithm starts by choosing some initial values for the parameters $\{\xi_n\}$, which we denote collectively by $\boldsymbol{\xi}^{\mathrm{old}}$. In the E step of the EM algorithm, we then use these parameter values to find the posterior distribution over $\mathbf{w}$, which is given by (10.156). In the M step, we then maximize the expected complete-data log likelihood which is given by

$$Q(\boldsymbol{\xi}, \boldsymbol{\xi}^{\text{old}}) = \mathbb{E}\left[\ln \{h(\mathbf{w}, \boldsymbol{\xi})p(\mathbf{w})\}\right] \tag{10.160}$$

where the expectation is taken with respect to the posterior distribution $q(\mathbf{w})$ evaluated using $\boldsymbol{\xi}^{\text{old}}$. Noting that $p(\mathbf{w})$ does not depend on $\boldsymbol{\xi}$, and substituting for $h(\mathbf{w}, \boldsymbol{\xi})$ we obtain

$$Q(\boldsymbol{\xi}, \boldsymbol{\xi}^{\text{old}}) = \sum_{n=1}^{N} \left\{ \ln \sigma(\xi_n) - \xi_n/2 - \lambda(\xi_n)(\boldsymbol{\phi}_n^{\mathrm{T}} \mathbb{E}[\mathbf{w}\mathbf{w}^{\mathrm{T}}]\boldsymbol{\phi}_n - \xi_n^2) \right\} + \text{const} \tag{10.161}$$

where 'const' denotes terms that are independent of $\boldsymbol{\xi}$. We now set the derivative with respect to ξ_n equal to zero. A few lines of algebra, making use of the definitions of $\sigma(\xi)$ and $\lambda(\xi)$, then gives

$$0 = \lambda'(\xi_n)(\boldsymbol{\phi}_n^{\mathrm{T}} \mathbb{E}[\mathbf{w}\mathbf{w}^{\mathrm{T}}]\boldsymbol{\phi}_n - \xi_n^2). \tag{10.162}$$

We now note that $\lambda'(\xi)$ is a monotonic function of ξ for $\xi \geqslant 0$, and that we can restrict attention to nonnegative values of ξ without loss of generality due to the symmetry of the bound around $\xi = 0$. Thus $\lambda'(\xi) \neq 0$, and hence we obtain the following re-estimation equations

Exercise 10.33

$$(\xi_n^{\text{new}})^2 = \boldsymbol{\phi}_n^{\mathrm{T}} \mathbb{E}[\mathbf{w}\mathbf{w}^{\mathrm{T}}]\boldsymbol{\phi}_n = \boldsymbol{\phi}_n^{\mathrm{T}} \left(\mathbf{S}_N + \mathbf{m}_N \mathbf{m}_N^{\mathrm{T}}\right) \boldsymbol{\phi}_n \tag{10.163}$$

where we have used (10.156).

Let us summarize the EM algorithm for finding the variational posterior distribution. We first initialize the variational parameters $\boldsymbol{\xi}^{\text{old}}$. In the E step, we evaluate the posterior distribution over $\mathbf{w}$ given by (10.156), in which the mean and covariance are defined by (10.157) and (10.158). In the M step, we then use this variational posterior to compute a new value for $\boldsymbol{\xi}$ given by (10.163). The E and M steps are repeated until a suitable convergence criterion is satisfied, which in practice typically requires only a few iterations.

An alternative approach to obtaining re-estimation equations for $\boldsymbol{\xi}$ is to note that in the integral over $\mathbf{w}$ in the definition (10.159) of the lower bound $\mathcal{L}(\boldsymbol{\xi})$, the integrand has a Gaussian-like form and so the integral can be evaluated analytically. Having evaluated the integral, we can then differentiate with respect to ξ_n. It turns out that this gives rise to exactly the same re-estimation equations as does the EM approach given by (10.163).

Exercise 10.34

As we have emphasized already, in the application of variational methods it is useful to be able to evaluate the lower bound $\mathcal{L}(\boldsymbol{\xi})$ given by (10.159). The integration over $\mathbf{w}$ can be performed analytically by noting that $p(\mathbf{w})$ is Gaussian and $h(\mathbf{w}, \boldsymbol{\xi})$ is the exponential of a quadratic function of $\mathbf{w}$. Thus, by completing the square and making use of the standard result for the normalization coefficient of a Gaussian distribution, we can obtain a closed form solution which takes the form

Exercise 10.35

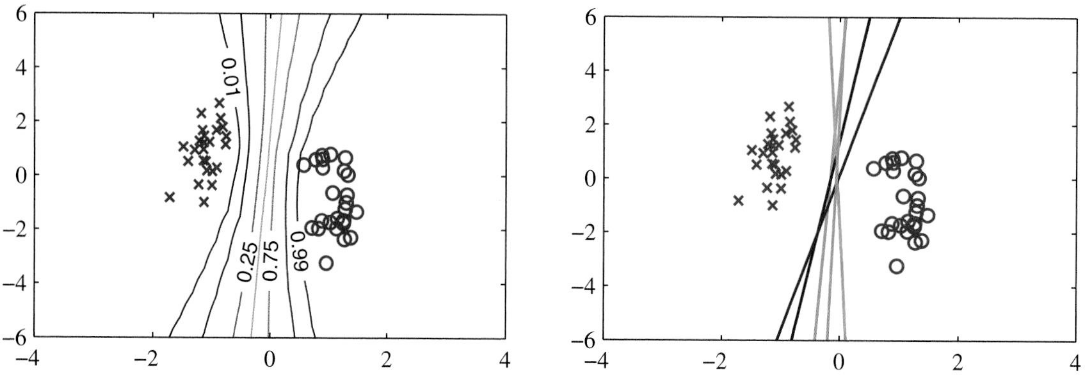

Figure 10.13 Illustration of the Bayesian approach to logistic regression for a simple linearly separable data set. The plot on the left shows the predictive distribution obtained using variational inference. We see that the decision boundary lies roughly mid way between the clusters of data points, and that the contours of the predictive distribution splay out away from the data reflecting the greater uncertainty in the classification of such regions. The plot on the right shows the decision boundaries corresponding to five samples of the parameter vector $\mathbf{w}$ drawn from the posterior distribution $p(\mathbf{w}|\mathbf{t})$.

$$\mathcal{L}(\boldsymbol{\xi}) = \frac{1}{2}\ln\frac{|\mathbf{S}_N|}{|\mathbf{S}_0|} + \frac{1}{2}\mathbf{m}_N^{\mathrm{T}}\mathbf{S}_N^{-1}\mathbf{m}_N - \frac{1}{2}\mathbf{m}_0^{\mathrm{T}}\mathbf{S}_0^{-1}\mathbf{m}_0 + \sum_{n=1}^{N}\left\{\ln\sigma(\xi_n) - \frac{1}{2}\xi_n + \lambda(\xi_n)\xi_n^2\right\}. \tag{10.164}$$

This variational framework can also be applied to situations in which the data is arriving sequentially (Jaakkola and Jordan, 2000). In this case we maintain a Gaussian posterior distribution over $\mathbf{w}$, which is initialized using the prior $p(\mathbf{w})$. As each data point arrives, the posterior is updated by making use of the bound (10.151) and then normalized to give an updated posterior distribution.

The predictive distribution is obtained by marginalizing over the posterior distribution, and takes the same form as for the Laplace approximation discussed in Section 4.5.2. Figure 10.13 shows the variational predictive distributions for a synthetic data set. This example provides interesting insights into the concept of 'large margin', which was discussed in Section 7.1 and which has qualitatively similar behaviour to the Bayesian solution.

10.6.3 Inference of hyperparameters

So far, we have treated the hyperparameter α in the prior distribution as a known constant. We now extend the Bayesian logistic regression model to allow the value of this parameter to be inferred from the data set. This can be achieved by combining the global and local variational approximations into a single framework, so as to maintain a lower bound on the marginal likelihood at each stage. Such a combined approach was adopted by Bishop and Svensén (2003) in the context of a Bayesian treatment of the hierarchical mixture of experts model.

Specifically, we consider once again a simple isotropic Gaussian prior distribution of the form

$$p(\mathbf{w}|\alpha) = \mathcal{N}(\mathbf{w}|\mathbf{0}, \alpha^{-1}\mathbf{I}). \tag{10.165}$$

Our analysis is readily extended to more general Gaussian priors, for instance if we wish to associate a different hyperparameter with different subsets of the parameters w_j. As usual, we consider a conjugate hyperprior over α given by a gamma distribution

$$p(\alpha) = \mathrm{Gam}(\alpha|a_0, b_0) \tag{10.166}$$

governed by the constants a_0 and b_0.

The marginal likelihood for this model now takes the form

$$p(\mathbf{t}) = \iint p(\mathbf{w}, \alpha, \mathbf{t})\,\mathrm{d}\mathbf{w}\,\mathrm{d}\alpha \tag{10.167}$$

where the joint distribution is given by

$$p(\mathbf{w}, \alpha, \mathbf{t}) = p(\mathbf{t}|\mathbf{w})p(\mathbf{w}|\alpha)p(\alpha). \tag{10.168}$$

We are now faced with an analytically intractable integration over $\mathbf{w}$ and α, which we shall tackle by using both the local and global variational approaches in the same model

To begin with, we introduce a variational distribution $q(\mathbf{w}, \alpha)$, and then apply the decomposition (10.2), which in this instance takes the form

$$\ln p(\mathbf{t}) = \mathcal{L}(q) + \mathrm{KL}(q\|p) \tag{10.169}$$

where the lower bound $\mathcal{L}(q)$ and the Kullback-Leibler divergence $\mathrm{KL}(q\|p)$ are defined by

$$\mathcal{L}(q) = \iint q(\mathbf{w}, \alpha) \ln \left\{ \frac{p(\mathbf{w}, \alpha, \mathbf{t})}{q(\mathbf{w}, \alpha)} \right\} \mathrm{d}\mathbf{w}\,\mathrm{d}\alpha \tag{10.170}$$

$$\mathrm{KL}(q\|p) = -\iint q(\mathbf{w}, \alpha) \ln \left\{ \frac{p(\mathbf{w}, \alpha|\mathbf{t}))}{q(\mathbf{w}, \alpha)} \right\} \mathrm{d}\mathbf{w}\,\mathrm{d}\alpha. \tag{10.171}$$

At this point, the lower bound $\mathcal{L}(q)$ is still intractable due to the form of the likelihood factor $p(\mathbf{t}|\mathbf{w})$. We therefore apply the local variational bound to each of the logistic sigmoid factors as before. This allows us to use the inequality (10.152) and place a lower bound on $\mathcal{L}(q)$, which will therefore also be a lower bound on the log marginal likelihood

$$\begin{aligned}\ln p(\mathbf{t}) &\geqslant \mathcal{L}(q) \geqslant \widetilde{\mathcal{L}}(q, \boldsymbol{\xi}) \\ &= \iint q(\mathbf{w}, \alpha) \ln \left\{ \frac{h(\mathbf{w}, \boldsymbol{\xi})p(\mathbf{w}|\alpha)p(\alpha)}{q(\mathbf{w}, \alpha)} \right\} \mathrm{d}\mathbf{w}\,\mathrm{d}\alpha. \end{aligned} \tag{10.172}$$

Next we assume that the variational distribution factorizes between parameters and hyperparameters so that

$$q(\mathbf{w}, \alpha) = q(\mathbf{w})q(\alpha). \tag{10.173}$$

With this factorization we can appeal to the general result (10.9) to find expressions for the optimal factors. Consider first the distribution $q(\mathbf{w})$. Discarding terms that are independent of $\mathbf{w}$, we have

$$\begin{aligned}\ln q(\mathbf{w}) &= \mathbb{E}_{\alpha}\left[\ln \{h(\mathbf{w}, \boldsymbol{\xi})p(\mathbf{w}|\alpha)p(\alpha)\}\right] + \text{const} \\ &= \ln h(\mathbf{w}, \boldsymbol{\xi}) + \mathbb{E}_{\alpha}\left[\ln p(\mathbf{w}|\alpha)\right] + \text{const.}\end{aligned}$$

We now substitute for $\ln h(\mathbf{w}, \boldsymbol{\xi})$ using (10.153), and for $\ln p(\mathbf{w}|\alpha)$ using (10.165), giving

$$\ln q(\mathbf{w}) = -\frac{\mathbb{E}[\alpha]}{2}\mathbf{w}^{\mathrm{T}}\mathbf{w} + \sum_{n=1}^{N}\left\{(t_n - 1/2)\mathbf{w}^{\mathrm{T}}\boldsymbol{\phi}_n - \lambda(\xi_n)\mathbf{w}^{\mathrm{T}}\boldsymbol{\phi}_n\boldsymbol{\phi}_n^{\mathrm{T}}\mathbf{w}\right\} + \text{const.}$$

We see that this is a quadratic function of $\mathbf{w}$ and so the solution for $q(\mathbf{w})$ will be Gaussian. Completing the square in the usual way, we obtain

$$q(\mathbf{w}) = \mathcal{N}(\mathbf{w}|\boldsymbol{\mu}_N, \boldsymbol{\Sigma}_N) \tag{10.174}$$

where we have defined

$$\boldsymbol{\Sigma}_N^{-1}\boldsymbol{\mu}_N = \sum_{n=1}^{N}(t_n - 1/2)\boldsymbol{\phi}_n \tag{10.175}$$

$$\boldsymbol{\Sigma}_N^{-1} = \mathbb{E}[\alpha]\mathbf{I} + 2\sum_{n=1}^{N}\lambda(\xi_n)\boldsymbol{\phi}_n\boldsymbol{\phi}_n^{\mathrm{T}}. \tag{10.176}$$

Similarly, the optimal solution for the factor $q(\alpha)$ is obtained from

$$\ln q(\alpha) = \mathbb{E}_{\mathbf{w}}\left[\ln p(\mathbf{w}|\alpha)\right] + \ln p(\alpha) + \text{const.}$$

Substituting for $\ln p(\mathbf{w}|\alpha)$ using (10.165), and for $\ln p(\alpha)$ using (10.166), we obtain

$$\ln q(\alpha) = \frac{M}{2}\ln\alpha - \frac{\alpha}{2}\mathbb{E}\left[\mathbf{w}^{\mathrm{T}}\mathbf{w}\right] + (a_0 - 1)\ln\alpha - b_0\alpha + \text{const.}$$

We recognize this as the log of a gamma distribution, and so we obtain

$$q(\alpha) = \text{Gam}(\alpha|a_N, b_N) = \frac{1}{\Gamma(a_N)}a_N^{b_N}\alpha^{a_N-1}e^{-b_N\alpha} \tag{10.177}$$

where

$$a_N = a_0 + \frac{M}{2} \tag{10.178}$$

$$b_N = b_0 + \frac{1}{2}\mathbb{E}_{\mathbf{w}}\left[\mathbf{w}^{\mathrm{T}}\mathbf{w}\right]. \tag{10.179}$$

We also need to optimize the variational parameters ξ_n, and this is also done by maximizing the lower bound $\widetilde{\mathcal{L}}(q, \boldsymbol{\xi})$. Omitting terms that are independent of $\boldsymbol{\xi}$, and integrating over α, we have

$$\widetilde{\mathcal{L}}(q, \boldsymbol{\xi}) = \int q(\mathbf{w}) \ln h(\mathbf{w}, \boldsymbol{\xi}) \, \mathrm{d}\mathbf{w} + \text{const.} \tag{10.180}$$

Note that this has precisely the same form as (10.160), and so we can again appeal to our earlier result (10.163), which can be obtained by direct optimization of the marginal likelihood function, leading to re-estimation equations of the form

$$(\xi_n^{\text{new}})^2 = \boldsymbol{\phi}_n^{\mathrm{T}} \left(\boldsymbol{\Sigma}_N + \boldsymbol{\mu}_N \boldsymbol{\mu}_N^{\mathrm{T}} \right) \boldsymbol{\phi}_n. \tag{10.181}$$

Appendix B

We have obtained re-estimation equations for the three quantities $q(\mathbf{w})$, $q(\alpha)$, and $\boldsymbol{\xi}$, and so after making suitable initializations, we can cycle through these quantities, updating each in turn. The required moments are given by

$$\mathbb{E}[\alpha] = \frac{a_N}{b_N} \tag{10.182}$$

$$\mathbb{E}\left[\mathbf{w}\mathbf{w}^{\mathrm{T}}\right] = \boldsymbol{\Sigma}_N + \boldsymbol{\mu}_N \boldsymbol{\mu}_N^{\mathrm{T}}. \tag{10.183}$$

10.7. Expectation Propagation

We conclude this chapter by discussing an alternative form of deterministic approximate inference, known as *expectation propagation* or *EP* (Minka, 2001a; Minka, 2001b). As with the variational Bayes methods discussed so far, this too is based on the minimization of a Kullback-Leibler divergence but now of the reverse form, which gives the approximation rather different properties.

Consider for a moment the problem of minimizing $\mathrm{KL}(p\|q)$ with respect to $q(\mathbf{z})$ when $p(\mathbf{z})$ is a fixed distribution and $q(\mathbf{z})$ is a member of the exponential family and so, from (2.194), can be written in the form

$$q(\mathbf{z}) = h(\mathbf{z}) g(\boldsymbol{\eta}) \exp\left\{\boldsymbol{\eta}^{\mathrm{T}} \mathbf{u}(\mathbf{z})\right\}. \tag{10.184}$$

As a function of $\boldsymbol{\eta}$, the Kullback-Leibler divergence then becomes

$$\mathrm{KL}(p\|q) = -\ln g(\boldsymbol{\eta}) - \boldsymbol{\eta}^{\mathrm{T}} \mathbb{E}_{p(\mathbf{z})}[\mathbf{u}(\mathbf{z})] + \text{const} \tag{10.185}$$

where the constant terms are independent of the natural parameters $\boldsymbol{\eta}$. We can minimize $\mathrm{KL}(p\|q)$ within this family of distributions by setting the gradient with respect to $\boldsymbol{\eta}$ to zero, giving

$$-\nabla \ln g(\boldsymbol{\eta}) = \mathbb{E}_{p(\mathbf{z})}[\mathbf{u}(\mathbf{z})]. \tag{10.186}$$

However, we have already seen in (2.226) that the negative gradient of $\ln g(\boldsymbol{\eta})$ is given by the expectation of $\mathbf{u}(\mathbf{z})$ under the distribution $q(\mathbf{z})$. Equating these two results, we obtain

$$\mathbb{E}_{q(\mathbf{z})}[\mathbf{u}(\mathbf{z})] = \mathbb{E}_{p(\mathbf{z})}[\mathbf{u}(\mathbf{z})]. \tag{10.187}$$

We see that the optimum solution simply corresponds to matching the expected sufficient statistics. So, for instance, if $q(\mathbf{z})$ is a Gaussian $\mathcal{N}(\mathbf{z}|\boldsymbol{\mu}, \boldsymbol{\Sigma})$ then we minimize the Kullback-Leibler divergence by setting the mean $\boldsymbol{\mu}$ of $q(\mathbf{z})$ equal to the mean of the distribution $p(\mathbf{z})$ and the covariance $\boldsymbol{\Sigma}$ equal to the covariance of $p(\mathbf{z})$. This is sometimes called *moment matching*. An example of this was seen in Figure 10.3(a).

Now let us exploit this result to obtain a practical algorithm for approximate inference. For many probabilistic models, the joint distribution of data $\mathcal{D}$ and hidden variables (including parameters) $\boldsymbol{\theta}$ comprises a product of factors in the form

$$p(\mathcal{D}, \boldsymbol{\theta}) = \prod_i f_i(\boldsymbol{\theta}). \tag{10.188}$$

This would arise, for example, in a model for independent, identically distributed data in which there is one factor $f_n(\boldsymbol{\theta}) = p(\mathbf{x}_n|\boldsymbol{\theta})$ for each data point $\mathbf{x}_n$, along with a factor $f_0(\boldsymbol{\theta}) = p(\boldsymbol{\theta})$ corresponding to the prior. More generally, it would also apply to any model defined by a directed probabilistic graph in which each factor is a conditional distribution corresponding to one of the nodes, or an undirected graph in which each factor is a clique potential. We are interested in evaluating the posterior distribution $p(\boldsymbol{\theta}|\mathcal{D})$ for the purpose of making predictions, as well as the model evidence $p(\mathcal{D})$ for the purpose of model comparison. From (10.188) the posterior is given by

$$p(\boldsymbol{\theta}|\mathcal{D}) = \frac{1}{p(\mathcal{D})} \prod_i f_i(\boldsymbol{\theta}) \tag{10.189}$$

and the model evidence is given by

$$p(\mathcal{D}) = \int \prod_i f_i(\boldsymbol{\theta}) \, \mathrm{d}\boldsymbol{\theta}. \tag{10.190}$$

Here we are considering continuous variables, but the following discussion applies equally to discrete variables with integrals replaced by summations. We shall suppose that the marginalization over $\boldsymbol{\theta}$, along with the marginalizations with respect to the posterior distribution required to make predictions, are intractable so that some form of approximation is required.

Expectation propagation is based on an approximation to the posterior distribution which is also given by a product of factors

$$q(\boldsymbol{\theta}) = \frac{1}{Z} \prod_i \widetilde{f}_i(\boldsymbol{\theta}) \tag{10.191}$$

in which each factor $\widetilde{f}_i(\boldsymbol{\theta})$ in the approximation corresponds to one of the factors $f_i(\boldsymbol{\theta})$ in the true posterior (10.189), and the factor $1/Z$ is the normalizing constant needed to ensure that the left-hand side of (10.191) integrates to unity. In order to obtain a practical algorithm, we need to constrain the factors $\widetilde{f}_i(\boldsymbol{\theta})$ in some way, and in particular we shall assume that they come from the exponential family. The product of the factors will therefore also be from the exponential family and so can be described by a finite set of sufficient statistics. For example, if each of the $\widetilde{f}_i(\boldsymbol{\theta})$ is a Gaussian, then the overall approximation $q(\boldsymbol{\theta})$ will also be Gaussian.

Ideally we would like to determine the $\widetilde{f}_i(\boldsymbol{\theta})$ by minimizing the Kullback-Leibler divergence between the true posterior and the approximation given by

$$\mathrm{KL}\,(p\|q) = \mathrm{KL}\left(\frac{1}{p(\mathcal{D})}\prod_i f_i(\boldsymbol{\theta}) \middle\| \frac{1}{Z}\prod_i \widetilde{f}_i(\boldsymbol{\theta})\right). \tag{10.192}$$

Note that this is the reverse form of KL divergence compared with that used in variational inference. In general, this minimization will be intractable because the KL divergence involves averaging with respect to the true distribution. As a rough approximation, we could instead minimize the KL divergences between the corresponding pairs $f_i(\boldsymbol{\theta})$ and $\widetilde{f}_i(\boldsymbol{\theta})$ of factors. This represents a much simpler problem to solve, and has the advantage that the algorithm is noniterative. However, because each factor is individually approximated, the product of the factors could well give a poor approximation.

Expectation propagation makes a much better approximation by optimizing each factor in turn in the context of all of the remaining factors. It starts by initializing the factors $\widetilde{f}_i(\boldsymbol{\theta})$, and then cycles through the factors refining them one at a time. This is similar in spirit to the update of factors in the variational Bayes framework considered earlier. Suppose we wish to refine factor $\widetilde{f}_j(\boldsymbol{\theta})$. We first remove this factor from the product to give $\prod_{i\neq j}\widetilde{f}_i(\boldsymbol{\theta})$. Conceptually, we will now determine a revised form of the factor $\widetilde{f}_j(\boldsymbol{\theta})$ by ensuring that the product

$$q^{\mathrm{new}}(\boldsymbol{\theta}) \propto \widetilde{f}_j(\boldsymbol{\theta})\prod_{i\neq j}\widetilde{f}_i(\boldsymbol{\theta}) \tag{10.193}$$

is as close as possible to

$$f_j(\boldsymbol{\theta})\prod_{i\neq j}\widetilde{f}_i(\boldsymbol{\theta}) \tag{10.194}$$

in which we keep fixed all of the factors $\widetilde{f}_i(\boldsymbol{\theta})$ for $i \neq j$. This ensures that the approximation is most accurate in the regions of high posterior probability as defined by the remaining factors. We shall see an example of this effect when we apply EP to the 'clutter problem'. To achieve this, we first remove the factor $\widetilde{f}_j(\boldsymbol{\theta})$ from the current approximation to the posterior by defining the unnormalized distribution

Section 10.7.1

$$q^{\backslash j}(\boldsymbol{\theta}) = \frac{q(\boldsymbol{\theta})}{\widetilde{f}_j(\boldsymbol{\theta})}. \tag{10.195}$$

Note that we could instead find $q^{\backslash j}(\boldsymbol{\theta})$ from the product of factors $i \neq j$, although in practice division is usually easier. This is now combined with the factor $f_j(\boldsymbol{\theta})$ to give a distribution

$$\frac{1}{Z_j}f_j(\boldsymbol{\theta})q^{\backslash j}(\boldsymbol{\theta}) \tag{10.196}$$

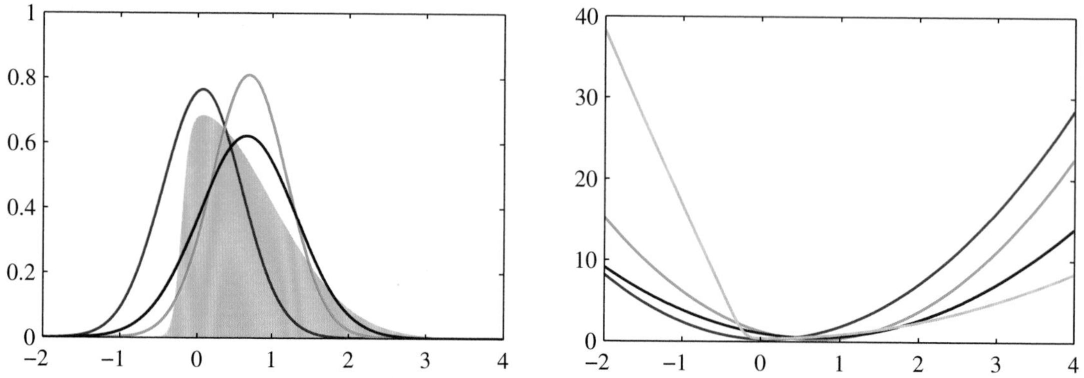

Figure 10.14 Illustration of the expectation propagation approximation using a Gaussian distribution for the example considered earlier in Figures 4.14 and 10.1. The left-hand plot shows the original distribution (yellow) along with the Laplace (red), global variational (green), and EP (blue) approximations, and the right-hand plot shows the corresponding negative logarithms of the distributions. Note that the EP distribution is broader than that obtained by variational inference, as a consequence of the different form of KL divergence.

where Z_j is the normalization constant given by

$$Z_j = \int f_j(\boldsymbol{\theta}) q^{\backslash j}(\boldsymbol{\theta}) \, \mathrm{d}\boldsymbol{\theta}. \tag{10.197}$$

We now determine a revised factor $\widetilde{f}_j(\boldsymbol{\theta})$ by minimizing the Kullback-Leibler divergence

$$\mathrm{KL}\left(\frac{f_j(\boldsymbol{\theta}) q^{\backslash j}(\boldsymbol{\theta})}{Z_j} \middle\| q^{\mathrm{new}}(\boldsymbol{\theta}) \right). \tag{10.198}$$

This is easily solved because the approximating distribution $q^{\mathrm{new}}(\boldsymbol{\theta})$ is from the exponential family, and so we can appeal to the result (10.187), which tells us that the parameters of $q^{\mathrm{new}}(\boldsymbol{\theta})$ are obtained by matching its expected sufficient statistics to the corresponding moments of (10.196). We shall assume that this is a tractable operation. For example, if we choose $q(\boldsymbol{\theta})$ to be a Gaussian distribution $\mathcal{N}(\boldsymbol{\theta}|\boldsymbol{\mu}, \boldsymbol{\Sigma})$, then $\boldsymbol{\mu}$ is set equal to the mean of the (unnormalized) distribution $f_j(\boldsymbol{\theta}) q^{\backslash j}(\boldsymbol{\theta})$, and $\boldsymbol{\Sigma}$ is set to its covariance. More generally, it is straightforward to obtain the required expectations for any member of the exponential family, provided it can be normalized, because the expected statistics can be related to the derivatives of the normalization coefficient, as given by (2.226). The EP approximation is illustrated in Figure 10.14.

From (10.193), we see that the revised factor $\widetilde{f}_j(\boldsymbol{\theta})$ can be found by taking $q^{\mathrm{new}}(\boldsymbol{\theta})$ and dividing out the remaining factors so that

$$\widetilde{f}_j(\boldsymbol{\theta}) = K \frac{q^{\mathrm{new}}(\boldsymbol{\theta})}{q^{\backslash j}(\boldsymbol{\theta})} \tag{10.199}$$

where we have used (10.195). The coefficient K is determined by multiplying both sides of (10.199) by $q^{\backslash j}(\boldsymbol{\theta})$ and integrating to give

$$K = \int \widetilde{f}_j(\boldsymbol{\theta}) q^{\backslash j}(\boldsymbol{\theta})\,\mathrm{d}\boldsymbol{\theta} \tag{10.200}$$

where we have used the fact that $q^{\mathrm{new}}(\boldsymbol{\theta})$ is normalized. The value of K can therefore be found by matching zeroth-order moments

$$\int \widetilde{f}_j(\boldsymbol{\theta}) q^{\backslash j}(\boldsymbol{\theta})\,\mathrm{d}\boldsymbol{\theta} = \int f_j(\boldsymbol{\theta}) q^{\backslash j}(\boldsymbol{\theta})\,\mathrm{d}\boldsymbol{\theta}. \tag{10.201}$$

Combining this with (10.197), we then see that $K = Z_j$ and so can be found by evaluating the integral in (10.197).

In practice, several passes are made through the set of factors, revising each factor in turn. The posterior distribution $p(\boldsymbol{\theta}|\mathcal{D})$ is then approximated using (10.191), and the model evidence $p(\mathcal{D})$ can be approximated by using (10.190) with the factors $f_i(\boldsymbol{\theta})$ replaced by their approximations $\widetilde{f}_i(\boldsymbol{\theta})$.

Expectation Propagation

We are given a joint distribution over observed data $\mathcal{D}$ and stochastic variables $\boldsymbol{\theta}$ in the form of a product of factors

$$p(\mathcal{D}, \boldsymbol{\theta}) = \prod_i f_i(\boldsymbol{\theta}) \tag{10.202}$$

and we wish to approximate the posterior distribution $p(\boldsymbol{\theta}|\mathcal{D})$ by a distribution of the form

$$q(\boldsymbol{\theta}) = \frac{1}{Z} \prod_i \widetilde{f}_i(\boldsymbol{\theta}). \tag{10.203}$$

We also wish to approximate the model evidence $p(\mathcal{D})$.

1. Initialize all of the approximating factors $\widetilde{f}_i(\boldsymbol{\theta})$.
2. Initialize the posterior approximation by setting

$$q(\boldsymbol{\theta}) \propto \prod_i \widetilde{f}_i(\boldsymbol{\theta}). \tag{10.204}$$

3. Until convergence:
   (a) Choose a factor $\widetilde{f}_j(\boldsymbol{\theta})$ to refine.
   (b) Remove $\widetilde{f}_j(\boldsymbol{\theta})$ from the posterior by division

$$q^{\backslash j}(\boldsymbol{\theta}) = \frac{q(\boldsymbol{\theta})}{\widetilde{f}_j(\boldsymbol{\theta})}. \tag{10.205}$$

(c) Evaluate the new posterior by setting the sufficient statistics (moments) of $q^{\text{new}}(\boldsymbol{\theta})$ equal to those of $q^{\setminus j}(\boldsymbol{\theta})f_j(\boldsymbol{\theta})$, including evaluation of the normalization constant

$$Z_j = \int q^{\setminus j}(\boldsymbol{\theta})f_j(\boldsymbol{\theta})\,\mathrm{d}\boldsymbol{\theta}. \tag{10.206}$$

(d) Evaluate and store the new factor

$$\widetilde{f}_j(\boldsymbol{\theta}) = Z_j \frac{q^{\text{new}}(\boldsymbol{\theta})}{q^{\setminus j}(\boldsymbol{\theta})}. \tag{10.207}$$

4. Evaluate the approximation to the model evidence

$$p(\mathcal{D}) \simeq \int \prod_i \widetilde{f}_i(\boldsymbol{\theta})\,\mathrm{d}\boldsymbol{\theta}. \tag{10.208}$$

A special case of EP, known as *assumed density filtering* (ADF) or *moment matching* (Maybeck, 1982; Lauritzen, 1992; Boyen and Koller, 1998; Opper and Winther, 1999), is obtained by initializing all of the approximating factors except the first to unity and then making one pass through the factors updating each of them once. Assumed density filtering can be appropriate for on-line learning in which data points are arriving in a sequence and we need to learn from each data point and then discard it before considering the next point. However, in a batch setting we have the opportunity to re-use the data points many times in order to achieve improved accuracy, and it is this idea that is exploited in expectation propagation. Furthermore, if we apply ADF to batch data, the results will have an undesirable dependence on the (arbitrary) order in which the data points are considered, which again EP can overcome.

One disadvantage of expectation propagation is that there is no guarantee that the iterations will converge. However, for approximations $q(\boldsymbol{\theta})$ in the exponential family, if the iterations do converge, the resulting solution will be a stationary point of a particular energy function (Minka, 2001a), although each iteration of EP does not necessarily decrease the value of this energy function. This is in contrast to variational Bayes, which iteratively maximizes a lower bound on the log marginal likelihood, in which each iteration is guaranteed not to decrease the bound. It is possible to optimize the EP cost function directly, in which case it is guaranteed to converge, although the resulting algorithms can be slower and more complex to implement.

Another difference between variational Bayes and EP arises from the form of KL divergence that is minimized by the two algorithms, because the former minimizes $\mathrm{KL}(q\|p)$ whereas the latter minimizes $\mathrm{KL}(p\|q)$. As we saw in Figure 10.3, for distributions $p(\boldsymbol{\theta})$ which are multimodal, minimizing $\mathrm{KL}(p\|q)$ can lead to poor approximations. In particular, if EP is applied to mixtures the results are not sensible because the approximation tries to capture all of the modes of the posterior distribution. Conversely, in logistic-type models, EP often out-performs both local variational methods and the Laplace approximation (Kuss and Rasmussen, 2006).

Figure 10.15 Illustration of the clutter problem for a data space dimensionality of $D = 1$. Training data points, denoted by the crosses, are drawn from a mixture of two Gaussians with components shown in red and green. The goal is to infer the mean of the green Gaussian from the observed data.

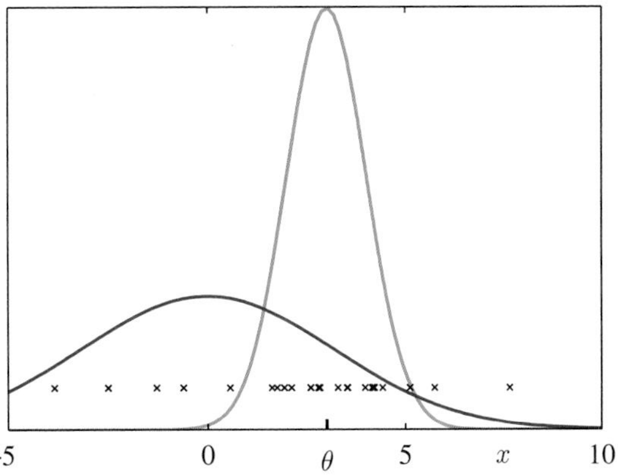

10.7.1 Example: The clutter problem

Following Minka (2001b), we illustrate the EP algorithm using a simple example in which the goal is to infer the mean $\boldsymbol{\theta}$ of a multivariate Gaussian distribution over a variable $\mathbf{x}$ given a set of observations drawn from that distribution. To make the problem more interesting, the observations are embedded in background clutter, which itself is also Gaussian distributed, as illustrated in Figure 10.15. The distribution of observed values $\mathbf{x}$ is therefore a mixture of Gaussians, which we take to be of the form

$$p(\mathbf{x}|\boldsymbol{\theta}) = (1-w)\mathcal{N}(\mathbf{x}|\boldsymbol{\theta}, \mathbf{I}) + w\mathcal{N}(\mathbf{x}|\mathbf{0}, a\mathbf{I}) \tag{10.209}$$

where w is the proportion of background clutter and is assumed to be known. The prior over $\boldsymbol{\theta}$ is taken to be Gaussian

$$p(\boldsymbol{\theta}) = \mathcal{N}(\boldsymbol{\theta}|\mathbf{0}, b\mathbf{I}) \tag{10.210}$$

and Minka (2001a) chooses the parameter values $a = 10$, $b = 100$ and $w = 0.5$. The joint distribution of N observations $\mathcal{D} = \{\mathbf{x}_1, \ldots, \mathbf{x}_N\}$ and $\boldsymbol{\theta}$ is given by

$$p(\mathcal{D}, \boldsymbol{\theta}) = p(\boldsymbol{\theta}) \prod_{n=1}^{N} p(\mathbf{x}_n|\boldsymbol{\theta}) \tag{10.211}$$

and so the posterior distribution comprises a mixture of 2^N Gaussians. Thus the computational cost of solving this problem exactly would grow exponentially with the size of the data set, and so an exact solution is intractable for moderately large N.

To apply EP to the clutter problem, we first identify the factors $f_0(\boldsymbol{\theta}) = p(\boldsymbol{\theta})$ and $f_n(\boldsymbol{\theta}) = p(\mathbf{x}_n|\boldsymbol{\theta})$. Next we select an approximating distribution from the exponential family, and for this example it is convenient to choose a spherical Gaussian

$$q(\boldsymbol{\theta}) = \mathcal{N}(\boldsymbol{\theta}|\mathbf{m}, v\mathbf{I}). \tag{10.212}$$

The factor approximations will therefore take the form of exponential-quadratic functions of the form

$$\widetilde{f}_n(\boldsymbol{\theta}) = s_n \mathcal{N}(\boldsymbol{\theta}|\mathbf{m}_n, v_n \mathbf{I}) \tag{10.213}$$

where $n = 1, \ldots, N$, and we set $\widetilde{f}_0(\boldsymbol{\theta})$ equal to the prior $p(\boldsymbol{\theta})$. Note that the use of $\mathcal{N}(\boldsymbol{\theta}|\cdot,\cdot)$ does not imply that the right-hand side is a well-defined Gaussian density (in fact, as we shall see, the variance parameter v_n can be negative) but is simply a convenient shorthand notation. The approximations $\widetilde{f}_n(\boldsymbol{\theta})$, for $n = 1, \ldots, N$, can be initialized to unity, corresponding to $s_n = (2\pi v_n)^{D/2}$, $v_n \to \infty$ and $\mathbf{m}_n = \mathbf{0}$, where D is the dimensionality of $\mathbf{x}$ and hence of $\boldsymbol{\theta}$. The initial $q(\boldsymbol{\theta})$, defined by (10.191), is therefore equal to the prior.

Exercise 10.37

We then iteratively refine the factors by taking one factor $f_n(\boldsymbol{\theta})$ at a time and applying (10.205), (10.206), and (10.207). Note that we do not need to revise the term $f_0(\boldsymbol{\theta})$ because an EP update will leave this term unchanged. Here we state the results and leave the reader to fill in the details.

Exercise 10.38

First we remove the current estimate $\widetilde{f}_n(\boldsymbol{\theta})$ from $q(\boldsymbol{\theta})$ by division using (10.205) to give $q^{\backslash n}(\boldsymbol{\theta})$, which has mean and inverse variance given by

$$\mathbf{m}^{\backslash n} = \mathbf{m} + v^{\backslash n} v_n^{-1}(\mathbf{m} - \mathbf{m}_n) \tag{10.214}$$

$$(v^{\backslash n})^{-1} = v^{-1} - v_n^{-1}. \tag{10.215}$$

Next we evaluate the normalization constant Z_n using (10.206) to give

$$Z_n = (1 - w)\mathcal{N}(\mathbf{x}_n|\mathbf{m}^{\backslash n}, (v^{\backslash n} + 1)\mathbf{I}) + w\mathcal{N}(\mathbf{x}_n|\mathbf{0}, a\mathbf{I}). \tag{10.216}$$

Exercise 10.39

Similarly, we compute the mean and variance of $q^{\text{new}}(\boldsymbol{\theta})$ by finding the mean and variance of $q^{\backslash n}(\boldsymbol{\theta}) f_n(\boldsymbol{\theta})$ to give

$$\mathbf{m} = \mathbf{m}^{\backslash n} + \rho_n \frac{v^{\backslash n}}{v^{\backslash n} + 1}(\mathbf{x}_n - \mathbf{m}^{\backslash n}) \tag{10.217}$$

$$v = v^{\backslash n} - \rho_n \frac{(v^{\backslash n})^2}{v^{\backslash n} + 1} + \rho_n(1 - \rho_n)\frac{(v^{\backslash n})^2 \|\mathbf{x}_n - \mathbf{m}^{\backslash n}\|^2}{D(v^{\backslash n} + 1)^2} \tag{10.218}$$

where the quantity

$$\rho_n = 1 - \frac{w}{Z_n}\mathcal{N}(\mathbf{x}_n|\mathbf{0}, a\mathbf{I}) \tag{10.219}$$

has a simple interpretation as the probability of the point $\mathbf{x}_n$ not being clutter. Then we use (10.207) to compute the refined factor $\widetilde{f}_n(\boldsymbol{\theta})$ whose parameters are given by

$$v_n^{-1} = (v^{\text{new}})^{-1} - (v^{\backslash n})^{-1} \tag{10.220}$$

$$\mathbf{m}_n = \mathbf{m}^{\backslash n} + (v_n + v^{\backslash n})(v^{\backslash n})^{-1}(\mathbf{m}^{\text{new}} - \mathbf{m}^{\backslash n}) \tag{10.221}$$

$$s_n = \frac{Z_n}{(2\pi v_n)^{D/2}\mathcal{N}(\mathbf{m}_n|\mathbf{m}^{\backslash n}, (v_n + v^{\backslash n})\mathbf{I})}. \tag{10.222}$$

This refinement process is repeated until a suitable termination criterion is satisfied, for instance that the maximum change in parameter values resulting from a complete

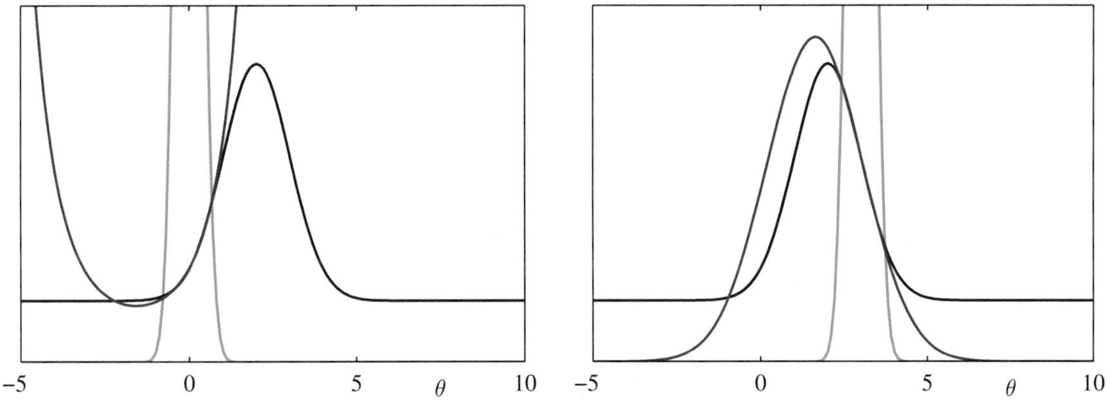

Figure 10.16 Examples of the approximation of specific factors for a one-dimensional version of the clutter problem, showing $f_n(\theta)$ in blue, $\widetilde{f}_n(\theta)$ in red, and $q^{\setminus n}(\theta)$ in green. Notice that the current form for $q^{\setminus n}(\theta)$ controls the range of θ over which $\widetilde{f}_n(\theta)$ will be a good approximation to $f_n(\theta)$.

pass through all factors is less than some threshold. Finally, we use (10.208) to evaluate the approximation to the model evidence, given by

$$p(\mathcal{D}) \simeq (2\pi v^{\text{new}})^{D/2} \exp(B/2) \prod_{n=1}^{N} \left\{ s_n (2\pi v_n)^{-D/2} \right\} \tag{10.223}$$

where

$$B = \frac{(\mathbf{m}^{\text{new}})^{\mathrm{T}} \mathbf{m}^{\text{new}}}{v} - \sum_{n=1}^{N} \frac{\mathbf{m}_n^{\mathrm{T}} \mathbf{m}_n}{v_n}. \tag{10.224}$$

Examples factor approximations for the clutter problem with a one-dimensional parameter space θ are shown in Figure 10.16. Note that the factor approximations can have infinite or even negative values for the 'variance' parameter v_n. This simply corresponds to approximations that curve upwards instead of downwards and are not necessarily problematic provided the overall approximate posterior $q(\boldsymbol{\theta})$ has positive variance. Figure 10.17 compares the performance of EP with variational Bayes (mean field theory) and the Laplace approximation on the clutter problem.

10.7.2 Expectation propagation on graphs

So far in our general discussion of EP, we have allowed the factors $f_i(\boldsymbol{\theta})$ in the distribution $p(\boldsymbol{\theta})$ to be functions of all of the components of $\boldsymbol{\theta}$, and similarly for the approximating factors $\widetilde{f}(\boldsymbol{\theta})$ in the approximating distribution $q(\boldsymbol{\theta})$. We now consider situations in which the factors depend only on subsets of the variables. Such restrictions can be conveniently expressed using the framework of probabilistic graphical models, as discussed in Chapter 8. Here we use a factor graph representation because this encompasses both directed and undirected graphs.

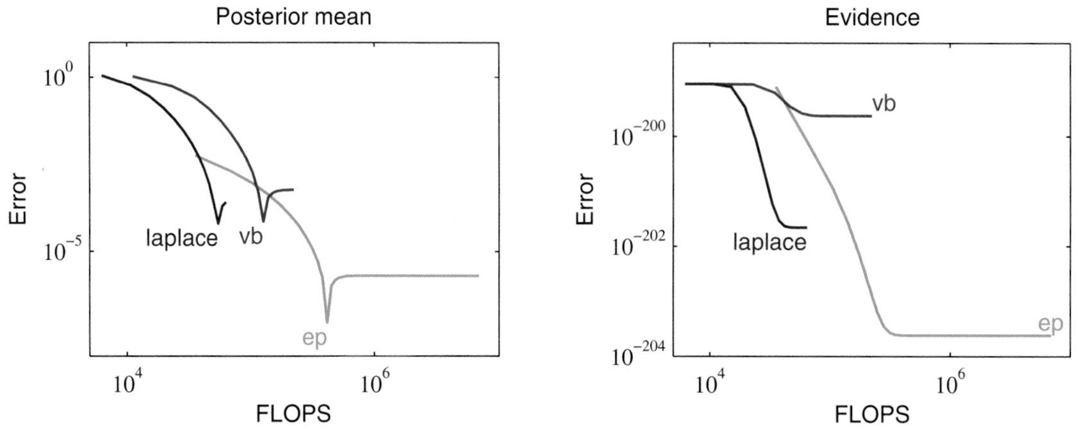

Figure 10.17 Comparison of expectation propagation, variational inference, and the Laplace approximation on the clutter problem. The left-hand plot shows the error in the predicted posterior mean versus the number of floating point operations, and the right-hand plot shows the corresponding results for the model evidence.

We shall focus on the case in which the approximating distribution is fully factorized, and we shall show that in this case expectation propagation reduces to loopy belief propagation (Minka, 2001a). To start with, we show this in the context of a simple example, and then we shall explore the general case.

First of all, recall from (10.17) that if we minimize the Kullback-Leibler divergence $\mathrm{KL}(p\|q)$ with respect to a factorized distribution q, then the optimal solution for each factor is simply the corresponding marginal of p.

Section 8.4.4

Now consider the factor graph shown on the left in Figure 10.18, which was introduced earlier in the context of the sum-product algorithm. The joint distribution is given by

$$p(\mathbf{x}) = f_a(x_1, x_2) f_b(x_2, x_3) f_c(x_2, x_4). \tag{10.225}$$

We seek an approximation $q(\mathbf{x})$ that has the same factorization, so that

$$q(\mathbf{x}) \propto \widetilde{f}_a(x_1, x_2) \widetilde{f}_b(x_2, x_3) \widetilde{f}_c(x_2, x_4). \tag{10.226}$$

Note that normalization constants have been omitted, and these can be re-instated at the end by local normalization, as is generally done in belief propagation. Now suppose we restrict attention to approximations in which the factors themselves factorize with respect to the individual variables so that

$$q(\mathbf{x}) \propto \widetilde{f}_{a1}(x_1) \widetilde{f}_{a2}(x_2) \widetilde{f}_{b2}(x_2) \widetilde{f}_{b3}(x_3) \widetilde{f}_{c2}(x_2) \widetilde{f}_{c4}(x_4) \tag{10.227}$$

which corresponds to the factor graph shown on the right in Figure 10.18. Because the individual factors are factorized, the overall distribution $q(\mathbf{x})$ is itself fully factorized.

Now we apply the EP algorithm using the fully factorized approximation. Suppose that we have initialized all of the factors and that we choose to refine factor

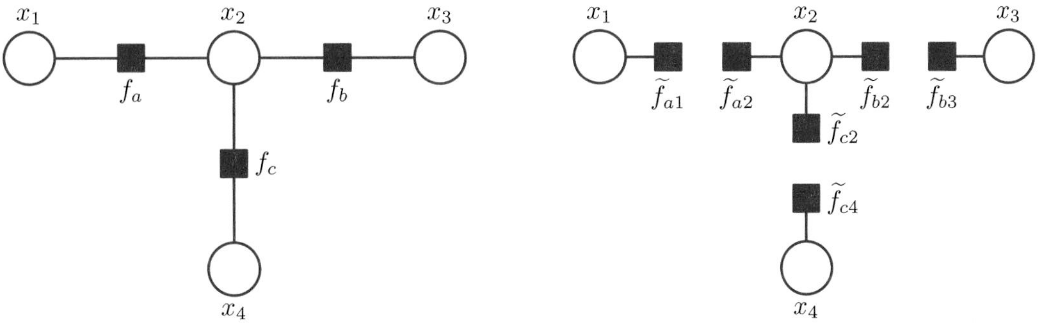

Figure 10.18 On the left is a simple factor graph from Figure 8.51 and reproduced here for convenience. On the right is the corresponding factorized approximation.

$\widetilde{f}_b(x_2, x_3) = \widetilde{f}_{b2}(x_2)\widetilde{f}_{b3}(x_3)$. We first remove this factor from the approximating distribution to give

$$q^{\backslash b}(\mathbf{x}) \propto \widetilde{f}_{a1}(x_1)\widetilde{f}_{a2}(x_2)\widetilde{f}_{c2}(x_2)\widetilde{f}_{c4}(x_4) \tag{10.228}$$

and we then multiply this by the exact factor $f_b(x_2, x_3)$ to give

$$\widehat{p}(\mathbf{x}) = q^{\backslash b}(\mathbf{x})f_b(x_2, x_3) = \widetilde{f}_{a1}(x_1)\widetilde{f}_{a2}(x_2)\widetilde{f}_{c2}(x_2)\widetilde{f}_{c4}(x_4)f_b(x_2, x_3). \tag{10.229}$$

We now find $q^{\text{new}}(\mathbf{x})$ by minimizing the Kullback-Leibler divergence $\text{KL}(\widehat{p}\|q^{\text{new}})$. The result, as noted above, is that $q^{\text{new}}(\mathbf{z})$ comprises the product of factors, one for each variable x_i, in which each factor is given by the corresponding marginal of $\widehat{p}(\mathbf{x})$. These four marginals are given by

$$\begin{aligned}
\widehat{p}(x_1) &\propto \widetilde{f}_{a1}(x_1) &(10.230)\\
\widehat{p}(x_2) &\propto \widetilde{f}_{a2}(x_2)\widetilde{f}_{c2}(x_2)\sum_{x_3} f_b(x_2, x_3) &(10.231)\\
\widehat{p}(x_3) &\propto \sum_{x_2}\left\{f_b(x_2, x_3)\widetilde{f}_{a2}(x_2)\widetilde{f}_{c2}(x_2)\right\} &(10.232)\\
\widehat{p}(x_4) &\propto \widetilde{f}_{c4}(x_4) &(10.233)
\end{aligned}$$

and $q^{\text{new}}(\mathbf{x})$ is obtained by multiplying these marginals together. We see that the only factors in $q(\mathbf{x})$ that change when we update $\widetilde{f}_b(x_2, x_3)$ are those that involve the variables in f_b namely x_2 and x_3. To obtain the refined factor $\widetilde{f}_b(x_2, x_3) = \widetilde{f}_{b2}(x_2)\widetilde{f}_{b3}(x_3)$ we simply divide $q^{\text{new}}(\mathbf{x})$ by $q^{\backslash b}(\mathbf{x})$, which gives

$$\begin{aligned}
\widetilde{f}_{b2}(x_2) &\propto \sum_{x_3} f_b(x_2, x_3) &(10.234)\\
\widetilde{f}_{b3}(x_3) &\propto \sum_{x_2}\left\{f_b(x_2, x_3)\widetilde{f}_{a2}(x_2)\widetilde{f}_{c2}(x_2)\right\}. &(10.235)
\end{aligned}$$

Section 8.4.4

These are precisely the messages obtained using belief propagation in which messages from variable nodes to factor nodes have been folded into the messages from factor nodes to variable nodes. In particular, $\widetilde{f}_{b2}(x_2)$ corresponds to the message $\mu_{f_b \to x_2}(x_2)$ sent by factor node f_b to variable node x_2 and is given by (8.81). Similarly, if we substitute (8.78) into (8.79), we obtain (10.235) in which $\widetilde{f}_{a2}(x_2)$ corresponds to $\mu_{f_a \to x_2}(x_2)$ and $\widetilde{f}_{c2}(x_2)$ corresponds to $\mu_{f_c \to x_2}(x_2)$, giving the message $\widetilde{f}_{b3}(x_3)$ which corresponds to $\mu_{f_b \to x_3}(x_3)$.

This result differs slightly from standard belief propagation in that messages are passed in both directions at the same time. We can easily modify the EP procedure to give the standard form of the sum-product algorithm by updating just one of the factors at a time, for instance if we refine only $\widetilde{f}_{b3}(x_3)$, then $\widetilde{f}_{b2}(x_2)$ is unchanged by definition, while the refined version of $\widetilde{f}_{b3}(x_3)$ is again given by (10.235). If we are refining only one term at a time, then we can choose the order in which the refinements are done as we wish. In particular, for a tree-structured graph we can follow a two-pass update scheme, corresponding to the standard belief propagation schedule, which will result in exact inference of the variable and factor marginals. The initialization of the approximation factors in this case is unimportant.

Now let us consider a general factor graph corresponding to the distribution

$$p(\boldsymbol{\theta}) = \prod_i f_i(\boldsymbol{\theta}_i) \tag{10.236}$$

where $\boldsymbol{\theta}_i$ represents the subset of variables associated with factor f_i. We approximate this using a fully factorized distribution of the form

$$q(\boldsymbol{\theta}) \propto \prod_i \prod_k \widetilde{f}_{ik}(\theta_k) \tag{10.237}$$

where θ_k corresponds to an individual variable node. Suppose that we wish to refine the particular term $\widetilde{f}_{jl}(\theta_l)$ keeping all other terms fixed. We first remove the term $\widetilde{f}_j(\boldsymbol{\theta}_j)$ from $q(\boldsymbol{\theta})$ to give

$$q^{\backslash j}(\boldsymbol{\theta}) \propto \prod_{i \neq j} \prod_k \widetilde{f}_{ik}(\theta_k) \tag{10.238}$$

and then multiply by the exact factor $f_j(\boldsymbol{\theta}_j)$. To determine the refined term $\widetilde{f}_{jl}(\theta_l)$, we need only consider the functional dependence on θ_l, and so we simply find the corresponding marginal of

$$q^{\backslash j}(\boldsymbol{\theta}) f_j(\boldsymbol{\theta}_j). \tag{10.239}$$

Up to a multiplicative constant, this involves taking the marginal of $f_j(\boldsymbol{\theta}_j)$ multiplied by any terms from $q^{\backslash j}(\boldsymbol{\theta})$ that are functions of any of the variables in $\boldsymbol{\theta}_j$. Terms that correspond to other factors $\widetilde{f}_i(\boldsymbol{\theta}_i)$ for $i \neq j$ will cancel between numerator and denominator when we subsequently divide by $q^{\backslash j}(\boldsymbol{\theta})$. We therefore obtain

$$\widetilde{f}_{jl}(\theta_l) \propto \sum_{\theta_{m \neq l} \in \boldsymbol{\theta}_j} f_j(\boldsymbol{\theta}_j) \prod_k \prod_{m \neq l} \widetilde{f}_{km}(\theta_m). \tag{10.240}$$

We recognize this as the sum-product rule in the form in which messages from variable nodes to factor nodes have been eliminated, as illustrated by the example shown in Figure 8.50. The quantity $\widetilde{f}_{jm}(\theta_m)$ corresponds to the message $\mu_{f_j \to \theta_m}(\theta_m)$, which factor node j sends to variable node m, and the product over k in (10.240) is over all factors that depend on the variables θ_m that have variables (other than variable θ_l) in common with factor $f_j(\boldsymbol{\theta}_j)$. In other words, to compute the outgoing message from a factor node, we take the product of all the incoming messages from other factor nodes, multiply by the local factor, and then marginalize.

Thus, the sum-product algorithm arises as a special case of expectation propagation if we use an approximating distribution that is fully factorized. This suggests that more flexible approximating distributions, corresponding to partially disconnected graphs, could be used to achieve higher accuracy. Another generalization is to group factors $f_i(\boldsymbol{\theta}_i)$ together into sets and to refine all the factors in a set together at each iteration. Both of these approaches can lead to improvements in accuracy (Minka, 2001b). In general, the problem of choosing the best combination of grouping and disconnection is an open research issue.

We have seen that variational message passing and expectation propagation optimize two different forms of the Kullback-Leibler divergence. Minka (2005) has shown that a broad range of message passing algorithms can be derived from a common framework involving minimization of members of the alpha family of divergences, given by (10.19). These include variational message passing, loopy belief propagation, and expectation propagation, as well as a range of other algorithms, which we do not have space to discuss here, such as *tree-reweighted message passing* (Wainwright *et al.*, 2005), *fractional belief propagation* (Wiegerinck and Heskes, 2003), and *power EP* (Minka, 2004).

Exercises

10.1 ($\star$) **www** Verify that the log marginal distribution of the observed data $\ln p(\mathbf{X})$ can be decomposed into two terms in the form (10.2) where $\mathcal{L}(q)$ is given by (10.3) and $\mathrm{KL}(q\|p)$ is given by (10.4).

10.2 ($\star$) Use the properties $\mathbb{E}[z_1] = m_1$ and $\mathbb{E}[z_2] = m_2$ to solve the simultaneous equations (10.13) and (10.15), and hence show that, provided the original distribution $p(\mathbf{z})$ is nonsingular, the unique solution for the means of the factors in the approximation distribution is given by $\mathbb{E}[z_1] = \mu_1$ and $\mathbb{E}[z_2] = \mu_2$.

10.3 ($\star\star$) **www** Consider a factorized variational distribution $q(\mathbf{Z})$ of the form (10.5). By using the technique of Lagrange multipliers, verify that minimization of the Kullback-Leibler divergence $\mathrm{KL}(p\|q)$ with respect to one of the factors $q_i(\mathbf{Z}_i)$, keeping all other factors fixed, leads to the solution (10.17).

10.4 ($\star\star$) Suppose that $p(\mathbf{x})$ is some fixed distribution and that we wish to approximate it using a Gaussian distribution $q(\mathbf{x}) = \mathcal{N}(\mathbf{x}|\boldsymbol{\mu}, \boldsymbol{\Sigma})$. By writing down the form of the KL divergence $\mathrm{KL}(p\|q)$ for a Gaussian $q(\mathbf{x})$ and then differentiating, show that minimization of $\mathrm{KL}(p\|q)$ with respect to $\boldsymbol{\mu}$ and $\boldsymbol{\Sigma}$ leads to the result that $\boldsymbol{\mu}$ is given by the expectation of $\mathbf{x}$ under $p(\mathbf{x})$ and that $\boldsymbol{\Sigma}$ is given by the covariance.

10.5 ($\star\star$) **www** Consider a model in which the set of all hidden stochastic variables, denoted collectively by $\mathbf{Z}$, comprises some latent variables $\mathbf{z}$ together with some model parameters $\boldsymbol{\theta}$. Suppose we use a variational distribution that factorizes between latent variables and parameters so that $q(\mathbf{z}, \boldsymbol{\theta}) = q_{\mathbf{z}}(\mathbf{z})q_{\boldsymbol{\theta}}(\boldsymbol{\theta})$, in which the distribution $q_{\boldsymbol{\theta}}(\boldsymbol{\theta})$ is approximated by a point estimate of the form $q_{\boldsymbol{\theta}}(\boldsymbol{\theta}) = \delta(\boldsymbol{\theta} - \boldsymbol{\theta}_0)$ where $\boldsymbol{\theta}_0$ is a vector of free parameters. Show that variational optimization of this factorized distribution is equivalent to an EM algorithm, in which the E step optimizes $q_{\mathbf{z}}(\mathbf{z})$, and the M step maximizes the expected complete-data log posterior distribution of $\boldsymbol{\theta}$ with respect to $\boldsymbol{\theta}_0$.

10.6 ($\star\star$) The alpha family of divergences is defined by (10.19). Show that the Kullback-Leibler divergence $\mathrm{KL}(p\|q)$ corresponds to $\alpha \to 1$. This can be done by writing $p^{\epsilon} = \exp(\epsilon \ln p) = 1 + \epsilon \ln p + O(\epsilon^2)$ and then taking $\epsilon \to 0$. Similarly show that $\mathrm{KL}(q\|p)$ corresponds to $\alpha \to -1$.

10.7 ($\star\star$) Consider the problem of inferring the mean and precision of a univariate Gaussian using a factorized variational approximation, as considered in Section 10.1.3. Show that the factor $q_{\mu}(\mu)$ is a Gaussian of the form $\mathcal{N}(\mu|\mu_N, \lambda_N^{-1})$ with mean and precision given by (10.26) and (10.27), respectively. Similarly show that the factor $q_{\tau}(\tau)$ is a gamma distribution of the form $\mathrm{Gam}(\tau|a_N, b_N)$ with parameters given by (10.29) and (10.30).

10.8 ($\star$) Consider the variational posterior distribution for the precision of a univariate Gaussian whose parameters are given by (10.29) and (10.30). By using the standard results for the mean and variance of the gamma distribution given by (B.27) and (B.28), show that if we let $N \to \infty$, this variational posterior distribution has a mean given by the inverse of the maximum likelihood estimator for the variance of the data, and a variance that goes to zero.

10.9 ($\star\star$) By making use of the standard result $\mathbb{E}[\tau] = a_N/b_N$ for the mean of a gamma distribution, together with (10.26), (10.27), (10.29), and (10.30), derive the result (10.33) for the reciprocal of the expected precision in the factorized variational treatment of a univariate Gaussian.

10.10 ($\star$) **www** Derive the decomposition given by (10.34) that is used to find approximate posterior distributions over models using variational inference.

10.11 ($\star\star$) **www** By using a Lagrange multiplier to enforce the normalization constraint on the distribution $q(m)$, show that the maximum of the lower bound (10.35) is given by (10.36).

10.12 ($\star\star$) Starting from the joint distribution (10.41), and applying the general result (10.9), show that the optimal variational distribution $q^{\star}(\mathbf{Z})$ over the latent variables for the Bayesian mixture of Gaussians is given by (10.48) by verifying the steps given in the text.

10.13 ($\star\star$) **www** Starting from (10.54), derive the result (10.59) for the optimum variational posterior distribution over $\boldsymbol{\mu}_k$ and $\boldsymbol{\Lambda}_k$ in the Bayesian mixture of Gaussians, and hence verify the expressions for the parameters of this distribution given by (10.60)–(10.63).

10.14 ($\star\star$) Using the distribution (10.59), verify the result (10.64).

10.15 ($\star$) Using the result (B.17), show that the expected value of the mixing coefficients in the variational mixture of Gaussians is given by (10.69).

10.16 ($\star\star$) **www** Verify the results (10.71) and (10.72) for the first two terms in the lower bound for the variational Gaussian mixture model given by (10.70).

10.17 ($\star\star\star$) Verify the results (10.73)–(10.77) for the remaining terms in the lower bound for the variational Gaussian mixture model given by (10.70).

10.18 ($\star\star\star$) In this exercise, we shall derive the variational re-estimation equations for the Gaussian mixture model by direct differentiation of the lower bound. To do this we assume that the variational distribution has the factorization defined by (10.42) and (10.55) with factors given by (10.48), (10.57), and (10.59). Substitute these into (10.70) and hence obtain the lower bound as a function of the parameters of the variational distribution. Then, by maximizing the bound with respect to these parameters, derive the re-estimation equations for the factors in the variational distribution, and show that these are the same as those obtained in Section 10.2.1.

10.19 ($\star\star$) Derive the result (10.81) for the predictive distribution in the variational treatment of the Bayesian mixture of Gaussians model.

10.20 ($\star\star$) **www** This exercise explores the variational Bayes solution for the mixture of Gaussians model when the size N of the data set is large and shows that it reduces (as we would expect) to the maximum likelihood solution based on EM derived in Chapter 9. Note that results from Appendix B may be used to help answer this exercise. First show that the posterior distribution $q^\star(\boldsymbol{\Lambda}_k)$ of the precisions becomes sharply peaked around the maximum likelihood solution. Do the same for the posterior distribution of the means $q^\star(\boldsymbol{\mu}_k|\boldsymbol{\Lambda}_k)$. Next consider the posterior distribution $q^\star(\boldsymbol{\pi})$ for the mixing coefficients and show that this too becomes sharply peaked around the maximum likelihood solution. Similarly, show that the responsibilities become equal to the corresponding maximum likelihood values for large N, by making use of the following asymptotic result for the digamma function for large x

$$\psi(x) = \ln x + O\left(1/x\right). \tag{10.241}$$

Finally, by making use of (10.80), show that for large N, the predictive distribution becomes a mixture of Gaussians.

10.21 ($\star$) Show that the number of equivalent parameter settings due to interchange symmetries in a mixture model with K components is $K!$.

10.22 ($\star\star$) We have seen that each mode of the posterior distribution in a Gaussian mixture model is a member of a family of $K!$ equivalent modes. Suppose that the result of running the variational inference algorithm is an approximate posterior distribution q that is localized in the neighbourhood of one of the modes. We can then approximate the full posterior distribution as a mixture of $K!$ such q distributions, once centred on each mode and having equal mixing coefficients. Show that if we assume negligible overlap between the components of the q mixture, the resulting lower bound differs from that for a single component q distribution through the addition of an extra term $\ln K!$.

10.23 ($\star\star$) **www** Consider a variational Gaussian mixture model in which there is no prior distribution over mixing coefficients $\{\pi_k\}$. Instead, the mixing coefficients are treated as parameters, whose values are to be found by maximizing the variational lower bound on the log marginal likelihood. Show that maximizing this lower bound with respect to the mixing coefficients, using a Lagrange multiplier to enforce the constraint that the mixing coefficients sum to one, leads to the re-estimation result (10.83). Note that there is no need to consider all of the terms in the lower bound but only the dependence of the bound on the $\{\pi_k\}$.

10.24 ($\star\star$) **www** We have seen in Section 10.2 that the singularities arising in the maximum likelihood treatment of Gaussian mixture models do not arise in a Bayesian treatment. Discuss whether such singularities would arise if the Bayesian model were solved using maximum posterior (MAP) estimation.

10.25 ($\star\star$) The variational treatment of the Bayesian mixture of Gaussians, discussed in Section 10.2, made use of a factorized approximation (10.5) to the posterior distribution. As we saw in Figure 10.2, the factorized assumption causes the variance of the posterior distribution to be under-estimated for certain directions in parameter space. Discuss qualitatively the effect this will have on the variational approximation to the model evidence, and how this effect will vary with the number of components in the mixture. Hence explain whether the variational Gaussian mixture will tend to under-estimate or over-estimate the optimal number of components.

10.26 ($\star\star\star$) Extend the variational treatment of Bayesian linear regression to include a gamma hyperprior $\text{Gam}(\beta|c_0, d_0)$ over β and solve variationally, by assuming a factorized variational distribution of the form $q(\mathbf{w})q(\alpha)q(\beta)$. Derive the variational update equations for the three factors in the variational distribution and also obtain an expression for the lower bound and for the predictive distribution.

10.27 ($\star\star$) By making use of the formulae given in Appendix B show that the variational lower bound for the linear basis function regression model can be written in the form (10.107) with the various terms defined by (10.108)–(10.112).

10.28 ($\star\star\star$) Rewrite the model for the Bayesian mixture of Gaussians, introduced in Section 10.2, as a conjugate model from the exponential family, as discussed in Section 10.4. Hence use the general results (10.115) and (10.119) to derive the specific results (10.48), (10.57), and (10.59).

10.29 ($\star$) **www** Show that the function $f(x) = \ln(x)$ is concave for $0 < x < \infty$ by computing its second derivative. Determine the form of the dual function $g(\eta)$ defined by (10.133), and verify that minimization of $\eta x - g(\eta)$ with respect to η according to (10.132) indeed recovers the function $\ln(x)$.

10.30 ($\star$) By evaluating the second derivative, show that the log logistic function $f(x) = -\ln(1 + e^{-x})$ is concave. Derive the variational upper bound (10.137) directly by making a first order Taylor expansion of the log logistic function around a point $x = \xi$.

10.31 ($\star\star$) By finding the second derivative with respect to x, show that the function $f(x) = -\ln(e^{x/2} + e^{-x/2})$ is a concave function of x. Now consider the second derivatives with respect to the variable x^2 and hence show that it is a convex function of x^2. Plot graphs of $f(x)$ against x and against x^2. Derive the lower bound (10.144) on the logistic sigmoid function directly by making a first order Taylor series expansion of the function $f(x)$ in the variable x^2 centred on the value ξ^2.

10.32 ($\star\star$) **www** Consider the variational treatment of logistic regression with sequential learning in which data points are arriving one at a time and each must be processed and discarded before the next data point arrives. Show that a Gaussian approximation to the posterior distribution can be maintained through the use of the lower bound (10.151), in which the distribution is initialized using the prior, and as each data point is absorbed its corresponding variational parameter ξ_n is optimized.

10.33 ($\star$) By differentiating the quantity $Q(\boldsymbol{\xi}, \boldsymbol{\xi}^{\text{old}})$ defined by (10.161) with respect to the variational parameter ξ_n show that the update equation for ξ_n for the Bayesian logistic regression model is given by (10.163).

10.34 ($\star\star$) In this exercise we derive re-estimation equations for the variational parameters $\boldsymbol{\xi}$ in the Bayesian logistic regression model of Section 4.5 by direct maximization of the lower bound given by (10.164). To do this set the derivative of $\mathcal{L}(\boldsymbol{\xi})$ with respect to ξ_n equal to zero, making use of the result (3.117) for the derivative of the log of a determinant, together with the expressions (10.157) and (10.158) which define the mean and covariance of the variational posterior distribution $q(\mathbf{w})$.

10.35 ($\star\star$) Derive the result (10.164) for the lower bound $\mathcal{L}(\boldsymbol{\xi})$ in the variational logistic regression model. This is most easily done by substituting the expressions for the Gaussian prior $q(\mathbf{w}) = \mathcal{N}(\mathbf{w}|\mathbf{m}_0, \mathbf{S}_0)$, together with the lower bound $h(\mathbf{w}, \boldsymbol{\xi})$ on the likelihood function, into the integral (10.159) which defines $\mathcal{L}(\boldsymbol{\xi})$. Next gather together the terms which depend on $\mathbf{w}$ in the exponential and complete the square to give a Gaussian integral, which can then be evaluated by invoking the standard result for the normalization coefficient of a multivariate Gaussian. Finally take the logarithm to obtain (10.164).

10.36 ($\star\star$) Consider the ADF approximation scheme discussed in Section 10.7, and show that inclusion of the factor $f_j(\boldsymbol{\theta})$ leads to an update of the model evidence of the form

$$p_j(\mathcal{D}) \simeq p_{j-1}(\mathcal{D})Z_j \tag{10.242}$$

where Z_j is the normalization constant defined by (10.197). By applying this result recursively, and initializing with $p_0(\mathcal{D}) = 1$, derive the result

$$p(\mathcal{D}) \simeq \prod_j Z_j. \tag{10.243}$$

10.37 ($\star$) **www** Consider the expectation propagation algorithm from Section 10.7, and suppose that one of the factors $f_0(\boldsymbol{\theta})$ in the definition (10.188) has the same exponential family functional form as the approximating distribution $q(\boldsymbol{\theta})$. Show that if the factor $\widetilde{f}_0(\boldsymbol{\theta})$ is initialized to be $f_0(\boldsymbol{\theta})$, then an EP update to refine $\widetilde{f}_0(\boldsymbol{\theta})$ leaves $\widetilde{f}_0(\boldsymbol{\theta})$ unchanged. This situation typically arises when one of the factors is the prior $p(\boldsymbol{\theta})$, and so we see that the prior factor can be incorporated once exactly and does not need to be refined.

10.38 ($\star\star\star$) In this exercise and the next, we shall verify the results (10.214)–(10.224) for the expectation propagation algorithm applied to the clutter problem. Begin by using the division formula (10.205) to derive the expressions (10.214) and (10.215) by completing the square inside the exponential to identify the mean and variance. Also, show that the normalization constant Z_n, defined by (10.206), is given for the clutter problem by (10.216). This can be done by making use of the general result (2.115).

10.39 ($\star\star\star$) Show that the mean and variance of $q^{\text{new}}(\boldsymbol{\theta})$ for EP applied to the clutter problem are given by (10.217) and (10.218). To do this, first prove the following results for the expectations of $\boldsymbol{\theta}$ and $\boldsymbol{\theta}\boldsymbol{\theta}^{\mathrm{T}}$ under $q^{\text{new}}(\boldsymbol{\theta})$

$$\mathbb{E}[\boldsymbol{\theta}] = \mathbf{m}^{\backslash n} + v^{\backslash n}\nabla_{\mathbf{m}^{\backslash n}} \ln Z_n \tag{10.244}$$

$$\mathbb{E}[\boldsymbol{\theta}^{\mathrm{T}}\boldsymbol{\theta}] = 2(v^{\backslash n})^2\nabla_{v^{\backslash n}} \ln Z_n + 2\mathbb{E}[\boldsymbol{\theta}]^{\mathrm{T}}\mathbf{m}^{\backslash n} - \|\mathbf{m}^{\backslash n}\|^2 + v^{\backslash n}D \tag{10.245}$$

and then make use of the result (10.216) for Z_n. Next, prove the results (10.220)–(10.222) by using (10.207) and completing the square in the exponential. Finally, use (10.208) to derive the result (10.223).

11 Sampling Methods

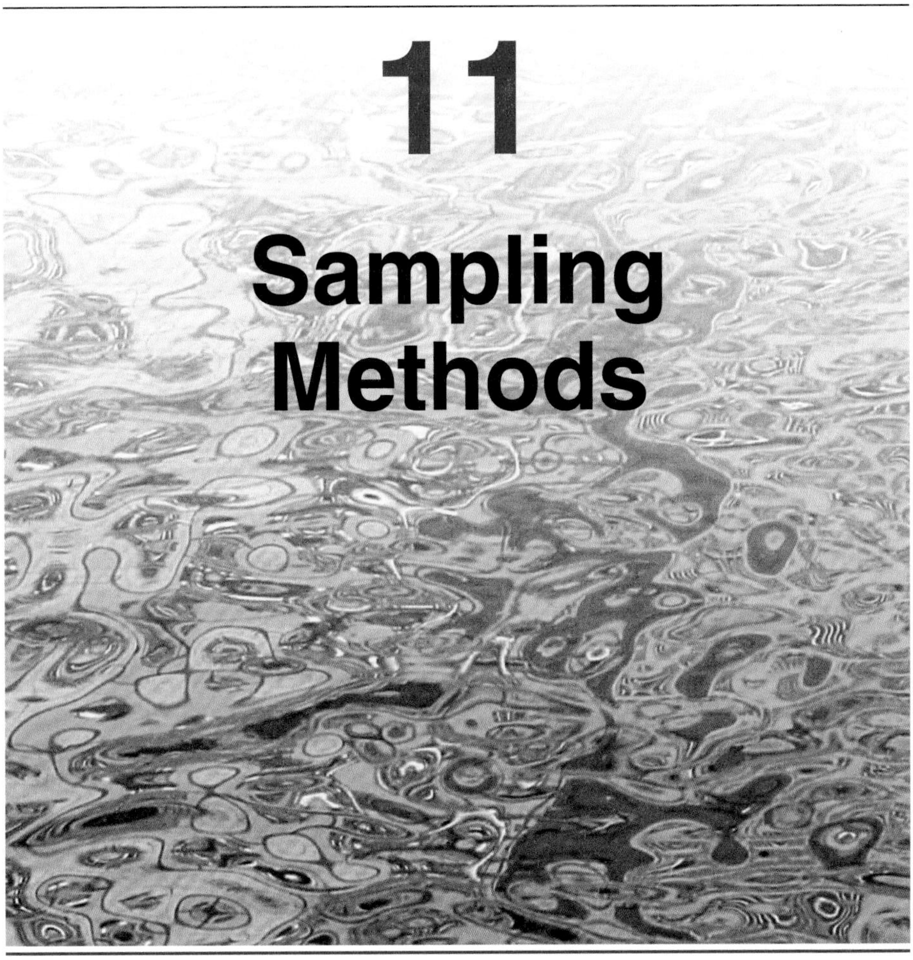

For most probabilistic models of practical interest, exact inference is intractable, and so we have to resort to some form of approximation. In Chapter 10, we discussed inference algorithms based on deterministic approximations, which include methods such as variational Bayes and expectation propagation. Here we consider approximate inference methods based on numerical sampling, also known as *Monte Carlo* techniques.

Although for some applications the posterior distribution over unobserved variables will be of direct interest in itself, for most situations the posterior distribution is required primarily for the purpose of evaluating expectations, for example in order to make predictions. The fundamental problem that we therefore wish to address in this chapter involves finding the expectation of some function $f(\mathbf{z})$ with respect to a probability distribution $p(\mathbf{z})$. Here, the components of $\mathbf{z}$ might comprise discrete or continuous variables or some combination of the two. Thus in the case of continuous

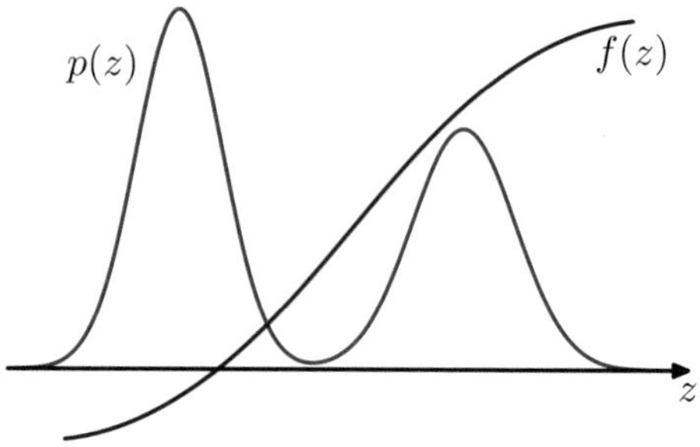

Figure 11.1 Schematic illustration of a function $f(z)$ whose expectation is to be evaluated with respect to a distribution $p(z)$.

variables, we wish to evaluate the expectation

$$\mathbb{E}[f] = \int f(\mathbf{z})p(\mathbf{z})\,\mathrm{d}\mathbf{z} \tag{11.1}$$

where the integral is replaced by summation in the case of discrete variables. This is illustrated schematically for a single continuous variable in Figure 11.1. We shall suppose that such expectations are too complex to be evaluated exactly using analytical techniques.

The general idea behind sampling methods is to obtain a set of samples $\mathbf{z}^{(l)}$ (where $l = 1, \ldots, L$) drawn independently from the distribution $p(\mathbf{z})$. This allows the expectation (11.1) to be approximated by a finite sum

$$\widehat{f} = \frac{1}{L}\sum_{l=1}^{L} f(\mathbf{z}^{(l)}). \tag{11.2}$$

As long as the samples $\mathbf{z}^{(l)}$ are drawn from the distribution $p(\mathbf{z})$, then $\mathbb{E}[\widehat{f}] = \mathbb{E}[f]$ and so the estimator $\widehat{f}$ has the correct mean. The variance of the estimator is given by

Exercise 11.1

$$\mathrm{var}[\widehat{f}] = \frac{1}{L}\mathbb{E}\left[(f - \mathbb{E}[f])^2\right] \tag{11.3}$$

is the variance of the function $f(\mathbf{z})$ under the distribution $p(\mathbf{z})$. It is worth emphasizing that the accuracy of the estimator therefore does not depend on the dimensionality of $\mathbf{z}$, and that, in principle, high accuracy may be achievable with a relatively small number of samples $\mathbf{z}^{(l)}$. In practice, ten or twenty independent samples may suffice to estimate an expectation to sufficient accuracy.

The problem, however, is that the samples $\{\mathbf{z}^{(l)}\}$ might not be independent, and so the effective sample size might be much smaller than the apparent sample size. Also, referring back to Figure 11.1, we note that if $f(\mathbf{z})$ is small in regions where $p(\mathbf{z})$ is large, and vice versa, then the expectation may be dominated by regions of small probability, implying that relatively large sample sizes will be required to achieve sufficient accuracy.

For many models, the joint distribution $p(\mathbf{z})$ is conveniently specified in terms of a graphical model. In the case of a directed graph with no observed variables, it is straightforward to sample from the joint distribution (assuming that it is possible to sample from the conditional distributions at each node) using the following *ancestral sampling* approach, discussed briefly in Section 8.1.2. The joint distribution is specified by

$$p(\mathbf{z}) = \prod_{i=1}^{M} p(\mathbf{z}_i|\mathrm{pa}_i) \tag{11.4}$$

where $\mathbf{z}_i$ are the set of variables associated with node i, and pa_i denotes the set of variables associated with the parents of node i. To obtain a sample from the joint distribution, we make one pass through the set of variables in the order $\mathbf{z}_1, \ldots, \mathbf{z}_M$ sampling from the conditional distributions $p(\mathbf{z}_i|\mathrm{pa}_i)$. This is always possible because at each step all of the parent values will have been instantiated. After one pass through the graph, we will have obtained a sample from the joint distribution.

Now consider the case of a directed graph in which some of the nodes are instantiated with observed values. We can in principle extend the above procedure, at least in the case of nodes representing discrete variables, to give the following *logic sampling* approach (Henrion, 1988), which can be seen as a special case of *importance sampling* discussed in Section 11.1.4. At each step, when a sampled value is obtained for a variable $\mathbf{z}_i$ whose value is observed, the sampled value is compared to the observed value, and if they agree then the sample value is retained and the algorithm proceeds to the next variable in turn. However, if the sampled value and the observed value disagree, then the whole sample so far is discarded and the algorithm starts again with the first node in the graph. This algorithm samples correctly from the posterior distribution because it corresponds simply to drawing samples from the joint distribution of hidden variables and data variables and then discarding those samples that disagree with the observed data (with the slight saving of not continuing with the sampling from the joint distribution as soon as one contradictory value is observed). However, the overall probability of accepting a sample from the posterior decreases rapidly as the number of observed variables increases and as the number of states that those variables can take increases, and so this approach is rarely used in practice.

In the case of probability distributions defined by an undirected graph, there is no one-pass sampling strategy that will sample even from the prior distribution with no observed variables. Instead, computationally more expensive techniques must be employed, such as Gibbs sampling, which is discussed in Section 11.3.

As well as sampling from conditional distributions, we may also require samples from a marginal distribution. If we already have a strategy for sampling from a joint distribution $p(\mathbf{u}, \mathbf{v})$, then it is straightforward to obtain samples from the marginal distribution $p(\mathbf{u})$ simply by ignoring the values for $\mathbf{v}$ in each sample.

There are numerous texts dealing with Monte Carlo methods. Those of particular interest from the statistical inference perspective include Chen *et al.* (2001), Gamerman (1997), Gilks *et al.* (1996), Liu (2001), Neal (1996), and Robert and Casella (1999). Also there are review articles by Besag *et al.* (1995), Brooks (1998), Diaconis and Saloff-Coste (1998), Jerrum and Sinclair (1996), Neal (1993), Tierney (1994), and Andrieu *et al.* (2003) that provide additional information on sampling methods for statistical inference.

Diagnostic tests for convergence of Markov chain Monte Carlo algorithms are summarized in Robert and Casella (1999).

11.1. Basic Sampling Algorithms

In this section, we consider some simple strategies for generating random samples from a given distribution. Because the samples will be generated by a computer algorithm they will in fact be *pseudo-random* numbers, that is, they will be deterministically calculated, but must nevertheless pass appropriate tests for randomness. Generating such numbers raises several subtleties (Press *et al.*, 1992) that lie outside the scope of this book. Here we shall assume that an algorithm has been provided that generates pseudo-random numbers distributed uniformly over $(0, 1)$, and indeed most software environments have such a facility built in.

11.1.1 Standard distributions

We first consider how to generate random numbers from simple nonuniform distributions, assuming that we already have available a source of uniformly distributed random numbers. Suppose that z is uniformly distributed over the interval $(0, 1)$, and that we transform the values of z using some function $f(\cdot)$ so that $y = f(z)$. The distribution of y will be governed by

$$p(y) = p(z) \left| \frac{\mathrm{d}z}{\mathrm{d}y} \right| \tag{11.5}$$

where, in this case, $p(z) = 1$. Our goal is to choose the function $f(z)$ such that the resulting values of y have some specific desired distribution $p(y)$. Integrating (11.5) we obtain

$$z = h(y) \equiv \int_{-\infty}^{y} p(\widehat{y})\,\mathrm{d}\widehat{y} \tag{11.6}$$

Exercise 11.2

which is the indefinite integral of $p(y)$. Thus, $y = h^{-1}(z)$, and so we have to transform the uniformly distributed random numbers using a function which is the inverse of the indefinite integral of the desired distribution. This is illustrated in Figure 11.2.

Consider for example the *exponential distribution*

$$p(y) = \lambda \exp(-\lambda y) \tag{11.7}$$

where $0 \leqslant y < \infty$. In this case the lower limit of the integral in (11.6) is 0, and so $h(y) = 1 - \exp(-\lambda y)$. Thus, if we transform our uniformly distributed variable z using $y = -\lambda^{-1} \ln(1 - z)$, then y will have an exponential distribution.

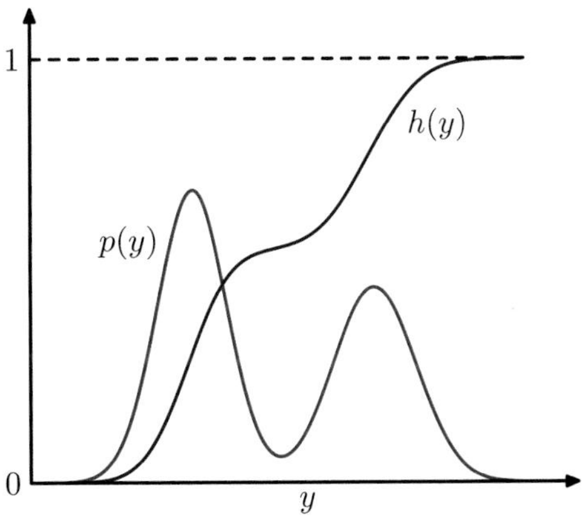

Figure 11.2 Geometrical interpretation of the transformation method for generating nonuniformly distributed random numbers. $h(y)$ is the indefinite integral of the desired distribution $p(y)$. If a uniformly distributed random variable z is transformed using $y = h^{-1}(z)$, then y will be distributed according to $p(y)$.

Another example of a distribution to which the transformation method can be applied is given by the Cauchy distribution

$$p(y) = \frac{1}{\pi} \frac{1}{1+y^2}. \tag{11.8}$$

In this case, the inverse of the indefinite integral can be expressed in terms of the 'tan' function.

Exercise 11.3

The generalization to multiple variables is straightforward and involves the Jacobian of the change of variables, so that

$$p(y_1, \ldots, y_M) = p(z_1, \ldots, z_M) \left| \frac{\partial(z_1, \ldots, z_M)}{\partial(y_1, \ldots, y_M)} \right|. \tag{11.9}$$

As a final example of the transformation method we consider the Box-Muller method for generating samples from a Gaussian distribution. First, suppose we generate pairs of uniformly distributed random numbers $z_1, z_2 \in (-1, 1)$, which we can do by transforming a variable distributed uniformly over $(0, 1)$ using $z \to 2z - 1$. Next we discard each pair unless it satisfies $z_1^2 + z_2^2 \leqslant 1$. This leads to a uniform distribution of points inside the unit circle with $p(z_1, z_2) = 1/\pi$, as illustrated in Figure 11.3. Then, for each pair z_1, z_2 we evaluate the quantities

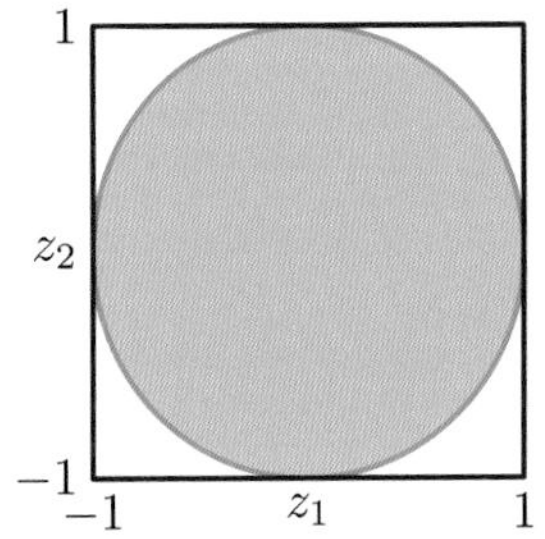

Figure 11.3 The Box-Muller method for generating Gaussian distributed random numbers starts by generating samples from a uniform distribution inside the unit circle.

$$y_1 = z_1 \left(\frac{-2\ln r^2}{r^2}\right)^{1/2} \tag{11.10}$$

$$y_2 = z_2 \left(\frac{-2\ln r^2}{r^2}\right)^{1/2} \tag{11.11}$$

Exercise 11.4

where $r^2 = z_1^2 + z_2^2$. Then the joint distribution of y_1 and y_2 is given by

$$\begin{aligned} p(y_1, y_2) &= p(z_1, z_2) \left| \frac{\partial(z_1, z_2)}{\partial(y_1, y_2)} \right| \\ &= \left[\frac{1}{\sqrt{2\pi}} \exp(-y_1^2/2)\right] \left[\frac{1}{\sqrt{2\pi}} \exp(-y_2^2/2)\right] \end{aligned} \tag{11.12}$$

and so y_1 and y_2 are independent and each has a Gaussian distribution with zero mean and unit variance.

If y has a Gaussian distribution with zero mean and unit variance, then $\sigma y + \mu$ will have a Gaussian distribution with mean μ and variance σ^2. To generate vector-valued variables having a multivariate Gaussian distribution with mean $\boldsymbol{\mu}$ and covariance $\boldsymbol{\Sigma}$, we can make use of the *Cholesky decomposition*, which takes the form $\boldsymbol{\Sigma} = \mathbf{L}\mathbf{L}^{\mathrm{T}}$ (Press *et al.*, 1992). Then, if $\mathbf{z}$ is a vector valued random variable whose components are independent and Gaussian distributed with zero mean and unit variance, then $\mathbf{y} = \boldsymbol{\mu} + \mathbf{L}\mathbf{z}$ will have mean $\boldsymbol{\mu}$ and covariance $\boldsymbol{\Sigma}$.

Exercise 11.5

Obviously, the transformation technique depends for its success on the ability to calculate and then invert the indefinite integral of the required distribution. Such operations will only be feasible for a limited number of simple distributions, and so we must turn to alternative approaches in search of a more general strategy. Here we consider two techniques called *rejection sampling* and *importance sampling*. Although mainly limited to univariate distributions and thus not directly applicable to complex problems in many dimensions, they do form important components in more general strategies.

11.1.2 Rejection sampling

The rejection sampling framework allows us to sample from relatively complex distributions, subject to certain constraints. We begin by considering univariate distributions and discuss the extension to multiple dimensions subsequently.

Suppose we wish to sample from a distribution $p(\mathbf{z})$ that is not one of the simple, standard distributions considered so far, and that sampling directly from $p(\mathbf{z})$ is difficult. Furthermore suppose, as is often the case, that we are easily able to evaluate $p(\mathbf{z})$ for any given value of $\mathbf{z}$, up to some normalizing constant Z, so that

$$p(z) = \frac{1}{Z_p}\widetilde{p}(z) \tag{11.13}$$

where $\widetilde{p}(z)$ can readily be evaluated, but Z_p is unknown.

In order to apply rejection sampling, we need some simpler distribution $q(z)$, sometimes called a *proposal distribution*, from which we can readily draw samples.

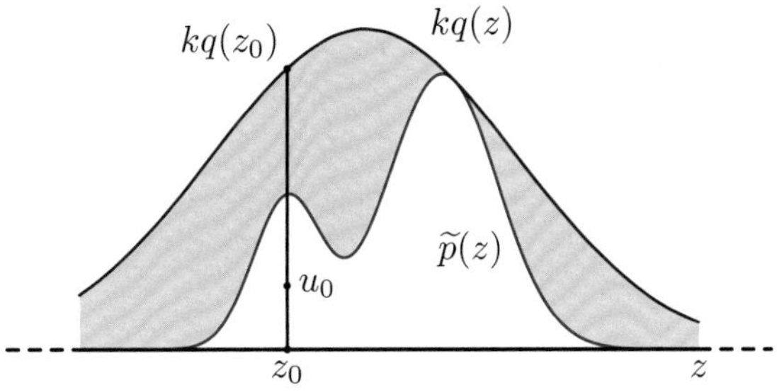

Figure 11.4 In the rejection sampling method, samples are drawn from a simple distribution $q(z)$ and rejected if they fall in the grey area between the unnormalized distribution $\widetilde{p}(z)$ and the scaled distribution $kq(z)$. The resulting samples are distributed according to $p(z)$, which is the normalized version of $\widetilde{p}(z)$.

We next introduce a constant k whose value is chosen such that $kq(z) \geqslant \widetilde{p}(z)$ for all values of z. The function $kq(z)$ is called the comparison function and is illustrated for a univariate distribution in Figure 11.4. Each step of the rejection sampler involves generating two random numbers. First, we generate a number z_0 from the distribution $q(z)$. Next, we generate a number u_0 from the uniform distribution over $[0, kq(z_0)]$. This pair of random numbers has uniform distribution under the curve of the function $kq(z)$. Finally, if $u_0 > \widetilde{p}(z_0)$ then the sample is rejected, otherwise u_0 is retained. Thus the pair is rejected if it lies in the grey shaded region in Figure 11.4. The remaining pairs then have uniform distribution under the curve of $\widetilde{p}(z)$, and hence the corresponding z values are distributed according to $p(z)$, as desired. *Exercise 11.6*

The original values of z are generated from the distribution $q(z)$, and these samples are then accepted with probability $\widetilde{p}(z)/kq(z)$, and so the probability that a sample will be accepted is given by

$$\begin{aligned} p(\text{accept}) &= \int \{\widetilde{p}(z)/kq(z)\}\, q(z)\, \mathrm{d}z \\ &= \frac{1}{k}\int \widetilde{p}(z)\, \mathrm{d}z. \end{aligned} \tag{11.14}$$

Thus the fraction of points that are rejected by this method depends on the ratio of the area under the unnormalized distribution $\widetilde{p}(z)$ to the area under the curve $kq(z)$. We therefore see that the constant k should be as small as possible subject to the limitation that $kq(z)$ must be nowhere less than $\widetilde{p}(z)$.

As an illustration of the use of rejection sampling, consider the task of sampling from the gamma distribution

$$\mathrm{Gam}(z|a, b) = \frac{b^a z^{a-1} \exp(-bz)}{\Gamma(a)} \tag{11.15}$$

which, for $a > 1$, has a bell-shaped form, as shown in Figure 11.5. A suitable proposal distribution is therefore the Cauchy (11.8) because this too is bell-shaped and because we can use the transformation method, discussed earlier, to sample from it. We need to generalize the Cauchy slightly to ensure that it nowhere has a smaller value than the gamma distribution. This can be achieved by transforming a uniform random variable y using $z = b \tan y + c$, which gives random numbers distributed according to *Exercise 11.7*

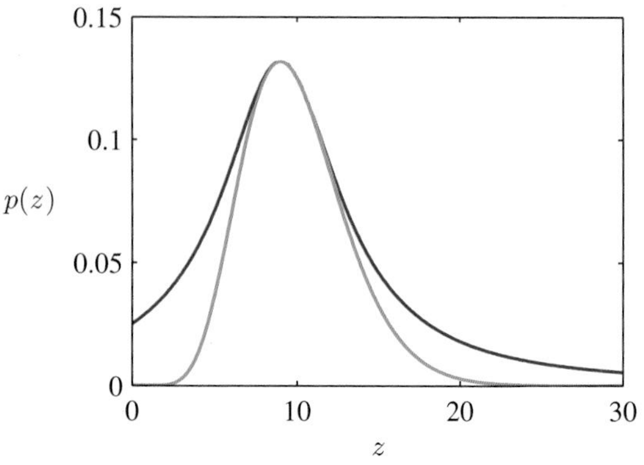

Figure 11.5 Plot showing the gamma distribution given by (11.15) as the green curve, with a scaled Cauchy proposal distribution shown by the red curve. Samples from the gamma distribution can be obtained by sampling from the Cauchy and then applying the rejection sampling criterion.

$$q(z) = \frac{k}{1 + (z - c)^2/b^2}. \tag{11.16}$$

The minimum reject rate is obtained by setting $c = a - 1$, $b^2 = 2a - 1$ and choosing the constant k to be as small as possible while still satisfying the requirement $kq(z) \geqslant \widetilde{p}(z)$. The resulting comparison function is also illustrated in Figure 11.5.

11.1.3 Adaptive rejection sampling

In many instances where we might wish to apply rejection sampling, it proves difficult to determine a suitable analytic form for the envelope distribution $q(z)$. An alternative approach is to construct the envelope function on the fly based on measured values of the distribution $p(z)$ (Gilks and Wild, 1992). Construction of an envelope function is particularly straightforward for cases in which $p(z)$ is log concave, in other words when $\ln p(z)$ has derivatives that are nonincreasing functions of z. The construction of a suitable envelope function is illustrated graphically in Figure 11.6.

The function $\ln p(z)$ and its gradient are evaluated at some initial set of grid points, and the intersections of the resulting tangent lines are used to construct the envelope function. Next a sample value is drawn from the envelope distribution. This is straightforward because the log of the envelope distribution is a succession

Exercise 11.9

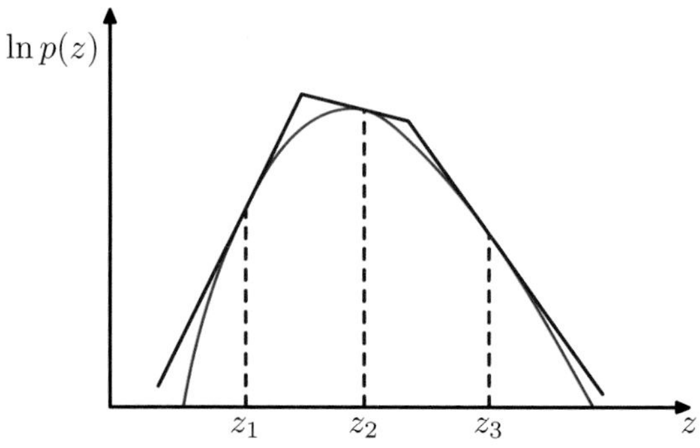

Figure 11.6 In the case of distributions that are log concave, an envelope function for use in rejection sampling can be constructed using the tangent lines computed at a set of grid points. If a sample point is rejected, it is added to the set of grid points and used to refine the envelope distribution.

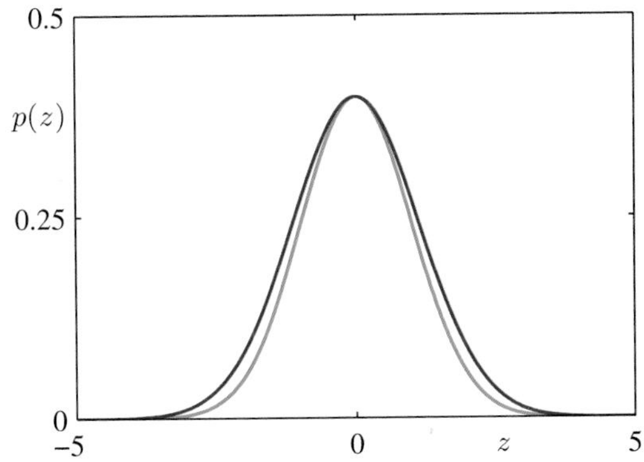

Figure 11.7 Illustrative example of rejection sampling involving sampling from a Gaussian distribution $p(z)$ shown by the green curve, by using rejection sampling from a proposal distribution $q(z)$ that is also Gaussian and whose scaled version $kq(z)$ is shown by the red curve.

of linear functions, and hence the envelope distribution itself comprises a piecewise exponential distribution of the form

$$q(z) = k_i \lambda_i \exp\left\{-\lambda_i (z - z_i)\right\} \qquad \widehat{z}_{i-1,i} < z \leqslant \widehat{z}_{i,i+1} \tag{11.17}$$

where $\widehat{z}_{i-1,i}$ is the point of intersection of the tangent lines at z_{i-1} and z_i, λ_i is the slope of the tangent at z_i and k_i accounts for the corresponding offset. Once a sample has been drawn, the usual rejection criterion can be applied. If the sample is accepted, then it will be a draw from the desired distribution. If, however, the sample is rejected, then it is incorporated into the set of grid points, a new tangent line is computed, and the envelope function is thereby refined. As the number of grid points increases, so the envelope function becomes a better approximation of the desired distribution $p(z)$ and the probability of rejection decreases.

A variant of the algorithm exists that avoids the evaluation of derivatives (Gilks, 1992). The adaptive rejection sampling framework can also be extended to distributions that are not log concave, simply by following each rejection sampling step with a Metropolis-Hastings step (to be discussed in Section 11.2.2), giving rise to *adaptive rejection Metropolis* sampling (Gilks *et al.*, 1995).

Clearly for rejection sampling to be of practical value, we require that the comparison function be close to the required distribution so that the rate of rejection is kept to a minimum. Now let us examine what happens when we try to use rejection sampling in spaces of high dimensionality. Consider, for the sake of illustration, a somewhat artificial problem in which we wish to sample from a zero-mean multivariate Gaussian distribution with covariance $\sigma_p^2\mathbf{I}$, where $\mathbf{I}$ is the unit matrix, by rejection sampling from a proposal distribution that is itself a zero-mean Gaussian distribution having covariance $\sigma_q^2\mathbf{I}$. Obviously, we must have $\sigma_q^2 \geqslant \sigma_p^2$ in order that there exists a k such that $kq(z) \geqslant p(z)$. In D-dimensions the optimum value of k is given by $k = (\sigma_q/\sigma_p)^D$, as illustrated for $D = 1$ in Figure 11.7. The acceptance rate will be the ratio of volumes under $p(z)$ and $kq(z)$, which, because both distributions are normalized, is just $1/k$. Thus the acceptance rate diminishes exponentially with dimensionality. Even if σ_q exceeds σ_p by just one percent, for $D = 1{,}000$ the acceptance ratio will be approximately $1/20{,}000$. In this illustrative example the comparison function is close to the required distribution. For more practical examples, where the desired distribution may be multimodal and sharply peaked, it will be extremely difficult to find a good proposal distribution and comparison function. Furthermore, the exponential decrease of acceptance rate with dimensionality is a generic feature of rejection sampling. Although rejection can be a useful technique in one or two dimensions it is unsuited to problems of high dimensionality. It can, however, play a role as a subroutine in more sophisticated algorithms for sampling in high dimensional spaces.

11.1.4 Importance sampling

One of the principal reasons for wishing to sample from complicated probability distributions is to be able to evaluate expectations of the form (11.1). The technique of *importance sampling* provides a framework for approximating expectations directly but does not itself provide a mechanism for drawing samples from distribution $p(\mathbf{z})$.

The finite sum approximation to the expectation, given by (11.2), depends on being able to draw samples from the distribution $p(\mathbf{z})$. Suppose, however, that it is impractical to sample directly from $p(\mathbf{z})$ but that we can evaluate $p(\mathbf{z})$ easily for any given value of $\mathbf{z}$. One simplistic strategy for evaluating expectations would be to discretize $\mathbf{z}$-space into a uniform grid and to evaluate the integrand as a sum of the form

$$\mathbb{E}[f] \simeq \sum_{l=1}^{L} p(\mathbf{z}^{(l)}) f(\mathbf{z}^{(l)}). \tag{11.18}$$

An obvious problem with this approach is that the number of terms in the summation grows exponentially with the dimensionality of $\mathbf{z}$. Furthermore, as we have already noted, the kinds of probability distributions of interest will often have much of their mass confined to relatively small regions of $\mathbf{z}$ space and so uniform sampling will be very inefficient because in high-dimensional problems, only a very small proportion of the samples will make a significant contribution to the sum. We would really like to choose the sample points to fall in regions where $p(\mathbf{z})$ is large, or ideally where the product $p(\mathbf{z})f(\mathbf{z})$ is large.

As in the case of rejection sampling, importance sampling is based on the use of a proposal distribution $q(\mathbf{z})$ from which it is easy to draw samples, as illustrated in Figure 11.8. We can then express the expectation in the form of a finite sum over samples $\{\mathbf{z}^{(l)}\}$ drawn from $q(\mathbf{z})$

$$\begin{aligned}
\mathbb{E}[f] &= \int f(\mathbf{z})p(\mathbf{z})\,\mathrm{d}\mathbf{z} \\
&= \int f(\mathbf{z})\frac{p(\mathbf{z})}{q(\mathbf{z})}q(\mathbf{z})\,\mathrm{d}\mathbf{z} \\
&\simeq \frac{1}{L}\sum_{l=1}^{L}\frac{p(\mathbf{z}^{(l)})}{q(\mathbf{z}^{(l)})}f(\mathbf{z}^{(l)}).
\end{aligned} \tag{11.19}$$

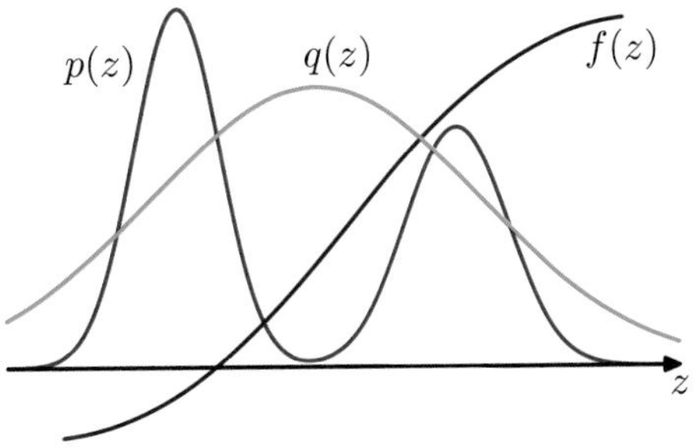

Figure 11.8 Importance sampling addresses the problem of evaluating the expectation of a function $f(z)$ with respect to a distribution $p(z)$ from which it is difficult to draw samples directly. Instead, samples $\{z^{(l)}\}$ are drawn from a simpler distribution $q(z)$, and the corresponding terms in the summation are weighted by the ratios $p(z^{(l)})/q(z^{(l)})$.

The quantities $r_l = p(\mathbf{z}^{(l)})/q(\mathbf{z}^{(l)})$ are known as *importance weights*, and they correct the bias introduced by sampling from the wrong distribution. Note that, unlike rejection sampling, all of the samples generated are retained.

It will often be the case that the distribution $p(\mathbf{z})$ can only be evaluated up to a normalization constant, so that $p(\mathbf{z}) = \widetilde{p}(\mathbf{z})/Z_p$ where $\widetilde{p}(\mathbf{z})$ can be evaluated easily, whereas Z_p is unknown. Similarly, we may wish to use an importance sampling distribution $q(\mathbf{z}) = \widetilde{q}(\mathbf{z})/Z_q$, which has the same property. We then have

$$\begin{aligned}\mathbb{E}[f] &= \int f(\mathbf{z})p(\mathbf{z})\,\mathrm{d}\mathbf{z} \\ &= \frac{Z_q}{Z_p}\int f(\mathbf{z})\frac{\widetilde{p}(\mathbf{z})}{\widetilde{q}(\mathbf{z})}q(\mathbf{z})\,\mathrm{d}\mathbf{z} \\ &\simeq \frac{Z_q}{Z_p}\frac{1}{L}\sum_{l=1}^{L}\widetilde{r}_l f(\mathbf{z}^{(l)}). \end{aligned} \tag{11.20}$$

where $\widetilde{r}_l = \widetilde{p}(\mathbf{z}^{(l)})/\widetilde{q}(\mathbf{z}^{(l)})$. We can use the same sample set to evaluate the ratio Z_p/Z_q with the result

$$\begin{aligned}\frac{Z_p}{Z_q} &= \frac{1}{Z_q}\int \widetilde{p}(\mathbf{z})\,\mathrm{d}\mathbf{z} = \int \frac{\widetilde{p}(\mathbf{z})}{\widetilde{q}(\mathbf{z})}q(\mathbf{z})\,\mathrm{d}\mathbf{z} \\ &\simeq \frac{1}{L}\sum_{l=1}^{L}\widetilde{r}_l \end{aligned} \tag{11.21}$$

and hence

$$\mathbb{E}[f] \simeq \sum_{l=1}^{L} w_l f(\mathbf{z}^{(l)}) \tag{11.22}$$

where we have defined

$$w_l = \frac{\widetilde{r}_l}{\sum_m \widetilde{r}_m} = \frac{\widetilde{p}(\mathbf{z}^{(l)})/q(\mathbf{z}^{(l)})}{\sum_m \widetilde{p}(\mathbf{z}^{(m)})/q(\mathbf{z}^{(m)})}. \tag{11.23}$$

As with rejection sampling, the success of the importance sampling approach depends crucially on how well the sampling distribution $q(\mathbf{z})$ matches the desired distribution $p(\mathbf{z})$. If, as is often the case, $p(\mathbf{z})f(\mathbf{z})$ is strongly varying and has a significant proportion of its mass concentrated over relatively small regions of $\mathbf{z}$ space, then the set of importance weights $\{r_l\}$ may be dominated by a few weights having large values, with the remaining weights being relatively insignificant. Thus the effective sample size can be much smaller than the apparent sample size L. The problem is even more severe if none of the samples falls in the regions where $p(\mathbf{z})f(\mathbf{z})$ is large. In that case, the apparent variances of r_l and $r_l f(\mathbf{z}^{(l)})$ may be small even though the estimate of the expectation may be severely wrong. Hence a major drawback of the importance sampling method is the potential to produce results that are arbitrarily in error and with no diagnostic indication. This also highlights a key requirement for the sampling distribution $q(\mathbf{z})$, namely that it should not be small or zero in regions where $p(\mathbf{z})$ may be significant.

For distributions defined in terms of a graphical model, we can apply the importance sampling technique in various ways. For discrete variables, a simple approach is called *uniform sampling*. The joint distribution for a directed graph is defined by (11.4). Each sample from the joint distribution is obtained by first setting those variables $\mathbf{z}_i$ that are in the evidence set equal to their observed values. Each of the remaining variables is then sampled independently from a uniform distribution over the space of possible instantiations. To determine the corresponding weight associated with a sample $\mathbf{z}^{(l)}$, we note that the sampling distribution $\widetilde{q}(\mathbf{z})$ is uniform over the possible choices for $\mathbf{z}$, and that $\widetilde{p}(\mathbf{z}|\mathbf{x}) = \widetilde{p}(\mathbf{z})$, where $\mathbf{x}$ denotes the subset of variables that are observed, and the equality follows from the fact that every sample $\mathbf{z}$ that is generated is necessarily consistent with the evidence. Thus the weights r_l are simply proportional to $p(\mathbf{z})$. Note that the variables can be sampled in any order. This approach can yield poor results if the posterior distribution is far from uniform, as is often the case in practice.

An improvement on this approach is called *likelihood weighted sampling* (Fung and Chang, 1990; Shachter and Peot, 1990) and is based on ancestral sampling of the variables. For each variable in turn, if that variable is in the evidence set, then it is just set to its instantiated value. If it is not in the evidence set, then it is sampled from the conditional distribution $p(\mathbf{z}_i|\mathrm{pa}_i)$ in which the conditioning variables are set to their currently sampled values. The weighting associated with the resulting sample $\mathbf{z}$ is then given by

$$r(\mathbf{z}) = \prod_{\mathbf{z}_i \notin \mathbf{e}} \frac{p(\mathbf{z}_i|\mathrm{pa}_i)}{p(\mathbf{z}_i|\mathrm{pa}_i)} \prod_{\mathbf{z}_i \in \mathbf{e}} \frac{p(\mathbf{z}_i|\mathrm{pa}_i)}{1} = \prod_{\mathbf{z}_i \in \mathbf{e}} p(\mathbf{z}_i|\mathrm{pa}_i). \tag{11.24}$$

This method can be further extended using *self-importance sampling* (Shachter and Peot, 1990) in which the importance sampling distribution is continually updated to reflect the current estimated posterior distribution.

11.1.5 Sampling-importance-resampling

The rejection sampling method discussed in Section 11.1.2 depends in part for its success on the determination of a suitable value for the constant k. For many pairs of distributions $p(\mathbf{z})$ and $q(\mathbf{z})$, it will be impractical to determine a suitable value for k in that any value that is sufficiently large to guarantee a bound on the desired distribution will lead to impractically small acceptance rates.

As in the case of rejection sampling, the *sampling-importance-resampling* (SIR) approach also makes use of a sampling distribution $q(\mathbf{z})$ but avoids having to determine the constant k. There are two stages to the scheme. In the first stage, L samples $\mathbf{z}^{(1)}, \ldots, \mathbf{z}^{(L)}$ are drawn from $q(\mathbf{z})$. Then in the second stage, weights $w_1, \ldots, w_L$ are constructed using (11.23). Finally, a second set of L samples is drawn from the discrete distribution $(\mathbf{z}^{(1)}, \ldots, \mathbf{z}^{(L)})$ with probabilities given by the weights $(w_1, \ldots, w_L)$.

The resulting L samples are only approximately distributed according to $p(\mathbf{z})$, but the distribution becomes correct in the limit $L \to \infty$. To see this, consider the univariate case, and note that the cumulative distribution of the resampled values is given by

$$\begin{aligned} p(z \leqslant a) &= \sum_{l:z^{(l)} \leqslant a} w_l \\ &= \frac{\sum_l I(z^{(l)} \leqslant a)\widetilde{p}(z^{(l)})/q(z^{(l)})}{\sum_l \widetilde{p}(z^{(l)})/q(z^{(l)})} \end{aligned} \tag{11.25}$$

where $I(\cdot)$ is the indicator function (which equals 1 if its argument is true and 0 otherwise). Taking the limit $L \to \infty$, and assuming suitable regularity of the distributions, we can replace the sums by integrals weighted according to the original sampling distribution $q(z)$

$$\begin{aligned} p(z \leqslant a) &= \frac{\displaystyle\int I(z \leqslant a)\,\{\widetilde{p}(z)/q(z)\}\,q(z)\,\mathrm{d}z}{\displaystyle\int \{\widetilde{p}(z)/q(z)\}\,q(z)\,\mathrm{d}z} \\ &= \frac{\displaystyle\int I(z \leqslant a)\widetilde{p}(z)\,\mathrm{d}z}{\displaystyle\int \widetilde{p}(z)\,\mathrm{d}z} \\ &= \int I(z \leqslant a)p(z)\,\mathrm{d}z \end{aligned} \tag{11.26}$$

which is the cumulative distribution function of $p(z)$. Again, we see that the normalization of $p(z)$ is not required.

For a finite value of L, and a given initial sample set, the resampled values will only approximately be drawn from the desired distribution. As with rejection sampling, the approximation improves as the sampling distribution $q(\mathbf{z})$ gets closer to the desired distribution $p(\mathbf{z})$. When $q(\mathbf{z}) = p(\mathbf{z})$, the initial samples $(\mathbf{z}^{(1)}, \ldots, \mathbf{z}^{(L)})$ have the desired distribution, and the weights $w_n = 1/L$ so that the resampled values also have the desired distribution.

If moments with respect to the distribution $p(\mathbf{z})$ are required, then they can be evaluated directly using the original samples together with the weights, because

$$\begin{aligned}
\mathbb{E}[f(\mathbf{z})] &= \int f(\mathbf{z})p(\mathbf{z})\,\mathrm{d}\mathbf{z} \\
&= \frac{\int f(\mathbf{z})[\widetilde{p}(\mathbf{z})/q(\mathbf{z})]q(\mathbf{z})\,\mathrm{d}\mathbf{z}}{\int [\widetilde{p}(\mathbf{z})/q(\mathbf{z})]q(\mathbf{z})\,\mathrm{d}\mathbf{z}} \\
&\simeq \sum_{l=1}^{L} w_l f(\mathbf{z}_l).
\end{aligned} \tag{11.27}$$

11.1.6 Sampling and the EM algorithm

In addition to providing a mechanism for direct implementation of the Bayesian framework, Monte Carlo methods can also play a role in the frequentist paradigm, for example to find maximum likelihood solutions. In particular, sampling methods can be used to approximate the E step of the EM algorithm for models in which the E step cannot be performed analytically. Consider a model with hidden variables $\mathbf{Z}$, visible (observed) variables $\mathbf{X}$, and parameters $\boldsymbol{\theta}$. The function that is optimized with respect to $\boldsymbol{\theta}$ in the M step is the expected complete-data log likelihood, given by

$$Q(\boldsymbol{\theta}, \boldsymbol{\theta}^{\text{old}}) = \int p(\mathbf{Z}|\mathbf{X}, \boldsymbol{\theta}^{\text{old}}) \ln p(\mathbf{Z}, \mathbf{X}|\boldsymbol{\theta})\,\mathrm{d}\mathbf{Z}. \tag{11.28}$$

We can use sampling methods to approximate this integral by a finite sum over samples $\{\mathbf{Z}^{(l)}\}$, which are drawn from the current estimate for the posterior distribution $p(\mathbf{Z}|\mathbf{X}, \boldsymbol{\theta}^{\text{old}})$, so that

$$Q(\boldsymbol{\theta}, \boldsymbol{\theta}^{\text{old}}) \simeq \frac{1}{L}\sum_{l=1}^{L} \ln p(\mathbf{Z}^{(l)}, \mathbf{X}|\boldsymbol{\theta}). \tag{11.29}$$

The Q function is then optimized in the usual way in the M step. This procedure is called the *Monte Carlo EM algorithm*.

It is straightforward to extend this to the problem of finding the mode of the posterior distribution over $\boldsymbol{\theta}$ (the MAP estimate) when a prior distribution $p(\boldsymbol{\theta})$ has been defined, simply by adding $\ln p(\boldsymbol{\theta})$ to the function $Q(\boldsymbol{\theta}, \boldsymbol{\theta}^{\text{old}})$ before performing the M step.

A particular instance of the Monte Carlo EM algorithm, called *stochastic EM*, arises if we consider a finite mixture model, and draw just one sample at each E step. Here the latent variable $\mathbf{Z}$ characterizes which of the K components of the mixture is responsible for generating each data point. In the E step, a sample of $\mathbf{Z}$ is taken from the posterior distribution $p(\mathbf{Z}|\mathbf{X}, \boldsymbol{\theta}^{\text{old}})$ where $\mathbf{X}$ is the data set. This effectively makes a hard assignment of each data point to one of the components in the mixture. In the M step, this sampled approximation to the posterior distribution is used to update the model parameters in the usual way.

Now suppose we move from a maximum likelihood approach to a full Bayesian treatment in which we wish to sample from the posterior distribution over the parameter vector $\boldsymbol{\theta}$. In principle, we would like to draw samples from the joint posterior $p(\boldsymbol{\theta}, \mathbf{Z}|\mathbf{X})$, but we shall suppose that this is computationally difficult. Suppose further that it is relatively straightforward to sample from the complete-data parameter posterior $p(\boldsymbol{\theta}|\mathbf{Z}, \mathbf{X})$. This inspires the *data augmentation* algorithm, which alternates between two steps known as the I-step (imputation step, analogous to an E step) and the P-step (posterior step, analogous to an M step).

IP Algorithm

I-step. We wish to sample from $p(\mathbf{Z}|\mathbf{X})$ but we cannot do this directly. We therefore note the relation

$$p(\mathbf{Z}|\mathbf{X}) = \int p(\mathbf{Z}|\boldsymbol{\theta}, \mathbf{X}) p(\boldsymbol{\theta}|\mathbf{X}) \, \mathrm{d}\boldsymbol{\theta} \tag{11.30}$$

and hence for $l = 1, \ldots, L$ we first draw a sample $\boldsymbol{\theta}^{(l)}$ from the current estimate for $p(\boldsymbol{\theta}|\mathbf{X})$, and then use this to draw a sample $\mathbf{Z}^{(l)}$ from $p(\mathbf{Z}|\boldsymbol{\theta}^{(l)}, \mathbf{X})$.

P-step. Given the relation

$$p(\boldsymbol{\theta}|\mathbf{X}) = \int p(\boldsymbol{\theta}|\mathbf{Z}, \mathbf{X}) p(\mathbf{Z}|\mathbf{X}) \, \mathrm{d}\mathbf{Z} \tag{11.31}$$

we use the samples $\{\mathbf{Z}^{(l)}\}$ obtained from the I-step to compute a revised estimate of the posterior distribution over $\boldsymbol{\theta}$ given by

$$p(\boldsymbol{\theta}|\mathbf{X}) \simeq \frac{1}{L} \sum_{l=1}^{L} p(\boldsymbol{\theta}|\mathbf{Z}^{(l)}, \mathbf{X}). \tag{11.32}$$

By assumption, it will be feasible to sample from this approximation in the I-step.

Note that we are making a (somewhat artificial) distinction between parameters $\boldsymbol{\theta}$ and hidden variables $\mathbf{Z}$. From now on, we blur this distinction and focus simply on the problem of drawing samples from a given posterior distribution.

11.2. Markov Chain Monte Carlo

In the previous section, we discussed the rejection sampling and importance sampling strategies for evaluating expectations of functions, and we saw that they suffer from severe limitations particularly in spaces of high dimensionality. We therefore turn in this section to a very general and powerful framework called Markov chain Monte Carlo (MCMC), which allows sampling from a large class of distributions, and which scales well with the dimensionality of the sample space. Markov chain Monte Carlo methods have their origins in physics (Metropolis and Ulam, 1949), and it was only towards the end of the 1980s that they started to have a significant impact in the field of statistics.

Section 11.2.1

As with rejection and importance sampling, we again sample from a proposal distribution. This time, however, we maintain a record of the current state $\mathbf{z}^{(\tau)}$, and the proposal distribution $q(\mathbf{z}|\mathbf{z}^{(\tau)})$ depends on this current state, and so the sequence of samples $\mathbf{z}^{(1)}, \mathbf{z}^{(2)}, \ldots$ forms a Markov chain. Again, if we write $p(\mathbf{z}) = \widetilde{p}(\mathbf{z})/Z_p$, we will assume that $\widetilde{p}(\mathbf{z})$ can readily be evaluated for any given value of $\mathbf{z}$, although the value of Z_p may be unknown. The proposal distribution itself is chosen to be sufficiently simple that it is straightforward to draw samples from it directly. At each cycle of the algorithm, we generate a candidate sample $\mathbf{z}^{\star}$ from the proposal distribution and then accept the sample according to an appropriate criterion.

In the basic *Metropolis* algorithm (Metropolis *et al.*, 1953), we assume that the proposal distribution is symmetric, that is $q(\mathbf{z}_A|\mathbf{z}_B) = q(\mathbf{z}_B|\mathbf{z}_A)$ for all values of $\mathbf{z}_A$ and $\mathbf{z}_B$. The candidate sample is then accepted with probability

$$A(\mathbf{z}^{\star}, \mathbf{z}^{(\tau)}) = \min\left(1, \frac{\widetilde{p}(\mathbf{z}^{\star})}{\widetilde{p}(\mathbf{z}^{(\tau)})}\right). \tag{11.33}$$

This can be achieved by choosing a random number u with uniform distribution over the unit interval $(0, 1)$ and then accepting the sample if $A(\mathbf{z}^{\star}, \mathbf{z}^{(\tau)}) > u$. Note that if the step from $\mathbf{z}^{(\tau)}$ to $\mathbf{z}^{\star}$ causes an increase in the value of $p(\mathbf{z})$, then the candidate point is certain to be kept.

If the candidate sample is accepted, then $\mathbf{z}^{(\tau+1)} = \mathbf{z}^{\star}$, otherwise the candidate point $\mathbf{z}^{\star}$ is discarded, $\mathbf{z}^{(\tau+1)}$ is set to $\mathbf{z}^{(\tau)}$ and another candidate sample is drawn from the distribution $q(\mathbf{z}|\mathbf{z}^{(\tau+1)})$. This is in contrast to rejection sampling, where rejected samples are simply discarded. In the Metropolis algorithm when a candidate point is rejected, the previous sample is included instead in the final list of samples, leading to multiple copies of samples. Of course, in a practical implementation, only a single copy of each retained sample would be kept, along with an integer weighting factor recording how many times that state appears. As we shall see, as long as $q(\mathbf{z}_A|\mathbf{z}_B)$ is positive for any values of $\mathbf{z}_A$ and $\mathbf{z}_B$ (this is a sufficient but not necessary condition), the distribution of $\mathbf{z}^{(\tau)}$ tends to $p(\mathbf{z})$ as $\tau \to \infty$. It should be emphasized, however, that the sequence $\mathbf{z}^{(1)}, \mathbf{z}^{(2)}, \ldots$ is not a set of independent samples from $p(\mathbf{z})$ because successive samples are highly correlated. If we wish to obtain independent samples, then we can discard most of the sequence and just retain every M^{th} sample. For M sufficiently large, the retained samples will for all practical purposes be independent. Figure 11.9 shows a simple illustrative example of sampling from a two-dimensional Gaussian distribution using the Metropolis algorithm in which the proposal distribution is an isotropic Gaussian.

Further insight into the nature of Markov chain Monte Carlo algorithms can be gleaned by looking at the properties of a specific example, namely a simple random

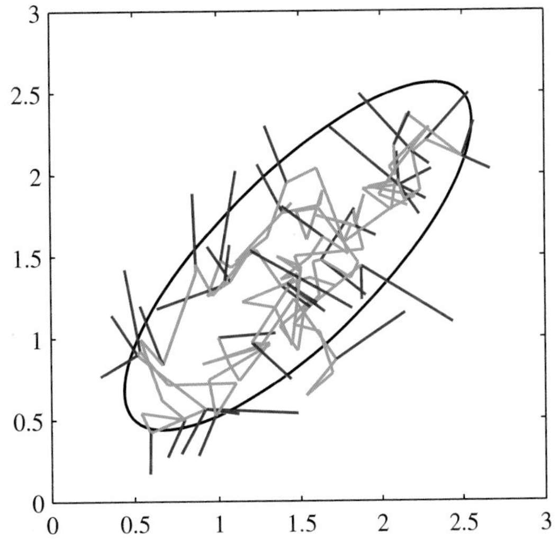

Figure 11.9 A simple illustration using Metropolis algorithm to sample from a Gaussian distribution whose one standard-deviation contour is shown by the ellipse. The proposal distribution is an isotropic Gaussian distribution whose standard deviation is 0.2. Steps that are accepted are shown as green lines, and rejected steps are shown in red. A total of 150 candidate samples are generated, of which 43 are rejected.

walk. Consider a state space z consisting of the integers, with probabilities

$$p(z^{(\tau+1)} = z^{(\tau)}) = 0.5 \tag{11.34}$$

$$p(z^{(\tau+1)} = z^{(\tau)} + 1) = 0.25 \tag{11.35}$$

$$p(z^{(\tau+1)} = z^{(\tau)} - 1) = 0.25 \tag{11.36}$$

where $z^{(\tau)}$ denotes the state at step τ. If the initial state is $z^{(0)} = 0$, then by symmetry the expected state at time τ will also be zero $\mathbb{E}[z^{(\tau)}] = 0$, and similarly it is easily seen that $\mathbb{E}[(z^{(\tau)})^2] = \tau/2$. Thus after τ steps, the random walk has only travelled a distance that on average is proportional to the square root of τ. This square root dependence is typical of random walk behaviour and shows that random walks are very inefficient in exploring the state space. As we shall see, a central goal in designing Markov chain Monte Carlo methods is to avoid random walk behaviour.

Exercise 11.10

11.2.1 Markov chains

Before discussing Markov chain Monte Carlo methods in more detail, it is useful to study some general properties of Markov chains in more detail. In particular, we ask under what circumstances will a Markov chain converge to the desired distribution. A first-order Markov chain is defined to be a series of random variables $\mathbf{z}^{(1)}, \ldots, \mathbf{z}^{(M)}$ such that the following conditional independence property holds for $m \in \{1, \ldots, M-1\}$

$$p(\mathbf{z}^{(m+1)}|\mathbf{z}^{(1)}, \ldots, \mathbf{z}^{(m)}) = p(\mathbf{z}^{(m+1)}|\mathbf{z}^{(m)}). \tag{11.37}$$

This of course can be represented as a directed graph in the form of a chain, an example of which is shown in Figure 8.38. We can then specify the Markov chain by giving the probability distribution for the initial variable $p(\mathbf{z}^{(0)})$ together with the conditional probabilities for subsequent variables in the form of *transition probabilities* $T_m(\mathbf{z}^{(m)}, \mathbf{z}^{(m+1)}) \equiv p(\mathbf{z}^{(m+1)}|\mathbf{z}^{(m)})$. A Markov chain is called *homogeneous* if the transition probabilities are the same for all m.

The marginal probability for a particular variable can be expressed in terms of the marginal probability for the previous variable in the chain in the form

$$p(\mathbf{z}^{(m+1)}) = \sum_{\mathbf{z}^{(m)}} p(\mathbf{z}^{(m+1)}|\mathbf{z}^{(m)})p(\mathbf{z}^{(m)}). \tag{11.38}$$

A distribution is said to be invariant, or stationary, with respect to a Markov chain if each step in the chain leaves that distribution invariant. Thus, for a homogeneous Markov chain with transition probabilities $T(\mathbf{z}', \mathbf{z})$, the distribution $p^\star(\mathbf{z})$ is invariant if

$$p^\star(\mathbf{z}) = \sum_{\mathbf{z}'} T(\mathbf{z}', \mathbf{z})p^\star(\mathbf{z}'). \tag{11.39}$$

Note that a given Markov chain may have more than one invariant distribution. For instance, if the transition probabilities are given by the identity transformation, then any distribution will be invariant.

A sufficient (but not necessary) condition for ensuring that the required distribution $p(\mathbf{z})$ is invariant is to choose the transition probabilities to satisfy the property of *detailed balance*, defined by

$$p^\star(\mathbf{z})T(\mathbf{z}, \mathbf{z}') = p^\star(\mathbf{z}')T(\mathbf{z}', \mathbf{z}) \tag{11.40}$$

for the particular distribution $p^\star(\mathbf{z})$. It is easily seen that a transition probability that satisfies detailed balance with respect to a particular distribution will leave that distribution invariant, because

$$\sum_{\mathbf{z}'} p^\star(\mathbf{z}')T(\mathbf{z}', \mathbf{z}) = \sum_{\mathbf{z}'} p^\star(\mathbf{z})T(\mathbf{z}, \mathbf{z}') = p^\star(\mathbf{z}) \sum_{\mathbf{z}'} p(\mathbf{z}'|\mathbf{z}) = p^\star(\mathbf{z}). \tag{11.41}$$

A Markov chain that respects detailed balance is said to be *reversible*.

Our goal is to use Markov chains to sample from a given distribution. We can achieve this if we set up a Markov chain such that the desired distribution is invariant. However, we must also require that for $m \to \infty$, the distribution $p(\mathbf{z}^{(m)})$ converges to the required invariant distribution $p^\star(\mathbf{z})$, irrespective of the choice of initial distribution $p(\mathbf{z}^{(0)})$. This property is called *ergodicity*, and the invariant distribution is then called the *equilibrium* distribution. Clearly, an ergodic Markov chain can have only one equilibrium distribution. It can be shown that a homogeneous Markov chain will be ergodic, subject only to weak restrictions on the invariant distribution and the transition probabilities (Neal, 1993).

In practice we often construct the transition probabilities from a set of 'base' transitions $B_1, \ldots, B_K$. This can be achieved through a mixture distribution of the form

$$T(\mathbf{z}', \mathbf{z}) = \sum_{k=1}^{K} \alpha_k B_k(\mathbf{z}', \mathbf{z}) \tag{11.42}$$

for some set of mixing coefficients $\alpha_1, \ldots, \alpha_K$ satisfying $\alpha_k \geqslant 0$ and $\sum_k \alpha_k = 1$. Alternatively, the base transitions may be combined through successive application, so that

$$T(\mathbf{z}', \mathbf{z}) = \sum_{\mathbf{z}_1} \ldots \sum_{\mathbf{z}_{K-1}} B_1(\mathbf{z}', \mathbf{z}_1) \ldots B_{K-1}(\mathbf{z}_{K-2}, \mathbf{z}_{K-1}) B_K(\mathbf{z}_{K-1}, \mathbf{z}). \quad (11.43)$$

If a distribution is invariant with respect to each of the base transitions, then obviously it will also be invariant with respect to either of the $T(\mathbf{z}', \mathbf{z})$ given by (11.42) or (11.43). For the case of the mixture (11.42), if each of the base transitions satisfies detailed balance, then the mixture transition T will also satisfy detailed balance. This does not hold for the transition probability constructed using (11.43), although by symmetrizing the order of application of the base transitions, in the form $B_1, B_2, \ldots, B_K, B_K, \ldots, B_2, B_1$, detailed balance can be restored. A common example of the use of composite transition probabilities is where each base transition changes only a subset of the variables.

11.2.2 The Metropolis-Hastings algorithm

Earlier we introduced the basic Metropolis algorithm, without actually demonstrating that it samples from the required distribution. Before giving a proof, we first discuss a generalization, known as the *Metropolis-Hastings* algorithm (Hastings, 1970), to the case where the proposal distribution is no longer a symmetric function of its arguments. In particular at step τ of the algorithm, in which the current state is $\mathbf{z}^{(\tau)}$, we draw a sample $\mathbf{z}^\star$ from the distribution $q_k(\mathbf{z}|\mathbf{z}^{(\tau)})$ and then accept it with probability $A_k(\mathbf{z}^\star, \mathbf{z}^{(\tau)})$ where

$$A_k(\mathbf{z}^\star, \mathbf{z}^{(\tau)}) = \min\left(1, \frac{\widetilde{p}(\mathbf{z}^\star) q_k(\mathbf{z}^{(\tau)}|\mathbf{z}^\star)}{\widetilde{p}(\mathbf{z}^{(\tau)}) q_k(\mathbf{z}^\star|\mathbf{z}^{(\tau)})}\right). \quad (11.44)$$

Here k labels the members of the set of possible transitions being considered. Again, the evaluation of the acceptance criterion does not require knowledge of the normalizing constant Z_p in the probability distribution $p(\mathbf{z}) = \widetilde{p}(\mathbf{z})/Z_p$. For a symmetric proposal distribution the Metropolis-Hastings criterion (11.44) reduces to the standard Metropolis criterion given by (11.33).

We can show that $p(\mathbf{z})$ is an invariant distribution of the Markov chain defined by the Metropolis-Hastings algorithm by showing that detailed balance, defined by (11.40), is satisfied. Using (11.44) we have

$$\begin{aligned} p(\mathbf{z}) q_k(\mathbf{z}'|\mathbf{z}) A_k(\mathbf{z}', \mathbf{z}) &= \min\left(p(\mathbf{z}) q_k(\mathbf{z}'|\mathbf{z}), p(\mathbf{z}') q_k(\mathbf{z}|\mathbf{z}')\right) \\ &= \min\left(p(\mathbf{z}') q_k(\mathbf{z}|\mathbf{z}'), p(\mathbf{z}) q_k(\mathbf{z}'|\mathbf{z})\right) \\ &= p(\mathbf{z}') q_k(\mathbf{z}|\mathbf{z}') A_k(\mathbf{z}, \mathbf{z}') \end{aligned} \quad (11.45)$$

as required.

The specific choice of proposal distribution can have a marked effect on the performance of the algorithm. For continuous state spaces, a common choice is a Gaussian centred on the current state, leading to an important trade-off in determining the variance parameter of this distribution. If the variance is small, then the

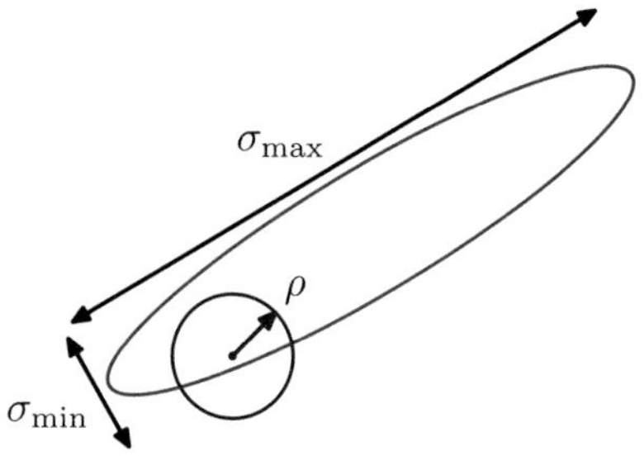

Figure 11.10 Schematic illustration of the use of an isotropic Gaussian proposal distribution (blue circle) to sample from a correlated multivariate Gaussian distribution (red ellipse) having very different standard deviations in different directions, using the Metropolis-Hastings algorithm. In order to keep the rejection rate low, the scale ρ of the proposal distribution should be on the order of the smallest standard deviation $\sigma_{\min}$, which leads to random walk behaviour in which the number of steps separating states that are approximately independent is of order $(\sigma_{\max}/\sigma_{\min})^2$ where $\sigma_{\max}$ is the largest standard deviation.

proportion of accepted transitions will be high, but progress through the state space takes the form of a slow random walk leading to long correlation times. However, if the variance parameter is large, then the rejection rate will be high because, in the kind of complex problems we are considering, many of the proposed steps will be to states for which the probability $p(\mathbf{z})$ is low. Consider a multivariate distribution $p(\mathbf{z})$ having strong correlations between the components of $\mathbf{z}$, as illustrated in Figure 11.10. The scale ρ of the proposal distribution should be as large as possible without incurring high rejection rates. This suggests that ρ should be of the same order as the smallest length scale $\sigma_{\min}$. The system then explores the distribution along the more extended direction by means of a random walk, and so the number of steps to arrive at a state that is more or less independent of the original state is of order $(\sigma_{\max}/\sigma_{\min})^2$. In fact in two dimensions, the increase in rejection rate as ρ increases is offset by the larger steps sizes of those transitions that are accepted, and more generally for a multivariate Gaussian the number of steps required to obtain independent samples scales like $(\sigma_{\max}/\sigma_2)^2$ where σ_2 is the second-smallest standard deviation (Neal, 1993). These details aside, it remains the case that if the length scales over which the distributions vary are very different in different directions, then the Metropolis Hastings algorithm can have very slow convergence.

11.3. Gibbs Sampling

Gibbs sampling (Geman and Geman, 1984) is a simple and widely applicable Markov chain Monte Carlo algorithm and can be seen as a special case of the Metropolis-Hastings algorithm.

Consider the distribution $p(\mathbf{z}) = p(z_1, \ldots, z_M)$ from which we wish to sample, and suppose that we have chosen some initial state for the Markov chain. Each step of the Gibbs sampling procedure involves replacing the value of one of the variables by a value drawn from the distribution of that variable conditioned on the values of the remaining variables. Thus we replace z_i by a value drawn from the distribution $p(z_i|\mathbf{z}_{\setminus i})$, where z_i denotes the ith component of $\mathbf{z}$, and $\mathbf{z}_{\setminus i}$ denotes $z_1, \ldots, z_M$ but with z_i omitted. This procedure is repeated either by cycling through the variables in some particular order or by choosing the variable to be updated at each step at random from some distribution.

For example, suppose we have a distribution $p(z_1, z_2, z_3)$ over three variables, and at step τ of the algorithm we have selected values $z_1^{(\tau)}, z_2^{(\tau)}$ and $z_3^{(\tau)}$. We first replace $z_1^{(\tau)}$ by a new value $z_1^{(\tau+1)}$ obtained by sampling from the conditional distribution

$$p(z_1|z_2^{(\tau)}, z_3^{(\tau)}). \tag{11.46}$$

Next we replace $z_2^{(\tau)}$ by a value $z_2^{(\tau+1)}$ obtained by sampling from the conditional distribution

$$p(z_2|z_1^{(\tau+1)}, z_3^{(\tau)}) \tag{11.47}$$

so that the new value for z_1 is used straight away in subsequent sampling steps. Then we update z_3 with a sample $z_3^{(\tau+1)}$ drawn from

$$p(z_3|z_1^{(\tau+1)}, z_2^{(\tau+1)}) \tag{11.48}$$

and so on, cycling through the three variables in turn.

Gibbs Sampling

1. Initialize $\{z_i : i = 1, \ldots, M\}$
2. For $\tau = 1, \ldots, T$:
   - Sample $z_1^{(\tau+1)} \sim p(z_1|z_2^{(\tau)}, z_3^{(\tau)}, \ldots, z_M^{(\tau)})$.
   - Sample $z_2^{(\tau+1)} \sim p(z_2|z_1^{(\tau+1)}, z_3^{(\tau)}, \ldots, z_M^{(\tau)})$.

     $\vdots$
   - Sample $z_j^{(\tau+1)} \sim p(z_j|z_1^{(\tau+1)}, \ldots, z_{j-1}^{(\tau+1)}, z_{j+1}^{(\tau)}, \ldots, z_M^{(\tau)})$.

     $\vdots$
   - Sample $z_M^{(\tau+1)} \sim p(z_M|z_1^{(\tau+1)}, z_2^{(\tau+1)}, \ldots, z_{M-1}^{(\tau+1)})$.

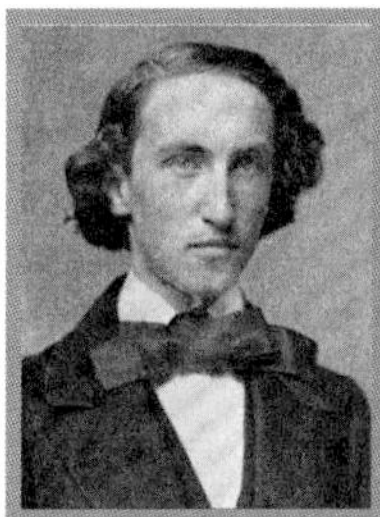

Josiah Willard Gibbs
1839–1903

Gibbs spent almost his entire life living in a house built by his father in New Haven, Connecticut. In 1863, Gibbs was granted the first PhD in engineering in the United States, and in 1871 he was appointed to the first chair of mathematical physics in the United States at Yale, a post for which he received no salary because at the time he had no publications. He developed the field of vector analysis and made contributions to crystallography and planetary orbits. His most famous work, entitled *On the Equilibrium of Heterogeneous Substances*, laid the foundations for the science of physical chemistry.

To show that this procedure samples from the required distribution, we first of all note that the distribution $p(\mathbf{z})$ is an invariant of each of the Gibbs sampling steps individually and hence of the whole Markov chain. This follows from the fact that when we sample from $p(z_i|\mathbf{z}_{\setminus i})$, the marginal distribution $p(\mathbf{z}_{\setminus i})$ is clearly invariant because the value of $\mathbf{z}_{\setminus i}$ is unchanged. Also, each step by definition samples from the correct conditional distribution $p(z_i|\mathbf{z}_{\setminus i})$. Because these conditional and marginal distributions together specify the joint distribution, we see that the joint distribution is itself invariant.

The second requirement to be satisfied in order that the Gibbs sampling procedure samples from the correct distribution is that it be ergodic. A sufficient condition for ergodicity is that none of the conditional distributions be anywhere zero. If this is the case, then any point in z space can be reached from any other point in a finite number of steps involving one update of each of the component variables. If this requirement is not satisfied, so that some of the conditional distributions have zeros, then ergodicity, if it applies, must be proven explicitly.

The distribution of initial states must also be specified in order to complete the algorithm, although samples drawn after many iterations will effectively become independent of this distribution. Of course, successive samples from the Markov chain will be highly correlated, and so to obtain samples that are nearly independent it will be necessary to subsample the sequence.

We can obtain the Gibbs sampling procedure as a particular instance of the Metropolis-Hastings algorithm as follows. Consider a Metropolis-Hastings sampling step involving the variable z_k in which the remaining variables $\mathbf{z}_{\setminus k}$ remain fixed, and for which the transition probability from $\mathbf{z}$ to $\mathbf{z}^\star$ is given by $q_k(\mathbf{z}^\star|\mathbf{z}) = p(z_k^\star|\mathbf{z}_{\setminus k})$. We note that $\mathbf{z}^\star_{\setminus k} = \mathbf{z}_{\setminus k}$ because these components are unchanged by the sampling step. Also, $p(\mathbf{z}) = p(z_k|\mathbf{z}_{\setminus k})p(\mathbf{z}_{\setminus k})$. Thus the factor that determines the acceptance probability in the Metropolis-Hastings (11.44) is given by

$$A(\mathbf{z}^\star, \mathbf{z}) = \frac{p(\mathbf{z}^\star)q_k(\mathbf{z}|\mathbf{z}^\star)}{p(\mathbf{z})q_k(\mathbf{z}^\star|\mathbf{z})} = \frac{p(z_k^\star|\mathbf{z}^\star_{\setminus k})p(\mathbf{z}^\star_{\setminus k})p(z_k|\mathbf{z}^\star_{\setminus k})}{p(z_k|\mathbf{z}_{\setminus k})p(\mathbf{z}_{\setminus k})p(z_k^\star|\mathbf{z}_{\setminus k})} = 1 \tag{11.49}$$

where we have used $\mathbf{z}^\star_{\setminus k} = \mathbf{z}_{\setminus k}$. Thus the Metropolis-Hastings steps are always accepted.

As with the Metropolis algorithm, we can gain some insight into the behaviour of Gibbs sampling by investigating its application to a Gaussian distribution. Consider a correlated Gaussian in two variables, as illustrated in Figure 11.11, having conditional distributions of width l and marginal distributions of width L. The typical step size is governed by the conditional distributions and will be of order l. Because the state evolves according to a random walk, the number of steps needed to obtain independent samples from the distribution will be of order $(L/l)^2$. Of course if the Gaussian distribution were uncorrelated, then the Gibbs sampling procedure would be optimally efficient. For this simple problem, we could rotate the coordinate system in order to decorrelate the variables. However, in practical applications it will generally be infeasible to find such transformations.

One approach to reducing random walk behaviour in Gibbs sampling is called *over-relaxation* (Adler, 1981). In its original form, this applies to problems for which

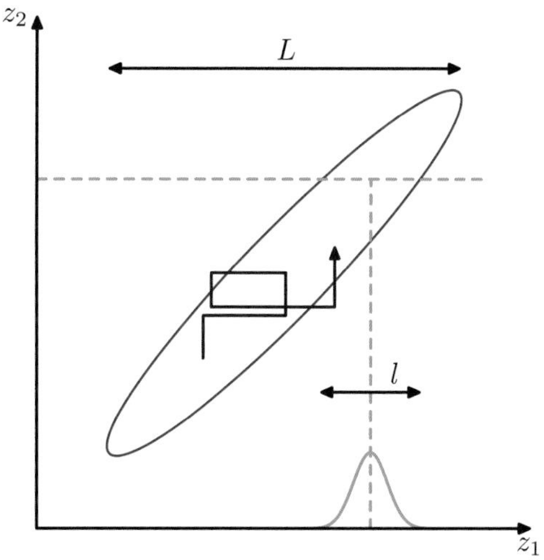

Figure 11.11 Illustration of Gibbs sampling by alternate updates of two variables whose distribution is a correlated Gaussian. The step size is governed by the standard deviation of the conditional distribution (green curve), and is $O(l)$, leading to slow progress in the direction of elongation of the joint distribution (red ellipse). The number of steps needed to obtain an independent sample from the distribution is $O((L/l)^2)$.

the conditional distributions are Gaussian, which represents a more general class of distributions than the multivariate Gaussian because, for example, the non-Gaussian distribution $p(z, y) \propto \exp(-z^2y^2)$ has Gaussian conditional distributions. At each step of the Gibbs sampling algorithm, the conditional distribution for a particular component z_i has some mean μ_i and some variance σ_i^2. In the over-relaxation framework, the value of z_i is replaced with

$$z_i' = \mu_i + \alpha(z_i - \mu_i) + \sigma_i(1 - \alpha^2)^{1/2}\nu \tag{11.50}$$

where ν is a Gaussian random variable with zero mean and unit variance, and α is a parameter such that $-1 < \alpha < 1$. For $\alpha = 0$, the method is equivalent to standard Gibbs sampling, and for $\alpha < 0$ the step is biased to the opposite side of the mean. This step leaves the desired distribution invariant because if z_i has mean μ_i and variance σ_i^2, then so too does z_i'. The effect of over-relaxation is to encourage directed motion through state space when the variables are highly correlated. The framework of *ordered over-relaxation* (Neal, 1999) generalizes this approach to non-Gaussian distributions.

The practical applicability of Gibbs sampling depends on the ease with which samples can be drawn from the conditional distributions $p(z_k|\mathbf{z}_{\setminus k})$. In the case of probability distributions specified using graphical models, the conditional distributions for individual nodes depend only on the variables in the corresponding Markov blankets, as illustrated in Figure 11.12. For directed graphs, a wide choice of conditional distributions for the individual nodes conditioned on their parents will lead to conditional distributions for Gibbs sampling that are log concave. The adaptive rejection sampling methods discussed in Section 11.1.3 therefore provide a framework for Monte Carlo sampling from directed graphs with broad applicability.

If the graph is constructed using distributions from the exponential family, and if the parent-child relationships preserve conjugacy, then the full conditional distributions arising in Gibbs sampling will have the same functional form as the orig-

Figure 11.12 The Gibbs sampling method requires samples to be drawn from the conditional distribution of a variable conditioned on the remaining variables. For graphical models, this conditional distribution is a function only of the states of the nodes in the Markov blanket. For an undirected graph this comprises the set of neighbours, as shown on the left, while for a directed graph the Markov blanket comprises the parents, the children, and the co-parents, as shown on the right.

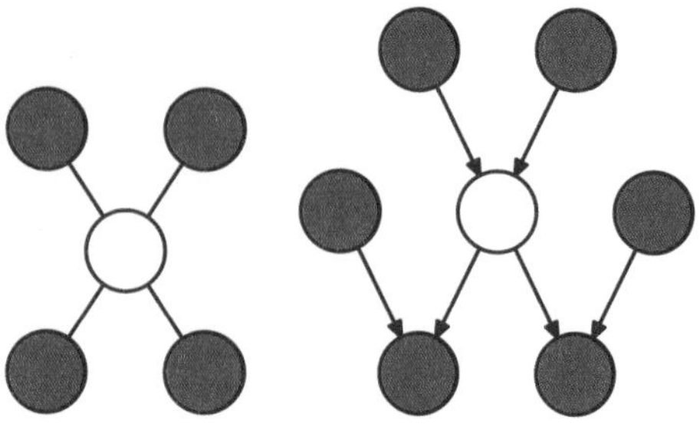

inal conditional distributions (conditioned on the parents) defining each node, and so standard sampling techniques can be employed. In general, the full conditional distributions will be of a complex form that does not permit the use of standard sampling algorithms. However, if these conditionals are log concave, then sampling can be done efficiently using adaptive rejection sampling (assuming the corresponding variable is a scalar).

If, at each stage of the Gibbs sampling algorithm, instead of drawing a sample from the corresponding conditional distribution, we make a point estimate of the variable given by the maximum of the conditional distribution, then we obtain the iterated conditional modes (ICM) algorithm discussed in Section 8.3.3. Thus ICM can be seen as a greedy approximation to Gibbs sampling.

Because the basic Gibbs sampling technique considers one variable at a time, there are strong dependencies between successive samples. At the opposite extreme, if we could draw samples directly from the joint distribution (an operation that we are supposing is intractable), then successive samples would be independent. We can hope to improve on the simple Gibbs sampler by adopting an intermediate strategy in which we sample successively from groups of variables rather than individual variables. This is achieved in the *blocking Gibbs* sampling algorithm by choosing blocks of variables, not necessarily disjoint, and then sampling jointly from the variables in each block in turn, conditioned on the remaining variables (Jensen *et al.*, 1995).

11.4. Slice Sampling

We have seen that one of the difficulties with the Metropolis algorithm is the sensitivity to step size. If this is too small, the result is slow decorrelation due to random walk behaviour, whereas if it is too large the result is inefficiency due to a high rejection rate. The technique of *slice sampling* (Neal, 2003) provides an adaptive step size that is automatically adjusted to match the characteristics of the distribution. Again it requires that we are able to evaluate the unnormalized distribution $\widetilde{p}(\mathbf{z})$.

Consider first the univariate case. Slice sampling involves augmenting z with an additional variable u and then drawing samples from the joint (z, u) space. We shall see another example of this approach when we discuss hybrid Monte Carlo in Section 11.5. The goal is to sample uniformly from the area under the distribution

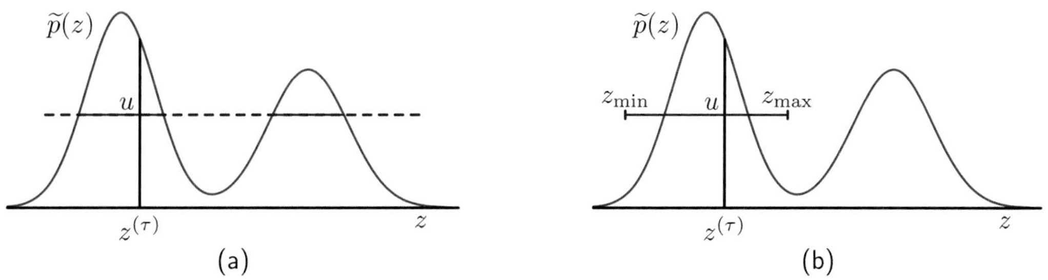

Figure 11.13 Illustration of slice sampling. (a) For a given value $z^{(\tau)}$, a value of u is chosen uniformly in the region $0 \leqslant u \leqslant \widetilde{p}(z^{(\tau)})$, which then defines a 'slice' through the distribution, shown by the solid horizontal lines. (b) Because it is infeasible to sample directly from a slice, a new sample of z is drawn from a region $z_{\min} \leqslant z \leqslant z_{\max}$, which contains the previous value $z^{(\tau)}$.

given by

$$\widehat{p}(z,u) = \begin{cases} 1/Z_p & \text{if } 0 \leqslant u \leqslant \widetilde{p}(z) \\ 0 & \text{otherwise} \end{cases} \tag{11.51}$$

where $Z_p = \int \widetilde{p}(z)\,\mathrm{d}z$. The marginal distribution over z is given by

$$\int \widehat{p}(z,u)\,\mathrm{d}u = \int_0^{\widetilde{p}(z)} \frac{1}{Z_p}\,\mathrm{d}u = \frac{\widetilde{p}(z)}{Z_p} = p(z) \tag{11.52}$$

and so we can sample from $p(z)$ by sampling from $\widehat{p}(z,u)$ and then ignoring the u values. This can be achieved by alternately sampling z and u. Given the value of z we evaluate $\widetilde{p}(z)$ and then sample u uniformly in the range $0 \leqslant u \leqslant \widetilde{p}(z)$, which is straightforward. Then we fix u and sample z uniformly from the 'slice' through the distribution defined by $\{z : \widetilde{p}(z) > u\}$. This is illustrated in Figure 11.13(a).

In practice, it can be difficult to sample directly from a slice through the distribution and so instead we define a sampling scheme that leaves the uniform distribution under $\widehat{p}(z,u)$ invariant, which can be achieved by ensuring that detailed balance is satisfied. Suppose the current value of z is denoted $z^{(\tau)}$ and that we have obtained a corresponding sample u. The next value of z is obtained by considering a region $z_{\min} \leqslant z \leqslant z_{\max}$ that contains $z^{(\tau)}$. It is in the choice of this region that the adaptation to the characteristic length scales of the distribution takes place. We want the region to encompass as much of the slice as possible so as to allow large moves in z space while having as little as possible of this region lying outside the slice, because this makes the sampling less efficient.

One approach to the choice of region involves starting with a region containing $z^{(\tau)}$ having some width w and then testing each of the end points to see if they lie within the slice. If either end point does not, then the region is extended in that direction by increments of value w until the end point lies outside the region. A candidate value z' is then chosen uniformly from this region, and if it lies within the slice, then it forms $z^{(\tau+1)}$. If it lies outside the slice, then the region is shrunk such that z' forms an end point and such that the region still contains $z^{(\tau)}$. Then another candidate point is drawn uniformly from this reduced region and so on, until a value of z is found that lies within the slice.

Slice sampling can be applied to multivariate distributions by repeatedly sampling each variable in turn, in the manner of Gibbs sampling. This requires that we are able to compute, for each component z_i, a function that is proportional to $p(z_i|\mathbf{z}_{\setminus i})$.

11.5. The Hybrid Monte Carlo Algorithm

As we have already noted, one of the major limitations of the Metropolis algorithm is that it can exhibit random walk behaviour whereby the distance traversed through the state space grows only as the square root of the number of steps. The problem cannot be resolved simply by taking bigger steps as this leads to a high rejection rate.

In this section, we introduce a more sophisticated class of transitions based on an analogy with physical systems and that has the property of being able to make large changes to the system state while keeping the rejection probability small. It is applicable to distributions over continuous variables for which we can readily evaluate the gradient of the log probability with respect to the state variables. We will discuss the dynamical systems framework in Section 11.5.1, and then in Section 11.5.2 we explain how this may be combined with the Metropolis algorithm to yield the powerful hybrid Monte Carlo algorithm. A background in physics is not required as this section is self-contained and the key results are all derived from first principles.

11.5.1 Dynamical systems

The dynamical approach to stochastic sampling has its origins in algorithms for simulating the behaviour of physical systems evolving under Hamiltonian dynamics. In a Markov chain Monte Carlo simulation, the goal is to sample from a given probability distribution $p(\mathbf{z})$. The framework of *Hamiltonian dynamics* is exploited by casting the probabilistic simulation in the form of a Hamiltonian system. In order to remain in keeping with the literature in this area, we make use of the relevant dynamical systems terminology where appropriate, which will be defined as we go along.

The dynamics that we consider corresponds to the evolution of the state variable $\mathbf{z} = \{z_i\}$ under continuous time, which we denote by τ. Classical dynamics is described by Newton's second law of motion in which the acceleration of an object is proportional to the applied force, corresponding to a second-order differential equation over time. We can decompose a second-order equation into two coupled first-order equations by introducing intermediate *momentum* variables $\mathbf{r}$, corresponding to the rate of change of the state variables $\mathbf{z}$, having components

$$r_i = \frac{\mathrm{d}z_i}{\mathrm{d}\tau} \tag{11.53}$$

where the z_i can be regarded as *position* variables in this dynamics perspective. Thus for each position variable there is a corresponding momentum variable, and the joint space of position and momentum variables is called *phase space*.

Without loss of generality, we can write the probability distribution $p(\mathbf{z})$ in the form

$$p(\mathbf{z}) = \frac{1}{Z_p} \exp\left(-E(\mathbf{z})\right) \tag{11.54}$$

where $E(\mathbf{z})$ is interpreted as the *potential energy* of the system when in state $\mathbf{z}$. The system acceleration is the rate of change of momentum and is given by the applied *force*, which itself is the negative gradient of the potential energy

$$\frac{\mathrm{d}r_i}{\mathrm{d}\tau} = -\frac{\partial E(\mathbf{z})}{\partial z_i}. \tag{11.55}$$

It is convenient to reformulate this dynamical system using the Hamiltonian framework. To do this, we first define the *kinetic energy* by

$$K(\mathbf{r}) = \frac{1}{2}\|\mathbf{r}\|^2 = \frac{1}{2}\sum_i r_i^2. \tag{11.56}$$

The total energy of the system is then the sum of its potential and kinetic energies

$$H(\mathbf{z}, \mathbf{r}) = E(\mathbf{z}) + K(\mathbf{r}) \tag{11.57}$$

where H is the *Hamiltonian* function. Using (11.53), (11.55), (11.56), and (11.57), we can now express the dynamics of the system in terms of the Hamiltonian equations given by

Exercise 11.15

$$\frac{\mathrm{d}z_i}{\mathrm{d}\tau} = \frac{\partial H}{\partial r_i} \tag{11.58}$$

$$\frac{\mathrm{d}r_i}{\mathrm{d}\tau} = -\frac{\partial H}{\partial z_i}. \tag{11.59}$$

William Hamilton
1805–1865

William Rowan Hamilton was an Irish mathematician and physicist, and child prodigy, who was appointed Professor of Astronomy at Trinity College, Dublin, in 1827, before he had even graduated. One of Hamilton's most important contributions was a new formulation of dynamics, which played a significant role in the later development of quantum mechanics.

His other great achievement was the development of *quaternions*, which generalize the concept of complex numbers by introducing three distinct square roots of minus one, which satisfy $i^2 = j^2 = k^2 = ijk = -1$. It is said that these equations occurred to him while walking along the Royal Canal in Dublin with his wife, on 16 October 1843, and he promptly carved the equations into the side of Broome bridge. Although there is no longer any evidence of the carving, there is now a stone plaque on the bridge commemorating the discovery and displaying the quaternion equations.

During the evolution of this dynamical system, the value of the Hamiltonian H is constant, as is easily seen by differentiation

$$\begin{aligned}\frac{\mathrm{d}H}{\mathrm{d}\tau} &= \sum_i \left\{ \frac{\partial H}{\partial z_i}\frac{\mathrm{d}z_i}{\mathrm{d}\tau} + \frac{\partial H}{\partial r_i}\frac{\mathrm{d}r_i}{\mathrm{d}\tau} \right\} \\ &= \sum_i \left\{ \frac{\partial H}{\partial z_i}\frac{\partial H}{\partial r_i} - \frac{\partial H}{\partial r_i}\frac{\partial H}{\partial z_i} \right\} = 0. \end{aligned} \tag{11.60}$$

A second important property of Hamiltonian dynamical systems, known as *Liouville's Theorem*, is that they preserve volume in phase space. In other words, if we consider a region within the space of variables $(\mathbf{z}, \mathbf{r})$, then as this region evolves under the equations of Hamiltonian dynamics, its shape may change but its volume will not. This can be seen by noting that the flow field (rate of change of location in phase space) is given by

$$\mathbf{V} = \left(\frac{\mathrm{d}\mathbf{z}}{\mathrm{d}\tau}, \frac{\mathrm{d}\mathbf{r}}{\mathrm{d}\tau} \right) \tag{11.61}$$

and that the divergence of this field vanishes

$$\begin{aligned}\operatorname{div} \mathbf{V} &= \sum_i \left\{ \frac{\partial}{\partial z_i}\frac{\mathrm{d}z_i}{\mathrm{d}\tau} + \frac{\partial}{\partial r_i}\frac{\mathrm{d}r_i}{\mathrm{d}\tau} \right\} \\ &= \sum_i \left\{ -\frac{\partial}{\partial z_i}\frac{\partial H}{\partial r_i} + \frac{\partial}{\partial r_i}\frac{\partial H}{\partial z_i} \right\} = 0. \end{aligned} \tag{11.62}$$

Now consider the joint distribution over phase space whose total energy is the Hamiltonian, i.e., the distribution given by

$$p(\mathbf{z}, \mathbf{r}) = \frac{1}{Z_H} \exp(-H(\mathbf{z}, \mathbf{r})). \tag{11.63}$$

Using the two results of conservation of volume and conservation of H, it follows that the Hamiltonian dynamics will leave $p(\mathbf{z}, \mathbf{r})$ invariant. This can be seen by considering a small region of phase space over which H is approximately constant. If we follow the evolution of the Hamiltonian equations for a finite time, then the volume of this region will remain unchanged as will the value of H in this region, and hence the probability density, which is a function only of H, will also be unchanged.

Although H is invariant, the values of $\mathbf{z}$ and $\mathbf{r}$ will vary, and so by integrating the Hamiltonian dynamics over a finite time duration it becomes possible to make large changes to $\mathbf{z}$ in a systematic way that avoids random walk behaviour.

Evolution under the Hamiltonian dynamics will not, however, sample ergodically from $p(\mathbf{z}, \mathbf{r})$ because the value of H is constant. In order to arrive at an ergodic sampling scheme, we can introduce additional moves in phase space that change the value of H while also leaving the distribution $p(\mathbf{z}, \mathbf{r})$ invariant. The simplest way to achieve this is to replace the value of $\mathbf{r}$ with one drawn from its distribution conditioned on $\mathbf{z}$. This can be regarded as a Gibbs sampling step, and hence from

Section 11.3 we see that this also leaves the desired distribution invariant. Noting that $\mathbf{z}$ and $\mathbf{r}$ are independent in the distribution $p(\mathbf{z}, \mathbf{r})$, we see that the conditional distribution $p(\mathbf{r}|\mathbf{z})$ is a Gaussian from which it is straightforward to sample.

Exercise 11.16

In a practical application of this approach, we have to address the problem of performing a numerical integration of the Hamiltonian equations. This will necessarily introduce numerical errors and so we should devise a scheme that minimizes the impact of such errors. In fact, it turns out that integration schemes can be devised for which Liouville's theorem still holds exactly. This property will be important in the hybrid Monte Carlo algorithm, which is discussed in Section 11.5.2. One scheme for achieving this is called the *leapfrog* discretization and involves alternately updating discrete-time approximations $\widehat{\mathbf{z}}$ and $\widehat{\mathbf{r}}$ to the position and momentum variables using

$$\widehat{r}_i(\tau + \epsilon/2) = \widehat{r}_i(\tau) - \frac{\epsilon}{2}\frac{\partial E}{\partial z_i}(\widehat{\mathbf{z}}(\tau)) \tag{11.64}$$

$$\widehat{z}_i(\tau + \epsilon) = \widehat{z}_i(\tau) + \epsilon\widehat{r}_i(\tau + \epsilon/2) \tag{11.65}$$

$$\widehat{r}_i(\tau + \epsilon) = \widehat{r}_i(\tau + \epsilon/2) - \frac{\epsilon}{2}\frac{\partial E}{\partial z_i}(\widehat{\mathbf{z}}(\tau + \epsilon)). \tag{11.66}$$

We see that this takes the form of a half-step update of the momentum variables with step size $\epsilon/2$, followed by a full-step update of the position variables with step size ϵ, followed by a second half-step update of the momentum variables. If several leapfrog steps are applied in succession, it can be seen that half-step updates to the momentum variables can be combined into full-step updates with step size ϵ. The successive updates to position and momentum variables then leapfrog over each other. In order to advance the dynamics by a time interval τ, we need to take τ/ϵ steps. The error involved in the discretized approximation to the continuous time dynamics will go to zero, assuming a smooth function $E(\mathbf{z})$, in the limit $\epsilon \to 0$. However, for a nonzero ϵ as used in practice, some residual error will remain. We shall see in Section 11.5.2 how the effects of such errors can be eliminated in the hybrid Monte Carlo algorithm.

In summary then, the Hamiltonian dynamical approach involves alternating between a series of leapfrog updates and a resampling of the momentum variables from their marginal distribution.

Note that the Hamiltonian dynamics method, unlike the basic Metropolis algorithm, is able to make use of information about the gradient of the log probability distribution as well as about the distribution itself. An analogous situation is familiar from the domain of function optimization. In most cases where gradient information is available, it is highly advantageous to make use of it. Informally, this follows from the fact that in a space of dimension D, the additional computational cost of evaluating a gradient compared with evaluating the function itself will typically be a fixed factor independent of D, whereas the D-dimensional gradient vector conveys D pieces of information compared with the one piece of information given by the function itself.

11.5.2 Hybrid Monte Carlo

As we discussed in the previous section, for a nonzero step size ϵ, the discretization of the leapfrog algorithm will introduce errors into the integration of the Hamiltonian dynamical equations. *Hybrid Monte Carlo* (Duane *et al.*, 1987; Neal, 1996) combines Hamiltonian dynamics with the Metropolis algorithm and thereby removes any bias associated with the discretization.

Specifically, the algorithm uses a Markov chain consisting of alternate stochastic updates of the momentum variable $\mathbf{r}$ and Hamiltonian dynamical updates using the leapfrog algorithm. After each application of the leapfrog algorithm, the resulting candidate state is accepted or rejected according to the Metropolis criterion based on the value of the Hamiltonian H. Thus if $(\mathbf{z}, \mathbf{r})$ is the initial state and $(\mathbf{z}^{\star}, \mathbf{r}^{\star})$ is the state after the leapfrog integration, then this candidate state is accepted with probability

$$\min\left(1, \exp\{H(\mathbf{z}, \mathbf{r}) - H(\mathbf{z}^{\star}, \mathbf{r}^{\star})\}\right). \tag{11.67}$$

If the leapfrog integration were to simulate the Hamiltonian dynamics perfectly, then every such candidate step would automatically be accepted because the value of H would be unchanged. Due to numerical errors, the value of H may sometimes decrease, and we would like the Metropolis criterion to remove any bias due to this effect and ensure that the resulting samples are indeed drawn from the required distribution. In order for this to be the case, we need to ensure that the update equations corresponding to the leapfrog integration satisfy detailed balance (11.40). This is easily achieved by modifying the leapfrog scheme as follows.

Before the start of each leapfrog integration sequence, we choose at random, with equal probability, whether to integrate forwards in time (using step size ϵ) or backwards in time (using step size $-\epsilon$). We first note that the leapfrog integration scheme (11.64), (11.65), and (11.66) is time-reversible, so that integration for L steps using step size $-\epsilon$ will exactly undo the effect of integration for L steps using step size ϵ. Next we show that the leapfrog integration preserves phase-space volume exactly. This follows from the fact that each step in the leapfrog scheme updates either a z_i variable or an r_i variable by an amount that is a function only of the other variable. As shown in Figure 11.14, this has the effect of shearing a region of phase space while not altering its volume.

Finally, we use these results to show that detailed balance holds. Consider a small region $\mathcal{R}$ of phase space that, under a sequence of L leapfrog iterations of step size ϵ, maps to a region $\mathcal{R}'$. Using conservation of volume under the leapfrog iteration, we see that if $\mathcal{R}$ has volume δV then so too will $\mathcal{R}'$. If we choose an initial point from the distribution (11.63) and then update it using L leapfrog interactions, the probability of the transition going from $\mathcal{R}$ to $\mathcal{R}'$ is given by

$$\frac{1}{Z_H} \exp(-H(\mathcal{R}))\delta V \frac{1}{2} \min\left\{1, \exp(H(\mathcal{R}) - H(\mathcal{R}'))\right\}. \tag{11.68}$$

where the factor of $1/2$ arises from the probability of choosing to integrate with a positive step size rather than a negative one. Similarly, the probability of starting in

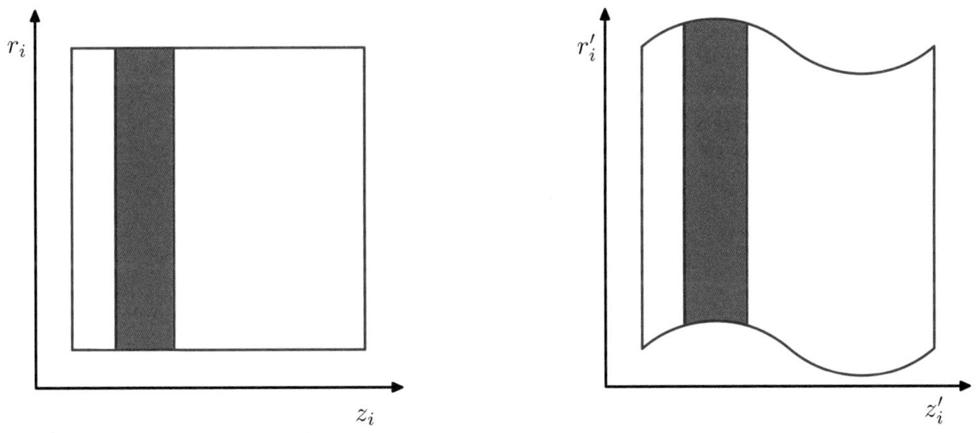

Figure 11.14 Each step of the leapfrog algorithm (11.64)–(11.66) modifies either a position variable z_i or a momentum variable r_i. Because the change to one variable is a function only of the other, any region in phase space will be sheared without change of volume.

region $\mathcal{R}'$ and integrating backwards in time to end up in region $\mathcal{R}$ is given by

$$\frac{1}{Z_H}\exp(-H(\mathcal{R}'))\delta V \frac{1}{2}\min\left\{1, \exp(H(\mathcal{R}') - H(\mathcal{R}))\right\}. \tag{11.69}$$

Exercise 11.17

It is easily seen that the two probabilities (11.68) and (11.69) are equal, and hence detailed balance holds. Note that this proof ignores any overlap between the regions $\mathcal{R}$ and $\mathcal{R}'$ but is easily generalized to allow for such overlap.

It is not difficult to construct examples for which the leapfrog algorithm returns to its starting position after a finite number of iterations. In such cases, the random replacement of the momentum values before each leapfrog integration will not be sufficient to ensure ergodicity because the position variables will never be updated. Such phenomena are easily avoided by choosing the magnitude of the step size at random from some small interval, before each leapfrog integration.

We can gain some insight into the behaviour of the hybrid Monte Carlo algorithm by considering its application to a multivariate Gaussian. For convenience, consider a Gaussian distribution $p(\mathbf{z})$ with independent components, for which the Hamiltonian is given by

$$H(\mathbf{z}, \mathbf{r}) = \frac{1}{2}\sum_i \frac{1}{\sigma_i^2} z_i^2 + \frac{1}{2}\sum_i r_i^2. \tag{11.70}$$

Our conclusions will be equally valid for a Gaussian distribution having correlated components because the hybrid Monte Carlo algorithm exhibits rotational isotropy. During the leapfrog integration, each pair of phase-space variables z_i, r_i evolves independently. However, the acceptance or rejection of the candidate point is based on the value of H, which depends on the values of all of the variables. Thus, a significant integration error in any one of the variables could lead to a high probability of rejection. In order that the discrete leapfrog integration be a reasonably good approximation to the true continuous-time dynamics, it is necessary for the leapfrog integration scale ϵ to be smaller than the shortest length-scale over which the potential is varying significantly. This is governed by the smallest value of σ_i, which we denote by $\sigma_{\min}$. Recall that the goal of the leapfrog integration in hybrid Monte Carlo is to move a substantial distance through phase space to a new state that is relatively independent of the initial state and still achieve a high probability of acceptance. In order to achieve this, the leapfrog integration must be continued for a number of iterations of order $\sigma_{\max}/\sigma_{\min}$.

By contrast, consider the behaviour of a simple Metropolis algorithm with an isotropic Gaussian proposal distribution of variance s^2, considered earlier. In order to avoid high rejection rates, the value of s must be of order $\sigma_{\min}$. The exploration of state space then proceeds by a random walk and takes of order $(\sigma_{\max}/\sigma_{\min})^2$ steps to arrive at a roughly independent state.

11.6. Estimating the Partition Function

As we have seen, most of the sampling algorithms considered in this chapter require only the functional form of the probability distribution up to a multiplicative constant. Thus if we write

$$p_E(\mathbf{z}) = \frac{1}{Z_E} \exp(-E(\mathbf{z})) \tag{11.71}$$

then the value of the normalization constant Z_E, also known as the partition function, is not needed in order to draw samples from $p(\mathbf{z})$. However, knowledge of the value of Z_E can be useful for Bayesian model comparison since it represents the model evidence (i.e., the probability of the observed data given the model), and so it is of interest to consider how its value might be obtained. We assume that direct evaluation by summing, or integrating, the function $\exp(-E(\mathbf{z}))$ over the state space of $\mathbf{z}$ is intractable.

For model comparison, it is actually the ratio of the partition functions for two models that is required. Multiplication of this ratio by the ratio of prior probabilities gives the ratio of posterior probabilities, which can then be used for model selection or model averaging.

One way to estimate a ratio of partition functions is to use importance sampling from a distribution with energy function $G(\mathbf{z})$

$$\begin{aligned}
\frac{Z_E}{Z_G} &= \frac{\sum_{\mathbf{z}} \exp(-E(\mathbf{z}))}{\sum_{\mathbf{z}} \exp(-G(\mathbf{z}))} \\
&= \frac{\sum_{\mathbf{z}} \exp(-E(\mathbf{z}) + G(\mathbf{z})) \exp(-G(\mathbf{z}))}{\sum_{\mathbf{z}} \exp(-G(\mathbf{z}))} \\
&= \mathbb{E}_{G(\mathbf{z})}[\exp(-E + G)] \\
&\simeq \frac{1}{L} \sum_{l} \exp(-E(\mathbf{z}^{(l)}) + G(\mathbf{z}^{(l)}))
\end{aligned} \tag{11.72}$$

where $\{\mathbf{z}^{(l)}\}$ are samples drawn from the distribution defined by $p_G(\mathbf{z})$. If the distribution p_G is one for which the partition function can be evaluated analytically, for example a Gaussian, then the absolute value of Z_E can be obtained.

This approach will only yield accurate results if the importance sampling distribution p_G is closely matched to the distribution p_E, so that the ratio p_E/p_G does not have wide variations. In practice, suitable analytically specified importance sampling distributions cannot readily be found for the kinds of complex models considered in this book.

An alternative approach is therefore to use the samples obtained from a Markov chain to define the importance-sampling distribution. If the transition probability for the Markov chain is given by $T(\mathbf{z}, \mathbf{z}')$, and the sample set is given by $\mathbf{z}^{(1)}, \ldots, \mathbf{z}^{(L)}$, then the sampling distribution can be written as

$$\frac{1}{Z_G} \exp\left(-G(\mathbf{z})\right) = \frac{1}{L} \sum_{l=1}^{L} T(\mathbf{z}^{(l)}, \mathbf{z}) \tag{11.73}$$

which can be used directly in (11.72).

Methods for estimating the ratio of two partition functions require for their success that the two corresponding distributions be reasonably closely matched. This is especially problematic if we wish to find the absolute value of the partition function for a complex distribution because it is only for relatively simple distributions that the partition function can be evaluated directly, and so attempting to estimate the ratio of partition functions directly is unlikely to be successful. This problem can be tackled using a technique known as *chaining* (Neal, 1993; Barber and Bishop, 1997), which involves introducing a succession of intermediate distributions $p_2, \ldots, p_{M-1}$ that interpolate between a simple distribution $p_1(\mathbf{z})$ for which we can evaluate the normalization coefficient Z_1 and the desired complex distribution $p_M(\mathbf{z})$. We then have

$$\frac{Z_M}{Z_1} = \frac{Z_2}{Z_1}\frac{Z_3}{Z_2} \cdots \frac{Z_M}{Z_{M-1}} \tag{11.74}$$

in which the intermediate ratios can be determined using Monte Carlo methods as discussed above. One way to construct such a sequence of intermediate systems is to use an energy function containing a continuous parameter $0 \leqslant \alpha \leqslant 1$ that interpolates between the two distributions

$$E_\alpha(\mathbf{z}) = (1-\alpha)E_1(\mathbf{z}) + \alpha E_M(\mathbf{z}). \tag{11.75}$$

If the intermediate ratios in (11.74) are to be found using Monte Carlo, it may be more efficient to use a single Markov chain run than to restart the Markov chain for each ratio. In this case, the Markov chain is run initially for the system p_1 and then after some suitable number of steps moves on to the next distribution in the sequence. Note, however, that the system must remain close to the equilibrium distribution at each stage.

Exercises

11.1 ($\star$) **www** Show that the finite sample estimator $\widehat{f}$ defined by (11.2) has mean equal to $\mathbb{E}[f]$ and variance given by (11.3).

11.2 ($\star$) Suppose that z is a random variable with uniform distribution over $(0, 1)$ and that we transform z using $y = h^{-1}(z)$ where $h(y)$ is given by (11.6). Show that y has the distribution $p(y)$.

11.3 ($\star$) Given a random variable z that is uniformly distributed over $(0, 1)$, find a transformation $y = f(z)$ such that y has a Cauchy distribution given by (11.8).

11.4 ($\star\star$) Suppose that z_1 and z_2 are uniformly distributed over the unit circle, as shown in Figure 11.3, and that we make the change of variables given by (11.10) and (11.11). Show that (y_1, y_2) will be distributed according to (11.12).

11.5 ($\star$) **www** Let $\mathbf{z}$ be a D-dimensional random variable having a Gaussian distribution with zero mean and unit covariance matrix, and suppose that the positive definite symmetric matrix $\boldsymbol{\Sigma}$ has the Cholesky decomposition $\boldsymbol{\Sigma} = \mathbf{L}\mathbf{L}^{\mathrm{T}}$ where $\mathbf{L}$ is a lower-triangular matrix (i.e., one with zeros above the leading diagonal). Show that the variable $\mathbf{y} = \boldsymbol{\mu} + \mathbf{L}\mathbf{z}$ has a Gaussian distribution with mean $\boldsymbol{\mu}$ and covariance $\boldsymbol{\Sigma}$. This provides a technique for generating samples from a general multivariate Gaussian using samples from a univariate Gaussian having zero mean and unit variance.

11.6 ($\star\star$) **www** In this exercise, we show more carefully that rejection sampling does indeed draw samples from the desired distribution $p(\mathbf{z})$. Suppose the proposal distribution is $q(\mathbf{z})$ and show that the probability of a sample value $\mathbf{z}$ being accepted is given by $\widetilde{p}(\mathbf{z})/kq(\mathbf{z})$ where $\widetilde{p}$ is any unnormalized distribution that is proportional to $p(\mathbf{z})$, and the constant k is set to the smallest value that ensures $kq(\mathbf{z}) \geqslant \widetilde{p}(\mathbf{z})$ for all values of $\mathbf{z}$. Note that the probability of drawing a value $\mathbf{z}$ is given by the probability of drawing that value from $q(\mathbf{z})$ times the probability of accepting that value given that it has been drawn. Make use of this, along with the sum and product rules of probability, to write down the normalized form for the distribution over $\mathbf{z}$, and show that it equals $p(\mathbf{z})$.

11.7 ($\star$) Suppose that z has a uniform distribution over the interval $[0, 1]$. Show that the variable $y = b \tan z + c$ has a Cauchy distribution given by (11.16).

11.8 ($\star\star$) Determine expressions for the coefficients k_i in the envelope distribution (11.17) for adaptive rejection sampling using the requirements of continuity and normalization.

11.9 ($\star\star$) By making use of the technique discussed in Section 11.1.1 for sampling from a single exponential distribution, devise an algorithm for sampling from the piecewise exponential distribution defined by (11.17).

11.10 ($\star$) Show that the simple random walk over the integers defined by (11.34), (11.35), and (11.36) has the property that $\mathbb{E}[(z^{(\tau)})^2] = \mathbb{E}[(z^{(\tau-1)})^2] + 1/2$ and hence by induction that $\mathbb{E}[(z^{(\tau)})^2] = \tau/2$.

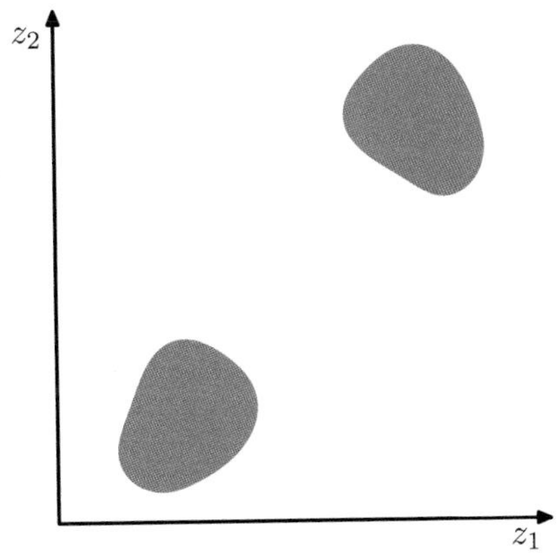

Figure 11.15 A probability distribution over two variables z_1 and z_2 that is uniform over the shaded regions and that is zero everywhere else.

11.11 ($\star\star$) **www** Show that the Gibbs sampling algorithm, discussed in Section 11.3, satisfies detailed balance as defined by (11.40).

11.12 ($\star$) Consider the distribution shown in Figure 11.15. Discuss whether the standard Gibbs sampling procedure for this distribution is ergodic, and therefore whether it would sample correctly from this distribution

11.13 ($\star\star$) Consider the simple 3-node graph shown in Figure 11.16 in which the observed node x is given by a Gaussian distribution $\mathcal{N}(x|\mu, \tau^{-1})$ with mean μ and precision τ. Suppose that the marginal distributions over the mean and precision are given by $\mathcal{N}(\mu|\mu_0, s_0)$ and $\mathrm{Gam}(\tau|a, b)$, where $\mathrm{Gam}(\cdot|\cdot, \cdot)$ denotes a gamma distribution. Write down expressions for the conditional distributions $p(\mu|x, \tau)$ and $p(\tau|x, \mu)$ that would be required in order to apply Gibbs sampling to the posterior distribution $p(\mu, \tau|x)$.

11.14 ($\star$) Verify that the over-relaxation update (11.50), in which z_i has mean μ_i and variance σ_i, and where ν has zero mean and unit variance, gives a value z_i' with mean μ_i and variance σ_i^2.

11.15 ($\star$) **www** Using (11.56) and (11.57), show that the Hamiltonian equation (11.58) is equivalent to (11.53). Similarly, using (11.57) show that (11.59) is equivalent to (11.55).

11.16 ($\star$) By making use of (11.56), (11.57), and (11.63), show that the conditional distribution $p(\mathbf{r}|\mathbf{z})$ is a Gaussian.

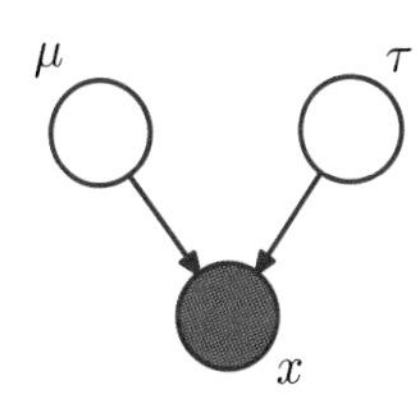

Figure 11.16 A graph involving an observed Gaussian variable x with prior distributions over its mean μ and precision τ.

11.17 ($\star$) **www** Verify that the two probabilities (11.68) and (11.69) are equal, and hence that detailed balance holds for the hybrid Monte Carlo algorithm.

12 Continuous Latent Variables

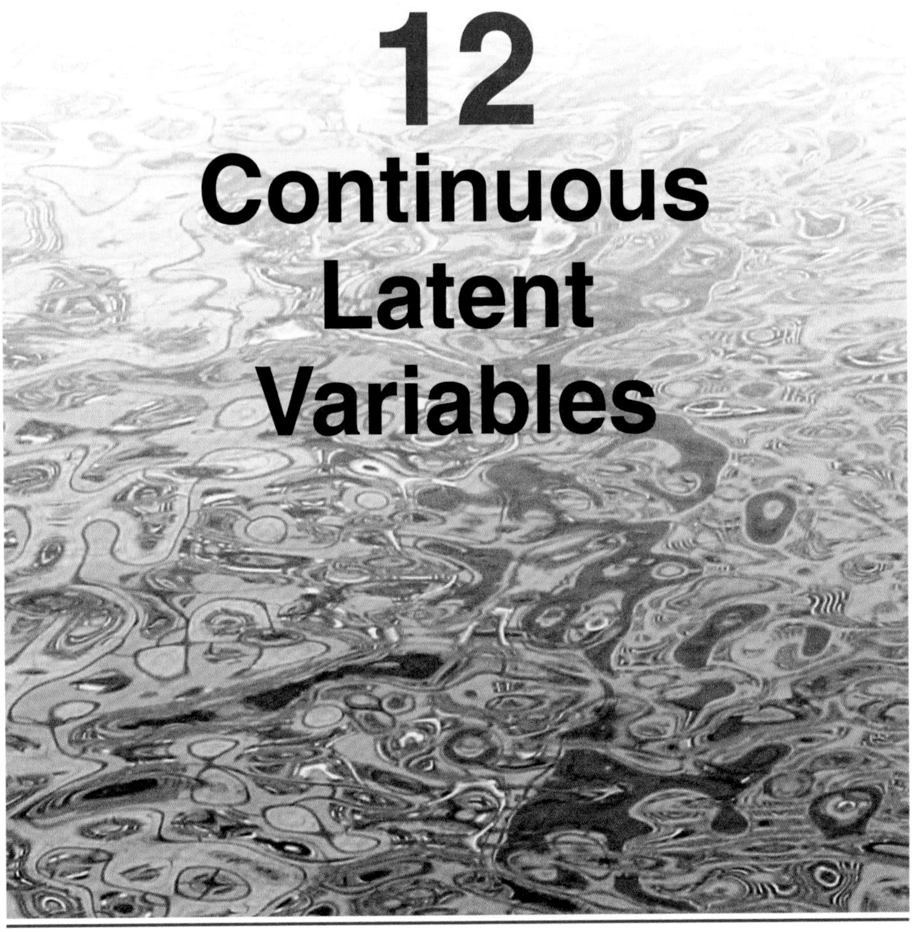

In Chapter 9, we discussed probabilistic models having discrete latent variables, such as the mixture of Gaussians. We now explore models in which some, or all, of the latent variables are continuous. An important motivation for such models is that many data sets have the property that the data points all lie close to a manifold of much lower dimensionality than that of the original data space. To see why this might arise, consider an artificial data set constructed by taking one of the off-line digits, represented by a 64×64 pixel grey-level image, and embedding it in a larger image of size 100×100 by padding with pixels having the value zero (corresponding to white pixels) in which the location and orientation of the digit is varied at random, as illustrated in Figure 12.1. Each of the resulting images is represented by a point in the $100 \times 100 = 10,000$-dimensional data space. However, across a data set of such images, there are only three *degrees of freedom* of variability, corresponding to the vertical and horizontal translations and the rotations. The data points will therefore live on a subspace of the data space whose *intrinsic dimensionality* is three. Note

Appendix A

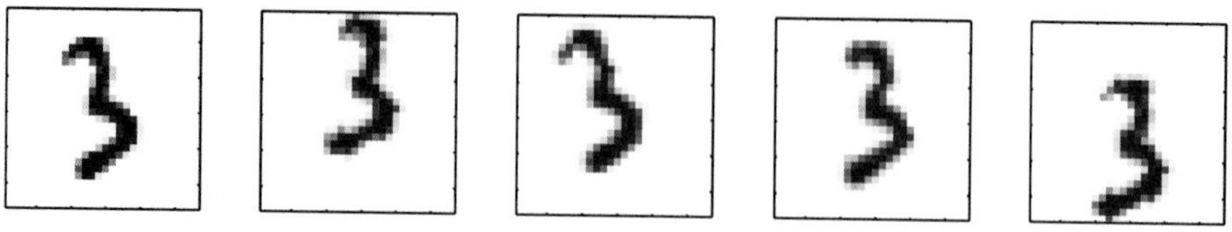

Figure 12.1 A synthetic data set obtained by taking one of the off-line digit images and creating multiple copies in each of which the digit has undergone a random displacement and rotation within some larger image field. The resulting images each have $100 \times 100 = 10,000$ pixels.

that the manifold will be nonlinear because, for instance, if we translate the digit past a particular pixel, that pixel value will go from zero (white) to one (black) and back to zero again, which is clearly a nonlinear function of the digit position. In this example, the translation and rotation parameters are latent variables because we observe only the image vectors and are not told which values of the translation or rotation variables were used to create them.

For real digit image data, there will be a further degree of freedom arising from scaling. Moreover there will be multiple additional degrees of freedom associated with more complex deformations due to the variability in an individual's writing as well as the differences in writing styles between individuals. Nevertheless, the number of such degrees of freedom will be small compared to the dimensionality of the data set.

Appendix A

Another example is provided by the oil flow data set, in which (for a given geometrical configuration of the gas, water, and oil phases) there are only two degrees of freedom of variability corresponding to the fraction of oil in the pipe and the fraction of water (the fraction of gas then being determined). Although the data space comprises 12 measurements, a data set of points will lie close to a two-dimensional manifold embedded within this space. In this case, the manifold comprises several distinct segments corresponding to different flow regimes, each such segment being a (noisy) continuous two-dimensional manifold. If our goal is data compression, or density modelling, then there can be benefits in exploiting this manifold structure.

In practice, the data points will not be confined precisely to a smooth low-dimensional manifold, and we can interpret the departures of data points from the manifold as 'noise'. This leads naturally to a generative view of such models in which we first select a point within the manifold according to some latent variable distribution and then generate an observed data point by adding noise, drawn from some conditional distribution of the data variables given the latent variables.

Section 8.1.4

The simplest continuous latent variable model assumes Gaussian distributions for both the latent and observed variables and makes use of a linear-Gaussian dependence of the observed variables on the state of the latent variables. This leads to a probabilistic formulation of the well-known technique of principal component analysis (PCA), as well as to a related model called factor analysis.

Section 12.1

In this chapter we will begin with a standard, nonprobabilistic treatment of PCA, and then we show how PCA arises naturally as the maximum likelihood solution to

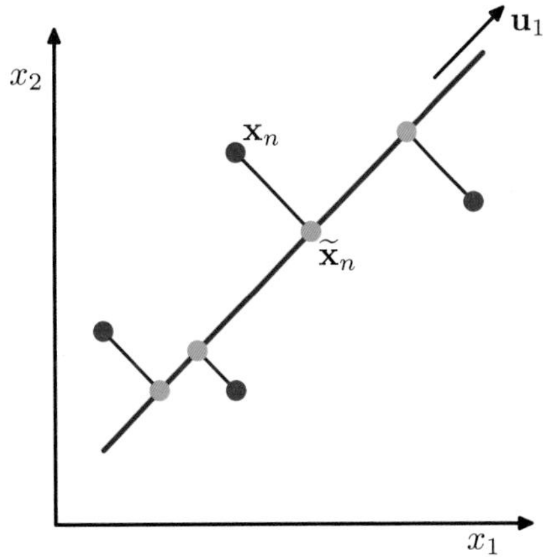

Figure 12.2 Principal component analysis seeks a space of lower dimensionality, known as the principal subspace and denoted by the magenta line, such that the orthogonal projection of the data points (red dots) onto this subspace maximizes the variance of the projected points (green dots). An alternative definition of PCA is based on minimizing the sum-of-squares of the projection errors, indicated by the blue lines.

Section 12.2

a particular form of linear-Gaussian latent variable model. This probabilistic reformulation brings many advantages, such as the use of EM for parameter estimation, principled extensions to mixtures of PCA models, and Bayesian formulations that allow the number of principal components to be determined automatically from the data. Finally, we discuss briefly several generalizations of the latent variable concept that go beyond the linear-Gaussian assumption including non-Gaussian latent variables, which leads to the framework of *independent component analysis*, as well as models having a nonlinear relationship between latent and observed variables.

Section 12.4

12.1. Principal Component Analysis

Principal component analysis, or PCA, is a technique that is widely used for applications such as dimensionality reduction, lossy data compression, feature extraction, and data visualization (Jolliffe, 2002). It is also known as the *Karhunen-Loève* transform.

There are two commonly used definitions of PCA that give rise to the same algorithm. PCA can be defined as the orthogonal projection of the data onto a lower dimensional linear space, known as the *principal subspace*, such that the variance of the projected data is maximized (Hotelling, 1933). Equivalently, it can be defined as the linear projection that minimizes the average projection cost, defined as the mean squared distance between the data points and their projections (Pearson, 1901). The process of orthogonal projection is illustrated in Figure 12.2. We consider each of these definitions in turn.

12.1.1 Maximum variance formulation

Consider a data set of observations $\{\mathbf{x}_n\}$ where $n = 1, \ldots, N$, and $\mathbf{x}_n$ is a Euclidean variable with dimensionality D. Our goal is to project the data onto a space having dimensionality $M < D$ while maximizing the variance of the projected data. For the moment, we shall assume that the value of M is given. Later in this chapter, we shall consider techniques to determine an appropriate value of M from the data.

To begin with, consider the projection onto a one-dimensional space ($M = 1$). We can define the direction of this space using a D-dimensional vector $\mathbf{u}_1$, which for convenience (and without loss of generality) we shall choose to be a unit vector so that $\mathbf{u}_1^{\mathrm{T}}\mathbf{u}_1 = 1$ (note that we are only interested in the direction defined by $\mathbf{u}_1$, not in the magnitude of $\mathbf{u}_1$ itself). Each data point $\mathbf{x}_n$ is then projected onto a scalar value $\mathbf{u}_1^{\mathrm{T}}\mathbf{x}_n$. The mean of the projected data is $\mathbf{u}_1^{\mathrm{T}}\overline{\mathbf{x}}$ where $\overline{\mathbf{x}}$ is the sample set mean given by

$$\overline{\mathbf{x}} = \frac{1}{N}\sum_{n=1}^{N}\mathbf{x}_n \tag{12.1}$$

and the variance of the projected data is given by

$$\frac{1}{N}\sum_{n=1}^{N}\left\{\mathbf{u}_1^{\mathrm{T}}\mathbf{x}_n - \mathbf{u}_1^{\mathrm{T}}\overline{\mathbf{x}}\right\}^2 = \mathbf{u}_1^{\mathrm{T}}\mathbf{S}\mathbf{u}_1 \tag{12.2}$$

where $\mathbf{S}$ is the data covariance matrix defined by

$$\mathbf{S} = \frac{1}{N}\sum_{n=1}^{N}(\mathbf{x}_n - \overline{\mathbf{x}})(\mathbf{x}_n - \overline{\mathbf{x}})^{\mathrm{T}}. \tag{12.3}$$

We now maximize the projected variance $\mathbf{u}_1^{\mathrm{T}}\mathbf{S}\mathbf{u}_1$ with respect to $\mathbf{u}_1$. Clearly, this has to be a constrained maximization to prevent $\|\mathbf{u}_1\| \to \infty$. The appropriate constraint comes from the normalization condition $\mathbf{u}_1^{\mathrm{T}}\mathbf{u}_1 = 1$. To enforce this constraint, we introduce a Lagrange multiplier that we shall denote by λ_1, and then make an unconstrained maximization of

Appendix E

$$\mathbf{u}_1^{\mathrm{T}}\mathbf{S}\mathbf{u}_1 + \lambda_1\left(1 - \mathbf{u}_1^{\mathrm{T}}\mathbf{u}_1\right). \tag{12.4}$$

By setting the derivative with respect to $\mathbf{u}_1$ equal to zero, we see that this quantity will have a stationary point when

$$\mathbf{S}\mathbf{u}_1 = \lambda_1\mathbf{u}_1 \tag{12.5}$$

which says that $\mathbf{u}_1$ must be an eigenvector of $\mathbf{S}$. If we left-multiply by $\mathbf{u}_1^{\mathrm{T}}$ and make use of $\mathbf{u}_1^{\mathrm{T}}\mathbf{u}_1 = 1$, we see that the variance is given by

$$\mathbf{u}_1^{\mathrm{T}}\mathbf{S}\mathbf{u}_1 = \lambda_1 \tag{12.6}$$

and so the variance will be a maximum when we set $\mathbf{u}_1$ equal to the eigenvector having the largest eigenvalue λ_1. This eigenvector is known as the first principal component.

We can define additional principal components in an incremental fashion by choosing each new direction to be that which maximizes the projected variance amongst all possible directions orthogonal to those already considered. If we consider the general case of an M-dimensional projection space, the optimal linear projection for which the variance of the projected data is maximized is now defined by the M eigenvectors $\mathbf{u}_1, \ldots, \mathbf{u}_M$ of the data covariance matrix $\mathbf{S}$ corresponding to the M largest eigenvalues $\lambda_1, \ldots, \lambda_M$. This is easily shown using proof by induction.

Exercise 12.1

To summarize, principal component analysis involves evaluating the mean $\overline{\mathbf{x}}$ and the covariance matrix $\mathbf{S}$ of the data set and then finding the M eigenvectors of $\mathbf{S}$ corresponding to the M largest eigenvalues. Algorithms for finding eigenvectors and eigenvalues, as well as additional theorems related to eigenvector decomposition, can be found in Golub and Van Loan (1996). Note that the computational cost of computing the full eigenvector decomposition for a matrix of size $D \times D$ is $O(D^3)$. If we plan to project our data onto the first M principal components, then we only need to find the first M eigenvalues and eigenvectors. This can be done with more efficient techniques, such as the *power method* (Golub and Van Loan, 1996), that scale like $O(MD^2)$, or alternatively we can make use of the EM algorithm.

Section 12.2.2

12.1.2 Minimum-error formulation

We now discuss an alternative formulation of PCA based on projection error minimization. To do this, we introduce a complete orthonormal set of D-dimensional basis vectors $\{\mathbf{u}_i\}$ where $i = 1, \ldots, D$ that satisfy

Appendix C

$$\mathbf{u}_i^{\mathrm{T}} \mathbf{u}_j = \delta_{ij}. \tag{12.7}$$

Because this basis is complete, each data point can be represented exactly by a linear combination of the basis vectors

$$\mathbf{x}_n = \sum_{i=1}^{D} \alpha_{ni} \mathbf{u}_i \tag{12.8}$$

where the coefficients α_{ni} will be different for different data points. This simply corresponds to a rotation of the coordinate system to a new system defined by the $\{\mathbf{u}_i\}$, and the original D components $\{x_{n1}, \ldots, x_{nD}\}$ are replaced by an equivalent set $\{\alpha_{n1}, \ldots, \alpha_{nD}\}$. Taking the inner product with $\mathbf{u}_j$, and making use of the orthonormality property, we obtain $\alpha_{nj} = \mathbf{x}_n^{\mathrm{T}} \mathbf{u}_j$, and so without loss of generality we can write

$$\mathbf{x}_n = \sum_{i=1}^{D} \left(\mathbf{x}_n^{\mathrm{T}} \mathbf{u}_i \right) \mathbf{u}_i. \tag{12.9}$$

Our goal, however, is to approximate this data point using a representation involving a restricted number $M < D$ of variables corresponding to a projection onto a lower-dimensional subspace. The M-dimensional linear subspace can be represented, without loss of generality, by the first M of the basis vectors, and so we approximate each data point $\mathbf{x}_n$ by

$$\widetilde{\mathbf{x}}_n = \sum_{i=1}^{M} z_{ni} \mathbf{u}_i + \sum_{i=M+1}^{D} b_i \mathbf{u}_i \tag{12.10}$$

where the $\{z_{ni}\}$ depend on the particular data point, whereas the $\{b_i\}$ are constants that are the same for all data points. We are free to choose the $\{\mathbf{u}_i\}$, the $\{z_{ni}\}$, and the $\{b_i\}$ so as to minimize the distortion introduced by the reduction in dimensionality. As our distortion measure, we shall use the squared distance between the original data point $\mathbf{x}_n$ and its approximation $\widetilde{\mathbf{x}}_n$, averaged over the data set, so that our goal is to minimize

$$J = \frac{1}{N} \sum_{n=1}^{N} \|\mathbf{x}_n - \widetilde{\mathbf{x}}_n\|^2. \tag{12.11}$$

Consider first of all the minimization with respect to the quantities $\{z_{ni}\}$. Substituting for $\widetilde{\mathbf{x}}_n$, setting the derivative with respect to z_{nj} to zero, and making use of the orthonormality conditions, we obtain

$$z_{nj} = \mathbf{x}_n^{\mathrm{T}} \mathbf{u}_j \tag{12.12}$$

where $j = 1, \ldots, M$. Similarly, setting the derivative of J with respect to b_i to zero, and again making use of the orthonormality relations, gives

$$b_j = \overline{\mathbf{x}}^{\mathrm{T}} \mathbf{u}_j \tag{12.13}$$

where $j = M+1, \ldots, D$. If we substitute for z_{ni} and b_i in (12.10), and make use of the general expansion (12.9), we obtain

$$\mathbf{x}_n - \widetilde{\mathbf{x}}_n = \sum_{i=M+1}^{D} \left\{ (\mathbf{x}_n - \overline{\mathbf{x}})^{\mathrm{T}} \mathbf{u}_i \right\} \mathbf{u}_i \tag{12.14}$$

from which we see that the displacement vector from $\mathbf{x}_n$ to $\widetilde{\mathbf{x}}_n$ lies in the space orthogonal to the principal subspace, because it is a linear combination of $\{\mathbf{u}_i\}$ for $i = M+1, \ldots, D$, as illustrated in Figure 12.2. This is to be expected because the projected points $\widetilde{\mathbf{x}}_n$ must lie within the principal subspace, but we can move them freely within that subspace, and so the minimum error is given by the orthogonal projection.

We therefore obtain an expression for the distortion measure J as a function purely of the $\{\mathbf{u}_i\}$ in the form

$$J = \frac{1}{N} \sum_{n=1}^{N} \sum_{i=M+1}^{D} \left(\mathbf{x}_n^{\mathrm{T}} \mathbf{u}_i - \overline{\mathbf{x}}^{\mathrm{T}} \mathbf{u}_i \right)^2 = \sum_{i=M+1}^{D} \mathbf{u}_i^{\mathrm{T}} \mathbf{S} \mathbf{u}_i. \tag{12.15}$$

There remains the task of minimizing J with respect to the $\{\mathbf{u}_i\}$, which must be a constrained minimization otherwise we will obtain the vacuous result $\mathbf{u}_i = 0$. The constraints arise from the orthonormality conditions and, as we shall see, the solution will be expressed in terms of the eigenvector expansion of the covariance matrix. Before considering a formal solution, let us try to obtain some intuition about the result by considering the case of a two-dimensional data space $D = 2$ and a one-dimensional principal subspace $M = 1$. We have to choose a direction $\mathbf{u}_2$ so as to minimize $J = \mathbf{u}_2^{\mathrm{T}}\mathbf{S}\mathbf{u}_2$, subject to the normalization constraint $\mathbf{u}_2^{\mathrm{T}}\mathbf{u}_2 = 1$. Using a Lagrange multiplier λ_2 to enforce the constraint, we consider the minimization of

$$\widetilde{J} = \mathbf{u}_2^{\mathrm{T}}\mathbf{S}\mathbf{u}_2 + \lambda_2\left(1 - \mathbf{u}_2^{\mathrm{T}}\mathbf{u}_2\right). \tag{12.16}$$

Setting the derivative with respect to $\mathbf{u}_2$ to zero, we obtain $\mathbf{S}\mathbf{u}_2 = \lambda_2\mathbf{u}_2$ so that $\mathbf{u}_2$ is an eigenvector of $\mathbf{S}$ with eigenvalue λ_2. Thus any eigenvector will define a stationary point of the distortion measure. To find the value of J at the minimum, we back-substitute the solution for $\mathbf{u}_2$ into the distortion measure to give $J = \lambda_2$. We therefore obtain the minimum value of J by choosing $\mathbf{u}_2$ to be the eigenvector corresponding to the smaller of the two eigenvalues. Thus we should choose the principal subspace to be aligned with the eigenvector having the *larger* eigenvalue. This result accords with our intuition that, in order to minimize the average squared projection distance, we should choose the principal component subspace to pass through the mean of the data points and to be aligned with the directions of maximum variance. For the case when the eigenvalues are equal, any choice of principal direction will give rise to the same value of J.

Exercise 12.2

The general solution to the minimization of J for arbitrary D and arbitrary $M < D$ is obtained by choosing the $\{\mathbf{u}_i\}$ to be eigenvectors of the covariance matrix given by

$$\mathbf{S}\mathbf{u}_i = \lambda_i\mathbf{u}_i \tag{12.17}$$

where $i = 1, \ldots, D$, and as usual the eigenvectors $\{\mathbf{u}_i\}$ are chosen to be orthonormal. The corresponding value of the distortion measure is then given by

$$J = \sum_{i=M+1}^{D} \lambda_i \tag{12.18}$$

which is simply the sum of the eigenvalues of those eigenvectors that are orthogonal to the principal subspace. We therefore obtain the minimum value of J by selecting these eigenvectors to be those having the $D - M$ smallest eigenvalues, and hence the eigenvectors defining the principal subspace are those corresponding to the M largest eigenvalues.

Although we have considered $M < D$, the PCA analysis still holds if $M = D$, in which case there is no dimensionality reduction but simply a rotation of the coordinate axes to align with principal components.

Finally, it is worth noting that there exists a closely related linear dimensionality reduction technique called *canonical correlation analysis*, or *CCA* (Hotelling, 1936; Bach and Jordan, 2002). Whereas PCA works with a single random variable, CCA considers two (or more) variables and tries to find a corresponding pair of linear subspaces that have high cross-correlation, so that each component within one of the subspaces is correlated with a single component from the other subspace. Its solution can be expressed in terms of a generalized eigenvector problem.

12.1.3 Applications of PCA

Appendix A

We can illustrate the use of PCA for data compression by considering the off-line digits data set, restricting our attention to images of the digit three. Because each

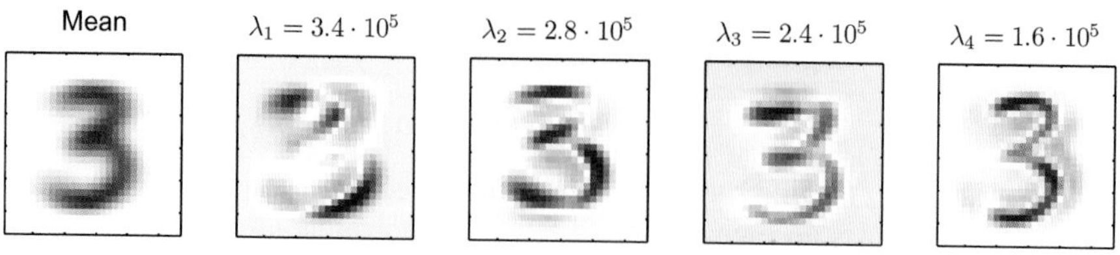

Figure 12.3 The mean vector $\overline{\mathbf{x}}$ along with the first four PCA eigenvectors $\mathbf{u}_1, \ldots, \mathbf{u}_4$ for the digit three from the off-line digits data set, together with the corresponding eigenvalues. Blue corresponds to positive values, white is zero and yellow corresponds to negative values.

eigenvector of the covariance matrix is a vector in the original D-dimensional space, we can represent the eigenvectors as images of the same size as the data points. The first four eigenvectors, along with the corresponding eigenvalues, are shown in Figure 12.3. A plot of the complete spectrum of eigenvalues, sorted into decreasing order, is shown in Figure 12.4(a). The distortion measure J associated with choosing a particular value of M is given by the sum of the eigenvalues from $M + 1$ up to D and is plotted for different values of M in Figure 12.4(b).

If we substitute (12.12) and (12.13) into (12.10), we can write the PCA approximation to a data vector $\mathbf{x}_n$ in the form

$$\widetilde{\mathbf{x}}_n = \sum_{i=1}^{M} (\mathbf{x}_n^{\mathrm{T}}\mathbf{u}_i)\mathbf{u}_i + \sum_{i=M+1}^{D} (\overline{\mathbf{x}}^{\mathrm{T}}\mathbf{u}_i)\mathbf{u}_i \tag{12.19}$$

$$= \overline{\mathbf{x}} + \sum_{i=1}^{M} \left(\mathbf{x}_n^{\mathrm{T}}\mathbf{u}_i - \overline{\mathbf{x}}^{\mathrm{T}}\mathbf{u}_i\right)\mathbf{u}_i \tag{12.20}$$

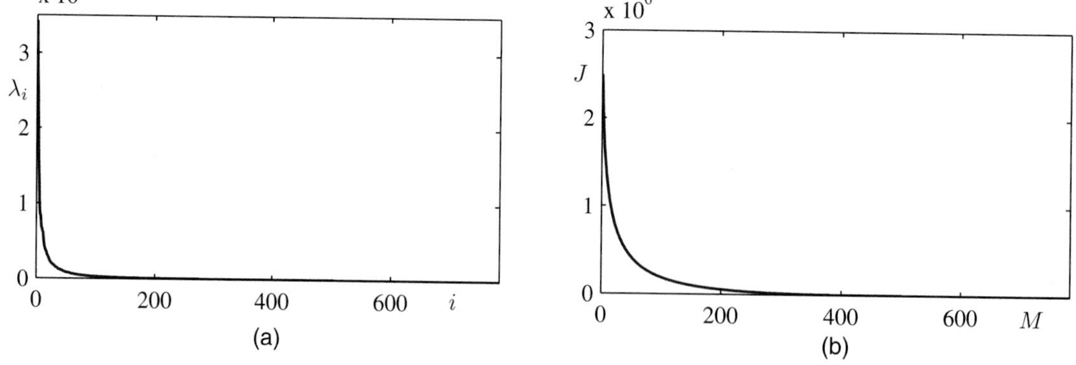

Figure 12.4 (a) Plot of the eigenvalue spectrum for the digit three from the off-line digits data set. (b) Plot of the sum of the discarded eigenvalues, which represents the sum-of-squares distortion J introduced by projecting the data onto a principal component subspace of dimensionality M.

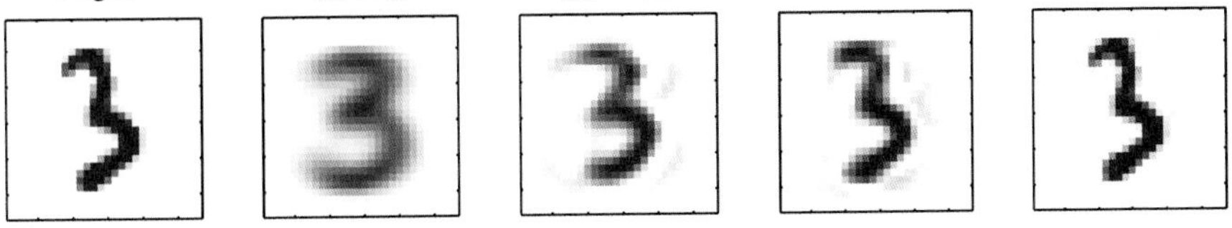

Figure 12.5 An original example from the off-line digits data set together with its PCA reconstructions obtained by retaining M principal components for various values of M. As M increases the reconstruction becomes more accurate and would become perfect when $M = D = 28 \times 28 = 784$.

where we have made use of the relation

$$\overline{\mathbf{x}} = \sum_{i=1}^{D} \left(\overline{\mathbf{x}}^{\mathrm{T}}\mathbf{u}_i\right)\mathbf{u}_i \tag{12.21}$$

which follows from the completeness of the $\{\mathbf{u}_i\}$. This represents a compression of the data set, because for each data point we have replaced the D-dimensional vector $\mathbf{x}_n$ with an M-dimensional vector having components $\left(\mathbf{x}_n^{\mathrm{T}}\mathbf{u}_i - \overline{\mathbf{x}}^{\mathrm{T}}\mathbf{u}_i\right)$. The smaller the value of M, the greater the degree of compression. Examples of reconstructions of a sample from the digit three data set are shown in Figure 12.5.

Another application of principal component analysis is to data pre-processing. In this case, the goal is not dimensionality reduction but rather the transformation of a data set in order to standardize certain of its properties. This can be important in allowing subsequent pattern recognition algorithms to be applied successfully to the data set. Typically, it is done when the original variables are measured in various different units or have significantly different variability. For instance in the Old Faithful data set, the time between eruptions is typically an order of magnitude greater than the duration of an eruption. When we applied the K-means algorithm to this data set, we first made a separate linear re-scaling of the individual variables such that each variable had zero mean and unit variance. This is known as *standardizing* the data, and the covariance matrix for the standardized data has components

Appendix A

Section 9.1

$$\rho_{ij} = \frac{1}{N}\sum_{n=1}^{N} \frac{(x_{ni} - \overline{x}_i)}{\sigma_i}\frac{(x_{nj} - \overline{x}_j)}{\sigma_j} \tag{12.22}$$

where σ_i is the standard deviation of x_i. This is known as the *correlation* matrix of the original data and has the property that if two components x_i and x_j of the data are perfectly correlated, then $\rho_{ij} = 1$, and if they are uncorrelated, then $\rho_{ij} = 0$.

However, using PCA we can make a more substantial normalization of the data to give it zero mean and unit covariance, so that different variables become decorrelated. To do this, we first write the eigenvector equation (12.17) in the form

$$\mathbf{SU} = \mathbf{UL} \tag{12.23}$$

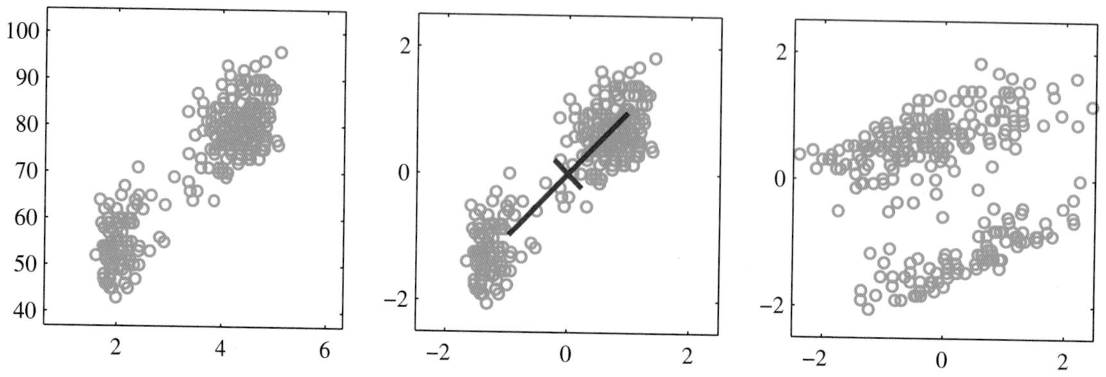

Figure 12.6 Illustration of the effects of linear pre-processing applied to the Old Faithful data set. The plot on the left shows the original data. The centre plot shows the result of standardizing the individual variables to zero mean and unit variance. Also shown are the principal axes of this normalized data set, plotted over the range $\pm\lambda_i^{1/2}$. The plot on the right shows the result of whitening of the data to give it zero mean and unit covariance.

where $\mathbf{L}$ is a $D \times D$ diagonal matrix with elements λ_i, and $\mathbf{U}$ is a $D \times D$ orthogonal matrix with columns given by $\mathbf{u}_i$. Then we define, for each data point $\mathbf{x}_n$, a transformed value given by

$$\mathbf{y}_n = \mathbf{L}^{-1/2}\mathbf{U}^{\mathrm{T}}(\mathbf{x}_n - \overline{\mathbf{x}}) \tag{12.24}$$

where $\overline{\mathbf{x}}$ is the sample mean defined by (12.1). Clearly, the set $\{\mathbf{y}_n\}$ has zero mean, and its covariance is given by the identity matrix because

$$\begin{aligned}\frac{1}{N}\sum_{n=1}^{N}\mathbf{y}_n\mathbf{y}_n^{\mathrm{T}} &= \frac{1}{N}\sum_{n=1}^{N}\mathbf{L}^{-1/2}\mathbf{U}^{\mathrm{T}}(\mathbf{x}_n - \overline{\mathbf{x}})(\mathbf{x}_n - \overline{\mathbf{x}})^{\mathrm{T}}\mathbf{U}\mathbf{L}^{-1/2} \\ &= \mathbf{L}^{-1/2}\mathbf{U}^{\mathrm{T}}\mathbf{S}\mathbf{U}\mathbf{L}^{-1/2} = \mathbf{L}^{-1/2}\mathbf{L}\mathbf{L}^{-1/2} = \mathbf{I}.\end{aligned} \tag{12.25}$$

Appendix A

This operation is known as *whitening* or *sphereing* the data and is illustrated for the Old Faithful data set in Figure 12.6.

It is interesting to compare PCA with the Fisher linear discriminant which was discussed in Section 4.1.4. Both methods can be viewed as techniques for linear dimensionality reduction. However, PCA is unsupervised and depends only on the values $\mathbf{x}_n$ whereas Fisher linear discriminant also uses class-label information. This difference is highlighted by the example in Figure 12.7.

Appendix A

Another common application of principal component analysis is to data visualization. Here each data point is projected onto a two-dimensional ($M = 2$) principal subspace, so that a data point $\mathbf{x}_n$ is plotted at Cartesian coordinates given by $\mathbf{x}_n^{\mathrm{T}}\mathbf{u}_1$ and $\mathbf{x}_n^{\mathrm{T}}\mathbf{u}_2$, where $\mathbf{u}_1$ and $\mathbf{u}_2$ are the eigenvectors corresponding to the largest and second largest eigenvalues. An example of such a plot, for the oil flow data set, is shown in Figure 12.8.

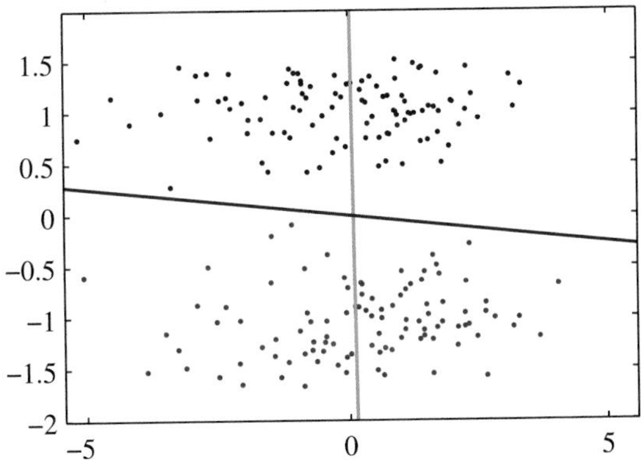

Figure 12.7 A comparison of principal component analysis with Fisher's linear discriminant for linear dimensionality reduction. Here the data in two dimensions, belonging to two classes shown in red and blue, is to be projected onto a single dimension. PCA chooses the direction of maximum variance, shown by the magenta curve, which leads to strong class overlap, whereas the Fisher linear discriminant takes account of the class labels and leads to a projection onto the green curve giving much better class separation.

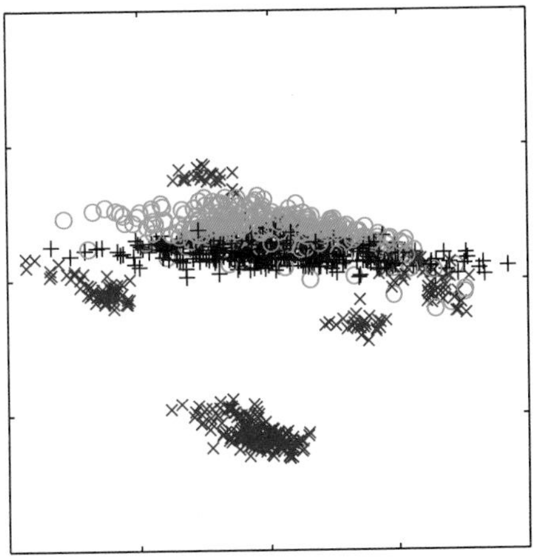

Figure 12.8 Visualization of the oil flow data set obtained by projecting the data onto the first two principal components. The red, blue, and green points correspond to the 'laminar', 'homogeneous', and 'annular' flow configurations respectively.

12.1.4 PCA for high-dimensional data

In some applications of principal component analysis, the number of data points is smaller than the dimensionality of the data space. For example, we might want to apply PCA to a data set of a few hundred images, each of which corresponds to a vector in a space of potentially several million dimensions (corresponding to three colour values for each of the pixels in the image). Note that in a D-dimensional space a set of N points, where $N < D$, defines a linear subspace whose dimensionality is at most $N - 1$, and so there is little point in applying PCA for values of M that are greater than $N - 1$. Indeed, if we perform PCA we will find that at least $D - N + 1$ of the eigenvalues are zero, corresponding to eigenvectors along whose directions the data set has zero variance. Furthermore, typical algorithms for finding the eigenvectors of a $D \times D$ matrix have a computational cost that scales like $O(D^3)$, and so for applications such as the image example, a direct application of PCA will be computationally infeasible.

We can resolve this problem as follows. First, let us define $\mathbf{X}$ to be the $(N \times D)$-

dimensional centred data matrix, whose n^{th} row is given by $(\mathbf{x}_n - \overline{\mathbf{x}})^{\mathrm{T}}$. The covariance matrix (12.3) can then be written as $\mathbf{S} = N^{-1}\mathbf{X}^{\mathrm{T}}\mathbf{X}$, and the corresponding eigenvector equation becomes

$$\frac{1}{N}\mathbf{X}^{\mathrm{T}}\mathbf{X}\mathbf{u}_i = \lambda_i \mathbf{u}_i. \tag{12.26}$$

Now pre-multiply both sides by $\mathbf{X}$ to give

$$\frac{1}{N}\mathbf{X}\mathbf{X}^{\mathrm{T}}(\mathbf{X}\mathbf{u}_i) = \lambda_i(\mathbf{X}\mathbf{u}_i). \tag{12.27}$$

If we now define $\mathbf{v}_i = \mathbf{X}\mathbf{u}_i$, we obtain

$$\frac{1}{N}\mathbf{X}\mathbf{X}^{\mathrm{T}}\mathbf{v}_i = \lambda_i \mathbf{v}_i \tag{12.28}$$

which is an eigenvector equation for the $N \times N$ matrix $N^{-1}\mathbf{X}\mathbf{X}^{\mathrm{T}}$. We see that this has the same $N-1$ eigenvalues as the original covariance matrix (which itself has an additional $D - N + 1$ eigenvalues of value zero). Thus we can solve the eigenvector problem in spaces of lower dimensionality with computational cost $O(N^3)$ instead of $O(D^3)$. In order to determine the eigenvectors, we multiply both sides of (12.28) by $\mathbf{X}^{\mathrm{T}}$ to give

$$\left(\frac{1}{N}\mathbf{X}^{\mathrm{T}}\mathbf{X}\right)(\mathbf{X}^{\mathrm{T}}\mathbf{v}_i) = \lambda_i(\mathbf{X}^{\mathrm{T}}\mathbf{v}_i) \tag{12.29}$$

from which we see that $(\mathbf{X}^{\mathrm{T}}\mathbf{v}_i)$ is an eigenvector of $\mathbf{S}$ with eigenvalue λ_i. Note, however, that these eigenvectors need not be normalized. To determine the appropriate normalization, we re-scale $\mathbf{u}_i \propto \mathbf{X}^{\mathrm{T}}\mathbf{v}_i$ by a constant such that $\|\mathbf{u}_i\| = 1$, which, assuming $\mathbf{v}_i$ has been normalized to unit length, gives

$$\mathbf{u}_i = \frac{1}{(N\lambda_i)^{1/2}}\mathbf{X}^{\mathrm{T}}\mathbf{v}_i. \tag{12.30}$$

In summary, to apply this approach we first evaluate $\mathbf{X}\mathbf{X}^{\mathrm{T}}$ and then find its eigenvectors and eigenvalues and then compute the eigenvectors in the original data space using (12.30).

12.2. Probabilistic PCA

The formulation of PCA discussed in the previous section was based on a linear projection of the data onto a subspace of lower dimensionality than the original data space. We now show that PCA can also be expressed as the maximum likelihood solution of a probabilistic latent variable model. This reformulation of PCA, known as *probabilistic PCA*, brings several advantages compared with conventional PCA:

- Probabilistic PCA represents a constrained form of the Gaussian distribution in which the number of free parameters can be restricted while still allowing the model to capture the dominant correlations in a data set.

- We can derive an EM algorithm for PCA that is computationally efficient in situations where only a few leading eigenvectors are required and that avoids having to evaluate the data covariance matrix as an intermediate step.

Section 12.2.2

- The combination of a probabilistic model and EM allows us to deal with missing values in the data set.

- Mixtures of probabilistic PCA models can be formulated in a principled way and trained using the EM algorithm.

- Probabilistic PCA forms the basis for a Bayesian treatment of PCA in which the dimensionality of the principal subspace can be found automatically from the data.

Section 12.2.3

- The existence of a likelihood function allows direct comparison with other probabilistic density models. By contrast, conventional PCA will assign a low reconstruction cost to data points that are close to the principal subspace even if they lie arbitrarily far from the training data.

- Probabilistic PCA can be used to model class-conditional densities and hence be applied to classification problems.

- The probabilistic PCA model can be run generatively to provide samples from the distribution.

This formulation of PCA as a probabilistic model was proposed independently by Tipping and Bishop (1997, 1999b) and by Roweis (1998). As we shall see later, it is closely related to *factor analysis* (Basilevsky, 1994).

Section 8.1.4

Probabilistic PCA is a simple example of the linear-Gaussian framework, in which all of the marginal and conditional distributions are Gaussian. We can formulate probabilistic PCA by first introducing an explicit latent variable $\mathbf{z}$ corresponding to the principal-component subspace. Next we define a Gaussian prior distribution $p(\mathbf{z})$ over the latent variable, together with a Gaussian conditional distribution $p(\mathbf{x}|\mathbf{z})$ for the observed variable $\mathbf{x}$ conditioned on the value of the latent variable. Specifically, the prior distribution over $\mathbf{z}$ is given by a zero-mean unit-covariance Gaussian

$$p(\mathbf{z}) = \mathcal{N}(\mathbf{z}|\mathbf{0}, \mathbf{I}). \tag{12.31}$$

Similarly, the conditional distribution of the observed variable $\mathbf{x}$, conditioned on the value of the latent variable $\mathbf{z}$, is again Gaussian, of the form

$$p(\mathbf{x}|\mathbf{z}) = \mathcal{N}(\mathbf{x}|\mathbf{W}\mathbf{z} + \boldsymbol{\mu}, \sigma^2\mathbf{I}) \tag{12.32}$$

in which the mean of $\mathbf{x}$ is a general linear function of $\mathbf{z}$ governed by the $D \times M$ matrix $\mathbf{W}$ and the D-dimensional vector $\boldsymbol{\mu}$. Note that this factorizes with respect to the elements of $\mathbf{x}$, in other words this is an example of the naive Bayes model. As we shall see shortly, the columns of $\mathbf{W}$ span a linear subspace within the data space that corresponds to the principal subspace. The other parameter in this model is the scalar σ^2 governing the variance of the conditional distribution. Note that there is no

Section 8.2.2

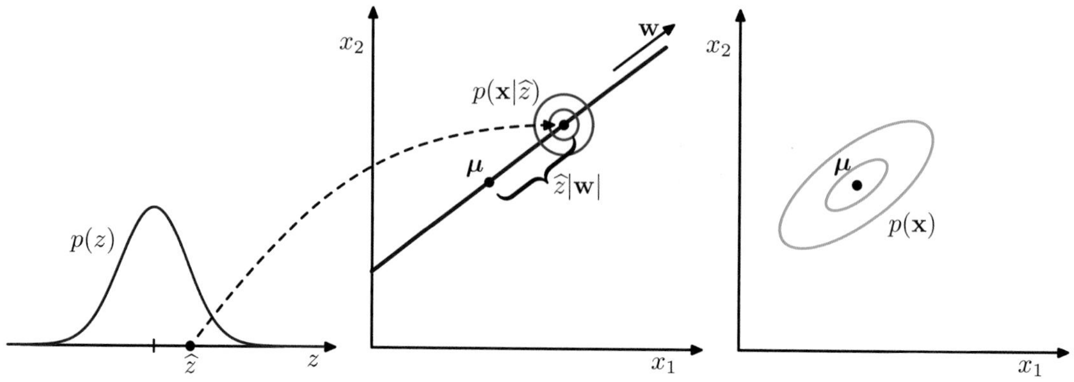

Figure 12.9 An illustration of the generative view of the probabilistic PCA model for a two-dimensional data space and a one-dimensional latent space. An observed data point $\mathbf{x}$ is generated by first drawing a value $\widehat{z}$ for the latent variable from its prior distribution $p(z)$ and then drawing a value for $\mathbf{x}$ from an isotropic Gaussian distribution (illustrated by the red circles) having mean $\mathbf{w}\widehat{z} + \boldsymbol{\mu}$ and covariance $\sigma^2\mathbf{I}$. The green ellipses show the density contours for the marginal distribution $p(\mathbf{x})$.

loss of generality in assuming a zero mean, unit covariance Gaussian for the latent distribution $p(\mathbf{z})$ because a more general Gaussian distribution would give rise to an equivalent probabilistic model.

Exercise 12.4

We can view the probabilistic PCA model from a generative viewpoint in which a sampled value of the observed variable is obtained by first choosing a value for the latent variable and then sampling the observed variable conditioned on this latent value. Specifically, the D-dimensional observed variable $\mathbf{x}$ is defined by a linear transformation of the M-dimensional latent variable $\mathbf{z}$ plus additive Gaussian 'noise', so that

$$\mathbf{x} = \mathbf{W}\mathbf{z} + \boldsymbol{\mu} + \boldsymbol{\epsilon} \tag{12.33}$$

where $\mathbf{z}$ is an M-dimensional Gaussian latent variable, and $\boldsymbol{\epsilon}$ is a D-dimensional zero-mean Gaussian-distributed noise variable with covariance $\sigma^2\mathbf{I}$. This generative process is illustrated in Figure 12.9. Note that this framework is based on a mapping from latent space to data space, in contrast to the more conventional view of PCA discussed above. The reverse mapping, from data space to the latent space, will be obtained shortly using Bayes' theorem.

Suppose we wish to determine the values of the parameters $\mathbf{W}$, $\boldsymbol{\mu}$ and σ^2 using maximum likelihood. To write down the likelihood function, we need an expression for the marginal distribution $p(\mathbf{x})$ of the observed variable. This is expressed, from the sum and product rules of probability, in the form

$$p(\mathbf{x}) = \int p(\mathbf{x}|\mathbf{z})p(\mathbf{z})\,\mathrm{d}\mathbf{z}. \tag{12.34}$$

Exercise 12.7

Because this corresponds to a linear-Gaussian model, this marginal distribution is again Gaussian, and is given by

$$p(\mathbf{x}) = \mathcal{N}(\mathbf{x}|\boldsymbol{\mu}, \mathbf{C}) \tag{12.35}$$

where the $D \times D$ covariance matrix $\mathbf{C}$ is defined by

$$\mathbf{C} = \mathbf{W}\mathbf{W}^{\mathrm{T}} + \sigma^2\mathbf{I}. \tag{12.36}$$

This result can also be derived more directly by noting that the predictive distribution will be Gaussian and then evaluating its mean and covariance using (12.33). This gives

$$\begin{aligned} \mathbb{E}[\mathbf{x}] &= \mathbb{E}[\mathbf{W}\mathbf{z} + \boldsymbol{\mu} + \boldsymbol{\epsilon}] = \boldsymbol{\mu} & (12.37)\\ \mathrm{cov}[\mathbf{x}] &= \mathbb{E}\left[(\mathbf{W}\mathbf{z} + \boldsymbol{\epsilon})(\mathbf{W}\mathbf{z} + \boldsymbol{\epsilon})^{\mathrm{T}}\right] \\ &= \mathbb{E}\left[\mathbf{W}\mathbf{z}\mathbf{z}^{\mathrm{T}}\mathbf{W}^{\mathrm{T}}\right] + \mathbb{E}[\boldsymbol{\epsilon}\boldsymbol{\epsilon}^{\mathrm{T}}] = \mathbf{W}\mathbf{W}^{\mathrm{T}} + \sigma^2\mathbf{I} & (12.38) \end{aligned}$$

where we have used the fact that $\mathbf{z}$ and $\boldsymbol{\epsilon}$ are independent random variables and hence are uncorrelated.

Intuitively, we can think of the distribution $p(\mathbf{x})$ as being defined by taking an isotropic Gaussian 'spray can' and moving it across the principal subspace spraying Gaussian ink with density determined by σ^2 and weighted by the prior distribution. The accumulated ink density gives rise to a 'pancake' shaped distribution representing the marginal density $p(\mathbf{x})$.

The predictive distribution $p(\mathbf{x})$ is governed by the parameters $\boldsymbol{\mu}$, $\mathbf{W}$, and σ^2. However, there is redundancy in this parameterization corresponding to rotations of the latent space coordinates. To see this, consider a matrix $\widetilde{\mathbf{W}} = \mathbf{W}\mathbf{R}$ where $\mathbf{R}$ is an orthogonal matrix. Using the orthogonality property $\mathbf{R}\mathbf{R}^{\mathrm{T}} = \mathbf{I}$, we see that the quantity $\widetilde{\mathbf{W}}\widetilde{\mathbf{W}}^{\mathrm{T}}$ that appears in the covariance matrix $\mathbf{C}$ takes the form

$$\widetilde{\mathbf{W}}\widetilde{\mathbf{W}}^{\mathrm{T}} = \mathbf{W}\mathbf{R}\mathbf{R}^{\mathrm{T}}\mathbf{W}^{\mathrm{T}} = \mathbf{W}\mathbf{W}^{\mathrm{T}} \tag{12.39}$$

and hence is independent of $\mathbf{R}$. Thus there is a whole family of matrices $\widetilde{\mathbf{W}}$ all of which give rise to the same predictive distribution. This invariance can be understood in terms of rotations within the latent space. We shall return to a discussion of the number of independent parameters in this model later.

When we evaluate the predictive distribution, we require $\mathbf{C}^{-1}$, which involves the inversion of a $D \times D$ matrix. The computation required to do this can be reduced by making use of the matrix inversion identity (C.7) to give

$$\mathbf{C}^{-1} = \sigma^{-2}\mathbf{I} - \sigma^{-2}\mathbf{W}\mathbf{M}^{-1}\mathbf{W}^{\mathrm{T}} \tag{12.40}$$

where the $M \times M$ matrix $\mathbf{M}$ is defined by

$$\mathbf{M} = \mathbf{W}^{\mathrm{T}}\mathbf{W} + \sigma^2\mathbf{I}. \tag{12.41}$$

Because we invert $\mathbf{M}$ rather than inverting $\mathbf{C}$ directly, the cost of evaluating $\mathbf{C}^{-1}$ is reduced from $O(D^3)$ to $O(M^3)$.

As well as the predictive distribution $p(\mathbf{x})$, we will also require the posterior distribution $p(\mathbf{z}|\mathbf{x})$, which can again be written down directly using the result (2.116) for linear-Gaussian models to give

Exercise 12.8

$$p(\mathbf{z}|\mathbf{x}) = \mathcal{N}\left(\mathbf{z}|\mathbf{M}^{-1}\mathbf{W}^{\mathrm{T}}(\mathbf{x} - \boldsymbol{\mu}), \sigma^2\mathbf{M}^{-1}\right). \tag{12.42}$$

Note that the posterior mean depends on $\mathbf{x}$, whereas the posterior covariance is independent of $\mathbf{x}$.

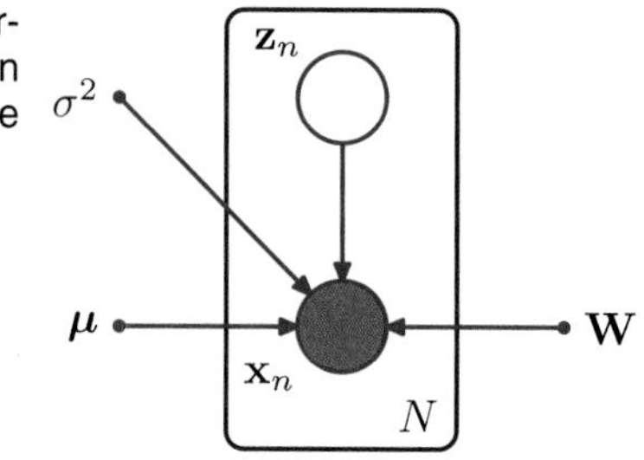

Figure 12.10 The probabilistic PCA model for a data set of N observations of $\mathbf{x}$ can be expressed as a directed graph in which each observation $\mathbf{x}_n$ is associated with a value $\mathbf{z}_n$ of the latent variable.

12.2.1 Maximum likelihood PCA

We next consider the determination of the model parameters using maximum likelihood. Given a data set $\mathbf{X} = \{\mathbf{x}_n\}$ of observed data points, the probabilistic PCA model can be expressed as a directed graph, as shown in Figure 12.10. The corresponding log likelihood function is given, from (12.35), by

$$\begin{aligned}\ln p(\mathbf{X}|\boldsymbol{\mu}, \mathbf{W}, \sigma^2) &= \sum_{n=1}^{N} \ln p(\mathbf{x}_n|\mathbf{W}, \boldsymbol{\mu}, \sigma^2) \\ &= -\frac{ND}{2}\ln(2\pi) - \frac{N}{2}\ln|\mathbf{C}| - \frac{1}{2}\sum_{n=1}^{N}(\mathbf{x}_n - \boldsymbol{\mu})^{\mathrm{T}}\mathbf{C}^{-1}(\mathbf{x}_n - \boldsymbol{\mu}). \qquad (12.43)\end{aligned}$$

Setting the derivative of the log likelihood with respect to $\boldsymbol{\mu}$ equal to zero gives the expected result $\boldsymbol{\mu} = \overline{\mathbf{x}}$ where $\overline{\mathbf{x}}$ is the data mean defined by (12.1). Back-substituting we can then write the log likelihood function in the form

$$\ln p(\mathbf{X}|\mathbf{W}, \boldsymbol{\mu}, \sigma^2) = -\frac{N}{2}\left\{D\ln(2\pi) + \ln|\mathbf{C}| + \mathrm{Tr}\left(\mathbf{C}^{-1}\mathbf{S}\right)\right\} \qquad (12.44)$$

where $\mathbf{S}$ is the data covariance matrix defined by (12.3). Because the log likelihood is a quadratic function of $\boldsymbol{\mu}$, this solution represents the unique maximum, as can be confirmed by computing second derivatives.

Maximization with respect to $\mathbf{W}$ and σ^2 is more complex but nonetheless has an exact closed-form solution. It was shown by Tipping and Bishop (1999b) that all of the stationary points of the log likelihood function can be written as

$$\mathbf{W}_{\mathrm{ML}} = \mathbf{U}_M(\mathbf{L}_M - \sigma^2\mathbf{I})^{1/2}\mathbf{R} \qquad (12.45)$$

where $\mathbf{U}_M$ is a $D \times M$ matrix whose columns are given by any subset (of size M) of the eigenvectors of the data covariance matrix $\mathbf{S}$, the $M \times M$ diagonal matrix $\mathbf{L}_M$ has elements given by the corresponding eigenvalues λ_i, and $\mathbf{R}$ is an arbitrary $M \times M$ orthogonal matrix.

Furthermore, Tipping and Bishop (1999b) showed that the *maximum* of the likelihood function is obtained when the M eigenvectors are chosen to be those whose eigenvalues are the M largest (all other solutions being saddle points). A similar result was conjectured independently by Roweis (1998), although no proof was given.

Again, we shall assume that the eigenvectors have been arranged in order of decreasing values of the corresponding eigenvalues, so that the M principal eigenvectors are $\mathbf{u}_1, \ldots, \mathbf{u}_M$. In this case, the columns of $\mathbf{W}$ define the principal subspace of standard PCA. The corresponding maximum likelihood solution for σ^2 is then given by

$$\sigma^2_{\mathrm{ML}} = \frac{1}{D-M} \sum_{i=M+1}^{D} \lambda_i \tag{12.46}$$

so that σ^2_{ML} is the average variance associated with the discarded dimensions.

Because $\mathbf{R}$ is orthogonal, it can be interpreted as a rotation matrix in the M-dimensional latent space. If we substitute the solution for $\mathbf{W}$ into the expression for $\mathbf{C}$, and make use of the orthogonality property $\mathbf{R}\mathbf{R}^{\mathrm{T}} = \mathbf{I}$, we see that $\mathbf{C}$ is independent of $\mathbf{R}$. This simply says that the predictive density is unchanged by rotations in the latent space as discussed earlier. For the particular case of $\mathbf{R} = \mathbf{I}$, we see that the columns of $\mathbf{W}$ are the principal component eigenvectors scaled by the square root of the variance parameters $\sqrt{\lambda_i - \sigma^2}$. The interpretation of these scaling factors is clear once we recognize that for a convolution of independent Gaussian distributions (in this case the latent space distribution and the noise model) the variances are additive. Thus the variance λ_i in the direction of an eigenvector $\mathbf{u}_i$ is composed of the sum of a contribution $\lambda_i - \sigma^2$ from the projection of the unit-variance latent space distribution into data space through the corresponding column of $\mathbf{W}$, plus an isotropic contribution of variance σ^2 which is added in all directions by the noise model.

It is worth taking a moment to study the form of the covariance matrix given by (12.36). Consider the variance of the predictive distribution along some direction specified by the unit vector $\mathbf{v}$, where $\mathbf{v}^{\mathrm{T}}\mathbf{v} = 1$, which is given by $\mathbf{v}^{\mathrm{T}}\mathbf{C}\mathbf{v}$. First suppose that $\mathbf{v}$ is orthogonal to the principal subspace, in other words it is given by some linear combination of the discarded eigenvectors. Then $\mathbf{v}^{\mathrm{T}}\mathbf{U} = \mathbf{0}$ and hence $\mathbf{v}^{\mathrm{T}}\mathbf{C}\mathbf{v} = \sigma^2$. Thus the model predicts a noise variance orthogonal to the principal subspace, which, from (12.46), is just the average of the discarded eigenvalues. Now suppose that $\mathbf{v} = \mathbf{u}_i$ where $\mathbf{u}_i$ is one of the retained eigenvectors defining the principal subspace. Then $\mathbf{v}^{\mathrm{T}}\mathbf{C}\mathbf{v} = (\lambda_i - \sigma^2) + \sigma^2 = \lambda_i$. In other words, this model correctly captures the variance of the data along the principal axes, and approximates the variance in all remaining directions with a single average value σ^2.

One way to construct the maximum likelihood density model would simply be to find the eigenvectors and eigenvalues of the data covariance matrix and then to evaluate $\mathbf{W}$ and σ^2 using the results given above. In this case, we would choose $\mathbf{R} = \mathbf{I}$ for convenience. However, if the maximum likelihood solution is found by numerical optimization of the likelihood function, for instance using an algorithm such as conjugate gradients (Fletcher, 1987; Nocedal and Wright, 1999; Bishop and Nabney, 2008) or through the EM algorithm, then the resulting value of $\mathbf{R}$ is essentially arbitrary. This implies that the columns of $\mathbf{W}$ need not be orthogonal. If an orthogonal basis is required, the matrix $\mathbf{W}$ can be post-processed appropriately (Golub and Van Loan, 1996). Alternatively, the EM algorithm can be modified in such a way as to yield orthonormal principal directions, sorted in descending order

Section 12.2.2

of the corresponding eigenvalues, directly (Ahn and Oh, 2003).

The rotational invariance in latent space represents a form of statistical nonidentifiability, analogous to that encountered for mixture models in the case of discrete latent variables. Here there is a continuum of parameters all of which lead to the same predictive density, in contrast to the discrete nonidentifiability associated with component re-labelling in the mixture setting.

If we consider the case of $M = D$, so that there is no reduction of dimensionality, then $\mathbf{U}_M = \mathbf{U}$ and $\mathbf{L}_M = \mathbf{L}$. Making use of the orthogonality properties $\mathbf{U}\mathbf{U}^{\mathrm{T}} = \mathbf{I}$ and $\mathbf{R}\mathbf{R}^{\mathrm{T}} = \mathbf{I}$, we see that the covariance $\mathbf{C}$ of the marginal distribution for $\mathbf{x}$ becomes

$$\mathbf{C} = \mathbf{U}(\mathbf{L} - \sigma^2\mathbf{I})^{1/2}\mathbf{R}\mathbf{R}^{\mathrm{T}}(\mathbf{L} - \sigma^2\mathbf{I})^{1/2}\mathbf{U}^{\mathrm{T}} + \sigma^2\mathbf{I} = \mathbf{U}\mathbf{L}\mathbf{U}^{\mathrm{T}} = \mathbf{S} \tag{12.47}$$

and so we obtain the standard maximum likelihood solution for an unconstrained Gaussian distribution in which the covariance matrix is given by the sample covariance.

Conventional PCA is generally formulated as a projection of points from the D-dimensional data space onto an M-dimensional linear subspace. Probabilistic PCA, however, is most naturally expressed as a mapping from the latent space into the data space via (12.33). For applications such as visualization and data compression, we can reverse this mapping using Bayes' theorem. Any point $\mathbf{x}$ in data space can then be summarized by its posterior mean and covariance in latent space. From (12.42) the mean is given by

$$\mathbb{E}[\mathbf{z}|\mathbf{x}] = \mathbf{M}^{-1}\mathbf{W}_{\mathrm{ML}}^{\mathrm{T}}(\mathbf{x} - \overline{\mathbf{x}}) \tag{12.48}$$

where $\mathbf{M}$ is given by (12.41). This projects to a point in data space given by

$$\mathbf{W}\mathbb{E}[\mathbf{z}|\mathbf{x}] + \boldsymbol{\mu}. \tag{12.49}$$

Section 3.3.1

Note that this takes the same form as the equations for regularized linear regression and is a consequence of maximizing the likelihood function for a linear Gaussian model. Similarly, the posterior covariance is given from (12.42) by $\sigma^2\mathbf{M}^{-1}$ and is independent of $\mathbf{x}$.

If we take the limit $\sigma^2 \to 0$, then the posterior mean reduces to

$$(\mathbf{W}_{\mathrm{ML}}^{\mathrm{T}}\mathbf{W}_{\mathrm{ML}})^{-1}\mathbf{W}_{\mathrm{ML}}^{\mathrm{T}}(\mathbf{x} - \overline{\mathbf{x}}) \tag{12.50}$$

Exercise 12.11

Exercise 12.12

which represents an orthogonal projection of the data point onto the latent space, and so we recover the standard PCA model. The posterior covariance in this limit is zero, however, and the density becomes singular. For $\sigma^2 > 0$, the latent projection is shifted towards the origin, relative to the orthogonal projection.

Section 2.3

Finally, we note that an important role for the probabilistic PCA model is in defining a multivariate Gaussian distribution in which the number of degrees of freedom, in other words the number of independent parameters, can be controlled whilst still allowing the model to capture the dominant correlations in the data. Recall that a general Gaussian distribution has $D(D+1)/2$ independent parameters in its covariance matrix (plus another D parameters in its mean). Thus the number of parameters scales quadratically with D and can become excessive in spaces of high dimensionality. If we restrict the covariance matrix to be diagonal, then it has only D independent parameters, and so the number of parameters now grows linearly with dimensionality. However, it now treats the variables as if they were independent and hence can no longer express any correlations between them. Probabilistic PCA provides an elegant compromise in which the M most significant correlations can be captured while still ensuring that the total number of parameters grows only linearly with D. We can see this by evaluating the number of degrees of freedom in the PPCA model as follows. The covariance matrix $\mathbf{C}$ depends on the parameters $\mathbf{W}$, which has size $D \times M$, and σ^2, giving a total parameter count of $DM+1$. However, we have seen that there is some redundancy in this parameterization associated with rotations of the coordinate system in the latent space. The orthogonal matrix $\mathbf{R}$ that expresses these rotations has size $M \times M$. In the first column of this matrix there are $M-1$ independent parameters, because the column vector must be normalized to unit length. In the second column there are $M-2$ independent parameters, because the column must be normalized and also must be orthogonal to the previous column, and so on. Summing this arithmetic series, we see that $\mathbf{R}$ has a total of $M(M-1)/2$ independent parameters. Thus the number of degrees of freedom in the covariance matrix $\mathbf{C}$ is given by

$$DM + 1 - M(M-1)/2. \tag{12.51}$$

The number of independent parameters in this model therefore only grows linearly with D, for fixed M. If we take $M = D - 1$, then we recover the standard result for a full covariance Gaussian. In this case, the variance along $D-1$ linearly independent directions is controlled by the columns of $\mathbf{W}$, and the variance along the remaining direction is given by σ^2. If $M = 0$, the model is equivalent to the isotropic covariance case.

Exercise 12.14

12.2.2 EM algorithm for PCA

As we have seen, the probabilistic PCA model can be expressed in terms of a marginalization over a continuous latent space $\mathbf{z}$ in which for each data point $\mathbf{x}_n$, there is a corresponding latent variable $\mathbf{z}_n$. We can therefore make use of the EM algorithm to find maximum likelihood estimates of the model parameters. This may seem rather pointless because we have already obtained an exact closed-form solution for the maximum likelihood parameter values. However, in spaces of high dimensionality, there may be computational advantages in using an iterative EM procedure rather than working directly with the sample covariance matrix. This EM procedure can also be extended to the factor analysis model, for which there is no closed-form solution. Finally, it allows missing data to be handled in a principled way.

Section 12.2.4

We can derive the EM algorithm for probabilistic PCA by following the general framework for EM. Thus we write down the complete-data log likelihood and take its expectation with respect to the posterior distribution of the latent variables evaluated using 'old' parameter values. Maximization of this expected complete-data log likelihood then yields the 'new' parameter values. Because the data points

Section 9.4

are assumed independent, the complete-data log likelihood function takes the form

$$\ln p\left(\mathbf{X}, \mathbf{Z}|\boldsymbol{\mu}, \mathbf{W}, \sigma^2\right) = \sum_{n=1}^{N} \{\ln p(\mathbf{x}_n|\mathbf{z}_n) + \ln p(\mathbf{z}_n)\} \tag{12.52}$$

where the nth row of the matrix $\mathbf{Z}$ is given by $\mathbf{z}_n$. We already know that the exact maximum likelihood solution for $\boldsymbol{\mu}$ is given by the sample mean $\overline{\mathbf{x}}$ defined by (12.1), and it is convenient to substitute for $\boldsymbol{\mu}$ at this stage. Making use of the expressions (12.31) and (12.32) for the latent and conditional distributions, respectively, and taking the expectation with respect to the posterior distribution over the latent variables, we obtain

$$\begin{aligned}\mathbb{E}[\ln p\left(\mathbf{X}, \mathbf{Z}|\boldsymbol{\mu}, \mathbf{W}, \sigma^2\right)] = -\sum_{n=1}^{N} \Bigg\{ \frac{D}{2}\ln(2\pi\sigma^2) + \frac{1}{2}\mathrm{Tr}\left(\mathbb{E}[\mathbf{z}_n\mathbf{z}_n^{\mathrm{T}}]\right) \\ + \frac{1}{2\sigma^2}\|\mathbf{x}_n - \boldsymbol{\mu}\|^2 - \frac{1}{\sigma^2}\mathbb{E}[\mathbf{z}_n]^{\mathrm{T}}\mathbf{W}^{\mathrm{T}}(\mathbf{x}_n - \boldsymbol{\mu}) \\ + \frac{1}{2\sigma^2}\mathrm{Tr}\left(\mathbb{E}[\mathbf{z}_n\mathbf{z}_n^{\mathrm{T}}]\mathbf{W}^{\mathrm{T}}\mathbf{W}\right) + \frac{M}{2}\ln(2\pi)\Bigg\}.\end{aligned} \tag{12.53}$$

Note that this depends on the posterior distribution only through the sufficient statistics of the Gaussian. Thus in the E step, we use the old parameter values to evaluate

$$\mathbb{E}[\mathbf{z}_n] = \mathbf{M}^{-1}\mathbf{W}^{\mathrm{T}}(\mathbf{x}_n - \overline{\mathbf{x}}) \tag{12.54}$$

$$\mathbb{E}[\mathbf{z}_n\mathbf{z}_n^{\mathrm{T}}] = \sigma^2\mathbf{M}^{-1} + \mathbb{E}[\mathbf{z}_n]\mathbb{E}[\mathbf{z}_n]^{\mathrm{T}} \tag{12.55}$$

which follow directly from the posterior distribution (12.42) together with the standard result $\mathbb{E}[\mathbf{z}_n\mathbf{z}_n^{\mathrm{T}}] = \mathrm{cov}[\mathbf{z}_n] + \mathbb{E}[\mathbf{z}_n]\mathbb{E}[\mathbf{z}_n]^{\mathrm{T}}$. Here $\mathbf{M}$ is defined by (12.41).

In the M step, we maximize with respect to $\mathbf{W}$ and σ^2, keeping the posterior statistics fixed. Maximization with respect to σ^2 is straightforward. For the maximization with respect to $\mathbf{W}$ we make use of (C.24), and obtain the M step equations

Exercise 12.15

$$\mathbf{W}_{\text{new}} = \left[\sum_{n=1}^{N}(\mathbf{x}_n - \overline{\mathbf{x}})\mathbb{E}[\mathbf{z}_n]^{\mathrm{T}}\right]\left[\sum_{n=1}^{N}\mathbb{E}[\mathbf{z}_n\mathbf{z}_n^{\mathrm{T}}]\right]^{-1} \tag{12.56}$$

$$\begin{aligned}\sigma^2_{\text{new}} = \frac{1}{ND}\sum_{n=1}^{N}\Big\{\|\mathbf{x}_n - \overline{\mathbf{x}}\|^2 - 2\mathbb{E}[\mathbf{z}_n]^{\mathrm{T}}\mathbf{W}_{\text{new}}^{\mathrm{T}}(\mathbf{x}_n - \overline{\mathbf{x}}) \\ +\mathrm{Tr}\left(\mathbb{E}[\mathbf{z}_n\mathbf{z}_n^{\mathrm{T}}]\mathbf{W}_{\text{new}}^{\mathrm{T}}\mathbf{W}_{\text{new}}\right)\Big\}.\end{aligned} \tag{12.57}$$

The EM algorithm for probabilistic PCA proceeds by initializing the parameters and then alternately computing the sufficient statistics of the latent space posterior distribution using (12.54) and (12.55) in the E step and revising the parameter values using (12.56) and (12.57) in the M step.

One of the benefits of the EM algorithm for PCA is computational efficiency for large-scale applications (Roweis, 1998). Unlike conventional PCA based on an eigenvector decomposition of the sample covariance matrix, the EM approach is iterative and so might appear to be less attractive. However, each cycle of the EM algorithm can be computationally much more efficient than conventional PCA in spaces of high dimensionality. To see this, we note that the eigendecomposition of the covariance matrix requires $O(D^3)$ computation. Often we are interested only in the first M eigenvectors and their corresponding eigenvalues, in which case we can use algorithms that are $O(MD^2)$. However, the evaluation of the covariance matrix itself takes $O(ND^2)$ computations, where N is the number of data points. Algorithms such as the snapshot method (Sirovich, 1987), which assume that the eigenvectors are linear combinations of the data vectors, avoid direct evaluation of the covariance matrix but are $O(N^3)$ and hence unsuited to large data sets. The EM algorithm described here also does not construct the covariance matrix explicitly. Instead, the most computationally demanding steps are those involving sums over the data set that are $O(NDM)$. For large D, and $M \ll D$, this can be a significant saving compared to $O(ND^2)$ and can offset the iterative nature of the EM algorithm.

Note that this EM algorithm can be implemented in an on-line form in which each D-dimensional data point is read in and processed and then discarded before the next data point is considered. To see this, note that the quantities evaluated in the E step (an M-dimensional vector and an $M \times M$ matrix) can be computed for each data point separately, and in the M step we need to accumulate sums over data points, which we can do incrementally. This approach can be advantageous if both N and D are large.

Because we now have a fully probabilistic model for PCA, we can deal with missing data, provided that it is missing at random, by marginalizing over the distribution of the unobserved variables. Again these missing values can be treated using the EM algorithm. We give an example of the use of this approach for data visualization in Figure 12.11.

Another elegant feature of the EM approach is that we can take the limit $\sigma^2 \to 0$, corresponding to standard PCA, and still obtain a valid EM-like algorithm (Roweis, 1998). From (12.55), we see that the only quantity we need to compute in the E step is $\mathbb{E}[\mathbf{z}_n]$. Furthermore, the M step is simplified because $\mathbf{M} = \mathbf{W}^{\mathrm{T}}\mathbf{W}$. To emphasize the simplicity of the algorithm, let us define $\widetilde{\mathbf{X}}$ to be a matrix of size $N \times D$ whose n^{th} row is given by the vector $\mathbf{x}_n - \overline{\mathbf{x}}$ and similarly define $\boldsymbol{\Omega}$ to be a matrix of size $M \times N$ whose n^{th} column is given by the vector $\mathbb{E}[\mathbf{z}_n]$. The E step (12.54) of the EM algorithm for PCA then becomes

$$\boldsymbol{\Omega} = (\mathbf{W}_{\text{old}}^{\mathrm{T}}\mathbf{W}_{\text{old}})^{-1}\mathbf{W}_{\text{old}}^{\mathrm{T}}\widetilde{\mathbf{X}}^{\mathrm{T}} \tag{12.58}$$

and the M step (12.56) takes the form

$$\mathbf{W}_{\text{new}} = \widetilde{\mathbf{X}}^{\mathrm{T}}\boldsymbol{\Omega}^{\mathrm{T}}(\boldsymbol{\Omega}\boldsymbol{\Omega}^{\mathrm{T}})^{-1}. \tag{12.59}$$

Again these can be implemented in an on-line form. These equations have a simple interpretation as follows. From our earlier discussion, we see that the E step involves an orthogonal projection of the data points onto the current estimate for the principal subspace. Correspondingly, the M step represents a re-estimation of the principal

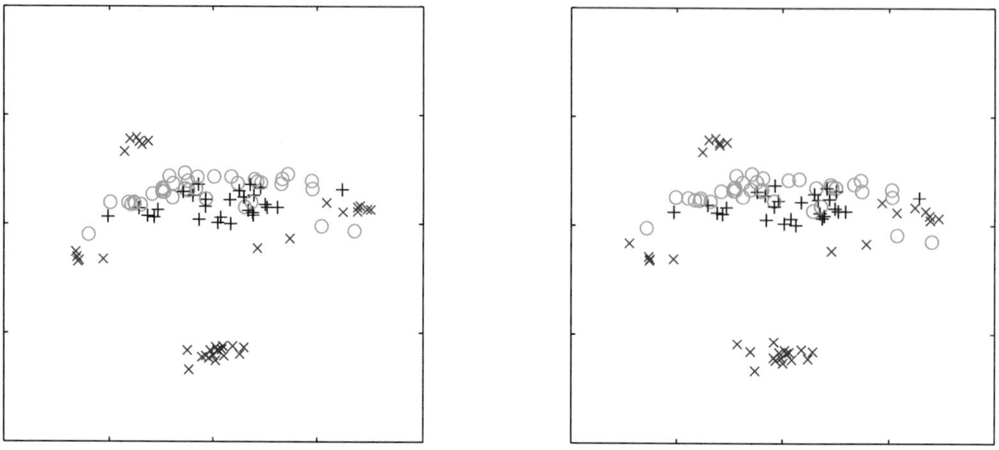

Figure 12.11 Probabilistic PCA visualization of a portion of the oil flow data set for the first 100 data points. The left-hand plot shows the posterior mean projections of the data points on the principal subspace. The right-hand plot is obtained by first randomly omitting 30% of the variable values and then using EM to handle the missing values. Note that each data point then has at least one missing measurement but that the plot is very similar to the one obtained without missing values.

subspace to minimize the squared reconstruction error in which the projections are fixed.

Exercise 12.17

We can give a simple physical analogy for this EM algorithm, which is easily visualized for $D = 2$ and $M = 1$. Consider a collection of data points in two dimensions, and let the one-dimensional principal subspace be represented by a solid rod. Now attach each data point to the rod via a spring obeying Hooke's law (stored energy is proportional to the square of the spring's length). In the E step, we keep the rod fixed and allow the attachment points to slide up and down the rod so as to minimize the energy. This causes each attachment point (independently) to position itself at the orthogonal projection of the corresponding data point onto the rod. In the M step, we keep the attachment points fixed and then release the rod and allow it to move to the minimum energy position. The E and M steps are then repeated until a suitable convergence criterion is satisfied, as is illustrated in Figure 12.12.

12.2.3 Bayesian PCA

So far in our discussion of PCA, we have assumed that the value M for the dimensionality of the principal subspace is given. In practice, we must choose a suitable value according to the application. For visualization, we generally choose $M = 2$, whereas for other applications the appropriate choice for M may be less clear. One approach is to plot the eigenvalue spectrum for the data set, analogous to the example in Figure 12.4 for the off-line digits data set, and look to see if the eigenvalues naturally form two groups comprising a set of small values separated by a significant gap from a set of relatively large values, indicating a natural choice for M. In practice, such a gap is often not seen.

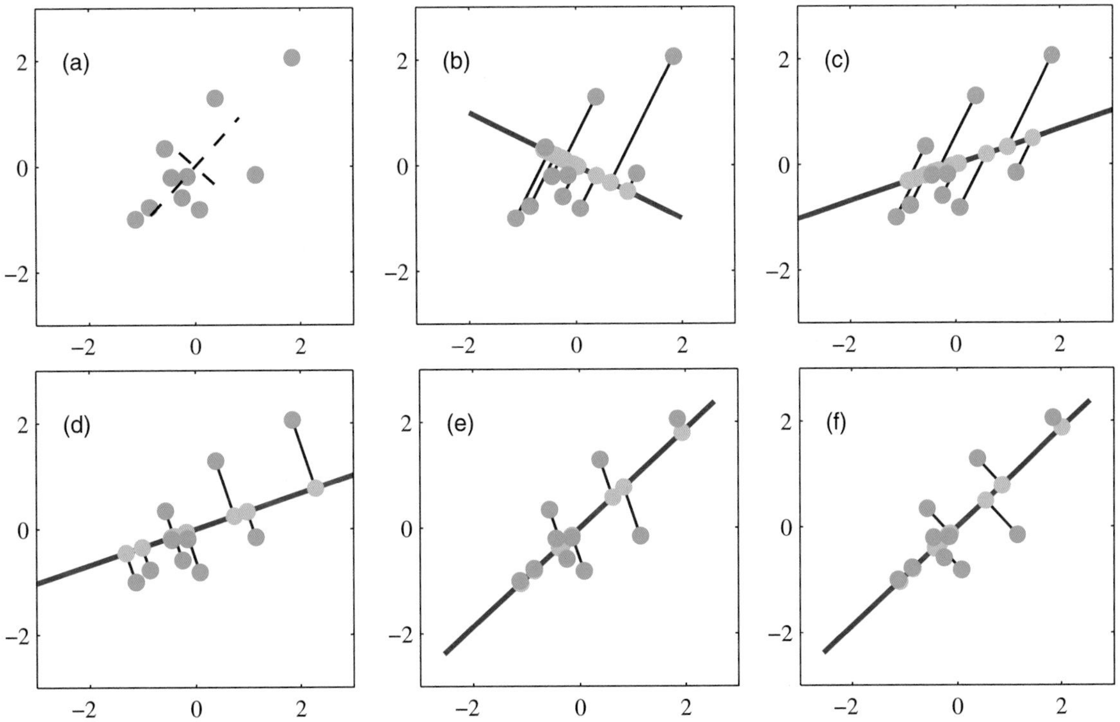

Figure 12.12 Synthetic data illustrating the EM algorithm for PCA defined by (12.58) and (12.59). (a) A data set $\mathbf{X}$ with the data points shown in green, together with the true principal components (shown as eigenvectors scaled by the square roots of the eigenvalues). (b) Initial configuration of the principal subspace defined by $\mathbf{W}$, shown in red, together with the projections of the latent points $\mathbf{Z}$ into the data space, given by $\mathbf{Z}\mathbf{W}^{\mathrm{T}}$, shown in cyan. (c) After one M step, the latent space has been updated with $\mathbf{Z}$ held fixed. (d) After the successive E step, the values of $\mathbf{Z}$ have been updated, giving orthogonal projections, with $\mathbf{W}$ held fixed. (e) After the second M step. (f) The converged solution.

Section 1.3

Because the probabilistic PCA model has a well-defined likelihood function, we could employ cross-validation to determine the value of dimensionality by selecting the largest log likelihood on a validation data set. Such an approach, however, can become computationally costly, particularly if we consider a probabilistic mixture of PCA models (Tipping and Bishop, 1999a) in which we seek to determine the appropriate dimensionality separately for each component in the mixture.

Given that we have a probabilistic formulation of PCA, it seems natural to seek a Bayesian approach to model selection. To do this, we need to marginalize out the model parameters $\boldsymbol{\mu}$, $\mathbf{W}$, and σ^2 with respect to appropriate prior distributions. This can be done by using a variational framework to approximate the analytically intractable marginalizations (Bishop, 1999b). The marginal likelihood values, given by the variational lower bound, can then be compared for a range of different values of M and the value giving the largest marginal likelihood selected.

Here we consider a simpler approach introduced by based on the *evidence ap-*

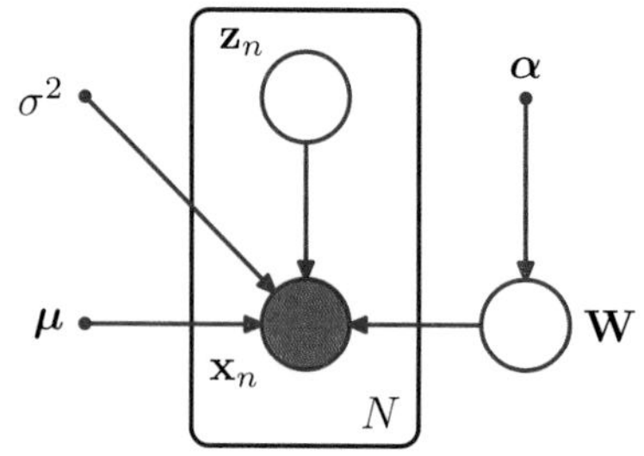

Figure 12.13 Probabilistic graphical model for Bayesian PCA in which the distribution over the parameter matrix $\mathbf{W}$ is governed by a vector $\boldsymbol{\alpha}$ of hyperparameters.

proximation, which is appropriate when the number of data points is relatively large and the corresponding posterior distribution is tightly peaked (Bishop, 1999a). It involves a specific choice of prior over $\mathbf{W}$ that allows surplus dimensions in the principal subspace to be pruned out of the model. This corresponds to an example of *automatic relevance determination*, or *ARD*, discussed in Section 7.2.2. Specifically, we define an independent Gaussian prior over each column of $\mathbf{W}$, which represent the vectors defining the principal subspace. Each such Gaussian has an independent variance governed by a precision hyperparameter α_i so that

$$p(\mathbf{W}|\boldsymbol{\alpha}) = \prod_{i=1}^{M} \left(\frac{\alpha_i}{2\pi}\right)^{D/2} \exp\left\{-\frac{1}{2}\alpha_i \mathbf{w}_i^{\mathrm{T}}\mathbf{w}_i\right\} \tag{12.60}$$

where $\mathbf{w}_i$ is the ith column of $\mathbf{W}$. The resulting model can be represented using the directed graph shown in Figure 12.13.

The values for α_i will be found iteratively by maximizing the marginal likelihood function in which $\mathbf{W}$ has been integrated out. As a result of this optimization, some of the α_i may be driven to infinity, with the corresponding parameters vector $\mathbf{w}_i$ being driven to zero (the posterior distribution becomes a delta function at the origin) giving a sparse solution. The effective dimensionality of the principal subspace is then determined by the number of finite α_i values, and the corresponding vectors $\mathbf{w}_i$ can be thought of as 'relevant' for modelling the data distribution. In this way, the Bayesian approach is automatically making the trade-off between improving the fit to the data, by using a larger number of vectors $\mathbf{w}_i$ with their corresponding eigenvalues λ_i each tuned to the data, and reducing the complexity of the model by suppressing some of the $\mathbf{w}_i$ vectors. The origins of this sparsity were discussed earlier in the context of relevance vector machines.

Section 7.2

The values of α_i are re-estimated during training by maximizing the marginal likelihood given by

$$p(\mathbf{X}|\boldsymbol{\alpha}, \boldsymbol{\mu}, \sigma^2) = \int p(\mathbf{X}|\mathbf{W}, \boldsymbol{\mu}, \sigma^2)p(\mathbf{W}|\boldsymbol{\alpha})\,\mathrm{d}\mathbf{W} \tag{12.61}$$

where the log of $p(\mathbf{X}|\mathbf{W}, \boldsymbol{\mu}, \sigma^2)$ is given by (12.43). Note that for simplicity we also treat $\boldsymbol{\mu}$ and σ^2 as parameters to be estimated, rather than defining priors over these parameters.

Section 4.4

Section 3.5.3

Because this integration is intractable, we make use of the Laplace approximation. If we assume that the posterior distribution is sharply peaked, as will occur for sufficiently large data sets, then the re-estimation equations obtained by maximizing the marginal likelihood with respect to α_i take the simple form

$$\alpha_i^{\text{new}} = \frac{D}{\mathbf{w}_i^{\mathrm{T}}\mathbf{w}_i} \tag{12.62}$$

which follows from (3.98), noting that the dimensionality of $\mathbf{w}_i$ is D. These re-estimations are interleaved with the EM algorithm updates for determining $\mathbf{W}$ and σ^2. The E-step equations are again given by (12.54) and (12.55). Similarly, the M-step equation for σ^2 is again given by (12.57). The only change is to the M-step equation for $\mathbf{W}$, which is modified to give

$$\mathbf{W}_{\text{new}} = \left[\sum_{n=1}^{N}(\mathbf{x}_n - \overline{\mathbf{x}})\mathbb{E}[\mathbf{z}_n]^{\mathrm{T}}\right]\left[\sum_{n=1}^{N}\mathbb{E}[\mathbf{z}_n\mathbf{z}_n^{\mathrm{T}}] + \sigma^2\mathbf{A}\right]^{-1} \tag{12.63}$$

where $\mathbf{A} = \text{diag}(\alpha_i)$. The value of $\boldsymbol{\mu}$ is given by the sample mean, as before.

If we choose $M = D - 1$ then, if all α_i values are finite, the model represents a full-covariance Gaussian, while if all the α_i go to infinity the model is equivalent to an isotropic Gaussian, and so the model can encompass all permissible values for the effective dimensionality of the principal subspace. It is also possible to consider smaller values of M, which will save on computational cost but which will limit the maximum dimensionality of the subspace. A comparison of the results of this algorithm with standard probabilistic PCA is shown in Figure 12.14.

Bayesian PCA provides an opportunity to illustrate the Gibbs sampling algorithm discussed in Section 11.3. Figure 12.15 shows an example of the samples from the hyperparameters $\ln \alpha_i$ for a data set in $D = 4$ dimensions in which the dimensionality of the latent space is $M = 3$ but in which the data set is generated from a probabilistic PCA model having one direction of high variance, with the remaining directions comprising low variance noise. This result shows clearly the presence of three distinct modes in the posterior distribution. At each step of the iteration, one of the hyperparameters has a small value and the remaining two have large values, so that two of the three latent variables are suppressed. During the course of the Gibbs sampling, the solution makes sharp transitions between the three modes.

The model described here involves a prior only over the matrix $\mathbf{W}$. A fully Bayesian treatment of PCA, including priors over $\boldsymbol{\mu}$, σ^2, and $\boldsymbol{\alpha}$, and solved using variational methods, is described in Bishop (1999b). For a discussion of various Bayesian approaches to determining the appropriate dimensionality for a PCA model, see Minka (2001c).

12.2.4 Factor analysis

Factor analysis is a linear-Gaussian latent variable model that is closely related to probabilistic PCA. Its definition differs from that of probabilistic PCA only in that the conditional distribution of the observed variable $\mathbf{x}$ given the latent variable $\mathbf{z}$ is

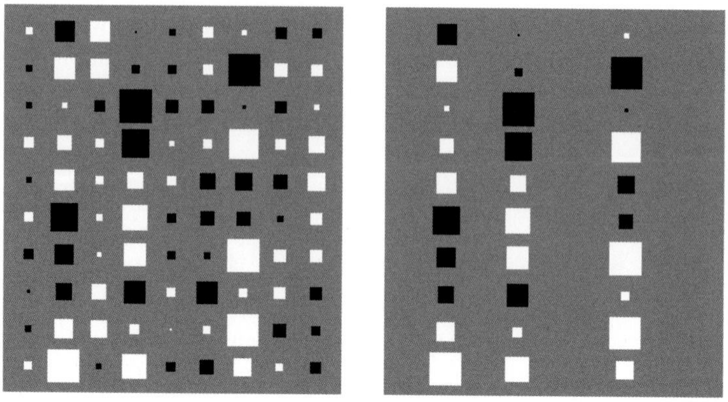

Figure 12.14 'Hinton' diagrams of the matrix $\mathbf{W}$ in which each element of the matrix is depicted as a square (white for positive and black for negative values) whose area is proportional to the magnitude of that element. The synthetic data set comprises 300 data points in $D = 10$ dimensions sampled from a Gaussian distribution having standard deviation 1.0 in 3 directions and standard deviation 0.5 in the remaining 7 directions for a data set in $D = 10$ dimensions having $M = 3$ directions with larger variance than the remaining 7 directions. The left-hand plot shows the result from maximum likelihood probabilistic PCA, and the right-hand plot shows the corresponding result from Bayesian PCA. We see how the Bayesian model is able to discover the appropriate dimensionality by suppressing the 6 surplus degrees of freedom.

taken to have a diagonal rather than an isotropic covariance so that

$$p(\mathbf{x}|\mathbf{z}) = \mathcal{N}(\mathbf{x}|\mathbf{W}\mathbf{z} + \boldsymbol{\mu}, \boldsymbol{\Psi}) \tag{12.64}$$

where $\boldsymbol{\Psi}$ is a $D \times D$ diagonal matrix. Note that the factor analysis model, in common with probabilistic PCA, assumes that the observed variables $x_1, \ldots, x_D$ are independent, given the latent variable $\mathbf{z}$. In essence, the factor analysis model is explaining the observed covariance structure of the data by representing the independent variance associated with each coordinate in the matrix $\boldsymbol{\Psi}$ and capturing the covariance between variables in the matrix $\mathbf{W}$. In the factor analysis literature, the columns of $\mathbf{W}$, which capture the correlations between observed variables, are called *factor loadings*, and the diagonal elements of $\boldsymbol{\Psi}$, which represent the independent noise variances for each of the variables, are called *uniquenesses*.

The origins of factor analysis are as old as those of PCA, and discussions of factor analysis can be found in the books by Everitt (1984), Bartholomew (1987), and Basilevsky (1994). Links between factor analysis and PCA were investigated by Lawley (1953) and Anderson (1963) who showed that at stationary points of the likelihood function, for a factor analysis model with $\boldsymbol{\Psi} = \sigma^2\mathbf{I}$, the columns of $\mathbf{W}$ are scaled eigenvectors of the sample covariance matrix, and σ^2 is the average of the discarded eigenvalues. Later, Tipping and Bishop (1999b) showed that the maximum of the log likelihood function occurs when the eigenvectors comprising $\mathbf{W}$ are chosen to be the principal eigenvectors.

Making use of (2.115), we see that the marginal distribution for the observed

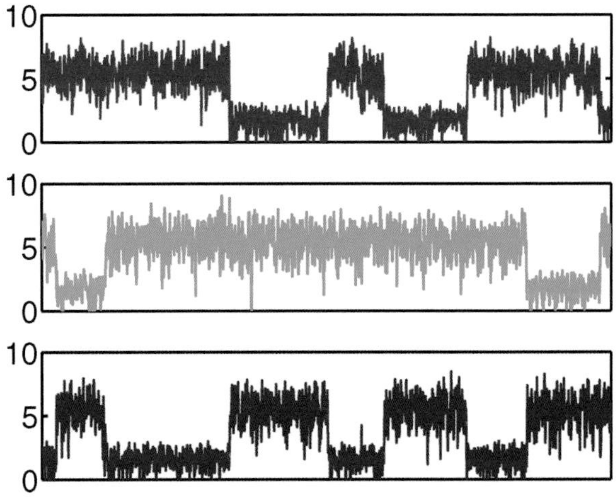

Figure 12.15 Gibbs sampling for Bayesian PCA showing plots of $\ln \alpha_i$ versus iteration number for three α values, showing transitions between the three modes of the posterior distribution.

variable is given by $p(\mathbf{x}) = \mathcal{N}(\mathbf{x}|\boldsymbol{\mu}, \mathbf{C})$ where now

$$\mathbf{C} = \mathbf{W}\mathbf{W}^{\mathrm{T}} + \boldsymbol{\Psi}. \tag{12.65}$$

Exercise 12.19

As with probabilistic PCA, this model is invariant to rotations in the latent space.

Historically, factor analysis has been the subject of controversy when attempts have been made to place an interpretation on the individual factors (the coordinates in $\mathbf{z}$-space), which has proven problematic due to the nonidentifiability of factor analysis associated with rotations in this space. From our perspective, however, we shall view factor analysis as a form of latent variable density model, in which the form of the latent space is of interest but not the particular choice of coordinates used to describe it. If we wish to remove the degeneracy associated with latent space rotations, we must consider non-Gaussian latent variable distributions, giving rise to independent component analysis (ICA) models.

Section 12.4

We can determine the parameters $\boldsymbol{\mu}$, $\mathbf{W}$, and $\boldsymbol{\Psi}$ in the factor analysis model by maximum likelihood. The solution for $\boldsymbol{\mu}$ is again given by the sample mean. However, unlike probabilistic PCA, there is no longer a closed-form maximum likelihood solution for $\mathbf{W}$, which must therefore be found iteratively. Because factor analysis is a latent variable model, this can be done using an EM algorithm (Rubin and Thayer, 1982) that is analogous to the one used for probabilistic PCA. Specifically, the E-step equations are given by

Exercise 12.21

$$\mathbb{E}[\mathbf{z}_n] = \mathbf{G}\mathbf{W}^{\mathrm{T}}\boldsymbol{\Psi}^{-1}(\mathbf{x}_n - \overline{\mathbf{x}}) \tag{12.66}$$

$$\mathbb{E}[\mathbf{z}_n\mathbf{z}_n^{\mathrm{T}}] = \mathbf{G} + \mathbb{E}[\mathbf{z}_n]\mathbb{E}[\mathbf{z}_n]^{\mathrm{T}} \tag{12.67}$$

where we have defined

$$\mathbf{G} = (\mathbf{I} + \mathbf{W}^{\mathrm{T}}\boldsymbol{\Psi}^{-1}\mathbf{W})^{-1}. \tag{12.68}$$

Note that this is expressed in a form that involves inversion of matrices of size $M \times M$ rather than $D \times D$ (except for the $D \times D$ diagonal matrix $\boldsymbol{\Psi}$ whose inverse is trivial to compute in $O(D)$ steps), which is convenient because often $M \ll D$. Similarly, *Exercise 12.22* the M-step equations take the form

$$\mathbf{W}_{\text{new}} = \left[\sum_{n=1}^{N}(\mathbf{x}_n - \overline{\mathbf{x}})\mathbb{E}[\mathbf{z}_n]^{\mathrm{T}}\right]\left[\sum_{n=1}^{N}\mathbb{E}[\mathbf{z}_n\mathbf{z}_n^{\mathrm{T}}]\right]^{-1} \tag{12.69}$$

$$\boldsymbol{\Psi}_{\text{new}} = \operatorname{diag}\left\{\mathbf{S} - \mathbf{W}_{\text{new}}\frac{1}{N}\sum_{n=1}^{N}\mathbb{E}[\mathbf{z}_n](\mathbf{x}_n - \overline{\mathbf{x}})^{\mathrm{T}}\right\} \tag{12.70}$$

where the 'diag' operator sets all of the nondiagonal elements of a matrix to zero. A Bayesian treatment of the factor analysis model can be obtained by a straightforward application of the techniques discussed in this book.

Another difference between probabilistic PCA and factor analysis concerns their *Exercise 12.25* different behaviour under transformations of the data set. For PCA and probabilistic PCA, if we rotate the coordinate system in data space, then we obtain exactly the same fit to the data but with the $\mathbf{W}$ matrix transformed by the corresponding rotation matrix. However, for factor analysis, the analogous property is that if we make a component-wise re-scaling of the data vectors, then this is absorbed into a corresponding re-scaling of the elements of $\boldsymbol{\Psi}$.

12.3. Kernel PCA

In Chapter 6, we saw how the technique of kernel substitution allows us to take an algorithm expressed in terms of scalar products of the form $\mathbf{x}^{\mathrm{T}}\mathbf{x}'$ and generalize that algorithm by replacing the scalar products with a nonlinear kernel. Here we apply this technique of kernel substitution to principal component analysis, thereby obtaining a nonlinear generalization called *kernel PCA* (Schölkopf *et al.*, 1998).

Consider a data set $\{\mathbf{x}_n\}$ of observations, where $n = 1, \ldots, N$, in a space of dimensionality D. In order to keep the notation uncluttered, we shall assume that we have already subtracted the sample mean from each of the vectors $\mathbf{x}_n$, so that $\sum_n \mathbf{x}_n = \mathbf{0}$. The first step is to express conventional PCA in such a form that the data vectors $\{\mathbf{x}_n\}$ appear only in the form of the scalar products $\mathbf{x}_n^{\mathrm{T}}\mathbf{x}_m$. Recall that the principal components are defined by the eigenvectors $\mathbf{u}_i$ of the covariance matrix

$$\mathbf{S}\mathbf{u}_i = \lambda_i \mathbf{u}_i \tag{12.71}$$

where $i = 1, \ldots, D$. Here the $D \times D$ sample covariance matrix $\mathbf{S}$ is defined by

$$\mathbf{S} = \frac{1}{N}\sum_{n=1}^{N}\mathbf{x}_n\mathbf{x}_n^{\mathrm{T}}, \tag{12.72}$$

and the eigenvectors are normalized such that $\mathbf{u}_i^{\mathrm{T}}\mathbf{u}_i = 1$.

Now consider a nonlinear transformation $\boldsymbol{\phi}(\mathbf{x})$ into an M-dimensional feature space, so that each data point $\mathbf{x}_n$ is thereby projected onto a point $\boldsymbol{\phi}(\mathbf{x}_n)$. We can

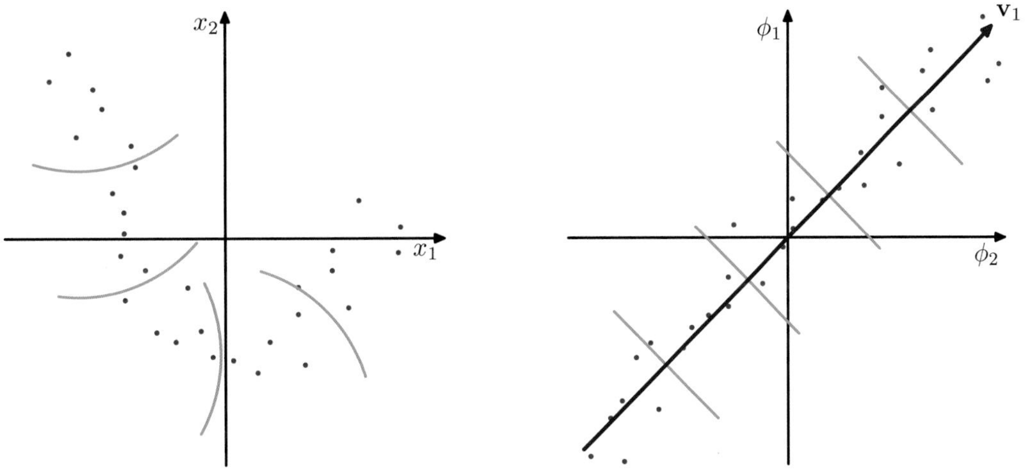

Figure 12.16 Schematic illustration of kernel PCA. A data set in the original data space (left-hand plot) is projected by a nonlinear transformation $\phi(\mathbf{x})$ into a feature space (right-hand plot). By performing PCA in the feature space, we obtain the principal components, of which the first is shown in blue and is denoted by the vector $\mathbf{v}_1$. The green lines in feature space indicate the linear projections onto the first principal component, which correspond to nonlinear projections in the original data space. Note that in general it is not possible to represent the nonlinear principal component by a vector in $\mathbf{x}$ space.

now perform standard PCA in the feature space, which implicitly defines a nonlinear principal component model in the original data space, as illustrated in Figure 12.16.

For the moment, let us assume that the projected data set also has zero mean, so that $\sum_n \phi(\mathbf{x}_n) = \mathbf{0}$. We shall return to this point shortly. The $M \times M$ sample covariance matrix in feature space is given by

$$\mathbf{C} = \frac{1}{N}\sum_{n=1}^{N} \phi(\mathbf{x}_n)\phi(\mathbf{x}_n)^{\mathrm{T}} \tag{12.73}$$

and its eigenvector expansion is defined by

$$\mathbf{C}\mathbf{v}_i = \lambda_i \mathbf{v}_i \tag{12.74}$$

$i = 1, \ldots, M$. Our goal is to solve this eigenvalue problem without having to work explicitly in the feature space. From the definition of $\mathbf{C}$, the eigenvector equations tells us that $\mathbf{v}_i$ satisfies

$$\frac{1}{N}\sum_{n=1}^{N} \phi(\mathbf{x}_n)\left\{\phi(\mathbf{x}_n)^{\mathrm{T}}\mathbf{v}_i\right\} = \lambda_i \mathbf{v}_i \tag{12.75}$$

and so we see that (provided $\lambda_i > 0$) the vector $\mathbf{v}_i$ is given by a linear combination of the $\phi(\mathbf{x}_n)$ and so can be written in the form

$$\mathbf{v}_i = \sum_{n=1}^{N} a_{in}\phi(\mathbf{x}_n). \tag{12.76}$$

Substituting this expansion back into the eigenvector equation, we obtain

$$\frac{1}{N}\sum_{n=1}^{N}\boldsymbol{\phi}(\mathbf{x}_n)\boldsymbol{\phi}(\mathbf{x}_n)^{\mathrm{T}}\sum_{m=1}^{N}a_{im}\boldsymbol{\phi}(\mathbf{x}_m) = \lambda_i\sum_{n=1}^{N}a_{in}\boldsymbol{\phi}(\mathbf{x}_n). \tag{12.77}$$

The key step is now to express this in terms of the kernel function $k(\mathbf{x}_n, \mathbf{x}_m) = \boldsymbol{\phi}(\mathbf{x}_n)^{\mathrm{T}}\boldsymbol{\phi}(\mathbf{x}_m)$, which we do by multiplying both sides by $\boldsymbol{\phi}(\mathbf{x}_l)^{\mathrm{T}}$ to give

$$\frac{1}{N}\sum_{n=1}^{N}k(\mathbf{x}_l, \mathbf{x}_n)\sum_{m=1}^{N}a_{im}k(\mathbf{x}_n, \mathbf{x}_m) = \lambda_i\sum_{n=1}^{N}a_{in}k(\mathbf{x}_l, \mathbf{x}_n). \tag{12.78}$$

This can be written in matrix notation as

$$\mathbf{K}^2\mathbf{a}_i = \lambda_i N\mathbf{K}\mathbf{a}_i \tag{12.79}$$

where $\mathbf{a}_i$ is an N-dimensional column vector with elements a_{in} for $n = 1, \ldots, N$. We can find solutions for $\mathbf{a}_i$ by solving the following eigenvalue problem

$$\mathbf{K}\mathbf{a}_i = \lambda_i N\mathbf{a}_i \tag{12.80}$$

in which we have removed a factor of $\mathbf{K}$ from both sides of (12.79). Note that the solutions of (12.79) and (12.80) differ only by eigenvectors of $\mathbf{K}$ having zero eigenvalues that do not affect the principal components projection.

Exercise 12.26

The normalization condition for the coefficients $\mathbf{a}_i$ is obtained by requiring that the eigenvectors in feature space be normalized. Using (12.76) and (12.80), we have

$$1 = \mathbf{v}_i^{\mathrm{T}}\mathbf{v}_i = \sum_{n=1}^{N}\sum_{m=1}^{N}a_{in}a_{im}\boldsymbol{\phi}(\mathbf{x}_n)^{\mathrm{T}}\boldsymbol{\phi}(\mathbf{x}_m) = \mathbf{a}_i^{\mathrm{T}}\mathbf{K}\mathbf{a}_i = \lambda_i N\mathbf{a}_i^{\mathrm{T}}\mathbf{a}_i. \tag{12.81}$$

Having solved the eigenvector problem, the resulting principal component projections can then also be cast in terms of the kernel function so that, using (12.76), the projection of a point $\mathbf{x}$ onto eigenvector i is given by

$$y_i(\mathbf{x}) = \boldsymbol{\phi}(\mathbf{x})^{\mathrm{T}}\mathbf{v}_i = \sum_{n=1}^{N}a_{in}\boldsymbol{\phi}(\mathbf{x})^{\mathrm{T}}\boldsymbol{\phi}(\mathbf{x}_n) = \sum_{n=1}^{N}a_{in}k(\mathbf{x}, \mathbf{x}_n) \tag{12.82}$$

and so again is expressed in terms of the kernel function.

In the original D-dimensional $\mathbf{x}$ space there are D orthogonal eigenvectors and hence we can find at most D linear principal components. The dimensionality M of the feature space, however, can be much larger than D (even infinite), and thus we can find a number of nonlinear principal components that can exceed D. Note, however, that the number of nonzero eigenvalues cannot exceed the number N of data points, because (even if $M > N$) the covariance matrix in feature space has rank at most equal to N. This is reflected in the fact that kernel PCA involves the eigenvector expansion of the $N \times N$ matrix $\mathbf{K}$.

So far we have assumed that the projected data set given by $\phi(\mathbf{x}_n)$ has zero mean, which in general will not be the case. We cannot simply compute and then subtract off the mean, since we wish to avoid working directly in feature space, and so again, we formulate the algorithm purely in terms of the kernel function. The projected data points after centralizing, denoted $\widetilde{\phi}(\mathbf{x}_n)$, are given by

$$\widetilde{\phi}(\mathbf{x}_n) = \phi(\mathbf{x}_n) - \frac{1}{N}\sum_{l=1}^{N}\phi(\mathbf{x}_l) \tag{12.83}$$

and the corresponding elements of the Gram matrix are given by

$$\begin{aligned}
\widetilde{K}_{nm} &= \widetilde{\phi}(\mathbf{x}_n)^{\mathrm{T}}\widetilde{\phi}(\mathbf{x}_m) \\
&= \phi(\mathbf{x}_n)^{\mathrm{T}}\phi(\mathbf{x}_m) - \frac{1}{N}\sum_{l=1}^{N}\phi(\mathbf{x}_n)^{\mathrm{T}}\phi(\mathbf{x}_l) \\
&\quad - \frac{1}{N}\sum_{l=1}^{N}\phi(\mathbf{x}_l)^{\mathrm{T}}\phi(\mathbf{x}_m) + \frac{1}{N^2}\sum_{j=1}^{N}\sum_{l=1}^{N}\phi(\mathbf{x}_j)^{\mathrm{T}}\phi(\mathbf{x}_l) \\
&= k(\mathbf{x}_n, \mathbf{x}_m) - \frac{1}{N}\sum_{l=1}^{N}k(\mathbf{x}_l, \mathbf{x}_m) \\
&\quad - \frac{1}{N}\sum_{l=1}^{N}k(\mathbf{x}_n, \mathbf{x}_l) + \frac{1}{N^2}\sum_{j=1}^{N}\sum_{l=1}^{N}k(\mathbf{x}_j, \mathbf{x}_l).
\end{aligned} \tag{12.84}$$

This can be expressed in matrix notation as

$$\widetilde{\mathbf{K}} = \mathbf{K} - \mathbf{1}_N\mathbf{K} - \mathbf{K}\mathbf{1}_N + \mathbf{1}_N\mathbf{K}\mathbf{1}_N \tag{12.85}$$

where $\mathbf{1}_N$ denotes the $N \times N$ matrix in which every element takes the value $1/N$. Thus we can evaluate $\widetilde{\mathbf{K}}$ using only the kernel function and then use $\widetilde{\mathbf{K}}$ to determine the eigenvalues and eigenvectors. Note that the standard PCA algorithm is recovered as a special case if we use a linear kernel $k(\mathbf{x}, \mathbf{x}') = \mathbf{x}^{\mathrm{T}}\mathbf{x}'$. Figure 12.17 shows an example of kernel PCA applied to a synthetic data set (Schölkopf *et al.*, 1998). Here a 'Gaussian' kernel of the form

Exercise 12.27

$$k(\mathbf{x}, \mathbf{x}') = \exp(-\|\mathbf{x} - \mathbf{x}'\|^2/0.1) \tag{12.86}$$

is applied to a synthetic data set. The lines correspond to contours along which the projection onto the corresponding principal component, defined by

$$\phi(\mathbf{x})^{\mathrm{T}}\mathbf{v}_i = \sum_{n=1}^{N} a_{in}k(\mathbf{x}, \mathbf{x}_n) \tag{12.87}$$

is constant.

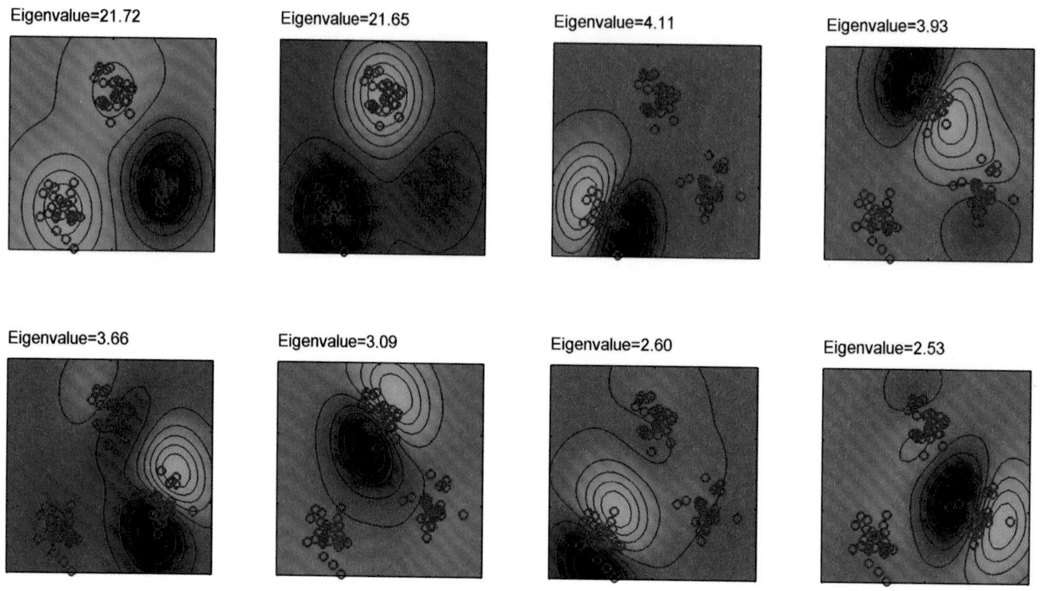

Figure 12.17 Example of kernel PCA, with a Gaussian kernel applied to a synthetic data set in two dimensions, showing the first eight eigenfunctions along with their eigenvalues. The contours are lines along which the projection onto the corresponding principal component is constant. Note how the first two eigenvectors separate the three clusters, the next three eigenvectors split each of the cluster into halves, and the following three eigenvectors again split the clusters into halves along directions orthogonal to the previous splits.

One obvious disadvantage of kernel PCA is that it involves finding the eigenvectors of the $N \times N$ matrix $\widetilde{\mathbf{K}}$ rather than the $D \times D$ matrix $\mathbf{S}$ of conventional linear PCA, and so in practice for large data sets approximations are often used.

Finally, we note that in standard linear PCA, we often retain some reduced number $L < D$ of eigenvectors and then approximate a data vector $\mathbf{x}_n$ by its projection $\widehat{\mathbf{x}}_n$ onto the L-dimensional principal subspace, defined by

$$\widehat{\mathbf{x}}_n = \sum_{i=1}^{L} \left(\mathbf{x}_n^{\mathrm{T}} \mathbf{u}_i\right) \mathbf{u}_i. \tag{12.88}$$

In kernel PCA, this will in general not be possible. To see this, note that the mapping $\boldsymbol{\phi}(\mathbf{x})$ maps the D-dimensional $\mathbf{x}$ space into a D-dimensional *manifold* in the M-dimensional feature space $\boldsymbol{\phi}$. The vector $\mathbf{x}$ is known as the *pre-image* of the corresponding point $\boldsymbol{\phi}(\mathbf{x})$. However, the projection of points in feature space onto the linear PCA subspace in that space will typically not lie on the nonlinear D-dimensional manifold and so will not have a corresponding pre-image in data space. Techniques have therefore been proposed for finding approximate pre-images (Bakir *et al.*, 2004).

12.4. Nonlinear Latent Variable Models

In this chapter, we have focussed on the simplest class of models having continuous latent variables, namely those based on linear-Gaussian distributions. As well as having great practical importance, these models are relatively easy to analyse and to fit to data and can also be used as components in more complex models. Here we consider briefly some generalizations of this framework to models that are either nonlinear or non-Gaussian, or both.

Exercise 12.28

In fact, the issues of nonlinearity and non-Gaussianity are related because a general probability density can be obtained from a simple fixed reference density, such as a Gaussian, by making a nonlinear change of variables. This idea forms the basis of several practical latent variable models as we shall see shortly.

12.4.1 Independent component analysis

We begin by considering models in which the observed variables are related linearly to the latent variables, but for which the latent distribution is non-Gaussian. An important class of such models, known as *independent component analysis*, or *ICA*, arises when we consider a distribution over the latent variables that factorizes, so that

$$p(\mathbf{z}) = \prod_{j=1}^{M} p(z_j). \tag{12.89}$$

To understand the role of such models, consider a situation in which two people are talking at the same time, and we record their voices using two microphones. If we ignore effects such as time delay and echoes, then the signals received by the microphones at any point in time will be given by linear combinations of the amplitudes of the two voices. The coefficients of this linear combination will be constant, and if we can infer their values from sample data, then we can invert the mixing process (assuming it is nonsingular) and thereby obtain two clean signals each of which contains the voice of just one person. This is an example of a problem called *blind source separation* in which 'blind' refers to the fact that we are given only the mixed data, and neither the original sources nor the mixing coefficients are observed (Cardoso, 1998).

This type of problem is sometimes addressed using the following approach (MacKay, 2003) in which we ignore the temporal nature of the signals and treat the successive samples as i.i.d. We consider a generative model in which there are two latent variables corresponding to the unobserved speech signal amplitudes, and there are two observed variables given by the signal values at the microphones. The latent variables have a joint distribution that factorizes as above, and the observed variables are given by a linear combination of the latent variables. There is no need to include a noise distribution because the number of latent variables equals the number of observed variables, and therefore the marginal distribution of the observed variables will not in general be singular, so the observed variables are simply deterministic linear combinations of the latent variables. Given a data set of observations, the likelihood function for this model is a function of the coefficients in the linear combination. The log likelihood can be maximized using gradient-based optimization giving rise to a particular version of independent component analysis.

The success of this approach requires that the latent variables have non-Gaussian distributions. To see this, recall that in probabilistic PCA (and in factor analysis) the latent-space distribution is given by a zero-mean isotropic Gaussian. The model therefore cannot distinguish between two different choices for the latent variables where these differ simply by a rotation in latent space. This can be verified directly by noting that the marginal density (12.35), and hence the likelihood function, is unchanged if we make the transformation $\mathbf{W} \rightarrow \mathbf{WR}$ where $\mathbf{R}$ is an orthogonal matrix satisfying $\mathbf{RR}^{\mathrm{T}} = \mathbf{I}$, because the matrix $\mathbf{C}$ given by (12.36) is itself invariant. Extending the model to allow more general Gaussian latent distributions does not change this conclusion because, as we have seen, such a model is equivalent to the zero-mean isotropic Gaussian latent variable model.

Another way to see why a Gaussian latent variable distribution in a linear model is insufficient to find independent components is to note that the principal components represent a rotation of the coordinate system in data space such as to diagonalize the covariance matrix, so that the data distribution in the new coordinates is then uncorrelated. Although zero correlation is a necessary condition for independence it is not, however, sufficient. In practice, a common choice for the latent-variable distribution is given by

Exercise 12.29

$$p(z_j) = \frac{1}{\pi \cosh(z_j)} = \frac{2}{\pi(e^{z_j} + e^{-z_j})} \tag{12.90}$$

which has heavy tails compared to a Gaussian, reflecting the observation that many real-world distributions also exhibit this property.

The original ICA model (Bell and Sejnowski, 1995) was based on the optimization of an objective function defined by information maximization. One advantage of a probabilistic latent variable formulation is that it helps to motivate and formulate generalizations of basic ICA. For instance, *independent factor analysis* (Attias, 1999a) considers a model in which the number of latent and observed variables can differ, the observed variables are noisy, and the individual latent variables have flexible distributions modelled by mixtures of Gaussians. The log likelihood for this model is maximized using EM, and the reconstruction of the latent variables is approximated using a variational approach. Many other types of model have been considered, and there is now a huge literature on ICA and its applications (Jutten and Herault, 1991; Comon *et al.*, 1991; Amari *et al.*, 1996; Pearlmutter and Parra, 1997; Hyvärinen and Oja, 1997; Hinton *et al.*, 2001; Miskin and MacKay, 2001; Hojen-Sorensen *et al.*, 2002; Choudrey and Roberts, 2003; Chan *et al.*, 2003; Stone, 2004).

12.4.2 Autoassociative neural networks

In Chapter 5 we considered neural networks in the context of supervised learning, where the role of the network is to predict the output variables given values

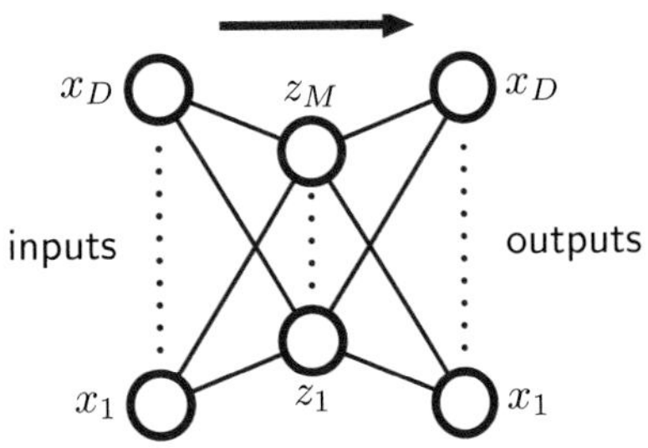

Figure 12.18 An autoassociative multilayer perceptron having two layers of weights. Such a network is trained to map input vectors onto themselves by minimization of a sum-of-squares error. Even with nonlinear units in the hidden layer, such a network is equivalent to linear principal component analysis. Links representing bias parameters have been omitted for clarity.

for the input variables. However, neural networks have also been applied to unsupervised learning where they have been used for dimensionality reduction. This is achieved by using a network having the same number of outputs as inputs, and optimizing the weights so as to minimize some measure of the reconstruction error between inputs and outputs with respect to a set of training data.

Consider first a multilayer perceptron of the form shown in Figure 12.18, having D inputs, D output units and M hidden units, with $M < D$. The targets used to train the network are simply the input vectors themselves, so that the network is attempting to map each input vector onto itself. Such a network is said to form an *autoassociative* mapping. Since the number of hidden units is smaller than the number of inputs, a perfect reconstruction of all input vectors is not in general possible. We therefore determine the network parameters $\mathbf{w}$ by minimizing an error function which captures the degree of mismatch between the input vectors and their reconstructions. In particular, we shall choose a sum-of-squares error of the form

$$E(\mathbf{w}) = \frac{1}{2}\sum_{n=1}^{N} \|\mathbf{y}(\mathbf{x}_n, \mathbf{w}) - \mathbf{x}_n\|^2. \tag{12.91}$$

If the hidden units have linear activation functions, then it can be shown that the error function has a unique global minimum, and that at this minimum the network performs a projection onto the M-dimensional subspace which is spanned by the first M principal components of the data (Bourlard and Kamp, 1988; Baldi and Hornik, 1989). Thus, the vectors of weights which lead into the hidden units in Figure 12.18 form a basis set which spans the principal subspace. Note, however, that these vectors need not be orthogonal or normalized. This result is unsurprising, since both principal component analysis and the neural network are using linear dimensionality reduction and are minimizing the same sum-of-squares error function.

It might be thought that the limitations of a linear dimensionality reduction could be overcome by using nonlinear (sigmoidal) activation functions for the hidden units in the network in Figure 12.18. However, even with nonlinear hidden units, the minimum error solution is again given by the projection onto the principal component subspace (Bourlard and Kamp, 1988). There is therefore no advantage in using two-layer neural networks to perform dimensionality reduction. Standard techniques for principal component analysis (based on singular value decomposition) are guaranteed to give the correct solution in finite time, and they also generate an ordered set of eigenvalues with corresponding orthonormal eigenvectors.

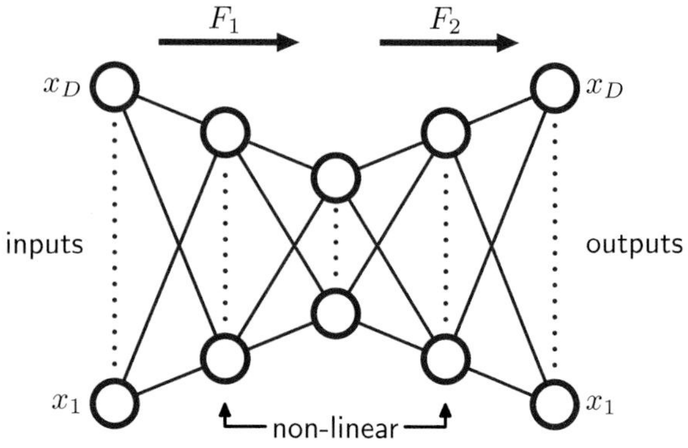

Figure 12.19 Addition of extra hidden layers of nonlinear units gives an autoassociative network which can perform a nonlinear dimensionality reduction.

The situation is different, however, if additional hidden layers are permitted in the network. Consider the four-layer autoassociative network shown in Figure 12.19. Again the output units are linear, and the M units in the second hidden layer can also be linear, however, the first and third hidden layers have sigmoidal nonlinear activation functions. The network is again trained by minimization of the error function (12.91). We can view this network as two successive functional mappings $\mathbf{F}_1$ and $\mathbf{F}_2$, as indicated in Figure 12.19. The first mapping $\mathbf{F}_1$ projects the original D-dimensional data onto an M-dimensional subspace $\mathcal{S}$ defined by the activations of the units in the second hidden layer. Because of the presence of the first hidden layer of nonlinear units, this mapping is very general, and in particular is not restricted to being linear. Similarly, the second half of the network defines an arbitrary functional mapping from the M-dimensional space back into the original D-dimensional input space. This has a simple geometrical interpretation, as indicated for the case $D = 3$ and $M = 2$ in Figure 12.20.

Such a network effectively performs a nonlinear principal component analysis.

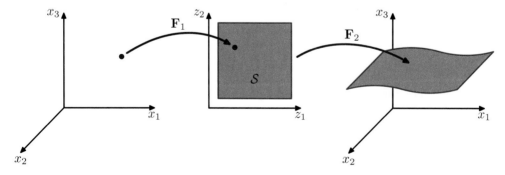

Figure 12.20 Geometrical interpretation of the mappings performed by the network in Figure 12.19 for the case of $D = 3$ inputs and $M = 2$ units in the middle hidden layer. The function $\mathbf{F}_2$ maps from an M-dimensional space $\mathcal{S}$ into a D-dimensional space and therefore defines the way in which the space $\mathcal{S}$ is embedded within the original $\mathbf{x}$-space. Since the mapping $\mathbf{F}_2$ can be nonlinear, the embedding of $\mathcal{S}$ can be nonplanar, as indicated in the figure. The mapping $\mathbf{F}_1$ then defines a projection of points in the original D-dimensional space into the M-dimensional subspace $\mathcal{S}$.

It has the advantage of not being limited to linear transformations, although it contains standard principal component analysis as a special case. However, training the network now involves a nonlinear optimization problem, since the error function (12.91) is no longer a quadratic function of the network parameters. Computationally intensive nonlinear optimization techniques must be used, and there is the risk of finding a suboptimal local minimum of the error function. Also, the dimensionality of the subspace must be specified before training the network.

12.4.3 Modelling nonlinear manifolds

As we have already noted, many natural sources of data correspond to low-dimensional, possibly noisy, nonlinear manifolds embedded within the higher dimensional observed data space. Capturing this property explicitly can lead to improved density modelling compared with more general methods. Here we consider briefly a range of techniques that attempt to do this.

One way to model the nonlinear structure is through a combination of linear models, so that we make a piece-wise linear approximation to the manifold. This can be obtained, for instance, by using a clustering technique such as K-means based on Euclidean distance to partition the data set into local groups with standard PCA applied to each group. A better approach is to use the reconstruction error for cluster assignment (Kambhatla and Leen, 1997; Hinton *et al.*, 1997) as then a common cost function is being optimized in each stage. However, these approaches still suffer from limitations due to the absence of an overall density model. By using probabilistic PCA it is straightforward to define a fully probabilistic model simply by considering a mixture distribution in which the components are probabilistic PCA models (Tipping and Bishop, 1999a). Such a model has both discrete latent variables, corresponding to the discrete mixture, as well as continuous latent variables, and the likelihood function can be maximized using the EM algorithm. A fully Bayesian treatment, based on variational inference (Bishop and Winn, 2000), allows the number of components in the mixture, as well as the effective dimensionalities of the individual models, to be inferred from the data. There are many variants of this model in which parameters such as the $\mathbf{W}$ matrix or the noise variances are tied across components in the mixture, or in which the isotropic noise distributions are replaced by diagonal ones, giving rise to a mixture of factor analysers (Ghahramani and Hinton, 1996a; Ghahramani and Beal, 2000). The mixture of probabilistic PCA models can also be extended hierarchically to produce an interactive data visualization algorithm (Bishop and Tipping, 1998).

An alternative to considering a mixture of linear models is to consider a single nonlinear model. Recall that conventional PCA finds a linear subspace that passes close to the data in a least-squares sense. This concept can be extended to one-dimensional nonlinear surfaces in the form of *principal curves* (Hastie and Stuetzle, 1989). We can describe a curve in a D-dimensional data space using a vector-valued function $\mathbf{f}(\lambda)$, which is a vector each of whose elements is a function of the scalar λ. There are many possible ways to parameterize the curve, of which a natural choice is the arc length along the curve. For any given point $\widehat{\mathbf{x}}$ in data space, we can find the point on the curve that is closest in Euclidean distance. We denote this point by

$\lambda = g_{\mathbf{f}}(\mathbf{x})$ because it depends on the particular curve $\mathbf{f}(\lambda)$. For a continuous data density $p(\mathbf{x})$, a principal curve is defined as one for which every point on the curve is the mean of all those points in data space that project to it, so that

$$\mathbb{E}\left[\mathbf{x}|g_{\mathbf{f}}(\mathbf{x}) = \lambda\right] = \mathbf{f}(\lambda). \tag{12.92}$$

For a given continuous density, there can be many principal curves. In practice, we are interested in finite data sets, and we also wish to restrict attention to smooth curves. Hastie and Stuetzle (1989) propose a two-stage iterative procedure for finding such principal curves, somewhat reminiscent of the EM algorithm for PCA. The curve is initialized using the first principal component, and then the algorithm alternates between a data projection step and curve re-estimation step. In the projection step, each data point is assigned to a value of λ corresponding to the closest point on the curve. Then in the re-estimation step, each point on the curve is given by a weighted average of those points that project to nearby points on the curve, with points closest on the curve given the greatest weight. In the case where the subspace is constrained to be linear, the procedure converges to the first principal component and is equivalent to the power method for finding the largest eigenvector of the covariance matrix. Principal curves can be generalized to multidimensional manifolds called *principal surfaces* although these have found limited use due to the difficulty of data smoothing in higher dimensions even for two-dimensional manifolds.

PCA is often used to project a data set onto a lower-dimensional space, for example two dimensional, for the purposes of visualization. Another linear technique with a similar aim is *multidimensional scaling*, or *MDS* (Cox and Cox, 2000). It finds a low-dimensional projection of the data such as to preserve, as closely as possible, the pairwise distances between data points, and involves finding the eigenvectors of the distance matrix. In the case where the distances are Euclidean, it gives equivalent results to PCA. The MDS concept can be extended to a wide variety of data types specified in terms of a similarity matrix, giving *nonmetric* MDS.

Two other nonprobabilistic methods for dimensionality reduction and data visualization are worthy of mention. *Locally linear embedding*, or *LLE* (Roweis and Saul, 2000) first computes the set of coefficients that best reconstructs each data point from its neighbours. These coefficients are arranged to be invariant to rotations, translations, and scalings of that data point and its neighbours, and hence they characterize the local geometrical properties of the neighbourhood. LLE then maps the high-dimensional data points down to a lower dimensional space while preserving these neighbourhood coefficients. If the local neighbourhood for a particular data point can be considered linear, then the transformation can be achieved using a combination of translation, rotation, and scaling, such as to preserve the angles formed between the data points and their neighbours. Because the weights are invariant to these transformations, we expect the same weight values to reconstruct the data points in the low-dimensional space as in the high-dimensional data space. In spite of the nonlinearity, the optimization for LLE does not exhibit local minima.

In *isometric feature mapping*, or *isomap* (Tenenbaum *et al.*, 2000), the goal is to project the data to a lower-dimensional space using MDS, but where the dissimilarities are defined in terms of the *geodesic distances* measured along the manifold. For instance, if two points lie on a circle, then the geodesic is the arc-length distance measured around the circumference of the circle not the straight line distance measured along the chord connecting them. The algorithm first defines the neighbourhood for each data point, either by finding the K nearest neighbours or by finding all points within a sphere of radius ϵ. A graph is then constructed by linking all neighbouring points and labelling them with their Euclidean distance. The geodesic distance between any pair of points is then approximated by the sum of the arc lengths along the shortest path connecting them (which itself is found using standard algorithms). Finally, metric MDS is applied to the geodesic distance matrix to find the low-dimensional projection.

Our focus in this chapter has been on models for which the observed variables are continuous. We can also consider models having continuous latent variables together with discrete observed variables, giving rise to *latent trait* models (Bartholomew, 1987). In this case, the marginalization over the continuous latent variables, even for a linear relationship between latent and observed variables, cannot be performed analytically, and so more sophisticated techniques are required. Tipping (1999) uses variational inference in a model with a two-dimensional latent space, allowing a binary data set to be visualized analogously to the use of PCA to visualize continuous data. Note that this model is the dual of the Bayesian logistic regression problem discussed in Section 4.5. In the case of logistic regression we have N observations of the feature vector ϕ_n which are parameterized by a single parameter vector $\mathbf{w}$, whereas in the latent space visualization model there is a single latent space variable $\mathbf{x}$ (analogous to ϕ) and N copies of the latent variable $\mathbf{w}_n$. A generalization of probabilistic latent variable models to general exponential family distributions is described in Collins *et al.* (2002).

We have already noted that an arbitrary distribution can be formed by taking a Gaussian random variable and transforming it through a suitable nonlinearity. This is exploited in a general latent variable model called a *density network* (MacKay, 1995; MacKay and Gibbs, 1999) in which the nonlinear function is governed by a multilayered neural network. If the network has enough hidden units, it can approximate a given nonlinear function to any desired accuracy. The downside of having such a flexible model is that the marginalization over the latent variables, required in order to obtain the likelihood function, is no longer analytically tractable. Instead, the likelihood is approximated using Monte Carlo techniques by drawing samples from the Gaussian prior. The marginalization over the latent variables then becomes a simple sum with one term for each sample. However, because a large number of sample points may be required in order to give an accurate representation of the marginal, this procedure can be computationally costly.

Chapter 5

Chapter 11

If we consider more restricted forms for the nonlinear function, and make an appropriate choice of the latent variable distribution, then we can construct a latent variable model that is both nonlinear and efficient to train. The *generative topographic mapping*, or *GTM* (Bishop *et al.*, 1996; Bishop *et al.*, 1997a; Bishop *et al.*, 1998b) uses a latent distribution that is defined by a finite regular grid of delta functions over the (typically two-dimensional) latent space. Marginalization over the latent space then simply involves summing over the contributions from each of the grid locations.

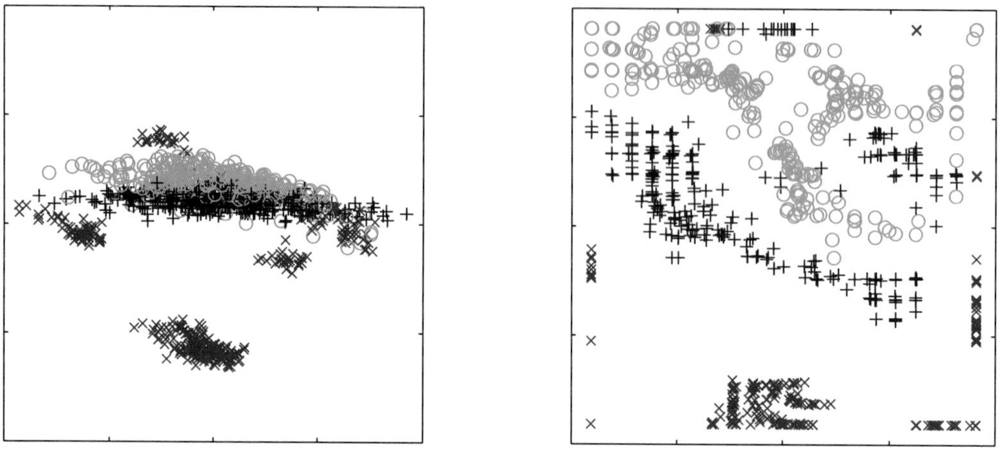

Figure 12.21 Plot of the oil flow data set visualized using PCA on the left and GTM on the right. For the GTM model, each data point is plotted at the mean of its posterior distribution in latent space. The nonlinearity of the GTM model allows the separation between the groups of data points to be seen more clearly.

Chapter 3

Section 1.4

The nonlinear mapping is given by a linear regression model that allows for general nonlinearity while being a linear function of the adaptive parameters. Note that the usual limitation of linear regression models arising from the curse of dimensionality does not arise in the context of the GTM since the manifold generally has two dimensions irrespective of the dimensionality of the data space. A consequence of these two choices is that the likelihood function can be expressed analytically in closed form and can be optimized efficiently using the EM algorithm. The resulting GTM model fits a two-dimensional nonlinear manifold to the data set, and by evaluating the posterior distribution over latent space for the data points, they can be projected back to the latent space for visualization purposes. Figure 12.21 shows a comparison of the oil data set visualized with linear PCA and with the nonlinear GTM.

The GTM can be seen as a probabilistic version of an earlier model called the *self organizing map*, or *SOM* (Kohonen, 1982; Kohonen, 1995), which also represents a two-dimensional nonlinear manifold as a regular array of discrete points. The SOM is somewhat reminiscent of the K-means algorithm in that data points are assigned to nearby prototype vectors that are then subsequently updated. Initially, the prototypes are distributed at random, and during the training process they 'self organize' so as to approximate a smooth manifold. Unlike K-means, however, the SOM is not optimizing any well-defined cost function (Erwin *et al.*, 1992) making it difficult to set the parameters of the model and to assess convergence. There is also no guarantee that the 'self-organization' will take place as this is dependent on the choice of appropriate parameter values for any particular data set.

By contrast, GTM optimizes the log likelihood function, and the resulting model defines a probability density in data space. In fact, it corresponds to a constrained mixture of Gaussians in which the components share a common variance, and the means are constrained to lie on a smooth two-dimensional manifold. This probabilistic foundation also makes it very straightforward to define generalizations of GTM (Bishop *et al.*, 1998a) such as a Bayesian treatment, dealing with missing values, a principled extension to discrete variables, the use of Gaussian processes to define the manifold, or a hierarchical GTM model (Tino and Nabney, 2002).

Section 6.4

Because the manifold in GTM is defined as a continuous surface, not just at the prototype vectors as in the SOM, it is possible to compute the *magnification factors* corresponding to the local expansions and compressions of the manifold needed to fit the data set (Bishop *et al.*, 1997b) as well as the *directional curvatures* of the manifold (Tino *et al.*, 2001). These can be visualized along with the projected data and provide additional insight into the model.

Exercises

12.1 ($\star\star$) **www** In this exercise, we use proof by induction to show that the linear projection onto an M-dimensional subspace that maximizes the variance of the projected data is defined by the M eigenvectors of the data covariance matrix $\mathbf{S}$, given by (12.3), corresponding to the M largest eigenvalues. In Section 12.1, this result was proven for the case of $M = 1$. Now suppose the result holds for some general value of M and show that it consequently holds for dimensionality $M + 1$. To do this, first set the derivative of the variance of the projected data with respect to a vector $\mathbf{u}_{M+1}$ defining the new direction in data space equal to zero. This should be done subject to the constraints that $\mathbf{u}_{M+1}$ be orthogonal to the existing vectors $\mathbf{u}_1, \ldots, \mathbf{u}_M$, and also that it be normalized to unit length. Use Lagrange multipliers to enforce these constraints. Then make use of the orthonormality properties of the vectors $\mathbf{u}_1, \ldots, \mathbf{u}_M$ to show that the new vector $\mathbf{u}_{M+1}$ is an eigenvector of $\mathbf{S}$. Finally, show that the variance is maximized if the eigenvector is chosen to be the one corresponding to eigenvector λ_{M+1} where the eigenvalues have been ordered in decreasing value.

Appendix E

12.2 ($\star\star$) Show that the minimum value of the PCA distortion measure J given by (12.15) with respect to the $\mathbf{u}_i$, subject to the orthonormality constraints (12.7), is obtained when the $\mathbf{u}_i$ are eigenvectors of the data covariance matrix $\mathbf{S}$. To do this, introduce a matrix $\mathbf{H}$ of Lagrange multipliers, one for each constraint, so that the modified distortion measure, in matrix notation reads

$$\widetilde{J} = \mathrm{Tr}\left\{\widehat{\mathbf{U}}^{\mathrm{T}}\mathbf{S}\widehat{\mathbf{U}}\right\} + \mathrm{Tr}\left\{\mathbf{H}(\mathbf{I} - \widehat{\mathbf{U}}^{\mathrm{T}}\widehat{\mathbf{U}})\right\} \tag{12.93}$$

where $\widehat{\mathbf{U}}$ is a matrix of dimension $D \times (D - M)$ whose columns are given by $\mathbf{u}_i$. Now minimize $\widetilde{J}$ with respect to $\widehat{\mathbf{U}}$ and show that the solution satisfies $\mathbf{S}\widehat{\mathbf{U}} = \widehat{\mathbf{U}}\mathbf{H}$. Clearly, one possible solution is that the columns of $\widehat{\mathbf{U}}$ are eigenvectors of $\mathbf{S}$, in which case $\mathbf{H}$ is a diagonal matrix containing the corresponding eigenvalues. To obtain the general solution, show that $\mathbf{H}$ can be assumed to be a symmetric matrix, and by using its eigenvector expansion show that the general solution to $\mathbf{S}\widehat{\mathbf{U}} = \widehat{\mathbf{U}}\mathbf{H}$ gives the same value for $\widetilde{J}$ as the specific solution in which the columns of $\widehat{\mathbf{U}}$ are the eigenvectors of $\mathbf{S}$. Because these solutions are all equivalent, it is convenient to choose the eigenvector solution.

12.3 $(\star)$ Verify that the eigenvectors defined by (12.30) are normalized to unit length, assuming that the eigenvectors $\mathbf{v}_i$ have unit length.

12.4 $(\star)$ **www** Suppose we replace the zero-mean, unit-covariance latent space distribution (12.31) in the probabilistic PCA model by a general Gaussian distribution of the form $\mathcal{N}(\mathbf{z}|\mathbf{m}, \boldsymbol{\Sigma})$. By redefining the parameters of the model, show that this leads to an identical model for the marginal distribution $p(\mathbf{x})$ over the observed variables for any valid choice of $\mathbf{m}$ and $\boldsymbol{\Sigma}$.

12.5 $(\star\star)$ Let $\mathbf{x}$ be a D-dimensional random variable having a Gaussian distribution given by $\mathcal{N}(\mathbf{x}|\boldsymbol{\mu}, \boldsymbol{\Sigma})$, and consider the M-dimensional random variable given by $\mathbf{y} = \mathbf{A}\mathbf{x} + \mathbf{b}$ where $\mathbf{A}$ is an $M \times D$ matrix. Show that $\mathbf{y}$ also has a Gaussian distribution, and find expressions for its mean and covariance. Discuss the form of this Gaussian distribution for $M < D$, for $M = D$, and for $M > D$.

12.6 $(\star)$ **www** Draw a directed probabilistic graph for the probabilistic PCA model described in Section 12.2 in which the components of the observed variable $\mathbf{x}$ are shown explicitly as separate nodes. Hence verify that the probabilistic PCA model has the same independence structure as the naive Bayes model discussed in Section 8.2.2.

12.7 $(\star\star)$ By making use of the results (2.270) and (2.271) for the mean and covariance of a general distribution, derive the result (12.35) for the marginal distribution $p(\mathbf{x})$ in the probabilistic PCA model.

12.8 $(\star\star)$ **www** By making use of the result (2.116), show that the posterior distribution $p(\mathbf{z}|\mathbf{x})$ for the probabilistic PCA model is given by (12.42).

12.9 $(\star)$ Verify that maximizing the log likelihood (12.43) for the probabilistic PCA model with respect to the parameter $\boldsymbol{\mu}$ gives the result $\boldsymbol{\mu}_{\mathrm{ML}} = \overline{\mathbf{x}}$ where $\overline{\mathbf{x}}$ is the mean of the data vectors.

12.10 $(\star\star)$ By evaluating the second derivatives of the log likelihood function (12.43) for the probabilistic PCA model with respect to the parameter $\boldsymbol{\mu}$, show that the stationary point $\boldsymbol{\mu}_{\mathrm{ML}} = \overline{\mathbf{x}}$ represents the unique maximum.

12.11 $(\star\star)$ **www** Show that in the limit $\sigma^2 \to 0$, the posterior mean for the probabilistic PCA model becomes an orthogonal projection onto the principal subspace, as in conventional PCA.

12.12 $(\star\star)$ For $\sigma^2 > 0$ show that the posterior mean in the probabilistic PCA model is shifted towards the origin relative to the orthogonal projection.

12.13 $(\star\star)$ Show that the optimal reconstruction of a data point under probabilistic PCA, according to the least squares projection cost of conventional PCA, is given by

$$\widetilde{\mathbf{x}} = \mathbf{W}_{\mathrm{ML}}(\mathbf{W}_{\mathrm{ML}}^{\mathrm{T}}\mathbf{W}_{\mathrm{ML}})^{-1}\mathbf{M}\mathbb{E}[\mathbf{z}|\mathbf{x}]. \tag{12.94}$$

12.14 ($\star$) The number of independent parameters in the covariance matrix for the probabilistic PCA model with an M-dimensional latent space and a D-dimensional data space is given by (12.51). Verify that in the case of $M = D - 1$, the number of independent parameters is the same as in a general covariance Gaussian, whereas for $M = 0$ it is the same as for a Gaussian with an isotropic covariance.

12.15 ($\star\star$) **www** Derive the M-step equations (12.56) and (12.57) for the probabilistic PCA model by maximization of the expected complete-data log likelihood function given by (12.53).

12.16 ($\star\star\star$) In Figure 12.11, we showed an application of probabilistic PCA to a data set in which some of the data values were missing at random. Derive the EM algorithm for maximizing the likelihood function for the probabilistic PCA model in this situation. Note that the $\{\mathbf{z}_n\}$, as well as the missing data values that are components of the vectors $\{\mathbf{x}_n\}$, are now latent variables. Show that in the special case in which all of the data values are observed, this reduces to the EM algorithm for probabilistic PCA derived in Section 12.2.2.

12.17 ($\star\star$) **www** Let $\mathbf{W}$ be a $D \times M$ matrix whose columns define a linear subspace of dimensionality M embedded within a data space of dimensionality D, and let $\boldsymbol{\mu}$ be a D-dimensional vector. Given a data set $\{\mathbf{x}_n\}$ where $n = 1, \ldots, N$, we can approximate the data points using a linear mapping from a set of M-dimensional vectors $\{\mathbf{z}_n\}$, so that $\mathbf{x}_n$ is approximated by $\mathbf{W}\mathbf{z}_n + \boldsymbol{\mu}$. The associated sum-of-squares reconstruction cost is given by

$$J = \sum_{n=1}^{N} \|\mathbf{x}_n - \boldsymbol{\mu} - \mathbf{W}\mathbf{z}_n\|^2. \tag{12.95}$$

First show that minimizing J with respect to $\boldsymbol{\mu}$ leads to an analogous expression with $\mathbf{x}_n$ and $\mathbf{z}_n$ replaced by zero-mean variables $\mathbf{x}_n - \overline{\mathbf{x}}$ and $\mathbf{z}_n - \overline{\mathbf{z}}$, respectively, where $\overline{\mathbf{x}}$ and $\overline{\mathbf{z}}$ denote sample means. Then show that minimizing J with respect to $\mathbf{z}_n$, where $\mathbf{W}$ is kept fixed, gives rise to the PCA E step (12.58), and that minimizing J with respect to $\mathbf{W}$, where $\{\mathbf{z}_n\}$ is kept fixed, gives rise to the PCA M step (12.59).

12.18 ($\star$) Derive an expression for the number of independent parameters in the factor analysis model described in Section 12.2.4.

12.19 ($\star\star$) **www** Show that the factor analysis model described in Section 12.2.4 is invariant under rotations of the latent space coordinates.

12.20 ($\star\star$) By considering second derivatives, show that the only stationary point of the log likelihood function for the factor analysis model discussed in Section 12.2.4 with respect to the parameter $\boldsymbol{\mu}$ is given by the sample mean defined by (12.1). Furthermore, show that this stationary point is a maximum.

12.21 ($\star\star$) Derive the formulae (12.66) and (12.67) for the E step of the EM algorithm for factor analysis. Note that from the result of Exercise 12.20, the parameter $\boldsymbol{\mu}$ can be replaced by the sample mean $\overline{\mathbf{x}}$.

12.22 ($\star\star$) Write down an expression for the expected complete-data log likelihood function for the factor analysis model, and hence derive the corresponding M step equations (12.69) and (12.70).

12.23 ($\star$) **www** Draw a directed probabilistic graphical model representing a discrete mixture of probabilistic PCA models in which each PCA model has its own values of $\mathbf{W}$, $\boldsymbol{\mu}$, and σ^2. Now draw a modified graph in which these parameter values are shared between the components of the mixture.

12.24 ($\star\star\star$) We saw in Section 2.3.7 that Student's t-distribution can be viewed as an infinite mixture of Gaussians in which we marginalize with respect to a continuous latent variable. By exploiting this representation, formulate an EM algorithm for maximizing the log likelihood function for a multivariate Student's t-distribution given an observed set of data points, and derive the forms of the E and M step equations.

12.25 ($\star\star$) **www** Consider a linear-Gaussian latent-variable model having a latent space distribution $p(\mathbf{z}) = \mathcal{N}(\mathbf{x}|\mathbf{0}, \mathbf{I})$ and a conditional distribution for the observed variable $p(\mathbf{x}|\mathbf{z}) = \mathcal{N}(\mathbf{x}|\mathbf{W}\mathbf{z} + \boldsymbol{\mu}, \boldsymbol{\Phi})$ where $\boldsymbol{\Phi}$ is an arbitrary symmetric, positive-definite noise covariance matrix. Now suppose that we make a nonsingular linear transformation of the data variables $\mathbf{x} \to \mathbf{A}\mathbf{x}$, where $\mathbf{A}$ is a $D \times D$ matrix. If $\boldsymbol{\mu}_{\mathrm{ML}}$, $\mathbf{W}_{\mathrm{ML}}$ and $\boldsymbol{\Phi}_{\mathrm{ML}}$ represent the maximum likelihood solution corresponding to the original untransformed data, show that $\mathbf{A}\boldsymbol{\mu}_{\mathrm{ML}}$, $\mathbf{A}\mathbf{W}_{\mathrm{ML}}$, and $\mathbf{A}\boldsymbol{\Phi}_{\mathrm{ML}}\mathbf{A}^{\mathrm{T}}$ will represent the corresponding maximum likelihood solution for the transformed data set. Finally, show that the form of the model is preserved in two cases: (i) $\mathbf{A}$ is a diagonal matrix and $\boldsymbol{\Phi}$ is a diagonal matrix. This corresponds to the case of factor analysis. The transformed $\boldsymbol{\Phi}$ remains diagonal, and hence factor analysis is *covariant* under component-wise re-scaling of the data variables; (ii) $\mathbf{A}$ is orthogonal and $\boldsymbol{\Phi}$ is proportional to the unit matrix so that $\boldsymbol{\Phi} = \sigma^2\mathbf{I}$. This corresponds to probabilistic PCA. The transformed $\boldsymbol{\Phi}$ matrix remains proportional to the unit matrix, and hence probabilistic PCA is covariant under a rotation of the axes of data space, as is the case for conventional PCA.

12.26 ($\star\star$) Show that any vector $\mathbf{a}_i$ that satisfies (12.80) will also satisfy (12.79). Also, show that for any solution of (12.80) having eigenvalue λ, we can add any multiple of an eigenvector of $\mathbf{K}$ having zero eigenvalue, and obtain a solution to (12.79) that also has eigenvalue λ. Finally, show that such modifications do not affect the principal-component projection given by (12.82).

12.27 ($\star\star$) Show that the conventional linear PCA algorithm is recovered as a special case of kernel PCA if we choose the linear kernel function given by $k(\mathbf{x}, \mathbf{x}') = \mathbf{x}^{\mathrm{T}}\mathbf{x}'$.

12.28 ($\star\star$) **www** Use the transformation property (1.27) of a probability density under a change of variable to show that any density $p(y)$ can be obtained from a fixed density $q(x)$ that is everywhere nonzero by making a nonlinear change of variable $y = f(x)$ in which $f(x)$ is a monotonic function so that $0 \leqslant f'(x) < \infty$. Write down the differential equation satisfied by $f(x)$ and draw a diagram illustrating the transformation of the density.

12.29 ($\star\star$) **www** Suppose that two variables z_1 and z_2 are independent so that $p(z_1, z_2) = p(z_1)p(z_2)$. Show that the covariance matrix between these variables is diagonal. This shows that independence is a sufficient condition for two variables to be uncorrelated. Now consider two variables y_1 and y_2 where y_1 is symmetrically distributed around 0 and $y_2 = y_1^2$. Write down the conditional distribution $p(y_2|y_1)$ and observe that this is dependent on y_1, showing that the two variables are not independent. Now show that the covariance matrix between these two variables is again diagonal. To do this, use the relation $p(y_1, y_2) = p(y_1)p(y_2|y_1)$ to show that the off-diagonal terms are zero. This counter-example shows that zero correlation is not a sufficient condition for independence.

13 Sequential Data

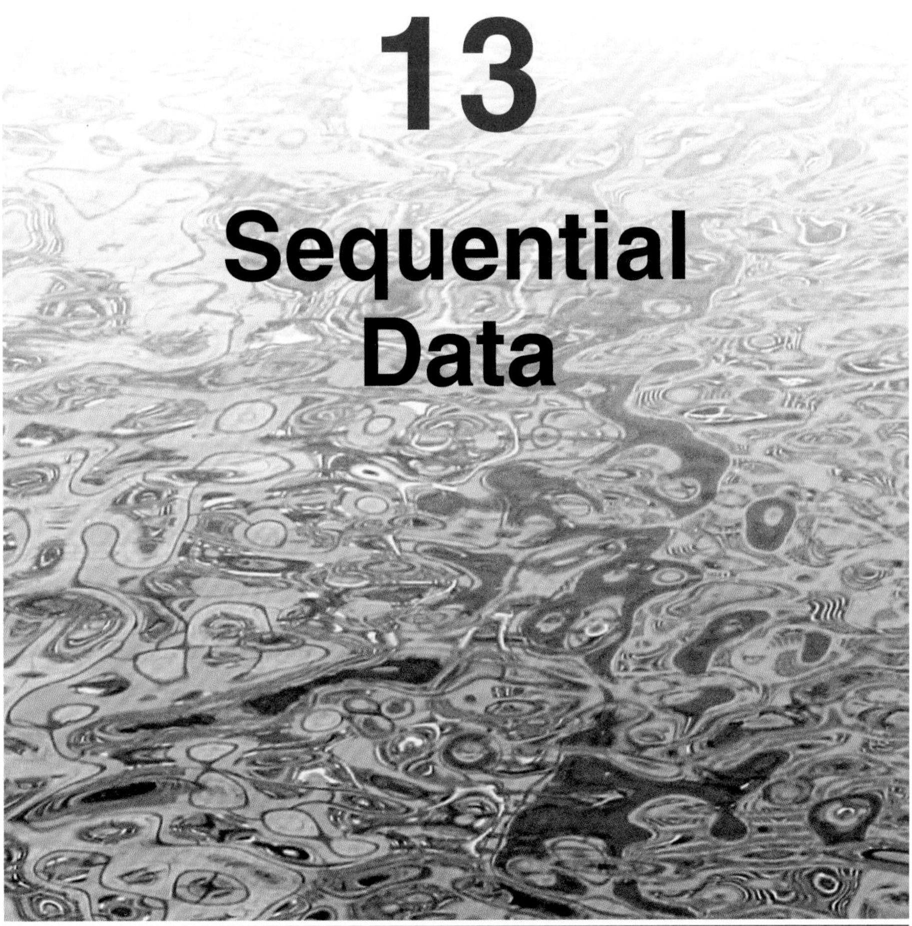

So far in this book, we have focussed primarily on sets of data points that were assumed to be independent and identically distributed (i.i.d.). This assumption allowed us to express the likelihood function as the product over all data points of the probability distribution evaluated at each data point. For many applications, however, the i.i.d. assumption will be a poor one. Here we consider a particularly important class of such data sets, namely those that describe sequential data. These often arise through measurement of time series, for example the rainfall measurements on successive days at a particular location, or the daily values of a currency exchange rate, or the acoustic features at successive time frames used for speech recognition. An example involving speech data is shown in Figure 13.1. Sequential data can also arise in contexts other than time series, for example the sequence of nucleotide base pairs along a strand of DNA or the sequence of characters in an English sentence. For convenience, we shall sometimes refer to 'past' and 'future' observations in a sequence. However, the models explored in this chapter are equally applicable to all

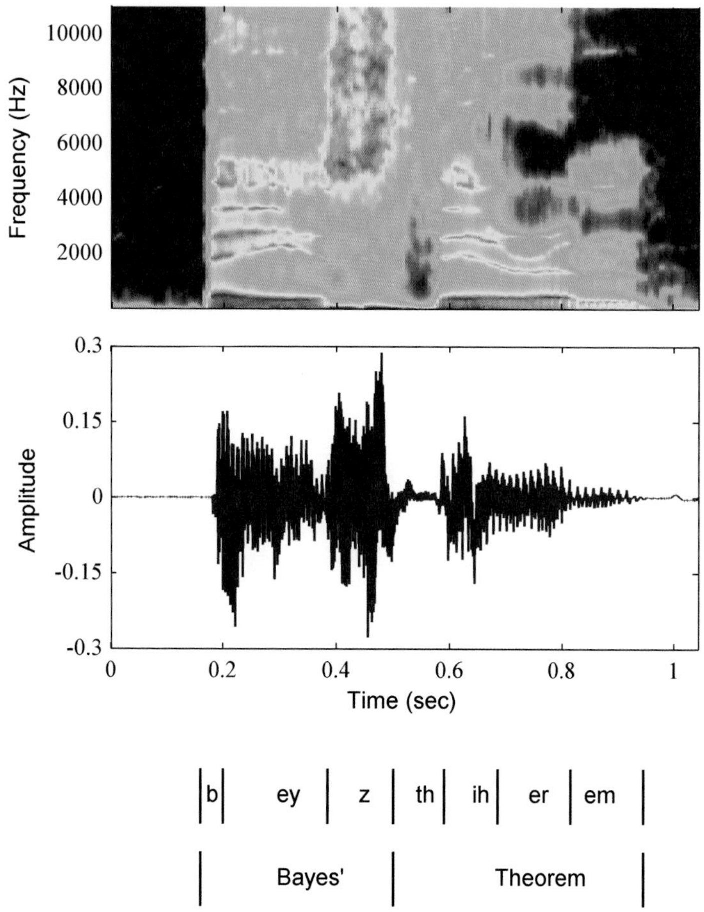

Figure 13.1 Example of a spectrogram of the spoken words "Bayes' theorem" showing a plot of the intensity of the spectral coefficients versus time index.

forms of sequential data, not just temporal sequences.

It is useful to distinguish between stationary and nonstationary sequential distributions. In the stationary case, the data evolves in time, but the distribution from which it is generated remains the same. For the more complex nonstationary situation, the generative distribution itself is evolving with time. Here we shall focus on the stationary case.

For many applications, such as financial forecasting, we wish to be able to predict the next value in a time series given observations of the previous values. Intuitively, we expect that recent observations are likely to be more informative than more historical observations in predicting future values. The example in Figure 13.1 shows that successive observations of the speech spectrum are indeed highly correlated. Furthermore, it would be impractical to consider a general dependence of future observations on all previous observations because the complexity of such a model would grow without limit as the number of observations increases. This leads us to consider *Markov models* in which we assume that future predictions are inde-

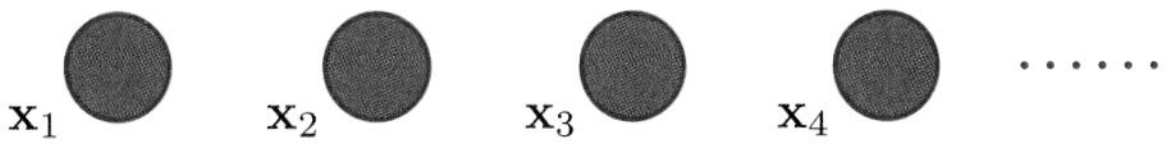

Figure 13.2 The simplest approach to modelling a sequence of observations is to treat them as independent, corresponding to a graph without links.

pendent of all but the most recent observations.

Although such models are tractable, they are also severely limited. We can obtain a more general framework, while still retaining tractability, by the introduction of latent variables, leading to *state space models*. As in Chapters 9 and 12, we shall see that complex models can thereby be constructed from simpler components (in particular, from distributions belonging to the exponential family) and can be readily characterized using the framework of probabilistic graphical models. Here we focus on the two most important examples of state space models, namely the *hidden Markov model*, in which the latent variables are discrete, and *linear dynamical systems*, in which the latent variables are Gaussian. Both models are described by directed graphs having a tree structure (no loops) for which inference can be performed efficiently using the sum-product algorithm.

13.1. Markov Models

The easiest way to treat sequential data would be simply to ignore the sequential aspects and treat the observations as i.i.d., corresponding to the graph in Figure 13.2. Such an approach, however, would fail to exploit the sequential patterns in the data, such as correlations between observations that are close in the sequence. Suppose, for instance, that we observe a binary variable denoting whether on a particular day it rained or not. Given a time series of recent observations of this variable, we wish to predict whether it will rain on the next day. If we treat the data as i.i.d., then the only information we can glean from the data is the relative frequency of rainy days. However, we know in practice that the weather often exhibits trends that may last for several days. Observing whether or not it rains today is therefore of significant help in predicting if it will rain tomorrow.

To express such effects in a probabilistic model, we need to relax the i.i.d. assumption, and one of the simplest ways to do this is to consider a *Markov model*. First of all we note that, without loss of generality, we can use the product rule to express the joint distribution for a sequence of observations in the form

$$p(\mathbf{x}_1, \ldots, \mathbf{x}_N) = \prod_{n=2}^{N} p(\mathbf{x}_n | \mathbf{x}_1, \ldots, \mathbf{x}_{n-1}). \tag{13.1}$$

If we now assume that each of the conditional distributions on the right-hand side is independent of all previous observations except the most recent, we obtain the *first-order Markov chain*, which is depicted as a graphical model in Figure 13.3. The

Figure 13.3 A first-order Markov chain of observations $\{\mathbf{x}_n\}$ in which the distribution $p(\mathbf{x}_n|\mathbf{x}_{n-1})$ of a particular observation $\mathbf{x}_n$ is conditioned on the value of the previous observation $\mathbf{x}_{n-1}$.

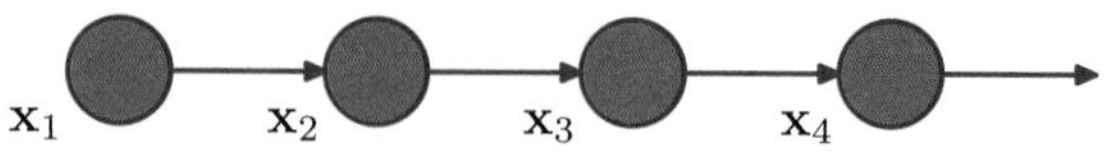

joint distribution for a sequence of N observations under this model is given by

$$p(\mathbf{x}_1, \ldots, \mathbf{x}_N) = p(\mathbf{x}_1) \prod_{n=2}^{N} p(\mathbf{x}_n|\mathbf{x}_{n-1}). \tag{13.2}$$

Section 8.2

From the d-separation property, we see that the conditional distribution for observation $\mathbf{x}_n$, given all of the observations up to time n, is given by

$$p(\mathbf{x}_n|\mathbf{x}_1, \ldots, \mathbf{x}_{n-1}) = p(\mathbf{x}_n|\mathbf{x}_{n-1}) \tag{13.3}$$

Exercise 13.1

which is easily verified by direct evaluation starting from (13.2) and using the product rule of probability. Thus if we use such a model to predict the next observation in a sequence, the distribution of predictions will depend only on the value of the immediately preceding observation and will be independent of all earlier observations.

In most applications of such models, the conditional distributions $p(\mathbf{x}_n|\mathbf{x}_{n-1})$ that define the model will be constrained to be equal, corresponding to the assumption of a stationary time series. The model is then known as a *homogeneous* Markov chain. For instance, if the conditional distributions depend on adjustable parameters (whose values might be inferred from a set of training data), then all of the conditional distributions in the chain will share the same values of those parameters.

Although this is more general than the independence model, it is still very restrictive. For many sequential observations, we anticipate that the trends in the data over several successive observations will provide important information in predicting the next value. One way to allow earlier observations to have an influence is to move to higher-order Markov chains. If we allow the predictions to depend also on the previous-but-one value, we obtain a second-order Markov chain, represented by the graph in Figure 13.4. The joint distribution is now given by

$$p(\mathbf{x}_1, \ldots, \mathbf{x}_N) = p(\mathbf{x}_1)p(\mathbf{x}_2|\mathbf{x}_1) \prod_{n=3}^{N} p(\mathbf{x}_n|\mathbf{x}_{n-1}, \mathbf{x}_{n-2}). \tag{13.4}$$

Again, using d-separation or by direct evaluation, we see that the conditional distribution of $\mathbf{x}_n$ given $\mathbf{x}_{n-1}$ and $\mathbf{x}_{n-2}$ is independent of all observations $\mathbf{x}_1, \ldots \mathbf{x}_{n-3}$.

Figure 13.4 A second-order Markov chain, in which the conditional distribution of a particular observation $\mathbf{x}_n$ depends on the values of the two previous observations $\mathbf{x}_{n-1}$ and $\mathbf{x}_{n-2}$.

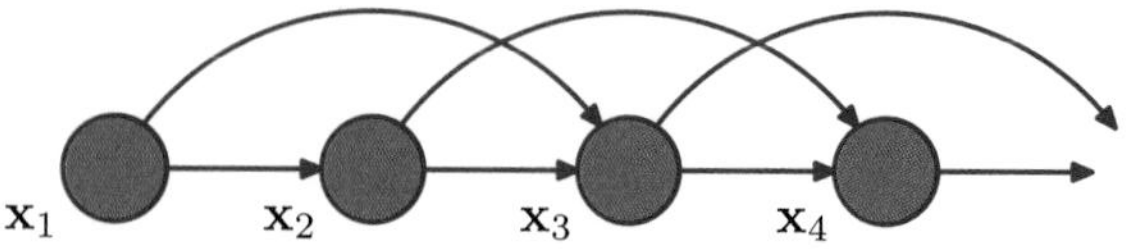

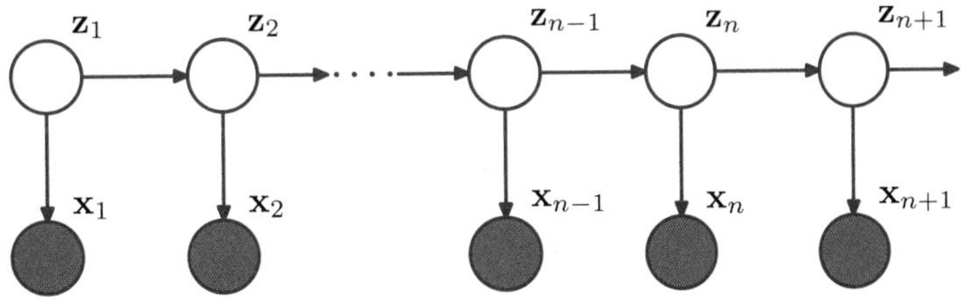

Figure 13.5 We can represent sequential data using a Markov chain of latent variables, with each observation conditioned on the state of the corresponding latent variable. This important graphical structure forms the foundation both for the hidden Markov model and for linear dynamical systems.

Each observation is now influenced by two previous observations. We can similarly consider extensions to an Mth order Markov chain in which the conditional distribution for a particular variable depends on the previous M variables. However, we have paid a price for this increased flexibility because the number of parameters in the model is now much larger. Suppose the observations are discrete variables having K states. Then the conditional distribution $p(\mathbf{x}_n|\mathbf{x}_{n-1})$ in a first-order Markov chain will be specified by a set of $K-1$ parameters for each of the K states of $\mathbf{x}_{n-1}$ giving a total of $K(K-1)$ parameters. Now suppose we extend the model to an Mth order Markov chain, so that the joint distribution is built up from conditionals $p(\mathbf{x}_n|\mathbf{x}_{n-M},\ldots,\mathbf{x}_{n-1})$. If the variables are discrete, and if the conditional distributions are represented by general conditional probability tables, then the number of parameters in such a model will have $K^M(K-1)$ parameters. Because this grows exponentially with M, it will often render this approach impractical for larger values of M.

For continuous variables, we can use linear-Gaussian conditional distributions in which each node has a Gaussian distribution whose mean is a linear function of its parents. This is known as an *autoregressive* or *AR* model (Box *et al.*, 1994; Thiesson *et al.*, 2004). An alternative approach is to use a parametric model for $p(\mathbf{x}_n|\mathbf{x}_{n-M},\ldots,\mathbf{x}_{n-1})$ such as a neural network. This technique is sometimes called a *tapped delay line* because it corresponds to storing (delaying) the previous M values of the observed variable in order to predict the next value. The number of parameters can then be much smaller than in a completely general model (for example it may grow linearly with M), although this is achieved at the expense of a restricted family of conditional distributions.

Suppose we wish to build a model for sequences that is not limited by the Markov assumption to any order and yet that can be specified using a limited number of free parameters. We can achieve this by introducing additional latent variables to permit a rich class of models to be constructed out of simple components, as we did with mixture distributions in Chapter 9 and with continuous latent variable models in Chapter 12. For each observation $\mathbf{x}_n$, we introduce a corresponding latent variable $\mathbf{z}_n$ (which may be of different type or dimensionality to the observed variable). We now assume that it is the latent variables that form a Markov chain, giving rise to the graphical structure known as a *state space model*, which is shown in Figure 13.5. It satisfies the key conditional independence property that $\mathbf{z}_{n-1}$ and $\mathbf{z}_{n+1}$ are independent given $\mathbf{z}_n$, so that

$$\mathbf{z}_{n+1} \perp\!\!\!\perp \mathbf{z}_{n-1} \mid \mathbf{z}_n. \tag{13.5}$$

The joint distribution for this model is given by

$$p(\mathbf{x}_1, \ldots, \mathbf{x}_N, \mathbf{z}_1, \ldots, \mathbf{z}_N) = p(\mathbf{z}_1) \left[\prod_{n=2}^{N} p(\mathbf{z}_n|\mathbf{z}_{n-1}) \right] \prod_{n=1}^{N} p(\mathbf{x}_n|\mathbf{z}_n). \tag{13.6}$$

Using the d-separation criterion, we see that there is always a path connecting any two observed variables $\mathbf{x}_n$ and $\mathbf{x}_m$ via the latent variables, and that this path is never blocked. Thus the predictive distribution $p(\mathbf{x}_{n+1}|\mathbf{x}_1, \ldots, \mathbf{x}_n)$ for observation $\mathbf{x}_{n+1}$ given all previous observations does not exhibit any conditional independence properties, and so our predictions for $\mathbf{x}_{n+1}$ depends on all previous observations. The observed variables, however, do not satisfy the Markov property at any order. We shall discuss how to evaluate the predictive distribution in later sections of this chapter.

There are two important models for sequential data that are described by this graph. If the latent variables are discrete, then we obtain the *hidden Markov model*, or *HMM* (Elliott *et al.*, 1995). Note that the observed variables in an HMM may be discrete or continuous, and a variety of different conditional distributions can be used to model them. If both the latent and the observed variables are Gaussian (with a linear-Gaussian dependence of the conditional distributions on their parents), then we obtain the *linear dynamical system*.

Section 13.2

Section 13.3

13.2. Hidden Markov Models

The hidden Markov model can be viewed as a specific instance of the state space model of Figure 13.5 in which the latent variables are discrete. However, if we examine a single time slice of the model, we see that it corresponds to a mixture distribution, with component densities given by $p(\mathbf{x}|\mathbf{z})$. It can therefore also be interpreted as an extension of a mixture model in which the choice of mixture component for each observation is not selected independently but depends on the choice of component for the previous observation. The HMM is widely used in speech recognition (Jelinek, 1997; Rabiner and Juang, 1993), natural language modelling (Manning and Schütze, 1999), on-line handwriting recognition (Nag *et al.*, 1986), and for the analysis of biological sequences such as proteins and DNA (Krogh *et al.*, 1994; Durbin *et al.*, 1998; Baldi and Brunak, 2001).

As in the case of a standard mixture model, the latent variables are the discrete multinomial variables $\mathbf{z}_n$ describing which component of the mixture is responsible for generating the corresponding observation $\mathbf{x}_n$. Again, it is convenient to use a 1-of-K coding scheme, as used for mixture models in Chapter 9. We now allow the probability distribution of $\mathbf{z}_n$ to depend on the state of the previous latent variable $\mathbf{z}_{n-1}$ through a conditional distribution $p(\mathbf{z}_n|\mathbf{z}_{n-1})$. Because the latent variables are K-dimensional binary variables, this conditional distribution corresponds to a table of numbers that we denote by $\mathbf{A}$, the elements of which are known as *transition probabilities*. They are given by $A_{jk} \equiv p(z_{nk} = 1|z_{n-1,j} = 1)$, and because they are probabilities, they satisfy $0 \leqslant A_{jk} \leqslant 1$ with $\sum_k A_{jk} = 1$, so that the matrix $\mathbf{A}$

Figure 13.6 Transition diagram showing a model whose latent variables have three possible states corresponding to the three boxes. The black lines denote the elements of the transition matrix A_{jk}.

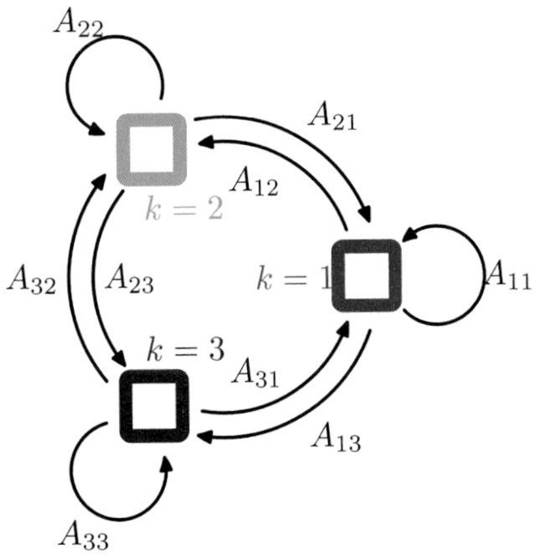

has $K(K-1)$ independent parameters. We can then write the conditional distribution explicitly in the form

$$p(\mathbf{z}_n|\mathbf{z}_{n-1}, \mathbf{A}) = \prod_{k=1}^{K}\prod_{j=1}^{K} A_{jk}^{z_{n-1,j}z_{nk}}. \tag{13.7}$$

The initial latent node $\mathbf{z}_1$ is special in that it does not have a parent node, and so it has a marginal distribution $p(\mathbf{z}_1)$ represented by a vector of probabilities $\boldsymbol{\pi}$ with elements $\pi_k \equiv p(z_{1k} = 1)$, so that

$$p(\mathbf{z}_1|\boldsymbol{\pi}) = \prod_{k=1}^{K} \pi_k^{z_{1k}} \tag{13.8}$$

where $\sum_k \pi_k = 1$.

The transition matrix is sometimes illustrated diagrammatically by drawing the states as nodes in a state transition diagram as shown in Figure 13.6 for the case of $K = 3$. Note that this does not represent a probabilistic graphical model, because the nodes are not separate variables but rather states of a single variable, and so we have shown the states as boxes rather than circles.

It is sometimes useful to take a state transition diagram, of the kind shown in Figure 13.6, and unfold it over time. This gives an alternative representation of the transitions between latent states, known as a *lattice* or *trellis* diagram, and which is shown for the case of the hidden Markov model in Figure 13.7.

Section 8.4.5

The specification of the probabilistic model is completed by defining the conditional distributions of the observed variables $p(\mathbf{x}_n|\mathbf{z}_n, \boldsymbol{\phi})$, where $\boldsymbol{\phi}$ is a set of parameters governing the distribution. These are known as *emission probabilities*, and might for example be given by Gaussians of the form (9.11) if the elements of $\mathbf{x}$ are continuous variables, or by conditional probability tables if $\mathbf{x}$ is discrete. Because $\mathbf{x}_n$ is observed, the distribution $p(\mathbf{x}_n|\mathbf{z}_n, \boldsymbol{\phi})$ consists, for a given value of $\boldsymbol{\phi}$, of a vector of K numbers corresponding to the K possible states of the binary vector $\mathbf{z}_n$.

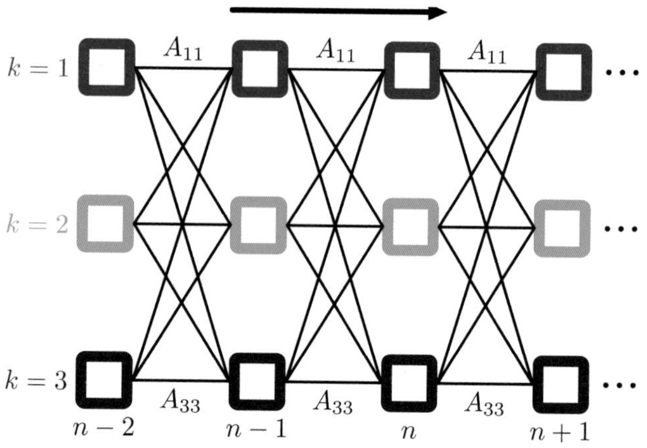

Figure 13.7 If we unfold the state transition diagram of Figure 13.6 over time, we obtain a lattice, or trellis, representation of the latent states. Each column of this diagram corresponds to one of the latent variables $\mathbf{z}_n$.

We can represent the emission probabilities in the form

$$p(\mathbf{x}_n|\mathbf{z}_n, \boldsymbol{\phi}) = \prod_{k=1}^{K} p(\mathbf{x}_n|\boldsymbol{\phi}_k)^{z_{nk}}. \tag{13.9}$$

We shall focus attention on *homogeneous* models for which all of the conditional distributions governing the latent variables share the same parameters $\mathbf{A}$, and similarly all of the emission distributions share the same parameters $\boldsymbol{\phi}$ (the extension to more general cases is straightforward). Note that a mixture model for an i.i.d. data set corresponds to the special case in which the parameters A_{jk} are the same for all values of j, so that the conditional distribution $p(\mathbf{z}_n|\mathbf{z}_{n-1})$ is independent of $\mathbf{z}_{n-1}$. This corresponds to deleting the horizontal links in the graphical model shown in Figure 13.5.

The joint probability distribution over both latent and observed variables is then given by

$$p(\mathbf{X}, \mathbf{Z}|\boldsymbol{\theta}) = p(\mathbf{z}_1|\boldsymbol{\pi}) \left[\prod_{n=2}^{N} p(\mathbf{z}_n|\mathbf{z}_{n-1}, \mathbf{A})\right] \prod_{m=1}^{N} p(\mathbf{x}_m|\mathbf{z}_m, \boldsymbol{\phi}) \tag{13.10}$$

where $\mathbf{X} = \{\mathbf{x}_1, \ldots, \mathbf{x}_N\}$, $\mathbf{Z} = \{\mathbf{z}_1, \ldots, \mathbf{z}_N\}$, and $\boldsymbol{\theta} = \{\boldsymbol{\pi}, \mathbf{A}, \boldsymbol{\phi}\}$ denotes the set of parameters governing the model. Most of our discussion of the hidden Markov model will be independent of the particular choice of the emission probabilities. Indeed, the model is tractable for a wide range of emission distributions including discrete tables, Gaussians, and mixtures of Gaussians. It is also possible to exploit discriminative models such as neural networks. These can be used to model the emission density $p(\mathbf{x}|\mathbf{z})$ directly, or to provide a representation for $p(\mathbf{z}|\mathbf{x})$ that can be converted into the required emission density $p(\mathbf{x}|\mathbf{z})$ using Bayes' theorem (Bishop *et al.*, 2004).

Exercise 13.4

We can gain a better understanding of the hidden Markov model by considering it from a generative point of view. Recall that to generate samples from a mixture of

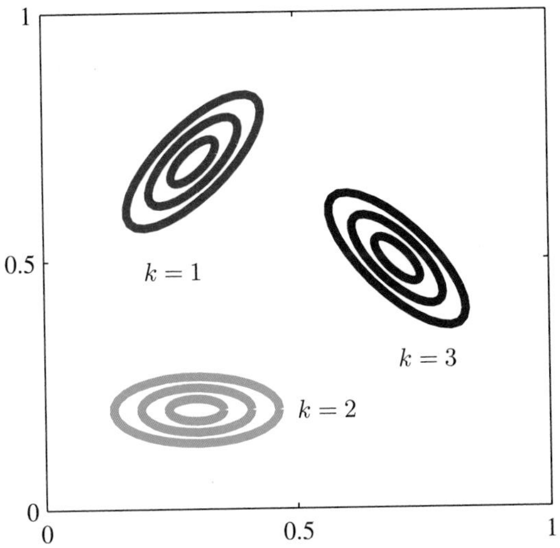

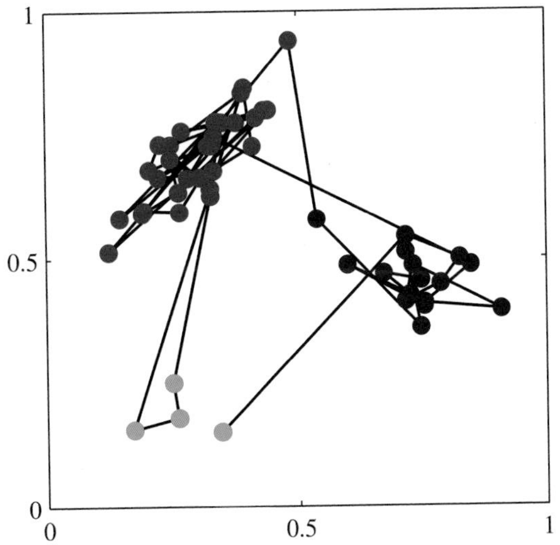

Figure 13.8 Illustration of sampling from a hidden Markov model having a 3-state latent variable $\mathbf{z}$ and a Gaussian emission model $p(\mathbf{x}|\mathbf{z})$ where $\mathbf{x}$ is 2-dimensional. (a) Contours of constant probability density for the emission distributions corresponding to each of the three states of the latent variable. (b) A sample of 50 points drawn from the hidden Markov model, colour coded according to the component that generated them and with lines connecting the successive observations. Here the transition matrix was fixed so that in any state there is a 5% probability of making a transition to each of the other states, and consequently a 90% probability of remaining in the same state.

Gaussians, we first chose one of the components at random with probability given by the mixing coefficients π_k and then generate a sample vector $\mathbf{x}$ from the corresponding Gaussian component. This process is repeated N times to generate a data set of N independent samples. In the case of the hidden Markov model, this procedure is modified as follows. We first choose the initial latent variable $\mathbf{z}_1$ with probabilities governed by the parameters π_k and then sample the corresponding observation $\mathbf{x}_1$. Now we choose the state of the variable $\mathbf{z}_2$ according to the transition probabilities $p(\mathbf{z}_2|\mathbf{z}_1)$ using the already instantiated value of $\mathbf{z}_1$. Thus suppose that the sample for $\mathbf{z}_1$ corresponds to state j. Then we choose the state k of $\mathbf{z}_2$ with probabilities A_{jk} for $k = 1, \ldots, K$. Once we know $\mathbf{z}_2$ we can draw a sample for $\mathbf{x}_2$ and also sample the next latent variable $\mathbf{z}_3$ and so on. This is an example of ancestral sampling for a directed graphical model. If, for instance, we have a model in which the diagonal transition elements A_{kk} are much larger than the off-diagonal elements, then a typical data sequence will have long runs of points generated from a single component, with infrequent transitions from one component to another. The generation of samples from a hidden Markov model is illustrated in Figure 13.8.

Section 8.1.2

There are many variants of the standard HMM model, obtained for instance by imposing constraints on the form of the transition matrix $\mathbf{A}$ (Rabiner, 1989). Here we mention one of particular practical importance called the *left-to-right* HMM, which is obtained by setting the elements A_{jk} of $\mathbf{A}$ to zero if $k < j$, as illustrated in the

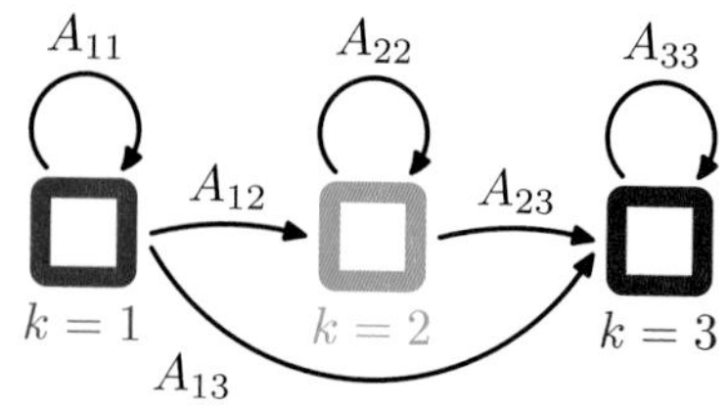

Figure 13.9 Example of the state transition diagram for a 3-state left-to-right hidden Markov model. Note that once a state has been vacated, it cannot later be re-entered.

state transition diagram for a 3-state HMM in Figure 13.9. Typically for such models the initial state probabilities for $p(\mathbf{z}_1)$ are modified so that $p(z_{11}) = 1$ and $p(z_{1j}) = 0$ for $j \neq 1$, in other words every sequence is constrained to start in state $j = 1$. The transition matrix may be further constrained to ensure that large changes in the state index do not occur, so that $A_{jk} = 0$ if $k > j + \Delta$. This type of model is illustrated using a lattice diagram in Figure 13.10.

Many applications of hidden Markov models, for example speech recognition, or on-line character recognition, make use of left-to-right architectures. As an illustration of the left-to-right hidden Markov model, we consider an example involving handwritten digits. This uses on-line data, meaning that each digit is represented by the trajectory of the pen as a function of time in the form of a sequence of pen coordinates, in contrast to the off-line digits data, discussed in Appendix A, which comprises static two-dimensional pixellated images of the ink. Examples of the on-line digits are shown in Figure 13.11. Here we train a hidden Markov model on a subset of data comprising 45 examples of the digit '2'. There are $K = 16$ states, each of which can generate a line segment of fixed length having one of 16 possible angles, and so the emission distribution is simply a 16×16 table of probabilities associated with the allowed angle values for each state index value. Transition probabilities are all set to zero except for those that keep the state index k the same or that increment it by 1, and the model parameters are optimized using 25 iterations of EM. We can gain some insight into the resulting model by running it generatively, as shown in Figure 13.11.

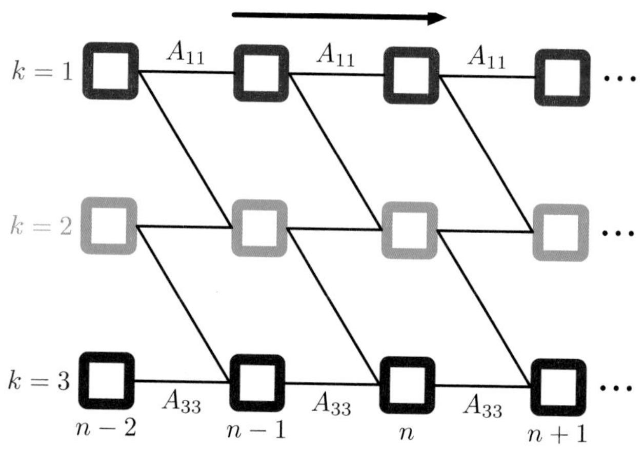

Figure 13.10 Lattice diagram for a 3-state left-to-right HMM in which the state index k is allowed to increase by at most 1 at each transition.

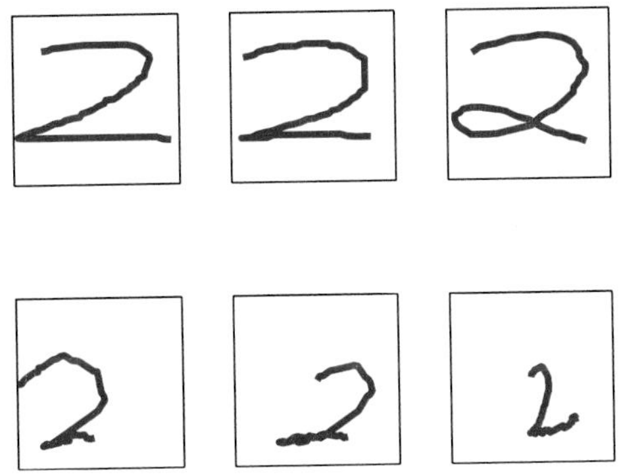

Figure 13.11 Top row: examples of on-line handwritten digits. Bottom row: synthetic digits sampled generatively from a left-to-right hidden Markov model that has been trained on a data set of 45 handwritten digits.

One of the most powerful properties of hidden Markov models is their ability to exhibit some degree of invariance to local warping (compression and stretching) of the time axis. To understand this, consider the way in which the digit '2' is written in the on-line handwritten digits example. A typical digit comprises two distinct sections joined at a cusp. The first part of the digit, which starts at the top left, has a sweeping arc down to the cusp or loop at the bottom left, followed by a second more-or-less straight sweep ending at the bottom right. Natural variations in writing style will cause the relative sizes of the two sections to vary, and hence the location of the cusp or loop within the temporal sequence will vary. From a generative perspective such variations can be accommodated by the hidden Markov model through changes in the number of transitions to the same state versus the number of transitions to the successive state. Note, however, that if a digit '2' is written in the reverse order, that is, starting at the bottom right and ending at the top left, then even though the pen tip coordinates may be identical to an example from the training set, the probability of the observations under the model will be extremely small. In the speech recognition context, warping of the time axis is associated with natural variations in the speed of speech, and again the hidden Markov model can accommodate such a distortion and not penalize it too heavily.

13.2.1 Maximum likelihood for the HMM

If we have observed a data set $\mathbf{X} = \{\mathbf{x}_1, \ldots, \mathbf{x}_N\}$, we can determine the parameters of an HMM using maximum likelihood. The likelihood function is obtained from the joint distribution (13.10) by marginalizing over the latent variables

$$p(\mathbf{X}|\boldsymbol{\theta}) = \sum_{\mathbf{Z}} p(\mathbf{X}, \mathbf{Z}|\boldsymbol{\theta}). \tag{13.11}$$

Because the joint distribution $p(\mathbf{X}, \mathbf{Z}|\boldsymbol{\theta})$ does not factorize over n (in contrast to the mixture distribution considered in Chapter 9), we cannot simply treat each of the summations over $\mathbf{z}_n$ independently. Nor can we perform the summations explicitly because there are N variables to be summed over, each of which has K states, resulting in a total of K^N terms. Thus the number of terms in the summation grows exponentially with the length of the chain. In fact, the summation in (13.11) corresponds to summing over exponentially many paths through the lattice diagram in Figure 13.7.

We have already encountered a similar difficulty when we considered the inference problem for the simple chain of variables in Figure 8.32. There we were able to make use of the conditional independence properties of the graph to re-order the summations in order to obtain an algorithm whose cost scales linearly, instead of exponentially, with the length of the chain. We shall apply a similar technique to the hidden Markov model.

A further difficulty with the expression (13.11) for the likelihood function is that, because it corresponds to a generalization of a mixture distribution, it represents a summation over the emission models for different settings of the latent variables. Direct maximization of the likelihood function will therefore lead to complex expressions with no closed-form solutions, as was the case for simple mixture models (recall that a mixture model for i.i.d. data is a special case of the HMM).

Section 9.2

We therefore turn to the expectation maximization algorithm to find an efficient framework for maximizing the likelihood function in hidden Markov models. The EM algorithm starts with some initial selection for the model parameters, which we denote by $\boldsymbol{\theta}^{\text{old}}$. In the E step, we take these parameter values and find the posterior distribution of the latent variables $p(\mathbf{Z}|\mathbf{X}, \boldsymbol{\theta}^{\text{old}})$. We then use this posterior distribution to evaluate the expectation of the logarithm of the complete-data likelihood function, as a function of the parameters $\boldsymbol{\theta}$, to give the function $Q(\boldsymbol{\theta}, \boldsymbol{\theta}^{\text{old}})$ defined by

$$Q(\boldsymbol{\theta}, \boldsymbol{\theta}^{\text{old}}) = \sum_{\mathbf{Z}} p(\mathbf{Z}|\mathbf{X}, \boldsymbol{\theta}^{\text{old}}) \ln p(\mathbf{X}, \mathbf{Z}|\boldsymbol{\theta}). \tag{13.12}$$

At this point, it is convenient to introduce some notation. We shall use $\gamma(\mathbf{z}_n)$ to denote the marginal posterior distribution of a latent variable $\mathbf{z}_n$, and $\xi(\mathbf{z}_{n-1}, \mathbf{z}_n)$ to denote the joint posterior distribution of two successive latent variables, so that

$$\gamma(\mathbf{z}_n) = p(\mathbf{z}_n|\mathbf{X}, \boldsymbol{\theta}^{\text{old}}) \tag{13.13}$$

$$\xi(\mathbf{z}_{n-1}, \mathbf{z}_n) = p(\mathbf{z}_{n-1}, \mathbf{z}_n|\mathbf{X}, \boldsymbol{\theta}^{\text{old}}). \tag{13.14}$$

For each value of n, we can store $\gamma(\mathbf{z}_n)$ using a set of K nonnegative numbers that sum to unity, and similarly we can store $\xi(\mathbf{z}_{n-1}, \mathbf{z}_n)$ using a $K \times K$ matrix of nonnegative numbers that again sum to unity. We shall also use $\gamma(z_{nk})$ to denote the conditional probability of $z_{nk} = 1$, with a similar use of notation for $\xi(z_{n-1,j}, z_{nk})$ and for other probabilistic variables introduced later. Because the expectation of a binary random variable is just the probability that it takes the value 1, we have

$$\gamma(z_{nk}) = \mathbb{E}[z_{nk}] = \sum_{\mathbf{z}_n} \gamma(\mathbf{z}) z_{nk} \tag{13.15}$$

$$\xi(z_{n-1,j}, z_{nk}) = \mathbb{E}[z_{n-1,j} z_{nk}] = \sum_{\mathbf{z}_{n-1}, \mathbf{z}_n} \gamma(\mathbf{z}) z_{n-1,j} z_{nk}. \tag{13.16}$$

If we substitute the joint distribution $p(\mathbf{X}, \mathbf{Z}|\boldsymbol{\theta})$ given by (13.10) into (13.12), and make use of the definitions of γ and ξ , we obtain

$$\begin{aligned}Q(\boldsymbol{\theta}, \boldsymbol{\theta}^{\text{old}}) &= \sum_{k=1}^{K} \gamma(z_{1k}) \ln \pi_k + \sum_{n=2}^{N} \sum_{j=1}^{K} \sum_{k=1}^{K} \xi(z_{n-1,j}, z_{nk}) \ln A_{jk} \\ &\quad + \sum_{n=1}^{N} \sum_{k=1}^{K} \gamma(z_{nk}) \ln p(\mathbf{x}_n | \boldsymbol{\phi}_k). \end{aligned} \tag{13.17}$$

The goal of the E step will be to evaluate the quantities $\gamma(\mathbf{z}_n)$ and $\xi(\mathbf{z}_{n-1}, \mathbf{z}_n)$ efficiently, and we shall discuss this in detail shortly.

In the M step, we maximize $Q(\boldsymbol{\theta}, \boldsymbol{\theta}^{\text{old}})$ with respect to the parameters $\boldsymbol{\theta} = \{\boldsymbol{\pi}, \mathbf{A}, \boldsymbol{\phi}\}$ in which we treat $\gamma(\mathbf{z}_n)$ and $\xi(\mathbf{z}_{n-1}, \mathbf{z}_n)$ as constant. Maximization with respect to $\boldsymbol{\pi}$ and $\mathbf{A}$ is easily achieved using appropriate Lagrange multipliers with the results

Exercise 13.5

$$\pi_k = \frac{\gamma(z_{1k})}{\sum_{j=1}^{K} \gamma(z_{1j})} \tag{13.18}$$

$$A_{jk} = \frac{\sum_{n=2}^{N} \xi(z_{n-1,j}, z_{nk})}{\sum_{l=1}^{K} \sum_{n=2}^{N} \xi(z_{n-1,j}, z_{nl})}. \tag{13.19}$$

The EM algorithm must be initialized by choosing starting values for $\boldsymbol{\pi}$ and $\mathbf{A}$, which should of course respect the summation constraints associated with their probabilistic interpretation. Note that any elements of $\boldsymbol{\pi}$ or $\mathbf{A}$ that are set to zero initially will remain zero in subsequent EM updates. A typical initialization procedure would involve selecting random starting values for these parameters subject to the summation and non-negativity constraints. Note that no particular modification to the EM results are required for the case of left-to-right models beyond choosing initial values for the elements A_{jk} in which the appropriate elements are set to zero, because these will remain zero throughout.

Exercise 13.6

To maximize $Q(\boldsymbol{\theta}, \boldsymbol{\theta}^{\text{old}})$ with respect to $\boldsymbol{\phi}_k$, we notice that only the final term in (13.17) depends on $\boldsymbol{\phi}_k$, and furthermore this term has exactly the same form as the data-dependent term in the corresponding function for a standard mixture distribution for i.i.d. data, as can be seen by comparison with (9.40) for the case of a Gaussian mixture. Here the quantities $\gamma(z_{nk})$ are playing the role of the responsibilities. If the parameters $\boldsymbol{\phi}_k$ are independent for the different components, then this term decouples into a sum of terms one for each value of k, each of which can be maximized independently. We are then simply maximizing the weighted log likelihood function for the emission density $p(\mathbf{x}|\boldsymbol{\phi}_k)$ with weights $\gamma(z_{nk})$. Here we shall suppose that this maximization can be done efficiently. For instance, in the case of

Gaussian emission densities we have $p(\mathbf{x}|\boldsymbol{\phi}_k) = \mathcal{N}(\mathbf{x}|\boldsymbol{\mu}_k, \boldsymbol{\Sigma}_k)$, and maximization of the function $Q(\boldsymbol{\theta}, \boldsymbol{\theta}^{\text{old}})$ then gives

$$\boldsymbol{\mu}_k = \frac{\sum_{n=1}^{N} \gamma(z_{nk})\mathbf{x}_n}{\sum_{n=1}^{N} \gamma(z_{nk})} \tag{13.20}$$

$$\boldsymbol{\Sigma}_k = \frac{\sum_{n=1}^{N} \gamma(z_{nk})(\mathbf{x}_n - \boldsymbol{\mu}_k)(\mathbf{x}_n - \boldsymbol{\mu}_k)^{\mathrm{T}}}{\sum_{n=1}^{N} \gamma(z_{nk})}. \tag{13.21}$$

For the case of discrete multinomial observed variables, the conditional distribution of the observations takes the form

$$p(\mathbf{x}|\mathbf{z}) = \prod_{i=1}^{D} \prod_{k=1}^{K} \mu_{ik}^{x_i z_k} \tag{13.22}$$

Exercise 13.8

and the corresponding M-step equations are given by

$$\mu_{ik} = \frac{\sum_{n=1}^{N} \gamma(z_{nk}) x_{ni}}{\sum_{n=1}^{N} \gamma(z_{nk})}. \tag{13.23}$$

An analogous result holds for Bernoulli observed variables.

The EM algorithm requires initial values for the parameters of the emission distribution. One way to set these is first to treat the data initially as i.i.d. and fit the emission density by maximum likelihood, and then use the resulting values to initialize the parameters for EM.

13.2.2 The forward-backward algorithm

Next we seek an efficient procedure for evaluating the quantities $\gamma(z_{nk})$ and $\xi(z_{n-1,j}, z_{nk})$, corresponding to the E step of the EM algorithm. The graph for the hidden Markov model, shown in Figure 13.5, is a tree, and so we know that the posterior distribution of the latent variables can be obtained efficiently using a two-stage message passing algorithm. In the particular context of the hidden Markov model, this is known as the *forward-backward* algorithm (Rabiner, 1989), or the *Baum-Welch* algorithm (Baum, 1972). There are in fact several variants of the basic algorithm, all of which lead to the exact marginals, according to the precise form of

Section 8.4

the messages that are propagated along the chain (Jordan, 2007). We shall focus on the most widely used of these, known as the alpha-beta algorithm.

As well as being of great practical importance in its own right, the forward-backward algorithm provides us with a nice illustration of many of the concepts introduced in earlier chapters. We shall therefore begin in this section with a 'conventional' derivation of the forward-backward equations, making use of the sum and product rules of probability, and exploiting conditional independence properties which we shall obtain from the corresponding graphical model using d-separation. Then in Section 13.2.3, we shall see how the forward-backward algorithm can be obtained very simply as a specific example of the sum-product algorithm introduced in Section 8.4.4.

It is worth emphasizing that evaluation of the posterior distributions of the latent variables is independent of the form of the emission density $p(\mathbf{x}|\mathbf{z})$ or indeed of whether the observed variables are continuous or discrete. All we require is the values of the quantities $p(\mathbf{x}_n|\mathbf{z}_n)$ for each value of $\mathbf{z}_n$ for every n. Also, in this section and the next we shall omit the explicit dependence on the model parameters $\boldsymbol{\theta}^{\text{old}}$ because these fixed throughout.

We therefore begin by writing down the following conditional independence properties (Jordan, 2007)

$$
\begin{aligned}
p(\mathbf{X}|\mathbf{z}_n) &= p(\mathbf{x}_1,\ldots,\mathbf{x}_n|\mathbf{z}_n) \\
&\qquad p(\mathbf{x}_{n+1},\ldots,\mathbf{x}_N|\mathbf{z}_n) && (13.24)\\
p(\mathbf{x}_1,\ldots,\mathbf{x}_{n-1}|\mathbf{x}_n,\mathbf{z}_n) &= p(\mathbf{x}_1,\ldots,\mathbf{x}_{n-1}|\mathbf{z}_n) && (13.25)\\
p(\mathbf{x}_1,\ldots,\mathbf{x}_{n-1}|\mathbf{z}_{n-1},\mathbf{z}_n) &= p(\mathbf{x}_1,\ldots,\mathbf{x}_{n-1}|\mathbf{z}_{n-1}) && (13.26)\\
p(\mathbf{x}_{n+1},\ldots,\mathbf{x}_N|\mathbf{z}_n,\mathbf{z}_{n+1}) &= p(\mathbf{x}_{n+1},\ldots,\mathbf{x}_N|\mathbf{z}_{n+1}) && (13.27)\\
p(\mathbf{x}_{n+2},\ldots,\mathbf{x}_N|\mathbf{z}_{n+1},\mathbf{x}_{n+1}) &= p(\mathbf{x}_{n+2},\ldots,\mathbf{x}_N|\mathbf{z}_{n+1}) && (13.28)\\
p(\mathbf{X}|\mathbf{z}_{n-1},\mathbf{z}_n) &= p(\mathbf{x}_1,\ldots,\mathbf{x}_{n-1}|\mathbf{z}_{n-1}) \\
&\qquad p(\mathbf{x}_n|\mathbf{z}_n)p(\mathbf{x}_{n+1},\ldots,\mathbf{x}_N|\mathbf{z}_n) && (13.29)\\
p(\mathbf{x}_{N+1}|\mathbf{X},\mathbf{z}_{N+1}) &= p(\mathbf{x}_{N+1}|\mathbf{z}_{N+1}) && (13.30)\\
p(\mathbf{z}_{N+1}|\mathbf{z}_N,\mathbf{X}) &= p(\mathbf{z}_{N+1}|\mathbf{z}_N) && (13.31)
\end{aligned}
$$

where $\mathbf{X} = \{\mathbf{x}_1,\ldots,\mathbf{x}_N\}$. These relations are most easily proved using d-separation. For instance in the second of these results, we note that every path from any one of the nodes $\mathbf{x}_1,\ldots,\mathbf{x}_{n-1}$ to the node $\mathbf{x}_n$ passes through the node $\mathbf{z}_n$, which is observed. Because all such paths are head-to-tail, it follows that the conditional independence property must hold. The reader should take a few moments to verify each of these properties in turn, as an exercise in the application of d-separation. These relations can also be proved directly, though with significantly greater effort, from the joint distribution for the hidden Markov model using the sum and product rules of probability.

Exercise 13.10

Let us begin by evaluating $\gamma(z_{nk})$. Recall that for a discrete multinomial random variable the expected value of one of its components is just the probability of that component having the value 1. Thus we are interested in finding the posterior distribution $p(\mathbf{z}_n|\mathbf{x}_1,\ldots,\mathbf{x}_N)$ of $\mathbf{z}_n$ given the observed data set $\mathbf{x}_1,\ldots,\mathbf{x}_N$. This represents a vector of length K whose entries correspond to the expected values of z_{nk}. Using Bayes' theorem, we have

$$\gamma(\mathbf{z}_n) = p(\mathbf{z}_n|\mathbf{X}) = \frac{p(\mathbf{X}|\mathbf{z}_n)p(\mathbf{z}_n)}{p(\mathbf{X})}. \tag{13.32}$$

Note that the denominator $p(\mathbf{X})$ is implicitly conditioned on the parameters $\boldsymbol{\theta}^{\text{old}}$ of the HMM and hence represents the likelihood function. Using the conditional independence property (13.24), together with the product rule of probability, we obtain

$$\gamma(\mathbf{z}_n) = \frac{p(\mathbf{x}_1,\ldots,\mathbf{x}_n,\mathbf{z}_n)p(\mathbf{x}_{n+1},\ldots,\mathbf{x}_N|\mathbf{z}_n)}{p(\mathbf{X})} = \frac{\alpha(\mathbf{z}_n)\beta(\mathbf{z}_n)}{p(\mathbf{X})} \tag{13.33}$$

where we have defined

$$\alpha(\mathbf{z}_n) \equiv p(\mathbf{x}_1,\ldots,\mathbf{x}_n,\mathbf{z}_n) \tag{13.34}$$

$$\beta(\mathbf{z}_n) \equiv p(\mathbf{x}_{n+1},\ldots,\mathbf{x}_N|\mathbf{z}_n). \tag{13.35}$$

The quantity $\alpha(\mathbf{z}_n)$ represents the joint probability of observing all of the given data up to time n and the value of $\mathbf{z}_n$, whereas $\beta(\mathbf{z}_n)$ represents the conditional probability of all future data from time $n+1$ up to N given the value of $\mathbf{z}_n$. Again, $\alpha(\mathbf{z}_n)$ and $\beta(\mathbf{z}_n)$ each represent a set of K numbers, one for each of the possible settings of the 1-of-K coded binary vector $\mathbf{z}_n$. We shall use the notation $\alpha(z_{nk})$ to denote the value of $\alpha(\mathbf{z}_n)$ when $z_{nk}=1$, with an analogous interpretation of $\beta(z_{nk})$.

We now derive recursion relations that allow $\alpha(\mathbf{z}_n)$ and $\beta(\mathbf{z}_n)$ to be evaluated efficiently. Again, we shall make use of conditional independence properties, in particular (13.25) and (13.26), together with the sum and product rules, allowing us to express $\alpha(\mathbf{z}_n)$ in terms of $\alpha(\mathbf{z}_{n-1})$ as follows

$$\begin{aligned}
\alpha(\mathbf{z}_n) &= p(\mathbf{x}_1,\ldots,\mathbf{x}_n,\mathbf{z}_n) \\
&= p(\mathbf{x}_1,\ldots,\mathbf{x}_n|\mathbf{z}_n)p(\mathbf{z}_n) \\
&= p(\mathbf{x}_n|\mathbf{z}_n)p(\mathbf{x}_1,\ldots,\mathbf{x}_{n-1}|\mathbf{z}_n)p(\mathbf{z}_n) \\
&= p(\mathbf{x}_n|\mathbf{z}_n)p(\mathbf{x}_1,\ldots,\mathbf{x}_{n-1},\mathbf{z}_n) \\
&= p(\mathbf{x}_n|\mathbf{z}_n)\sum_{\mathbf{z}_{n-1}} p(\mathbf{x}_1,\ldots,\mathbf{x}_{n-1},\mathbf{z}_{n-1},\mathbf{z}_n) \\
&= p(\mathbf{x}_n|\mathbf{z}_n)\sum_{\mathbf{z}_{n-1}} p(\mathbf{x}_1,\ldots,\mathbf{x}_{n-1},\mathbf{z}_n|\mathbf{z}_{n-1})p(\mathbf{z}_{n-1}) \\
&= p(\mathbf{x}_n|\mathbf{z}_n)\sum_{\mathbf{z}_{n-1}} p(\mathbf{x}_1,\ldots,\mathbf{x}_{n-1}|\mathbf{z}_{n-1})p(\mathbf{z}_n|\mathbf{z}_{n-1})p(\mathbf{z}_{n-1}) \\
&= p(\mathbf{x}_n|\mathbf{z}_n)\sum_{\mathbf{z}_{n-1}} p(\mathbf{x}_1,\ldots,\mathbf{x}_{n-1},\mathbf{z}_{n-1})p(\mathbf{z}_n|\mathbf{z}_{n-1})
\end{aligned}$$

Making use of the definition (13.34) for $\alpha(\mathbf{z}_n)$, we then obtain

$$\alpha(\mathbf{z}_n) = p(\mathbf{x}_n|\mathbf{z}_n)\sum_{\mathbf{z}_{n-1}} \alpha(\mathbf{z}_{n-1})p(\mathbf{z}_n|\mathbf{z}_{n-1}). \tag{13.36}$$

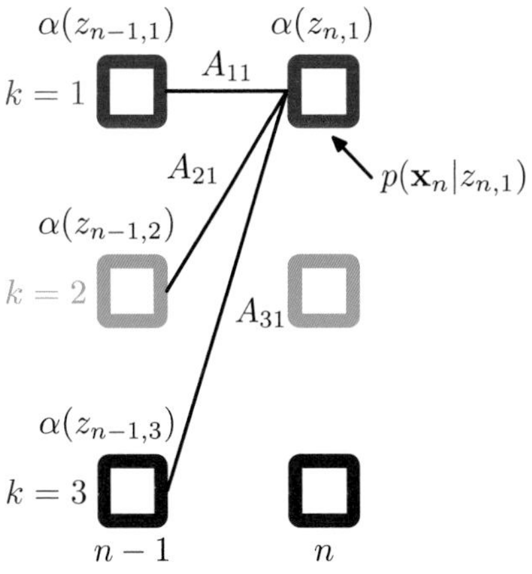

Figure 13.12 Illustration of the forward recursion (13.36) for evaluation of the α variables. In this fragment of the lattice, we see that the quantity $\alpha(z_{n,1})$ is obtained by taking the elements $\alpha(z_{n-1,j})$ of $\alpha(\mathbf{z}_{n-1})$ at step $n-1$ and summing them up with weights given by A_{j1}, corresponding to the values of $p(\mathbf{z}_n|\mathbf{z}_{n-1})$, and then multiplying by the data contribution $p(\mathbf{x}_n|z_{n1})$.

It is worth taking a moment to study this recursion relation in some detail. Note that there are K terms in the summation, and the right-hand side has to be evaluated for each of the K values of $\mathbf{z}_n$ so each step of the α recursion has computational cost that scaled like $O(K^2)$. The forward recursion equation for $\alpha(\mathbf{z}_n)$ is illustrated using a lattice diagram in Figure 13.12.

In order to start this recursion, we need an initial condition that is given by

$$\alpha(\mathbf{z}_1) = p(\mathbf{x}_1, \mathbf{z}_1) = p(\mathbf{z}_1)p(\mathbf{x}_1|\mathbf{z}_1) = \prod_{k=1}^{K} \{\pi_k p(\mathbf{x}_1|\boldsymbol{\phi}_k)\}^{z_{1k}} \tag{13.37}$$

which tells us that $\alpha(z_{1k})$, for $k = 1, \ldots, K$, takes the value $\pi_k p(\mathbf{x}_1|\boldsymbol{\phi}_k)$. Starting at the first node of the chain, we can then work along the chain and evaluate $\alpha(\mathbf{z}_n)$ for every latent node. Because each step of the recursion involves multiplying by a $K \times K$ matrix, the overall cost of evaluating these quantities for the whole chain is of $O(K^2N)$.

We can similarly find a recursion relation for the quantities $\beta(\mathbf{z}_n)$ by making use of the conditional independence properties (13.27) and (13.28) giving

$$\begin{aligned}
\beta(\mathbf{z}_n) &= p(\mathbf{x}_{n+1}, \ldots, \mathbf{x}_N|\mathbf{z}_n) \\
&= \sum_{\mathbf{z}_{n+1}} p(\mathbf{x}_{n+1}, \ldots, \mathbf{x}_N, \mathbf{z}_{n+1}|\mathbf{z}_n) \\
&= \sum_{\mathbf{z}_{n+1}} p(\mathbf{x}_{n+1}, \ldots, \mathbf{x}_N|\mathbf{z}_n, \mathbf{z}_{n+1})p(\mathbf{z}_{n+1}|\mathbf{z}_n) \\
&= \sum_{\mathbf{z}_{n+1}} p(\mathbf{x}_{n+1}, \ldots, \mathbf{x}_N|\mathbf{z}_{n+1})p(\mathbf{z}_{n+1}|\mathbf{z}_n) \\
&= \sum_{\mathbf{z}_{n+1}} p(\mathbf{x}_{n+2}, \ldots, \mathbf{x}_N|\mathbf{z}_{n+1})p(\mathbf{x}_{n+1}|\mathbf{z}_{n+1})p(\mathbf{z}_{n+1}|\mathbf{z}_n).
\end{aligned}$$

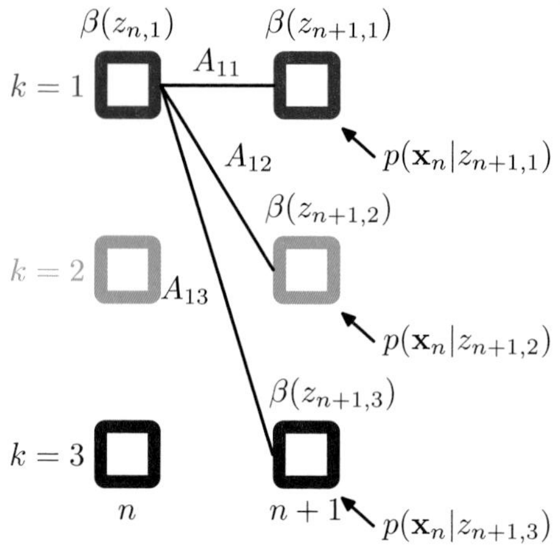

Figure 13.13 Illustration of the backward recursion (13.38) for evaluation of the β variables. In this fragment of the lattice, we see that the quantity $\beta(z_{n,1})$ is obtained by taking the components $\beta(z_{n+1,k})$ of $\beta(\mathbf{z}_{n+1})$ at step $n+1$ and summing them up with weights given by the products of A_{1k}, corresponding to the values of $p(\mathbf{z}_{n+1}|\mathbf{z}_n)$ and the corresponding values of the emission density $p(\mathbf{x}_n|z_{n+1,k})$.

Making use of the definition (13.35) for $\beta(\mathbf{z}_n)$, we then obtain

$$\beta(\mathbf{z}_n) = \sum_{\mathbf{z}_{n+1}} \beta(\mathbf{z}_{n+1}) p(\mathbf{x}_{n+1}|\mathbf{z}_{n+1}) p(\mathbf{z}_{n+1}|\mathbf{z}_n). \tag{13.38}$$

Note that in this case we have a backward message passing algorithm that evaluates $\beta(\mathbf{z}_n)$ in terms of $\beta(\mathbf{z}_{n+1})$. At each step, we absorb the effect of observation $\mathbf{x}_{n+1}$ through the emission probability $p(\mathbf{x}_{n+1}|\mathbf{z}_{n+1})$, multiply by the transition matrix $p(\mathbf{z}_{n+1}|\mathbf{z}_n)$, and then marginalize out $\mathbf{z}_{n+1}$. This is illustrated in Figure 13.13.

Again we need a starting condition for the recursion, namely a value for $\beta(\mathbf{z}_N)$. This can be obtained by setting $n = N$ in (13.33) and replacing $\alpha(\mathbf{z}_N)$ with its definition (13.34) to give

$$p(\mathbf{z}_N|\mathbf{X}) = \frac{p(\mathbf{X}, \mathbf{z}_N)\beta(\mathbf{z}_N)}{p(\mathbf{X})} \tag{13.39}$$

which we see will be correct provided we take $\beta(\mathbf{z}_N) = 1$ for all settings of $\mathbf{z}_N$.

In the M step equations, the quantity $p(\mathbf{X})$ will cancel out, as can be seen, for instance, in the M-step equation for $\boldsymbol{\mu}_k$ given by (13.20), which takes the form

$$\boldsymbol{\mu}_k = \frac{\sum_{n=1}^{n} \gamma(z_{nk})\mathbf{x}_n}{\sum_{n=1}^{n} \gamma(z_{nk})} = \frac{\sum_{n=1}^{n} \alpha(z_{nk})\beta(z_{nk})\mathbf{x}_n}{\sum_{n=1}^{n} \alpha(z_{nk})\beta(z_{nk})}. \tag{13.40}$$

However, the quantity $p(\mathbf{X})$ represents the likelihood function whose value we typically wish to monitor during the EM optimization, and so it is useful to be able to evaluate it. If we sum both sides of (13.33) over $\mathbf{z}_n$, and use the fact that the left-hand side is a normalized distribution, we obtain

$$p(\mathbf{X}) = \sum_{\mathbf{z}_n} \alpha(\mathbf{z}_n)\beta(\mathbf{z}_n). \tag{13.41}$$

Thus we can evaluate the likelihood function by computing this sum, for any convenient choice of n. For instance, if we only want to evaluate the likelihood function, then we can do this by running the α recursion from the start to the end of the chain, and then use this result for $n = N$, making use of the fact that $\beta(\mathbf{z}_N)$ is a vector of 1s. In this case no β recursion is required, and we simply have

$$p(\mathbf{X}) = \sum_{\mathbf{z}_N} \alpha(\mathbf{z}_N). \tag{13.42}$$

Let us take a moment to interpret this result for $p(\mathbf{X})$. Recall that to compute the likelihood we should take the joint distribution $p(\mathbf{X}, \mathbf{Z})$ and sum over all possible values of $\mathbf{Z}$. Each such value represents a particular choice of hidden state for every time step, in other words every term in the summation is a path through the lattice diagram, and recall that there are exponentially many such paths. By expressing the likelihood function in the form (13.42), we have reduced the computational cost from being exponential in the length of the chain to being linear by swapping the order of the summation and multiplications, so that at each time step n we sum the contributions from all paths passing through each of the states z_{nk} to give the intermediate quantities $\alpha(\mathbf{z}_n)$.

Next we consider the evaluation of the quantities $\xi(\mathbf{z}_{n-1}, \mathbf{z}_n)$, which correspond to the values of the conditional probabilities $p(\mathbf{z}_{n-1}, \mathbf{z}_n|\mathbf{X})$ for each of the $K \times K$ settings for $(\mathbf{z}_{n-1}, \mathbf{z}_n)$. Using the definition of $\xi(\mathbf{z}_{n-1}, \mathbf{z}_n)$, and applying Bayes' theorem, we have

$$\begin{aligned}
\xi(\mathbf{z}_{n-1}, \mathbf{z}_n) &= p(\mathbf{z}_{n-1}, \mathbf{z}_n|\mathbf{X}) \\
&= \frac{p(\mathbf{X}|\mathbf{z}_{n-1}, \mathbf{z}_n)p(\mathbf{z}_{n-1}, \mathbf{z}_n)}{p(\mathbf{X})} \\
&= \frac{p(\mathbf{x}_1, \ldots, \mathbf{x}_{n-1}|\mathbf{z}_{n-1})p(\mathbf{x}_n|\mathbf{z}_n)p(\mathbf{x}_{n+1}, \ldots, \mathbf{x}_N|\mathbf{z}_n)p(\mathbf{z}_n|\mathbf{z}_{n-1})p(\mathbf{z}_{n-1})}{p(\mathbf{X})} \\
&= \frac{\alpha(\mathbf{z}_{n-1})p(\mathbf{x}_n|\mathbf{z}_n)p(\mathbf{z}_n|\mathbf{z}_{n-1})\beta(\mathbf{z}_n)}{p(\mathbf{X})}
\end{aligned} \tag{13.43}$$

where we have made use of the conditional independence property (13.29) together with the definitions of $\alpha(\mathbf{z}_n)$ and $\beta(\mathbf{z}_n)$ given by (13.34) and (13.35). Thus we can calculate the $\xi(\mathbf{z}_{n-1}, \mathbf{z}_n)$ directly by using the results of the α and β recursions.

Let us summarize the steps required to train a hidden Markov model using the EM algorithm. We first make an initial selection of the parameters $\boldsymbol{\theta}^{\text{old}}$ where $\boldsymbol{\theta} \equiv (\boldsymbol{\pi}, \mathbf{A}, \boldsymbol{\phi})$. The $\mathbf{A}$ and $\boldsymbol{\pi}$ parameters are often initialized either uniformly or randomly from a uniform distribution (respecting their non-negativity and summation constraints). Initialization of the parameters $\boldsymbol{\phi}$ will depend on the form of the distribution. For instance in the case of Gaussians, the parameters $\boldsymbol{\mu}_k$ might be initialized by applying the K-means algorithm to the data, and $\boldsymbol{\Sigma}_k$ might be initialized to the covariance matrix of the corresponding K means cluster. Then we run both the forward α recursion and the backward β recursion and use the results to evaluate $\gamma(\mathbf{z}_n)$ and $\xi(\mathbf{z}_{n-1}, \mathbf{z}_n)$. At this stage, we can also evaluate the likelihood function.

This completes the E step, and we use the results to find a revised set of parameters $\boldsymbol{\theta}^{\text{new}}$ using the M-step equations from Section 13.2.1. We then continue to alternate between E and M steps until some convergence criterion is satisfied, for instance when the change in the likelihood function is below some threshold.

Note that in these recursion relations the observations enter through conditional distributions of the form $p(\mathbf{x}_n|\mathbf{z}_n)$. The recursions are therefore independent of the type or dimensionality of the observed variables or the form of this conditional distribution, so long as its value can be computed for each of the K possible states of $\mathbf{z}_n$.

We have seen in earlier chapters that the maximum likelihood approach is most effective when the number of data points is large in relation to the number of parameters. Here we note that a hidden Markov model can be trained effectively, using maximum likelihood, provided the training sequence is sufficiently long. Alternatively, we can make use of multiple shorter sequences, which requires a straightforward modification of the hidden Markov model EM algorithm. In the case of left-to-right models, this is particularly important because, in a given observation sequence, a given state transition corresponding to a nondiagonal element of $\mathbf{A}$ will seen at most once.

Exercise 13.12

Another quantity of interest is the predictive distribution, in which the observed data is $\mathbf{X} = \{\mathbf{x}_1, \ldots, \mathbf{x}_N\}$ and we wish to predict $\mathbf{x}_{N+1}$, which would be important for real-time applications such as financial forecasting. Again we make use of the sum and product rules together with the conditional independence properties (13.30) and (13.31) giving

$$\begin{aligned}
p(\mathbf{x}_{N+1}|\mathbf{X}) &= \sum_{\mathbf{z}_{N+1}} p(\mathbf{x}_{N+1}, \mathbf{z}_{N+1}|\mathbf{X}) \\
&= \sum_{\mathbf{z}_{N+1}} p(\mathbf{x}_{N+1}|\mathbf{z}_{N+1})p(\mathbf{z}_{N+1}|\mathbf{X}) \\
&= \sum_{\mathbf{z}_{N+1}} p(\mathbf{x}_{N+1}|\mathbf{z}_{N+1}) \sum_{\mathbf{z}_N} p(\mathbf{z}_{N+1}, \mathbf{z}_N|\mathbf{X}) \\
&= \sum_{\mathbf{z}_{N+1}} p(\mathbf{x}_{N+1}|\mathbf{z}_{N+1}) \sum_{\mathbf{z}_N} p(\mathbf{z}_{N+1}|\mathbf{z}_N)p(\mathbf{z}_N|\mathbf{X}) \\
&= \sum_{\mathbf{z}_{N+1}} p(\mathbf{x}_{N+1}|\mathbf{z}_{N+1}) \sum_{\mathbf{z}_N} p(\mathbf{z}_{N+1}|\mathbf{z}_N)\frac{p(\mathbf{z}_N, \mathbf{X})}{p(\mathbf{X})} \\
&= \frac{1}{p(\mathbf{X})} \sum_{\mathbf{z}_{N+1}} p(\mathbf{x}_{N+1}|\mathbf{z}_{N+1}) \sum_{\mathbf{z}_N} p(\mathbf{z}_{N+1}|\mathbf{z}_N)\alpha(\mathbf{z}_N) \qquad (13.44)
\end{aligned}$$

which can be evaluated by first running a forward α recursion and then computing the final summations over $\mathbf{z}_N$ and $\mathbf{z}_{N+1}$. The result of the first summation over $\mathbf{z}_N$ can be stored and used once the value of $\mathbf{x}_{N+1}$ is observed in order to run the α recursion forward to the next step in order to predict the subsequent value $\mathbf{x}_{N+2}$.

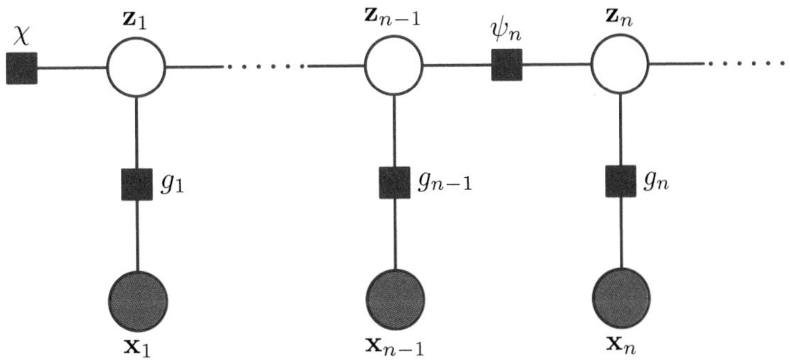

Figure 13.14 A fragment of the factor graph representation for the hidden Markov model.

Note that in (13.44), the influence of all data from $\mathbf{x}_1$ to $\mathbf{x}_N$ is summarized in the K values of $\alpha(\mathbf{z}_N)$. Thus the predictive distribution can be carried forward indefinitely using a fixed amount of storage, as may be required for real-time applications.

Here we have discussed the estimation of the parameters of an HMM using maximum likelihood. This framework is easily extended to regularized maximum likelihood by introducing priors over the model parameters $\boldsymbol{\pi}$, $\mathbf{A}$ and $\boldsymbol{\phi}$ whose values are then estimated by maximizing their posterior probability. This can again be done using the EM algorithm in which the E step is the same as discussed above, and the M step involves adding the log of the prior distribution $p(\boldsymbol{\theta})$ to the function $Q(\boldsymbol{\theta}, \boldsymbol{\theta}^{\text{old}})$ before maximization and represents a straightforward application of the techniques developed at various points in this book. Furthermore, we can use variational methods to give a fully Bayesian treatment of the HMM in which we marginalize over the parameter distributions (MacKay, 1997). As with maximum likelihood, this leads to a two-pass forward-backward recursion to compute posterior probabilities.

Section 10.1

13.2.3 The sum-product algorithm for the HMM

The directed graph that represents the hidden Markov model, shown in Figure 13.5, is a tree and so we can solve the problem of finding local marginals for the hidden variables using the sum-product algorithm. Not surprisingly, this turns out to be equivalent to the forward-backward algorithm considered in the previous section, and so the sum-product algorithm therefore provides us with a simple way to derive the alpha-beta recursion formulae.

Section 8.4.4

We begin by transforming the directed graph of Figure 13.5 into a factor graph, of which a representative fragment is shown in Figure 13.14. This form of the factor graph shows all variables, both latent and observed, explicitly. However, for the purpose of solving the inference problem, we shall always be conditioning on the variables $\mathbf{x}_1, \ldots, \mathbf{x}_N$, and so we can simplify the factor graph by absorbing the emission probabilities into the transition probability factors. This leads to the simplified factor graph representation in Figure 13.15, in which the factors are given by

$$h(\mathbf{z}_1) = p(\mathbf{z}_1)p(\mathbf{x}_1|\mathbf{z}_1) \tag{13.45}$$

$$f_n(\mathbf{z}_{n-1}, \mathbf{z}_n) = p(\mathbf{z}_n|\mathbf{z}_{n-1})p(\mathbf{x}_n|\mathbf{z}_n). \tag{13.46}$$

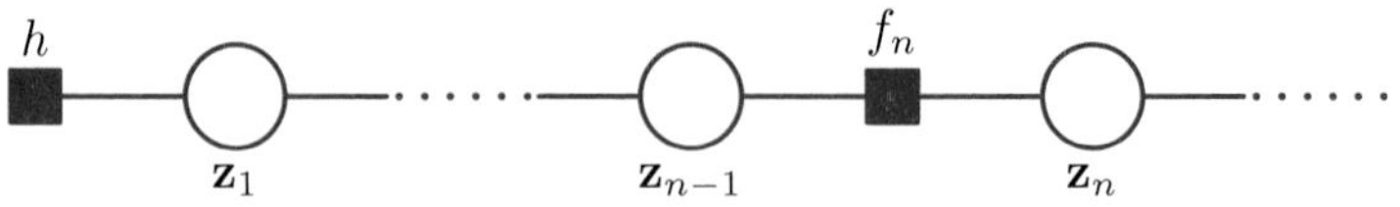

Figure 13.15 A simplified form of factor graph to describe the hidden Markov model.

To derive the alpha-beta algorithm, we denote the final hidden variable $\mathbf{z}_N$ as the root node, and first pass messages from the leaf node h to the root. From the general results (8.66) and (8.69) for message propagation, we see that the messages which are propagated in the hidden Markov model take the form

$$\mu_{\mathbf{z}_{n-1}\rightarrow f_n}(\mathbf{z}_{n-1}) = \mu_{f_{n-1}\rightarrow \mathbf{z}_{n-1}}(\mathbf{z}_{n-1}) \tag{13.47}$$

$$\mu_{f_n\rightarrow \mathbf{z}_n}(\mathbf{z}_n) = \sum_{\mathbf{z}_{n-1}} f_n(\mathbf{z}_{n-1}, \mathbf{z}_n)\mu_{\mathbf{z}_{n-1}\rightarrow f_n}(\mathbf{z}_{n-1}) \tag{13.48}$$

These equations represent the propagation of messages forward along the chain and are equivalent to the alpha recursions derived in the previous section, as we shall now show. Note that because the variable nodes $\mathbf{z}_n$ have only two neighbours, they perform no computation.

We can eliminate $\mu_{\mathbf{z}_{n-1}\rightarrow f_n}(\mathbf{z}_{n-1})$ from (13.48) using (13.47) to give a recursion for the $f \rightarrow \mathbf{z}$ messages of the form

$$\mu_{f_n\rightarrow \mathbf{z}_n}(\mathbf{z}_n) = \sum_{\mathbf{z}_{n-1}} f_n(\mathbf{z}_{n-1}, \mathbf{z}_n)\mu_{f_{n-1}\rightarrow \mathbf{z}_{n-1}}(\mathbf{z}_{n-1}). \tag{13.49}$$

If we now recall the definition (13.46), and if we define

$$\alpha(\mathbf{z}_n) = \mu_{f_n\rightarrow \mathbf{z}_n}(\mathbf{z}_n) \tag{13.50}$$

then we obtain the alpha recursion given by (13.36). We also need to verify that the quantities $\alpha(\mathbf{z}_n)$ are themselves equivalent to those defined previously. This is easily done by using the initial condition (8.71) and noting that $\alpha(\mathbf{z}_1)$ is given by $h(\mathbf{z}_1) = p(\mathbf{z}_1)p(\mathbf{x}_1|\mathbf{z}_1)$ which is identical to (13.37). Because the initial α is the same, and because they are iteratively computed using the same equation, all subsequent α quantities must be the same.

Next we consider the messages that are propagated from the root node back to the leaf node. These take the form

$$\mu_{f_{n+1}\rightarrow f_n}(\mathbf{z}_n) = \sum_{\mathbf{z}_{n+1}} f_{n+1}(\mathbf{z}_n, \mathbf{z}_{n+1})\mu_{f_{n+2}\rightarrow f_{n+1}}(\mathbf{z}_{n+1}) \tag{13.51}$$

where, as before, we have eliminated the messages of the type $\mathbf{z} \rightarrow f$ since the variable nodes perform no computation. Using the definition (13.46) to substitute for $f_{n+1}(\mathbf{z}_n, \mathbf{z}_{n+1})$, and defining

$$\beta(\mathbf{z}_n) = \mu_{f_{n+1}\rightarrow \mathbf{z}_n}(\mathbf{z}_n) \tag{13.52}$$

we obtain the beta recursion given by (13.38). Again, we can verify that the beta variables themselves are equivalent by noting that (8.70) implies that the initial message send by the root variable node is $\mu_{\mathbf{z}_N \to f_N}(\mathbf{z}_N) = 1$, which is identical to the initialization of $\beta(\mathbf{z}_N)$ given in Section 13.2.2.

The sum-product algorithm also specifies how to evaluate the marginals once all the messages have been evaluated. In particular, the result (8.63) shows that the local marginal at the node $\mathbf{z}_n$ is given by the product of the incoming messages. Because we have conditioned on the variables $\mathbf{X} = \{\mathbf{x}_1, \ldots, \mathbf{x}_N\}$, we are computing the joint distribution

$$p(\mathbf{z}_n, \mathbf{X}) = \mu_{f_n \to \mathbf{z}_n}(\mathbf{z}_n)\mu_{f_{n+1} \to \mathbf{z}_n}(\mathbf{z}_n) = \alpha(\mathbf{z}_n)\beta(\mathbf{z}_n). \tag{13.53}$$

Dividing both sides by $p(\mathbf{X})$, we then obtain

$$\gamma(\mathbf{z}_n) = \frac{p(\mathbf{z}_n, \mathbf{X})}{p(\mathbf{X})} = \frac{\alpha(\mathbf{z}_n)\beta(\mathbf{z}_n)}{p(\mathbf{X})} \tag{13.54}$$

Exercise 13.11

in agreement with (13.33). The result (13.43) can similarly be derived from (8.72).

13.2.4 Scaling factors

There is an important issue that must be addressed before we can make use of the forward backward algorithm in practice. From the recursion relation (13.36), we note that at each step the new value $\alpha(\mathbf{z}_n)$ is obtained from the previous value $\alpha(\mathbf{z}_{n-1})$ by multiplying by quantities $p(\mathbf{z}_n|\mathbf{z}_{n-1})$ and $p(\mathbf{x}_n|\mathbf{z}_n)$. Because these probabilities are often significantly less than unity, as we work our way forward along the chain, the values of $\alpha(\mathbf{z}_n)$ can go to zero exponentially quickly. For moderate lengths of chain (say 100 or so), the calculation of the $\alpha(\mathbf{z}_n)$ will soon exceed the dynamic range of the computer, even if double precision floating point is used.

In the case of i.i.d. data, we implicitly circumvented this problem with the evaluation of likelihood functions by taking logarithms. Unfortunately, this will not help here because we are forming sums of products of small numbers (we are in fact implicitly summing over all possible paths through the lattice diagram of Figure 13.7). We therefore work with re-scaled versions of $\alpha(\mathbf{z}_n)$ and $\beta(\mathbf{z}_n)$ whose values remain of order unity. As we shall see, the corresponding scaling factors cancel out when we use these re-scaled quantities in the EM algorithm.

In (13.34), we defined $\alpha(\mathbf{z}_n) = p(\mathbf{x}_1, \ldots, \mathbf{x}_n, \mathbf{z}_n)$ representing the joint distribution of all the observations up to $\mathbf{x}_n$ and the latent variable $\mathbf{z}_n$. Now we define a normalized version of α given by

$$\widehat{\alpha}(\mathbf{z}_n) = p(\mathbf{z}_n|\mathbf{x}_1, \ldots, \mathbf{x}_n) = \frac{\alpha(\mathbf{z}_n)}{p(\mathbf{x}_1, \ldots, \mathbf{x}_n)} \tag{13.55}$$

which we expect to be well behaved numerically because it is a probability distribution over K variables for any value of n. In order to relate the scaled and original alpha variables, we introduce scaling factors defined by conditional distributions over the observed variables

$$c_n = p(\mathbf{x}_n|\mathbf{x}_1, \ldots, \mathbf{x}_{n-1}). \tag{13.56}$$

From the product rule, we then have

$$p(\mathbf{x}_1, \ldots, \mathbf{x}_n) = \prod_{m=1}^{n} c_m \tag{13.57}$$

and so

$$\alpha(\mathbf{z}_n) = p(\mathbf{z}_n|\mathbf{x}_1, \ldots, \mathbf{x}_n)p(\mathbf{x}_1, \ldots, \mathbf{x}_n) = \left(\prod_{m=1}^{n} c_m\right)\widehat{\alpha}(\mathbf{z}_n). \tag{13.58}$$

We can then turn the recursion equation (13.36) for α into one for $\widehat{\alpha}$ given by

$$c_n\widehat{\alpha}(\mathbf{z}_n) = p(\mathbf{x}_n|\mathbf{z}_n)\sum_{\mathbf{z}_{n-1}}\widehat{\alpha}(\mathbf{z}_{n-1})p(\mathbf{z}_n|\mathbf{z}_{n-1}). \tag{13.59}$$

Note that at each stage of the forward message passing phase, used to evaluate $\widehat{\alpha}(\mathbf{z}_n)$, we have to evaluate and store c_n, which is easily done because it is the coefficient that normalizes the right-hand side of (13.59) to give $\widehat{\alpha}(\mathbf{z}_n)$.

We can similarly define re-scaled variables $\widehat{\beta}(\mathbf{z}_n)$ using

$$\beta(\mathbf{z}_n) = \left(\prod_{m=n+1}^{N} c_m\right)\widehat{\beta}(\mathbf{z}_n) \tag{13.60}$$

which will again remain within machine precision because, from (13.35), the quantities $\widehat{\beta}(\mathbf{z}_n)$ are simply the ratio of two conditional probabilities

$$\widehat{\beta}(\mathbf{z}_n) = \frac{p(\mathbf{x}_{n+1}, \ldots, \mathbf{x}_N|\mathbf{z}_n)}{p(\mathbf{x}_{n+1}, \ldots, \mathbf{x}_N|\mathbf{x}_1, \ldots, \mathbf{x}_n)}. \tag{13.61}$$

The recursion result (13.38) for β then gives the following recursion for the re-scaled variables

$$c_{n+1}\widehat{\beta}(\mathbf{z}_n) = \sum_{\mathbf{z}_{n+1}}\widehat{\beta}(\mathbf{z}_{n+1})p(\mathbf{x}_{n+1}|\mathbf{z}_{n+1})p(\mathbf{z}_{n+1}|\mathbf{z}_n). \tag{13.62}$$

In applying this recursion relation, we make use of the scaling factors c_n that were previously computed in the α phase.

From (13.57), we see that the likelihood function can be found using

$$p(\mathbf{X}) = \prod_{n=1}^{N} c_n. \tag{13.63}$$

Exercise 13.15

Similarly, using (13.33) and (13.43), together with (13.63), we see that the required marginals are given by

$$\gamma(\mathbf{z}_n) = \widehat{\alpha}(\mathbf{z}_n)\widehat{\beta}(\mathbf{z}_n) \tag{13.64}$$

$$\xi(\mathbf{z}_{n-1}, \mathbf{z}_n) = c_n^{-1}\widehat{\alpha}(\mathbf{z}_{n-1})p(\mathbf{x}_n|\mathbf{z}_n)p(\mathbf{z}_n|\mathbf{z}_{n-1})\widehat{\beta}(\mathbf{z}_n). \tag{13.65}$$

Finally, we note that there is an alternative formulation of the forward-backward algorithm (Jordan, 2007) in which the backward pass is defined by a recursion based on the quantities $\gamma(\mathbf{z}_n) = \widehat{\alpha}(\mathbf{z}_n)\widehat{\beta}(\mathbf{z}_n)$ instead of using $\widehat{\beta}(\mathbf{z}_n)$. This α–γ recursion requires that the forward pass be completed first so that all the quantities $\widehat{\alpha}(\mathbf{z}_n)$ are available for the backward pass, whereas the forward and backward passes of the α–β algorithm can be done independently. Although these two algorithms have comparable computational cost, the α–β version is the most commonly encountered one in the case of hidden Markov models, whereas for linear dynamical systems a recursion analogous to the α–γ form is more usual.

Section 13.3

13.2.5 The Viterbi algorithm

In many applications of hidden Markov models, the latent variables have some meaningful interpretation, and so it is often of interest to find the most probable sequence of hidden states for a given observation sequence. For instance in speech recognition, we might wish to find the most probable phoneme sequence for a given series of acoustic observations. Because the graph for the hidden Markov model is a directed tree, this problem can be solved exactly using the max-sum algorithm. We recall from our discussion in Section 8.4.5 that the problem of finding the most probable sequence of latent states is not the same as that of finding the set of states that are individually the most probable. The latter problem can be solved by first running the forward-backward (sum-product) algorithm to find the latent variable marginals $\gamma(\mathbf{z}_n)$ and then maximizing each of these individually (Duda *et al.*, 2001). However, the set of such states will not, in general, correspond to the most probable sequence of states. In fact, this set of states might even represent a sequence having zero probability, if it so happens that two successive states, which in isolation are individually the most probable, are such that the transition matrix element connecting them is zero.

In practice, we are usually interested in finding the most probable *sequence* of states, and this can be solved efficiently using the max-sum algorithm, which in the context of hidden Markov models is known as the *Viterbi* algorithm (Viterbi, 1967). Note that the max-sum algorithm works with log probabilities and so there is no need to use re-scaled variables as was done with the forward-backward algorithm. Figure 13.16 shows a fragment of the hidden Markov model expanded as lattice diagram. As we have already noted, the number of possible paths through the lattice grows exponentially with the length of the chain. The Viterbi algorithm searches this space of paths efficiently to find the most probable path with a computational cost that grows only linearly with the length of the chain.

As with the sum-product algorithm, we first represent the hidden Markov model as a factor graph, as shown in Figure 13.15. Again, we treat the variable node $\mathbf{z}_N$ as the root, and pass messages to the root starting with the leaf nodes. Using the results (8.93) and (8.94), we see that the messages passed in the max-sum algorithm are given by

$$\mu_{\mathbf{z}_n \to f_{n+1}}(\mathbf{z}_n) = \mu_{f_n \to \mathbf{z}_n}(\mathbf{z}_n) \tag{13.66}$$

$$\mu_{f_{n+1} \to \mathbf{z}_{n+1}}(\mathbf{z}_{n+1}) = \max_{\mathbf{z}_n} \left\{ \ln f_{n+1}(\mathbf{z}_n, \mathbf{z}_{n+1}) + \mu_{\mathbf{z}_n \to f_{n+1}}(\mathbf{z}_n) \right\}. \tag{13.67}$$

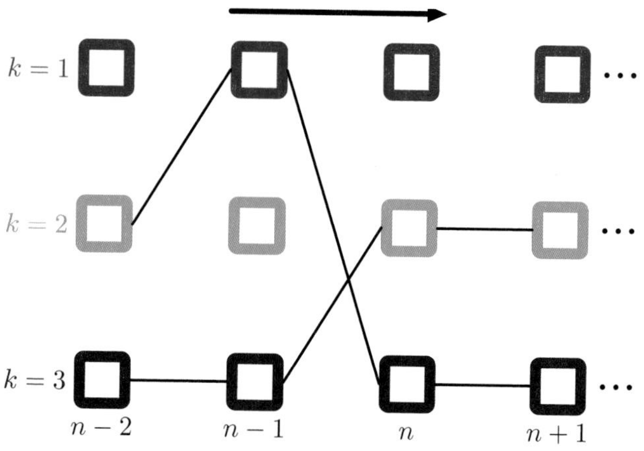

Figure 13.16 A fragment of the HMM lattice showing two possible paths. The Viterbi algorithm efficiently determines the most probable path from amongst the exponentially many possibilities. For any given path, the corresponding probability is given by the product of the elements of the transition matrix A_{jk}, corresponding to the probabilities $p(\mathbf{z}_{n+1}|\mathbf{z}_n)$ for each segment of the path, along with the emission densities $p(\mathbf{x}_n|k)$ associated with each node on the path.

If we eliminate $\mu_{\mathbf{z}_n \to f_{n+1}}(\mathbf{z}_n)$ between these two equations, and make use of (13.46), we obtain a recursion for the $f \to \mathbf{z}$ messages of the form

$$\omega(\mathbf{z}_{n+1}) = \ln p(\mathbf{x}_{n+1}|\mathbf{z}_{n+1}) + \max_{\mathbf{z}_n} \{\ln p(\mathbf{z}_{n+1}|\mathbf{z}_n) + \omega(\mathbf{z}_n)\} \tag{13.68}$$

where we have introduced the notation $\omega(\mathbf{z}_n) \equiv \mu_{f_n \to \mathbf{z}_n}(\mathbf{z}_n)$.

From (8.95) and (8.96), these messages are initialized using

$$\omega(\mathbf{z}_1) = \ln p(\mathbf{z}_1) + \ln p(\mathbf{x}_1|\mathbf{z}_1). \tag{13.69}$$

where we have used (13.45). Note that to keep the notation uncluttered, we omit the dependence on the model parameters $\boldsymbol{\theta}$ that are held fixed when finding the most probable sequence.

The Viterbi algorithm can also be derived directly from the definition (13.6) of the joint distribution by taking the logarithm and then exchanging maximizations and summations. It is easily seen that the quantities $\omega(\mathbf{z}_n)$ have the probabilistic interpretation

Exercise 13.16

$$\omega(\mathbf{z}_n) = \max_{\mathbf{z}_1,\ldots,\mathbf{z}_{n-1}} \ln p(\mathbf{x}_1,\ldots,\mathbf{x}_n,\mathbf{z}_1,\ldots,\mathbf{z}_n). \tag{13.70}$$

Once we have completed the final maximization over $\mathbf{z}_N$, we will obtain the value of the joint distribution $p(\mathbf{X}, \mathbf{Z})$ corresponding to the most probable path. We also wish to find the sequence of latent variable values that corresponds to this path. To do this, we simply make use of the back-tracking procedure discussed in Section 8.4.5. Specifically, we note that the maximization over $\mathbf{z}_n$ must be performed for each of the K possible values of $\mathbf{z}_{n+1}$. Suppose we keep a record of the values of $\mathbf{z}_n$ that correspond to the maxima for each value of the K values of $\mathbf{z}_{n+1}$. Let us denote this function by $\psi(k_n)$ where $k \in \{1,\ldots,K\}$. Once we have passed messages to the end of the chain and found the most probable state of $\mathbf{z}_N$, we can then use this function to backtrack along the chain by applying it recursively

$$k_{n-1}^{\max} = \psi(k_n^{\max}). \tag{13.71}$$

Intuitively, we can understand the Viterbi algorithm as follows. Naively, we could consider explicitly all of the exponentially many paths through the lattice, evaluate the probability for each, and then select the path having the highest probability. However, we notice that we can make a dramatic saving in computational cost as follows. Suppose that for each path we evaluate its probability by summing up products of transition and emission probabilities as we work our way forward along each path through the lattice. Consider a particular time step n and a particular state k at that time step. There will be many possible paths converging on the corresponding node in the lattice diagram. However, we need only retain that particular path that so far has the highest probability. Because there are K states at time step n, we need to keep track of K such paths. At time step $n+1$, there will be K^2 possible paths to consider, comprising K possible paths leading out of each of the K current states, but again we need only retain K of these corresponding to the best path for each state at time $n+1$. When we reach the final time step N we will discover which state corresponds to the overall most probable path. Because there is a unique path coming into that state we can trace the path back to step $N-1$ to see what state it occupied at that time, and so on back through the lattice to the state $n=1$.

13.2.6 Extensions of the hidden Markov model

The basic hidden Markov model, along with the standard training algorithm based on maximum likelihood, has been extended in numerous ways to meet the requirements of particular applications. Here we discuss a few of the more important examples.

We see from the digits example in Figure 13.11 that hidden Markov models can be quite poor generative models for the data, because many of the synthetic digits look quite unrepresentative of the training data. If the goal is sequence classification, there can be significant benefit in determining the parameters of hidden Markov models using discriminative rather than maximum likelihood techniques. Suppose we have a training set of R observation sequences $\mathbf{X}_r$, where $r = 1, \ldots, R$, each of which is labelled according to its class m, where $m = 1, \ldots, M$. For each class, we have a separate hidden Markov model with its own parameters $\boldsymbol{\theta}_m$, and we treat the problem of determining the parameter values as a standard classification problem in which we optimize the cross-entropy

$$\sum_{r=1}^{R} \ln p(m_r|\mathbf{X}_r). \tag{13.72}$$

Using Bayes' theorem this can be expressed in terms of the sequence probabilities associated with the hidden Markov models

$$\sum_{r=1}^{R} \ln \left\{ \frac{p(\mathbf{X}_r|\boldsymbol{\theta}_r)p(m_r)}{\sum_{l=1}^{M} p(\mathbf{X}_r|\boldsymbol{\theta}_l)p(l_r)} \right\} \tag{13.73}$$

where $p(m)$ is the prior probability of class m. Optimization of this cost function is more complex than for maximum likelihood (Kapadia, 1998), and in particular

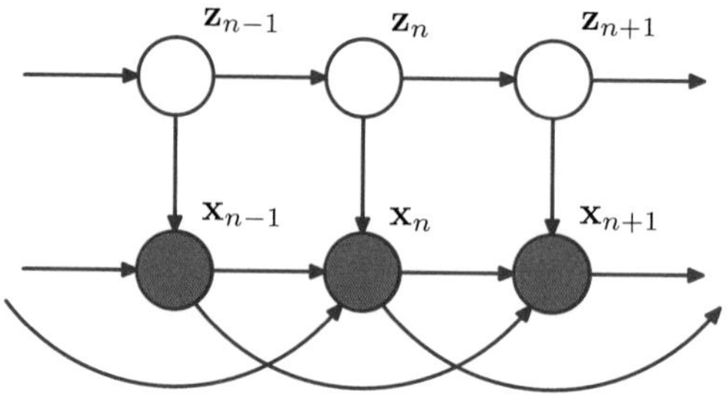

Figure 13.17 Section of an autoregressive hidden Markov model, in which the distribution of the observation $\mathbf{x}_n$ depends on a subset of the previous observations as well as on the hidden state $\mathbf{z}_n$. In this example, the distribution of $\mathbf{x}_n$ depends on the two previous observations $\mathbf{x}_{n-1}$ and $\mathbf{x}_{n-2}$.

requires that every training sequence be evaluated under each of the models in order to compute the denominator in (13.73). Hidden Markov models, coupled with discriminative training methods, are widely used in speech recognition (Kapadia, 1998).

A significant weakness of the hidden Markov model is the way in which it represents the distribution of times for which the system remains in a given state. To see the problem, note that the probability that a sequence sampled from a given hidden Markov model will spend precisely T steps in state k and then make a transition to a different state is given by

$$p(T) = (A_{kk})^T(1 - A_{kk}) \propto \exp(T \ln A_{kk}) \tag{13.74}$$

and so is an exponentially decaying function of T. For many applications, this will be a very unrealistic model of state duration. The problem can be resolved by modelling state duration directly in which the diagonal coefficients A_{kk} are all set to zero, and each state k is explicitly associated with a probability distribution $p(T|k)$ of possible duration times. From a generative point of view, when a state k is entered, a value T representing the number of time steps that the system will remain in state k is then drawn from $p(T|k)$. The model then emits T values of the observed variable $\mathbf{x}_t$, which are generally assumed to be independent so that the corresponding emission density is simply $\prod_{t=1}^{T} p(\mathbf{x}_t|k)$. This approach requires some straightforward modifications to the EM optimization procedure (Rabiner, 1989).

Another limitation of the standard HMM is that it is poor at capturing long-range correlations between the observed variables (i.e., between variables that are separated by many time steps) because these must be mediated via the first-order Markov chain of hidden states. Longer-range effects could in principle be included by adding extra links to the graphical model of Figure 13.5. One way to address this is to generalize the HMM to give the *autoregressive hidden Markov model* (Ephraim *et al.*, 1989), an example of which is shown in Figure 13.17. For discrete observations, this corresponds to expanded tables of conditional probabilities for the emission distributions. In the case of a Gaussian emission density, we can use the linear-Gaussian framework in which the conditional distribution for $\mathbf{x}_n$ given the values of the previous observations, and the value of $\mathbf{z}_n$, is a Gaussian whose mean is a linear combination of the values of the conditioning variables. Clearly the number of additional links in the graph must be limited to avoid an excessive number of free parameters. In the example shown in Figure 13.17, each observation depends on

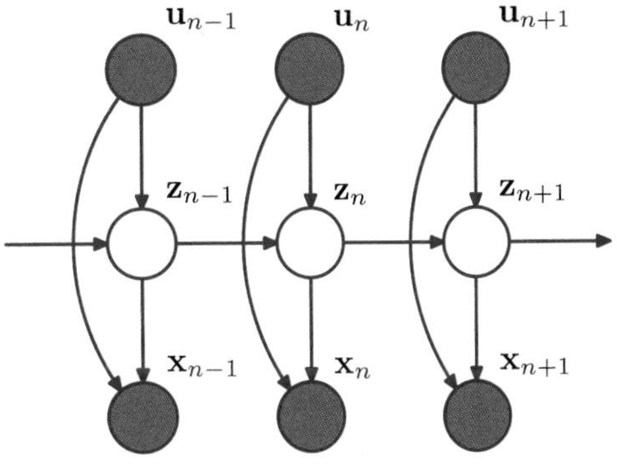

Figure 13.18 Example of an input-output hidden Markov model. In this case, both the emission probabilities and the transition probabilities depend on the values of a sequence of observations $\mathbf{u}_1, \ldots, \mathbf{u}_N$.

the two preceding observed variables as well as on the hidden state. Although this graph looks messy, we can again appeal to d-separation to see that in fact it still has a simple probabilistic structure. In particular, if we imagine conditioning on $\mathbf{z}_n$ we see that, as with the standard HMM, the values of $\mathbf{z}_{n-1}$ and $\mathbf{z}_{n+1}$ are independent, corresponding to the conditional independence property (13.5). This is easily verified by noting that every path from node $\mathbf{z}_{n-1}$ to node $\mathbf{z}_{n+1}$ passes through at least one observed node that is head-to-tail with respect to that path. As a consequence, we can again use a forward-backward recursion in the E step of the EM algorithm to determine the posterior distributions of the latent variables in a computational time that is linear in the length of the chain. Similarly, the M step involves only a minor modification of the standard M-step equations. In the case of Gaussian emission densities this involves estimating the parameters using the standard linear regression equations, discussed in Chapter 3.

We have seen that the autoregressive HMM appears as a natural extension of the standard HMM when viewed as a graphical model. In fact the probabilistic graphical modelling viewpoint motivates a plethora of different graphical structures based on the HMM. Another example is the *input-output* hidden Markov model (Bengio and Frasconi, 1995), in which we have a sequence of observed variables $\mathbf{u}_1, \ldots, \mathbf{u}_N$, in addition to the output variables $\mathbf{x}_1, \ldots, \mathbf{x}_N$, whose values influence either the distribution of latent variables or output variables, or both. An example is shown in Figure 13.18. This extends the HMM framework to the domain of supervised learning for sequential data. It is again easy to show, through the use of the d-separation criterion, that the Markov property (13.5) for the chain of latent variables still holds. To verify this, simply note that there is only one path from node $\mathbf{z}_{n-1}$ to node $\mathbf{z}_{n+1}$ and this is head-to-tail with respect to the observed node $\mathbf{z}_n$. This conditional independence property again allows the formulation of a computationally efficient learning algorithm. In particular, we can determine the parameters $\boldsymbol{\theta}$ of the model by maximizing the likelihood function $L(\boldsymbol{\theta}) = p(\mathbf{X}|\mathbf{U}, \boldsymbol{\theta})$ where $\mathbf{U}$ is a matrix whose rows are given by $\mathbf{u}_n^{\mathrm{T}}$. As a consequence of the conditional independence property (13.5) this likelihood function can be maximized efficiently using an EM algorithm in which the E step involves forward and backward recursions.

Exercise 13.18

Another variant of the HMM worthy of mention is the *factorial hidden Markov model* (Ghahramani and Jordan, 1997), in which there are multiple independent

Figure 13.19 A factorial hidden Markov model comprising two Markov chains of latent variables. For continuous observed variables $\mathbf{x}$, one possible choice of emission model is a linear-Gaussian density in which the mean of the Gaussian is a linear combination of the states of the corresponding latent variables.

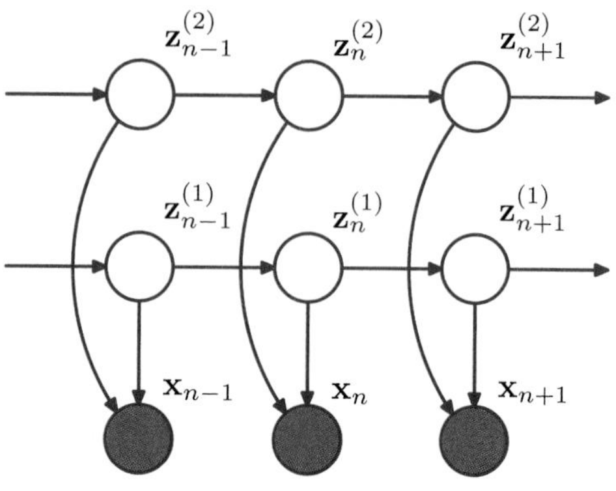

Markov chains of latent variables, and the distribution of the observed variable at a given time step is conditional on the states of all of the corresponding latent variables at that same time step. Figure 13.19 shows the corresponding graphical model. The motivation for considering factorial HMM can be seen by noting that in order to represent, say, 10 bits of information at a given time step, a standard HMM would need $K = 2^{10} = 1024$ latent states, whereas a factorial HMM could make use of 10 binary latent chains. The primary disadvantage of factorial HMMs, however, lies in the additional complexity of training them. The M step for the factorial HMM model is straightforward. However, observation of the $\mathbf{x}$ variables introduces dependencies between the latent chains, leading to difficulties with the E step. This can be seen by noting that in Figure 13.19, the variables $\mathbf{z}_n^{(1)}$ and $\mathbf{z}_n^{(2)}$ are connected by a path which is head-to-head at node $\mathbf{x}_n$ and hence they are not d-separated. The exact E step for this model does *not* correspond to running forward and backward recursions along the M Markov chains independently. This is confirmed by noting that the key conditional independence property (13.5) is not satisfied for the individual Markov chains in the factorial HMM model, as is shown using d-separation in Figure 13.20. Now suppose that there are M chains of hidden nodes and for simplicity suppose that all latent variables have the same number K of states. Then one approach would be to note that there are K^M combinations of latent variables at a given time step

Figure 13.20 Example of a path, highlighted in green, which is head-to-head at the observed nodes $\mathbf{x}_{n-1}$ and $\mathbf{x}_{n+1}$, and head-to-tail at the unobserved nodes $\mathbf{z}_{n-1}^{(2)}$, $\mathbf{z}_n^{(2)}$ and $\mathbf{z}_{n+1}^{(2)}$. Thus the path is not blocked and so the conditional independence property (13.5) does not hold for the individual latent chains of the factorial HMM model. As a consequence, there is no efficient exact E step for this model.

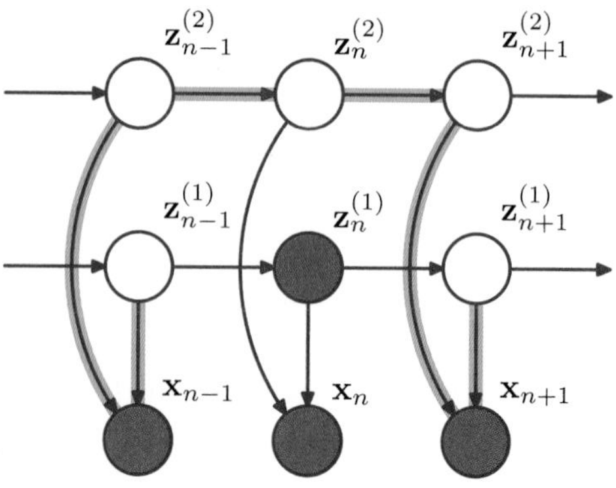

and so we can transform the model into an equivalent standard HMM having a single chain of latent variables each of which has K^M latent states. We can then run the standard forward-backward recursions in the E step. This has computational complexity $O(NK^{2M})$ that is exponential in the number M of latent chains and so will be intractable for anything other than small values of M. One solution would be to use sampling methods (discussed in Chapter 11). As an elegant deterministic alternative, Ghahramani and Jordan (1997) exploited variational inference techniques to obtain a tractable algorithm for approximate inference. This can be done using a simple variational posterior distribution that is fully factorized with respect to the latent variables, or alternatively by using a more powerful approach in which the variational distribution is described by independent Markov chains corresponding to the chains of latent variables in the original model. In the latter case, the variational inference algorithms involves running independent forward and backward recursions along each chain, which is computationally efficient and yet is also able to capture correlations between variables within the same chain.

Section 10.1

Clearly, there are many possible probabilistic structures that can be constructed according to the needs of particular applications. Graphical models provide a general technique for motivating, describing, and analysing such structures, and variational methods provide a powerful framework for performing inference in those models for which exact solution is intractable.

13.3. Linear Dynamical Systems

In order to motivate the concept of linear dynamical systems, let us consider the following simple problem, which often arises in practical settings. Suppose we wish to measure the value of an unknown quantity $\mathbf{z}$ using a noisy sensor that returns a observation $\mathbf{x}$ representing the value of $\mathbf{z}$ plus zero-mean Gaussian noise. Given a single measurement, our best guess for $\mathbf{z}$ is to assume that $\mathbf{z} = \mathbf{x}$. However, we can improve our estimate for $\mathbf{z}$ by taking lots of measurements and averaging them, because the random noise terms will tend to cancel each other. Now let's make the situation more complicated by assuming that we wish to measure a quantity $\mathbf{z}$ that is changing over time. We can take regular measurements of $\mathbf{x}$ so that at some point in time we have obtained $\mathbf{x}_1, \ldots, \mathbf{x}_N$ and we wish to find the corresponding values $\mathbf{z}_1, \ldots, \mathbf{z}_N$. If we simply average the measurements, the error due to random noise will be reduced, but unfortunately we will just obtain a single averaged estimate, in which we have averaged over the changing value of $\mathbf{z}$, thereby introducing a new source of error.

Intuitively, we could imagine doing a bit better as follows. To estimate the value of $\mathbf{z}_N$, we take only the most recent few measurements, say $\mathbf{x}_{N-L}, \ldots, \mathbf{x}_N$ and just average these. If $\mathbf{z}$ is changing slowly, and the random noise level in the sensor is high, it would make sense to choose a relatively long window of observations to average. Conversely, if the signal is changing quickly, and the noise levels are small, we might be better just to use $\mathbf{x}_N$ directly as our estimate of $\mathbf{z}_N$. Perhaps we could do even better if we take a weighted average, in which more recent measurements make a greater contribution than less recent ones.

Although this sort of intuitive argument seems plausible, it does not tell us how to form a weighted average, and any sort of hand-crafted weighing is hardly likely to be optimal. Fortunately, we can address problems such as this much more systematically by defining a probabilistic model that captures the time evolution and measurement processes and then applying the inference and learning methods developed in earlier chapters. Here we shall focus on a widely used model known as a *linear dynamical system*.

As we have seen, the HMM corresponds to the state space model shown in Figure 13.5 in which the latent variables are discrete but with arbitrary emission probability distributions. This graph of course describes a much broader class of probability distributions, all of which factorize according to (13.6). We now consider extensions to other distributions for the latent variables. In particular, we consider continuous latent variables in which the summations of the sum-product algorithm become integrals. The general form of the inference algorithms will, however, be the same as for the hidden Markov model. It is interesting to note that, historically, hidden Markov models and linear dynamical systems were developed independently. Once they are both expressed as graphical models, however, the deep relationship between them immediately becomes apparent.

One key requirement is that we retain an efficient algorithm for inference which is linear in the length of the chain. This requires that, for instance, when we take a quantity $\widehat{\alpha}(\mathbf{z}_{n-1})$, representing the posterior probability of $\mathbf{z}_{n-1}$ given observations $\mathbf{x}_1, \ldots, \mathbf{x}_{n-1}$, and multiply by the transition probability $p(\mathbf{z}_n|\mathbf{z}_{n-1})$ and the emission probability $p(\mathbf{x}_n|\mathbf{z}_n)$ and then marginalize over $\mathbf{z}_{n-1}$, we obtain a distribution over $\mathbf{z}_n$ that is of the same functional form as that over $\widehat{\alpha}(\mathbf{z}_{n-1})$. That is to say, the distribution must not become more complex at each stage, but must only change in its parameter values. Not surprisingly, the only distributions that have this property of being closed under multiplication are those belonging to the exponential family.

Here we consider the most important example from a practical perspective, which is the Gaussian. In particular, we consider a linear-Gaussian state space model so that the latent variables $\{\mathbf{z}_n\}$, as well as the observed variables $\{\mathbf{x}_n\}$, are multivariate Gaussian distributions whose means are linear functions of the states of their parents in the graph. We have seen that a directed graph of linear-Gaussian units is equivalent to a joint Gaussian distribution over all of the variables. Furthermore, marginals such as $\widehat{\alpha}(\mathbf{z}_n)$ are also Gaussian, so that the functional form of the messages is preserved and we will obtain an efficient inference algorithm. By contrast, suppose that the emission densities $p(\mathbf{x}_n|\mathbf{z}_n)$ comprise a mixture of K Gaussians each of which has a mean that is linear in $\mathbf{z}_n$. Then even if $\widehat{\alpha}(\mathbf{z}_1)$ is Gaussian, the quantity $\widehat{\alpha}(\mathbf{z}_2)$ will be a mixture of K Gaussians, $\widehat{\alpha}(\mathbf{z}_3)$ will be a mixture of K^2 Gaussians, and so on, and exact inference will not be of practical value.

We have seen that the hidden Markov model can be viewed as an extension of the mixture models of Chapter 9 to allow for sequential correlations in the data. In a similar way, we can view the linear dynamical system as a generalization of the continuous latent variable models of Chapter 12 such as probabilistic PCA and factor analysis. Each pair of nodes $\{\mathbf{z}_n, \mathbf{x}_n\}$ represents a linear-Gaussian latent variable model for that particular observation. However, the latent variables $\{\mathbf{z}_n\}$ are no longer treated as independent but now form a Markov chain.

Because the model is represented by a tree-structured directed graph, inference problems can be solved efficiently using the sum-product algorithm. The forward recursions, analogous to the α messages of the hidden Markov model, are known as the *Kalman filter* equations (Kalman, 1960; Zarchan and Musoff, 2005), and the backward recursions, analogous to the β messages, are known as the *Kalman smoother* equations, or the *Rauch-Tung-Striebel* (RTS) equations (Rauch *et al.*, 1965). The Kalman filter is widely used in many real-time tracking applications.

Because the linear dynamical system is a linear-Gaussian model, the joint distribution over all variables, as well as all marginals and conditionals, will be Gaussian. It follows that the sequence of individually most probable latent variable values is the same as the most probable latent sequence. There is thus no need to consider the analogue of the Viterbi algorithm for the linear dynamical system.

Exercise 13.19

Because the model has linear-Gaussian conditional distributions, we can write the transition and emission distributions in the general form

$$p(\mathbf{z}_n|\mathbf{z}_{n-1}) = \mathcal{N}(\mathbf{z}_n|\mathbf{A}\mathbf{z}_{n-1}, \mathbf{\Gamma}) \tag{13.75}$$

$$p(\mathbf{x}_n|\mathbf{z}_n) = \mathcal{N}(\mathbf{x}_n|\mathbf{C}\mathbf{z}_n, \mathbf{\Sigma}). \tag{13.76}$$

The initial latent variable also has a Gaussian distribution which we write as

$$p(\mathbf{z}_1) = \mathcal{N}(\mathbf{z}_1|\boldsymbol{\mu}_0, \mathbf{P}_0). \tag{13.77}$$

Note that in order to simplify the notation, we have omitted additive constant terms from the means of the Gaussians. In fact, it is straightforward to include them if desired. Traditionally, these distributions are more commonly expressed in an equivalent form in terms of noisy linear equations given by

Exercise 13.24

$$\mathbf{z}_n = \mathbf{A}\mathbf{z}_{n-1} + \mathbf{w}_n \tag{13.78}$$

$$\mathbf{x}_n = \mathbf{C}\mathbf{z}_n + \mathbf{v}_n \tag{13.79}$$

$$\mathbf{z}_1 = \boldsymbol{\mu}_0 + \mathbf{u} \tag{13.80}$$

where the noise terms have the distributions

$$\mathbf{w} \sim \mathcal{N}(\mathbf{w}|\mathbf{0}, \mathbf{\Gamma}) \tag{13.81}$$

$$\mathbf{v} \sim \mathcal{N}(\mathbf{v}|\mathbf{0}, \mathbf{\Sigma}) \tag{13.82}$$

$$\mathbf{u} \sim \mathcal{N}(\mathbf{u}|\mathbf{0}, \mathbf{P}_0). \tag{13.83}$$

The parameters of the model, denoted by $\boldsymbol{\theta} = \{\mathbf{A}, \mathbf{\Gamma}, \mathbf{C}, \mathbf{\Sigma}, \boldsymbol{\mu}_0, \mathbf{P}_0\}$, can be determined using maximum likelihood through the EM algorithm. In the E step, we need to solve the inference problem of determining the local posterior marginals for the latent variables, which can be solved efficiently using the sum-product algorithm, as we discuss in the next section.

13.3.1 Inference in LDS

We now turn to the problem of finding the marginal distributions for the latent variables conditional on the observation sequence. For given parameter settings, we also wish to make predictions of the next latent state $\mathbf{z}_n$ and of the next observation $\mathbf{x}_n$ conditioned on the observed data $\mathbf{x}_1, \ldots, \mathbf{x}_{n-1}$ for use in real-time applications. These inference problems can be solved efficiently using the sum-product algorithm, which in the context of the linear dynamical system gives rise to the Kalman filter and Kalman smoother equations.

It is worth emphasizing that because the linear dynamical system is a linear-Gaussian model, the joint distribution over all latent and observed variables is simply a Gaussian, and so in principle we could solve inference problems by using the standard results derived in previous chapters for the marginals and conditionals of a multivariate Gaussian. The role of the sum-product algorithm is to provide a more efficient way to perform such computations.

Linear dynamical systems have the identical factorization, given by (13.6), to hidden Markov models, and are again described by the factor graphs in Figures 13.14 and 13.15. Inference algorithms therefore take precisely the same form except that summations over latent variables are replaced by integrations. We begin by considering the forward equations in which we treat $\mathbf{z}_N$ as the root node, and propagate messages from the leaf node $h(\mathbf{z}_1)$ to the root. From (13.77), the initial message will be Gaussian, and because each of the factors is Gaussian, all subsequent messages will also be Gaussian. By convention, we shall propagate messages that are normalized marginal distributions corresponding to $p(\mathbf{z}_n|\mathbf{x}_1, \ldots, \mathbf{x}_n)$, which we denote by

$$\widehat{\alpha}(\mathbf{z}_n) = \mathcal{N}(\mathbf{z}_n|\boldsymbol{\mu}_n, \mathbf{V}_n). \tag{13.84}$$

This is precisely analogous to the propagation of scaled variables $\widehat{\alpha}(\mathbf{z}_n)$ given by (13.59) in the discrete case of the hidden Markov model, and so the recursion equation now takes the form

$$c_n\widehat{\alpha}(\mathbf{z}_n) = p(\mathbf{x}_n|\mathbf{z}_n) \int \widehat{\alpha}(\mathbf{z}_{n-1})p(\mathbf{z}_n|\mathbf{z}_{n-1})\,\mathrm{d}\mathbf{z}_{n-1}. \tag{13.85}$$

Substituting for the conditionals $p(\mathbf{z}_n|\mathbf{z}_{n-1})$ and $p(\mathbf{x}_n|\mathbf{z}_n)$, using (13.75) and (13.76), respectively, and making use of (13.84), we see that (13.85) becomes

$$\begin{aligned} c_n\mathcal{N}(\mathbf{z}_n|\boldsymbol{\mu}_n, \mathbf{V}_n) &= \mathcal{N}(\mathbf{x}_n|\mathbf{C}\mathbf{z}_n, \boldsymbol{\Sigma}) \\ &\int \mathcal{N}(\mathbf{z}_n|\mathbf{A}\mathbf{z}_{n-1}, \boldsymbol{\Gamma})\mathcal{N}(\mathbf{z}_{n-1}|\boldsymbol{\mu}_{n-1}, \mathbf{V}_{n-1})\,\mathrm{d}\mathbf{z}_{n-1}. \end{aligned} \tag{13.86}$$

Here we are supposing that $\boldsymbol{\mu}_{n-1}$ and $\mathbf{V}_{n-1}$ are known, and by evaluating the integral in (13.86), we wish to determine values for $\boldsymbol{\mu}_n$ and $\mathbf{V}_n$. The integral is easily evaluated by making use of the result (2.115), from which it follows that

$$\begin{aligned} &\int \mathcal{N}(\mathbf{z}_n|\mathbf{A}\mathbf{z}_{n-1}, \boldsymbol{\Gamma})\mathcal{N}(\mathbf{z}_{n-1}|\boldsymbol{\mu}_{n-1}, \mathbf{V}_{n-1})\,\mathrm{d}\mathbf{z}_{n-1} \\ &\qquad = \mathcal{N}(\mathbf{z}_n|\mathbf{A}\boldsymbol{\mu}_{n-1}, \mathbf{P}_{n-1}) \end{aligned} \tag{13.87}$$

where we have defined

$$\mathbf{P}_{n-1} = \mathbf{A}\mathbf{V}_{n-1}\mathbf{A}^{\mathrm{T}} + \mathbf{\Gamma}. \tag{13.88}$$

We can now combine this result with the first factor on the right-hand side of (13.86) by making use of (2.115) and (2.116) to give

$$\boldsymbol{\mu}_n = \mathbf{A}\boldsymbol{\mu}_{n-1} + \mathbf{K}_n(\mathbf{x}_n - \mathbf{C}\mathbf{A}\boldsymbol{\mu}_{n-1}) \tag{13.89}$$

$$\mathbf{V}_n = (\mathbf{I} - \mathbf{K}_n\mathbf{C})\mathbf{P}_{n-1} \tag{13.90}$$

$$c_n = \mathcal{N}(\mathbf{x}_n|\mathbf{C}\mathbf{A}\boldsymbol{\mu}_{n-1}, \mathbf{C}\mathbf{P}_{n-1}\mathbf{C}^{\mathrm{T}} + \boldsymbol{\Sigma}). \tag{13.91}$$

Here we have made use of the matrix inverse identities (C.5) and (C.7) and also defined the *Kalman gain matrix*

$$\mathbf{K}_n = \mathbf{P}_{n-1}\mathbf{C}^{\mathrm{T}}\left(\mathbf{C}\mathbf{P}_{n-1}\mathbf{C}^{\mathrm{T}} + \boldsymbol{\Sigma}\right)^{-1}. \tag{13.92}$$

Thus, given the values of $\boldsymbol{\mu}_{n-1}$ and $\mathbf{V}_{n-1}$, together with the new observation $\mathbf{x}_n$, we can evaluate the Gaussian marginal for $\mathbf{z}_n$ having mean $\boldsymbol{\mu}_n$ and covariance $\mathbf{V}_n$, as well as the normalization coefficient c_n.

The initial conditions for these recursion equations are obtained from

$$c_1\widehat{\alpha}(\mathbf{z}_1) = p(\mathbf{z}_1)p(\mathbf{x}_1|\mathbf{z}_1). \tag{13.93}$$

Because $p(\mathbf{z}_1)$ is given by (13.77), and $p(\mathbf{x}_1|\mathbf{z}_1)$ is given by (13.76), we can again make use of (2.115) to calculate c_1 and (2.116) to calculate $\boldsymbol{\mu}_1$ and $\mathbf{V}_1$ giving

$$\boldsymbol{\mu}_1 = \boldsymbol{\mu}_0 + \mathbf{K}_1(\mathbf{x}_1 - \mathbf{C}\boldsymbol{\mu}_0) \tag{13.94}$$

$$\mathbf{V}_1 = (\mathbf{I} - \mathbf{K}_1\mathbf{C})\mathbf{P}_0 \tag{13.95}$$

$$c_1 = \mathcal{N}(\mathbf{x}_1|\mathbf{C}\boldsymbol{\mu}_0, \mathbf{C}\mathbf{P}_0\mathbf{C}^{\mathrm{T}} + \boldsymbol{\Sigma}) \tag{13.96}$$

where

$$\mathbf{K}_1 = \mathbf{P}_0\mathbf{C}^{\mathrm{T}}\left(\mathbf{C}\mathbf{P}_0\mathbf{C}^{\mathrm{T}} + \boldsymbol{\Sigma}\right)^{-1}. \tag{13.97}$$

Similarly, the likelihood function for the linear dynamical system is given by (13.63) in which the factors c_n are found using the Kalman filtering equations.

We can interpret the steps involved in going from the posterior marginal over $\mathbf{z}_{n-1}$ to the posterior marginal over $\mathbf{z}_n$ as follows. In (13.89), we can view the quantity $\mathbf{A}\boldsymbol{\mu}_{n-1}$ as the prediction of the mean over $\mathbf{z}_n$ obtained by simply taking the mean over $\mathbf{z}_{n-1}$ and projecting it forward one step using the transition probability matrix $\mathbf{A}$. This predicted mean would give a predicted observation for $\mathbf{x}_n$ given by $\mathbf{C}\mathbf{A}\mathbf{z}_{n-1}$ obtained by applying the emission probability matrix $\mathbf{C}$ to the predicted hidden state mean. We can view the update equation (13.89) for the mean of the hidden variable distribution as taking the predicted mean $\mathbf{A}\boldsymbol{\mu}_{n-1}$ and then adding a correction that is proportional to the error $\mathbf{x}_n - \mathbf{C}\mathbf{A}\mathbf{z}_{n-1}$ between the predicted observation and the actual observation. The coefficient of this correction is given by the Kalman gain matrix. Thus we can view the Kalman filter as a process of making successive predictions and then correcting these predictions in the light of the new observations. This is illustrated graphically in Figure 13.21.

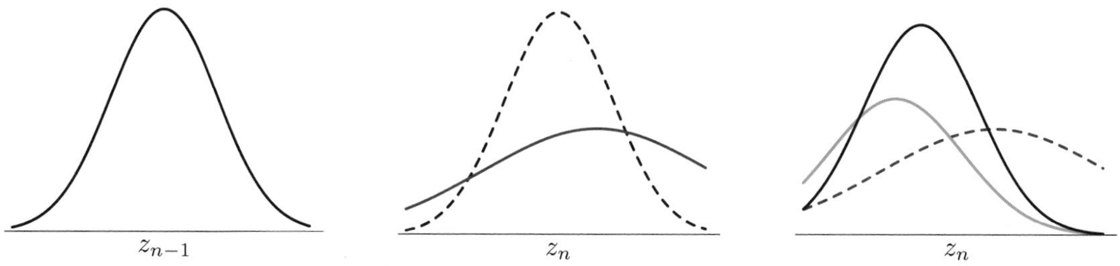

Figure 13.21 The linear dynamical system can be viewed as a sequence of steps in which increasing uncertainty in the state variable due to diffusion is compensated by the arrival of new data. In the left-hand plot, the blue curve shows the distribution $p(\mathbf{z}_{n-1}|\mathbf{x}_1, \ldots, \mathbf{x}_{n-1})$, which incorporates all the data up to step $n-1$. The diffusion arising from the nonzero variance of the transition probability $p(\mathbf{z}_n|\mathbf{z}_{n-1})$ gives the distribution $p(\mathbf{z}_n|\mathbf{x}_1, \ldots, \mathbf{x}_{n-1})$, shown in red in the centre plot. Note that this is broader and shifted relative to the blue curve (which is shown dashed in the centre plot for comparison). The next data observation $\mathbf{x}_n$ contributes through the emission density $p(\mathbf{x}_n|\mathbf{z}_n)$, which is shown as a function of $\mathbf{z}_n$ in green on the right-hand plot. Note that this is not a density with respect to $\mathbf{z}_n$ and so is not normalized to one. Inclusion of this new data point leads to a revised distribution $p(\mathbf{z}_n|\mathbf{x}_1, \ldots, \mathbf{x}_n)$ for the state density shown in blue. We see that observation of the data has shifted and narrowed the distribution compared to $p(\mathbf{z}_n|\mathbf{x}_1, \ldots, \mathbf{x}_{n-1})$ (which is shown in dashed in the right-hand plot for comparison).

If we consider a situation in which the measurement noise is small compared to the rate at which the latent variable is evolving, then we find that the posterior distribution for $\mathbf{z}_n$ depends only on the current measurement $\mathbf{x}_n$, in accordance with the intuition from our simple example at the start of the section. Similarly, if the latent variable is evolving slowly relative to the observation noise level, we find that the posterior mean for $\mathbf{z}_n$ is obtained by averaging all of the measurements obtained up to that time.

Exercise 13.27

Exercise 13.28

One of the most important applications of the Kalman filter is to tracking, and this is illustrated using a simple example of an object moving in two dimensions in Figure 13.22.

So far, we have solved the inference problem of finding the posterior marginal for a node $\mathbf{z}_n$ given observations from $\mathbf{x}_1$ up to $\mathbf{x}_n$. Next we turn to the problem of finding the marginal for a node $\mathbf{z}_n$ given all observations $\mathbf{x}_1$ to $\mathbf{x}_N$. For temporal data, this corresponds to the inclusion of future as well as past observations. Although this cannot be used for real-time prediction, it plays a key role in learning the parameters of the model. By analogy with the hidden Markov model, this problem can be solved by propagating messages from node $\mathbf{x}_N$ back to node $\mathbf{x}_1$ and combining this information with that obtained during the forward message passing stage used to compute the $\widehat{\alpha}(\mathbf{z}_n)$.

In the LDS literature, it is usual to formulate this backward recursion in terms of $\gamma(\mathbf{z}_n) = \widehat{\alpha}(\mathbf{z}_n)\widehat{\beta}(\mathbf{z}_n)$ rather than in terms of $\widehat{\beta}(\mathbf{z}_n)$. Because $\gamma(\mathbf{z}_n)$ must also be Gaussian, we write it in the form

$$\gamma(\mathbf{z}_n) = \widehat{\alpha}(\mathbf{z}_n)\widehat{\beta}(\mathbf{z}_n) = \mathcal{N}(\mathbf{z}_n|\widehat{\boldsymbol{\mu}}_n, \widehat{\mathbf{V}}_n). \tag{13.98}$$

To derive the required recursion, we start from the backward recursion (13.62) for

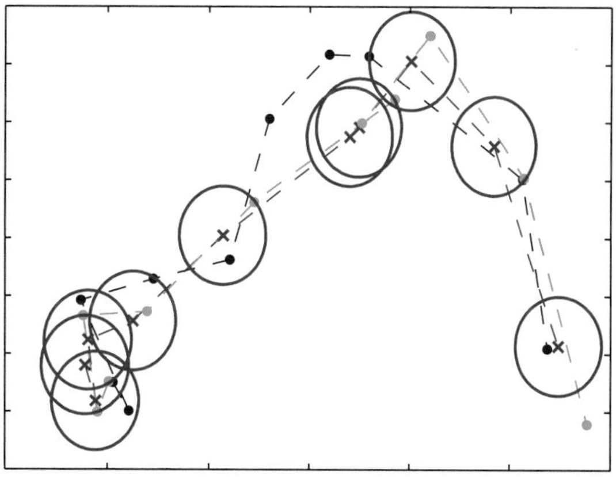

Figure 13.22 An illustration of a linear dynamical system being used to track a moving object. The blue points indicate the true positions of the object in a two-dimensional space at successive time steps, the green points denote noisy measurements of the positions, and the red crosses indicate the means of the inferred posterior distributions of the positions obtained by running the Kalman filtering equations. The covariances of the inferred positions are indicated by the red ellipses, which correspond to contours having one standard deviation.

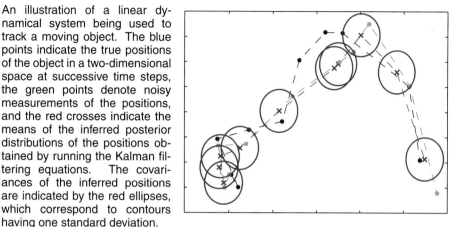

$\widehat{\beta}(\mathbf{z}_n)$, which, for continuous latent variables, can be written in the form

$$c_{n+1}\widehat{\beta}(\mathbf{z}_n) = \int \widehat{\beta}(\mathbf{z}_{n+1})p(\mathbf{x}_{n+1}|\mathbf{z}_{n+1})p(\mathbf{z}_{n+1}|\mathbf{z}_n)\,\mathrm{d}\mathbf{z}_{n+1}. \tag{13.99}$$

Exercise 13.29

We now multiply both sides of (13.99) by $\widehat{\alpha}(\mathbf{z}_n)$ and substitute for $p(\mathbf{x}_{n+1}|\mathbf{z}_{n+1})$ and $p(\mathbf{z}_{n+1}|\mathbf{z}_n)$ using (13.75) and (13.76). Then we make use of (13.89), (13.90) and (13.91), together with (13.98), and after some manipulation we obtain

$$\widehat{\boldsymbol{\mu}}_n = \boldsymbol{\mu}_n + \mathbf{J}_n\left(\widehat{\boldsymbol{\mu}}_{n+1} - \mathbf{A}\boldsymbol{\mu}_n\right) \tag{13.100}$$

$$\widehat{\mathbf{V}}_n = \mathbf{V}_n + \mathbf{J}_n\left(\widehat{\mathbf{V}}_{n+1} - \mathbf{P}_n\right)\mathbf{J}_n^{\mathrm{T}} \tag{13.101}$$

where we have defined

$$\mathbf{J}_n = \mathbf{V}_n\mathbf{A}^{\mathrm{T}}\left(\mathbf{P}_n\right)^{-1} \tag{13.102}$$

and we have made use of $\mathbf{A}\mathbf{V}_n = \mathbf{P}_n\mathbf{J}_n^{\mathrm{T}}$. Note that these recursions require that the forward pass be completed first so that the quantities $\boldsymbol{\mu}_n$ and $\mathbf{V}_n$ will be available for the backward pass.

For the EM algorithm, we also require the pairwise posterior marginals, which can be obtained from (13.65) in the form

$$\begin{aligned}\xi(\mathbf{z}_{n-1}, \mathbf{z}_n) &= (c_n)^{-1}\,\widehat{\alpha}(\mathbf{z}_{n-1})p(\mathbf{x}_n|\mathbf{z}_n)p(\mathbf{z}_n|\mathbf{z}_{n-1})\widehat{\beta}(\mathbf{z}_n) \\ &= \frac{\mathcal{N}(\mathbf{z}_{n-1}|\boldsymbol{\mu}_{n-1}, \mathbf{V}_{n-1})\mathcal{N}(\mathbf{z}_n|\mathbf{A}\mathbf{z}_{n-1}, \boldsymbol{\Gamma})\mathcal{N}(\mathbf{x}_n|\mathbf{C}\mathbf{z}_n, \boldsymbol{\Sigma})\mathcal{N}(\mathbf{z}_n|\widehat{\boldsymbol{\mu}}_n, \widehat{\mathbf{V}}_n)}{c_n\widehat{\alpha}(\mathbf{z}_n)}.\end{aligned} \tag{13.103}$$

Exercise 13.31

Substituting for $\widehat{\alpha}(\mathbf{z}_n)$ using (13.84) and rearranging, we see that $\xi(\mathbf{z}_{n-1}, \mathbf{z}_n)$ is a Gaussian with mean given by $[\widehat{\boldsymbol{\mu}}_{n-1}, \widehat{\boldsymbol{\mu}}_n]^{\mathrm{T}}$ and a covariance between $\mathbf{z}_n$ and $\mathbf{z}_{n-1}$ given by

$$\mathrm{cov}[\mathbf{z}_{n-1}, \mathbf{z}_n] = \mathbf{J}_{n-1}\widehat{\mathbf{V}}_n. \tag{13.104}$$

13.3.2 Learning in LDS

So far, we have considered the inference problem for linear dynamical systems, assuming that the model parameters $\boldsymbol{\theta} = \{\mathbf{A}, \boldsymbol{\Gamma}, \mathbf{C}, \boldsymbol{\Sigma}, \boldsymbol{\mu}_0, \mathbf{P}_0\}$ are known. Next, we consider the determination of these parameters using maximum likelihood (Ghahramani and Hinton, 1996b). Because the model has latent variables, this can be addressed using the EM algorithm, which was discussed in general terms in Chapter 9.

We can derive the EM algorithm for the linear dynamical system as follows. Let us denote the estimated parameter values at some particular cycle of the algorithm by $\boldsymbol{\theta}^{\text{old}}$. For these parameter values, we can run the inference algorithm to determine the posterior distribution of the latent variables $p(\mathbf{Z}|\mathbf{X}, \boldsymbol{\theta}^{\text{old}})$, or more precisely those local posterior marginals that are required in the M step. In particular, we shall require the following expectations

$$\mathbb{E}[\mathbf{z}_n] = \widehat{\boldsymbol{\mu}}_n \tag{13.105}$$

$$\mathbb{E}\left[\mathbf{z}_n \mathbf{z}_{n-1}^{\mathrm{T}}\right] = \widehat{\mathbf{V}}_n \mathbf{J}_{n-1}^{\mathrm{T}} + \widehat{\boldsymbol{\mu}}_n \widehat{\boldsymbol{\mu}}_{n-1}^{\mathrm{T}} \tag{13.106}$$

$$\mathbb{E}\left[\mathbf{z}_n \mathbf{z}_n^{\mathrm{T}}\right] = \widehat{\mathbf{V}}_n + \widehat{\boldsymbol{\mu}}_n \widehat{\boldsymbol{\mu}}_n^{\mathrm{T}} \tag{13.107}$$

where we have used (13.104).

Now we consider the complete-data log likelihood function, which is obtained by taking the logarithm of (13.6) and is therefore given by

$$\begin{aligned}\ln p(\mathbf{X}, \mathbf{Z}|\boldsymbol{\theta}) &= \ln p(\mathbf{z}_1|\boldsymbol{\mu}_0, \mathbf{P}_0) + \sum_{n=2}^{N} \ln p(\mathbf{z}_n|\mathbf{z}_{n-1}, \mathbf{A}, \boldsymbol{\Gamma}) \\ &\quad + \sum_{n=1}^{N} \ln p(\mathbf{x}_n|\mathbf{z}_n, \mathbf{C}, \boldsymbol{\Sigma})\end{aligned} \tag{13.108}$$

in which we have made the dependence on the parameters explicit. We now take the expectation of the complete-data log likelihood with respect to the posterior distribution $p(\mathbf{Z}|\mathbf{X}, \boldsymbol{\theta}^{\text{old}})$ which defines the function

$$Q(\boldsymbol{\theta}, \boldsymbol{\theta}^{\text{old}}) = \mathbb{E}_{\mathbf{Z}|\boldsymbol{\theta}^{\text{old}}}\left[\ln p(\mathbf{X}, \mathbf{Z}|\boldsymbol{\theta})\right]. \tag{13.109}$$

In the M step, this function is maximized with respect to the components of $\boldsymbol{\theta}$.

Consider first the parameters $\boldsymbol{\mu}_0$ and $\mathbf{P}_0$. If we substitute for $p(\mathbf{z}_1|\boldsymbol{\mu}_0, \mathbf{P}_0)$ in (13.108) using (13.77), and then take the expectation with respect to $\mathbf{Z}$, we obtain

$$Q(\boldsymbol{\theta}, \boldsymbol{\theta}^{\text{old}}) = -\frac{1}{2}\ln|\mathbf{P}_0| - \mathbb{E}_{\mathbf{Z}|\boldsymbol{\theta}^{\text{old}}}\left[\frac{1}{2}(\mathbf{z}_1 - \boldsymbol{\mu}_0)^{\mathrm{T}}\mathbf{P}_0^{-1}(\mathbf{z}_1 - \boldsymbol{\mu}_0)\right] + \text{const}$$

where all terms not dependent on $\boldsymbol{\mu}_0$ or $\mathbf{P}_0$ have been absorbed into the additive constant. Maximization with respect to $\boldsymbol{\mu}_0$ and $\mathbf{P}_0$ is easily performed by making use of the maximum likelihood solution for a Gaussian distribution discussed in Section 2.3.4, giving

Exercise 13.32

$$\boldsymbol{\mu}_0^{\text{new}} = \mathbb{E}[\mathbf{z}_1] \tag{13.110}$$

$$\mathbf{P}_0^{\text{new}} = \mathbb{E}[\mathbf{z}_1\mathbf{z}_1^{\mathrm{T}}] - \mathbb{E}[\mathbf{z}_1]\mathbb{E}[\mathbf{z}_1^{\mathrm{T}}]. \tag{13.111}$$

Similarly, to optimize $\mathbf{A}$ and $\boldsymbol{\Gamma}$, we substitute for $p(\mathbf{z}_n|\mathbf{z}_{n-1}, \mathbf{A}, \boldsymbol{\Gamma})$ in (13.108) using (13.75) giving

$$Q(\boldsymbol{\theta}, \boldsymbol{\theta}^{\text{old}}) = -\frac{N-1}{2}\ln|\boldsymbol{\Gamma}| \\ -\mathbb{E}_{\mathbf{Z}|\boldsymbol{\theta}^{\text{old}}}\left[\frac{1}{2}\sum_{n=2}^{N}(\mathbf{z}_n - \mathbf{A}\mathbf{z}_{n-1})^{\mathrm{T}}\boldsymbol{\Gamma}^{-1}(\mathbf{z}_n - \mathbf{A}\mathbf{z}_{n-1})\right] + \text{const} \tag{13.112}$$

Exercise 13.33

in which the constant comprises terms that are independent of $\mathbf{A}$ and $\boldsymbol{\Gamma}$. Maximizing with respect to these parameters then gives

$$\mathbf{A}^{\text{new}} = \left(\sum_{n=2}^{N}\mathbb{E}\left[\mathbf{z}_n\mathbf{z}_{n-1}^{\mathrm{T}}\right]\right)\left(\sum_{n=2}^{N}\mathbb{E}\left[\mathbf{z}_{n-1}\mathbf{z}_{n-1}^{\mathrm{T}}\right]\right)^{-1} \tag{13.113}$$

$$\boldsymbol{\Gamma}^{\text{new}} = \frac{1}{N-1}\sum_{n=2}^{N}\left\{\mathbb{E}\left[\mathbf{z}_n\mathbf{z}_n^{\mathrm{T}}\right] - \mathbf{A}^{\text{new}}\mathbb{E}\left[\mathbf{z}_{n-1}\mathbf{z}_n^{\mathrm{T}}\right]\right. \\ \left. -\mathbb{E}\left[\mathbf{z}_n\mathbf{z}_{n-1}^{\mathrm{T}}\right](\mathbf{A}^{\text{new}})^{\mathrm{T}} + \mathbf{A}^{\text{new}}\mathbb{E}\left[\mathbf{z}_{n-1}\mathbf{z}_{n-1}^{\mathrm{T}}\right](\mathbf{A}^{\text{new}})^{\mathrm{T}}\right\}. \tag{13.114}$$

Note that $\mathbf{A}^{\text{new}}$ must be evaluated first, and the result can then be used to determine $\boldsymbol{\Gamma}^{\text{new}}$.

Finally, in order to determine the new values of $\mathbf{C}$ and $\boldsymbol{\Sigma}$, we substitute for $p(\mathbf{x}_n|\mathbf{z}_n, \mathbf{C}, \boldsymbol{\Sigma})$ in (13.108) using (13.76) giving

$$Q(\boldsymbol{\theta}, \boldsymbol{\theta}^{\text{old}}) = -\frac{N}{2}\ln|\boldsymbol{\Sigma}| \\ -\mathbb{E}_{\mathbf{Z}|\boldsymbol{\theta}^{\text{old}}}\left[\frac{1}{2}\sum_{n=1}^{N}(\mathbf{x}_n - \mathbf{C}\mathbf{z}_n)^{\mathrm{T}}\boldsymbol{\Sigma}^{-1}(\mathbf{x}_n - \mathbf{C}\mathbf{z}_n)\right] + \text{const.}$$

Exercise 13.34

Maximizing with respect to $\mathbf{C}$ and $\boldsymbol{\Sigma}$ then gives

$$\mathbf{C}^{\text{new}} = \left(\sum_{n=1}^{N}\mathbf{x}_n\mathbb{E}\left[\mathbf{z}_n^{\mathrm{T}}\right]\right)\left(\sum_{n=1}^{N}\mathbb{E}\left[\mathbf{z}_n\mathbf{z}_n^{\mathrm{T}}\right]\right)^{-1} \tag{13.115}$$

$$\boldsymbol{\Sigma}^{\text{new}} = \frac{1}{N}\sum_{n=1}^{N}\left\{\mathbf{x}_n\mathbf{x}_n^{\mathrm{T}} - (\mathbf{C}^{\text{new}})^{\mathrm{T}}\mathbb{E}\left[\mathbf{z}_n\right]\mathbf{x}_n^{\mathrm{T}}\right. \\ \left. -\mathbf{x}_n\mathbb{E}\left[\mathbf{z}_n^{\mathrm{T}}\right]\mathbf{C}^{\text{new}} + (\mathbf{C}^{\text{new}})^{\mathrm{T}}\mathbb{E}\left[\mathbf{z}_n\mathbf{z}_n^{\mathrm{T}}\right]\mathbf{C}^{\text{new}}\right\}. \tag{13.116}$$

We have approached parameter learning in the linear dynamical system using maximum likelihood. Inclusion of priors to give a MAP estimate is straightforward, and a fully Bayesian treatment can be found by applying the analytical approximation techniques discussed in Chapter 10, though a detailed treatment is precluded here due to lack of space.

13.3.3 Extensions of LDS

As with the hidden Markov model, there is considerable interest in extending the basic linear dynamical system in order to increase its capabilities. Although the assumption of a linear-Gaussian model leads to efficient algorithms for inference and learning, it also implies that the marginal distribution of the observed variables is simply a Gaussian, which represents a significant limitation. One simple extension of the linear dynamical system is to use a Gaussian mixture as the initial distribution for $\mathbf{z}_1$. If this mixture has K components, then the forward recursion equations (13.85) will lead to a mixture of K Gaussians over each hidden variable $\mathbf{z}_n$, and so the model is again tractable.

For many applications, the Gaussian emission density is a poor approximation. If instead we try to use a mixture of K Gaussians as the emission density, then the posterior $\widehat{\alpha}(\mathbf{z}_1)$ will also be a mixture of K Gaussians. However, from (13.85) the posterior $\widehat{\alpha}(\mathbf{z}_2)$ will comprise a mixture of K^2 Gaussians, and so on, with $\widehat{\alpha}(\mathbf{z}_n)$ being given by a mixture of K^n Gaussians. Thus the number of components grows exponentially with the length of the chain, and so this model is impractical.

More generally, introducing transition or emission models that depart from the linear-Gaussian (or other exponential family) model leads to an intractable inference problem. We can make deterministic approximations such as assumed density filtering or expectation propagation, or we can make use of sampling methods, as discussed in Section 13.3.4. One widely used approach is to make a Gaussian approximation by linearizing around the mean of the predicted distribution, which gives rise to the *extended Kalman filter* (Zarchan and Musoff, 2005).

Chapter 10

As with hidden Markov models, we can develop interesting extensions of the basic linear dynamical system by expanding its graphical representation. For example, the *switching state space model* (Ghahramani and Hinton, 1998) can be viewed as a combination of the hidden Markov model with a set of linear dynamical systems. The model has multiple Markov chains of continuous linear-Gaussian latent variables, each of which is analogous to the latent chain of the linear dynamical system discussed earlier, together with a Markov chain of discrete variables of the form used in a hidden Markov model. The output at each time step is determined by stochastically choosing one of the continuous latent chains, using the state of the discrete latent variable as a switch, and then emitting an observation from the corresponding conditional output distribution. Exact inference in this model is intractable, but variational methods lead to an efficient inference scheme involving forward-backward recursions along each of the continuous and discrete Markov chains independently. Note that, if we consider multiple chains of discrete latent variables, and use one as the switch to select from the remainder, we obtain an analogous model having only discrete latent variables known as the *switching hidden Markov model*.

13.3.4 Particle filters

Chapter 11

For dynamical systems which are not linear-Gaussian, for example, if they use a non-Gaussian emission density, we can turn to sampling methods in order to find a tractable inference algorithm. In particular, we can apply the sampling-importance-resampling formalism of Section 11.1.5 to obtain a sequential Monte Carlo algorithm known as the particle filter.

Consider the class of distributions represented by the graphical model in Figure 13.5, and suppose we are given the observed values $\mathbf{X}_n = (\mathbf{x}_1, \ldots, \mathbf{x}_n)$ and we wish to draw L samples from the posterior distribution $p(\mathbf{z}_n|\mathbf{X}_n)$. Using Bayes' theorem, we have

$$\begin{aligned}
\mathbb{E}[f(\mathbf{z}_n)] &= \int f(\mathbf{z}_n) p(\mathbf{z}_n|\mathbf{X}_n)\,\mathrm{d}\mathbf{z}_n \\
&= \int f(\mathbf{z}_n) p(\mathbf{z}_n|\mathbf{x}_n, \mathbf{X}_{n-1})\,\mathrm{d}\mathbf{z}_n \\
&= \frac{\displaystyle\int f(\mathbf{z}_n) p(\mathbf{x}_n|\mathbf{z}_n) p(\mathbf{z}_n|\mathbf{X}_{n-1})\,\mathrm{d}\mathbf{z}_n}{\displaystyle\int p(\mathbf{x}_n|\mathbf{z}_n) p(\mathbf{z}_n|\mathbf{X}_{n-1})\,\mathrm{d}\mathbf{z}_n} \\
&\simeq \sum_{l=1}^{L} w_n^{(l)} f(\mathbf{z}_n^{(l)})
\end{aligned} \tag{13.117}$$

where $\{\mathbf{z}_n^{(l)}\}$ is a set of samples drawn from $p(\mathbf{z}_n|\mathbf{X}_{n-1})$ and we have made use of the conditional independence property $p(\mathbf{x}_n|\mathbf{z}_n, \mathbf{X}_{n-1}) = p(\mathbf{x}_n|\mathbf{z}_n)$, which follows from the graph in Figure 13.5. The sampling weights $\{w_n^{(l)}\}$ are defined by

$$w_n^{(l)} = \frac{p(\mathbf{x}_n|\mathbf{z}_n^{(l)})}{\sum_{m=1}^{L} p(\mathbf{x}_n|\mathbf{z}_n^{(m)})} \tag{13.118}$$

where the same samples are used in the numerator as in the denominator. Thus the posterior distribution $p(\mathbf{z}_n|\mathbf{x}_n)$ is represented by the set of samples $\{\mathbf{z}_n^{(l)}\}$ together with the corresponding weights $\{w_n^{(l)}\}$. Note that these weights satisfy $0 \leqslant w_n^{(l)} \leqslant 1$ and $\sum_l w_n^{(l)} = 1$.

Because we wish to find a sequential sampling scheme, we shall suppose that a set of samples and weights have been obtained at time step n, and that we have subsequently observed the value of $\mathbf{x}_{n+1}$, and we wish to find the weights and samples at time step $n+1$. We first sample from the distribution $p(\mathbf{z}_{n+1}|\mathbf{X}_n)$. This is straightforward since, again using Bayes' theorem

$$\begin{aligned} p(\mathbf{z}_{n+1}|\mathbf{X}_n) &= \int p(\mathbf{z}_{n+1}|\mathbf{z}_n, \mathbf{X}_n)p(\mathbf{z}_n|\mathbf{X}_n)\,\mathrm{d}\mathbf{z}_n \\ &= \int p(\mathbf{z}_{n+1}|\mathbf{z}_n)p(\mathbf{z}_n|\mathbf{X}_n)\,\mathrm{d}\mathbf{z}_n \\ &= \int p(\mathbf{z}_{n+1}|\mathbf{z}_n)p(\mathbf{z}_n|\mathbf{x}_n, \mathbf{X}_{n-1})\,\mathrm{d}\mathbf{z}_n \\ &= \frac{\int p(\mathbf{z}_{n+1}|\mathbf{z}_n)p(\mathbf{x}_n|\mathbf{z}_n)p(\mathbf{z}_n|\mathbf{X}_{n-1})\,\mathrm{d}\mathbf{z}_n}{\int p(\mathbf{x}_n|\mathbf{z}_n)p(\mathbf{z}_n|\mathbf{X}_{n-1})\,\mathrm{d}\mathbf{z}_n} \\ &\simeq \sum_l w_n^{(l)} p(\mathbf{z}_{n+1}|\mathbf{z}_n^{(l)}) \end{aligned} \tag{13.119}$$

where we have made use of the conditional independence properties

$$p(\mathbf{z}_{n+1}|\mathbf{z}_n, \mathbf{X}_n) = p(\mathbf{z}_{n+1}|\mathbf{z}_n) \tag{13.120}$$

$$p(\mathbf{x}_n|\mathbf{z}_n, \mathbf{X}_{n-1}) = p(\mathbf{x}_n|\mathbf{z}_n) \tag{13.121}$$

which follow from the application of the d-separation criterion to the graph in Figure 13.5. The distribution given by (13.119) is a mixture distribution, and samples can be drawn by choosing a component l with probability given by the mixing coefficients $w^{(l)}$ and then drawing a sample from the corresponding component.

In summary, we can view each step of the particle filter algorithm as comprising two stages. At time step n, we have a sample representation of the posterior distribution $p(\mathbf{z}_n|\mathbf{X}_n)$ expressed as samples $\{\mathbf{z}_n^{(l)}\}$ with corresponding weights $\{w_n^{(l)}\}$. This can be viewed as a mixture representation of the form (13.119). To obtain the corresponding representation for the next time step, we first draw L samples from the mixture distribution (13.119), and then for each sample we use the new observation $\mathbf{x}_{n+1}$ to evaluate the corresponding weights $w_{n+1}^{(l)} \propto p(\mathbf{x}_{n+1}|\mathbf{z}_{n+1}^{(l)})$. This is illustrated, for the case of a single variable z, in Figure 13.23.

The particle filtering, or sequential Monte Carlo, approach has appeared in the literature under various names including the *bootstrap filter* (Gordon *et al.*, 1993), *survival of the fittest* (Kanazawa *et al.*, 1995), and the *condensation* algorithm (Isard and Blake, 1998).

Exercises

13.1 ($\star$) **www** Use the technique of d-separation, discussed in Section 8.2, to verify that the Markov model shown in Figure 13.3 having N nodes in total satisfies the conditional independence properties (13.3) for $n = 2, \ldots, N$. Similarly, show that a model described by the graph in Figure 13.4 in which there are N nodes in total

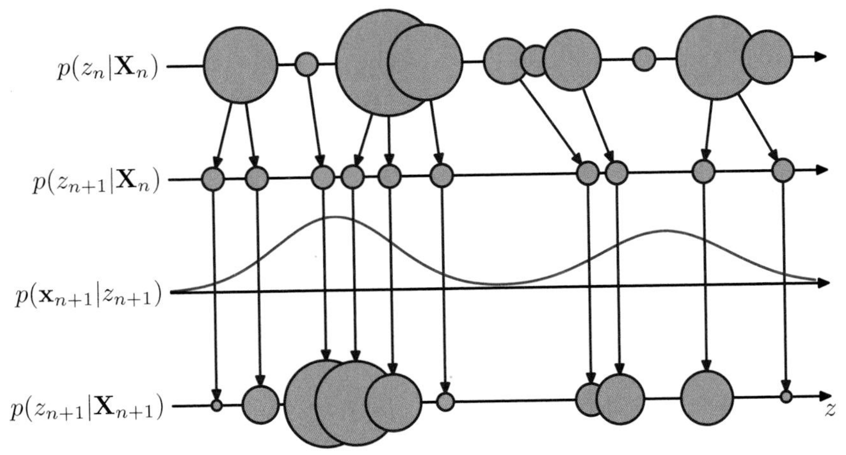

Figure 13.23 Schematic illustration of the operation of the particle filter for a one-dimensional latent space. At time step n, the posterior $p(z_n|\mathbf{x}_n)$ is represented as a mixture distribution, shown schematically as circles whose sizes are proportional to the weights $w_n^{(l)}$. A set of L samples is then drawn from this distribution and the new weights $w_{n+1}^{(l)}$ evaluated using $p(\mathbf{x}_{n+1}|\mathbf{z}_{n+1}^{(l)})$.

satisfies the conditional independence properties

$$p(\mathbf{x}_n|\mathbf{x}_1, \ldots, \mathbf{x}_{n-1}) = p(\mathbf{x}_n|\mathbf{x}_{n-1}, \mathbf{x}_{n-2}) \tag{13.122}$$

for $n = 3, \ldots, N$.

13.2 ($\star\star$) Consider the joint probability distribution (13.2) corresponding to the directed graph of Figure 13.3. Using the sum and product rules of probability, verify that this joint distribution satisfies the conditional independence property (13.3) for $n = 2, \ldots, N$. Similarly, show that the second-order Markov model described by the joint distribution (13.4) satisfies the conditional independence property

$$p(\mathbf{x}_n|\mathbf{x}_1, \ldots, \mathbf{x}_{n-1}) = p(\mathbf{x}_n|\mathbf{x}_{n-1}, \mathbf{x}_{n-2}) \tag{13.123}$$

for $n = 3, \ldots, N$.

13.3 ($\star$) By using d-separation, show that the distribution $p(\mathbf{x}_1, \ldots, \mathbf{x}_N)$ of the observed data for the state space model represented by the directed graph in Figure 13.5 does not satisfy any conditional independence properties and hence does not exhibit the Markov property at any finite order.

13.4 ($\star\star$) **www** Consider a hidden Markov model in which the emission densities are represented by a parametric model $p(\mathbf{x}|\mathbf{z}, \mathbf{w})$, such as a linear regression model or a neural network, in which $\mathbf{w}$ is a vector of adaptive parameters. Describe how the parameters $\mathbf{w}$ can be learned from data using maximum likelihood.

13.5 ($\star\star$) Verify the M-step equations (13.18) and (13.19) for the initial state probabilities and transition probability parameters of the hidden Markov model by maximization of the expected complete-data log likelihood function (13.17), using appropriate Lagrange multipliers to enforce the summation constraints on the components of $\boldsymbol{\pi}$ and $\mathbf{A}$.

13.6 ($\star$) Show that if any elements of the parameters $\boldsymbol{\pi}$ or $\mathbf{A}$ for a hidden Markov model are initially set to zero, then those elements will remain zero in all subsequent updates of the EM algorithm.

13.7 ($\star$) Consider a hidden Markov model with Gaussian emission densities. Show that maximization of the function $Q(\boldsymbol{\theta}, \boldsymbol{\theta}^{\text{old}})$ with respect to the mean and covariance parameters of the Gaussians gives rise to the M-step equations (13.20) and (13.21).

13.8 ($\star\star$) **www** For a hidden Markov model having discrete observations governed by a multinomial distribution, show that the conditional distribution of the observations given the hidden variables is given by (13.22) and the corresponding M step equations are given by (13.23). Write down the analogous equations for the conditional distribution and the M step equations for the case of a hidden Markov with multiple binary output variables each of which is governed by a Bernoulli conditional distribution. Hint: refer to Sections 2.1 and 2.2 for a discussion of the corresponding maximum likelihood solutions for i.i.d. data if required.

13.9 ($\star\star$) **www** Use the d-separation criterion to verify that the conditional independence properties (13.24)–(13.31) are satisfied by the joint distribution for the hidden Markov model defined by (13.6).

13.10 ($\star\star\star$) By applying the sum and product rules of probability, verify that the conditional independence properties (13.24)–(13.31) are satisfied by the joint distribution for the hidden Markov model defined by (13.6).

13.11 ($\star\star$) Starting from the expression (8.72) for the marginal distribution over the variables of a factor in a factor graph, together with the results for the messages in the sum-product algorithm obtained in Section 13.2.3, derive the result (13.43) for the joint posterior distribution over two successive latent variables in a hidden Markov model.

13.12 ($\star\star$) Suppose we wish to train a hidden Markov model by maximum likelihood using data that comprises R independent sequences of observations, which we denote by $\mathbf{X}^{(r)}$ where $r = 1, \ldots, R$. Show that in the E step of the EM algorithm, we simply evaluate posterior probabilities for the latent variables by running the α and β recursions independently for each of the sequences. Also show that in the M step, the initial probability and transition probability parameters are re-estimated using modified forms of (13.18) and (13.19) given by

$$\pi_k = \frac{\sum_{r=1}^{R} \gamma(z_{1k}^{(r)})}{\sum_{r=1}^{R}\sum_{j=1}^{K} \gamma(z_{1j}^{(r)})} \tag{13.124}$$

$$A_{jk} = \frac{\sum_{r=1}^{R}\sum_{n=2}^{N} \xi(z_{n-1,j}^{(r)}, z_{n,k}^{(r)})}{\sum_{r=1}^{R}\sum_{l=1}^{K}\sum_{n=2}^{N} \xi(z_{n-1,j}^{(r)}, z_{n,l}^{(r)})} \tag{13.125}$$

where, for notational convenience, we have assumed that the sequences are of the same length (the generalization to sequences of different lengths is straightforward). Similarly, show that the M-step equation for re-estimation of the means of Gaussian emission models is given by

$$\boldsymbol{\mu}_k = \frac{\sum_{r=1}^{R}\sum_{n=1}^{N} \gamma(z_{nk}^{(r)})\mathbf{x}_n^{(r)}}{\sum_{r=1}^{R}\sum_{n=1}^{N} \gamma(z_{nk}^{(r)})}. \tag{13.126}$$

Note that the M-step equations for other emission model parameters and distributions take an analogous form.

13.13 ($\star\star$) **www** Use the definition (8.64) of the messages passed from a factor node to a variable node in a factor graph, together with the expression (13.6) for the joint distribution in a hidden Markov model, to show that the definition (13.50) of the alpha message is the same as the definition (13.34).

13.14 ($\star\star$) Use the definition (8.67) of the messages passed from a factor node to a variable node in a factor graph, together with the expression (13.6) for the joint distribution in a hidden Markov model, to show that the definition (13.52) of the beta message is the same as the definition (13.35).

13.15 ($\star\star$) Use the expressions (13.33) and (13.43) for the marginals in a hidden Markov model to derive the corresponding results (13.64) and (13.65) expressed in terms of re-scaled variables.

13.16 ($\star\star\star$) In this exercise, we derive the forward message passing equation for the Viterbi algorithm directly from the expression (13.6) for the joint distribution. This involves maximizing over all of the hidden variables $\mathbf{z}_1, \ldots, \mathbf{z}_N$. By taking the logarithm and then exchanging maximizations and summations, derive the recursion

(13.68) where the quantities $\omega(\mathbf{z}_n)$ are defined by (13.70). Show that the initial condition for this recursion is given by (13.69).

13.17 ($\star$) **www** Show that the directed graph for the input-output hidden Markov model, given in Figure 13.18, can be expressed as a tree-structured factor graph of the form shown in Figure 13.15 and write down expressions for the initial factor $h(\mathbf{z}_1)$ and for the general factor $f_n(\mathbf{z}_{n-1}, \mathbf{z}_n)$ where $2 \leqslant n \leqslant N$.

13.18 ($\star\star\star$) Using the result of Exercise 13.17, derive the recursion equations, including the initial conditions, for the forward-backward algorithm for the input-output hidden Markov model shown in Figure 13.18.

13.19 ($\star$) **www** The Kalman filter and smoother equations allow the posterior distributions over individual latent variables, conditioned on all of the observed variables, to be found efficiently for linear dynamical systems. Show that the sequence of latent variable values obtained by maximizing each of these posterior distributions individually is the same as the most probable sequence of latent values. To do this, simply note that the joint distribution of all latent and observed variables in a linear dynamical system is Gaussian, and hence all conditionals and marginals will also be Gaussian, and then make use of the result (2.98).

13.20 ($\star\star$) **www** Use the result (2.115) to prove (13.87).

13.21 ($\star\star$) Use the results (2.115) and (2.116), together with the matrix identities (C.5) and (C.7), to derive the results (13.89), (13.90), and (13.91), where the Kalman gain matrix $\mathbf{K}_n$ is defined by (13.92).

13.22 ($\star\star$) **www** Using (13.93), together with the definitions (13.76) and (13.77) and the result (2.115), derive (13.96).

13.23 ($\star\star$) Using (13.93), together with the definitions (13.76) and (13.77) and the result (2.116), derive (13.94), (13.95) and (13.97).

13.24 ($\star\star$) **www** Consider a generalization of (13.75) and (13.76) in which we include constant terms $\mathbf{a}$ and $\mathbf{c}$ in the Gaussian means, so that

$$p(\mathbf{z}_n|\mathbf{z}_{n-1}) = \mathcal{N}(\mathbf{z}_n|\mathbf{A}\mathbf{z}_{n-1} + \mathbf{a}, \mathbf{\Gamma}) \tag{13.127}$$

$$p(\mathbf{x}_n|\mathbf{z}_n) = \mathcal{N}(\mathbf{x}_n|\mathbf{C}\mathbf{z}_n + \mathbf{c}, \mathbf{\Sigma}). \tag{13.128}$$

Show that this extension can be re-cast in the framework discussed in this chapter by defining a state vector $\mathbf{z}$ with an additional component fixed at unity, and then augmenting the matrices $\mathbf{A}$ and $\mathbf{C}$ using extra columns corresponding to the parameters $\mathbf{a}$ and $\mathbf{c}$.

13.25 ($\star\star$) In this exercise, we show that when the Kalman filter equations are applied to independent observations, they reduce to the results given in Section 2.3 for the maximum likelihood solution for a single Gaussian distribution. Consider the problem of finding the mean μ of a single Gaussian random variable x, in which we are given a set of independent observations $\{x_1, \ldots, x_N\}$. To model this we can use a linear dynamical system governed by (13.75) and (13.76), with latent variables $\{z_1, \ldots, z_N\}$ in which $\mathbf{C} = 1$, $\mathbf{A} = 1$ and $\boldsymbol{\Gamma} = 0$. Let the parameters $\boldsymbol{\mu}_0$ and $\mathbf{P}_0$ of the initial state be denoted by μ_0 and σ_0^2, respectively, and suppose that $\boldsymbol{\Sigma}$ becomes σ^2. Write down the corresponding Kalman filter equations starting from the general results (13.89) and (13.90), together with (13.94) and (13.95). Show that these are equivalent to the results (2.141) and (2.142) obtained directly by considering independent data.

13.26 $(\star\star\star)$ Consider a special case of the linear dynamical system of Section 13.3 that is equivalent to probabilistic PCA, so that the transition matrix $\mathbf{A} = \mathbf{0}$, the covariance $\boldsymbol{\Gamma} = \mathbf{I}$, and the noise covariance $\boldsymbol{\Sigma} = \sigma^2\mathbf{I}$. By making use of the matrix inversion identity (C.7) show that, if the emission density matrix $\mathbf{C}$ is denoted $\mathbf{W}$, then the posterior distribution over the hidden states defined by (13.89) and (13.90) reduces to the result (12.42) for probabilistic PCA, assuming $\boldsymbol{\mu} = \mathbf{0}$ in (12.42).

13.27 $(\star)$ **www** Consider a linear dynamical system of the form discussed in Section 13.3 in which the amplitude of the observation noise goes to zero, so that $\boldsymbol{\Sigma} = \mathbf{0}$. Show that, in the case $\mathbf{C} = \mathbf{I}$, the posterior distribution for $\mathbf{z}_n$ has mean $\mathbf{x}_n$ and zero variance. This accords with our intuition that if there is no noise, we should just use the current observation $\mathbf{x}_n$ to estimate the state variable $\mathbf{z}_n$ and ignore all previous observations.

13.28 $(\star\star\star)$ Consider a special case of the linear dynamical system of Section 13.3 in which the state variable $\mathbf{z}_n$ is constrained to be equal to the previous state variable, which corresponds to $\mathbf{A} = \mathbf{I}$ and $\boldsymbol{\Gamma} = \mathbf{0}$. For simplicity, assume also that $\mathbf{C} = \mathbf{I}$ and that $\mathbf{P}_0 \to \infty$ so that the initial conditions for $\mathbf{z}$ are unimportant, and the predictions are determined purely by the data. Use proof by induction to show that the posterior mean for state $\mathbf{z}_n$ is determined by the average of $\mathbf{x}_1, \ldots, \mathbf{x}_n$. This corresponds to the intuitive result that if the state variable is constant, our best estimate is obtained by averaging the observations.

13.29 $(\star\star\star)$ Starting from the backwards recursion equation (13.99), derive the RTS smoothing equations (13.100) and (13.101) for the Gaussian linear dynamical system.

13.30 $(\star\star)$ Starting from the result (13.65) for the pairwise posterior marginal in a state space model, derive the specific form (13.103) for the case of the Gaussian linear dynamical system.

13.31 $(\star\star)$ Starting from the result (13.103) and by substituting for $\widehat{\alpha}(\mathbf{z}_n)$ using (13.84), verify the result (13.104) for the covariance between $\mathbf{z}_n$ and $\mathbf{z}_{n-1}$.

13.32 $(\star\star)$ **www** Verify the results (13.110) and (13.111) for the M-step equations for $\boldsymbol{\mu}_0$ and $\mathbf{P}_0$ in the linear dynamical system.

13.33 $(\star\star)$ Verify the results (13.113) and (13.114) for the M-step equations for $\mathbf{A}$ and $\boldsymbol{\Gamma}$ in the linear dynamical system.

13.34 $(\star\star)$ Verify the results (13.115) and (13.116) for the M-step equations for $\mathbf{C}$ and $\boldsymbol{\Sigma}$ in the linear dynamical system.

14 Combining Models

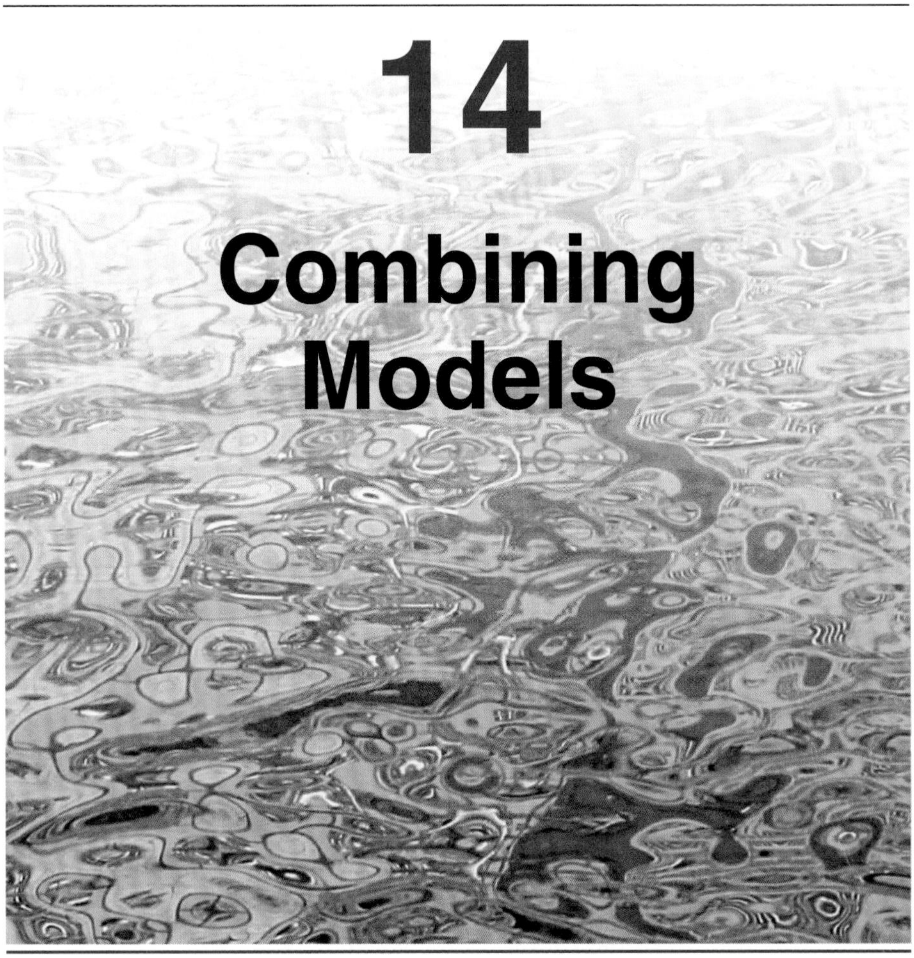

In earlier chapters, we have explored a range of different models for solving classification and regression problems. It is often found that improved performance can be obtained by combining multiple models together in some way, instead of just using a single model in isolation. For instance, we might train L different models and then make predictions using the average of the predictions made by each model. Such combinations of models are sometimes called *committees*. In Section 14.2, we discuss ways to apply the committee concept in practice, and we also give some insight into why it can sometimes be an effective procedure.

One important variant of the committee method, known as *boosting*, involves training multiple models in sequence in which the error function used to train a particular model depends on the performance of the previous models. This can produce substantial improvements in performance compared to the use of a single model and is discussed in Section 14.3.

Instead of averaging the predictions of a set of models, an alternative form of model combination is to select one of the models to make the prediction, in which the choice of model is a function of the input variables. Thus different models become responsible for making predictions in different regions of input space. One widely used framework of this kind is known as a *decision tree* in which the selection process can be described as a sequence of binary selections corresponding to the traversal of a tree structure and is discussed in Section 14.4. In this case, the individual models are generally chosen to be very simple, and the overall flexibility of the model arises from the input-dependent selection process. Decision trees can be applied to both classification and regression problems.

One limitation of decision trees is that the division of input space is based on hard splits in which only one model is responsible for making predictions for any given value of the input variables. The decision process can be softened by moving to a probabilistic framework for combining models, as discussed in Section 14.5. For example, if we have a set of K models for a conditional distribution $p(t|\mathbf{x}, k)$ where $\mathbf{x}$ is the input variable, t is the target variable, and $k = 1, \ldots, K$ indexes the model, then we can form a probabilistic mixture of the form

$$p(t|\mathbf{x}) = \sum_{k=1}^{K} \pi_k(\mathbf{x}) p(t|\mathbf{x}, k) \tag{14.1}$$

in which $\pi_k(\mathbf{x}) = p(k|\mathbf{x})$ represent the input-dependent mixing coefficients. Such models can be viewed as mixture distributions in which the component densities, as well as the mixing coefficients, are conditioned on the input variables and are known as *mixtures of experts*. They are closely related to the mixture density network model discussed in Section 5.6.

14.1. Bayesian Model Averaging

Section 9.2

It is important to distinguish between model combination methods and Bayesian model averaging, as the two are often confused. To understand the difference, consider the example of density estimation using a mixture of Gaussians in which several Gaussian components are combined probabilistically. The model contains a binary latent variable $\mathbf{z}$ that indicates which component of the mixture is responsible for generating the corresponding data point. Thus the model is specified in terms of a joint distribution

$$p(\mathbf{x}, \mathbf{z}) \tag{14.2}$$

and the corresponding density over the observed variable $\mathbf{x}$ is obtained by marginalizing over the latent variable

$$p(\mathbf{x}) = \sum_{\mathbf{z}} p(\mathbf{x}, \mathbf{z}). \tag{14.3}$$

In the case of our Gaussian mixture example, this leads to a distribution of the form

$$p(\mathbf{x}) = \sum_{k=1}^{K} \pi_k \mathcal{N}(\mathbf{x}|\boldsymbol{\mu}_k, \boldsymbol{\Sigma}_k) \tag{14.4}$$

with the usual interpretation of the symbols. This is an example of model combination. For independent, identically distributed data, we can use (14.3) to write the marginal probability of a data set $\mathbf{X} = \{\mathbf{x}_1, \ldots, \mathbf{x}_N\}$ in the form

$$p(\mathbf{X}) = \prod_{n=1}^{N} p(\mathbf{x}_n) = \prod_{n=1}^{N} \left[\sum_{\mathbf{z}_n} p(\mathbf{x}_n, \mathbf{z}_n) \right]. \tag{14.5}$$

Thus we see that each observed data point $\mathbf{x}_n$ has a corresponding latent variable $\mathbf{z}_n$.

Now suppose we have several different models indexed by $h = 1, \ldots, H$ with prior probabilities $p(h)$. For instance one model might be a mixture of Gaussians and another model might be a mixture of Cauchy distributions. The marginal distribution over the data set is given by

$$p(\mathbf{X}) = \sum_{h=1}^{H} p(\mathbf{X}|h)p(h). \tag{14.6}$$

This is an example of Bayesian model averaging. The interpretation of this summation over h is that just one model is responsible for generating the whole data set, and the probability distribution over h simply reflects our uncertainty as to which model that is. As the size of the data set increases, this uncertainty reduces, and the posterior probabilities $p(h|\mathbf{X})$ become increasingly focussed on just one of the models.

This highlights the key difference between Bayesian model averaging and model combination, because in Bayesian model averaging the whole data set is generated by a single model. By contrast, when we combine multiple models, as in (14.5), we see that different data points within the data set can potentially be generated from different values of the latent variable $\mathbf{z}$ and hence by different components.

Although we have considered the marginal probability $p(\mathbf{X})$, the same considerations apply for the predictive density $p(\mathbf{x}|\mathbf{X})$ or for conditional distributions such as $p(\mathbf{t}|\mathbf{x}, \mathbf{X}, \mathbf{T})$.

Exercise 14.1

14.2. Committees

The simplest way to construct a committee is to average the predictions of a set of individual models. Such a procedure can be motivated from a frequentist perspective by considering the trade-off between bias and variance, which decomposes the error due to a model into the bias component that arises from differences between the model and the true function to be predicted, and the variance component that represents the sensitivity of the model to the individual data points. Recall from Figure 3.5

Section 3.2

that when we trained multiple polynomials using the sinusoidal data, and then averaged the resulting functions, the contribution arising from the variance term tended to cancel, leading to improved predictions. When we averaged a set of low-bias models (corresponding to higher order polynomials), we obtained accurate predictions for the underlying sinusoidal function from which the data were generated.

In practice, of course, we have only a single data set, and so we have to find a way to introduce variability between the different models within the committee. One approach is to use *bootstrap* data sets, discussed in Section 1.2.3. Consider a regression problem in which we are trying to predict the value of a single continuous variable, and suppose we generate M bootstrap data sets and then use each to train a separate copy $y_m(\mathbf{x})$ of a predictive model where $m = 1, \ldots, M$. The committee prediction is given by

$$y_{\text{COM}}(\mathbf{x}) = \frac{1}{M} \sum_{m=1}^{M} y_m(\mathbf{x}). \tag{14.7}$$

This procedure is known as bootstrap aggregation or *bagging* (Breiman, 1996).

Suppose the true regression function that we are trying to predict is given by $h(\mathbf{x})$, so that the output of each of the models can be written as the true value plus an error in the form

$$y_m(\mathbf{x}) = h(\mathbf{x}) + \epsilon_m(\mathbf{x}). \tag{14.8}$$

The average sum-of-squares error then takes the form

$$\mathbb{E}_{\mathbf{x}}\left[\{y_m(\mathbf{x}) - h(\mathbf{x})\}^2\right] = \mathbb{E}_{\mathbf{x}}\left[\epsilon_m(\mathbf{x})^2\right] \tag{14.9}$$

where $\mathbb{E}_{\mathbf{x}}[\cdot]$ denotes a frequentist expectation with respect to the distribution of the input vector $\mathbf{x}$. The average error made by the models acting individually is therefore

$$E_{\text{AV}} = \frac{1}{M} \sum_{m=1}^{M} \mathbb{E}_{\mathbf{x}}\left[\epsilon_m(\mathbf{x})^2\right]. \tag{14.10}$$

Similarly, the expected error from the committee (14.7) is given by

$$\begin{aligned} E_{\text{COM}} &= \mathbb{E}_{\mathbf{x}}\left[\left\{\frac{1}{M}\sum_{m=1}^{M} y_m(\mathbf{x}) - h(\mathbf{x})\right\}^2\right] \\ &= \mathbb{E}_{\mathbf{x}}\left[\left\{\frac{1}{M}\sum_{m=1}^{M} \epsilon_m(\mathbf{x})\right\}^2\right] \end{aligned} \tag{14.11}$$

If we assume that the errors have zero mean and are uncorrelated, so that

$$\mathbb{E}_{\mathbf{x}}\left[\epsilon_m(\mathbf{x})\right] = 0 \tag{14.12}$$

$$\mathbb{E}_{\mathbf{x}}\left[\epsilon_m(\mathbf{x})\epsilon_l(\mathbf{x})\right] = 0, \qquad m \neq l \tag{14.13}$$

Exercise 14.2

then we obtain

$$E_{\mathrm{COM}} = \frac{1}{M} E_{\mathrm{AV}}. \tag{14.14}$$

This apparently dramatic result suggests that the average error of a model can be reduced by a factor of M simply by averaging M versions of the model. Unfortunately, it depends on the key assumption that the errors due to the individual models are uncorrelated. In practice, the errors are typically highly correlated, and the reduction in overall error is generally small. It can, however, be shown that the expected committee error will not exceed the expected error of the constituent models, so that $E_{\mathrm{COM}} \leqslant E_{\mathrm{AV}}$. In order to achieve more significant improvements, we turn to a more sophisticated technique for building committees, known as boosting.

Exercise 14.3

14.3. Boosting

Boosting is a powerful technique for combining multiple 'base' classifiers to produce a form of committee whose performance can be significantly better than that of any of the base classifiers. Here we describe the most widely used form of boosting algorithm called *AdaBoost*, short for 'adaptive boosting', developed by Freund and Schapire (1996). Boosting can give good results even if the base classifiers have a performance that is only slightly better than random, and hence sometimes the base classifiers are known as *weak learners*. Originally designed for solving classification problems, boosting can also be extended to regression (Friedman, 2001).

The principal difference between boosting and the committee methods such as bagging discussed above, is that the base classifiers are trained in sequence, and each base classifier is trained using a weighted form of the data set in which the weighting coefficient associated with each data point depends on the performance of the previous classifiers. In particular, points that are misclassified by one of the base classifiers are given greater weight when used to train the next classifier in the sequence. Once all the classifiers have been trained, their predictions are then combined through a weighted majority voting scheme, as illustrated schematically in Figure 14.1.

Consider a two-class classification problem, in which the training data comprises input vectors $\mathbf{x}_1, \ldots, \mathbf{x}_N$ along with corresponding binary target variables $t_1, \ldots, t_N$ where $t_n \in \{-1, 1\}$. Each data point is given an associated weighting parameter w_n, which is initially set $1/N$ for all data points. We shall suppose that we have a procedure available for training a base classifier using weighted data to give a function $y(\mathbf{x}) \in \{-1, 1\}$. At each stage of the algorithm, AdaBoost trains a new classifier using a data set in which the weighting coefficients are adjusted according to the performance of the previously trained classifier so as to give greater weight to the misclassified data points. Finally, when the desired number of base classifiers have been trained, they are combined to form a committee using coefficients that give different weight to different base classifiers. The precise form of the AdaBoost algorithm is given below.

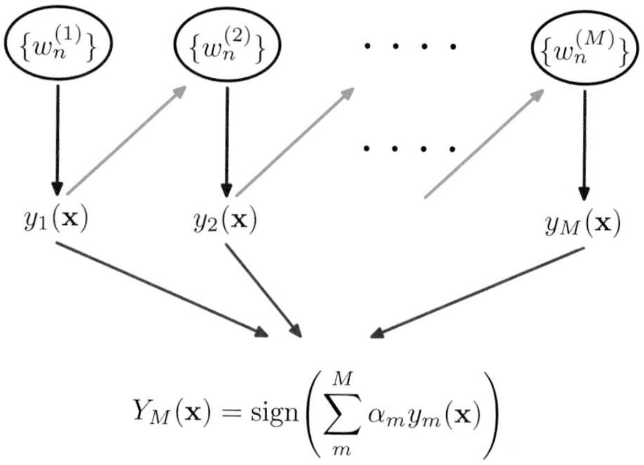

Figure 14.1 Schematic illustration of the boosting framework. Each base classifier $y_m(\mathbf{x})$ is trained on a weighted form of the training set (blue arrows) in which the weights $w_n^{(m)}$ depend on the performance of the previous base classifier $y_{m-1}(\mathbf{x})$ (green arrows). Once all base classifiers have been trained, they are combined to give the final classifier $Y_M(\mathbf{x})$ (red arrows).

AdaBoost

1. Initialize the data weighting coefficients $\{w_n\}$ by setting $w_n^{(1)} = 1/N$ for $n = 1, \ldots, N$.
2. For $m = 1, \ldots, M$:
   (a) Fit a classifier $y_m(\mathbf{x})$ to the training data by minimizing the weighted error function

$$J_m = \sum_{n=1}^{N} w_n^{(m)} I(y_m(\mathbf{x}_n) \neq t_n) \tag{14.15}$$

   where $I(y_m(\mathbf{x}_n) \neq t_n)$ is the indicator function and equals 1 when $y_m(\mathbf{x}_n) \neq t_n$ and 0 otherwise.

   (b) Evaluate the quantities

$$\epsilon_m = \frac{\sum_{n=1}^{N} w_n^{(m)} I(y_m(\mathbf{x}_n) \neq t_n)}{\sum_{n=1}^{N} w_n^{(m)}} \tag{14.16}$$

   and then use these to evaluate

$$\alpha_m = \ln\left\{\frac{1-\epsilon_m}{\epsilon_m}\right\}. \tag{14.17}$$

   (c) Update the data weighting coefficients

$$w_n^{(m+1)} = w_n^{(m)} \exp\left\{\alpha_m I(y_m(\mathbf{x}_n) \neq t_n)\right\} \tag{14.18}$$

3. Make predictions using the final model, which is given by

$$Y_M(\mathbf{x}) = \operatorname{sign}\left(\sum_{m=1}^{M} \alpha_m y_m(\mathbf{x})\right). \qquad (14.19)$$

We see that the first base classifier $y_1(\mathbf{x})$ is trained using weighting coefficients $w_n^{(1)}$ that are all equal, which therefore corresponds to the usual procedure for training a single classifier. From (14.18), we see that in subsequent iterations the weighting coefficients $w_n^{(m)}$ are increased for data points that are misclassified and unchanged for data points that are correctly classified. Successive classifiers are therefore forced to place greater emphasis on points that have been misclassified by previous classifiers, and data points that continue to be misclassified by successive classifiers receive ever greater weight. The quantities ϵ_m represent weighted measures of the error rates of each of the base classifiers on the data set. We therefore see that the weighting coefficients α_m defined by (14.17) give greater weight to the more accurate classifiers when computing the overall output given by (14.19).

The AdaBoost algorithm is illustrated in Figure 14.2, using a subset of 30 data points taken from the toy classification data set shown in Figure A.7. Here each base learners consists of a threshold on one of the input variables. This simple classifier corresponds to a form of decision tree known as a 'decision stumps', i.e., a decision tree with a single node. Thus each base learner classifies an input according to whether one of the input features exceeds some threshold and therefore simply partitions the space into two regions separated by a linear decision surface that is perpendicular to one of the axes.

Section 14.4

14.3.1 Minimizing exponential error

Boosting was originally motivated using statistical learning theory, leading to upper bounds on the generalization error. However, these bounds turn out to be too loose to have practical value, and the actual performance of boosting is much better than the bounds alone would suggest. Friedman *et al.* (2000) gave a different and very simple interpretation of boosting in terms of the sequential minimization of an exponential error function.

Consider the exponential error function defined by

$$E = \sum_{n=1}^{N} \exp\left\{-t_n f_m(\mathbf{x}_n)\right\} \qquad (14.20)$$

where $f_m(\mathbf{x})$ is a classifier defined in terms of a linear combination of base classifiers $y_l(\mathbf{x})$ of the form

$$f_m(\mathbf{x}) = \frac{1}{2}\sum_{l=1}^{m} \alpha_l y_l(\mathbf{x}) \qquad (14.21)$$

and $t_n \in \{-1, 1\}$ are the training set target values. Our goal is to minimize E with respect to both the weighting coefficients α_l and the parameters of the base classifiers $y_l(\mathbf{x})$.

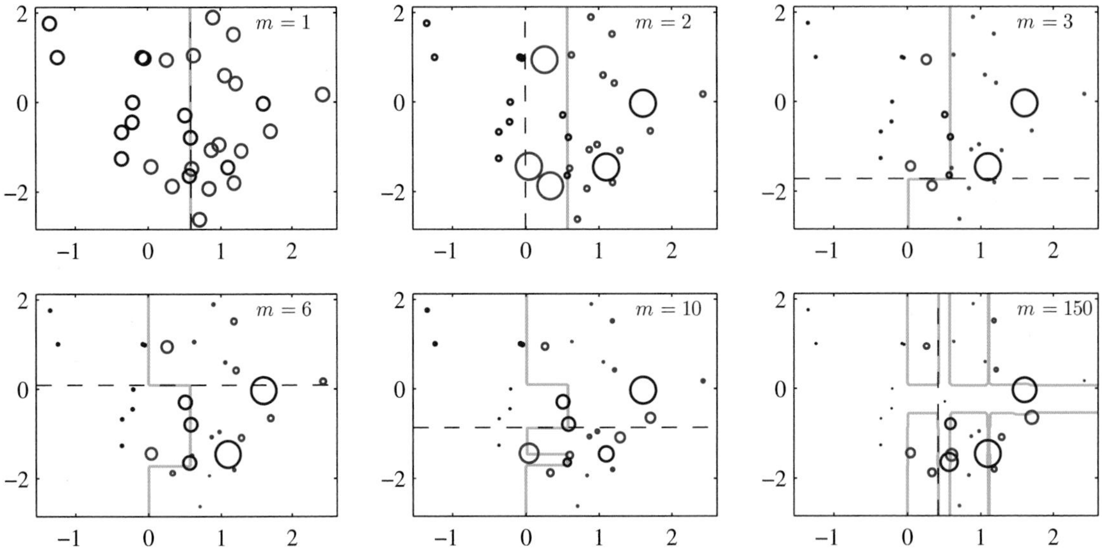

Figure 14.2 Illustration of boosting in which the base learners consist of simple thresholds applied to one or other of the axes. Each figure shows the number m of base learners trained so far, along with the decision boundary of the most recent base learner (dashed black line) and the combined decision boundary of the ensemble (solid green line). Each data point is depicted by a circle whose radius indicates the weight assigned to that data point when training the most recently added base learner. Thus, for instance, we see that points that are misclassified by the $m = 1$ base learner are given greater weight when training the $m = 2$ base learner.

Instead of doing a global error function minimization, however, we shall suppose that the base classifiers $y_1(\mathbf{x}), \ldots, y_{m-1}(\mathbf{x})$ are fixed, as are their coefficients $\alpha_1, \ldots, \alpha_{m-1}$, and so we are minimizing only with respect to α_m and $y_m(\mathbf{x})$. Separating off the contribution from base classifier $y_m(\mathbf{x})$, we can then write the error function in the form

$$\begin{aligned} E &= \sum_{n=1}^{N} \exp\left\{-t_n f_{m-1}(\mathbf{x}_n) - \frac{1}{2} t_n \alpha_m y_m(\mathbf{x}_n)\right\} \\ &= \sum_{n=1}^{N} w_n^{(m)} \exp\left\{-\frac{1}{2} t_n \alpha_m y_m(\mathbf{x}_n)\right\} \end{aligned} \tag{14.22}$$

where the coefficients $w_n^{(m)} = \exp\{-t_n f_{m-1}(\mathbf{x}_n)\}$ can be viewed as constants because we are optimizing only α_m and $y_m(\mathbf{x})$. If we denote by $\mathcal{T}_m$ the set of data points that are correctly classified by $y_m(\mathbf{x})$, and if we denote the remaining misclassified points by $\mathcal{M}_m$, then we can in turn rewrite the error function in the form

$$\begin{aligned} E &= e^{-\alpha_m/2} \sum_{n \in \mathcal{T}_m} w_n^{(m)} + e^{\alpha_m/2} \sum_{n \in \mathcal{M}_m} w_n^{(m)} \\ &= (e^{\alpha_m/2} - e^{-\alpha_m/2}) \sum_{n=1}^{N} w_n^{(m)} I(y_m(\mathbf{x}_n) \neq t_n) + e^{-\alpha_m/2} \sum_{n=1}^{N} w_n^{(m)}. \end{aligned} \tag{14.23}$$

When we minimize this with respect to $y_m(\mathbf{x})$, we see that the second term is constant, and so this is equivalent to minimizing (14.15) because the overall multiplicative factor in front of the summation does not affect the location of the minimum. Similarly, minimizing with respect to α_m, we obtain (14.17) in which ϵ_m is defined by (14.16).

Exercise 14.6

From (14.22) we see that, having found α_m and $y_m(\mathbf{x})$, the weights on the data points are updated using

$$w_n^{(m+1)} = w_n^{(m)} \exp\left\{-\frac{1}{2} t_n \alpha_m y_m(\mathbf{x}_n)\right\}. \tag{14.24}$$

Making use of the fact that

$$t_n y_m(\mathbf{x}_n) = 1 - 2I(y_m(\mathbf{x}_n) \neq t_n) \tag{14.25}$$

we see that the weights $w_n^{(m)}$ are updated at the next iteration using

$$w_n^{(m+1)} = w_n^{(m)} \exp(-\alpha_m/2) \exp\left\{\alpha_m I(y_m(\mathbf{x}_n) \neq t_n)\right\}. \tag{14.26}$$

Because the term $\exp(-\alpha_m/2)$ is independent of n, we see that it weights all data points by the same factor and so can be discarded. Thus we obtain (14.18).

Finally, once all the base classifiers are trained, new data points are classified by evaluating the sign of the combined function defined according to (14.21). Because the factor of $1/2$ does not affect the sign it can be omitted, giving (14.19).

14.3.2 Error functions for boosting

The exponential error function that is minimized by the AdaBoost algorithm differs from those considered in previous chapters. To gain some insight into the nature of the exponential error function, we first consider the expected error given by

$$\mathbb{E}_{\mathbf{x},t}\left[\exp\{-ty(\mathbf{x})\}\right] = \sum_t \int \exp\{-ty(\mathbf{x})\} p(t|\mathbf{x}) p(\mathbf{x})\,\mathrm{d}\mathbf{x}. \tag{14.27}$$

If we perform a variational minimization with respect to all possible functions $y(\mathbf{x})$, we obtain

Exercise 14.7

$$y(\mathbf{x}) = \frac{1}{2} \ln\left\{\frac{p(t=1|\mathbf{x})}{p(t=-1|\mathbf{x})}\right\} \tag{14.28}$$

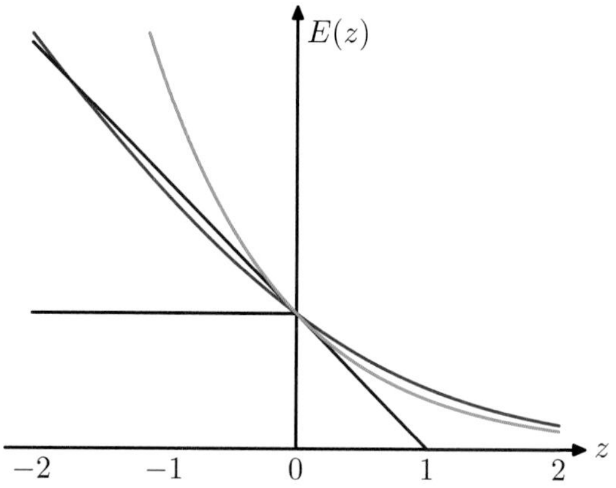

Figure 14.3 Plot of the exponential (green) and rescaled cross-entropy (red) error functions along with the hinge error (blue) used in support vector machines, and the misclassification error (black). Note that for large negative values of $z = ty(\mathbf{x})$, the cross-entropy gives a linearly increasing penalty, whereas the exponential loss gives an exponentially increasing penalty.

which is half the log-odds. Thus the AdaBoost algorithm is seeking the best approximation to the log odds ratio, within the space of functions represented by the linear combination of base classifiers, subject to the constrained minimization resulting from the sequential optimization strategy. This result motivates the use of the sign function in (14.19) to arrive at the final classification decision.

We have already seen that the minimizer $y(\mathbf{x})$ of the cross-entropy error (4.90) for two-class classification is given by the posterior class probability. In the case of a target variable $t \in \{-1, 1\}$, we have seen that the error function is given by $\ln(1 + \exp(-yt))$. This is compared with the exponential error function in Figure 14.3, where we have divided the cross-entropy error by a constant factor $\ln(2)$ so that it passes through the point $(0, 1)$ for ease of comparison. We see that both can be seen as continuous approximations to the ideal misclassification error function. An advantage of the exponential error is that its sequential minimization leads to the simple AdaBoost scheme. One drawback, however, is that it penalizes large negative values of $ty(\mathbf{x})$ much more strongly than cross-entropy. In particular, we see that for large negative values of ty, the cross-entropy grows linearly with $|ty|$, whereas the exponential error function grows exponentially with $|ty|$. Thus the exponential error function will be much less robust to outliers or misclassified data points. Another important difference between cross-entropy and the exponential error function is that the latter cannot be interpreted as the log likelihood function of any well-defined probabilistic model. Furthermore, the exponential error does not generalize to classification problems having $K > 2$ classes, again in contrast to the cross-entropy for a probabilistic model, which is easily generalized to give (4.108).

Section 7.1.2

Exercise 14.8

Section 4.3.4

The interpretation of boosting as the sequential optimization of an additive model under an exponential error (Friedman *et al.*, 2000) opens the door to a wide range of boosting-like algorithms, including multiclass extensions, by altering the choice of error function. It also motivates the extension to regression problems (Friedman, 2001). If we consider a sum-of-squares error function for regression, then sequential minimization of an additive model of the form (14.21) simply involves fitting each new base classifier to the residual errors $t_n - f_{m-1}(\mathbf{x}_n)$ from the previous model. As we have noted, however, the sum-of-squares error is not robust to outliers, and this

Exercise 14.9

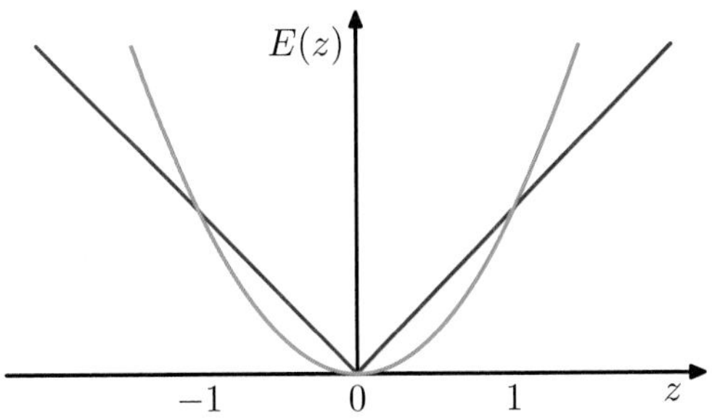

Figure 14.4 Comparison of the squared error (green) with the absolute error (red) showing how the latter places much less emphasis on large errors and hence is more robust to outliers and mislabelled data points.

can be addressed by basing the boosting algorithm on the absolute deviation $|y - t|$ instead. These two error functions are compared in Figure 14.4.

14.4. Tree-based Models

There are various simple, but widely used, models that work by partitioning the input space into cuboid regions, whose edges are aligned with the axes, and then assigning a simple model (for example, a constant) to each region. They can be viewed as a model combination method in which only one model is responsible for making predictions at any given point in input space. The process of selecting a specific model, given a new input $\mathbf{x}$, can be described by a sequential decision making process corresponding to the traversal of a binary tree (one that splits into two branches at each node). Here we focus on a particular tree-based framework called *classification and regression trees*, or *CART* (Breiman *et al.*, 1984), although there are many other variants going by such names as ID3 and C4.5 (Quinlan, 1986; Quinlan, 1993).

Figures 14.5 and 14.6 show an illustration of a recursive binary partitioning of the input space, along with the corresponding tree structure. In this example, the first

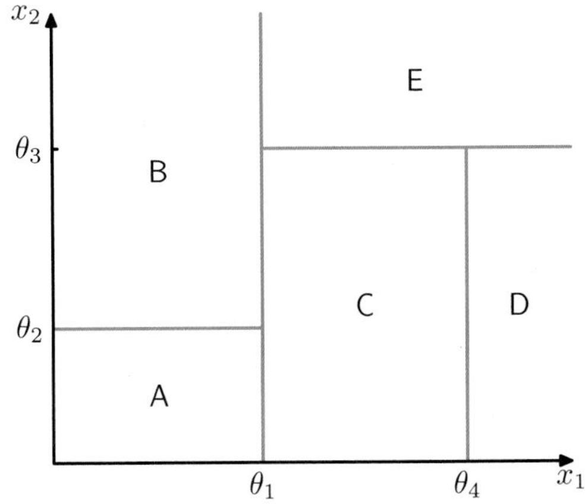

Figure 14.5 Illustration of a two-dimensional input space that has been partitioned into five regions using axis-aligned boundaries.

Figure 14.6 Binary tree corresponding to the partitioning of input space shown in Figure 14.5.

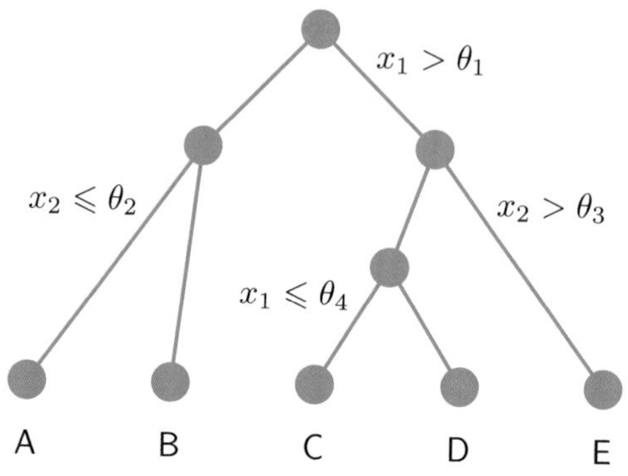

step divides the whole of the input space into two regions according to whether $x_1 \leqslant \theta_1$ or $x_1 > \theta_1$ where θ_1 is a parameter of the model. This creates two subregions, each of which can then be subdivided independently. For instance, the region $x_1 \leqslant \theta_1$ is further subdivided according to whether $x_2 \leqslant \theta_2$ or $x_2 > \theta_2$, giving rise to the regions denoted A and B. The recursive subdivision can be described by the traversal of the binary tree shown in Figure 14.6. For any new input $\mathbf{x}$, we determine which region it falls into by starting at the top of the tree at the root node and following a path down to a specific leaf node according to the decision criteria at each node. Note that such decision trees are not probabilistic graphical models.

Within each region, there is a separate model to predict the target variable. For instance, in regression we might simply predict a constant over each region, or in classification we might assign each region to a specific class. A key property of tree-based models, which makes them popular in fields such as medical diagnosis, for example, is that they are readily interpretable by humans because they correspond to a sequence of binary decisions applied to the individual input variables. For instance, to predict a patient's disease, we might first ask "is their temperature greater than some threshold?". If the answer is yes, then we might next ask "is their blood pressure less than some threshold?". Each leaf of the tree is then associated with a specific diagnosis.

In order to learn such a model from a training set, we have to determine the structure of the tree, including which input variable is chosen at each node to form the split criterion as well as the value of the threshold parameter θ_i for the split. We also have to determine the values of the predictive variable within each region.

Consider first a regression problem in which the goal is to predict a single target variable t from a D-dimensional vector $\mathbf{x} = (x_1, \ldots, x_D)^{\mathrm{T}}$ of input variables. The training data consists of input vectors $\{\mathbf{x}_1, \ldots, \mathbf{x}_N\}$ along with the corresponding continuous labels $\{t_1, \ldots, t_N\}$. If the partitioning of the input space is given, and we minimize the sum-of-squares error function, then the optimal value of the predictive variable within any given region is just given by the average of the values of t_n for those data points that fall in that region.

Exercise 14.10

Now consider how to determine the structure of the decision tree. Even for a fixed number of nodes in the tree, the problem of determining the optimal structure (including choice of input variable for each split as well as the corresponding thresholds) to minimize the sum-of-squares error is usually computationally infeasible due to the combinatorially large number of possible solutions. Instead, a greedy optimization is generally done by starting with a single root node, corresponding to the whole input space, and then growing the tree by adding nodes one at a time. At each step there will be some number of candidate regions in input space that can be split, corresponding to the addition of a pair of leaf nodes to the existing tree. For each of these, there is a choice of which of the D input variables to split, as well as the value of the threshold. The joint optimization of the choice of region to split, and the choice of input variable and threshold, can be done efficiently by exhaustive search noting that, for a given choice of split variable and threshold, the optimal choice of predictive variable is given by the local average of the data, as noted earlier. This is repeated for all possible choices of variable to be split, and the one that gives the smallest residual sum-of-squares error is retained.

Given a greedy strategy for growing the tree, there remains the issue of when to stop adding nodes. A simple approach would be to stop when the reduction in residual error falls below some threshold. However, it is found empirically that often none of the available splits produces a significant reduction in error, and yet after several more splits a substantial error reduction is found. For this reason, it is common practice to grow a large tree, using a stopping criterion based on the number of data points associated with the leaf nodes, and then prune back the resulting tree. The pruning is based on a criterion that balances residual error against a measure of model complexity. If we denote the starting tree for pruning by T_0, then we define $T \subset T_0$ to be a subtree of T_0 if it can be obtained by pruning nodes from T_0 (in other words, by collapsing internal nodes by combining the corresponding regions). Suppose the leaf nodes are indexed by $\tau = 1, \ldots, |T|$, with leaf node τ representing a region $\mathcal{R}_\tau$ of input space having N_τ data points, and $|T|$ denoting the total number of leaf nodes. The optimal prediction for region $\mathcal{R}_\tau$ is then given by

$$y_\tau = \frac{1}{N_\tau} \sum_{\mathbf{x}_n \in \mathcal{R}_\tau} t_n \tag{14.29}$$

and the corresponding contribution to the residual sum-of-squares is then

$$Q_\tau(T) = \sum_{\mathbf{x}_n \in \mathcal{R}_\tau} \{t_n - y_\tau\}^2 . \tag{14.30}$$

The pruning criterion is then given by

$$C(T) = \sum_{\tau=1}^{|T|} Q_\tau(T) + \lambda |T| \tag{14.31}$$

The regularization parameter λ determines the trade-off between the overall residual sum-of-squares error and the complexity of the model as measured by the number $|T|$ of leaf nodes, and its value is chosen by cross-validation.

For classification problems, the process of growing and pruning the tree is similar, except that the sum-of-squares error is replaced by a more appropriate measure of performance. If we define $p_{\tau k}$ to be the proportion of data points in region $\mathcal{R}_\tau$ assigned to class k, where $k = 1, \ldots, K$, then two commonly used choices are the negative cross-entropy

$$Q_\tau(T) = \sum_{k=1}^{K} p_{\tau k} \ln p_{\tau k} \tag{14.32}$$

and the *Gini index*

$$Q_\tau(T) = \sum_{k=1}^{K} p_{\tau k} \left(1 - p_{\tau k}\right). \tag{14.33}$$

These both vanish for $p_{\tau k} = 0$ and $p_{\tau k} = 1$ and have a maximum at $p_{\tau k} = 0.5$. They encourage the formation of regions in which a high proportion of the data points are assigned to one class. The cross entropy and the Gini index are better measures than the misclassification rate for growing the tree because they are more sensitive to the node probabilities. Also, unlike misclassification rate, they are differentiable and hence better suited to gradient based optimization methods. For subsequent pruning of the tree, the misclassification rate is generally used.

Exercise 14.11

The human interpretability of a tree model such as CART is often seen as its major strength. However, in practice it is found that the particular tree structure that is learned is very sensitive to the details of the data set, so that a small change to the training data can result in a very different set of splits (Hastie *et al.*, 2001).

There are other problems with tree-based methods of the kind considered in this section. One is that the splits are aligned with the axes of the feature space, which may be very suboptimal. For instance, to separate two classes whose optimal decision boundary runs at 45 degrees to the axes would need a large number of axis-parallel splits of the input space as compared to a single non-axis-aligned split. Furthermore, the splits in a decision tree are hard, so that each region of input space is associated with one, and only one, leaf node model. The last issue is particularly problematic in regression where we are typically aiming to model smooth functions, and yet the tree model produces piecewise-constant predictions with discontinuities at the split boundaries.

14.5. Conditional Mixture Models

We have seen that standard decision trees are restricted by hard, axis-aligned splits of the input space. These constraints can be relaxed, at the expense of interpretability, by allowing soft, probabilistic splits that can be functions of all of the input variables, not just one of them at a time. If we also give the leaf models a probabilistic interpretation, we arrive at a fully probabilistic tree-based model called the *hierarchical mixture of experts*, which we consider in Section 14.5.3.

An alternative way to motivate the hierarchical mixture of experts model is to start with a standard probabilistic mixtures of unconditional density models such as Gaussians and replace the component densities with conditional distributions. Here we consider mixtures of linear regression models (Section 14.5.1) and mixtures of

Chapter 9

logistic regression models (Section 14.5.2). In the simplest case, the mixing coefficients are independent of the input variables. If we make a further generalization to allow the mixing coefficients also to depend on the inputs then we obtain a *mixture of experts* model. Finally, if we allow each component in the mixture model to be itself a mixture of experts model, then we obtain a hierarchical mixture of experts.

14.5.1 Mixtures of linear regression models

One of the many advantages of giving a probabilistic interpretation to the linear regression model is that it can then be used as a component in more complex probabilistic models. This can be done, for instance, by viewing the conditional distribution representing the linear regression model as a node in a directed probabilistic graph. Here we consider a simple example corresponding to a mixture of linear regression models, which represents a straightforward extension of the Gaussian mixture model discussed in Section 9.2 to the case of conditional Gaussian distributions.

We therefore consider K linear regression models, each governed by its own weight parameter $\mathbf{w}_k$. In many applications, it will be appropriate to use a common noise variance, governed by a precision parameter β, for all K components, and this is the case we consider here. We will once again restrict attention to a single target variable t, though the extension to multiple outputs is straightforward. If we denote the mixing coefficients by π_k, then the mixture distribution can be written

Exercise 14.12

$$p(t|\boldsymbol{\theta}) = \sum_{k=1}^{K} \pi_k \mathcal{N}(t|\mathbf{w}_k^{\mathrm{T}}\boldsymbol{\phi}, \beta^{-1}) \tag{14.34}$$

where $\boldsymbol{\theta}$ denotes the set of all adaptive parameters in the model, namely $\mathbf{W} = \{\mathbf{w}_k\}$, $\boldsymbol{\pi} = \{\pi_k\}$, and β. The log likelihood function for this model, given a data set of observations $\{\boldsymbol{\phi}_n, t_n\}$, then takes the form

$$\ln p(\mathbf{t}|\boldsymbol{\theta}) = \sum_{n=1}^{N} \ln \left(\sum_{k=1}^{K} \pi_k \mathcal{N}(t_n|\mathbf{w}_k^{\mathrm{T}}\boldsymbol{\phi}_n, \beta^{-1}) \right) \tag{14.35}$$

where $\mathbf{t} = (t_1, \ldots, t_N)^{\mathrm{T}}$ denotes the vector of target variables.

In order to maximize this likelihood function, we can once again appeal to the EM algorithm, which will turn out to be a simple extension of the EM algorithm for unconditional Gaussian mixtures of Section 9.2. We can therefore build on our experience with the unconditional mixture and introduce a set $\mathbf{Z} = \{\mathbf{z}_n\}$ of binary latent variables where $z_{nk} \in \{0, 1\}$ in which, for each data point n, all of the elements $k = 1, \ldots, K$ are zero except for a single value of 1 indicating which component of the mixture was responsible for generating that data point. The joint distribution over latent and observed variables can be represented by the graphical model shown in Figure 14.7.

Exercise 14.13

The complete-data log likelihood function then takes the form

$$\ln p(\mathbf{t}, \mathbf{Z}|\boldsymbol{\theta}) = \sum_{n=1}^{N} \sum_{k=1}^{K} z_{nk} \ln \left\{ \pi_k \mathcal{N}(t_n|\mathbf{w}_k^{\mathrm{T}}\boldsymbol{\phi}_n, \beta^{-1}) \right\}. \tag{14.36}$$

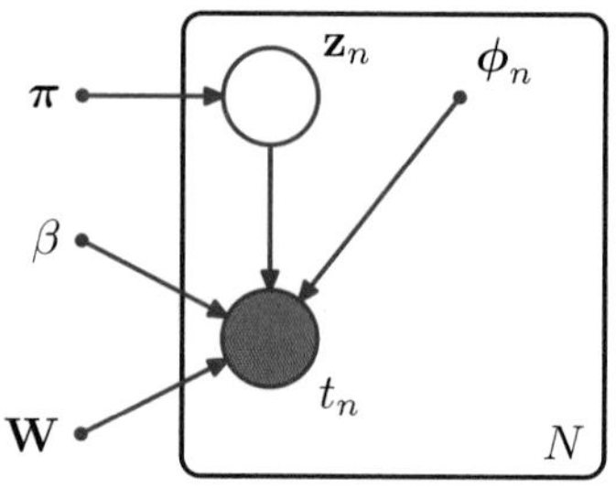

Figure 14.7 Probabilistic directed graph representing a mixture of linear regression models, defined by (14.35).

The EM algorithm begins by first choosing an initial value $\boldsymbol{\theta}^{\text{old}}$ for the model parameters. In the E step, these parameter values are then used to evaluate the posterior probabilities, or responsibilities, of each component k for every data point n given by

$$\gamma_{nk} = \mathbb{E}[z_{nk}] = p(k|\boldsymbol{\phi}_n, \boldsymbol{\theta}^{\text{old}}) = \frac{\pi_k \mathcal{N}(t_n|\mathbf{w}_k^{\text{T}}\boldsymbol{\phi}_n, \beta^{-1})}{\sum_j \pi_j \mathcal{N}(t_n|\mathbf{w}_j^{\text{T}}\boldsymbol{\phi}_n, \beta^{-1})}. \tag{14.37}$$

The responsibilities are then used to determine the expectation, with respect to the posterior distribution $p(\mathbf{Z}|\mathbf{t}, \boldsymbol{\theta}^{\text{old}})$, of the complete-data log likelihood, which takes the form

$$Q(\boldsymbol{\theta}, \boldsymbol{\theta}^{\text{old}}) = \mathbb{E}_{\mathbf{Z}}\left[\ln p(\mathbf{t}, \mathbf{Z}|\boldsymbol{\theta})\right] = \sum_{n=1}^{N}\sum_{k=1}^{K} \gamma_{nk}\left\{\ln \pi_k + \ln \mathcal{N}(t_n|\mathbf{w}_k^{\text{T}}\boldsymbol{\phi}_n, \beta^{-1})\right\}.$$

In the M step, we maximize the function $Q(\boldsymbol{\theta}, \boldsymbol{\theta}^{\text{old}})$ with respect to $\boldsymbol{\theta}$, keeping the γ_{nk} fixed. For the optimization with respect to the mixing coefficients π_k we need to take account of the constraint $\sum_k \pi_k = 1$, which can be done with the aid of a Lagrange multiplier, leading to an M-step re-estimation equation for π_k in the form

Exercise 14.14

$$\pi_k = \frac{1}{N}\sum_{n=1}^{N}\gamma_{nk}. \tag{14.38}$$

Note that this has exactly the same form as the corresponding result for a simple mixture of unconditional Gaussians given by (9.22).

Next consider the maximization with respect to the parameter vector $\mathbf{w}_k$ of the kth linear regression model. Substituting for the Gaussian distribution, we see that the function $Q(\boldsymbol{\theta}, \boldsymbol{\theta}^{\text{old}})$, as a function of the parameter vector $\mathbf{w}_k$, takes the form

$$Q(\boldsymbol{\theta}, \boldsymbol{\theta}^{\text{old}}) = \sum_{n=1}^{N}\gamma_{nk}\left\{-\frac{\beta}{2}\left(t_n - \mathbf{w}_k^{\text{T}}\boldsymbol{\phi}_n\right)^2\right\} + \text{const} \tag{14.39}$$

where the constant term includes the contributions from other weight vectors $\mathbf{w}_j$ for $j \neq k$. Note that the quantity we are maximizing is similar to the (negative of the) standard sum-of-squares error (3.12) for a single linear regression model, but with the inclusion of the responsibilities γ_{nk}. This represents a *weighted least squares* problem, in which the term corresponding to the nth data point carries a weighting coefficient given by $\beta\gamma_{nk}$, which could be interpreted as an effective precision for each data point. We see that each component linear regression model in the mixture, governed by its own parameter vector $\mathbf{w}_k$, is fitted separately to the whole data set in the M step, but with each data point n weighted by the responsibility γ_{nk} that model k takes for that data point. Setting the derivative of (14.39) with respect to $\mathbf{w}_k$ equal to zero gives

$$0 = \sum_{n=1}^{N} \gamma_{nk} \left(t_n - \mathbf{w}_k^{\mathrm{T}} \boldsymbol{\phi}_n \right) \boldsymbol{\phi}_n \tag{14.40}$$

which we can write in matrix notation as

$$0 = \boldsymbol{\Phi}^{\mathrm{T}} \mathbf{R}_k (\mathbf{t} - \boldsymbol{\Phi} \mathbf{w}_k) \tag{14.41}$$

where $\mathbf{R}_k = \mathrm{diag}(\gamma_{nk})$ is a diagonal matrix of size $N \times N$. Solving for $\mathbf{w}_k$, we obtain

$$\mathbf{w}_k = \left(\boldsymbol{\Phi}^{\mathrm{T}} \mathbf{R}_k \boldsymbol{\Phi} \right)^{-1} \boldsymbol{\Phi}^{\mathrm{T}} \mathbf{R}_k \mathbf{t}. \tag{14.42}$$

This represents a set of modified normal equations corresponding to the weighted least squares problem, of the same form as (4.99) found in the context of logistic regression. Note that after each E step, the matrix $\mathbf{R}_k$ will change and so we will have to solve the normal equations afresh in the subsequent M step.

Finally, we maximize $Q(\boldsymbol{\theta}, \boldsymbol{\theta}^{\mathrm{old}})$ with respect to β. Keeping only terms that depend on β, the function $Q(\boldsymbol{\theta}, \boldsymbol{\theta}^{\mathrm{old}})$ can be written

$$Q(\boldsymbol{\theta}, \boldsymbol{\theta}^{\mathrm{old}}) = \sum_{n=1}^{N} \sum_{k=1}^{K} \gamma_{nk} \left\{ \frac{1}{2} \ln \beta - \frac{\beta}{2} \left(t_n - \mathbf{w}_k^{\mathrm{T}} \boldsymbol{\phi}_n \right)^2 \right\}. \tag{14.43}$$

Setting the derivative with respect to β equal to zero, and rearranging, we obtain the M-step equation for β in the form

$$\frac{1}{\beta} = \frac{1}{N} \sum_{n=1}^{N} \sum_{k=1}^{K} \gamma_{nk} \left(t_n - \mathbf{w}_k^{\mathrm{T}} \boldsymbol{\phi}_n \right)^2. \tag{14.44}$$

In Figure 14.8, we illustrate this EM algorithm using the simple example of fitting a mixture of two straight lines to a data set having one input variable x and one target variable t. The predictive density (14.34) is plotted in Figure 14.9 using the converged parameter values obtained from the EM algorithm, corresponding to the right-hand plot in Figure 14.8. Also shown in this figure is the result of fitting a single linear regression model, which gives a unimodal predictive density. We see that the mixture model gives a much better representation of the data distribution, and this is reflected in the higher likelihood value. However, the mixture model also assigns significant probability mass to regions where there is no data because its predictive distribution is bimodal for all values of x. This problem can be resolved by extending the model to allow the mixture coefficients themselves to be functions of x, leading to models such as the mixture density networks discussed in Section 5.6, and hierarchical mixture of experts discussed in Section 14.5.3.

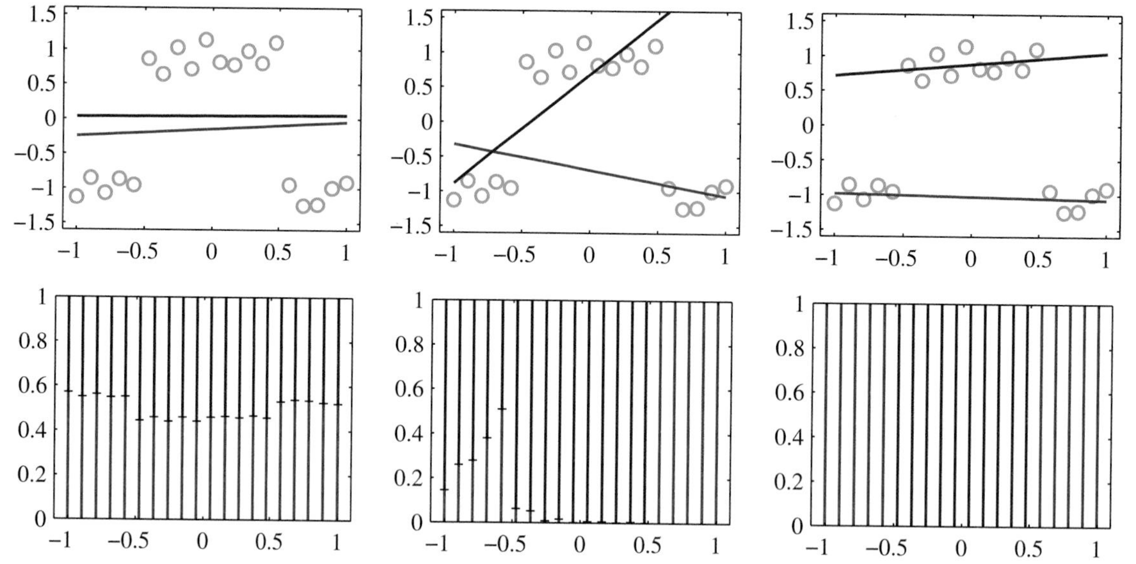

Figure 14.8 Example of a synthetic data set, shown by the green points, having one input variable x and one target variable t, together with a mixture of two linear regression models whose mean functions $y(x, \mathbf{w}_k)$, where $k \in \{1, 2\}$, are shown by the blue and red lines. The upper three plots show the initial configuration (left), the result of running 30 iterations of EM (centre), and the result after 50 iterations of EM (right). Here β was initialized to the reciprocal of the true variance of the set of target values. The lower three plots show the corresponding responsibilities plotted as a vertical line for each data point in which the length of the blue segment gives the posterior probability of the blue line for that data point (and similarly for the red segment).

14.5.2 Mixtures of logistic models

Because the logistic regression model defines a conditional distribution for the target variable, given the input vector, it is straightforward to use it as the component distribution in a mixture model, thereby giving rise to a richer family of conditional distributions compared to a single logistic regression model. This example involves a straightforward combination of ideas encountered in earlier sections of the book and will help consolidate these for the reader.

The conditional distribution of the target variable, for a probabilistic mixture of K logistic regression models, is given by

$$p(t|\boldsymbol{\phi}, \boldsymbol{\theta}) = \sum_{k=1}^{K} \pi_k y_k^t \left[1 - y_k\right]^{1-t} \tag{14.45}$$

where $\boldsymbol{\phi}$ is the feature vector, $y_k = \sigma\left(\mathbf{w}_k^{\mathrm{T}}\boldsymbol{\phi}\right)$ is the output of component k, and $\boldsymbol{\theta}$ denotes the adjustable parameters namely $\{\pi_k\}$ and $\{\mathbf{w}_k\}$.

Now suppose we are given a data set $\{\boldsymbol{\phi}_n, t_n\}$. The corresponding likelihood

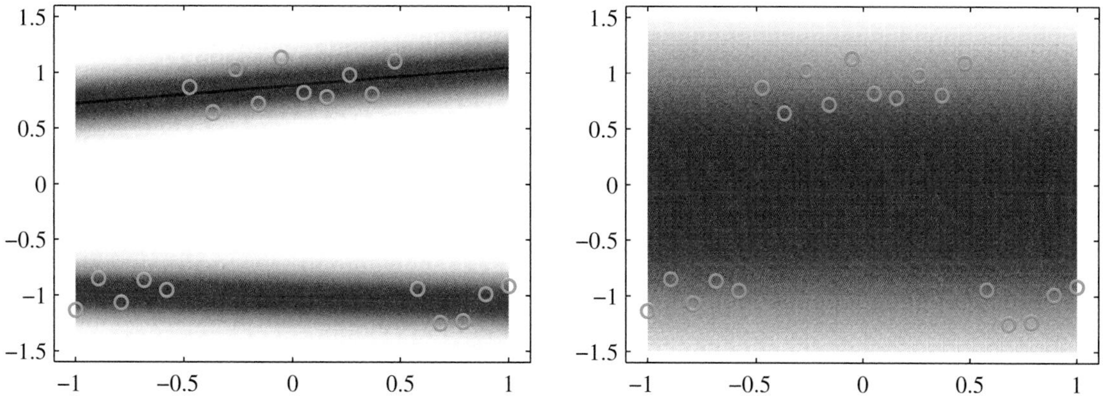

Figure 14.9 The left plot shows the predictive conditional density corresponding to the converged solution in Figure 14.8. This gives a log likelihood value of -3.0. A vertical slice through one of these plots at a particular value of x represents the corresponding conditional distribution $p(t|x)$, which we see is bimodal. The plot on the right shows the predictive density for a single linear regression model fitted to the same data set using maximum likelihood. This model has a smaller log likelihood of -27.6.

function is then given by

$$p(\mathbf{t}|\boldsymbol{\theta}) = \prod_{n=1}^{N} \left(\sum_{k=1}^{K} \pi_k y_{nk}^{t_n} \left[1 - y_{nk}\right]^{1-t_n} \right) \tag{14.46}$$

where $y_{nk} = \sigma(\mathbf{w}_k^{\mathrm{T}} \boldsymbol{\phi}_n)$ and $\mathbf{t} = (t_1, \ldots, t_N)^{\mathrm{T}}$. We can maximize this likelihood function iteratively by making use of the EM algorithm. This involves introducing latent variables z_{nk} that correspond to a 1-of-K coded binary indicator variable for each data point n. The complete-data likelihood function is then given by

$$p(\mathbf{t}, \mathbf{Z}|\boldsymbol{\theta}) = \prod_{n=1}^{N} \prod_{k=1}^{K} \left\{ \pi_k y_{nk}^{t_n} \left[1 - y_{nk}\right]^{1-t_n} \right\}^{z_{nk}} \tag{14.47}$$

where $\mathbf{Z}$ is the matrix of latent variables with elements z_{nk}. We initialize the EM algorithm by choosing an initial value $\boldsymbol{\theta}^{\mathrm{old}}$ for the model parameters. In the E step, we then use these parameter values to evaluate the posterior probabilities of the components k for each data point n, which are given by

$$\gamma_{nk} = \mathbb{E}[z_{nk}] = p(k|\boldsymbol{\phi}_n, \boldsymbol{\theta}^{\mathrm{old}}) = \frac{\pi_k y_{nk}^{t_n} \left[1 - y_{nk}\right]^{1-t_n}}{\sum_j \pi_j y_{nj}^{t_n} \left[1 - y_{nj}\right]^{1-t_n}}. \tag{14.48}$$

These responsibilities are then used to find the expected complete-data log likelihood as a function of $\boldsymbol{\theta}$, given by

$$\begin{aligned} Q(\boldsymbol{\theta}, \boldsymbol{\theta}^{\mathrm{old}}) &= \mathbb{E}_{\mathbf{Z}} \left[\ln p(\mathbf{t}, \mathbf{Z}|\boldsymbol{\theta}) \right] \\ &= \sum_{n=1}^{N} \sum_{k=1}^{K} \gamma_{nk} \left\{ \ln \pi_k + t_n \ln y_{nk} + (1 - t_n) \ln (1 - y_{nk}) \right\}. \end{aligned} \tag{14.49}$$

The M step involves maximization of this function with respect to $\boldsymbol{\theta}$, keeping $\boldsymbol{\theta}^{\text{old}}$, and hence γ_{nk}, fixed. Maximization with respect to π_k can be done in the usual way, with a Lagrange multiplier to enforce the summation constraint $\sum_k \pi_k = 1$, giving the familiar result

$$\pi_k = \frac{1}{N} \sum_{n=1}^{N} \gamma_{nk}. \tag{14.50}$$

To determine the $\{\mathbf{w}_k\}$, we note that the $Q(\boldsymbol{\theta}, \boldsymbol{\theta}^{\text{old}})$ function comprises a sum over terms indexed by k each of which depends only on one of the vectors $\mathbf{w}_k$, so that the different vectors are decoupled in the M step of the EM algorithm. In other words, the different components interact only via the responsibilities, which are fixed during the M step. Note that the M step does not have a closed-form solution and must be solved iteratively using, for instance, the iterative reweighted least squares (IRLS) algorithm. The gradient and the Hessian for the vector $\mathbf{w}_k$ are given by

Section 4.3.3

$$\nabla_k Q = \sum_{n=1}^{N} \gamma_{nk}(t_n - y_{nk})\boldsymbol{\phi}_n \tag{14.51}$$

$$\mathbf{H}_k = -\nabla_k \nabla_k Q = \sum_{n=1}^{N} \gamma_{nk} y_{nk}(1 - y_{nk})\boldsymbol{\phi}_n \boldsymbol{\phi}_n^{\mathrm{T}} \tag{14.52}$$

where ∇_k denotes the gradient with respect to $\mathbf{w}_k$. For fixed γ_{nk}, these are independent of $\{\mathbf{w}_j\}$ for $j \neq k$ and so we can solve for each $\mathbf{w}_k$ separately using the IRLS algorithm. Thus the M-step equations for component k correspond simply to fitting a single logistic regression model to a weighted data set in which data point n carries a weight γ_{nk}. Figure 14.10 shows an example of the mixture of logistic regression models applied to a simple classification problem. The extension of this model to a mixture of softmax models for more than two classes is straightforward.

Section 4.3.3

Exercise 14.16

14.5.3 Mixtures of experts

In Section 14.5.1, we considered a mixture of linear regression models, and in Section 14.5.2 we discussed the analogous mixture of linear classifiers. Although these simple mixtures extend the flexibility of linear models to include more complex (e.g., multimodal) predictive distributions, they are still very limited. We can further increase the capability of such models by allowing the mixing coefficients themselves to be functions of the input variable, so that

$$p(\mathbf{t}|\mathbf{x}) = \sum_{k=1}^{K} \pi_k(\mathbf{x}) p_k(\mathbf{t}|\mathbf{x}). \tag{14.53}$$

This is known as a *mixture of experts* model (Jacobs *et al.*, 1991) in which the mixing coefficients $\pi_k(\mathbf{x})$ are known as *gating* functions and the individual component densities $p_k(\mathbf{t}|\mathbf{x})$ are called *experts*. The notion behind the terminology is that different components can model the distribution in different regions of input space (they

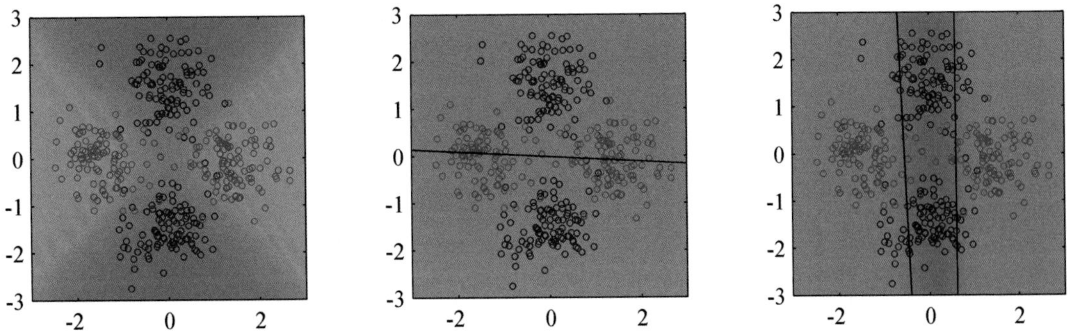

Figure 14.10 Illustration of a mixture of logistic regression models. The left plot shows data points drawn from two classes denoted red and blue, in which the background colour (which varies from pure red to pure blue) denotes the true probability of the class label. The centre plot shows the result of fitting a single logistic regression model using maximum likelihood, in which the background colour denotes the corresponding probability of the class label. Because the colour is a near-uniform purple, we see that the model assigns a probability of around 0.5 to each of the classes over most of input space. The right plot shows the result of fitting a mixture of two logistic regression models, which now gives much higher probability to the correct labels for many of the points in the blue class.

are 'experts' at making predictions in their own regions), and the gating functions determine which components are dominant in which region.

The gating functions $\pi_k(\mathbf{x})$ must satisfy the usual constraints for mixing coefficients, namely $0 \leqslant \pi_k(\mathbf{x}) \leqslant 1$ and $\sum_k \pi_k(\mathbf{x}) = 1$. They can therefore be represented, for example, by linear softmax models of the form (4.104) and (4.105). If the experts are also linear (regression or classification) models, then the whole model can be fitted efficiently using the EM algorithm, with iterative reweighted least squares being employed in the M step (Jordan and Jacobs, 1994).

Such a model still has significant limitations due to the use of linear models for the gating and expert functions. A much more flexible model is obtained by using a multilevel gating function to give the *hierarchical mixture of experts*, or *HME* model (Jordan and Jacobs, 1994). To understand the structure of this model, imagine a mixture distribution in which each component in the mixture is itself a mixture distribution. For simple unconditional mixtures, this hierarchical mixture is trivially equivalent to a single flat mixture distribution. However, when the mixing coefficients are input dependent, this hierarchical model becomes nontrivial. The HME model can also be viewed as a probabilistic version of *decision trees* discussed in Section 14.4 and can again be trained efficiently by maximum likelihood using an EM algorithm with IRLS in the M step. A Bayesian treatment of the HME has been given by Bishop and Svensén (2003) based on variational inference.

Exercise 14.17

Section 4.3.3

We shall not discuss the HME in detail here. However, it is worth pointing out the close connection with the *mixture density network* discussed in Section 5.6. The principal advantage of the mixtures of experts model is that it can be optimized by EM in which the M step for each mixture component and gating model involves a convex optimization (although the overall optimization is nonconvex). By contrast, the advantage of the mixture density network approach is that the component densities and the mixing coefficients share the hidden units of the neural network. Furthermore, in the mixture density network, the splits of the input space are further relaxed compared to the hierarchical mixture of experts in that they are not only soft, and not constrained to be axis aligned, but they can also be nonlinear.

Exercises

14.1 ($\star\star$) **www** Consider a set models of the form $p(\mathbf{t}|\mathbf{x}, \mathbf{z}_h, \boldsymbol{\theta}_h, h)$ in which $\mathbf{x}$ is the input vector, $\mathbf{t}$ is the target vector, h indexes the different models, $\mathbf{z}_h$ is a latent variable for model h, and $\boldsymbol{\theta}_h$ is the set of parameters for model h. Suppose the models have prior probabilities $p(h)$ and that we are given a training set $\mathbf{X} = \{\mathbf{x}_1, \ldots, \mathbf{x}_N\}$ and $\mathbf{T} = \{\mathbf{t}_1, \ldots, \mathbf{t}_N\}$. Write down the formulae needed to evaluate the predictive distribution $p(\mathbf{t}|\mathbf{x}, \mathbf{X}, \mathbf{T})$ in which the latent variables and the model index are marginalized out. Use these formulae to highlight the difference between Bayesian averaging of different models and the use of latent variables within a single model.

14.2 ($\star$) The expected sum-of-squares error E_{AV} for a simple committee model can be defined by (14.10), and the expected error of the committee itself is given by (14.11). Assuming that the individual errors satisfy (14.12) and (14.13), derive the result (14.14).

14.3 ($\star$) **www** By making use of Jensen's inequality (1.115), for the special case of the convex function $f(x) = x^2$, show that the average expected sum-of-squares error E_{AV} of the members of a simple committee model, given by (14.10), and the expected error E_{COM} of the committee itself, given by (14.11), satisfy

$$E_{\mathrm{COM}} \leqslant E_{\mathrm{AV}}. \tag{14.54}$$

14.4 ($\star\star$) By making use of Jensen's in equality (1.115), show that the result (14.54) derived in the previous exercise holds for any error function $E(y)$, not just sum-of-squares, provided it is a convex function of y.

14.5 ($\star\star$) **www** Consider a committee in which we allow unequal weighting of the constituent models, so that

$$y_{\mathrm{COM}}(\mathbf{x}) = \sum_{m=1}^{M} \alpha_m y_m(\mathbf{x}). \tag{14.55}$$

In order to ensure that the predictions $y_{\mathrm{COM}}(\mathbf{x})$ remain within sensible limits, suppose that we require that they be bounded at each value of $\mathbf{x}$ by the minimum and maximum values given by any of the members of the committee, so that

$$y_{\min}(\mathbf{x}) \leqslant y_{\mathrm{COM}}(\mathbf{x}) \leqslant y_{\max}(\mathbf{x}). \tag{14.56}$$

Show that a necessary and sufficient condition for this constraint is that the coefficients α_m satisfy

$$\alpha_m \geqslant 0, \qquad \sum_{m=1}^{M} \alpha_m = 1. \tag{14.57}$$

14.6 ($\star$) **www** By differentiating the error function (14.23) with respect to α_m, show that the parameters α_m in the AdaBoost algorithm are updated using (14.17) in which ϵ_m is defined by (14.16).

14.7 ($\star$) By making a variational minimization of the expected exponential error function given by (14.27) with respect to all possible functions $y(\mathbf{x})$, show that the minimizing function is given by (14.28).

14.8 ($\star$) Show that the exponential error function (14.20), which is minimized by the AdaBoost algorithm, does not correspond to the log likelihood of any well-behaved probabilistic model. This can be done by showing that the corresponding conditional distribution $p(t|\mathbf{x})$ cannot be correctly normalized.

14.9 ($\star$) **www** Show that the sequential minimization of the sum-of-squares error function for an additive model of the form (14.21) in the style of boosting simply involves fitting each new base classifier to the residual errors $t_n - f_{m-1}(\mathbf{x}_n)$ from the previous model.

14.10 ($\star$) Verify that if we minimize the sum-of-squares error between a set of training values $\{t_n\}$ and a single predictive value t, then the optimal solution for t is given by the mean of the $\{t_n\}$.

14.11 ($\star\star$) Consider a data set comprising 400 data points from class $\mathcal{C}_1$ and 400 data points from class $\mathcal{C}_2$. Suppose that a tree model A splits these into $(300, 100)$ assigned to the first leaf node (predicting $\mathcal{C}_1$) and $(100, 300)$ assigned to the second leaf node (predicting $\mathcal{C}_2$), where (n, m) denotes that n points come from class $\mathcal{C}_1$ and m points come from class $\mathcal{C}_2$. Similarly, suppose that a second tree model B splits them into $(200, 400)$ and $(200, 0)$, respectively. Evaluate the misclassification rates for the two trees and hence show that they are equal. Similarly, evaluate the pruning criterion (14.31) for the cross-entropy case (14.32) and the Gini index case (14.33) for the two trees and show that they are both lower for tree B than for tree A.

14.12 ($\star\star$) Extend the results of Section 14.5.1 for a mixture of linear regression models to the case of multiple target values described by a vector $\mathbf{t}$. To do this, make use of the results of Section 3.1.5.

14.13 ($\star$) **www** Verify that the complete-data log likelihood function for the mixture of linear regression models is given by (14.36).

14.14 ($\star$) Use the technique of Lagrange multipliers (Appendix E) to show that the M-step re-estimation equation for the mixing coefficients in the mixture of linear regression models trained by maximum likelihood EM is given by (14.38).

14.15 ($\star$) **www** We have already noted that if we use a squared loss function in a regression problem, the corresponding optimal prediction of the target variable for a new input vector is given by the conditional mean of the predictive distribution. Show that the conditional mean for the mixture of linear regression models discussed in Section 14.5.1 is given by a linear combination of the means of each component distribution. Note that if the conditional distribution of the target data is multimodal, the conditional mean can give poor predictions.

14.16 ($\star\star\star$) Extend the logistic regression mixture model of Section 14.5.2 to a mixture of softmax classifiers representing $C \geqslant 2$ classes. Write down the EM algorithm for determining the parameters of this model through maximum likelihood.

14.17 ($\star\star$) **www** Consider a mixture model for a conditional distribution $p(t|\mathbf{x})$ of the form

$$p(t|\mathbf{x}) = \sum_{k=1}^{K} \pi_k \psi_k(t|\mathbf{x}) \tag{14.58}$$

in which each mixture component $\psi_k(t|\mathbf{x})$ is itself a mixture model. Show that this two-level hierarchical mixture is equivalent to a conventional single-level mixture model. Now suppose that the mixing coefficients in both levels of such a hierarchical model are arbitrary functions of $\mathbf{x}$. Again, show that this hierarchical model is again equivalent to a single-level model with $\mathbf{x}$-dependent mixing coefficients. Finally, consider the case in which the mixing coefficients at both levels of the hierarchical mixture are constrained to be linear classification (logistic or softmax) models. Show that the hierarchical mixture cannot in general be represented by a single-level mixture having linear classification models for the mixing coefficients. Hint: to do this it is sufficient to construct a single counter-example, so consider a mixture of two components in which one of those components is itself a mixture of two components, with mixing coefficients given by linear-logistic models. Show that this cannot be represented by a single-level mixture of 3 components having mixing coefficients determined by a linear-softmax model.

Appendix A. Data Sets

In this appendix, we give a brief introduction to the data sets used to illustrate some of the algorithms described in this book. Detailed information on file formats for these data sets, as well as the data files themselves, can be obtained from the book web site:

http://research.microsoft.com/∼cmbishop/PRML

Handwritten Digits

The digits data used in this book is taken from the MNIST data set (LeCun *et al.*, 1998), which itself was constructed by modifying a subset of the much larger data set produced by NIST (the National Institute of Standards and Technology). It comprises a training set of $60,000$ examples and a test set of $10,000$ examples. Some of the data was collected from Census Bureau employees and the rest was collected from high-school children, and care was taken to ensure that the test examples were written by different individuals to the training examples.

The original NIST data had binary (black or white) pixels. To create MNIST, these images were size normalized to fit in a 20×20 pixel box while preserving their aspect ratio. As a consequence of the anti-aliasing used to change the resolution of the images, the resulting MNIST digits are grey scale. These images were then centred in a 28×28 box. Examples of the MNIST digits are shown in Figure A.1.

Error rates for classifying the digits range from 12% for a simple linear classifier, through 0.56% for a carefully designed support vector machine, to 0.4% for a convolutional neural network (LeCun *et al.*, 1998).

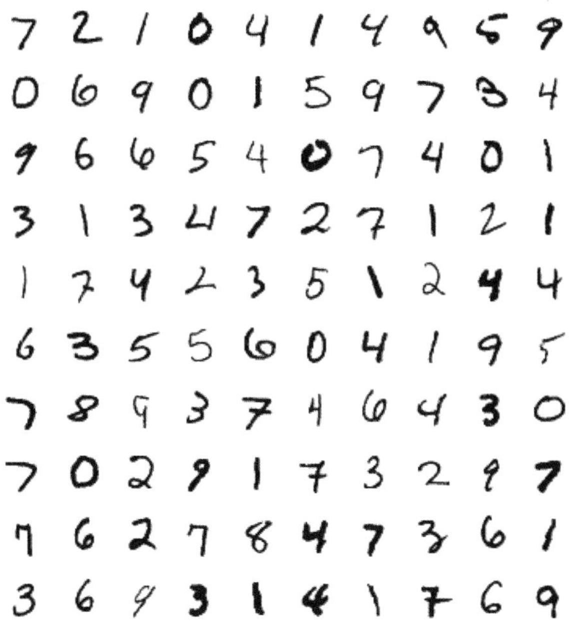

Figure A.1 One hundred examples of the MNIST digits chosen at random from the training set.

Oil Flow

This is a synthetic data set that arose out of a project aimed at measuring noninvasively the proportions of oil, water, and gas in North Sea oil transfer pipelines (Bishop and James, 1993). It is based on the principle of *dual-energy gamma densitometry*. The ideas is that if a narrow beam of gamma rays is passed through the pipe, the attenuation in the intensity of the beam provides information about the density of material along its path. Thus, for instance, the beam will be attenuated more strongly by oil than by gas.

A single attenuation measurement alone is not sufficient because there are two degrees of freedom corresponding to the fraction of oil and the fraction of water (the fraction of gas is redundant because the three fractions must add to one). To address this, two gamma beams of different energies (in other words different frequencies or wavelengths) are passed through the pipe along the same path, and the attenuation of each is measured. Because the absorbtion properties of different materials vary differently as a function of energy, measurement of the attenuations at the two energies provides two independent pieces of information. Given the known absorbtion properties of oil, water, and gas at the two energies, it is then a simple matter to calculate the average fractions of oil and water (and hence of gas) measured *along the path* of the gamma beams.

There is a further complication, however, associated with the motion of the materials along the pipe. If the flow velocity is small, then the oil floats on top of the water with the gas sitting above the oil. This is known as a *laminar* or *stratified* flow

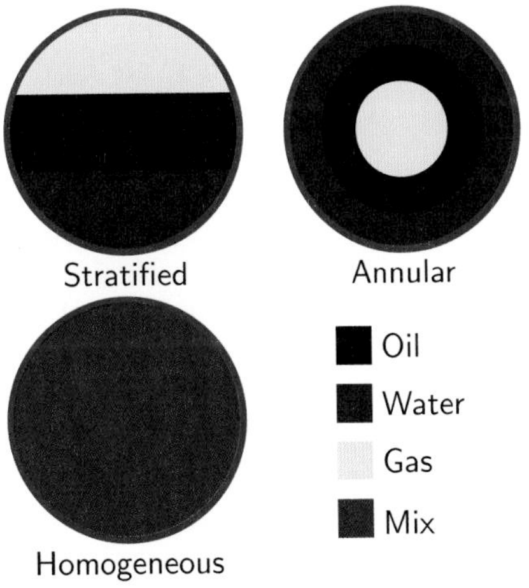

Figure A.2 The three geometrical configurations of the oil, water, and gas phases used to generate the oil-flow data set. For each configuration, the proportions of the three phases can vary.

configuration and is illustrated in Figure A.2. As the flow velocity is increased, more complex geometrical configurations of the oil, water, and gas can arise. For the purposes of this data set, two specific idealizations are considered. In the *annular* configuration the oil, water, and gas form concentric cylinders with the water around the outside and the gas in the centre, whereas in the *homogeneous* configuration the oil, water and gas are assumed to be intimately mixed as might occur at high flow velocities under turbulent conditions. These configurations are also illustrated in Figure A.2.

We have seen that a single dual-energy beam gives the oil and water fractions measured along the path length, whereas we are interested in the volume fractions of oil and water. This can be addressed by using multiple dual-energy gamma densitometers whose beams pass through different regions of the pipe. For this particular data set, there are six such beams, and their spatial arrangement is shown in Figure A.3. A single observation is therefore represented by a 12-dimensional vector comprising the fractions of oil and water measured along the paths of each of the beams. We are, however, interested in obtaining the overall volume fractions of the three phases in the pipe. This is much like the classical problem of tomographic reconstruction, used in medical imaging for example, in which a two-dimensional dis-

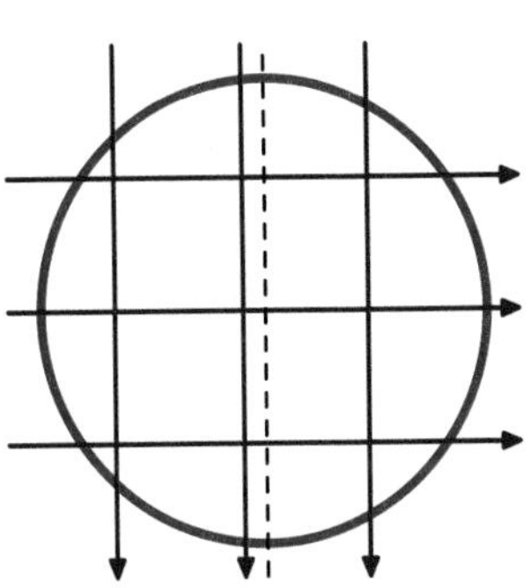

Figure A.3 Cross section of the pipe showing the arrangement of the six beam lines, each of which comprises a single dual-energy gamma densitometer. Note that the vertical beams are asymmetrically arranged relative to the central axis (shown by the dotted line).

tribution is to be reconstructed from an number of one-dimensional averages. Here there are far fewer line measurements than in a typical tomography application. On the other hand the range of geometrical configurations is much more limited, and so the configuration, as well as the phase fractions, can be predicted with reasonable accuracy from the densitometer data.

For safety reasons, the intensity of the gamma beams is kept relatively weak and so to obtain an accurate measurement of the attenuation, the measured beam intensity is integrated over a specific time interval. For a finite integration time, there are random fluctuations in the measured intensity due to the fact that the gamma beams comprise discrete packets of energy called photons. In practice, the integration time is chosen as a compromise between reducing the noise level (which requires a long integration time) and detecting temporal variations in the flow (which requires a short integration time). The oil flow data set is generated using realistic known values for the absorption properties of oil, water, and gas at the two gamma energies used, and with a specific choice of integration time (10 seconds) chosen as characteristic of a typical practical setup.

Each point in the data set is generated independently using the following steps:

1. Choose one of the three phase configurations at random with equal probability.

2. Choose three random numbers f_1, f_2 and f_3 from the uniform distribution over $(0, 1)$ and define

$$f_{\text{oil}} = \frac{f_1}{f_1 + f_2 + f_3}, \qquad f_{\text{water}} = \frac{f_2}{f_1 + f_2 + f_3}. \tag{A.1}$$

   This treats the three phases on an equal footing and ensures that the volume fractions add to one.

3. For each of the six beam lines, calculate the effective path lengths through oil and water for the given phase configuration.

4. Perturb the path lengths using the Poisson distribution based on the known beam intensities and integration time to allow for the effect of photon statistics.

Each point in the data set comprises the 12 path length measurements, together with the fractions of oil and water and a binary label describing the phase configuration. The data set is divided into training, validation, and test sets, each of which comprises $1,000$ independent data points. Details of the data format are available from the book web site.

In Bishop and James (1993), statistical machine learning techniques were used to predict the volume fractions and also the geometrical configuration of the phases shown in Figure A.2, from the 12-dimensional vector of measurements. The 12-dimensional observation vectors can also be used to test data visualization algorithms.

This data set has a rich and interesting structure, as follows. For any given configuration there are two degrees of freedom corresponding to the fractions of oil and water, and so for infinite integration time the data will locally live on a two-dimensional manifold. For a finite integration time, the individual data points will be perturbed away from the manifold by the photon noise. In the homogeneous phase configuration, the path lengths in oil and water are linearly related to the fractions of oil and water, and so the data points lie close to a linear manifold. For the annular configuration, the relationship between phase fraction and path length is nonlinear and so the manifold will be nonlinear. In the case of the laminar configuration the situation is even more complex because small variations in the phase fractions can cause one of the horizontal phase boundaries to move across one of the horizontal beam lines leading to a discontinuous jump in the 12-dimensional observation space. In this way, the two-dimensional nonlinear manifold for the laminar configuration is broken into ten distinct segments. Note also that some of the manifolds for different phase configurations meet at specific points, for example if the pipe is filled entirely with oil, it corresponds to specific instances of the laminar, annular, and homogeneous configurations.

Old Faithful

Old Faithful, shown in Figure A.4, is a hydrothermal geyser in Yellowstone National Park in the state of Wyoming, U.S.A., and is a popular tourist attraction. Its name stems from the supposed regularity of its eruptions.

The data set comprises 272 observations, each of which represents a single eruption and contains two variables corresponding to the duration in minutes of the eruption, and the time until the next eruption, also in minutes. Figure A.5 shows a plot of the time to the next eruption versus the duration of the eruptions. It can be seen that the time to the next eruption varies considerably, although knowledge of the duration of the current eruption allows it to be predicted more accurately. Note that there exist several other data sets relating to the eruptions of Old Faithful.

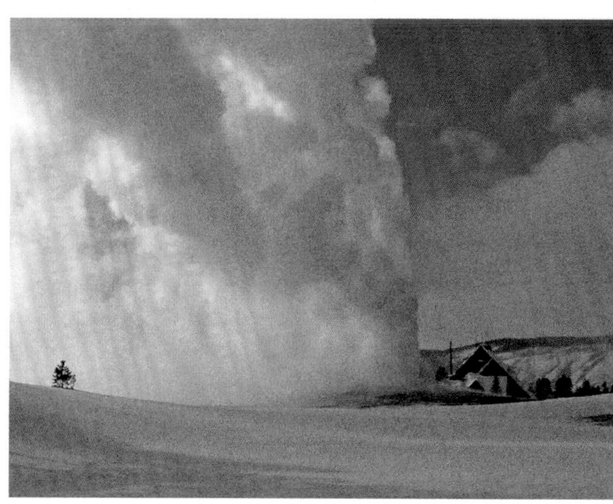

Figure A.4 The Old Faithful geyser in Yellowstone National Park. ©Bruce T. Gourley www.brucegourley.com.

Figure A.5 Plot of the time to the next eruption in minutes (vertical axis) versus the duration of the eruption in minutes (horizontal axis) for the Old Faithful data set.

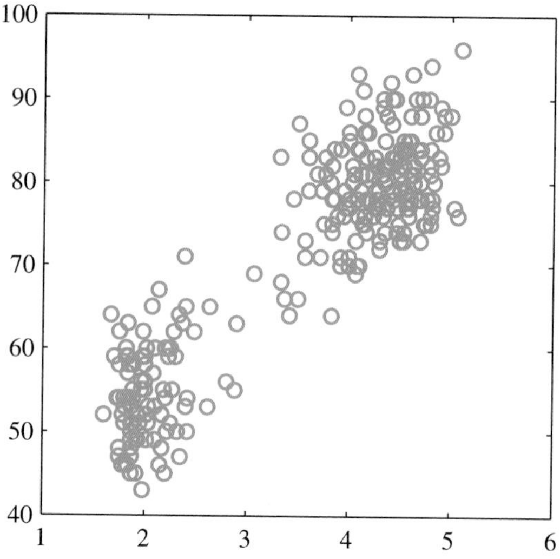

Synthetic Data

Throughout the book, we use two simple synthetic data sets to illustrate many of the algorithms. The first of these is a regression problem, based on the sinusoidal function, shown in Figure A.6. The input values $\{x_n\}$ are generated uniformly in range $(0, 1)$, and the corresponding target values $\{t_n\}$ are obtained by first computing the corresponding values of the function $\sin(2\pi x)$, and then adding random noise with a Gaussian distribution having standard deviation 0.3. Various forms of this data set, having different numbers of data points, are used in the book.

The second data set is a classification problem having two classes, with equal prior probabilities, and is shown in Figure A.7. The blue class is generated from a single Gaussian while the red class comes from a mixture of two Gaussians. Because we know the class priors and the class-conditional densities, it is straightforward to evaluate and plot the true posterior probabilities as well as the minimum misclassification-rate decision boundary, as shown in Figure A.7.

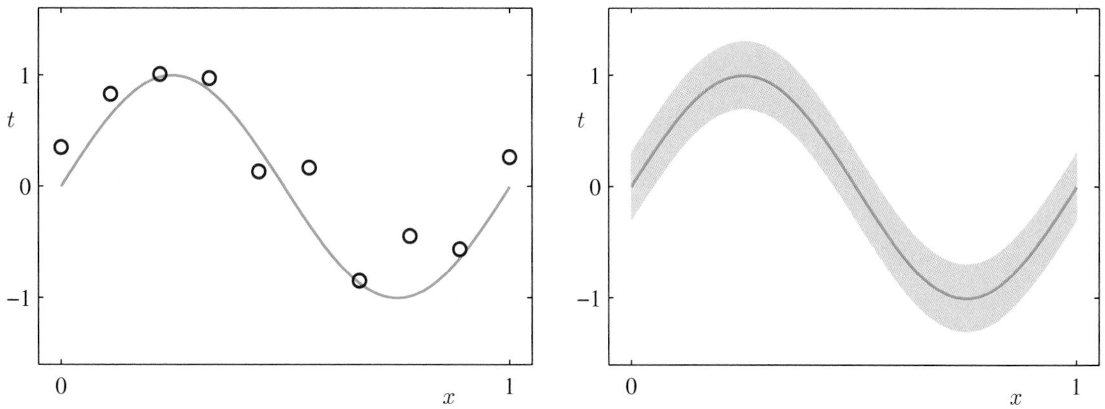

Figure A.6 The left-hand plot shows the synthetic regression data set along with the underlying sinusoidal function from which the data points were generated. The right-hand plot shows the true conditional distribution $p(t|x)$ from which the labels are generated, in which the green curve denotes the mean, and the shaded region spans one standard deviation on each side of the mean.

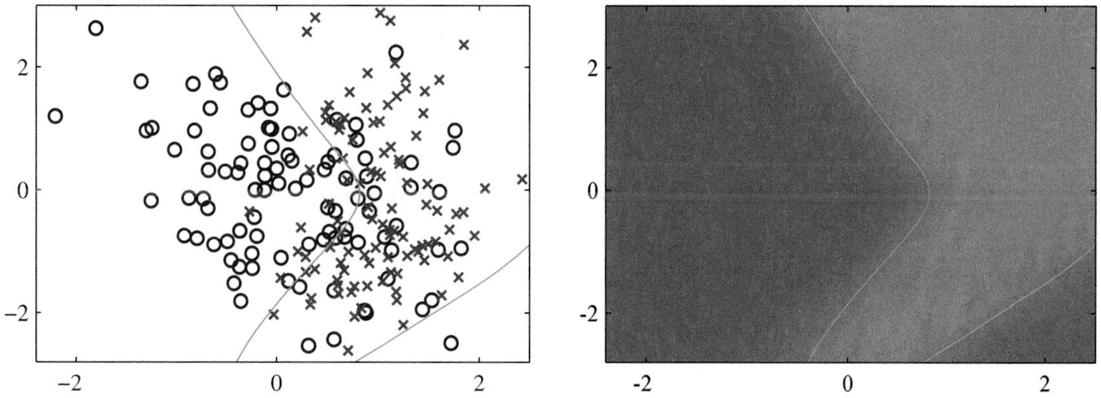

Figure A.7 The left plot shows the synthetic classification data set with data from the two classes shown in red and blue. On the right is a plot of the true posterior probabilities, shown on a colour scale going from pure red denoting probability of the red class is 1 to pure blue denoting probability of the red class is 0. Because these probabilities are known, the optimal decision boundary for minimizing the misclassification rate (which corresponds to the contour along which the posterior probabilities for each class equal 0.5) can be evaluated and is shown by the green curve. This decision boundary is also plotted on the left-hand figure.

Appendix B. Probability Distributions

In this appendix, we summarize the main properties of some of the most widely used probability distributions, and for each distribution we list some key statistics such as the expectation $\mathbb{E}[\mathbf{x}]$, the variance (or covariance), the mode, and the entropy $\mathrm{H}[\mathbf{x}]$. All of these distributions are members of the exponential family and are widely used as building blocks for more sophisticated probabilistic models.

Bernoulli

This is the distribution for a single binary variable $x \in \{0, 1\}$ representing, for example, the result of flipping a coin. It is governed by a single continuous parameter $\mu \in [0, 1]$ that represents the probability of $x = 1$.

$$\begin{aligned} \mathrm{Bern}(x|\mu) &= \mu^x (1-\mu)^{1-x} && \text{(B.1)} \\ \mathbb{E}[x] &= \mu && \text{(B.2)} \\ \mathrm{var}[x] &= \mu(1-\mu) && \text{(B.3)} \\ \mathrm{mode}[x] &= \begin{cases} 1 & \text{if } \mu \geqslant 0.5, \\ 0 & \text{otherwise} \end{cases} && \text{(B.4)} \\ \mathrm{H}[x] &= -\mu \ln \mu - (1-\mu)\ln(1-\mu). && \text{(B.5)} \end{aligned}$$

The Bernoulli is a special case of the binomial distribution for the case of a single observation. Its conjugate prior for μ is the beta distribution.

Beta

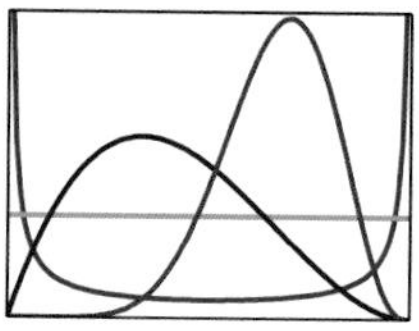

This is a distribution over a continuous variable $\mu \in [0, 1]$, which is often used to represent the probability for some binary event. It is governed by two parameters a and b that are constrained by $a > 0$ and $b > 0$ to ensure that the distribution can be normalized.

$$\mathrm{Beta}(\mu|a, b) = \frac{\Gamma(a + b)}{\Gamma(a)\Gamma(b)} \mu^{a-1}(1 - \mu)^{b-1} \tag{B.6}$$

$$\mathbb{E}[\mu] = \frac{a}{a + b} \tag{B.7}$$

$$\mathrm{var}[\mu] = \frac{ab}{(a + b)^2(a + b + 1)} \tag{B.8}$$

$$\mathrm{mode}[\mu] = \frac{a - 1}{a + b - 2}. \tag{B.9}$$

The beta is the conjugate prior for the Bernoulli distribution, for which a and b can be interpreted as the effective prior number of observations of $x = 1$ and $x = 0$, respectively. Its density is finite if $a \geqslant 1$ and $b \geqslant 1$, otherwise there is a singularity at $\mu = 0$ and/or $\mu = 1$. For $a = b = 1$, it reduces to a uniform distribution. The beta distribution is a special case of the K-state Dirichlet distribution for $K = 2$.

Binomial

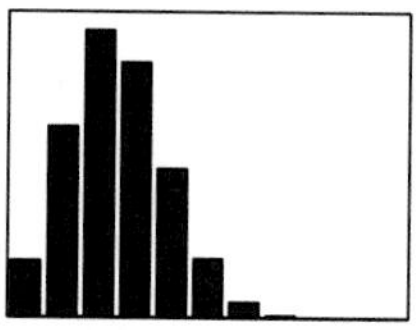

The binomial distribution gives the probability of observing m occurrences of $x = 1$ in a set of N samples from a Bernoulli distribution, where the probability of observing $x = 1$ is $\mu \in [0, 1]$.

$$\mathrm{Bin}(m|N, \mu) = \binom{N}{m} \mu^m (1 - \mu)^{N-m} \tag{B.10}$$

$$\mathbb{E}[m] = N\mu \tag{B.11}$$

$$\mathrm{var}[m] = N\mu(1 - \mu) \tag{B.12}$$

$$\mathrm{mode}[m] = \lfloor (N + 1)\mu \rfloor \tag{B.13}$$

where $\lfloor (N + 1)\mu \rfloor$ denotes the largest integer that is less than or equal to $(N + 1)\mu$, and the quantity

$$\binom{N}{m} = \frac{N!}{m!(N - m)!} \tag{B.14}$$

denotes the number of ways of choosing m objects out of a total of N identical objects. Here $m!$, pronounced 'factorial m', denotes the product $m \times (m - 1) \times \ldots, \times 2 \times 1$. The particular case of the binomial distribution for $N = 1$ is known as the Bernoulli distribution, and for large N the binomial distribution is approximately Gaussian. The conjugate prior for μ is the beta distribution.

Dirichlet

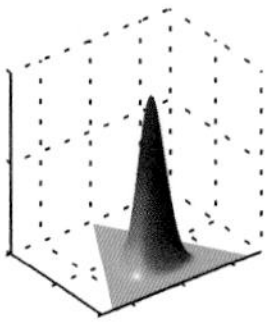

The Dirichlet is a multivariate distribution over K random variables $0 \leqslant \mu_k \leqslant 1$, where $k = 1, \ldots, K$, subject to the constraints

$$0 \leqslant \mu_k \leqslant 1, \qquad \sum_{k=1}^{K} \mu_k = 1. \tag{B.15}$$

Denoting $\boldsymbol{\mu} = (\mu_1, \ldots, \mu_K)^{\mathrm{T}}$ and $\boldsymbol{\alpha} = (\alpha_1, \ldots, \alpha_K)^{\mathrm{T}}$, we have

$$\mathrm{Dir}(\boldsymbol{\mu}|\boldsymbol{\alpha}) = C(\boldsymbol{\alpha}) \prod_{k=1}^{K} \mu_k^{\alpha_k - 1} \tag{B.16}$$

$$\mathbb{E}[\mu_k] = \frac{\alpha_k}{\widehat{\alpha}} \tag{B.17}$$

$$\mathrm{var}[\mu_k] = \frac{\alpha_k(\widehat{\alpha} - \alpha_k)}{\widehat{\alpha}^2(\widehat{\alpha} + 1)} \tag{B.18}$$

$$\mathrm{cov}[\mu_j \mu_k] = -\frac{\alpha_j \alpha_k}{\widehat{\alpha}^2(\widehat{\alpha} + 1)} \tag{B.19}$$

$$\mathrm{mode}[\mu_k] = \frac{\alpha_k - 1}{\widehat{\alpha} - K} \tag{B.20}$$

$$\mathbb{E}[\ln \mu_k] = \psi(\alpha_k) - \psi(\widehat{\alpha}) \tag{B.21}$$

$$\mathrm{H}[\boldsymbol{\mu}] = -\sum_{k=1}^{K} (\alpha_k - 1)\left\{\psi(\alpha_k) - \psi(\widehat{\alpha})\right\} - \ln C(\boldsymbol{\alpha}) \tag{B.22}$$

where

$$C(\boldsymbol{\alpha}) = \frac{\Gamma(\widehat{\alpha})}{\Gamma(\alpha_1) \cdots \Gamma(\alpha_K)} \tag{B.23}$$

and

$$\widehat{\alpha} = \sum_{k=1}^{K} \alpha_k. \tag{B.24}$$

Here

$$\psi(a) \equiv \frac{d}{da} \ln \Gamma(a) \tag{B.25}$$

is known as the *digamma* function (Abramowitz and Stegun, 1965). The parameters α_k are subject to the constraint $\alpha_k > 0$ in order to ensure that the distribution can be normalized.

The Dirichlet forms the conjugate prior for the multinomial distribution and represents a generalization of the beta distribution. In this case, the parameters α_k can be interpreted as effective numbers of observations of the corresponding values of the K-dimensional binary observation vector $\mathbf{x}$. As with the beta distribution, the Dirichlet has finite density everywhere provided $\alpha_k \geqslant 1$ for all k.

Gamma

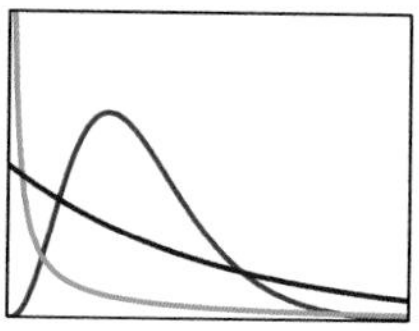

The Gamma is a probability distribution over a positive random variable $\tau > 0$ governed by parameters a and b that are subject to the constraints $a > 0$ and $b > 0$ to ensure that the distribution can be normalized.

$$\mathrm{Gam}(\tau|a,b) = \frac{1}{\Gamma(a)} b^a \tau^{a-1} e^{-b\tau} \tag{B.26}$$

$$\mathbb{E}[\tau] = \frac{a}{b} \tag{B.27}$$

$$\mathrm{var}[\tau] = \frac{a}{b^2} \tag{B.28}$$

$$\mathrm{mode}[\tau] = \frac{a-1}{b} \quad \text{for } a \geqslant 1 \tag{B.29}$$

$$\mathbb{E}[\ln \tau] = \psi(a) - \ln b \tag{B.30}$$

$$\mathrm{H}[\tau] = \ln \Gamma(a) - (a-1)\psi(a) - \ln b + a \tag{B.31}$$

where $\psi(\cdot)$ is the digamma function defined by (B.25). The gamma distribution is the conjugate prior for the precision (inverse variance) of a univariate Gaussian. For $a \geqslant 1$ the density is everywhere finite, and the special case of $a = 1$ is known as the *exponential* distribution.

Gaussian

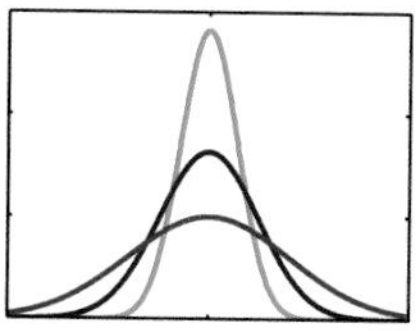

The Gaussian is the most widely used distribution for continuous variables. It is also known as the *normal* distribution. In the case of a single variable $x \in (-\infty, \infty)$ it is governed by two parameters, the mean $\mu \in (-\infty, \infty)$ and the variance $\sigma^2 > 0$.

$$\mathcal{N}(x|\mu,\sigma^2) = \frac{1}{(2\pi\sigma^2)^{1/2}} \exp\left\{-\frac{1}{2\sigma^2}(x-\mu)^2\right\} \tag{B.32}$$

$$\mathbb{E}[x] = \mu \tag{B.33}$$

$$\mathrm{var}[x] = \sigma^2 \tag{B.34}$$

$$\mathrm{mode}[x] = \mu \tag{B.35}$$

$$\mathrm{H}[x] = \frac{1}{2}\ln \sigma^2 + \frac{1}{2}(1 + \ln(2\pi)). \tag{B.36}$$

The inverse of the variance $\tau = 1/\sigma^2$ is called the precision, and the square root of the variance σ is called the standard deviation. The conjugate prior for μ is the Gaussian, and the conjugate prior for τ is the gamma distribution. If both μ and τ are unknown, their joint conjugate prior is the Gaussian-gamma distribution.

For a D-dimensional vector $\mathbf{x}$, the Gaussian is governed by a D-dimensional mean vector $\boldsymbol{\mu}$ and a $D \times D$ covariance matrix $\boldsymbol{\Sigma}$ that must be symmetric and positive-definite.

$$\mathcal{N}(\mathbf{x}|\boldsymbol{\mu}, \boldsymbol{\Sigma}) = \frac{1}{(2\pi)^{D/2}} \frac{1}{|\boldsymbol{\Sigma}|^{1/2}} \exp\left\{-\frac{1}{2}(\mathbf{x}-\boldsymbol{\mu})^{\mathrm{T}}\boldsymbol{\Sigma}^{-1}(\mathbf{x}-\boldsymbol{\mu})\right\} \quad (B.37)$$

$$\mathbb{E}[\mathbf{x}] = \boldsymbol{\mu} \quad (B.38)$$

$$\operatorname{cov}[\mathbf{x}] = \boldsymbol{\Sigma} \quad (B.39)$$

$$\operatorname{mode}[\mathbf{x}] = \boldsymbol{\mu} \quad (B.40)$$

$$\mathrm{H}[\mathbf{x}] = \frac{1}{2}\ln|\boldsymbol{\Sigma}| + \frac{D}{2}(1+\ln(2\pi)). \quad (B.41)$$

The inverse of the covariance matrix $\boldsymbol{\Lambda} = \boldsymbol{\Sigma}^{-1}$ is the precision matrix, which is also symmetric and positive definite. Averages of random variables tend to a Gaussian, by the central limit theorem, and the sum of two Gaussian variables is again Gaussian. The Gaussian is the distribution that maximizes the entropy for a given variance (or covariance). Any linear transformation of a Gaussian random variable is again Gaussian. The marginal distribution of a multivariate Gaussian with respect to a subset of the variables is itself Gaussian, and similarly the conditional distribution is also Gaussian. The conjugate prior for $\boldsymbol{\mu}$ is the Gaussian, the conjugate prior for $\boldsymbol{\Lambda}$ is the Wishart, and the conjugate prior for $(\boldsymbol{\mu}, \boldsymbol{\Lambda})$ is the Gaussian-Wishart.

If we have a marginal Gaussian distribution for $\mathbf{x}$ and a conditional Gaussian distribution for $\mathbf{y}$ given $\mathbf{x}$ in the form

$$p(\mathbf{x}) = \mathcal{N}(\mathbf{x}|\boldsymbol{\mu}, \boldsymbol{\Lambda}^{-1}) \quad (B.42)$$

$$p(\mathbf{y}|\mathbf{x}) = \mathcal{N}(\mathbf{y}|\mathbf{A}\mathbf{x}+\mathbf{b}, \mathbf{L}^{-1}) \quad (B.43)$$

then the marginal distribution of $\mathbf{y}$, and the conditional distribution of $\mathbf{x}$ given $\mathbf{y}$, are given by

$$p(\mathbf{y}) = \mathcal{N}(\mathbf{y}|\mathbf{A}\boldsymbol{\mu}+\mathbf{b}, \mathbf{L}^{-1}+\mathbf{A}\boldsymbol{\Lambda}^{-1}\mathbf{A}^{\mathrm{T}}) \quad (B.44)$$

$$p(\mathbf{x}|\mathbf{y}) = \mathcal{N}(\mathbf{x}|\boldsymbol{\Sigma}\{\mathbf{A}^{\mathrm{T}}\mathbf{L}(\mathbf{y}-\mathbf{b})+\boldsymbol{\Lambda}\boldsymbol{\mu}\}, \boldsymbol{\Sigma}) \quad (B.45)$$

where

$$\boldsymbol{\Sigma} = (\boldsymbol{\Lambda} + \mathbf{A}^{\mathrm{T}}\mathbf{L}\mathbf{A})^{-1}. \quad (B.46)$$

If we have a joint Gaussian distribution $\mathcal{N}(\mathbf{x}|\boldsymbol{\mu}, \boldsymbol{\Sigma})$ with $\boldsymbol{\Lambda} \equiv \boldsymbol{\Sigma}^{-1}$ and we define the following partitions

$$\mathbf{x} = \begin{pmatrix}\mathbf{x}_a \\ \mathbf{x}_b\end{pmatrix}, \quad \boldsymbol{\mu} = \begin{pmatrix}\boldsymbol{\mu}_a \\ \boldsymbol{\mu}_b\end{pmatrix} \quad (B.47)$$

$$\boldsymbol{\Sigma} = \begin{pmatrix}\boldsymbol{\Sigma}_{aa} & \boldsymbol{\Sigma}_{ab} \\ \boldsymbol{\Sigma}_{ba} & \boldsymbol{\Sigma}_{bb}\end{pmatrix}, \quad \boldsymbol{\Lambda} = \begin{pmatrix}\boldsymbol{\Lambda}_{aa} & \boldsymbol{\Lambda}_{ab} \\ \boldsymbol{\Lambda}_{ba} & \boldsymbol{\Lambda}_{bb}\end{pmatrix} \quad (B.48)$$

then the conditional distribution $p(\mathbf{x}_a|\mathbf{x}_b)$ is given by

$$p(\mathbf{x}_a|\mathbf{x}_b) = \mathcal{N}(\mathbf{x}|\boldsymbol{\mu}_{a|b}, \boldsymbol{\Lambda}_{aa}^{-1}) \quad (B.49)$$

$$\boldsymbol{\mu}_{a|b} = \boldsymbol{\mu}_a - \boldsymbol{\Lambda}_{aa}^{-1}\boldsymbol{\Lambda}_{ab}(\mathbf{x}_b - \boldsymbol{\mu}_b) \quad (B.50)$$

and the marginal distribution $p(\mathbf{x}_a)$ is given by

$$p(\mathbf{x}_a) = \mathcal{N}(\mathbf{x}_a|\boldsymbol{\mu}_a, \boldsymbol{\Sigma}_{aa}). \tag{B.51}$$

Gaussian-Gamma

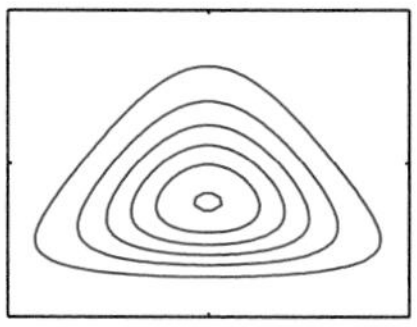

This is the conjugate prior distribution for a univariate Gaussian $\mathcal{N}(x|\mu, \lambda^{-1})$ in which the mean μ and the precision λ are both unknown and is also called the *normal-gamma* distribution. It comprises the product of a Gaussian distribution for μ, whose precision is proportional to λ, and a gamma distribution over λ.

$$p(\mu, \lambda|\mu_0, \beta, a, b) = \mathcal{N}\left(\mu|\mu_o, (\beta\lambda)^{-1}\right) \mathrm{Gam}(\lambda|a, b). \tag{B.52}$$

Gaussian-Wishart

This is the conjugate prior distribution for a multivariate Gaussian $\mathcal{N}(\mathbf{x}|\boldsymbol{\mu}, \boldsymbol{\Lambda})$ in which both the mean $\boldsymbol{\mu}$ and the precision $\boldsymbol{\Lambda}$ are unknown, and is also called the normal-Wishart distribution. It comprises the product of a Gaussian distribution for $\boldsymbol{\mu}$, whose precision is proportional to $\boldsymbol{\Lambda}$, and a Wishart distribution over $\boldsymbol{\Lambda}$.

$$p(\boldsymbol{\mu}, \boldsymbol{\Lambda}|\boldsymbol{\mu}_0, \beta, \mathbf{W}, \nu) = \mathcal{N}\left(\boldsymbol{\mu}|\boldsymbol{\mu}_0, (\beta\boldsymbol{\Lambda})^{-1}\right) \mathcal{W}(\boldsymbol{\Lambda}|\mathbf{W}, \nu). \tag{B.53}$$

For the particular case of a scalar x, this is equivalent to the Gaussian-gamma distribution.

Multinomial

If we generalize the Bernoulli distribution to an K-dimensional binary variable $\mathbf{x}$ with components $x_k \in \{0, 1\}$ such that $\sum_k x_k = 1$, then we obtain the following discrete distribution

$$p(\mathbf{x}) = \prod_{k=1}^{K} \mu_k^{x_k} \tag{B.54}$$

$$\mathbb{E}[x_k] = \mu_k \tag{B.55}$$

$$\mathrm{var}[x_k] = \mu_k(1 - \mu_k) \tag{B.56}$$

$$\mathrm{cov}[x_j x_k] = -\mu_j \mu_k, j \neq k \tag{B.57}$$

$$\mathrm{H}[\mathbf{x}] = -\sum_{k=1}^{M} \mu_k \ln \mu_k \tag{B.58}$$

where I_{jk} is the j, k element of the identity matrix. Because $p(x_k = 1) = \mu_k$, the parameters must satisfy $0 \leqslant \mu_k \leqslant 1$ and $\sum_k \mu_k = 1$.

The multinomial distribution is a multivariate generalization of the binomial and gives the distribution over counts m_k for a K-state discrete variable to be in state k given a total number of observations N.

$$\begin{aligned}
\mathrm{Mult}(m_1, m_2, \ldots, m_K|\boldsymbol{\mu}, N) &= \binom{N}{m_1 m_2 \ldots m_K} \prod_{k=1}^{K} \mu_k^{m_k} &\text{(B.59)}\\
\mathbb{E}[m_k] &= N\mu_k &\text{(B.60)}\\
\mathrm{var}[m_k] &= N\mu_k(1-\mu_k) &\text{(B.61)}\\
\mathrm{cov}[m_j m_k] &= -N\mu_j\mu_k, j \neq k &\text{(B.62)}
\end{aligned}$$

where $\boldsymbol{\mu} = (\mu_1, \ldots, \mu_K)^{\mathrm{T}}$, and the quantity

$$\binom{N}{m_1 m_2 \ldots m_K} = \frac{N!}{m_1! \ldots m_K!} \tag{B.63}$$

gives the number of ways of taking N identical objects and assigning m_k of them to bin k for $k = 1, \ldots, K$. The value of μ_k gives the probability of the random variable taking state k, and so these parameters are subject to the constraints $0 \leqslant \mu_k \leqslant 1$ and $\sum_k \mu_k = 1$. The conjugate prior distribution for the parameters $\{\mu_k\}$ is the Dirichlet.

Normal

The normal distribution is simply another name for the Gaussian. In this book, we use the term Gaussian throughout, although we retain the conventional use of the symbol $\mathcal{N}$ to denote this distribution. For consistency, we shall refer to the normal-gamma distribution as the Gaussian-gamma distribution, and similarly the normal-Wishart is called the Gaussian-Wishart.

Student's t

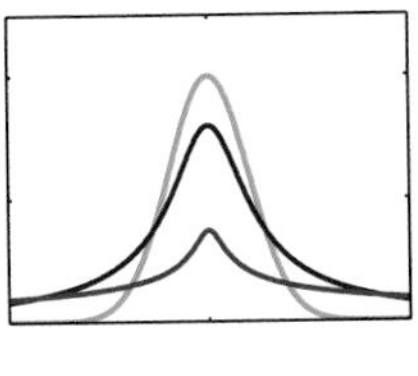

This distribution was published by William Gosset in 1908, but his employer, Guiness Breweries, required him to publish under a pseudonym, so he chose 'Student'. In the univariate form, Student's t-distribution is obtained by placing a conjugate gamma prior over the precision of a univariate Gaussian distribution and then integrating out the precision variable. It can therefore be viewed as an infinite mixture of Gaussians having the same mean but different variances.

$$\mathrm{St}(x|\mu,\lambda,\nu) = \frac{\Gamma(\nu/2+1/2)}{\Gamma(\nu/2)}\left(\frac{\lambda}{\pi\nu}\right)^{1/2}\left[1+\frac{\lambda(x-\mu)^2}{\nu}\right]^{-\nu/2-1/2} \quad \text{(B.64)}$$

$$\mathbb{E}[x] = \mu \quad \text{for } \nu > 1 \quad \text{(B.65)}$$

$$\mathrm{var}[x] = \frac{1}{\lambda}\frac{\nu}{\nu-2} \quad \text{for } \nu > 2 \quad \text{(B.66)}$$

$$\mathrm{mode}[x] = \mu. \quad \text{(B.67)}$$

Here $\nu > 0$ is called the number of degrees of freedom of the distribution. The particular case of $\nu = 1$ is called the *Cauchy* distribution.

For a D-dimensional variable $\mathbf{x}$, Student's t-distribution corresponds to marginalizing the precision matrix of a multivariate Gaussian with respect to a conjugate Wishart prior and takes the form

$$\mathrm{St}(\mathbf{x}|\boldsymbol{\mu},\boldsymbol{\Lambda},\nu) = \frac{\Gamma(\nu/2+D/2)}{\Gamma(\nu/2)}\frac{|\boldsymbol{\Lambda}|^{1/2}}{(\nu\pi)^{D/2}}\left[1+\frac{\Delta^2}{\nu}\right]^{-\nu/2-D/2} \quad \text{(B.68)}$$

$$\mathbb{E}[\mathbf{x}] = \boldsymbol{\mu} \quad \text{for } \nu > 1 \quad \text{(B.69)}$$

$$\mathrm{cov}[\mathbf{x}] = \frac{\nu}{\nu-2}\boldsymbol{\Lambda}^{-1} \quad \text{for } \nu > 2 \quad \text{(B.70)}$$

$$\mathrm{mode}[\mathbf{x}] = \boldsymbol{\mu} \quad \text{(B.71)}$$

where Δ^2 is the squared Mahalanobis distance defined by

$$\Delta^2 = (\mathbf{x}-\boldsymbol{\mu})^{\mathrm{T}}\boldsymbol{\Lambda}(\mathbf{x}-\boldsymbol{\mu}). \quad \text{(B.72)}$$

In the limit $\nu \to \infty$, the t-distribution reduces to a Gaussian with mean μ and precision $\boldsymbol{\Lambda}$. Student's t-distribution provides a generalization of the Gaussian whose maximum likelihood parameter values are robust to outliers.

Uniform

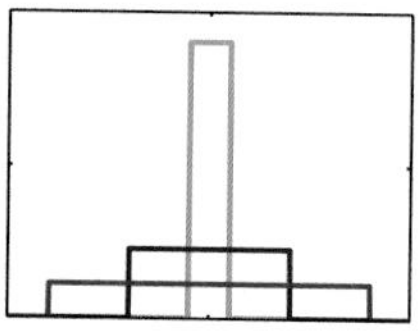

This is a simple distribution for a continuous variable x defined over a finite interval $x \in [a, b]$ where $b > a$.

$$\mathrm{U}(x|a,b) = \frac{1}{b-a} \quad \text{(B.73)}$$

$$\mathbb{E}[x] = \frac{(b+a)}{2} \quad \text{(B.74)}$$

$$\mathrm{var}[x] = \frac{(b-a)^2}{12} \quad \text{(B.75)}$$

$$\mathrm{H}[x] = \ln(b-a). \quad \text{(B.76)}$$

If x has distribution $\mathrm{U}(x|0,1)$, then $a + (b-a)x$ will have distribution $\mathrm{U}(x|a,b)$.

Von Mises

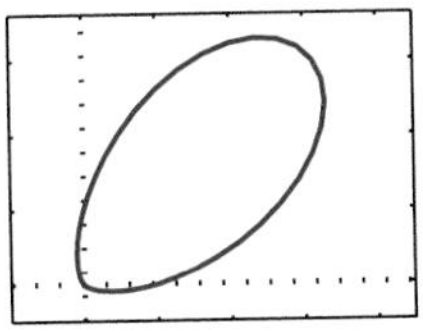

The von Mises distribution, also known as the circular normal or the circular Gaussian, is a univariate Gaussian-like periodic distribution for a variable $\theta \in [0, 2\pi)$.

$$p(\theta|\theta_0, m) = \frac{1}{2\pi I_0(m)} \exp\{m\cos(\theta - \theta_0)\} \tag{B.77}$$

where $I_0(m)$ is the zeroth-order Bessel function of the first kind. The distribution has period 2π so that $p(\theta + 2\pi) = p(\theta)$ for all θ. Care must be taken in interpreting this distribution because simple expectations will be dependent on the (arbitrary) choice of origin for the variable θ. The parameter θ_0 is analogous to the mean of a univariate Gaussian, and the parameter $m > 0$, known as the *concentration* parameter, is analogous to the precision (inverse variance). For large m, the von Mises distribution is approximately a Gaussian centred on θ_0.

Wishart

The Wishart distribution is the conjugate prior for the precision matrix of a multivariate Gaussian.

$$\mathcal{W}(\mathbf{\Lambda}|\mathbf{W}, \nu) = B(\mathbf{W}, \nu)|\mathbf{\Lambda}|^{(\nu-D-1)/2} \exp\left(-\frac{1}{2}\mathrm{Tr}(\mathbf{W}^{-1}\mathbf{\Lambda})\right) \tag{B.78}$$

where

$$B(\mathbf{W}, \nu) \equiv |\mathbf{W}|^{-\nu/2}\left(2^{\nu D/2}\,\pi^{D(D-1)/4}\prod_{i=1}^{D}\Gamma\left(\frac{\nu+1-i}{2}\right)\right)^{-1} \tag{B.79}$$

$$\mathbb{E}[\mathbf{\Lambda}] = \nu\mathbf{W} \tag{B.80}$$

$$\mathbb{E}[\ln|\mathbf{\Lambda}|] = \sum_{i=1}^{D}\psi\left(\frac{\nu+1-i}{2}\right) + D\ln 2 + \ln|\mathbf{W}| \tag{B.81}$$

$$\mathrm{H}[\mathbf{\Lambda}] = -\ln B(\mathbf{W}, \nu) - \frac{(\nu-D-1)}{2}\mathbb{E}[\ln|\mathbf{\Lambda}|] + \frac{\nu D}{2} \tag{B.82}$$

where $\mathbf{W}$ is a $D \times D$ symmetric, positive definite matrix, and $\psi(\cdot)$ is the digamma function defined by (B.25). The parameter ν is called the *number of degrees of freedom* of the distribution and is restricted to $\nu > D - 1$ to ensure that the Gamma function in the normalization factor is well-defined. In one dimension, the Wishart reduces to the gamma distribution $\mathrm{Gam}(\lambda|a, b)$ given by (B.26) with parameters $a = \nu/2$ and $b = 1/2W$.

Appendix C. Properties of Matrices

In this appendix, we gather together some useful properties and identities involving matrices and determinants. This is not intended to be an introductory tutorial, and it is assumed that the reader is already familiar with basic linear algebra. For some results, we indicate how to prove them, whereas in more complex cases we leave the interested reader to refer to standard textbooks on the subject. In all cases, we assume that inverses exist and that matrix dimensions are such that the formulae are correctly defined. A comprehensive discussion of linear algebra can be found in Golub and Van Loan (1996), and an extensive collection of matrix properties is given by Lütkepohl (1996). Matrix derivatives are discussed in Magnus and Neudecker (1999).

Basic Matrix Identities

A matrix $\mathbf{A}$ has elements A_{ij} where i indexes the rows, and j indexes the columns. We use $\mathbf{I}_N$ to denote the $N \times N$ identity matrix (also called the unit matrix), and where there is no ambiguity over dimensionality we simply use $\mathbf{I}$. The transpose matrix $\mathbf{A}^{\mathrm{T}}$ has elements $(\mathbf{A}^{\mathrm{T}})_{ij} = A_{ji}$. From the definition of transpose, we have

$$(\mathbf{AB})^{\mathrm{T}} = \mathbf{B}^{\mathrm{T}}\mathbf{A}^{\mathrm{T}} \tag{C.1}$$

which can be verified by writing out the indices. The inverse of $\mathbf{A}$, denoted $\mathbf{A}^{-1}$, satisfies

$$\mathbf{A}\mathbf{A}^{-1} = \mathbf{A}^{-1}\mathbf{A} = \mathbf{I}. \tag{C.2}$$

Because $\mathbf{ABB}^{-1}\mathbf{A}^{-1} = \mathbf{I}$, we have

$$(\mathbf{AB})^{-1} = \mathbf{B}^{-1}\mathbf{A}^{-1}. \tag{C.3}$$

Also we have

$$\left(\mathbf{A}^{\mathrm{T}}\right)^{-1} = \left(\mathbf{A}^{-1}\right)^{\mathrm{T}} \tag{C.4}$$

which is easily proven by taking the transpose of (C.2) and applying (C.1).

A useful identity involving matrix inverses is the following

$$(\mathbf{P}^{-1} + \mathbf{B}^{\mathrm{T}}\mathbf{R}^{-1}\mathbf{B})^{-1}\mathbf{B}^{\mathrm{T}}\mathbf{R}^{-1} = \mathbf{P}\mathbf{B}^{\mathrm{T}}(\mathbf{B}\mathbf{P}\mathbf{B}^{\mathrm{T}} + \mathbf{R})^{-1}. \tag{C.5}$$

which is easily verified by right multiplying both sides by $(\mathbf{B}\mathbf{P}\mathbf{B}^{\mathrm{T}} + \mathbf{R})$. Suppose that $\mathbf{P}$ has dimensionality $N \times N$ while $\mathbf{R}$ has dimensionality $M \times M$, so that $\mathbf{B}$ is $M \times N$. Then if $M \ll N$, it will be much cheaper to evaluate the right-hand side of (C.5) than the left-hand side. A special case that sometimes arises is

$$(\mathbf{I} + \mathbf{A}\mathbf{B})^{-1}\mathbf{A} = \mathbf{A}(\mathbf{I} + \mathbf{B}\mathbf{A})^{-1}. \tag{C.6}$$

Another useful identity involving inverses is the following:

$$(\mathbf{A} + \mathbf{B}\mathbf{D}^{-1}\mathbf{C})^{-1} = \mathbf{A}^{-1} - \mathbf{A}^{-1}\mathbf{B}(\mathbf{D} + \mathbf{C}\mathbf{A}^{-1}\mathbf{B})^{-1}\mathbf{C}\mathbf{A}^{-1} \tag{C.7}$$

which is known as the *Woodbury identity* and which can be verified by multiplying both sides by $(\mathbf{A} + \mathbf{B}\mathbf{D}^{-1}\mathbf{C})$. This is useful, for instance, when $\mathbf{A}$ is large and diagonal, and hence easy to invert, while $\mathbf{B}$ has many rows but few columns (and conversely for $\mathbf{C}$) so that the right-hand side is much cheaper to evaluate than the left-hand side.

A set of vectors $\{\mathbf{a}_1, \ldots, \mathbf{a}_N\}$ is said to be *linearly independent* if the relation $\sum_n \alpha_n \mathbf{a}_n = 0$ holds only if all $\alpha_n = 0$. This implies that none of the vectors can be expressed as a linear combination of the remainder. The rank of a matrix is the maximum number of linearly independent rows (or equivalently the maximum number of linearly independent columns).

Traces and Determinants

Trace and determinant apply to square matrices. The trace $\mathrm{Tr}(\mathbf{A})$ of a matrix $\mathbf{A}$ is defined as the sum of the elements on the leading diagonal. By writing out the indices, we see that

$$\mathrm{Tr}(\mathbf{A}\mathbf{B}) = \mathrm{Tr}(\mathbf{B}\mathbf{A}). \tag{C.8}$$

By applying this formula multiple times to the product of three matrices, we see that

$$\mathrm{Tr}(\mathbf{A}\mathbf{B}\mathbf{C}) = \mathrm{Tr}(\mathbf{C}\mathbf{A}\mathbf{B}) = \mathrm{Tr}(\mathbf{B}\mathbf{C}\mathbf{A}) \tag{C.9}$$

which is known as the *cyclic* property of the trace operator and which clearly extends to the product of any number of matrices. The determinant $|\mathbf{A}|$ of an $N \times N$ matrix $\mathbf{A}$ is defined by

$$|\mathbf{A}| = \sum (\pm 1) A_{1i_1} A_{2i_2} \cdots A_{Ni_N} \tag{C.10}$$

in which the sum is taken over all products consisting of precisely one element from each row and one element from each column, with a coefficient $+1$ or -1 according to whether the permutation $i_1 i_2 \ldots i_N$ is even or odd, respectively. Note that $|\mathbf{I}| = 1$. Thus, for a 2×2 matrix, the determinant takes the form

$$|\mathbf{A}| = \begin{vmatrix} a_{11} & a_{12} \\ a_{21} & a_{22} \end{vmatrix} = a_{11}a_{22} - a_{12}a_{21}. \tag{C.11}$$

The determinant of a product of two matrices is given by

$$|\mathbf{AB}| = |\mathbf{A}||\mathbf{B}| \tag{C.12}$$

as can be shown from (C.10). Also, the determinant of an inverse matrix is given by

$$\left|\mathbf{A}^{-1}\right| = \frac{1}{|\mathbf{A}|} \tag{C.13}$$

which can be shown by taking the determinant of (C.2) and applying (C.12).

If $\mathbf{A}$ and $\mathbf{B}$ are matrices of size $N \times M$, then

$$\left|\mathbf{I}_N + \mathbf{A}\mathbf{B}^{\mathrm{T}}\right| = \left|\mathbf{I}_M + \mathbf{A}^{\mathrm{T}}\mathbf{B}\right|. \tag{C.14}$$

A useful special case is

$$\left|\mathbf{I}_N + \mathbf{a}\mathbf{b}^{\mathrm{T}}\right| = 1 + \mathbf{a}^{\mathrm{T}}\mathbf{b} \tag{C.15}$$

where $\mathbf{a}$ and $\mathbf{b}$ are N-dimensional column vectors.

Matrix Derivatives

Sometimes we need to consider derivatives of vectors and matrices with respect to scalars. The derivative of a vector $\mathbf{a}$ with respect to a scalar x is itself a vector whose components are given by

$$\left(\frac{\partial \mathbf{a}}{\partial x}\right)_i = \frac{\partial a_i}{\partial x} \tag{C.16}$$

with an analogous definition for the derivative of a matrix. Derivatives with respect to vectors and matrices can also be defined, for instance

$$\left(\frac{\partial x}{\partial \mathbf{a}}\right)_i = \frac{\partial x}{\partial a_i} \tag{C.17}$$

and similarly

$$\left(\frac{\partial \mathbf{a}}{\partial \mathbf{b}}\right)_{ij} = \frac{\partial a_i}{\partial b_j}. \tag{C.18}$$

The following is easily proven by writing out the components

$$\frac{\partial}{\partial \mathbf{x}}\left(\mathbf{x}^{\mathrm{T}}\mathbf{a}\right) = \frac{\partial}{\partial \mathbf{x}}\left(\mathbf{a}^{\mathrm{T}}\mathbf{x}\right) = \mathbf{a}. \tag{C.19}$$

Similarly

$$\frac{\partial}{\partial x}(\mathbf{AB}) = \frac{\partial \mathbf{A}}{\partial x}\mathbf{B} + \mathbf{A}\frac{\partial \mathbf{B}}{\partial x}. \tag{C.20}$$

The derivative of the inverse of a matrix can be expressed as

$$\frac{\partial}{\partial x}\left(\mathbf{A}^{-1}\right) = -\mathbf{A}^{-1}\frac{\partial \mathbf{A}}{\partial x}\mathbf{A}^{-1} \tag{C.21}$$

as can be shown by differentiating the equation $\mathbf{A}^{-1}\mathbf{A} = \mathbf{I}$ using (C.20) and then right multiplying by $\mathbf{A}^{-1}$. Also

$$\frac{\partial}{\partial x}\ln|\mathbf{A}| = \mathrm{Tr}\left(\mathbf{A}^{-1}\frac{\partial \mathbf{A}}{\partial x}\right) \tag{C.22}$$

which we shall prove later. If we choose x to be one of the elements of $\mathbf{A}$, we have

$$\frac{\partial}{\partial A_{ij}}\mathrm{Tr}\,(\mathbf{AB}) = B_{ji} \tag{C.23}$$

as can be seen by writing out the matrices using index notation. We can write this result more compactly in the form

$$\frac{\partial}{\partial \mathbf{A}}\mathrm{Tr}\,(\mathbf{AB}) = \mathbf{B}^{\mathrm{T}}. \tag{C.24}$$

With this notation, we have the following properties

$$\frac{\partial}{\partial \mathbf{A}}\mathrm{Tr}\left(\mathbf{A}^{\mathrm{T}}\mathbf{B}\right) = \mathbf{B} \tag{C.25}$$

$$\frac{\partial}{\partial \mathbf{A}}\mathrm{Tr}(\mathbf{A}) = \mathbf{I} \tag{C.26}$$

$$\frac{\partial}{\partial \mathbf{A}}\mathrm{Tr}(\mathbf{ABA}^{\mathrm{T}}) = \mathbf{A}(\mathbf{B}+\mathbf{B}^{\mathrm{T}}) \tag{C.27}$$

which can again be proven by writing out the matrix indices. We also have

$$\frac{\partial}{\partial \mathbf{A}}\ln|\mathbf{A}| = \left(\mathbf{A}^{-1}\right)^{\mathrm{T}} \tag{C.28}$$

which follows from (C.22) and (C.24).

Eigenvector Equation

For a square matrix $\mathbf{A}$ of size $M \times M$, the eigenvector equation is defined by

$$\mathbf{A}\mathbf{u}_i = \lambda_i \mathbf{u}_i \tag{C.29}$$

for $i = 1, \ldots, M$, where $\mathbf{u}_i$ is an *eigenvector* and λ_i is the corresponding *eigenvalue*. This can be viewed as a set of M simultaneous homogeneous linear equations, and the condition for a solution is that

$$|\mathbf{A} - \lambda_i \mathbf{I}| = 0 \tag{C.30}$$

which is known as the *characteristic equation*. Because this is a polynomial of order M in λ_i, it must have M solutions (though these need not all be distinct). The rank of $\mathbf{A}$ is equal to the number of nonzero eigenvalues.

Of particular interest are symmetric matrices, which arise as covariance matrices, kernel matrices, and Hessians. Symmetric matrices have the property that $A_{ij} = A_{ji}$, or equivalently $\mathbf{A}^{\mathrm{T}} = \mathbf{A}$. The inverse of a symmetric matrix is also symmetric, as can be seen by taking the transpose of $\mathbf{A}^{-1}\mathbf{A} = \mathbf{I}$ and using $\mathbf{A}\mathbf{A}^{-1} = \mathbf{I}$ together with the symmetry of $\mathbf{I}$.

In general, the eigenvalues of a matrix are complex numbers, but for symmetric matrices the eigenvalues λ_i are real. This can be seen by first left multiplying (C.29) by $(\mathbf{u}_i^{\star})^{\mathrm{T}}$, where $\star$ denotes the complex conjugate, to give

$$(\mathbf{u}_i^{\star})^{\mathrm{T}} \mathbf{A}\mathbf{u}_i = \lambda_i (\mathbf{u}_i^{\star})^{\mathrm{T}} \mathbf{u}_i. \tag{C.31}$$

Next we take the complex conjugate of (C.29) and left multiply by $\mathbf{u}_i^{\mathrm{T}}$ to give

$$\mathbf{u}_i^{\mathrm{T}} \mathbf{A}\mathbf{u}_i^{\star} = \lambda_i^{\star} \mathbf{u}_i^{\mathrm{T}} \mathbf{u}_i^{\star}. \tag{C.32}$$

where we have used $\mathbf{A}^{\star} = \mathbf{A}$ because we consider only real matrices $\mathbf{A}$. Taking the transpose of the second of these equations, and using $\mathbf{A}^{\mathrm{T}} = \mathbf{A}$, we see that the left-hand sides of the two equations are equal, and hence that $\lambda_i^{\star} = \lambda_i$ and so λ_i must be real.

The eigenvectors $\mathbf{u}_i$ of a real symmetric matrix can be chosen to be orthonormal (i.e., orthogonal and of unit length) so that

$$\mathbf{u}_i^{\mathrm{T}} \mathbf{u}_j = I_{ij} \tag{C.33}$$

where I_{ij} are the elements of the identity matrix $\mathbf{I}$. To show this, we first left multiply (C.29) by $\mathbf{u}_j^{\mathrm{T}}$ to give

$$\mathbf{u}_j^{\mathrm{T}} \mathbf{A}\mathbf{u}_i = \lambda_i \mathbf{u}_j^{\mathrm{T}} \mathbf{u}_i \tag{C.34}$$

and hence, by exchange of indices, we have

$$\mathbf{u}_i^{\mathrm{T}} \mathbf{A}\mathbf{u}_j = \lambda_j \mathbf{u}_i^{\mathrm{T}} \mathbf{u}_j. \tag{C.35}$$

We now take the transpose of the second equation and make use of the symmetry property $\mathbf{A}^{\mathrm{T}} = \mathbf{A}$, and then subtract the two equations to give

$$(\lambda_i - \lambda_j)\, \mathbf{u}_i^{\mathrm{T}} \mathbf{u}_j = 0. \tag{C.36}$$

Hence, for $\lambda_i \neq \lambda_j$, we have $\mathbf{u}_i^{\mathrm{T}} \mathbf{u}_j = 0$, and hence $\mathbf{u}_i$ and $\mathbf{u}_j$ are orthogonal. If the two eigenvalues are equal, then any linear combination $\alpha\mathbf{u}_i + \beta\mathbf{u}_j$ is also an eigenvector with the same eigenvalue, so we can select one linear combination arbitrarily, and then choose the second to be orthogonal to the first (it can be shown that the degenerate eigenvectors are never linearly dependent). Hence the eigenvectors can be chosen to be orthogonal, and by normalizing can be set to unit length. Because there are M eigenvalues, the corresponding M orthogonal eigenvectors form a complete set and so any M-dimensional vector can be expressed as a linear combination of the eigenvectors.

We can take the eigenvectors $\mathbf{u}_i$ to be the columns of an $M \times M$ matrix $\mathbf{U}$, which from orthonormality satisfies

$$\mathbf{U}^{\mathrm{T}}\mathbf{U} = \mathbf{I}. \tag{C.37}$$

Such a matrix is said to be *orthogonal*. Interestingly, the rows of this matrix are also orthogonal, so that $\mathbf{U}\mathbf{U}^{\mathrm{T}} = \mathbf{I}$. To show this, note that (C.37) implies $\mathbf{U}^{\mathrm{T}}\mathbf{U}\mathbf{U}^{-1} = \mathbf{U}^{-1} = \mathbf{U}^{\mathrm{T}}$ and so $\mathbf{U}\mathbf{U}^{-1} = \mathbf{U}\mathbf{U}^{\mathrm{T}} = \mathbf{I}$. Using (C.12), it also follows that $|\mathbf{U}| = 1$.

The eigenvector equation (C.29) can be expressed in terms of $\mathbf{U}$ in the form

$$\mathbf{A}\mathbf{U} = \mathbf{U}\boldsymbol{\Lambda} \tag{C.38}$$

where $\boldsymbol{\Lambda}$ is an $M \times M$ diagonal matrix whose diagonal elements are given by the eigenvalues λ_i.

If we consider a column vector $\mathbf{x}$ that is transformed by an orthogonal matrix $\mathbf{U}$ to give a new vector

$$\widetilde{\mathbf{x}} = \mathbf{U}\mathbf{x} \tag{C.39}$$

then the length of the vector is preserved because

$$\widetilde{\mathbf{x}}^{\mathrm{T}}\widetilde{\mathbf{x}} = \mathbf{x}^{\mathrm{T}}\mathbf{U}^{\mathrm{T}}\mathbf{U}\mathbf{x} = \mathbf{x}^{\mathrm{T}}\mathbf{x} \tag{C.40}$$

and similarly the angle between any two such vectors is preserved because

$$\widetilde{\mathbf{x}}^{\mathrm{T}}\widetilde{\mathbf{y}} = \mathbf{x}^{\mathrm{T}}\mathbf{U}^{\mathrm{T}}\mathbf{U}\mathbf{y} = \mathbf{x}^{\mathrm{T}}\mathbf{y}. \tag{C.41}$$

Thus, multiplication by $\mathbf{U}$ can be interpreted as a rigid rotation of the coordinate system.

From (C.38), it follows that

$$\mathbf{U}^{\mathrm{T}}\mathbf{A}\mathbf{U} = \boldsymbol{\Lambda} \tag{C.42}$$

and because $\boldsymbol{\Lambda}$ is a diagonal matrix, we say that the matrix $\mathbf{A}$ is *diagonalized* by the matrix $\mathbf{U}$. If we left multiply by $\mathbf{U}$ and right multiply by $\mathbf{U}^{\mathrm{T}}$, we obtain

$$\mathbf{A} = \mathbf{U}\boldsymbol{\Lambda}\mathbf{U}^{\mathrm{T}} \tag{C.43}$$

Taking the inverse of this equation, and using (C.3) together with $\mathbf{U}^{-1} = \mathbf{U}^{\mathrm{T}}$, we have

$$\mathbf{A}^{-1} = \mathbf{U}\boldsymbol{\Lambda}^{-1}\mathbf{U}^{\mathrm{T}}. \tag{C.44}$$

These last two equations can also be written in the form

$$\mathbf{A} = \sum_{i=1}^{M} \lambda_i \mathbf{u}_i \mathbf{u}_i^{\mathrm{T}} \tag{C.45}$$

$$\mathbf{A}^{-1} = \sum_{i=1}^{M} \frac{1}{\lambda_i} \mathbf{u}_i \mathbf{u}_i^{\mathrm{T}}. \tag{C.46}$$

If we take the determinant of (C.43), and use (C.12), we obtain

$$|\mathbf{A}| = \prod_{i=1}^{M} \lambda_i. \tag{C.47}$$

Similarly, taking the trace of (C.43), and using the cyclic property (C.8) of the trace operator together with $\mathbf{U}^{\mathrm{T}}\mathbf{U} = \mathbf{I}$, we have

$$\mathrm{Tr}(\mathbf{A}) = \sum_{i=1}^{M} \lambda_i. \tag{C.48}$$

We leave it as an exercise for the reader to verify (C.22) by making use of the results (C.33), (C.45), (C.46), and (C.47).

A matrix $\mathbf{A}$ is said to be *positive definite*, denoted by $\mathbf{A} \succ 0$, if $\mathbf{w}^{\mathrm{T}}\mathbf{A}\mathbf{w} > 0$ for all non-zero values of the vector $\mathbf{w}$. Equivalently, a positive definite matrix has $\lambda_i > 0$ for all of its eigenvalues (as can be seen by setting $\mathbf{w}$ to each of the eigenvectors in turn, and by noting that an arbitrary vector can be expanded as a linear combination of the eigenvectors). Note that positive definite is not the same as all the elements being positive. For example, the matrix

$$\begin{pmatrix} 1 & 2 \\ 3 & 4 \end{pmatrix} \tag{C.49}$$

has eigenvalues $\lambda_1 \simeq 5.37$ and $\lambda_2 \simeq -0.37$. A matrix is said to be *positive semi-definite* if $\mathbf{w}^{\mathrm{T}}\mathbf{A}\mathbf{w} \geqslant 0$ holds for all values of $\mathbf{w}$, which is denoted $\mathbf{A} \succeq 0$, and is equivalent to $\lambda_i \geqslant 0$.

Appendix D. Calculus of Variations

We can think of a function $y(x)$ as being an operator that, for any input value x, returns an output value y. In the same way, we can define a *functional* $F[y]$ to be an operator that takes a function $y(x)$ and returns an output value F. An example of a functional is the length of a curve drawn in a two-dimensional plane in which the path of the curve is defined in terms of a function. In the context of machine learning, a widely used functional is the entropy $\mathrm{H}[x]$ for a continuous variable x because, for any choice of probability density function $p(x)$, it returns a scalar value representing the entropy of x under that density. Thus the entropy of $p(x)$ could equally well have been written as $\mathrm{H}[p]$.

A common problem in conventional calculus is to find a value of x that maximizes (or minimizes) a function $y(x)$. Similarly, in the calculus of variations we seek a function $y(x)$ that maximizes (or minimizes) a functional $F[y]$. That is, of all possible functions $y(x)$, we wish to find the particular function for which the functional $F[y]$ is a maximum (or minimum). The calculus of variations can be used, for instance, to show that the shortest path between two points is a straight line or that the maximum entropy distribution is a Gaussian.

If we weren't familiar with the rules of ordinary calculus, we could evaluate a conventional derivative $\mathrm{d}y/\,\mathrm{d}x$ by making a small change ϵ to the variable x and then expanding in powers of ϵ, so that

$$y(x+\epsilon) = y(x) + \frac{\mathrm{d}y}{\mathrm{d}x}\epsilon + O(\epsilon^2) \tag{D.1}$$

and finally taking the limit $\epsilon \to 0$. Similarly, for a function of several variables $y(x_1, \ldots, x_D)$, the corresponding partial derivatives are defined by

$$y(x_1+\epsilon_1, \ldots, x_D+\epsilon_D) = y(x_1, \ldots, x_D) + \sum_{i=1}^{D} \frac{\partial y}{\partial x_i}\epsilon_i + O(\epsilon^2). \tag{D.2}$$

The analogous definition of a functional derivative arises when we consider how much a functional $F[y]$ changes when we make a small change $\epsilon\eta(x)$ to the function

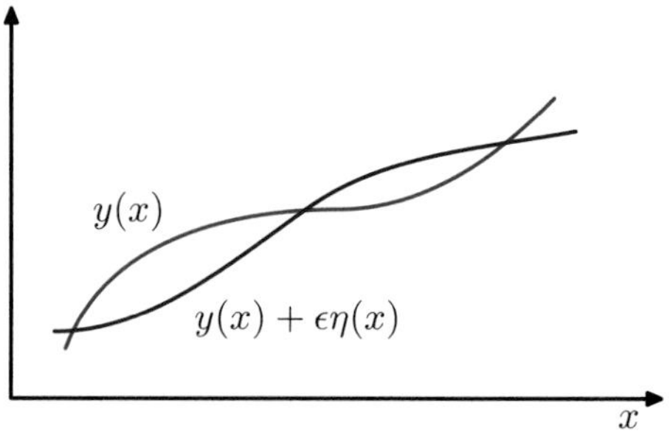

Figure D.1 A functional derivative can be defined by considering how the value of a functional $F[y]$ changes when the function $y(x)$ is changed to $y(x) + \epsilon\eta(x)$ where $\eta(x)$ is an arbitrary function of x.

$y(x)$, where $\eta(x)$ is an arbitrary function of x, as illustrated in Figure D.1. We denote the functional derivative of $F[y]$ with respect to $y(x)$ by $\delta F/\delta y(x)$, and define it by the following relation:

$$F[y(x) + \epsilon\eta(x)] = F[y(x)] + \epsilon \int \frac{\delta F}{\delta y(x)} \eta(x)\, \mathrm{d}x + O(\epsilon^2). \tag{D.3}$$

This can be seen as a natural extension of (D.2) in which $F[y]$ now depends on a continuous set of variables, namely the values of y at all points x. Requiring that the functional be stationary with respect to small variations in the function $y(x)$ gives

$$\int \frac{\delta F}{\delta y(x)} \eta(x)\, \mathrm{d}x = 0. \tag{D.4}$$

Because this must hold for an arbitrary choice of $\eta(x)$, it follows that the functional derivative must vanish. To see this, imagine choosing a perturbation $\eta(x)$ that is zero everywhere except in the neighbourhood of a point $\widehat{x}$, in which case the functional derivative must be zero at $x = \widehat{x}$. However, because this must be true for every choice of $\widehat{x}$, the functional derivative must vanish for all values of x.

Consider a functional that is defined by an integral over a function $G(y, y', x)$ that depends on both $y(x)$ and its derivative $y'(x)$ as well as having a direct dependence on x

$$F[y] = \int G\left(y(x), y'(x), x\right)\, \mathrm{d}x \tag{D.5}$$

where the value of $y(x)$ is assumed to be fixed at the boundary of the region of integration (which might be at infinity). If we now consider variations in the function $y(x)$, we obtain

$$F[y(x) + \epsilon\eta(x)] = F[y(x)] + \epsilon \int \left\{ \frac{\partial G}{\partial y}\eta(x) + \frac{\partial G}{\partial y'}\eta'(x) \right\} \mathrm{d}x + O(\epsilon^2). \tag{D.6}$$

We now have to cast this in the form (D.3). To do so, we integrate the second term by parts and make use of the fact that $\eta(x)$ must vanish at the boundary of the integral (because $y(x)$ is fixed at the boundary). This gives

$$F[y(x) + \epsilon\eta(x)] = F[y(x)] + \epsilon \int \left\{ \frac{\partial G}{\partial y} - \frac{\mathrm{d}}{\mathrm{d}x}\left(\frac{\partial G}{\partial y'}\right) \right\} \eta(x)\, \mathrm{d}x + O(\epsilon^2) \tag{D.7}$$

from which we can read off the functional derivative by comparison with (D.3). Requiring that the functional derivative vanishes then gives

$$\frac{\partial G}{\partial y} - \frac{\mathrm{d}}{\mathrm{d}x}\left(\frac{\partial G}{\partial y'}\right) = 0 \tag{D.8}$$

which are known as the *Euler-Lagrange* equations. For example, if

$$G = y(x)^2 + (y'(x))^2 \tag{D.9}$$

then the Euler-Lagrange equations take the form

$$y(x) - \frac{\mathrm{d}^2 y}{\mathrm{d}x^2} = 0. \tag{D.10}$$

This second order differential equation can be solved for $y(x)$ by making use of the boundary conditions on $y(x)$.

Often, we consider functionals defined by integrals whose integrands take the form $G(y, x)$ and that do not depend on the derivatives of $y(x)$. In this case, stationarity simply requires that $\partial G/\partial y(x) = 0$ for all values of x.

Appendix E

If we are optimizing a functional with respect to a probability distribution, then we need to maintain the normalization constraint on the probabilities. This is often most conveniently done using a Lagrange multiplier, which then allows an unconstrained optimization to be performed.

The extension of the above results to a multidimensional variable $\mathbf{x}$ is straightforward. For a more comprehensive discussion of the calculus of variations, see Sagan (1969).

Appendix E. Lagrange Multipliers

Lagrange multipliers, also sometimes called *undetermined multipliers*, are used to find the stationary points of a function of several variables subject to one or more constraints.

Consider the problem of finding the maximum of a function $f(x_1, x_2)$ subject to a constraint relating x_1 and x_2, which we write in the form

$$g(x_1, x_2) = 0. \tag{E.1}$$

One approach would be to solve the constraint equation (E.1) and thus express x_2 as a function of x_1 in the form $x_2 = h(x_1)$. This can then be substituted into $f(x_1, x_2)$ to give a function of x_1 alone of the form $f(x_1, h(x_1))$. The maximum with respect to x_1 could then be found by differentiation in the usual way, to give the stationary value $x_1^\star$, with the corresponding value of x_2 given by $x_2^\star = h(x_1^\star)$.

One problem with this approach is that it may be difficult to find an analytic solution of the constraint equation that allows x_2 to be expressed as an explicit function of x_1. Also, this approach treats x_1 and x_2 differently and so spoils the natural symmetry between these variables.

A more elegant, and often simpler, approach is based on the introduction of a parameter λ called a Lagrange multiplier. We shall motivate this technique from a geometrical perspective. Consider a D-dimensional variable $\mathbf{x}$ with components $x_1, \ldots, x_D$. The constraint equation $g(\mathbf{x}) = 0$ then represents a $(D-1)$-dimensional surface in $\mathbf{x}$-space as indicated in Figure E.1.

We first note that at any point on the constraint surface the gradient $\nabla g(\mathbf{x})$ of the constraint function will be orthogonal to the surface. To see this, consider a point $\mathbf{x}$ that lies on the constraint surface, and consider a nearby point $\mathbf{x} + \boldsymbol{\epsilon}$ that also lies on the surface. If we make a Taylor expansion around $\mathbf{x}$, we have

$$g(\mathbf{x} + \boldsymbol{\epsilon}) \simeq g(\mathbf{x}) + \boldsymbol{\epsilon}^{\mathrm{T}} \nabla g(\mathbf{x}). \tag{E.2}$$

Because both $\mathbf{x}$ and $\mathbf{x} + \boldsymbol{\epsilon}$ lie on the constraint surface, we have $g(\mathbf{x}) = g(\mathbf{x} + \boldsymbol{\epsilon})$ and hence $\boldsymbol{\epsilon}^{\mathrm{T}} \nabla g(\mathbf{x}) \simeq 0$. In the limit $\|\boldsymbol{\epsilon}\| \to 0$ we have $\boldsymbol{\epsilon}^{\mathrm{T}} \nabla g(\mathbf{x}) = 0$, and because $\boldsymbol{\epsilon}$ is

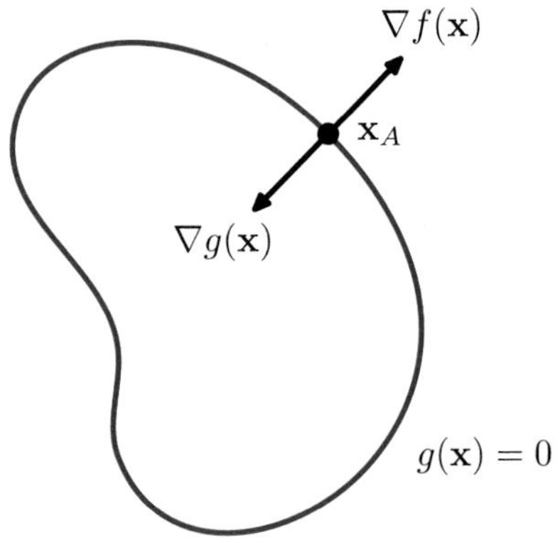

Figure E.1 A geometrical picture of the technique of Lagrange multipliers in which we seek to maximize a function $f(\mathbf{x})$, subject to the constraint $g(\mathbf{x}) = 0$. If $\mathbf{x}$ is D dimensional, the constraint $g(\mathbf{x}) = 0$ corresponds to a subspace of dimensionality $D - 1$, indicated by the red curve. The problem can be solved by optimizing the Lagrangian function $L(\mathbf{x}, \lambda) = f(\mathbf{x}) + \lambda g(\mathbf{x})$.

then parallel to the constraint surface $g(\mathbf{x}) = 0$, we see that the vector ∇g is normal to the surface.

Next we seek a point $\mathbf{x}^\star$ on the constraint surface such that $f(\mathbf{x})$ is maximized. Such a point must have the property that the vector $\nabla f(\mathbf{x})$ is also orthogonal to the constraint surface, as illustrated in Figure E.1, because otherwise we could increase the value of $f(\mathbf{x})$ by moving a short distance along the constraint surface. Thus ∇f and ∇g are parallel (or anti-parallel) vectors, and so there must exist a parameter λ such that

$$\nabla f + \lambda \nabla g = 0 \tag{E.3}$$

where $\lambda \neq 0$ is known as a *Lagrange multiplier*. Note that λ can have either sign.

At this point, it is convenient to introduce the *Lagrangian* function defined by

$$L(\mathbf{x}, \lambda) \equiv f(\mathbf{x}) + \lambda g(\mathbf{x}). \tag{E.4}$$

The constrained stationarity condition (E.3) is obtained by setting $\nabla_{\mathbf{x}} L = 0$. Furthermore, the condition $\partial L/\partial \lambda = 0$ leads to the constraint equation $g(\mathbf{x}) = 0$.

Thus to find the maximum of a function $f(\mathbf{x})$ subject to the constraint $g(\mathbf{x}) = 0$, we define the Lagrangian function given by (E.4) and we then find the stationary point of $L(\mathbf{x}, \lambda)$ with respect to both $\mathbf{x}$ and λ. For a D-dimensional vector $\mathbf{x}$, this gives $D+1$ equations that determine both the stationary point $\mathbf{x}^\star$ and the value of λ. If we are only interested in $\mathbf{x}^\star$, then we can eliminate λ from the stationarity equations without needing to find its value (hence the term 'undetermined multiplier').

As a simple example, suppose we wish to find the stationary point of the function $f(x_1, x_2) = 1 - x_1^2 - x_2^2$ subject to the constraint $g(x_1, x_2) = x_1 + x_2 - 1 = 0$, as illustrated in Figure E.2. The corresponding Lagrangian function is given by

$$L(\mathbf{x}, \lambda) = 1 - x_1^2 - x_2^2 + \lambda(x_1 + x_2 - 1). \tag{E.5}$$

The conditions for this Lagrangian to be stationary with respect to x_1, x_2, and λ give the following coupled equations:

$$-2x_1 + \lambda = 0 \tag{E.6}$$

$$-2x_2 + \lambda = 0 \tag{E.7}$$

$$x_1 + x_2 - 1 = 0. \tag{E.8}$$

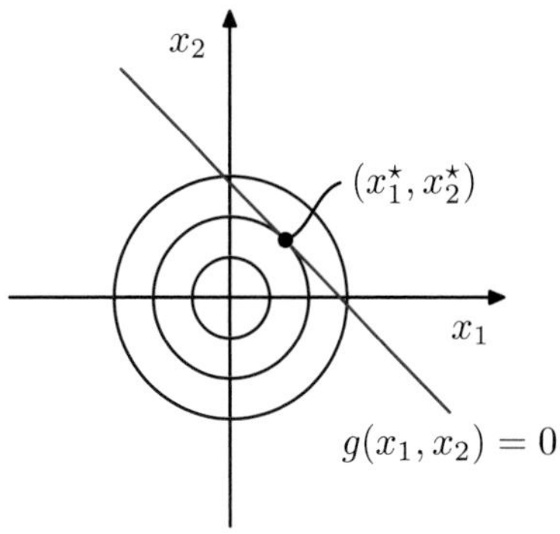

Figure E.2 A simple example of the use of Lagrange multipliers in which the aim is to maximize $f(x_1, x_2) = 1 - x_1^2 - x_2^2$ subject to the constraint $g(x_1, x_2) = 0$ where $g(x_1, x_2) = x_1 + x_2 - 1$. The circles show contours of the function $f(x_1, x_2)$, and the diagonal line shows the constraint surface $g(x_1, x_2) = 0$.

Solution of these equations then gives the stationary point as $(x_1^\star, x_2^\star) = (\frac{1}{2}, \frac{1}{2})$, and the corresponding value for the Lagrange multiplier is $\lambda = 1$.

So far, we have considered the problem of maximizing a function subject to an *equality constraint* of the form $g(\mathbf{x}) = 0$. We now consider the problem of maximizing $f(\mathbf{x})$ subject to an *inequality constraint* of the form $g(\mathbf{x}) \geqslant 0$, as illustrated in Figure E.3.

There are now two kinds of solution possible, according to whether the constrained stationary point lies in the region where $g(\mathbf{x}) > 0$, in which case the constraint is *inactive*, or whether it lies on the boundary $g(\mathbf{x}) = 0$, in which case the constraint is said to be *active*. In the former case, the function $g(\mathbf{x})$ plays no role and so the stationary condition is simply $\nabla f(\mathbf{x}) = 0$. This again corresponds to a stationary point of the Lagrange function (E.4) but this time with $\lambda = 0$. The latter case, where the solution lies on the boundary, is analogous to the equality constraint discussed previously and corresponds to a stationary point of the Lagrange function (E.4) with $\lambda \neq 0$. Now, however, the sign of the Lagrange multiplier is crucial, because the function $f(\mathbf{x})$ will only be at a maximum if its gradient is oriented away from the region $g(\mathbf{x}) > 0$, as illustrated in Figure E.3. We therefore have $\nabla f(\mathbf{x}) = -\lambda \nabla g(\mathbf{x})$ for some value of $\lambda > 0$.

For either of these two cases, the product $\lambda g(\mathbf{x}) = 0$. Thus the solution to the

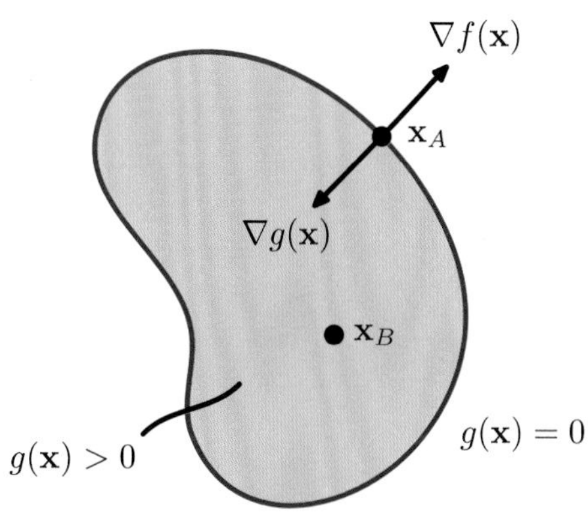

Figure E.3 Illustration of the problem of maximizing $f(\mathbf{x})$ subject to the inequality constraint $g(\mathbf{x}) \geqslant 0$.

problem of maximizing $f(\mathbf{x})$ subject to $g(\mathbf{x}) \geqslant 0$ is obtained by optimizing the Lagrange function (E.4) with respect to $\mathbf{x}$ and λ subject to the conditions

$$g(\mathbf{x}) \geqslant 0 \tag{E.9}$$

$$\lambda \geqslant 0 \tag{E.10}$$

$$\lambda g(\mathbf{x}) = 0 \tag{E.11}$$

These are known as the *Karush-Kuhn-Tucker* (KKT) conditions (Karush, 1939; Kuhn and Tucker, 1951).

Note that if we wish to minimize (rather than maximize) the function $f(\mathbf{x})$ subject to an inequality constraint $g(\mathbf{x}) \geqslant 0$, then we minimize the Lagrangian function $L(\mathbf{x}, \lambda) = f(\mathbf{x}) - \lambda g(\mathbf{x})$ with respect to $\mathbf{x}$, again subject to $\lambda \geqslant 0$.

Finally, it is straightforward to extend the technique of Lagrange multipliers to the case of multiple equality and inequality constraints. Suppose we wish to maximize $f(\mathbf{x})$ subject to $g_j(\mathbf{x}) = 0$ for $j = 1, \ldots, J$, and $h_k(\mathbf{x}) \geqslant 0$ for $k = 1, \ldots, K$. We then introduce Lagrange multipliers $\{\lambda_j\}$ and $\{\mu_k\}$, and then optimize the Lagrangian function given by

$$L(\mathbf{x}, \{\lambda_j\}, \{\mu_k\}) = f(\mathbf{x}) + \sum_{j=1}^{J} \lambda_j g_j(\mathbf{x}) + \sum_{k=1}^{K} \mu_k h_k(\mathbf{x}) \tag{E.12}$$

subject to $\mu_k \geqslant 0$ and $\mu_k h_k(\mathbf{x}) = 0$ for $k = 1, \ldots, K$. Extensions to constrained functional derivatives are similarly straightforward. For a more detailed discussion of the technique of Lagrange multipliers, see Nocedal and Wright (1999).

Appendix D

References

Abramowitz, M. and I. A. Stegun (1965). *Handbook of Mathematical Functions*. Dover.

Adler, S. L. (1981). Over-relaxation method for the Monte Carlo evaluation of the partition function for multiquadratic actions. *Physical Review D* **23**, 2901–2904.

Ahn, J. H. and J. H. Oh (2003). A constrained EM algorithm for principal component analysis. *Neural Computation* **15**(1), 57–65.

Aizerman, M. A., E. M. Braverman, and L. I. Rozonoer (1964). The probability problem of pattern recognition learning and the method of potential functions. *Automation and Remote Control* **25**, 1175–1190.

Akaike, H. (1974). A new look at statistical model identification. *IEEE Transactions on Automatic Control* **19**, 716–723.

Ali, S. M. and S. D. Silvey (1966). A general class of coefficients of divergence of one distribution from another. *Journal of the Royal Statistical Society, B* **28**(1), 131–142.

Allwein, E. L., R. E. Schapire, and Y. Singer (2000). Reducing multiclass to binary: a unifying approach for margin classifiers. *Journal of Machine Learning Research* **1**, 113–141.

Amari, S. (1985). *Differential-Geometrical Methods in Statistics*. Springer.

Amari, S. (1998). Natural gradient works efficiently in learning. *Neural Computation* **10**, 251–276.

Amari, S., A. Cichocki, and H. H. Yang (1996). A new learning algorithm for blind signal separation. In D. S. Touretzky, M. C. Mozer, and M. E. Hasselmo (Eds.), *Advances in Neural Information Processing Systems*, Volume 8, pp. 757–763. MIT Press.

Anderson, J. A. and E. Rosenfeld (Eds.) (1988). *Neurocomputing: Foundations of Research*. MIT Press.

Anderson, T. W. (1963). Asymptotic theory for principal component analysis. *Annals of Mathematical Statistics* **34**, 122–148.

Andrieu, C., N. de Freitas, A. Doucet, and M. I. Jordan (2003). An introduction to MCMC for machine learning. *Machine Learning* **50**, 5–43.

Anthony, M. and N. Biggs (1992). *An Introduction to Computational Learning Theory*. Cambridge University Press.

Attias, H. (1999a). Independent factor analysis. *Neural Computation* **11**(4), 803–851.

Attias, H. (1999b). Inferring parameters and structure of latent variable models by variational Bayes. In K. B. Laskey and H. Prade (Eds.), *Uncertainty in Artificial Intelligence: Proceed-*

ings of the Fifth Conference, pp. 21–30. Morgan Kaufmann.

Bach, F. R. and M. I. Jordan (2002). Kernel independent component analysis. *Journal of Machine Learning Research* **3**, 1–48.

Bakir, G. H., J. Weston, and B. Schölkopf (2004). Learning to find pre-images. In S. Thrun, L. K. Saul, and B. Schölkopf (Eds.), *Advances in Neural Information Processing Systems*, Volume 16, pp. 449–456. MIT Press.

Baldi, P. and S. Brunak (2001). *Bioinformatics: The Machine Learning Approach* (Second ed.). MIT Press.

Baldi, P. and K. Hornik (1989). Neural networks and principal component analysis: learning from examples without local minima. *Neural Networks* **2**(1), 53–58.

Barber, D. and C. M. Bishop (1997). Bayesian model comparison by Monte Carlo chaining. In M. Mozer, M. Jordan, and T. Petsche (Eds.), *Advances in Neural Information Processing Systems*, Volume 9, pp. 333–339. MIT Press.

Barber, D. and C. M. Bishop (1998a). Ensemble learning for multi-layer networks. In M. I. Jordan, K. J. Kearns, and S. A. Solla (Eds.), *Advances in Neural Information Processing Systems*, Volume 10, pp. 395–401.

Barber, D. and C. M. Bishop (1998b). Ensemble learning in Bayesian neural networks. In C. M. Bishop (Ed.), *Generalization in Neural Networks and Machine Learning*, pp. 215–237. Springer.

Bartholomew, D. J. (1987). *Latent Variable Models and Factor Analysis*. Charles Griffin.

Basilevsky, A. (1994). *Statistical Factor Analysis and Related Methods: Theory and Applications*. Wiley.

Bather, J. (2000). *Decision Theory: An Introduction to Dynamic Programming and Sequential Decisions*. Wiley.

Baudat, G. and F. Anouar (2000). Generalized discriminant analysis using a kernel approach. *Neural Computation* **12**(10), 2385–2404.

Baum, L. E. (1972). An inequality and associated maximization technique in statistical estimation of probabilistic functions of Markov processes. *Inequalities* **3**, 1–8.

Becker, S. and Y. Le Cun (1989). Improving the convergence of back-propagation learning with second order methods. In D. Touretzky, G. E. Hinton, and T. J. Sejnowski (Eds.), *Proceedings of the 1988 Connectionist Models Summer School*, pp. 29–37. Morgan Kaufmann.

Bell, A. J. and T. J. Sejnowski (1995). An information maximization approach to blind separation and blind deconvolution. *Neural Computation* **7**(6), 1129–1159.

Bellman, R. (1961). *Adaptive Control Processes: A Guided Tour*. Princeton University Press.

Bengio, Y. and P. Frasconi (1995). An input output HMM architecture. In G. Tesauro, D. S. Touretzky, and T. K. Leen (Eds.), *Advances in Neural Information Processing Systems*, Volume 7, pp. 427–434. MIT Press.

Bennett, K. P. (1992). Robust linear programming discrimination of two linearly separable sets. *Optimization Methods and Software* **1**, 23–34.

Berger, J. O. (1985). *Statistical Decision Theory and Bayesian Analysis* (Second ed.). Springer.

Bernardo, J. M. and A. F. M. Smith (1994). *Bayesian Theory*. Wiley.

Berrou, C., A. Glavieux, and P. Thitimajshima (1993). Near Shannon limit error-correcting coding and decoding: Turbo-codes (1). In *Proceedings ICC'93*, pp. 1064–1070.

Besag, J. (1974). On spatio-temporal models and Markov fields. In *Transactions of the 7th Prague Conference on Information Theory, Statistical Decision Functions and Random Processes*, pp. 47–75. Academia.

Besag, J. (1986). On the statistical analysis of dirty pictures. *Journal of the Royal Statistical Society* **B-48**, 259–302.

Besag, J., P. J. Green, D. Hidgon, and K. Megersen (1995). Bayesian computation and stochastic systems. *Statistical Science* **10**(1), 3–66.

Bishop, C. M. (1991). A fast procedure for retraining the multilayer perceptron. *International Journal of Neural Systems* **2**(3), 229–236.

Bishop, C. M. (1992). Exact calculation of the Hessian matrix for the multilayer perceptron. *Neural Computation* **4**(4), 494–501.

Bishop, C. M. (1993). Curvature-driven smoothing: a learning algorithm for feedforward networks. *IEEE Transactions on Neural Networks* **4**(5), 882–884.

Bishop, C. M. (1994). Novelty detection and neural network validation. *IEE Proceedings: Vision, Image and Signal Processing* **141**(4), 217–222. Special issue on applications of neural networks.

Bishop, C. M. (1995a). *Neural Networks for Pattern Recognition*. Oxford University Press.

Bishop, C. M. (1995b). Training with noise is equivalent to Tikhonov regularization. *Neural Computation* **7**(1), 108–116.

Bishop, C. M. (1999a). Bayesian PCA. In M. S. Kearns, S. A. Solla, and D. A. Cohn (Eds.), *Advances in Neural Information Processing Systems*, Volume 11, pp. 382–388. MIT Press.

Bishop, C. M. (1999b). Variational principal components. In *Proceedings Ninth International Conference on Artificial Neural Networks, ICANN'99*, Volume 1, pp. 509–514. IEE.

Bishop, C. M. and G. D. James (1993). Analysis of multiphase flows using dual-energy gamma densitometry and neural networks. *Nuclear Instruments and Methods in Physics Research* **A327**, 580–593.

Bishop, C. M. and I. T. Nabney (1996). Modelling conditional probability distributions for periodic variables. *Neural Computation* **8**(5), 1123–1133.

Bishop, C. M. and I. T. Nabney (2008). *Optimization Algorithms for Machine Learning*. In preparation.

Bishop, C. M., D. Spiegelhalter, and J. Winn (2003). VIBES: A variational inference engine for Bayesian networks. In S. Becker, S. Thrun, and K. Obermeyer (Eds.), *Advances in Neural Information Processing Systems*, Volume 15, pp. 793–800. MIT Press.

Bishop, C. M. and M. Svensén (2003). Bayesian hierarchical mixtures of experts. In U. Kjaerulff and C. Meek (Eds.), *Proceedings Nineteenth Conference on Uncertainty in Artificial Intelligence*, pp. 57–64. Morgan Kaufmann.

Bishop, C. M., M. Svensén, and G. E. Hinton (2004). Distinguishing text from graphics in online handwritten ink. In F. Kimura and H. Fujisawa (Eds.), *Proceedings Ninth International Workshop on Frontiers in Handwriting Recognition, IWFHR-9*, Tokyo, Japan, pp. 142–147.

Bishop, C. M., M. Svensén, and C. K. I. Williams (1996). EM optimization of latent variable density models. In D. S. Touretzky, M. C. Mozer, and M. E. Hasselmo (Eds.), *Advances in Neural Information Processing Systems*, Volume 8, pp. 465–471. MIT Press.

Bishop, C. M., M. Svensén, and C. K. I. Williams (1997a). GTM: a principled alternative to the Self-Organizing Map. In M. C. Mozer, M. I. Jordan, and T. Petche (Eds.), *Advances in Neural Information Processing Systems*, Volume 9, pp. 354–360. MIT Press.

Bishop, C. M., M. Svensén, and C. K. I. Williams (1997b). Magnification factors for the GTM algorithm. In *Proceedings IEE Fifth International Conference on Artificial Neural Networks, Cambridge, U.K.*, pp. 64–69. Institute of Electrical Engineers.

Bishop, C. M., M. Svensén, and C. K. I. Williams (1998a). Developments of the Generative Topographic Mapping. *Neurocomputing* **21**, 203–224.

Bishop, C. M., M. Svensén, and C. K. I. Williams (1998b). GTM: the Generative Topographic Mapping. *Neural Computation* **10**(1), 215–234.

Bishop, C. M. and M. E. Tipping (1998). A hierarchical latent variable model for data visualization. *IEEE Transactions on Pattern Analysis and Machine Intelligence* **20**(3), 281–293.

Bishop, C. M. and J. Winn (2000). Non-linear Bayesian image modelling. In *Proceedings Sixth European Conference on Computer Vision, Dublin*, Volume 1, pp. 3–17. Springer.

Blei, D. M., M. I. Jordan, and A. Y. Ng (2003). Hierarchical Bayesian models for applications in information retrieval. In J. M. Bernardo et al. (Eds.), *Bayesian Statistics, 7*, pp. 25–43. Oxford University Press.

Block, H. D. (1962). The perceptron: a model for brain functioning. *Reviews of Modern Physics* **34**(1), 123–135. Reprinted in Anderson and Rosenfeld (1988).

Blum, J. A. (1965). Multidimensional stochastic approximation methods. *Annals of Mathematical Statistics* **25**, 737–744.

Bodlaender, H. (1993). A tourist guide through treewidth. *Acta Cybernetica* **11**, 1–21.

Boser, B. E., I. M. Guyon, and V. N. Vapnik (1992). A training algorithm for optimal margin classifiers. In D. Haussler (Ed.), *Proceedings Fifth Annual Workshop on Computational Learning Theory (COLT)*, pp. 144–152. ACM.

Bourlard, H. and Y. Kamp (1988). Auto-association by multilayer perceptrons and singular value decomposition. *Biological Cybernetics* **59**, 291–294.

Box, G. E. P., G. M. Jenkins, and G. C. Reinsel (1994). *Time Series Analysis*. Prentice Hall.

Box, G. E. P. and G. C. Tiao (1973). *Bayesian Inference in Statistical Analysis*. Wiley.

Boyd, S. and L. Vandenberghe (2004). *Convex Optimization*. Cambridge University Press.

Boyen, X. and D. Koller (1998). Tractable inference for complex stochastic processes. In G. F. Cooper and S. Moral (Eds.), *Proceedings 14th Annual Conference on Uncertainty in Artificial Intelligence (UAI)*, pp. 33–42. Morgan Kaufmann.

Boykov, Y., O. Veksler, and R. Zabih (2001). Fast approximate energy minimization via graph cuts. *IEEE Transactions on Pattern Analysis and Machine Intelligence* **23**(11), 1222–1239.

Breiman, L. (1996). Bagging predictors. *Machine Learning* **26**, 123–140.

Breiman, L., J. H. Friedman, R. A. Olshen, and P. J. Stone (1984). *Classification and Regression Trees*. Wadsworth.

Brooks, S. P. (1998). Markov chain Monte Carlo method and its application. *The Statistician* **47**(1), 69–100.

Broomhead, D. S. and D. Lowe (1988). Multivariable functional interpolation and adaptive networks. *Complex Systems* **2**, 321–355.

Buntine, W. and A. Weigend (1991). Bayesian backpropagation. *Complex Systems* **5**, 603–643.

Buntine, W. L. and A. S. Weigend (1993). Computing second derivatives in feed-forward networks: a review. *IEEE Transactions on Neural Networks* **5**(3), 480–488.

Burges, C. J. C. (1998). A tutorial on support vector machines for pattern recognition. *Knowledge Discovery and Data Mining* **2**(2), 121–167.

Cardoso, J.-F. (1998). Blind signal separation: statistical principles. *Proceedings of the IEEE* **9**(10), 2009–2025.

Casella, G. and R. L. Berger (2002). *Statistical Inference* (Second ed.). Duxbury.

Castillo, E., J. M. Gutiérrez, and A. S. Hadi (1997). *Expert Systems and Probabilistic Network Models*. Springer.

Chan, K., T. Lee, and T. J. Sejnowski (2003). Variational Bayesian learning of ICA with missing data. *Neural Computation* **15**(8), 1991–2011.

Chen, A. M., H. Lu, and R. Hecht-Nielsen (1993). On the geometry of feedforward neural network error surfaces. *Neural Computation* **5**(6), 910–927.

Chen, M. H., Q. M. Shao, and J. G. Ibrahim (Eds.) (2001). *Monte Carlo Methods for Bayesian Computation*. Springer.

Chen, S., C. F. N. Cowan, and P. M. Grant (1991). Orthogonal least squares learning algorithm for radial basis function networks. *IEEE Transactions on Neural Networks* **2**(2), 302–309.

Choudrey, R. A. and S. J. Roberts (2003). Variational mixture of Bayesian independent component analyzers. *Neural Computation* **15**(1), 213–252.

Clifford, P. (1990). Markov random fields in statistics. In G. R. Grimmett and D. J. A. Welsh (Eds.), *Disorder in Physical Systems. A Volume in Honour of John M. Hammersley*, pp. 19–32. Oxford University Press.

Collins, M., S. Dasgupta, and R. E. Schapire (2002). A generalization of principal component analysis to the exponential family. In T. G. Dietterich, S. Becker, and Z. Ghahramani (Eds.), *Advances in Neural Information Processing Systems*, Volume 14, pp. 617–624. MIT Press.

Comon, P., C. Jutten, and J. Herault (1991). Blind source separation, 2: problems statement. *Signal Processing* **24**(1), 11–20.

Corduneanu, A. and C. M. Bishop (2001). Variational Bayesian model selection for mixture distributions. In T. Richardson and T. Jaakkola (Eds.), *Proceedings Eighth International Conference on Artificial Intelligence and Statistics*, pp. 27–34. Morgan Kaufmann.

Cormen, T. H., C. E. Leiserson, R. L. Rivest, and C. Stein (2001). *Introduction to Algorithms* (Second ed.). MIT Press.

Cortes, C. and V. N. Vapnik (1995). Support vector networks. *Machine Learning* **20**, 273–297.

Cotter, N. E. (1990). The Stone-Weierstrass theorem and its application to neural networks. *IEEE Transactions on Neural Networks* **1**(4), 290–295.

Cover, T. and P. Hart (1967). Nearest neighbor pattern classification. *IEEE Transactions on Information Theory* **IT-11**, 21–27.

Cover, T. M. and J. A. Thomas (1991). *Elements of Information Theory*. Wiley.

Cowell, R. G., A. P. Dawid, S. L. Lauritzen, and D. J. Spiegelhalter (1999). *Probabilistic Networks and Expert Systems*. Springer.

Cox, R. T. (1946). Probability, frequency and reasonable expectation. *American Journal of Physics* **14**(1), 1–13.

Cox, T. F. and M. A. A. Cox (2000). *Multidimensional Scaling* (Second ed.). Chapman and Hall.

Cressie, N. (1993). *Statistics for Spatial Data*. Wiley.

Cristianini, N. and J. Shawe-Taylor (2000). *Support vector machines and other kernel-based learning methods*. Cambridge University Press.

Csató, L. and M. Opper (2002). Sparse on-line Gaussian processes. *Neural Computation* **14**(3), 641–668.

Csiszàr, I. and G. Tusnàdy (1984). Information geometry and alternating minimization procedures. *Statistics and Decisions* **1**(1), 205–237.

Cybenko, G. (1989). Approximation by superpositions of a sigmoidal function. *Mathematics of Control, Signals and Systems* **2**, 304–314.

Dawid, A. P. (1979). Conditional independence in statistical theory (with discussion). *Journal of the Royal Statistical Society, Series B* **4**, 1–31.

Dawid, A. P. (1980). Conditional independence for statistical operations. *Annals of Statistics* **8**, 598–617.

deFinetti, B. (1970). *Theory of Probability*. Wiley and Sons.

Dempster, A. P., N. M. Laird, and D. B. Rubin (1977). Maximum likelihood from incomplete data via the EM algorithm. *Journal of the Royal Statistical Society, B* **39**(1), 1–38.

Denison, D. G. T., C. C. Holmes, B. K. Mallick, and A. F. M. Smith (2002). *Bayesian Methods for Nonlinear Classification and Regression*. Wiley.

Diaconis, P. and L. Saloff-Coste (1998). What do we know about the Metropolis algorithm? *Journal of Computer and System Sciences* **57**, 20–36.

Dietterich, T. G. and G. Bakiri (1995). Solving multiclass learning problems via error-correcting output codes. *Journal of Artificial Intelligence Research* **2**, 263–286.

Duane, S., A. D. Kennedy, B. J. Pendleton, and D. Roweth (1987). Hybrid Monte Carlo. *Physics Letters B* **195**(2), 216–222.

Duda, R. O. and P. E. Hart (1973). *Pattern Classification and Scene Analysis*. Wiley.

Duda, R. O., P. E. Hart, and D. G. Stork (2001). *Pattern Classification* (Second ed.). Wiley.

Durbin, R., S. Eddy, A. Krogh, and G. Mitchison (1998). *Biological Sequence Analysis*. Cambridge University Press.

Dybowski, R. and S. Roberts (2005). An anthology of probabilistic models for medical informatics. In D. Husmeier, R. Dybowski, and S. Roberts (Eds.), *Probabilistic Modeling in Bioinformatics and Medical Informatics*, pp. 297–349. Springer.

Efron, B. (1979). Bootstrap methods: another look at the jackknife. *Annals of Statistics* **7**, 1–26.

Elkan, C. (2003). Using the triangle inequality to accelerate k-means. In *Proceedings of the Twelfth International Conference on Machine Learning*, pp. 147–153. AAAI.

Elliott, R. J., L. Aggoun, and J. B. Moore (1995). *Hidden Markov Models: Estimation and Control*. Springer.

Ephraim, Y., D. Malah, and B. H. Juang (1989). On the application of hidden Markov models for enhancing noisy speech. *IEEE Transactions on Acoustics, Speech and Signal Processing* **37**(12), 1846–1856.

Erwin, E., K. Obermayer, and K. Schulten (1992). Self-organizing maps: ordering, convergence properties and energy functions. *Biological Cybernetics* **67**, 47–55.

Everitt, B. S. (1984). *An Introduction to Latent Variable Models*. Chapman and Hall.

Faul, A. C. and M. E. Tipping (2002). Analysis of sparse Bayesian learning. In T. G. Dietterich, S. Becker, and Z. Ghahramani (Eds.), *Advances in Neural Information Processing Systems*, Volume 14, pp. 383–389. MIT Press.

Feller, W. (1966). *An Introduction to Probability Theory and its Applications* (Second ed.), Volume 2. Wiley.

Feynman, R. P., R. B. Leighton, and M. Sands (1964). *The Feynman Lectures of Physics*, Volume Two. Addison-Wesley. Chapter 19.

Fletcher, R. (1987). *Practical Methods of Optimization* (Second ed.). Wiley.

Forsyth, D. A. and J. Ponce (2003). *Computer Vision: A Modern Approach*. Prentice Hall.

Freund, Y. and R. E. Schapire (1996). Experiments with a new boosting algorithm. In L. Saitta (Ed.), *Thirteenth International Conference on Machine Learning*, pp. 148–156. Morgan Kaufmann.

Frey, B. J. (1998). *Graphical Models for Machine Learning and Digital Communication*. MIT Press.

Frey, B. J. and D. J. C. MacKay (1998). A revolution: Belief propagation in graphs with cycles. In M. I. Jordan, M. J. Kearns, and S. A. Solla (Eds.), *Advances in Neural Information Processing Systems*, Volume 10. MIT Press.

Friedman, J. H. (2001). Greedy function approximation: a gradient boosting machine. *Annals of Statistics* **29**(5), 1189–1232.

Friedman, J. H., T. Hastie, and R. Tibshirani (2000). Additive logistic regression: a statistical view of boosting. *Annals of Statistics* **28**, 337–407.

Friedman, N. and D. Koller (2003). Being Bayesian about network structure: A Bayesian approach to structure discovery in Bayesian networks. *Machine Learning* **50**, 95–126.

Frydenberg, M. (1990). The chain graph Markov property. *Scandinavian Journal of Statistics* **17**, 333–353.

Fukunaga, K. (1990). *Introduction to Statistical Pattern Recognition* (Second ed.). Academic Press.

Funahashi, K. (1989). On the approximate realization of continuous mappings by neural networks. *Neural Networks* **2**(3), 183–192.

Fung, R. and K. C. Chang (1990). Weighting and integrating evidence for stochastic simulation in Bayesian networks. In P. P. Bonissone, M. Henrion, L. N. Kanal, and J. F. Lemmer (Eds.), *Uncertainty in Artificial Intelligence*, Volume 5, pp. 208–219. Elsevier.

Gallager, R. G. (1963). *Low-Density Parity-Check Codes*. MIT Press.

Gamerman, D. (1997). *Markov Chain Monte Carlo: Stochastic Simulation for Bayesian Inference.* Chapman and Hall.

Gelman, A., J. B. Carlin, H. S. Stern, and D. B. Rubin (2004). *Bayesian Data Analysis* (Second ed.). Chapman and Hall.

Geman, S. and D. Geman (1984). Stochastic relaxation, Gibbs distributions, and the Bayesian restoration of images. *IEEE Transactions on Pattern Analysis and Machine Intelligence* **6**(1), 721–741.

Ghahramani, Z. and M. J. Beal (2000). Variational inference for Bayesian mixtures of factor analyzers. In S. A. Solla, T. K. Leen, and K. R. Müller (Eds.), *Advances in Neural Information Processing Systems*, Volume 12, pp. 449–455. MIT Press.

Ghahramani, Z. and G. E. Hinton (1996a). The EM algorithm for mixtures of factor analyzers. Technical Report CRG-TR-96-1, University of Toronto.

Ghahramani, Z. and G. E. Hinton (1996b). Parameter estimation for linear dynamical systems. Technical Report CRG-TR-96-2, University of Toronto.

Ghahramani, Z. and G. E. Hinton (1998). Variational learning for switching state-space models. *Neural Computation* **12**(4), 963–996.

Ghahramani, Z. and M. I. Jordan (1994). Supervised learning from incomplete data via an EM appproach. In J. D. Cowan, G. T. Tesauro, and J. Alspector (Eds.), *Advances in Neural Information Processing Systems*, Volume 6, pp. 120–127. Morgan Kaufmann.

Ghahramani, Z. and M. I. Jordan (1997). Factorial hidden Markov models. *Machine Learning* **29**, 245–275.

Gibbs, M. N. (1997). *Bayesian Gaussian processes for regression and classification*. Phd thesis, University of Cambridge.

Gibbs, M. N. and D. J. C. MacKay (2000). Variational Gaussian process classifiers. *IEEE Transactions on Neural Networks* **11**, 1458–1464.

Gilks, W. R. (1992). Derivative-free adaptive rejection sampling for Gibbs sampling. In J. Bernardo, J. Berger, A. P. Dawid, and A. F. M. Smith (Eds.), *Bayesian Statistics*, Volume 4. Oxford University Press.

Gilks, W. R., N. G. Best, and K. K. C. Tan (1995). Adaptive rejection Metropolis sampling. *Applied Statistics* **44**, 455–472.

Gilks, W. R., S. Richardson, and D. J. Spiegelhalter (Eds.) (1996). *Markov Chain Monte Carlo in Practice*. Chapman and Hall.

Gilks, W. R. and P. Wild (1992). Adaptive rejection sampling for Gibbs sampling. *Applied Statistics* **41**, 337–348.

Gill, P. E., W. Murray, and M. H. Wright (1981). *Practical Optimization*. Academic Press.

Goldberg, P. W., C. K. I. Williams, and C. M. Bishop (1998). Regression with input-dependent noise: A Gaussian process treatment. In *Advances in Neural Information Processing Systems*, Volume 10, pp. 493–499. MIT Press.

Golub, G. H. and C. F. Van Loan (1996). *Matrix Computations* (Third ed.). John Hopkins University Press.

Good, I. (1950). *Probability and the Weighing of Evidence*. Hafners.

Gordon, N. J., D. J. Salmond, and A. F. M. Smith (1993). Novel approach to nonlinear/non-Gaussian Bayesian state estimation. *IEE Proceedings-F* **140**(2), 107–113.

Graepel, T. (2003). Solving noisy linear operator equations by Gaussian processes: Application to ordinary and partial differential equations. In *Proceedings of the Twentieth International Conference on Machine Learning*, pp. 234–241.

Greig, D., B. Porteous, and A. Seheult (1989). Exact maximum a-posteriori estimation for binary images. *Journal of the Royal Statistical Society, Series B* **51**(2), 271–279.

Gull, S. F. (1989). Developments in maximum entropy data analysis. In J. Skilling (Ed.), *Maximum Entropy and Bayesian Methods*, pp. 53–71. Kluwer.

Hassibi, B. and D. G. Stork (1993). Second order derivatives for network pruning: optimal brain surgeon. In S. J. Hanson, J. D. Cowan, and C. L. Giles (Eds.), *Advances in Neural Information Processing Systems*, Volume 5, pp. 164–171. Morgan Kaufmann.

Hastie, T. and W. Stuetzle (1989). Principal curves. *Journal of the American Statistical Association* **84**(106), 502–516.

Hastie, T., R. Tibshirani, and J. Friedman (2001). *The Elements of Statistical Learning*. Springer.

Hastings, W. K. (1970). Monte Carlo sampling methods using Markov chains and their applications. *Biometrika* **57**, 97–109.

Hathaway, R. J. (1986). Another interpretation of the EM algorithm for mixture distributions. *Statistics and Probability Letters* **4**, 53–56.

Haussler, D. (1999). Convolution kernels on discrete structures. Technical Report UCSC-CRL-99-10, University of California, Santa Cruz, Computer Science Department.

Henrion, M. (1988). Propagation of uncertainty by logic sampling in Bayes' networks. In J. F. Lemmer and L. N. Kanal (Eds.), *Uncertainty in Artificial Intelligence*, Volume 2, pp. 149–164. North Holland.

Herbrich, R. (2002). *Learning Kernel Classifiers*. MIT Press.

Hertz, J., A. Krogh, and R. G. Palmer (1991). *Introduction to the Theory of Neural Computation*. Addison Wesley.

Hinton, G. E., P. Dayan, and M. Revow (1997). Modelling the manifolds of images of handwritten digits. *IEEE Transactions on Neural Networks* **8**(1), 65–74.

Hinton, G. E. and D. van Camp (1993). Keeping neural networks simple by minimizing the description length of the weights. In *Proceedings of the Sixth Annual Conference on Computational Learning Theory*, pp. 5–13. ACM.

Hinton, G. E., M. Welling, Y. W. Teh, and S. Osindero (2001). A new view of ICA. In *Proceedings of the International Conference on Independent Component Analysis and Blind Signal Separation*, Volume 3.

Hodgson, M. E. (1998). Reducing computational requirements of the minimum-distance classifier. *Remote Sensing of Environments* **25**, 117–128.

Hoerl, A. E. and R. Kennard (1970). Ridge regression: biased estimation for nonorthogonal problems. *Technometrics* **12**, 55–67.

Hofmann, T. (2000). Learning the similarity of documents: an information-geometric approach to document retrieval and classification. In S. A. Solla, T. K. Leen, and K. R. Müller (Eds.), *Advances in Neural Information Processing Systems*, Volume 12, pp. 914–920. MIT Press.

Hojen-Sorensen, P. A., O. Winther, and L. K. Hansen (2002). Mean field approaches to independent component analysis. *Neural Computation* **14**(4), 889–918.

Hornik, K. (1991). Approximation capabilities of multilayer feedforward networks. *Neural Networks* **4**(2), 251–257.

Hornik, K., M. Stinchcombe, and H. White (1989). Multilayer feedforward networks are universal approximators. *Neural Networks* **2**(5), 359–366.

Hotelling, H. (1933). Analysis of a complex of statistical variables into principal components. *Journal of Educational Psychology* **24**, 417–441.

Hotelling, H. (1936). Relations between two sets of variables. *Biometrika* **28**, 321–377.

Hyvärinen, A. and E. Oja (1997). A fast fixed-point algorithm for independent component analysis. *Neural Computation* **9**(7), 1483–1492.

Isard, M. and A. Blake (1998). CONDENSATION – conditional density propagation for visual tracking. *International Journal of Computer Vision* **29**(1), 5–18.

Ito, Y. (1991). Representation of functions by superpositions of a step or sigmoid function and their applications to neural network theory. *Neural Networks* **4**(3), 385–394.

Jaakkola, T. and M. I. Jordan (2000). Bayesian parameter estimation via variational methods. *Statistics and Computing* **10**, 25–37.

Jaakkola, T. S. (2001). Tutorial on variational approximation methods. In M. Opper and D. Saad (Eds.), *Advances in Mean Field Methods*, pp. 129–159. MIT Press.

Jaakkola, T. S. and D. Haussler (1999). Exploiting generative models in discriminative classifiers. In M. S. Kearns, S. A. Solla, and D. A. Cohn (Eds.), *Advances in Neural Information Processing Systems*, Volume 11. MIT Press.

Jacobs, R. A., M. I. Jordan, S. J. Nowlan, and G. E. Hinton (1991). Adaptive mixtures of local experts. *Neural Computation* **3**(1), 79–87.

Jaynes, E. T. (2003). *Probability Theory: The Logic of Science*. Cambridge University Press.

Jebara, T. (2004). *Machine Learning: Discriminative and Generative*. Kluwer.

Jeffreys, H. (1946). An invariant form for the prior probability in estimation problems. *Pro. Roy. Soc. AA* **186**, 453–461.

Jelinek, F. (1997). *Statistical Methods for Speech Recognition*. MIT Press.

Jensen, C., A. Kong, and U. Kjaerulff (1995). Blocking gibbs sampling in very large probabilistic expert systems. *International Journal of Human Computer Studies. Special Issue on Real-World Applications of Uncertain Reasoning.* **42**, 647–666.

Jensen, F. V. (1996). *An Introduction to Bayesian Networks*. UCL Press.

Jerrum, M. and A. Sinclair (1996). The Markov chain Monte Carlo method: an approach to approximate counting and integration. In D. S. Hochbaum (Ed.), *Approximation Algorithms for NP-Hard Problems*. PWS Publishing.

Jolliffe, I. T. (2002). *Principal Component Analysis* (Second ed.). Springer.

Jordan, M. I. (1999). *Learning in Graphical Models*. MIT Press.

Jordan, M. I. (2007). *An Introduction to Probabilistic Graphical Models*. In preparation.

Jordan, M. I., Z. Ghahramani, T. S. Jaakkola, and L. K. Saul (1999). An introduction to variational methods for graphical models. In M. I. Jordan (Ed.), *Learning in Graphical Models*, pp. 105–162. MIT Press.

Jordan, M. I. and R. A. Jacobs (1994). Hierarchical mixtures of experts and the EM algorithm. *Neural Computation* **6**(2), 181–214.

Jutten, C. and J. Herault (1991). Blind separation of sources, 1: An adaptive algorithm based on neuromimetic architecture. *Signal Processing* **24**(1), 1–10.

Kalman, R. E. (1960). A new approach to linear filtering and prediction problems. *Transactions of the American Society for Mechanical Engineering, Series D, Journal of Basic Engineering* **82**, 35–45.

Kambhatla, N. and T. K. Leen (1997). Dimension reduction by local principal component analysis. *Neural Computation* **9**(7), 1493–1516.

Kanazawa, K., D. Koller, and S. Russel (1995). Stochastic simulation algorithms for dynamic probabilistic networks. In *Uncertainty in Artificial Intelligence*, Volume 11. Morgan Kaufmann.

Kapadia, S. (1998). *Discriminative Training of Hidden Markov Models*. Phd thesis, University of Cambridge, U.K.

Kapur, J. (1989). *Maximum entropy methods in science and engineering*. Wiley.

Karush, W. (1939). Minima of functions of several variables with inequalities as side constraints. Master's thesis, Department of Mathematics, University of Chicago.

Kass, R. E. and A. E. Raftery (1995). Bayes factors. *Journal of the American Statistical Association* **90**, 377–395.

Kearns, M. J. and U. V. Vazirani (1994). *An Introduction to Computational Learning Theory*. MIT Press.

Kindermann, R. and J. L. Snell (1980). *Markov Random Fields and Their Applications*. American Mathematical Society.

Kittler, J. and J. Föglein (1984). Contextual classification of multispectral pixel data. *Image and Vision Computing* **2**, 13–29.

Kohonen, T. (1982). Self-organized formation of topologically correct feature maps. *Biological Cybernetics* **43**, 59–69.

Kohonen, T. (1995). *Self-Organizing Maps*. Springer.

Kolmogorov, V. and R. Zabih (2004). What energy functions can be minimized via graph cuts? *IEEE Transactions on Pattern Analysis and Machine Intelligence* **26**(2), 147–159.

Kreinovich, V. Y. (1991). Arbitrary nonlinearity is sufficient to represent all functions by neural networks: a theorem. *Neural Networks* **4**(3), 381–383.

Krogh, A., M. Brown, I. S. Mian, K. Sjölander, and D. Haussler (1994). Hidden Markov models in computational biology: Applications to protein modelling. *Journal of Molecular Biology* **235**, 1501–1531.

Kschischnang, F. R., B. J. Frey, and H. A. Loeliger (2001). Factor graphs and the sum-product algorithm. *IEEE Transactions on Information Theory* **47**(2), 498–519.

Kuhn, H. W. and A. W. Tucker (1951). Nonlinear programming. In *Proceedings of the 2nd Berkeley Symposium on Mathematical Statistics and Probabilities*, pp. 481–492. University of California Press.

Kullback, S. and R. A. Leibler (1951). On information and sufficiency. *Annals of Mathematical Statistics* **22**(1), 79–86.

Kůrková, V. and P. C. Kainen (1994). Functionally equivalent feed-forward neural networks. *Neural Computation* **6**(3), 543–558.

Kuss, M. and C. Rasmussen (2006). Assessing approximations for Gaussian process classification. In *Advances in Neural Information Processing Systems*, Number 18. MIT Press. in press.

Lasserre, J., C. M. Bishop, and T. Minka (2006). Principled hybrids of generative and discriminative models. In *Proceedings 2006 IEEE Conference on Computer Vision and Pattern Recognition, New York*.

Lauritzen, S. and N. Wermuth (1989). Graphical models for association between variables, some of which are qualitative some quantitative. *Annals of Statistics* **17**, 31–57.

Lauritzen, S. L. (1992). Propagation of probabilities, means and variances in mixed graphical association models. *Journal of the American Statistical Association* **87**, 1098–1108.

Lauritzen, S. L. (1996). *Graphical Models*. Oxford University Press.

Lauritzen, S. L. and D. J. Spiegelhalter (1988). Local computations with probabailities on graphical structures and their application to expert systems. *Journal of the Royal Statistical Society* **50**, 157–224.

Lawley, D. N. (1953). A modified method of estimation in factor analysis and some large sample results. In *Uppsala Symposium on Psychological Factor Analysis*, Number 3 in Nordisk Psykologi Monograph Series, pp. 35–42. Uppsala: Almqvist and Wiksell.

Lawrence, N. D., A. I. T. Rowstron, C. M. Bishop, and M. J. Taylor (2002). Optimising synchronisation times for mobile devices. In T. G. Dietterich, S. Becker, and Z. Ghahramani (Eds.), *Advances in Neural Information Processing Systems*, Volume 14, pp. 1401–1408. MIT Press.

Lazarsfeld, P. F. and N. W. Henry (1968). *Latent Structure Analysis*. Houghton Mifflin.

Le Cun, Y., B. Boser, J. S. Denker, D. Henderson, R. E. Howard, W. Hubbard, and L. D. Jackel (1989). Backpropagation applied to handwritten zip code recognition. *Neural Computation* **1**(4), 541–551.

Le Cun, Y., J. S. Denker, and S. A. Solla (1990). Optimal brain damage. In D. S. Touretzky (Ed.),

Advances in Neural Information Processing Systems, Volume 2, pp. 598–605. Morgan Kaufmann.

LeCun, Y., L. Bottou, Y. Bengio, and P. Haffner (1998). Gradient-based learning applied to document recognition. *Proceedings of the IEEE* **86**, 2278–2324.

Lee, Y., Y. Lin, and G. Wahba (2001). Multicategory support vector machines. Technical Report 1040, Department of Statistics, University of Madison, Wisconsin.

Leen, T. K. (1995). From data distributions to regularization in invariant learning. *Neural Computation* **7**, 974–981.

Lindley, D. V. (1982). Scoring rules and the inevitability of probability. *International Statistical Review* **50**, 1–26.

Liu, J. S. (Ed.) (2001). *Monte Carlo Strategies in Scientific Computing*. Springer.

Lloyd, S. P. (1982). Least squares quantization in PCM. *IEEE Transactions on Information Theory* **28**(2), 129–137.

Lütkepohl, H. (1996). *Handbook of Matrices*. Wiley.

MacKay, D. J. C. (1992a). Bayesian interpolation. *Neural Computation* **4**(3), 415–447.

MacKay, D. J. C. (1992b). The evidence framework applied to classification networks. *Neural Computation* **4**(5), 720–736.

MacKay, D. J. C. (1992c). A practical Bayesian framework for back-propagation networks. *Neural Computation* **4**(3), 448–472.

MacKay, D. J. C. (1994). Bayesian methods for backprop networks. In E. Domany, J. L. van Hemmen, and K. Schulten (Eds.), *Models of Neural Networks, III*, Chapter 6, pp. 211–254. Springer.

MacKay, D. J. C. (1995). Bayesian neural networks and density networks. *Nuclear Instruments and Methods in Physics Research, A* **354**(1), 73–80.

MacKay, D. J. C. (1997). Ensemble learning for hidden Markov models. Unpublished manuscript, Department of Physics, University of Cambridge.

MacKay, D. J. C. (1998). Introduction to Gaussian processes. In C. M. Bishop (Ed.), *Neural Networks and Machine Learning*, pp. 133–166. Springer.

MacKay, D. J. C. (1999). Comparison of approximate methods for handling hyperparameters. *Neural Computation* **11**(5), 1035–1068.

MacKay, D. J. C. (2003). *Information Theory, Inference and Learning Algorithms*. Cambridge University Press.

MacKay, D. J. C. and M. N. Gibbs (1999). Density networks. In J. W. Kay and D. M. Titterington (Eds.), *Statistics and Neural Networks: Advances at the Interface*, Chapter 5, pp. 129–145. Oxford University Press.

MacKay, D. J. C. and R. M. Neal (1999). Good error-correcting codes based on very sparse matrices. *IEEE Transactions on Information Theory* **45**, 399–431.

MacQueen, J. (1967). Some methods for classification and analysis of multivariate observations. In L. M. LeCam and J. Neyman (Eds.), *Proceedings of the Fifth Berkeley Symposium on Mathematical Statistics and Probability*, Volume I, pp. 281–297. University of California Press.

Magnus, J. R. and H. Neudecker (1999). *Matrix Differential Calculus with Applications in Statistics and Econometrics*. Wiley.

Mallat, S. (1999). *A Wavelet Tour of Signal Processing* (Second ed.). Academic Press.

Manning, C. D. and H. Schütze (1999). *Foundations of Statistical Natural Language Processing*. MIT Press.

Mardia, K. V. and P. E. Jupp (2000). *Directional Statistics*. Wiley.

Maybeck, P. S. (1982). *Stochastic models, estimation and control*. Academic Press.

McAllester, D. A. (2003). PAC-Bayesian stochastic model selection. *Machine Learning* **51**(1), 5–21.

McCullagh, P. and J. A. Nelder (1989). *Generalized Linear Models* (Second ed.). Chapman and Hall.

McCulloch, W. S. and W. Pitts (1943). A logical calculus of the ideas immanent in nervous activity. *Bulletin of Mathematical Biophysics* **5**, 115–133. Reprinted in Anderson and Rosenfeld (1988).

McEliece, R. J., D. J. C. MacKay, and J. F. Cheng (1998). Turbo decoding as an instance of Pearl's 'Belief Ppropagation' algorithm. *IEEE Journal on Selected Areas in Communications* **16**, 140–152.

McLachlan, G. J. and K. E. Basford (1988). *Mixture Models: Inference and Applications to Clustering*. Marcel Dekker.

McLachlan, G. J. and T. Krishnan (1997). *The EM Algorithm and its Extensions*. Wiley.

McLachlan, G. J. and D. Peel (2000). *Finite Mixture Models*. Wiley.

Meng, X. L. and D. B. Rubin (1993). Maximum likelihood estimation via the ECM algorithm: a general framework. *Biometrika* **80**, 267–278.

Metropolis, N., A. W. Rosenbluth, M. N. Rosenbluth, A. H. Teller, and E. Teller (1953). Equation of state calculations by fast computing machines. *Journal of Chemical Physics* **21**(6), 1087–1092.

Metropolis, N. and S. Ulam (1949). The Monte Carlo method. *Journal of the American Statistical Association* **44**(247), 335–341.

Mika, S., G. Rätsch, J. Weston, and B. Schölkopf (1999). Fisher discriminant analysis with kernels. In Y. H. Hu, J. Larsen, E. Wilson, and S. Douglas (Eds.), *Neural Networks for Signal Processing IX*, pp. 41–48. IEEE.

Minka, T. (1998). Inferring a Gaussian distribution. Media Lab note, MIT. Available from http://research.microsoft.com/$\sim$minka/.

Minka, T. (2001a). Expectation propagation for approximate Bayesian inference. In J. Breese and D. Koller (Eds.), *Proceedings of the Seventeenth Conference on Uncertainty in Artificial Intelligence*, pp. 362–369. Morgan Kaufmann.

Minka, T. (2001b). *A family of approximate algorithms for Bayesian inference*. Ph. D. thesis, MIT.

Minka, T. (2004). Power EP. Technical Report MSR-TR-2004-149, Microsoft Research Cambridge.

Minka, T. (2005). Divergence measures and message passing. Technical Report MSR-TR-2005-173, Microsoft Research Cambridge.

Minka, T. P. (2001c). Automatic choice of dimensionality for PCA. In T. K. Leen, T. G. Dietterich, and V. Tresp (Eds.), *Advances in Neural Information Processing Systems*, Volume 13, pp. 598–604. MIT Press.

Minsky, M. L. and S. A. Papert (1969). *Perceptrons*. MIT Press. Expanded edition 1990.

Miskin, J. W. and D. J. C. MacKay (2001). Ensemble learning for blind source separation. In S. J. Roberts and R. M. Everson (Eds.), *Independent Component Analysis: Principles and Practice*. Cambridge University Press.

Møller, M. (1993). Efficient Training of Feed-Forward Neural Networks. Ph. D. thesis, Aarhus University, Denmark.

Moody, J. and C. J. Darken (1989). Fast learning in networks of locally-tuned processing units. *Neural Computation* **1**(2), 281–294.

Moore, A. W. (2000). The anchors hierarch: using the triangle inequality to survive high dimensional data. In *Proceedings of the Twelfth Conference on Uncertainty in Artificial Intelligence*, pp. 397–405.

Müller, K. R., S. Mika, G. Rätsch, K. Tsuda, and B. Schölkopf (2001). An introduction to kernel-based learning algorithms. *IEEE Transactions on Neural Networks* **12**(2), 181–202.

Müller, P. and F. A. Quintana (2004). Nonparametric Bayesian data analysis. *Statistical Science* **19**(1), 95–110.

Nabney, I. T. (2002). *Netlab: Algorithms for Pattern Recognition*. Springer.

Nadaraya, É. A. (1964). On estimating regression. *Theory of Probability and its Applications* **9**(1), 141–142.

Nag, R., K. Wong, and F. Fallside (1986). Script recognition using hidden markov models. In *ICASSP86*, pp. 2071–2074. IEEE.

Neal, R. M. (1993). Probabilistic inference using Markov chain Monte Carlo methods. Technical Report CRG-TR-93-1, Department of Computer Science, University of Toronto, Canada.

Neal, R. M. (1996). *Bayesian Learning for Neural Networks*. Springer. Lecture Notes in Statistics 118.

Neal, R. M. (1997). Monte Carlo implementation of Gaussian process models for Bayesian regression and classification. Technical Report 9702, Department of Computer Statistics, University of Toronto.

Neal, R. M. (1999). Suppressing random walks in Markov chain Monte Carlo using ordered overrelaxation. In M. I. Jordan (Ed.), *Learning in Graphical Models*, pp. 205–228. MIT Press.

Neal, R. M. (2000). Markov chain sampling for Dirichlet process mixture models. *Journal of Computational and Graphical Statistics* **9**, 249–265.

Neal, R. M. (2003). Slice sampling. *Annals of Statistics* **31**, 705–767.

Neal, R. M. and G. E. Hinton (1999). A new view of the EM algorithm that justifies incremental and other variants. In M. I. Jordan (Ed.), *Learning in Graphical Models*, pp. 355–368. MIT Press.

Nelder, J. A. and R. W. M. Wedderburn (1972). Generalized linear models. *Journal of the Royal Statistical Society, A* **135**, 370–384.

Nilsson, N. J. (1965). *Learning Machines*. McGraw-Hill. Reprinted as *The Mathematical Foundations of Learning Machines*, Morgan Kaufmann, (1990).

Nocedal, J. and S. J. Wright (1999). *Numerical Optimization*. Springer.

Nowlan, S. J. and G. E. Hinton (1992). Simplifying neural networks by soft weight sharing. *Neural Computation* **4**(4), 473–493.

Ogden, R. T. (1997). *Essential Wavelets for Statistical Applications and Data Analysis*. Birkhäuser.

Opper, M. and O. Winther (1999). A Bayesian approach to on-line learning. In D. Saad (Ed.), *On-Line Learning in Neural Networks*, pp. 363–378. Cambridge University Press.

Opper, M. and O. Winther (2000a). Gaussian processes and SVM: mean field theory and leave-one-out. In A. J. Smola, P. L. Bartlett, B. Schölkopf, and D. Shuurmans (Eds.), *Advances in Large Margin Classifiers*, pp. 311–326. MIT Press.

Opper, M. and O. Winther (2000b). Gaussian processes for classification. *Neural Computation* **12**(11), 2655–2684.

Osuna, E., R. Freund, and F. Girosi (1996). Support vector machines: training and applications. A.I. Memo AIM-1602, MIT.

Papoulis, A. (1984). *Probability, Random Variables, and Stochastic Processes* (Second ed.). McGraw-Hill.

Parisi, G. (1988). *Statistical Field Theory*. Addison-Wesley.

Pearl, J. (1988). *Probabilistic Reasoning in Intelligent Systems*. Morgan Kaufmann.

Pearlmutter, B. A. (1994). Fast exact multiplication by the Hessian. *Neural Computation* **6**(1), 147–160.

Pearlmutter, B. A. and L. C. Parra (1997). Maximum likelihood source separation: a context-sensitive generalization of ICA. In M. C. Mozer, M. I. Jordan, and T. Petsche (Eds.), *Advances in Neural Information Processing Systems*, Volume 9, pp. 613–619. MIT Press.

Pearson, K. (1901). On lines and planes of closest fit to systems of points in space. *The London, Edin-*

burgh and Dublin Philosophical Magazine and Journal of Science, Sixth Series **2**, 559–572.

Platt, J. C. (1999). Fast training of support vector machines using sequential minimal optimization. In B. Schölkopf, C. J. C. Burges, and A. J. Smola (Eds.), *Advances in Kernel Methods – Support Vector Learning*, pp. 185–208. MIT Press.

Platt, J. C. (2000). Probabilities for SV machines. In A. J. Smola, P. L. Bartlett, B. Schölkopf, and D. Shuurmans (Eds.), *Advances in Large Margin Classifiers*, pp. 61–73. MIT Press.

Platt, J. C., N. Cristianini, and J. Shawe-Taylor (2000). Large margin DAGs for multiclass classification. In S. A. Solla, T. K. Leen, and K. R. Müller (Eds.), *Advances in Neural Information Processing Systems*, Volume 12, pp. 547–553. MIT Press.

Poggio, T. and F. Girosi (1990). Networks for approximation and learning. *Proceedings of the IEEE* **78**(9), 1481–1497.

Powell, M. J. D. (1987). Radial basis functions for multivariable interpolation: a review. In J. C. Mason and M. G. Cox (Eds.), *Algorithms for Approximation*, pp. 143–167. Oxford University Press.

Press, W. H., S. A. Teukolsky, W. T. Vetterling, and B. P. Flannery (1992). *Numerical Recipes in C: The Art of Scientific Computing* (Second ed.). Cambridge University Press.

Qazaz, C. S., C. K. I. Williams, and C. M. Bishop (1997). An upper bound on the Bayesian error bars for generalized linear regression. In S. W. Ellacott, J. C. Mason, and I. J. Anderson (Eds.), *Mathematics of Neural Networks: Models, Algorithms and Applications*, pp. 295–299. Kluwer.

Quinlan, J. R. (1986). Induction of decision trees. *Machine Learning* **1**(1), 81–106.

Quinlan, J. R. (1993). *C4.5: Programs for Machine Learning*. Morgan Kaufmann.

Rabiner, L. and B. H. Juang (1993). *Fundamentals of Speech Recognition*. Prentice Hall.

Rabiner, L. R. (1989). A tutorial on hidden Markov models and selected applications in speech recognition. *Proceedings of the IEEE* **77**(2), 257–285.

Ramasubramanian, V. and K. K. Paliwal (1990). A generalized optimization of the k-d tree for fast nearest-neighbour search. In *Proceedings Fourth IEEE Region 10 International Conference (TENCON'89)*, pp. 565–568.

Ramsey, F. (1931). Truth and probability. In R. Braithwaite (Ed.), *The Foundations of Mathematics and other Logical Essays*. Humanities Press.

Rao, C. R. and S. K. Mitra (1971). *Generalized Inverse of Matrices and Its Applications*. Wiley.

Rasmussen, C. E. (1996). *Evaluation of Gaussian Processes and Other Methods for Non-Linear Regression*. Ph. D. thesis, University of Toronto.

Rasmussen, C. E. and J. Quiñonero-Candela (2005). Healing the relevance vector machine by augmentation. In L. D. Raedt and S. Wrobel (Eds.), *Proceedings of the 22nd International Conference on Machine Learning*, pp. 689–696.

Rasmussen, C. E. and C. K. I. Williams (2006). *Gaussian Processes for Machine Learning*. MIT Press.

Rauch, H. E., F. Tung, and C. T. Striebel (1965). Maximum likelihood estimates of linear dynamical systems. *AIAA Journal* **3**, 1445–1450.

Ricotti, L. P., S. Ragazzini, and G. Martinelli (1988). Learning of word stress in a sub-optimal second order backpropagation neural network. In *Proceedings of the IEEE International Conference on Neural Networks*, Volume 1, pp. 355–361. IEEE.

Ripley, B. D. (1996). *Pattern Recognition and Neural Networks*. Cambridge University Press.

Robbins, H. and S. Monro (1951). A stochastic approximation method. *Annals of Mathematical Statistics* **22**, 400–407.

Robert, C. P. and G. Casella (1999). *Monte Carlo Statistical Methods*. Springer.

Rockafellar, R. (1972). *Convex Analysis*. Princeton University Press.

Rosenblatt, F. (1962). *Principles of Neurodynamics: Perceptrons and the Theory of Brain Mechanisms*. Spartan.

Roth, V. and V. Steinhage (2000). Nonlinear discriminant analysis using kernel functions. In S. A. Solla, T. K. Leen, and K. R. Müller (Eds.), *Advances in Neural Information Processing Systems*, Volume 12. MIT Press.

Roweis, S. (1998). EM algorithms for PCA and SPCA. In M. I. Jordan, M. J. Kearns, and S. A. Solla (Eds.), *Advances in Neural Information Processing Systems*, Volume 10, pp. 626–632. MIT Press.

Roweis, S. and Z. Ghahramani (1999). A unifying review of linear Gaussian models. *Neural Computation* **11**(2), 305–345.

Roweis, S. and L. Saul (2000, December). Nonlinear dimensionality reduction by locally linear embedding. *Science* **290**, 2323–2326.

Rubin, D. B. (1983). Iteratively reweighted least squares. In *Encyclopedia of Statistical Sciences*, Volume 4, pp. 272–275. Wiley.

Rubin, D. B. and D. T. Thayer (1982). EM algorithms for ML factor analysis. *Psychometrika* **47**(1), 69–76.

Rumelhart, D. E., G. E. Hinton, and R. J. Williams (1986). Learning internal representations by error propagation. In D. E. Rumelhart, J. L. McClelland, and the PDP Research Group (Eds.), *Parallel Distributed Processing: Explorations in the Microstructure of Cognition*, Volume 1: Foundations, pp. 318–362. MIT Press. Reprinted in Anderson and Rosenfeld (1988).

Rumelhart, D. E., J. L. McClelland, and the PDP Research Group (Eds.) (1986). *Parallel Distributed Processing: Explorations in the Microstructure of Cognition*, Volume 1: Foundations. MIT Press.

Sagan, H. (1969). *Introduction to the Calculus of Variations*. Dover.

Savage, L. J. (1961). The subjective basis of statistical practice. Technical report, Department of Statistics, University of Michigan, Ann Arbor.

Schölkopf, B., J. Platt, J. Shawe-Taylor, A. Smola, and R. C. Williamson (2001). Estimating the support of a high-dimensional distribution. *Neural Computation* **13**(7), 1433–1471.

Schölkopf, B., A. Smola, and K.-R. Müller (1998). Nonlinear component analysis as a kernel eigenvalue problem. *Neural Computation* **10**(5), 1299–1319.

Schölkopf, B., A. Smola, R. C. Williamson, and P. L. Bartlett (2000). New support vector algorithms. *Neural Computation* **12**(5), 1207–1245.

Schölkopf, B. and A. J. Smola (2002). *Learning with Kernels*. MIT Press.

Schwarz, G. (1978). Estimating the dimension of a model. *Annals of Statistics* **6**, 461–464.

Schwarz, H. R. (1988). *Finite element methods*. Academic Press.

Seeger, M. (2003). *Bayesian Gaussian Process Models: PAC-Bayesian Generalization Error Bounds and Sparse Approximations*. Ph. D. thesis, University of Edinburg.

Seeger, M., C. K. I. Williams, and N. Lawrence (2003). Fast forward selection to speed up sparse Gaussian processes. In C. M. Bishop and B. Frey (Eds.), *Proceedings Ninth International Workshop on Artificial Intelligence and Statistics, Key West, Florida*.

Shachter, R. D. and M. Peot (1990). Simulation approaches to general probabilistic inference on belief networks. In P. P. Bonissone, M. Henrion, L. N. Kanal, and J. F. Lemmer (Eds.), *Uncertainty in Artificial Intelligence*, Volume 5. Elsevier.

Shannon, C. E. (1948). A mathematical theory of communication. *The Bell System Technical Journal* **27**(3), 379–423 and 623–656.

Shawe-Taylor, J. and N. Cristianini (2004). *Kernel Methods for Pattern Analysis*. Cambridge University Press.

Sietsma, J. and R. J. F. Dow (1991). Creating artificial neural networks that generalize. *Neural Networks* **4**(1), 67–79.

Simard, P., Y. Le Cun, and J. Denker (1993). Efficient pattern recognition using a new transformation distance. In S. J. Hanson, J. D. Cowan, and C. L. Giles (Eds.), *Advances in Neural Information Processing Systems*, Volume 5, pp. 50–58. Morgan Kaufmann.

Simard, P., B. Victorri, Y. Le Cun, and J. Denker (1992). Tangent prop – a formalism for specifying selected invariances in an adaptive network. In J. E. Moody, S. J. Hanson, and R. P. Lippmann (Eds.), *Advances in Neural Information Processing Systems*, Volume 4, pp. 895–903. Morgan Kaufmann.

Simard, P. Y., D. Steinkraus, and J. Platt (2003). Best practice for convolutional neural networks applied to visual document analysis. In *Proceedings International Conference on Document Analysis and Recognition (ICDAR)*, pp. 958–962. IEEE Computer Society.

Sirovich, L. (1987). Turbulence and the dynamics of coherent structures. *Quarterly Applied Mathematics* **45**(3), 561–590.

Smola, A. J. and P. Bartlett (2001). Sparse greedy Gaussian process regression. In T. K. Leen, T. G. Dietterich, and V. Tresp (Eds.), *Advances in Neural Information Processing Systems*, Volume 13, pp. 619–625. MIT Press.

Spiegelhalter, D. and S. Lauritzen (1990). Sequential updating of conditional probabilities on directed graphical structures. *Networks* **20**, 579–605.

Stinchecombe, M. and H. White (1989). Universal approximation using feed-forward networks with non-sigmoid hidden layer activation functions. In *International Joint Conference on Neural Networks*, Volume 1, pp. 613–618. IEEE.

Stone, J. V. (2004). *Independent Component Analysis: A Tutorial Introduction*. MIT Press.

Sung, K. K. and T. Poggio (1994). Example-based learning for view-based human face detection. A.I. Memo 1521, MIT.

Sutton, R. S. and A. G. Barto (1998). *Reinforcement Learning: An Introduction*. MIT Press.

Svensén, M. and C. M. Bishop (2004). Robust Bayesian mixture modelling. *Neurocomputing* **64**, 235–252.

Tarassenko, L. (1995). Novelty detection for the identification of masses in mamograms. In *Proceedings Fourth IEE International Conference on Artificial Neural Networks*, Volume 4, pp. 442–447. IEE.

Tax, D. and R. Duin (1999). Data domain description by support vectors. In M. Verleysen (Ed.), *Proceedings European Symposium on Artificial Neural Networks, ESANN*, pp. 251–256. D. Facto Press.

Teh, Y. W., M. I. Jordan, M. J. Beal, and D. M. Blei (2006). Hierarchical Dirichlet processes. *Journal of the Americal Statistical Association*. to appear.

Tenenbaum, J. B., V. de Silva, and J. C. Langford (2000, December). A global framework for non-linear dimensionality reduction. *Science* **290**, 2319–2323.

Tesauro, G. (1994). TD-Gammon, a self-teaching backgammon program, achieves master-level play. *Neural Computation* **6**(2), 215–219.

Thiesson, B., D. M. Chickering, D. Heckerman, and C. Meek (2004). ARMA time-series modelling with graphical models. In M. Chickering and J. Halpern (Eds.), *Proceedings of the Twentieth Conference on Uncertainty in Artificial Intelligence, Banff, Canada*, pp. 552–560. AUAI Press.

Tibshirani, R. (1996). Regression shrinkage and selection via the lasso. *Journal of the Royal Statistical Society, B* **58**, 267–288.

Tierney, L. (1994). Markov chains for exploring posterior distributions. *Annals of Statistics* **22**(4), 1701–1762.

Tikhonov, A. N. and V. Y. Arsenin (1977). *Solutions of Ill-Posed Problems*. V. H. Winston.

Tino, P. and I. T. Nabney (2002). Hierarchical GTM: constructing localized non-linear projection manifolds in a principled way. *IEEE Trans-*

actions on Pattern Analysis and Machine Intelligence **24**(5), 639–656.

Tino, P., I. T. Nabney, and Y. Sun (2001). Using directional curvatures to visualize folding patterns of the GTM projection manifolds. In G. Dorffner, H. Bischof, and K. Hornik (Eds.), *Artificial Neural Networks – ICANN 2001*, pp. 421–428. Springer.

Tipping, M. E. (1999). Probabilistic visualisation of high-dimensional binary data. In M. S. Kearns, S. A. Solla, and D. A. Cohn (Eds.), *Advances in Neural Information Processing Systems*, Volume 11, pp. 592–598. MIT Press.

Tipping, M. E. (2001). Sparse Bayesian learning and the relevance vector machine. *Journal of Machine Learning Research* **1**, 211–244.

Tipping, M. E. and C. M. Bishop (1997). Probabilistic principal component analysis. Technical Report NCRG/97/010, Neural Computing Research Group, Aston University.

Tipping, M. E. and C. M. Bishop (1999a). Mixtures of probabilistic principal component analyzers. *Neural Computation* **11**(2), 443–482.

Tipping, M. E. and C. M. Bishop (1999b). Probabilistic principal component analysis. *Journal of the Royal Statistical Society, Series B* **21**(3), 611–622.

Tipping, M. E. and A. Faul (2003). Fast marginal likelihood maximization for sparse Bayesian models. In C. M. Bishop and B. Frey (Eds.), *Proceedings Ninth International Workshop on Artificial Intelligence and Statistics, Key West, Florida*.

Tong, S. and D. Koller (2000). Restricted Bayes optimal classifiers. In *Proceedings 17th National Conference on Artificial Intelligence*, pp. 658–664. AAAI.

Tresp, V. (2001). Scaling kernel-based systems to large data sets. *Data Mining and Knowledge Discovery* **5**(3), 197–211.

Uhlenbeck, G. E. and L. S. Ornstein (1930). On the theory of Brownian motion. *Phys. Rev.* **36**, 823–841.

Valiant, L. G. (1984). A theory of the learnable. *Communications of the Association for Computing Machinery* **27**, 1134–1142.

Vapnik, V. N. (1982). *Estimation of dependences based on empirical data*. Springer.

Vapnik, V. N. (1995). *The nature of statistical learning theory*. Springer.

Vapnik, V. N. (1998). *Statistical learning theory*. Wiley.

Veropoulos, K., C. Campbell, and N. Cristianini (1999). Controlling the sensitivity of support vector machines. In *Proceedings of the International Joint Conference on Artificial Intelligence (IJCAI99), Workshop ML3*, pp. 55–60.

Vidakovic, B. (1999). *Statistical Modelling by Wavelets*. Wiley.

Viola, P. and M. Jones (2004). Robust real-time face detection. *International Journal of Computer Vision* **57**(2), 137–154.

Viterbi, A. J. (1967). Error bounds for convolutional codes and an asymptotically optimum decoding algorithm. *IEEE Transactions on Information Theory* **IT-13**, 260–267.

Viterbi, A. J. and J. K. Omura (1979). *Principles of Digital Communication and Coding*. McGraw-Hill.

Wahba, G. (1975). A comparison of GCV and GML for choosing the smoothing parameter in the generalized spline smoothing problem. *Numerical Mathematics* **24**, 383–393.

Wainwright, M. J., T. S. Jaakkola, and A. S. Willsky (2005). A new class of upper bounds on the log partition function. *IEEE Transactions on Information Theory* **51**, 2313–2335.

Walker, A. M. (1969). On the asymptotic behaviour of posterior distributions. *Journal of the Royal Statistical Society, B* **31**(1), 80–88.

Walker, S. G., P. Damien, P. W. Laud, and A. F. M. Smith (1999). Bayesian nonparametric inference for random distributions and related functions (with discussion). *Journal of the Royal Statistical Society, B* **61**(3), 485–527.

Watson, G. S. (1964). Smooth regression analysis. *Sankhyā: The Indian Journal of Statistics. Series A* **26**, 359–372.

Webb, A. R. (1994). Functional approximation by feed-forward networks: a least-squares approach to generalisation. *IEEE Transactions on Neural Networks* **5**(3), 363–371.

Weisstein, E. W. (1999). *CRC Concise Encyclopedia of Mathematics*. Chapman and Hall, and CRC.

Weston, J. and C. Watkins (1999). Multi-class support vector machines. In M. Verlysen (Ed.), *Proceedings ESANN'99, Brussels*. D-Facto Publications.

Whittaker, J. (1990). *Graphical Models in Applied Multivariate Statistics*. Wiley.

Widrow, B. and M. E. Hoff (1960). Adaptive switching circuits. In *IRE WESCON Convention Record*, Volume 4, pp. 96–104. Reprinted in Anderson and Rosenfeld (1988).

Widrow, B. and M. A. Lehr (1990). 30 years of adaptive neural networks: perceptron, madeline, and backpropagation. *Proceedings of the IEEE* **78**(9), 1415–1442.

Wiegerinck, W. and T. Heskes (2003). Fractional belief propagation. In S. Becker, S. Thrun, and K. Obermayer (Eds.), *Advances in Neural Information Processing Systems*, Volume 15, pp. 455–462. MIT Press.

Williams, C. K. I. (1998). Computation with infinite neural networks. *Neural Computation* **10**(5), 1203–1216.

Williams, C. K. I. (1999). Prediction with Gaussian processes: from linear regression to linear prediction and beyond. In M. I. Jordan (Ed.), *Learning in Graphical Models*, pp. 599–621. MIT Press.

Williams, C. K. I. and D. Barber (1998). Bayesian classification with Gaussian processes. *IEEE Transactions on Pattern Analysis and Machine Intelligence* **20**, 1342–1351.

Williams, C. K. I. and M. Seeger (2001). Using the Nystrom method to speed up kernel machines. In T. K. Leen, T. G. Dietterich, and V. Tresp (Eds.), *Advances in Neural Information Processing Systems*, Volume 13, pp. 682–688. MIT Press.

Williams, O., A. Blake, and R. Cipolla (2005). Sparse Bayesian learning for efficient visual tracking. *IEEE Transactions on Pattern Analysis and Machine Intelligence* **27**(8), 1292–1304.

Williams, P. M. (1996). Using neural networks to model conditional multivariate densities. *Neural Computation* **8**(4), 843–854.

Winn, J. and C. M. Bishop (2005). Variational message passing. *Journal of Machine Learning Research* **6**, 661–694.

Zarchan, P. and H. Musoff (2005). *Fundamentals of Kalman Filtering: A Practical Approach* (Second ed.). AIAA.

Index

Page numbers in **bold** indicate the primary source of information for the corresponding topic.